JACARANDA NATURE OF BIOLOGY 2

VCE UNITS 3 AND 4 | SIXTH EDITION

JUDITH KINNEAR

MARJORY MARTIN

LUCY CASSAR

ELISE MEEHAN

RITU TYAGI

CONTRIBUTING AUTHORS

Stephanie Young | Sacha O'Connor-Price | Quinn Grace Arnold

jacaranda
A Wiley Brand

Sixth edition published 2021 by
John Wiley & Sons Australia, Ltd
42 McDougall Street, Milton, Qld 4064

First edition published 1992
Second edition published 2000
Third edition published 2006
Fourth edition published 2013
Fifth edition published 2017

Typeset in 11/14 pt Times Ltd Std

ISBN: 978-0-7303-7126-7

Cover image: © apple2499/Shutterstock

Illustrated by Harry Slaghekke (p.716), various artists, diacriTech and Wiley Composition Services

Typeset in India by diacriTech

A catalogue record for this book is available from the National Library of Australia

All activities have been written with the safety of both teacher and student in mind. Some, however, involve physical activity or the use of equipment or tools. **All due care should be taken when performing such activities.** Neither the publisher nor the authors can accept responsibility for any injury that may be sustained when completing activities described in this textbook.

This suite of print and digital resources may contain images of, or references to, members of Aboriginal and Torres Strait Islander communities who are, or may be, deceased. These images and references have been included to help Australian students from all cultural backgrounds develop a better understanding of Aboriginal and Torres Strait Islander peoples' history, culture and lived experience.

Printed in Singapore
M WEP295519 110724

Contents

About this resource

Jacaranda Nature of Biology 2 Sixth Edition has been revised and reimagined to provide students and teachers with the most relevant and comprehensive resource on the market. This engaging and purposeful suite of resources is fully aligned to the VCE Biology Study Design (2022–2026).

Formats

Jacaranda Nature of Biology is now available in digital and print formats:

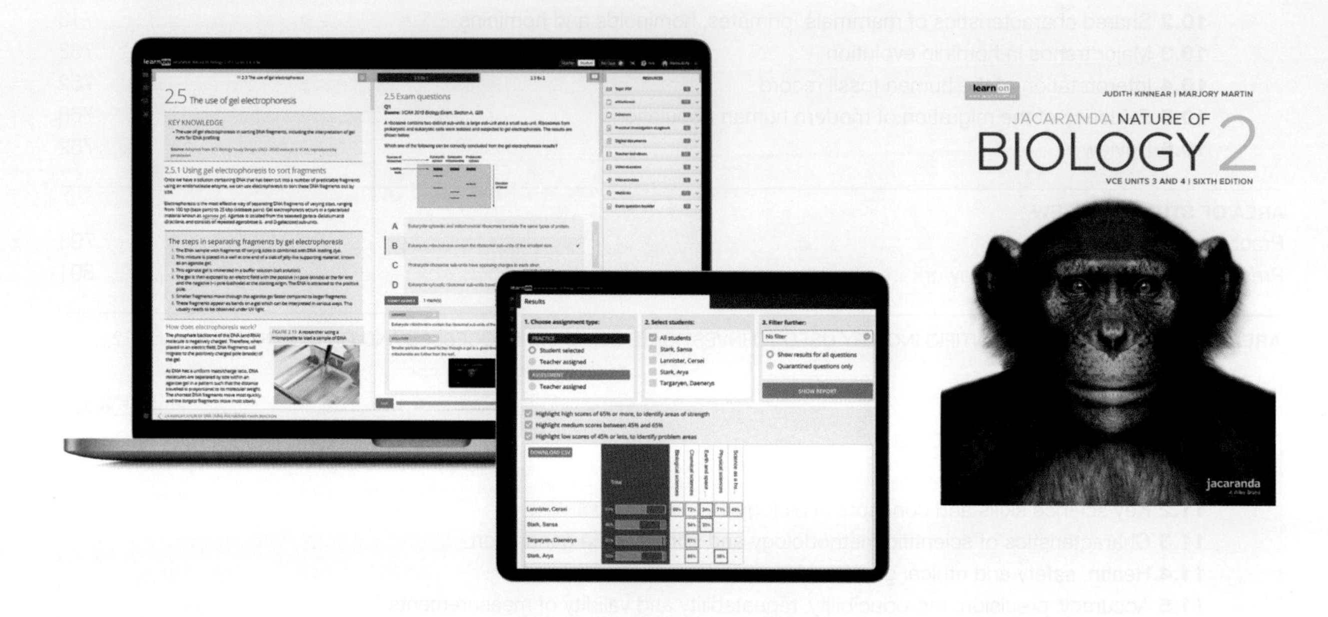

Fully aligned to the VCE Biology Study Design

Have confidence that you are covering the entire VCE Biology Study Design (2022–2026), with:

- key knowledge stated at the start of every topic and subtopic
- explicit support through dedicated topic for key science skills
- tailored exercise sets at the end of every subtopic, including past VCAA exam questions
- additional background information, case studies and extension easily distinguished from curriculum content
- comprehensive topic reviews including a summary flowchart, plus topic review exercises including exam questions
- practice SACs and exams for each Area of Study outcome
- glossary boxes to target key biological literacy
- key ideas at the end of each subtopic to help summarise important points
- suggested practical investigations to support VCAA requirements
- onResources boxes highlighting additional online resources

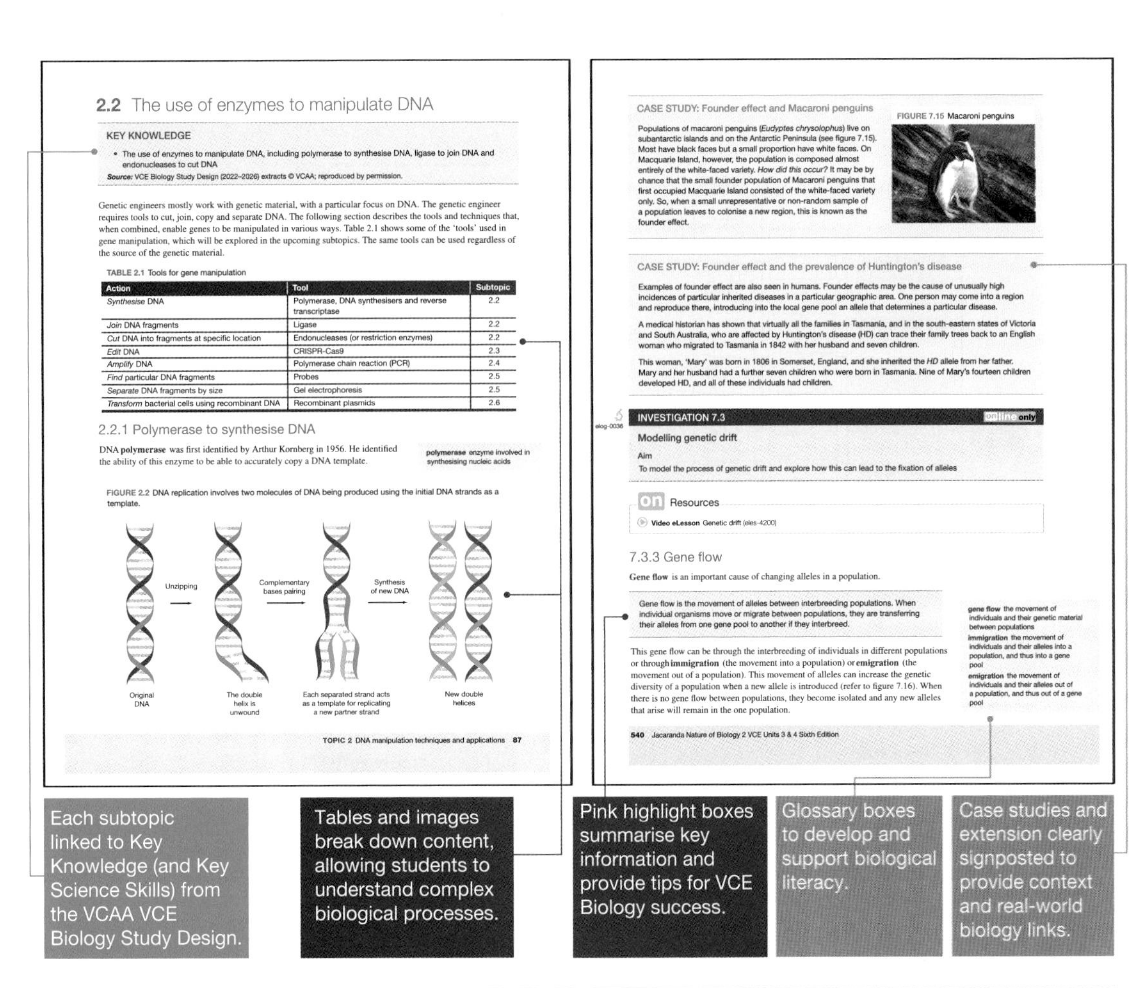

2.2 The use of enzymes to manipulate DNA

KEY KNOWLEDGE

- The use of enzymes to manipulate DNA, including polymerase to synthesise DNA, ligase to join DNA and endonucleases to cut DNA

Source: VCE Biology Study Design (2022–2026) extracts © VCAA; reproduced by permission.

Genetic engineers mostly work with genetic material, with a particular focus on DNA. The genetic engineer requires tools to cut, join, copy and separate DNA. The following section describes the tools and techniques that, when combined, enable genes to be manipulated in various ways. Table 2.1 shows some of the 'tools' used in gene manipulation, which will be explored in the upcoming subtopics. The same tools can be used regardless of the source of the genetic material.

TABLE 2.1 Tools for gene manipulation

Action	Tool	Subtopic
Synthesise DNA	Polymerase, DNA synthesisers and reverse transcriptase	2.2
Join DNA fragments	Ligase	2.2
Cut DNA into fragments at specific location	Endonucleases (or restriction enzymes)	2.2
Edit DNA	CRISPR-Cas9	2.3
Amplify DNA	Polymerase chain reaction (PCR)	2.4
Find particular DNA fragments	Probes	2.5
Separate DNA fragments by size	Gel electrophoresis	2.5
Transform bacterial cells using recombinant DNA	Recombinant plasmids	2.6

2.2.1 Polymerase to synthesise DNA

DNA **polymerase** was first identified by Arthur Kornberg in 1956. He identified the ability of this enzyme to be able to accurately copy a DNA template.

polymerase enzyme involved in synthesising nucleic acids

FIGURE 2.2 DNA replication involves two molecules of DNA being produced using the initial DNA strands as a template.

TOPIC 2 DNA manipulation techniques and applications 87

CASE STUDY: Founder effect and Macaroni penguins

Populations of macaroni penguins (*Eudyptes chrysolophus*) live on subantarctic islands and on the Antarctic Peninsula (see figure 7.15). Most have black faces but a small proportion have white faces. On Macquarie Island, however, the population is composed almost entirely of the white-faced variety. *How did this occur?* It may be by chance that the small founder population of Macaroni penguins that first occupied Macquarie Island consisted of the white-faced variety only. So, when a small unrepresentative or non-random sample of a population leaves to colonise a new region, this is known as the founder effect.

FIGURE 7.15 Macaroni penguins

CASE STUDY: Founder effect and the prevalence of Huntington's disease

Examples of founder effect are also seen in humans. Founder effects may be the cause of unusually high incidences of particular inherited diseases in a particular geographic area. One person may come into a region and reproduce there, introducing into the local gene pool an allele that determines a particular disease.

A medical historian has shown that virtually all the families in Tasmania, and in the south-eastern states of Victoria and South Australia, who are affected by Huntington's disease (HD) can trace their family trees back to an English woman who migrated to Tasmania in 1842 with her husband and seven children.

This woman, 'Mary' was born in 1806 in Somerset, England, and she inherited the *HD* allele from her father. Mary and her husband had a further seven children who were born in Tasmania. Nine of Mary's fourteen children developed HD, and all of these individuals had children.

elog-0036

INVESTIGATION 7.3 online only

Modelling genetic drift

Aim

To model the process of genetic drift and explore how this can lead to the fixation of alleles

on Resources

Video eLesson Genetic drift (eles-4200)

7.3.3 Gene flow

Gene flow is an important cause of changing alleles in a population.

Gene flow is the movement of alleles between interbreeding populations. When individual organisms move or migrate between populations, they are transferring their alleles from one gene pool to another if they interbreed.

This gene flow can be through the interbreeding of individuals in different populations or through **immigration** (the movement into a population) or **emigration** (the movement out of a population). This movement of alleles can increase the genetic diversity of a population when a new allele is introduced (refer to figure 7.16). When there is no gene flow between populations, they become isolated and any new alleles that arise will remain in the one population.

gene flow the movement of individuals and their genetic material between populations

immigration the movement of individuals and their alleles into a population, and thus into a gene pool

emigration the movement of individuals and their alleles out of a population, and thus out of a gene pool

540 Jacaranda Nature of Biology 2 VCE Units 3 & 4 Sixth Edition

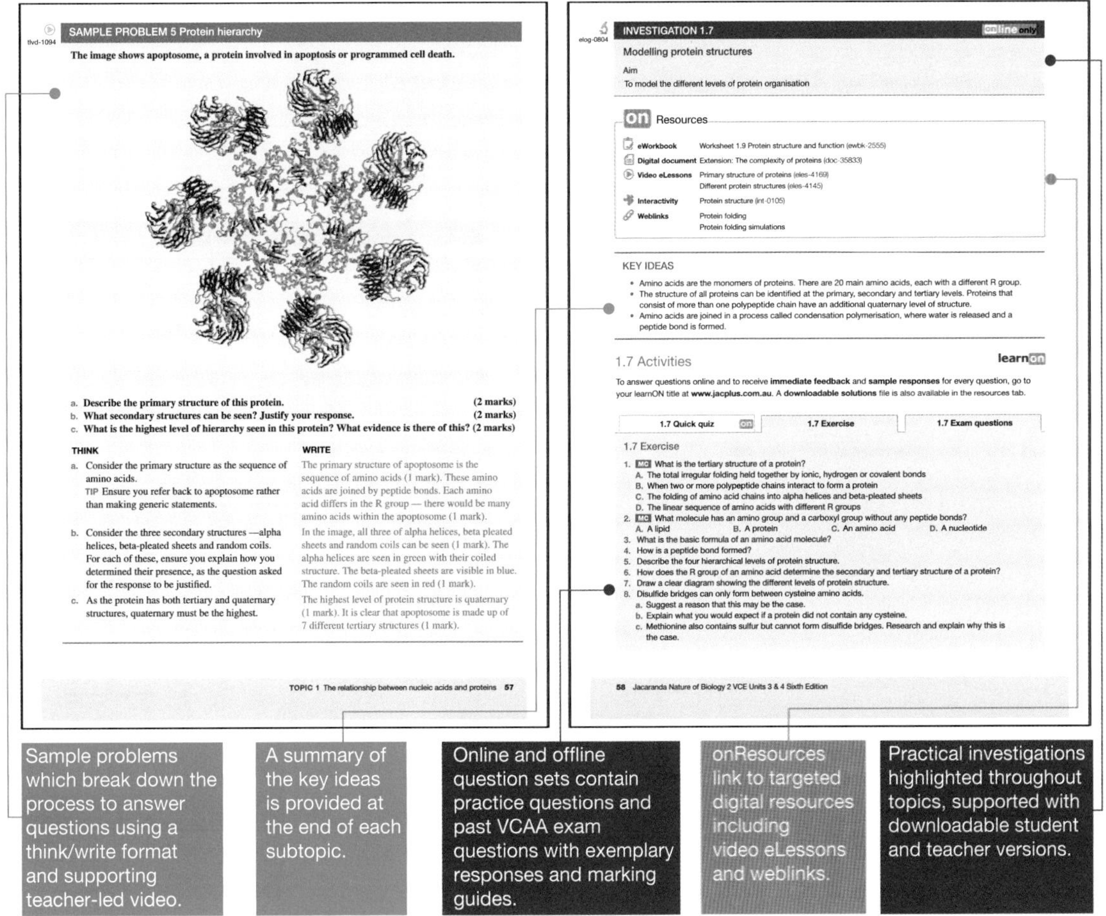

tlvd-1094

SAMPLE PROBLEM 5 Protein hierarchy

The image shows apoptosome, a protein involved in apoptosis or programmed cell death.

a. **Describe the primary structure of this protein.** **(2 marks)**
b. **What secondary structures can be seen? Justify your response.** **(2 marks)**
c. **What is the highest level of hierarchy seen in this protein? What evidence is there of this?** **(2 marks)**

THINK	WRITE
a. Consider the primary structure as the sequence of amino acids. TIP Ensure you refer back to apoptosome rather than making generic statements.	The primary structure of apoptosome is the sequence of amino acids (1 mark). These amino acids are joined by peptide bonds. Each amino acid differs in the R group — there would be many amino acids within the apoptosome (1 mark).
b. Consider the three secondary structures —alpha helices, beta-pleated sheets and random coils. For each of these, ensure you explain how you determined their presence, as the question asked for the response to be justified.	In the image, all three of alpha helices, beta pleated sheets and random coils can be seen (1 mark). The alpha helices are seen in green with their coiled structure. The beta-pleated sheets are visible in blue. The random coils are seen in red (1 mark).
c. As the protein has both tertiary and quaternary structures, quaternary must be the highest.	The highest level of protein structure is quaternary (1 mark). It is clear that apoptosome is made up of 7 different tertiary structures (1 mark).

TOPIC 1 The relationship between nucleic acids and proteins 57

elog-0804

INVESTIGATION 1.7 online only

Modelling protein structures

Aim

To model the different levels of protein organisation

on Resources

eWorkbook Worksheet 1.9 Protein structure and function (ewbk-2555)
Digital document Extension: The complexity of proteins (doc-35833)
Video eLessons Primary structure of proteins (eles-4169)
Different protein structures (eles-4145)
Interactivity Protein structure (int-0105)
Weblinks Protein folding
Protein folding simulations

KEY IDEAS

- Amino acids are the monomers of proteins. There are 20 main amino acids, each with a different R group.
- The structure of all proteins can be identified at the primary, secondary and tertiary levels. Proteins that consist of more than one polypeptide chain have an additional quaternary level of structure.
- Amino acids are joined in a process called condensation polymerisation, where water is released and a peptide bond is formed.

1.7 Activities

learnON

To answer questions online and to receive **immediate feedback** and **sample responses** for every question, go to your learnON title at **www.jacplus.com.au**. A **downloadable solutions** file is also available in the resources tab.

1.7 Quick quiz | 1.7 Exercise | 1.7 Exam questions

1.7 Exercise

1. MC What is the tertiary structure of a protein?
   A. The total irregular folding held together by ionic, hydrogen or covalent bonds
   B. When two or more polypeptide chains interact to form a protein
   C. The folding of amino acid chains into alpha helices and beta-pleated sheets
   D. The linear sequence of amino acids with different R groups
2. MC What molecule has an amino group and a carboxyl group without any peptide bonds?
   A. A lipid B. A protein C. An amino acid D. A nucleotide
3. What is the basic formula of an amino acid molecule?
4. How is a peptide bond formed?
5. Describe the four hierarchical levels of protein structure.
6. How does the R group of an amino acid determine the secondary and tertiary structure of a protein?
7. Draw a clear diagram showing the different levels of protein structure.
8. Disulfide bridges can only form between cysteine amino acids.
   a. Suggest a reason that this may be the case.
   b. Explain what you would expect if a protein did not contain any cysteine.
   c. Methionine also contains sulfur but cannot form disulfide bridges. Research and explain why this is the case.

58 Jacaranda Nature of Biology 2 VCE Units 3 & 4 Sixth Edition

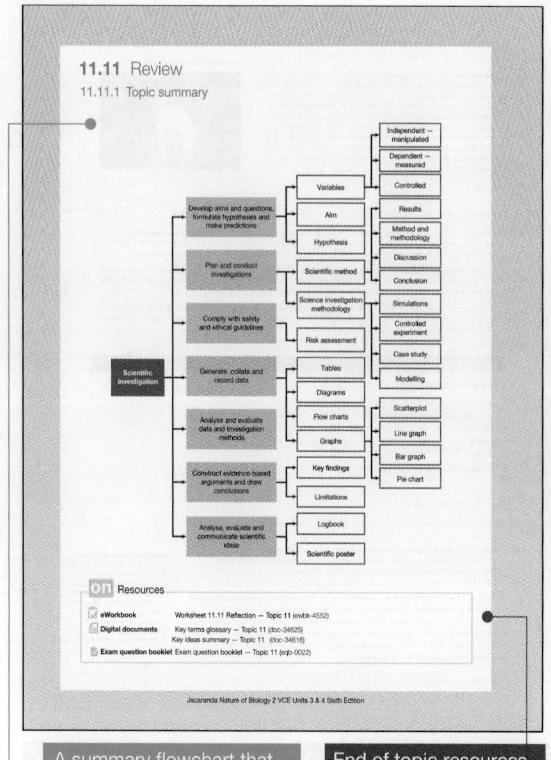

11.11 Review

11.11.1 Topic summary

Resources		
eWorkbook	Worksheet 11.11 Reflection — Topic 11 (ewbk-4552)
Digital documents	Key terms glossary — Topic 11 (doc-34625) Key ideas summary — Topic 11 (doc-34616)
Exam question booklet	Exam question booklet — Topic 11 (eqb-0022)

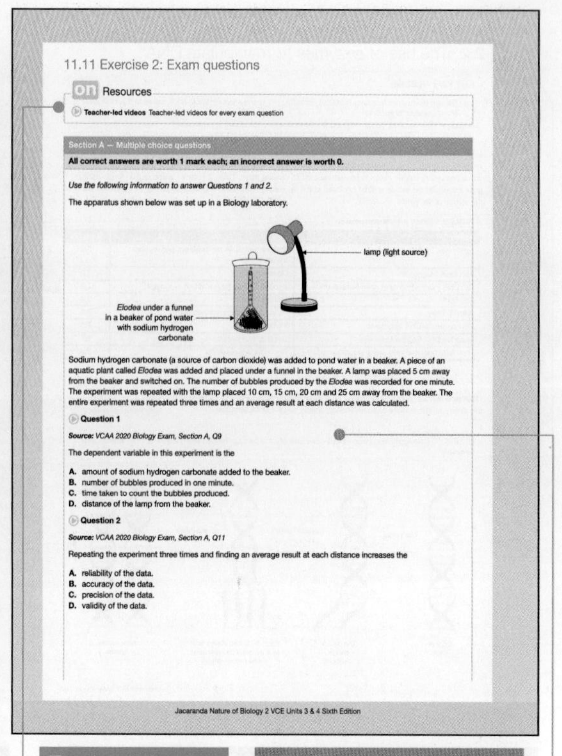

11.11 Exercise 2: Exam questions

Resources

Teacher-led videos Teacher-led videos for every exam question

Section A — Multiple choice questions

All correct answers are worth 1 mark each; an incorrect answer is worth 0.

Use the following information to answer Questions 1 and 2.

The apparatus shown below was set up in a Biology laboratory.

Sodium hydrogen carbonate (a source of carbon dioxide) was added to pond water in a beaker. A piece of an aquatic plant called *Elodea* was added and placed under a funnel in the beaker. A lamp was placed 5 cm away from the beaker and switched on. The number of bubbles produced by the *Elodea* was recorded for one minute. The experiment was repeated with the lamp placed 10 cm, 15 cm, 20 cm and 25 cm away from the beaker. The entire experiment was repeated three times and an average result at each distance was calculated.

Question 1

Source: VCAA 2020 Biology Exam, Section A, Q9

The dependent variable in this experiment is the

A. amount of sodium hydrogen carbonate added to the beaker.
B. number of bubbles produced in one minute.
C. time taken to count the bubbles produced.
D. distance of the lamp from the beaker.

Question 2

Source: VCAA 2020 Biology Exam, Section A, Q11

Repeating the experiment three times and finding an average result at each distance increases the

A. reliability of the data.
B. accuracy of the data.
C. precision of the data.
D. validity of the data.

Jacaranda Nature of Biology 2 VCE Units 3 & 4 Sixth Edition

A summary flowchart that shows the interrelationship between the main ideas of the topic. This includes links to both Key Knowledge and Key Science Skills.

End of topic resources including a reflection, key terms glossary and key ideas summary.

Teacher-led videos for every VCAA exam question through the topic.

End of topic exam questions, containing both multiple choice and short answer VCAA question questions that link together various concepts.

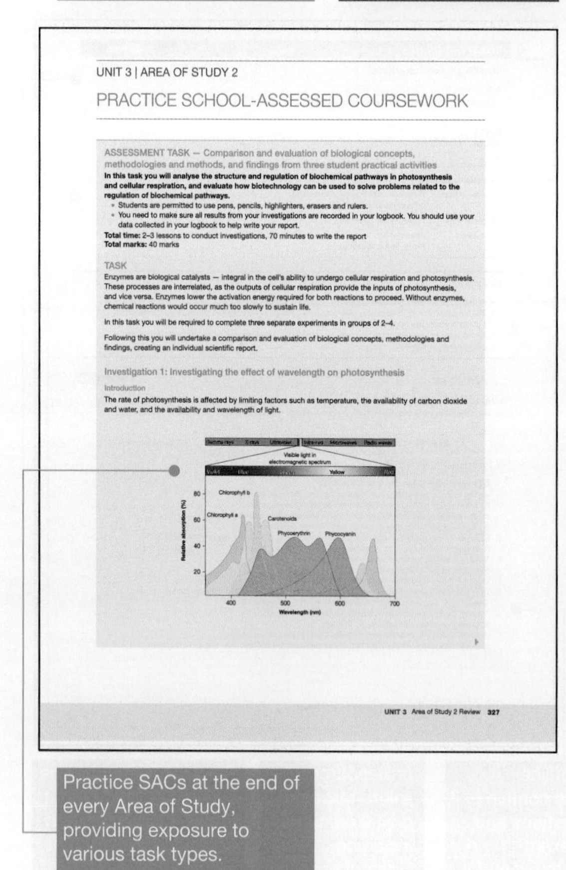

UNIT 3 | AREA OF STUDY 2

PRACTICE SCHOOL-ASSESSED COURSEWORK

ASSESSMENT TASK — Comparison and evaluation of biological concepts, methodologies and methods, and findings from three student practical activities

In this task you will analyse the structure and regulation of biochemical pathways in photosynthesis and cellular respiration, and evaluate how biotechnology can be used to solve problems related to the regulation of biochemical pathways.

- Students are permitted to use pens, pencils, highlighters, erasers and rulers.
- You need to make sure all results from your investigations are recorded in your logbook. You should use your data collected in your logbook to help write your report.

Total time: 2–3 lessons to conduct investigations, 70 minutes to write the report
Total marks: 40 marks

TASK

Enzymes are biological catalysts — integral in the cell's ability to undergo cellular respiration and photosynthesis. These processes are interrelated, as the outputs of cellular respiration provide the inputs of photosynthesis, and vice versa. Enzymes lower the activation energy required for both reactions to proceed. Without enzymes, chemical reactions would occur much too slowly to sustain life.

In this task you will be required to complete three separate experiments in groups of 2–4.

Following this you will undertake a comparison and evaluation of biological concepts, methodologies and findings, creating an individual scientific report.

Investigation 1: Investigating the effect of wavelength on photosynthesis

Introduction

The rate of photosynthesis is affected by limiting factors such as temperature, the availability of carbon dioxide and water, and the availability and wavelength of light.

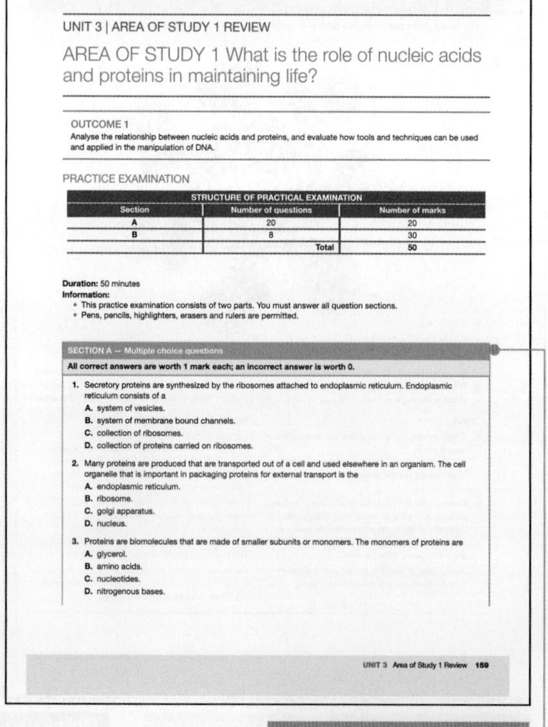

UNIT 3 | AREA OF STUDY 1 REVIEW

AREA OF STUDY 1 What is the role of nucleic acids and proteins in maintaining life?

OUTCOME 1

Analyse the relationship between nucleic acids and proteins, and evaluate how tools and techniques can be used and applied in the manipulation of DNA.

PRACTICE EXAMINATION

STRUCTURE OF PRACTICAL EXAMINATION		
Section	Number of questions	Number of marks
A	20	20
B	8	30
	Total	50

Duration: 50 minutes
Information:
- This practice examination consists of two parts. You must answer all question sections.
- Pens, pencils, highlighters, erasers and rulers are permitted.

SECTION A — Multiple choice questions

All correct answers are worth 1 mark each; an incorrect answer is worth 0.

1. Secretory proteins are synthesized by the ribosomes attached to endoplasmic reticulum. Endoplasmic reticulum consists of a
   A. system of vesicles.
   B. system of membrane bound channels.
   C. collection of ribosomes.
   D. collection of proteins carried on ribosomes.
2. Many proteins are produced that are transported out of a cell and used elsewhere in an organism. The cell organelle that is important in packaging proteins for external transport is the
   A. endoplasmic reticulum.
   B. ribosome.
   C. golgi apparatus.
   D. nucleus.
3. Proteins are biomolecules that are made of smaller subunits or monomers. The monomers of proteins are
   A. glycerol.
   B. amino acids.
   C. nucleotides.
   D. nitrogenous bases.

UNIT 3 Area of Study 1 Review 159

Practice SACs at the end of every Area of Study, providing exposure to various task types.

Practice exams at the end of every Area of Study to further support student learning across a broader range of concepts.

Accessing online resources

The power of learnON

The *Jacaranda Nature of Biology* series is now available on learnON, immersive digital learning platform that provides teachers with valuable insights into their students' learning and engagement. It's so much more than a textbook! It empowers students to be independent learners and allows teachers to assign, mark and track student work in real time.

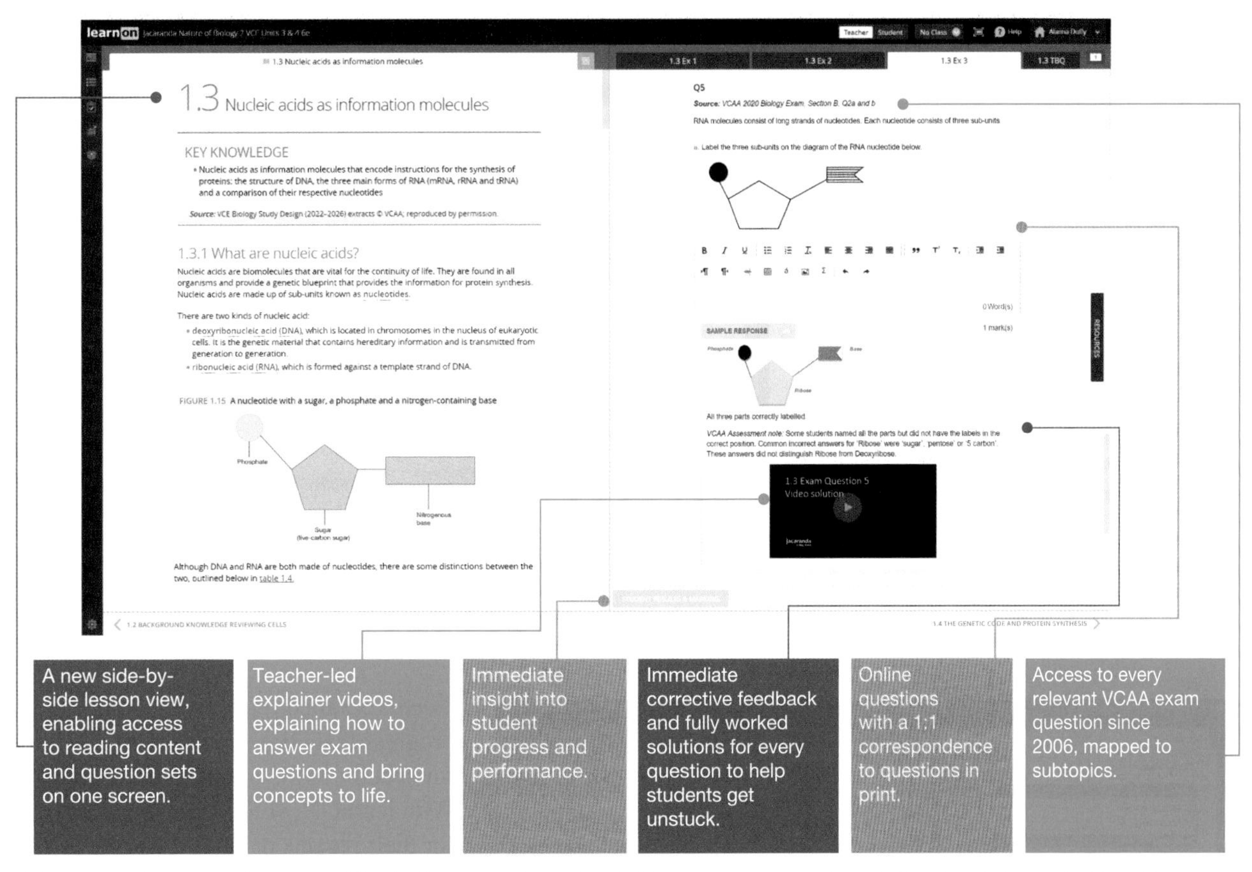

Alongside the powerful learnON platforms, a downloadable and customisable eWorkbook is avaliable. This is further supplemented by a practical investigation eLogbook, complete with risk assessments, to support all aspects of VCE Biology.

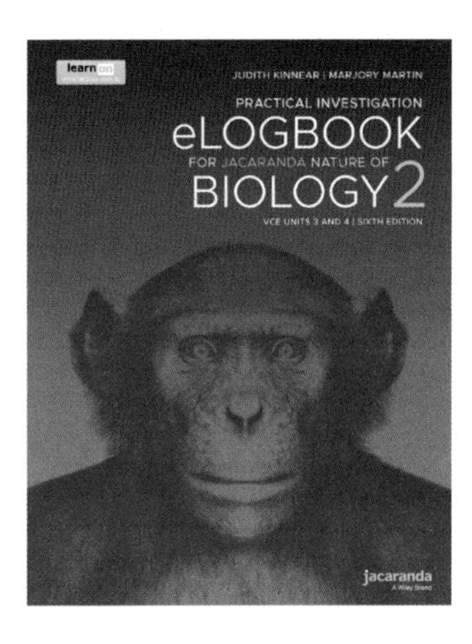

A wealth of teacher resources

Nature of Biology empowers teachers to teach their class their way, with an extensive range of teacher resources including:

- comprehensive support material, including work programs and teaching advice
- quarantined practice SACs and topic tests with answers and exemplary responses
- an easy-to-navigate marking interface that allows teachers to see student responses, add comments and mark their work
- the ability to create custom tests for your class from the entire question pool — including all subtopic, topic review and past VCAA exam questions.
- customisable course content, giving teachers more flexibility to create their own course
- the ability to separate a class into subgroups, making differentiation simpler
- dashboards to track progress and insight to students' strengths and weaknesses.

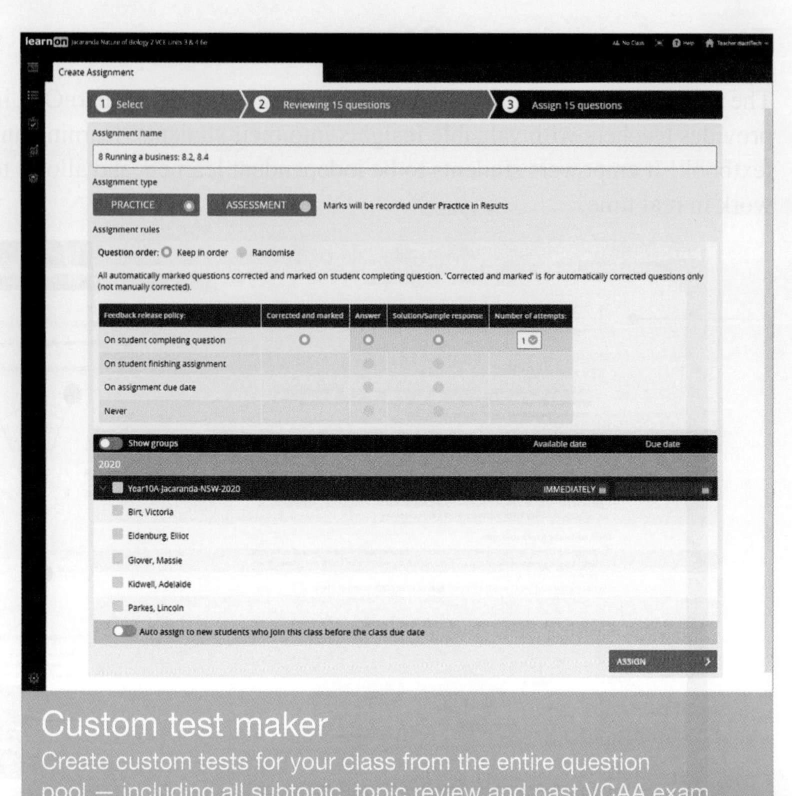

Custom test maker
Create custom tests for your class from the entire question pool — including all subtopic, topic review and past VCAA exam questions

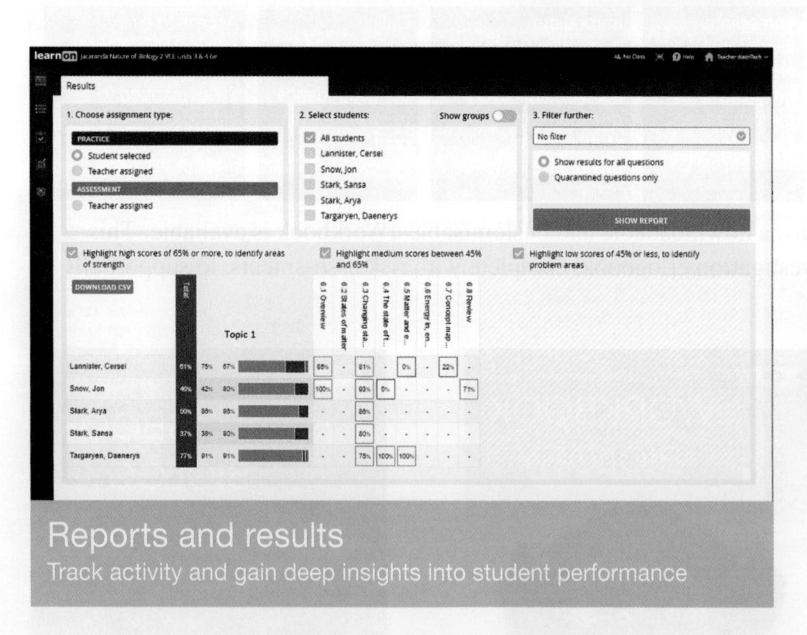

Reports and results
Track activity and gain deep insights into student performance

Access detailed reports on student progress that allow you to filter results for specific skills or question types. With learnON, you can show students (or their parents or carers) their own assessment data in fine detail. You can filter their results to identify areas of strength or weakness. Results are also colourcoded to help students understand their strengths and weaknesses at a glance.

Acknowledgements

The authors would like to thank those people who have played a key role in the production of this text. Their families and friends were always patient and supportive, especially when deadlines were imminent. This project was greatly enhanced by the generous cooperation of many academic colleagues and friends.

The staff of Wiley are also deserving of the highest praise. Their professionalism and expertise are greatly admired and appreciated. We build on each other's work.

The authors and publisher would like to thank the following copyright holders, organisations and individuals for their assistance and for permission to reproduce copyright material in this book.

Selected extracts from the VCE Biology Study Design (2022–2026) are copyright Victorian Curriculum and Assessment Authority (VCAA), reproduced by permission. VCE® is a registered trademark of the VCAA. The VCAA does not endorse this product and makes no warranties regarding the correctness or accuracy of its content. To the extent permitted by law, the VCAA excludes all liability for any loss or damage suffered or incurred as a result of accessing, using or relying on the content. Current VCE Study Designs and related content can be accessed directly at www.vcaa.vic.edu.au. Teachers are advised to check the VCAA Bulletin for updates.

Images

• Alamy Stock Photo: **760**, **763** • Designua / Shutterstock: Kateryna Kon / Shutterstock, **6**. Dr Klaus Boller / Science Photo Library **61** • Getty Images/SPL Creative RM: **376**, **728**, **756** • JOHN READER/SCIENCE PHOTO LIBRAR: **757** • KEITH CHAMBERS/SCIENCE PHOTO LIB: **1** • P.PLAILLY/E.DAYNES/SCIENCE PHOTO: **712** • PASCAL GOETGHELUCK/SCIENCE PHOTO: **734** • SCIENCE PHOTO LIBRARY: **329** • Shutterstock / CherylRamalho: **718** • Shutterstock / Convit: **730** • Shutterstock / Eric Isselee: **798** • Shutterstock / Heiti Paves: **161** • Shutterstock / jamesteohart: **798** • Shutterstock / Juan Gaertner: **403** • Shutterstock / Oksana Shufrych: **805** • Shutterstock / Ondrej Prosicky: **715** • Shutterstock / Rattana: **796** • Shutterstock / Rawpixel.com: **746** • Shutterstock / StudioMolekuul: **163** • Shutterstock / tristan tan: **711** • Shutterstock / Valentyna Chukhly: **798** • SPUTNIK/SCIENCE PHOTO LIBRARY: **761** • ZEPHYR/SCIENCE PHOTO LIBRARY: **353** • © 2005 - 2019 The Regents of the University of Michigan: **62** • © Adalbert Dragon / Shutterstock: **650** • © Aiempp147 / Shutterstock: **631** • © AJCespedes / Shutterstock: **347** • © Aldona Griskeviciene / Shutterstock: **7 (upper)**, **663** • © ALFRED PASIEKA / SCIENCE PHOTO LIBRARY: **447** • © Alhovik / Shutterstock: **200** • © Alila Medical Media / Shutterstock: **382**, **404**, **405**, **424**, **429** • © alinabel / Shutterstock: **604** • © alvindom / Shutterstock: **306** • © Amy Kerkemeyer / Shutterstock: **356** • © annalisa e marina durante / Shutterstock: **693** • © anyaivanova/ Shutterstock: **128** • © apple2499 / Shutterstock: **4**, **8**, **10**, **12**, **14–22**, **24**, **26**, **27**, **31**, **32**, **34–36**, **40–42**, **44–48**, **50–55**, **57–60**, **63**, **65–67**, **71**, **72**, **76–81**, **83**, **86–89**, **91–96**, **98**, **101**, **105**, **107**, **109**, **111–118**, **120–128**, **130–132**, **134**, **137**, **139**, **140**, **142**, **143**, **145**, **146**, **148–156**, **170**, **171**, **173–177**, **180**, **181**, **183–188**, **191–194**, **197**, **198**, **200–205**, **207–224**, **226–234**, **236–244**, **248**, **253**, **255–257**, **259**, **260**, **262–264**, **266–268**, **270–272**, **276**, **279**, **281**, **283**, **284**, **289–291**, **295**, **296**, **300**, **305**, **309–314**, **316–318**, **320**, **336–340**, **345**, **346**, **348**, **350**, **352**, **357–359**, **361**, **362**, **365**, **367**, **373–376**, **379**, **380**, **383–388**, **391**, **394**, **397–400**, **402**, **403**, **405–407**, **409**, **410**, **414–417**, **419–423**, **425**, **428**, **430**, **435**, **436**, **438**, **440–443**, **445**, **446**, **448**, **452**, **454**, **455**, **458**, **460–463**, **465–467**, **471**, **474**, **475**, **477–479**, **482**, **485**, **489**, **491**, **497–500**, **502**, **504–507**, **511**, **514–516**, **525–530**, **533–539**, **541–543**, **546–554**, **557–560**, **562**, **564**, **568**, **571–573**, **576**, **577**, **580–584**, **587–591**, **594–597**, **600**, **602–611**, **614–617**, **619–623**, **625**, **626**, **628–631**, **633**, **634**, **636**, **637**, **640–642**, **645–647**, **649**, **651–653**, **656**, **657**, **659**, **661**, **664**, **666–669**, **671–675**, **677**, **678**, **683**, **684**, **689–693**, **695–706**, **709** • © Barbol / Shutterstock: **370** • © Based on data from AIHW: **508** • © Bertold Werkmann / Shutterstock: **305** • © Bettmann / Getty Images Australia: **350** • © BigBearCamera / Shutterstock: **189** • © BlueRingMedia / Shutterstock: **228** • © Breck P. Kent / Shutterstock: **609** • © Centers for Disease Control and Prevention: **346**, **460**, **469** • © Central Historic Books / Alamy Stock Photo: **258** • © Choksawatdikorn / Shutterstock: **110** • © Courtesy: National Human Genome Research Institute: **30** • © Created by Dr. Newcomb. Rights held by the Board of Regents of the University of Wisconsin System University of Wisconsin-Madison: Department of Botany, **230** • © Cultura Creative RF / Alamy Stock Photo: **148** • © Design_Cells / Shutterstock: **335** • © Designua / Shutterstock: **5**, **198**, **235**, **349**, **419** • © Dizfoto / Shutterstock: **138** • © Dmitry Lobanov / Shutterstock: **429** • © DR DAN KALMAN / SCIENCE PHOTO LIBRARY: **463** • © DR ELENA KISELEVA / SCIENCE PHOTO LIBRARY: **24** • © DR KARI LOUNATMAA / SCIENCE PHOTO LIBRARY: **225** • © Dr Olivier Schwartz: Institute Pasteur / Science Photo Library, **377** • © Driscoll: Youngquist & Baldeschwieler, Caltech/Science Photo Library, **19** • © Ed Reschke / Getty Images Australia: **229** • © effe45 / Shutterstock: **185** • © ellepigrafica/Shutterstock: **450** • © Emre Terim / Shutterstock: **54 (lower)** • © extender_01 / Shutterstock: **110**

(upper) • © FCG / Shutterstock: **389** • © FLHC 76 / Alamy Stock Photo: **670** • © Francis Leroy: Biocosmos / Science Photo Library, **679** • © Frank Fichtmueller / Shutterstock: **568** • © fStop Images GmbH / Alamy Australia Pty Ltd: **256** • © Golden Rice Humanitarian Board: **142 (lower)** • © Gunilla Elam / Science Photo Library: **99** • © Helen Sushitskaya / Shutterstock: **170** • © http://www.jmol.org / RCSB PCB: **192** • © Hunter: B. & Carmody, J. 2015. Estimating the Aboriginal Population in Early Colonial Australia: The Role of Chickenpox Reconsidered. Australian Economic History Review 552. Wiley., **347, 456** • © Hybrid Medical Animation / Science Photo Library: **12, 16** • © I. ANDERSSON: OXFORD MOLECULAR BIOPHYSICS LABORATORY / SCIENCE PHOTO LIBRARY, **169** • © Image courtesy of Dr Fred Cohen: University of California, San Franscisco, **351** • © isak55 / Shutterstock: **665** • © James King-Holmes / Science Photo Library: **632** • © Jarun Ontakrai / Shutterstock: **463** • © jeka84 / Shutterstock: **233** • © JI de Wet / Shutterstock: **282** • © jomphong / Shutterstock: **431** • © Jose Antonio Penas / Science Photo Library: **689** • © Jose Calvo/Science Photo Library: **13** • © Jose Luis Calvo / Shutterstock: **367, 378** • © Josh Anon / Shutterstock: **638** • © Juan Gaertner / Shutterstock: **4, 97** • © Kateryna Kon / Shutterstock: **6, 286, 347, 367, 377, 378** • © Kondor83 / Shutterstock: **373, 391** • © LAGUNA DESIGN / SCIENCE PHOTO LIBRARY: **246** • © Lawrence Berkeley Laboratory: **55 (lower)** • © LindseyRN / Shutterstock: **381** • © Linnas / Shutterstock: **614** • © Lourens Smak / Alamy Stock Photo: **627** • © M. PATTHAWEE / Shutterstock: **181** • © Magicleaf / Shutterstock: **694** • © Manjurul Haque / Shutterstock: **104** • © MAPgraphics Pty Ltd: Brisbane, **655** • © mariait / Shutterstock: **673** • © Martina Simonazzi / Alamy Stock Photo: **285** • © Maurizio De Mattei / Shutterstock: **643** • © MAURO FERMARIELLO / SCIENCE PHOTO LIBRARY: **467** • © Merlin74 / Shutterstock: **599** • © Michael Marschall / Newspix: **113** • © Natali_ Mis / Shutterstock: **86** • © National Science Foundation / Science Photo Library: **654** • © Nika_Akin / Pixabay: **370** • © NoPainNoGain / Shutterstock: **626** • © Papadopulos: A., Igea, J., Dunning, L., Osborne, O., Quan, X., Pellicer, J., Turnbull, C., Hutton, I., Baker, W., Butlin, R. & Savolainen, V. 2019. Ecological speciation in sympatric palms: **2019** . Genetic map reveals genomic islands underlying species divergence in Howea. International Journal of Evolution. 739, **1986** -1995., **648** • © Pascal Geotgheluck / Science Photo Library: **110** • © Pascal Goetgheluck / Science Photo Library: **625** • © Permission to use granted by June Kwak. Department of New Biology: DGIST, Republic of Korea, **246** • © Peter Etchells / Shutterstock: **640** • © Peter Hermes Furian / Shutterstock: **644** • © Philippe Plailly / Science Photo Library: **85** • © Photo©Bluestar®: **68 (lower)** • © physicsgirl / pixabay: **671** • © PjrStudio / Alamy Stock Photo: **612** • © podsy / Shutterstock: **488** • © Professor P. Motta & T. Naguro / Science Photo Library: **67** • © Public Domain: **688** • © Public Domain/Louisa Howard: **66** • © Raimundo79 / Shutterstock: **678** • © Rena Schild / Shutterstock: **280** • © Rokas Tenys / Shutterstock: **304** • © Sajee Rod / Shutterstock: **432** • © Sakurra / Shutterstock: **396** • © Sasi Ponchaisang / EyeEm / Getty Images Australia: **100** • © Schira / Shutterstock: **42** • © Science History Images / Alamy Stock Photo: **550** • © SCIENCE PHOTO LIBRARY: **249** • © Science Photo Library / Getty Images: **56** • © Sebastian Kaulitzki / Shutterstock: **392** • © Secchi-Lecaque / Roussel-Uclaf / CNRI / Science Photo Library: **413** • © Sheila Say / Shutterstock: **642** • © Siberian Art / Shutterstock: **368** • © SingjaiStocker / Shutterstock: **397** • © Soleil Nordic / Shutterstock: **106** • © SPUTNIK / SCIENCE PHOTO LIBRARY: **448** • © Sputnik / Science Photo Library: **655** • © STEFAN DILLER / SCIENCE PHOTO LIBRARY: **258** • © stihii / Shutterstock: **612, 675** • © StudioMolekuul / Shutterstock: **104** • © T-flex / Shutterstock: **119** • © Tetyana Dotsenko / Shutterstock: **540** • © Thomas Shafee: **680** • © TIM VERNON: LTH NHS TRUST / SCIENCE PHOTO LIBRARY, **464** • © TOKYO GAS Co.: Ltd. Sourced from https://www.tokyo-gas.co.jp/, **304** • © Tom Mchugh / Science Photo Library: **600** • © Universal Images Group North America LLC/Alamy Stock Photo: **643** • © US AGENCY FOR INTERNATIONAL DEVELOPMENT / DVIDS: US DEPARTMENT OF DEFENSE / SCIENCE PHOTO LIBRARY, **487** • © Valentijn: K.M., F.D. Gumkowski and J.D. Jamieson. 1999. The subapical actin cytoskeleton regulates secretion and membrane retrieval in pancreatic acinar cells. J. Cell Science. 112:**81–96.**, **69** • © Vecton / Shutterstock: **369** • © VectorMine / Shutterstock: **247, 274, 281, 395** • © VectorMine/Shutterstock: **390** • © Victoria Antonova / Shutterstock: **432** • © vitstudio / Shutterstock: **3** • © Volker Steger / Science Photo Library: **133 (upper and lower)** • © Wikimedia Commons: **672**

Every effort has been made to trace the ownership of copyright material. Information that will enable the publisher to rectify any error or omission in subsequent reprints will be welcome. In such cases, please contact the Permissions Section of John Wiley & Sons Australia, Ltd.

UNIT

3 How do cells maintain life?

AREA OF STUDY 1

What is the role of nucleic acids and proteins in maintaining life?

OUTCOME 1

Analyse the relationship between nucleic acids and proteins, and evaluate how tools and techniques can be used and applied in the manipulation of DNA.

AREA OF STUDY 2

How are biochemical pathways regulated?

OUTCOME 2

Analyse the structure and regulation of biochemical pathways in photosynthesis and cellular respiration, and evaluate how biotechnology can be used to solve problems related to the regulation of biochemical pathways.

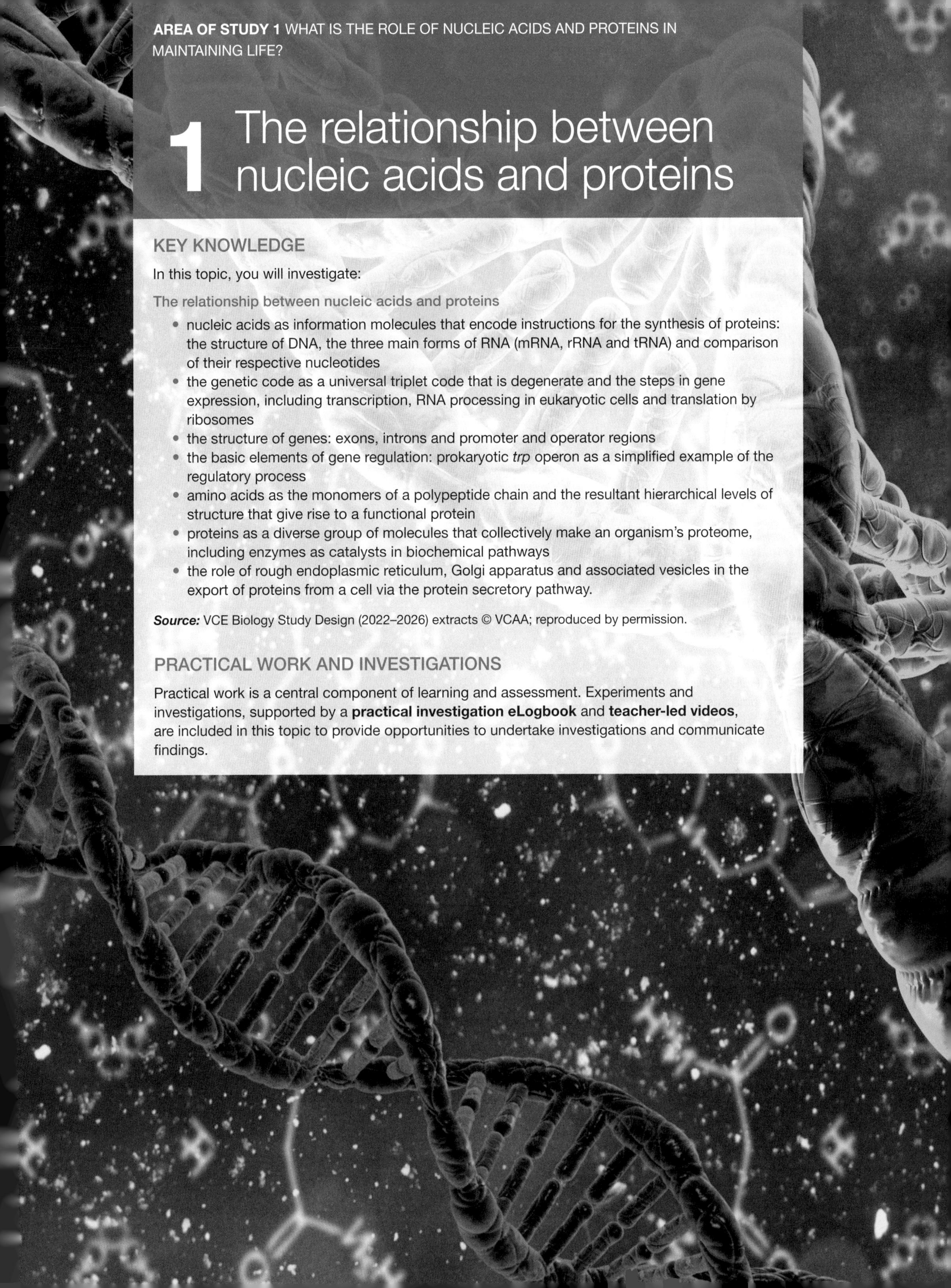

1 The relationship between nucleic acids and proteins

KEY KNOWLEDGE

In this topic, you will investigate:

The relationship between nucleic acids and proteins

- nucleic acids as information molecules that encode instructions for the synthesis of proteins: the structure of DNA, the three main forms of RNA (mRNA, rRNA and tRNA) and comparison of their respective nucleotides
- the genetic code as a universal triplet code that is degenerate and the steps in gene expression, including transcription, RNA processing in eukaryotic cells and translation by ribosomes
- the structure of genes: exons, introns and promoter and operator regions
- the basic elements of gene regulation: prokaryotic *trp* operon as a simplified example of the regulatory process
- amino acids as the monomers of a polypeptide chain and the resultant hierarchical levels of structure that give rise to a functional protein
- proteins as a diverse group of molecules that collectively make an organism's proteome, including enzymes as catalysts in biochemical pathways
- the role of rough endoplasmic reticulum, Golgi apparatus and associated vesicles in the export of proteins from a cell via the protein secretory pathway.

PRACTICAL WORK AND INVESTIGATIONS

Practical work is a central component of learning and assessment. Experiments and investigations, supported by a **practical investigation eLogbook** and **teacher-led videos**, are included in this topic to provide opportunities to undertake investigations and communicate findings.

1.1 Overview

Numerous **videos** and **interactivities** are available just where you need them, at the point of learning, in your digital formats, learnON and eBookPLUS, and at **www.jacplus.com.au**.

1.1.1 Introduction

Cells are what define us as living organisms, allowing us to reproduce, adapt, survive and grow. Understanding the structure and components of cells is fundamental to understanding life itself. How do we control what enters and leaves our cells? How do we synthesise the proteins that allow us to thrive?

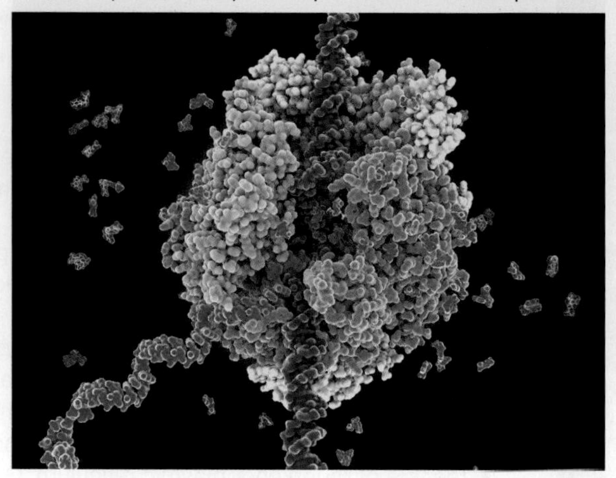

FIGURE 1.1 RNA polymerase unwinding a DNA strand (seen in violet) and building a new RNA strand (seen in red) in the process of transcription

Our cells are incredible in their ability not only to synthesise proteins, but also to regulate their production so they are only made where and when they are required. Proteins are fundamental to our survival, forming diverse structures with a variety of functions and making up everything from our enzymes to our fibrous tissues.

In this topic, we will examine our cells' amazing ability to make proteins from a DNA blueprint. DNA is the basis of the incredible diversity of life on Earth.

LEARNING SEQUENCE

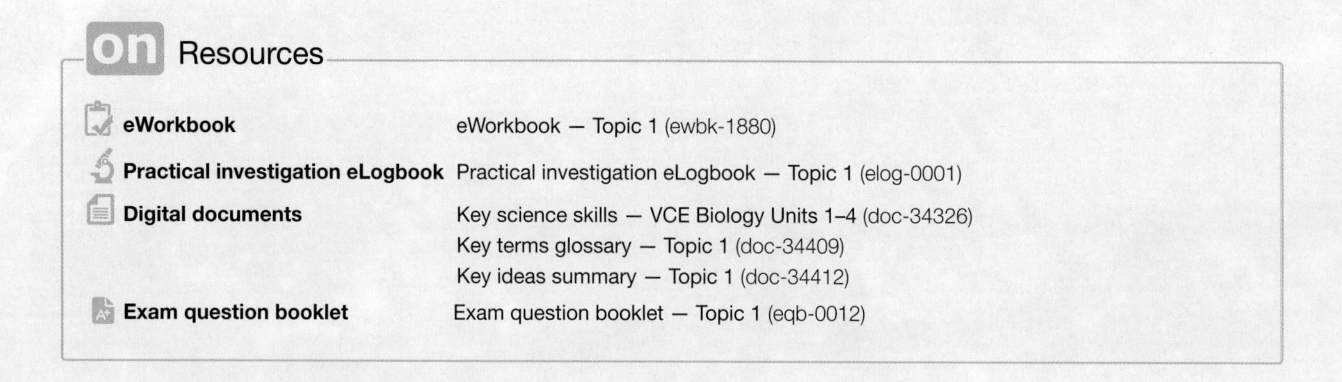

on Resources

eWorkbook	eWorkbook — Topic 1 (ewbk-1880)
Practical investigation eLogbook	Practical investigation eLogbook — Topic 1 (elog-0001)
Digital documents	Key science skills — VCE Biology Units 1–4 (doc-34326) Key terms glossary — Topic 1 (doc-34409) Key ideas summary — Topic 1 (doc-34412)
Exam question booklet	Exam question booklet — Topic 1 (eqb-0012)

1.2 BACKGROUND KNOWLEDGE Reviewing cells

BACKGROUND KNOWLEDGE

This subtopic will review concepts from Units 1 and 2 that will help you understand various key knowledge points of the Study Design for Units 3 and 4. This content is not examinable in Units 3 and 4.

- Cells as the basic structural feature of life on Earth, including the distinction between prokaryotic and eukaryotic cells
- The structure and specialisation of plant and animal cell organelles for distinct functions
- The structure and function of the plasma membrane in the passage of water, hydrophobic and hydrophilic substances via osmosis, facilitated diffusion and active transport

1.2.1 What are cells?

Cells are the basic structural and functional units of life, and all living organisms are built of one or more cells. Cells, with only a very few exceptions, are too small to be seen with an unaided eye. Throughout Units 3 and 4, you will learn about the structure and function of various types of cells.

FIGURE 1.2 Comparing the size of various cells, organelles and non-cellular pathogens. Each diagram zooms in on the previous diagram.

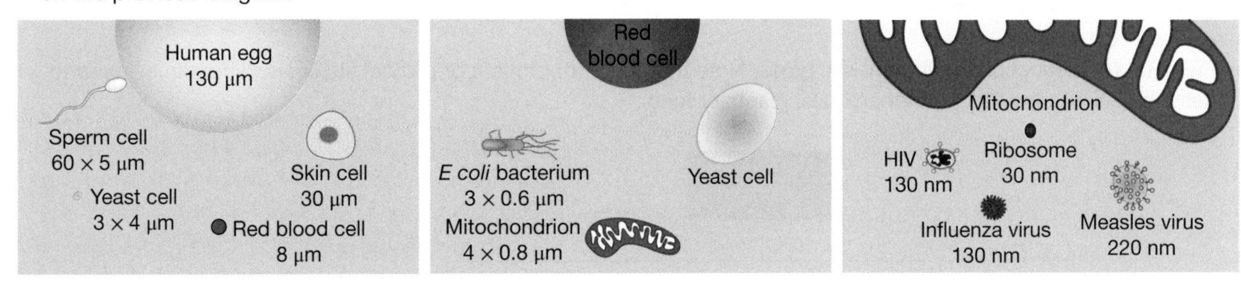

How big are cells?

Cells and organisms vary greatly in size. Some examples of these are compared in figure 1.2.

- Most animal cells fall within the size range of 10 to 40 μm. Among the smallest human cells are red blood cells, with diameters for normal cells in the range of 6 to 8 μm. Only a few single cells are large enough to see with the unaided human eye, for example human egg cells with diameters around 0.1 mm.
- The common amoeba (*Amoeba proteus*) is a unicellular organism with an average size ranging from 0.25 to 0.75 mm. (You would see an amoeba as about the size of a full stop.)
- Microbial cells are on average 10 times smaller than plant and animal cells, with diameters in the range of 0.4 to 2.0 μm and lengths in the range of 0.5 to 5 μm. (However, the smallest bacterium, *Pelagibacter ubique*, consists of a cell just 0.2 μm in diameter.)

Cells constantly need to transport materials as they exchange ions with the extracellular environment, gain nutrients and remove wastes. Cells need to be small in order to maximise their surface area to volume ratio, allowing for the movement of ions, nutrients and wastes to occur quickly. Without a large surface area to volume ratio, cells will not survive.

There is no fixed shape for cells. Cells vary in shape and their shapes often reflect their functions. For example, immune cells such as dendritic cells and macrophages have very different cellular shapes compared to other cells such as red blood cells.

FIGURE 1.3 a. Various shapes of some different cells b. A 3D representation of a dendritic cell

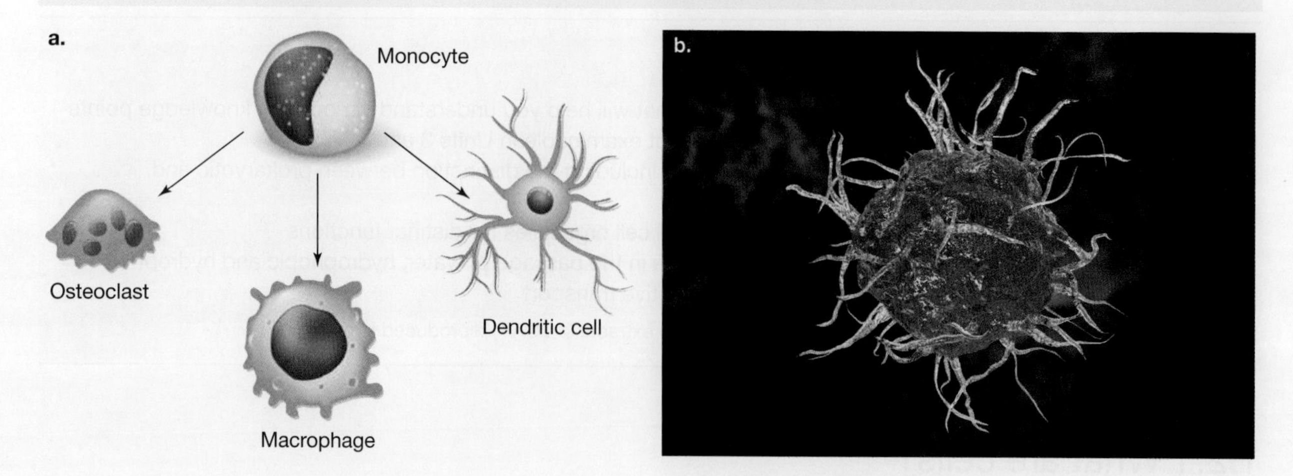

1.2.2 Comparing eukaryotic and prokaryotic cells

Prokaryotes and eukaryotes

Although cells vary greatly in complexity, they can be distinguished into two main types — **prokaryotic cells** and **eukaryotic cells**.

FIGURE 1.4 Distinction of different cell types. Note that the term protista or protoctista is also sometimes used to describe eukaryotes that are not animals, plants or fungi.

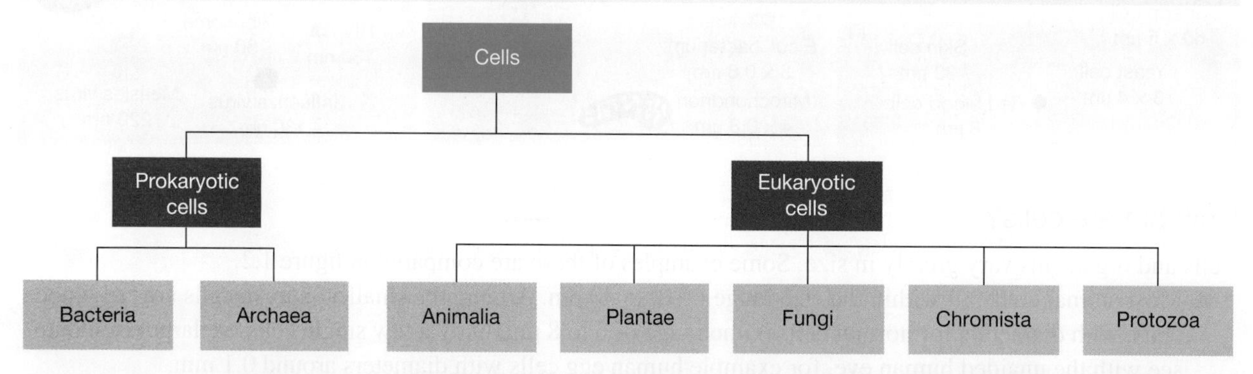

The microscopically tiny creatures that we call 'microbes' are a diverse group of organisms. The microbes comprise two different classification groups, namely bacteria and archaea. The cells of all these microbes can be distinguished from the cells of the other major groups of living organisms: fungi, plants and animals. The key distinguishing feature of archaea and bacteria is that their cells lack a membrane-bound nucleus (see figures 1.5 and 1.6). Cells with this characteristic are described as prokaryotic cells, and organisms displaying this feature are called **prokaryotes**. Like all other kinds of organism, archaea and bacteria have DNA in their cells, but the DNA in prokaryotic cells is dispersed, not enclosed within a separate membrane-bound compartment.

In contrast, the cells of all other organisms have a definite nucleus (see figure 1.5) bordered by a double membrane. Organisms with these characteristics are called **eukaryotes**. The nucleus of a eukaryotic cell contains DNA, the genetic material of cells. Eukaryotic cells contain many membrane-bound cell organelles that are not present in prokaryotic cells.

prokaryotic cells cells within prokaryotes that lack a membrane-bound nucleus

eukaryotic cells cells within eukaryotes that have a membrane-bound nucleus and other membrane-bound organelles

prokaryotes any cells or organisms without a membrane-bound nucleus

eukaryotes any cells or organisms with a membrane-bound nucleus

FIGURE 1.5 Comparing the structure of a typical prokaryotic cell with that of a eukaryotic cell. Note that a prokaryotic cell has a simple architecture in contrast to a eukaryotic cell.

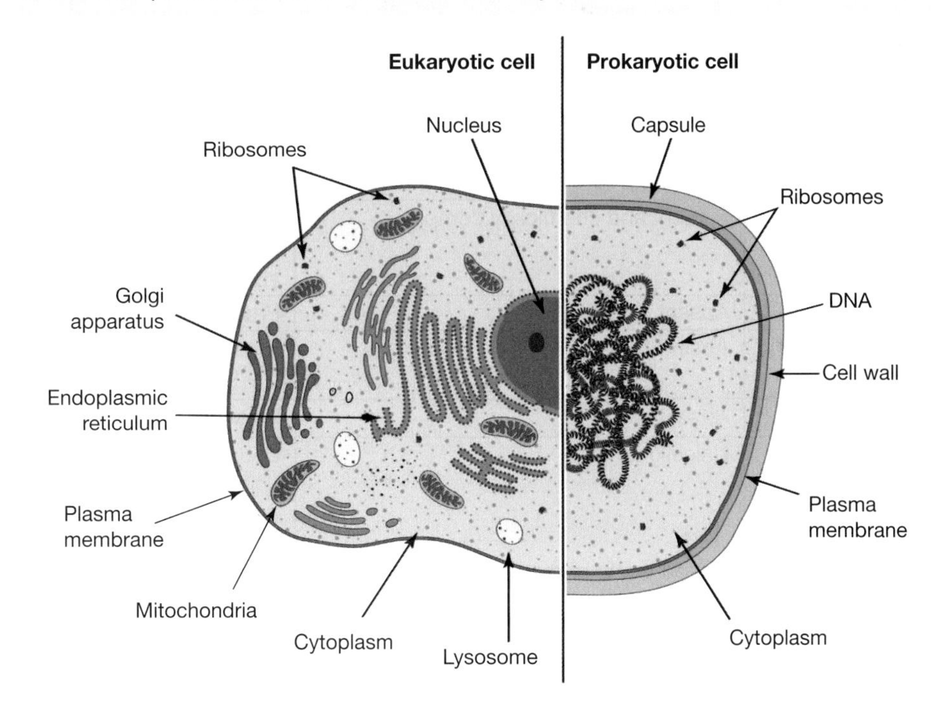

FIGURE 1.6 a. A prokaryotic cell in the process of dividing. Note the dispersed genetic material (stained red). b. A confocal fluorescence microscope view of eukaryotic cells (human breast cells). Note the discrete rounded nuclei (shown in blue). (Image courtesy of Leigh Ackland)

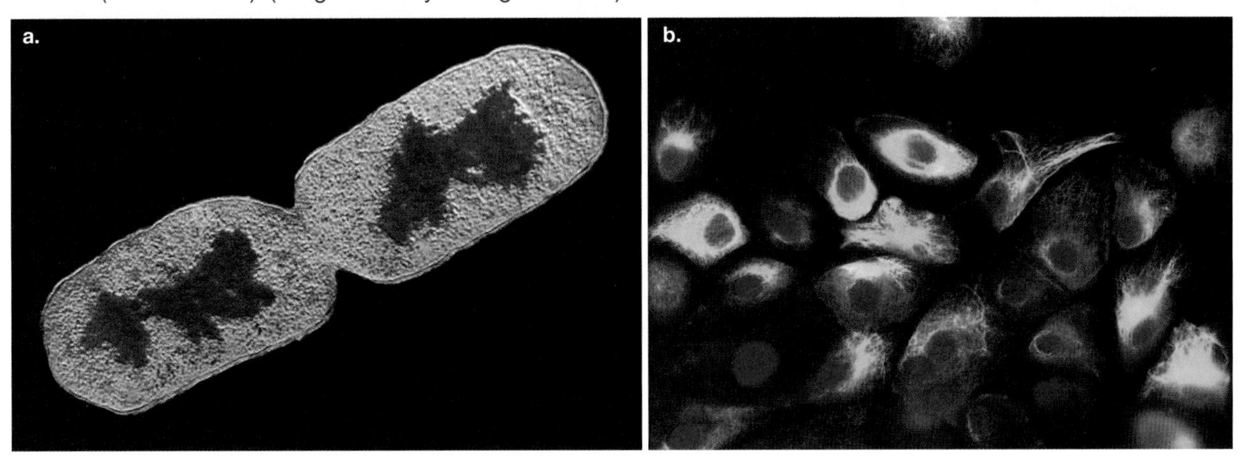

The distinguishing feature between a prokaryotic cell and a eukaryotic cell is the absence of a membrane-bound nucleus in prokaryotes. This can be linked back to the naming of prokaryotes and eukaryotes:

pro = before + *karyon* = nut, kernel, nucleus

eu = well (= true) + *karyon* = nut, kernel, nucleus.

TABLE 1.1 Comparison of prokaryotic and eukaryotic cells

Feature	Prokaryote	Eukaryote
Size	Small: typically 12 μm diameter	Larger: typically in range 10–100 μm
Chromosomes	Present as single circular DNA molecule	Present as multiple linear DNA molecules
Ribosomes	Present; small size (20 nm or 70S)*	Present; large size (25–30 nm or 80S)*
Cell membrane	Present	Present
Cell wall	Present and chemically complex	Present in plants, fungi, and some protists, but chemically simple; absent in animal cells
Membrane-bound nucleus	Absent	Present
Membrane-bound cell organelles	Absent	Present, e.g. lysosomes, mitochondria
Cytoskeleton	Absent	Present
Number of cells	Unicellular	Usually multicellular but can be unicellular (e.g. protists such as *Amoeba* and *Euglena*, algae such as *Chlorella* and yeasts)

* S denotes Svedberg units, which measure the time it takes for a particle to settle to the bottom of a solution. For ribosomes, this time can be correlated to particle size.

Although there are some differences in aspects of the structure of eukaryotic and prokaryotic cells, there are many similarities in their structures and functioning.

Both prokaryotic and eukaryotic cells:

- have DNA as their genetic material
- have cell membranes that selectively control the entry and exit of dissolved materials into and out of the cell
- use the same chemical building blocks, including carbon, nitrogen, oxygen, hydrogen and phosphorus, to build the organic molecules that form their structure and enable their function
- produce proteins through the same mechanism (transcription and translation)
- use ATP as their source of energy to drive the energy-requiring activities of their cells.

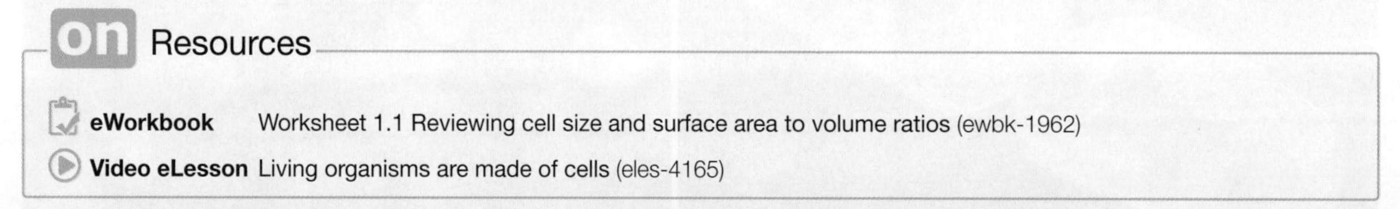

1.2.3 Organelles

Eukaryotic cells are organised internally into various compartments, each enclosed by a membrane. These compartments are known as organelles and give eukaryotes a much more complex structure than prokaryotes.

Compartmentalisation in eukaryotic cells is about efficiency. Separating the cell into specific components allows for the creation of specific microenvironments within a cell. That way, each organelle can have all the advantages it needs to perform to the best of its ability.

Think about how a house is subdivided into rooms to support different functions: you shower in the bathroom, not in the kitchen; the stove is in the kitchen, not in the bedroom, and so on. A eukaryotic cell can be likened to a house — its various compartments are like different rooms where different tasks are carried out. The conditions (such as pH and ion concentration) within the different kinds of compartment can vary from each other and from the cellular environment in which they are found.

Some of the main organelles in eukaryotes are explored in figure 1.7 and table 1.2. When writing organelle names, you should try to write the full name of the organelle to reduce the chance for confusion. Many of these organelles will be further explored in subtopic 1.9.

FIGURE 1.7 Two examples of eukaryotic cells: **a.** a generalised animal cell; **b.** a generalised plant cell

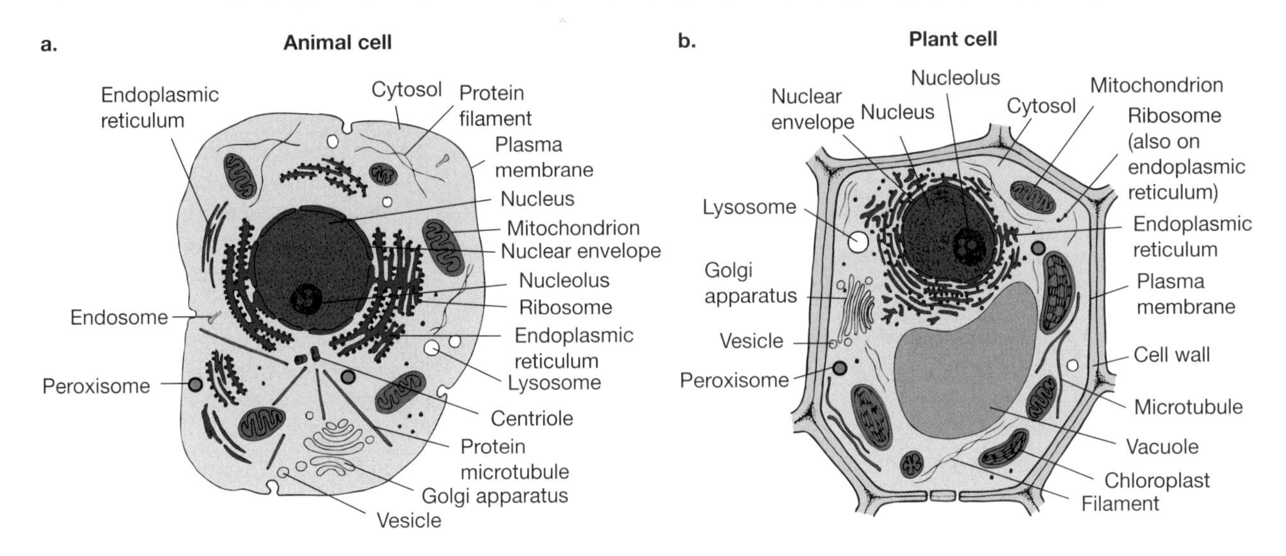

TABLE 1.2 Summary of the structure and function of various organelles

General role	Organelle	Structure	Function
Storage and transcription of genetic information	**Nucleus**	• Membrane surrounded by the nuclear envelope (double membrane), perforated by nuclear pores • The nuclear envelope is continuous with the endoplasmic reticulum.	• Houses chromosomes, made of chromatin (DNA, the genetic material and proteins) • Contains the nucleolus, where ribosomal sub-units are synthesised and assembled • Nuclear pores regulate entry and exit of materials including proteins and RNA.
	Ribosomes	• Consist of two major components: the small ribosomal sub-units and the large sub-units. • Ribosomes can be located within the cytosol or bound to the endoplasmic reticulum.	• Protein synthesis • The small ribosomal subunits read the mRNA, and the large subunits join amino acids to form a polypeptide chain.
Endomembrane system regulates the synthesis and transport of specific proteins	**Endoplasmic reticulum**	• Part of an interconnected network of flattened, membrane-enclosed sacs or tube-like structures • The membranes of the endoplasmic reticulum are continuous with the outer nuclear membrane.	• Smooth endoplasmic reticulum is involved in the synthesis of lipids, metabolism of carbohydrates, calcium storage, and detoxication of drugs and poisons. • Rough endoplasmic reticulum is involved in the synthesis of specific proteins from bound ribosomes.
	Golgi apparatus	Consists of a collection of fused, flattened sacs, enclosed in a single membrane	Modifies, sorts, tags, packages and distributes proteins to be secreted via vesicles.

(continued)

TABLE 1.2 Summary of the structure and function of various organelles *(continued)*

General role	Organelle	Structure	Function
Digestion, breakdown and storage	**Lysosomes**	Membrane-enclosed sacs of hydrolytic enzymes (found only in animal cells)	Breaks down ingested substances, cell macromolecules and damaged organelles for recycling.
	Vacuoles	Large membrane-bound vesicles (found only in plant cells)	Digestion, storage, waste disposal, water balance, cell growth and protection
	Peroxisomes	Metabolic compartment bound by a single membrane	An enzyme that transfers hydrogen to water, producing hydrogen peroxide (H_2O_2) as a by-product, which is converted to water by other enzymes in the peroxisomes
Conversion of inorganic to organic compounds	**Mitochondria**	Bound by a double membrane; the inner membrane has a series of folds (cristae) containing enzymes for ATP synthesis.	The site of cellular respiration, where ATP synthesis occurs for the cell
	Chloroplast	Double membrane around fluid stroma, which contains membranous thylakoids stacks (sacs) in the grana (found only in plant cells)	Conducts photosynthesis, a process by which inorganic compounds are converted to chemical energy, resulting in the production of oxygen and energy-rich organic compounds (simple and complex sugars).
Controlling the entry and exit of substances	**Plasma or cell membrane**	Consists of a phospholipid bilayer with transport and receptor proteins, enclosed by a single membrane	The structural boundary that controls the entry of raw materials into the cell, such as amino acids, the building blocks of proteins

elog-0003

INVESTIGATION 1.1

online only

Viewing and staining cells

Aim

To describe the microscopic structure of a variety of cells and show the effect of staining on distinguishing organelles

on Resources

eWorkbook Worksheet 1.2 Labelling organelles (ewbk-1964)

Weblink Cells resources

1.2.4 The plasma membrane

The cells of all living organisms have boundaries that separate their internal environments from their surroundings. From single-celled organisms, such as amoebae or bacteria, to multicellular organisms, such as mushrooms, palm trees and human beings, each of their living cells has an active boundary called the **plasma membrane**, also known as the cell membrane.

> The plasma membrane is the active boundary around all living cells, consisting of a phospholipid bilayer and associated proteins, that separates the cell contents from their external environment.

The function of the membrane

The plasma membrane boundary can be thought of as a busy gatekeeper selectively controlling the entry and exit of materials into and out of cells. As such, the plasma membrane is said to be **semi-permeable** or **selectively permeable**, meaning that it allows only some substances to cross it — in or out. It can exclude some substances from entering the cell while permitting entry of other substances and allowing for the elimination of certain substances. Without such a boundary, life could not exist, and indeed could not have evolved.

> The plasma membrane carries out several important functions for a cell. The plasma membrane:
> - is an active and selective boundary
> - denotes cell identity (which is vital in the immune response)
> - receives external signals
> - transports materials.

The plasma membrane forms the active boundary of a cell, separating the cell from its external environment and other cells. The plasma membrane forms the boundary of a compartment in which the cytosol — the internal environment of a living cell — can be held within a narrow range of conditions that are different from those of the external environment.

Within the cell, similar membranes form the active boundaries of cell organelles, including the nucleus, the **endoplasmic reticulum**, the **Golgi apparatus** and **lysosomes**. In other cell organelles, such as mitochondria and chloroplasts, membranes form both the external boundary and part of the internal structure. Because of the presence of their membrane boundaries, membrane-bound cell organelles can maintain internal environments that differ from those in the surrounding cytosol and so can perform different functions.

Transporting materials

As mentioned, the membrane is vital to the transportation of materials, acting as a semi-permeable barrier. Various factors affect a substance's ability to cross a membrane, including:

- molecular size
- charge (positive or negative)
- solubility in aqueous solution (**hydrophobic**/nonpolar, or **hydrophilic**/polar)
- concentration gradient.

> Hydrophilic (water-loving) molecules dissolve readily in water.
>
> Hydrophobic (water-fearing) molecules are usually lipophilic (lipid-loving) and dissolve readily in organic solvents such as benzene.

plasma membrane partially permeable boundary of a cell controlling entry to and exit of substances from a cell

semi-permeable allows only certain molecules to cross by diffusion

selectively permeable another term for semi-permeable, where only particular molecules can pass through

endoplasmic reticulum cell organelle consisting of a system of membrane-bound channels that transport substances within the cell

Golgi apparatus organelle that packages material into vesicles for export from a cell (also known as Golgi complex or Golgi body)

lysosomes membrane-bound vesicles containing digestive enzymes

hydrophobic substances that tend to be insoluble in water; also termed nonpolar

hydrophilic substances that dissolves easily in water; also termed polar

FIGURE 1.8 Diagram showing the semi-permeable nature of a phospholipid bilayer membrane

Gases	Small uncharged polar molecules	Nonpolar molecules	Large uncharged polar molecules	Ions	Charged polar molecules
CO_2	Ethanol	Oestrogen	Glucose	K^+	Amino acids
N_2	H_2O	Benzene	Sucrose	Mg^{2+}	ATP
O_2				Ca^{2+}	

Substances can cross a membrane by several different methods. These can be passive (not requiring energy) or active (requiring energy).

Passive methods include the following:

- **Simple diffusion** is the means of transport of small lipophilic substances. Water can also move across the plasma membrane by diffusion; this is a special case of diffusion known as **osmosis**.
- **Facilitated diffusion** involves protein transporters and is the means of transport of dissolved hydrophilic substances down their concentration gradients.

Active methods include the following:

- **Active transport** involves protein transporters known as pumps and is the means of transport of dissolved hydrophilic substances against their concentration gradients.
- Bulk transport of macromolecules and fluid includes:
  - **endocytosis** (movement into the cell)
  - **exocytosis** (movement out of the cell).

These are summarised in table 1.3.

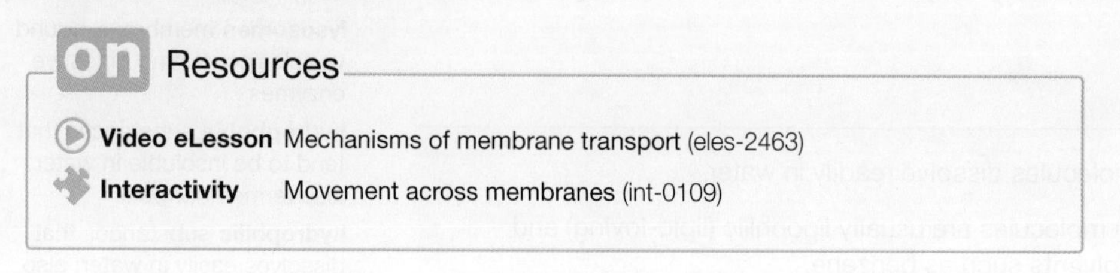

simple diffusion the movement of substances from a region of higher concentration to one of lower concentration of that substance; that is, *down* its concentration gradient

osmosis a specialised process of passive transport in which water molecules move across a partially permeable membrane from an area of high water (low solute) to an area of low water (high solute)

facilitated diffusion form of diffusion involving a specific carrier molecule for the substance

active transport net movement of dissolved substances across a cell membrane by an energy-requiring process that moves substances against a concentration gradient from a region of lower to higher concentration

endocytosis an energy-requiring process of bulk transport, in which solids or liquids move into the cell by engulfment

exocytosis an energy-requiring process of bulk transport, in which solids or liquids move out of the cell via vesicles

TABLE 1.3 Summary of modes of transport

	Simple diffusion	Osmosis	Facilitated diffusion	Active transport	Bulk transport
Direction of movement	High to low concentration of solute	High to low concentration of water (high water = low dissolved solute)	High to low concentration of solute	Low to high concentration of solute	High to low concentration of solute
Is energy required?	No	No	No	Yes	Yes
Extra requirements	None	None	Protein transporter	Protein pumps	Vesicles
Use	Nonpolar molecules and small polar molecules	Water	Large polar and charged molecules	Movement of various molecules against the concentration gradient	Bulk movement of various molecules, such as the transport of synthesised proteins from a cell via vesicles

The structure of the membrane

Small, but vitally important, the plasma membrane is just 8 nanometres (nm) wide and so is only visible using a transmission electron microscope (TEM). A TEM image of the plasma membrane (as seen in figure 1.9) has a 'train track' appearance with two dark lines separated by a more lightly stained region. These images were important clues in elucidating the structure of the plasma membrane.

The plasma membrane has the following major components, which can be observed in figure 1.10:

- **phospholipids**. Various kinds of phospholipids are the main structural components of the plasma membrane. They are organised as two layers (leaflets).
- **proteins**. Some proteins are embedded in the plasma membrane; others are attached at the membrane surfaces.
- **carbohydrate groups**. These are attached to some lipids, forming glycolipids, and to some proteins, forming glycoproteins. Both of these occur at the membrane surfaces.

FIGURE 1.9 Coloured transmission electron micrograph (TEM) showing the plasma membrane (red) of an intestinal brush border. In the inset, the structure of the membrane is evident.

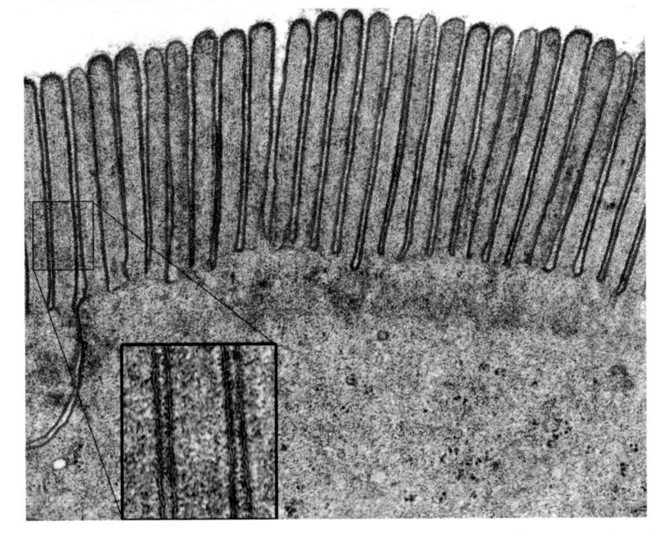

phospholipids major type of lipid found in plasma membranes and the main structural component of plasma membranes

proteins macromolecules built of amino acid sub-units and linked by peptide bonds to form a chain, sometimes termed a polypeptide

carbohydrate groups molecules that are associated with the plasma membrane and are associated with cell to cell communication and signalling

The fluid mosaic model

The **fluid mosaic model** describes the structure of the plasma membrane. This model also applies to the membranes that form the outer boundary of cell organelles, such as the membranes that surround the cell nucleus and other cell organelles.

The fluid mosaic model proposes that the plasma membrane and other intracellular membranes should be considered as fluid layers in which proteins are embedded.

The term 'fluid' comes from the fact that the fatty chains of the phospholipids are like a thick oily fluid, and the term 'mosaic' comes from the fact that the external surface (when viewed from above) has the appearance of a mosaic because of the various embedded proteins set in a uniform background.

FIGURE 1.10 Diagram showing the fluid mosaic model of membrane structure

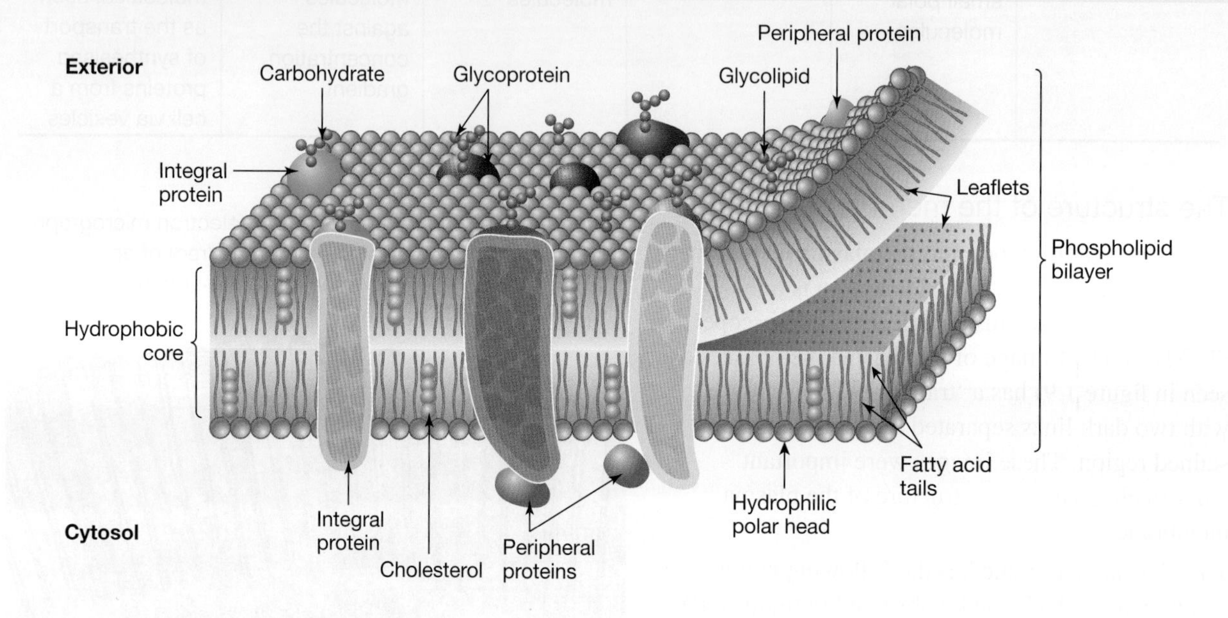

Phospholipids

The plasma membrane consists of a double layer (bilayer) of phospholipids. Each phospholipid molecule consists of two fatty acid chains joined to a phosphate-containing group. The phosphate-containing group forms the water-loving (hydrophilic or polar) head of the molecule. The fatty acid chains constitute the water-fearing (hydrophobic or nonpolar) tail of each phospholipid molecule. As seen in figures 1.10 and 1.11, the two layers of phospholipids in a plasma membrane are arranged so that the hydrophilic polar heads are exposed to both the external environment of the cell and the cytosol (the internal environment of the cell). In contrast, the two layers of hydrophobic nonpolar tails face each other in the central region of the plasma membrane.

fluid mosaic model a model which proposes that the plasma membrane and other intracellular membranes should be considered as two-dimensional fluids in which proteins are embedded

FIGURE 1.11 a. Chemical structure of a phospholipid b. Diagram showing part of the bilayer of phospholipid molecules in the plasma membrane

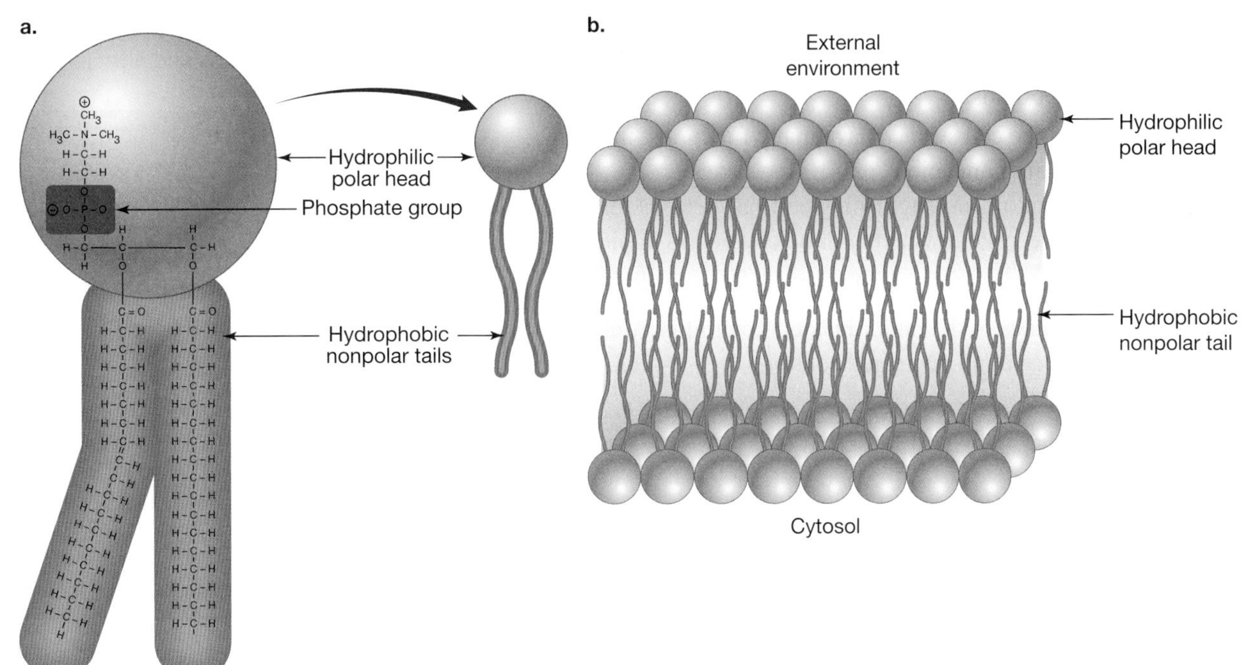

Proteins

Proteins form the second essential part of the structure of the plasma membrane. Many different kinds of protein make up the plasma membrane. They can be broadly grouped into:

- integral proteins
- peripheral proteins.

Integral proteins, as their name implies, are fundamental components of the plasma membrane. These proteins are embedded in the phospholipid bilayer. Typically, they span the width of the plasma membrane, with part of the protein being exposed on both sides of the membrane; these proteins are described as being trans-membrane.

Trans-membrane proteins serve many functions, including as transporters, receptors, channels and carriers. Integral proteins can be separated from the plasma membrane only by harsh treatments that disrupt the phospholipid bilayer, such as treatment with strong detergents.

Figure 1.12 shows a ribbon model of a trans-membrane protein that functions as an acid-sensing (H^+) ion channel. The red and blue lines are not part of the protein but indicate the extracellular (red) and the cytoplasmic (blue) sides of the plasma membrane. Most of the protein extends into the extracellular space, some is embedded in the plasma membrane, and a small part lies within the cell.

Peripheral proteins are either anchored to the exterior of the plasma membrane through bonding with lipids, or indirectly associated with the plasma membrane through interactions with integral proteins in the membrane.

FIGURE 1.12 Ribbon model of a trans-membrane protein

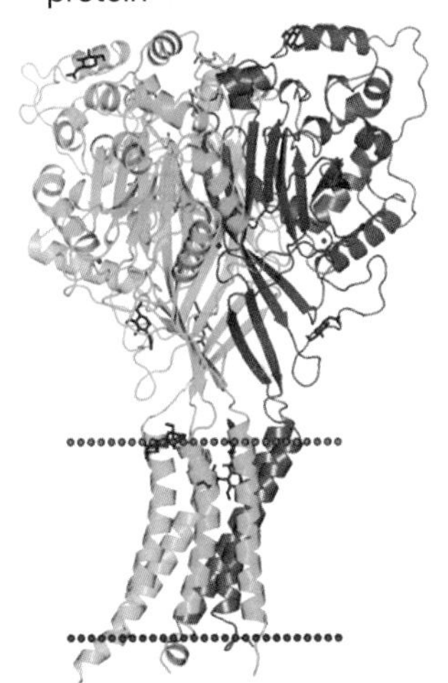

integral proteins proteins that are embedded in the phospholipid bilayer

trans-membrane proteins proteins that are embedded within and span the plasma membrane, allowing them to have parts exposed to both the intracellular and extracellular environment

peripheral proteins proteins that are anchored to the exterior of the plasma membrane through bonding with either lipids or integral proteins

Carbohydrates

In many cases, carbohydrate groups, such as sugars, are attached to the exposed parts of proteins on the outer side of the membrane, creating combinations called glycoproteins (see figure 1.13). Some carbohydrates instead covalently link directly to the lipids in the membrane; these are referred to as glycolipids.

Carbohydrates on the cell surface have many functions, including:

- cell-to-cell communication
- acting as receptors, distinguishing cells as 'self' (a feature that is vital in the immune system, which will be covered in Topic 5).

Glycoproteins in particular are of vital importance in immune recognition and include molecules of the major histocompatibility complex, which is found on the surface of all nucleated cells.

> The prefix 'glyco' means sugar.
>
> Sugars attached to a protein = glycoprotein
>
> Sugars attached to a lipid = glycolipid

FIGURE 1.13 A computer generated image of the plasma membrane. Glycoproteins and glycolipids are shown in red.

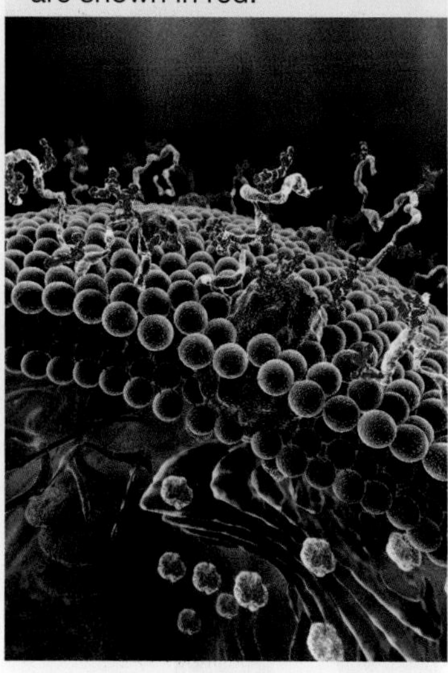

FIGURE 1.14 Diagram showing two integral proteins embedded in and spanning the plasma membrane. The left image shows associated carbohydrates.

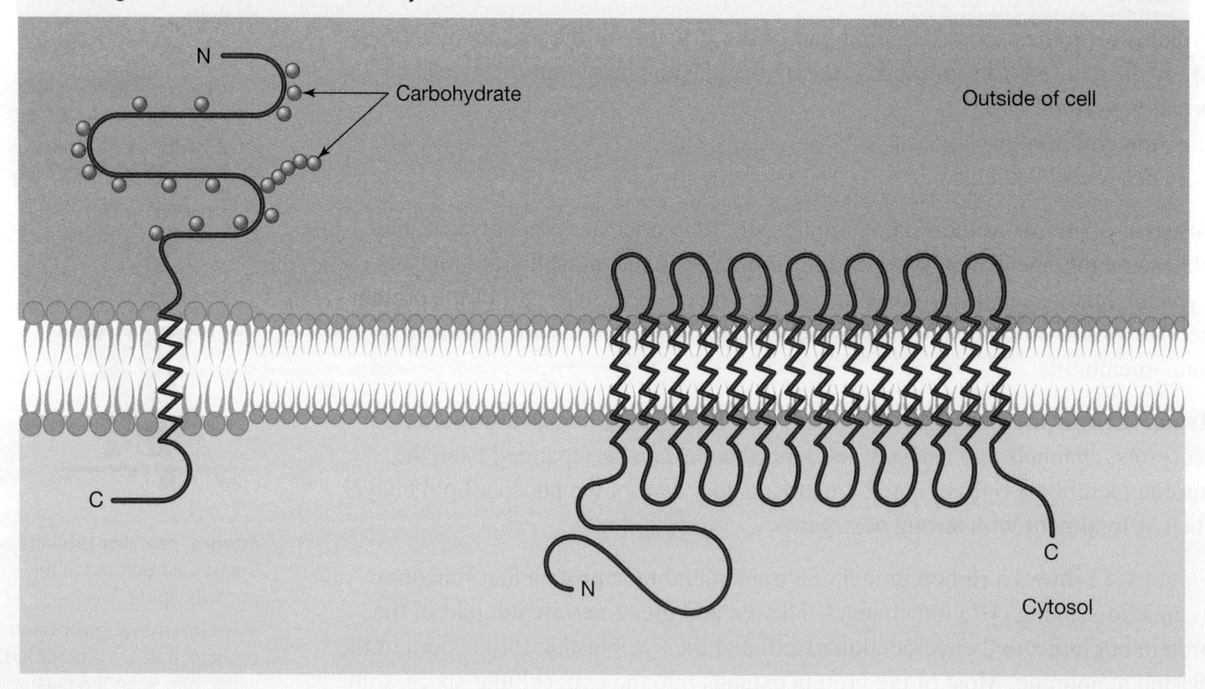

Cholesterol

In animal cells only, cholesterol is an essential component of the plasma membrane, acting in a fluid manner similar to an iceberg. It makes up about 20 percent of the membrane by mass. Cholesterol molecules are inserted alongside phospholipid molecules in both leaflets of the membrane.

Cholesterol acts on the plasma membrane in several ways:

- At low temperatures, cholesterol molecules maintain the fluidity of the membrane by keeping phospholipid molecules separated and preventing the membrane from become too stiff.
- At high temperatures, cholesterol stabilises the membrane by raising its melting point and preventing it from becoming excessively fluid.

elog-0005

INVESTIGATION 1.2

Membrane transport across a semi-permeable membrane

Aim

To observe the semi-permeability of an artificial membrane and relate this to plasma membranes

Resources

eWorkbook Worksheet 1.3 Structure of the membrane and membrane transport (ewbk-1966)

KEY IDEAS

- Cells are the major structural unit of life. They can vary in size and shape.
- Prokaryotic cells do not contain membrane-bound organelles (such as nuclei). Eukaryotic cells contain membrane-bound organelles, with DNA situated within the nucleus.
- Organelles are compartments that carry out specific functions.
- The major structural component of the plasma membrane is a bilayer of phospholipid molecules, each with a hydrophilic head and hydrophobic tail.
- A major role of the plasma membrane of a cell is to act as a gatekeeper that controls the entry and exit of materials into and out of the cell.
- Passive methods of membrane transport include facilitated diffusion, simple diffusion and osmosis. Active methods include active transport, exocytosis and endocytosis.

1.2 Activities

learnon

To answer questions online and to receive **immediate feedback** and **sample responses** for every question, go to your learnON title at **www.jacplus.com.au**. A **downloadable solutions** file is also available in the resources tab.

1.2 Quick quiz 	**1.2 Exercise**

1.2 Exercise

1. Compare and contrast eukaryotic and prokaryotic cells.
2. Outline the role of the two main organelles in the endomembrane system that regulate and transport specific proteins.
3. Describe the function(s) of the plasma membrane in a eukaryotic cell.
4. The plasma membrane is often described as a 'fluid mosaic'. Why?
5. Describe the importance of the polar head and nonpolar tails in the plasma membrane.
6. Explain the importance of cholesterol as a component of animal cell membranes.
7. Specifically identify the substances that can pass easily through a plasma membrane (including charge) and the substances that cannot pass through a plasma membrane (including charge).

1.3 Nucleic acids as information molecules

KEY KNOWLEDGE

- Nucleic acids as information molecules that encode instructions for the synthesis of proteins: the structure of DNA, the three main forms of RNA (mRNA, rRNA and tRNA) and a comparison of their respective nucleotides

Source: VCE Biology Study Design (2022–2026) extracts © VCAA; reproduced by permission.

1.3.1 What are nucleic acids?

Nucleic acids are biomolecules that are vital for the continuity of life. They are found in all organisms and provide a genetic blueprint that provides the information for protein synthesis. Nucleic acids are made up of sub-units known as **nucleotides**.

There are two kinds of nucleic acid:

- **deoxyribonucleic acid (DNA)**, which is located in chromosomes in the nucleus of eukaryotic cells. It is the genetic material that contains hereditary information and is transmitted from generation to generation.
- **ribonucleic acid (RNA)**, which is formed against a template strand of DNA.

nucleotides basic building blocks or sub-units of DNA and RNA consisting of a phosphate group, a base and a five-carbon sugar

deoxyribonucleic acid (DNA) nucleic acid consisting of nucleotide sub-units that contain the sugar deoxyribose and the bases A, C, G and T; DNA forms the major component of chromosomes

ribonucleic acid (RNA) nucleic acid consisting of a single chain of nucleotide sub-units that contain the sugar ribose and the bases A, U, C and G; RNA

All nucleic acids are polymers made up of sub-units (or monomers) known as nucleotides. Each nucleotide has:

- a 5-carbon (pentose) sugar
- a phosphate
- a nitrogenous base.

FIGURE 1.15 A nucleotide with a sugar, a phosphate and a nitrogen-containing base

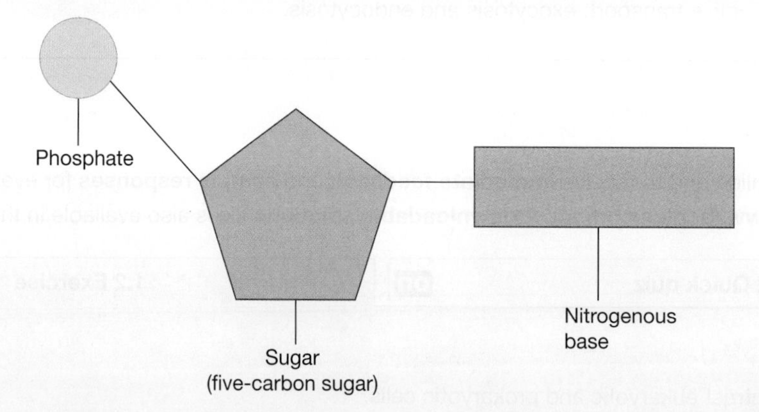

Although DNA and RNA are both made of nucleotides, there are some distinctions between the two, outlined below in table 1.4.

TABLE 1.4 Comparison of DNA and RNA

	DNA	RNA
Type of sugar	Deoxyribose	Ribose
Nitrogen-containing bases	Adenine, cytosine, guanine and thymine	Adenine, cytosine, guanine and uracil
Number of strands	Two	One

TIP: Be careful when spelling biological terminology in your responses. If the spelling means that your term will be confused with something else, it will not be accepted. For example, when spelling cytosine, cytosin or cytocine would be accepted, but cystine would not be accepted (as it is too close to the amino acid cysteine). In some cases, the symbol for the nucleotide may be appropriate (A, C, G, T, U).

1.3.2 The structure of DNA

Our DNA is what makes every single person unique. DNA is vital to code for all the proteins in our bodies — from the melanin that determines our skin colour, to enzymes such as lactase, which allows us to break down lactose. Differences in our genetic code lead to the production of different proteins that allow us to have different traits.

Each DNA molecule is made of two **complementary** chains of nucleotides that run anti-parallel (in opposite directions). Each of the chains of nucleotides is made up of a sugar-phosphate backbone (bonded through phosphodiester bonds). The terms 3′ and 5′ are very important in understanding the direction of the each chain of the nucleic acid. The 5′ end is the phosphate end, which is attached to the 5′ carbon of the sugar. The 3′ end is the hydroxyl end of the sugar, which is associated with the 3′ carbon. One strand runs 3′ to 5′ and the opposite strand runs 5′ to 3′.

FIGURE 1.16 Part of a DNA double helix revealed through scanning tunnelling microscopy

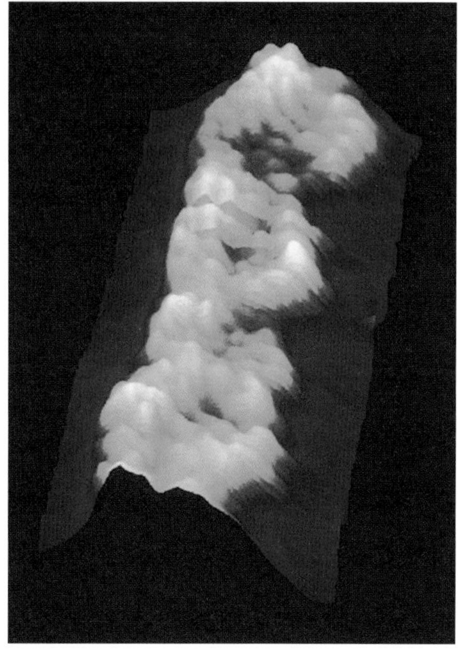

on Resources

Video eLesson DNA structure (eles-4211)

FIGURE 1.17 DNA made up of monomers of nucleotides

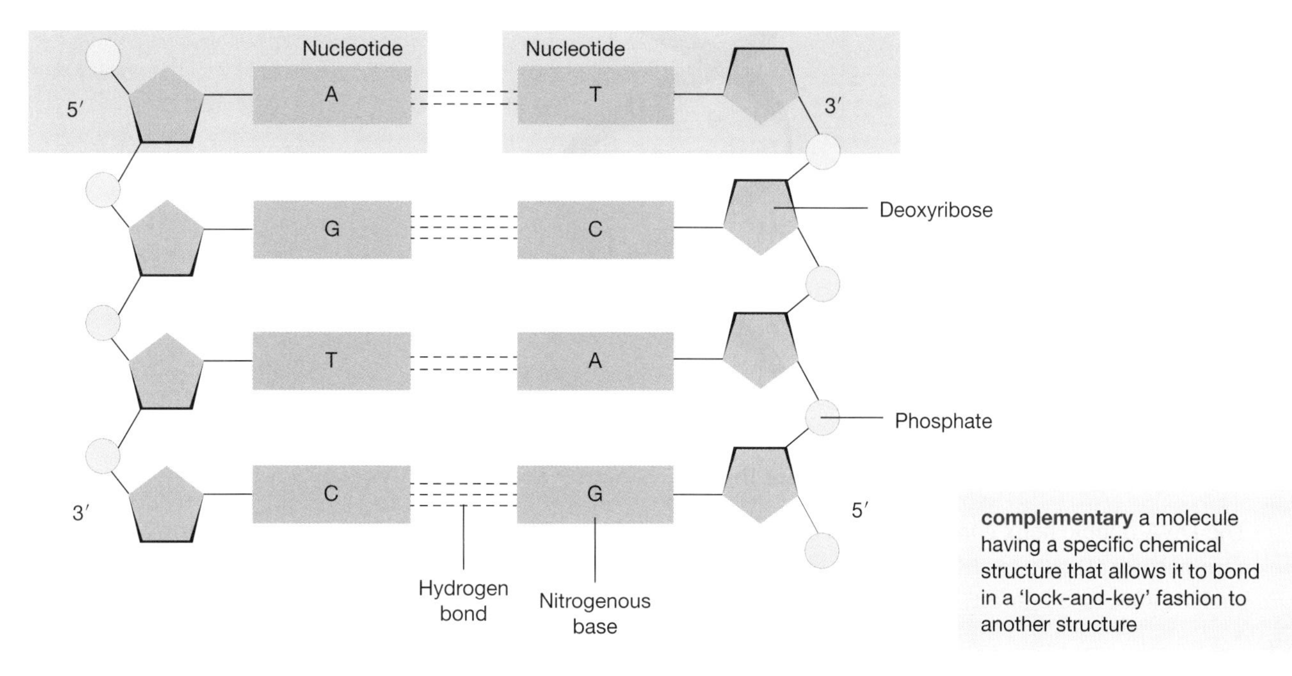

complementary a molecule having a specific chemical structure that allows it to bond in a 'lock-and-key' fashion to another structure

It can be observed that the sugar (deoxyribose) and phosphate parts are the same in each nucleotide. However, there are four different kinds of nucleotides because four different kinds of nitrogen-containing bases are involved: adenine (A), thymine (T), cytosine (C) and guanine (G). Hydrogen bonds form between complementary nitrogenous bases on opposite strands. The two chains form a double-helical molecule of DNA.

The base pairs between the two strands, namely A with T and C with G, are said to be complementary pairs. A and T bond with 2 hydrogen bonds, and C and G bond with 3 hydrogen bonds (as seen in figure 1.18).

FIGURE 1.18 The base pairing rules in DNA

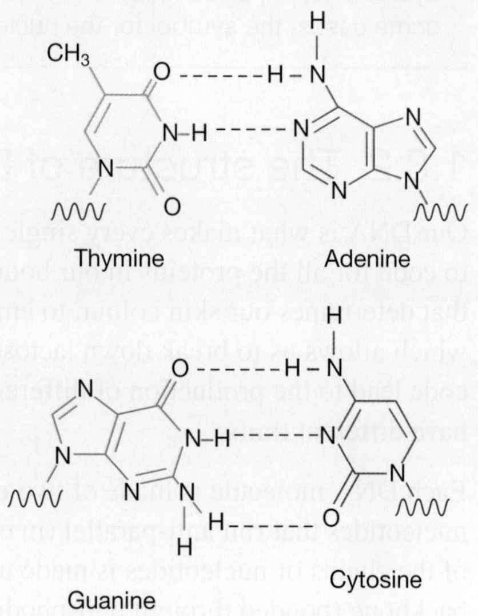

Thymine
Adenine
Guanine
Cytosine

Base pairing rules in DNA

A pairs with T (**a**rrow in the **t**arget).

C pairs with G (**c**ar in the **g**arage).

int-0133

FIGURE 1.19 The double helix structure of DNA. The two chains are held together by hydrogen bonds between complementary bases.

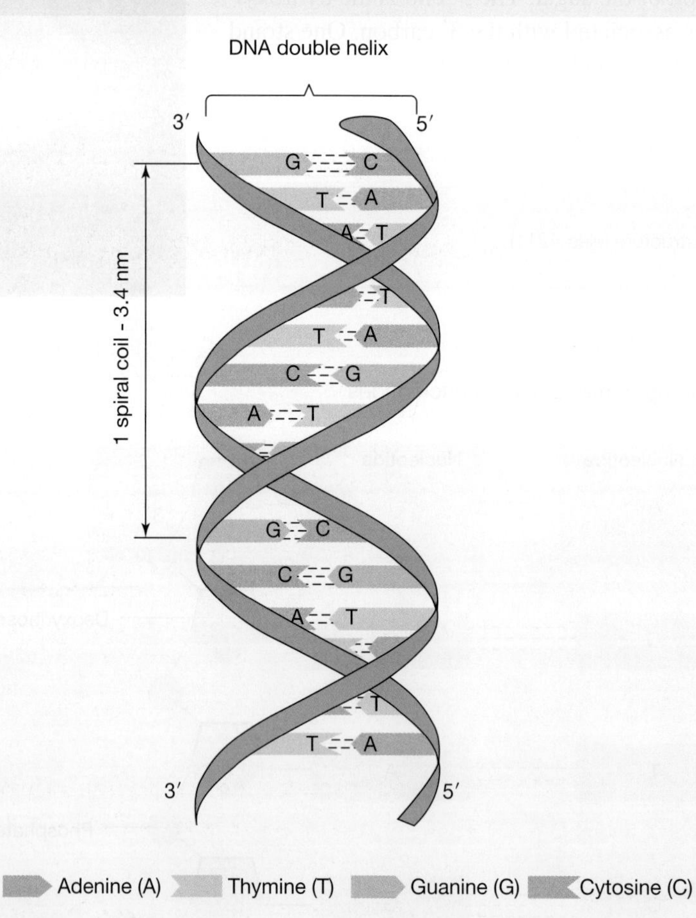

The DNA double helix combines with certain proteins, in particular histones, as it condenses to form a chromosome (see figure 1.20a). As the DNA winds around clusters of histone proteins, it forms structures called nucleosomes (see figure 1.20b).

FIGURE 1.20 a. Diagram showing the coiling and supercoiling of one molecule of a DNA double helix to form a eukaryotic chromosome b. A model of a nucleosome showing the DNA double helix (grey) coiling around a cluster of histone proteins (shown in colours) (Image b. courtesy of Dr Song Tan, Pennsylvania State University)

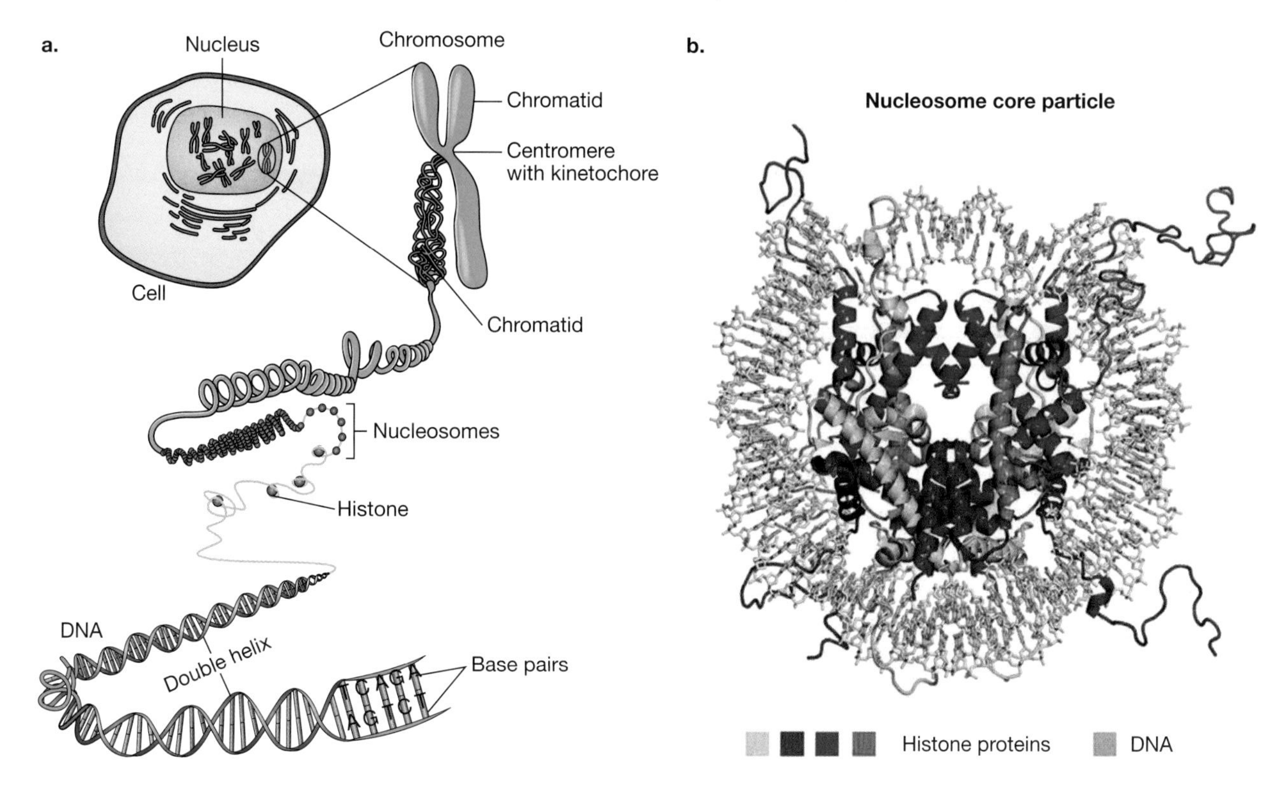

The total length of the DNA double helix molecule in an 'average' human chromosome is about five centimetres. By coiling and supercoiling, this long DNA molecule becomes compressed into a microscopic chromosome. In one cell alone, the length of all the DNA is around two metres. In the entire human body, the length of all DNA is trillions of metres long.

elog-0007

INVESTIGATION 1.3

online only

Extraction of DNA

Aim

To extract and observe DNA from the nucleus of a cell

Representing DNA

There are many ways we can represent a DNA strand. Part of a single chain of DNA could be shown as follows:

… -nucleotide-nucleotide-nucleotide-nucleotide-nucleotide- …

OR it could be shown as:

```
... – P – sugar – P – sugar – P – sugar – P – sugar – P – sugar – ...
           |             |             |             |             |
          base          base          base          base          base
```

OR the specific bases in the nucleotides in one chain could be shown as:

5′ ... A T T A G C T T G A G G C G ... 3′

DNA is not always represented in diagrams as a double helix. Figure 1.21 shows some of the many ways of representing DNA. The representation used will depend on the purpose of the diagram. Each provides different information about DNA.

FIGURE 1.21 Different ways of representing DNA

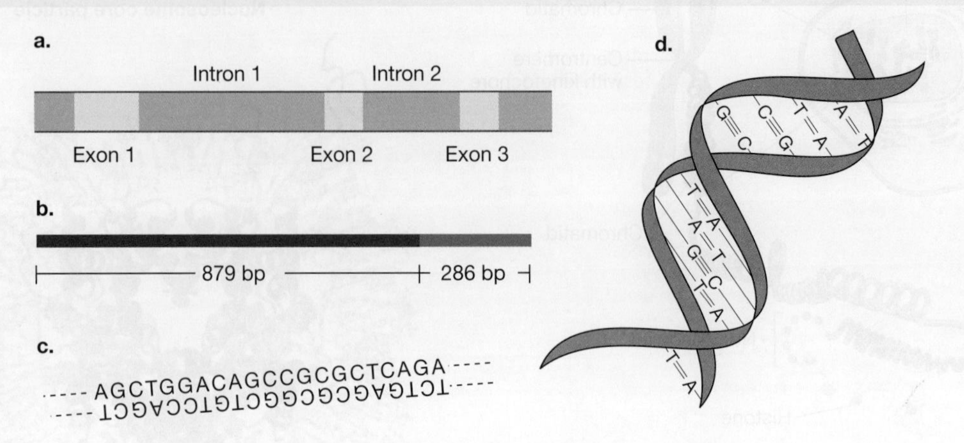

EXTENSION: Mitochondrial DNA

When we consider DNA, our minds often jump straight to linear nuclear DNA. However, DNA is also found within our mitochondria. This type of DNA is circular and is referred to as mitochondrial DNA (mtDNA). Like nuclear DNA, it is made up of nucleotides joined with phosphodiester bonds. mtDNA contains around 40 genes that code for proteins involved in cellular respiration.

Mitochondrial DNA is an extension concept for this subtopic but will be explored further in Unit 4 in Topic 10.

To access more information on this extension concept, please download the digital document.

FIGURE 1.22 Mitochondrial DNA within the mitochondrion

on Resources

Digital document Extension: Mitochondrial DNA (doc-35832)

1.3.3 Forms of RNA

Ribonucleic acid (RNA) is also a polymer of nucleotides (see figure 1.23). It differs from DNA in that it:

- is an unpaired chain of nucleotides
- contains the sugar ribose
- contains uracil rather than thymine.

FIGURE 1.23 The four nucleotide sub-units, uracil, adenine, guanine and cytosine, from which the three forms of RNA are constructed

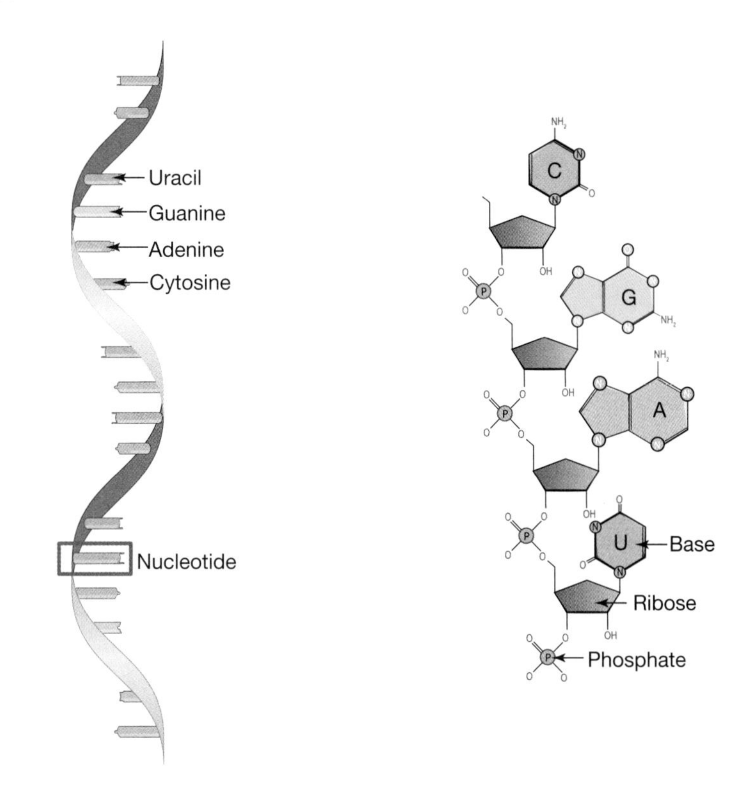

RNA has three main forms with different functions and structures.

The three different forms of RNA are all produced in the nucleus using DNA as a template:

- **messenger RNA (mRNA)**, which carries the genetic message from the DNA within the nucleus to the **ribosomes**, where the message is translated into a particular protein. Each group of three nucleotides in mRNA (known as a codon) provides the information for the addition of one amino acid. A special form of mRNA known as pre-mRNA is made through transcription in the nucleus.
- **ribosomal RNA (rRNA)**, which, together with particular proteins, makes the ribosomes found in cytosol
- **transfer RNA (tRNA)**, molecules that carry amino acids to ribosomes that are free in the cytoplasm, where they are used to construct proteins. An **anticodon** (a set of three nucleotides) binds to the complementary codon on mRNA.

In each of the forms of RNA, the strand of nucleotides is folded in a different way (see figure 1.24).

messenger RNA (mRNA) form of RNA synthesised by the transcription of a DNA template strand in the nucleus; mRNA carries a copy of the genetic information into the cytoplasm

ribosomes organelles that are major sites of protein production in cells in both eukaryotes and prokaryotes

ribosomal RNA (rRNA) stable form of RNA found in ribosomes

transfer RNA (tRNA) form of RNA that can attach to specific amino acids and carry them to a ribosome during translation

anticodon sequence of three bases in a transfer RNA molecule that can pair with the complementary codon of a messenger RNA molecule

FIGURE 1.24 The different forms of RNA

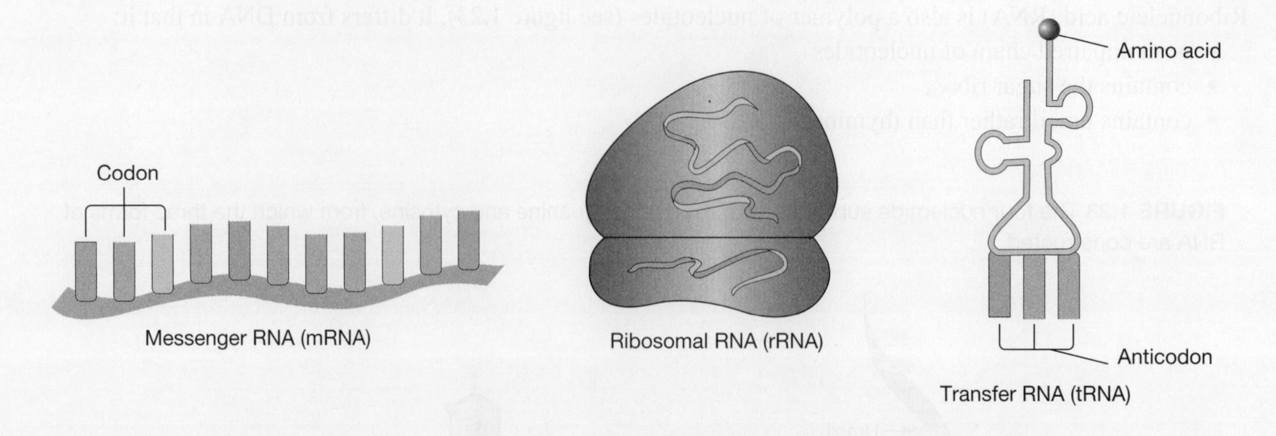

In figure 1.25, ribosomes (blue) attach to the mRNA strand (pink). A tRNA molecule carrying a corresponding amino acid binds to the ribosome. As the ribosome moves onto the next base along the mRNA, a protein (green) grows from the ribosome.

FIGURE 1.25 A transmission electron micrograph (TEM) of a fragment of an mRNA translation unit from the salivary gland cell of a midge (*Chironomus* sp.)

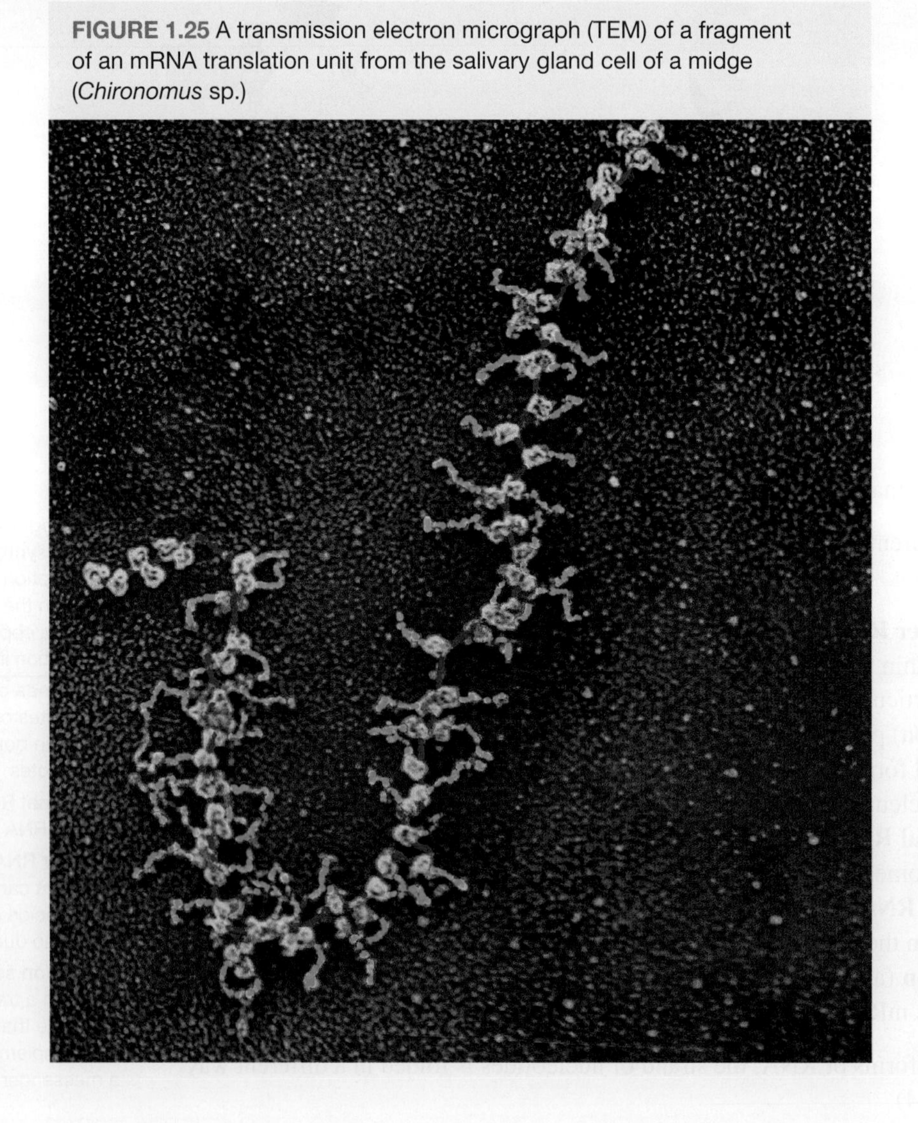

tlvd-1090

SAMPLE PROBLEM 1 Comparing DNA and RNA

Compare and contrast the monomers of DNA and RNA. **(2 marks)**

THINK	WRITE
1. In compare and contrast questions, both similarities and differences must be addressed.	
2. Note down similarities.	DNA and RNA are both made up of nucleotide monomers, which contain phosphate, a sugar and a nitrogenous base (1 mark).
3. Note down differences. **TIP** As this question is asking about monomers, stating one is double-stranded and one is single-stranded is incorrect.	DNA contains thymine as a base instead of uracil, which is in RNA. DNA contains the deoxyribose sugar, whereas RNA contains the ribose sugar (1 mark).

elog-0009

INVESTIGATION 1.4

online only

Building a model of DNA and RNA

Aim

To construct a model to examine the structure of DNA and RNA and understand differences between these two nucleic acids

Resources

eWorkbook Worksheet 1.4 Comparing DNA and RNA (ewbk-1968)

Interactivity RNA structure (int-0111)

Weblink DNA and genes resources

KEY IDEAS

- Proteins, polysaccharides and nucleic acids are polymeric organic molecules built out of a very large number of repeating sub-units.
- The nucleic acids, double-helical DNA and single-stranded RNA, are built out of a very large number of repeating sub-units called nucleotides.
- Each nucleotide consists of a sugar, a phosphate and a nitrogen-containing base, with the sugar in DNA being deoxyribose and that in RNA being ribose.
- Each DNA molecule consists of two chains of nucleotides that are complementary to each other and held together by hydrogen bonds.
- In DNA the nitrogenous bases are adenine, cytosine, guanine and thymine. In RNA, thymine is replaced with uracil.
- Each RNA molecule consists of a single strand of nucleotides.
- There are three main types of RNA. Messenger RNA (mRNA) carries the genetic material in DNA from the nucleus to ribosomes. Ribosomal RNA (rRNA) makes up ribosomes. Transfer RNA (tRNA) carries amino acids to ribosomes.

1.3 Activities

learnON

To answer questions online and to receive **immediate feedback** and **sample responses** for every question, go to your learnON title at **www.jacplus.com.au**. A **downloadable solutions** file is also available in the resources tab.

1.3 Quick quiz on	1.3 Exercise	1.3 Exam questions

1.3 Exercise

1. MC Nucleic acids come in two main forms — deoxyribonucleic acid (DNA) and ribonucleic acid (RNA). Which of the following is a similarity between polymers of DNA and RNA?
   A. They have the same number of oxygen atoms.
   B. They contain phosphodiester bonds.
   C. They contain an identical sugar.
   D. They contain the nitrogenous base thymine.
2. **a.** Draw and label a double stranded of DNA with four nucleotides.
   b. DNA is a double helix. How does a double helix form? Identify the bonds that hold the nucleotides in position.
3. Within a DNA strand, which would be harder to separate into two strands: DNA composed predominantly of AT base pairs, or DNA composed predominantly of GC base pairs? Why?
4. How is pre-mRNA made? What is the name of the process? Where does this happen?
5. Describe the roles messenger RNA (mRNA) and transfer RNA (tRNA) perform in protein synthesis.
6. If a double-stranded DNA molecule contains 13.5% cytosine, what would be the percentages of the other three nitrogenous bases?
7. **a.** A strand of DNA has the sequence 3′ ATGCCGGATA 5′. What would be the sequence of the other strand?
   b. What is meant by 3′ and 5′ in regards to nucleic acids?

1.3 Exam questions

Use the following information to answer Questions 1 and 2.

The diagram below represents part of a DNA molecule.

Question 1 (1 mark)

***Source:** VCAA 2015 Biology Exam, Section A, Q3*

MC A single DNA nucleotide is shown by sub-unit(s)

A. X alone.
B. X and Y together.
C. Y and Z together.
D. X, Y and Z together.

Question 2 (1 mark)

***Source:** Adapted from VCAA 2015 Biology Exam, Section A, Q4*

MC A feature of DNA that can be seen in the diagram is

A. the anti-parallel arrangement of the two strands of nucleotides.
B. the process of semi-conservative replication.
C. its ribose sugar–phosphate backbone.
D. its double-helix structure.

Question 3 (1 mark)

***Source:** Adapted from VCAA 2016 Biology Exam, Section A, Q4*

MC A portion of one strand of a macromolecule has the sequence -CGATTCGGTTAA-.

The complementary strand would be

A. -CGATTCGGTTAA-
B. -AATTGGCTTAGC-
C. -GCTAAGCCAATT-
D. -GCUAAGCCAAUU-

Question 4 (4 marks)

***Source:** Adapted from VCAA 2011 Biology Exam 1, Section B, Q1b*

The following figure represents a portion of a plant cell.

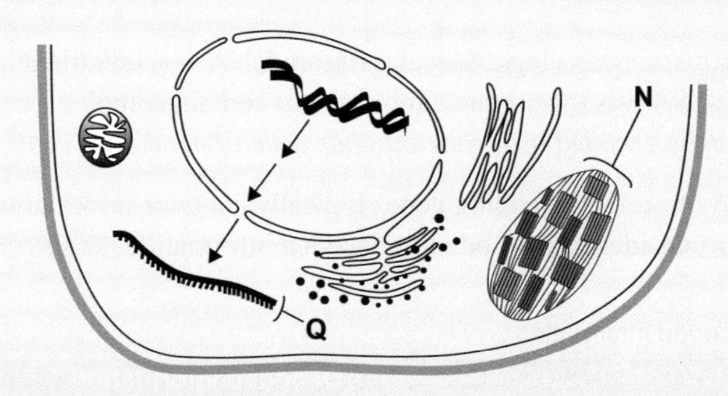

Examine the figure above and answer the following.

a. i. Identify the type of nucleic acid found in Structure N. **1 mark**
 ii. Outline the specific function of this nucleic acid. **1 mark**

b. i. Identify the type of nucleic acid found in Structure Q. **1 mark**
 ii. Outline the specific function of this nucleic acid. **1 mark**

Question 5 (3 marks)

***Source:** Adapted from VCAA 2020 Biology Exam, Section B, Q2a and b*

RNA molecules consist of long strands of nucleotides. Each nucleotide consists of three sub-units.

a. Label the three sub-units on the diagram of the RNA nucleotide below. **1 mark**

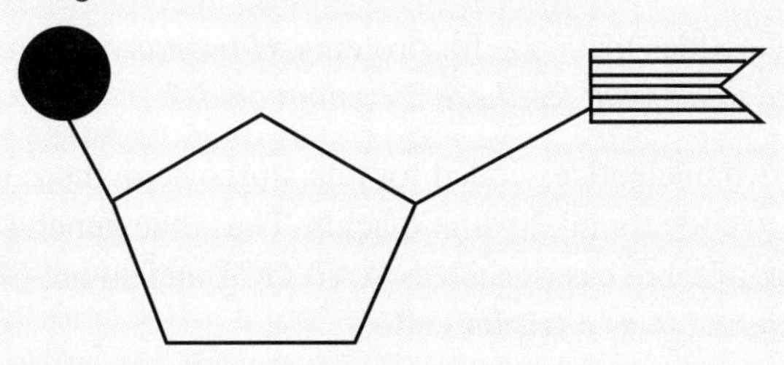

b. Describe the role in a cell of the two types of RNA listed:
 i. tRNA **1 mark**
 ii. mRNA **1 mark**

More exam questions are available in your learnON title.

1.4 The genetic code and protein synthesis

KEY KNOWLEDGE

- The genetic code as a universal triplet code that is degenerate and the steps in gene expression including transcription, RNA processing in eukaryotic cells and translation by ribosomes

DNA is an information molecule that provides the code to produce proteins. How does DNA control functions within cells? All metabolic reactions in cells are controlled by enzymes that, almost without exception, are proteins built of one or more polypeptide chains — that is, chains of amino acids. Hence, the DNA in the nucleus of a eukaryotic cell controls all the metabolic processes in a cell through the polypeptide chains for which the DNA dictates production.

1.4.1 Features of the genetic code

The genetic instructions for all organisms are written in a code that uses an 'alphabet' of four letters only, namely A, T, C and G.

The DNA of genes is an information-carrying molecule. Before DNA was identified as the genetic material, many biologists thought that DNA was too simple a molecule to contain complex genetic instructions. *How can a large amount of information be encoded by a code that has a small number of elements?*

The **genetic code** in the DNA of protein-encoding genes typically contains information for joining amino acids to form polypeptides. We can consider this as coded or decoded information, as shown in table 1.5.

TABLE 1.5 Coded and decoded information

Coded information	Decoded information
Nucleotide sequences in DNA template strand	Order of amino acids in polypeptides

The genetic code as a triplet code

Consider two observations:

1. Genes typically contain coded information for assembling amino acids to form polypeptides.
2. Polypeptides are made of combinations of 20 different amino acid sub-units.

From these observations, it may be inferred that the genetic code must have at least 20 different instructions or pieces of information.

There are only 4 possible nucleotides that code for proteins. *How many different nucleotides would be required to account for these 20 amino acids?*

If a sequence of only one or two nucleotides coded for one amino acid, there would not be enough combinations to code for all 20 amino acids. Thus, one genetic instruction consists of a group of three bases, such as AAT, GCT and so on. Because of this, the genetic code is referred to as a **triplet code**.

genetic code representation of genetic information through a non-overlapping series of groups of three bases (triplets) in a DNA template chain

triplet code the idea that the genetic code consists of triplets or three-base sequences

TABLE 1.6 Comparing the number of instructions coded for by different numbers of nucleotides

Number of nucleotides in one instruction	Total number of different instructions possible
1 (e.g. T)	4
2 (e.g. AA, AT, GA)	16
3 (e.g. TTA, GCC, AAA)	64
4 (e.g. GGGA, TGCA, AATG)	256

This code is non-overlapping. So, a fragment of DNA consisting of 12 bases contains four pieces of information or instructions, with each triplet leading to the addition of a single amino acid.

EXTENSION: Breaking the code

The genetic code was originally unknown and had to be broken. In 1961, the first piece of the genetic code was broken when the three-base triplet 'AAA' in DNA was decoded as 'Add the amino acid *phe* into a protein being constructed'. We will see later in this topic that the translation of each DNA triplet occurs through an intermediate molecule, messenger RNA (mRNA). By 1966, all 64 pieces of the genetic code had been deciphered.

The different amino acids that each triplet codes for are shown in table 1.7. Note that some combinations lead to instructions 'START adding amino acids' and 'STOP adding amino acids'. Refer to table 1.12 in section 1.7.1 to learn more about the different amino acids. Refer to Appendix: Amino acid data to see the structures of the amino acids and their three-letter and single-letter abbreviations.

The genetic code as a universal code

The code is essentially the same in bacteria, in plants and in animals — it is said to be **universal**. The same sequence of nucleotides codes for the same amino acid (for example, CCA codes for proline in plants, animals and bacteria).

universal the property of the genetic code in which the code is essentially the same across all organisms

The information in the DNA template strand also includes a START instruction (TAC) and three STOP instructions (ATT, ATC or ACT).

TABLE 1.7 Triplets of DNA and the corresponding amino acids (and in three cases, the STOP signal). Each amino acid is represented with a different colour. Amino acids of similar colours (blue, pink, purple, orange or green) have similar properties.

FIRST BASE	SECOND BASE							
	T		C		A		G	
T	TTT	*Phe*	TCT	*Ser*	TAT	*Tyr*	TGT	*Cys*
T	TTC	*Phe*	TCC	*Ser*	TAC	*Tyr*	TGC	*Cys*
T	TTA	*Leu*	TCA	*Ser*	TAA	*STOP*	TGA	*STOP*
T	TTG	*Leu*	TCG	*Ser*	TAG	*STOP*	TGG	*Trp*
C	CTT	*Leu*	CCT	*Pro*	CAT	*His*	CGT	*Arg*
C	CTC	*Leu*	CCC	*Pro*	CAC	*His*	CGC	*Arg*
C	CTA	*Leu*	CCA	*Pro*	CAA	*Gln*	CGA	*Arg*
C	CTG	*Leu*	CCG	*Pro*	CAG	*Gln*	CGG	*Arg*
A	ATT	*Ile*	ACT	*Thr*	AAT	*Asn*	AGT	*Ser*
A	ATC	*Ile*	ACC	*Thr*	AAC	*Asn*	AGC	*Ser*
A	ATA	*Ile*	ACA	*Thr*	AAA	*Lys*	AGA	*Arg*
A	ATG	*Met**	ACG	*Thr*	AAG	*Lys*	AGG	*Arg*
G	GTT	*Val*	GCT	*Ala*	GAT	*Asp*	GGT	*Gly*
G	GTC	*Val*	GCC	*Ala*	GAC	*Asp*	GGC	*Gly*
G	GTA	*Val*	GCA	*Ala*	GAA	*Glu*	GGA	*Gly*
G	GTG	*Val*	GCG	*Ala*	GAG	*Glu*	GGG	*Gly*

* *Met* is the amino acid methionine and is the first amino acid added (the START instruction).

The genetic code is degenerate

An important feature of the genetic code is that it is **degenerate** or **redundant**.

In many cases, more than one triplet of bases codes for one amino acid, as shown in table 1.8. Multiple triplets in DNA can lead to the addition of the same amino acid. For example, if an individual has a mutation where CCA is changed to CCG, there will be no impact on the final protein produced, as this mutation is silent.

TABLE 1.8 A small section of the amino acid coding chart

CCT	*Pro*
CCC	*Pro*
CCA	*Pro*
CCG	*Pro*

This idea of having a degenerate code is important as it allows us to be more tolerant to mutations. This is especially important in genes that are vital for function and survival. Two individuals may produce the exact same functional protein, despite having slight variations in their genetic code.

1.4.2 Steps in gene expression

The genetic instructions for producing proteins are found within the DNA of chromosomes. A gene is a segment of DNA that codes for a protein. In eukaryotic organisms, how do genetic instructions get from the nucleus to the ribosomes?

Gene expression involves protein synthesis, which can be summarised in two main steps:

1. **Transcription:** DNA to mRNA
2. **Translation:** mRNA to protein.

Another step, known as RNA processing, occurs between transcription and translation, in which pre-mRNA produced in transcription is processed and made into mature mRNA to be used in translation.

A gene consists of a particular part of a double-helical molecule of DNA that codes for a particular protein. Only one of the two chains is used during protein synthesis; this is called the **template strand**. The complementary chain to this is sometimes called the non-template strand or the **coding strand**.

degenerate the property of the genetic code in which more than one triplet of bases can code or one amino acid

redundant see 'degenerate'

transcription process of copying the genetic instructions present in DNA to messenger RNA

translation process of decoding the genetic instructions in mRNA into a protein (polypeptide chain) built of amino acids

template strand one strand of a DNA double helix that is used to produce a complementary mRNA strand during transcription; sometimes called the sense strand

coding strand one strand of a DNA double helix that is complementary to the template strand

FIGURE 1.26 The two main steps in protein synthesis. A third step, known as RNA processing, occurs after transcription and before translation.

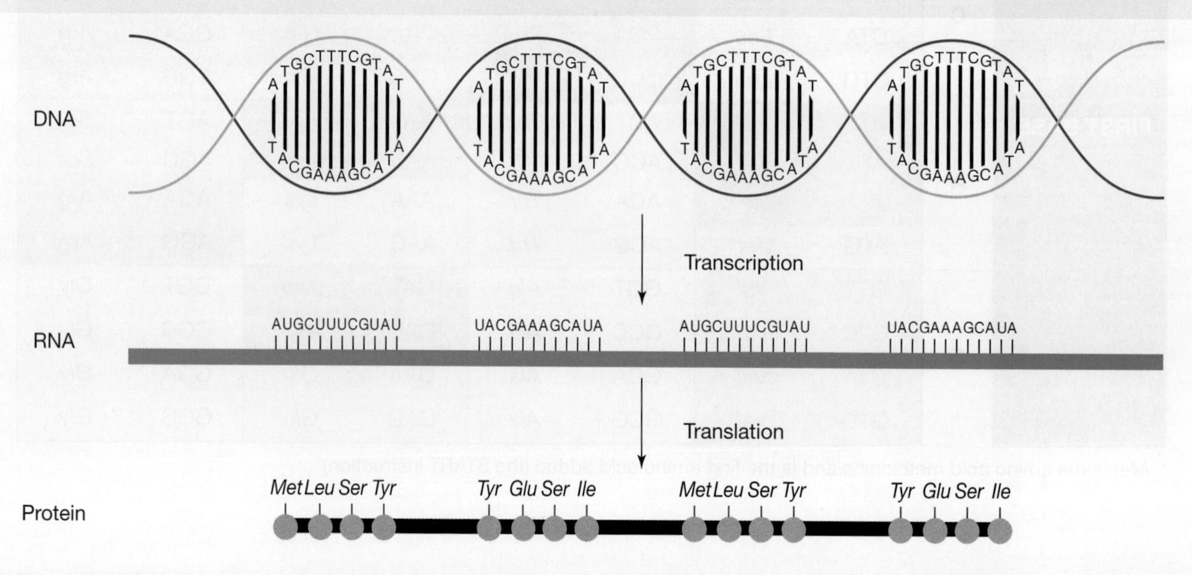

When a gene becomes active, it first makes a mobile copy of the coded instruction (known as mRNA) that it contains. This occurs by a process known as transcription. This mobile copy of a genetic instruction leaves the nucleus and moves to the cytoplasm, where the instruction is decoded. This occurs by a process known as translation.

For example, the production of a particular protein, such as beta chains of haemoglobin, starts in the cell nucleus. It is here that pre-mRNA molecules are produced by transcription from a template DNA chain. The mRNA leaves the nucleus and moves to the cytosol, where it becomes attached to ribosomes. It is here that protein chains are formed by translation of mRNA. This process can be seen in figure 1.27.

FIGURE 1.27 Representation of the processes of transcription and translation in a eukaryote cell. (Image courtesy of the National Human Genome Research Institute)

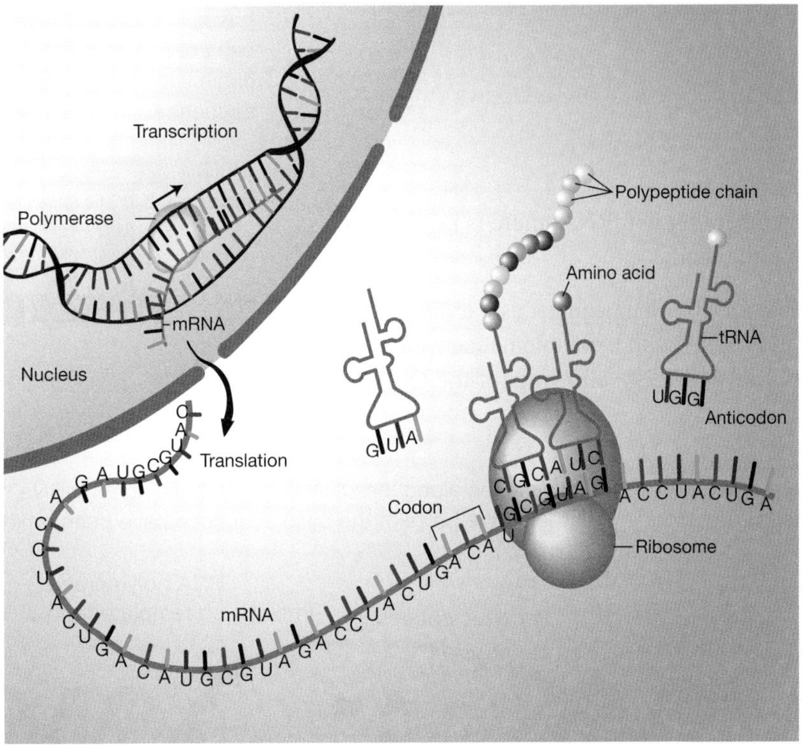

1.4.3 Transcription

The first stage of protein synthesis is transcription, which occurs in the nucleus.

> Transcription is the process in which the genetic instructions from DNA are copied into a form that is able to leave the nucleus, known as messenger RNA (mRNA).

The nucleus of a eukaryotic cell is like a safe that contains the genetic master plan in the form of DNA. The genetic master plan contains the entire set of instructions for an organism — it is like the complete plan for the construction of a complex structure, such as a jumbo jet. One gene or instruction for a protein is like the plan for making one component of the jet, such as a wing flap.

The workers at the site where the wing flaps are made do not work directly from the complete master plan; instead, they have copies of the relevant section of the plan. Likewise, before a genetic instruction in DNA is decoded, that instruction is copied (transcribed) from the genetic master plan, which remains in the nucleus. This copy is encoded in a different nucleic acid called ribonucleic acid (RNA).

From DNA to mRNA: step by step

1. An enzyme, known as **RNA polymerase**, attaches to a specific promoter sequence of DNA in the upstream region of the template strand. The double-stranded DNA of the gene unwinds and exposes the bases of the template strand.
2. The base sequence of the DNA template guides the building of a complementary copy of the mRNA sequence. The RNA polymerase enzyme moves along the DNA template in a 3′ to 5′ direction. As it moves, complementary nucleotides are brought into place and, one by one, are joined to form an RNA chain. These new nucleotides are added onto the growing 3′ end of the mRNA strand.
3. After the RNA polymerase moves past the coding region and into the downstream region of the gene, transcription stops and the mRNA molecule (pre-mRNA) is released from the template.

Pairing or hybridisation can occur between the bases in one DNA strand and complementary bases in an RNA strand as follows:

- A pairs with U
- T pairs with A
- C pairs with G
- G pairs with C.

It is important to remember that in mRNA, there is no thymine (T). This is replaced with uracil.

This pairing means that a DNA chain can act as a template to guide the construction of RNA with a complementary base sequence (see figures 1.28 and 1.29), allowing for the DNA to be accurately copied.

FIGURE 1.28 Pairing of complementary bases in DNA and RNA. One DNA chain can act as a template to build an RNA chain with a predictable nucleotide sequence.

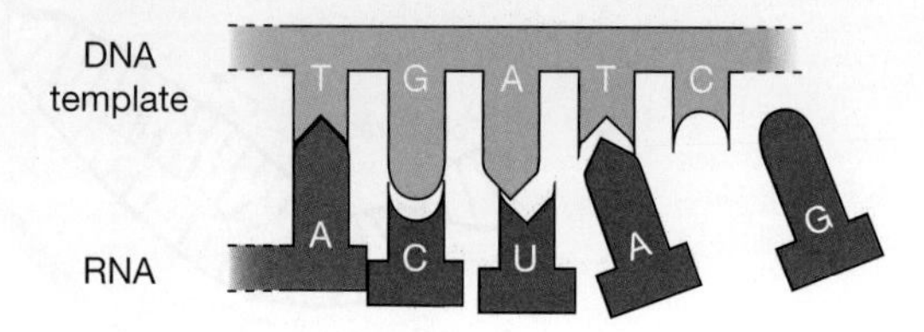

FIGURE 1.29 The enzyme RNA polymerase moves along the DNA template building an mRNA molecule at the rate of about 30 bases per second. Note that some parts of the base sequence have been omitted for simplicity.

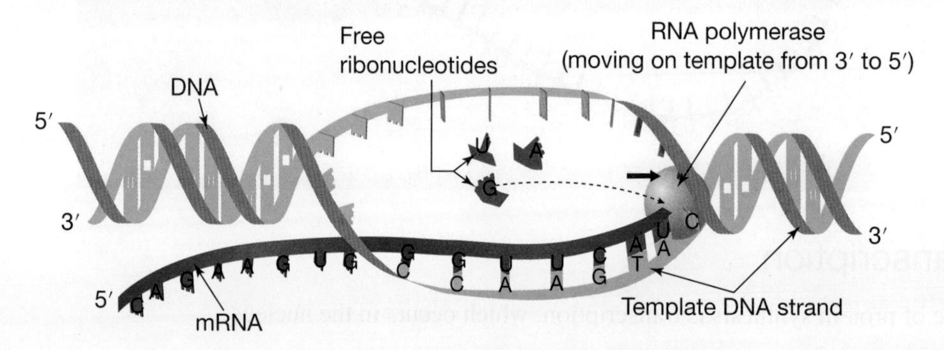

The result of this process is a single-stranded molecule, called pre-mRNA. The base sequence in the pre-mRNA molecule is complementary to the base sequence of the DNA of the template strand.

The base sequence of the mRNA primary transcript is not identical to that of the template DNA strand; instead, the mRNA has a complementary sequence. However, the mRNA base sequence matches that of the complementary non-template DNA strand, except that U replaces A. For this reason, the non-template DNA strand is also called the coding DNA strand.

RNA polymerase an enzyme that controls the synthesis of an RNA strand from a DNA template during transcription

TABLE 1.9 Comparing the nucleotides in non-template, template and mRNA stands

Non-template DNA	T	A	C	G	G	A	C	T	T	A
Template DNA	A	T	G	C	C	T	G	A	A	T
mRNA transcript	U	A	C	G	G	A	C	U	U	A

TIP: Ensure you check if you have been given the template or non-template (coding) strand before you transcribe your sequence, as this affects the mRNA sequence. If you are given a DNA segment to transcribe and the strand is not specified, you can assume it is the template.

tlvd-1091

SAMPLE PROBLEM 2 Applying transcription

A template sequence of DNA was found to be ATGCCTGAA. Provide the mRNA for this strand. **(1 mark)**

THINK	WRITE
1. Divide the template DNA sequence into triplets. (*Hint:* number each triplet to prevent the code from getting lost) This DNA can act as a template to guide the formation of an RNA molecule.	Original strand: ATGCCTGAA Separated strand: ATG CCT GAA 1 2 3
2. Transcribe the complementary base sequence to (pre)mRNA.	Transcribed strand: UAC GGA CUU (1 mark) 1 2 3

on Resources

Video eLesson Transcription (eles-4167)

Interactivity Transcription (int-8125)

1.4.4 RNA processing in eukaryotic cells

The primary product of transcription is pre-mRNA, also known as the primary transcript. The sequence of bases in the pre-mRNA is complementary to all the DNA bases of a gene, both **introns** and **exons** (these will be further explored in section 1.5.1). The primary mRNA transcript then undergoes a process termed **RNA processing** or **post-transcription modification** (see figure 1.30).

RNA processing: step by step

RNA processing occurs in the nucleus and includes the following processes:

1. *Capping:* The 5′ end of the pre-mRNA is capped with an altered guanine (G) base (methyl guanosine). The methyl cap protects the pre-mRNA from enzyme attack and contributes to its stability, helping it attach to the ribosome.
2. *Adding a tail:* The primary transcript is clipped at a specific point downstream of the coding region and a poly-adenine (A) tail, with up to 250 As, is then added at the 3′ end. The poly A tail contributes to the stability of the mRNA and facilitates mRNA export from the nucleus.
3. *Splicing:* The regions in the pre-mRNA that correspond to the introns are spliced and the remaining exons are joined together. This cutting and splicing is done by **spliceosomes**, which recognise specific base sequences at the ends of the introns: GU at the 5′ end and AG at the 3′ end.

introns parts of the coding region of a gene that are transcribed but not translated

exons parts of the coding region of a gene that are both transcribed and translated

RNA processing occurs after transcription and involves modifying pre-mRNA to form mature mRNA; also known as post-transcription modification

post-transcription modification process occurring after transcription in which pre-mRNA is altered to become mature mRNA

spliceosomes complex molecules present in the nucleus that remove introns from the pre-mRNA transcript

The final mRNA product (referred to as mature mRNA) now moves across the nuclear membrane into the cytosol, carrying with it a copy of the information originally encoded in the DNA of the gene.

FIGURE 1.30 RNA processing of pre-mRNA to form mRNA

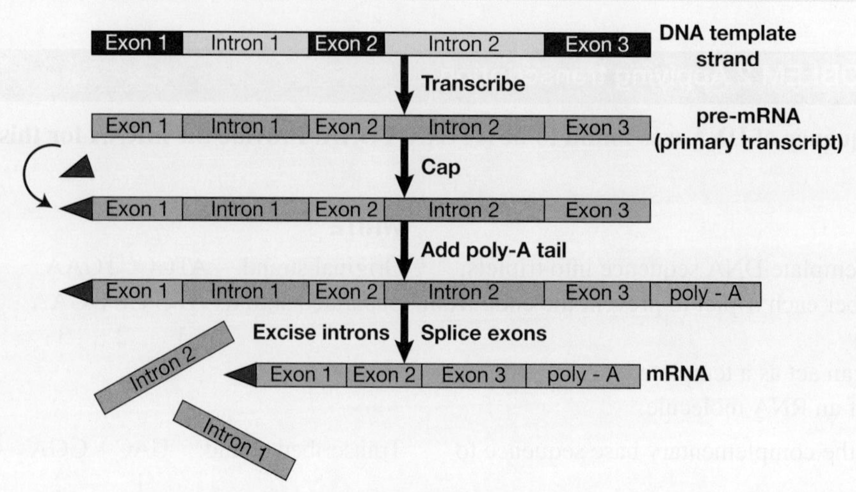

This process is carefully controlled. Mistakes in RNA processing can lead to different proteins being produced in translation.

However, there is added complexity to this as in some cases additional exons are removed deliberately by the spliceosomes. This is known as alternative splicing and adds further complexity to RNA processing and gene regulation, enabling one gene to produce a multitude of proteins if required.

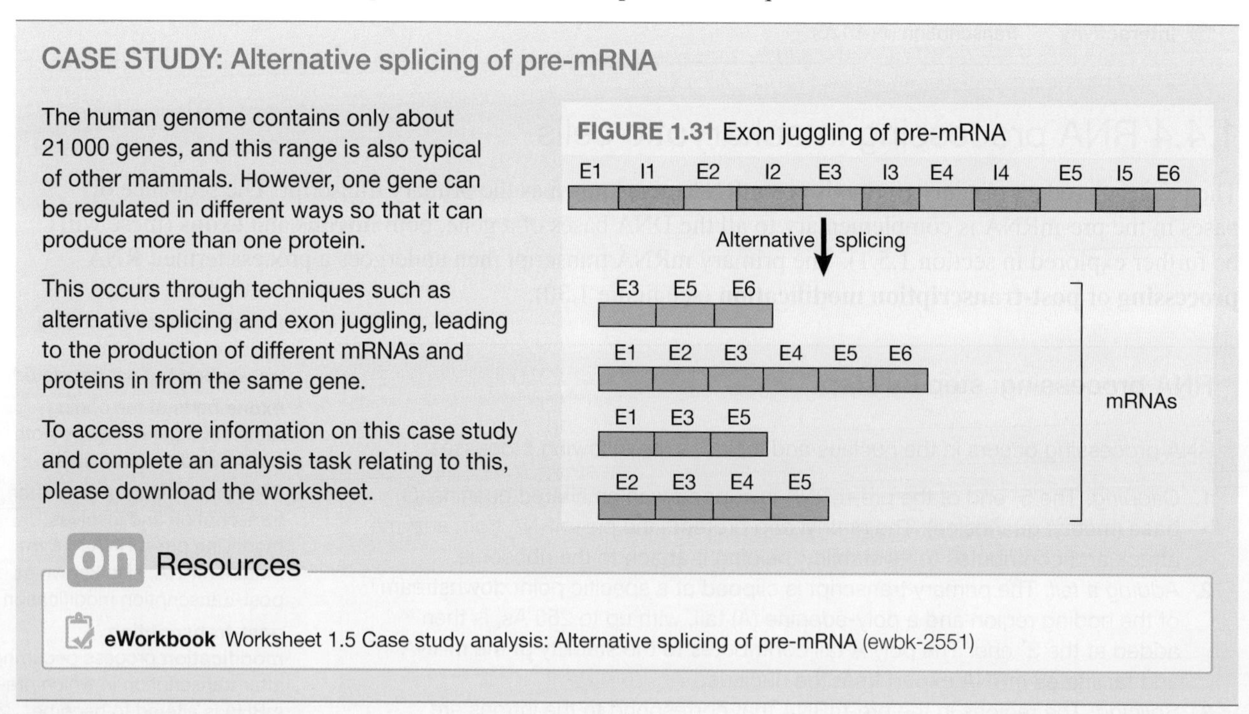

CASE STUDY: Alternative splicing of pre-mRNA

The human genome contains only about 21 000 genes, and this range is also typical of other mammals. However, one gene can be regulated in different ways so that it can produce more than one protein.

This occurs through techniques such as alternative splicing and exon juggling, leading to the production of different mRNAs and proteins in from the same gene.

To access more information on this case study and complete an analysis task relating to this, please download the worksheet.

FIGURE 1.31 Exon juggling of pre-mRNA

on Resources

eWorkbook Worksheet 1.5 Case study analysis: Alternative splicing of pre-mRNA (ewbk-2551)

1.4.5 Translation

The decoding of the genetic instructions occurs through the process of translation, which takes place in the cytoplasm. By the end of this process, the genetic instructions carried in mRNA have been decoded and translated into a protein chain built of amino acids. This is known as the **primary structure** of the protein, which will be explored further in subtopic 1.7. The coded instruction in the mRNA is not changed in this process, just as the plan of a jumbo jet part is unaltered after the part is made.

primary structure the specific linear sequence of amino acids in a protein

Translation is the process in which the sequence mRNA is decoded and translated into a protein chain by the addition of corresponding amino acids.

Translation involves the combined action of several agents (see table 1.10).

TABLE 1.10 Important components of translation

Agents	Analogy
DNA in the nucleus	Master plan with complete set of instructions
mRNA	Working copy of one instruction
Ribosomes	Construction site
tRNA	Carriers of raw material
Amino acids	Raw material
Protein chain	End product

The mRNA moves from the nucleus to the cytoplasm, where it attaches to submicroscopic organelles known as ribosomes (see figure 1.32).

The construction of a protein according to the coded instructions in mRNA involves the assembly of amino acid sub-units. The various **amino acids** are present in solution in the cytosol.

The information in mRNA is present in coded form as sets of three bases or triplets. These triplets, such as AGG and UCU, are called **codons**.

Most codons contain the information to add one specific amino acid to a protein chain. In addition, one codon (AUG) is a *start translation* instruction, and three different codons (UAA, UAG and UGA) are *stop translation* instructions (see table 1.11).

FIGURE 1.32 In this electron photomicrograph, ribosomes appear as dark dots located on the endoplasmic reticulum. (Image courtesy of Dr Maret Vesk)

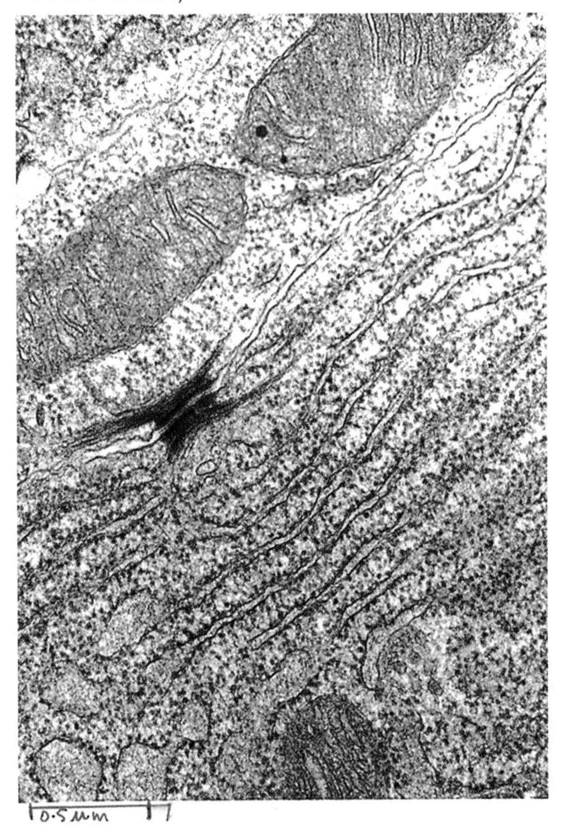

From mRNA to protein: step by step

1. mRNA moves to the ribosome, where it is read in groups of three known as codons. Translation begins at the 'start adding amino acids' signal (AUG codon) (see figure 1.33a).
2. Each amino acid is brought to the mRNA on the ribosomes by a carrier molecule called transfer RNA (tRNA). At one end of each tRNA molecule are three bases that make up an anticodon. At the other end of a tRNA molecule is a region that attaches to one specific amino acid. The first amino acid to be added is methionine (*met*).
3. The ribosome continues to move along the mRNA and tRNA molecules to deliver the appropriate amino acid, determined by the code seen in table 1.11. As amino acids are added, they are joined by peptide bonds.
4. A codon representing STOP is reached (UAA, UAG or UGA) and the polypeptide is released from the ribosome.

amino acids basic building blocks or sub-units of polypeptide chains and proteins

codons sequences of three bases in a messenger RNA molecule that contain information either to bring amino acids into place in a polypeptide chain or to start or stop this process

FIGURE 1.33 **a.** The mRNA molecule attaches to a ribosome. In turn, as the ribosome moves along the mRNA molecule, each codon pairs with the tRNA with the complementary anticodon. **b.** The amino acids carried by each tRNA molecule are joined to form a chain. The final product is a protein (or polypeptide) consisting of a chain of amino acid sub-units.

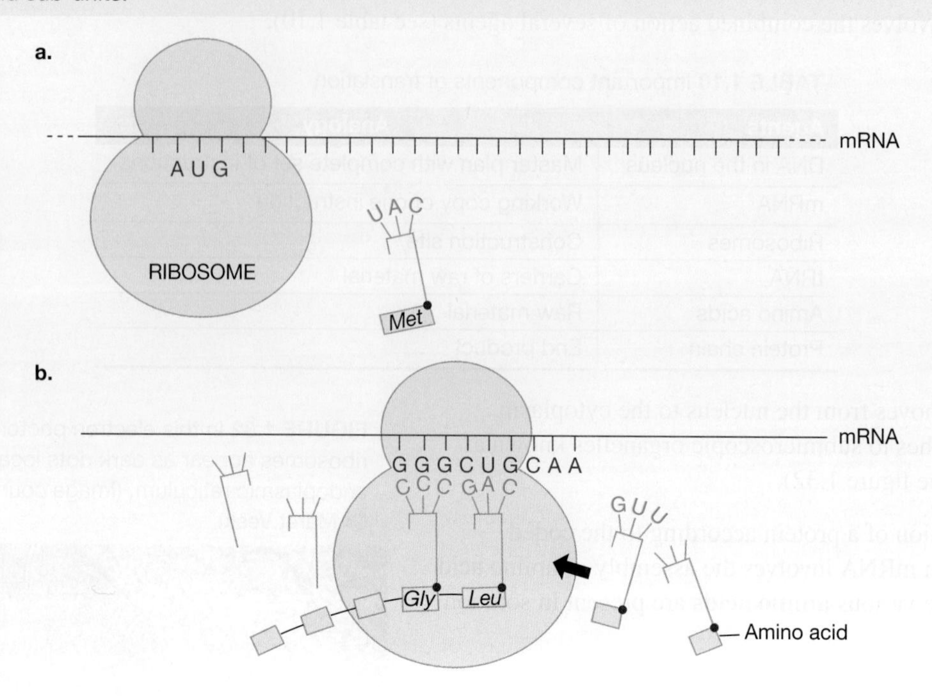

Table 1.11 shows the chart of amino acids. Refer to the Appendix (Amino acid data) for the full names of the amino acids and their chemical properties. Each amino acid is shown using a specific colour. Amino acids with similar colours have similar chemical properties (for example *his*, *lys* and *arg* are all shades of green as they are all have positively charged side chains.).

TABLE 1.11 Genetic code shown as the 64 mRNA codons and the amino acids they specify

FIRST BASE		SECOND BASE							
		U		C		A		G	
	U	UUU	*Phe*	UCU	*Ser*	UAU	*Tyr*	UGU	*Cys*
		UUC	*Phe*	UCC	*Ser*	UAC	*Tyr*	UGC	*Cys*
		UUA	*Leu*	UCA	*Ser*	UAA	*STOP*	UGA	*STOP*
		UUG	*Leu*	UCG	*Ser*	UAG	*STOP*	UGG	*Trp*
	C	CUU	*Leu*	CCU	*Pro*	CAU	*His*	CGU	*Arg*
		CUC	*Leu*	CCC	*Pro*	CAC	*His*	CGC	*Arg*
		CUA	*Leu*	CCA	*Pro*	CAA	*Gln*	CGA	*Arg*
		CUG	*Leu*	CCG	*Pro*	CAG	*Gln*	CGG	*Arg*
	A	AUU	*Ile*	ACU	*Thr*	AAU	*Asn*	AGU	*Ser*
		AUC	*Ile*	ACC	*Thr*	AAC	*Asn*	AGC	*Ser*
		AUA	*Ile*	ACA	*Thr*	AAA	*Lys*	AGA	*Arg*
		AUG	*Met**	ACG	*Thr*	AAG	*Lys*	AGG	*Arg*
	G	GUU	*Val*	GCU	*Ala*	GAU	*Asp*	GGU	*Gly*
		GUC	*Val*	GCC	*Ala*	GAC	*Asp*	GGC	*Gly*
		GUA	*Val*	GCA	*Ala*	GAA	*Glu*	GGA	*Gly*
		GUG	*Val*	GCG	*Ala*	GAG	*Glu*	GGG	*Gly*

* *Met* is the amino acid methionine and is the first amino acid added (the start instruction).

Transfer RNA

As mentioned previously, tRNA is a vital molecule involved in translation. Each tRNA molecule consists of a single strand of 76 nucleotides coiled and paired with themselves. At one end of each tRNA molecule are three bases that make up an anticodon. At the other end of a tRNA molecule is a region that attaches to one specific amino acid (see figure 1.34). An enzyme, amino acyl tRNA synthetase, catalyses the linking of each amino acid to its specific tRNA carrier.

FIGURE 1.34 a. Transfer RNA (tRNA) molecule. At one end is an amino acid attachment site; at the opposite end is an anticodon region. **b.** A simplified diagram of tRNA showing the anticodon

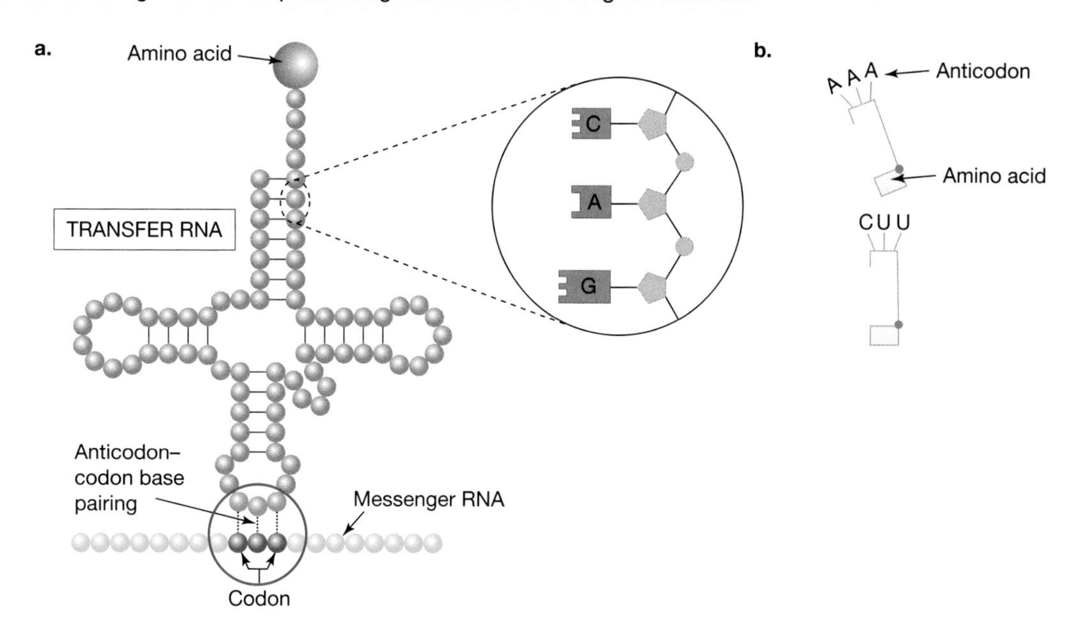

As each codon is translated, the tRNA molecule with the complementary anticodon pairs momentarily with the mRNA. The pairing between bases in codons and the complementary bases in an anticodon is as follows:

- A pairs with U
- U pairs with A
- C pairs with G
- G pairs with C.

So, when the mRNA codon UUU is reached, the tRNA carrier molecule that has the anticodon AAA comes into place with its specific cargo of the amino acid *phe*. The amino acid carried by that tRNA is brought into the correct position to be joined into the growing protein chain. Amino acids continue to be added until a STOP signal is reached, which stops the addition of amino acids to the protein chain.

Messenger RNA (mRNA) formed during gene transcription has a short life. This contrasts with ribosomal RNA (rRNA), which forms part of the ribosomes and is very stable.

tlvd-1092

SAMPLE PROBLEM 3 Combining transcription and translation

A section of coding DNA was sequenced and found to be as follows:

TACGGACTT

a. **What would the complementary template sequence be for this strand?** **(1 mark)**
b. **What would this sequence be transcribed to?** **(1 mark)**
c. **Show the amino acid that would be produced for this process.** **(1 mark)**

THINK	WRITE
a. The complementary strand will show base pairing and be opposite to the coding strand (A pairs with T and C with G).	ATG CCT GAA (1 mark)
b. The template strand is used to make the mRNA that is complementary.	UAC GGA CUU (1 mark)
c. Use the coding chart to determine the amino acids to be added based on the codons. The first codon is UAC, which corresponds to *tyr*. The second codon is GGA, which corresponds to *gly*. The third codon is CUU, which corresponds to *leu*.	*tyr – gly – leu* (1 mark)

elog-0011

INVESTIGATION 1.5 online only

Using the genetic code to model protein synthesis

Aim

To model the stages of protein synthesis and understand the importance of the genetic code

on Resources

eWorkbook Worksheet 1.6 Gene expression and protein synthesis (ewbk-1970)

Digital document Codon and triplet charts (doc-36165)

Video eLesson Translation (eles-4168)

Interactivities Protein synthesis (int-0112)
Comparing transcription and translation (int-0113)
Translation (int-8126)

Weblinks Transcription and translation practice
DNA to RNA to protein simulations
Examining protein synthesis in fireflies

KEY IDEAS

- Genes contain coded instructions for joining specific amino acids into polypeptides.
- The genetic code in DNA is a non-overlapping triplet code consisting of groups of three bases.
- The genetic code is degenerate and in many cases, more than one triplet of bases codes for one particular amino acid.
- One chain of the double stranded DNA is known as the template strand; its complementary chain is called the non-template or coding strand.
- One piece of genetic code typically contains the information to add one amino acid to a protein. This is referred to as the gene.
- During transcription, the information in the template strand of the DNA of a gene is copied into a pre-RNA molecule.
- The final mRNA molecule is produced when post-transcription modification is complete, including intron excision and exon splicing.
- The mRNA moves to the protein and is translated into a protein.
- During translation, amino acids are transported by tRNA molecules to the ribosomes to build the protein.

1.4 Activities

To answer questions online and to receive **immediate feedback** and **sample responses** for every question, go to your learnON title at **www.jacplus.com.au**. A **downloadable solutions** file is also available in the resources tab.

1.4 Quick quiz on	1.4 Exercise	1.4 Exam questions

1.4 Exercise

1. MC Which does **not** occur in RNA processing in eukaryotic cells?
   A. Capping of the 5′end
   B. Splicing and removal of introns
   C. Deletion of exons interfering in the sequence
   D. Addition of a poly tail at the 3′end
2. Explain the steps involved in DNA transcription.
3. Use a flowchart to show the steps involved in the process of translation.
4. Explain the relationship between genes and polypeptides.
5. A section of mRNA has the following sequence:

   A U G C A G G A G G C U U A A

   a. Write down the DNA sequence that was a template leading to the production for this section of mRNA.
   b. Write down the amino acid sequence created by this piece of mRNA.
   c. A mutation occurs which changes the fourth base to U. Describe the effect this mutation will have.
6. Explain the advantages and disadvantages of the universality of the genetic code to humans.

1.4 Exam questions

Question 1 (1 mark)

Source: *Adapted from VCAA 2020 Biology Exam, Section A, Q7*

MC The codon chart in figure 1.11 can be used to determine amino acids coded for by a nucleotide sequence.

It is correct to state that

A. identical amino acid sequences are found in all organisms.
B. the genetic coding is degenerate with respect to Met.
C. the codon GGU adds Trp to a polypeptide chain.
D. the DNA template sequence GAA codes for Leu.

Question 2 (1 mark)

Source: *VCAA 2018 Biology Exam, Section A, Q4*

MC The genetic code is described as a degenerate code.

This means that

A. in almost all organisms the same DNA triplet is translated to the same amino acid.
B. some amino acids may be encoded by more than one codon.
C. a single nucleotide cannot be part of two adjacent codons.
D. three codons are needed to specify one amino acid.

Question 3 (1 mark)

Source: *Adapted from VCAA 2015 Biology Exam, Section A, Q23*

MC The following is a sequence of amino acids located within a polypeptide: – Asn – Gly – Pro – Arg – Ser –

Using table 1.11, the DNA template sequence that could code for this amino acid sequence is

A. TTG / CCC / GGT / GCT / TCG
B. TTG / GTT / GGT / GCT / TCG
C. TTG / CCC / GGT / GCT / TCT
D. UUG / CCC / GGU / CGU / UGC

Question 4 (4 marks)

Source: *VCAA 2020 Biology Exam, Section A, Q2c*

Outline **two** events that occur during RNA processing and the importance of each event in gene expression.

Question 5 (7 marks)

Source: *Adapted from VCAA 2016 Biology Exam, Section B, Q6*

The hormone insulin is a relatively small protein. Researchers studying the production of insulin in the cells of the pancreas noted that one of the early steps in this process was the formation of a polypeptide called preproinsulin.

Researchers noted that the formation of this polypeptide required repeated use of different types of Molecule X, as shown.

Molecule X

anticodon

a. i. What is the name of Molecule X? **1 mark**

ii. How does Molecule X play a role in the production of preproinsulin? **3 marks**

b. The coding information in the DNA molecule for preproinsulin is initially transferred to another molecule (Molecule W). However, Molecule W has a different nucleotide sequence from the coding section of the DNA molecule.

Describe how Molecule W is synthesised. **3 marks**

More exam questions are available in your learnON title.

1.5 The structure of genes

KEY KNOWLEDGE

- The structure of genes: exons, introns and promoter and operator regions

1.5.1 The structure of genes

The part of a gene that contains the coded information for making a protein is called the **coding region** of a gene. The regions on either side of the coding region of a gene are called **flanking regions**. The flanking region before the start of the coding region is called the upstream region. The flanking region after the end of the coding region is called the downstream region.

coding region part of a gene that contains the coded information for making a polypeptide chain

flanking regions regions located either downstream or upstream of the coding region of a gene

FIGURE 1.35 A simplified diagram outlining the key components in a eukaryotic gene

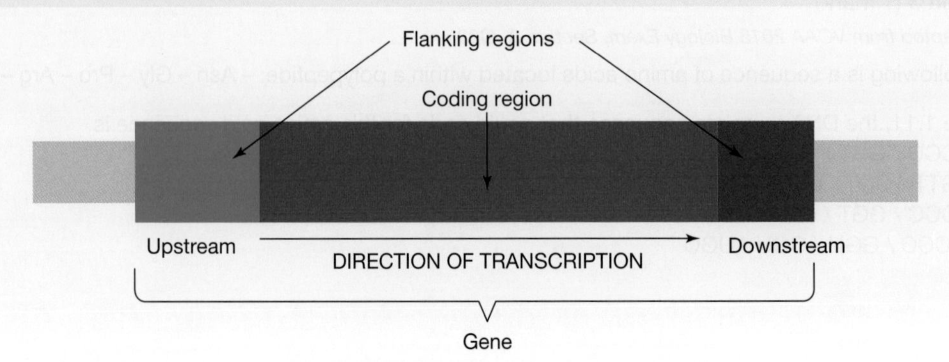

1.5.2 Exons and introns

The coding region of a gene is the segment of DNA double helix that includes the DNA template strand, which encodes the information that will later be translated into the amino acid sequence of a polypeptide. This region of a DNA template strand begins with a start signal (TAC) and, some distance away, there is a stop signal (ATT or ATC or ACT).

An unexpected discovery about the genes in eukaryotes was made in 1977. Until then, the coding region of a gene was thought to be continuous. The coding region is made of two main parts — exons and introns.

Exons contain the instructions for the synthesis of the protein and are both transcribed and translated. They provide the instructions that code for the amino acids in the produced protein.

FIGURE 1.36 A coding region is made up of exons broken into segments by noncoding segments known as introns.

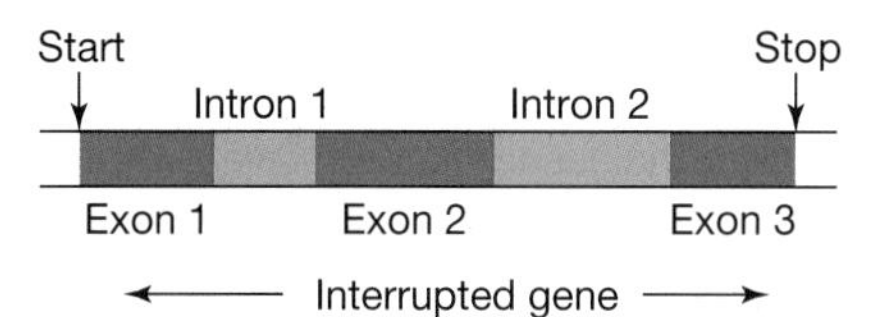

The exons are separated by lengths of DNA that do not contain instructions relating to the protein chain. These interrupting segments are called introns. They are transcribed in the nucleus but are cut out in RNA processing and therefore are not translated to form the polypeptide product.

TIP: To avoid confusing the two terms *exon* and *intron*, think about **IN**terruption and **IN**tron to help remember the functional difference. Introns are *transcribed* but not *translated*.

The number of exons and introns in genes varies. The DNA making up the *HBB* gene, which controls the production of one chain of haemoglobin molecules, consists of three exons and two introns. The *F8C* gene, which controls the production of factor VIII, which assists in blood clotting, consists of 26 exons and 25 introns. Often, larger genes tend to have more exons and introns.

So, eukaryote genes are not like nursery rhymes in a book, where the reader starts at the beginning and reads through to the end. The information in genes is broken up into segments, and the sections in between are filled with other printed material that is unrelated.

1.5.3 Promoter regions — upstream

The region of DNA on the template strand upstream from the coding region contains some particular base sequences. These can be seen in figure 1.38. One part of the upstream region is rich in As and Ts and is often called the **TATA box**, because the sequence TATA AA (or similar) occurs there. This is located around 25 to 35 base pairs from the transcription start site. This region is known as the **promoter**. The CAT or CAAT box, located around 60 to 100 bases upstream of the transcription start site, is also part of some promoter regions.

The promoter is where transcription factors and RNA polymerase binds to initiate transcription. Without a functioning promoter region, transcription cannot be properly initiated.

Observations related to the upstream region and promoter are as follows:

- Upstream sequences are invariably found in all organisms. It is reasonable to suggest that these upstream sequences serve an important function since they have been maintained during evolution.
- If upstream sequences are altered by mutation, the activity of the coding region of the gene may be reduced or even become inactive. The absence of the correct upstream signal is a cause of some inherited human diseases. One form of thalassaemia is due to a missing TATA group in the upstream region of the DNA of both copies of the specific gene in the people concerned.
- The upstream region includes segments of DNA to which hormones can attach. The fact that some hormones can bind to DNA provides one clue as to how hormones can influence the action of genes.

TATA box short base sequence consistently found in the upstream flanking region of the coding region of genes of many different species

promoter part of the upstream flanking region of a gene where RNA polymerase binds that contains base sequences that control the activity of that gene

These observations support the conclusion that specific DNA sequences upstream of the coding region of a gene initiate transcription, the process by which the encoded information in the DNA coding region is transcribed into mRNA. Promoters also act as sites where proteins called transcription factors can bind and regulate the expression of genes.

CASE STUDY: Beta thalassaemia

Issues in the flanking regions can have drastic effects on the final protein product formed.

An example of this occurs in beta thalassaemia. Individuals with beta thalassaemia can have symptoms including slow growth, fatigue, anemia, a pale or jaundiced appearance and a swollen abdomen. This is due to a mutation that affects the beta chain of haemoglobin coded for by the *HBB* gene on chromosome 11.

Most often, this mutation occurs in the promoter region preceding the beta-globin genes, leading to issues with producing the protein through transcription and translation.

FIGURE 1.37 Individuals with thalassaemia have abnormal red blood cells, as seen in the below smear.

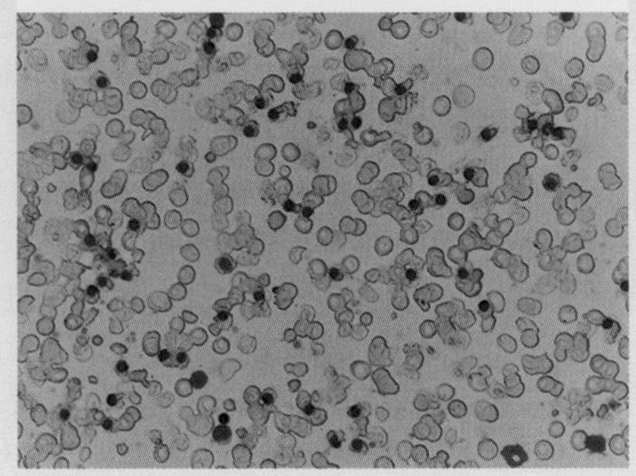

1.5.4 Leader region

Another region that is upstream of the coding region of some genes is the leader region. In prokaryotes, it tends to be small, but can be much longer in eukaryotes.

Leader regions contain sections known as attenuators and as such are involved in a process of gene regulation called attenuation. This involves the formation of hairpin loops and the stalling and detachment of the ribosome and in turn RNA polymerase when the structural genes do not need to be transcribed and translated. This process is explored in a digital document in subtopic 1.6.

EXTENSION: The downstream region and parts of a gene

The DNA following the end of the coding region is referred to as the 'downstream' region (see figure 1.38). About 20 bases downstream, the sequence AATAAA is usually found. If this sequence is altered, the gene action is altered. The downstream region includes an 'end transcription signal', which terminates the process of transcription of mRNA from the DNA template.

FIGURE 1.38 Regions of the template strand of a typical gene.

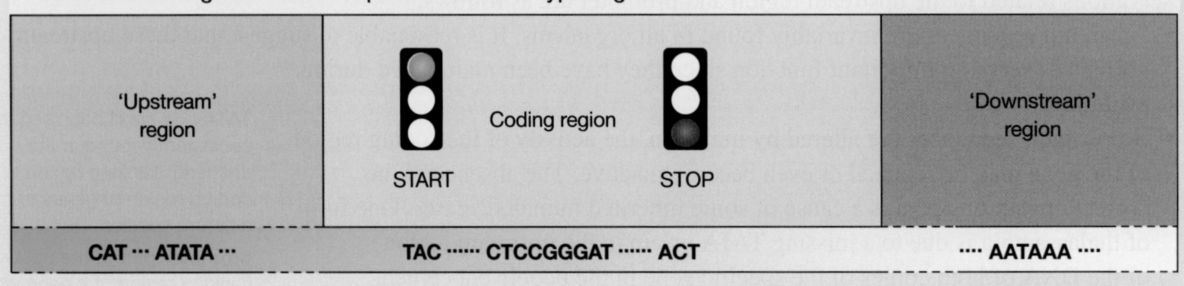

1.5.5 Operator regions

Operator regions are special sections found in prokaryotic genes. An operator is found between the promoter and the gene being transcribed. It is the binding site for **repressor** proteins. When a repressor (produced in another gene) binds to the operator, it prevents the RNA polymerase binding to the promoter, and thus transcription cannot be initiated. A specific operator region involved in the *trp* operon will be explored in subtopic 1.6.

operator a region found in an operon where a repressor is able to bind

repressor a protein produced by a regulatory gene that can bind to DNA and prevent transcription

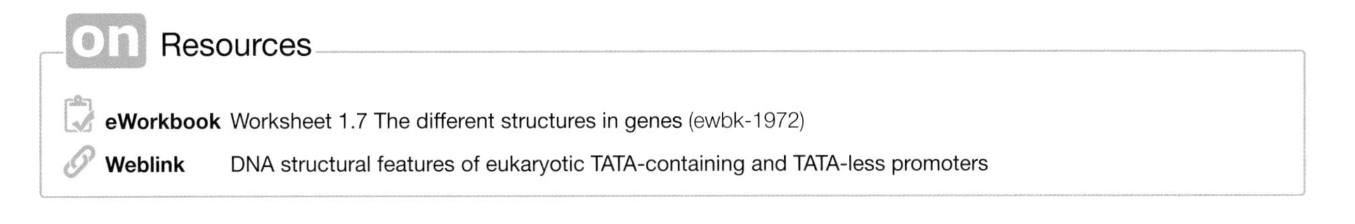

KEY IDEAS

- Eukaryotic genes consist of a coding region and flanking regions.
- The coding region contains exons, which are translated into proteins, and introns, which are spliced out before translation.
- One region in the flanking region is the upstream region. This contains the promoter region, where RNA polymerase binds to commence transcription.
- The downstream region contains signals to stop transcription.
- The operator is a region in prokaryotes in which a repressor can bind to prevent transcription.

1.5 Activities

learnon

To answer questions online and to receive **immediate feedback** and **sample responses** for every question, go to your learnON title at **www.jacplus.com.au**. A **downloadable solutions** file is also available in the resources tab.

1.5 Quick quiz on	1.5 Exercise	1.5 Exam questions

1.5 Exercise

1. **MC** Which of the following are found in eukaryotic genes?
   A. Regions where hormones can attach in the upstream region
   B. Non-coding segments known as exons, which are not translated in a protein product
   C. A TATA box found in the downstream region for binding RNA polymerase
   D. Flanking regions that are translated into regulatory proteins
2. Using words or diagrams, distinguish between the members of each of the following pairs:
   a. intron and exon
   b. coding region and flanking region.
3. List one difference between the promoter and operator regions.
4. The structure of genes in eukaryotic cells includes promoter regions, exons and introns. Explain their role and function.
5. What would happen to a protein if introns were not removed?
6. Predict the relationship between gene size and number of introns.

1.5 Exam questions

Question 1 (1 mark)

Source: *VCAA 2017 Biology Exam, Section A, Q2*

MC The *lac* operon was originally identified in *Escherichia coli*. The *lac* operon has three structural genes: *lacZ*, *lacY* and *lacA*. The *lacZ* gene codes for the production of the enzyme β-galactosidase, which catalyses the breakdown of lactose into glucose and galactose. Below is a diagram that shows the order of the genes found in the lac operon. The dots represent the DNA nucleotides between the genes.

...	regulatory gene	...	promoter gene	operator gene	*lacZ*	*lacY*	*lacA*	...

To begin transcription of the three structural genes, RNA polymerase needs to bind to the
- A. operator gene.
- B. promoter gene.
- C. regulatory gene.
- D. structural genes

Question 2 (1 mark)

Source: *VCAA 2018 Biology Exam, Section A, Q22*

In humans, Duchenne muscular dystrophy (DMD) is caused by mutations in the dystrophin-encoding DMD gene.

The DMD gene contains 79 exons. In some patients, duplication of one exon occurs.

MC If the number of nucleotides in the duplicated exon is divisible by three, the
- A. transcribed mRNA will contain many stop codons.
- B. length of each of the 79 exons will increase by three nucleotides.
- C. translated protein will be longer than the dystrophin protein found in a person without DMD.
- D. dystrophin of these patients will show one amino acid change in the sequence compared to normal dystrophin.

Question 3 (1 mark)

Source: *VCAA 2006 Biology Exam 2, Section A, Q1*

MC In eukaryotic organisms genes are
- A. composed of DNA.
- B. alternative forms of an allele.
- C. composed of DNA and protein.
- D. the same length as a chromosome.

Question 4 (3 marks)

Source: *VCAA 2017 Biology Exam, Section B, Q1c*

Different cells within an organism have different proteins. In some cases different proteins can be coded for by the same gene.

Explain how the expression of a single gene can lead to the production of different proteins.

Question 5 (1 mark)

MC Which statement is correct regarding introns and exons?
- A. Introns are sections of pre-mRNA retained in the mRNA when it leaves the nucleus.
- B. Exons are sections of pre-mRNA that code for the translated polypeptide.
- C. After the introns are removed from pre-mRNA, the exons are also removed.
- D. Exons are the non-coding sections of the pre-mRNA.

More exam questions are available in your learnON title.

1.6 Gene regulation

KEY KNOWLEDGE

- The basic elements of gene regulation: prokaryotic *trp* operon as a simplified example of the regulatory process

The ability to regulate genes is important. There is a time and a place for every gene to be active — some genes are on all the time, others are only on during development, and others are only on in certain tissues. There are many ways to regulate genes. It is important to understand the types of genes to help understand this concept.

1.6.1 Types of genes

Eukaryotic cells have genes that contain a great level of complexity. Humans have around 21 000 genes, so the accurate transcription and translation of these genes is vital for the concise production of proteins. Therefore, there is a set structure for all genes in eukaryotic cells.

Genes vary in the functions that they carry out in the cells of an organism. Some genes produce proteins that become part of the structure and the functioning of the organism. These genes are termed **structural genes**.

Some genes produce proteins that control the action of other genes. These genes are termed **regulator genes**, and their actions determine whether other genes are active ('on') or not ('off') and, if active, the rate at which their products are made. Many of these act as DNA binding proteins, binding directly to sections on the DNA. Other act as signalling molecules, binding to receptors on the cell surface.

structural genes genes that produce proteins that contribute to the structure or functioning of an organism

regulator genes genes that produce proteins that control the activity of other genes

CASE STUDY: Homeotic genes

Regulatory genes that control embryonic development are known as homeotic genes. These are a type of master control gene. Homeotic genes control the action of hundreds of other genes that are needed to build the various parts of an animal body in their correct locations.

FIGURE 1.39 Fly with mutations in homeotic genes that result in the appearance of a second pair of wings instead of halteres and legs instead of antennae

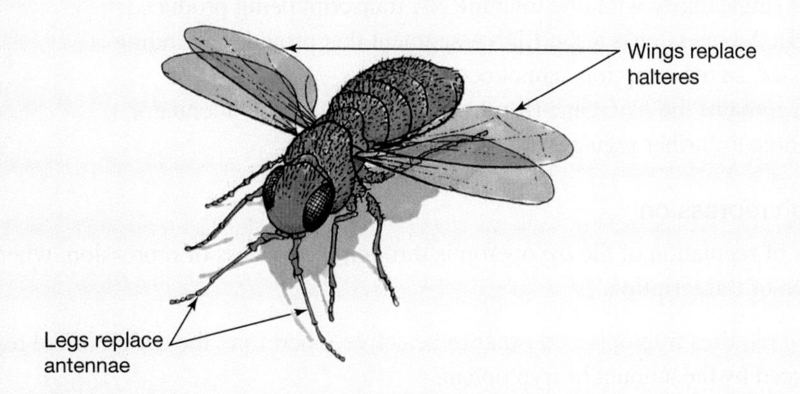

Mutations in 'master' or homeotic genes result in the appearance of body parts in unexpected locations.

To access more information on this case study, please download the digital document.

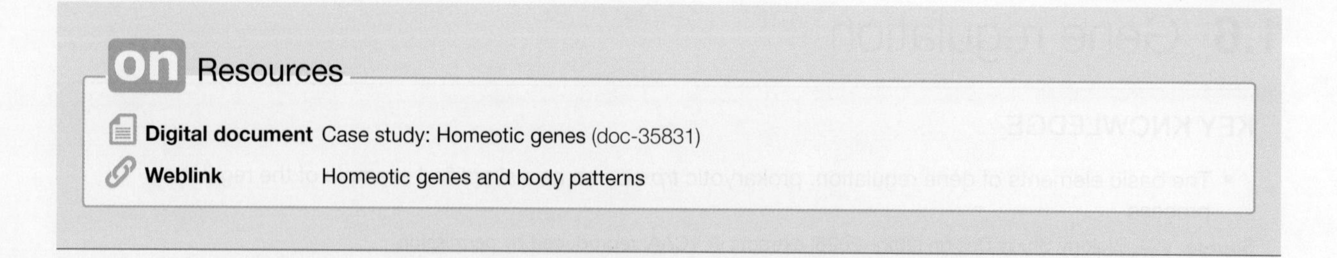

1.6.2 The *trp* operon

Tryptophan is an amino acid that the bacterium *Escherichia coli* (*E. coli*) is able to ingest from the surrounding environment. *E. coli* is also able to synthesise tryptophan using enzymes encoded by five genes. However, as to not waste energy, these enzymes are not produced in conditions when tryptophan is present. This is a basic example of gene regulation, where the genes corresponding to the enzymes to synthesise tryptophan are only active when required. This is controlled by the ***trp* operon.**

The tryptophan (*trp*) **operon** contains five structural genes that encode for the enzymes involved in the synthesis of tryptophan (such as tryptophan synthetase) with an upstream promoter (P_{trp}) and the tryptophan operator sequence (O_{trp}). These five structural genes are *trpA*, *trpB*, *trpC*, *trpD* and *trpE*. Another unique aspect to the operon is that the *trp* operator region partly overlaps the *trp* promoter. The operon regulates transcription when tryptophan is in low abundances in the cell.

An operon is group of linked structural genes with a common promoter and operator that is transcribed as a single unit.

The *trp* operon is comprised of the following components:

- **structural genes:** a gene that codes for any RNA or protein product other than a regulator (*trpA*, *trpB*, *trpC*, *trpD*, and *trpE*, which encodes tryptophan synthetase)
- a **regulatory gene:** a gene that codes for a product (typically protein) that controls the expression of other genes (usually at the level of transcription). In this case, the regulatory gene encodes for a repressor, which when active, binds to DNA and thus regulates the expression of genes by decreasing the rate of transcription.
- a **promoter (P_{trp})**: The promoter is a short DNA segment where RNA polymerase can attach and start transcription of the structural genes. The *trp* genes are transcribed as a single entity with one long mRNA transcript being produced.
- an **operator (O)**: An operator is a short DNA segment that provides a binding site for a repressor, so transcription cannot occur.
- a **leader** which contains the *trpL* gene (for the leader peptide) and attenuator section, that is able to further regulate transcription.

***trp* operon** a collection of adjacent genes in bacteria that code for the enzymes needed in the production of tryptophan

operon a cluster of adjacent structural genes in bacteria controlled by a single promoter and operating as a coordinated unit

Regulation through repression

One of the key ways of regulation of the *trp* operon is through the process of repression, where a repressor prevents the initiation of transcription.

The repressor protein requires tryptophan in order to be active. Therefore, the action of the repressor and in turn the operon is influenced by the amount of tryptophan.

When tryptophan is present, the operon is switched OFF and tryptophan synthetase is not produced.

When tryptophan is absent, the operon is switched ON and tryptophan synthetase is produced.

Presence of tryptophan

- When tryptophan is present, it binds to the repressor protein causing a configurational change in its shape, allowing it to be active.
- This allows the repressor to bind at the operator.
- Therefore, RNA polymerase is unable to bind to the promoter and transcription does not occur. Thus the operon is OFF (see figure 1.40a).

Absence of tryptophan

- When tryptophan is not present, the repressor is unable to bind to the operator (as it is still in an inactive form).
- This means that RNA polymerase can bind to the promoter and start transcription of the structural genes; thus, the operon is ON (see figure 1.40b).

FIGURE 1.40 Structure and regulation of the tryptophan operon in bacteria a. in the presence of tryptophan and b. in the absence of tryptophan

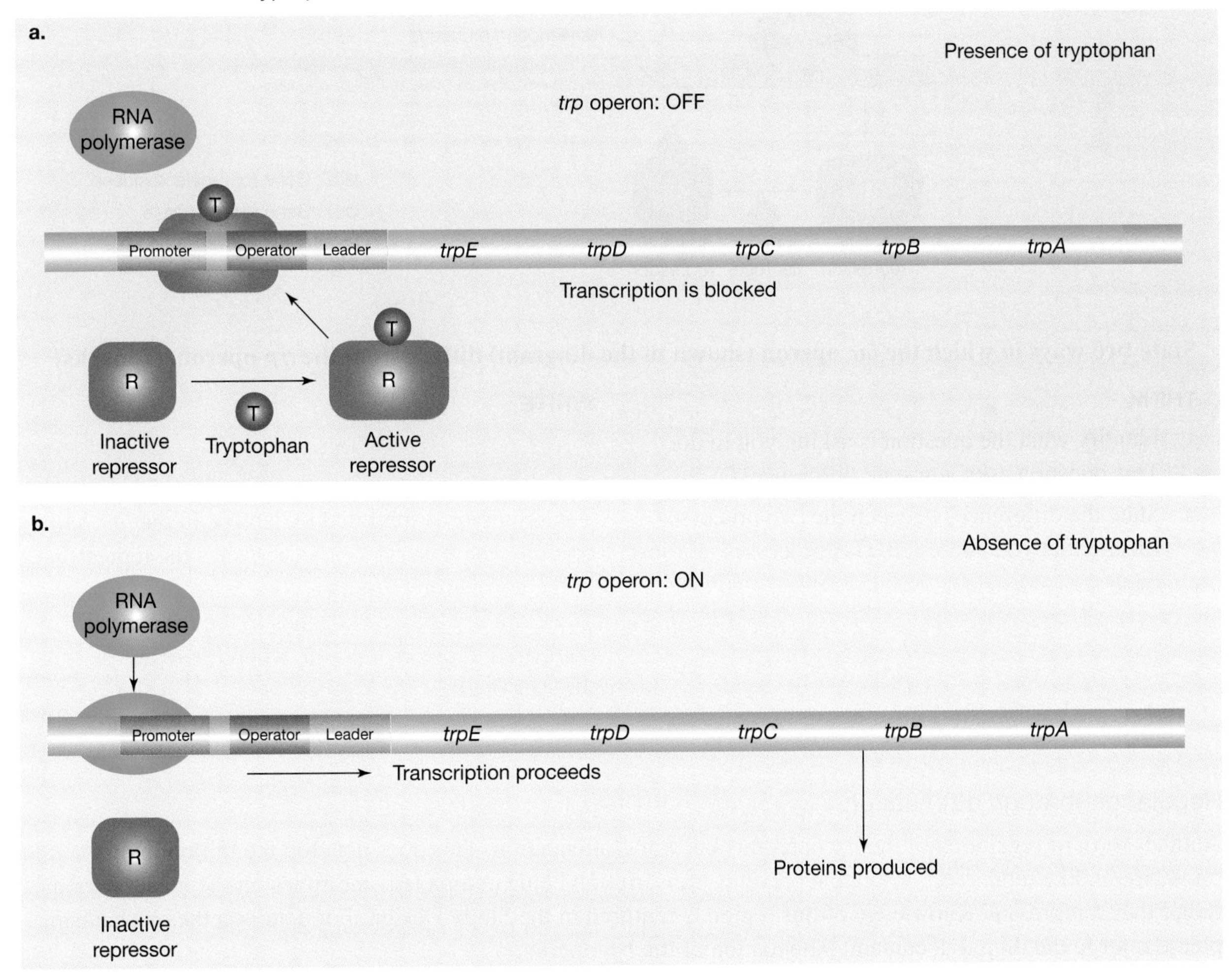

tlvd-1093

SAMPLE PROBLEM 4 Comparing operons

The diagram below shows another gene regulation system in bacteria, known as the *lac* operon.

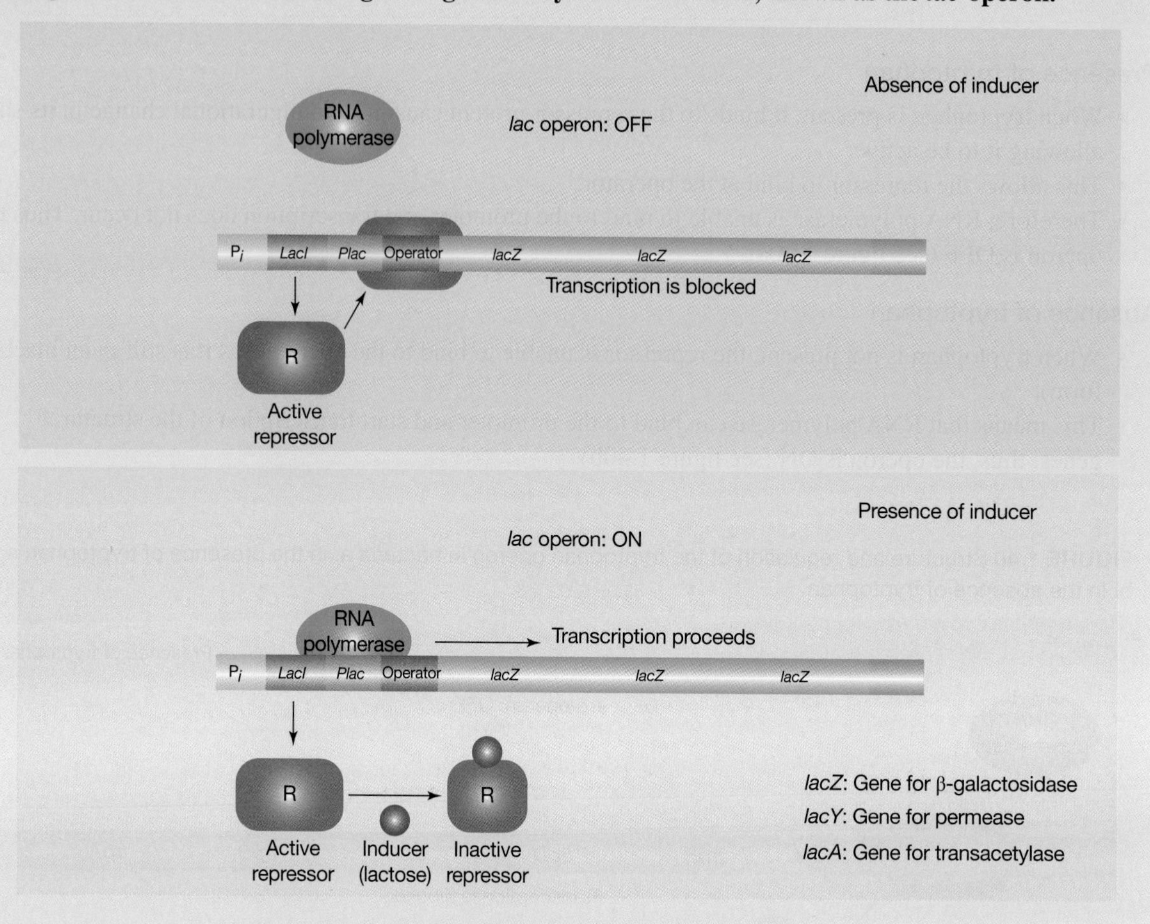

State two ways in which the *lac* operon (shown in the diagram) differs from the *trp* operon. (2 marks)

THINK	WRITE
1. Identify what the question is asking you to do. This question asks you state differences.	
2. State the differences between the two processes. Ensure that you address BOTH processes, not just one.	The repressor is inhibited by lactose in the *lac* operon, whereas tryptophan activates the repressor in the *trp* operon (1 mark). There are five structural genes in the *trp* operon compared to the *lac* operon, where there are three (1 mark).

Regulation through attenuation

Another form of regulation of the *trp* operon is known as attenuation. This does not stop the initiation of transcription like repression. Instead, it prevents transcription being completed, and relies on the leader region, rather than a repressor. Knowledge of this region is required in the Study Design. For detail on the mechanism, please refer to the digital document: Attenuation of the *trp* operon.

on Resources

eWorkbook Worksheet 1.8 Exploring the *trp* operon (ewbk-1974)

Video eLesson The *trp* operon (eles-5079)

Digital document Attenuation of the *trp* operon (doc-39265)

KEY IDEAS

- Genes can be classified as structural genes or as regulator genes.
- Regulator genes control the activity of other genes, switching them on or off, either directly through the action of DNA-binding proteins or indirectly through the action of signalling proteins.
- The *trp* operon is found in many bacteria to help synthesise tryptophan. An operon is a cluster of adjacent structural genes controlled by a single promoter.
- The structural genes in the *trp* operon are only expressed when tryptophan is absent.
- When tryptophan is present, it activates the repressor, which binds to a site on the operon known as an operator, preventing RNA polymerase from transcribing the gene.

1.6 Activities

learnon

To answer questions online and to receive **immediate feedback** and **sample responses** for every question, go to your learnON title at **www.jacplus.com.au**. A **downloadable solutions** file is also available in the resources tab.

1.6 Quick quiz on	1.6 Exercise	1.6 Exam questions

1.6 Exercise

1. Explain what is meant by the term operon.
2. How is the structure of an operon different to eukaryotic genes?
3. Why is it important for bacteria to only have *trp* genes transcribed in the absence of tryptophan?
4. If there is a mutation in the operator of the operon and the repressor was unable to bind, what would you expect:
   a. when tryptophan is present
   b. when tryptophan is absent?
5. Using the *trp* operon in *E. coli* as an example, explain how gene regulation by transcriptional factors expressed by regulatory genes can occur.
6. In what two ways is gene regulation in eukaryotes different from gene regulation in prokaryotes?
7. Define the functional difference between structural genes and regulatory genes.

1.6 Exam questions

Question 1 (1 mark)

Source: *Adapted from VCAA 2020 Biology Exam, Section A, Q26*

MC The *trp* operon in prokaryotes illustrates the switching off and on of genes.

The operator within the *trp* operon

A. is a regulatory gene.
B. attaches RNA polymerase.
C. codes for the production of an enzyme.
D. is the binding site for the repressor protein.

Question 2 (1 mark)

Source: *VCAA 2016 Biology Exam, Section A, Q32*

MC Which one of the following statements about gene regulation is correct?

A. Regulator genes are composed of mRNA.
B. Gene regulation is expressed only during the process of meiosis.
C. Regulator genes produce factors that alter the expression of another gene.
D. Gene regulation is not affected by environmental factors external to the cell.

Question 3 (1 mark)

***Source:** VCAA 2009 Biology Exam 2, Section A, Q20*

MC Regulatory and structural genes differ in their arrangement in the genomes of prokaryotic and eukaryotic cells. In prokaryotic cells, regulatory genes are arranged side by side. This arrangement is known as an operon. In eukaryotic cells, the regulatory genes and structural genes may be located on different chromosomes.

Therefore, it would be reasonable to say that

A. eukaryotic stem cells have all genes switched on.
B. the environment has no impact on whether a gene is switched on or off.
C. all bacterial operons are located on a large circular chromosome within the cell.
D. mutations in distant regulatory genes will have no effect on their related structural genes in eukaryotic cells.

Question 4 (4 marks)

***Source:** VCAA 2006 Biology Exam 2, Section B, Q7*

Organisms can regulate the expression of their genes in a number of ways.

a. Suggest why an organism regulates the expression of its genes. **1 mark**

One example in bacteria is the regulation of the expression of a gene which produces an enzyme (enzyme X) involved in the metabolism of the amino acid tryptophan. Enzyme X is only produced when tryptophan is in high concentration. This gene regulation involves several genes. Two of the genes include a gene for the production of enzyme X and an operator gene. If a protein, called a repressor protein, binds to the operator gene, transcription of the gene for enzyme X is stopped. If no repressor protein is bound to the operator, transcription of the gene for enzyme X occurs. A summary of this regulation is shown in Figure 1.

FIGURE 1

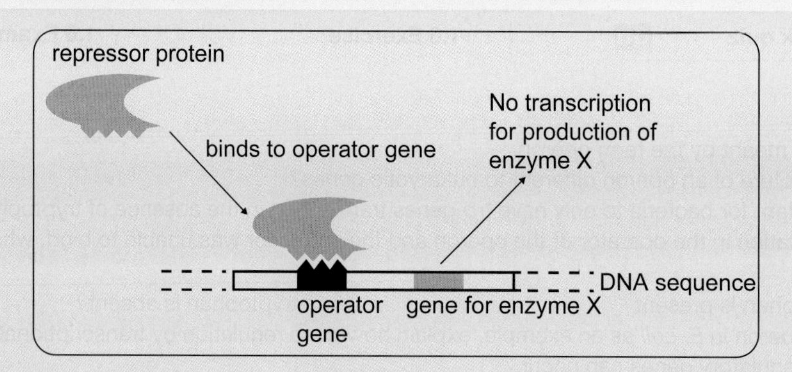

b. The gene coding for enzyme X is not transcribed when the repressor protein binds to the operator gene. What enzyme is prevented from functioning during this binding? **1 mark**

When tryptophan binds to the repressor protein, the repressor protein can no longer bind to the operator gene. (See Figure 2.)

FIGURE 2

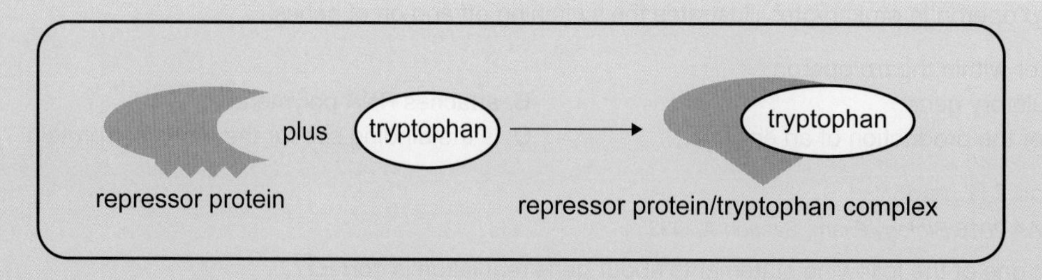

c. When tryptophan binds to the repressor protein what will happen to the production of enzyme X? **1 mark**
d. Based on Figure 2, suggest how tryptophan prevents repressor protein function. **1 mark**

Question 5 (5 marks)

Source: *VCAA 2007 Biology Exam 2, Section B, Q3*

Bacteria require amino acids to produce proteins. For example, bacteria in a human intestine may absorb amino acids from digested food, but at times there may be a deficiency of a particular amino acid. If this is the case, the bacteria will produce the necessary amino acid themselves.

The diagram below is a regulation system in a bacterial cell involving the production of the amino acid tryptophan. Note that there are two pathways (X and Y). Tryptophan is the regulatory compound in these two pathways and acts as a repressor in both.

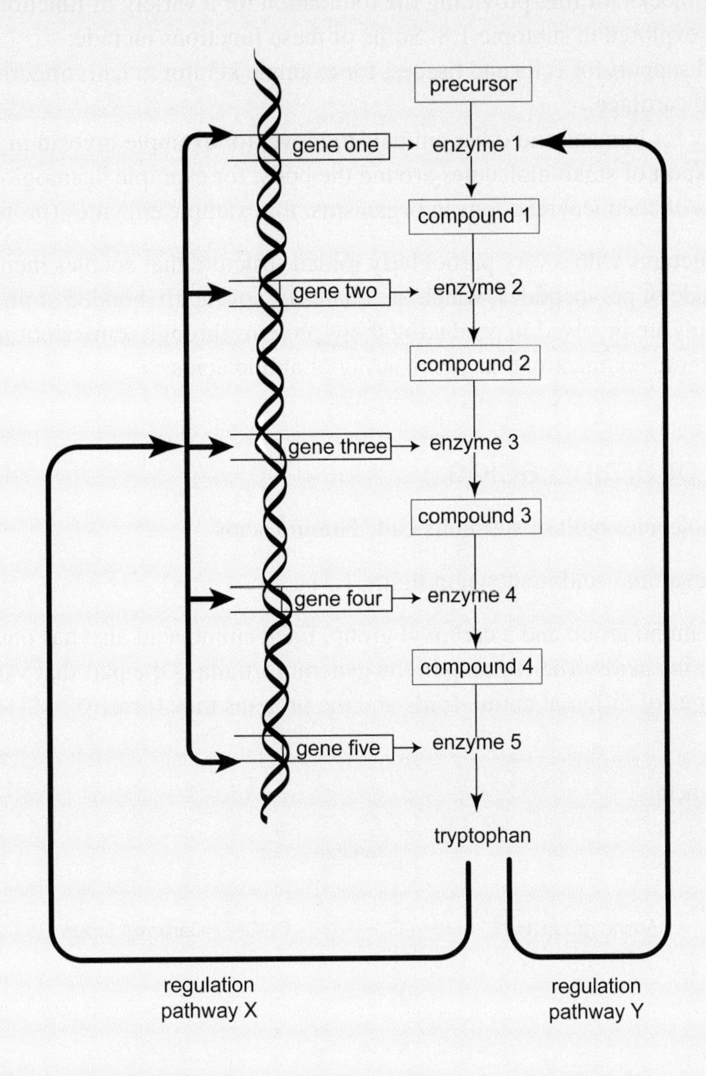

a. Explain what is meant by gene regulation. **1 mark**

b. Describe the immediate outcome when tryptophan activates pathway X. **1 mark**

c. Describe the immediate outcome when tryptophan activates pathway Y. **1 mark**

d. Suggest how the action of tryptophan as a repressor in this system could be of selective advantage to a bacterial cell in the digestive tract. **2 marks**

More exam questions are available in your learnON title.

1.7 Amino acids and polypeptides

KEY KNOWLEDGE

- Amino acids as the monomers of a polypeptide chain and the resultant hierarchical levels of structure that give rise to a functional protein

Source: VCE Biology Study Design (2022–2026) extracts © VCAA; reproduced by permission.

Proteins are the building blocks for life, providing the foundation for a variety of functions and structures. These functions will be further explored in subtopic 1.8. Some of these functions include:

- providing structural support for cells and tissues, for example keratin in hair, fingernails and skin, and collagen in skin and cartilage.
- providing the ability for humans and other animals to move, for example myosin in muscle.
- facilitating the transport of small molecules around the body, for example haemoglobin.
- controlling the rates of chemical reactions in organisms, for example enzymes (protease lipase, amylase).

Proteins are complex structures with a very particularly folded structure that enables them to achieve their function. Proteins are made of polypeptides, which are chains of covalently bonded amino acids. We have explored how nucleic acids are involved in producing these proteins through transcription and translation. However, proteins have more complexity than just a string of amino acids.

1.7.1 Structure of amino acids

Polypeptides are large molecules built of sub-units called amino acids.

The general formula of an amino acid is shown in figure 1.41.

All amino acids have an amino group and a carboxyl group. Each amino acid also has one part of its molecule that differs from other amino acids. The R group in the general formula is the part that varies. This causes great differences in the properties of different amino acids and the proteins they form.

FIGURE 1.41 An amino acid

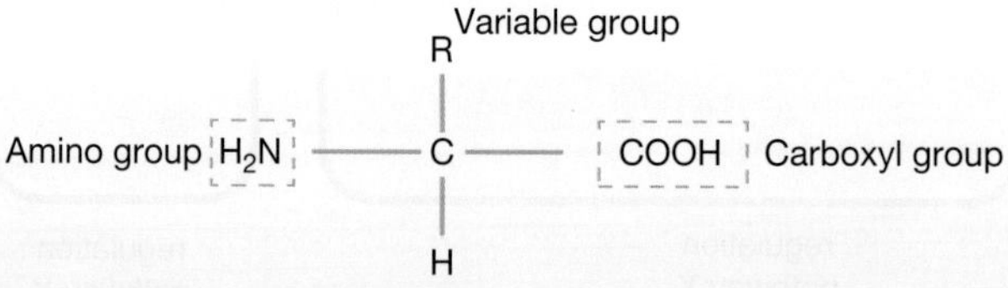

TABLE 1.12 A list of the 20 amino acids, colour-coded according to their properties: hydrophobic (blue), hydrophilic (pink), positively charged (green), negatively charged (orange) and hydrophobic and aromatic (purple)

Alanine	Arginine	Asparagine	Aspartic acid (or aspartate)	Cysteine
Glutamine	Glutamic acid (or glutamate)	Glycine	Histidine	Isoleucine
Leucine	Lysine	Methionine	Phenylalanine	Proline
Serine	Threonine	Tryptophan	Tyrosine	Valine

elog-0013

INVESTIGATION 1.6

online only

What's in a protein

Aim

To identify the presence of proteins and digest this protein into its monomers

EXTENSION: Where do amino acids come from?

There are 20 naturally occurring amino acids as listed in table 1.12 (refer to the Appendix for their structure). Humans are unable to make all 20 amino acids and must rely on their food for the nine they are unable to make. These include: histidine, isoleucine, leucine, lysine, methionine, phenylalanine, threonine, tryptophan and valine.

Not all plants can make all 20 amino acids, so a vegetarian diet should be well planned to ensure a balanced intake of appropriate amino acids. Generally, animal proteins are a better source of amino acids for humans because animal protein is more like that of humans.

1.7.2 Formation of a polypeptide chain

Proteins are assembled from amino acids that are joined by peptide bonds. Each peptide bond forms by the linkage of an amino group from one amino acid and a carboxyl group of another amino acid (refer to figure 1.42). A number of amino acids joined by peptide bonds form a **polypeptide** chain. This process requires an input of energy.

FIGURE 1.42 Proteins are assembled from amino acids that are joined by peptide bonds. Note that the different R groups are represented by different colours. The remainder of each amino acid molecule is identical.

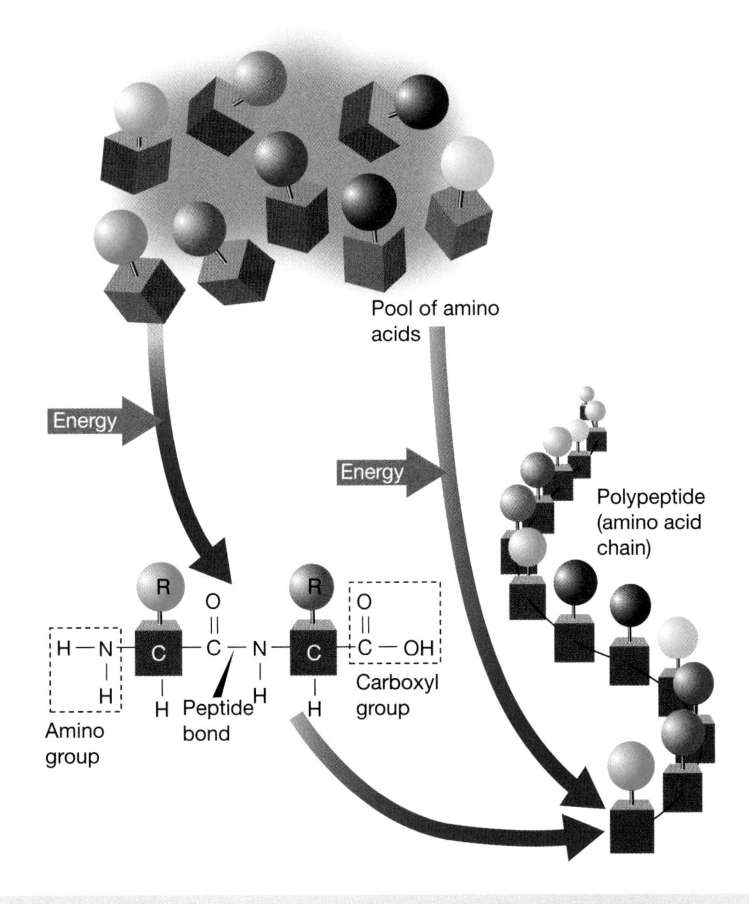

Condensation polymerisation to form polypeptides

The reaction that forms peptide bonds is an example of condensation polymerisation. Condensation refers to the fact that water is released. (In hydrolysis, which occurs during the breakdown of polypeptides, water is required.) The term polymerisation refers to monomers being bonded together to form a polymer (see figure 1.43).

This process requires energy in the form of ATP to allow for the formation of a peptide bond between amino acids.

FIGURE 1.43 Two monomers with different R groups come together by forming a peptide bond (CONH).

$$H_2N-CH_2-COOH + H_2N-CH(CH_3)-COOH \rightarrow H_2N-CH_2-CO-NH-CH(CH_3)-COOH + H_2O$$

Peptide bond

ε

glycine + alanine ⟶ dipeptide + water

1.7.3 Hierarchical structure of a functional protein

Polypeptides fold and organise into different shapes to form proteins. This folding depends on the amino acid sequence. Protein structure is described at four different levels of organisation (see figure 1.44).

The four levels of protein structure are:

- primary — the order of amino acids in the molecule
- secondary — folding of some portion of the amino acid chain (note the three different modes)
- tertiary — the shape of the entire polypeptide chain
- quaternary — some proteins comprise a number of polypeptide chains.

FIGURE 1.44 The four levels of protein structure. The structures are clearly interrelated.

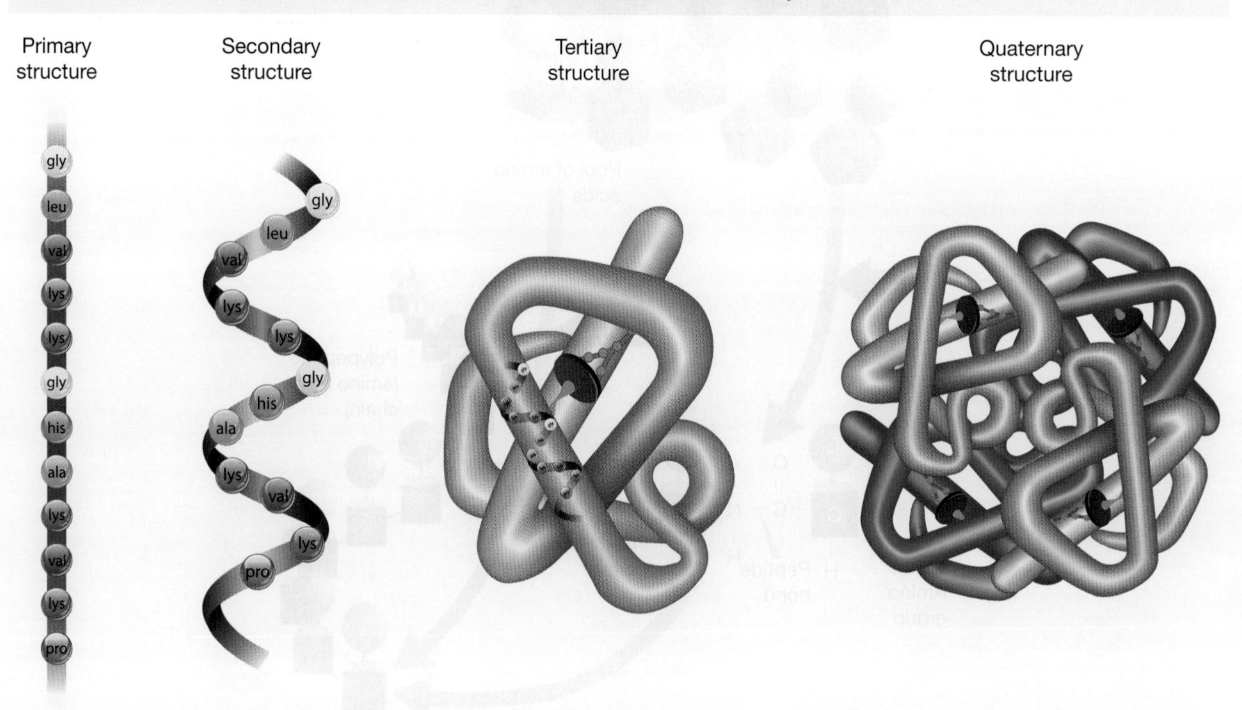

Primary structure

The primary structure of a protein is the specific linear sequence of amino acids in the protein (figure 1.45). Different proteins have different primary structures and hence have different functions. The sequence of amino acids in a protein is determined by the genetic material in the nucleus. These are brought together as a polypeptide during translation in the ribosome.

FIGURE 1.45 The primary structure of a protein

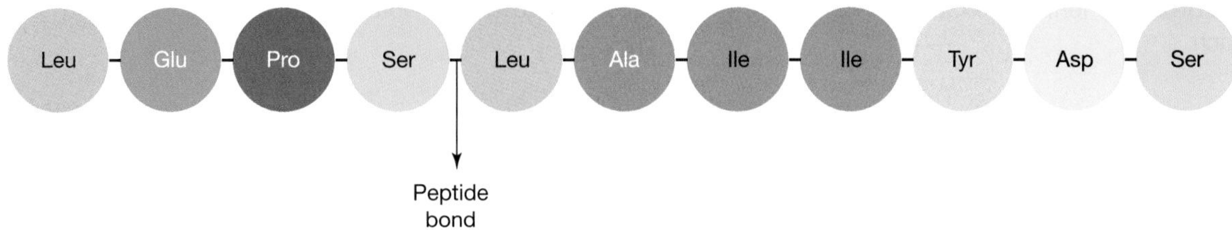

Secondary structure

The next level of protein structure is the **secondary structure**, where three different folds can occur in amino acid chains, depending on the R groups in the different amino acids (see figure 1.46).

Hydrogen bonds form between segments of the folded chain that have come close together and help stabilise the three-dimensional shape of the protein. The following are some examples of secondary structure.

1. **Alpha helix**: the major proteins of wool are keratins that have a spiral secondary structure, known as an alpha helix (see figure 1.46a). If the fibre is stretched and the hydrogen bonds are broken, the fibre becomes extended. If the fibre is then 'let go', the hydrogen bonds reform and the fibre returns to its original length. The secondary structure of myoglobin, the oxygen-binding protein of muscle, consists mainly (75%) of a coiled alpha helix structure.
2. **Beta-pleated sheet**: the major protein of silk is fibroin, which is fully extended and lacks the coiling found in the structure of wool. The silk molecules form a beta-pleated sheet (see figure 1.46b). The polypeptide chains of silk are already extended and cannot be extended further.
3. **Random coiling**: the secondary structure of portions of a protein is called random coiling if the portions do not conform to the shape of an alpha helix or a beta-pleated sheet (see figure 1.46c).

secondary structure a type of protein structure where three different folds of alpha helices, beta-pleated sheets and random coils can occur in amino acid chains, depending on the R groups in the different amino acids

alpha helix a type of secondary structure in proteins that appears as a tight twist

beta-pleated sheet a type of secondary structure in proteins that appears as folded sheets, with a change in direction of the polypeptide chain

random coiling a type of secondary structure in proteins that does not fit in as either a alpha helix or beta-pleated sheet

FIGURE 1.46 Examples of some different types of secondary structures: a. alpha helices seen in keratin; b. beta-pleated sheets found in silk fibroin; c. random coils seen in lactoferrin

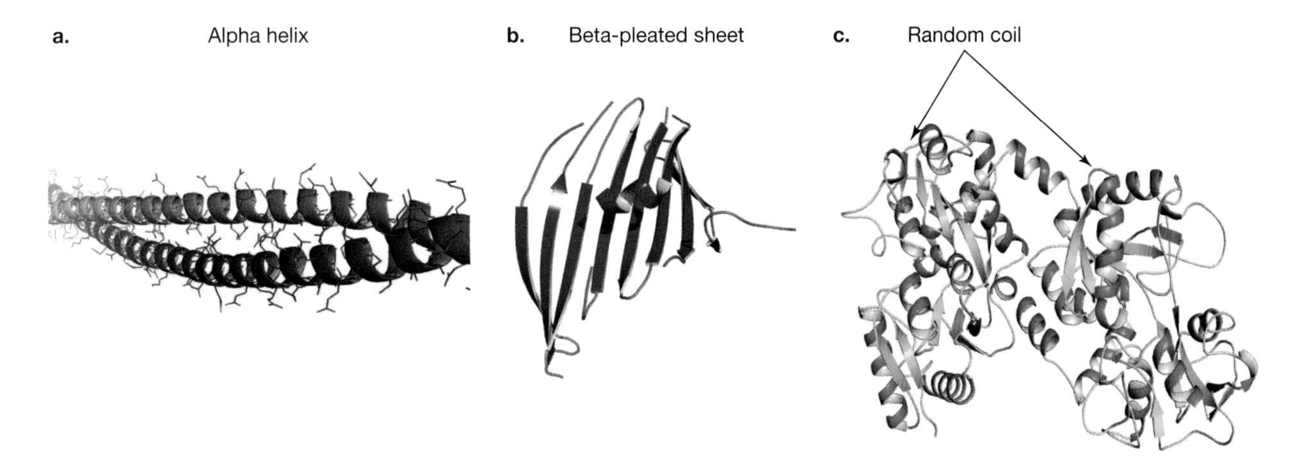

Tertiary structure

The **tertiary structure** refers to the total irregular 3D folding held together by ionic or hydrogen bonds forming a complex shape, such as that of myoglobin. The bonds form between side chains of amino acids to form a complex structure. The tertiary structure depends on both the primary and the secondary structures.

The 3D shape that constitutes the tertiary structure of a protein is critical for its function. For example, if the shape of an enzyme is changed, particularly at its active site, the protein can no longer function as an enzyme.

FIGURE 1.47 Tertiary structures of various proteins: **a.** thrombin; **b.** tubulin. Note the secondary structures in each.

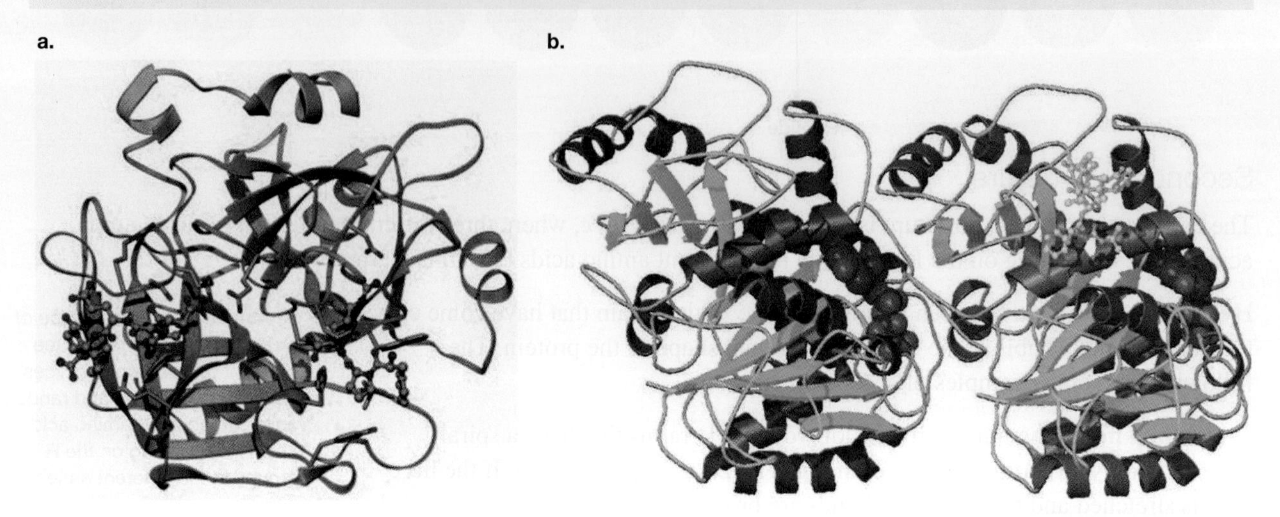

The forces that maintain the tertiary structure of proteins are:
- hydrogen bonds
- ionic attractions between charged R groups
- interactions between hydrophobic R groups in the protein interior
- covalent disulfide cross links.

These bonds differ in their strength and frequency. Disulfide bridges are the strongest bonds that can form in a tertiary structure. These are only able to occur between two cysteine amino acids, between the sulfur in their R groups.

tertiary structure the total irregular 3D folding of a protein held together by various bonds forming a complex shape

quaternary structure the final level of protein structure in which multiple polypeptides join together to form a protein complex

Quaternary structure

The **quaternary structure** describes a structure in which two or more polypeptide chains interact to form a protein. The resulting structure can be, for example, globular as in haemoglobin (figure 1.48) or fibrous as in collagen, the most common protein in skin, bone and cartilage.

In haemoglobin, the quaternary molecular structure comprises four chains: two alpha chains and two beta chains. The amino acid sequence in a protein is important. If the order of amino acids in either chain is altered, a defective chain results. An individual inherits, from each parent, the DNA that encodes the beta chain. If a defect in this DNA is inherited from both parents, an individual is unable to produce any normal haemoglobin and has the genetic disorder beta thalassaemia.

FIGURE 1.48 The secondary and tertiary structures seen within a quaternary haemoglobin structure

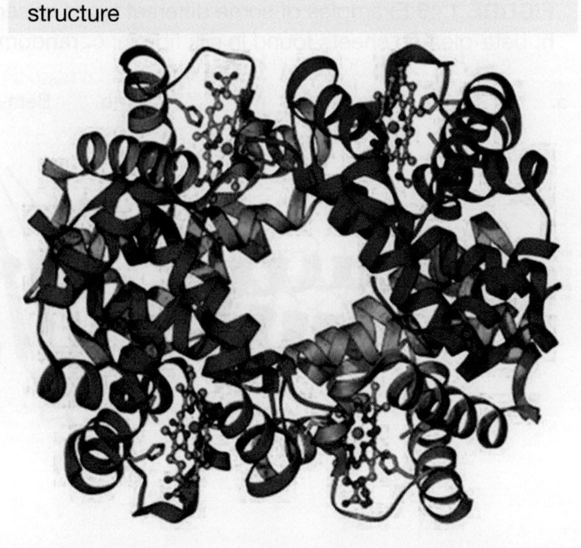

tlvd-1094

SAMPLE PROBLEM 5 Protein hierarchy

The image shows apoptosome, a protein involved in apoptosis or programmed cell death.

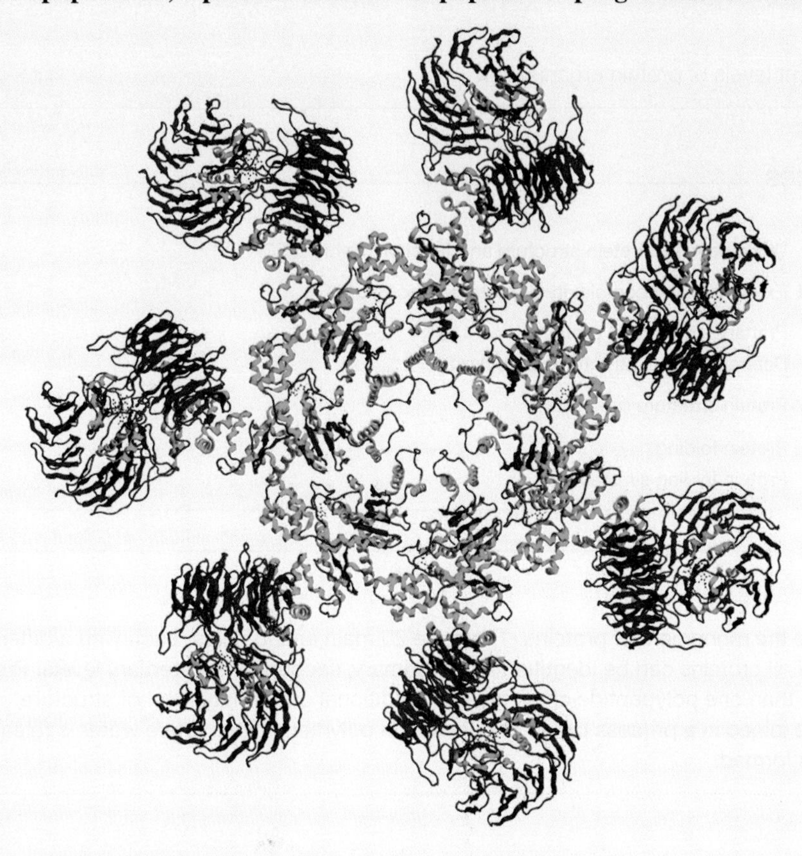

a. **Describe the primary structure of this protein.** **(2 marks)**
b. **What secondary structures can be seen? Justify your response.** **(2 marks)**
c. **What is the highest level of hierarchy seen in this protein? What evidence is there of this? (2 marks)**

THINK	WRITE
a. Consider the primary structure as the sequence of amino acids. TIP Ensure you refer back to apoptosome rather than making generic statements.	The primary structure of apoptosome is the sequence of amino acids (1 mark). These amino acids are joined by peptide bonds. Each amino acid differs in the R group — there would be many amino acids within the apoptosome (1 mark).
b. Consider the three secondary structures —alpha helices, beta-pleated sheets and random coils. For each of these, ensure you explain how you determined their presence, as the question asked for the response to be justified.	In the image, all three of alpha helices, beta pleated sheets and random coils can be seen (1 mark). The alpha helices are seen in green with their coiled structure. The beta-pleated sheets are visible in blue. The random coils are seen in red (1 mark).
c. As the protein has both tertiary and quaternary structures, quaternary must be the highest.	The highest level of protein structure is quaternary (1 mark). It is clear that apoptosome is made up of 7 different tertiary structures (1 mark).

elog-0804

INVESTIGATION 1.7

online only

Modelling protein structures

Aim

To model the different levels of protein organisation

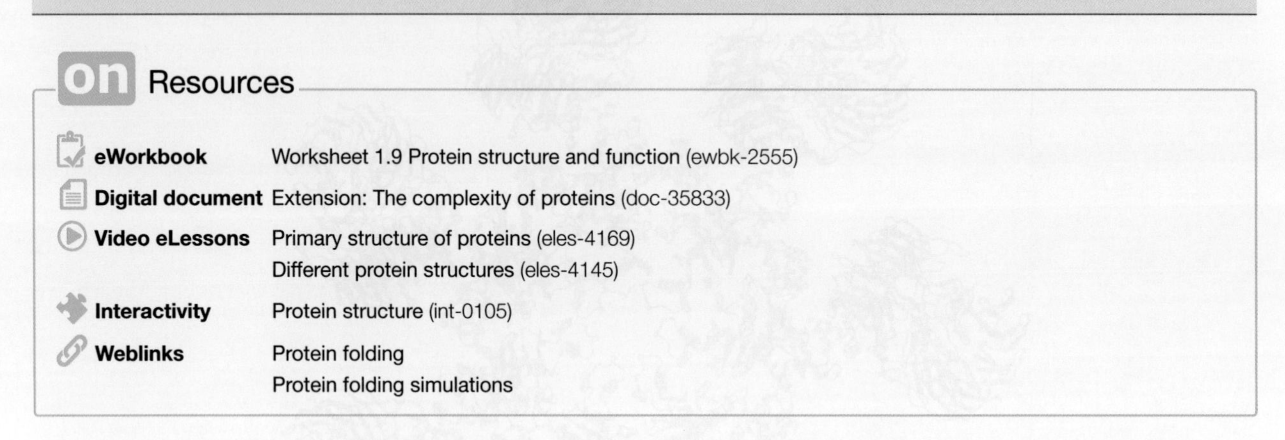

on Resources

eWorkbook	Worksheet 1.9 Protein structure and function (ewbk-2555)
Digital document	Extension: The complexity of proteins (doc-35833)
Video eLessons	Primary structure of proteins (eles-4169) Different protein structures (eles-4145)
Interactivity	Protein structure (int-0105)
Weblinks	Protein folding Protein folding simulations

KEY IDEAS

- Amino acids are the monomers of proteins. There are 20 main amino acids, each with a different R group.
- The structure of all proteins can be identified at the primary, secondary and tertiary levels. Proteins that consist of more than one polypeptide chain have an additional quaternary level of structure.
- Amino acids are joined in a process called condensation polymerisation, where water is released and a peptide bond is formed.

1.7 Activities

learn on

To answer questions online and to receive **immediate feedback** and **sample responses** for every question, go to your learnON title at **www.jacplus.com.au**. A **downloadable solutions** file is also available in the resources tab.

1.7 Quick quiz **on**	1.7 Exercise	1.7 Exam questions

1.7 Exercise

1. **MC** What is the tertiary structure of a protein?
   A. The total irregular folding held together by ionic, hydrogen or covalent bonds
   B. When two or more polypeptide chains interact to form a protein
   C. The folding of amino acid chains into alpha helices and beta-pleated sheets
   D. The linear sequence of amino acids with different R groups
2. **MC** What molecule has an amino group and a carboxyl group without any peptide bonds?
   A. A lipid B. A protein C. An amino acid D. A nucleotide
3. What is the basic formula of an amino acid molecule?
4. How is a peptide bond formed?
5. Describe the four hierarchical levels of protein structure.
6. How does the R group of an amino acid determine the secondary and tertiary structure of a protein?
7. Draw a clear diagram showing the different levels of protein structure.
8. Disulfide bridges can only form between cysteine amino acids.
   a. Suggest a reason that this may be the case.
   b. Explain what you would expect if a protein did not contain any cysteine.
   c. Methionine also contains sulfur but cannot form disulfide bridges. Research and explain why this is the case.

1.7 Exam questions

Question 1 (1 mark)

Source: *Adapted from VCAA 2018 Biology Exam, Section A, Q3*

MC The diagram represents adjacent amino acids being joined together.

The joining of adjacent amino acids

A. results in the formation of a nucleic acid.
B. is an energy-releasing reaction.
C. is catalysed by DNA ligase.
D. is a condensation reaction.

Question 2 (1 mark)

Source: *VCAA 2017 Biology Exam, Section A, Q1*

MC Consider the structure and functional importance of proteins.

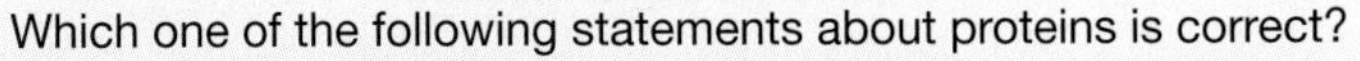

Which one of the following statements about proteins is correct?

A. A change in the tertiary structure of a protein may result in the protein becoming biologically inactive.
B. Proteins with a quaternary structure will be more active than proteins without a quaternary structure.
C. Two different proteins with the same number of amino acids will have identical functions.
D. Denaturation will alter the primary structure of a protein.

Question 3 (1 mark)

Source: *VCAA 2008 Biology Exam 1, Section A, Q12*

MC Insulin is a complex protein that is said to have a quaternary structure.

This means that insulin

A. cannot be denatured.
B. lacks disulphide bridges.
C. contains all the known amino acids.
D. has more than one polypeptide chain.

Question 4 (2 marks)

Source: *VCAA 2015 Biology Exam, Section B, Q1a*

The diagrams below represent examples of three levels of structure with respect to the folding and assembly of a protein. The diagrams are not to scale.

A.

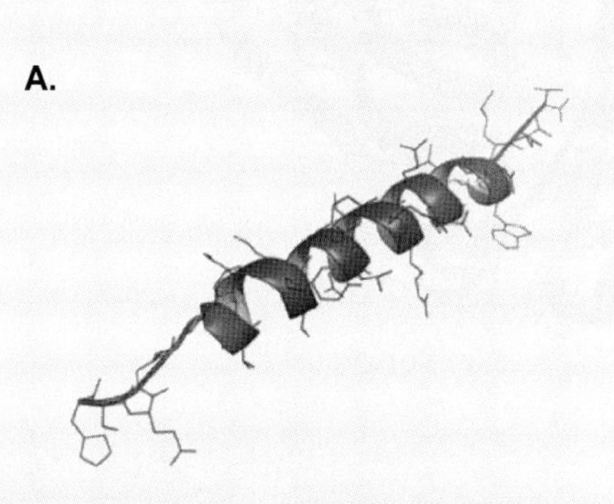

alpha-helix

B.

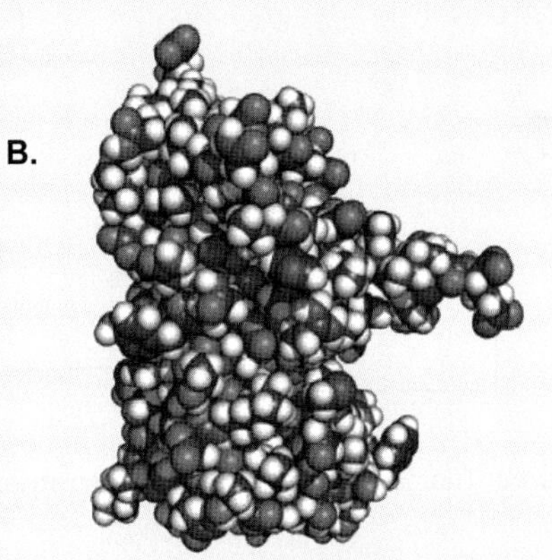

human interferon

C.

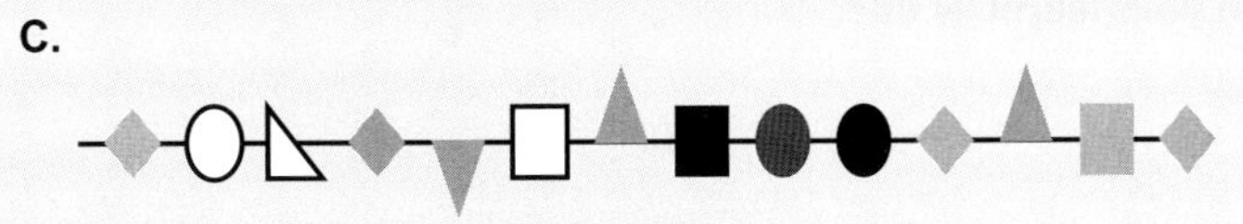

a polypeptide sequence

i. Complete the table below to indicate the diagram that represents the structural level of the protein given. **1 mark**

Structural level of protein	Diagram (A., B. or C.)
primary	
secondary	
tertiary	

ii. Name the molecular sub-unit of a protein. **1 mark**

Question 5 (5 marks)

Source: *VCAA 2016 Biology Exam, Section B, Q1*

a. Immunoglobulins, or antibody molecules, have an important role in the immune system. They are made up of two heavy chains and two light chains. Name the molecular monomer of these chains. **1 mark**

Part of a mouse immunoglobulin molecule bound to an antigen is shown in the diagram below.

Two arrows point to two different types of secondary structures of the immunoglobulin molecule.

b. Give the name of each structure in the boxes provided. **2 marks**

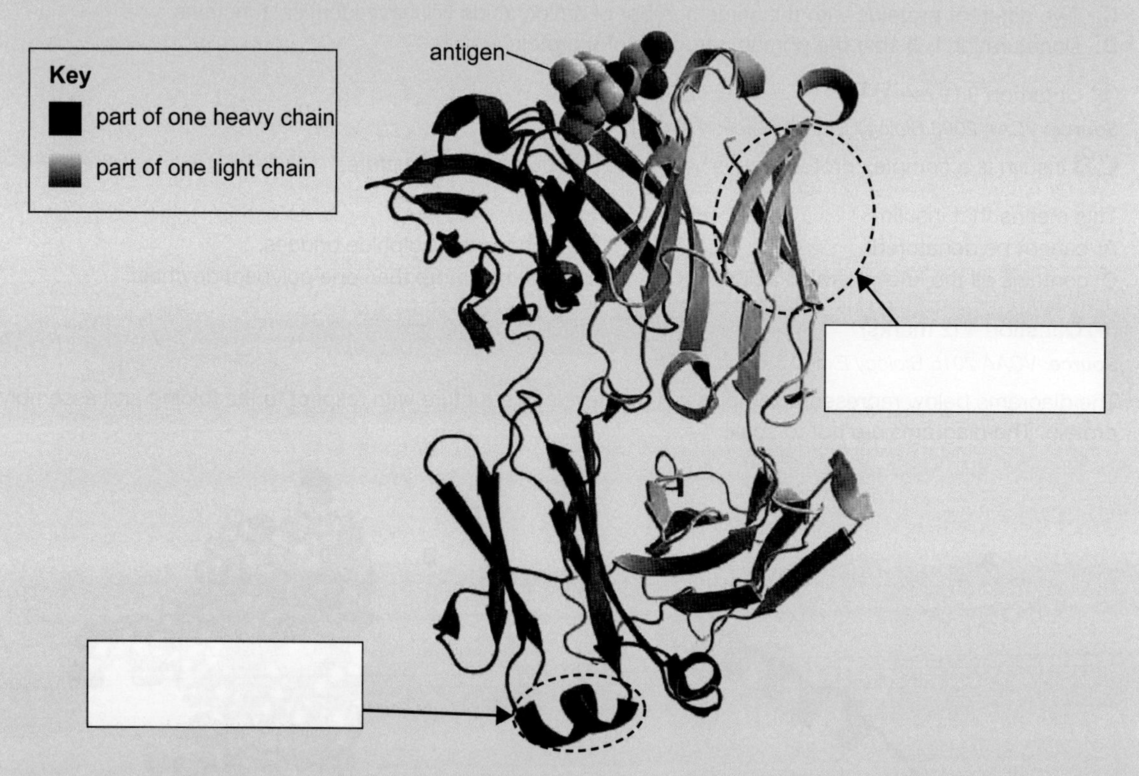

Source: Thomas Splettstoesser (www.scistyle.com)

c. Immunoglobulin molecules also display a tertiary structure and a quaternary structure. Referring to the diagram, explain what 'quaternary' means. **2 marks**

More exam questions are available in your learnON title.

1.8 The proteome

KEY KNOWLEDGE

- Proteins as a diverse group of molecules that collectively make an organism's proteome, including enzymes as catalysts in biochemical pathways.

1.8.1 The diversity of proteins

Although water is the main compound in living cells, about 18 per cent of living cells is protein. There are thousands of different proteins in each cell, and many of these control the metabolic processes. These proteins are made up of one or more polypeptide molecules, which in turn are made up of amino acid monomers.

A polypeptide on its own may be a functional protein; in other cases, a functional protein may be built of several different polypeptide chains.

Proteins are very diverse polymers, greatly differing in their structure and their role. Table 1.13 lists a number of different types of protein and their functions. Some examples of proteins are also shown in figure 1.49.

TABLE 1.13 Examples of proteins and their functions.

Type of protein	Function	Example
Structural	Fibrous support tissue in skin, bone, tendons, cartilage, blood vessels, heart valves, and cornea of the eye	Collagen, keratin
Enzyme	Catalyse reactions	ATP synthase
Contractile	Muscle movement	Myosin, actin
Immunoglobulin	Defence against disease	Antibodies
Hormone	Regulate body activity	Insulin, growth hormones
Receptor	Respond to stimuli	Insulin receptors
Transport	Carry other molecules	Haemoglobin

FIGURE 1.49 a. A scanning electron micrograph (SEM) of collagen bundles from connective tissue that wraps around and supports nerve fibres (notice the characteristic banding of collagen fibres) **b.** An SEM of skeletal muscle fibre showing the thick filaments that are made up of myosin **c.** A transmission electron micrograph (TEM) showing Y-shaped antibodies (yellow)

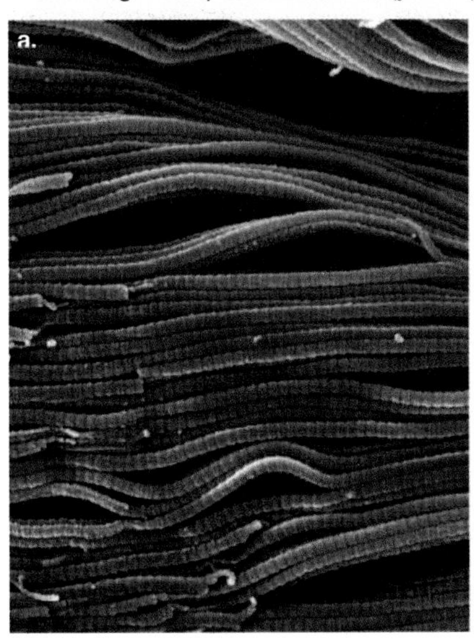

The role of enzymes as catalysts

Catalysts are substances that speed up the rate of chemical reactions without the catalyst itself being used up in the reactions. The basic function of enzymes is to increase the rate of almost all the chemical reactions in living organisms and to do this within the prevailing conditions of temperature and pH within cells. In Topic 3, we will explore how enzymes speed up the rates of chemical reactions in cells.

Consider the following scenario. The steak that you recently ate was broken down in the acidic conditions of your stomach at body temperature over a period of several hours. This breakdown was the result of the catalytic action of digestive enzymes such as pepsin and trypsin acting on the protein of the steak. To do the same without enzymes also requires acidic conditions, but would require a few days (not hours) and a temperature of 100 °C (not body temperature). Without enzymes, the speed of biochemical reactions would be far too slow to sustain the living state.

FIGURE 1.50 Different enzymes involved in digestion

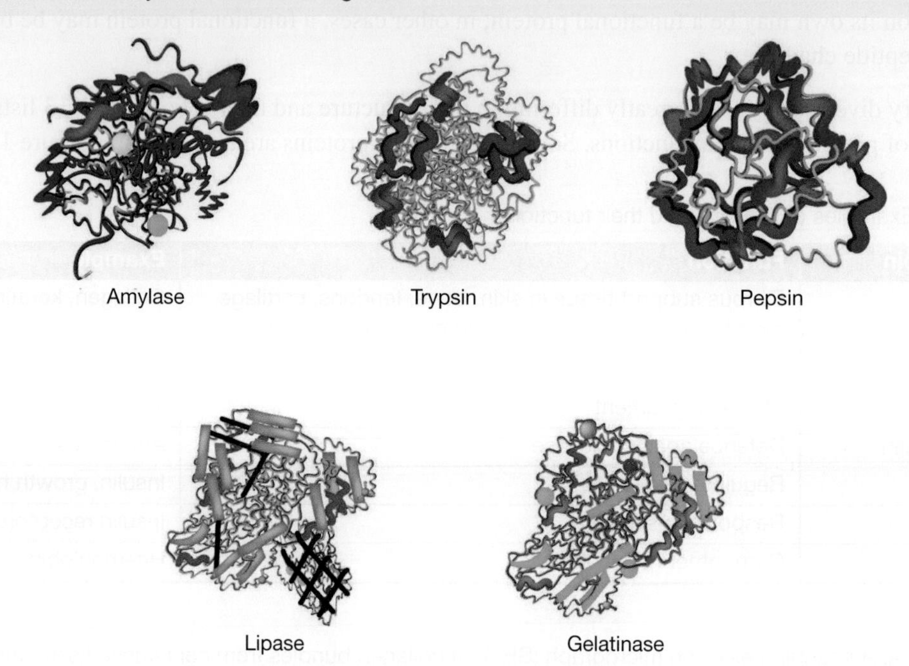

EXTENSION: Are all enzymes proteins?

The exception to the statement 'All enzymes are proteins' was discovered by Thomas Cech (b. 1947) and Sidney Altman (b. 1939), who discovered ribozymes. Ribozymes are catalytic RNA molecules that can cut themselves out of long RNA sequences.

1.8.2 What is the proteome?

In living organisms, proteins are involved in one way or another in virtually every chemical reaction. They may be the enzymes involved, they may be the reactants or the products, or they may be all three. The complete array of proteins produced by a single cell or organism in a particular environment is called the **proteome** of the cell or organism. The study of the proteome is called **proteomics**.

Scientists are moving away from investigating single proteins as no protein acts in isolation from other proteins. They are now exploring the total pattern of proteins produced by a cell and analysing these patterns to compare them with patterns from different kinds of cells.

proteome the complete array of proteins produced by a single cell or an organism in a particular environment

proteomics the study of the proteome, the complete array of proteins produced by an organism

Some questions when exploring various aspects of the proteome include:

- What are the differences?
- What are the similarities?
- What is the proteome profile of diseased tissue or even the fluids surrounding the tissue?
- In what ways do they differ from the healthy state?

Knowing that a protein exists is different from knowing how that protein operates. Understanding the structure and function of various proteins enables us to make testable predictions about the roles of other proteins on the basis of their structures. Knowledge of three-dimensional structures of well-known proteins may give insights into their functions.

FIGURE 1.51 The proteome has a huge level of complexity compared to the genome and transcriptome.

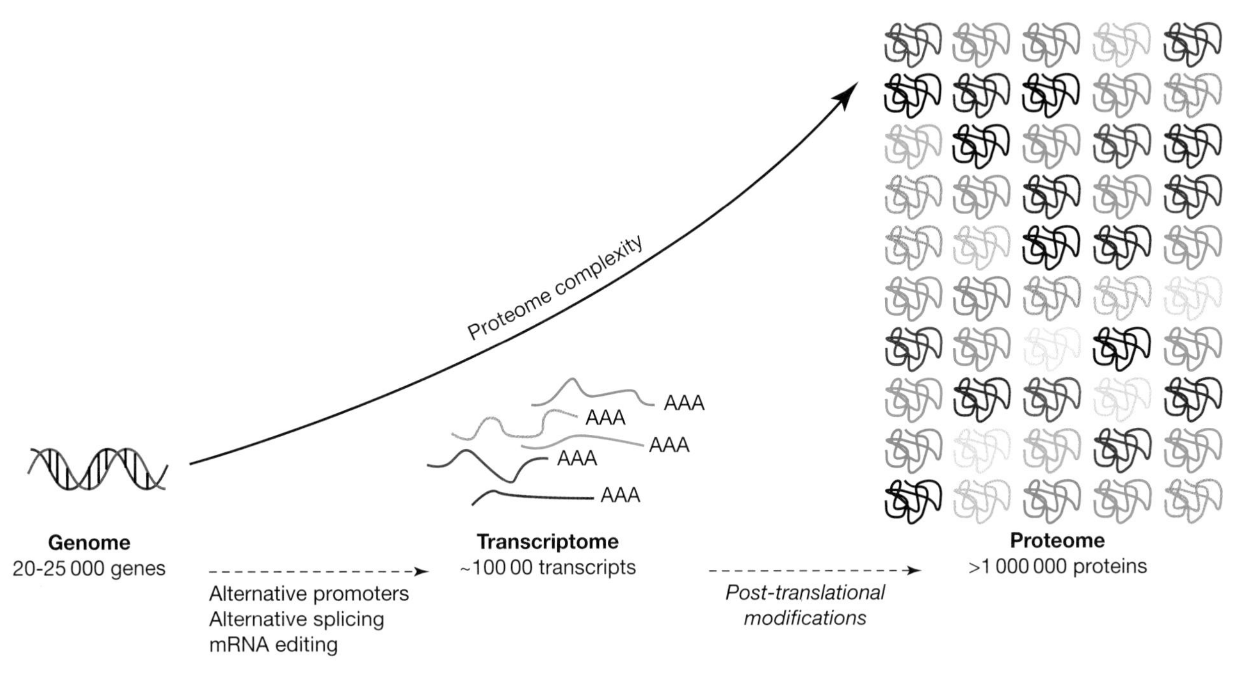

KEY IDEAS

- Proteins are very diverse, having functions involving structure, regulation and transport.
- A key role of proteins is to act as biological catalysts to speed up reactions. These proteins are known as enzymes.
- The complement of proteins is known as the proteome. It is useful to study proteins together as many influence other proteins.

1.8 Activities

learnON

To answer questions online and to receive **immediate feedback** and **sample responses** for every question, go to your learnON title at **www.jacplus.com.au**. A **downloadable solutions** file is also available in the resources tab.

1.8 Quick quiz on	1.8 Exercise	1.8 Exam questions

1.8 Exercise

1. MC What is the proteome?
   A. All the genetic material in a cell that codes for proteins
   B. The complete array of proteins produced by a single cell or organism
   C. The examination of proteins with quaternary structures
   D. The monomer of polypeptides
2. Give an example of:
   a. a structural protein
   b. a contractile protein
   c. a conjugated protein.
3. a. What is a proteome?
   b. What is the difference between the proteome and the genome of a species?
   c. What are two advantages of studying the proteome over the genome?
4. Why is proteomics considered important?
5. An individual has a deficiency in an enzyme that catalyses the breakdown of lactose.
   a. What are the possible consequences of this?
   b. With reference to genes and proteins, explain how this may occur.

1.8 Exam questions

Question 1 (1 mark)

Source: *VCAA 2019 Biology Exam, Section A, Q5*

MC Which one of the following statements about proteins is correct?

A. The activity of a protein may be affected by the temperature and pH of its environment.
B. The primary structure of a protein refers to its three-dimensional protein shape.
C. Proteins are not involved in the human immune response.
D. A protein with a quaternary structure will be an enzyme.

Question 2 (1 mark)

Source: *VCAA 2018 Biology Exam, Section A, Q2*

MC The proteome is

A. the total DNA content that is present within one cell of an organism.
B. a complete set of chromosomes found inside a cell of an organism.
C. the entire set of proteins expressed by an organism at a given time.
D. the four hierarchical levels of protein structure.

Question 3 (1 mark)

Source: *VCAA 2012 Biology Exam 1, Section A, Q8*

MC Protein forms part of the structure of

A. polysaccharides.
B. transfer RNA.
C. phospholipids.
D. haemoglobin.

Question 4 (1 mark)

Source: *VCAA 2011 Biology Exam 1, Section A, Q1*

MC The term used to indicate all proteins in an organism is

A. protozoa.
B. protease.
C. proteome.
D. proterozoic.

Question 5 (5 marks)

Source: *Adapted from VCAA 2006 Biology Exam 1, Section B, Q1a*

Scientists are now turning to the study of the proteome (all of the proteins) of an organism rather than the study of single proteins.

a. Briefly outline **one** reason why the emphasis is now on the study of all the proteins of an organism rather than on one protein at a time. **1 mark**

Protein molecules come in many shapes and forms that can be classified into primary, secondary, tertiary and quaternary. The secondary, tertiary and quaternary shapes arise as a result of different kinds of folding of a primary structure. One kind of secondary structure is a pleated sheet where the primary molecule extends along the folded sheet. The primary structures in the layers are held together by hydrogen bonding.

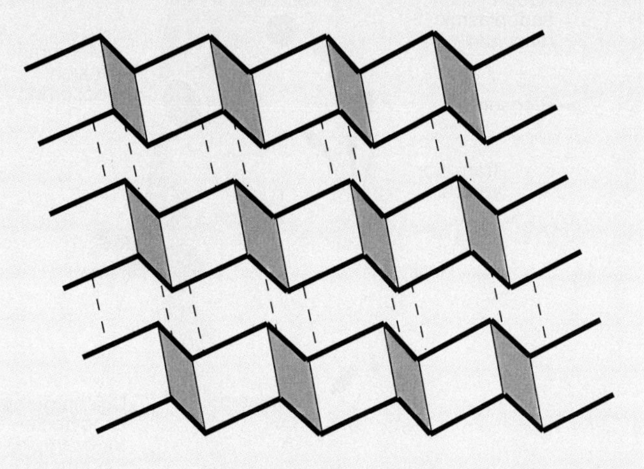

b. Explain why such a structure may be important in the function of a particular protein. **1 mark**

c. Proteins can also be classified on the basis of their general function. Three of these functions are listed below.

- structural
- transport
- regulatory

Give an example of a protein for each of the functions listed. **3 marks**

More exam questions are available in your learnON title.

1.9 Organelles involved in the protein secretory pathway

KEY KNOWLEDGE

- The role of the rough endoplasmic reticulum, Golgi apparatus and associated vesicles in the export of proteins from a cell via the protein secretory pathway

Source: VCE Biology Study Design (2022–2026) extracts © VCAA; reproduced by permission.

Cells make a range of proteins for many purposes, such as the manufacture of haemoglobin, an oxygen-transporting protein by developing human red blood cells in the bone marrow; the manufacture of the contractile proteins actin and myosin by the muscle cells; and the manufacture of the hormone insulin and digestive enzymes including lipases by different cells of the pancreas. Insulin is an example of a protein that is exported from the cell in which it was manufactured.

Proteins produced in ribosomes at the **rough endoplasmic reticulum** that are intended for export must be transferred to the Golgi apparatus and then to secretory **vesicles**. As there is no direct connection between the membranes of the endoplasmic reticulum and the Golgi apparatus, the proteins are shuttled from the rough endoplasmic reticulum to the Golgi apparatus in membrane-bound transition vesicles. Once there, the proteins are taken into the Golgi apparatus, where they are concentrated and packaged into secretory vesicles. These vesicles with their protein cargo move to the plasma membrane of the cell, merge with it and discharge their contents.

FIGURE 1.52 The secretory export pathway for proteins

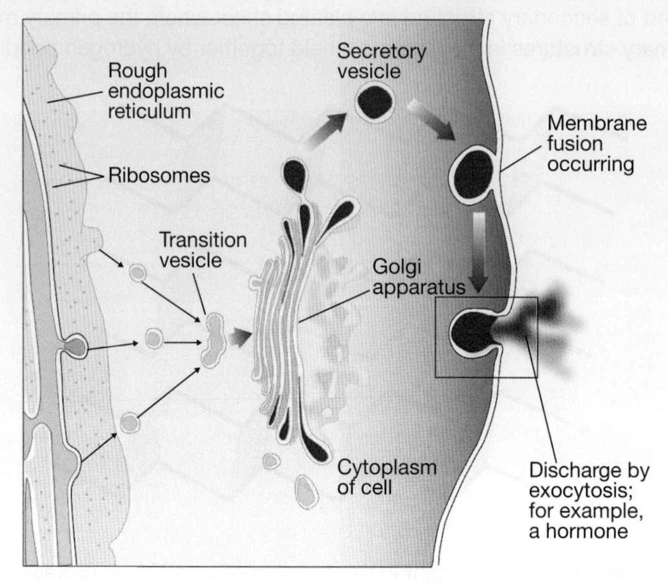

1.9.1 Ribosomes

Ribosomes are the sites in a cell where proteins are made. Eukaryotic ribosomes are composed of strands of rRNA (ribosomal RNA) merged with about 80 ribosomal proteins (r-proteins). It is on the ribosomes that amino acids are assembled and joined into polypeptide chains or proteins.

rough endoplasmic reticulum endoplasmic reticulum with ribosomes attached

vesicles membrane-bound sacs found within a cell, such as secretory vesicles, which are involved in the export of proteins

Ribosomes are the sites of translation, in which amino acids are joined to form polypeptides.

To produce any protein, its specific sequence of amino acids must be assembled, one by one, at the ribosome and joined by peptide bonds. Ribosomes can join amino acids into a protein chain at the rate of about 200 per minute.

The diameter of a ribosome is only about 0.03 μm. Because of their very small size, ribosomes can be seen only through use of an electron microscope (see figure 1.54). However, ribosomes are present in very large numbers in a cell. Unlike many organelles, they are not enclosed within a membrane.

Ribosomes were discovered in 1955 by George Palade (1912–2008) and were originally called Palade particles. Palade was awarded a Nobel Prize in 1974.

Within a cell, many ribosomes are attached to membranous channels known as the endoplasmic reticulum. An estimated 13 million ribosomes are attached to the rough endoplasmic reticulum in a typical

FIGURE 1.53 Ribosomes

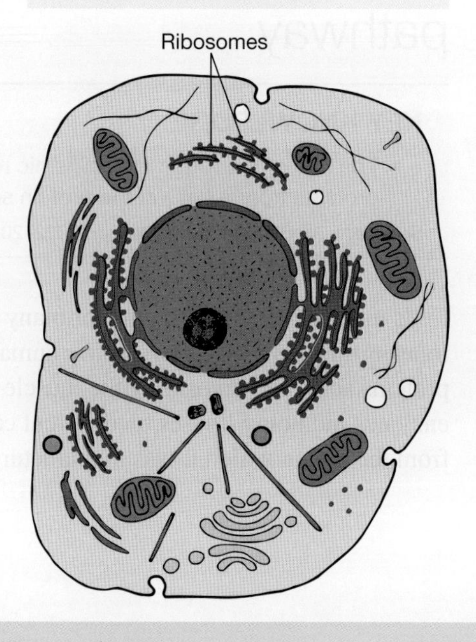

human liver cell. Other ribosomes are found free in the cytosol. Proteins made by 'free' ribosomes are for local use within the cell.

FIGURE 1.54 a. TEM image of a section of cell showing the rough endoplasmic reticulum (er) with ribosomes (ri) and b. a 3D representation of ribosomes on the rough endoplasmic reticulum

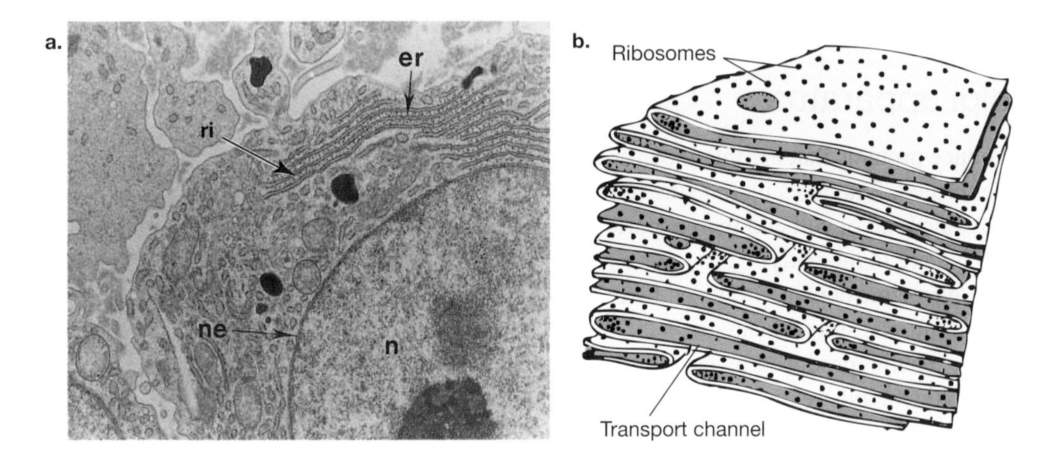

1.9.2 Rough endoplasmic reticulum

The endoplasmic reticulum is an interconnected system of membrane-enclosed flattened channels. Figure 1.55 shows part of the channels of the endoplasmic reticulum in a eukaryotic cell. When the endoplasmic reticulum has ribosomes attached to the outer surface of its channels, it is known as rough endoplasmic reticulum. Proteins produced by ribosomes on the endoplasmic reticulum are generally exported from the cell.

Through its network of channels, the rough endoplasmic reticulum is involved in transporting some of the proteins to various sites within a cell.

FIGURE 1.55 a. Rough endoplasmic reticulum b. False-coloured scanning electron micrograph of part of a eukaryotic cell showing channels of rough endoplasmic reticulum

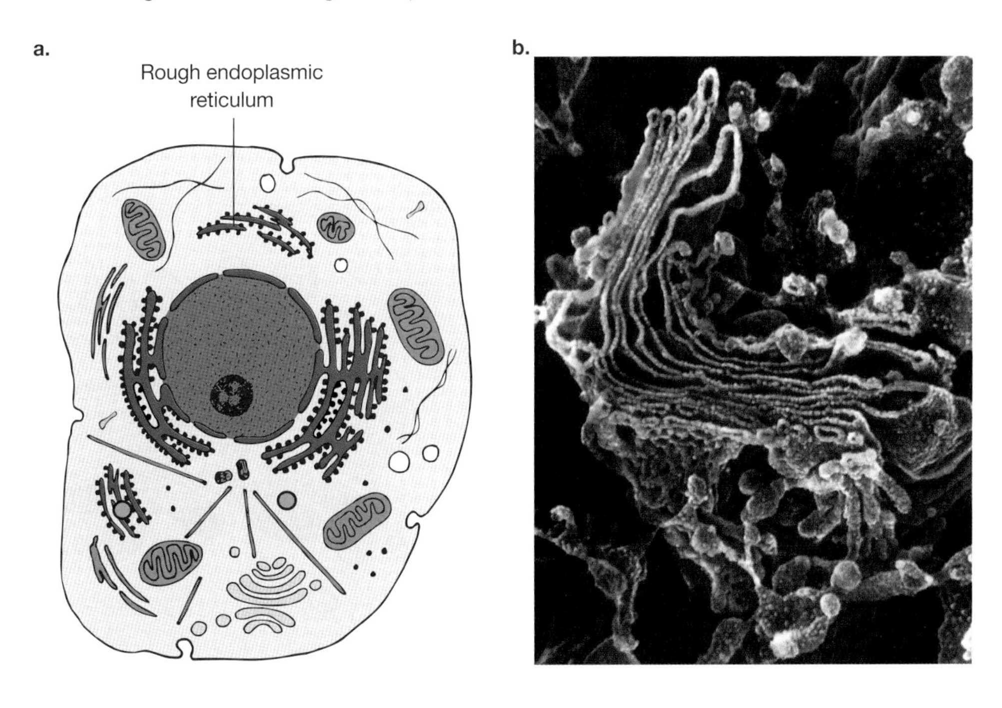

Proteins delivered from the ribosomes into the channels of the rough endoplasmic reticulum are also processed before they are transported. The processing of proteins within the rough endoplasmic reticulum includes:

- attaching sugar groups to some proteins to form glycoproteins
- folding proteins into their correct functional shape or conformation
- assembling complex proteins by linking together several polypeptide chains, such as the four polypeptide chains that comprise the haemoglobin protein.

1.9.3 Golgi apparatus

Some cells produce proteins that are intended for use outside the cells where they are formed. Examples include the following proteins that are produced by one kind of cell and then exported (secreted) by those cells for use elsewhere in the body:

- the digestive enzyme pepsin, produced by cells lining the stomach and secreted into the stomach cavity
- the hormone insulin, produced by cells in the pancreas and secreted into the bloodstream
- protein antibodies, produced in special lymphocytes and secreted at an area of infection.

How do these substances get exported from cells? The cell organelle responsible for packaging substances for export out of cells is the Golgi apparatus, also known as the Golgi complex.

The Golgi apparatus has a multi-layered structure composed of stacks of membrane-lined channels (see figures 1.56 and 1.57). The Golgi apparatus is named after Camillo Golgi, who in 1898 first identified this cell organelle.

FIGURE 1.56 The Golgi apparatus or Golgi complex

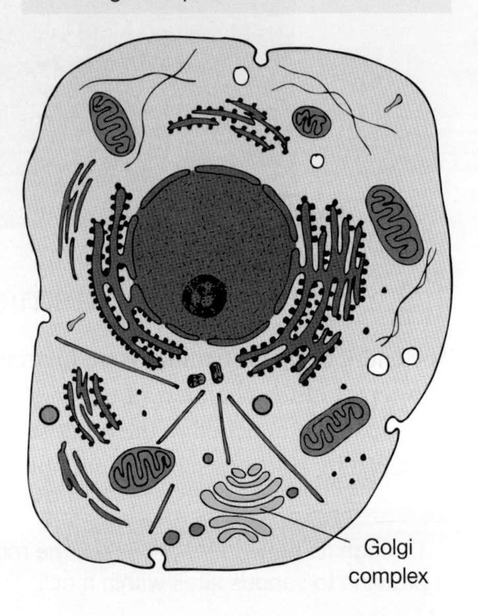

The Golgi apparatus packages proteins into vesicles for export from the cell through exocytosis.

FIGURE 1.57 a. False coloured TEM image of the Golgi apparatus (orange). Note the stacks of flattened membrane-lined channels with their wider ends that can break free as separate vesicles. **b.** 3D representation of the Golgi apparatus

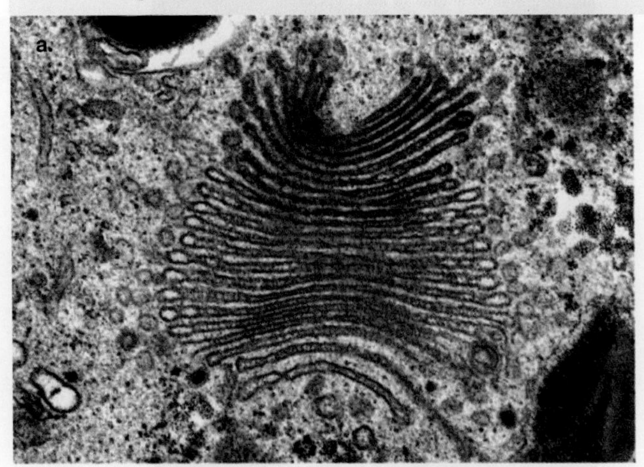

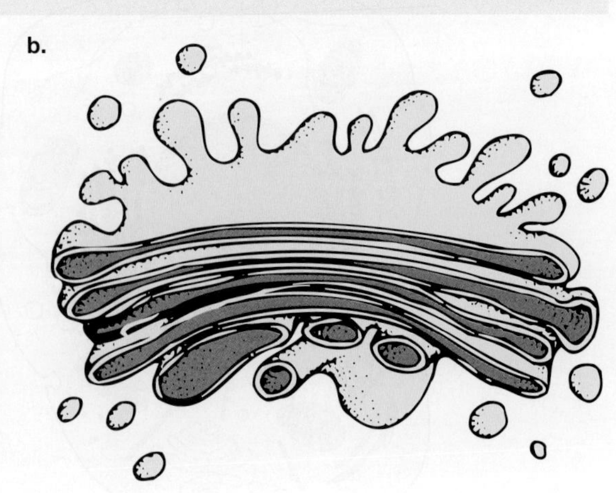

1.9.4 Vesicles

The secretory vesicles with their protein cargo break free from the Golgi apparatus. These membrane-bound vesicles move to the plasma membrane of the cell, where they fuse with it and discharge their protein contents to the exterior through exocytosis.

FIGURE 1.58 a. Vesicles associated with the Golgi apparatus b. Transmission electron micrograph of a small part of a pancreatic ascinar cell that exports the pancreatic proenzymes it synthesises

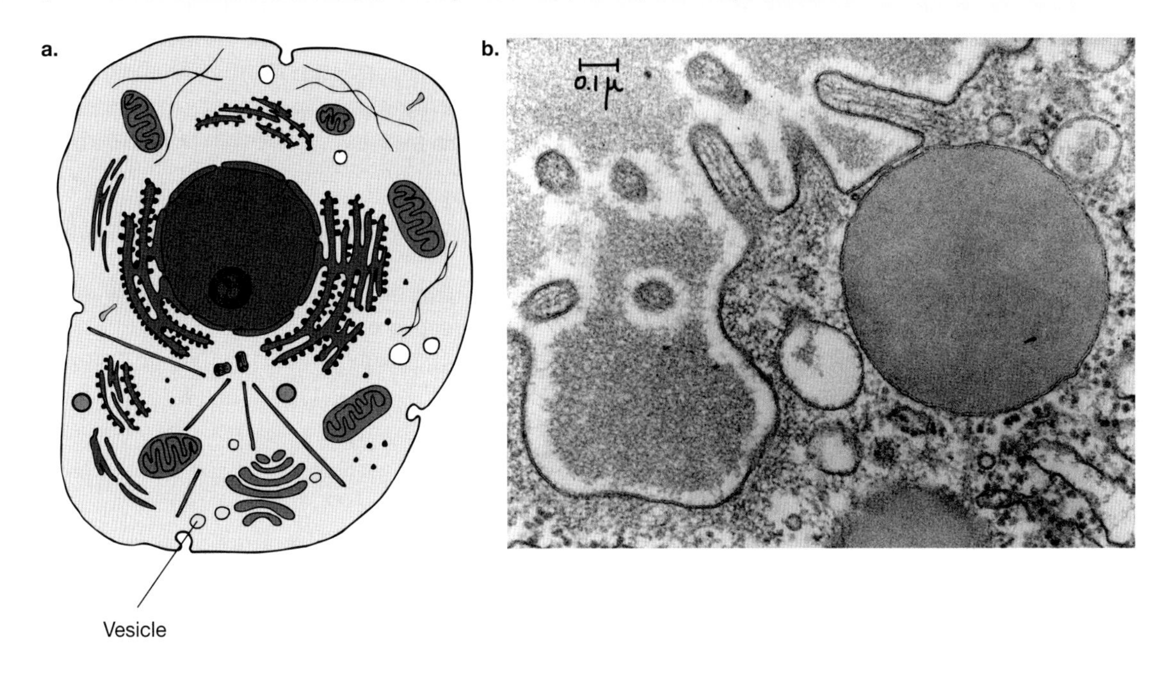

Resources

eWorkbook Worksheet 1.10 Exploring organelles in protein synthesis (ewbk-2557)

KEY IDEAS

- Ribosomes are cell organelles where proteins are manufactured during translation.
- The endoplasmic reticulum is made of a series of membrane-bound channels.
- The rough endoplasmic reticulum is so named because of the presence of ribosomes on the external surface of its membranes.
- Rough endoplasmic reticulum is involved in the processing of proteins and in their transport.
- The Golgi apparatus packages substances into vesicles for export from a cell.
- Vesicles move towards the plasma membrane and expel contents using exocytosis.

1.9 Activities

To answer questions online and to receive **immediate feedback** and **sample responses** for every question, go to your learnON title at **www.jacplus.com.au**. A **downloadable solutions** file is also available in the resources tab.

1.9 Quick quiz on	**1.9 Exercise**	**1.9 Exam questions**

1.9 Exercise

1. Identify whether each of the following is true or false. Justify your response.
   a. The RNA of the ribosomes is made in the nucleolus.
   b. The folding of a protein into its functional 3D shape takes place on the ribosomes.
   c. Ribosomes are membrane-bound organelles that form part of the cell cytoplasm.
   d. The channels of the Golgi apparatus are connected to those of the ER.
2. A scientist wishes to examine ribosomes in a liver cell.
   a. Where should the scientist look: in the nucleus or the cytoplasm?
   b. What kind of microscope is likely to be used by the scientist: a light microscope or a transmission electron microscope? Explain.
3. In the synthesis of proteins for secretion, explain how vesicles are used in rough endoplasmic reticulum and the Golgi apparatus, including the way in which they form and are reabsorbed.
4. Explain how secretory organelles including vesicles export a protein product from the cell through exocytosis.

1.9 Exam questions

Question 1 (1 mark)

Source: *VCAA 2019 Biology Exam, Section A, Q2*

MC Which one of the following organelles has the role of synthesising proteins from their monomers?

A. Golgi apparatus
B. ribosomes
C. vesicles
D. nucleus

Question 2 (1 mark)

Source: *VCAA 2015 Biology Exam, Section A, Q5*

MC All specialised cells that secrete protein molecules

A. have a rigid cell wall.
B. contain numerous lysosomes.
C. contain functional chloroplasts.
D. have an extensive Golgi apparatus.

Question 3 (1 mark)

Source: *VCAA 2016 Biology Exam, Section A, Q7*

MC In animal cells, tight junctions are multi-protein complexes that mediate cell-to-cell adhesion and regulate transport through the extracellular matrix.

Proteins that form these complexes are made within the cell.

One pathway for the production of protein for these junctions is

A. nucleus – ribosome – Golgi apparatus – vesicle – endoplasmic reticulum.
B. nucleus – ribosome – endoplasmic reticulum – vesicle – Golgi apparatus.
C. nucleus – vesicle – endoplasmic reticulum – Golgi apparatus – ribosome.
D. nucleus – vesicle – Golgi apparatus – ribosome – endoplasmic reticulum.

Question 4 (1 mark)

Source: *VCAA 2010 Biology Exam 1, Section A, Q19*

MC At extremely low oxygen levels, important cellular proteins unravel. Another intracellular protein, Hsp60, has been discovered in a wide range of prokaryotic and eukaryotic cells. This protein may bind to other proteins and prevent their unravelling.

It would be reasonable to infer that Hsp60

A. is manufactured at free ribosomes.
B. has a structure which is secondary.
C. is produced continuously by a cell.
D. is produced in equal amounts by living cells in a multicellular organism.

Question 5 (2 marks)

Source: *VCAA 2009 Biology Exam 1, Section B, Q5e and f*

The following diagram shows some detail of a cell from a thyroid gland.

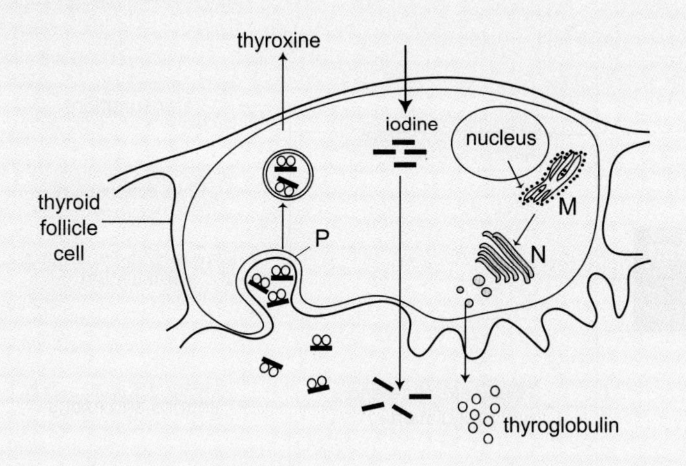

a. Name structure M. **1 mark**
b. Name structure N. **1 mark**

More exam questions are available in your learnON title.

1.10 Review

1.10.1 Topic Summary

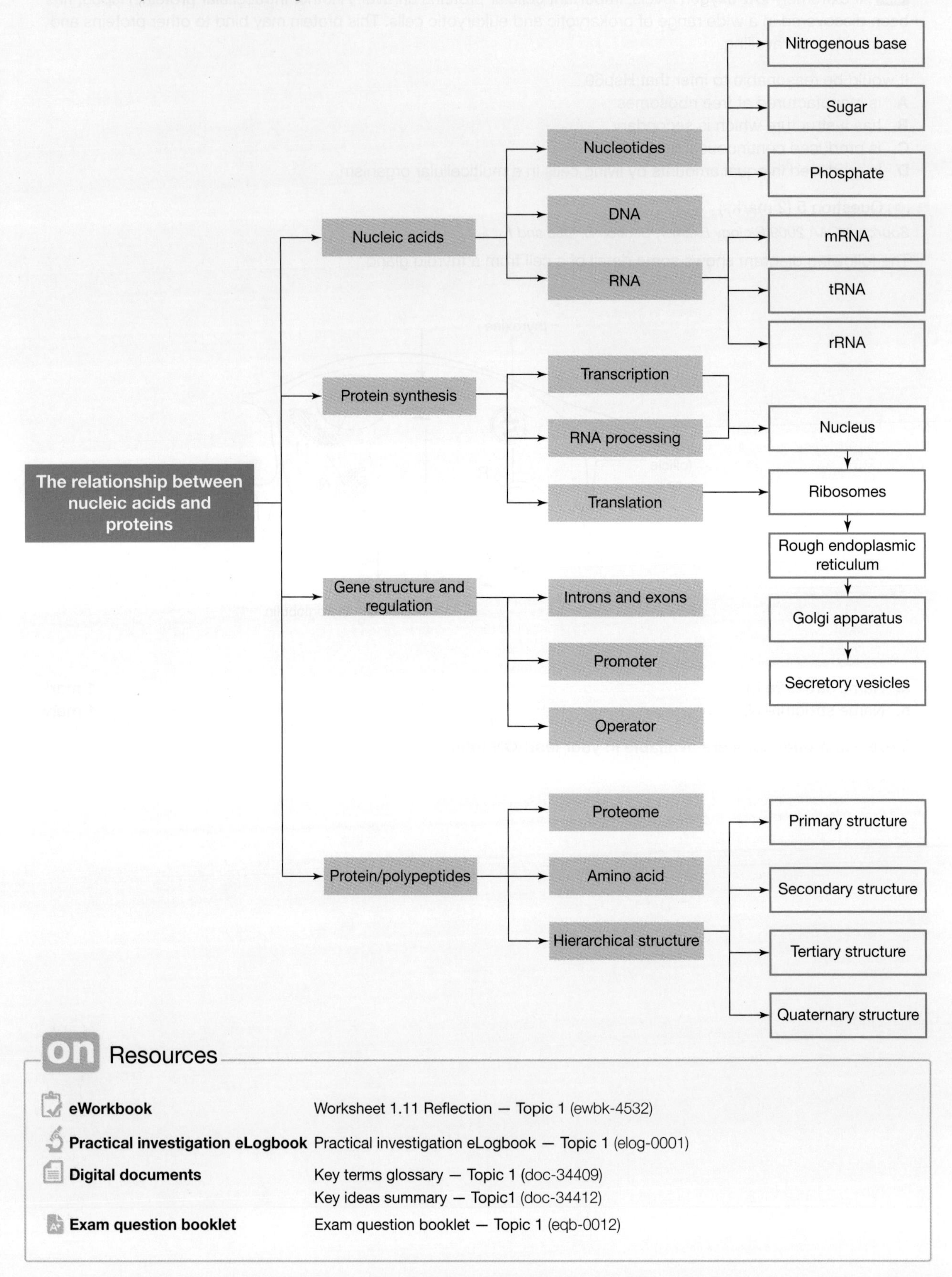

on Resources

eWorkbook	Worksheet 1.11 Reflection — Topic 1 (ewbk-4532)
Practical investigation eLogbook	Practical investigation eLogbook — Topic 1 (elog-0001)
Digital documents	Key terms glossary — Topic 1 (doc-34409) Key ideas summary — Topic1 (doc-34412)
Exam question booklet	Exam question booklet — Topic 1 (eqb-0012)

1.10 Exercises

To answer questions online and to receive **immediate feedback** and **sample responses** for every question, go to your learnON title at **www.jacplus.com.au**. A **downloadable solutions** file is also available in the resources tab.

1.10 Exercise 1: Review questions

1. a. When two monomers such as amino acids join together, a molecule of water is produced. Use a diagram with two amino acid molecules to explain where this water comes from.
   b. How many water molecules would be required to completely hydrolyse a protein polymer that contained 100 monomers? (*Note:* Hydrolysing involves breaking a polymer down into its monomers.)
2. Before the introduction of genetically engineered insulin for use by people with diabetes, the protein hormone was extracted from beef or pig pancreas. Explain how you would expect the sequence of amino acids in the beef and pig insulin to compare with that in humans.
3. A particular small polypeptide contains nine amino acids. The polypeptide has been fragmented in various experiments by breaking particular peptide bonds. The fragments obtained were:
   - *ser – cys – his – pro – arg – cys*
   - *pro – arg – cys*
   - X – *gly – met – cys*
   - *his – pro – arg – cys*
   - X – *gly – met – cys – ser – cys.*

   X is known to be the first amino acid in the polypeptide. What is the primary structure of the polypeptide?
4. Assume that the nitrogenous base sequence in the coding strand in a DNA molecule is:

   A T C G A C A T G G A A T A C C T C

   a. What is the base sequence in the complementary template strand?
   b. How many amino acids does this piece of DNA code for?
5. The following is part of the nitrogenous base sequence in the template strand of part of a gene:

   ... T A T G G G C A T G T A A T G G G C ...

   a. Identify the base sequence in each of the following:
      i. the complementary DNA strand
      ii. the mRNA that would be transcribed from this template.
   b. How many codons are present in this mRNA?
   c. List the anticodons that correspond with each codon.
6. Refer to the genetic code (see table 1.11) and answer the following questions.
   a. Which codons in mRNA control the addition of the amino acid *gly*?
   b. How many codons contain the information to add the amino acid *lys* to a protein? For each codon, write the complementary anticodon.
   c. When the mRNA codon UUU is translated, which amino acid is added to a protein chain?
7. A protein includes the following amino acids in part of its structure:

   ... – *val – thr – lys – pro* – ...

   a. How many codons are needed for the instruction to put these amino acids into place?
   b. Write this instruction in genetic code, as it would appear in mRNA. Would you predict that your code would be identical to that written by all your fellow students? Explain.
8. A mutation in one gene (*G1*) affects just one kind of protein produced by a cell. A mutation in another gene (*G2*) affects a large number of different kinds of proteins produced by that cell. One of these genes carries the coded instructions to make one kind of tRNA. The other carries coded information to make one kind of salivary enzyme.

   Which is more likely to be the gene for the tRNA: *G1* or *G2*? Explain.

9. A segment of mRNA has the base sequence:

...C A U A A G A A U C U U G C...

a. Write the base sequence of the DNA template strand.
b. Write the amino acids that would be translated from this mRNA segment.
c. Assume that a base substitution occurs in the original DNA so that the third base (U) of the mRNA is replaced by a G:

...C A G* A A G A A U C U U G C...

Write the amino acid sequence that would result from this change.
d. Assume that a base addition occurs in the original DNA so that a G is added between the third and fourth bases, which is shown in the mRNA below:

...C A U G*A A G A A U C U U G C...

Write the amino acid sequence that would result from this change.
e. On the basis of this information, what kind of change in DNA has a more extensive effect on the protein resulting from gene translation — a base substitution or a base addition? Explain.

10. a. Where would you find the following in eukaryotes?
i. A codon
ii. Gene transcription in action
iii. An anticodon
iv. Gene translation in action

b. Suggest possible identities for the following.
i. A STOP codon
ii. An amino acid that has six codons coding for it
iii. A self-replicating molecule that carries information in coded form
iv. The cell organelle to which an mRNA molecule attaches for translation

11. Undertake some online research to find the gene that causes cystic fibrosis (the search term OMIM may be useful). Answer the following questions about the gene that causes cystic fibrosis.
a. How many exons are there in the DNA of this gene?
b. The pre-mRNA is longer than the mRNA. Explain.
c. The coding sequence within the mRNA is shorter than the mRNA. Explain.
d. What relationship exists between the number of bases in the coding sequence of the mRNA and the number of amino acids in the protein product?

12. *E. coli* bacteria have a requirement for amino acids, including tryptophan (*trp*). These bacteria can take up *trp* from their environment, but if it is not available, *E. coli* can synthesise this amino acid. The five genes involved in the synthesis of tryptophan are part of a system called the *trp* operon.
a. As well as the five structural genes needed to synthesise *trp*, what other DNA segments form part of the *trp* operon?
b. Draw a rough line diagram showing the essential components of the *trp* operon.

Consider a situation in which tryptophan is present in the environment in which the *E. coli* bacteria are growing.

c. Under these conditions, do the *E. coli* need to synthesise *trp*?
d. Under these conditions, would you expect that the *trp* operon would be repressed or be activated?

If tryptophan is *not* available from the environment in which the bacteria are growing, they will manufacture it themselves.

e. Under these conditions, would you expect that the *trp* operon would be repressed or be active?

When the *trp* operon is repressed, the structural genes that encode the various enzymes needed are silent.

f. Identify a possible means by which the *trp* operon might be repressed.

When the *trp* operon is activated, the structural genes that encode the various enzymes needed for its synthesis and transport from cells are transcribed and translated.

g. Identify a possible means by which the *trp* operon might be activated.

13. Haemoglobin is a protein that is vital in carrying oxygen around the blood.
    a. Draw a clearly labelled monomer of haemoglobin.
    b. Explain the steps involved in producing a polypeptide of haemoglobin through protein synthesis.
    c. Describe the process in which monomers of haemoglobin are joined together. Where does this process occur?
    d. Haemoglobin has both a specific secondary structure and a tertiary structure. Identify the difference between these two terms.

    Haemoglobin is made of 4 different sub-units: two alpha and two beta sub-units. This forms the quaternary structure of haemoglobin. One disorder that affects haemoglobin is thalassaemia, which is a genetically inherited disease. It can vary in severity, depending on how many sub-units are affected.

    One form of thalassaemia, known as beta thalassaemia, is prevalent amongst Mediterranean populations.

    e. Mutations in those with beta thalassaemia are often found to have occurred in the upstream region of the *HBB* (haemoglobin beta) gene rather than in the coding region of chromosome 11. Explain how this can prevent the production of a protein.
    f. Two siblings were tested for thalassaemia. When their DNA code was examined at the *HBB* gene, it was found that they had differences in their genetic code, but they produced the exact same primary structure of haemoglobin. Referring to features of the genetic code, explain how this occurred.
14. Nucleotides are the monomers of nucleic acids.
    a. Which part of a nucleotide would be the same between polymers of DNA and RNA?
    b. RNA exists in three main forms. Explain how each of these forms is involved in protein synthesis.

    Some relevant anticodons for amino acids are shown in the table below.

Anticodon	Amino acid
ACC	*trp*
GAG	*leu*
GUA	*his*
AUA	*tyr*
CGA	*ala*
UAA	*ile*

    A section of protein was examined and found to have the following amino acid sequence:

    ala – tyr– ile – his

    c. What base sequence would be found on the initial DNA template strand, assuming that only the anticodons in the above table were used?
    d. It was found that the protein above was coded for by a regulatory gene. What are regulatory genes and how do they function?

1.10 Exercise 2: Exam questions

on Resources

Teacher-led videos Teacher-led videos for every exam question

Section A — Multiple choice questions

All correct answers are worth 1 mark each; an incorrect answer is worth 0.

Question 1

Source: *VCAA 2012 Biology Exam 1, Section A, Q5*

A particular DNA double helix is 100 nucleotide pairs long and contains 25 adenine bases.

The number of guanine bases in this DNA double helix would be

A. 25
B. 50
C. 75
D. 100

Question 2

Source: *VCAA 2019 Biology Exam, Section A, Q11*

Two different cells taken from the same human were viewed using a microscope.

The diagrams below show the structure of the two cells, not drawn to the same scale.

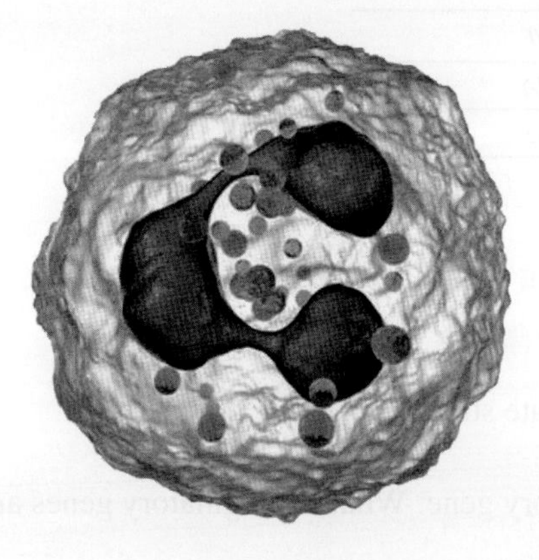

a neutrophil

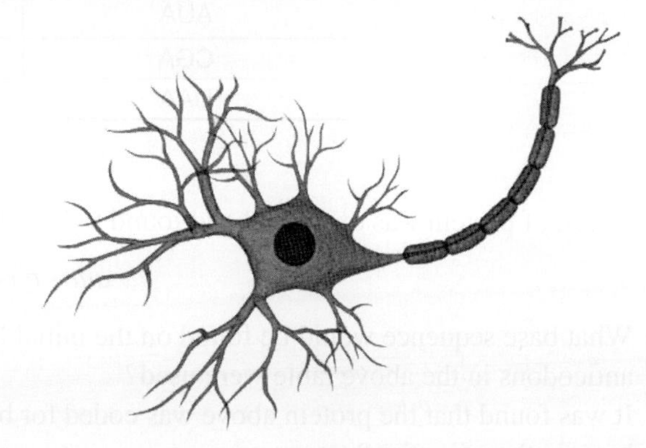

a neuron

Sources (from left): KaterynaKon/Shutterstock.com; SebastianKaulitzki/Shutterstock.com

Which one of the following is a correct conclusion to reach when comparing the two cells?

A. At any given time, the genes expressed in each cell may be different.
B. All proteins in each cell will have similar tertiary structures.
C. The two cells have the same proteome.
D. The two cells have different genomes.

Question 3

Source: *Adapted from VCAA 2018 Biology Exam, Section A, Q20*

Consider the following sequence of six amino acids that make up part of a polypeptide.

---- phe ---- leu ---- pro ---- val ---- tyr ---- ala ----

A mutation within the gene coding for this sequence of six amino acids resulted in the following six amino acids in the same position.

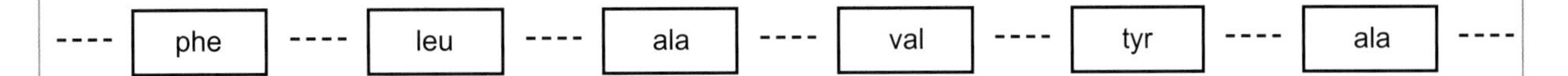

This change in the sequence of amino acids was caused by

A. a deletion of a nucleotide.
B. an insertion of a nucleotide.
C. a substitution of a nucleotide.
D. an inversion of adjacent nucleotides.

Question 4

Source: *VCAA 2006 Biology Exam 1, Section A, Q11*

There are 4 polypeptide chains in a human haemoglobin molecule. The monomers in a small section of each of the 4 chains is shown.

chain 1... leu-ser-pro-ala-asp-lys-thr-asn-val-lys...

chain 2... leu-thr-pro-glu-glu-lys-ser-ala-val-thr...

chain 3... leu-ser-pro-ala-asp-lys-thr-asn-val-lys...

chain 4... leu-thr-pro-glu-glu-lys-ser-ala-val-thr...

Consider the sections of the chains shown.

The information given suggests that

A. each of the chains is the result of the same DNA sequence.
B. each total chain contains the same number of monomers.
C. adjacent monomers are linked by a peptide bond.
D. each monomer is specified by a nucleotide.

Question 5

Source: *Adapted from VCAA 2019 Biology Exam, Section A, Q4*

Structural genes can be switched off and turned on by transcriptional factors expressed by regulatory genes. In prokaryotes, a group of genes associated with the synthesis of tryptophan is grouped together in a single operon called the *trp* operon.

Transcription of the structural genes within the *trp* operon results in the production of molecules of

A. a transcription factor.
B. a repressor protein.
C. tryptophan.
D. mRNA.

Question 6

Source: *VCAA 2011 Biology Exam 1, Section A, Q10*

Consider the following cell.

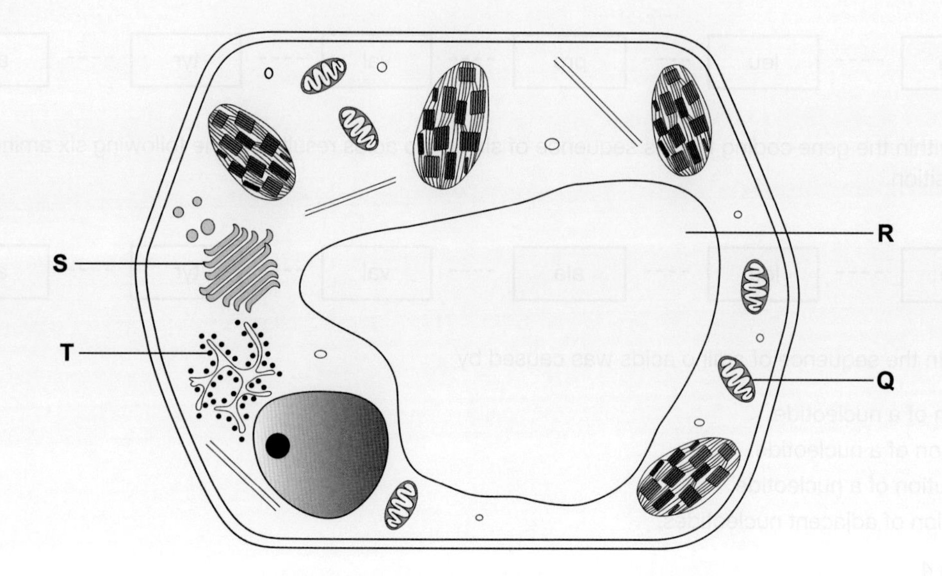

The synthesis of

A. glucose occurs in structure Q.
B. DNA occurs in structure R.
C. RNA occurs at structure S.
D. protein occurs at structure T.

Question 7

A mutation in the regulatory gene for the *trp* operon has led to the production of a repressor that is unable to bind to tryptophan. Which of the following statements is correct?

A. RNA polymerase would be prevented from transcribing the structural genes.
B. Tryptophan would never be produced by this bacterium.
C. The repressor would be permanently active.
D. The structural genes in the *trp* operon would be constantly on.

Question 8

Geneticists studying the nematode worm *Caenorhablditis elegans* found a gene that codes for a 22-nucleotide strand of 'microRNA'. The microRNA binds to a complementary sequence in a particular target mRNA in *C. elegans*, thus preventing the translation of the target mRNA into a protein.

What term describes the type of gene that codes for the 'microRNA' in *C. elegans*?

A. A structural gene
B. A regulator gene
C. A promoter sequence
D. A transcription unit

Question 9

Source: *VCAA 2009 Biology Exam 1, Section A, Q25*

The proteome is the complete collection of proteins in any given cell or organism.

The study of proteomics is concerned with the systematic, large-scale analysis of a proteome. A range of techniques are used to identify all proteins and their components, and the amount of each present. The information is stored in a database.

The following diagram outlines one approach to a study of the proteins in a cell. Steps T and R are alternative steps that can be taken.

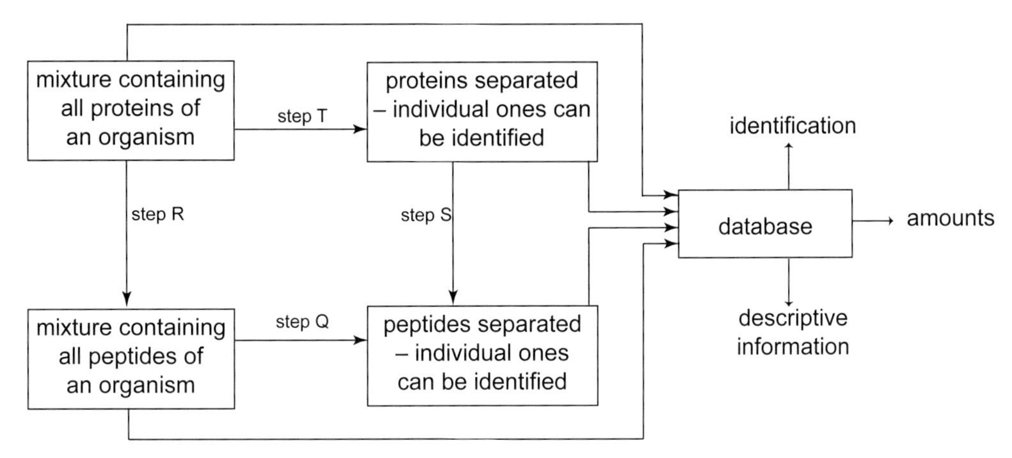

From this information it is reasonable to conclude that

A. step Q involves the use of peptidases.
B. step R involves the joining of proteins to form polypeptides.
C. children with the same parents would have identical proteins.
D. the number of amino acids in the mixture containing all proteins is the same as the amino acids in the separated peptides.

Question 10

A cell is found to have a large number of ribosomes, rough endoplasmic reticulum and Golgi apparatuses. What is the likely identity of this cell?

A. A muscle cell found in the triceps of a body builder
B. A pancreatic cell secreting the hormone insulin into the bloodstream
C. A red blood cell circulating in the bloodstream
D. A phagocytic cell that engulfs foreign matter such as bacteria

Section B — Short answer questions

Question 11 (6 marks)

Source: *Adapted from VCAA 2018 Biology Exam, Section B, Q1*

Tryptase is an enzyme that is released, along with histamine and other chemicals, from human mast cells. Nucleic acids encode instructions for the synthesis of tryptase in a mast cell.

a. Outline the steps of translation in the synthesis of tryptase. **3 marks**

b. After being synthesised, tryptase is released from mast cells via exocytosis.
Name three different organelles directly associated with the transport of the synthesised tryptase within or from mast cells and state the role of each organelle in this process. **3 marks**

Question 12 (6 marks)

Source: *Adapted from VCAA 2012 Biology Exam 1, Section B, Q3*

Human insulin is a macromolecule composed of two amino acid chains. The chains are connected by disulfide bonds.

a. To what group of macromolecules does insulin belong? **1 mark**

Insulin found in other animals varies from human insulin. The following table compares all the differences seen in the primary structure of human, cow, pig and sheep insulin.

	Amino acid position number within	
	Alpha chain	**Beta chain**
	-8 - 9 - 10-	**-30-**
human	-thr - ser - ile-	thr
cow	-ala - ser - val-	ala
pig	-thr - ser - ile-	ala
sheep	-ala - gly - val-	ala

b. What is meant by the term 'primary structure' of the insulin macromolecule? **1 mark**

c. Humans with diabetes take insulin injections to maintain their health.
If supplies of human insulin were not available, which one of the other three animals listed in the table would be the best source of insulin?
Explain your reason for choosing this particular animal. **2 marks**

d. Refer back to table 1.11, which contains the genetic code for protein production.
Use the information in the table **to explain**

i. the different sequence of nucleotides in humans and cows with respect to the DNA coding for the amino acid at position 30. **1 mark**

ii. whether the sequence of nucleotides in DNA coding for the amino acid at position 30 will be identical in cows, pigs and sheep. **1 mark**

Question 13 (3 marks)

Source: *VCAA 2018 Biology Exam, Section B, Q6*

a. Describe the functional difference between a structural gene and a regulatory gene. **2 marks**

The *Hox* genes are master regulatory genes that influence cells in a particular location of an animal embryo in order to develop structures for that part of the body.

In the brine shrimp, *Artemia*, the expression of the *Hox* genes *Ubx* and *Scr* results in the growth of either a swimming appendage or a feeding appendage, depending on whether the genes are expressed in cells that are in the mid-region of the body or that are near the mouth. These specialised appendages are labelled in the diagram below.

Adult brine shrimp (*Artemia*)

b. Describe **one** way that genes are regulated so that the same genes can produce different appendages when the genes are expressed in different locations in the *Artemia* embryo **1 mark**

Question 14 (4 marks)

Source: *VCAA 2019 Biology Exam, Section B, Q1*

Diagrams of two molecules that are required for the production of proteins within a cell are shown below.

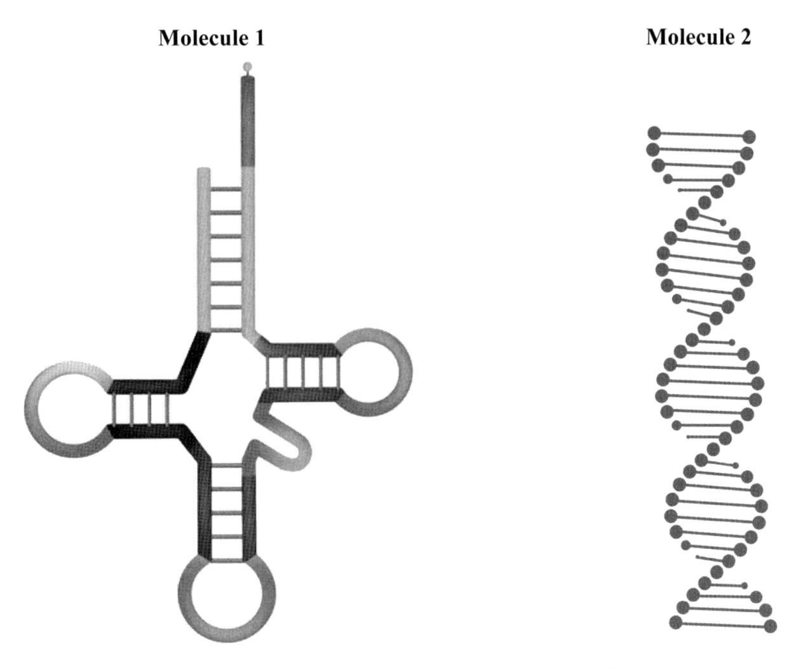

Sources (from left): Designua/Shutterstock.com; PannaKotta/Shutterstock.com

a. Complete the table below to describe two differences between the monomers of the two molecules. **2 marks**

	Molecule 1	Molecule 2
Difference 1		
Difference 2		

b. Ten amino acids that form part of a protein are shown below.

-phe-val-asn-gln-his-leu-cys-gly-ser-his-

The section of an RNA molecule found in the nucleus of the cell associated with the translation of these 10 amino acids was found to contain over 300 monomers.
Explain how there can be over 300 monomers in this section of the RNA molecule but only 10 amino acids translated. **2 marks**

Question 15 (5 marks)

The *trp* operon is a section of the chromosome in the bacterium *Escherichia coli.* The *trp* operon controls the production of enzymes involved in the synthesis of tryptophan.

The *trp* operon contains:

- five structural genes
- a promoter sequence
- an operator sequence.

a. Define the term operon. **1 mark**

b. The operon is controlled by a regulator gene known as *trp*, which produces a repressor.
Explain the mechanism of action of this repressor if tryptophan was present. **2 marks**

c. A mutation occurs in the repressor, where it is unable to bind to tryptophan. Explain what would occur in the bacteria when tryptophan was present. **2 marks**

1.10 Exercise 3: Biochallenge online only

on Resources

eWorkbook Biochallenge — Topic 1 (ewbk-8082)

Solutions Solutions — Topic 1 (sol-0657)

Past VCAA examinations

Sit past VCAA examinations and receive immediate feedback, marking guides and examiner's report notes.
Access Course Content and select 'Past VCAA examinations' to sit the examinations online or offline.

teachon

Test maker
Create unique tests and exams from our extensive range of questions, including past VCAA questions.
Access the assignments section in learnON to begin creating and assigning assessments to students.

Online Resources

Below is a full list of **rich resources** available online for this topic. These resources are designed to bring ideas to life, to promote deep and lasting learning and to support the different learning needs of each individual.

eWorkbook

- 1.1 eWorkbook — Topic 1 (ewbk-1880) ☐
- 1.2 Worksheet 1.1 Reviewing cell size and surface area to volume ratios (ewbk-1962) ☐
  - Worksheet 1.2 Labelling organelles (ewbk-1964) ☐
  - Worksheet 1.3 Structure of the membrane and membrane transport (ewbk-1966) ☐
- 1.3 Worksheet 1.4 Comparing DNA and RNA (ewbk-1968) ☐
- 1.4 Worksheet 1.5 Case study analysis: Alternative splicing of pre-mRNA (ewbk-2551) ☐
  - Worksheet 1.6 Gene expression and protein synthesis (ewbk-1970) ☐
- 1.5 Worksheet 1.7 The different structures in genes (ewbk-1972) ☐
- 1.6 Worksheet 1.8 Exploring the *trp* operon (ewbk-1974) ☐
- 1.7 Worksheet 1.9 Protein structure and function (ewbk-2555) ☐
- 1.9 Worksheet 1.10 Exploring organelles in protein synthesis (ewbk-2557) ☐
- 1.10 Worksheet 1.11 Reflection — Topic 1 (ewbk-4532) ☐
  - Biochallenge — Topic 1 (ewbk-8082) ☐

Solutions

- 1.10 Solutions — Topic 1 (sol-0657) ☐

Practical investigation eLogbook

- 1.1 Practical investigation eLogbook — Topic 1 (elog-0001) ☐
- 1.2 Investigation 1.1 Viewing and staining cells (elog-0003) ☐
  - Investigation 1.2 Membrane transport across a semi-permeable membrane (elog-0005) ☐
- 1.3 Investigation 1.3 Extraction of DNA (elog-0007) ☐
  - Investigation 1.4 Building a model of DNA and RNA (elog-0009) ☐
- 1.4 Investigation 1.5 Using the genetic code to model protein synthesis (elog-0011) ☐
- 1.7 Investigation 1.6 What's in a protein (elog-0013) ☐
  - Investigation 1.7 Modelling protein structures (elog-0804) ☐

Digital documents

- 1.1 Key science skills — VCE Biology Units 1–4 (doc-34326) ☐
  - Key terms glossary — Topic 1 (doc-34409) ☐
  - Key ideas summary — Topic 1 (doc-34412) ☐
- 1.3 Extension: Mitochondrial DNA (doc-35832) ☐
- 1.4 Codon and triplet charts (doc-36165) ☐
- 1.6 Case study: Homeotic genes (doc-35831) ☐
  - Attenuation of the *trp* operon (doc-39265) ☐
- 1.7 Extension: The complexity of proteins (doc-35833) ☐

Teacher-led videos

- Exam questions — Topic 1 ☐
- 1.3 Sample problem 1 Comparing DNA and RNA (tlvd-1090) ☐
- 1.4 Sample problem 2 Applying transcription (tlvd-1091) ☐
  - Sample problem 3 Combining transcription and translation (tlvd-1092) ☐
- 1.6 Sample problem 4 Comparing operons (tlvd-1093) ☐
- 1.7 Sample problem 5 Protein hierarchy (tlvd-1094) ☐

Video eLessons

- 1.2 Living organisms are made of cells (eles-4165) ☐
  - Mechanisms of membrane transport (eles-2463) ☐
- 1.3 DNA structure (eles-4211) ☐
- 1.4 Transcription (eles-4167) ☐
  - Translation (eles-4168) ☐
- 1.6 The *trp* operon (eles-5079) ☐
- 1.7 Primary structure of proteins (eles-4169) ☐
  - Different protein structures (eles-4145) ☐

Interactivities

- 1.2 Movement across membranes (int-0109) ☐
- 1.3 Complementary DNA (int-0133) ☐
  - RNA structure (int-0111) ☐
- 1.4 Transcription (int-8125) ☐
  - Protein synthesis (int-0112) ☐
  - Comparing transcription and translation (int-0113) ☐
  - Translation (int-8126) ☐
- 1.7 Protein structure (int-0105) ☐

Weblinks

- 1.2 Cells resources ☐
- 1.3 DNA and genes resources ☐
- 1.4 Transcription and translation practice ☐
  - DNA to RNA to protein simulations ☐
  - Examining protein synthesis in fireflies ☐
- 1.5 DNA structural features of eukaryotic TATA-containing and TATA-less promoters ☐
- 1.6 Homeotic genes and body patterns ☐
- 1.7 Protein folding ☐
  - Protein folding simulations ☐
- 1.8 Types of proteins ☐

Exam question booklet

- 1.1 Exam question booklet — Topic 1 (eqb-0012) ☐

Teacher resources

There are many resources available exclusively for teachers online.

To access these online resources, log on to **www.jacplus.com.au**

2 DNA manipulation techniques and applications

KEY KNOWLEDGE

In this topic, you will investigate:

DNA manipulation techniques and applications

- the use of enzymes to manipulate DNA, including polymerase to synthesise DNA, ligase to join DNA and endonucleases to cut DNA
- the function of CRISPR-Cas9 in bacteria and the application of this function in editing an organism's genome
- amplification of DNA using polymerase chain reaction and the use of gel electrophoresis in sorting DNA fragments, including the interpretation of gel runs for DNA profiling
- the use of recombinant plasmids as vectors to transform bacterial cells as demonstrated by the production of human insulin
- the use of genetically modified and transgenic organisms in agriculture to increase crop productivity and to provide resistance to disease.

PRACTICAL WORK AND INVESTIGATIONS

Practical work is a central component of learning and assessment. Experiments and investigations, supported by a **practical investigation eLogbook** and **teacher-led videos**, are included in this topic to provide opportunities to undertake investigations and communicate findings.

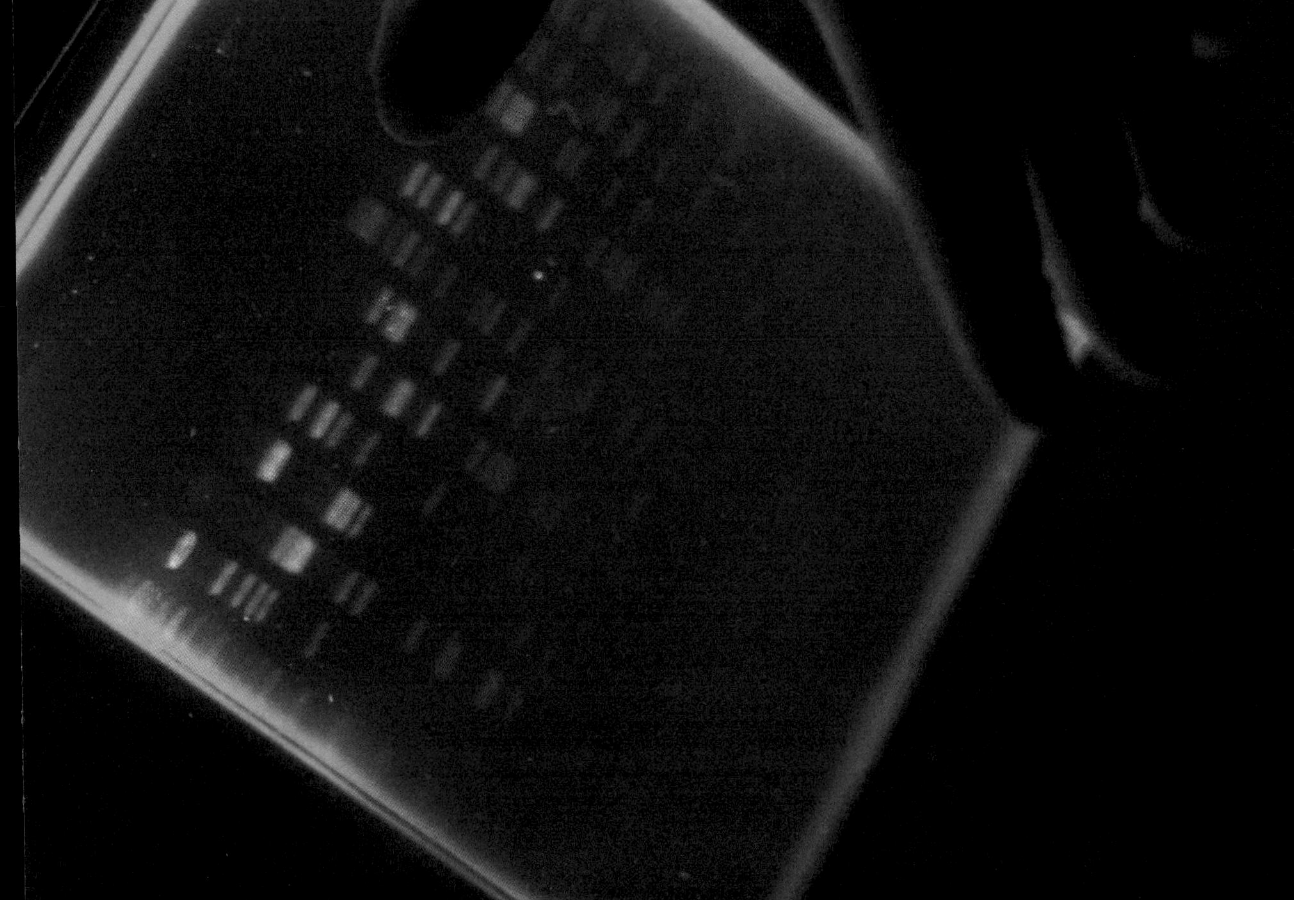

2.1 Overview

Numerous **videos** and **interactivities** are available just where you need them, at the point of learning, in your digital formats, learnON and eBookPLUS, and at **www.jacplus.com.au**.

2.1.1 Introduction

DNA is fundamental to who we are, providing a genetic blueprint that makes us unique. Our DNA is responsible for many traits, including everything from our eye colour to the inheritance of genetic disease. All individuals have mutations, with new mutations accumulated each generation. Many of these mutations are silent, lurking unknown in our genome. However, some mutations are deleterious, causing a negative effect on the lives of individuals who possess them, and even shortening life expectancies for individuals.

FIGURE 2.1 The possibilities around DNA manipulation techniques and applications are exciting and endless.

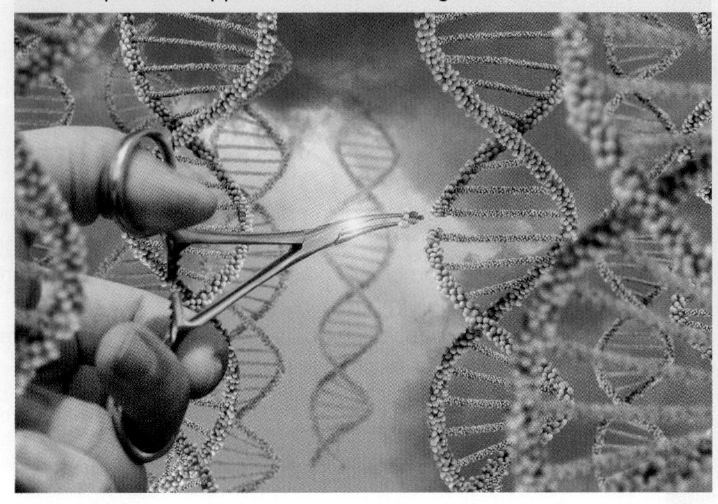

What if we could manipulate our genes? Scientists who use gene manipulation technology or 'genetic engineers' do this every day. A genetic engineer requires tools to cut, join, amplify, copy and separate DNA. This topic will explore techniques to manipulate DNA and applications of DNA, such as its use in forensics and DNA profiling.

LEARNING SEQUENCE

on Resources

eWorkbook	eWorkbook — Topic 2 (ewbk-1881)
Practical investigation eLogbook	Practical investigation eLogbook — Topic 2 (elog-0015)
Digital documents	Key science skills — VCE Biology Units 1–4 (doc-34326)
	Key terms glossary — Topic 2 (doc-34414)
	Key ideas summary — Topic 2 (doc-34417)
Exam question booklet	Exam question booklet — Topic 2 (eqb-0013)

2.2 The use of enzymes to manipulate DNA

KEY KNOWLEDGE

- The use of enzymes to manipulate DNA, including polymerase to synthesise DNA, ligase to join DNA and endonucleases to cut DNA

Genetic engineers mostly work with genetic material, with a particular focus on DNA. The genetic engineer requires tools to cut, join, copy and separate DNA. The following section describes the tools and techniques that, when combined, enable genes to be manipulated in various ways. Table 2.1 shows some of the 'tools' used in gene manipulation, which will be explored in the upcoming subtopics. The same tools can be used regardless of the source of the genetic material.

TABLE 2.1 Tools for gene manipulation

Action	Tool	Subtopic
Synthesise DNA	Polymerase, DNA synthesisers and reverse transcriptase	2.2
Join DNA fragments	Ligase	2.2
Cut DNA into fragments at specific location	Endonucleases (or restriction enzymes)	2.2
Edit DNA	CRISPR-Cas9	2.3
Amplify DNA	Polymerase chain reaction (PCR)	2.4
Find particular DNA fragments	Probes	2.5
Separate DNA fragments by size	Gel electrophoresis	2.5
Transform bacterial cells using recombinant DNA	Recombinant plasmids	2.6

2.2.1 Polymerase to synthesise DNA

DNA **polymerase** was first identified by Arthur Kornberg in 1956. He identified the ability of this enzyme to be able to accurately copy a DNA template.

polymerase enzyme involved in synthesising nucleic acids

FIGURE 2.2 DNA replication involves two molecules of DNA being produced using the initial DNA strands as a template.

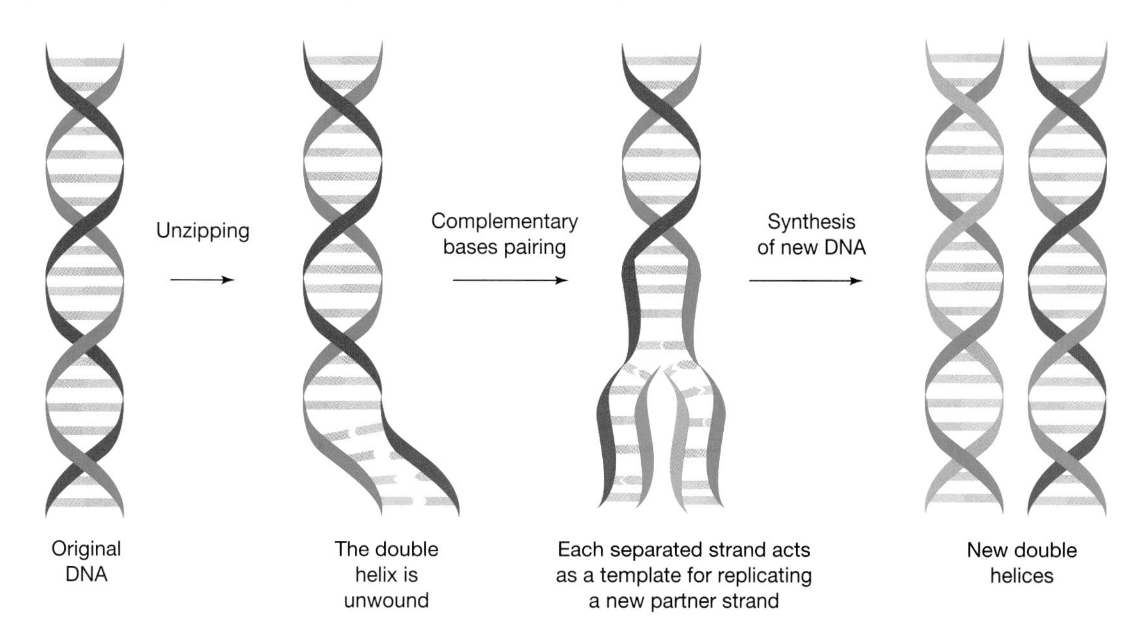

Interestingly, the first DNA polymerase to be identified (DNA polymerase I) is not the major enzyme responsible for replication. Both prokaryotic and eukaryotic cells contain several different DNA polymerases that have distinct roles in the replication and repair of DNA.

In order for DNA polymerase to synthesise DNA, the double helix needs to be unzipped and made single stranded. DNA polymerase can then work alongside other enzymes to add free nucleotides to the DNA template as shown in figures 2.2 and 2.3.

FIGURE 2.3 DNA replication involves numerous enzymes, including DNA polymerase.

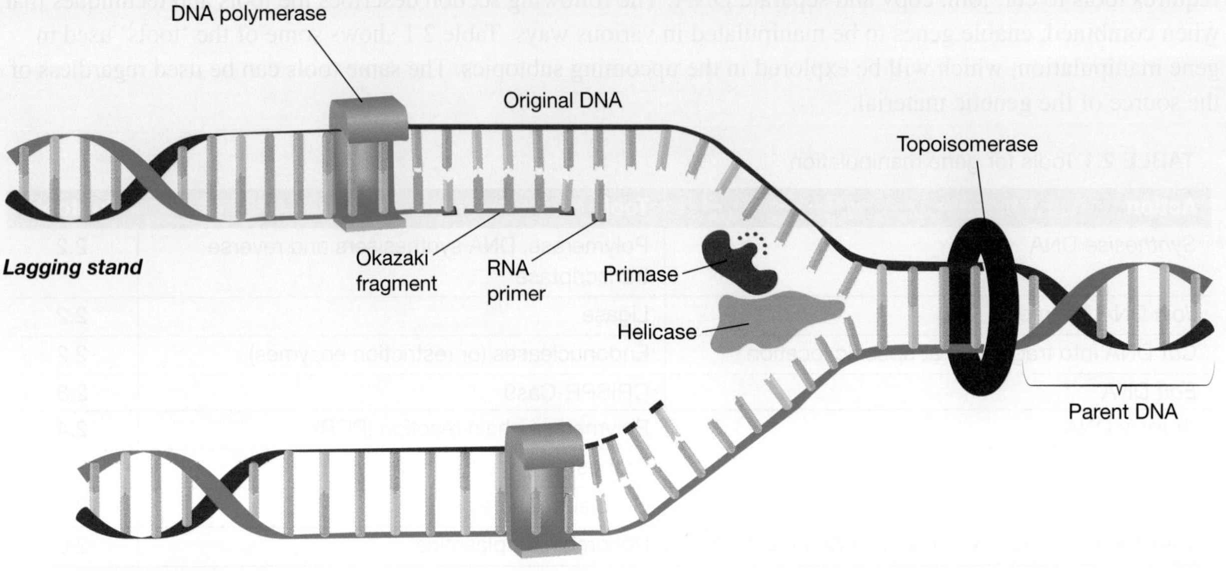

Another special type of polymerase, known as *Taq* DNA polymerase, is used in a process known as the **polymerase chain reaction**, which is explored in subtopic 2.4.

Other methods to synthesise DNA

Sometimes, when only specific genes or small sections of DNA are required to be synthesised, using the mRNA as a template can be more useful. In order to make a copy of DNA using an mRNA template, mRNA needs to isolated from the specific cells in which the gene concerned is active. The enzyme **reverse transcriptase** uses the mRNA as a template to build a single-stranded DNA with a complementary base sequence. Polymerase then builds the second strand. This DNA is known as **complementary DNA (cDNA)**. The reverse transcriptase procedure has been used successfully to make copies of the human growth hormone gene and also the gene for tissue plasminogen activator (t-PA), an enzyme that dissolves blood clots and is used in the treatment of some forms of heart attack. This process is shown in figure 2.4.

Another method to make DNA is through the use of DNA synthesisers to synthesise DNA from nucleotide building blocks. To use this method, the base sequence of the required DNA must be known. Instruments called DNA synthesisers can join nucleotide sub-units in a pre-defined order to produce DNA segments with lengths greater than 100 bases. The chemical synthesis of DNA does not require a template strand nor does it require the enzyme DNA polymerase.

polymerase chain reaction a technique used to amplify a segment of DNA

reverse transcriptase enzyme that directs the formation of copy DNA from a messenger RNA template through reverse transcription

complementary DNA (cDNA) a strand of DNA that has complementary bases to the opposite strand and is usually produced through reverse transcription

FIGURE 2.4 Reverse transcription

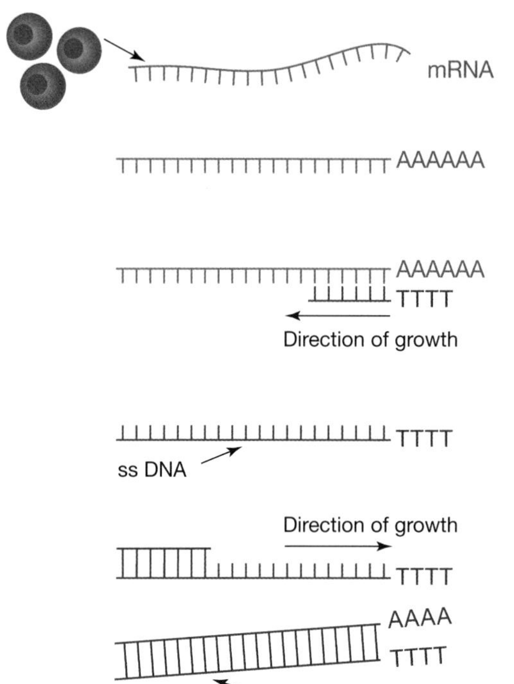

1. mRNA isolated from cystol of specific cells.
2. Poly-A tail added to this mRNA. (This segment provides an anchor to which a primer can attach at the 3′ end.)
3. Oligo-dT primer added and binds to poly-A mRNA tail. Reverse transcriptase enzyme added. DNA lengthens by addition of nucleotides. The order is controlled by the sequence in the mRNA.
4. When DNA chain is complete, mRNA is removed by alkali treatment.
5. Polymerase enzyme added. This catalyses the building of complementary DNA strand.
6. Final double-stranded DNA product (cDNA)

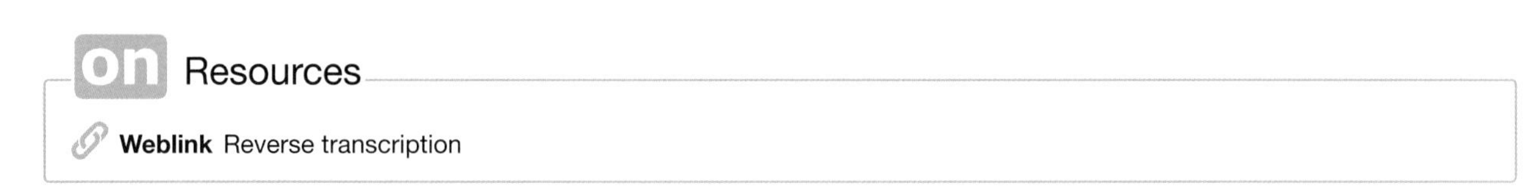

2.2.2 Endonucleases to cut DNA

Restriction enzymes (**endonucleases**) are often referred to as 'molecular scissors'. They were discovered and characterised by molecular biologists Werner Arber, Hamilton Smith and Daniel Nathans in the late 1960s–early 1970s. Endonucleases are found in nature. They derive their name from the genus, species and strain designations of the bacteria that produce them; for example, *Eco*RI is produced by *Escherichia coli* strain RY13, and *Bgl*I is produced by *Bacillus globigii* (and was the first endonuclease discovered from that species).

Endonucleases cleave DNA at specific recognition sequences known as restriction sites, splitting DNA into smaller fragments.

Endonucleases make one incision on each of the two complementary strands of DNA. The ability of the enzymes to cut DNA at precise locations enabled researchers to isolate gene-containing fragments and recombine them with other molecules of DNA, such as in gene cloning.

The position where a cutting enzyme can snip double-stranded DNA is the recognition sequence of that restriction enzyme. A restriction site is a particular order of nucleotides (refer to table 2.2). The number of nucleotide base pairs (bp) in a restriction site varies from four to about eight base pairs (bp). This length determines how frequently a restriction enzyme is likely to cut a random sequence of DNA or, in other words, the average distance between cuts.

endonucleases enzymes, also known as restriction enzymes, that cut at specific sites within DNA molecules

TABLE 2.2 Endonucleases in common use and their recognition sites. The arrowheads indicate the site of the cut in each strand of the double-stranded DNA molecule. Note that in some resources, italics are not used when naming the endonuclease.

Enzyme	Source	Recognition site
*Alu*I	*Arthrobacter luteus*	AG$^{\vee}$CT TC^GA
*Bam*HI	*Bacillus amyloliquefaciens H*	G$^{\vee}$GATC C C CTAG^G
*Bcl*I	*Bacillus caldolyticus*	T$^{\vee}$GATC A A CTAG^T
*Bgl*II	*Bacillus globigii*	A$^{\vee}$GATC T T CTAG^A
*Eco*RI	*Escherichia coli R factor*	G$^{\vee}$AATT C C TTAA^G
*Hae*III	*Hemophilus aegyptus*	GG$^{\vee}$CC CC^GG
*Hind*III	*Hemophilus influenza Rd*	A$^{\vee}$AGCT T T TCGA^A
*Kpn*I	*Klebsiella pneumoniae*	G GTAC$^{\vee}$C C^CATG G
*Not*I	*Norcadia otitidis-caviarum*	GC$^{\vee}$GGCC GC CG CCGG^CG
*Pst*I	*Providencia stuartii*	C TGCA$^{\vee}$G G^ACGT C
*Sac*I	*Streptomyces achromogenes*	GAG$^{\vee}$CTC CTC^GAG
*Taq*I	*Thermos aquaticus*	T$^{\vee}$CG A A GC^T

TIP: When identifying restriction sites, it is important to ensure you are looking at the correct strand of DNA. The direction is also important (one strand runs 5′ to 3′ and the other strand runs 3′ to 5′). (For example, if the restriction site for the top strand is TCGA, it must be in that direction. It will not cut AGCT on the same strand of DNA).

Blunt and sticky ends

There are two ways in which a DNA strand may be cut:

- Some endonucleases cut the two strands of a DNA molecule at points directly opposite each other to produce cut ends that are **'blunt ends'**.
  - In 'blunt end' ligation, the DNA fragments are joined directly together through the use of DNA **ligase**. Scientists have less control over the orientation of the resultant insertion.
- Other endonucleases cut one strand at one point but cut the second strand at a point that is not directly opposite. The overhanging cut ends made by these cutting enzymes are called **'sticky ends'**. These sticky ends are complementary.
  - The DNA ligase enzyme connects the single-stranded DNA together via the sugar-phosphate backbones. Through careful selection of restriction enzymes to create the sticky ends, scientists can exercise greater control over the site of ligation.

blunt ends ends of a DNA fragment with no overhanging bases after being cut by an endonuclease

ligase an enzyme that catalyses the joining of two double-stranded DNA fragments

sticky ends ends of a DNA fragment with overhanging bases after being cut by an endonuclease

FIGURE 2.5 a. 'Blunt ends' are produced when the two DNA strands are produced with no overhang on one or the other strand. b. 'Sticky ends' are produced when each of the strands extend beyond the complementary region of the strand pair.

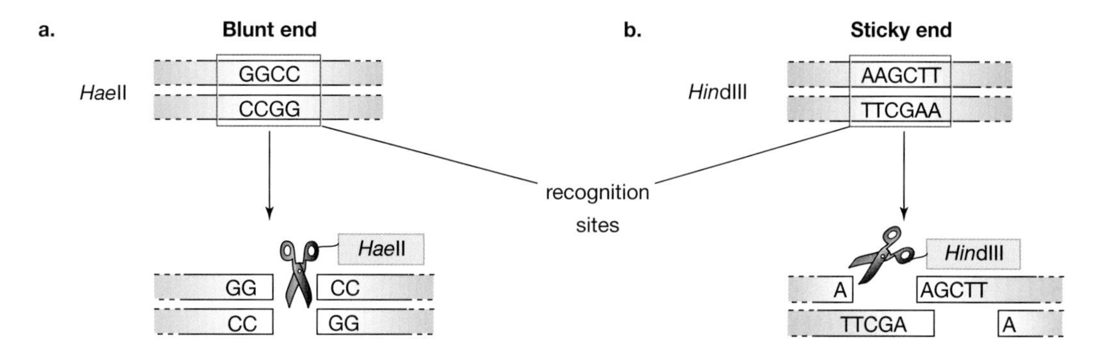

For cutting, a sample of DNA is dissolved and the particular endonuclease is added. Provided recognition sites are present in the DNA sample, the DNA molecules will be cut into two or more fragments. The lengths of the fragments depend on the relative positions of the recognition sites (see figure 2.6).

FIGURE 2.6 Each cutting enzyme snips DNA at a particular site. Would a different cutting enzyme be expected to produce three fragments of the same sizes as those produced by *Bgl*II? (Note that kbp = kilobase pairs.)

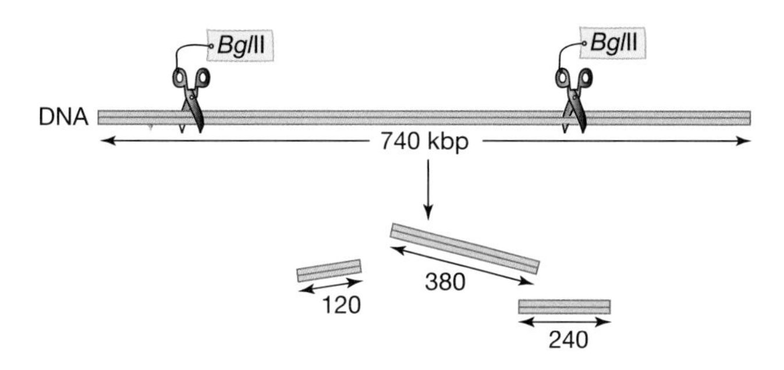

TIP: Remember to consider if DNA is circular or linear when determining the number of fragments that will be produced.

- If a linear piece of DNA is cut twice, three fragments are produced.
- If a circular piece of DNA is cut twice, only two fragments are produced.

tlvd-1095

SAMPLE PROBLEM 1 Examining fragments and endonucleases

A piece of double-stranded DNA is shown below.

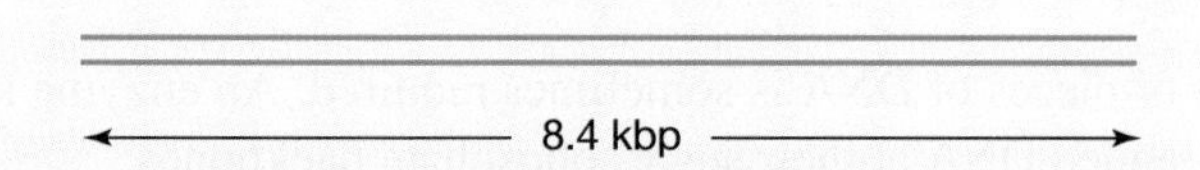

A sample of this DNA was treated with the cutting enzyme *Alu*I. The cutting site for this enzyme is AGCT. It was run on an agarose gel to determine the relative sizes of the resulting DNA fragments, which are shown in the diagram provided.

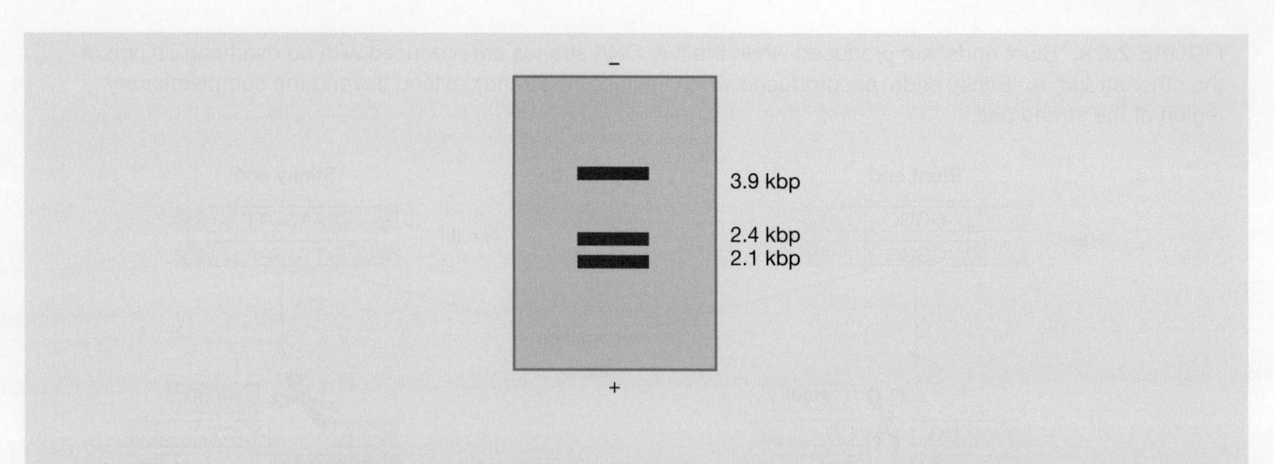

a. **The numbers denote the sizes of the fragments in kbp. What conclusion may be drawn about the recognition sites in the DNA fragment? (1 mark)**
b. **A second sample of this DNA was treated with a different cutting enzyme, *Hin*dIII. The cutting site for this enzyme is AAGCTT. The result after electrophoresis revealed two DNA fragments of sizes 5.4 kbp and 3.0 kbp. What conclusion may be drawn? (1 mark)**

THINK	WRITE
a. Examine how many bands there are. In this case there are three bands, so the DNA must have been cut twice.	It can be concluded that the DNA fragment contains two recognition sites of AGCT for the *Alu*I endonuclease (1 mark).
b. Carefully read the question and determine that in this case, only two fragments were formed; thus there must only be one cut made by the endonuclease.	This fragment of DNA contains one recognition site of AAGCTT for the *Hin*dIII endonuclease (1 mark).

Several thousand endonucleases have now been isolated, mainly from bacteria but some from archaea. About 300 endonucleases are available from commercial suppliers. Genetic engineers simply buy a quantity of the required enzyme in highly purified form from a commercial supplier.

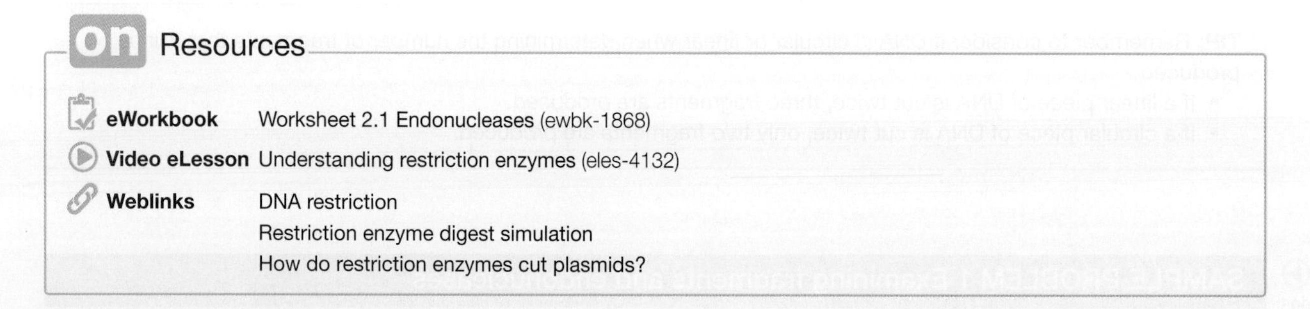
on Resources

eWorkbook Worksheet 2.1 Endonucleases (ewbk-1868)

Video eLesson Understanding restriction enzymes (eles-4132)

Weblinks DNA restriction
Restriction enzyme digest simulation
How do restriction enzymes cut plasmids?

2.2.3 Ligase to join DNA

In gene manipulation, the joining of pieces of DNA is sometimes required. An enzyme known as ligase catalyses the joining of pieces of double-stranded DNA at their sugar–phosphate backbones.

The bonds that form in this case are strong phosphodiester (covalent) bonds. The joining can produce one long piece of DNA (see figure 2.7a) or it can produce a circular molecule of DNA if the fragments have complementary ‘sticky ends’ at both ends (see figure 2.7b).

Within molecular biology, DNA ligase, can be used to insert genes of interest into plasmid vectors. DNA ligase is able to be used on different lengths of DNA which have a 'blunt' or 'sticky' end following the use of endonucleases.

FIGURE 2.7 Joining DNA fragments with ligase **a.** The joining can produce one longer piece of DNA. **b.** The joining can produce a circular molecule of DNA if the fragments have complementary 'sticky ends' at both ends.

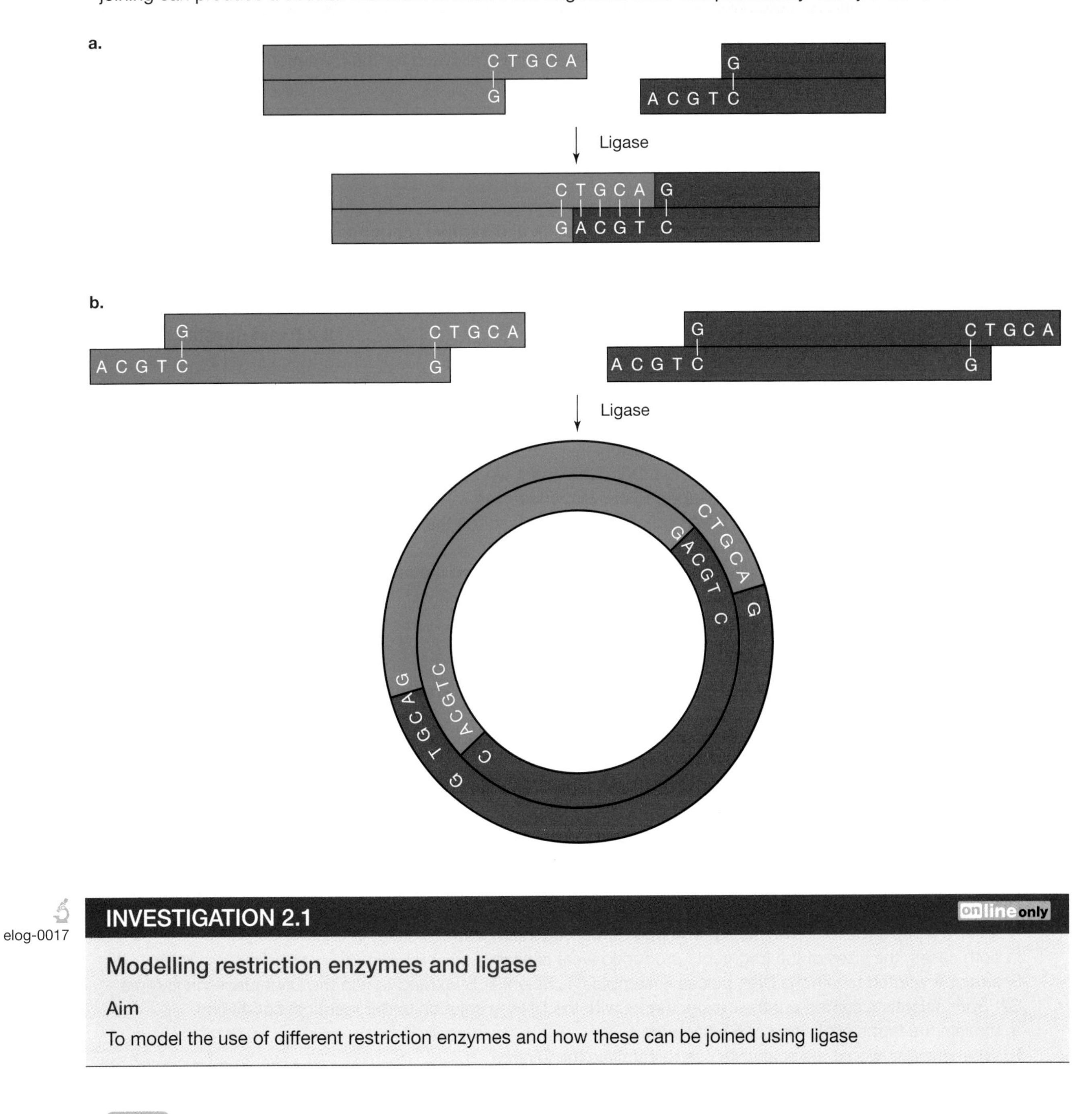

elog-0017

INVESTIGATION 2.1

online only

Modelling restriction enzymes and ligase

Aim

To model the use of different restriction enzymes and how these can be joined using ligase

on Resources

Weblink The role of DNA ligase

KEY IDEAS

- DNA can be manipulated in various ways, including by being cut into fragments, and DNA fragments can be sorted and joined.
- Cutting or restriction enzymes (endonucleases) cut DNA into fragments at specific recognition sites.
- Cuts of double-stranded DNA by some restriction enzymes produce 'sticky' ends in which there are overhanging nucleotides, and some produce blunt ends.
- Polymerases are enzymes that allow for the synthesis of nucleic acids such as DNA.
- Ligases are enzymes that allow for DNA fragments to be joined, allowing for the formation of bonds in the sugar–phosphate backbone.

2.2 Activities

learnon

To answer questions online and to receive **immediate feedback** and **sample responses** for every question, go to your learnON title at **www.jacplus.com.au**. A **downloadable solutions** file is also available in the resources tab.

2.2 Quick quiz on	2.2 Exercise	2.2 Exam questions

2.2 Exercise

1. Describe the role of DNA polymerase in the synthesis of a new DNA strand.
2. Explain the role of DNA ligase in DNA manipulation.
3. In a long DNA sequence, it is reasonable to predict that there would be more cut sites for *Hae*III than for *Hin*dIII? Explain why. Use table 2.2 to support your response.
4. Could endonuclease act on human DNA and mouse DNA? Describe the specificity of endonuclease across different species.
5. Restriction enzymes are extensively used in molecular biology. Using the recognition sites of two of these enzymes, *Bam*HI and *Bcl*I, answer the following questions.
   You are given the DNA shown below.

   5′ ATTGAGGATCCGTAATGTGTCCTGATCACGCTCCACG 3′
   3′ TAACTCCTAGGCATTACACAGGACTAGTGCGAGGTGC 5′

   a. If this DNA was cut with *Bam*HI, how many DNA fragments would you expect? The sequence needs to represent the double-stranded DNA fragments.
   b. If the DNA shown above was cut with *Bcl*I, how many DNA fragments would you expect? The sequence needs to represent the double-stranded DNA fragments.
   c. You can ligate the smaller restriction fragment produced in part **a** to the smaller restriction fragment produced in part **b**. Write out the sequence of the resulting recombinant fragment.
6. Describe one means by which a scientist can obtain a copy of a gene.
7. One sample (S1) of DNA was cut into fragments with blunt ends.
   A second sample (S2) of DNA was cut into fragments with sticky ends.
   In both cases, the sizes of the fragments produced were similar.
   Scientist A wanted to join the DNA pieces in sample S1. Scientist B wanted to join the DNA pieces in sample S2. Both scientists carried out their experiments with the DNA in solution under identical conditions.
   a. Explain the terms *sticky end* and *blunt end.*
   b. What enzyme would the scientists use to catalyse the joining?
   c. The scientists observed that the pieces of DNA from one sample joined more quickly than those in the other sample. In which sample would the faster rate of joining be observed? Justify your choice.

2.2 Exam questions

Question 1 (1 mark)

Source: *VCAA 2019 Biology Exam, Section A, Q39*

MC The diagram below is a map of a bacterial plasmid showing ORI, the origin of DNA replication, and selected restriction endonuclease sites.

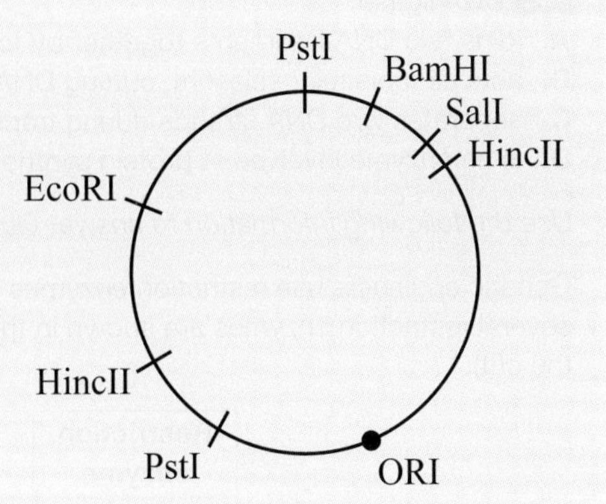

One plasmid was mixed with the restriction enzymes EcoRI, BamHI and HincII.

Which of the following shows the number of restriction sites that have been cut and the resulting number of DNA fragments produced?

	Number of restriction sites cut	Number of DNA fragments produced
A.	3	3
B.	3	4
C.	4	4
D.	4	5

Question 2 (1 mark)

Source: *VCAA 2008 Biology Exam 2, Section A, Q20*

MC The genome of a small virus is depicted below, showing the positions of cutting sites (P and Q) for two restriction enzymes.

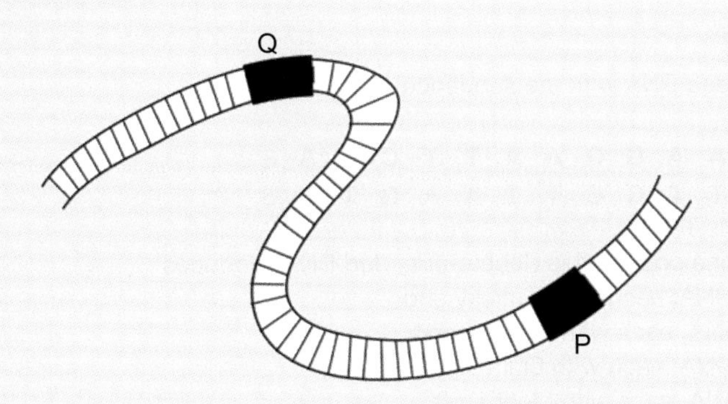

The length of DNA fragments obtained when using these restriction enzymes is shown in the table below.

Cutting site	Restriction enzyme used	Length of DNA fragments obtained (kB)
Q	EcoR1	3, 7
P	BamH1	8, 2

If both EcoR1 and BamH1 are used together on this viral DNA, the length of fragments obtained would be

A. 3, 8, 5, 2
B. 7, 2, 1
C. 3, 5, 2
D. 3, 7, 8, 2

Question 3 (1 mark)

Source: *VCAA 2019 Biology Exam, Section A, Q38*

MC DNA ligase

A. joins two DNA fragments together by forming phosphodiester bonds between the two fragments.
B. acts as molecular scissors, cutting DNA molecules at specific nucleotide sequences.
C. separates two DNA strands during transcription so that a copy can be made.
D. is an enzyme involved in protein synthesis.

Use the following information to answer Questions 4 and 5.

Genetic engineers use restriction enzymes to cut DNA into smaller lengths. The recognition sequences of several restriction enzymes are shown in the table below. The symbol * denotes the restriction site (position of the cut).

Restriction enzyme	Recognition sequence (read in 5' to 3' direction)
EcoRI	G* A A T T C C T T A A *G
HindIII	A* A G C T T T T C G A *A
AluI	A G* C T T C* G A
HaeIII	G G* C C C C* G G

Question 4 (1 mark)

Source: *VCAA 2013 Biology Section A, Q29*

MC Consider a length of double-stranded DNA with the sequence

5' **T T A A G G A A T T C A A** 3'
3' **A A T T C C T T A A G T T** 5'

Adding EcoRI to a solution containing one copy of this double-stranded DNA produces

A. two fragments of double-stranded DNA, each with a sticky end.
B. four fragments of single-stranded DNA, each with a sticky end.
C. two fragments of double-stranded DNA, each with blunt ends.
D. four fragments of single-stranded DNA, each with blunt ends.

Question 5 (1 mark)

Source: *VCAA 2013 Biology Section A, Q30*

MC Now consider a different length of double-stranded DNA with the sequence

5' **C T T A A G C T T C C A A A T T A C C G A** 3'
3' **G A A T T C G A A G G T T T A A T G G C T** 5'

Which enzyme(s) will cut this piece of DNA?

A. EcoRI only
B. HindIII only
C. AluI and HindIII only
D. AluI, HindIII and HaeIII only

More exam questions are available in your learnON title.

2.3 CRISPR-Cas9

KEY KNOWLEDGE

- The function of CRISPR-Cas9 in bacteria and the application of this function in editing an organism's genome

2.3.1 The function of CRISPR-Cas9 in bacteria

The first attempts at genetic manipulation in the 1970s involved adding DNA to the plant and animal genomes.

EXTENSION: Early techniques used for genetic manipulation

Early techniques used to add DNA to target cells included:

- the use of so-called 'gene guns' in which gold atoms coated with DNA were blasted into cells, a technique used mainly to insert genes for pesticide or herbicide resistance into plant cells
- the use of modified viruses to act as vectors to carry a functional gene into cells with a defective gene. Viruses used as vectors include retroviruses and adenoviruses.

These techniques were clearly very limited because there was no control over where any added DNA would be inserted into the genome of the target cells. Instead, DNA was simply added randomly into the genome. What would be expected to happen if the added DNA was by chance inserted into a gene that was essential for cell survival, causing the gene to become non-functional? What would be expected to happen if the added DNA was by chance inserted into a cancer-suppressing gene so that this gene was disabled?

Another major limitation was that, although these early techniques could carry a functional replacement copy of a gene into cells with a defective gene, these techniques were not able to mend a defective gene by editing or disabling it. Editing is a process of correction, such as occurs when an editor of a manuscript corrects a misspelt word, adds or replaces a word or phrase, or deletes a sentence.

What is CRISPR-Cas9?

CRISPR-Cas9 (*Cluster regularly interspaced short palindromic repeats-CRISPR-associated protein 9*) was first identified by a team of scientists in Japan in 1987. The scientists noticed an unusual genetic sequence composed of alternative repeating and non-repeating sequences. The function and importance of these sequences was not identified until 2007. In 2007, Rodolphe Barrangou and his team identified that CRISPR-Cas9 was essential in the adaptive immunity in most bacterial and archaeal organisms in the defensive mechanism against the invasion of bacteriophage infection. The addition and removal of the phage's genomic sequences led to a diverse phage resistance phenotype of the bacteria. Within the same study, researchers also noticed near the CRISPR sites a particular set of CRISPR-associated (*Cas*) genes that coded for Cas proteins. Cas proteins are vital in the immunological defences against DNA virus in prokaryotic organisms. Cas proteins were also identified as being relevant to the CRISPR-mediated immunity because they repressed the expression of these genes that disrupted CRISPR function.

FIGURE 2.8 The Cas9 enzymes in complex with RNA (yellow) and DNA (violet)

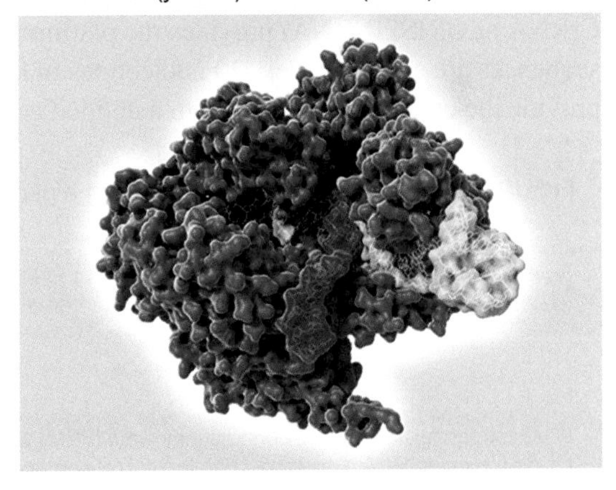

CRISPR are short repeated segments of DNA, with each repeated segment being separated by a length of spacer DNA. Cas9 is a specific enyzme (specifically an endonuclease) that can cut DNA associated with CRISPR.

CRISPR-Cas9 a tool for precise and targeted genome editing that uses specific RNA sequences to guide an endonuclease, Cas9, to cut DNA at the required positions

What was most significant was the recognition that CRISPR-Cas9 technology could be readily adapted to provide an inexpensive and easy-to-use means of **genome editing** for use in an endless range of genes from any organism.

In 2012, Dr Jennifer Doudna and Dr Emmanuelle Charpentier (shown in figure 2.9) expanded on the previous understanding of CRISPR-Cas9. They were co-inventors of the technology that allowed CRISPR-Cas9 to be used for programmed and targeted gene-editing.

FIGURE 2.9 Jennifer Doudna (left) and Emmanuelle Charpentier (right)

This tool, borrowed and adapted from the bacterial adaptive immune system, is now being put to use in editing faulty genes and in silencing genes in plants and animals. Many applications in agriculture are being explored (which will be further discussed in subtopic 4.8). Research is already underway on the use of the CRISPR technology as a safe and reliable means of editing the defective alleles responsible for human diseases, such as cystic fibrosis and sickle-cell anaemia.

How CRISPR is used in the editing of an organism's genome

The CRISPR gene-editing technique simply requires molecular 'scissors' to cut the target DNA and a 'guide' to direct the 'scissors' to the site where the cuts in the DNA will be made. The guide RNA is typically denoted by gRNA or sgRNA, where sg means single or synthetic guide. SgRNA (also referred to as CrRNA or CRISPR-RNA) provides the platform to allow the Cas protein to bind. These sequences are approximately 20 nucleotides in length and are defined as 'spacers'; these provide the cellular machinery with a defined genomic target to be modified.

genome editing a process by which changes are made to the nucleic acid sequence of genes; also termed gene editing

FIGURE 2.10 The steps involved in editing a genome using CRISPR-Cas9

Scientist creates a complex consisting of a short RNA sequence (CrRNA or sgRNA) and CRISPR-associated endonucleases (Cas protein). The RNA sequence is complementary to the target DNA sequence.

↓

Cas9 recognises a protospacer adjacent motif (PAM) sequence, a very short nucleotide sequence adjacent to the target spacer. Using the guide RNA, Cas9 identifies the corresponding DNA sequence within the host cell's genome and cuts both strands of DNA.

↓

Scientists can insert a new piece of DNA (donor DNA) at the site of the cut. They may also remove or replace sections of DNA, or add and delete single nucleotides.

↓

The cell detects and repairs the broken strand of DNA. When the DNA is repaired, any changes made are integrated into the genomic DNA.

FIGURE 2.11 Biological process in gene targeting using CRISPR-Cas9

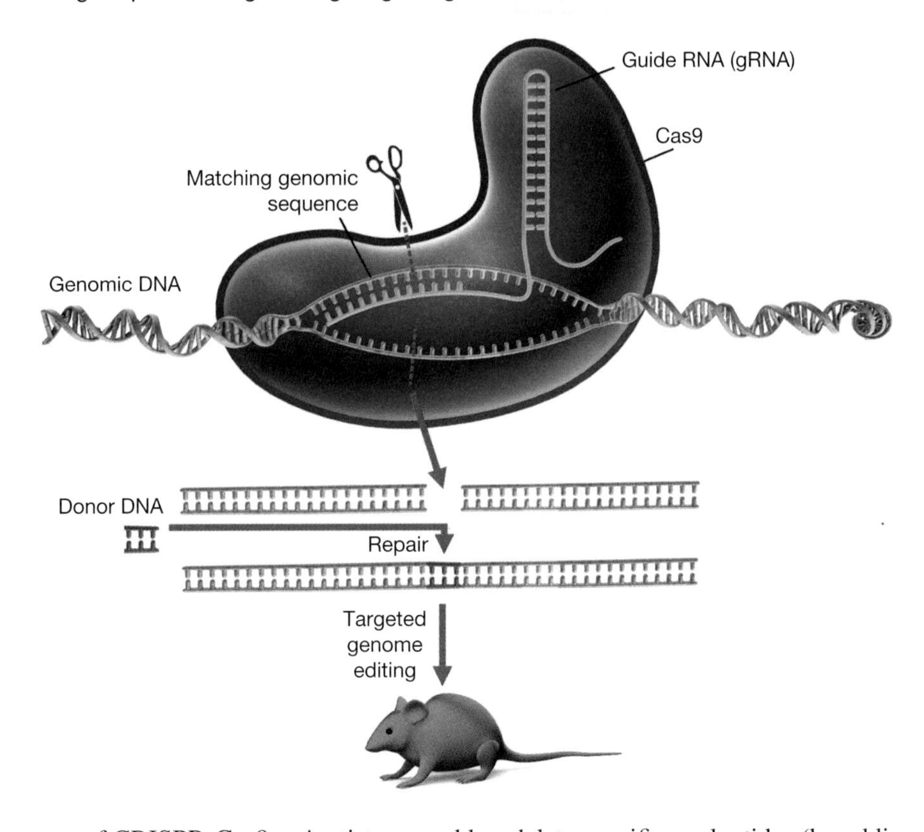

Through the process of CRISPR-Cas9, scientists can add or delete specific nucleotides (by adding template strands that contain these changes), delete sections of DNA, or introduce a new gene altogether (for example correcting a disease-causing mutation).

tlvd-1096

SAMPLE PROBLEM 2 Defining the role of CRISPR

a. Define the role of CRISPR-Cas9 in gene editing and comment on the importance of this process. (2 marks)

b. Outline the two major features of how the process occurs. (2 marks)

THINK	WRITE
a. 1. Identify what the question is asking you to do. You are being specifically asked for two things — a definition of the role of CRISPR-Cas9 and a comment on the importance of this.	
2. As the question is worth two marks, two statements important to the process are required. TIP: • Identify what the statement is asking before commencing your answer. • Each mark awarded equates to one statement.	• CRISPR-Cas9 is a unique technology enabling scientists and researchers to edit parts of the genome by removing, adding or altering sections of the DNA sequence (1 mark). • The CRISPR-Cas9 system is currently the gold standard in gene editing, as it is the fastest, cheapest and most reliable system currently available (1 mark).

b. 1. This question is asking you to *outline*, which requires a brief statement on each feature.

2. As the question is worth two marks, two statements important to the process are required. Each statement should focus on a major feature of the CRISPR process.

The CRISPR-Cas9 system consists of two key molecules that introduce a change (mutation) into the DNA.

- The enzyme Cas9. This acts as a pair of 'molecular scissors' that cuts the two strands of DNA at a specific location in the genome. This allows for the addition or removal of DNA within the genome (1 mark).
- The guide RNA (sgRNA). This consists of a small piece of pre-designed RNA sequence (about 20 bases long) located within the RNA scaffold. In turn, the scaffold binds to DNA and the pre-designed sequence 'guides' Cas9 at the correct location within the genome, ensuring the Cas9 enzyme cuts at the correct point in the genome (1 mark).

2.3.2 Application of CRISPR-Cas9

CRISPR-Cas9 technology has extraordinary potential for widespread use in clinical, agricultural and research settings, for example:

- for a myriad of research purposes:
  - to 'knock out' genes, one at a time, in order to identify their function
  - to introduce specific mutations in a DNA sequence.
- to edit a faulty allele of a gene in a person with a severe inherited disease
- to snip out the faulty segment of a gene and replace it with a working copy
- to activate or to repress a gene
- to add a new gene to the genome.
- to advance animal welfare (e.g. hornless dairy cows)
- to modify crops to increase nutritional value.

FIGURE 2.12 Concerns exist around gene editing. Chinese scientists used the CRISPR technique to confer HIV resistance on twin girls named 'Lulu' and 'Nana', who were born in 2018. Other scientists and ethicists criticised this use of the technique as unnecessary and irresponsible.

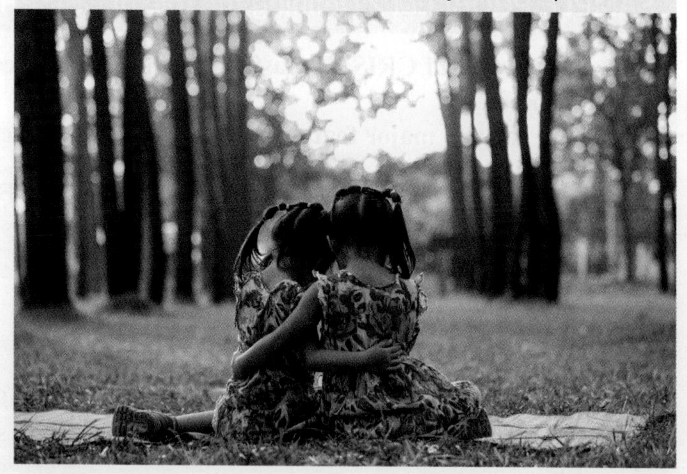

This potential is highlighted by the fact that major pharmaceutical companies have made significant investments in CRISPR-Cas9 technologies. For example, in October 2015, Vertex Pharmaceuticals paid in excess of US$100 million to use the gene-editing technology of CRISPR Therapeutics. Their intention is to develop treatments for cystic fibrosis and sickle-cell diseases. More recently, Bayer Pharmaceuticals and CRISPR Therapeutics set up a joint venture aimed at developing therapies to treat blood disorders, blindness and congenital heart disease.

FIGURE 2.13 Future applications of CRISPR

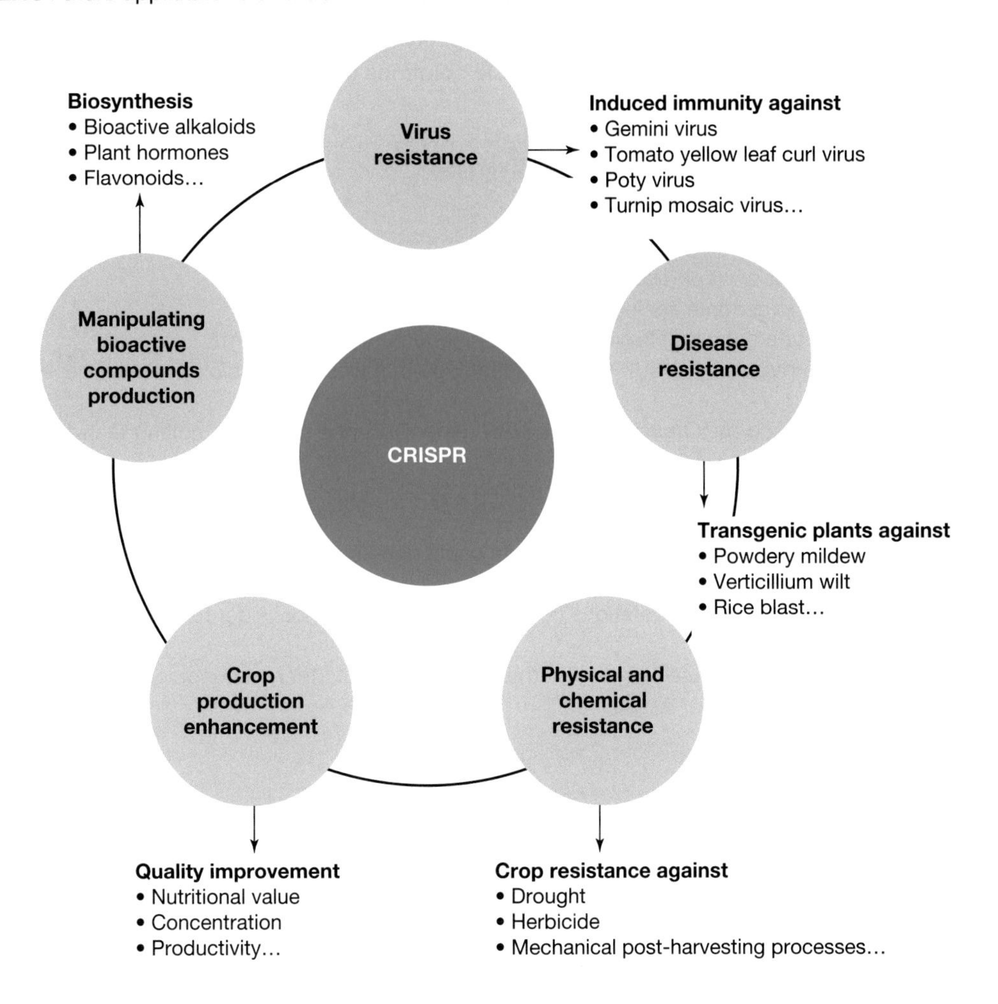

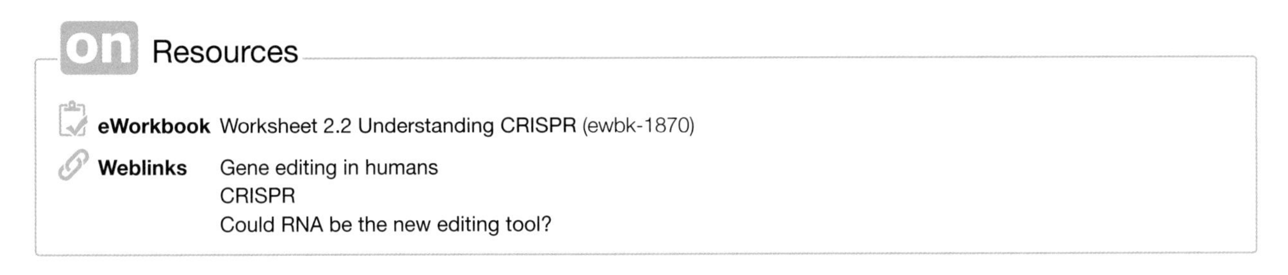

KEY IDEAS

- CRISPR-Cas9 technology is an efficient, easy-to-use, and inexpensive tool for precise gene editing.
- CRISPR-Cas9 technology can be applied to every eukaryotic species, including humans.
- The key components of CRISPR-Cas9 technology are a guide RNA and an endonuclease enzyme, Cas9.
- Early gene therapy techniques added DNA segments to genomes but at unpredictable locations.
- CRISPR-Cas9 technology enables precise genomic editing.
- CRISPR technology can not only add DNA but can also edit, modify or disable or delete DNA from the genome.
- Like any new technology with so many potential applications, aspects of the use of CRISPR technology raise ethical considerations.

2.3 Activities

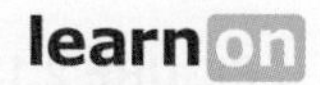

To answer questions online and to receive **immediate feedback** and **sample responses** for every question, go to your learnON title at **www.jacplus.com.au**. A **downloadable solutions** file is also available in the resources tab.

2.3 Quick quiz on	**2.3 Exercise**	**2.3 Exam questions**

2.3 Exercise

1. 'CRISPR' is an acronym for what phrase?
2. What is the role of Cas9 in genome editing?
3. In a cell, what happens to a gene that has been cut by a genome editing system?
4. Identify one capability provided by CRISPR-Cas9 technology that was not possible in earlier gene therapy techniques.
5. Sickle cell anaemia is caused by a mutation in the *HBB* gene. Describe how gene editing technology could be used to correct this mutation.
6. Research the use of CRISPR-Cas9 and the gene editing used on 'Lulu' and 'Nana' (figure 2.12). Use this information to answer the following questions.
   a. Identify and explain two biological considerations in using gene sequencing to identify potential diseases.
   b. Identify and explain one advantage of genetic testing specifically for children like 'Lulu' and 'Nana'.
   c. Identify and explain the legal considerations countries would need to consider to ensure the privacy of individuals.
   d. Outline how CRISPR could be used to prevent the development of other immunological diseases.

2.3 Exam questions

Question 1 (1 mark)

MC Which of the following is **NOT** an application of CRISPR-Cas9?

A. Adding a new gene to a genome
B. Replacing entire chromosomes
C. Deleting a single base from a genome
D. To repress specific genes in a genome

Question 2 (4 marks)

Duchenne Muscular Dystrophy (DMD) is a genetic disorder that affects the protein dystrophin, an important protein that connects the cytoskeleton to muscle fibres. In DMD, the muscle cells are non-functioning due to the lack of functional dystrophin, and thus muscles weaken and die. By age 10, most individuals with DMD are wheelchair-bound.

The most common cause of DMD is a deletion in one or more exons in the gene that codes for dystrophin.

Outline how CRISPR-Cas9 may be used in the future in individuals with DMD.

Question 3 (6 marks)

In 2020, the Nobel Prize in Chemistry was awarded to Jennifer Doudna and Emmanuelle Charpentier. This was historic as it was the first time that two women were jointly awarded this prize.

Doudna and Charpentier were the first to propose the use of CRISPR-Cas9 for the editing of genomes.

a. As part of their research, Doudna and Charpentier had to comment on the ethical implications of the use of CRISPR-Cas9. Outline two ethical considerations that relate to this technology. **2 marks**
b. Prior to the discoveries by Doudna and Charpentier, gene editing was usually achieved through the use of gene guns or modified viruses. These technologies cause DNA to randomly integrate into the target genome. Describe two advantages that CRISPR-Cas9 has over these technologies. **2 marks**

The CRISPR-Cas9 system was first discovered in bacteria in 1987. The function of this system was not known until 2007, when Rodolphe Barrangou and his team identified the system as essential in the adaptive immune system of bacteria. It was found to mainly act against DNA viruses that can infect and integrate into the bacterial chromosome.

c. Explain how this system would be used by prokaryotic organisms as a defence against viruses. **2 marks**

Question 4 (1 mark)

MC CRISPR-Cas9 is a technology used to edit genomes. The steps taken to edit the genome using CRISPR-Cas9 are summarised below. The order of the steps has been mixed up.

S. Scientists insert a new piece of DNA at the site of the cut.
T. A short guide RNA sequence is produced that is complementary to the target DNA sequence and Cas protein.
U. The guide RNA identifies the appropriate location and Cas9 cuts the DNA in the host at that location.
V. The cell detects and repairs the DNA, integrating the new piece of DNA.

The correct order of these steps is:

A. S, V, T, U **B.** T, U, S, V **C.** T, U, V, S **D.** U, T, S, V

Question 5 (9 marks)

Gene editing from CRISPR-Cas9 made news in 2018, where two embryos in China had their genomes edited. Lulu and Nana were therefore the first babies born whose cells had undergone CRISPR-Cas9 treatment.

In this case, CRISPR-Cas9 was used to provide HIV resistance in Lulu and Nana. A certain gene, *CCR5*, produces the protein that sits on the cell surface of immune cells and acts as a receptor. Although this protein has some uses in the immune system, it is also targeted by HIV to allow it to invade cells.

Some individuals have a variant of this gene known as *CCR5Δ32*, in which 32 base pairs are deleted compared to the functional gene. This leads to a truncated protein that does not reach the surface of the immune cells, reducing the chance an individual is infected with HIV. Individuals with two copies of this variant are thought to be highly resistant to HIV.

a. One technique scientists used was to delete some of the bases from *CCR5* to produce the non-functional *CCR5Δ32* variant. Draw a clear diagram outlining how the CRISPR-Cas9 system could be used to do this. **3 marks**

b. Suggest two reasons why CRISPR-Cas9 was performed on embryos (with only around 200 cells) rather than after the twins were born. **2 marks**

c. For Lulu and Nana, this editing was not entirely successful. Firstly, not all cells had the altered *CCR5Δ32* gene. The use of CRISPR-Cas9 in this case posed many problems, particularly around future implications of editing the genome and possible misuse of the technology. As a result, conversations about the regulation of these technologies, particularly in embryos (and germline cells that may be inherited by future generation) have increased. Identify two possible regulations that may be important around the use of CRISPR-Cas9. Justify your response. **4 marks**

More exam questions are available in your learnON title.

2.4 Amplification of DNA using polymerase chain reaction

KEY KNOWLEDGE

- Amplification of DNA using polymerase chain reaction

Source: Adapted from VCE Biology Study Design (2022–2026) extracts © VCAA; reproduced by permission

Polymerase chain reaction (PCR) is a technique used to amplify a segment of DNA accurately and quickly. Before the development of PCR, the amplification of recombinant DNA fragments was laborious and time-consuming.

PCR enables scientists to obtain large quantities of a specific DNA sequence from a small initial sample.

This amplification (which can be achieved with as little as one initial cell) is required for various applications, including molecular biology, forensic analysis, evolutionary biology and medical diagnostics. For example, tiny bone and tissue fragments on ancient stone tools are a source of DNA. This DNA has been amplified using PCR to identify the species from which it came.

PCR was developed in 1983 by Kary B. Mullis, an American biochemist who won the Nobel Prize for Chemistry in 1993 for his work.

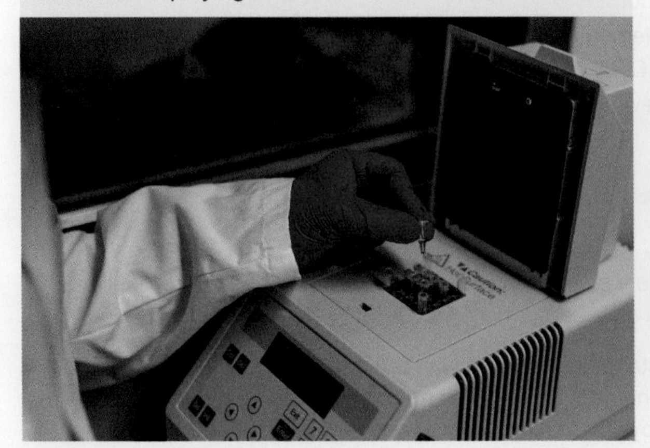

FIGURE 2.14 The polymerase chain reaction involves amplifying DNA from a small initial amount.

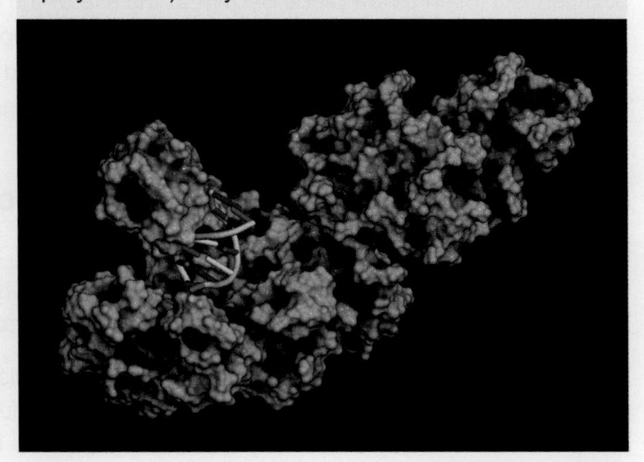

FIGURE 2.15 *Taq* polymerase (*Thermus aquaticus* polymerase) enzyme bound to DNA

The polymerase chain reaction depends on the enzyme DNA polymerase (as shown in figure 2.15) to amplify or make multiple copies of a sample of DNA. The polymerase enzyme comes from a bacterial species (*Thermus aquaticus*) that lives in hot springs. Its optimal living conditions are environments around 70 °C. This enzyme is known as ***Taq* polymerase**.

When you are discussing PCR (whether in a SAC or an exam), ensure you refer specifically to *Taq* polymerase and not just DNA polymerase.

For PCR to occur, there are many components required. Table 2.3 outlines the components required for PCR.

***Taq* polymerase** an enzyme used in PCR that adds free nucleotides to the single stranded DNA in order to synthesise a new strand

denaturing the first stage in PCR, in which a double-stranded piece of DNA is heated and separated into single-stranded DNA

annealing the second stage in PCR, in which primers attach to the single-stranded DNA

extension the third stage in PCR, in which the *Taq* polymerase enzyme synthesises a new strand of DNA by adding free nucleotides

TABLE 2.3 Components required for PCR

PCR component	Purpose
DNA sample	To provide a template to produce copies of in PCR
Primers	To bind to the single-stranded DNA and to provide a point in which DNA synthesis can be initiated and designate the sequence to be copied.
Taq polymerase	To make multiple copies of the DNA strand by adding nucleotides
Free nucleotides (dNTPs — deoxyribonucleotide triphosphate)	To be added by *Taq* polymerase to produce the new DNA strand
Mix buffer	To provide a suitable chemical environment for the activity of *Taq* polymerase by maintaining the appropriate pH and providing any required salts
PCR tube	To provide a vessel for the PCR reaction to occur. The tube will contain the DNA sample, polymerase, primers, nucleotides and buffer.

PCR is a repeating process involving three main steps. These steps are referred to as **denaturing**, **annealing** and **extension**.

Steps in PCR

1. *Denaturing*
   - The DNA sample needs to be separated into two separate strands.
   - The separation happens by raising the temperature of the PCR mixture, causing the hydrogen bonds between the complementary DNA strands to break
   - This occurs at approximately 94 °C for one minute.
2. *Annealing*
   - Short segments of single-stranded DNA, known as primers (forward and reverse sequences), are added.
   - Primers bind to the target DNA sequences (at 3′ ends) and initiate DNA synthesis (polymerisation). These primers can bind to their complementary sequences on the single-stranded template DNA.
   - This occurs at approximately 55 °C for two minutes.
3. *Extension*
   - The polymerase enzyme (*Taq* polymerase) uses the primers as a starting point and extending the primers, synthesising the new strands of DNA by adding nucleotides.
   - A supply of 'free' nucleotides must be available in order for the DNA polymerase to create a new strand of 'complementary DNA' to each of the single template strands.
   - This occurs at 72 °C for one minute (the optimal temperature for *Taq* polymerase).

About five minutes is required for each cycle of 'denature — anneal — extend'. This process is repeated, doubling the DNA each time.

FIGURE 2.16 The steps involved in PCR. A range of temperatures for the first two steps is acceptable.

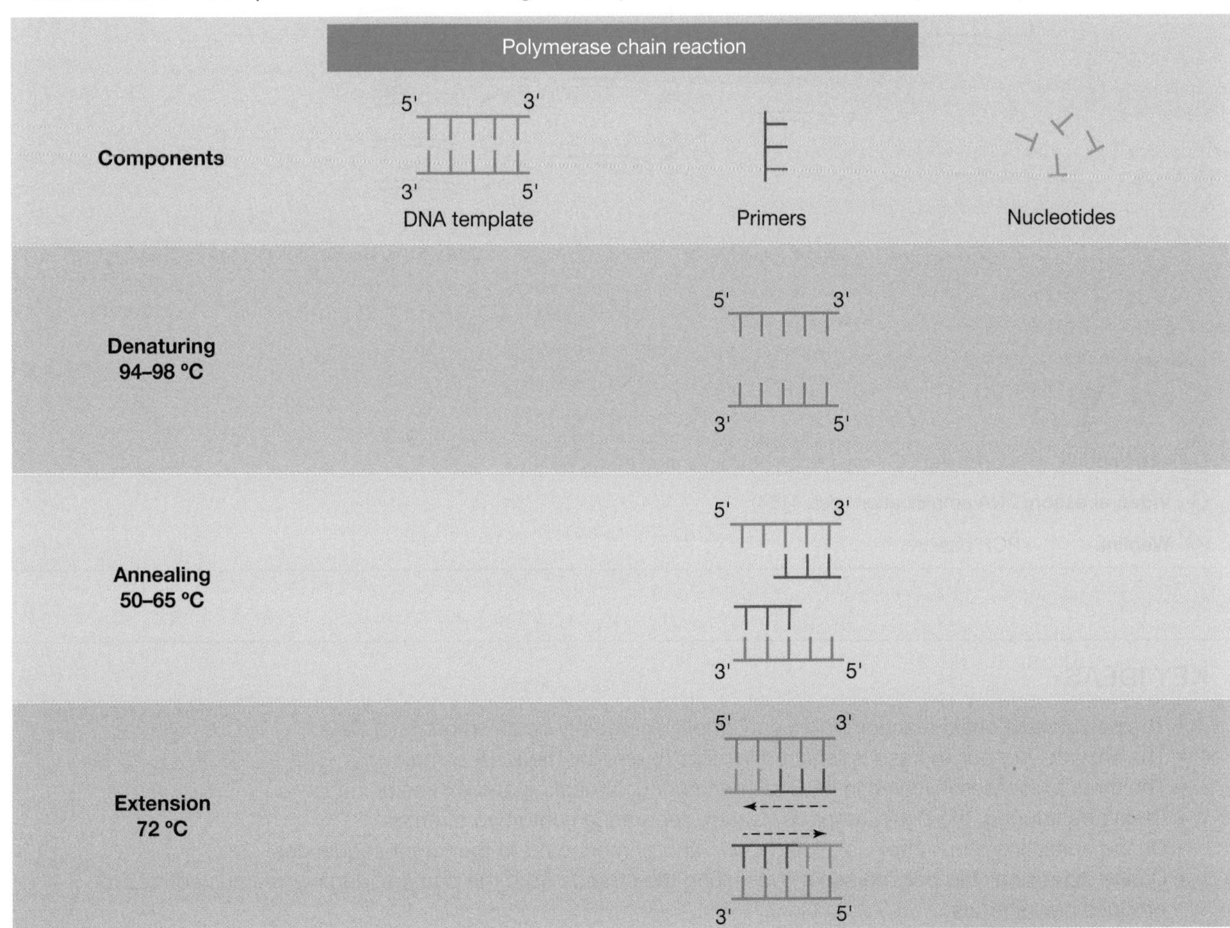

Over time, an exponential growth of the DNA from the two original strands occurs (with the number of DNA copies doubling each cycle). After around 35 cycles an exponentially number of the exact copies of the target DNA is produced.

FIGURE 2.17 During PCR, DNA strands are doubled each cycle. How many copies would be present after 35 cycles?

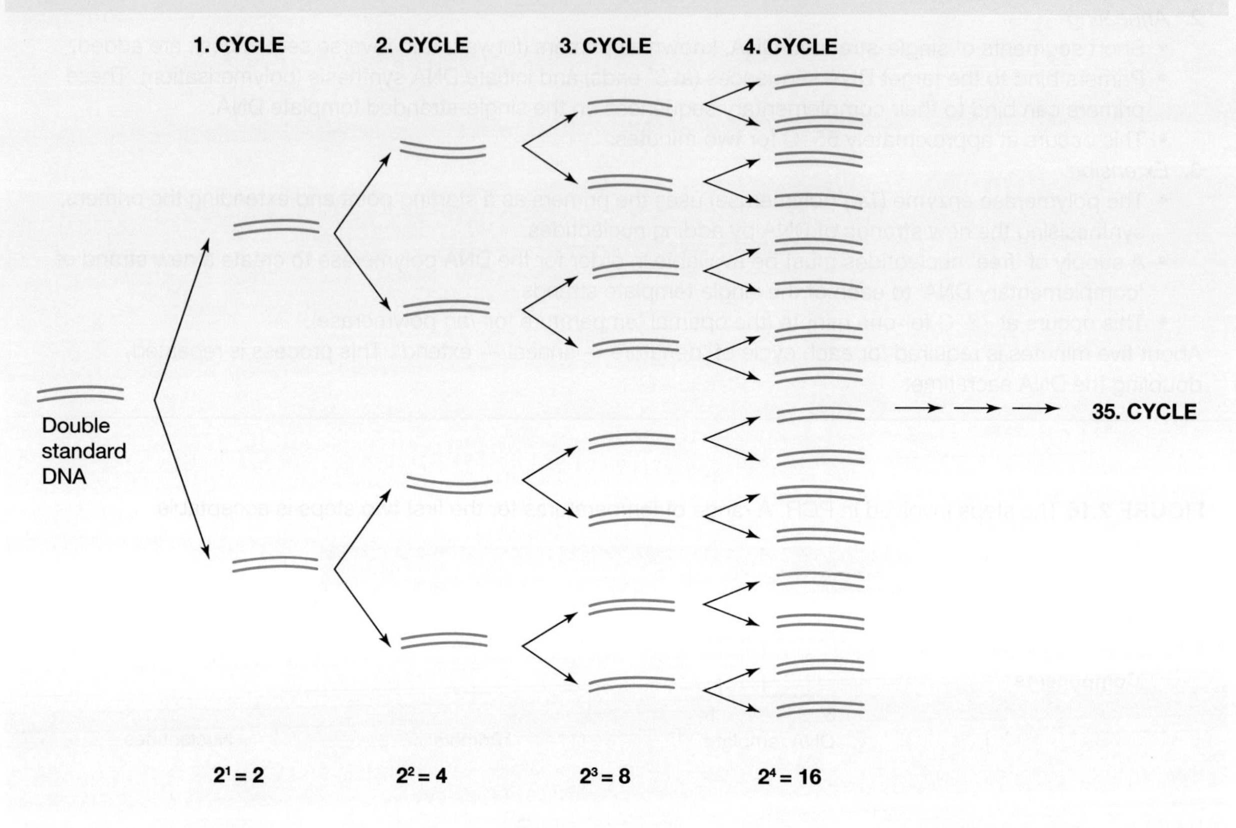

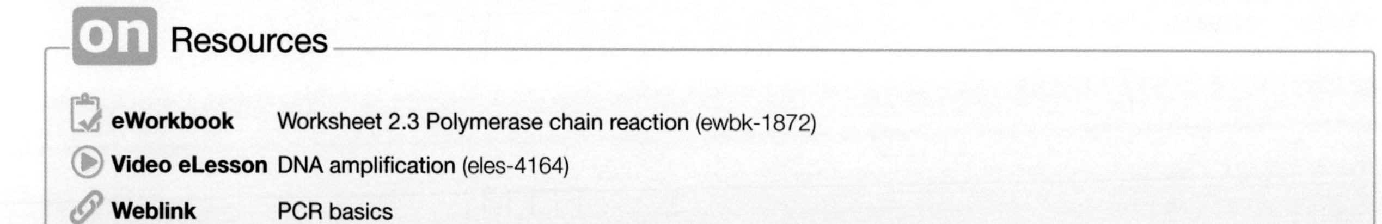

KEY IDEAS

- The polymerase chain reaction (PCR) can amplify incredibly small amounts of DNA.
- The enzyme *Taq* polymerase is required in order to amplify the DNA sample.
- The three processes involved in PCR are denaturing, annealing and extension.
- During denaturing, the DNA sample is heated, separating it into two strands.
- During annealing, the sample is cooled down and primers bind to the target sequences.
- During extension, *Taq* polymerase moves along the strands from the primers, adding free nucleotides to produce new strands.
- The process is repeated, allowing for DNA to double every cycle.

2.4 Activities

To answer questions online and to receive **immediate feedback** and **sample responses** for every question, go to your learnON title at **www.jacplus.com.au**. A **downloadable solutions** file is also available in the resources tab.

2.4 Quick quiz on	2.4 Exercise	2.4 Exam questions

2.4 Exercise

1. List two applications of DNA amplification.
2. DNA amplification can be achieved through polymerase chain reaction (PCR) relying on the enzyme polymerases. Compare DNA polymerases and RNA polymerases.
3. a. List the reagents found in a PCR mixture.
   b. Outline the main steps involved in the polymerase chain reaction.
4. What is meant by the phrase 'denature — anneal — extend'?
5. Explain the purpose of using *Taq* polymerase for PCR rather than DNA polymerase from *E. coli*.
6. If you started a PCR cycle with a double-stranded molecule of DNA, how many double-stranded molecules would there be after 25 PCR cycles?

2.4 Exam questions

Question 1 (1 mark)

Source: *VCAA 2020 Biology Exam, Section A, Q29*

MC A small sample of DNA was obtained from a fossil. Polymerase chain reaction (PCR) was used to amplify the amount of DNA obtained from the sample.

Which one of the following is a correct statement regarding the PCR process?

A. DNA polymerase catalyses the pairing of primers with complementary nucleotides.
B. RNA polymerase catalyses the additions of nucleotides to a DNA strand.
C. Annealing and extension of the DNA occur at different temperatures.
D. The number of copies of the DNA is quadrupled in each cycle.

Question 2 (3 marks)

Source: *Adapted from VCAA 2019 Biology Exam, Section B, Q8b*

The Genomics Health Futures Mission will run a $32 million trial, starting in 2019, to screen over 10 000 couples who are in early pregnancy or who are planning to have a baby. Using a blood test, individuals will be screened for 500 severe or deadly recessive gene mutations.

Couples will be told they have a genetic mutation if both individuals in the couple carry the same mutation. The trial may lead to a population-wide carrier screening program. The researchers will evaluate cost effectiveness, psychological impact, ethics and barriers to screening. It is anticipated that future tests will be free of charge.

The blood sample from an individual will provide researchers with only a small amount of DNA. Polymerase chain reaction (PCR) will be used to amplify the DNA.

Describe what happens in each of the three stages of PCR. The stages must be given in the order in which they occur.

Question 3 (1 mark)

Source: *VCAA 2017 Biology Exam, Section A, Q34*

MC The process known as polymerase chain reaction (PCR) involves repeated cycles made up of several steps.

During PCR the

A. first step in each cycle is to anneal primers to the DNA at a low temperature.
B. temperature must be lowered to 37 °C before the beginning of each cycle.
C. second step in each cycle is to heat the DNA to a high temperature.
D. final step of each cycle involves the use of DNA polymerase.

Question 4 (1 mark)

Source: *VCAA 2006 Biology Exam 2, Section A, Q11*

MC Amplification of DNA in the polymerase chain reaction requires
A. nucleotides of uracil.
B. DNA polymerase.
C. amino acids.
D. ribose sugar.

Question 5 (1 mark)

Source: *VCAA 2010 Biology Exam 2, Section B, Q8b*

A young man, Ben, wants to find out more about his genetic ancestry. He sends a sample of cells, obtained from a swab of his mouth, to a laboratory. On receipt of the sample, the laboratory treats the cells to release the DNA to enable identification of STR markers.

Name the process used to produce many copies of the STR markers.

More exam questions are available in your learnON title.

2.5 The use of gel electrophoresis

KEY KNOWLEDGE

- The use of gel electrophoresis in sorting DNA fragments, including the interpretation of gel runs for DNA profiling

Source: Adapted from VCE Biology Study Design (2022–2026) extracts © VCAA; reproduced by permission

Being able to sort fragments by size has many vital scientific applications. This can be done using a process known as **gel electrophoresis**. Gel electrophoresis has a variety of applications; for example, it is used in **DNA fingerprinting** and the detection of genetic variants and proteins involved in health and disease. It is used in profiling to help solve crimes, determine paternity and identify bodies in events such as natural disasters. It is also used to aid in the detection of pathogens (disease-causing organisms) that may be present in blood or other tissues, or in sources such as food. In many instances, nucleic acids or proteins that are detected and purified with gel electrophoresis are investigated further by means of **DNA sequencing**.

Gel electrophoresis is one of several techniques used to separate macromolecules such as DNA, RNA, or protein on the basis of their size or electric charge (for example, using the negative phosphate–sugar backbone present in DNA and RNA).

2.5.1 Using gel electrophoresis to sort fragments

Once we have a solution containing DNA that has been cut into a number of predictable fragments using an endonuclease enzyme, we can use electrophoresis to sort these DNA fragments out by size.

Electrophoresis is the most effective way of separating DNA fragments of varying sizes, ranging from 100 bp (base pairs) to 25 kbp (kilobase pairs). Gel electrophoresis occurs in a specialised material known as **agarose gel**. Agarose is isolated from the seaweed genera *Gelidium* and *Gracilaria*, and consists of repeated agarobiose (L- and D-galactose) sub-units.

gel electrophoresis a technique for sorting a mixture of DNA fragments (and other molecules with a net charge) through an electric field on the basis of different fragment lengths

DNA fingerprinting technique for identifying DNA from different individuals based on variable numbers of tandem repeats of short DNA segments near the ends of chromosomes

DNA sequencing identification of the order or sequence of bases along a DNA strand

agarose gel a special type of porous gel used in gel electrophoresis, which allows for DNA to be separated by fragment size

The steps in separating fragments by gel electrophoresis

1. The DNA sample with fragments of varying sizes is combined with DNA loading dye.
2. This mixture is placed in a well at one end of a slab of jelly-like supporting material, known as an agarose gel.
3. This agarose gel is immersed in a buffer solution (salt solution).
4. The gel is then exposed to an electric field with the positive (+) pole (anode) at the far end and the negative (–) pole (cathode) at the starting origin. The DNA is attracted to the positive pole.
5. Smaller fragments move through the agarose gel faster compared to larger fragments.
6. These fragments appear as bands on a gel which can be interpreted in various ways. This usually needs to be observed under UV light.

FIGURE 2.18 The equipment used for gel electrophoresis and the observations that occur

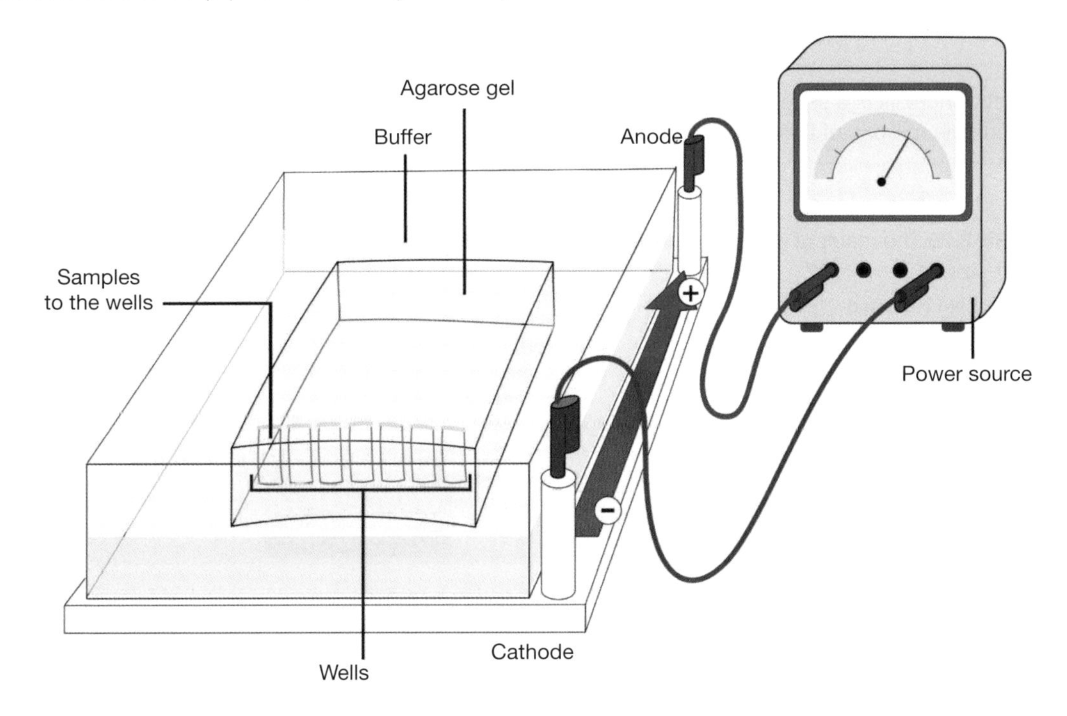

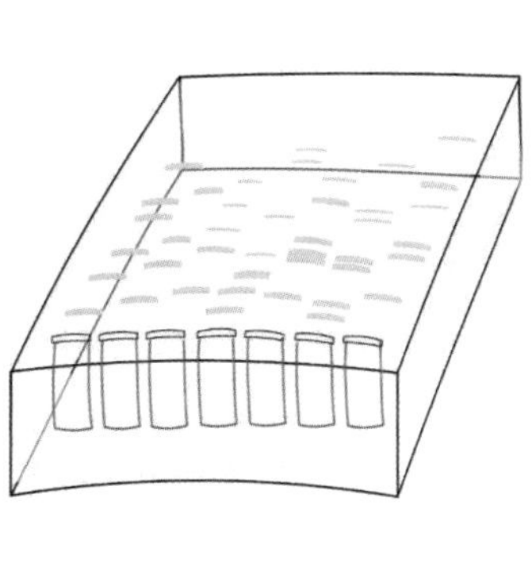

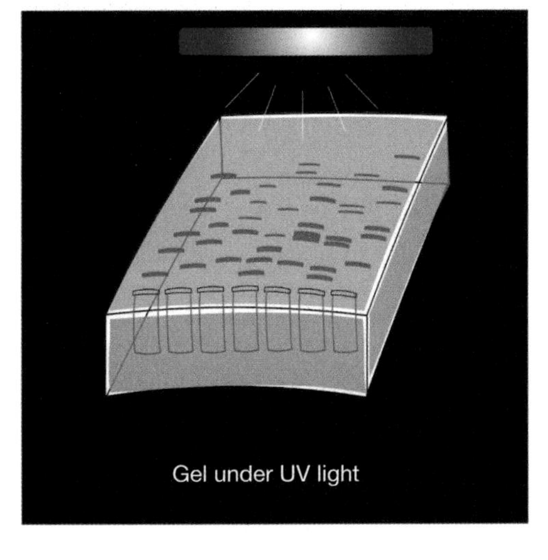

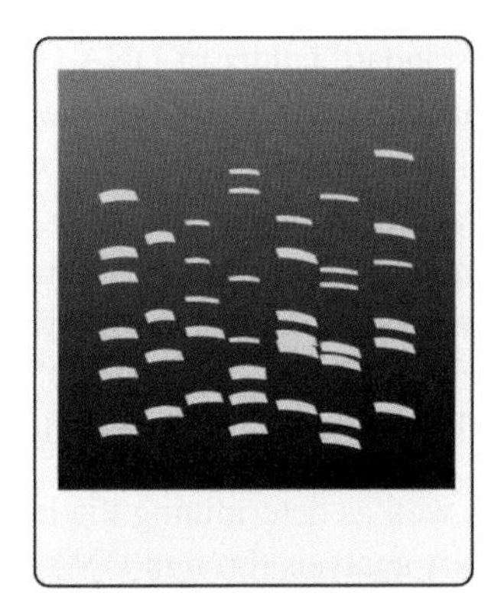

How does electrophoresis work?

The phosphate backbone of the DNA (and RNA) molecule is negatively charged. Therefore, when placed in an electric field, DNA fragments will migrate to the positively charged pole (anode) of the gel.

As DNA has a uniform mass/charge ratio, DNA molecules are separated by size within an agarose gel in a pattern such that the distance travelled is proportional to its molecular weight. The shortest DNA fragments move most quickly, and the longest fragments move most slowly.

The distance the band travels is also influenced by the concentration or viscosity of the agarose and the specific voltage or power used.

FIGURE 2.19 A researcher using a micropipette to load a sample of DNA

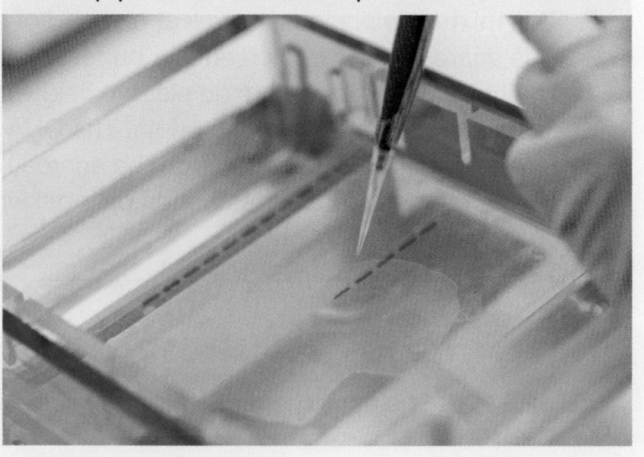

The result of electrophoresis is a series of parallel bands of DNA fragments at differing distances down the gel. Each band can contain millions of DNA molecules of the same size.

FIGURE 2.20 The result of gel electrophoresis where the fragments are separated according to their sizes

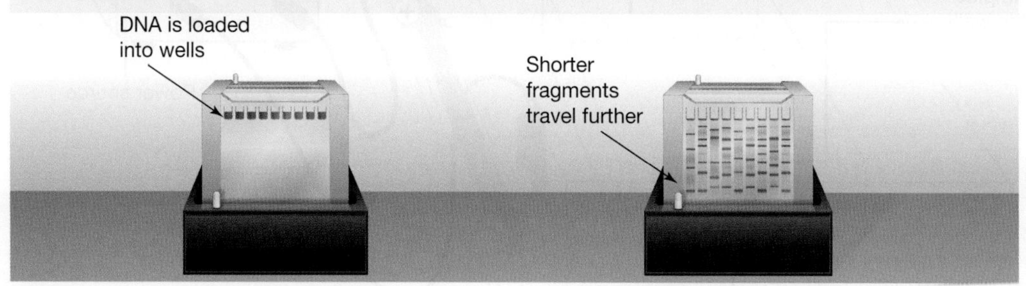

Once the gel run is complete, the separated DNA bands must be made visible either through the use of a dye or a labelled **probe** (that binds to a target sequence). One common dye is **ethidium bromide** (EtBr), which illuminates under ultraviolet light; the DNA bound to EtBr will fluoresce pale pink (see figure 2.21).

A standard ladder of DNA fragments of known sizes is usually run through the gel at the same time as the unknown DNA samples. In some cases, this is referred to as an allele ladder. The sizes of unknown DNA fragments can be approximated by comparing the positions of their bands with those of the known standards. Figures 2.21 and 2.22 show a ladder of DNA markers of known sizes in lane 1, at the left-hand side of the gel.

FIGURE 2.21 Gel stained with ethidium bromide after completion of electrophoresis

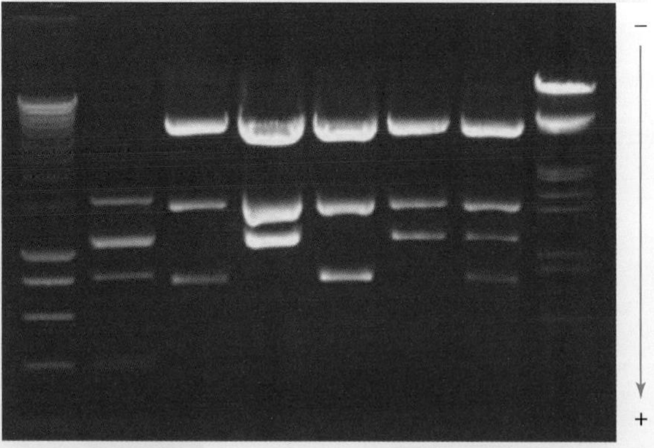

As well as determining the lengths of fragments, gel electrophoresis can be used for DNA analysis through DNA fingerprinting. This can be used in forensic investigations (for example, by comparing the DNA found at a crime scene to that of a suspect); in examining disease; or in paternity disputes (by examining the DNA of the child in comparison to the DNA of their parents). This will be further explored in section 2.5.2.

probe single-stranded segment of DNA (or RNA) carrying a radioactive or fluorescent label with a base sequence complementary to that in a target strand of DNA

ethidium bromide a dye that binds to DNA and illuminates under UV light

FIGURE 2.22 Bands are separated by the size of their fragments. The unknown bands can be approximated to be 5000, 3300, 1500 and 600 base pairs long.

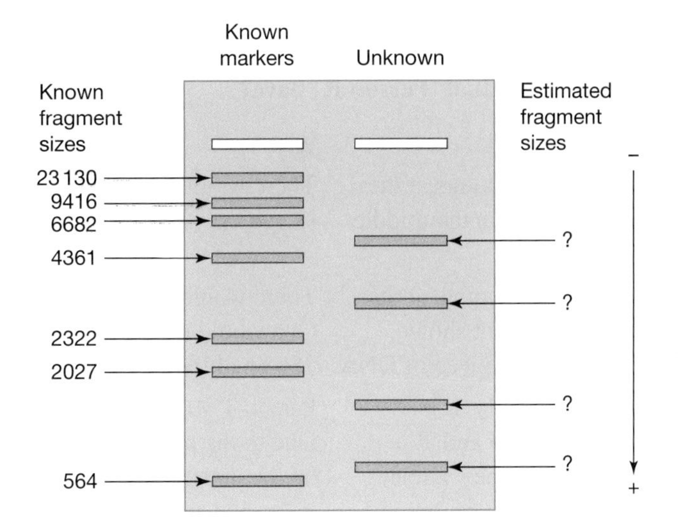

on Resources

Video eLesson Gel electrophoresis (eles-4133)

tlvd-1097

SAMPLE PROBLEM 3 Analysing gels

Samples of the DNA of the *F8C* gene, which controls blood clotting, were collected from several different people, two males (J and T) and one female (K). Defects in this gene can lead to haemophilia.

The *F8C* gene is located on the X chromosome and has two alleles: the *F* allele that determines normal blood clotting and the *f* allele that determines defective blood clotting (haemophilia). Males, only having one X chromosome, only require one version of the faulty *F8C* allele to have haemophilia. Females need two copies of the faulty allele to havehaemophilia.

Samples were taken from all three individuals and treated with the endonuclease *Bcl*I. The result of *Bcl*I treatment is below, with the blue arrows showing the cutting sites for the restriction enzyme.

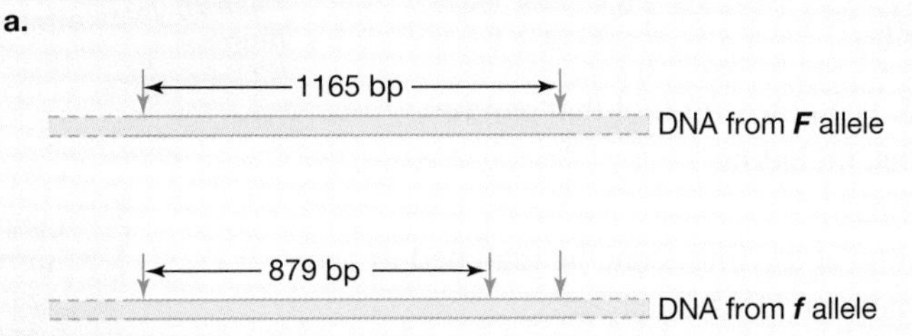

The DNA samples were then subjected to electrophoresis and the positions of the bands were identified using a fluorescent probe. The flanking sections outside the cuts (shown with the dotted lines) were not run.

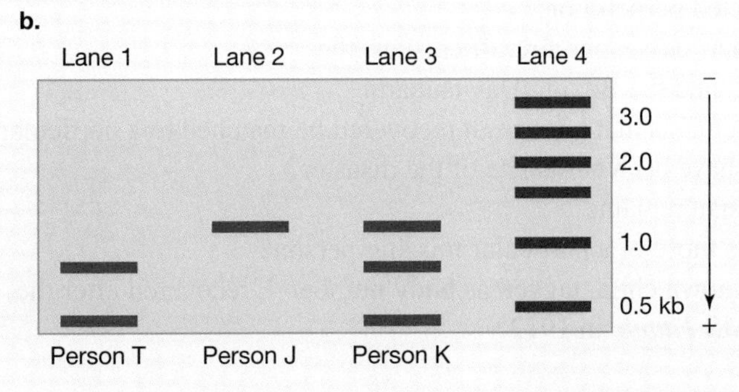

a. **How many fragments would result from *Bcl*I digestion of the *F* allele?** **(1 mark)**
b. **How many fragments would be produced by *Bcl*I treatment of the *f* allele? What are their expected sizes?** **(2 marks)**
c. **Which male will have haemophilia? Provide evidence for your response.** **(2 marks)**
d. **What alleles does the female individual (Person K) have?** **(1 mark)**

THINK	WRITE
a. Examine the DNA — there are two points at the end that were cut. There are no cuts in the middle of the 1165 bp piece.	There would be one fragment produced of 1165 bp (1 mark).
b. Examine the DNA — there are two points at the end that were cut to form the fragment shown. There is one cut in the middle of the piece of DNA.	There would be two fragments produced (1 mark). One would be 879 bp and the other would be 286 bp (1165 – 879 = 286) (1 mark).
c. Ensure you read the question carefully. It is specifically asking about the males: J and T. To be affected, an individual needs the *f* allele, which, as answered in part **b**, has two fragments.	Person T will have haemophilia (1 mark). Due to the presence of two bands at 879 bp and 286 bp, he must have the *f* allele and thus must have the disease (1 mark).
d. Examine Person K on the gel. She has three bands at around 1165 bp, 879 bp and 286 bp.	Person K has both the *f* and *F* alleles, as she has the 1165 bp band from the *F* allele and the 879 bp and 286 bp fragments from the *f* allele (1 mark).

2.5.2 Why do we conduct DNA profiling?

Individuals can be identified through a technique known as **DNA profiling**. The amount of DNA needed for DNA profiling is very small, because DNA can be amplified through the polymerase chain reaction (PCR) (refer to subtopic 2.4). A person's DNA profile is constant, regardless of the type of cell used to prepare the profile. A DNA profile prepared from a person's white blood cells is identical to that prepared from the same person's skin cells or other somatic cells.

FIGURE 2.23 Headlines signalling the power of DNA in identification

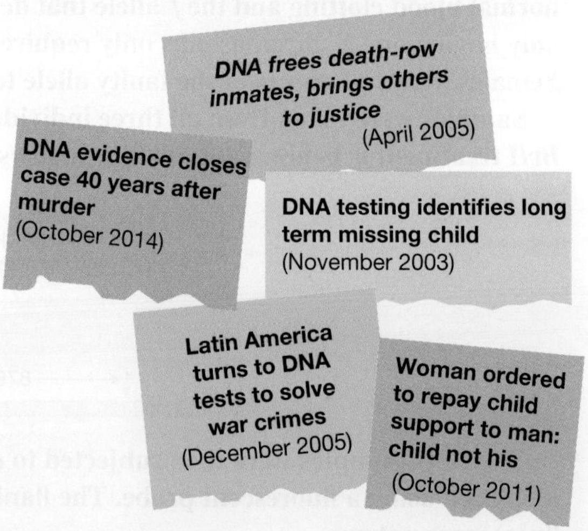

Identification using DNA is a powerful tool that can be applied in many situations, including:

- forensic investigations
  - Can the DNA found at a crime scene be matched to a person on the national DNA database?
  - Is this blood spot from the victim or from the possible assailant?
  - In a sexual assault case, is this semen from a previously convicted perpetrator?
- mass disasters, such as passenger aircraft crashes, the 9/11 terrorist attacks and the Boxing Day tsunami
  - Can the various remains that have been recovered be matched to a particular person known to have been at the site of the disaster?
- identification of human remains
  - Are these remains those of a particular missing person?
  - Who was the unknown child, tagged as body number 4, recovered after the sinking of the RMS *Titanic* in 1912?

DNA profiling technique for identifying DNA from different individuals based on variable regions known as short tandem repeats (STRs) or microsatellites

CASE STUDY: DNA profiling in forensic investigations

Because DNA molecules are only slowly degraded, DNA profiling can be carried out on biological samples from crime scenes from years ago. This profiling has led to many 'cold cases' worldwide being solved, such as the murder of a 22-year-old female bank clerk by Gerald Hyland in 1980.

DNA profiling has not just been used to convict those who are guilty, but in many cases has been used to exonerate innocent individuals and pardon them of any wrongdoing.

To access more information on various case studies and complete an analysis task relating to these, please download the worksheet.

FIGURE 2.24 Senior biologist Kamen Eckhoff holds up a t-shirt worn by the travelling companion of a victim

 Resources

 **eWorkbook** Worksheet 2.4 Case study analysis: DNA profiling in forensic investigations (ewbk-1875)

CASE STUDY: DNA profiling in identifying remains

DNA samples are not just used to solve crimes; they are also used to identify human remains. This may include remains from natural disasters or wars, or from ancient burial sites. These samples can be used to produce DNA profiles, such as with the identification of the lost soldiers of the battle of Fromelles.

To access more information on various case studies, please download the digital document.

FIGURE 2.25 This cemetery in France is the last resting place of some Australian soldiers killed in Fromelles.

 Resources

 **Digital document** Case study: DNA profiling in identifying remains (doc-36161)

 Resources

 **Weblinks** Solving a 28-year-old cold case
Police cracking cold cases
The Fromelles Project

2.5.3 Which DNA is used for identification?

Depending on the purpose and circumstances of the identification, the DNA used comes from either the chromosomes (nuclear DNA) or from mitochondria (mtDNA). Some important differences between nuclear DNA and mtDNA include:

1. mtDNA is almost exclusively inherited via the maternal line — that is, it is matrilineal; nuclear DNA is inherited from both parents. (New research is refining this idea about the inheritance of mtDNA.)
2. mtDNA does not undergo recombination and any changes in mtDNA must arise by mutation; nuclear DNA undergoes recombination, meaning that it is shuffled around in each generation by crossing over during meiosis.
3. mtDNA in each somatic cell is present in high numbers of copies; nuclear DNA is only present within the single nucleus. For this reason, mtDNA is the DNA that can be recovered from ancient specimens, such as bones and teeth.

FIGURE 2.26 Nuclear versus mitochondrial DNA

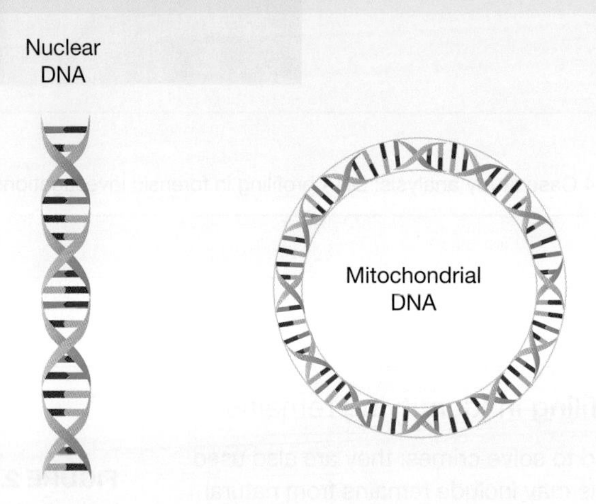

In both nuclear DNA and mtDNA, however, their use in identification depends on the existence of segments of DNA that vary greatly between individuals. Such regions of DNA are termed hypervariable.

Hypervariable regions of DNA that are currently used for identification are:

- **short tandem repeats** (STRs) in the nuclear DNA, also known as microsatellites. A large number of STRs are present on different human chromosomes. On its own, DNA from STRs can identify one person uniquely (apart from identical siblings). DNA samples from relatives are *not* required. So, when there is a need to match a DNA sample from a crime scene to just one particular person, STRs are used.
- hypervariable regions (HVRs) in the noncoding region of mtDNA. It should be noted that mtDNA identification is less precise because individuals from the same maternal line have almost identical mtDNA profiles. mtDNA is used, for example, when nuclear DNA cannot be recovered or when nuclear DNA is degraded because of age.

During the late 1980s and early 1990s, new and highly sensitive methods of visualising DNA were developed that used fluorescent labelling of DNA. This, alongside the use of STRs, allowed DNA profiling to be come the technique of choice for identification today.

hypervariable regions regions in DNA that are highly polymorphic

short tandem repeats (STRs) chromosomal sites where many copies of a short DNA sequence are joined end-to-end; the number of repeats is variable between unrelated people

These developments, combined with the use of the polymerase chain reaction (PCR) to amplify DNA, produced the powerful technique of DNA profiling that has applications in so many areas, including forensics and medicine.

What is an STR?

STRs are found within chromosomes (usually in the non-coding introns) where sequences of just two to seven base pairs are repeated over and over. These regions are very common and hundreds are scattered throughout the human chromosomes. STRs are termed 'short' because the repeat sequences are only two to five base pairs long, and 'tandem' because the repeats occur one after the other (see figure 2.27).

However, the number of repeats of an STR marker can vary between people and each variation is a distinct allele. The number of repeats of a 4-base pair sequence of one STR marker on the number-5 chromosome ranges from 7 to 15. In most cases, the alleles at an STR locus on a human chromosome are named according to the number of repeats and so are identified as allele 7, allele 8 and so on.

FIGURE 2.27 Diagrammatic representation of an STR showing one allele with 5 repeats of a 4-base sequence and the other allele with 7 repeats on homologous chromosomes.

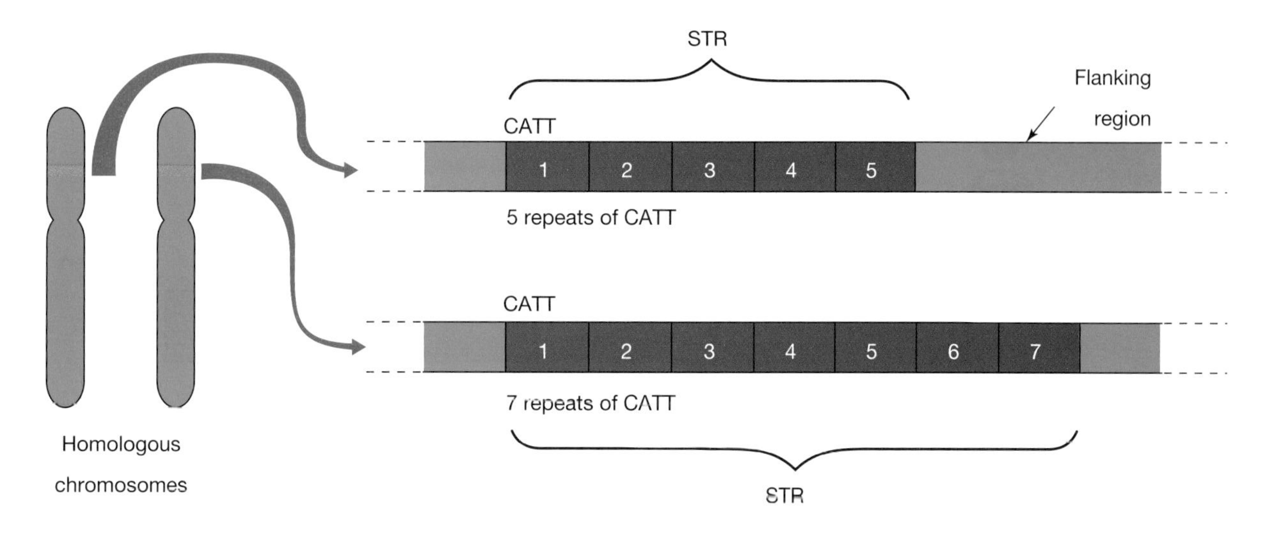

At each STR locus, an individual is either homozygous or heterozygous and so can have a maximum of just two different alleles. These alleles are inherited in a Mendelian fashion, and are co-dominant; that is, where two different alleles of the same STR marker are present, both can be detected.

BACKGROUND KNOWLEDGE: Reviewing alleles

An allele is a variant form of a gene. Homozygous means you have two identical copies of an allele (one on each chromosome). Heterozygous means you have two different alleles. Figure 2.27 shows a person who is heterozygous (5/7) at one particular STR locus and has one allele with 5 repeats and the other allele with 7 repeats.

STRs in paternity testing

In DNA paternity tests, the markers used are STRs, short pieces of DNA that occur in highly differential repeat patterns among individuals. Each individual's DNA contains two copies of these markers — one copy inherited from the father and one from the mother. Within a population, the markers at each person's DNA location could differ in length and sometimes sequence, depending on the markers inherited from the parents.

FIGURE 2.28 Individuals can have large variations in STRs across various sites.

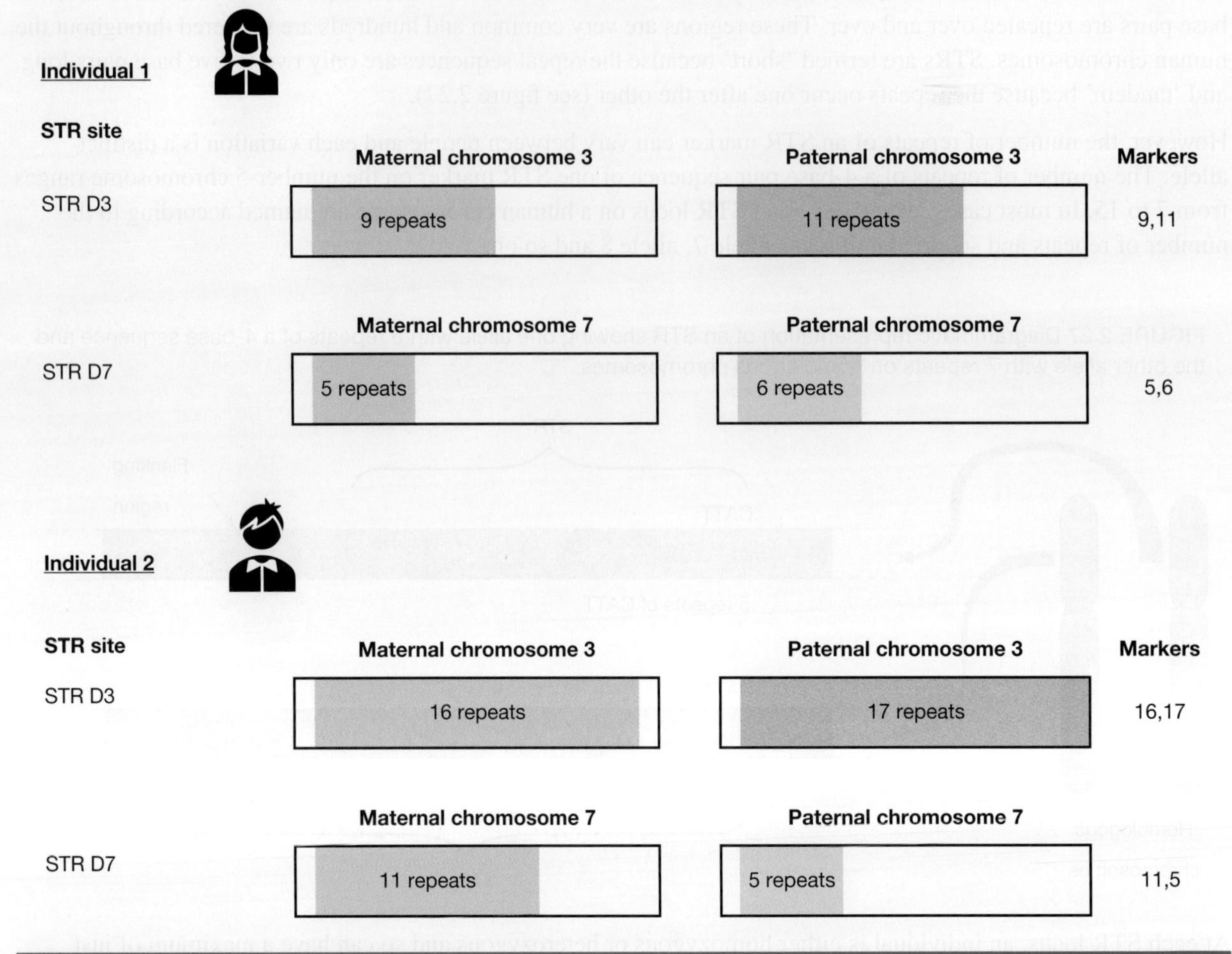

tlvd-1098

SAMPLE PROBLEM 4 Analysing STRs

The STR profiles of a mother, father and three children were examined across 4 STR loci. The results for the number of repeats at each marker is shown below.

STR locus	Mother	Father	Child 1	Child 2	Child 3
D3	14, 18	16, 17	14, 17	17, 18	14, 16
D8	10, 15	12, 13	13, 15	10, 13	10, 15
D21	28, 30	31, 32	28, 31	28, 32	30, 30
D7	8, 10	7, 13	7, 8	10, 13	7, 10

a. **Identify which parent each marker came from for Child 1, 2 and 3 at STR locus D3.** **(3 marks)**

b. **What conclusion can you draw about Child 3? Provide evidence for your response.** **(2 marks)**

THINK	WRITE
a. Examine each child individually at D3.	
1. Child 1 has the STRs 14 and 17.	Child 1 received the 14 repeat allele from the mother and the 17 repeat allele from the father (1 mark).
2. Child 2 has the STRs 18 and 17.	Child 2 received the 18 repeat allele from the mother and the 17 repeat allele from the father (1 mark).
3. Child 3 has the STRs 14 and 16.	Child 3 received the 14 repeat allele from the mother and the 16 repeat allele from the father (1 mark).

b. Examine each STR locus for Child 3. D3 was examined in part **a**. At D8, they have 10 and 15 repeats. The suspected father has neither of those, even though you must get an STR from each parent. At D21, they have 30 and 30 repeats so must have inherited one of these from their father. The suspected father does not have this. *Note:* As you have determined that the father is not the biological father of Child 3, you only need to provide evidence that shows this. Therefore, you do not need to refer to D7, as it does not provide evidence to support your response.	The father is not the biological father of Child 3 (1 mark). Evidence to support this is as follows: • At D8, the father does not have either 10 or 15 repeat alleles • At D21, the father does not have the 30 repeat allele that the child has. • As some of the STRs do not come from the suspected father, another individual must be the biological father of Child 3 (1 mark).

STRs in crime scenes

In other cases, such as crime scenes, DNA identified through STRs may be a full or partial match. A full match suggests a high likelihood that the DNA belongs to that individual. Partial matches imply that the DNA belongs to a close relative. This is shown in figure 2.29.

FIGURE 2.29 Matches between an individual's STRs and those at a crime scene may be **a.** a full match or **b.** a partial match.

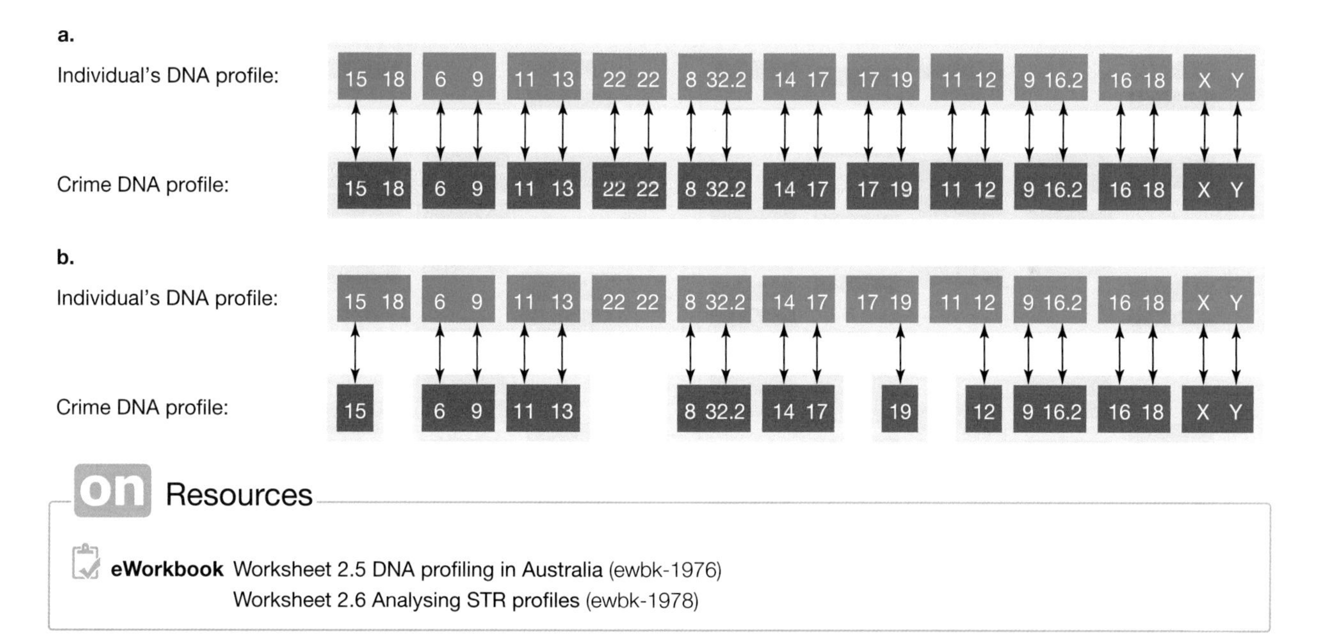

on Resources

eWorkbook Worksheet 2.5 DNA profiling in Australia (ewbk-1976)
Worksheet 2.6 Analysing STR profiles (ewbk-1978)

2.5.4 Interpretation of gel runs for DNA profiling

The combination of marker sizes found in each person makes up his or her unique genetic profile. When determining the relationship between two individuals, their genetic profiles are compared to see if they share the same inheritance patterns at a statistically conclusive rate (refer to figure 2.30a).

These profiles are often produced using STRs. Within a population there are usually several different versions of the DNA at an STR locus with different repeat lengths. Each STR generated is essential in order to generate match probabilities. If only one STR was analysed, many people would have the same alleles only by chance.

It is therefore necessary to analyse a number of different STR loci to ensure that the chance of two unrelated people having matching DNA profiles is very small. The STRs used in Australia are explored in the worksheet DNA profiling in Australia.

Steps involved in generating and analysing a DNA gel

Example: Paternity case

Step 1: To identify an unknown child, firstly identify the DNA bands (size and location) from the mothers that appears in the child's DNA profile.

Step 2: Match the remaining bands with one of the unknown samples from the different male possibilities. As 50 percent of the DNA inherited in the child will come from the mother and 50 percent of the DNA will come from the father, the remaining bands should match with the unknown father's sample. Colours can be used to match the child's remaining bands to the correct father's band.

An example of this is shown in figure 2.30.

Note: A similar technique can be used in forensic cases by using the victim's blood and the possible suspects to match an unknown sample found at a crime scene. This can be used to obtain partial or complete matches.

FIGURE 2.30 Paternity testing (comparing DNA of offspring against potential fathers) **a.** Initial gel **b.** Analysis of the gel showing the source of each band in the child's gel

a. Initial gel

Mother | Possible father 1 | Possible father 2 | Possible father 3 | Child

b. Analysis

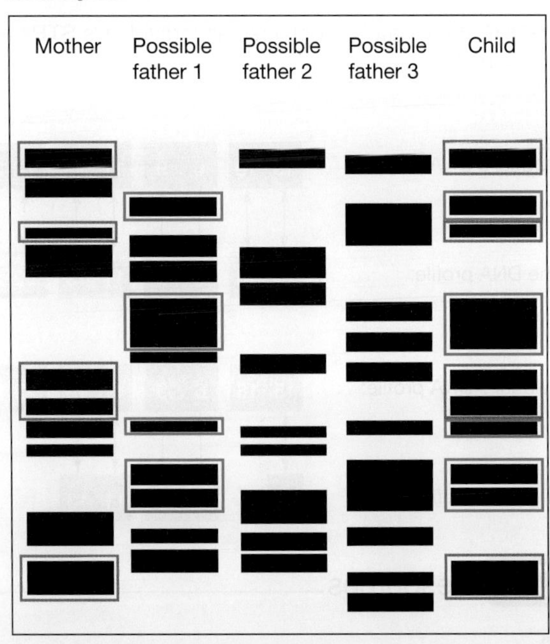

elog-0019

INVESTIGATION 2.2

online only

Using gel electrophoresis for DNA profiling

Aim

To undertake PCR and gel electrophoresis to produce a specific banding pattern and conduct DNA profiling

tlvd-1099

SAMPLE PROBLEM 5 Analysing DNA profiles from a crime scene

DNA was collected from a robbery from a splatter of blood on the window, which had been smashed during the intrusion. The DNA profiles of eight suspects are shown below.

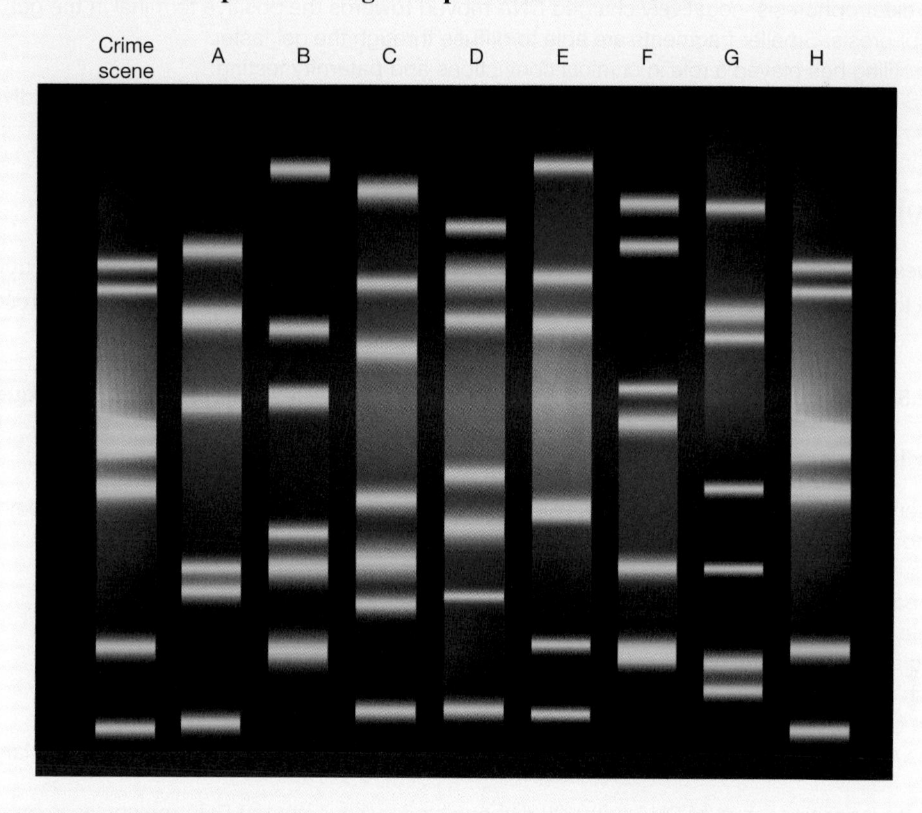

a. **Using the DNA profiles shown, who was the likely perpetrator of the crime? (1 mark)**

b. **Does this provide evidence that the individual with the DNA match was the only person involved in the crime? Justify your response. (2 marks)**

THINK	WRITE
a. Examine the gels for the DNA that matches the crime scene.	Suspect H likely committed the crime (1 mark).
b. Consider the evidence; is it the only evidence from the crime scene?	While the results provide evidence that the individual was present at the scene of the crime, it does not prove he was the only individual present (1 mark).
As this is a two mark question, ensure you justify your response.	Only one sample of DNA was collected from the crime scene. There may be DNA at other points, or the other individuals may have been completely covered to reduce the risk of DNA evidence being found (1 mark).

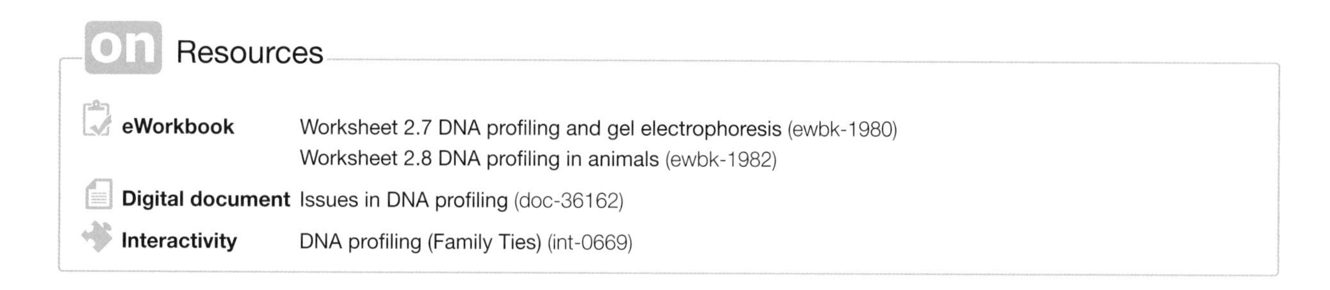

on Resources

eWorkbook Worksheet 2.7 DNA profiling and gel electrophoresis (ewbk-1980)
Worksheet 2.8 DNA profiling in animals (ewbk-1982)

Digital document Issues in DNA profiling (doc-36162)

Interactivity DNA profiling (Family Ties) (int-0669)

KEY IDEAS

- Electrophoresis sorts DNA fragments according to their lengths, measured in base pairs.
- During electrophoresis, negatively charged DNA moved towards the positive terminal in the gel electrophoresis. Smaller fragments are able to diffuse through the gel faster.
- DNA profiling has played a role in criminal convictions and paternity testing.
- Often, STRs are used in DNA profiling, as there is a much higher level of variation between individuals.

2.5 Activities

learnon

To answer questions online and to receive **immediate feedback** and **sample responses** for every question, go to your learnON title at **www.jacplus.com.au**. A **downloadable solutions** file is also available in the resources tab.

2.5 Quick quiz on | **2.5 Exercise** | **2.5 Exam questions**

2.5 Exercise

1. **MC** Identify which of the following DNA will migrate faster. The condition is the molecular weight of the following is equal.
   A. Supercoiled circular DNA
   B. Nicked circular DNA (open circular)
   C. Single stranded DNA
   D. Double stranded DNA
2. a. Which macromolecules can be separated using electrophoresis?
   b. State two applications of gel electrophoresis.
3. Explain why in electrophoresis, shorter fragments of DNA will move further from the well than larger segments of DNA.
4. Explain the specific feature of DNA primarily responsible for movement of DNA molecules in an electrical field?
5. Explain how fluorescent dye, such as ethidium bromide, is used for visualising DNA. How does ethidium bromide bind to DNA?
6. A student performed a gel electrophoresis experiment, and the results are shown in the diagram. The student used a circular plasmid that had been separately digested with three different restriction enzymes (lanes 2, 3 and 4). The uncut sample is in lane 1 and the DNA ladder is in lane 5. In some of the lanes, the student made mistakes with the DNA they loaded.

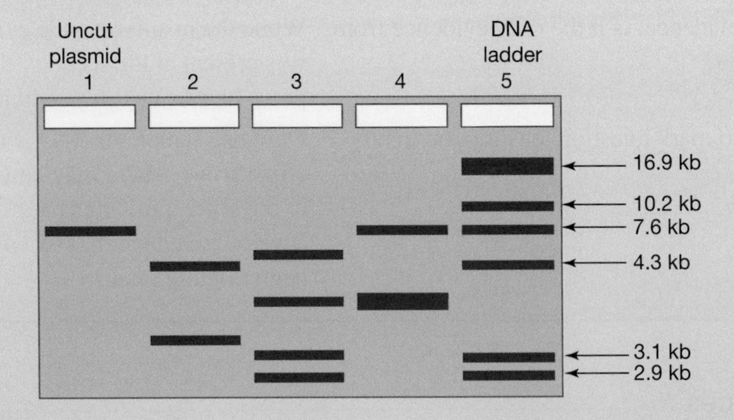

   a. Calculate the size of the uncut plasmid.
   b. In lane 2, how many recognition sites for the enzyme were in the plasmid? Justify your response.
   c. Explain the DNA fragment in lane 4.
   d. Suggest an explanation for the results in lane 3.

7. A gel was run that contained five different linear DNA samples and a lane with samples of known size. These DNA samples were exposed to the endonuclease *Bgl*II. The results are shown in the diagram.

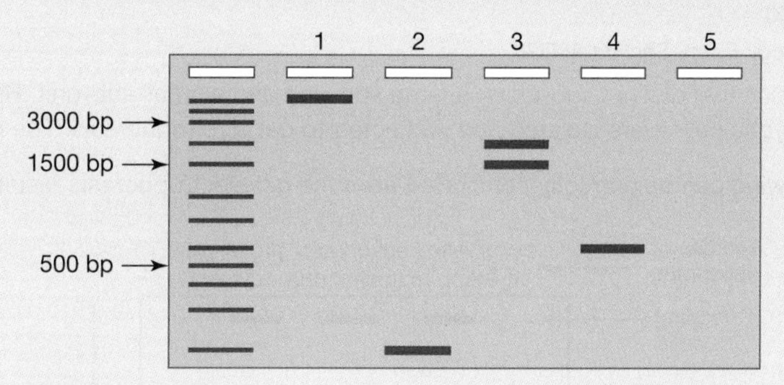

a. Identify the lane that contains:
   i. the shortest fragments of DNA
   ii. DNA fragments that include one restriction site for *Bgl*II and were treated with that restriction enzyme
   iii. DNA fragments with an approximate size of 1500 bps
   iv. DNA fragments with a size in excess of 3000 bps.
b. If this same DNA that was treated with *Bgl*II had instead been treated with *Hin*dIII, would the pattern of bands obtained in electrophoresis have been the same? Explain.
c. A different sample of DNA was treated with a particular restriction enzyme that produced three fragments of sizes 400, 750 and 2500 bps. Consider that this DNA mixture was loaded into lane 5 of the gel shown in the figure above. Draw the expected pattern on completion of electrophoresis of this DNA in lane 5.

8. A piece of double-stranded DNA is shown below.

A sample of this DNA was treated with the endonuclease *Alu*I. The cutting site for this enzyme is AGCT.

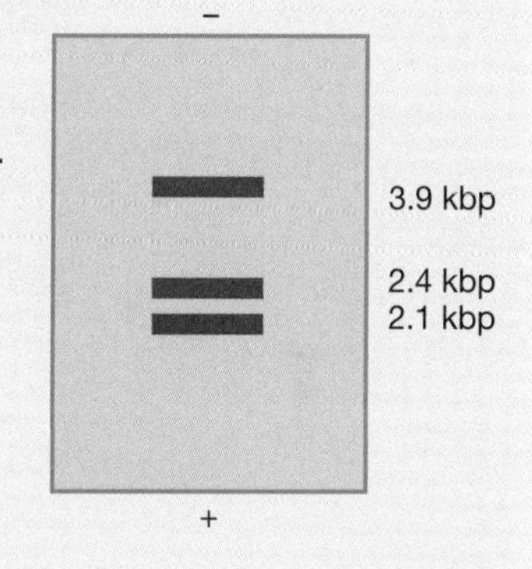

a. After this treatment, the sample was subjected to electrophoresis. The result is shown in the figure. The numbers denote the sizes of the fragments in kbp. What conclusion may be drawn?
b. A second sample of this DNA was treated with a different endonuclease, *Hin*dIII. The cutting site for this enzyme is AAGCTT. The result after electrophoresis revealed two DNA fragments of sizes 5.4 kbp and 3.0 kbp. What conclusion may be drawn?
c. A third sample of this DNA was treated with yet another endonuclease, *Not*I. The result after electrophoresis revealed one fragment of 8.4 kbp. What conclusion may be drawn?

9. Consider the following information: Daniel Fitzgerald was convicted of a murder after DNA evidence linked him to a didgeridoo found at a crime scene. There was no other evidence linking this man to the crime.
   a. If you had been a member of the jury involved in this case, would you have agreed with the verdict? Give a reason for your decision.
   b. If you were one of the High Court judges hearing this appeal, would you uphold the appeal against the conviction or reject the appeal?
   c. The High Court judges ruled that the recovery of the appellant's DNA from the didgeridoo did not raise any inference about the *time when* or *circumstances in which the DNA was deposited* there. Suggest two possible ways in which Fitzgerald's DNA might have been transferred to the didgeridoo.

2.5 Exam questions

Question 1 (1 mark)

Source: *VCAA 2015 Biology Exam, Section A, Q28*

MC A ribosome contains two distinct sub-units: a large sub-unit and a small sub-unit. Ribosomes from prokaryotic and eukaryotic cells were isolated and subjected to gel electrophoresis. The results are shown below.

Which one of the following can be correctly concluded from the gel electrophoresis results?

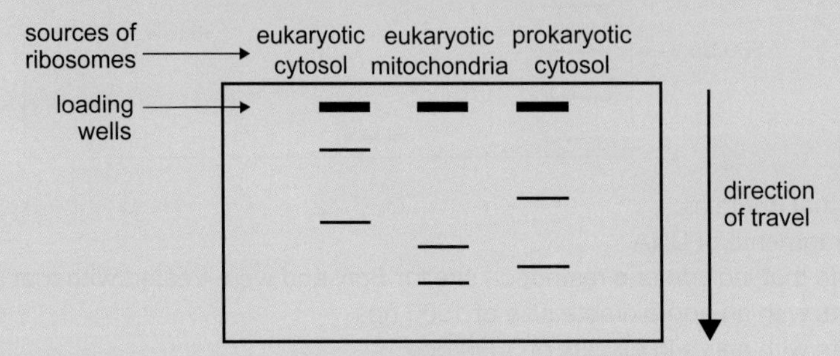

A. Eukaryote cytosolic and mitochondrial ribosomes translate the same types of protein.
B. Eukaryote mitochondria contain the ribosomal sub-units of the smallest size.
C. Prokaryote ribosomal sub-units have opposing charges to each other.
D. Eukaryote cytosolic ribosomal sub-units travel at the greatest speeds.

Use the following information to answer Questions 2 and 3.

Four samples of DNA were loaded into four different wells in lanes W, X, Y and Z. A standard ladder was loaded into the well in lane S. The results of gel electrophoresis are shown below.

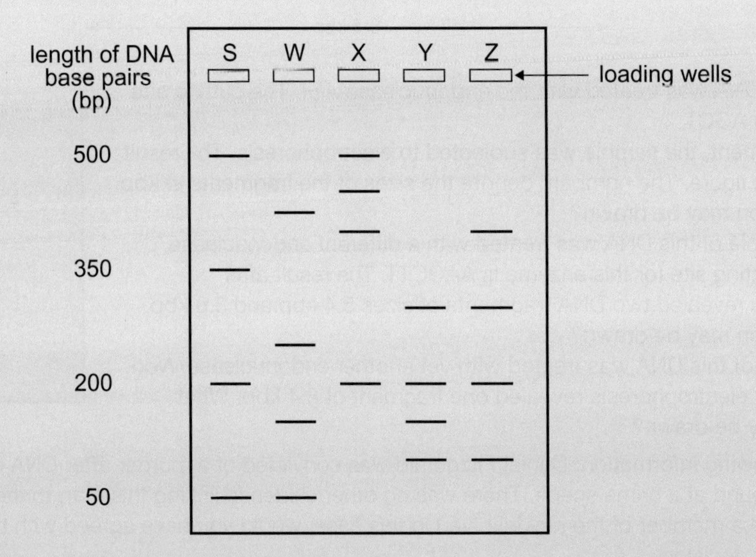

Question 2 (1 mark)

Source: *VCAA 2018 Biology Exam, Section A, Q30*

MC Which lane represents a sample that was loaded with DNA fragments of four different lengths: 100 bp, 150 bp, 200 bp and 300 bp?

A. W B. X C. Y D. Z

Question 3 (1 mark)

Source: *VCAA 2018 Biology Exam, Section A, Q31*

MC Which lane contains the band that is closest to the negative electrode?

A. W B. X C. Y D. Z

Question 4 (1 mark)

***Source:** VCAA 2013 Biology Section A, Q28*

MC During a fight between a number of people, one was seriously injured. Blood samples were taken from the victim, the crime scene and four suspects.

DNA was extracted from white blood cells in each of the blood samples and electrophoresis of the samples was carried out.

The results are shown in the following diagram.

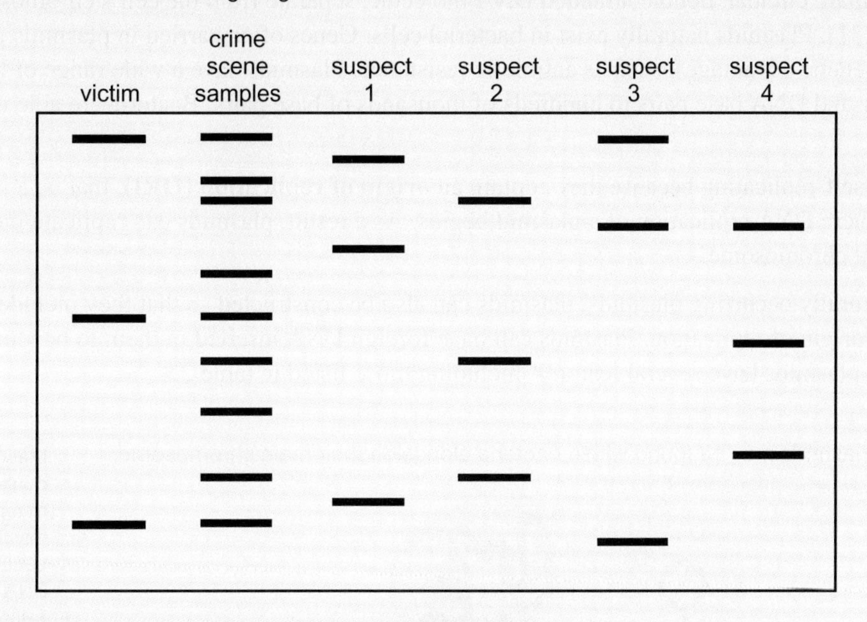

The person most likely to have been at the crime scene is suspect

A. 1. **B.** 2. **C.** 3. **D.** 4.

Question 5 (2 marks)

***Source:** VCAA 2010 Biology Exam 2, Section B, Q8c*

Each of the STR markers produced is labelled with a dye and subjected to gel electrophoresis. Five of Ben's STR markers were compared with three family groups who have the same surname as him. The following gels resulted.

Ben family X family Y family Z

Explain which family is Ben's most recent common ancestor.

More exam questions are available in your learnON title.

2.6 Recombinant plasmids

KEY KNOWLEDGE

- The use of recombinant plasmids as vectors to transform bacterial cells as demonstrated by the production of human insulin

A **plasmid** is a small, circular, double-stranded DNA molecule, separate from the cell's chromosomal DNA (refer to figure 2.31). Plasmids naturally exist in bacterial cells. Genes often carried in plasmids provide the bacterium with genetic advantages, such as antibiotic resistance. Plasmids have a wide range of lengths, from roughly one thousand DNA base pairs to hundreds of thousands of base pairs. Bacteria are able to carry multiple plasmids.

All plasmids are self-replicating because they contain an **origin of replication (ORI)**, that is, a specific DNA base sequence where DNA replication of a plasmid begins. As a result, plasmids can replicate independently of the main bacterial chromosome.

In addition to naturally occurring plasmids, plasmids can also be constructed so that they include certain features in addition to an origin of replication. Plasmids can have foreign DNA inserted in them to become **recombinant plasmids**. These plasmids have several features, including those listed in table 2.4.

FIGURE 2.31 Plasmids may be found within bacteria alongside their main chromosome.

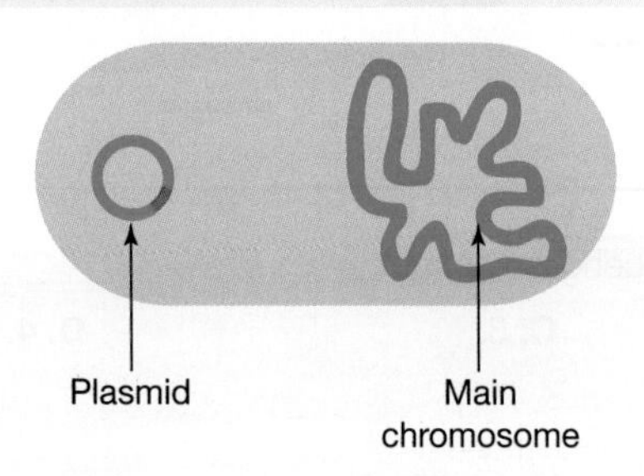

plasmid a small circular piece of double-stranded DNA that is able to reproduce independently and may be taken up by cells (usually bacteria) in addition to chromosomal DNA

origin of replication (ORI) a DNA base sequence in a plasmid in which DNA replication begins

recombinant plasmids plasmids that carry foreign DNA

TABLE 2.4 Features found in a recombinant plasmid

Plasmid features	Function
Origin of replication (ORI)	Section of DNA sequence which is recognised by a cell's DNA replication proteins, allowing initiation of new DNA synthesis.
Antibiotic resistance gene	The marker gene is often a gene governing antibiotic resistance, with the particular form of the gene being the allele for resistance to an antibiotic, such as ampicillin (Amp^R). Bacteria containing the plasmid with the marker can be selected from among other bacteria by exposure to the antibiotic.
Multiple cloning site (MCS) (or polylinker region)	A DNA region within a plasmid that contains multiple unique endonuclease cut sites. Plasmids are very useful as they allow foreign DNA to be easily inserted. This enables the plasmid to act as a vector to insert DNA into another cell — for example, to create a transgenic organism or for use in gene therapy.
Promoter region	Promoters are found upstream of their target genes. The sequence of the promoter region controls the binding of the RNA polymerase; therefore, promoters play a large role in determining where and when your gene of interest will be expressed.
Selectable marker	The antibiotic resistance gene allows for selection in bacteria. However, many plasmids also have selectable markers for use in other cell types; for example, green fluorescent protein (GFP) makes cells glow green under UV light.
Screening marker	A screening marker allows for the confirmation that a plasmid is actually recombinant. An example of this is with the *lacZ* screening marker, which is explored in the case study box in section 2.6.2.

Figure 2.32 shows two of many plasmids that can be used as **vectors** to transport foreign DNA into bacterial cells to transform them. These are termed vectors as they are used as vehicles to transfer genetic material into other cells.

Both plasmids have the required origin of replication (ORI) and also have a **selectable marker** (Amp^{R}) for the antibiotic ampicillin. In addition, the pUC19 plasmid has a second marker (Tet^{R}) for a second antibiotic, tetracycline.

vector an agent or vehicle used to transfer pathogens or genes between cells and organisms
selectable marker genes carried by plasmids for certain traits, often for antibiotic resistance

FIGURE 2.32 a. Plasmid pBR322 with many cloning sites present throughout the plasmid (the numbers represent the distance clockwise from the origin in base pairs). **b.** Plasmid pUC19 has a cluster of many restriction enzyme recognition sites in the sector of the plasmid labelled as polylinker (or multiple cloning site).

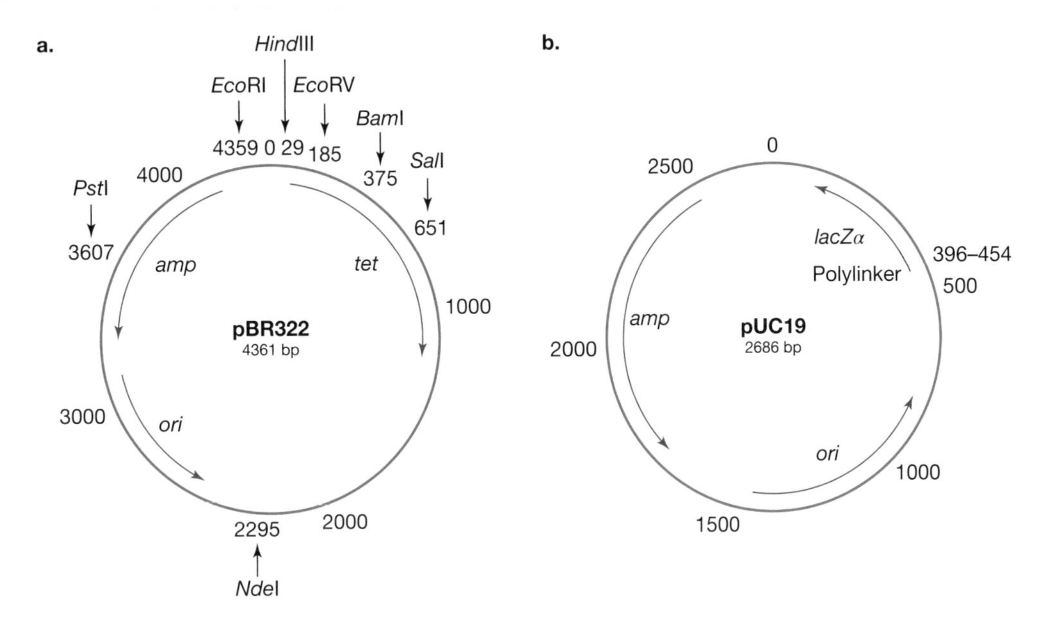

2.6.1 Making recombinant plasmids

Recombinant plasmids are plasmids that carry foreign DNA.

Making recombinant plasmids

1. The DNA of the plasmid is cut using a specific cutting enzyme (endonuclease) in order to create sticky ends. This changes the plasmid from circular to linear.
2. The foreign DNA fragments are prepared using the same endonuclease so that the foreign DNA has sticky ends complementary to the cut plasmid. Often, the process of reverse transcription is used to create these foreign fragments (refer to section 2.2.1) to ensure that non-coding introns are not included.
3. The foreign DNA fragments and the plasmids are mixed, and, in some cases, their 'sticky ends' pair by using weak hydrogen bonds. A recombinant plasmid has been created. (Other pairings will also occur, such as cut plasmids resealing themselves so that they are not recombinant plasmids).
4. The joining enzyme, ligase, is added and this makes the joins permanent through covalent bonding.

Figure 2.33 shows the process by which recombinant plasmids are formed.

FIGURE 2.33 The steps in the formation of a recombinant plasmid

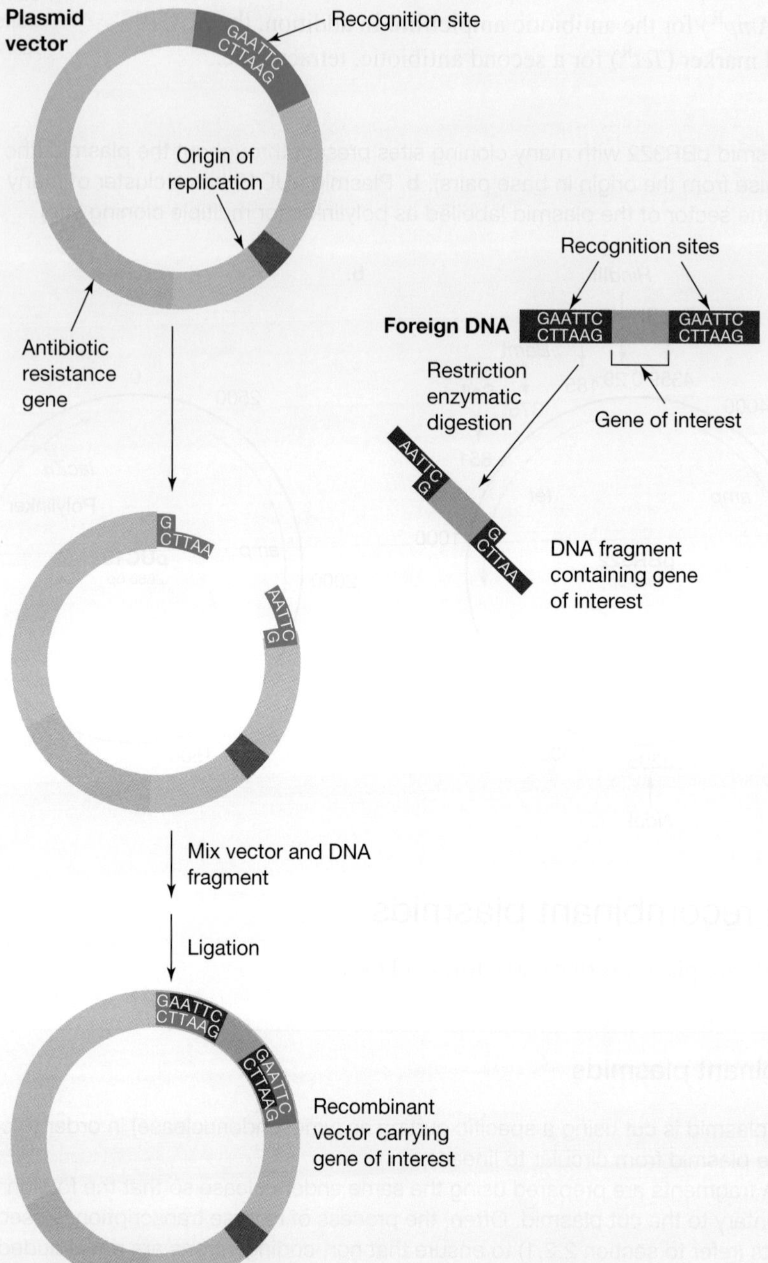

2.6.2 Getting plasmids into bacterial cells

Once recombinant plasmids have been created, the challenge is to transfer them into bacterial cells. This process is known as **bacterial transformation**. It is important that the bacterium selected is appropriate. The bacterium needs to be able to reproduce quickly and be harmless to individuals working with it. A harmless strain of *E. coli* is commonly used for this purpose. However, yeast cells and other eukaryotic cells can be used as host cells when glycoproteins are being produced.

bacterial transformation process in which bacterial cells take up segments of foreign DNA that become part of their genetic make-up

In general, the success rate for the transfer of recombinant plasmids into bacterial cells is low, but the success rate can be increased through various techniques:

- One technique is termed **electroporation**. Cells are briefly placed in an electric field that shocks them and appears to create holes in their plasma membranes, allowing the plasmid entry.
- Another method is **heat shock**. The bacterial cells are suspended in an ice-cold salt solution and then transferred to 42 °C for less than one minute. This treatment appears to increase the fluidity of the plasma membranes of the bacterial cells and increases the chance of uptake of plasmids.

electroporation a technique that uses brief exposure of host cells to an electric field to enable the entry of segments of foreign DNA into the cells

heat shock a technique to transform bacteria in which cells are suspended in a ice cold solution and then moved into a warm solution to increase plasma membrane fluidity

An example of the heat-shock treatment is shown in figure 2.34. The starting culture of bacterial cells is sensitive to the antibiotic tetracycline, and is denoted Tet^S.

Steps in heat shock

1. The bacterial culture is placed in an ice bath and chilled.
2. Recombinant plasmids with the tetracycline resistance allele Tet^R are added to the bacterial culture and chilled.
3. The bacteria and plasmid mix are placed in hot water at 42 °C for 50 seconds, producing a heat shock. This is the stage when the plasma membranes of the bacterial cells are altered, increasing the chance of uptake of plasmids by the cells.
4. The mix is returned to an ice bath for two minutes.
5. The bacteria are plated on an agar plate containing the antibiotic tetracycline and incubated at 37 °C overnight. Bacteria that have not taken up the plasmids are killed by the tetracycline. Bacterial cells that have taken up the plasmids will be selected as they will survive and replicate.

FIGURE 2.34 Stages in the uptake of plasmids by bacterial calls, resulting in their transformation

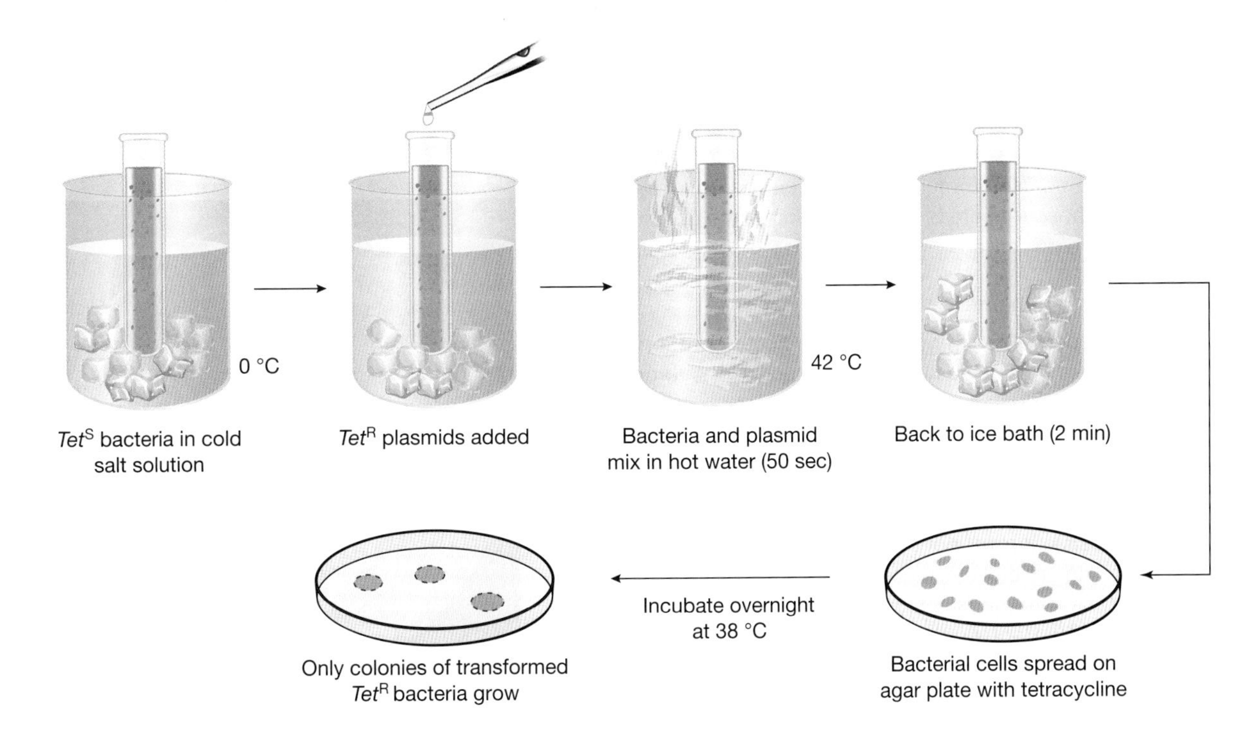

The importance of recombinant plasmids is that the foreign DNA they incorporate can come from any source: it may be a human gene, a plant gene, a jellyfish gene, a yeast gene and so on. The importance of transformed bacteria is that they express the foreign gene in their phenotype. We will see in section 2.6.3 how these transformed bacteria can be factories to produce the protein products of these foreign genes, such as human insulin.

Using selectable and screening markers

Transforming bacteria using recombinant plasmids is not a perfect process. Sometimes the gene is not incorporated in the plasmid or the plasmid is not taken up by the bacteria. There are many techniques that allow us to quickly screen bacteria to ensure that it has taken up the recombinant plasmid.

Checking that the plasmid has the gene

One common technique to ensure that the gene of interest that has been inserted is using a screening marker such as *lacZ*. *LacZ* codes for an enzyme known as β-galactosidase (beta-galactosidase), which converts a colourless substrate (X-gal) to a blue product.

The plasmid below (pUC18) contains the gene *lacZ*. If the gene of interest is incorporated into the plasmid at the *Bam*HI recognition site, the enzyme that makes the blue product is lost because the DNA of this segment is disrupted. If the gene of interest is not present, the enzyme is produced and the blue colour can be produced.

FIGURE 2.35 A plasmid pUC18 with both selectable and screening markers

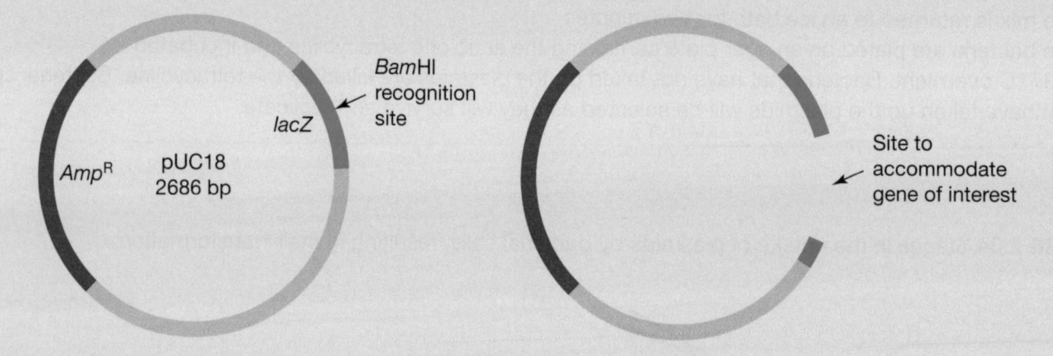

Therefore, when the bacteria are grown on an agar plate containing X-gal, those containing the gene of interest appear white and can be used for further replication, as observed in figure 2.36. Those that are blue do not contain the gene of interest, so are discarded.

FIGURE 2.36 An agar plate with bacterial colonies

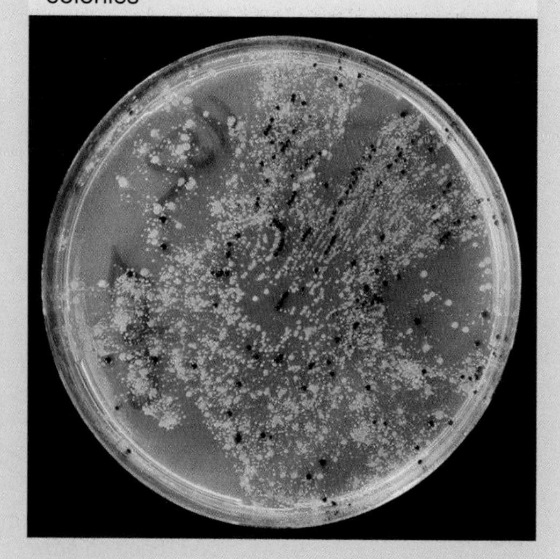

Checking that the bacteria contains the plasmid

Transforming bacteria with heat shock or electroporation is not always successful. One easy way to examine this is through the use of a plasmid with an antibiotic resistance gene. Often a plasmid with resistance to multiple antibiotics is used to reduce the chance of a bacteria having natural resistance.

Usually, under a specific antibiotic, bacteria without a plasmid conferring resistance will not survive. By ensuring that the plasmid has an antibiotic resistance gene within it, it is easy to select bacteria with the plasmids. A common example of this is with ampicillin. Bacteria that have not taken up the plasmid with *Amp*^{R} will not grow on an agar plate with ampicillin present. However, those bacteria that have taken up the plasmid will. This is a quick and easy way to check that the bacteria being selected have taken up the plasmid.

tlvd-1100

SAMPLE PROBLEM 6 Describing bacterial transformation

A plasmid with genes for ampicillin and tetracycline resistance was inserted into a bacterium. Within the plasmid, a foreign gene, known as green fluorescent protein (GFP), was inserted into the plasmid to make it recombinant.

a. **Outline the process to insert the foreign gene into the plasmid.** **(2 marks)**

b. **Identify two ways in which the plasmid is taken up by the bacterium.** **(1 mark)**

After transformation, the bacteria were plated upon the three following plates.

1. **Nutrient agar**
2. **Nutrient agar + ampicillin**
3. **Nutrient agar + ampicillin + tetracyline**

The bacteria were allowed to grow, and the plates were explored under UV light to see if they fluoresced green.

c. **Which plate would you expect to have the greatest amount of bacterial growth?** **(1 mark)**

d. **Why might Plate 2 have had more growth than Plate 3?** **(2 marks)**

e. **Some bacteria on Plate 3 were able to grow but did not fluoresce green. Explain the reasons for this observation.** **(2 marks)**

THINK	WRITE
a. Carefully read what the question is asking you to do. As there are two marks for this question you should outline two points: the use of endonucleases and the use of ligase.	• Both ends of the gene and the plasmid are cut with the same endonuclease to produce sticky ends (1 mark). • These sticky ends are joined using the ligase enzyme (1 mark).
b. This is a one-part question and you just need to identify the processes.	The plasmid can be taken up using either heat shock or electroporation (1 mark).
c. Consider the three plates: one plate has no antibiotics, so this would grow both resistant and sensitive bacteria.	Plate 1 would have the most growth, as bacteria with and without the plasmid would be able to grow (1 mark).
d. As the question is worth two marks, you need to make two clear points. One way to do this is to: • outline why there might be less growth on plate 3 • outline why there might be more growth on plate 2 **TIP** If you choose, you may write your two key points as bullets. This is accepted by VCAA in exams.	• The only bacteria likely to grow on Plate 3 were those that took up the plasmid, as this provided them resistance to both antibiotics. This resistance to both antibiotics is unlikely to be common in other bacteria (1 mark). • There is a chance that some bacteria were resistant to the ampicillin beforehand without the recombinant plasmid. Therefore, they would be able to grow on Plate 2, alongside any bacteria that had taken up the plasmid. It is more likely that a bacteria is already resistant to only one antibiotic and not two; therefore, more bacteria could grow on Plate 2 compared to Plate 3 (1 mark).
e. Refer back to the question, which requires you to explain your response. Consider reasons bacteria can grow on Plate 3 (presence of the plasmid with antibiotic resistances) and what would be required to fluoresce (presence of the gene). Use this to explain why bacteria may not fluoresce green.	• In order to fluoresce, the *GFP* gene needs to have been incorporated into the plasmid. There is a chance this didn't occur, and the plasmid just religated (1 mark). • This plasmid may have been taken up by a bacterium. Therefore, bacteria may have been able to grow as they had the resistance to ampicillin and tetracycline, but could not glow as the GFP gene was not present (1 mark).

elog-0021

INVESTIGATION 2.3

online only

Transformation using the pGlo plasmid

Aim

To transform *E. coli* using a plasmid containing the *pGlo* gene

Resources

Video eLesson Gene cloning (eles-4134)

Weblink Transforming bacteria

2.6.3 Producing human insulin

One of the most common uses of recombinant proteins is through the synthesis of human **insulin**, a life-saving hormone used in the treatment of diabetes. The mass production of insulin for therapeutic use first began in 1923, saving millions of lives.

Producing recombinant insulin

To understand the process of recombinant insulin, it is important to review some related key ideas.

TABLE 2.5 Important concepts related to recombination

Gene cloning	The process by which a *gene* of interest is located and cloned (copied) to produce multiple copies
Recombinant DNA	DNA that is formed by combining DNA from different sources, often from different kinds of organisms
Recombinant proteins	Proteins that are expressed by recombinant DNA present in an organism. A protein obtained by introducing recombinant DNA into a host (such as a bacterial cell or yeast cell) produces the gene product. A cloned gene encodes the amino acid sequence of such a protein.

In the 1980s, a major breakthrough occurred when the American drug company Eli Lilly used gene cloning and expression of the copied genes to produce recombinant human insulin, known as Humulin® (see figure 2.37a). In late October 1982, the Food and Drug Administration (FDA) of the United States approved Humulin® for human use, and it has now replaced the insulin prepared from non-human sources, such as from the pancreatic tissue of cows and pigs.

insulin a hormone that allows for glucose to enter cells, reducing blood glucose levels

FIGURE 2.37 a. Humulin® or recombinant insulin produced by gene cloning **b.** An insulin pen that can be conveniently carried and used to administer insulin

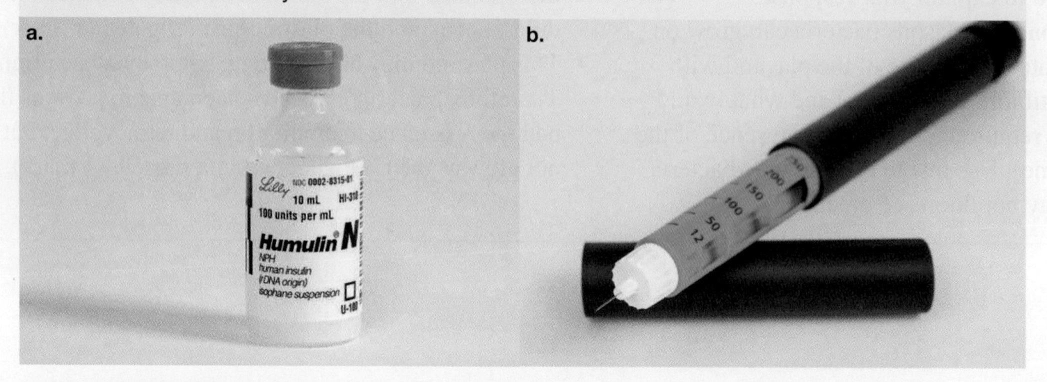

Today, if the beta cells of the human pancreas fail to produce insulin, recombinant human insulin can substitute for the missing hormone. Following the development of Humulin®, other recombinant insulins have been developed with slightly different properties, including rapid-acting forms (e.g. generic name: Insulin lispro) and long-acting forms (e.g. generic name: Insulin glargine). Insulin is now supplied in vials, as pumps and as pens (see figure 2.37b).

FIGURE 2.38 Time course of action of various forms of recombinant insulin on blood glucose levels

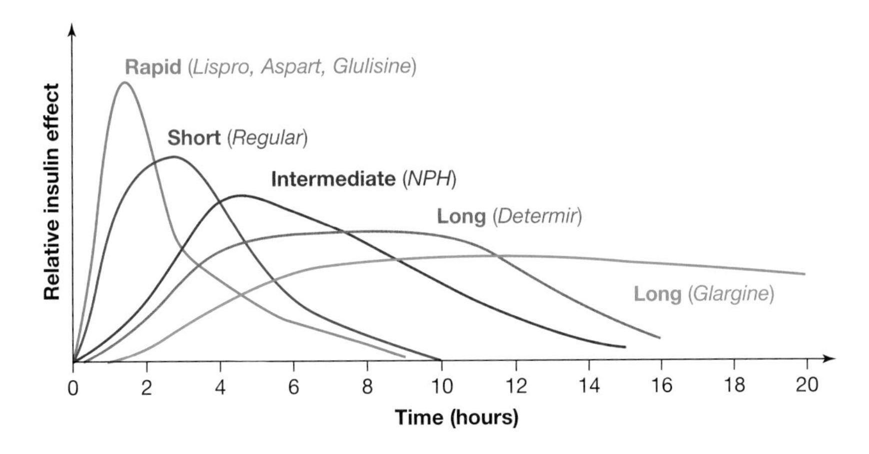

Advantages of producing proteins through expression of cloned genes as compared to extraction from other biological sources include:

- high levels of purity
- reliability of supply
- reduced chance of side effects such as allergy (compared to pig or cow derived insulin)
- consistency of quality between batches.

Steps involved in producing recombinant insulin

Insulin is a protein consisting of two separate chains of amino acids, amino acid *chain A* above an amino acid *chain B*. Before becoming an active insulin protein, insulin is first produced as **pre-proinsulin**. This is one single long protein chain with the A and B chains not yet separated. The pre-proinsulin chain evolves into **proinsulin**, and finally the proinsulin is modified to become the active protein insulin. At each step, the protein needs specific enzymes to produce the next form of insulin. One way to overcome this is by producing chain A and B separately.

pre-proinsulin an inactive precursor molecule of insulin that includes a signal peptide

proinsulin a modified form of pre-proinsulin and the precursor to insulin, in which the signal peptide has been removed and disulfide bridges have formed between chains

FIGURE 2.39 Primary structure of human insulin showing chain A and chain B

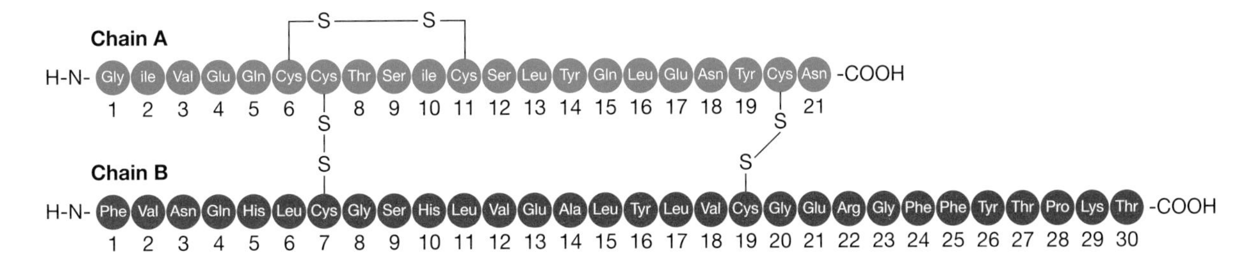

Producing the two insulin chains separately

Step 1: One method used to manufacture insulin is to produce the two insulin chains separately. As the amino acid sequence is known for each of these chains, this can be used to determine the DNA sequence that will lead to the production of insulin.

Step 2: The DNA is often produced using reverse transcription. The mature mRNA strands that code for the insulin chains are used as a template. Mature mRNA is used as it no longer contains non-coding introns. With the aid of the reverse transcriptase enzyme, the DNA segments (or the genes) coding for insulin can be produced. Alternatively, DNA synthesisers may also be used.

Step 3: The insulin gene and plasmid are cut with an endonuclease (usually *Bam*HI), forming sticky ends. The two DNA molecules are then inserted into plasmids using ligase (usually in different plasmids).

Step 4: The plasmids are then inserted into the bacterium *E. coli.* The plasmid contains the selectable marker the *lacZ* gene. *LacZ* encodes for *β*-galactosidase (the use of selectable markers was covered in section 2.6.2).

Step 5: The recombinant plasmids are then mixed with the bacterial cells. Plasmids enter the bacteria in a process called transformation (using either heat shock or electroporation). The bacteria culture is incubated, allowing for the replication of bacteria and plasmids. (See figure 2.40.)

Step 6: Bacteria are spread over culture plates containing ampicillin and X-gal. Bacteria with the recombinant plasmid are able to grow in the presence of ampicillin, and as they can't express *β*-galactosidase, they will be colourless or white in the presence of X-gal. Bacteria are also often plated on control plates (either with no ampicillin or with ampicillin only) at the same time.

FIGURE 2.40 Growing cultures with the insulin gene

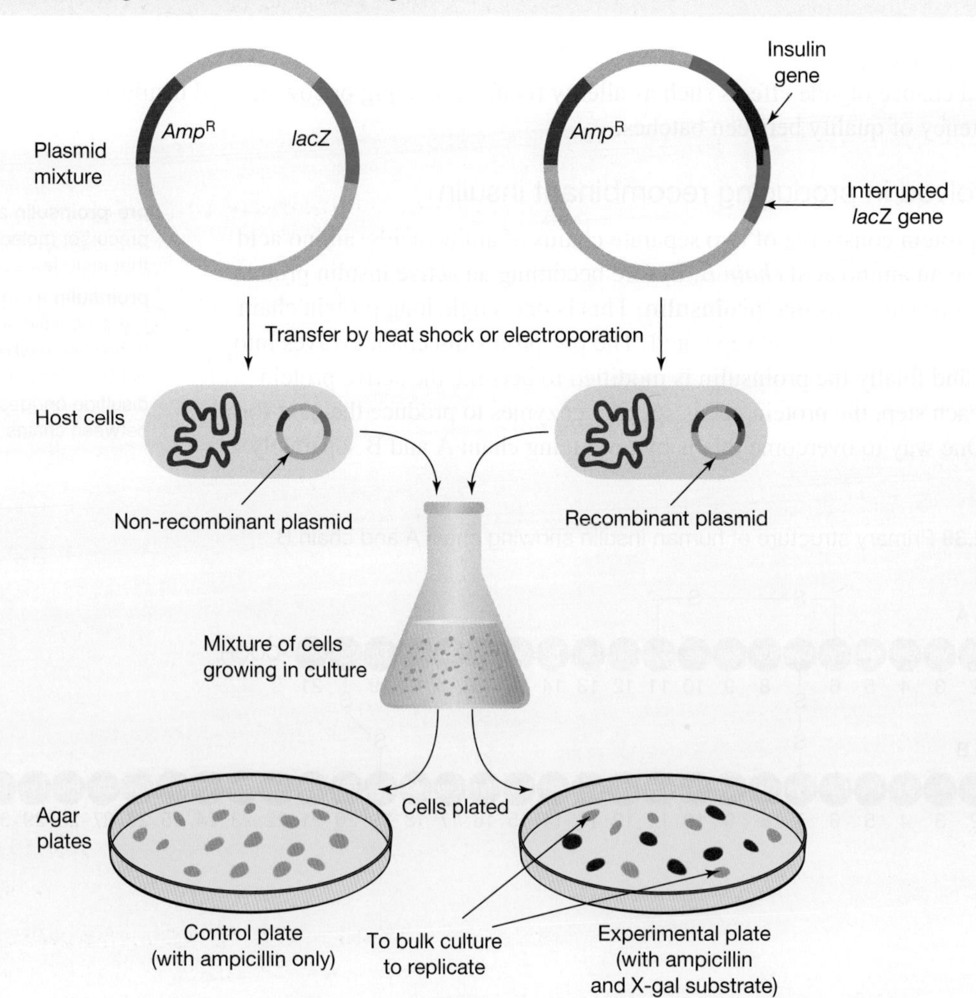

Step 7: The bacteria with the recombinant plasmid undergo a fermentation process. The millions of bacteria replicate roughly every 20 minutes, expressing the insulin gene.

Step 8: Cells are then lysed using lysosomes, where the the expressed protein is able to be extracted. The protein is then treated with cyanogen bromide, separating the insulin chains from other proteins that were also translated.

Step 9: The purified samples can be tested to ensure none of the bacteria's *E. coli* proteins are mixed in with the insulin. This is done by using a marker protein to detect the *E. coli* DNA. Scientists are then able to determine whether the purification process successfully removed the *E. coli.*

Step 10: The two chains are then mixed together and re-joined by forming disulfide bonds (or bridges).

Step 11: At the end of the manufacturing process, ingredients are also added to intermediate and long-acting insulin to produce the desired duration type of insulin.

Part of this process is summarised in figure 2.43.

FIGURE 2.41 Purifying a sample using affinity chromatography

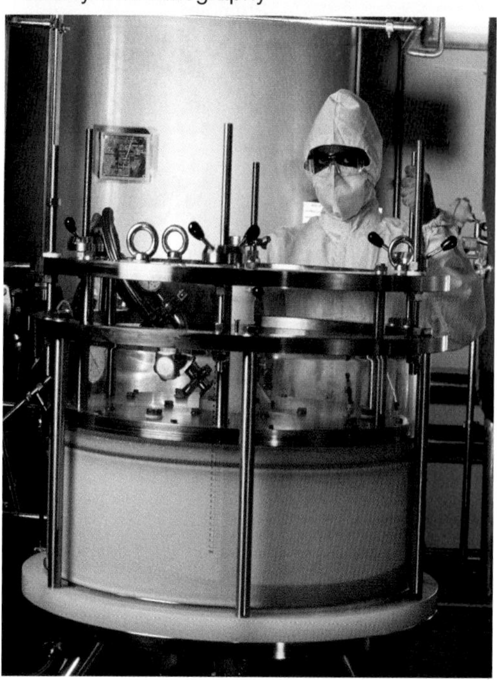

FIGURE 2.42 Stainless steel fermentation vats in which mass culturing of recombinant *E. coli* bacterial cells takes place and where those cells express the insulin gene

FIGURE 2.43 Flow diagram outlining the steps involved in the manufacture of recombinant insulin

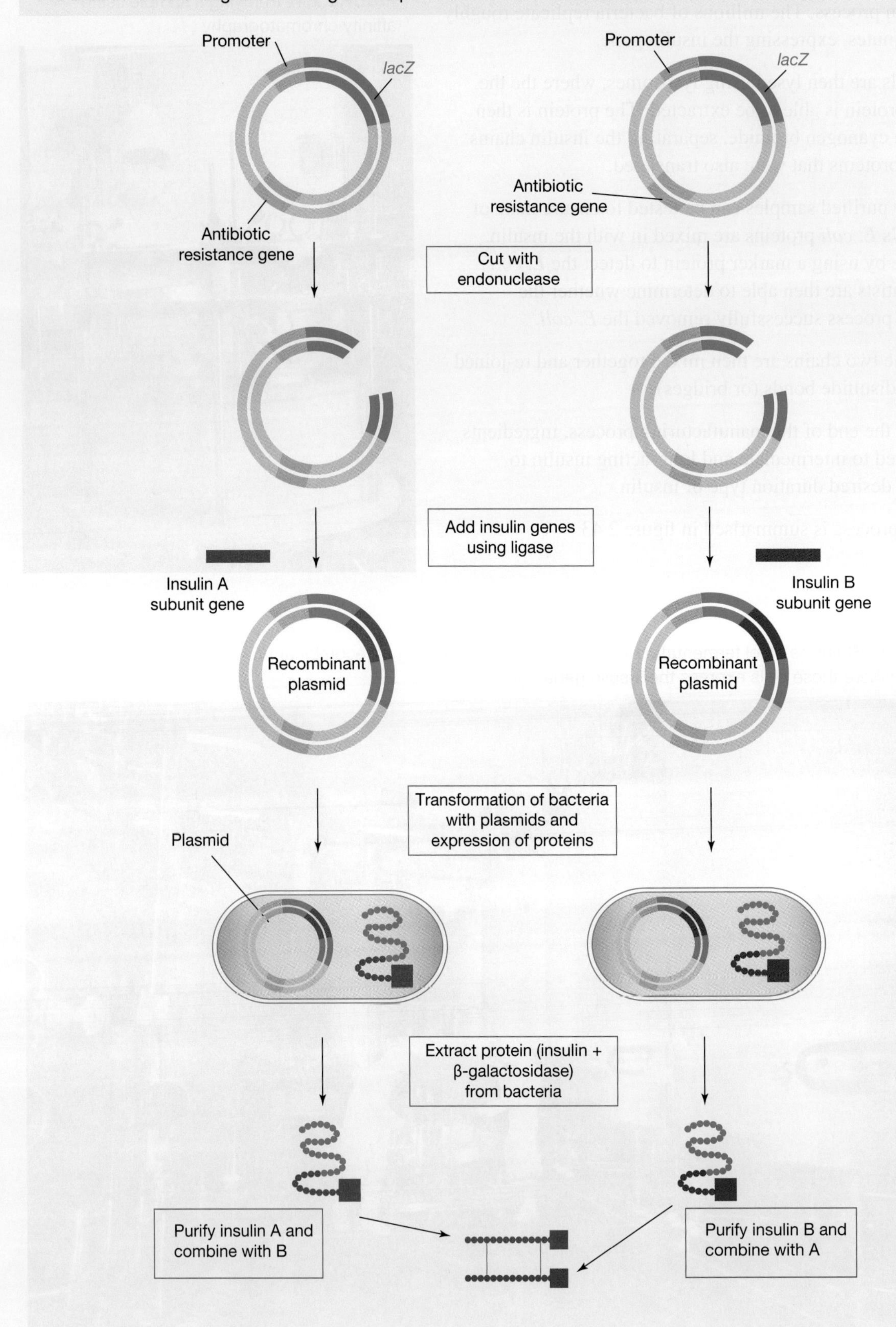

Producing proinsulin

The steps described previously showed one method of insulin production in which the two chains of insulin are produced and combined.

There are other ways insulin may be produced. For example, rather than producing the two separate insulin chains, proinsulin may instead be produced. To do this, the purified sequence coding for proinsulin is inserted into the *E. coli* bacterium. The bacteria go through the fermentation process (through mass culturing) where they reproduce and produce proinsulin. An enzyme is then used to modify the proinsulin sequence, resulting in the purified insulin molecule.

elog-0023

INVESTIGATION 2.4

online only

Modelling the production of recombinant insulin

Aim

To model the process in which recombinant plasmids are used to create human insulin

EXTENSION: Other useful recombinant proteins

There are various useful recombinant plasmids that can be used to produce other proteins that have therapeutic uses.

TABLE 2.6 Examples of recombinant proteins

Example of rec protein	Host cell tye	Use of rec protein
Humulin® (rec insulin)	*E. coli* bacteria or yeast cells	Used in treatment of type 1 diabetes
Humatrope® (rec human growth hormone)	*E. coli* bacteria or yeast cells	Used in treatment of growth deficiency in children
Puregon® (rec human follicle-stimulating hormone)	Hamster ovary cells	Promotes ovulation and is used in the treatment of infertility
Recombinate® (rec factor 8)	Hamster ovary cells	Blood clotting factor used to control bleeding episodes in haemophilia
Erythropoietin (rec hEPO)	Hamster ovary cells	Used in treatment of severe anaemia
Interferons (recIF)	*E. coli* bacteria	Used as an anti-viral treatment

on Resources

eWorkbook Worksheet 2.9 Recombinant plasmids and cloning genes (ewbk-1986)

Digital document Case study: The history of insulin (doc-36163)

Weblink The history of cloning insulin

KEY IDEAS

- To make a recombinant plasmid, genes and plasmids need to be cut with endonucleases and ligated together using ligase.
- This plasmid can be taken up by bacterium using heat shock or electroporation.
- To test for bacteria containing the recombinant plasmid, selectable and screening markers are used.
- Insulin can be incorporated into a recombinant plasmid. This allows the bacteria to take this up, replicate and express insulin, which can be purified and used for humans.

2.6 Activities

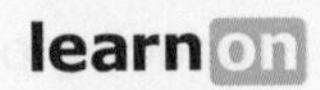

To answer questions online and to receive **immediate feedback** and **sample responses** for every question, go to your learnON title at **www.jacplus.com.au**. A **downloadable solutions** file is also available in the resources tab.

2.6 Quick quiz on	2.6 Exercise	2.6 Exam questions

2.6 Exercise

1. **MC** What is a plasmid?
   A. A linear section of DNA that contains the gene of interest
   B. A small circular piece of DNA that is self-replicating
   C. A bacterium that has contains an inserted gene
   D. A specific sequence of bases that is cut by a restriction enzyme
2. Explain why the same recognition sites must be present on the ends of the gene of interest and at a known site in the selected plasmid.
3. Bacterial colonies with cells containing a recombinant plasmid do not appear blue when plated on agar containing X-gal. Explain, with reference to the *lacZ* gene, why this is the case.
4. Identify two procedures that can increase the chance of uptake of plasmids by bacteria.
5. Explain why insulin was called 'a miracle drug' when it first became available.
6. Provide a brief description of how bacteria can be used to produce human insulin.
7. Suppose that you are required to produce a recombinant protein. Some of the steps that you will perform are identified as follows, but they are not in the correct order:
   Copy recombinant DNA, identify relevant gene, screen transformed cells, transfer recombinant DNA into host cells, construct recombinant DNA vector.
   a. List these steps in the appropriate order.
   b. For each step, briefly describe its outcome.

2.6 Exam questions

Use the following information to answer Questions 1 and 2.

A bacterial plasmid was modified by inserting a gene for an enzyme that provides resistance to the antibiotic ampicillin. A nutrient solution containing cells of the bacterium *Escherichia coli* was obtained. *E. coli* is naturally sensitive to the antibiotic ampicillin. The solution was divided into two equal volumes. The bacteria in one half of the solution were left untreated. Plasmids were added to the other half of the solution and the bacteria were treated to increase their chance of taking up the plasmids.

The next day, the bacterial cells were spread on agar plates as follows:

- Plate 1 — Untreated bacterial cells on nutrient agar
- Plate 2 — Untreated bacterial cells on nutrient agar with ampicillin
- Plate 3 — Treated bacterial cells on nutrient agar with ampicillin
- Plate 4 — Treated bacterial cells on nutrient agar

The plates were incubated overnight.

Question 1 (1 mark)

Source: *VCAA 2020 Biology Exam, Section A, Q30*

MC In order to collect only bacterial cells that had taken up the plasmid successfully, a sample should be taken from

A. Plate 1.
B. Plate 2.
C. Plate 3.
D. Plate 4.

Question 2 (1 mark)

Source: *VCAA 2020 Biology Exam, Section A, Q31*

MC The process in which the bacterial cell takes up the plasmid is called

A. translation.
B. transcription.
C. translocation.
D. transformation.

Question 3 (1 mark)

Source: *VCAA 2020 Biology Exam, Section A, Q32*

MC Target DNA is to be inserted into a plasmid.

For a recombinant plasmid to be produced

A. the plasmid sections and the target DNA must have blunt ends.
B. the target DNA must come from the same species as the bacteria.
C. the plasmid and the target DNA must be cut by a polymerase.
D. DNA ligase is used to rejoin the sugar–phosphate sections of the plasmid and the target DNA.

Question 4 (4 marks)

Source: *VCAA 2020 Biology Exam, Section B, Q6*

Gene cloning has allowed the pharmaceutical industry to manufacture large quantities of proteins at a low cost. These proteins are produced by bacteria and are used to treat certain health conditions.

In the past, before the development of DNA technology, proteins for treating certain health conditions could be obtained only from animals, such as cattle and pigs, or from human corpses.

a. State **two** advantages of using gene cloning to manufacture pharmaceutical proteins rather than sourcing the proteins from animals or human corpses. **2 marks**

b. Outline **one** ethical issue associated with the use of gene cloning in the manufacture of a pharmaceutical product. **2 marks**

Question 5 (6 marks)

Source: *VCAA 2006 Biology Exam 2, Section B, Q4*

a. Describe the appearance of a bacterial plasmid. **1 mark**

A bacterial plasmid was modified in the laboratory so that it contained a gene for an enzyme which provided resistance to the antibiotic tetracycline.

Bacterial cells, which in their natural environment were sensitive to the antibiotic tetracycline, were mixed with the modified plasmid. The bacterial cells were treated so that they could take up the plasmid.

b. What is the name of the process in which a bacterial cell takes up a plasmid and expresses the genes of the plasmid? **1 mark**

The outcome of an experiment is shown below.

A
bacterial cells only, spread on agar

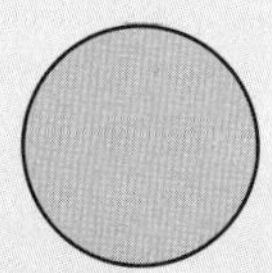

B
bacterial cells only, spread on agar with tetracycline

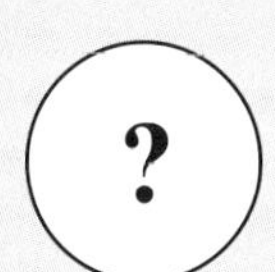

C
bacterial cells exposed to the plasmid, spread on agar with tetracycline

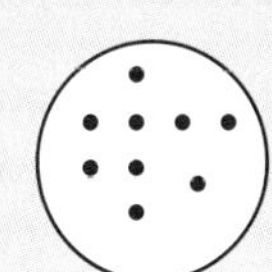

c. With respect to the growth of bacteria the results of plates A and C are shown. On plate A there is a continuous growth of bacteria over the surface of the agar. On plate C the colonies are distinguishable from each other.
   i. What result would you expect on plate B with respect to the growth of the bacteria? **1 mark**
   ii. Explain your answer to cii. **1 mark**

d. Explain why there is a difference in the way the bacteria have grown on plates A and C. **2 marks**

More exam questions are available in your learnON title.

2.7 The use of genetically modified and transgenic organisms

KEY KNOWLEDGE

- The use of genetically modified and transgenic organisms in agriculture to increase crop productivity and to provide resistance to disease.

2.7.1 Distinction between genetically modified and transgenic organisms

Genetically modified organisms (GMOs) are those organisms whose genomes have been altered using genetic engineering technology. GMOs have novel combinations of genes that do not occur in natural populations of the species concerned but are created through the means of genetic engineering. The genetic alteration achieved through genetic engineering may involve:

- the addition of a gene or a segment of DNA
- the silencing of a gene so that its function is lost.

For an organism to be identified as a GMO, any gene or DNA segment that is added through genetic engineering should be heritable; that is, it should be able to be passed on to the next generation.

Transgenic organisms comprise a subgroup of GMOs that includes those GMOs in which the alteration to the genome involves the genetic material from a different species. So, it follows that all transgenic organisms are GMOs, but not all GMOs are transgenic.

genetically modified organisms (GMOs) organisms whose genomes are altered through the use of genetic engineering technology

transgenic organisms organisms that carry in their genomes one or more genes artificially introduced from another species

Genetically modified organisms are those whose genomes have been altered. Transgenic organisms are specific types of genetically modified organisms that contain genes from other species.

The question to ask in order to identify a transgenic organism is: *'Does the gene that is added to the genome come from the same species or does it come from a different species?'*

If the gene is from the same species, the organism is termed a GMO. If the gene comes from a different species, the recipient organism is described as being a transgenic organism and, at the same time, a GMO.

For example, GloFish that have incorporated genes from a sea anemone into their genomes are transgenic as well as GMOs. An organism whose genome has *not* been subjected to change by genetic engineering is termed a non-GMO.

FIGURE 2.44 GloFish (Tetra): a transgenic genetically modified fish

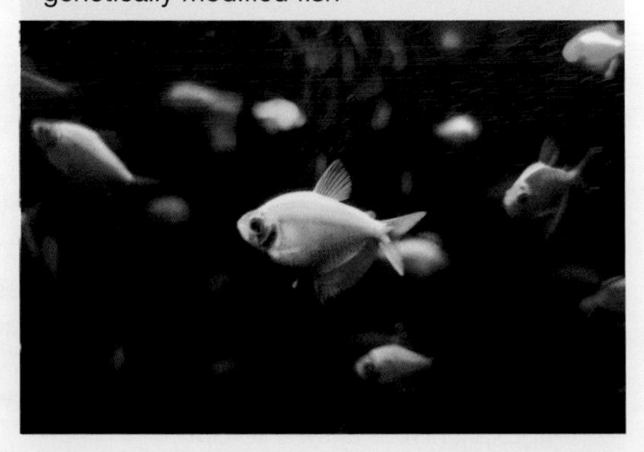

2.7.2 Applications in agriculture

Foreign genes that confer resistance to insect pests and viruses, and that confer tolerance to various herbicides, have been engineered into major food crops with high market value. Currently, examples of genetic modified crops include pest-resistant maize, cotton, potato and rice, virus-resistant squash and papaya, and herbicide-tolerant cotton, maize, soybean and canola. In addition, some crops, such as corn and cotton, have been engineered to carry the combined ('stacked') traits of insect resistance and herbicide tolerance.

TABLE 2.7 Examples of intended purposes of genetic manipulation of plant crops

Transgenic trait	Crops
Insect (pest) resistance	Cotton, corn, potato, tomato
Herbicide tolerance	Canola, cotton, corn, rice, flax, sugar beet
Virus resistance	Papaya, squash, potato
Delayed ripening	Tomato
Tolerance to environmental stress: • drought tolerance • flood tolerance • salt tolerance	 Rice Rice Rice, wheat, barley, tomato
Altered oil composition	Canola, soybean
Enhanced nutritional value	Rice, wheat
Improved post-harvest shelf life	Mango, papaya, tomato, pineapple

GM crops worldwide

Worldwide in 1996, an initial area of 1.7 million hectares was planted with genetically modified (GM) crops. This had increased by 2018 to nearly 190 million hectares, with the main GM crops under commercial cultivation being herbicide-tolerant and insect-resistant cotton, canola, corn and soybean. This number has continued to increase, closing in on 200 million hectares globally in 2020.

In total, about 30 countries have GM crops under cultivation. However, in 2015 just 10 of these countries accounted for 98 per cent of the total GM area under cultivation. Countries with the largest areas under cultivation with GM crops are shown in figure 2.45. The United States has the largest area, followed by Brazil and Argentina. The group identified as 'Others' includes Australia and eight European Union nations, including Sweden and Germany (who mostly plant GM potatoes engineered to have altered starch content). In 2020, USA, Brazil and Argentina still lead the way with the cultivation of GM crops.

FIGURE 2.45 Pie chart showing the worldwide areas of cultivation of GM crops in 2015. These trends are still seen today.

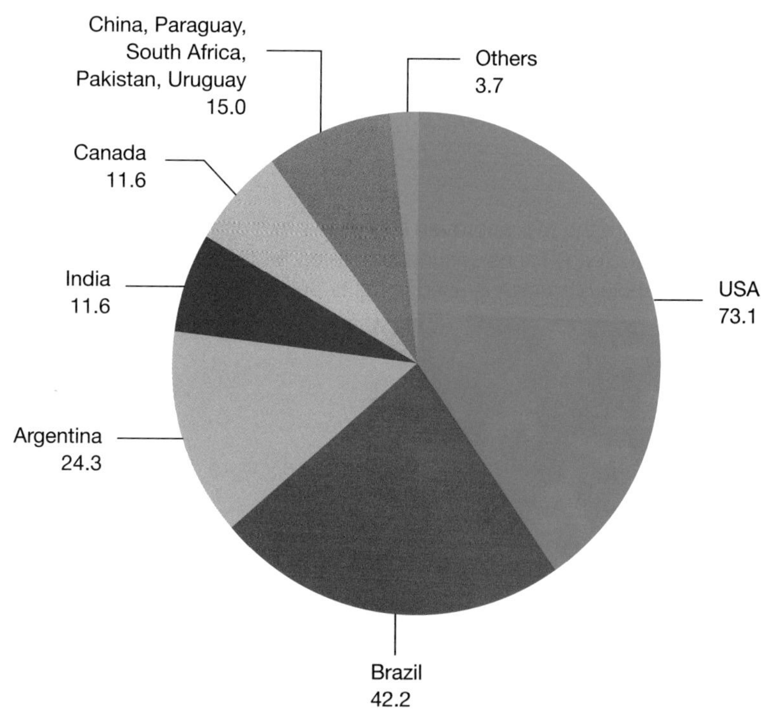

Note: Total area of GM crops in 2015 was 181.5 million hectares.

Source: ISAAA 2015

As well as the largest area of GM plantings, the United States has the widest range of GM crops, either in commercial plantings or under development. The commercial plantings of approved GM food crops in the USA include soybean, corn, cotton, apple, canola, chicory, flax and linseed, melon, papaya, potato, rice, squash, sugar beet, tomato and tobacco.

Worldwide, however, the majority of GM crops under cultivation include four main crops: soybean (51 per cent), maize (corn) (30 per cent), cotton (13 per cent) and canola (5 per cent). All other GM crops constitute just 1 per cent of the total GM plantings (see figure 2.46). This trend continues in 2020.

FIGURE 2.46 GM crops as a percentage of total area under GM cultivation. Just four crops dominate worldwide. Plantings of all other GM crops constitute just 1 per cent.

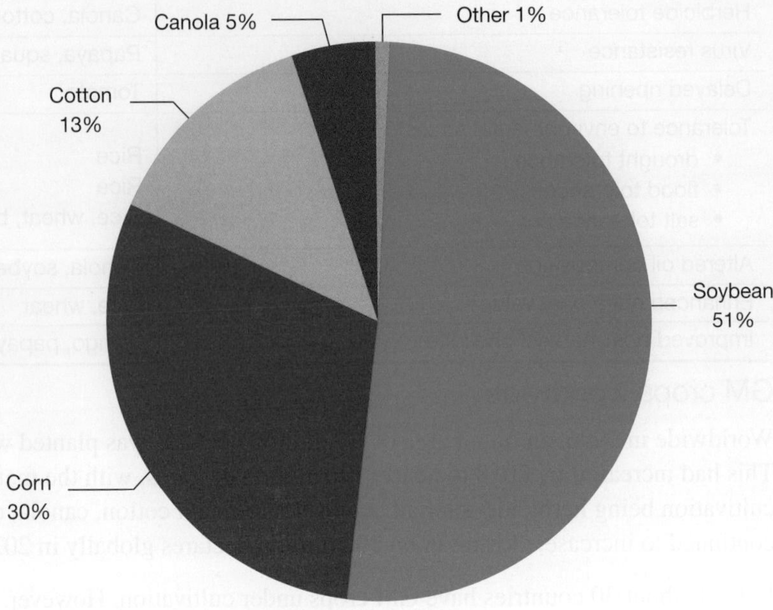

Source: CBAN, 2016

GMOs in Australia

In Australia, two genetically engineered crops grown on a large scale are cotton (*Gossypium hirsutum*) and canola (*Brassica napus*).

GM cotton

The fruit of the cotton plant is a structure called the cotton boll (see figure 2.47a). Inside each boll is the lint that forms the cotton fibres that are spun into the cotton fabrics of clothing, towels, furniture covers and the like. The cotton boll also contains a number of seeds, which are a source of cottonseed oil.

GM canola

In spring, crops of canola form vivid displays of bright yellow flowers (see figure 2.47b) that are a valuable source of oilseed. Herbicide-tolerant genetically modified canola is grown across approximately half a million hectares in New South Wales, Victoria and Western Australia. Around 20 per cent of all canola grown in Australia is genetically modified.

FIGURE 2.47 Cotton and canola are the two approved GM plants that can be grown commercially in Australia. **a.** Crop of cotton plants, with fruits ready for harvesting in autumn. **b.** Canola plants (*Brassica napus*) in flower. Canola seeds are a source of canola oil, which is used in many processed food products.

The regulation of genetically modified organisms in Australia

The Australian Government's Office of the Gene Technology Regulator (OGTR) is responsible for regulating all GMOs in Australia, ensuring genetic modification is done safely and ethically. However, it is up to state governments on how to govern this.

As of 2019, both Tasmania and South Australia banned GMOs. In 2019, this ban in Tasmania was extended to at least the year 2029. In 2020, the South Australian government changed the laws around the ban of GMOs, leading to the lifting of restrictions — particularly around GM crops. However, local councils were given the power to decide if they want their region to remain GM free.

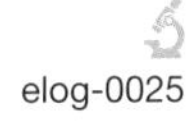
elog-0025

INVESTIGATION 2.5 online only

Investigating genetically modified and non-genetically modified foods

Aim

To compare genetically modified and non-genetically modified foods

2.7.3 Increased productivity and resistance to disease

Most existing genetically modified crops have been developed to improve yield through the introduction of resistance to plant diseases or increased tolerance of herbicides. There are two collections of genetically modified crops that are widely grown around the world and in Australia.

Group 1: Herbicide-resistant crops

- These crops were first altered so that they were not affected by the herbicide glyphosate (active ingredient in the commercially purchased herbicide Roundup), allowing farmers to eliminate weeds without harming their crops.
- Glyphosate-resistant crops can increase crop efficiency while eliminating weeds.
- An example of a herbicide resistant variety of canola being grown in Australia is Roundup Ready® canola, developed by Monsanto, was approved for commercial production in Australia in 2003. The *gox* gene, from the bacterial species *Ochrobactrum anthropi*, was engineered into canola plants using *Agrobacterium* as a vector. The *gox* gene encodes an enzyme that breaks down glyphosate, the active ingredient of Roundup® herbicide, into a harmless product. Spraying canola crops with this herbicide kills weeds and other plants that lack the *gox* gene, but leaves GM canola plants unaffected.

Group 2: Natural pesticides and insecticides

- These crops are altered to produce their own natural pesticides.
- Protect the crops against insect infestation, as the pesticide is located within the plant and therefore needs to be ingested by the insects to have an effect.
- The use of natural pesticides is environmentally friendly as it eliminates the use of sprays that could be harmful and toxic to other organisms.
- Many insecticides, such as endosulfan and broad-spectrum organophosphates, are highly toxic to people, pets, livestock on farms, wildlife, and they kill beneficial insects as well as their major target pests.

Examples of natural pesticide crops of cotton being grown in Australia include the following:

- In 1996, Ingard® cotton became the first genetically modified field crop to be grown in Australia. These cotton plants were genetically engineered to include the *cry1Ac* gene from the bacterial species *Bacillus thuringiensis*. The *cry1Ac* gene encodes a protein, known as Bt, that acts as an insecticide. Bt insecticides are highly effective against the leaf-feeding larvae (caterpillars) of the cotton bollworm (*Helicoverpa armigera*), which are major pests of cotton crops (see figure 2.48).

- In 2003, Bollgard II® cotton was developed by CSIRO in collaboration with the US-based multinational company Monsanto, and was released for commercial use. A second gene, the *cry2Ab* gene, that encodes a toxin that acts against different insect pests was engineered into Bollgard II cotton plants.
- In recent years, Bollgard II was replaced with Bollgard 3®. Currently Bollgard 3 is the main Bt cotton grown into Australia, containing both the the *cry1Ac* and *cry2Ab* genes, as well as a third gene, *vip3A,* which produces an insecticide.

FIGURE 2.48 Larva of *Helicoverpa armigera*, known as the bollworm, eating a cotton boll. *H. armigera* and other *Helicoverpa* species are the most serious insect pests of cotton and are also pests of other crops, including corn. (Courtesy of Judith Kinnear.)

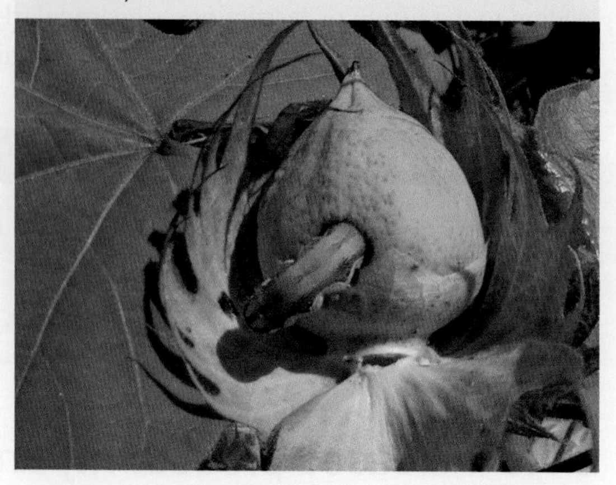

Increasing crop productivity

As well as being modified to introduce resistance, many crops have been modified to increase productivity. Examples of such modifications include improved photosynthetic efficiency, greater crop yields and faster growth rates. This includes the use of CRISPR-Cas9 technologies to modify plant genomes. This will be further explored in subtopic 4.8.

CASE STUDY: Genetically modified Golden Rice

The World Health Organization (WHO) has identified vitamin A deficiency as one of the major causes of preventable blindness in children. Vitamin A deficiency also increases the risk of disease and death from infections, particularly in the poorest segments of populations in low- and middle-income countries. Vitamin A deficiency is the lack of an adequate intake of vitamin A in the diet. The direct sources of vitamin A in the human diet are animal products, such as egg yolk, butter, cream, cod liver oil and liver. However, plants, such as carrots, leafy green vegetables, sweet potatoes and cantaloupes, are also the source of beta-carotene, which, in the human body, is converted to vitamin A.

FIGURE 2.49 Standard rice (left) and Golden Rice (at right)

The normal role of the carotenoids is that of accessory photosynthetic pigments found in the leaves of green plants. Rice plants produce beta-carotene in their photosynthetic leaf cells, but the genes concerned are not active in the cells of rice seeds (grains). This means that a serve of rice does not provide any vitamin A precursor to the person eating it.

Golden Rice 2 is a strain of rice (*Oryza sativa*) that has been genetically modified to produce beta-carotene, a precursor of vitamin A, in the endosperm of the rice grain. Golden Rice is so named because it has a distinctive golden colour that is due to the presence of high levels of beta-carotene. Eating a serve of Golden Rice 2 provides beta-carotene, which can be converted to vitamin A. The development of Golden Rice does not involve any multinational companies; it has been developed as a public-good project.

The genome of the Golden Rice variety has been genetically engineered by the insertion into the rice genome of two additional genes in the pathway for the biosynthesis of carotene and the addition of a promoter sequence that keeps this pathway active in the cells of the rice endosperm.

The two genes added are:

1. the *PSY* gene from daffodils (*Narcissus pseudonarcissus*), which encodes the enzyme phytoene synthase
2. the *CRTI* gene from the soil bacterium *Pantoea ananatis*, which encodes the enzyme phytoene desaturase; this enzyme catalyses multiple steps in the synthesis of carotenoids up to lycopene.

To access more information on additional case studies of genetically modified plants, please download the digital document.

 Resources

 **Digital document** Case study: Other examples of genetically modified plants (doc-35872)

CASE STUDY: Genetically modified animals

It is not only plants that are able to be genetically modified. Animals may also be genetically modified. In subtopic 2.3, we explored some uses of gene editing through CRISPR-Cas9. Various techniques may be used to modify genes in animals such as mice, salmon, cattle and hamsters.

To access more information on various case studies and complete an analysis task relating to these, please download the worksheet.

FIGURE 2.50 Salmon may be genetically modified to reach a larger size faster.

 Resources

 **eWorkbook** Worksheet 2.10 Case study analysis: Genetically modified animals (ewbk-1878)

 Resources

 **eWorkbook** Worksheet 2.11 Manipulating genes in organisms (ewbk-1988)

 **Weblinks** Natural fluorescence in fish
OGTR website
Regulation of genetically modified crops in Australia
Database of GM approvals worldwide

KEY IDEAS

- Gene manipulation may be carried out for a variety of purposes.
- GM plants containing particular genes that are silenced have been produced using antisense mRNA and RNA interference techniques.
- GM animals have been produced through the insertion of DNA constructs carrying a structural gene of interest and relevant promoter genes.
- Issues, such as social, ecological, and legal ones, have arisen in relation to various aspects of genetic engineering.

2.7 Activities

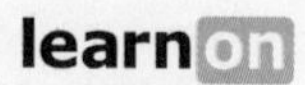

To answer questions online and to receive **immediate feedback** and **sample responses** for every question, go to your learnON title at **www.jacplus.com.au**. A **downloadable solutions** file is also available in the resources tab.

2.7 Quick quiz on	2.7 Exercise	2.7 Exam questions

2.7 Exercise

1. Describe the difference between genetically modified and transgenic organisms. Provide an example of each of these.
2. In Australia could you purchase a GloFish from your local pet store? Explain.
3. Would Australia be considered one of the top ten countries in terms of area of GM crops under cultivation? Justify your response.
4. What are genetically modified foods? Explain the two common genetic modifications found to increase yields and protect from disease.

2.7 Exam questions

Question 1 (1 mark)

Source: *VCAA 2019 Biology Exam, Section A, Q36*

MC Certain yeast (*Saccharomyces cerevisiae*) can be modified and made to express a human gene, resulting in the production of insulin.

S. cerevisiae can most accurately be described as a

A. transgenic organism.
B. yeast–human hybrid.
C. genetically mutated organism.
D. laboratory-produced organism.

Question 2 (1 mark)

Source: *VCAA 2017 Biology Exam, Section A, Q40*

MC Plant viruses are a major problem for farmers growing crops. A particular plant virus can infect many different plant species.

Scientists are trialling a spray treatment on tobacco crops. The treatment does not alter the DNA of the tobacco plants.

During this treatment, tobacco plants are sprayed with clay nanoparticles containing double-stranded RNA (dsRNA). The dsRNA released from each of the clay nanoparticles enters the plant cells. Inside each cell the dsRNA silences a gene from the virus by causing viral RNA to break down.

In this technique the

A. dsRNA would have a nucleotide sequence complementary to a section of DNA nucleotides in the tobacco plants.
B. dsRNA would silence a gene from the virus by initiating changes that prevent translation of the viral gene.
C. spray treatment would be effective only on tobacco plants and not on other plant species.
D. sprayed tobacco plants would be regarded as transgenic organisms.

Question 3 (2 marks)

Scientists achieved some success in transferring a specific jellyfish gene into fertilised mouse eggs.

The gene concerned controls production of a fluorescent protein. After the mouse eggs developed into baby mice, those mice carrying the gene were readily identified because they showed a green fluorescence when placed in ultraviolet (UV) light.

a. Explain whether or not the production of these fluorescent mice is an example of genetic transformation. **1 mark**
b. Suggest a possible reason for the scientists' choice of this particular gene for their experiment. **1 mark**

Question 4 (8 marks)

***Source:** Adapted from VCAA 2018 Biology Exam, Section B, Q10*

Should we grow GM crops?

by Mary Nguyen

More than 25 years after genetically modified (GM) food first appeared, growing GM crops remains a hotly debated topic. Some people argue that GM crops are the only way to feed the growing world population and to minimise environmental harm. Other people express different views.

Bt cotton is a type of cotton that contains two genes from a soil bacterium, *Bacillus thuringiensis*, enabling it to produce insect-resistant proteins. Australian farmers of Bt cotton use only 15% of the quantity of the insecticide that was once needed to protect their cotton crops*. However, Bt cotton is not as resistant to the main insect pest of cotton crops, *Helicoverpa*, as it has been in the past*.

In Australia, Bt cotton is picked by machine, but in India, it is picked by hand. Workers in India have developed skin allergies, which have been attributed to Bt cotton proteins. Traditionally, farmers have saved money by keeping seed from one year's crop to plant the following year. However, it is illegal for farmers to keep Bt cotton seeds because these seeds have been declared the legal property of the company Monsanto. Every year, cotton farmers must buy more seeds from Monsanto.

Unlike Monsanto, the company that produces the GM food crop Golden Rice allows farmers to replant the rice they harvested the previous year. By inserting a gene from the bacteria *Erwinia uredovora* and another from a daffodil, *Narcissus pseudonarcissus*, into white rice, scientists produced Golden Rice – a rice variety containing higher levels of vitamin A†. People who eat Golden Rice avoid vitamin A deficiency. Trials conducted in several countries have shown that Golden Rice is safe to eat‡.

References: *CSIRO, 'Cotton pest management', case study, <www.csiro.au>; †JA Paine et al., 'Improving the nutritional value of Golden Rice through increased pro-vitamin A content', *Nature Biotechnology* , 23, 27 March 2005, pp. 482–487; ‡A Coghlan, 'Golden Rice gets approval in the US', *NewScientist*, magazine issue 3180, 2 June 2018

a. Bt cotton and Golden Rice are genetically modified organisms but are they also transgenic organisms? Support your response with evidence from the article above. **3 marks**

b. How can planting a Bt cotton crop lead to an increase in crop yield? **1 mark**

c. Using information from the article, describe one social and one biological implication relevant to the use of Bt cotton and one social and one biological implication relevant to the use Golden Rice. The same implication should **not** be used twice. **4 marks**

Question 5 (1 mark)

***Source:** Adapted from VCAA 2020 Biology Exam, Section A, Q37*

MC Rice (*Oryza sativa*) is a staple food for billions of people worldwide, particularly in Asia. Although rice supplies energy, it is low in micronutrients, such as iron and zinc. Australian scientists created a strain of biofortified rice that has been trialled in the Philippines and has been recently introduced to Bangladesh.

The biofortified rice was created when two particular genes were inserted into normal rice. The biofortified rice plants responded as if they were iron deficient by permanently 'switching on' another gene to take up iron and zinc from the soil. Details of the two inserted genes are given in the table below.

Inserted gene	Protein function	Source of gene
rice nicotianamine synthase (OsNAS2)	assists iron uptake by roots of rice plants	rice plants
soybean ferritin (Sfer-HI)	binds and stores large amounts of iron	soybean plants

Which one of the following is the best description for this strain of biofortified rice?

A. genetically screened
B. genetically modified and transgenic
C. genetically transformed by gene silencing
D. genetically engineered by adding iron and zinc

More exam questions are available in your learnON title.

2.8 Review

2.8.1 Topic summary

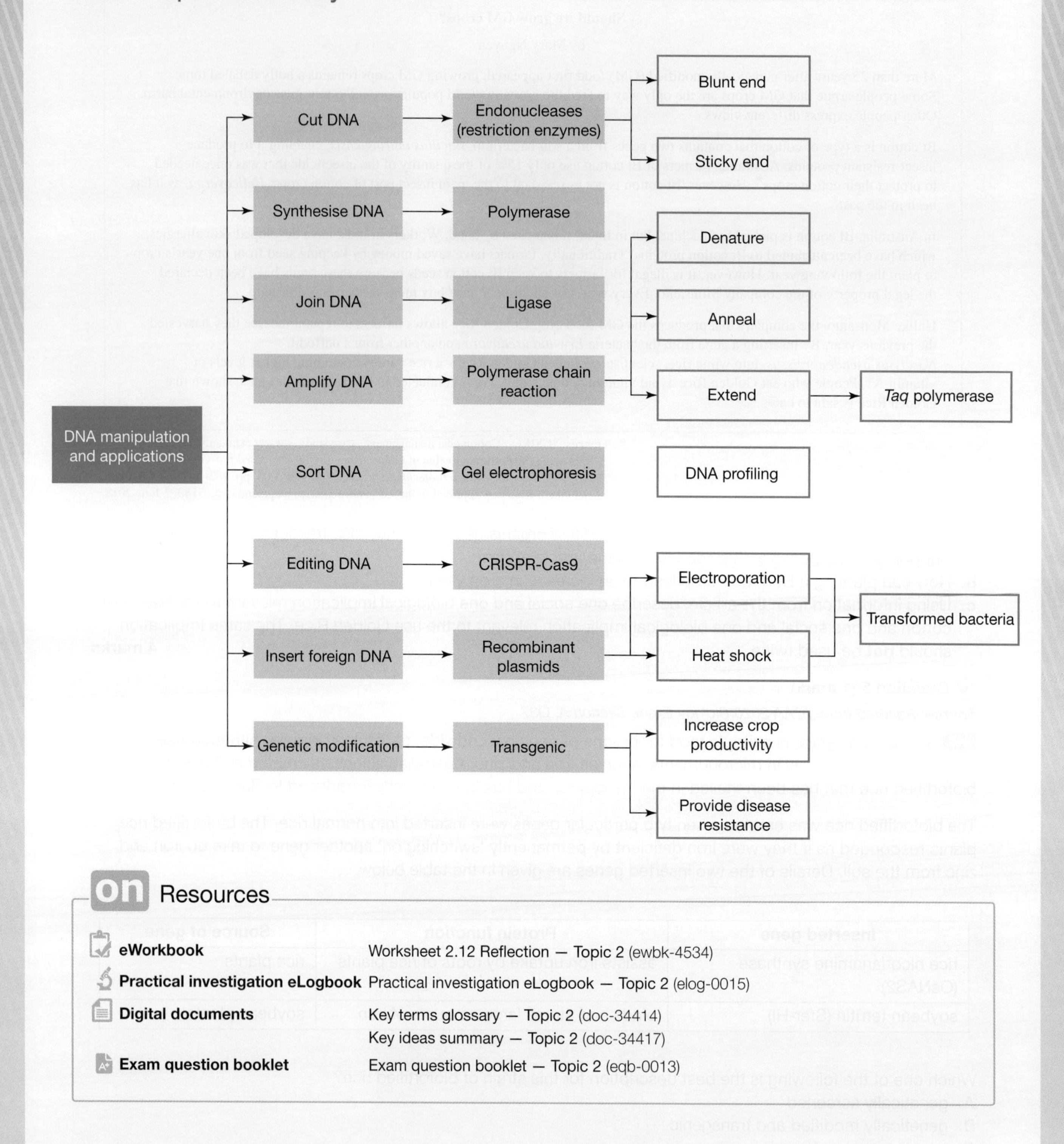

on Resources

eWorkbook Worksheet 2.12 Reflection — Topic 2 (ewbk-4534)

Practical investigation eLogbook Practical investigation eLogbook — Topic 2 (elog-0015)

Digital documents Key terms glossary — Topic 2 (doc-34414)
Key ideas summary — Topic 2 (doc-34417)

Exam question booklet Exam question booklet — Topic 2 (eqb-0013)

2.8 Exercises

To answer questions online and to receive **immediate feedback** and **sample responses** for every question, go to your learnON title at **www.jacplus.com.au**. A **downloadable solutions** file is also available in the resources tab.

2.8 Exercise 1: Review questions

1. A solution contains many copies of a piece of linear DNA with three cutting sites for a particular restriction enzyme. This solution of DNA is treated with the restriction enzyme under controlled conditions.
   a. What is the expected result of this treatment?
   b. Is this change in DNA an example of a chemical or a physical change?
2. The action of the endonuclease *Pst*I is as follows:

   CTGCAG → CTGCA G
   GACGTC G ACGTC

   a. What kinds of ends does this enzyme produce?
   Copies of one long fragment of DNA were treated with *Pst*I and three smaller fragments were produced:
   - fragment 1 2.3 kbp
   - fragment 2 8.1 kbp
   - fragment 3 4.4 kbp

   b. How long was the original DNA fragment?
   c. How many recognition sites for this enzyme were present?
   d. If the three fragments were subjected to electrophoresis, draw the expected outcome.
3. A young woman from a family with a history of haemophilia (an X-linked genetic disorder) had a test that revealed that she was a carrier of the haemophilia *c* allele. This woman later became pregnant and found out that her fetus was male. She wished to know if her fetus would be affected by haemophilia.
   a. Carefully explain if and how an answer might be given to her.
   One DNA test for haemophilia involves the use of the restriction enzyme *Bcl*I.
   b. For this test, could *Bcl*I be replaced with another restriction enzyme, such as *Sac*I? Explain your decision.
4. CRISPR-Cas9 technology allows for gene editing. An issue that is the subject of debate is: Should gene editing be allowed on human cells, such as eggs, sperm and embryos, whose DNA can pass to future generations? Or, should gene editing be restricted to somatic cells where that DNA is not passed to the next generation? Construct arguments that supports and refutes the influence of social, economic, legal and ethical factors relevant to CRISPR-Cas9 in editing the human genome.
5. *E. coli* cells carrying the foreign gene for human growth hormone are to be grown in mass culture in 50 000 litre fermenting tanks over 2 days. The recombinant plasmid present in these *E. coli* cells is pUC18.
   a. Person A suggested that 0.5 gram of these bacteria would be sufficient to start this mass culture, but person B said 'No way! You will need lots more'. Indicate with which person you agree and give a reason for your choice.
   b. Outline a procedure by which the *E. coli* bacteria with recombinant plasmids might be separated from bacteria not possessing recombinant plasmids.
   c. The culture fluid in which the *E. coli* bacteria cells will grow provides the water, glucose, salts and nitrates required for their growth, but it will also support the growth of many other kinds of bacteria. Identify a possible measure that could be taken to prevent the growth of contaminating microbes.
6. What is the essential difference between the members of the following pairs?
   a. A transgenic organism and a genetically modified organism
   b. Insect resistance and herbicide tolerance
7. a. What is the role of DNA polymerase in the polymerase chain reaction (PCR)?
   b. Outline the steps involved in the PCR process.
   c. What purpose does heating the DNA strand during the first step of the PCR process serve?
   d. What are the advantages of using PCR?

8. Detection and amplification of DNA has become much more sensitive, so that DNA can now be isolated from just a small number of cells, estimated at 5 to 20 cells. This fact has given rise to the concept of 'touch DNA', that is DNA that may be recovered from skin cells left behind on any object that a person has touched or handled.

   a. Suggest why the technician in this figure is gloved and masked.
   b. Suggest three other possible sources of touch DNA at a crime scene.

   The DNA profile from touch DNA found on a door handle at a crime scene was found to match that of the next-door neighbour of the victim.

   c. Would this evidence alone justify the immediate arrest of that neighbour for the crime concerned? Explain your answer.
9. Consider the following observations and give an explanation to account for each.
   a. Inserting a foreign gene into the genome of an organism may unintentionally cause unexpected harmful effects.
   b. Individuals who were allergic to Brazil nuts suffered an allergic reaction when they ate GM soybeans into which a Brazil nut gene had been incorporated.
   c. GM crops are more extensively tested than non-GM varieties before release, both for their environmental effects and as foods.
   d. DNA constructs used to genetically engineer an organism would be expected to include the gene encoding the protein of interest, a promoter and a terminator.

2.8 Exercise 2: Exam questions

on Resources

Teacher-led videos Teacher-led videos for every exam question

Section A — Multiple choice questions

All correct answers are worth 1 mark each; an incorrect answer is worth 0.

Question 1

Source: *VCAA 2017 Biology Exam, Section A, Q38*

The diagram below represents a DNA molecule and the position of the recognition sites for the restriction enzymes BamHI, EcoRI, HaeIII and SalI.

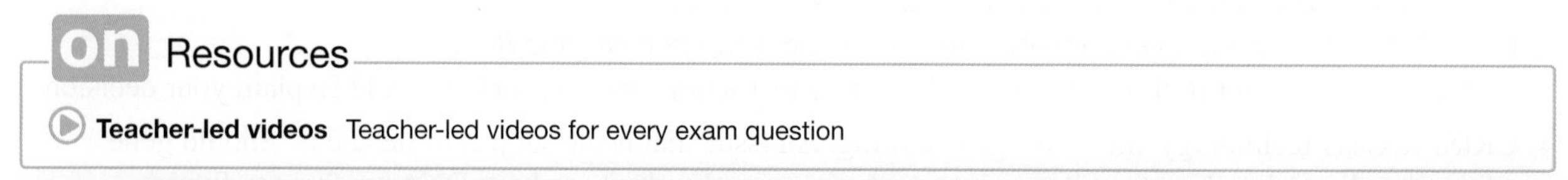

Also shown is a diagram of an electrophoresis gel in which the lanes R, S, T and U show the separation of DNA segments resulting from digestion of the molecule with one of the restriction enzymes.

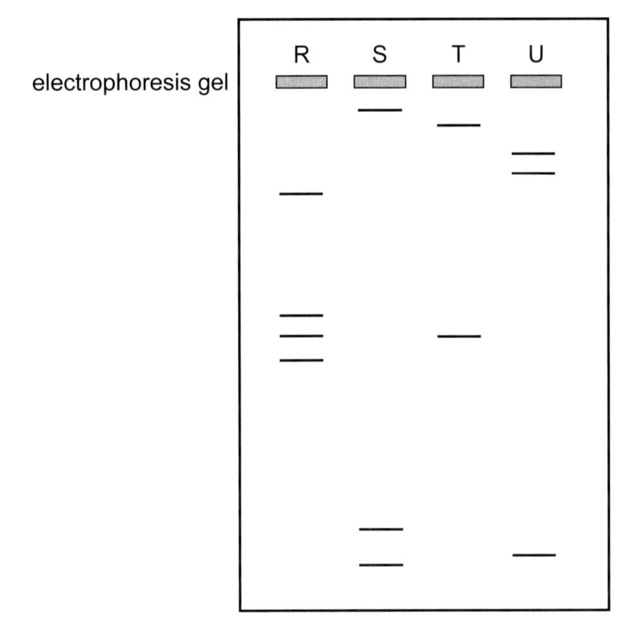

Which of the following shows the correct match between the lane and the restriction enzyme used to digest the DNA molecule?

	R	S	T	U
A.	Sa1I	EcoRI	HaeIII	BamHI
B.	EcoRI	BamHI	HaeIII	Sa1I
C.	EcoRI	BamHI	Sa1I	HaeIII
D.	HaeIII	Sa1I	BamHI	EcoRI

Question 2

Source: *VCAA 2018 Biology Exam, Section A, Q28*

The diagram below represents a method of DNA manipulation.

The method represented is

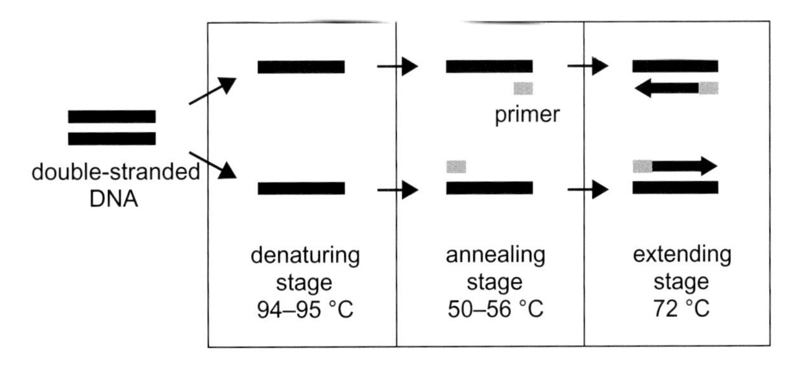

Source: Genome Research Limited, in Your Genome, <www.yourgenome.org>

A. gel electrophoresis.

B. DNA transformation.

C. bacterial transformation.

D. polymerase chain reaction.

Question 3

Source: *VCAA 2014 Biology Exam, Section A, Q25*

Plasmids of bacteria are used to transfer selected genes from one species to another.

The process can be represented as follows.

bacterial plasmid cut → foreign gene and plasmid mixed → plasmid with inserted foreign gene

Enzymes are used to facilitate several of these steps.

Which one of the following shows the enzymes required for the first and last steps of process?

	Cuts plasmid	Insert genes
A.	restriction enzyme	DNA ligase
B.	restriction enzyme	DNA polymerase
C.	DNA ligase	DNA polymerase
D.	DNA polymerase	DNA ligase

Question 4

Source: *VCAA 2018 Biology Exam, Section A, Q34*

Genetic testing can be used to test for the allele for Huntington's disease (HD). The onset of HD predominantly occurs in adulthood.

Eight individual family members were tested for the HD allele. The diagram below shows the electrophoresis gel results of a test for the presence of the allele. Individuals 4 and 8 have been diagnosed with the disease.

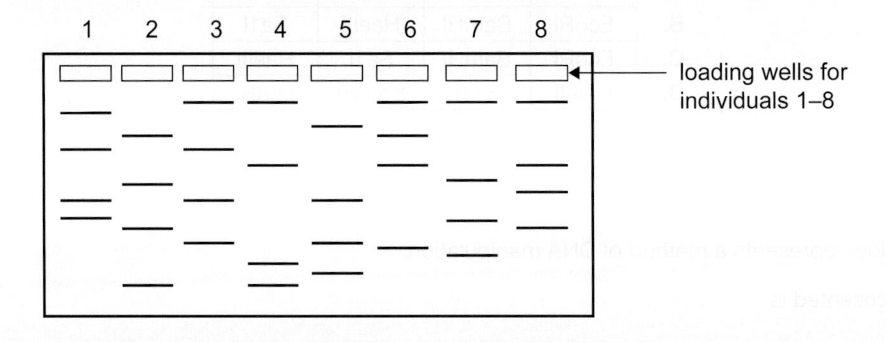

Which other individual is likely to suffer from HD now or in the future?

A. 1
B. 2
C. 5
D. 6

Use the following information to answer Questions 5 and 6.

Question 5

Source: *VCAA 2013 Biology Section A, Q34*

Bacteria can be transformed with an artificial insulin gene and cultured to make insulin in commercial quantities. The steps taken to produce genetically engineered insulin are summarised below. The order of the steps has been mixed up.

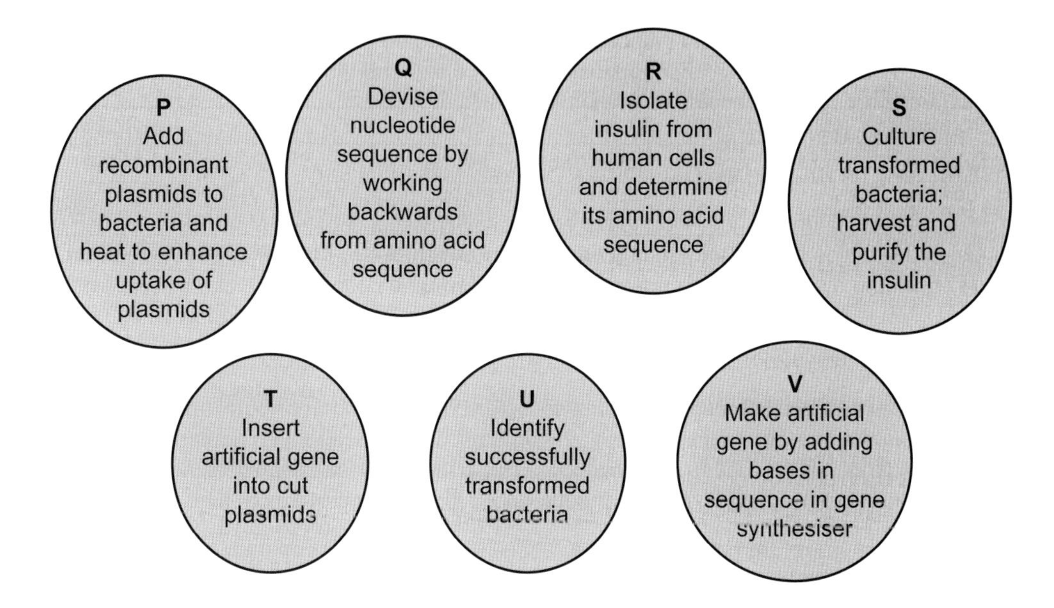

The correct sequence of steps when producing the insulin is

A. V, P, T, S, U, R, Q.
B. V, T, P, U, S, Q, R.
C. R, Q, V, T, P, U, S.
D. R, V, Q, T, P, S, U.

Question 6

Source: *VCAA 2013 Biology Section A, Q35*

The tool used for joining the artificial gene to plasmid DNA at step T is

A. a primer.
B. DNA ligase.
C. DNA polymerase.
D. gel electrophoresis.

Question 7

Which of the following is correct regarding gel electrophoresis?

A. DNA moves towards the negative end as it is positively charged.
B. Smaller fragments of DNA move faster down the gel compared to larger fragments.
C. Fragments can be easily observed without the use of a dye.
D. No more than five samples can be run per gel.

Question 8

Source: *VCAA 2017 Biology Exam, Section A, Q39*

DNA profiling, using short tandem repeats (STR) within a person's DNA, helps to determine the genetic relationship between individuals.

DNA profiles based on four STRs for five individuals are shown below. The results of a gender identifier are also shown.

STR	Individual 1	Individual 2	Individual 3	Individual 4	Individual 5
CSF1PO	7,14	7,11	8,13	7,14	7,14
TPOX	6,10	10,12	6,9	10,12	10,10
D21S11	27,30	29,32	27,27	29,30	27,28
D8S1179	9,11	12,13	17,17	11,12	9,11

	Individual 1	Individual 2	Individual 3	Individual 4	Individual 5
Gender identifier	**male**	**female**	**female**	**male**	**male**

Which one of the following conclusions can be made using the information given?

A. Individual 4 is the father of Individual 5.
B. Individual 3 is the mother of Individual 4.
C. Individual 5 could be the child of Individual 1 and Individual 3.
D. The parents of Individual 4 could be Individual 1 and Individual 2.

Question 9

The recognition sites of various endonucleases are shown below.

Enzyme	Source	Recognition site
*Bam*HI	*Bacillus amyloliquefaciens H*	**G$^{\vee}$GATC C** **C CTAG$^{\wedge}$G**
*Bgl*II	*Bacillus globigii*	**A$^{\vee}$GATC T** **T CTAG$^{\wedge}$A**
*Hae*III	*Hemophilus aegyptus*	**GG$^{\vee}$CC** **CC$^{\wedge}$GG**
*Taq*I	*Thermus aquaticus*	**T$^{\vee}$CG A** **A GC$^{\wedge}$T**

The following double-stranded sequence of DNA contains flanking regions as well as a desired gene. The desired gene is shown in bold.

G G C C T C G A A **C G A G A T C T C A C A** G G C C G G A T C C G G T T C G A T T
C C G G A G C T T **G C T C T A G A G T G T** C C G G C C T A G G C C A A G C T A A

Which restriction enzyme would be the best choice to use to cut out the gene to be incorporated into a recombinant plasmid?

A. *Hae*III
B. *Bgl*II
C. *Bam*HI
D. *Taq*I

Question 10

Which of the following would NOT be true for bacterial transformation?

A. Antibiotic sensitivity genes are included in the plasmid to differentiate between bacteria that have taken up the plasmid and those that have not.
B. Bacteria may be subject to heat shock therapy to introduce the plasmid into the bacterial cells.
C. The ligase enzyme must be used to allow for the gene of interest and the plasmid to bond.
D. Bacterial plasmids are cut with the same restriction enzyme as used for the gene of interest.

Section B — Short answer questions

Question 11 (7 marks)

Source: *Adapted from VCAA 2017 Biology Exam, Section B, Q9*

Scientists use recombinant bacterial plasmids as vectors to transform bacteria for a range of purposes in research and biotechnology.

a. What is meant by the term 'vector' in the context given? **1 mark**

A particular bacterial plasmid contains recognition sites for the restriction enzymes EcoRI, HindIII and BamHI, along with two antibiotic-resistant genes, ampicillin resistance (amp) and tetracycline resistance (tcl), and an origin of replication (ORI).

The diagram shows the positions of these recognition sites and antibiotic-resistant genes as well as the position of the origin of replication within this plasmid.

One purpose of using recombinant bacterial plasmids is to produce bacteria capable of synthesising human protein.

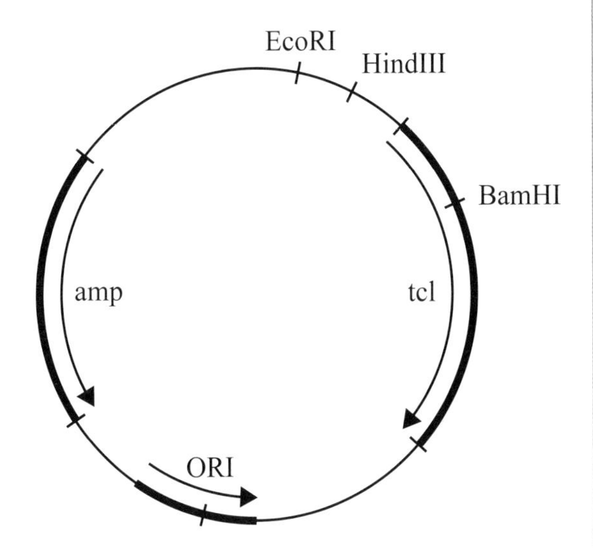

b. The restriction enzyme BamHI was used to help insert a gene coding for a human protein into this plasmid.

i. Describe how restriction enzymes such as BamHI are used to help insert a gene coding for a human protein into this plasmid. **2 marks**

ii. Draw and label a diagram to show the position of the human gene in this plasmid when BamHI is used. Include the position of the recognition sites for the restriction enzymes EcoRI, HindIII and BamHI on the plasmid. **1 mark**

c. After the scientists had carried out the steps required to make plasmids with the inserted human gene, these plasmids were mixed with a culture of bacteria. This mixture was treated so that these plasmids would move into the bacterial cells. Not all bacteria took up these plasmids.

Explain how scientists use antibiotics to identify which of the bacterial cells have been successfully transformed with plasmids carrying the human gene. **3 marks**

Question 12 (9 marks)

Source: *VCAA 2007 Biology Exam 2, Section B, Q2*

Victoria Police forensic scientists conduct DNA profiling using samples taken from crime scenes. Traces of DNA of less than 1 nanogram can be amplified and then profiled.

a. Name the process which is used to amplify the DNA. **1 mark**

Below is a diagram showing part of this process.

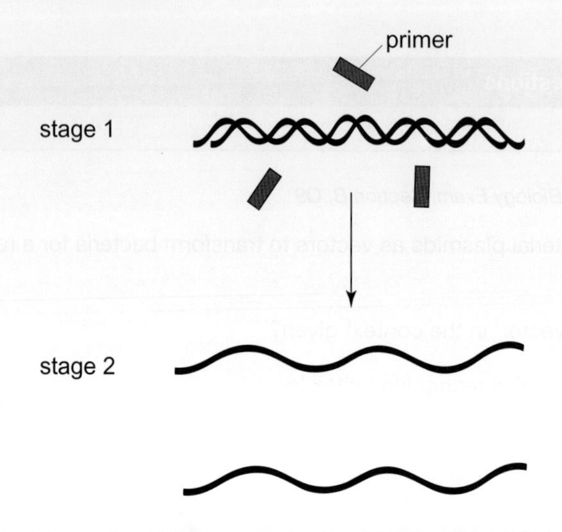

b. What must be done between stages 1 and 2 to separate the strands of the DNA molecule? **1 mark**

c. Complete and label the diagram at stage 2. **2 marks**

Small pieces of DNA of differing length can be compared to determine whether or not a sample could have come from a particular person. In a case, samples of DNA from the victim and the crime scene were compared with samples from two suspects.

The DNA samples were treated with restriction enzymes, amplified and run through gel electrophoresis. The results for one gene locus are shown in the diagram below.

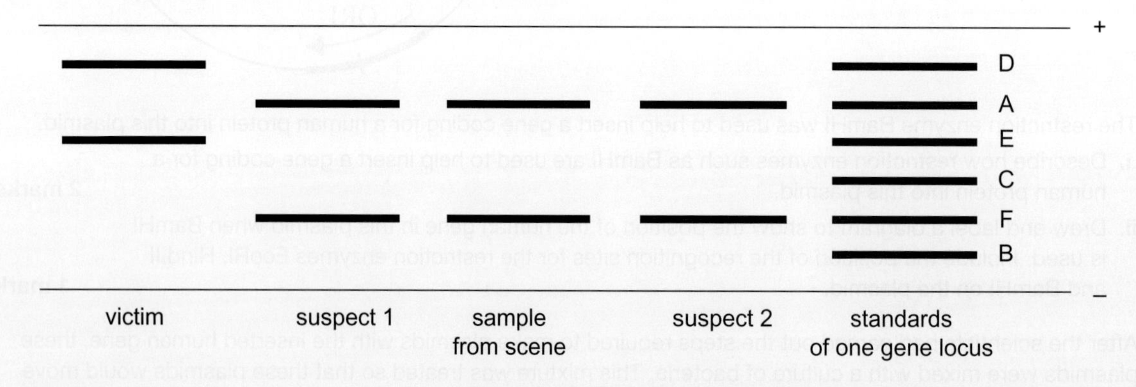

d. Draw an arrow on the right-hand side of the diagram to indicate the direction of movement of the DNA fragments. **1 mark**

e. What do the standards consist of, and what is their purpose? **2 marks**

f. From these results, give a conclusion which could be drawn about the sample taken from the crime scene. **1 mark**

g. What further action would you recommend to the forensic scientists investigating this case? **1 mark**

Question 13 (11 marks)

Source: *VCAA 2020 Biology Exam, Section B, Q11*

A student wanted to investigate the effect of two different endonucleases (restriction enzymes) on a linear DNA fragment.

The student used three tubes containing a buffered solution of linear DNA fragments, each fragment being 9500 base pairs in length.

Two different endonucleases were available: BamHI and HindIII.

The student followed the steps below.

Step 1 – 2 μL of BamHI was added to the sample in Tube 1.

Step 2 – 2 μL of HindIII was added to the sample in Tube 2.

Step 3 – 2 μL of HindIII and 2 μL of BamHI were added to the sample in Tube 3.

Step 4 – All three tubes were incubated for one hour at a constant temperature of 37 °C.

Step 5 – A 1% agarose gel was placed into an electrophoresis chamber and the gel was covered with buffer solution.

Step 6 – 40 μL of a DNA ladder with fragments of known sizes was added to the first well of the 1% agarose gel. The known sizes of the fragments were 10 000 bp, 8000 bp, 6000 bp, 5000 bp, 4000 bp, 3000 bp, 2000 bp, 1500 bp, 1000 bp, 500 bp and 250 bp.

Step 7 – 40 μL of the contents of each of the tubes was loaded into three separate wells of the1% agarose gel.

Step 8 – An electric current of 100 V was run through the gel for 45 minutes.

After 45 minutes the student obtained the results shown below.

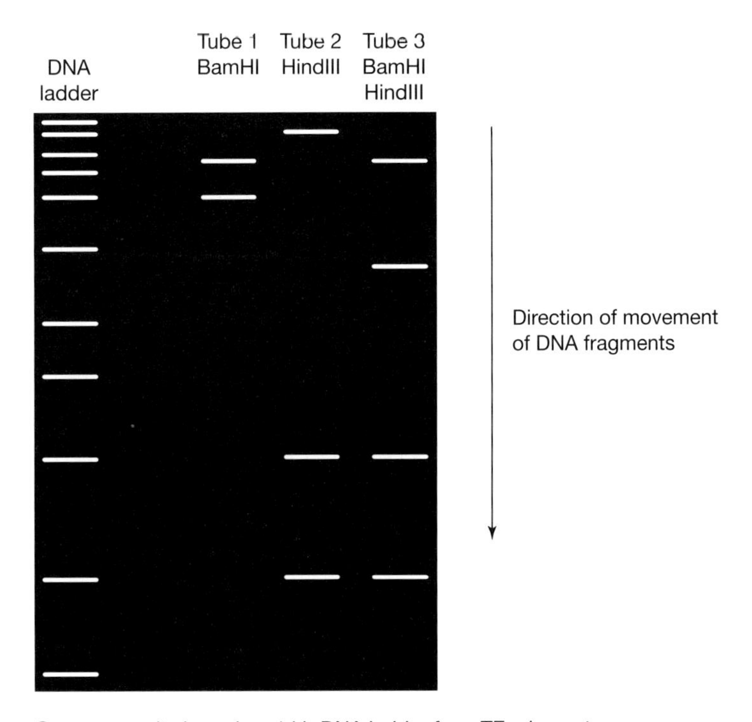

Source: results based on 1 kb DNA ladder from TEquipment, <www.tequipment.net>

a. Analyse the results of the experiment performed by the student. **5 marks**

The student repeated the experiment the next day and obtained the following results.

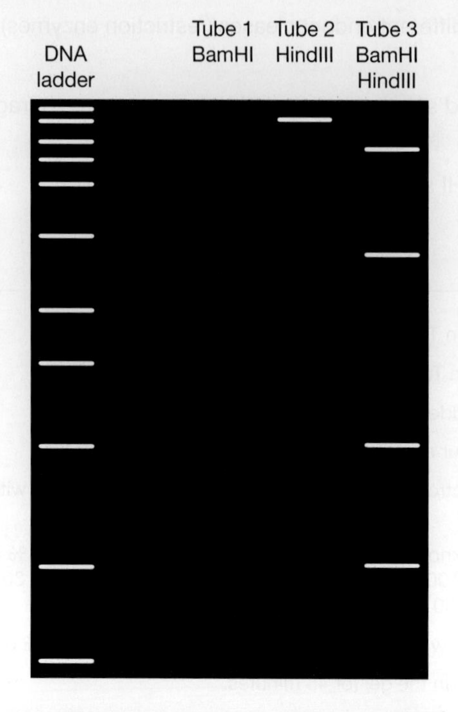

a. Identify **one** difference between the new results and the previous results, and suggest a possible reason for this difference. **2 marks**

b. State **two** factors that will have an impact on the rate of movement of the DNA fragments through the agarose gel. **2 marks**

c. Outline **two** safety guidelines that should have been followed by the student. **2 marks**

Question 14 (8 marks)

a. Describe the function of endonucleases in producing a recombinant plasmid. **1 mark**

b. The restriction enzyme *Bam*HI cuts DNA between the two Gs when it encounters the base sequence.

GGATCC
CCTAGG

Mark the recognition sites on the segment of DNA when the restriction enzyme *Bam*HI is used. **2 marks**

TACGGATCCTAGGGCATAGCTCAGGATCCCGTCAATGGGGATCCC
ATGCCTAGGATCCCGTATCGAGTCCTAGGGCAGTTACCCCTAGGG

c. Outline the role bacteria play in producing recombinant human insulin. **2 marks**

d. Antigens are proteins that can trigger an immune response. Vaccines contain these proteins in order for the human body to provide immunity from such pathogens, such as bacteria that harbour harmful antigens. Describe how bacteria could benefit humans in terms of antigen production. **3 marks**

Question 15 (10 marks)

Genetically modified organisms are commonly seen in society, in both agriculture and animal industries.

a. Define the term genetically modified organisms (GMO). **1 mark**

b. List four advantages farmers would desire in GMO plants. **2 marks**

c. Explain why the *Bacillus thuringiensis* (Bt) gene is added to corn genomes. **2 marks**

d. What are the core issues of concern for human health? **2 marks**

e. Describe the processes used to place a new gene into a corn plant. **3 marks**

2.8 Exercise 3: Biochallenge online only

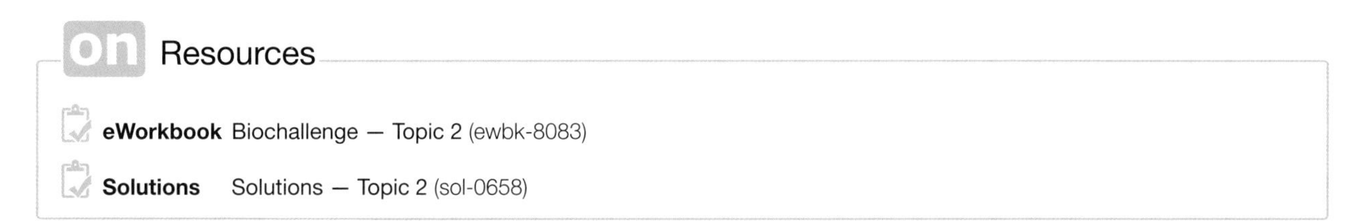

on Resources

eWorkbook Biochallenge — Topic 2 (ewbk-8083)

Solutions Solutions — Topic 2 (sol-0658)

Past VCAA examinations

Sit past VCAA examinations and receive immediate feedback, marking guides and examiner's report notes.
Access Course Content and select 'Past VCAA examinations' to sit the examinations online or offline.

teachon

Test maker

Create unique tests and exams from our extensive range of questions, including past VCAA questions.
Access the assignments section in learnON to begin creating and assigning assessments to students.

Online Resources

Below is a full list of **rich resources** available online for this topic. These resources are designed to bring ideas to life, to promote deep and lasting learning and to support the different learning needs of each individual.

eWorkbook

- **2.1** eWorkbook — Topic 2 (ewbk-1881) ☐
- **2.2** Worksheet 2.1 Endonucleases (ewbk-1868) ☐
- **2.3** Worksheet 2.2 Understanding CRISPR (ewbk-1870) ☐
- **2.4** Worksheet 2.3 Polymerase chain reaction (ewbk-1872) ☐
- **2.5** Worksheet 2.4 Case study analysis: DNA profiling in forensic investigations (ewbk-1875) ☐
  - Worksheet 2.5 DNA profiling in Australia (ewbk-1976) ☐
  - Worksheet 2.6 Analysing STR profiles (ewbk-1978) ☐
  - Worksheet 2.7 DNA profiling and gel electrophoresis (ewbk-1980) ☐
  - Worksheet 2.8 DNA profiling in animals (ewbk-1982) ☐
- **2.6** Worksheet 2.9 Recombinant plasmids and cloning genes (ewbk-1986) ☐
- **2.7** Worksheet 2.10 Case study analysis: Genetically modified animals (ewbk-1878) ☐
  - Worksheet 2.11 Manipulating genes in organisms (ewbk-1988) ☐
- **2.8** Worksheet 2.12 Reflection — Topic 2 (ewbk-4534) ☐
  - Biochallenge — Topic 2 (ewbk-8083) ☐

Solutions

- **2.8** Solutions — Topic 2 (sol-0658) ☐

Practical investigation eLogbook

- **2.1** Practical investigation eLogbook — Topic 2 (elog-0015) ☐
- **2.2** Investigation 2.1 Modelling restriction enzymes and ligase (elog-0017) ☐
- **2.5** Investigation 2.2 Using gel electrophoresis for DNA profiling (elog-0019) ☐
- **2.6** Investigation 2.3 Transformation using the pGlo plasmid (elog-0021) ☐
  - Investigation 2.4 Modelling the production of recombinant insulin (elog-0023) ☐
- **2.7** Investigation 2.5 Investigating genetically modified and non-genetically modified foods (elog-0025) ☐

Digital documents

- **2.1** Key science skills — VCE Biology Units 1–4 (doc-34326) ☐
  - Key terms glossary — Topic 2 (doc-34414) ☐
  - Key ideas summary — Topic 2 (doc-34417) ☐
- **2.5** Case study: DNA profiling in identifying remains (doc-36161) ☐
  - Issues in DNA profiling (doc-36162) ☐
- **2.6** Case study: The history of insulin (doc-36163) ☐
- **2.7** Case study: Other examples of genetically modified plants (doc-35872) ☐

Teacher-led videos

Exam questions — Topic 2

- **2.2** Sample problem 1 Examining fragments and endonucleases (tlvd-1095) ☐
- **2.3** Sample problem 2 Defining the role of CRISPR (tlvd-1096) ☐
- **2.5** Sample problem 3 Analysing gels (tlvd-1097) ☐
  - Sample problem 4 Analysing STRs (tlvd-1098) ☐
  - Sample problem 5 Analysing DNA profiles from a crime scene (tlvd-1099) ☐
- **2.6** Sample problem 6 Describing bacterial transformation (tlvd-1100) ☐

Video eLessons

- **2.2** Understanding restriction enzymes (eles-4132) ☐
- **2.4** DNA amplification (eles-4164) ☐
- **2.5** Gel electrophoresis (eles-4133) ☐
- **2.6** Gene cloning (eles-4134) ☐

Interactivities

- **2.5** DNA profiling (int-0669) ☐

Weblinks

- **2.2** Reverse transcription ☐
  - DNA restriction ☐
  - Restriction enzyme digest simulation ☐
  - How do restriction enzymes cut plasmids? ☐
  - The role of DNA ligase ☐
- **2.3** Gene editing in humans ☐
  - CRISPR ☐
  - Could RNA be the new editing tool?
- **2.4** PCR basics ☐
- **2.5** Solving a 28-year-old cold case ☐
  - Police cracking cold cases ☐
  - The Fromelles Project ☐
- **2.6** Transforming bacteria ☐
  - The history of cloning insulin ☐
- **2.7** Natural fluorescence in fish ☐
  - OGTR website ☐
  - Regulation of genetically modified crops in Australia ☐
  - Database of GM approvals worldwide ☐

Exam question booklet

- **2.1** Exam question booklet — Topic 2 (eqb-0013) ☐

Teacher resources

There are many resources available exclusively for teachers online.

To access these online resources, log on to **www.jacplus.com.au**

AREA OF STUDY 1 What is the role of nucleic acids and proteins in maintaining life?

OUTCOME 1

Analyse the relationship between nucleic acids and proteins, and evaluate how tools and techniques can be used and applied in the manipulation of DNA.

PRACTICE EXAMINATION

STRUCTURE OF PRACTICAL EXAMINATION		
Section	**Number of questions**	**Number of marks**
A	20	20
B	8	30
	Total	**50**

Duration: 50 minutes

Information:

- This practice examination consists of two parts. You must answer all question sections.
- Pens, pencils, highlighters, erasers and rulers are permitted.

SECTION A — Multiple choice questions

All correct answers are worth 1 mark each; an incorrect answer is worth 0.

1. Secretory proteins are synthesized by the ribosomes attached to endoplasmic reticulum. Endoplasmic reticulum consists of a
   A. system of vesicles.
   B. system of membrane bound channels.
   C. collection of ribosomes.
   D. collection of proteins carried on ribosomes.

2. Many proteins are produced that are transported out of a cell and used elsewhere in an organism. The cell organelle that is important in packaging proteins for external transport is the
   A. endoplasmic reticulum.
   B. ribosome.
   C. Golgi apparatus.
   D. nucleus.

3. Proteins are biomolecules that are made of smaller subunits or monomers. The monomers of proteins are
   A. glycerol.
   B. amino acids.
   C. nucleotides.
   D. nitrogenous bases.

4. A sequence of three nucleotide bases that are found in a DNA molecule and which codes for a particular amino acid is called a(n)
   A. codon.
   B. anticodon.
   C. triplet.
   D. sequence.

5. In the diagram shown, process 1 is

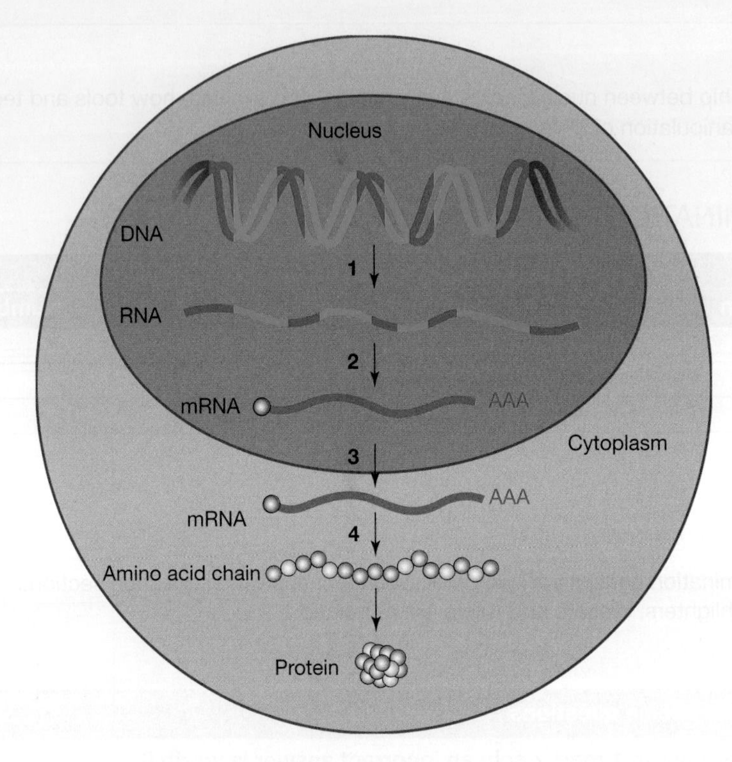

   A. exocytosis.
   B. RNA processing.
   C. transcription.
   D. translation.

6. Consider the template DNA sequence: AAT CGC TTT ATC. The mRNA transcript from this DNA is
   A. AAT CGC TTT ATC
   B. TTA GCG AAA TAG
   C. AAU CGC UUU AUC
   D. UUA GCG AAA UAG

7. In eukaryotes, RNA polymerase helps in the transcription of DNA into mRNA. For this process to take place, RNA polymerase must be able to bind to the
   A. operator.
   B. promoter.
   C. regulator.
   D. enhancer.

8. Geneticists studying the nematode worm *Caenorhabiditis elegans* found a gene that codes for a 22-nucleotide strand of 'microRNA'. The microRNA binds to a complementary sequence in a particular target mRNA in *C. elegans*, thus preventing the translation of the target mRNA into a protein.

The correct term to describe the type of gene that codes for the 'microRNA' in *C. elegans* is a

A. regulatory gene.
B. structural gene.
C. promoter sequence.
D. transcription unit.

9. The three dimensional shape of a protein can be seen in the

A. primary structure.
B. secondary structure.
C. tertiary structure.
D. quaternary structure.

10. The function of transfer RNA in the cell is to

A. carry information from the nucleus to the ribosome.
B. carry specific amino acids to the ribosomes.
C. complement the structure of the ribosomes.
D. relay information within the nucleus.

11. Each endonuclease cuts DNA at a specific sequence of bases called the

A. transcription unit.
B. operator sequence.
C. promoter sequence.
D. recognition sequence.

12. The technique of gel electrophoresis can be used to separate a mixture of DNA fragments. This technique separates various DNA fragments into discrete bands because the DNA fragments in a sample

A. differ in size.
B. carry different charges.
C. carry different number of positive charges.
D. are differentially repelled by the gel molecules.

13. How many cycles are needed to generate 64 copies of DNA in the polymerase chain reaction?
A. 3
B. 4
C. 5
D. 6

14. CRISPR refers to repeated sequences located in
A. bacterial DNA.
B. fungal DNA.
C. viral DNA.
D. viral RNA.

15. Cas9, a bacterial enzyme that is used in the CRISPR gene-editing technique. It is reasonable to state that
A. Cas9 is an endonuclease enzyme.
B. Cas9 is an exonuclease enzyme.
C. Cas9 is used to cut RNA strands at a precise location.
D. Cas9 joins DNA strands.

16. When a specific gene from an organism is cloned, this means that
A. the DNA profile of the organism concerned is identified.
B. the specific gene is sequenced.
C. the genome of the organism concerned is replicated.
D. multiple copies of the specific gene are made.

17. Assume that a particular kind of bacterial cell makes a total of 4 copies of its plasmids as it matures. In addition, this bacterial cell undergoes binary fission every 30 minutes. After two hours, the number of plasmids in the bacterial culture would be equal to
A. 4
B. 16
C. 64
D. 120

18. Genetic transformation of an organism (or cell) occurs when foreign DNA is incorporated into the DNA of that organism (or cell), as, for example, when the human insulin gene (*INS*) is introduced into *E. coli* bacterial cells. It may reasonably be concluded that
A. the human *INS* gene has been transformed.
B. genetic transformation must involve bacterial species.
C. the *E. coli* bacterial cells have been transformed.
D. genetic transformation is restricted to animal cells.

19. The image shows two mice born from the same litter at 24 weeks of age. The mouse on the right had the gene for human growth hormone (GH1) successfully inserted into its DNA when it was a newly fertilised egg.

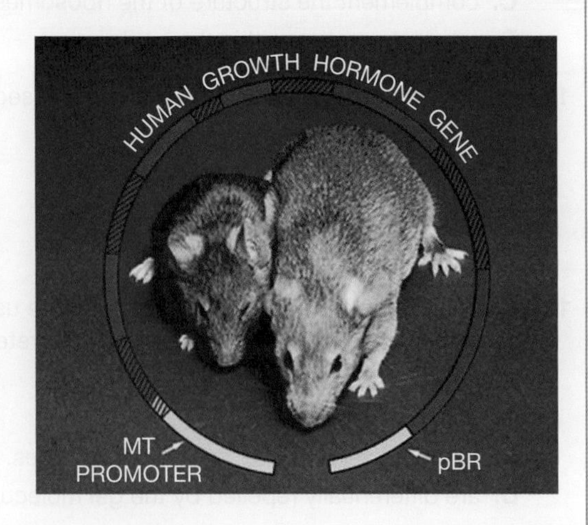

What term describes the mouse on the right?
A. Vector
B. Clone
C. Transgenic
D. Growth hormone 1

20. In gel electrophoresis
A. smaller fragments travel faster in the gel.
B. larger fragments travel faster in the gel.
C. positively charged particles travel faster than negatively charged particles.
D. neutral charged particles travel faster than negatively charged particles.

SECTION B — Short answer questions

Question 21 (4 marks)

The diagram illustrates a type of bulk transport.

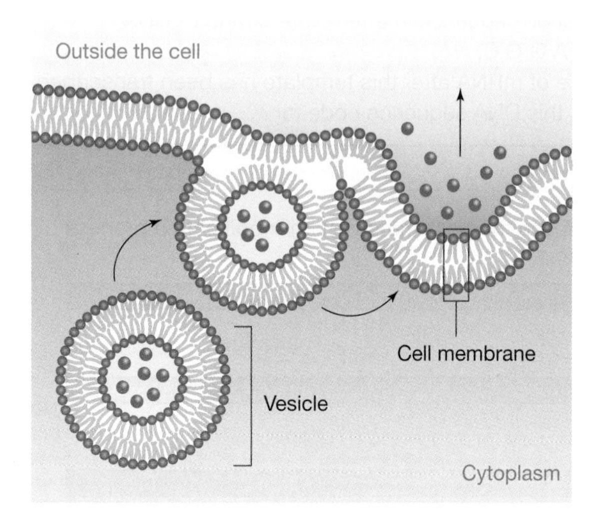

a. Name the process illustrated in the diagram. **1 mark**

b. Insulin is an example of a secretory protein that is exported from the cell in which it was manufactured. Name three different organelles directly associated with the transport of the insulin from the location where it is produced inside the pancreatic cell to the extracellular environment. State the role of each organelle in this process. **3 marks**

Question 22 (4 marks)

Two different proteins are shown in the diagrams. Part of a mouse protein is shown in figure 1 below. Two different types of secondary structures are shown.

The haemoglobin molecule comprises four chains: two alpha chains and two beta chains, shown in figure 2.

FIGURE 1

FIGURE 2

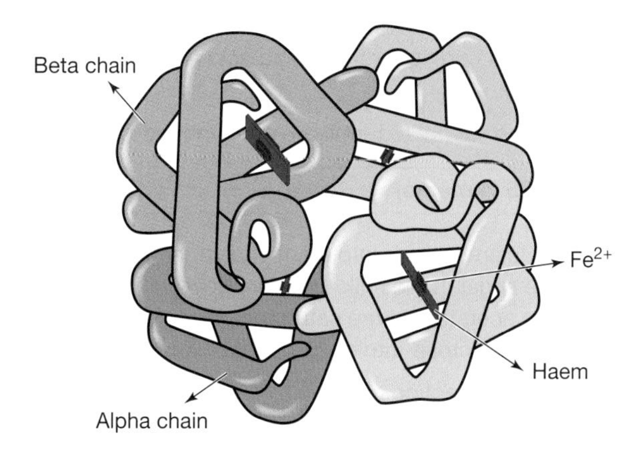

a. Name the monomers of the chains of the two proteins shown. **1 mark**

b. The haemoglobin molecule also display a quaternary structure. Referring to the diagram, explain what 'quaternary' means. **1 mark**

c. Give the names of the secondary structures shown in red and the secondary structure shown in green in figure 1. **1 mark**

d. Different proteins can have different primary structures. How can the primary structure vary? **1 mark**

Question 23 (4 marks)

a. Nucleic acids encode instructions for the synthesis of insulin in pancreatic cells. Outline the steps of insulin production that occur at the ribosome. **2 marks**

b. Consider the following nucleotide sequence in a template strand of DNA:
A A T G G C T A T A C C T T A G G C.
Write the nucleotide sequence of mRNA after this template has been transcribed. **1 mark**

c. How many amino acids does this DNA sequence code for? **1 mark**

Question 24 (5 marks)

The diagram summarises the steps in the post-transcription modification process of the *HBB* gene.

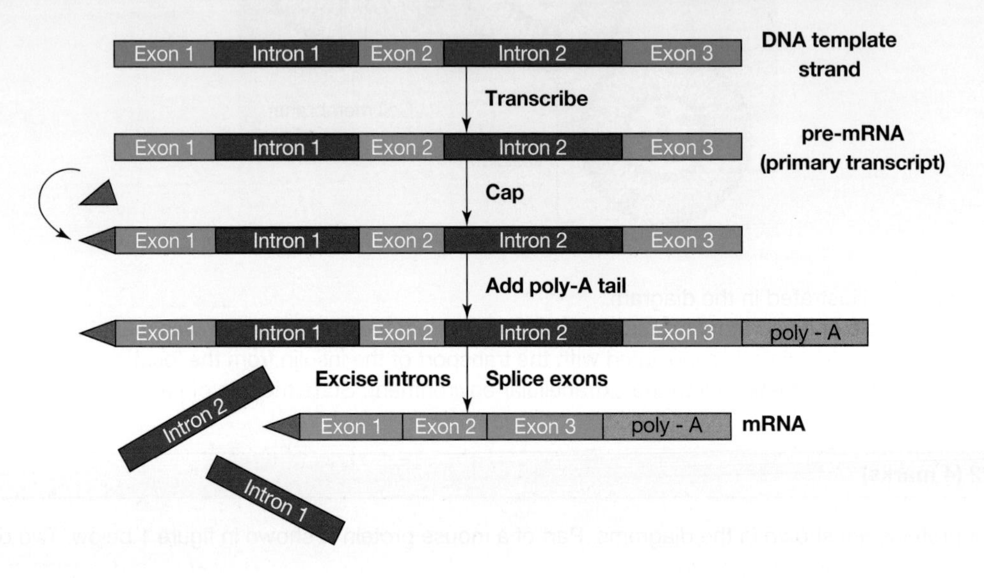

a. How many coding sections are present in the *HBB* gene? **1 mark**

b. How many non-coding sections of pre-mRNA are removed before the final mRNA leaves the nucleus? **1 mark**

c. A gene consists of its coding region, and its activity is regulated by regions upstream and downstream of the gene. In which of these regions would you expect to find:

i. a triplet coding for the amino acid Arg? **1 mark**

ii. a TATA box? **1 mark**

iii. a 'stop' signal to terminate a polypeptide chain? **1 mark**

Question 25 (4 marks)

Some adults are lactose intolerant and cannot drink ordinary cow's milk because they react adversely to the lactose sugar in the milk. Ordinary cow's milk can be passed down a special column filled with fibres of cellulase acetate that are saturated with an enzyme. After passing through the column, the milk can be consumed by lactose-intolerant people.

a. What is the function of this enzyme? **1 mark**

b. Enzymes are protein catalysts that have been folded in a particular way in order to bind to a specific substrate. Identify what organelle this folding would occur in. **1 mark**

c. In individuals who are not lactose intolerant, lactase is produced in the cells in the small intestine. Explain why lactase isn't produced in all cells and outline how this occurs. **2 marks**

Question 26 (4 marks)

a. Consider the endonuclease *AluI* obtained from *Arthrobacter luteus*. *AluI* cuts the DNA sequence at 5' AG*CT 3'. (*Note:* The symbol * denotes the position of the cut.)

  i. This cut produces blunt ends. Note down the other DNA strand (including the 3' and 5') and show the location of the cut. **1 mark**

  ii. *AluI* was added to a DNA segment with the sequence 5' AAATTTCCGAGCTGCGCTAA 3'. How many fragments of DNA would be produced? **1 mark**

b. The action of the cutting enzyme *PstI* is as follows:

```
C T G C A G     C T G C A | G
             →  ┌─────────┘
G A C G T C     G | A C G T C
```

  i. What kinds of ends does this enzyme produce? **1 mark**

  ii. Copies of one long fragment of DNA were treated with *PstI* and three smaller fragments were produced:

  - fragment 1: 2.3 kbp
  - fragment 2: 8.1 kbp
  - fragment 3: 4.4 kbp.

  (One kbp denotes one thousand (kilo) base pairs.)
  How long was the original DNA fragment? **1 mark**

Question 27 (3 marks)

One method of DNA amplification is the polymerase chain reaction (PCR), which uses a heat-resistant DNA polymerase enzyme (known as *Taq* polymerase) to make multiple copies of a DNA segment. Typically, PCR is a cyclic process that is repeated up to 28 times. The polymerase chain reaction is often described as a series of 'denature–anneal–extend' cycles.

a. Explain why the DNA polymerase enzyme used is heat resistant. **1 mark**

b. The polymerase chain reaction is often described as a series of 'denature–anneal–extend' cycles. Why is it necessary to denature the double-stranded DNA fragment that is to be amplified? **1 mark**

c. Describe what occurs during the 'anneal' phase and the 'extend' phase. **1 mark**

Question 28 (2 marks)

When making recombinant DNA, molecular biologists use the same restriction enzyme to cut DNA fragments from different sources. The fragments are then joined together.

a. Explain why the same restriction enzyme must be used to cut the fragments. **1 mark**

b. Bacterial plasmids can be used in the cloning of foreign genes. Outline an advantage of using plasmids for this purpose. **1 mark**

END OF EXAMINATION

PRACTICE SCHOOL-ASSESSED COURSEWORK

ASSESSMENT TASK — Analysis and evaluation of a contemporary bioethical issue

In this task you will analyse the relationship between nucleic acids and proteins, and evaluate how tools and techniques can be used and applied in the manipulation of DNA.

- Students are permitted to use pens, pencils, highlighters, erasers and rulers.
- Students should research their bioethical issue of focus BEFORE completing the task.
- You need to select one of the two bioethical issues as a point of focus, then answer the associated questions.

Total time: 50 minutes
Total marks: 40 marks
Select either Bioethical issue 1 or Bioethical issue 2 for this assessment task.

BIOETHICAL ISSUE 1: Using familial DNA to convict criminals

One of the most infamous criminals in the United States was known as the Golden State Killer. Between 1974 and 1986, they were connected with 13 murders and over 50 sexual assaults.

At the start of 2018, the identity of the Golden State Killer had not been discovered, with investigators having little evidence and little idea of who committed the crimes. They had DNA evidence from the suspected killer, but they were never able to match it to any criminals who were in their DNA database. How was the identity of the Golden State Killer determined?

In recent times, many individuals have used online sites to trace their genealogy and do DNA testing to find their risk for various diseases. Sites such as 23andMe and GEDmatch provide information on personal genomics.

In early 2018, detectives and FBI agents sent their DNA evidence into GEDmatch in the hope that they might catch the Golden State Killer. Unsurprisingly, there were no direct matches. However, around 20 distant relatives of the killer were tracked down, allowing for the eventual identification of the suspect. The Golden State Killer was found to be Joseph James DeAngelo, a former police office. He pleaded guilty in June 2020 and was sentenced in August 2020 to life imprisonment.

1. Describe the structure of DNA and draw a clear diagram showing a DNA monomer. **2 marks**
2. The type of DNA that assisted in determining the culprit of these crimes was short tandem repeats or STRs. These STRs are highly variable between individuals.
   a. Explain how STRs would have been used in this scenario and why they are useful. **2 marks**
   b. STRs are more commonly located in introns, rather than in exons. Explain the difference between introns and exons, and explain why STRs are more often found in introns rather than exons. **2 marks**
3. Identify and explain the purpose of PCR in solving this crime. **2 marks**
4. As part of determining the STRs present in an individual, it is important to use endonucleases before analysis. Why are endonucleases important in cases such as this one? **2 marks**
5. Describe how DNA from distant relatives would have assisted in solving the case. **2 marks**
6. The STRs across three sites in three family members are shown below. The values show the number of repeats within the STR. The first value is the number of repeats on the maternally inherited chromosome and the second value is the number of repeats on paternally inherited chromosome.

TABLE Repeats at three different STR locations

STR	Individual 1	Individual 2	Individual 3
STR 1	14, 12	6, 8	10, 10
STR 2	13, 11	8, 8	12, 13
STR 3	14, 19	7, 6	13, 9

a. The DNA of the individuals is run on a gel to separate the fragment by size. What is the name of this process? **1 mark**
b. Draw a clearly labelled gel showing the bands present for all individuals. **3 marks**
c. Explain how you determined the location of each of the bands. **1 mark**

7. Why did the investigators still need to collect DNA from Joseph James DeAngelo to ensure a conviction? **2 marks**
8. Many ethical factors need to be taken into consideration when solving crimes.
   a. Outline three bioethical considerations that need to be made when disclosing the results to the individuals. **3 marks**
   b. Two ethical concepts that need to be considered when judging bioethics are integrity (ensuring honesty in reporting and gathering information) and justice (ensuring there is no unfair burden on a particular group). Describe how each of these factors need be taken into consideration when conducting familial testing. **4 marks**
9. How long should the DNA of the family members used to help with the conviction of Joseph James DeAngelo be kept on file? Outline your thoughts on this in detail and justify your response. (You should ensure that your response provides at least five reasons to support your thoughts.) **6 marks**
10. Many individuals believe the DNA of all individuals should be kept in a DNA database. Other individuals are greatly opposed to this. Construct an argument with clear reasons and outline two pros and two cons about the use of DNA databases. **8 marks**

BIOETHICAL ISSUE 2: The right to choose — predictive testing

Genetic counsellors have important roles in supporting individuals with genetic diagnoses. There are many reasons an individual person might seek or be offered genetic testing. For example, someone might want to find out whether they possess a variant (allele) of a particular gene that causes a genetic disorder, significantly increases their susceptibility to a disease, or would allow them to benefit from a particular drug.

Types of genetic testing include the following:

- adult screening for increased risk of a disease — determining if an individual is at an increased risk for disease
- carrier detection — determining if an individual carries a defective allele
- prenatal screening and diagnosis — testing unborn babies for genetic disease
- predictive screening or presymptomatic testing — detecting mutations that cause disease later in life
- embryo biopsy or pre-implantation genetic screening — testing embryos for genetic disease before implantation and in vitro fertilisation (IVF).

One disease that can be tested for is Huntington's disease, which can be identified through presymptomatic testing. Children of a parent with Huntington's disease have a 50 : 50 chance of inheriting the defective *H* allele from that parent and are said to be 'at risk'. Typically, those children who inherit the *H* allele show the first signs of this disorder only in mid adulthood.

The *H* allele that causes Huntington's disease can be distinguished from the normal allele (*h*) by the presence in the coding region of the gene, typically, of 40 or more copies of a trinucleotide repeat, CAG.

In people who are *not* at risk of Huntington's disease, the number of repeats of this trinucleotide is in the range of 6 to 35.

The *HD* gene contains the instructions for the protein huntingtin. The presence of the extra trinucleotide repeats in the *H* allele means that this protein has additional *gln* amino acids in its structure. The abnormal huntingtin protein produces neurological symptoms, typically in middle age. This is an incurable neurological disease.

1. Outline how Huntington's disease can affect the primary, secondary and tertiary structure of the huntingtin protein. **3 marks**
2. Using your knowledge of CRISPR-Cas9, explain how this may be examined as a possible mechanism to treat Huntington's disease. Draw a clear diagram to assist in your response. **3 marks**
3. Why is PCR important in the process of diagnosing Huntington's disease? **2 marks**
4. Two parents are examined for the Huntington's allele. The mother, who has Huntington's disease, has one allele with 48 repeats and a second allele with 18 repeats. The father has one allele with 19 repeats and a second allele with 17 repeats.

a. The DNA of both parents is run on a gel to separate the fragments by size. What is the name of this process? **1 mark**
b. Draw a clear labelled gel showing the bands present for both parents. **2 marks**
c. Explain how you determined the location of each of the bands. **1 mark**

5. These two individuals have three children, none of whom show symptoms of the disease. All of the children were born before the mother was diagnosed. The children had their DNA tested for HD, and their results are shown below.

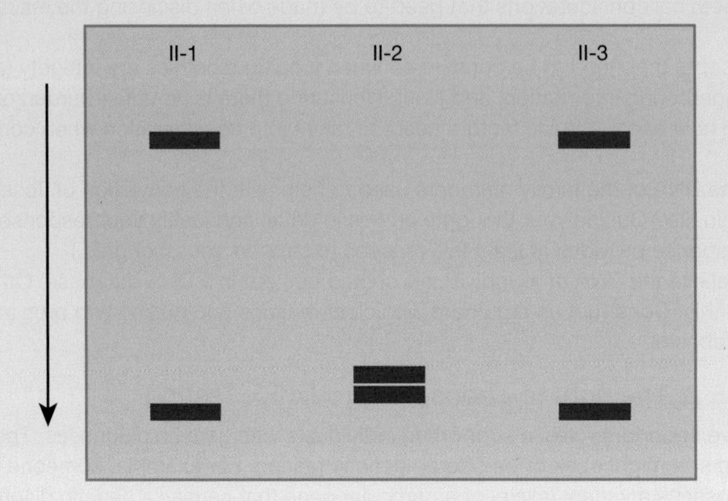

Which child or children will likely develop Huntington's disease? How do you know? **3 marks**

6. Of the three children, the oldest child (II-1) is 19 years old. The second child (II-2) is 14 years old. The youngest child (II-3) is only 5 years old. While testing has occurred, none of the children have found out their results.
   a. Outline three bioethical considerations that need to be made when disclosing the results with the individuals. **3 marks**
   b. Two ethical concepts that need to be considered when judging bioethics are respect (ensuring autonomy) and non-maleficence (avoiding causations of harm). Describe how each of these factors need to be taken into account in the testing of the children for Huntington's disease. **4 marks**
7. The second child chose to never find out if they would develop Huntington's disease. Describe two reasons that may have motivated them to make this decision. **2 marks**
8. When the second child has their own child, their child (at age 16) chooses to have testing for Huntington's disease. What implications may these results have for their parent? **2 marks**
9. With whom, if anyone, should a symptomless person whose test results show that she or he possesses the *H* allele share the test result: a fiancé, an employer, an insurance agent, or a living parent who does not show the HD symptoms? Outline your thoughts on this in details and justify your response. (You should ensure that your response provides at least five reasons to support your thoughts.) **6 marks**
10. Should there be an age limit for presymptomatic testing, or should it be up to the parents? Construct a detailed argument with clear reasons and outline two pros and two cons about having age limits for presymptomatic testing. **8 marks**

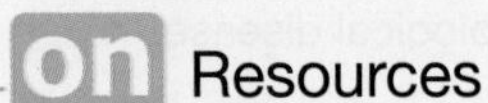

Resources

Digital document Unit 3 AOS 1 School-assessed coursework (doc-34406)

3 Regulation of biochemical pathways

KEY KNOWLEDGE

In this topic you will investigate:

Regulation of biochemical pathways

- the general structure of the biochemical pathways in photosynthesis and cellular respiration from initial reactant to final product
- the general role of enzymes and coenzymes in facilitating steps in photosynthesis and cellular respiration
- the general factors that impact on enzyme function in relation to photosynthesis and cellular respiration: changes in temperature, pH, concentration, competitive and non-competitive enzyme inhibitors.

PRACTICAL WORK AND INVESTIGATIONS

Practical work is a central component of learning and assessment. Experiments and investigations, supported by a **practical investigation eLogbook** and **teacher-led videos**, are included in this topic to provide opportunities to undertake investigations and communicate findings.

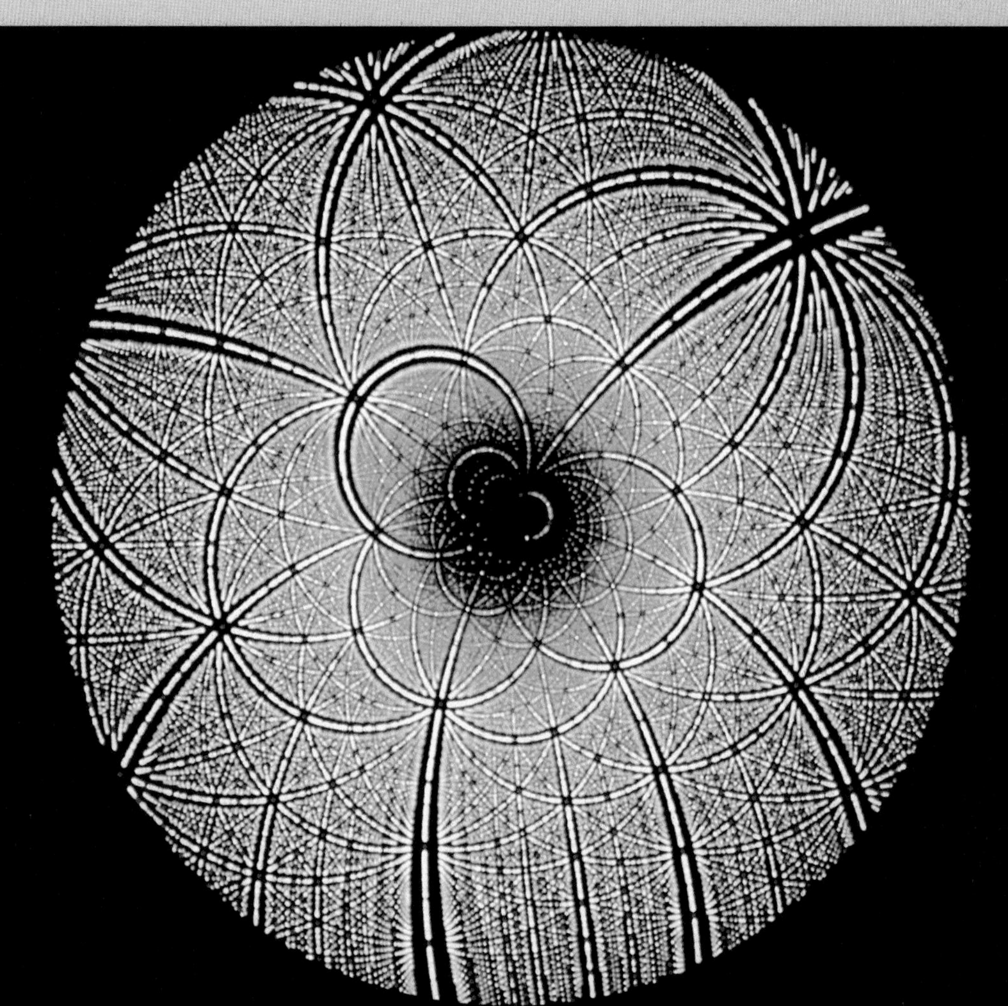

3.1 Overview

Numerous **videos** and **interactivities** are available just where you need them, at the point of learning, in your digital formats, learnON and eBookPLUS at **www.jacplus.com.au**.

3.1.1 Introduction

The bloodspot screening program shown in figure 3.1 is a familiar one around Australia; it is conducted on nearly every baby within two to three days after its birth. Here, a tiny sample of blood has been taken by pricking the baby's heel. Each bloodspot is screened for the activity of several key enzymes. Serious medical disorders can result if the activity of any of these enzymes is missing or is greatly reduced.

FIGURE 3.1 A blood sample from this baby (taken from the heel) will be screened for various rare but serious disorders.

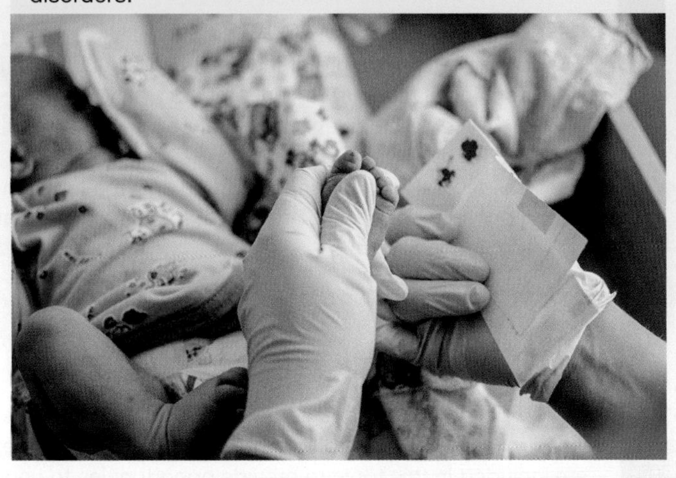

One enzyme that is screened for in newborns is the enzyme phenylalanine hydroxylase (PAH). A deficiency of this enzyme causes phenylketonuria (PKU), a disorder in which severe and irreversible damage to the brain occurs, resulting in intellectual disability. Symptoms of this disease can be prevented through early dietary modification.

Enzymes are not just important for humans, but for all organisms. The opening image for this topic shows the diffraction pattern of a crystal of the enzyme Rubisco from a plant. This enzyme is thought to be the most abundant protein in nature and is vital for photosynthesis in plants. Without this enzyme, life on Earth would unlikely exist. In this topic you will explore enzymes and learn how enzymes are critical for cells to carry out a network of interconnected biochemical pathways that maintain the living state.

LEARNING SEQUENCE

on Resources

eWorkbook	eWorkbook — Topic 3 (ewbk-1882)
Practical investigation eLogbook	Practical investigation eLogbook — Topic 3 (elog-0183)
Digital documents	Key science skills — VCE Biology Units 1–4 (doc-34326) Key terms glossary — Topic 3 (doc-34617) Key ideas summary — Topic 3 (doc-34608)
Exam question booklet	Exam question booklet — Topic 3 (eqb-0014)

3.2 Biochemical pathways

KEY KNOWLEDGE

- The general structure of the biochemical pathways in photosynthesis and cellular respiration from initial reactant to final product

Source: VCE Biology Study Design (2022–2026) extracts © VCAA; reproduced by permission.

3.2.1 What are biochemical pathways?

Cells of living organisms are also constantly carrying out chemical reactions. These reactions are referred to as **biochemical reactions**. At every moment of life, the cells of your tissues and organs are carrying out a multitude of biochemical reactions in response to stimuli received from both outside and from within your body. These biochemical reactions are not isolated from each other, but instead are interconnected or linked to form multi-step **biochemical pathways**. The total activity of the reactions of all biochemical pathways in a living organism is its **metabolism**.

- A biochemical pathway is a series of linked biochemical reactions that start with an initial reactant that is converted in a stepwise fashion to a final product.
- The product of the one reaction becomes the starting reactant for the next step, until the final product is reached.
- Each step in a biochemical pathway requires the activity of a specific enzyme.

Figure 3.2 shows a highly simplified representation of two biochemical pathways in which letters denote molecules and E1–E6 denote enzymes. Every step is catalysed by specific enzymes.

FIGURE 3.2 A simple representation of two biochemical pathways. E1 through E6 denote different enzymes.

Pathway 1

$$A \xrightarrow{E1} B \xrightarrow{E2} C \xrightarrow{E3} D$$

Pathway 2

$$J \xrightarrow{E4} K \xrightarrow{E5} L \xrightarrow{E6} M + N$$

biochemical reactions reactions occurring in cells that lead to the formation of a product from a reactant

biochemical pathways a series of linked biochemical reactions

metabolism total of all chemical reactions occurring in an organism

enzyme a protein that acts as a biological catalyst, speeding up reactions without being used up

Table 3.1 identifies some major biochemical pathways and a key **enzyme** involved. Each enzyme identified catalyses one particular reaction in the pathway. The photosynthesis pathway occurs in green plants and algae, but cellular respiration and nucleic acid processes occur in both plants and animals.

TABLE 3.1 Examples of biochemical pathways and some of the enzymes that play a role in them

Biochemical pathway	Enzyme
Cellular respiration	
Glycolysis	Phosphofructokinase (PFK)
Krebs cycle	Malate dehydrogenase
Electron transport chain	Cytochrome *c* oxidase
Nucleic acid pathways	
DNA replication	DNA polymerase
Transcription of mRNA	RNA polymerase
mRNA translation to protein	Aminoacyl tRNA synthetase

(continued)

TABLE 3.1 Examples of biochemical pathways and some of the enzymes that play a role in them *(continued)*

Biochemical pathway	Enzyme
Photosynthesis	
Light-dependent stage	Water-splitting enzyme (Photosystem II)
Light-independent stage	Rubisco*

Note: *Rubisco = **ribul**ose-1, 5-**bis**phosphate **c**arboxylase/**o**xygenase

The number of steps in a biochemical pathway can vary from just a few to a large number. For example:

- The first stage of cellular respiration (glycolysis) is a 10-step biochemical pathway, in which one molecule of glucose (with six C atoms) is broken down to two molecules of pyruvate (with three C atoms).
- The second stage of photosynthesis (light-independent stage) is a 9-step biochemical circular pathway in which glucose is built from carbon dioxide. A different enzyme is required for each step in these pathways.

From initial reactant to final product

In reactions, **reactants** (the starting molecules) are chemically changed to form **products**. In biochemical pathways the product formed by one step of the pathway becomes the reactant for the next step of the pathway and so on. Since each reaction involves the action of a specific enzyme, reactant molecules in each step can also be called the **substrates** for their particular enzymes.

Refer back to figure 3.2. In pathway 1, the initial reactant is molecule A and is the substrate for enzyme E1. This leads to a product, molecule B. Molecule B then acts as the reactant to produce molecule C. The final product of the biochemical pathway is molecule D.

reactant a substance that is changed during a chemical reaction

product the compound that is produced in a reaction

substrate a compound on which an enzyme acts

anabolic chemical reaction in a cell in which complex molecules are built from simple molecules

endergonic a chemical reaction that is energy-requiring

3.2.2 Types of biochemical pathways

Biochemical pathways do not just differ in their substrates, enzymes and products. Pathways catalysed by enzymes can be identified in terms of whether they build complex molecules from smaller ones and require energy or whether they break down complex molecules into smaller ones and release energy.

BACKGROUND KNOWLEDGE: Biochemical pathways — build up or break down?

Anabolic pathways

Anabolic pathways assemble simple molecules into more complex molecules. Building complex molecules from simple ones requires a net input of energy, so anabolic pathways are energy-requiring or **endergonic**.

Photosynthesis is an example of an anabolic pathway, where glucose molecules are synthesised from carbon dioxide and water using radiant energy from the Sun (figure 3.3). Other examples are the pathways that assemble amino acids into proteins, build macromolecules of glycogen from glucose, build starch from sugars and build DNA from nucleotides.

FIGURE 3.3 Simplified representation of photosynthesis showing only the initial reactants (at left) and the end products (at right). This is an energy-requiring anabolic pathway.

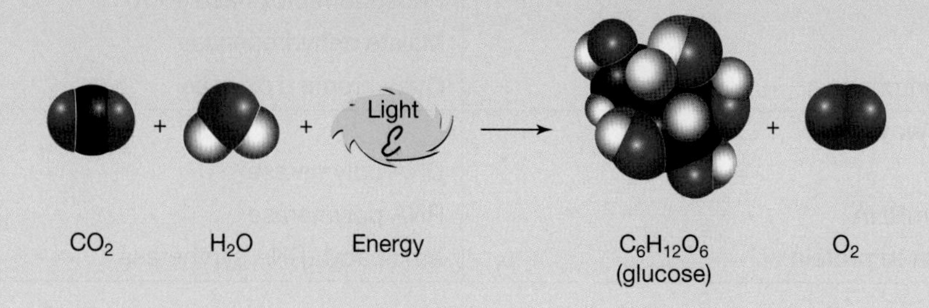

Catabolic pathways

Catabolic pathways break down complex molecules into more simple molecules. Catabolic pathways are energy-releasing or **exergonic**.

Aerobic cellular respiration is an example of a catabolic pathway, where glucose molecules are broken down into carbon dioxide and water molecules in (figure 3.4). Another example is the breakdown of fatty acids to carbon dioxide.

FIGURE 3.4 Simplified representation of cellular respiration showing only the initial reactants (at left) and the end products (at right). Can you identify the colour of the hydrogen atoms?

Comparing anabolic and catabolic reactions

Catabolic reactions produce a net release of energy as the energy level of the initial reactants is higher than that of the final products (figure 3.5a). Anabolic reactions require energy, so the energy level of the initial reactants is lower than that of the final products (figure 3.5b).

FIGURE 3.5 Energy change in **a.** anabolic (endergonic) reactions **b.** catabolic (exergonic) reactions

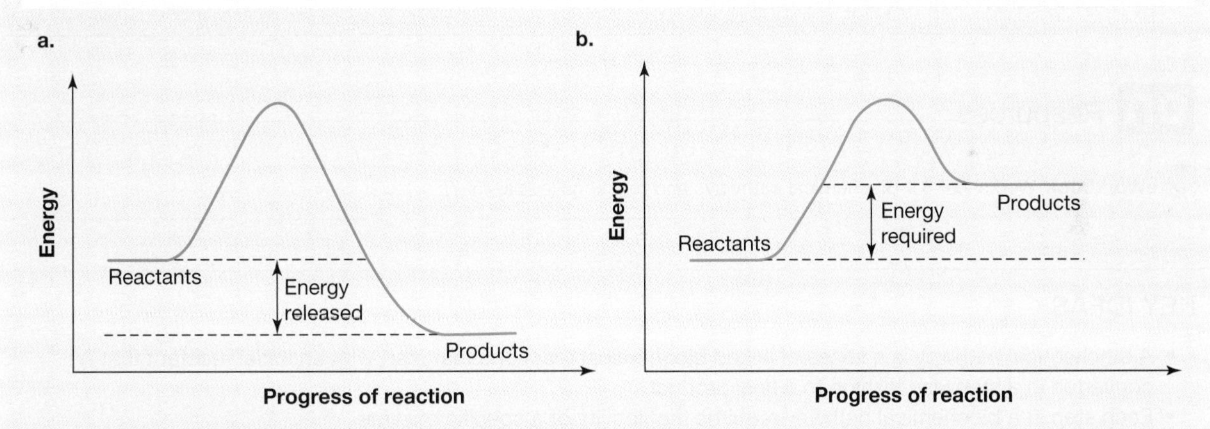

In cells, catabolic and anabolic reactions are constantly operating, but do not operate in isolation. Instead, reactions are coupled so that the energy released by catabolic reactions can be transferred to drive energy-requiring anabolic pathways. Note that the energy released by a catabolic pathway is transferred as chemical energy in ATP and often used to drive the anabolic pathway.

TIP: A good way to remember the differences between endergonic and exergonic is to think that **ex**ergonic reactions have energy **ex**it the system and **en**dergonic reactions have energy **en**ter the system.

catabolic chemical reaction in a cell in which complex molecules are broken down into simple molecules

exergonic a chemical reaction that is energy-releasing

tlvd-1764

SAMPLE PROBLEM 1 Comparing and contrasting exergonic and endergonic reactions

Compare and contrast exergonic and endergonic reactions. **(2 marks)**

THINK	WRITE
1. Identify what the question is asking you to do. This question asks you to both compare (find the similarities) and contrast (find the differences).	
2. Compare the two reactions.	Both exergonic and endergonic reactions are involved in metabolism and catalysed by enzymes. An initial activation energy is required in order for these reactions to occur (1 mark).
3. Contrast the two reactions. When you contrast, ensure that you address BOTH reactions, not just one.	Exergonic reactions are catabolic and involve the breakdown of a complex molecule to release energy. Endergonic reactions, on the other hand, are anabolic, and involve building up a complex molecule from simple subunits with the input of energy (1 mark).

elog-0818

INVESTIGATION 3.1

online only

Endergonic and exergonic reactions

Aim

To compare endergonic and exergonic reactions

Resources

eWorkbook Worksheet 3.1 Biochemical pathways and reactions (ewbk-7527)

KEY IDEAS

- A biochemical pathway is a series of linked biochemical reactions that start with an initial reactant that is converted in a stepwise fashion to a final product.
- Each step in a biochemical pathway requires the activity of a specific enzyme.
- The total activity of the reactions of all biochemical pathways in a living organism is its metabolism.
- Anabolic pathways assemble simple molecules into more complex molecules.
- Anabolic pathways are endergonic (require energy).
- Photosynthesis is an anabolic pathway.
- Catabolic pathways break down complex molecules into more simple molecules.
- Catabolic pathways are exergonic (release energy).
- Cellular respiration is an exergonic pathway.

3.2 Activities

To answer questions online and to receive **immediate feedback** and **sample responses** for every question, go to your learnON title at **www.jacplus.com.au**. A **downloadable solutions** file is also available in the resources tab.

3.2 Quick quiz on	3.2 Exercise	3.2 Exam questions

3.2 Exercise

1. Give an example of a biochemical pathway (and the enzyme involved) that
   a. occurs in green plants and algae only
   b. is involved in cellular respiration
   c. requires an overall input of energy to proceed
   d. involves the transformation of the radiant energy of sunlight to the chemical energy of sugars.
2. What is a key difference between the members of the following pairs?
   a. Anabolism and catabolism
   b. Endergonic and exergonic reactions
   c. An enzyme-catalysed reaction and an uncatalysed reaction
   d. A biochemical pathway and a biochemical reaction
3. Identify the following statements as true or false and, if false, rewrite it as a true statement.
   a. All enzymes have a similar reaction rate.
   b. Exergonic reactions include photosynthesis.
   c. The reactant is the final molecule produced in a biochemical pathway.
   d. Some enzymes are only active in specific cells.
4. Briefly explain in your own words the following observation:
   'An anabolic reaction can only proceed if it is coupled to a catabolic reaction.'
5. Photosynthesis uses energy from the Sun to allow for the production of glucose. Cellular respiration breaks down glucose to release energy in the form of ATP. Identify and explain which process is exergonic and which is endergonic.

3.2 Exam questions

Question 1 (1 mark)

Source: *VCAA 2017 Biology Exam, Section A, Q5*

MC The biochemical pathway of glycolysis involves nine intermediate reaction steps.

One of these steps is represented in the diagram below.

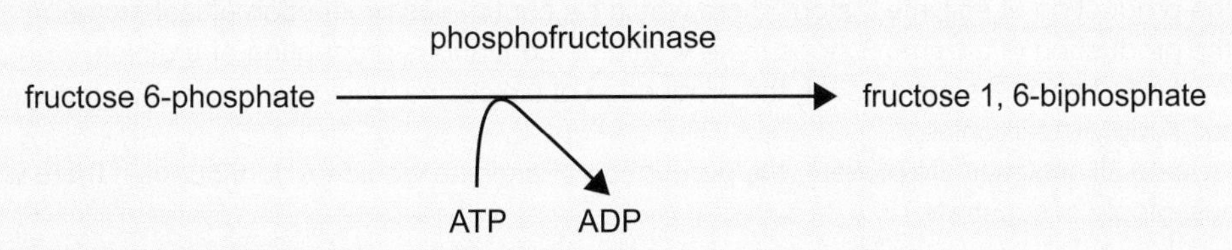

It is correct to state that, in this reaction, phosphofructokinase

A. acts as a coenzyme.
B. increases the rate of reaction.
C. is the substrate for the reaction.
D. releases energy in the form of ADP.

Question 2 (1 mark)

Source: *VCAA 2015 Biology Exam, Section A, Q6*

MC An experiment was conducted to investigate enzyme activity. A small quantity of amylase solution was added to a solution of starch dissolved in water at 35 °C. It was observed that maltose was produced.

Which one of the following is the substrate in this reaction?

A. water
B. starch
C. maltose
D. amylase

Question 3 (1 mark)

Source: *VCAA 2008 Biology Exam 1, Section A, Q18*

MC Sucrose (cane sugar) is a disaccharide used by plants as a transport molecule. Sucrose is formed in the following reaction

$$\text{glucose + fructose} \xrightarrow{\text{enzyme}} \text{sucrose}$$

With reference to this process it can be stated that

A. glucose and fructose are polysaccharides.
B. the production of sucrose is an endergonic reaction.
C. sucrose is a reactant and glucose is a product of the reaction.
D. a molecule of fructose contains more stored energy than a molecule of sucrose.

Question 4 (1 mark)

Source: *VCAA 2006 Biology Exam 1, Section A, Q25*

MC In the production of isoleucine from threonine in bacteria (Biochemical Pathway 1 [**BP 1**]), the end product acts as an inhibitor of the first enzyme in the pathway. In the production of arginine (Biochemical Pathway 2 [**BP 2**]), the end product has no influence on other enzymes in the pathway.

Biochemical Pathway 1 (BP 1)

inhibits enzyme 1

threonine → (enzyme 1) compound X → (enzyme 2) compound Y → (enzyme 3) isoleucine

Biochemical Pathway 2 (BP 2)

substrate → (enzyme 1) compound M → (enzyme 2) compound N → (enzyme 3) arginine

It is reasonable to conclude that in

A. BP 1, if the production of enzyme 3 stops there would be continuous production of isoleucine.
B. BP 2, if the production of enzyme 3 stops there would be continuous production of arginine.
C. BP 1, providing all enzymes are present, the production of isoleucine would be continuous if there was a continuous supply of threonine.
D. BP 2, providing all enzymes are present, the production of arginine would be continuous if there was a continuous supply of substrate.

Question 5 (1 mark)

Source: *Adapted from VCAA 2007 Biology Exam 1, Section B, Q3a*

Many living cells produce hydrogen peroxide as a by-product of some metabolic reactions. Hydrogen peroxide is a poisonous substance for these cells and is immediately decomposed into water and oxygen by an enzyme called catalase.

The reaction is represented by the equation

$$2H_2O_2 \xrightarrow{\text{catalase}} 2H_2O_2 + O_2$$

Which is the substrate in this chemical reaction?

More exam questions are available in your learnON title.

3.3 The role of enzymes in biochemical pathways

KEY KNOWLEDGE

- The general role of enzymes in facilitating steps in photosynthesis and cellular respiration

3.3.1 The role of enzymes

At any given moment within a cell, many biochemical pathways are proceeding — molecules are being oxidised or reduced, chemical bonds in molecules are being broken or formed, atoms are being rearranged in molecules, molecules are being joined or split apart. Each step or reaction in a biochemical pathway involves enzyme action.

Enzymes are composed of proteins and their key role in cells is to act as catalysts of the reactions in biochemical pathways.

Catalysts speed up the rate of a reaction. Each different kind of reaction is catalysed by a specific enzyme. Enzymes speed up the rates at which the products of reactions are formed by lowering the **activation energy** needed for reactions. Enzymes are *not* themselves used up or changed in reactions, so enzymes at the end of a reaction can be reused for another reaction.

Enzymes and activation energy

All chemical reactions require an input of energy to get started. For example, a match can burn but it will not do so spontaneously. However, when struck, the match head heats up and provides the input of energy to start the match burning (the activation energy).

In the presence of an enzyme, the activation energy needed to start a reaction is much lower than that for an uncatalysed reaction (see figure 3.6). Since less energy is required for an enzyme-catalysed reaction to get started, more substrate molecules and enzymes will have sufficient energy to react when they collide. So, enzymes lower the energy level that is needed to activate reactant molecules. This lowering of activation energy occurs in both catabolic reactions (shown in figure 3.6) and anabolic reactions (refer back to figure 3.5a).

catalyst a factor that causes an increase in the rate of a reaction
activation energy minimum amount of energy required to initiate a chemical reaction

FIGURE 3.6 The different 'routes' taken by an enzyme-catalysed reaction (pink line) as compared with the non-catalysed reaction (blue line) in a catabolic reaction. Enzyme-catalysed reactions have a lower activation energy than the corresponding reaction in the absence of enzymes. This also occurs in anabolic reactions.

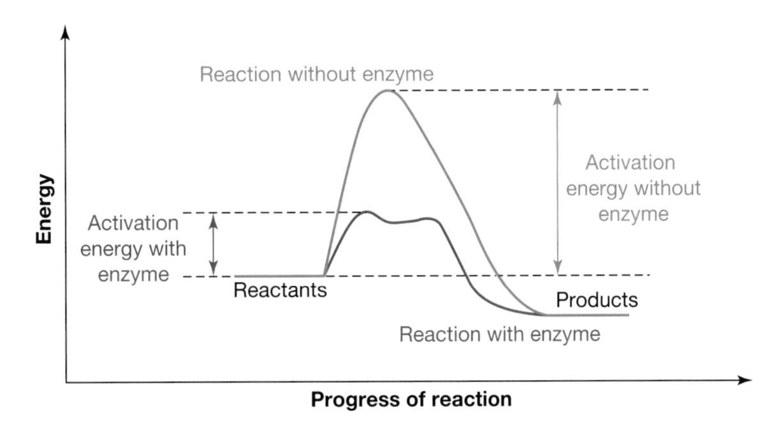

Enzymes change the rate of biochemical reactions

Biochemical reactions catalysed by enzymes occur at rates that can be thousands of times faster than the same reactions in the absence of enzymes. Action rates of enzymes vary over a wide range, for example:

- the cellular respiration enzyme, triose phosphate isomerase, can act on more than 10 000 substrate molecules per second
- Rubisco, a key enzyme of the photosynthesis pathway, is a relatively slow enzyme, acting on just 3 to 10 substrate molecules per second. Green plant cells compensate for this by having a very high concentration of this enzyme in their chloroplasts.

At the prevailing temperature and pH of the internal cellular environment, biochemical reactions occur rapidly in the presence of enzymes. However, in the absence of enzymes, these reactions would proceed at rates so incredibly slow that years may pass before the product appeared. One estimate is that, at room temperature, in the absence of an endopeptidase enzyme, the reaction to break a peptide bond of a protein spontaneously would take 400 years. Clearly, this situation could not support the living state.

elog-0820

INVESTIGATION 3.2

online only

Enzymes in plants

Aim

To investigate the action of enzymes within fresh pineapple on jelly

Resources

Video eLesson Enzymes as catalysts (eles-4191)

tlvd-1765

SAMPLE PROBLEM 2 Analysing the effect of enzymes on biochemical pathways

When alcohol (ethanol) is consumed, enzymes break down ethanol into less harmful substances. These enzymes are found in the liver. This is a similar situation in organisms that produce ethanol by anaerobic fermentation.

Ethanol is converted into aldetaldehyde by the alcohol dehydrogenase enzyme (ADH). Another enzyme, acetaldehyde dehydrogenase 2 (ALDH2), breaks down acetaldehyde into acetate (or acetic acid), a non-toxic substance. Both ethanol and aldetaldehyde are toxic so must be broken down. The accumulation of ethanol leads to cellular death and the accumulation of acetaldehyde can cause skin to flush and the release of histamines, leading to inflammation.

This is simplified in the following pathway:

$$\text{Ethanol} \xrightarrow{\text{ADH}} \text{Acetaldehyde} \xrightarrow{\text{ALDH2}} \text{Acetate}$$

a. **Some individuals have alcohol intolerance, in which drinking small amounts of alcohol leads to discomfort, skin redness, swelling and rashes. Explain, with reference to the metabolic pathway, which enzyme would not be functioning in an individual for them to suffer from these symptoms.** **(2 marks)**

Two drugs, formepizole and disulfirum act as inhibitors of the above pathway. Inhibitors prevent the action of enzymes. Formepizole inhibits ADH and disulfirum inhibits ALDH-2.

b. **Suggest one symptom that could be caused by formepizole if an individual consumed this alongside alcohol.** **(1 mark)**

c. **One species of yeast, *Sacchharomyces cerevisiae,* also uses the above pathway to break down ethanol into acetate. What would be the effect of exposing *S. cerevisiae* to disulfirum?** **(2 marks)**

THINK	WRITE
a. 1. Identify what the question is asking you to do. In an *explain* question you need to account for the reasons something occurs. The question also clearly states that you need to refer to the metabolic pathways.	
2. Look at the information in the question and stimulus material for what causes redness, swelling and rashes, and link this back to the metabolic pathway.	From the symptoms of skin redness, swelling and rashes, there is clearly a buildup in acetaldehyde (1 mark). As ALDH-2 converts acetaldehyde to acetate, the lack of a functioning form of this enzyme would result in a build up of acetaldehyde (1 mark). Therefore, those with alcohol intolerance would lack this enzyme.
b. In a *suggest* question, you need to be able to apply a solution to what may be an unfamiliar concept. The inactivity of formepizole will prevent ethanol being converted into acetaldehyde.	Formepizole, when consumed with alcohol, would lead to the buildup in ethanol, which may lead to organ damage and death (1 mark).
c. 1. Identify what the question is asking you to do. In this case, you need to explore the effect of a change in the metabolic pathway. This question is worth two marks, so two clear points should be addressed.	
2. Explore the process that would be affected by the inhibition of ALDH-2 — the conversion of acetaldehyde to acetate. Use this to formulate your response.	Disulfirum inhibits acetaldehyde dehydrogenase. The ethanol produced from fermentation would be able to be broken down in acetaldehyde, but this would not be converted to acetate. Therefore, there would be an excess in acetaldehyde (1 mark) and a deficiency in acetate (1 mark) in the yeast.

3.3.2 The features of enzymes

Enzymes have many features that allow them to act as biological catalysts and speed up biochemical reactions. Their ability to speed up reactions is vital for the functioning and survival of organisms.

BACKGROUND KNOWLEDGE: Reviewing the features of enzymes

Enzymes are proteins

Apart from a few catalytic RNA molecules (ribozymes), all the enzymes present in living cells are soluble globular proteins. Enzymes typically can be recognised by the suffix -ase in their recommended names: for example, hexokinase, isomerase and succinate dehydrogenase. The name of an enzyme can often give a clue to its activity; for example, a dehydrogenase enzyme removes hydrogen atoms from substrate molecules.

Enzymes consist of one or more polypeptide chains

A minority of enzymes consist of a single polypeptide chain formed of amino acids joined by peptide bonds. The number of amino acid residues in the single polypeptide chain of an enzyme ranges from 100 to 300. Single-chain enzymes include trypsin (in the gastric juices of the stomach), ribonuclease and lysozyme (figure 3.7 which represents amino acids as different colours). Lysozyme (also known are muramidase) has antimicrobial properties and is present in saliva, mucus and tears, as well as in some cells of the immune system.

Most enzymes consist of two or more polypeptide chain sub-units. One example of a multi-chain enzyme is phosphofructokinase (PFK), an enzyme that is active in the glycolysis stage of cellular respiration.

▶

PFK has four identical sub-units, which are distinguished by four different colours in figure 3.8. PFK is involved in regulation of the cellular respiration pathway.

FIGURE 3.7 Ribbon model of lysozyme (muramidase), a single-chain enzyme, showing the overall globular shape of this enzyme

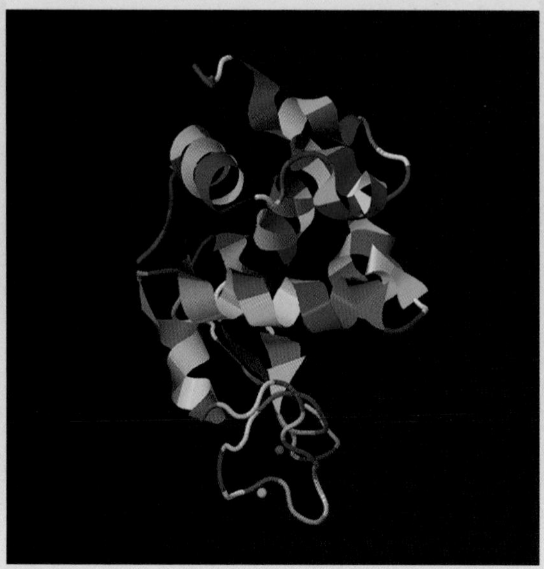

Source: Tsujino, S., Tomizaki, T., RCSB Protein Data Bank

FIGURE 3.8 Ribbon model of the enzyme phosphofructokinase (PFK) showing its four identical sub-units

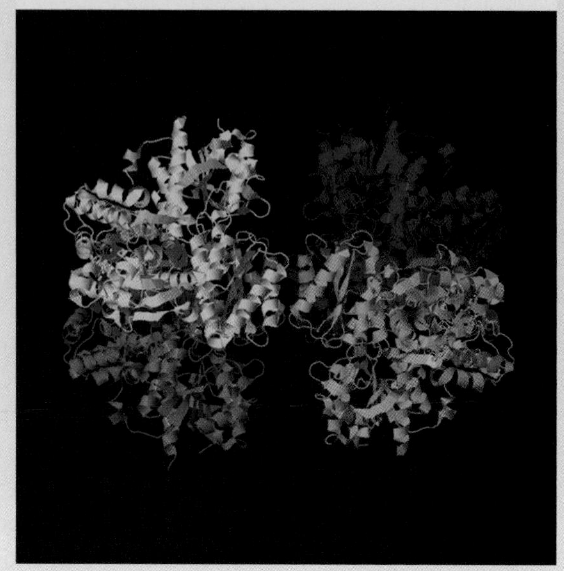

Source: Kloos, M., Straeter, N., RCSB Protein Data Bank

Every enzyme has an active site

Typically, the **active site** of an enzyme is a small part of its structure that has a unique 3D shape. The shape is complementary to that of its specific substrate molecule. The active site is where a substrate (S) binds to the enzyme at the enzyme binding site and forms a temporary **enzyme–substrate complex** (E–S), and it is here where the catalytic action of an enzyme occurs (figure 3.9).

active site region of an enzyme that binds temporarily with the specific substrate of the enzyme

enzyme–substrate complex transient compound produced by the bonding of an enzyme with its specific substrate

At an active site, substrate molecules can be split apart into smaller molecules, they can be rearranged or they can be assembled into more complex molecules. Once the product of the enzyme action is formed and released, the unchanged enzyme is now free to act on other substrate molecules.

Figure 3.10 shows a highly simplified outline of enzyme action. The active site of an enzyme and its substrate do not fit together tightly in a rigid lock-and-key arrangement. Instead, the shape of the active site of an enzyme is not rigid and it can adjust to the shape of its substrate — in some cases, some enzymes even shift their shapes to enclose their substrate; for example, the Rubisco enzyme forms a series of loops that close over its substrate to enfold and capture it.

This is known as the induced fit model of enzyme activity and it has replaced the old lock-and-key model of enzyme action.

FIGURE 3.9 Representation of a substrate (S; in black) at the active site, which includes an enzyme binding site (blue) and the catalytic site (red) where the reaction occurs

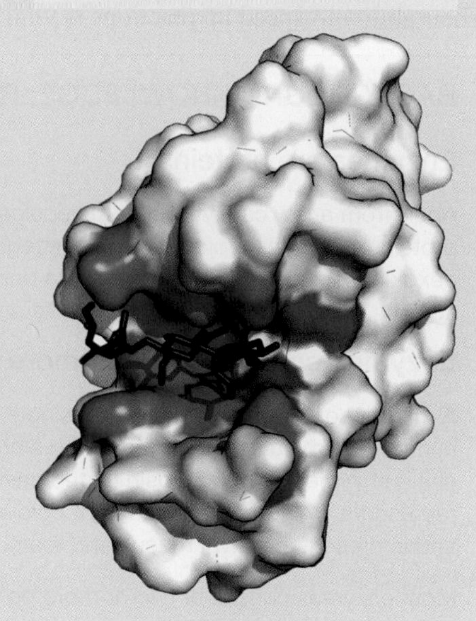

FIGURE 3.10 Simplified representation of the stages in an enzyme-catalysed catabolic reaction through induced fit. Note that the enzyme shifts the shape of the active site slightly to enclose the substrate.

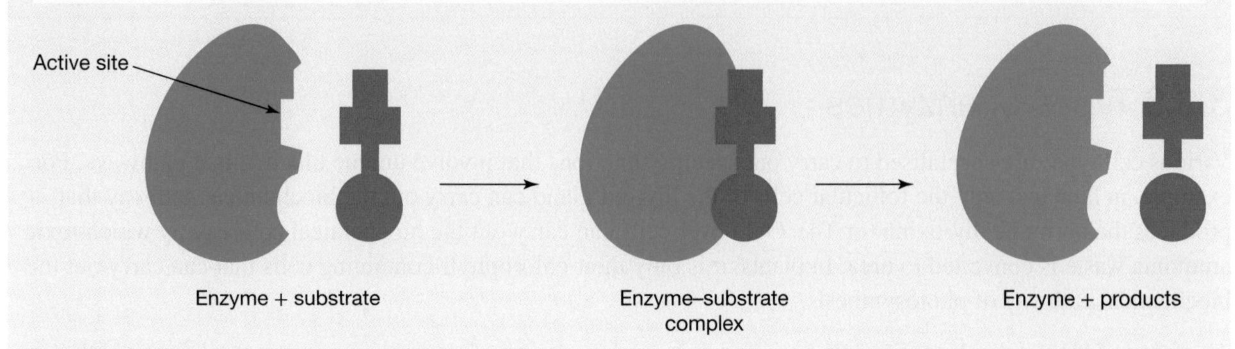

Consider figure 3.10:

1. The enzyme and substrate join to form a transient structure, the E–S complex, when a substrate binds to the active site of its enzyme.
2. The substrate forms weak bonds with particular amino acid residues at the active site of the enzyme (figure 3.11a). The bonds that temporarily hold the E–S complex together are weak bonds that can be easily broken. The E–S complex exists for just a short period before it is rapidly converted to the products of the reaction.
3. The enzyme is not consumed in the reaction. Once the product is formed, enzyme molecules are free to catalyse more substrate molecules. Because of this reuse of an enzyme, a very small amount of enzyme can produce large quantities of a product.

FIGURE 3.11 **a.** 2-D representation of a substrate molecule bound to the active site of an enzyme to form a transient E–S complex, **b.** Various bonds, such as hydrogen and ionic bonds form between a substrate and enzyme within the active site.

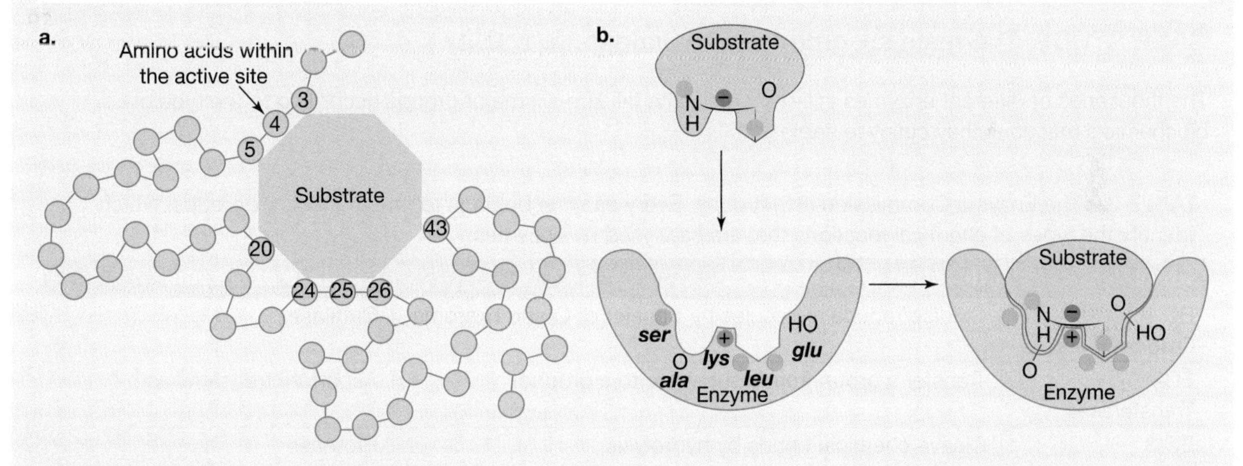

Enzymes have a specific activity

The specificity of enzymes varies, Some enzymes have:

- *absolute or substrate specificity*: they catalyse one particular substrate only, such as the enzyme lactase can only hydrolyse the milk sugar (lactose), breaking it apart to form glucose and galactose
- *bond specificity*: they act only on one kind of chemical bond, such as the peptidase enzymes that act specifically on the peptide bonds that join amino acids in a polypeptide chain
- *group specificity*: they act only on particular functional groups present as part of the substrate molecule, such as a carboxyl (-COOH) group or a hydroxyl (-OH) group.

on Resources

Video eLesson Structure and specificity of enzymes (eles-4192)

3.3.3 Types of enzymes

Various cell types are specialised to carry out specific functions that involve unique biochemical pathways. For example, in humans, only the follicular cells of the thyroid gland can carry out the biochemical pathway that produces the hormone thyroxine (or T4). Only liver cells can carry out the biochemical pathway by which toxic ammonia waste is converted to urea. In plants, it is only their chlorophyll-containing cells that can carry out the biochemical pathway of photosynthesis.

The types of biochemical reactions that can occur depend on the set of enzymes present in the specific cell. Some enzymes are only present in particular cells, like those involved in the production of thyroxine. Other enzymes, such as those involved in cellular respiration, are present in all types of cells.

A single human cell typically contains an estimated 1300 different enzymes, while all the cells of the human body, in total, contain several thousand different kinds of enzyme. The single-celled bacterium, *E. coli*, (part of your gut microflora) possesses about 1000 different enzymes and depend on their catalytic actions to maintain its living state.

Life requires the constant activity of catabolic pathways that break down glucose in cellular respiration and supply the energy needs of cells. The living state also requires the constant activity of anabolic pathways to build macromolecules needed for growth, replacement and repair. Each of these types of biochemical pathways can only proceed in the presence of enzymes. In effect, *the catalytic properties of enzymes are essential for life.*

CASE STUDY: The main six groups of enzymes

The thousands of different enzymes in living organisms fall into six major groups according to the type of biochemical reactions they catalyse (table 3.2).

TABLE 3.2 The Enzyme Commission (EC) groups. Every enzyme belongs to one of these six groups, which identify the types of chemical reactions that are catalysed by enzymes.

Group	Reaction type catalysed	Example of enzyme
EC 1 Oxidoreductases	Oxidise/reduce molecules by transfer of O and H atoms	Catalase
EC2 Transferases	Transfer a group from a substrate to a product	Aspartate transaminase
EC 3 Hydrolases	Cleave chemical bonds by hydrolysis	Protease
EC 4 Lyases	Cleave bonds in a substrate (not by hydrolysis)	Aldolase
EC 5 Isomerases	Change arrangement of atoms in a molecule	Glucose isomerase
EC 6 Ligases	Join two molecules by forming a new bond	Glutamate-cysteine ligase

elog-0822

INVESTIGATION 3.3 online only

The actions of the catalase enzyme

Aim

To investigate how catalase breaks down hydrogen peroxide

3.3.4 Enzymes involved in photosynthesis

The greatest source of free energy on planet Earth is sunlight. Green plants (and some bacteria) can capture the radiant energy of sunlight and transform it to the chemical energy of glucose molecules using the biochemical pathway of photosynthesis. The transformation of sunlight into glucose depends on the presence of the green pigment chlorophyll in plant cells and on the availability of carbon dioxide from the air and water, most commonly from the soil.

Overall, photosynthesis is an energy-hungry anabolic pathway that can be shown in its most simple form as:

$6CO_2$	$+ 6H_2O +$	ε	$\rightarrow C_6H_{12}O_6$	$+ 6O_2$
carbon dioxide	water	sunlight	glucose	oxygen

Note that sunlight can be placed above the arrow rather than next to it. As for any biochemical pathway, each step of photosynthesis depends on the presence of an active specific enzyme.

Included among the many enzymes involved in photosynthesis are:

- the water-oxidising enzyme that is critical in the first stage of photosynthesis — this enzyme splits water molecules, releasing hydrogen ions and electrons that will be built in energy-rich molecules. Oxygen is released as a side (waste) product.
- the ATP synthase enzyme, also active in the first stage of photosynthesis, generates energy-rich ATP for use in energy-hungry second stage of photosynthesis
- the Rubisco enzyme at the heart of photosynthesis (figure 3.12; see also opening image of this topic). Rubisco captures inorganic carbon dioxide molecules from the air and catalyses their conversion to organic 3-carbon molecules that can then be assembled into sugar molecules. The Rubisco enzyme is built of 16 polypeptide chains (8 large and 8 small).

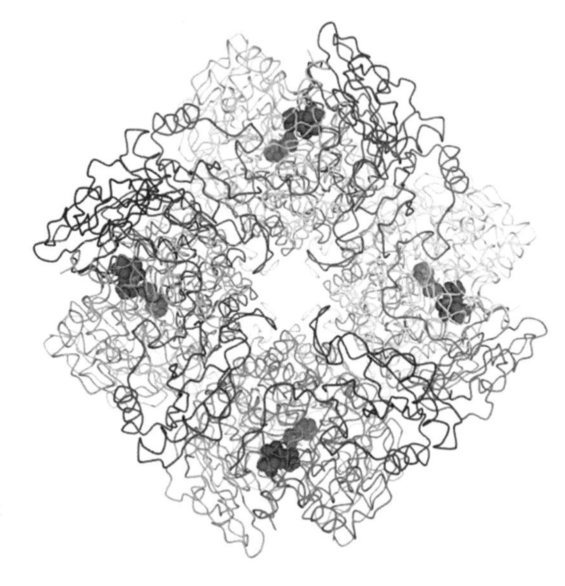

FIGURE 3.12 Model of the Rubisco enzyme. Four substrate molecules (shown as spacefill models) can be seen within the enzyme structure.

There are many other enzymes involved in the processes, of a variety of enzyme types:

- oxidoreductases, such as glyceraldehyde-3-phosphate dehydrogenase
- transferases, such as phosphoglycerate kinase
- hydrolases, such as fructose 1,6-biphosphatase
- lyases, such as aldolase
- isomerases, such as triose-phosphate isomerase.

Details of photosynthesis are discussed further in topic 4.

on Resources

Weblink Enzyme activity in photosynthesis

3.3.5 Enzymes involved in cellular respiration

Cellular respiration is occurring at every moment of an organism's life — whether that organism is a human or other animal, plant, fungus or a microbe. The essence of cellular respiration is the generation of energy in a form that is useable by cells, that is, ATP (adenosine triphosphate). In cellular respiration, energy is generated by a complex biochemical pathway in which glucose molecules are broken down to carbon dioxide and water.

Overall, cellular respiration is an energy-releasing catabolic pathway that can be shown in its most simple form as:

$$\underset{\text{glucose}}{C_6H_{12}O_6} + \underset{\text{oxygen}}{6O_2} \rightarrow \underset{\text{carbon dioxide}}{6CO_2} + \underset{\text{water}}{6H_2O} + \underset{\text{ATP}}{\varepsilon}$$

In reality, the cellular respiration pathway is a complex multi-step process and each step requires the action of a specific enzyme. Examples of some key enzymes include:

- *phosphofructokinase enzyme (PFK).* This enzyme catalyses an early step in the cellular respiration pathway and is a control point where the reaction rate can be regulated.
- *dehydrogenase enzymes.* These enzymes, as their name suggests, catalyse the removal of hydrogen atoms from substrates. Consuming one glucose ($C_6H_{12}O_6$) molecule to produce six carbon dioxide (CO_2) molecules involves the removal of 12 hydrogen atoms. Examples including malate dehydrogenase, pyruvate dehydrogenase (figure 3.13), and alpha ketoglutarate dehydrogenase.

Some other enzymes in cellular respiration include:

- oxidoreductases, such as pyruvate dehydrogenase
- transferases, such as citrate synthase and hexokinase
- lyases, such as fumarase (fumarate hydratase)
- isomerases, such as glucose-6-phosphate isomerase
- ligases, such as succinyl coenzyme A synthetase.

FIGURE 3.13 Ribbon model of the pyruvate dehydrogenase enzyme

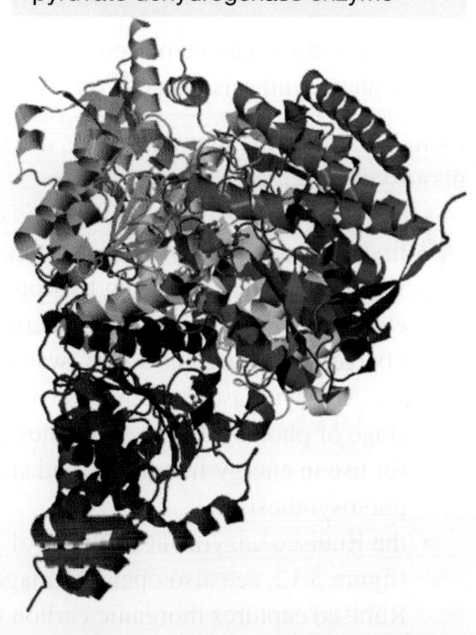

TIP You don't need to know the names of individual enzymes, but you should recognise that enzymes end in -ase and that a variety of different enzymes are used in both cellular respiration and photosynthesis.

tlvd-1766

SAMPLE PROBLEM 3 Exploring the role of enzymes in cellular respiration

Fructose-bisphosphate aldolase is an important enzyme in the body involved in the first step of cellular respiration known as glycolysis. This enzyme breaks down an intermediate product of glycolysis, fructose 1, 6-bisphosphate, into two products: glyceraldehyde-3-phosphate and dihydroxyacetone phosphate.

Use labelled diagrams to show the formation of the enzyme–substrate complex and the release of the products of this reaction. (3 marks)

THINK	WRITE
1. Show the enzyme and substrate, clearly labelling the active site, enzyme and substrate. Make sure that you use the substrate and enzyme name provided in the question stem. You should show the induced fit model of enzyme action, as this is the more supported model.	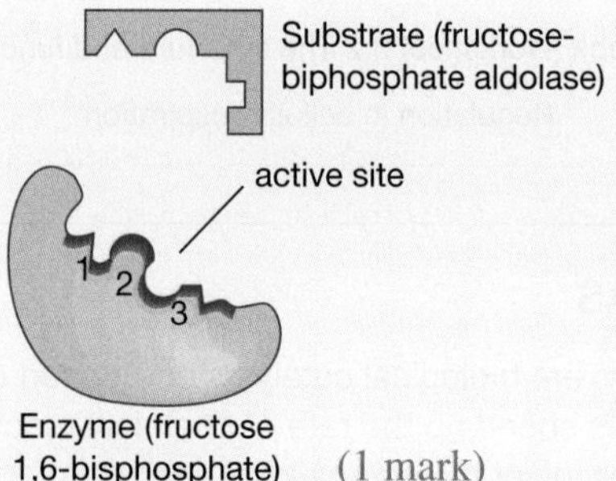 (1 mark)
2. Show the formation of the enzyme–substrate complex.	1 2 3 Enzyme–substrate complex (1 mark)
3. Show the formation of the products, ensuring you use the names provided in the question stem.	Products (glyceraldehyde-3-phosphate and dihydroxyacetone phosphate) 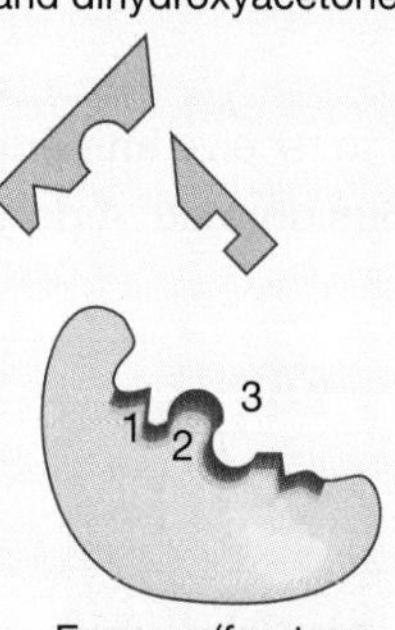 Enzyme (fructose 1,6-bisphosphate) (1 mark)

3.3.6 Case studies of enzymes

CASE STUDY: The plague of the sea

Scurvy is a deficiency disease, resulting from a lack of vitamin C (ascorbic acid). It was often referred to as 'the plague of the sea', and rose to prominence in the fifteenth century when explorers first ventured on extended ocean voyages.

Many animals can produce their own vitamin C through various biochemical pathways, which humans cannot do due to the lack of a particular enzyme (known as L-gulonolactone oxidase or the GULO enzyme).

To access more information on scurvy and the biochemical pathways related to this, please download the digital document.

FIGURE 3.14 Examples of foods that are rich sources of vitamin C (ascorbic acid)

on Resources

Digital document Case study: The plague of the sea (doc-35827)

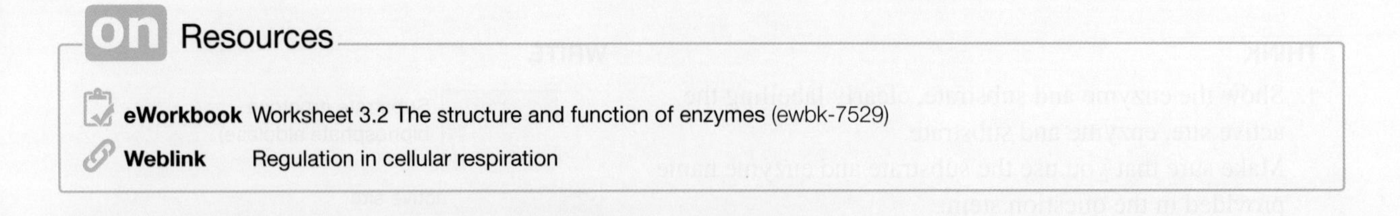

KEY IDEAS

- Enzymes are biological catalysts composed of proteins, each consisting of one or more polypeptide chains.
- Catalysts speed up the rate of chemical reactions.
- Enzymes lower the energy level needed to activate reactant molecules as compared with the non-enzyme catalysed situation.
- Each different biochemical reaction requires a different enzyme, including those involved in both photosynthesis and cellular respiration.
- Some examples of enzymes involved in photosynthesis are Rubisco, PEP carboxylase, ATP synthase, phosphoglycerate kinase and transketolase.

3.3 Activities

learnon

To answer questions online and to receive **immediate feedback** and **sample responses** for every question, go to your learnON title at **www.jacplus.com.au**. A **downloadable solutions** file is also available in the resources tab.

3.3 Quick quiz on | **3.3 Exercise** | **3.3 Exam questions**

3.3 Exercise

1. MC Enzymes are unable to
   A. lower the activation energy needed to activate reactant molecules.
   B. increase the amount of product produced in a reaction.
   C. consist of more than one polypeptide chain.
   D. form weak bonds with the substrate.
2. Give an example of an enzyme that
   a. transfers a group from a substrate to a product
   b. acts on the peptide bonds joining amino acids, cleaving them through hydrolysis
   c. changes the arrangement of atoms in a molecule
   d. can join two molecules by forming a new bond.
3. Identify the following statements as true or false and, if false, rewrite it as a true statement.
   a. All enzymes consist of a single polypeptide chain formed of amino acids joined by peptide bonds.
   b. Enzymes speed up the rate of the reactions that they catalyse.
   c. After an enzyme-catalysed reaction is completed, only enzyme and substrate molecules will be present.
   d. Bacterial cells do not contain any enzymes.
   e. In catabolic reactions, the energy level of the reactants is greater than that of the products.
4. Provide brief answers to the following.
   a. In which biochemical pathway is the Rubisco enzyme active?
   b. What is the role of the Rubisco enzyme?
   c. What is the structure of the Rubisco enzyme?
   d. At what rate does Rubisco act?

5. Give a brief scientific explanation for the following observations.
   a. Only a relatively few enzyme molecules are needed to catalyse a very large number of the substrate molecules of that enzyme.
   b. Pancreatic cells can produce proinsulin, but not growth hormone through biochemical pathways, while pituitary gland cells can produce growth hormone, but not proinsulin through biochemical pathways.
6. Explain the following in your own words.
   a. Why enzymes are important in the cellular respiration pathway
   b. The meaning of the term *enzyme specificity*.

3.3 Exam questions

Question 1 (1 mark)

Source: *VCAA 2008 Biology Exam 1, Section A, Q19*

MC Activation energy in a biological reaction

A. increases in the presence of an enzyme.
B. increases with an increase in temperature.
C. is the energy required to start the reaction.
D. is involved in the formation of complex molecules only.

Question 2 (1 mark)

Source: *VCAA 2011 Biology Exam 1, Section A, Q23*

MC The following graphs depict two different reactions.

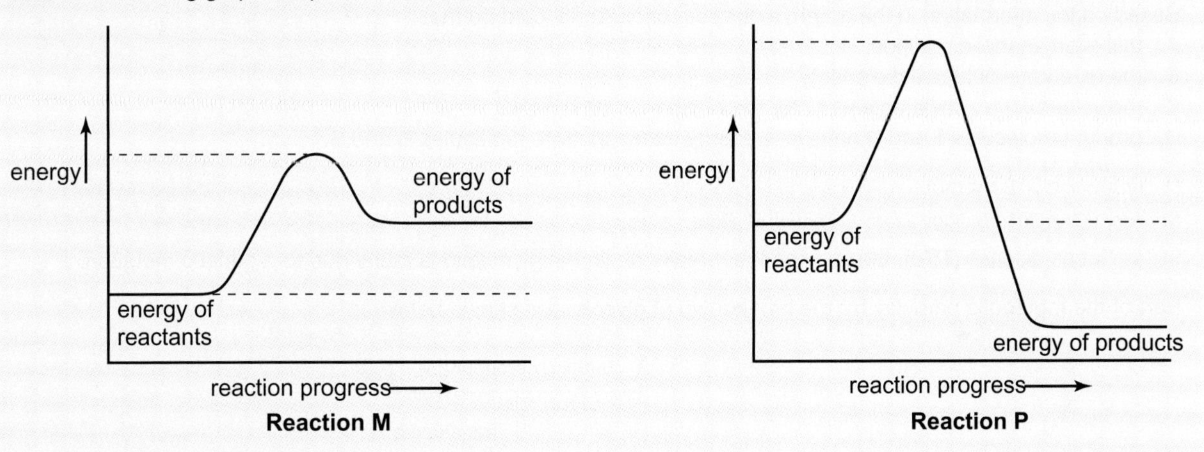

From the two graphs, it is reasonable to conclude that

A. in reaction **P**, the energy level of the products is greater than that of the reactants.
B. activation energy of reaction **M** is greater than that of reaction **P**.
C. both graphs **M** and **P** represent endothermic reactions.
D. energy is released in reaction **P** only.

Question 3 (1 mark)

Source: *VCAA 2008 Biology Exam 1, Section B, Q4a and b*

Living organisms cannot survive without the presence of enzymes.

a. Explain why enzymes are necessary in living organisms. **1 mark**
b. Describe the 'active site' of an enzyme and explain its role. **1 mark**

Question 4 (1 mark)

Source: *VCAA 2013 Biology Section A, Q8*

MC Consider the following reaction in which substrate molecule R and substrate molecule S are converted into product molecule T and product molecule U.

$$\text{R and S} \rightarrow \text{T and U}$$

The following graph shows the energy available in the molecules against time.

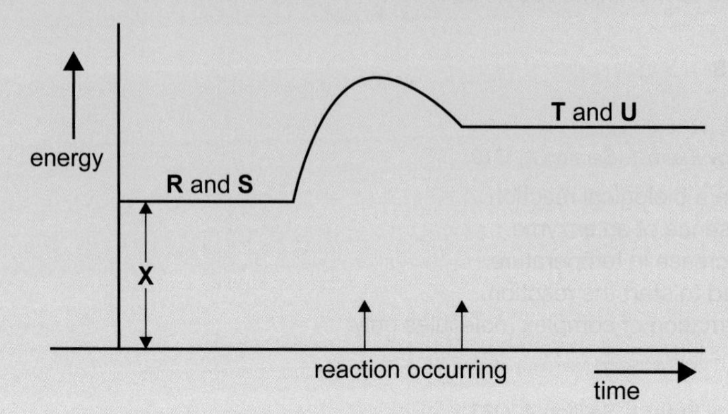

Based on the information in the graph, a correct conclusion would be that

A. this is an anabolic reaction.
B. the reaction would release energy.
C. the value of the activation energy for the reaction is shown by X.
D. product molecules T and U have less energy than substrate molecules R and S.

Question 5 (2 marks)

Source: *VCAA 2010 Biology Exam 1, Section B, Q6c*

In Tay Sachs disease (TSD), the enzyme that breaks down the glycolipid is faulty due to a genetic mutation.

Part of the metabolic pathway for the breakdown of glycolipid is shown below.

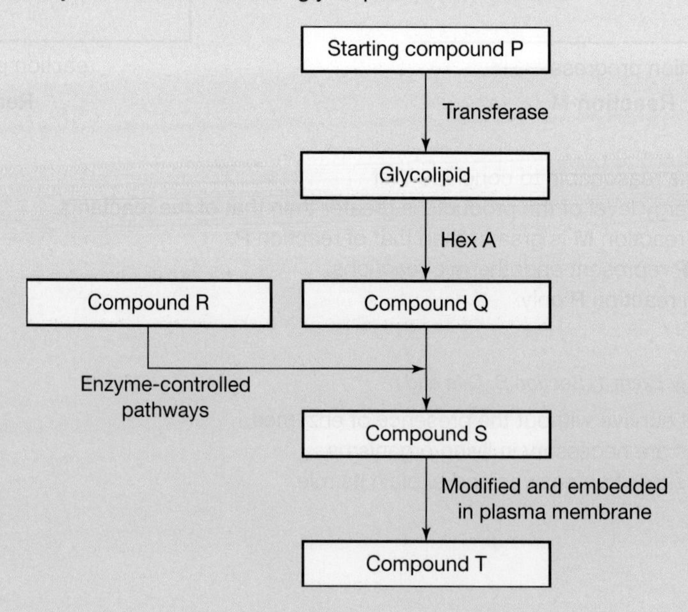

Explain which enzyme–substrate complex fails to form correctly in sufferers of TSD.

More exam questions are available in your learnON title.

3.4 The role of coenzymes in biochemical pathways

KEY KNOWLEDGE

- The general role of coenzymes in facilitating steps in photosynthesis and cellular respiration

Source: Adapted from VCE Biology Study Design (2022–2026) extracts © VCAA; reproduced by permission

3.4.1 What are coenzymes?

Enzymes are composed of protein and, for many enzymes, that is all that is required for their activity. However, many enzymes have additional non-protein components that are necessary for their activity and stability. These additional non-protein groups that are needed for enzymes to function are termed **cofactors**. Cofactors include organic coenzymes, such as those used in cellular respiration and photosynthesis.

Inorganic and organic cofactors

Inorganic cofactors

Inorganic cofactors do not contain carbon and include metal ions such as magnesium (Mg^{2+}), copper (Cu^{2+}), manganese (Mn^{2+}), calcium (Ca^{2+}) or clusters of several ions. The inorganic cofactor of the water-splitting enzyme of photosynthesis is a cluster of four Mn^{2+} ions and one Ca^{2+} ion. Many enzymes include a metal cofactor (see table 3.4). Metal cofactors form bonds at the active site of their enzyme and also bond with the substrate — as a result, the substrate is held temporarily in the position required for enzyme action. For example, the enzyme carboxypeptidase requires zinc ions (Zn^{2+}) as a cofactor to stabilise its substrate in position. Note that you do not need to memorise these cofactors.

cofactor a non-protein molecule or ions that is essential for the normal functioning of some enzymes

TABLE 3.4 Examples of enzymes requiring involvement of metal ions for their activity

Metal ions	Enzymes that require ion
Copper Cu^{+}/Cu^{2+}	Cytochrome oxidase
Magnesium Mg^{2+}	Glucose-6-phosphatase Hexokinase Pyruvate kinase Rubisco
Iron Fe^{2+}/Fe^{3+}	Cytochrome oxidase Catalase
Zinc Zn^{2+}	Carboxypeptidase Alcohol dehydrogenase

FIGURE 3.15 A simplified diagram showing how a cofactor can bind to an enzyme to allow for it to function

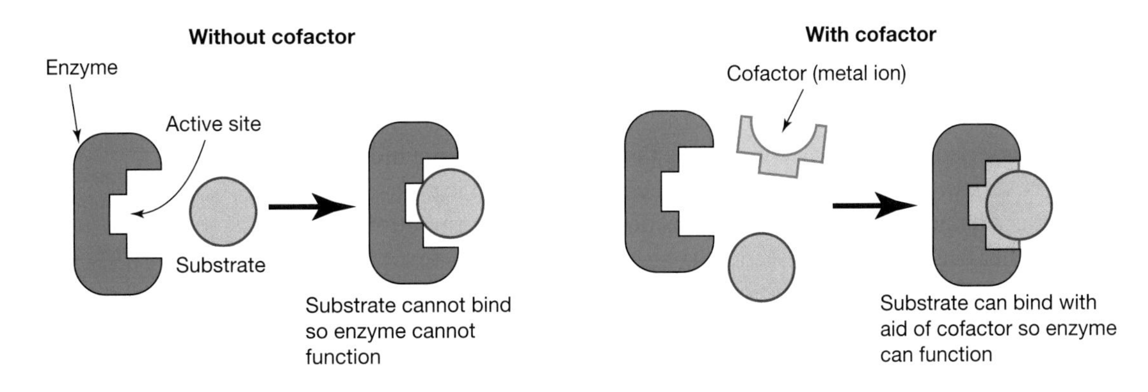

Organic cofactors and coenzymes

Organic cofactors are small non-protein organic molecules that are essential for the function of particular enzymes. They comprise:

1. prosthetic groups: cofactors that are tightly bound to an enzyme and are essential for it to function as a catalyst. The enzyme, cytochrome *c*, contains four heme molecules as an essential part of its structure.
2. **coenzymes**: cofactors that loosely bound to their enzymes only when the enzyme is acting on a substrate. (*Note:* Coenzymes are a key focus of the Study Design.)

Major coenzymes involved in photosynthesis and cellular respiration

- NAD (nicotinamide adenine dinucleotide)
- NADP (nicotinamide adenine dinucleotide phosphate)
- FAD (flavin adenine dinucleotide)
- ATP (adenosine triphosphate)
- Coenzyme A (CoA).

Many coenzymes are derived from vitamins; for example, nicotinamide is derived from the vitamin niacin (vitamin B3), flavin from the vitamin riboflavin (vitamin B2), and coenzyme A is from the vitamin pantothenic acid (vitamin B5).

tlvd-1767

SAMPLE PROBLEM 4 Defining coenzymes

Define and provide an example of a coenzyme. **(2 marks)**

THINK	WRITE
1. Identify what the question is asking you to do. This question asks you to both *define* and *provide an example*.	
2. Write a clear definition, which shows a clear statement about what a coenzyme is.	A coenzyme is an organic non-protein organic compound that binds loosely with an enzyme at its active site, assisting the catalytic function of the enzyme (1 mark).
3. Provide an example.	An example of a coenzyme is ATP (adenosine triphosphate) (1 mark).

3.4.2 The role of coenzymes in biochemical pathways

Coenzymes play helper roles that are essential for certain classes of enzyme to perform their catalytic actions. The roles by which coenzymes help enzymes include:

- transfers of atoms or groups of atoms, such as hydrogens, phosphate groups and acetyl groups
- energy transfers.

Coenzymes exist in two inter-convertible forms:

- a high energy form that is loaded with a group that can be transferred
- a lower energy form that is unloaded.

These forms can be demonstrated by considering the coenzyme nicotinamide adenine dinucleotide (NAD). In its two forms, **loaded** and **unloaded**, NAD is a versatile coenzyme that is a helper for enzymes in key reactions in cellular respiration, which is discussed in further detail in the next section.

- In its unloaded form, NAD^+, it accepts hydrogen ions and their associated electrons.
- In its loaded form, NADH, it can donate hydrogen ions and electrons and transfer energy.

coenzyme an organic molecule that acts with an enzyme to alter the rate of a reaction

loaded the form of coenzymes that can act as electron donors

unloaded a form of coenzymes that can act as electron acceptors

Whichever form it is in, either loaded or unloaded, this coenzyme is in constant use, allowing for a variety of reactions to occur that are catalysed by enzymes.

FIGURE 3.16 The two forms of NAD: the unloaded form, NAD^+, that is a hydrogen acceptor (left) and the loaded form, NADH, that is a hydrogen donor (right)

Unloaded form | **Loaded form**

Reduction / Oxidation

$$\mathbf{NAD^+ + H^+ + 2e^- \rightleftharpoons NADH}$$

3.4.3 Coenzymes involved in cellular respiration

Cellular respiration (explored further in subtopic 4.5) involves many reactions that require the actions of various coenzymes and enzymes.

NAD: Nicotinamide adenine dinucleotide

As seen in figure 3.17, NAD exists in two forms — an unloaded NAD^+ and a loaded NADH. These are outlined in table 3.5.

TABLE 3.5 Inter-convertible forms of the coenzyme NAD that are helpers for enzymes involved in hydrogen transfers (redox reactions) and energy transfers. Hydrogen comprises a H^+ ion and an electron (e^-).

Coenzyme	Alternate forms of coenzyme	Role of coenzyme	Energy level	Redox state
NAD (Nicotinamide adenine dinucleotide)	Unloaded NAD^+	Hydrogen acceptor	Low	Oxidised
	Loaded NADH	Hydrogen donor	High	Reduced

FIGURE 3.17 Simplified representation of the helper role of the NAD^+ coenzyme as a hydrogen acceptor for dehydrogenase enzymes.

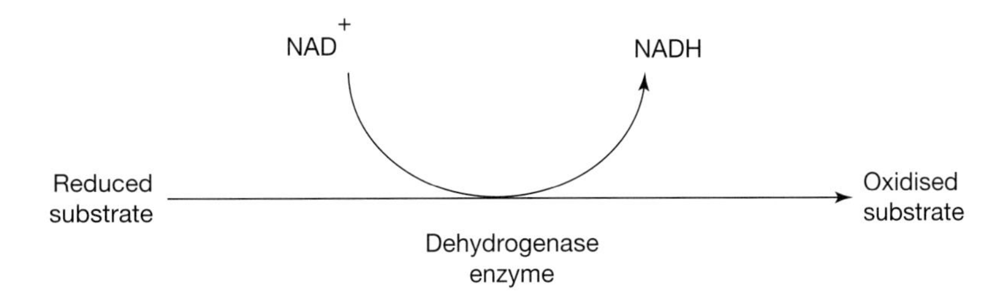

Unloaded NAD$^+$

In its unloaded form, **NAD$^+$**, is a receiver of electrons and hydrogen ions. When a substrate needs to give up electrons and hydrogen ions (or is 'oxidised'), NAD$^+$ is able to accept these, becoming NADH. As NAD transitions between its loaded and unloaded forms, it switches from being a helper for the group of enzymes that catalyse reactions where electrons are lost ('reduction') to being a helper for those enzymes catalysing reactions where electrons are gained ('oxidation').

Unloaded NAD$^+$ is an essential coenzyme for several dehydrogenase enzymes that catalyse hydrogen transfers in the cellular respiration pathway. These enzymes, such as malate dehydrogenase, are associated with unloaded NAD$^+$ coenzymes that act as hydrogen acceptors for hydrogen stripped from their substrates. As this happens, the coenzymes are converted to the high-energy loaded form of NADH. Figure 3.18 shows the enzyme malate dehydrogenase associated with its coenzyme NAD$^+$ (shown as ball-and-stick models) that acts as an acceptor of hydrogens stripped from the enzyme's substrate.

FIGURE 3.18 Ribbon model of malate dehydrogenase, an enzyme of the cellular respiration pathway. The zoomed sections shows where NAD$^+$ associates with malate dehydrogenase and accepts electrons and hydrogen ions.

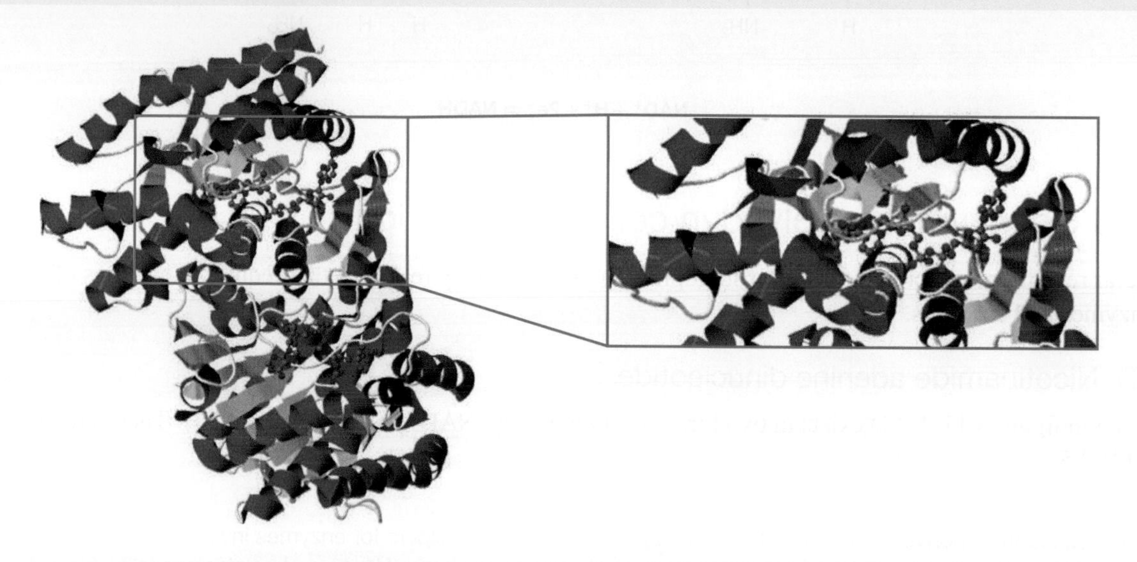

Loaded NADH

Loaded **NADH** coenzymes are important helpers in cellular respiration. NADH coenzymes transfer their hydrogens — in the form of hydrogen ions (H^+) and electrons (e^-) — in the last stages of this pathway. It catalyses reactions in which the substrate is reduced and needs to gain hydrogen ions and electrons. When NADH unloads its electrons the energy released is transferred via a number of steps and used in the production of ATP (figure 3.19).

FIGURE 3.19 Simplified representation showing the release of energy when NADH is converted to NAD$^+$ and its use in ATP production

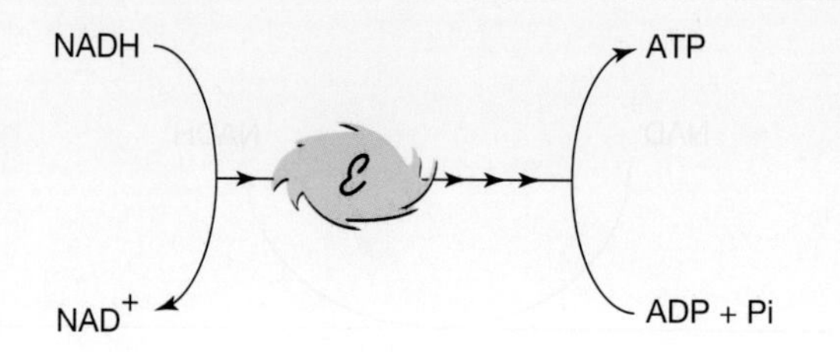

NAD$^+$ the unloaded form of NADH, which can accept hydrogen ions and electrons during cellular respiration

NADH the loaded form of NAD$^+$ which can donate hydrogen ions and electrons during cellular respiration

FAD: Flavin adenine dinucleotide

FAD is very similar in function to NAD. In its loaded and unloaded forms, **$FADH_2$** can act as a hydrogen and electron donor and FAD as a hydrogen and electron acceptor:

$$FAD + 2H^+ + 2e^- \leftrightharpoons FADH_2$$

The unloaded form of this coenzyme, FAD, plays a role in the second stage of cellular respiration (the Krebs cycle) and accepts high-energy electrons. The loaded form, $FADH_2$, plays a role in the last stage of the cellular respiration pathway (the electron transport chain) as a donor of these high-energy electrons into a chain of electron acceptors.

ATP: adenosine triphosphate

ATP, with its three phosphate groups, is energy rich and it is a major player in transferring energy within cells. This energy is made available when the last phosphate group of ATP is removed, creating **ADP** (adenosine diphosphate) at a lower energy level (see figure 3.20). ATP can be regenerated from ADP in an energy-requiring reaction with the addition of a phosphate (Pi).

$$ADP + Pi \leftrightharpoons ATP$$

FIGURE 3.20 Diagram showing the interconversion of ATP and ADP

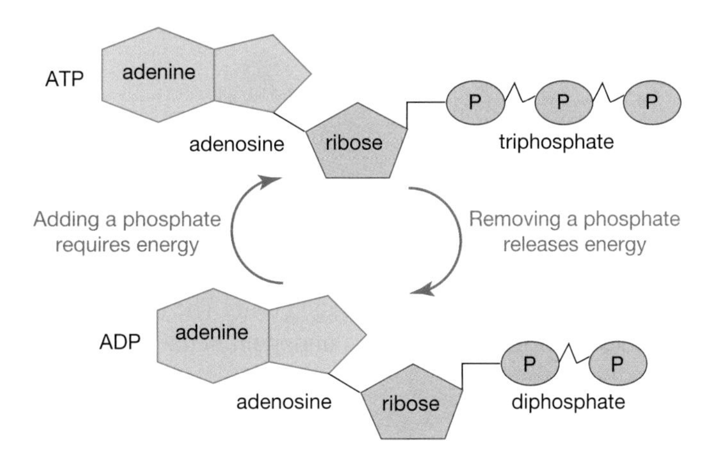

ATP is a coenzyme helper for several enzymes that catalyse energy-requiring reactions in the cellular respiration pathway. The major role of ATP is the transfer of its energy to drive endergonic reactions in biochemical pathways. In some reactions, ATP transfers both energy and donates a phosphate group to substrates (figure 3.21a).

ADP is a coenzyme helper to some enzymes when it acts as an acceptor of a phosphate group from the substrate of an enzyme to form ATP (see figure 3.21b).

FAD a coenzyme with a similar function to NAD, accepting hydrogen ions and electrons during cellular respiration

$FADH_2$ the loaded form of FAD, which can donate hydrogen ions and electrons during cellular respiration

ATP adenosine triphosphate; the common source of chemical energy for cells

ADP adenosine diphosphate; a coenzyme that accepts a phosphate group to form ATP

FIGURE 3.21 Examples of the role of ATP and ADP as coenzymes in **a.** an endergonic reaction, where coenzyme ATP transfers both energy and a phosphate group to assist the action of the hexokinase enzyme on its substrate, glucose to form glucose-6-phosphate **b.** an exergonic reaction, where the pyruvate kinase enzyme is assisted by an ADP coenzyme that accepts a phosphate group and energy from PEP, allowing for pyruvate and ATP to be formed.

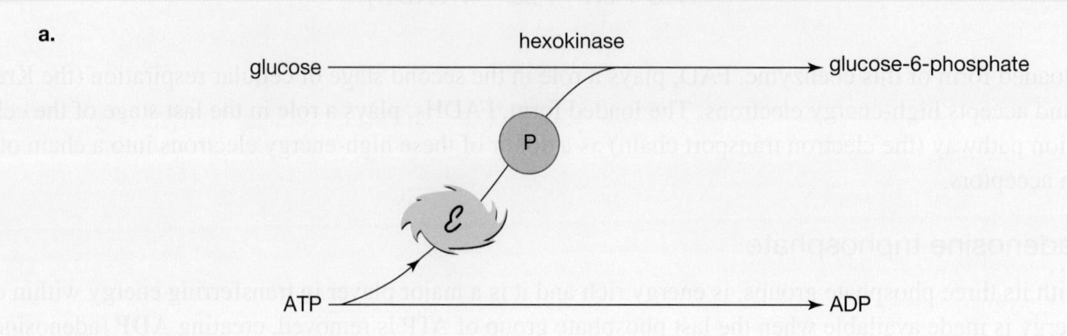

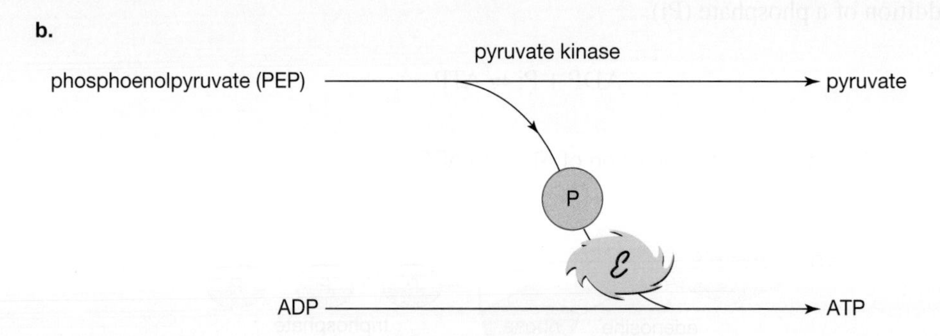

Coenzyme A (CoA)

Coenzyme A is involved in the cellular respiration pathway as a helper to the pyruvate decarboxylase enzyme in the mitochondria. This enzyme strips one carbon dioxide molecule from the 3-C pyruvate molecule formed in glycolysis., leaving behind a 2-C fragment, called an acetyl group. The role of CoA is to accept the acetyl group and transfer it to another acceptor molecule. This is important in the second stage of cellular respiration, the Krebs cycle.

$$\text{pyruvate (3C)} \xrightarrow{\text{CoA}} \text{acetyl CoA (2C)}$$

coenzyme A a coenzyme that aids pyruvate decarboxylase during cellular respiration, accepting an acetyl group

3.4.4 Coenzymes involved in photosynthesis

Photosynthesis (explored further in subtopic 4.2) also involves many reactions that require the actions of various coenzymes and enzymes.

NADP: nicotinamide adenine dinucleotide phosphate

NADP exists in two forms: an unloaded $NADP^+$ and a loaded NADPH. These are outlined in table 3.6.

TABLE 3.6 The two forms of the coenzyme $NADP^+$ assist enzymes involved in hydrogen transfer reactions.

Coenzyme	Alternate forms of coenzyme	Role of coenzyme	Energy level	Redox state
NADP (nicotinamide adenine dinucleotide phosphate)	$NADP^+$	Hydrogen acceptor	Low	Oxidised
	NADPH	Hydrogen donor	High	Reduced

In its loaded form, **NADPH**, this coenzyme plays an important role in anabolic pathways in which small molecules are built into larger energy-rich ones. NADPH coenzymes assist the enzymes involved in these pathways by acting as hydrogen donors and in the transfer of energy.

The most biologically significant anabolic pathway is the assembly of low-energy inorganic carbon dioxide (CO_2) into energy-rich sugars, such as glucose ($C_6H_{12}O_6$). This pathway involves adding hydrogen to form glucose and the enzymes involved depend on coenzymes to act as hydrogen donors.

In photosynthesis, NAPDH is formed from **NADP⁺** using high-energy electrons initially produced by uptake of the radiant energy of sunlight as shown below:

$$NADP^+ + H^+ + 2e^- \leftrightharpoons NAPDH$$

The high-energy NADPH coenzymes are used in the second stage of the photosynthesis pathway when they donate electrons and transfer energy required by enzymes in this pathway as sugars are built. In so doing, NADPH reverts to $NADP^+$.

TIP: To avoid getting confused between NADP and NAD, remember that the extra **P** can remind us that it is used for **p**hotosynthesis.

ATP

ATP is also an essential coenzyme for reactions in the photosynthesis pathway.

In photosynthesis, high-energy ATP is formed from ADP (refer to figure 3.20) and the energy required is initially derived from the radiant energy of sunlight. ATP can then release its energy to help enzymes involved in energy-requiring reactions that build sugar molecules from carbon dioxide.

NADPH the loaded form of $NADP^+$, which can donate hydrogen ions and electrons during photosynthesis

$NADP^+$ the unloaded form of NADPH, which can accept hydrogen ions and electrons during photosynthesis

eWorkbook Worksheet 3.3 Investigating coenzymes (ewbk-7531)

Video eLesson ATP is the energy currency of cells (eles-4193)
Coenzymes (eles-3904)

KEY IDEAS

- Coenzymes are organic cofactors that are non-protein molecules that contribute to the activity and stability of enzymes.
- One key group of cofactors includes the coenzymes NAD, NADP and ATP, which move groups (such as hydrogen and phosphate) and energy between metabolic pathways in cells.
- NAD is used in cellular respiration and exists in two forms — the loaded NAD^+ and the unloaded NADH.
- FAD is used in cellular respiration and exists in two forms — the loaded FAD and the unloaded $FADH_2$.
- Coenzyme A is used in cellular respiration, accepting an acetyl group to transfer.
- NADP is used in photosynthesis and exists in two forms — the loaded $NADP^+$ and the unloaded NADPH.
- ATP helps to catalyse energy requiring reactions. The loaded ATP can release a phosphate group and energy to form the unloaded ADP.

3.4 Activities

To answer questions online and to receive **immediate feedback** and **sample responses** for every question, go to your learnON title at **www.jacplus.com.au**. A **downloadable solutions** file is also available in the resources tab.

3.4 Quick quiz on	3.4 Exercise	3.4 Exam questions

3.4 Exercise

1. **MC** Which of the following is correct regarding ATP?
   A. It is the unloaded form of ADP.
   B. It contains one less phosphate compared to ADP.
   C. It cannot be converted back to ADP.
   D. It is the direct usable source of energy for cells.
2. What is a key difference between the members of the following pair?
   a. An enzyme and a coenzyme
   b. Loaded NAD and unloaded NAD
3. Give an example of
   a. a coenzyme that can act as an acceptor of hydrogens and their electrons
   b. a coenzyme that can accept a phosphate group
   c. a high-energy coenzyme than can transfer energy to an enzyme-catalysed reaction
   d. the reaction for which acetyl CoA is the helper.
4. Identify the following statements as true or false and, if false, rewrite it as a true statement
   a. In the cellular respiration pathway, the malate dehydrogenase enzyme can only function if loaded NADH coenzyme is available to it.
   b. In the photosynthesis pathway, the coenzyme $NADP^+$ is an essential helper to several different enzymes.
   c. A coenzyme forms a permanent bond with the enzyme for which it is a helper.
   d. After they are used in reactions, coenzymes such as NAD^+ and NADH can be regenerated for reuse in another reaction.
   e. All enzymes require a coenzyme for their action.
5. Briefly describe the role of the following as enzyme helpers
   a. a metal cofactor, such as Zn^{2+}
   b. a coenzyme of your choice.
6. Give a brief scientific explanation for the following observation
   'In the absence of its coenzyme, an enzyme cannot carry out its catalytic activity.'

3.4 Exam questions

Question 1 (1 mark)

Source: *VCAA 2016 Biology Exam, Section A, Q9*

MC ATP is important in living cells as it
A. is required for osmosis.
B. provides a supply of usable energy for the cell.
C. provides one of the building blocks for lipid synthesis.
D. is an important structural component of the plasma membrane.

Question 2 (1 mark)

Source: *VCAA 2017 Biology Exam, Section A, Q16*

MC A molecule that takes part in many biochemical reactions is $NADP^+$.

It is correct to state that
A. $NADP^+$ becomes NADH when it is loaded.
B. $NADP^+$ has a higher energy level when it is unloaded.
C. energy is released when $NADP^+$ is converted to NADPH.
D. $NADP^+$ carries additional energy when protons and electrons are added to it.

Question 3 (1 mark)

Source: *VCAA 2018 Biology Exam, Section A, Q10*

MC Which pair of molecules contains the greatest amount of stored energy?

A. NADH and ATP
B. NAD^+ and ATP
C. NAD^+ and ADP
D. NADH and ADP

Question 4 (3 marks)

The diagram represents a molecule of ATP (adenosine triphosphate). (*Note:* P = phosphate)

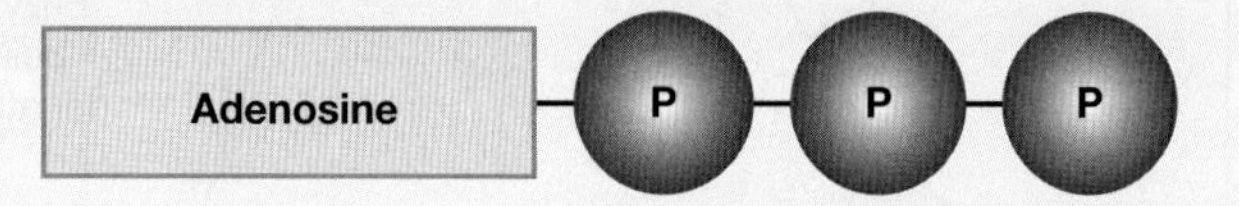

Explain how an ATP releases energy directly to a cellular reaction and what is formed as a result.

Question 5 (2 marks)

NAD^+ and $NADP^+$ are coenzymes.

a. Define the term coenzyme. **1 mark**
b. Name two cellular processes that involve NAD^+ or $NADP^+$. **1 mark**

More exam questions are available in your learnON title.

3.5 Factors that impact enzyme function

KEY KNOWLEDGE

- The general factors that impact on enzyme function in relation to photosynthesis and cellular respiration: changes in temperature, pH, concentration, competitive and non-competitive enzyme inhibitors

Various factors can affect enzyme activity. These include:

- temperature
- pH
- substrate concentration.

3.5.1 Effect of temperature on enzyme activity

Every enzyme has an optimum temperature at which it operates most efficiently and at that point, the enzyme is operating at its maximum rate. The **optimum temperature** varies for enzymes from different organisms.

Figure 3.22 shows a graph of the reaction rate (enzyme activity) against temperature for a typical human enzyme. Note that this enzyme has its fastest reaction rate at a temperature of about 37 °C or normal body temperature, and this is its optimum temperature.

optimum temperature the temperature at which the rate of reaction catalysed by an enzyme is at its highest

FIGURE 3.22 Graph showing the effect of temperature on the rate of a reaction catalysed by an intracellular human enzyme. Note that reaction rate is a measure of enzyme activity.

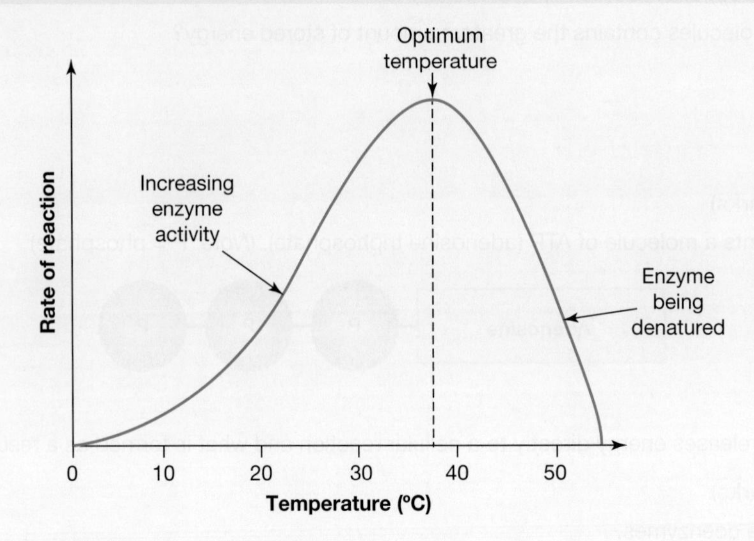

It can be seen that when temperature first increases, the reaction rate also increases. This is because the reactants absorb heat energy and move faster, allowing them to collide and overcome the activation energy. However, in enzyme-catalysed reactions, this increase only occurs up to the optimum temperature, after which the rate decreases, due to denaturation and the change in enzyme structure.

On either side of the optimum is a range of temperatures at which enzymes will still operate, but at decreasing rates of reaction, and for different reasons.

Temperature decrease from optimum

As the temperature drops *below* the optimum, the rate of reaction reduces progressively. As the temperature falls below optimum, molecular movements slow, resulting in fewer collisions between substrates and enzyme, and fewer molecules have sufficient energy to interact. However, while low temperatures can inactivate enzymes, this is not a permanent change and cold-inactivated enzymes can be reactivated with an increase in temperature. So, enzymes are tolerant of low temperatures and can be stored in laboratories in a solution of glycerol at temperatures as low as –35 °C.

Temperature increase from optimum

As the temperature increases *above* the optimum, the reaction rate reduces sharply. This occurs because enzymes are proteins and when temperatures exceed the optimum, heat **denaturation** of the protein occurs, as shown in figure 3.23. This involves the breaking of bonds and the loss of the secondary, tertiary and quaternary structure in the enzymes. Denaturation is irreversible. As the catalytic properties of enzymes depend upon the particular shape of their active sites, denaturation changes their 3D shape, permanently inactivating them.

FIGURE 3.23 The denaturation of an enzyme occurs when the temperature is too high, leading to the loss of the active site due to the breaking of bonds in the protein structure.

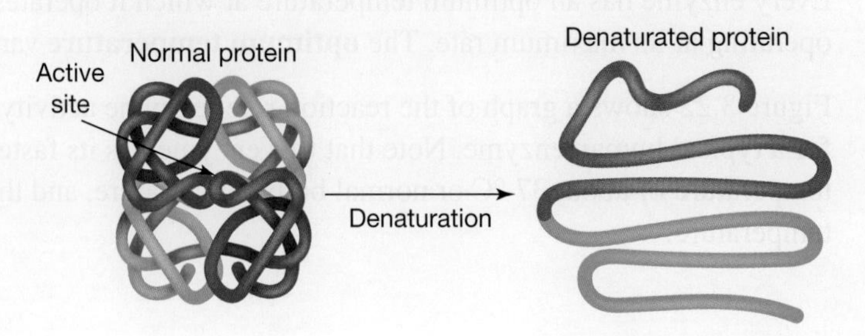

Heat denaturation may be seen when an egg is boiled — it causes the albumin proteins of the egg white to solidify. As the egg white cannot be returned to its original state — it is an irreversible change.

denaturation the loss of enzyme structure due to the breaking of bonds upon heating, irreversibly changing the shape of the active site

tlvd-1768

SAMPLE PROBLEM 5 Designing an experiment on factors affecting rate of reaction

Pyruvate decarboxyase is an enzyme that breaks down pyruvate into acetaldehyde and carbon dioxide in anaerobic respiration. The amount of carbon dioxide produced can be detected using phenol red, which turns pink when carbon dioxide levels are high, and yellow when carbon dioxide levels are low.
Design an investigation to determine the optimum temperature of pyruvate decarboxylase. **(6 marks)**

THINK	WRITE
1. Determine the independent and dependent variables in the investigation.	Independent variable: Temperature Dependent variable: Amount of carbon dioxide (1 mark)
2. Determine how you will measure your dependent variable.	The amount of carbon dioxide produced will be determined by recording the colour of the solution when tested with phenol red (1 mark).
3. Determine variables that need to be controlled.	Variables that will need to be controlled across various temperatures include the amount and concentration of the initial pyruvate to be tested, the amount of pyruvate decarboxylase and phenol red used, the pH conditions at the start of the investigation and the time for the reaction to occur (1 mark).
4. Write a clear experimental method that is reproducible, with identifiable quantities and an obvious control of variables. Ensure you include a large sample size (or multiple trials), information about collecting data and a method that will allow for valid results to be obtained.	1. Label 6 test tubes with the following: S1, S2, S3, S4, S5 and S6. 2. Place 10 mL of pyruvate in each of the 6 labelled test tubes. 3. Label 6 more test tubes with the following: A1, A2, A3, A4, A5 and A6. 4. Place 10 mL of pyruvate decarboxylase solution in each of the 6 labelled test tubes. 5. Place test tubes labelled 1 in a water bath set to 10 °C, those labelled 2 in a water bath set to 20 °C, 3 in a water bath set to 30 °C, 4 at 40 °C, 5 at 50 °C and 6 at 60 °C. 6. Leave these for 5 minutes. 7. Place the pyruvate decarboxylase solution in the respective test tube of pyruvate solution and leave at the specified temperature for a further 10 minutes. 8. During this stage, place 20 mL of water into one test tube and 20 mL of pyruvate into another test tubes and label these as control groups. 9. Add in 3 drops of phenol red into each test tube (both control and experimental). 10. Record the colour of each test tube. The ones that turn the darkest shade of pink had the greatest rate of enzyme activity. (1 mark for reproducible method including quantities, 1 mark for control of variables, 1 mark for large sample size)

3.5.2 Effect of pH on enzyme activity

BACKGROUND KNOWLEDGE: Reviewing pH

The pH (hydrogen ion or H^+ concentration) is a measure of acidity and alkalinity. Acidic solutions have a low pH and alkaline or basic solutions have a high pH. The pH scale is shown in figure 3.24.

FIGURE 3.24 Review of the pH scale. Acids have a pH less than 7 and bases, which are alkaline, have a pH above 7.

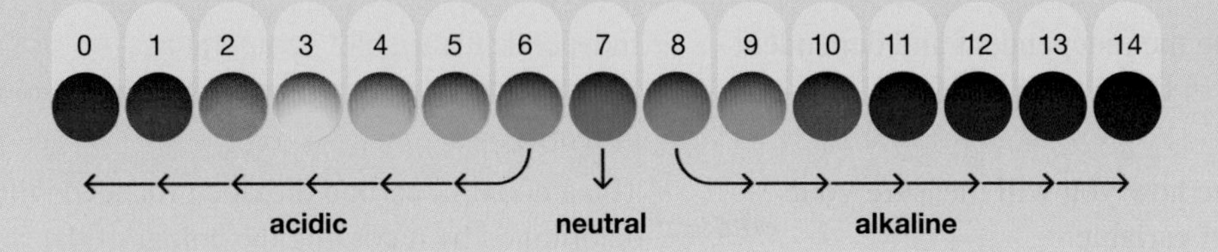

The level of acidity at which enzymes can operate varies, typically according to the environment in which the enzyme normally operates. Different enzymes have corresponding optimal pH. Figure 3.25 shows the activity rate for three human enzymes against pH. Enzymes that act within human cells typically have an optimal pH of about 7.0–7.4. The extracellular environment of various compartments of the human body have different prevailing pH values, such as the stomach with a pH of 2 to 3, and the duodenum with a pH around 8.5.

Enzymes can operate as catalysts over a limited pH range. The optimum pH for an enzyme is the pH at which the enzyme displays its highest activity, measured as the reaction rate (substrate molecules catalysed per unit of time). One kind of bond that contributes to the shape of enzymes is an ionic bond,which is an attraction between positively and negatively charged groups. As the pH moves progressively from the optimum value — either up or down — various ionic bonds that contribute to the shape of the enzyme can be altered, and the bonds that temporarily hold a substrate in place in the active site can be altered. If this happens, the enzyme is less able to combine with its substrate, and its activity is reduced or shut down.

FIGURE 3.25 Graph showing the effect of pH on the reaction rate of three human enzymes. Pepsin is a stomach enzyme, trypsin is found in the duodenum and carbonic anhydrase is found in various locations, such as red blood cells.

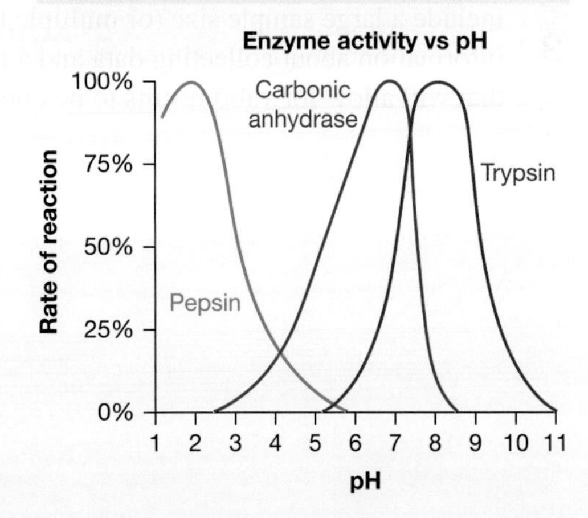

As pH increases or decreases from optimum pH, the rate of reaction and the activity of an enzyme decreases.

Figure 3.26 shows that at the pH optimum of the enzyme, the substrate is held in place in the position needed for the enzyme to act by two ionic bonds (red dotted lines) each between a positively-charged amino group (NH_3^+) and a negatively-charged carboxyl group (COO^-).

When the pH becomes too acidic or too basic (below or above the optimum range for the enzyme), the groups between the active site of the enzyme and the substrate are no longer able to form an ionic bond, causing enzyme activity to decline and eventually stop. This is due to the loss of the charge on COO^- in acidic conditions and the loss of the charge on NH_3^+ in alkaline conditions.

FIGURE 3.26 A substrate molecule in the active site of an enzyme at its pH optimum of 7.2

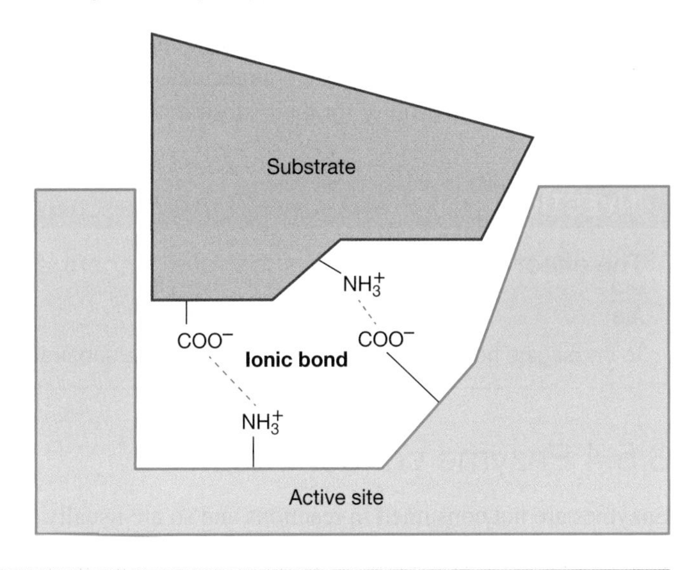

elog-0824

INVESTIGATION 3.4

online only

The effect of temperature and pH on enzyme activity

Aim

To investigate how changing the temperature and pH affects the activity of amylase

3.5.3 Substrate concentration

Initially, increasing the concentration of a substrate would be expected to result in an increase in the rate of an enzyme-catalysed reaction, as shown in figure 3.27. This happens when some enzymes have unoccupied active sites that are free to bind to substrate molecules and form E–S complexes. However, as only a fixed amount of enzyme is present, the rate of the reaction progressively tapers off until all the active sites of the enzyme molecules become occupied. The E–S complexes must dissociate before the active sites of the enzyme are free to accommodate more substrate molecules, which limits the rate of reaction.

For a given enzyme concentration, the rate of reaction increases with increasing substrate concentration — but only up to a point. Beyond this, any further increase in substrate concentration produces no significant change in reaction rate, as all the active sites of the enzyme molecules at any given moment are occupied by substrate molecules.

FIGURE 3.27 Graph showing reaction rate plotted against increasing substrate concentration

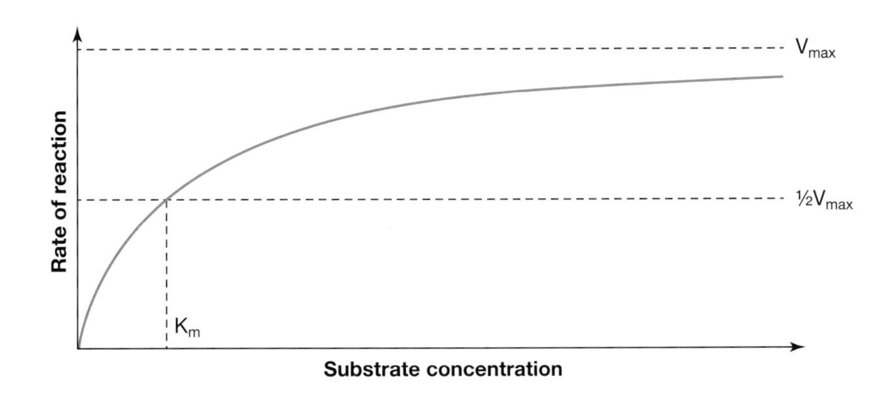

Figure 3.27 shows the idealised reaction rate of an enzyme against its substrate concentration. Note that at low substrate concentrations, the rate of reaction increases rapidly because the active sites of some enzymes are not occupied. However, the curve begins to flatten out as the substrate concentration reaches higher levels. Eventually, the maximum reaction rate (V_{max}) is reached and, at any time after that, the active sites of all enzyme molecules are saturated by substrate molecules. Enzymes differ in their ability or affinity to bind substrates. Enzymes with a high affinity for their substrate reach their maximum rate faster compared to those with a low affinity.

INVESTIGATION 3.5 online only

The effect of changing the substrate concentration on enzyme activity

Aim

To investigate how changing the concentration of a substrate impacts the activity of an enzyme

3.5.4 Enzyme concentration

Enzymes are not consumed in reactions and so are usually needed in small amounts only. For this reason, the effect of increasing enzyme concentration on reaction rate is not a common concern. Provided that the substrate concentration is high and that temperature and pH are kept constant, the rate of reaction is proportional to the enzyme concentration, as shown in figure 3.28.

FIGURE 3.28 Provided substrate concentrations are high, increases in enzyme concentration are expected to produce a continual increase in reaction rate. This may level off if substrate concentration becomes a limiting factor.

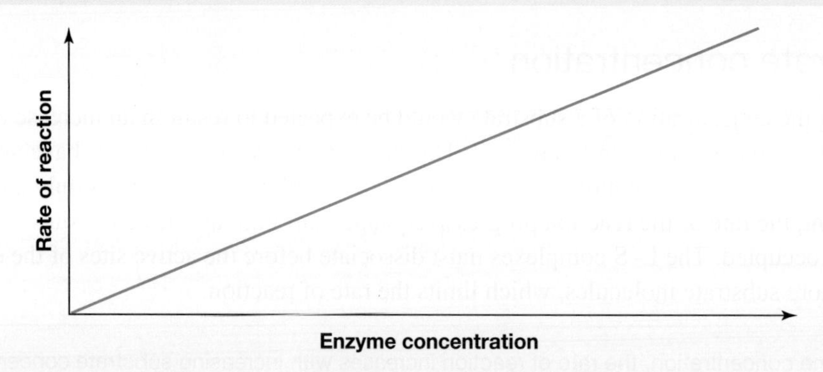

In the most unusual situation of a very high concentration of enzyme, the substrate concentration may become rate-limiting and, if so, the reaction rate will stop increasing and will flatten out, as there is not enough substrate to fill the active sites of the enzymes. This will lead to a graph with a trend similar to that shown back in figure 3.27.

INVESTIGATION 3.6 online only

Modelling the effect of substrate and enzyme concentration on reaction rate

Aim

To model the effect of increasing substrate and enzyme concentrations

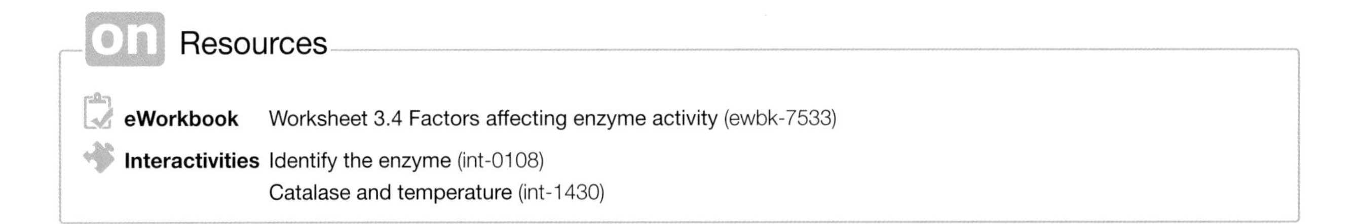

3.5.5 Competitive inhibitors

A **competitive inhibitor** of an enzyme is a molecule that contains a region with a shape that is similar to that in the usual substrate molecule (figure 3.29a). As a result, these inhibitor molecules prevent the formation of the E–S complexes. Over a given period of time, fewer substrate molecules will be able to bind to the active site of the enzyme so that the rate of the reaction is decreased (see figure 3.29b).

Competitive inhibitors bind to the active site of an enzyme, preventing substrates from binding.

FIGURE 3.29 a. In competitive inhibition, inhibitor molecules compete with the substrate molecules for access and bind to the active site of the enzyme. **b.** Graph showing the rate of enzyme action versus substrate concentration in the absence of a competitive inhibitor and in the presence of a competitive inhibitor

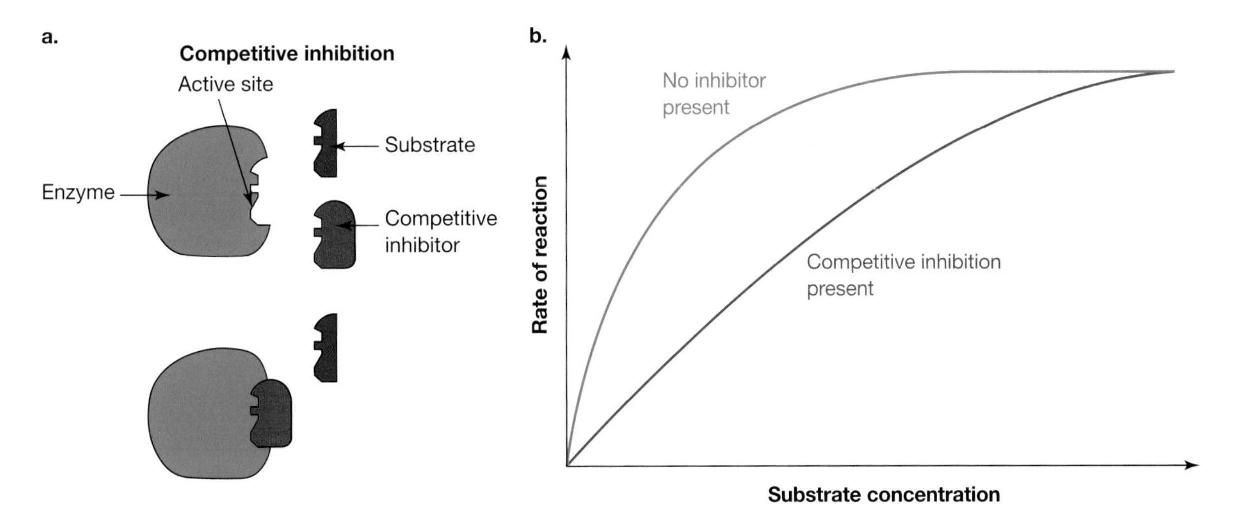

Factors affecting competitive inhibition

Because substrate and competitive inhibitor molecules compete for the same active sites on an enzyme, the degree of inhibition depends on the relative concentrations of substrate and inhibitor molecules, as shown in figure 3.29b. Adding more substrate molecules increases the chance that a random collision with the active site will be with a substrate molecule rather than a molecule of the competitive inhibitor.

So, the amount of enzyme inhibition depends upon:

- the concentration of the inhibitor
- the substrate concentration
- the relative affinities of enzyme for the inhibitor and substrate.

competitive inhibition inhibition in which a molecule binds to the active site of a molecule instead of the usual substrate

3.5.6 Non-competitive inhibitors

In **non-competitive inhibition**, the inhibitor molecule binds with the enzyme but at a site that is NOT the active site — this region is termed an **allosteric site** (figure 3.30a).

Non-competitive inhibitors bind to an allosteric site, causing a conformational change in an enzyme so it cannot properly bind with the substrate and produce a product.

The substrate and the non-competitive inhibitor are not in competition for the active site. This means that increasing the substrate concentration will not remove the effect of the non-competitive inhibition. As a result, a fixed percentage of the enzyme molecules are always inactivated by a non-competitive inhibitor.

In non-competitive inhibition, the enzyme-catalysed reaction still occurs, but the maximum rate (V_{max}) of the reaction is decreased. Figure 3.30b shows the effect on reaction rate in the presence of a non-competitive inhibitor of the enzyme involved.

non-competitive inhibition inhibition in which a molecule binds to the allosteric site of an enzyme causing a conformation change in

allosteric site location on an enzyme molecule where a compound can bind and alter the shape of the enzyme

FIGURE 3.30 a. Enzyme with an allosteric site and an active site **b.** Graph of rates of enzyme action versus substrate concentration in the absence of inhibitor (blue line) and in the presence of a non-competitive inhibitor (pink line)

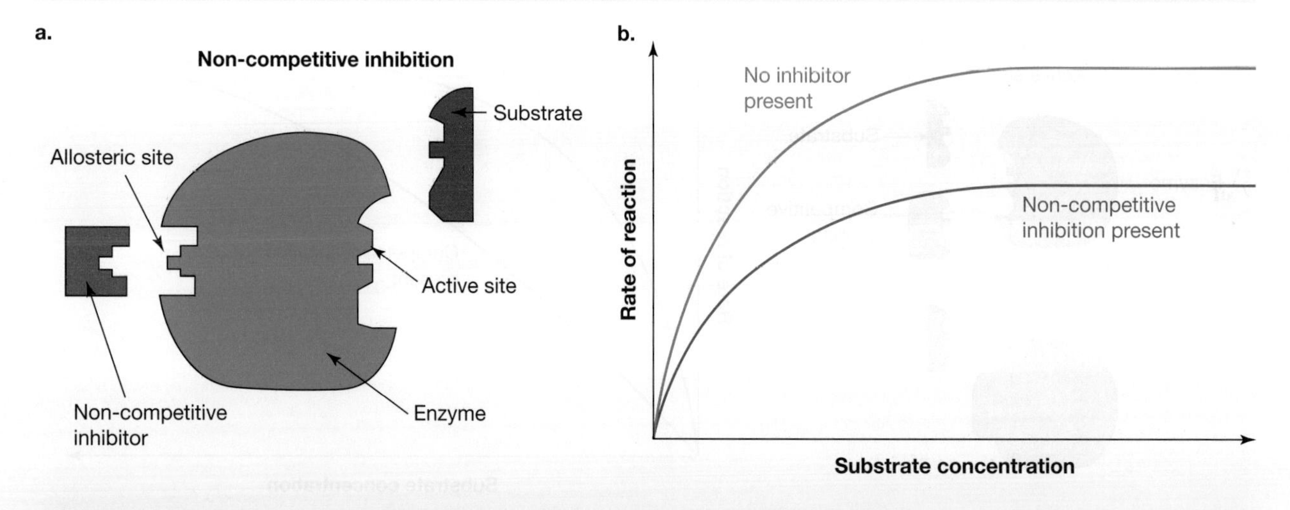

elog-0830

INVESTIGATION 3.7

online only

Inhibitors and enzyme activity

Aim

To investigate the impact of competitive and non-competitive inhibitors of enzyme activity

Irreversible inhibition

Both competitive and non-competitive inhibition of enzymes are processes that can be reversed. However, an *irreversible* inhibition of enzymes is also possible. This type of inhibition occurs when a specific molecule can form a strong covalent bond with an enzyme at its active site, so that the normal substrate is permanently blocked from accessing the active site. Not surprisingly, compounds that are irreversible inhibitors of enzymes are termed poisons and can result in death, for example:

- Exposure to DIFP (Diisopropyl fluorophosphate), a neurotoxin, produces irreversible inhibition of the enzyme acetylcholine esterase (ACE). ACE normally deactivates neurotransmitters, such as acetylcholine, after a nerve impulse has passed. DIFP binds tightly to an amino acid at the enzyme's active site and so blocks access by the normal substrate. Poisoning by DIFP disables the ACE enzyme resulting in continuous muscle contractions, constriction of the bronchial tubes of the lungs, seizures and death.

FIGURE 3.31 Diagram showing the irreversible inhibition of the enzyme acetylcholine esterase (ACE) resulting from the tight covalent binding of the poison DIFP to an amino acid residue (ser) at the active site of the enzyme

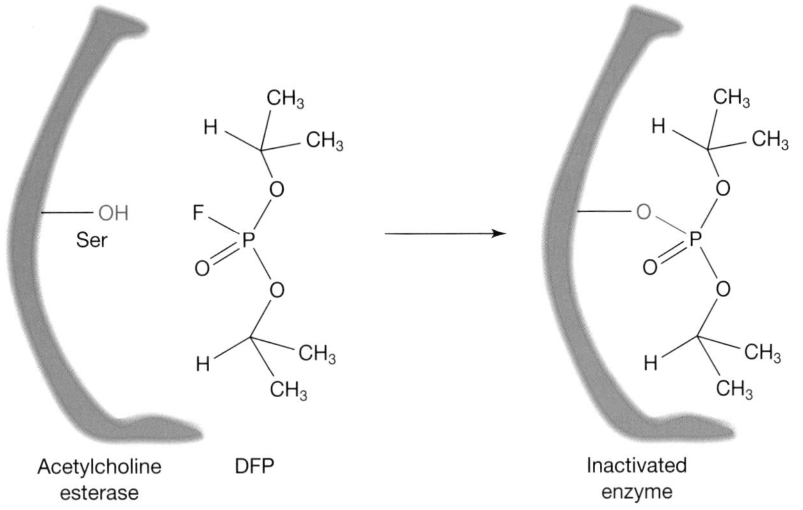

tlvd-1769

SAMPLE PROBLEM 6 Describing the action of inhibitors on enzymes

Cyanide is a toxic chemical that acts as an inhibitor on cytochrome *c* oxidase, a key enzyme in the process of cellular respiration that allows for the formation of ATP and water. Cyanide binds covalently to the enzyme and has a remarkably different structure from the usual substrate (oxygen) of cytochrome *c* oxidase.

Describe the mechanism of action of cyanide and explain how it leads to death. (2 marks)

THINK	WRITE
1. Identify what the question is asking you to do. The first part of the question is asking you to *describe*. In a describe question, you need to clearly show factual recall and communicate a concept.	
2. Consider what key factors are required to describe the mechanism of action. Is it competitive or non-competitive?	Cyanide binds to the allosteric site of cytochrome *c* oxidase, acting as a non-competitive inhibitor, as it is unable to bind to the active site (1 mark).
3. Provide an answer to the second part of the question, which asks you to *explain* how it leads to death. When being asked to explain you need to account clearly for the reasons something occurs.	As cyanide changes a key enzyme involved in cellular respiration and the production of energy, energy cannot be produced in adequate amounts to sustain life, which leads to death (1 mark).

on Resources

eWorkbook Worksheet 3.5 Inhibitors of enzyme activity (ewbk-7535)

Video eLesson Inhibition of enzymes (eles-3905)

3.5.7 Regulation of biochemical pathways

Biochemical pathways in cells do not operate continuously at maximum rates. Instead, biochemical pathways are precisely coordinated and regulated so that a balance is established between energy production and energy needs of cells. Controlling enzyme activity is a major means by which regulation of biochemical pathways is achieved.

Regulation of the rates of reaction of the steps in biochemical pathway is necessary:

- to prevent waste, such as would occur if pathway products were made in excess of cell requirements. For example, the ATP produced by cellular respiration cannot be stored and, because it is unstable, any excess ATP loses its energy and is wasted. The production of ATP is tightly regulated so that the rate of supply meets the cell demand. At times of high demand, the rate ATP production is stepped up and at times of lower demand, the rate is slowed.
- to prevent the build-up in cells of products to potentially harmful levels
- to prevent depletion of substrates. If the supply of an initial reactant is continually used at high rates in biochemical reactions it will deplete or exhaust the supply of the substrate, causing possible problems.

How are biochemical pathways regulated?

Let's look at how a biochemical pathway may be regulated.

- Pathways may be down-regulated by slowing or stopping the activity of specific enzymes in the pathway.
- Pathways may be up-regulated by increasing the activity of specific enzymes.

Control of a pathway depends on regulating the activity of one or more enzymes in the pathway.

Means of regulation enzyme activity include:

- **allosteric regulation**
- **feedback inhibition.**

Allosteric regulation

Allosteric regulation of one or more enzymes in the pathway is controlled through conformational changes in enzymes. This occurs when a regulator molecule binds to a specific site on an enzyme (but not its active site) and this binding produces a change in the enzyme shape that affects its activity. Binding by regulator molecules to enzymes is not permanent and can be reversed.

Regulator molecules may be:

- **allosteric inhibitors**: their binding produces a change of shape in the enzyme that stops enzyme activity; they act like an OFF switch.
- **allosteric activators**: the shape change resulting from the binding produces an increase in enzyme activity; they act like an ON switch (see figure 3.32).

allosteric regulation the control of the reaction rate of enzymes through conformational changes in enzymes

feedback inhibition inhibition occurs when the end product of a pathway inhibits an enzymes earlier in the pathway as a negative feedback mechanism; also known as end-product inhibition

allosteric inhibitors molecules that bind to the allosteric site of an enzyme and stop enzyme activity

allosteric activators molecules that bind to the allosteric site of an enzyme and increase enzyme activity

FIGURE 3.32 Shape changes in an enzyme from the binding of an allosteric inhibitor (at left) prevents its catalytic activity, preventing a substrate binding. The binding of an allosteric activator (right) causes shape changes that activate an enzyme, allowing the substrate to bind.

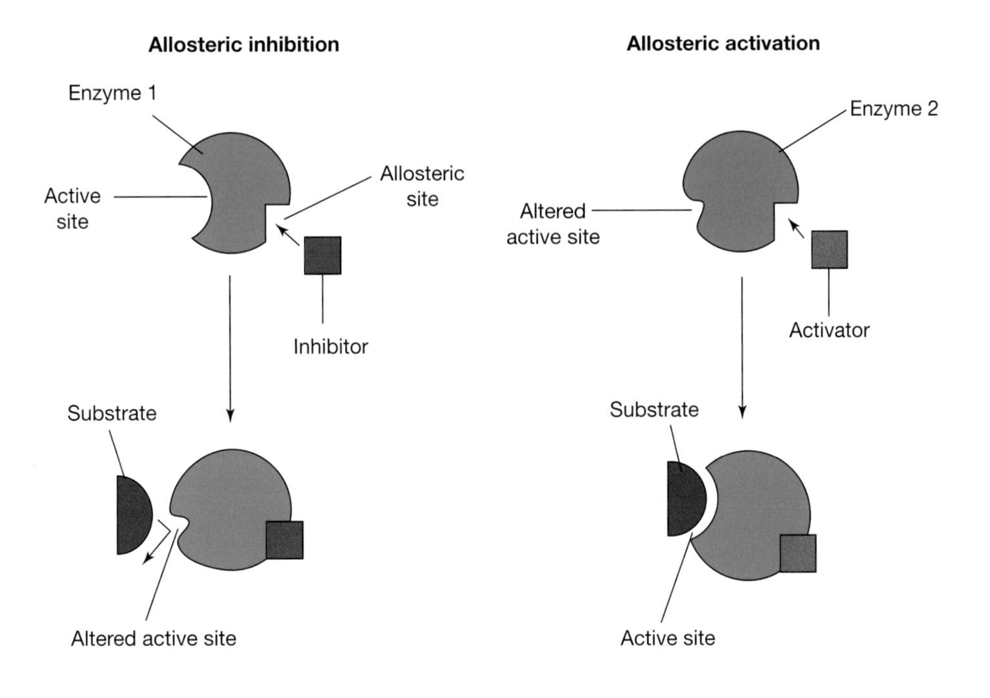

CASE STUDY: Examples of allosteric regulation

Allosteric inhibition of pyruvate kinase

During cellular respiration, an substrate known as phosphoenol pyruvate (PEP) is converted to pyruvate. This pathway leads to the production of ATP. A highly simplified version of this pathway is shown below:

$$\text{Phosphoenol pyruvate} + \text{ADP} + \text{Pi} \xrightarrow{\text{PK}} \text{Pyruvate} + \text{ATP}$$

ATP also acts as a non-competitive inhibitor for the pyruvate kinase (PK) enzyme that catalyses this step. This inhibition by ATP at high levels stops the production of pyruvate from phosphoenol pyruvate (PEP) by allosteric inhibition. This inhibition also prevents the continued production of ATP that would otherwise be in excess of requirements and ATP cannot be stored. When ATP levels fall, this inhibition is lifted and the energy production can resume.

Allosteric regulation of phosphofructokinase

ATP is also an allosteric inhibitor of the phosphofructokinase (PFK) enzyme that catalyses early steps of the cellular respiration pathway. This inhibition only occurs when ATP supplies are plentiful.

AMP (adenosine monophosphate) is an allosteric activator of the PFK enzyme. This makes sense since AMP is generated particularly when ATP supplies are limited and cells are beginning to be starved of energy. The presence of AMP increases the activity of PFK, until the ATP levels are back to normal.

Other enzymes in later steps of the cellular respiration pathways are also subject to allosteric inhibition by ATP and also NADH, as well as allosteric activation by AMP and ADP.

Feedback inhibition

Regulation of biochemical pathways can also occur through feedback inhibition, also known as **end-product inhibition**. In this process, the end product of a metabolic pathway acts as an inhibitor of the key enzyme that catalyses the first step in a pathway (figure 3.33). Feedback inhibition comes into operation when the end product is plentiful and abundant. This leads to the first enzyme being inhibited, leading to a reduction in the production of the end product (and other products in the pathway). When the end product falls to low levels, the inhibition ceases and the biochemical pathway resumes, replenishing the supply of the end product.

end-product inhibition inhibition of the early stage of a multi-step pathway by the final product of that pathway

FIGURE 3.33 In feedback inhibition, the abundance of the end product acts as an inhibitor of the key enzyme of the first step in the pathway

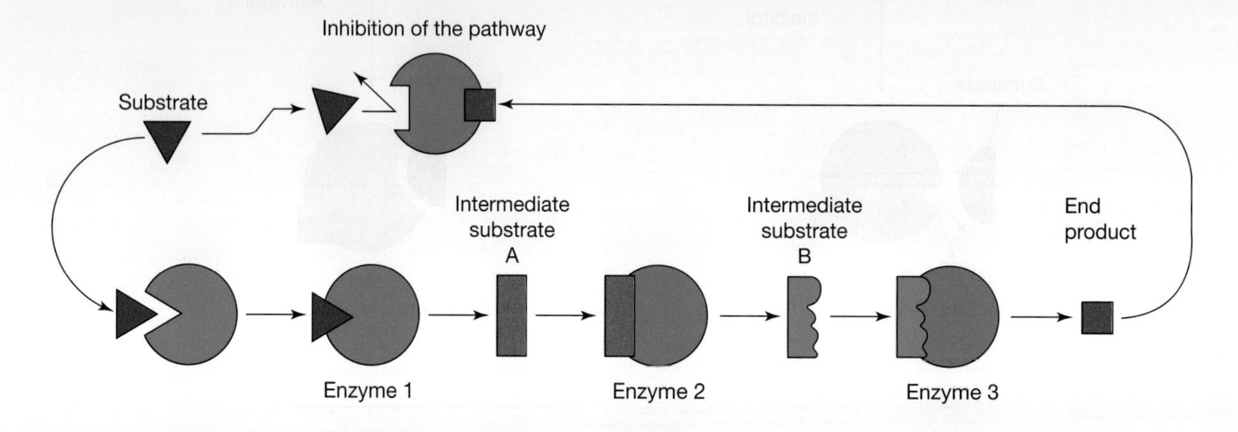

on Resources

eWorkbook Worksheet 3.6 Regulation of biochemical pathways (ewbk-7537)

KEY IDEAS

- The rate of an enzyme-catalysed reaction is affected by several factors, mainly temperature, pH and concentration.
- Increasing temperature causes an initial increase in enzyme activity due to increased collisions between molecules and substrates. Eventually, the enzyme (being protein based) begins to denature, causing enzyme activity to decrease.
- Enzymes have an optimal pH. A pH that is too high (too alkaline) or too low (too acidic) causes the substrate to not be able to bind properly to the enzyme's active site.
- Increasing both enzyme and substrate concentration causes an increase in enzyme rate. However, this eventually levels off if another factor is limiting the reaction rate.
- Enzymes can be inhibited and prevented from catalysing reactions.
- Reversible inhibition includes competitive and non-competitive inhibition.
- End-product or feedback inhibition is a form of non-competitive inhibition that is used to regulate the operation of a metabolic pathway.

3.5 Activities

To answer questions online and to receive **immediate feedback** and **sample responses** for every question, go to your learnON title at **www.jacplus.com.au**. A **downloadable solutions** file is also available in the resources tab.

3.5 Quick quiz	3.5 Exercise	3.5 Exam questions

3.5 Exercise

1. Identify the following statements as true or false and, if false, rewrite it as a true statement
   a. Regulation of biochemical pathways can be achieved by regulating the activity of rate-determining enzymes in the pathway.
   b. All enzyme molecules are saturated with substrate molecules when the maximum reaction rate of an enzyme is reached.
   c. Inhibitors always bind to the active site.
   d. To regulate a particular pathway, every enzyme involved in the pathway must be regulated.
   e. Enzymes in people would be expected to have a temperature optimum that varies according to the external ambient temperature.
2. Using figure 3.25, give an example of a human enzyme with
   a. an optimum pH of about 2.0
   b. an optimum pH of about 8.0.
3. **a.** What is a key difference between competitive and non-competitive inhibition of enzymes?
   b. Construct a graph to compare the rate of reaction in the presence of a competitive inhibitor, non-competitive inhibitor and no inhibitor as the level of a substrate increases.
4. Identify if each of the following statements applies to a competitive or non-competitive inhibitor.
   a. Its inhibition can be reduced adding more substrate.
   b. It binds reversibly to the enzyme, but not to the active site.
   c. Its structure is similar to that of the normal substrate.
   d. It can form a covalent link at the enzyme's active site, permanently excluding the substrate.
   e. It causes a conformational change in an enzyme, changing its shape
5. Suggest a possible explanation for the following observations
   a. Salivary amylase enzyme breaks down the starch in food to its glucose sub-units. This glucose is then used for cellular respiration. This breakdown starts in the mouth, but when the food reaches the stomach, the rate of enzyme action drops to zero.
   b. Enzyme J and enzyme K are both exposed to the same competitive inhibitor. Only the functionality of enzyme J is affected.
   c. Above a critical temperature, there is a rapid rate of loss of enzyme activity.
6. The normal substrate of the enzyme succinate dehydrogenase is succinic acid. However, malonic molecules can also bind to this enzyme. The structure of these two molecules are

```
 O    H   H    O          O    H    O
  \\  |   |   //           \\  |   //
   C—C—C—C                  C—C—C
  /   |   |   \            /   |   \
HO    H   H    OH        HO    H    OH
```

Succinic acid **Malonic acid**

   Given this information, what prediction might reasonably be made about the kind of interaction that takes place between malonic acid and succinic acid relative to the succinate dehydrogenase enzyme? Give a reason for your answer.
7. During cellular respiration, hydrogen peroxide may be generated. This is broken down by catalase into water and oxygen. Design an experiment to show the effect of catalase concentration on the breakdown of hydrogen peroxide.
8. Glycogen phosphorylase is an enzyme that converts glycogen into glucose so it can be used to produce energy through cellular respiration. When glucose levels are low, glycogen phosphorylase is active, and glycogen binds to its active site. When glucose levels are high, the enzyme is inactive. Describe the mechanism of action of glucose and explain how it is important in its own regulation.

9. The graph below shows the effect of pH on the activity of an enzyme with the same function isolated from two locations in the body.

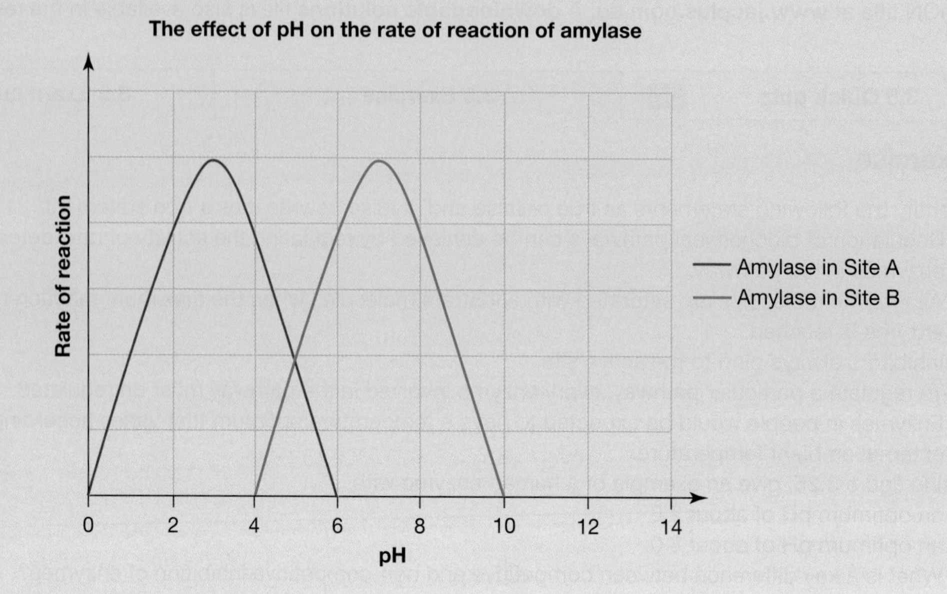

a. What prediction might be made about the sites in which these enzymes act? Give a brief explanation for your prediction.
b. One of these enzymes acts in the small intestine and one acts in the mouth. Which would you expect to be which? Justify your response.
c. What is the optimum pH for the amylase:
   i. at site A
   ii. at site B?
d. Why does the rate of reaction increase on the upside (left-hand side) of the curves?
e. Why does the rate of reaction decrease on the downside (right-hand side) of the curves?
f. Estimate the pH at which the amylase at site B would operate at about half its maximum rate.

10. Give two reasons why regulation of biochemical pathways is necessary.

11. If a person suffers an injury causing bleeding, an interconnected pathway is initiated to stop the bleeding by clot formation. Some of the reactions in this process are as follows:
- Prothrombin (an inactive enzyme) is converted to thrombin, an active enzyme; this reaction is catalysed by the enzyme, prothrombinase, also known as factor Xa.
- Soluble fibrinogen is converted to insoluble fibrin that forms a clot that stops the bleeding; this reaction is catalysed by the thrombin enzyme.

a. Show the word equations showing the relationship between these two reactions.
b. Biting blood-feeding invertebrates, including leeches, ticks and tse-tse flies, produce anti-coagulants in their saliva that prevent the blood of the mammal or bird that they have bitten from clotting. Based on the information provided, suggest a possible mode of action of this anti-coagulant.

3.5 Exam questions

Question 1 (3 marks)

Source: *Adapted from VCAA 2015, Biology Exam, Section B, Q1d*

Enzymes are protein catalysts.

Use labelled diagrams to illustrate both enzyme denaturation and enzyme inhibition. Include both the enzyme and substrate in your diagrams.

Question 2 (1 mark)

Source: *VCAA 2018 Biology Exam, Section A, Q7*

MC Four students performed a series of experiments to investigate the effect of four different variables on the rate of an enzyme-catalysed reaction. In each experiment the students changed one of the following variables: substrate concentration, pH, temperature and enzyme concentration. After recording their data, the students displayed their results in a series of graphs, as shown below. Each graph is a line of best fit for their data.

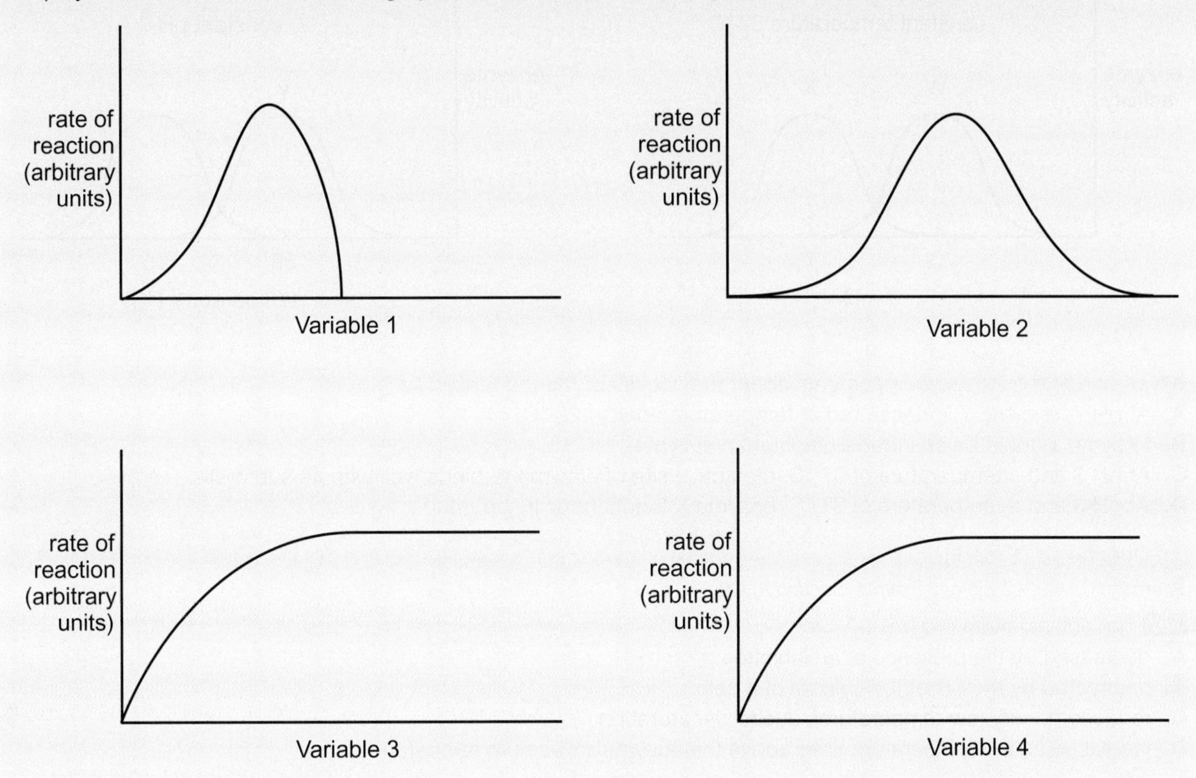

The students did not label the horizontal axis on any of their graphs. The next day, the students could not agree on which variable should be labelled on the horizontal axis of each graph. The students made the following suggestions as to what each variable could be.

Student	Variable 1	Valuable 2	Variable 3	Valuable 4
Marcus	substrate concentration	temperature	pH	enzyme concentration
Billy	temperature	substrate concentration	enzyme concentration	pH
Voula	enzyme concentration	temperature	substrate concentration	pH
Sheena	temperature	pH	enzyme concentration	substrate concentration

Which student correctly identified all four variables on the horizontal axes?

A. Marcus
B. Billy
C. Voula
D. Sheena

Question 3 (1 mark)

Source: *VCAA 2014 Biology Exam, Section A, Q13*

MC The following graphs show the way four enzymes, W, X, Y and Z, change their activity in different pH and temperature situations.

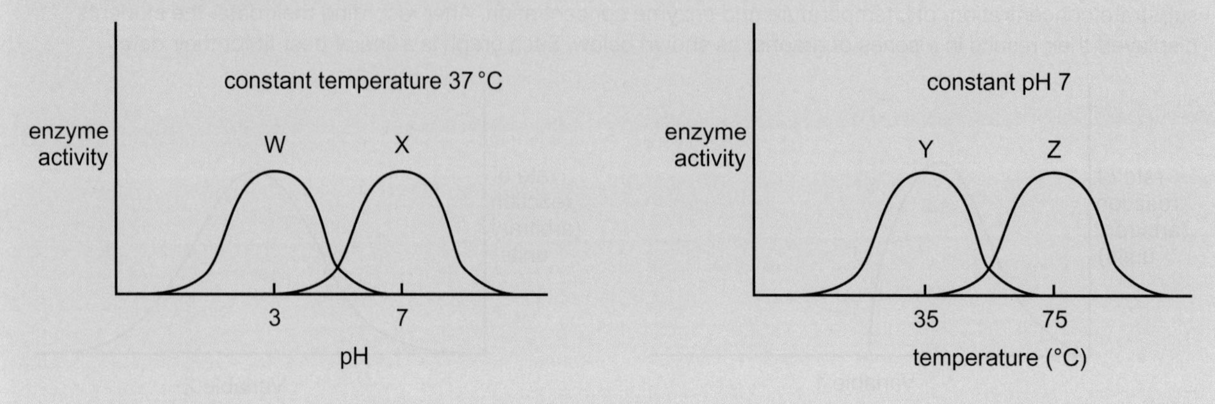

Which one of the following statements about the activity of the four enzymes is true?

A. At pH 7, enzyme Y is denatured at temperatures below 20 °C.
B. Enzyme Z could be an intracellular human enzyme.
C. At pH 3 and a temperature of 37 °C, the active site of enzyme W binds well with its substrate.
D. At pH 3 and a temperature of 37 °C, enzyme X functions at its optimum.

Question 4 (1 mark)

Source: *VCAA 2012 Biology Exam 1, Section A, Q2*

MC The activity of an enzyme is

A. decreased by the presence of an inhibitor.
B. unaffected by the pH of the cytosol of a cell.
C. reduced at very low temperatures due to denaturation.
D. increased as the temperature rises above the enzyme's optimum temperature.

Question 5 (2 marks)

Source: *VCAA 2016 Biology Exam, Section B, Q2d*

Plant materials containing cellulose and other polysaccharides are reacted with acids to break them down to produce glucose. This glucose is then used by yeast cells for fermentation.

A by-product of the acid treatment of plant materials is a group of chemical compounds called furans. It has been observed that as the concentration of furans increases, the rate of fermentation decreases. The enzyme alcohol dehydrogenase is required for the process of fermentation.

Scientists have proposed that furfural is a competitive inhibitor of the enzyme alcohol dehydrogenase.

Explain how furfural could act as a competitive inhibitor of the enzyme alcohol dehydrogenase.

More exam questions are available in your learnON title.

3.6 Review

3.6.1 Topic summary

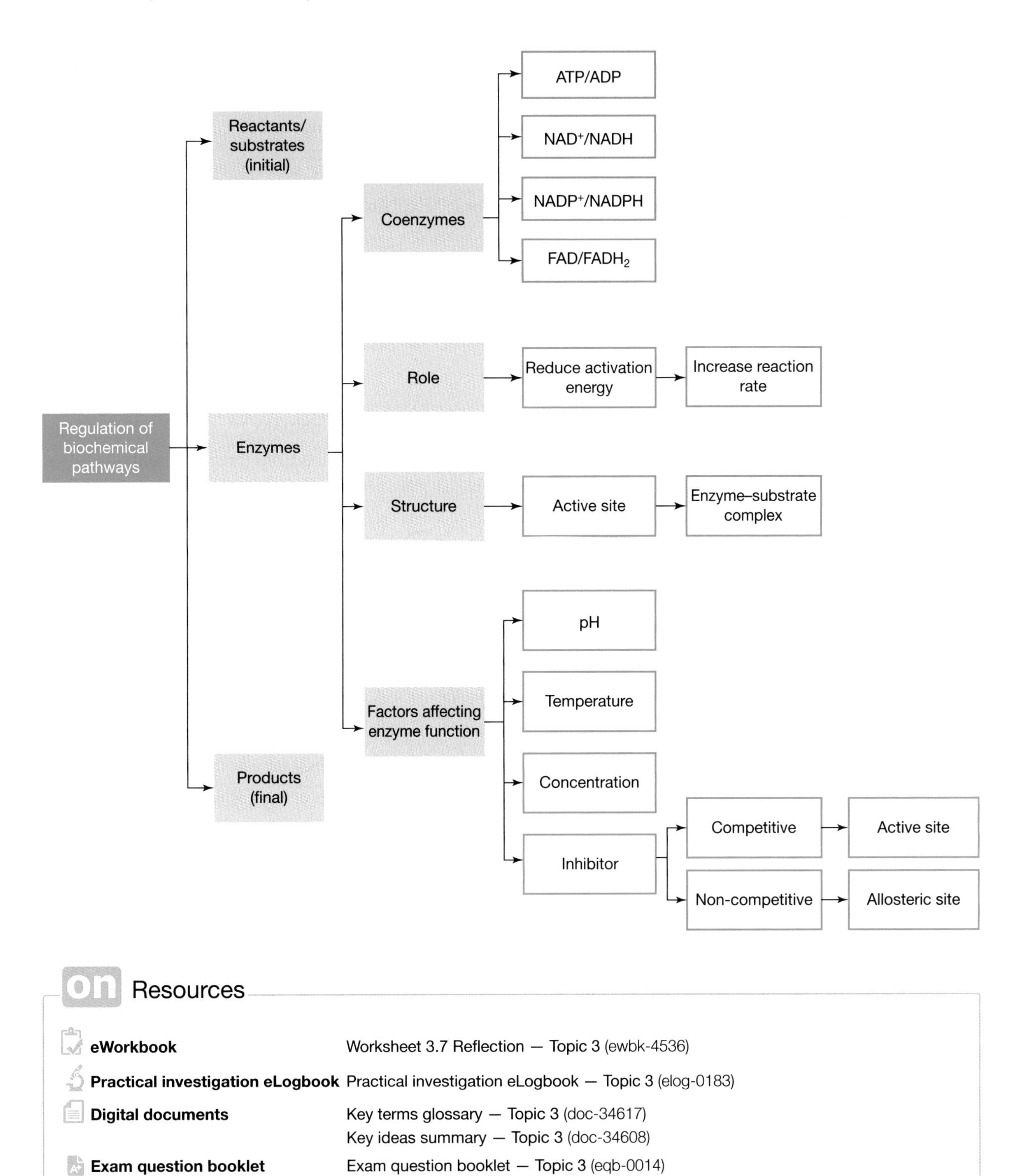

on Resources

eWorkbook	Worksheet 3.7 Reflection — Topic 3 (ewbk-4536)
Practical investigation eLogbook	Practical investigation eLogbook — Topic 3 (elog-0183)
Digital documents	Key terms glossary — Topic 3 (doc-34617) Key ideas summary — Topic 3 (doc-34608)
Exam question booklet	Exam question booklet — Topic 3 (eqb-0014)

3.6 Exercises

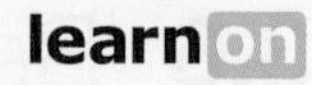

To answer questions online and to receive **immediate feedback** and **sample responses** for every question, go to your learnON title at **www.jacplus.com.au**. A **downloadable solutions** file is also available in the resources tab.

3.6 Exercise 1: Review questions

1. Examine the figure shown. Compound A is a type of protein.
   a. What is the general name given to the type of protein represented by compound A?
   b. What are the general names given to each of the parts labelled B, C, D and E?
   c. Name the type of reaction shown in this diagram — is it catabolic or anabolic? Explain.
   d. Is this reaction more likely to be endergonic or exergonic? Explain.

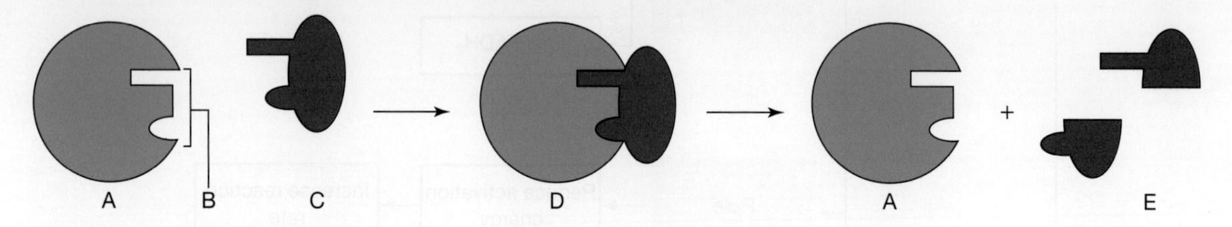

2. Explain in your own words the term end-product inhibition (or feedback inhibition).
3. The graph below shows the effect of temperature on the activity of an enzyme from three different organisms.

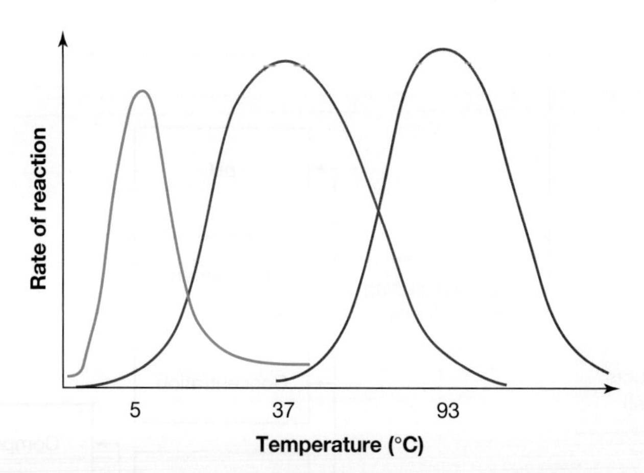

   a. What predictions, if any, might be made about the organisms that were the sources of these enzymes? Give a brief rationale for any predictions that you make.
   b. Note that the shapes of the curves are different. Suggest a reasonable explanation for these differences.
4. Amino acids in a polypeptide chain are often numbered, starting from the amino terminal; for example, the third amino acid might be Val-3 and the tenth in the sequence might be Leu-10, and so on . . .
   The active site of the enzyme trypsin, a digestive enzyme, consists of three amino acid as follows: Asp-102, His-57 and Ser-195; that is, these three amino acids are the 57th, the 102nd and the 195th amino acids in the primary structure of this enzyme.
   a. Identify two events that could occur at the active site of an enzyme.
   b. What stabilises a substrate in the active site of an enzyme?
   One student suggested that the three amino acids at the active site of trypsin had been misnumbered. The student said that there is quite a distance separating them so that they cannot be close together at the active site.
   c. Do you agree with the student? Give a clear explanation for the decision you have made.

5. Identify a key difference between the members of the following pairs
   a. substrate and product
   b. loaded NAD and unloaded NAD
   c. competitive inhibition and feedback inhibition
   d. catabolism and anabolism.
6. Examine this figure. Explain what this diagram is showing.

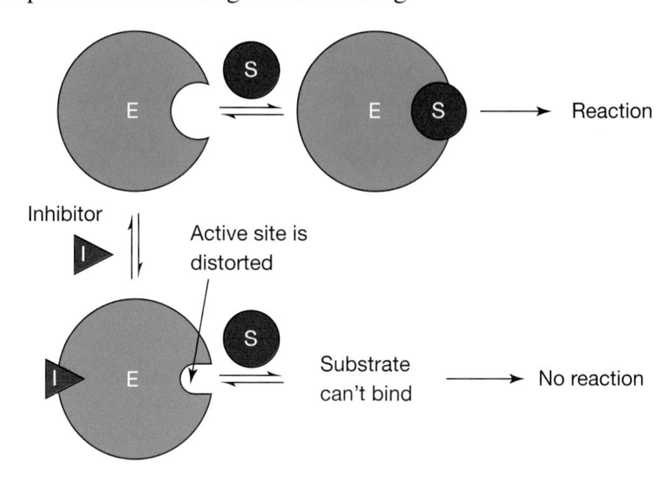

7. Consider the following figure, which shows the total amount of product formed during an enzyme-catalysed reaction.

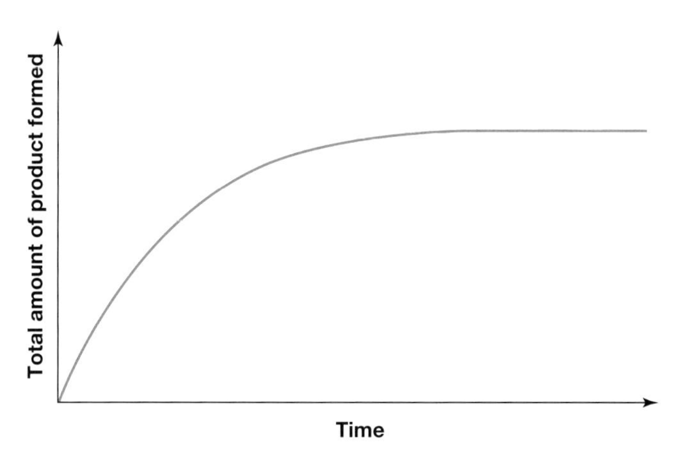

   Explain the following features of this graph:
   a. the initial steep slope
   b. the slowdown in the rate of product formation
   c. the flattening of the curve to form a plateau.
8. Consider each of the following observations and formulate an explanation for the observation.
   a. Salivary amylase enzyme breaks down starch. The process begins when saliva mixes with food in the mouth but, when it reaches the stomach, the amylase enzyme no longer functions.
   b. The jelly-like white of a raw hen's egg consists principally of the protein albumin but, after it is cooked, it forms a white solid.
   c. When an inhibitor binds to an enzyme at a site other than the active site, this inhibition cannot be overcome by increasing the concentration of substrate molecules.
   d. Enzyme-catalysed reactions operate at much faster rates than non-catalysed reactions.

9. Protein X (shown in the figure) is an enzyme.

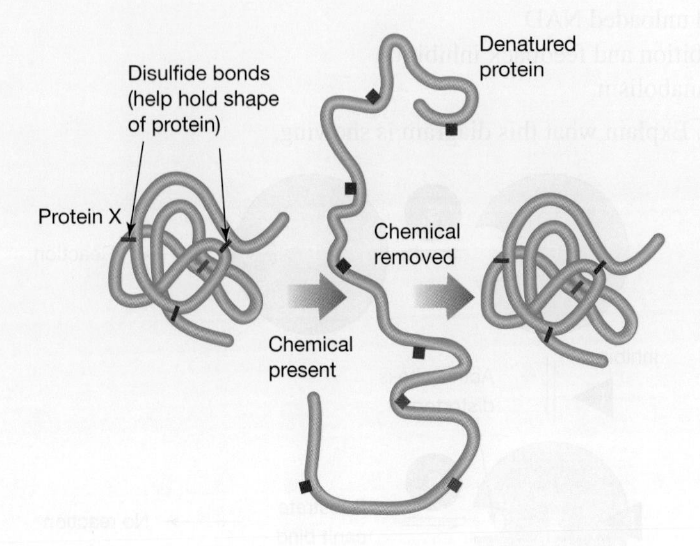

a. What is the general structure of protein X (as shown in the left-hand image)?
b. When treated with a particular chemical, the shape of protein X was radically changed, as shown in the middle image. What specific effects has the chemical had on protein X?
c. Describe what happened when the chemical was removed.
d. Identify the action of the chemical on protein X.
e. How, if at all, would the ability of protein X to catalyse its specific reaction be affected when it was in the 'long snake form'?

10. The enzyme succinate dehydrogenase is active in the cellular respiration pathway where it catalyses the reaction below and requires the coenzyme NAD^+ for its action:

$$C_4H_6O_4 \rightarrow C_4H_4O_4$$

succinic acid → fumaric acid

a. What is the product of the action of this enzyme on succinic acid?
b. What is the role of the NAD^+ coenzyme in this reaction?
c. Which molecule of NAD is in the unloaded form and which molecule is in the loaded form?
d. Would the rate of cellular respiration be expected to increase, decrease or stay the same if fumaric acid was added to cells undergoing cellular respiration?

11. The main target for regulation of a biochemical pathway is a key enzyme in that pathway.
a. Are enzymes good targets for regulation of biochemical pathways?
b. Identify two means by which enzymes can be regulated.
c. A student suggested that pathways could be regulated by reducing or increasing the concentration of the initial substrate in the cell. Do you agree or disagree with this student? Explain your decision.

The enzyme phosphofrutokinase (PFK) catalyses an early reaction in the 10-step glycolysis pathway that is the initial stage of the cellular respiration pathway. The glycolysis pathway produces ATP.
d. How might regulation of the PFK enzyme prevent the excessive production of ATP?
e. When and by what means would this regulation stop?

3.6 Exercise 2: Exam questions

on Resources

Teacher-led videos Teacher-led videos for every exam question

Section A — Multiple choice questions

All correct answers are worth 1 mark each; an incorrect answer is worth 0.

Question 1

Source: *VCAA 2020 Biology Exam, Section A, Q4*

A group of Biology students set up an experiment with three test tubes. Each test tube contained the same enzyme and was kept under the same experimental conditions. All concentrations and volumes of substrate and enzyme placed in the test tubes were kept the same. The following diagram shows the initial contents of each test tube and the final contents of test tubes 1 and 2.

	Initial contents of each test tube				**Final contents of each test tube**		
Test tube 1	Enzyme	+	Substrate A	⟶	Enzyme	+	Product
Test tube 2	Enzyme	+	Substrate B	⟶	Enzyme	+	Substrate B
Test tube 3	Enzyme	+	Substrate A + Substrate B	⟶	**?**		

An enzyme–substrate complex was formed in each of the three test tubes.

When compared to the final contents of Test tube 1, the concentration of product contained in the final contents of Test tube 3 will

A. be the same because both Substrate A and Substrate B are able to bind to the enzyme at the same time.
B. increase because Substrate A is a substrate that can bind to the enzyme's active site.
C. decrease due to competitive reversible inhibition by Substrate B.
D. be zero due to the presence of an irreversible inhibitor.

Question 2

Coenzymes such as NADPH and ATP are

A. enzymes that require a cofactor to function.
B. the substrate that binds in the active site.
C. inorganic substances that also include metal ions.
D. organic compounds that act with an enzyme to increase the rate of a reaction.

Use the following information to answer Questions 3 and 4

The enzyme lactase digests lactose.

$$\text{lactose} \xrightarrow{\text{lactase}} \text{glucose + galactose}$$

Two test tubes where set up using 5 mL of lactose syrup and 0.5 mL of lactase. Test tube **one** was incubated at 37 °C, while test tube **two** was incubated at 15 °C. Both tubes were incubated for 10 minutes.

Question 3

Source: *VCAA 2009 Biology Exam 1, Section A, Q7*

At the end of 10 minutes, the amount of glucose produced in test tube **two** when compared to test tube **one** would be

A. lower as the enzyme's active site would have denatured at this temperature.
B. equal as lowering the temperature does not affect digestion of lactose.
C. lower as there would be fewer collisions between the substrate and the enzyme.
D. equal as the two test tubes contained the same amount of lactose and lactase enzyme.

Question 4

Source: *VCAA 2009 Biology Exam 1, Section A, Q8*

In another experiment, test tube **three** was compared with test tube **four**. Each tube contained 5 mL of lactose syrup. Tube **three** contained 0.5 mL of lactase and tube **four** contained 0.25 mL of lactase. The two tubes were incubated at 15 °C and monitored for 10 minutes.

The result for test tube **three** is shown below

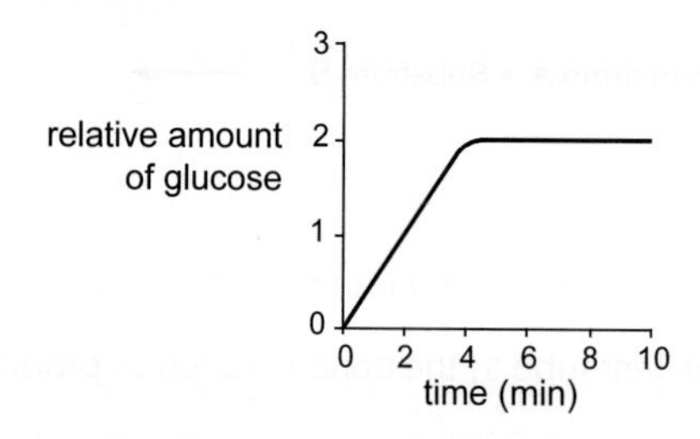

The graph of results for tube **four** would resemble

A.
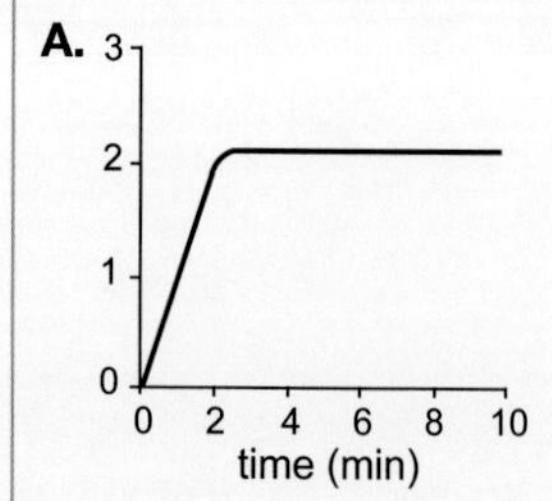

B.
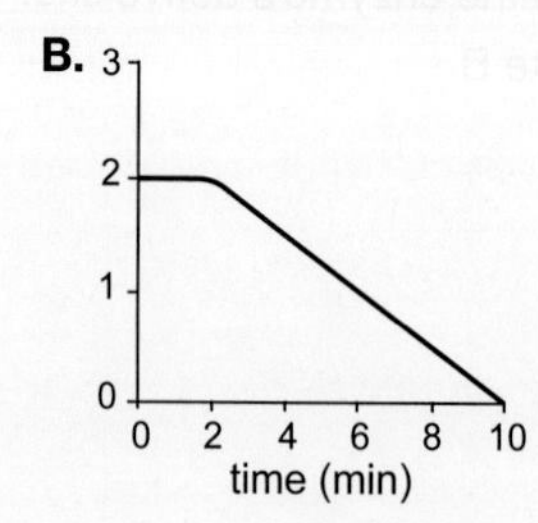

C.
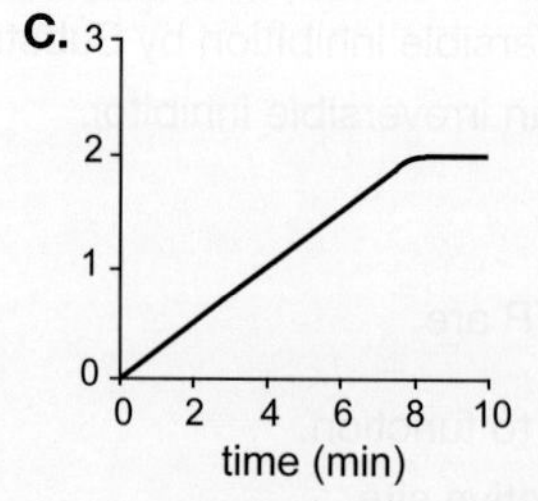

D.
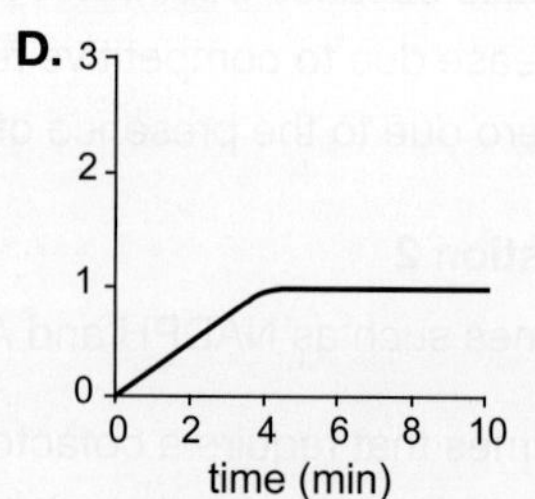

Use the following information to answer Questions 5 and 6

Hydrogen peroxide is a toxic by-product of many biochemical reactions. Cells break down hydrogen peroxide into water and oxygen gas with the help of the intracellular enzyme catalase. The optimum pH of catalase is 7.

A Biology student measured the activity of catalase by recording the volume of oxygen gas produced from the decomposition of hydrogen peroxide when a catalase suspension was added to it. The catalase suspension was made from ground, raw potato mixed with distilled water. The student performed two tests and graphed the results.

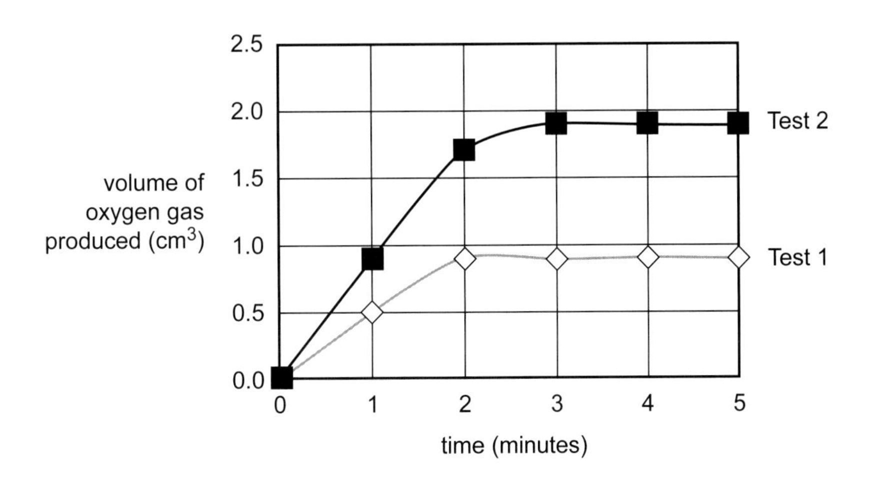

Test 1 used 5 mL of 3% hydrogen peroxide solution and 0.5 mL of catalase suspension, and was conducted at 20 °C in a buffer solution of pH 7.

Test 2 was carried out under identical conditions to Test 1, except for one factor that the student changed.

Question 5

***Source:** VCAA 2017 Biology Exam, Section A, Q7*

An explanation for the results of Test 2 would be that the student

A. increased the concentration of catalase by adding less water to the ground potato.
B. increased the temperature by placing the test tube in a water bath set at 30 °C.
C. used a hydrogen peroxide solution with a higher concentration.
D. added a catalase suspension made from a cooked potato chip.

Question 6

***Source:** VCAA 2017 Biology Exam, Section A, Q8*

The student then performed more tests by varying the pH of the buffer solution.

It is expected that

A. at pH 6 the reaction will cease.
B. at pH 9 the reaction will be faster.
C. at pH 2 and pH 10 very little oxygen will be produced.
D. a greater volume of oxygen will be produced each time the pH is increased.

Question 7

***Source:** VCAA 2019 Biology Exam, Section A, Q6*

Acetylcholinesterase is an enzyme that catalyses the breakdown of the neurotransmitter acetylcholine into acetate and choline.

Some chemicals used to kill insects contain aldicarb.

Aldicarb is a reversible inhibitor of acetylcholinesterase that

A. permanently blocks the active site of acetylcholinesterase.
B. acts by strongly binding to the active site of acetylcholinesterase.
C. increases the rate at which acetylcholine is broken down into acetate and choline.
D. reduces acetylcholinesterase activity by reducing the number of active sites for acetylcholine to bind to.

Question 8

***Source:** VCAA 2020 Biology Exam, Section A, Q13*

Celluclast® is an enzyme. The activity of Celluclast® at a range of temperatures and at a pH of 5 was measured. The experiment was repeated five times. The relative activity (%) of Celluclast® was calculated and plotted on a graph, as shown below. The range of the calculated measurements at each temperature is shown as an error bar on the graph.

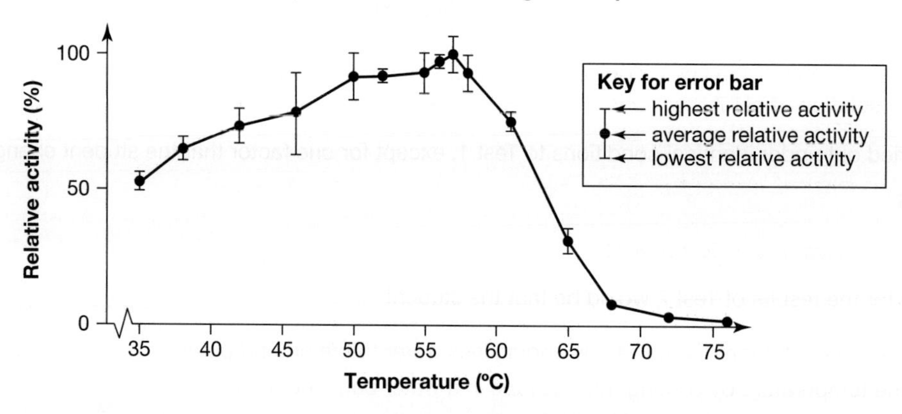

It is reasonable to conclude that

A. Celluclast® is inactive at 61 °C.
B. Celluclast® is denatured at 35 °C.
C. the optimum pH for Celluclast® is pH 5.
D. the optimum temperature for Celluclast® is around 57 °C.

Question 9

Isocitrate dehydrogenase is an enzyme used during the process of cellular respiration. Which of the following is correct regarding the activity of isocitrate dehydrogenase in humans?

A. At 45 °C, the peptide bonds in the primary structure of isocitrate would be destroyed.
B. At 10 °C, the number of collisions between isocitrate dehydrogenase and the substrate would be reduced.
C. At 37 °C, isocitrate and substrate molecules would not have sufficient energy to react.
D. At 65 °C, the denaturation of isocitrate is reversible.

Question 10

Many animals can produce their own vitamin C in their liver or kidneys using glucose as an input. The following biochemical pathway shows the multi-step conversion of glucose to vitamin C.

Glucose →(E1) →(E2) →(E3) Glucuronate →(E4) →(E5) Gulonate →(E6) Gulono-1, 4-lactone →(GULO enzyme) Ascorbic acid (vitamin C)

In order for the levels of vitamin C to be regulated, it is likely that

A. gulonate binds irreversibly to E4 to stop the metabolic pathway.
B. gulono-1, 4-lactone will be converted into glucose when vitamin C is too high.
C. The GULO enzyme is permanently denatured as vitamin C increases.
D. vitamin C acts as a reversible inhibitor E1.

Section B — Short answer questions

Question 11 (6 marks)

Source: *VCAA 2014 Biology Exam, Section B, Q1*

CTP is a substance used by cells to make RNA. The cell initially synthesises CTP using a metabolic pathway starting with the amino acid aspartane (A) and another complex molecule (B).

The pathway for making CTP is represented below. The enzyme involved in the first step of the pathway is called ATCase.

A + B →(ATCase) D →(Enzyme X) intermediate reactions not shown →(Enzyme Y) N →(Enzyme Z) CTP

a. What is the role of ATCase? Explain how it performs this role. **2 marks**

The graph below shows the change in the rate of production of D in solutions with different concentrations of CTP, keeping all other variables constant

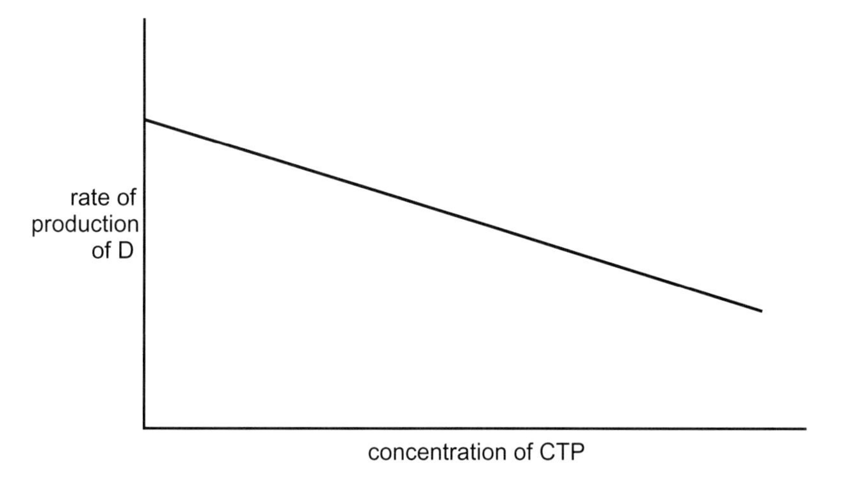

b. i. Using the information in the graph, state what is happening to the rate of production of D as the concentration of CTP increases. **1 mark**

ii. CTP changes the quaternary structure of the enzyme ATCase.

Use this information to explain at a molecular level how the production of CTP is regulated within a cell. **3 marks**

Question 12 (5 marks)

Source: *VCAA 2018 Biology Exam, Section B, Q8a and b*

Methotrexate is a drug used in the treatment of some cancers and an autoimmune disease, psoriasis, which affects the skin and sometimes the joints. In both of these diseases, cells grow rapidly. Methotrexate is structurally very similar to folic acid. Methotrexate works by inhibiting an enzyme that catalyses the change of an inactive form of folic acid into an active form. The active form of folic acid is needed for DNA production.

a. Based on the information given, explain how methotrexate could be acting as a competitive inhibitor of the enzyme. **2 marks**

b. Explain why methotrexate affects cell growth. **3 marks**

Question 13 (3 marks)

Source: *VCAA 2017 Biology Exam, Section B, Q3b*

Many people in Australia have high blood cholesterol levels. High blood cholesterol levels are linked to coronary heart disease. Cholesterol is produced by a series of chemical reactions. One of the reactions in the series is shown below.

$$\text{HMG-CoA} \xrightarrow{\text{HMG-CoA reductase}} \text{mevalonate}$$

Lovastatin is used to treat high blood cholesterol. When administered, the lovastatin is converted to an active form that is a reversible competitive inhibitor of HMG-CoA reductase.

i. What does this suggest about the structure of the active form of lovastatin? **1 mark**

ii. To successfully treat high levels of cholesterol, lovastatin must be taken every day at regular time intervals. Considering the mode of action of lovastatin, why is it important to prevent the blood concentration of lovastatin from becoming too low? **2 marks**

Question 14 (10 marks)

Galactosemia is a metabolic disorder in which the enzyme galactose-1-phosphate uridylyltransferase (GALT) is missing. Individuals with galactosemia are unable to break down galactose, a sugar found in a variety of foods. This leads to a build up of galactose, causing a variety of symptoms such as learning difficulties, lethargy and neurological issues. It is usually inherited due to a gene mutation passed on from the parents of affected children.

Babies are checked for galactosemia shortly after birth and, if they are found to be missing the key enzymes involved in galactose metabolism, early intervention is quickly begun. Individuals must avoid galactose in their diet to reduce symptoms of the disease.

a. GALT leads to the conversion of galactose into glucose. Explain why this enzyme only acts on galactose and not on other sugars. **1 mark**

b. Lactose is a disaccharide that is hydrolysed in the body into galactose and glucose. Explain why individuals with galactosemia must also avoid lactose in their diet. **1 mark**

c. Identify if the breakdown of lactose into galactose and glucose is an endergonic or an exergonic reaction. Justify your response. **2 marks**

d. GALT was extracted from an individual unaffected with galactosemia to examine factors that affect its mechanism of action. It was examined under a variety of temperatures. Draw a clear graph showing reaction rate versus temperature to show the expected results of this experiment, including the expected optimal temperature. **3 marks**

e. Explain, on a molecular level, what is happening to the GALT enzyme at extremely high temperatures. **2 marks**

f. It is important that individuals with galactosemia obtain adequate amounts of glucose. In relation to cellular respiration, outline the important of glucose. **1 mark**

Question 15 (6 marks)

Hexokinase is an enzyme used in glycolysis, the first step in cellular respiration. A molecular model of this enzyme is shown.

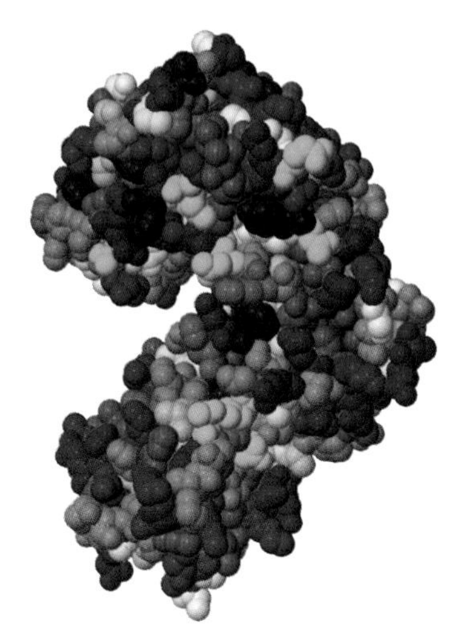

Hexokinase allows for the glucose to be converted to glucose-6-phosphate in an endergonic reaction.

a. Describe how hexokinase increases the rate of the conversion of glucose to glucose-6-phosphate. **2 marks**

b. ATP is used in this reaction and is known as a loaded form. What does this term mean? **1 mark**

c. Explain the role of the coenzyme ATP in this reaction. **1 mark**

d. If the concentration of the glucose substrate was increased, explain what would happen to the rate of the reaction. **2 marks**

3.6 Exercise 3: Biochallenge online only

Resources

eWorkbook Biochallenge — Topic 3 (ewbk-8084)

Solutions Solutions — Topic 3 (sol-0659)

Past VCAA examinations

Sit past VCAA examinations and receive immediate feedback, marking guides and examiner's report notes.
Access Course Content and select 'Past VCAA examinations' to sit the examinations online or offline.

teachon

Test maker
Create unique tests and exams from our extensive range of questions, including past VCAA questions.
Access the assignments section in learnON to begin creating and assigning assessments to students.

Online Resources

Below is a full list of **rich resources** available online for this topic. These resources are designed to bring ideas to life, to promote deep and lasting learning and to support the different learning needs of each individual.

eWorkbook

- 3.1 eWorkbook — Topic 3 (ewbk-1882) ☐
- 3.2 Worksheet 3.1 Biochemical pathways and reactions (ewbk-7527) ☐
- 3.3 Worksheet 3.2 The structure and function of enzymes (ewbk-7529) ☐
- 3.4 Worksheet 3.3 Investigating coenzymes (ewbk-7531) ☐
- 3.5 Worksheet 3.4 Factors affecting enzyme activity (ewbk-7533) ☐
- Worksheet 3.5 Inhibitors of enzyme activity (ewbk-7535) ☐
- Worksheet 3.6 Regulation of biochemical pathways (ewbk-7537) ☐
- 3.6 Worksheet 3.7 Reflection — Topic 3 (ewbk-4536) ☐
- Worksheet 3.8 Biochallenge — Topic 3 (ewbk-8084) ☐

Solutions

- 3.6 Solutions — Topic 3 (sol-0659)

Practical investigation eLogbook

- 3.1 Practical investigation eLogbook — Topic 3 (elog-0183) ☐
- 3.2 Investigation 3.1 Endergonic and exergonic reactions (elog-0818) ☐
- 3.3 Investigation 3.2 Enzymes in plants (elog-0820) ☐
- Investigation 3.3 The actions of the catalase enzyme (elog-0822) ☐
- 3.5 Investigation 3.4 The effect of temperature and pH on enzyme activity (elog-0824) ☐
- Investigation 3.5 The effect of changing the substrate concentration on enzyme activity (elog-0826) ☐
- Investigation 3.6 Modelling the effect of substrate and enzyme concentration on reaction rate (elog-0828) ☐
- Investigation 3.7 Inhibitors and enzyme activity (elog-0830) ☐

Digital documents

- 3.1 Key science skills — VCE Biology Units 1–4 (doc-34326) ☐
- Key terms glossary — Topic 3 (doc-34617) ☐
- Key ideas summary — Topic 3 (doc-34608) ☐
- 3.3 Case study: The plague of the sea (doc-35827) ☐

Teacher-led videos

- Exam questions — Topic 3 ☐
- 3.2 Sample problem 1 Comparing and contrasting exergonic and endergonic reactions (tlvd-1764) ☐
- 3.3 Sample problem 2 Analysing the effect of enzymes on biochemical pathways (tlvd-1765) ☐
- Sample problem 3 Exploring the role of enzymes in cellular respiration (tlvd-1766) ☐
- 3.4 Sample problem 4 Defining coenzymes (tlvd-1767) ☐
- 3.5 Sample problem 5 Designing an experiment on factors affecting rate of reaction (tlvd-1768) ☐
- Sample problem 6 Describing the action of inhibitors on enzymes (tlvd-1769) ☐

Video eLessons

- 3.3 Enzymes as catalysts (eles-4191) ☐
- Structure and specificity of enzymes (eles-4192) ☐
- 3.4 ATP is the energy currency of cells (eles-4193) ☐
- Coenzymes (eles-3904) ☐
- 3.5 Inhibition of enzymes (eles-3905) ☐

Interactivities

- 3.5 Identify the enzyme (int-0108) ☐
- Catalase and temperature (int-1430) ☐

Weblinks

- 3.3 Enzyme activity in photosynthesis ☐
- Regulation of cellular respiration ☐

Exam question booklet

- 3.1 Exam question booklet — Topic 3 (eqb-0014) ☐

Teacher resources

There are many resources available exclusively for teachers online.

To access these online resources, log on to **www.jacplus.com.au**

4 Photosynthesis, cellular respiration and biotechnological applications

KEY KNOWLEDGE

In this topic, you will investigate:

Photosynthesis as an example of biochemical pathways

- inputs, outputs and locations of the light-dependent and light-independent stages of photosynthesis in C_3 plants (details of biochemical pathway mechanisms are not required)
- the role of Rubisco in photosynthesis, including adaptations of C_3, C_4 and CAM plants to maximise the efficiency of photosynthesis
- the factors that affect the rate of photosynthesis: light availability, water availability, temperature and carbon dioxide concentration

Cellular respiration as an example of biochemical pathways

- the main inputs, outputs and locations of glycolysis, Krebs cycle and electron transport chain including ATP yield (details of biochemical pathway mechanisms are not required)
- the location of inputs and the difference in outputs of anaerobic fermentation in animals and yeasts
- the factors that affect the rate of cellular respiration: temperature, glucose availability and oxygen concentration

Biotechnological applications of biochemical pathways

- potential uses and applications of CRISPR-Cas9 technologies to improve photosynthetic efficiencies and crop yields
- uses and applications of anaerobic fermentation of biomass for biofuel production.

PRACTICAL WORK AND INVESTIGATIONS

Practical work is a central component of learning and assessment. Experiments and investigations, supported by a **practical investigation eLogbook** and **teacher-led videos**, are included in this topic to provide opportunities to undertake investigations and communicate findings.

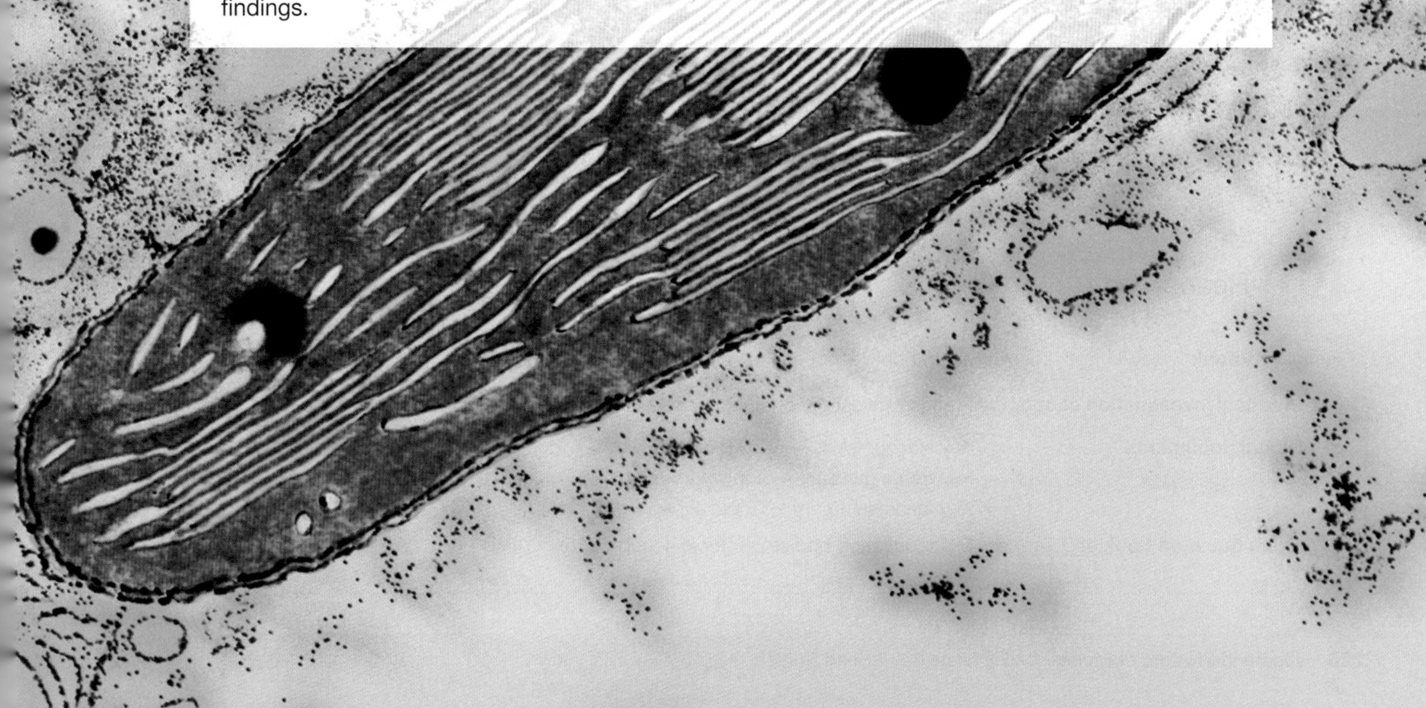

4.1 Overview

Numerous **videos** and **interactivities** are available just where you need them, at the point of learning, in your digital formats, learnON and eBookPLUS and at **www.jacplus.com.au**.

4.1.1 Introduction

Being alive requires a constant input of energy. Every animal, plant, fungus and microbe must have access to energy to drive the biochemical reactions that enable life to function. The first challenge for living organisms is to capture energy for their use from an external source in their environment. A second challenge is to convert that energy into a form that can be used by cells.

Every day, a massive amount of radiant energy reaches Earth. Many of Earth's life forms have evolved mechanisms to exploit the energy provided by our Sun. These organisms — green plants, algae and cyanobacteria — can just remain fixed in one spot or float on water, exposing their surface area to the Sun, capturing its radiant energy, and transforming it to the chemical energy of organic molecules for their energy needs. These organisms are said to be autotrophs or 'self feeders'.

FIGURE 4.1 Humans capture the chemical energy from organic molecules (food)

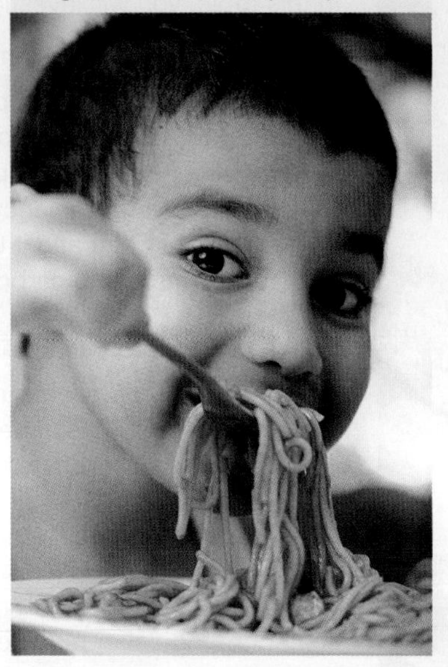

In contrast, animals, fungi and most microbes must capture the energy needed for living from a different source. Organisms of this type capture energy from their environment in the form of the chemical energy of organic molecules in their food. 'Food' consists of the organic molecules of other organisms, living or dead, or their products. In every case, 'food' is any organic substance that is a source of chemical energy which can be absorbed and used by an organism to provide energy for its cellular functions. Organisms that gain their energy for living in this manner are termed heterotrophs or 'other feeders'.

LEARNING SEQUENCE

on Resources

eWorkbook	eWorkbook — Topic 4 (ewbk-1883)
Practical investigation eLogbook	Practical investigation eLogbook — Topic 4 (elog-0185)
Digital documents	Key science skills — VCE Biology Units 1–4 (doc-34326)
	Key terms glossary — Topic 4 (doc-34618)
	Key ideas summary — Topic 4 (doc-34609)
Exam question booklet	Exam question booklet — Topic 4 (eqb-0015)

4.2 Photosynthesis

KEY KNOWLEDGE

- Inputs, outputs and locations of the light-dependent and light-independent stages of photosynthesis in C_3 plants (details of biochemical pathway mechanisms are not required)

Source: VCE Biology Study Design (2022–2026) extracts © VCAA; reproduced by permission.

4.2.1 Role of photosynthesis

Photosynthesis is a remarkable process. It is the only biochemical pathway that uses radiant energy captured from beyond Earth. However, the sunlight captured by **autotrophs** such as plants and algae is *not* a useful form of energy for cells. The radiant energy of sunlight is a diffuse form of energy that cannot be used directly for cellular processes; it cannot be transported, and it cannot be stored (see figure 4.2). Instead, photosynthesis transforms this radiant energy into chemical energy in the form of sugars such as glucose.

In contrast, the chemical energy of organic molecules, such as glucose, is a dense form of energy that can be used as an energy source for cells and can be stored in the form of either starch (in plants) and glycogen (in animals). Both plants and animals can also store energy as fats or oils.

FIGURE 4.2 The radiant energy of sunlight is a diffuse form of energy. In that form, sunlight can neither be transported nor stored.

The essential purpose of photosynthesis is to capture sunlight and transform this energy into the concentrated chemical energy of organic sugar molecules — that is, to make sugars, such as glucose, from sunlight.

4.2.2 Biochemical pathway of photosynthesis

We can show, in simple form, the making of sugars, such as glucose, from sunlight in several ways:

- As a word equation:

$$\text{carbon dioxide} + \text{water} \xrightarrow[\text{chlorophyll}]{\text{light}} \text{glucose} + \text{oxygen}$$

- As a balanced equation showing inputs and outputs:

$$6CO_2 + 12H_2O \xrightarrow[\text{chlorophyll}]{\text{light}} C_6H_{12}O_6 + 6O_2 + 6H_2O$$

that can be simplified showing water as an input only:

$$6CO_2 + 6H_2O \xrightarrow[\text{chlorophyll}]{\text{light}} C_6H_{12}O_6 + 6O_2$$

photosynthesis process by which plants use the radiant energy of sunlight trapped by chlorophyll to build carbohydrates from carbon dioxide and water

autotrophs organisms that, when given a source of energy, produce their own food from simple inorganic substances

TIP: When writing out the equation for photosynthesis, ensure that you include sunlight (or light or radiant energy) and chlorophyll above and below the line. These are not used up in the process, so should not be placed with carbon dioxide and water as inputs. You may lose marks in assessments if you do not include these aspects or include them in the wrong location.

- As a figure with a focus on inputs and outputs:

eles-4332

FIGURE 4.3 The inputs and outputs of photosynthesis shown diagrammatically. Glucose is not released from the leaf but is translocated as sucrose throughout the plant.

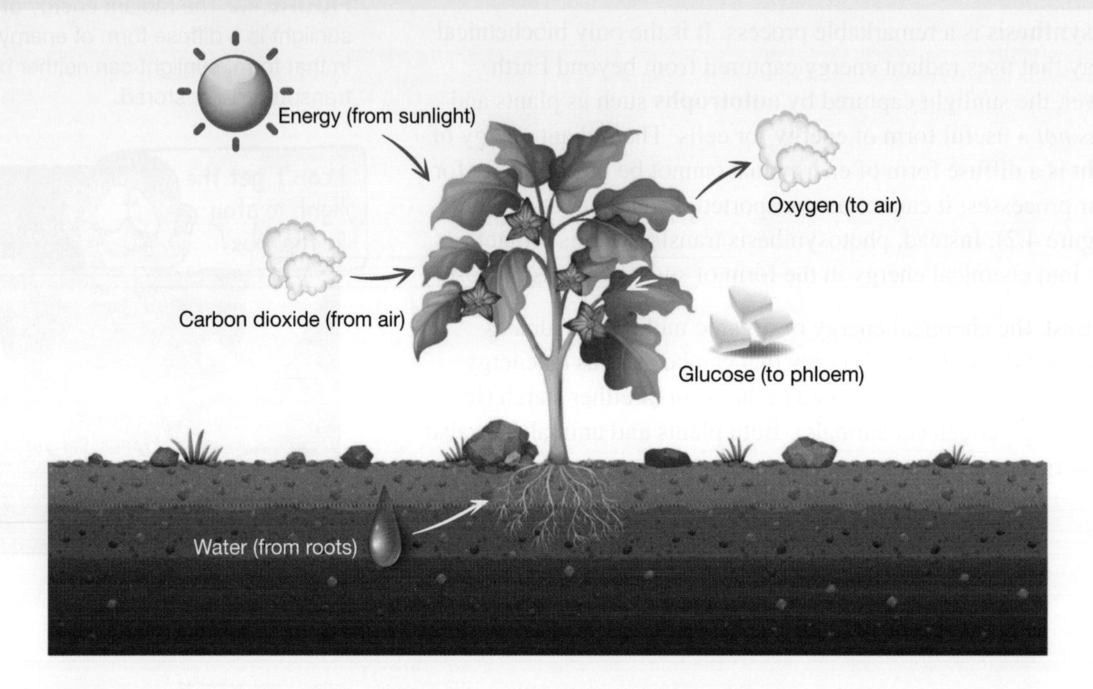

These simple representations of photosynthesis identify the initial inputs and the final outputs of this biochemical pathway. However, these representations conceal the intricacy of the photosynthesis pathway that occurs within the chloroplasts and they do not identify some other players, such as coenzymes, that are essential cofactors for enzymes in the photosynthesis pathway.

Photosynthesis involves two distinct interconnected stages, namely:

- the **light-dependent stage**
- the **light-independent stage**.

These stages will be explored in detail in sections 4.2.4 and 4.2.5.

light-dependent stage the first stage of photosynthesis where light energy is trapped by chlorophyll

light-independent stage the second stage of photosynthesis, in which glucose is produced

The key events in these two stages are:

- the light-dependent stage, which can only proceed in the presence of light. The function of the light-dependent stage is to transform sunlight energy that is captured by chlorophyll into the chemical energy of loaded coenzymes.
- the light-independent stage, which is also known as the Calvin cycle, and can only proceed in the presence of high-energy coenzymes. The function of this stage is to assemble simple inorganic carbon dioxide molecules into more complex organic glucose molecules.

on Resources

Interactivity Photosynthesis summary equation (int-0107)

4.2.3 The chloroplast: the location of photosynthesis

Chloroplasts are the cell organelles where all the action of the photosynthesis pathway occurs. The green pigment, **chlorophyll**, an essential player in the photosynthesis pathway, is embedded in the internal membranes of a special cell organelle, the chloroplast.

Chloroplasts are present in high numbers in the cytosol of particular cells of photosynthetic plants, typically ranging from about 40 to 200 chloroplasts per cell (see figure 4.4a).

Chlorophyll enables plants to capture the radiant energy of sunlight, bringing it into cells as the starting point of photosynthesis.

Accessory pigments, such as carotenoids, are also present in chloroplasts — they can also capture sunlight energy and transfer the radiant energy they absorb to the chlorophylls.

Not all the cells of a green leaf contain chloroplasts. Almost exclusively, it is the mesophyll cell layers of a leaf where the chloroplast-containing cells are located. This can be observed in figure 4.4, where the cells with chloroplasts are those of the palisade mesophyll (cells arranged like pickets in a fence) and the spongy mesophyll (cells loosely packed).

FIGURE 4.4 a. Photomicrograph of cells of leaf tissue showing multiple green chloroplasts present in the cells b. Stylised diagram of a transverse section through a leaf showing the chloroplast-containing cells

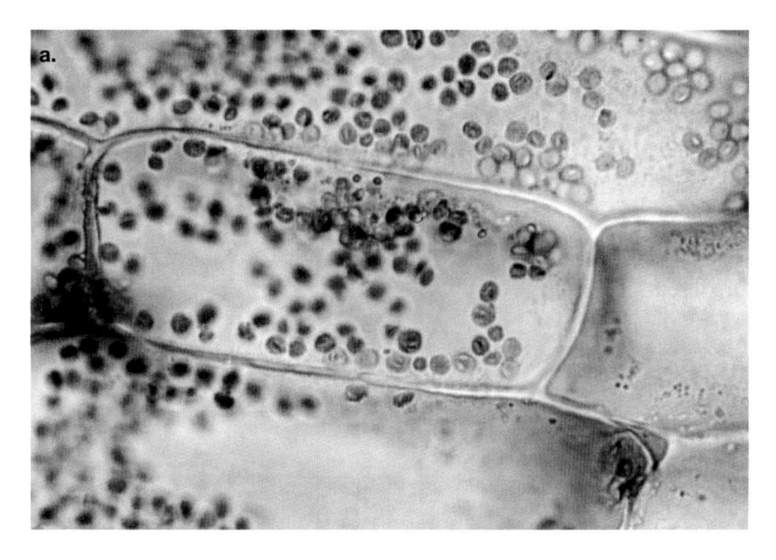

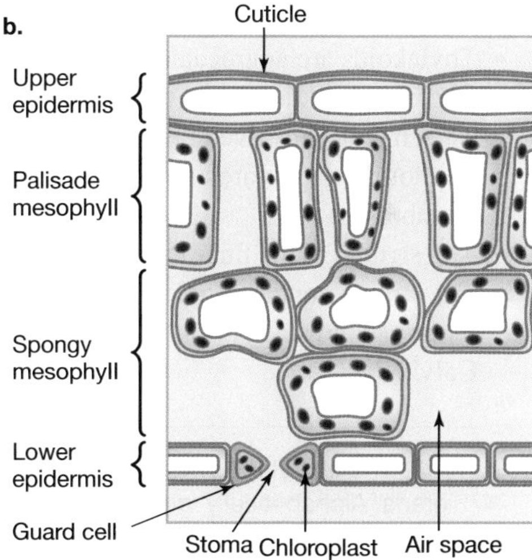

The structure of the chloroplast

Let's look at the structure of chloroplasts: the site of the photosynthesis pathway.

Figure 4.5a is a diagrammatic representation of the structure of a chloroplast, showing the various membranes that are part of its structure.

chloroplast chlorophyll-containing organelle that is the site of photosynthesis

chlorophyll green pigment that traps the radiant energy of light

FIGURE 4.5 a. Stylised diagram of a single chloroplast showing its main features b. Transmission electron micrograph of a single chloroplast. Note the many dark regions where thylakoids are piled into stacks called grana.

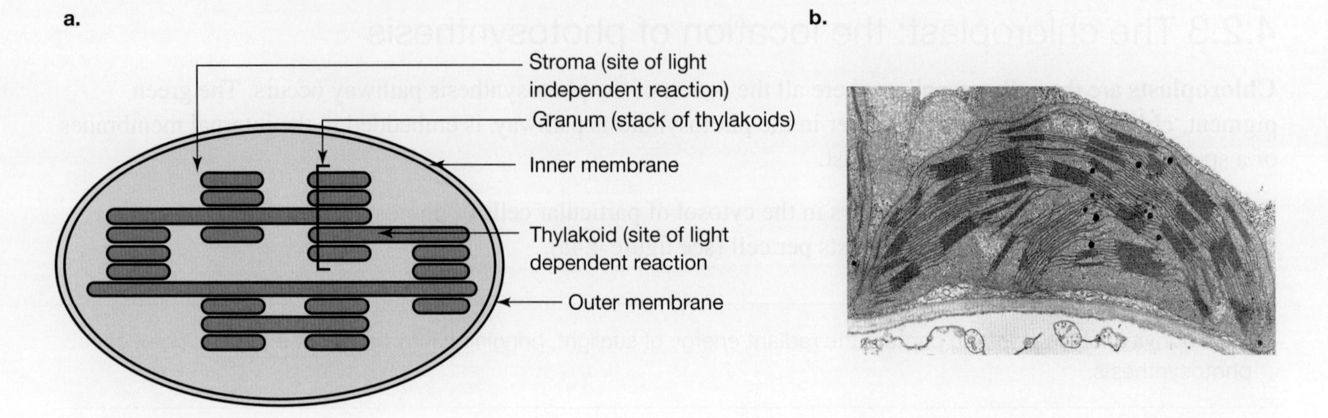

Note the following:

- Each chloroplast is enclosed in an envelope that is formed by an outer membrane and an inner membrane.
- Within this envelope are many membrane-bound discs known as **thylakoids** (from the Greek *thylakos:* pouch or sac).
- The light-trapping pigment, chlorophyll (see figure 4.6), is embedded in the thylakoid membranes. The thylakoids provide a large surface area for the capture of sunlight.
- Thylakoids are aggregated into stacks (a bit like stacks of pancakes) known as **grana** (singular: granum).
- The enzymes involved in the light-dependent stage of photosynthesis are also located in the thylakoid membranes.
- The **stroma** is the fluid inside the chloroplast.
- The stroma contains the enzymes that are involved in the second stage of photosynthesis, known as the Calvin cycle.

TIP: The first stage of photosynthesis occurs in the grana. Alphabetically 'grana' comes before 'stroma' and, therefore, the stage in the grana occurs first.

Remember that **s**troma is the '**s**olution' in chloroplasts and the **s**econd stage of photosynthesis occurs there.

FIGURE 4.6 Structural formula of the light-trapping molecules, chlorophyll *a* and *b*. The hydrophilic head is where the sunlight energy is trapped.

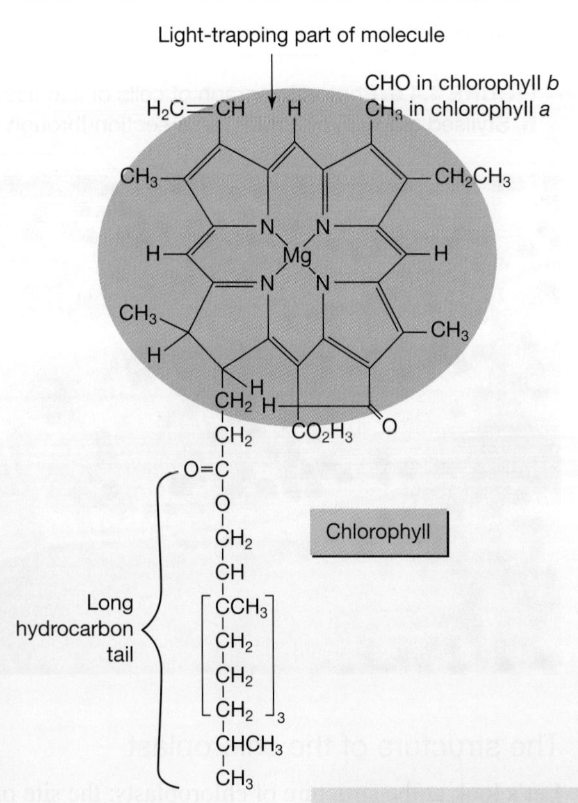

thylakoid flattened membranous sacs in chloroplasts that contain chlorophyll

grana stacks of flattened thylakoids; singular: granum

stroma in chloroplasts, the semi-fluid substance which contains enzymes for some of the reactions of photosynthesis

4.2.4 Light-dependent stage of photosynthesis

The light-dependent stage of photosynthesis is involved in the capture of sunlight and the transformation of its energy to the chemical energy of loaded coenzymes, NADPH and ATP. This process occurs through a pathway that takes place within the grana, in the thylakoid membranes of chloroplasts.

FIGURE 4.7 Electrons become 'excited' when they are moved to a higher energy level by absorbing light energy. Excited electrons are able to move between acceptor molecules, releasing energy as they go.

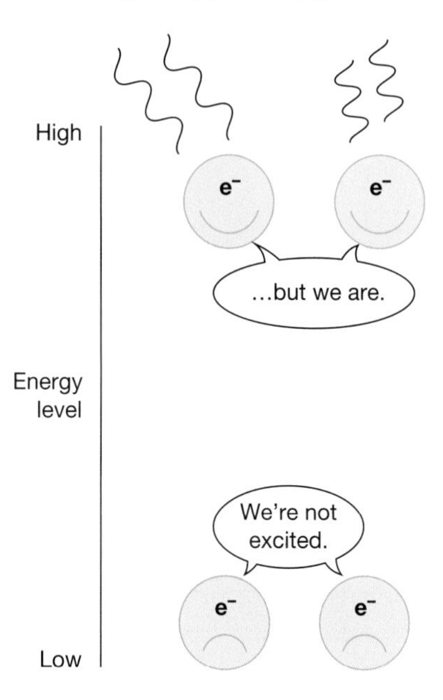

The key events in the light-dependent stage

- The capture of the radiant energy of sunlight by chlorophyll molecules in the thylakoids
- The absorption of this energy by electrons in the chlorophyll to become high-energy or 'excited' electrons (see figure 4.7)
- The splitting of water molecules, that produces electrons, hydrogen ions (H^+) (also known as protons) and oxygen
- The passage of high-energy of electrons down a chain of electron acceptors during which electrons release their energy
- The loading of electrons and hydrogen ions onto $NADP^+$ to form NADPH
  ($NADP^+ + H^+ + 2e^- \rightarrow NADPH$)
- Use of this energy to pump protons from the stroma to inside the thylakoid, creating a proton gratient
- Passive movement of protons down this gradient back into the stroma produces kinetic energy that is used by the ATP synthase enzyme to produce ATP from ADP and Pi.

The inputs and outputs of the light-dependent stage

The inputs and outputs of the light-dependent stage of photosynthesis are shown in simple form in figure 4.8.

FIGURE 4.8 Simple diagram showing the inputs and outputs of the light-dependent stage of photosynthesis. Note that $NADP^+$ is loaded with hydrogen ions and electrons to form NADPH.

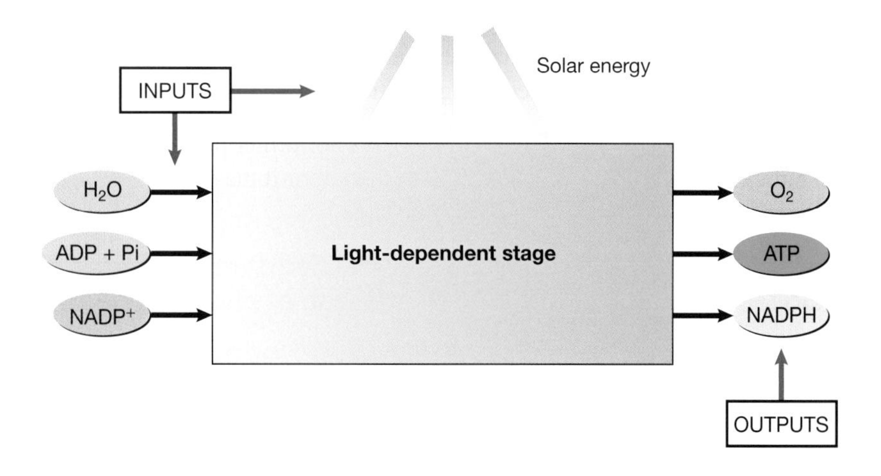

The inputs to the light-dependent stage and their roles are shown in table 4.1. Note the following:

- sunlight energy: this energy is needed to generate 'excited' electrons that are involved in energy transfers
- water: water molecules are split to produce a supply of electrons and hydrogen ions for the light-dependent stage
- $NADP^+$: accepts two electrons and a hydrogen ion and is converted to loaded NADPH
- ADP + Pi: uses energy released to produce ATP.

FIGURE 4.9 Diagram showing the reversible reactions between unloaded $NADP^+$ and loaded NADPH (only the part of the molecule that changes is shown — the rest of the molecule is represented as R)

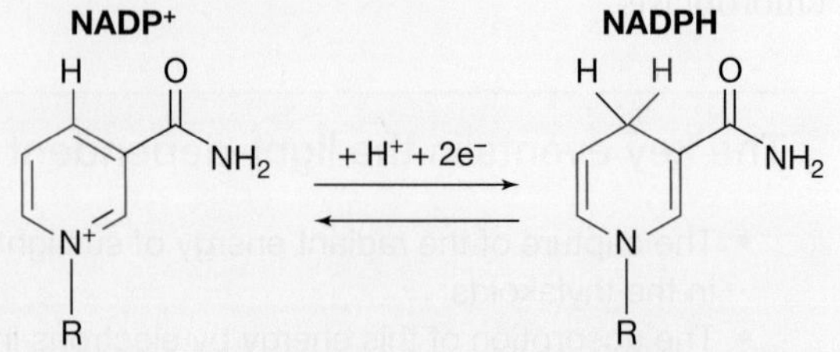

TABLE 4.1 Table showing the inputs and outputs of the light-dependent stage of photosynthesis, their roles and the reactions involved

Inputs	Role	Outputs	Role	Reaction
Sunlight	Initial input of energy to chlorophyll			
$NADP^+$	Unloaded coenzyme and acceptor of hydrogen ions and electrons	**NADPH**	Loaded coenzyme and donor of hydrogen ions and electrons	$NADP^+ + H^+ + 2e^- \rightarrow$ **NADPH**
ADP	Unloaded coenzyme	**ATP**	Loaded coenzyme and energy supplier	ADP + Pi $\rightarrow$ **ATP**
Water	Supplier of electrons and hydrogen ions	**Oxygen**	By-product of splitting water	$H_2O \rightarrow 2e^- + 2H^+ + \frac{1}{2}O_2$

Note: Pi represents the phosphate group, $-PO_4^{3-}$

At the end of the light-dependent stage of photosynthesis, chloroplasts have:

- a supply of high-energy loaded **ATP** molecules that can unload and transfer their energy to drive the energy-requiring anabolic reactions of the second stage of photosynthesis, the Calvin cycle (or light-independent stage)
- a supply of high-energy loaded **NADPH** coenzymes that can act as donors of hydrogen ions and electrons in the energy-requiring reactions of the Calvin cycle.

ATP adenine triphosphate; common source of chemical energy for cells

NADPH the loaded form of $NADP^+$, which can donate hydrogen and electrons during photosynthesis

Note that the oxygen produced in the light-dependent stage plays no further part in photosynthesis. It will either be released from the leaves into the air, or it may be used as an input to cellular respiration in the plant.

EXTENSION: A closer look at the light-dependent stage of photosynthesis

Events in the light-dependent stage of photosynthesis are as follows:

- Radiant energy of sunlight is captured by chlorophyll molecules.
- Absorption of radiant energy produces 'excited' (high energy) electrons in the chlorophyll.
- Excited electrons from the chlorophyll pass down a chain of protein complexes embedded in the thylakoid membrane. These protein complexes, in turn, accept and then donate electrons; they form an electron transport chain.
- Energy becomes available as the electrons move along this chain.
- This energy is used to produce ATP and NADPH.

The energy of the electrons can be used in two different ways:

1. *Production of ATP.* ATP is formed from ADP and Pi using energy related from transfer of electrons down the electron transport chain. Note that the electrons complete their cyclic path by returning to the chlorophyll (see figure 4.10).
2. *Production of loaded NADPH.* At the end of the electron transport chain, unloaded $NADP^+$ can accept electrons and convert to loaded NADPH. Note that the chlorophyll now has a shortage of electrons and these are replaced by electrons produced by splitting water molecules.

FIGURE 4.10 Diagram showing a highly simplified path of excited electrons down an electron transport chain and the pumping of protons.

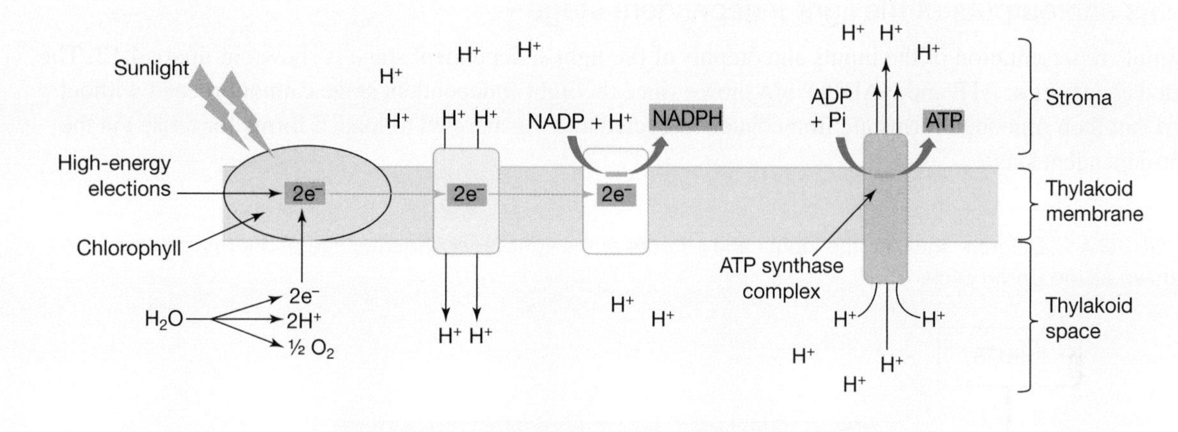

4.2.5 Light-independent stage or Calvin cycle

The light-independent stage of photosynthesis (or the **Calvin cycle**) is the stage of photosynthesis during which inorganic carbon dioxide molecules (CO_2) are built into energy-rich reduced organic molecules, such as glucose ($C_6H_{12}O_6$). The enzymes required for this stage are in solution in the stroma. The main enzyme in C3 plants is known as **Rubisco**, which is vital in **carbon fixation**.

As well as essential enzymes to catalyse each step of the cycle, other key requirements are the supply of:

- NADPH to donate hydrogens and electrons
- ATP as an energy source.

FIGURE 4.11 A wheat crop ready for harvest. This crop is a physical expression of the outcome of photosynthesis — glucose production. The starch molecules stored in wheat grains are formed from glucose molecules joined into long or branched chains.

Calvin cycle cycle of reactions occurring in the stroma of chloroplasts in the light-independent stage of photosynthesis

Rubisco an important enzyme involved in the process of carbon fixation

carbon fixation process by which atmospheric carbon dioxide is incorporated into organic molecules such as sugars

The key events in the light-independent stage

- Inorganic CO_2 is converted into the carbon in organic molecules, a process termed carbon fixation.
- Carbon dioxide molecules are accepted into the Calvin cycle by organic 5C acceptor molecules.
- Loaded NADPH coenzymes donate hydrogens and electrons as molecules are reduced to higher energy levels.
- ATP supplies energy for the anabolic steps of this cycle.
- Glucose is formed as an output.

Inputs and outputs of the light-independent stage

A simple representation of the inputs and outputs of the light-independent stage is shown in figure 4.12. The loaded coenzymes, ATP and NADPH, are shown since the light-independent stage cannot proceed without them, but their unloaded forms are immediately recycled and regenerated to loaded forms for reuse via the light-dependent stage.

FIGURE 4.12 Diagram showing the inputs and outputs of the light-independent stage of photosynthesis, also known as the Calvin cycle

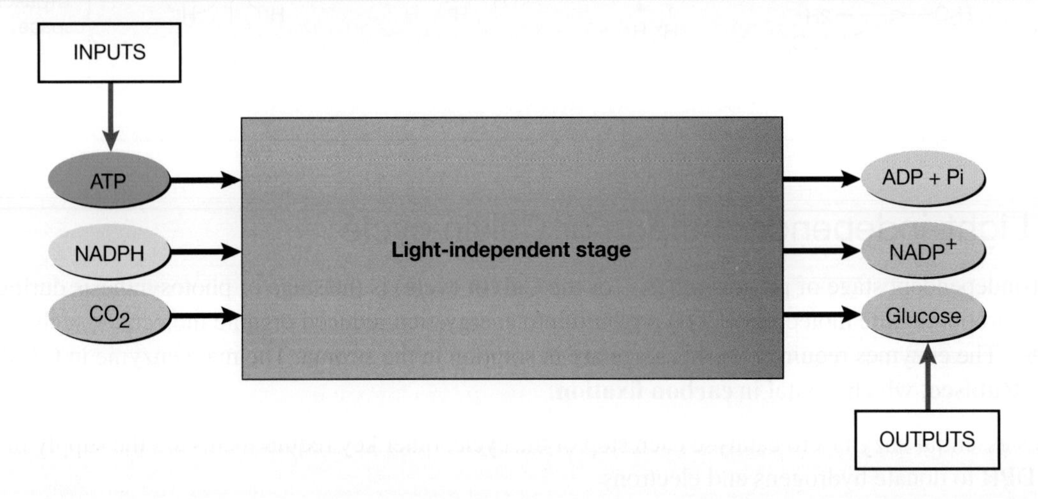

TABLE 4.2 Table showing the inputs and outputs of the light-independent stage of photosynthesis, their roles and the reactions involved

Inputs	Role	Outputs	Role	Reaction
NADPH	Loaded coenzyme and donor of hydrogens and electrons	**$NADP^+$**	Unloaded coenzyme and acceptor of hydrogens and electrons	**$NADPH \rightarrow NADP^+$** $+ H^- + 2e^-$
ATP	Loaded coenzyme and energy supplier	**ADP**	Unloaded coenzyme	**ATP** $\rightarrow$ **ADP** + Pi
Carbon dioxide	Supplier of carbon and oxygen atoms	**Glucose**	Final product in photosynthesis	**$CO_2 + 2e^- + 2H^+ \rightarrow C_6H_{12}O_6$** (Note this is not balanced, but rather a simplification of the process involved.)

EXTENSION: Exploring the Calvin cycle

Let's explore the Calvin cycle. Check out figure 4.13, which shows that the Calvin cycle is composed of three major steps that take place in the chloroplast stroma.

FIGURE 4.13 A highly simplified representation of the Calvin cycle. The black circles represent carbon atoms and the purple circles represent other parts of the molecule.

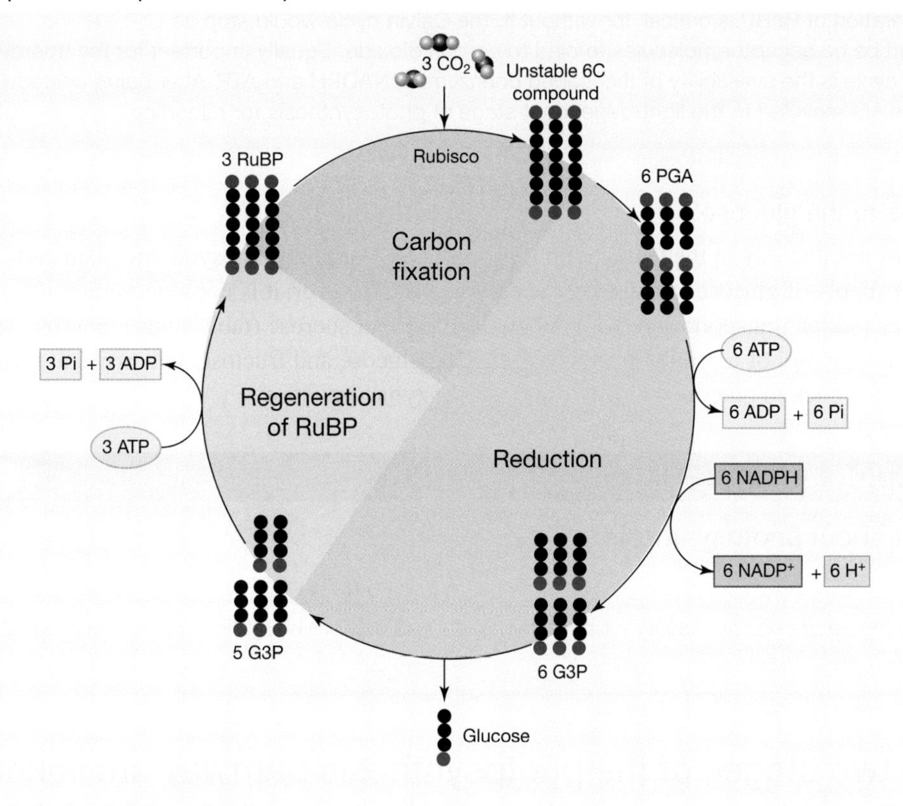

The three steps in the Calvin cycle are:

1. *a carbon fixation step.* In this step, inorganic carbon dioxide molecules are converted to organic molecules. The CO_2 is picked up by an acceptor, RuBP (ribulose bisphosphate), a five-carbon molecule. This reaction, catalysed by the Rubisco enzyme, produces an unstable six-carbon molecule that immediately breaks down to form two three-carbon molecules. So, the first stable product of the Calvin cycle is PGA (phosphoglyceric acid).

   The key reaction that starts the Calvin cycle is the conversion of inorganic carbon dioxide from the air into an organic molecule (PGA). Note that figure 4.13 shows this reaction for three molecules of RuBP and carbon dioxide.

$C_5H_6O_5$	+	CO_2	+	H_2O	+	Pi	→	$2C_3H_4O_4$~Pi
RuBP		carbon dioxide		water				PGA

   Rubisco is the world's busiest enzyme and key to starting the process of converting carbon dioxide (a low-energy inorganic form of carbon) to glucose (a high-energy form of carbon). The worldwide conversion of carbon dioxide to an organic form (carbon fixation) by the Rubisco enzyme is estimated to be occurring at the rate of about 2×10^{23} carbon dioxide molecules per second. The enzyme itself is relatively slow but there is a very high concentration of Rubisco in chloroplasts enabling this high rate.

2. *a reduction step.* In this step, PGA molecules are reduced to another three-carbon molecule, G3P (glyceraldehyde phosphate) by accepting hydrogens from loaded NADPH coenzymes. The energy needed to drive this step is provided by ATP. The G3P molecules are the output of the Calvin cycle, with two GPA of these molecules needed to form a six-carbon glucose.
3. ***an acceptor regeneration** step.* The RuBP acceptor that is used up in the first step is re-formed in a reaction in which five G3P molecules are reassembled to form three RuBP molecules. Note that this anabolic reaction requires energy that is supplied by ATP.
   The regeneration of RuBP is critical, for without it, the Calvin cycle would stop as there would be no acceptor molecules to bind to carbon dioxide. Equally important for the operation of the Calvin cycle is the availability of the loaded coenzymes, NADPH and ATP. After being unloaded, these coenzymes are recycled to the light-dependent stage of photosynthesis for reloading.

What happens to the glucose?

The glucose does not just stay in the leaves after it is produced. All non-photosynthetic plant cells must also have access to a supply of glucose to meet their energy needs. However, it is not glucose that is translocated in the phloem tissue (which transports small organic molecules), but sucrose (table sugar). Sucrose, a disaccharide, is assembled in the cytosol of mesophyll cells from linking glucose and fructose. On arrival at the cell walls, sucrose is converted back to glucose and fructose by the enzyme, invertase.

elog-0262

INVESTIGATION 4.1 online only

Finding out about photosynthesis

Aim

To investigate the conditions necessary for photosynthesis and explore how this leads to the production of carbohydrates

4.2.6 The two stages of the photosynthesis pathway are linked

The light-dependent stage and the light-independent stage (Calvin cycle) are tightly linked, with some of the outputs of each stage becoming inputs to the other stage, so that the two stages depend on each other.

Loaded coenzymes that are outputs from the first stage are the inputs to the second stage. Unloaded coenzymes produced in the second stage are recycled back to the first stage for reloading. If the supply and recycling of either coenzyme is interrupted, photosynthesis stops. This recycling is shown in figure 4.14.

acceptor regeneration a process where acceptor molecules are reformed, allowing for them to be recycled in biochemical pathways

TABLE 4.3 Many outputs of the first stage of photosynthesis are the inputs for the second stage (shown in green). Many of the outputs for the second stage (shown in red) are inputs for the first stage as they are recycled. The other inputs and outputs are not recycled, and instead form the primary inputs and outputs as seen in the general photosynthesis equation.

	Inputs	Outputs
Light-dependent stage	H_2O ADP + Pi $NADP^+$ Sunlight	O_2 ATP NADPH
Light-independent stage	CO_2 ATP NADPH	$C_6H_{12}O_6$ ADP + Pi $NADP^+$

FIGURE 4.14 Simplified diagram showing how the outputs from the light-dependent reactions in photosynthesis (ATP and NADPH) are used to drive the energy requiring light-independent stage

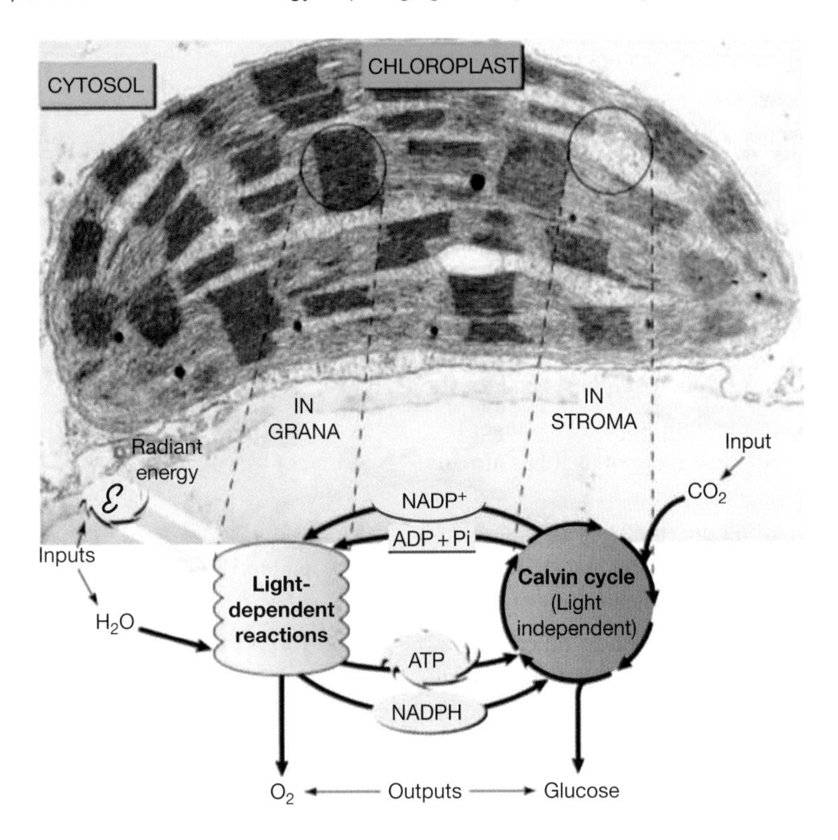

tlvd-1424

SAMPLE PROBLEM 1 Exploring the stages of photosynthesis

A student set up an experiment exploring the photosynthesis of plants in the light and the dark. They placed one piece of *Elodea*, a type of water plant, in two separate beakers and filled this with pond water. They placed two drops of phenol red in each test tube. They placed one test tube in the dark and the other in the light.

a. **In which test tube would oxygen be released? (1 mark)**
b. **At what stage of photosynthesis is oxygen produced, and at what location does this stage occur? (2 marks)**
c. **Identify two other outputs of this stage. (2 marks)**

When carbon dioxide dissolves in water, carbonic acid is produced. When phenol red reacts with carbonic acid, the solution is yellow. When there is no carbonic acid present, the solution is pink.

d. **Which test tube would you expect to have yellow coloured water and which would you expect to have pink coloured water? Justify your response. (4 marks)**
e. **The locations of the tube were swapped, with the tube in the light being moved to the dark and vice versa. The student noticed that carbon dioxide was still being used up in the tube in the dark, despite there being no sunlight. Explain why this is the case. (3 marks)**

THINK	WRITE
a. Determine when oxygen is released in photosynthesis. This usually occurs in the first stage (light-dependent stage).	
Examine the two tubes and determine which tube will have oxygen produced. In this case it is the one that is having an input of radiant energy (light).	The *Elodea* in the test tube that was placed in the light will produce oxygen (1 mark).
b. 1. Carefully read the question and determine what needs to be included in the answer. In this case, there are two questions around the stage and location.	Oxygen is produced in the light-dependent stage (1 mark) which occurs in the thylakoid (or grana) in the chloroplast (1 mark).
2. Review the stages of photosynthesis and locations.	
c. State two outputs of the light-dependent stage. It is important you do not state oxygen as it has already been mentioned.	Other outputs include NADPH (1 mark) and ATP (1 mark).
d. 1. Recall when carbon dioxide is used up — in the light-independent stage. Determine if the light-independent stage can occur in each tube.	Carbon dioxide is used up in the light-independent stage (1 mark).
2. Link this to the results. If photosynthesis is occurring, carbon dioxide is used up, resulting in a pink solution. If it is not occurring, it is likely being produced instead, so the solution is pink.	Therefore, if this stage is occurring, there is less CO_2 and the water will appear pink (1 mark). Photosynthesis will occur in the tube in the light, so it will appear pink (1 mark).
3. Ensure you justify your response.	While CO_2 is produced during the light-independent stage, inputs from the light-dependent stage (such as NADPH and ATP) are required for it to work. Therefore, the tube in the dark would not undergo photosynthesis so will appear yellow (1 mark).
e. 1. Consider when CO_2 is used up — the light-independent stage (so it can occur without an impact of sunlight).	In order for CO_2 to be used up, the light-independent stage needs to occur. For this to occur, an input of CO_2, ATP and NADPH is required (1 mark).
2. Think about what else is needed for this stage: NADPH and ATP from the light-dependent stage. Link this to your observations.	As the tube was in the light earlier, ATP and NADPH would be available for use in the light-independent stage (1 mark). Therefore, some CO_2 can be used to produce glucose until ATP and NAPH is depleted (1 mark).

on Resources

eWorkbook Worksheet 4.1 The stages of photosynthesis (ewbk-7575)

Digital document Extensions of photosynthesis (doc-36166)

KEY IDEAS

- Photosynthesis is split into two stages: the light-dependent and the light-independent stages.
- The overall equation for photosynthesis is:

$$6CO_2 + 6H_2O \xrightarrow[\text{chlorophyll}]{\text{light}} 6O_2 + C_6H_{12}O_6$$

- The light-dependent stage occurs in the grana on the thylakoid membranes in the chloroplast.
- The light-independent stage occurs in the stroma of the chloroplast.
- Inputs to the light-dependent stage are sunlight, $NADP^+$, ADP and Pi and H_2O.
- Outputs of the light-dependent stage are NADPH, ATP and O_2.
- Inputs to the light-independent stage are NADPH, ATP and CO_2.
- Outputs of the light-independent stage are $NADP^+$, ADP & Pi and glucose ($C_6H_{12}O_6$).

4.2 Activities

learnON

To answer questions online and to receive **immediate feedback** and **sample responses** for every question, go to your learnON title at **www.jacplus.com.au**. A **downloadable solutions** file is also available in the resources tab.

4.2 Quick quiz on	**4.2 Exercise**	**4.2 Exam questions**

4.2 Exercise

1. MC The light-independent reaction of photosynthesis
   A. uses the coenzymes NADH and ATP.
   B. uses the carbon dioxide produced during the light-dependent reaction.
   C. leads to the formation of a carbohydrate.
   D. occurs only at night-time.
2. Identify the location within a chloroplast where light-trapping activity occurs.
3. Examine the image below and identify the numbered structures.

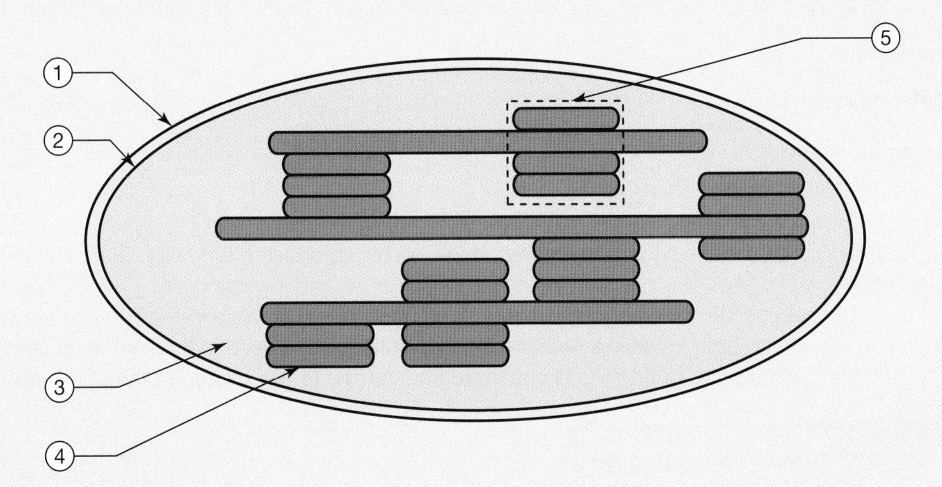

4. Identify the following statements as true or false and justify your response.
   a. The energy of sunlight is used directly to make sugars in photosynthesis.
   b. The electron acceptor in the light-dependent stage of photosynthesis is $NADP^+$.
   c. ATP formed in the light-dependent stage is a by-product of the splitting of water.
   d. The outputs of the light-dependent stage of photosynthesis are the inputs to the light-independent stage.
   e. Chlorophyll molecules absorb radiant light and transfer the energy to excited electrons.

5. Briefly explain the difference between the light-dependent and light-independent stages of photosynthesis.
6. The light-dependent and light-independent stages of photosynthesis are often described as linked processes. Explain this, and use a diagram to support your response.
7. A student wished to explore the process of photosynthesis to see the impact of light on the light-dependent and light-independent stages. They kept the plant in sunlight for 3 hours and then moved into darkness for 2 hours. Throughout the process, they measured the production of O_2 and glucose produced during that time. This is shown in the following graph:

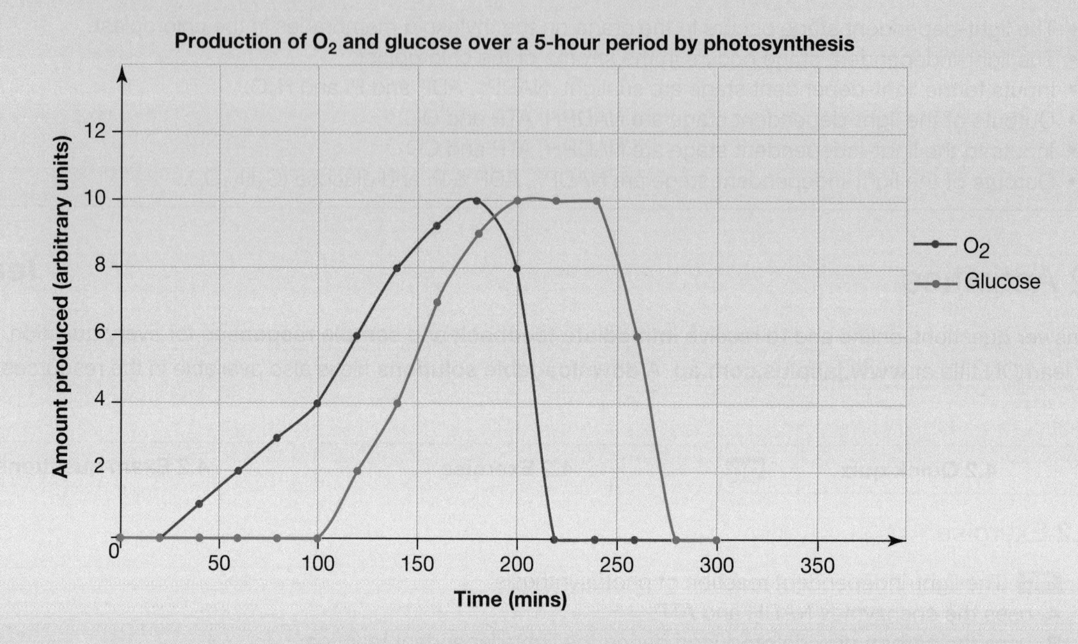

a. Why does the production of oxygen not start until 20 minutes into the photosynthetic process?
b. Explain what is occurring in the graph at 100 minutes.
c. Why does the amount of glucose begin to decrease after 240 minutes, despite being formed in the light-independent stage?
d. Why does the amount of O_2 produced not immediately drop to 0 when sunlight is no longer being provided?

4.2 Exam questions

Question 1 (1 mark)

Source: *VCAA 2020 Biology Exam, Section A, Q12*

MC During photosynthesis

A. ATP and NADH created in the light independent stage are transported to the chloroplasts' thylakoid membranes.
B. ADP and NADH are used in the electron transport chain after being created in the light-dependent stage.
C. ATP and NADPH are created in the grana of the chloroplasts and are used in the light-independent stage.
D. ADP and NADPH are created during the Krebs cycle and carried to the stroma of the chloroplasts.

Question 2 (1 mark)

Source: *VCAA 2016 Biology Exam, Section A, Q10*

MC Plants grown in light were supplied with water containing radioactive oxygen atoms. After four hours, an analysis of the chemicals in and around the plants was undertaken.

Which one of the following would contain the radioactive oxygen atoms after four hours?

A. protein
B. glucose
C. oxygen gas
D. carbon dioxide gas

Question 3 (4 marks)

Source: *VCAA 2020 Biology Exam, Section B, Q3a*

Greenhouses have been used to generate higher crop yields than open-field agriculture. To encourage plant growth in greenhouses, the conditions required for photosynthesis are controlled. Commercial greenhouses, like the ones shown, often use a lot of energy for heating, ventilation, lighting and water.

a. Consider the reactions of photosynthesis. Why would it be important to maintain the temperature within narrow limits in a commercial greenhouse? Justify your answer. **2 marks**

b. Scientists are developing a new material to cover greenhouses, which can split incoming light and convert the rays from green wavelengths into red wavelengths. Explain how this new material increases crop yields. **2 marks**

Source: SUPEE PURATO/Shutterstock.com

Question 4 (1 mark)

Source: *VCAA 2018 Biology Exam, Section A, Q15*

MC Which one of the following diagrams correctly represents the inputs and outputs of photosynthesis?

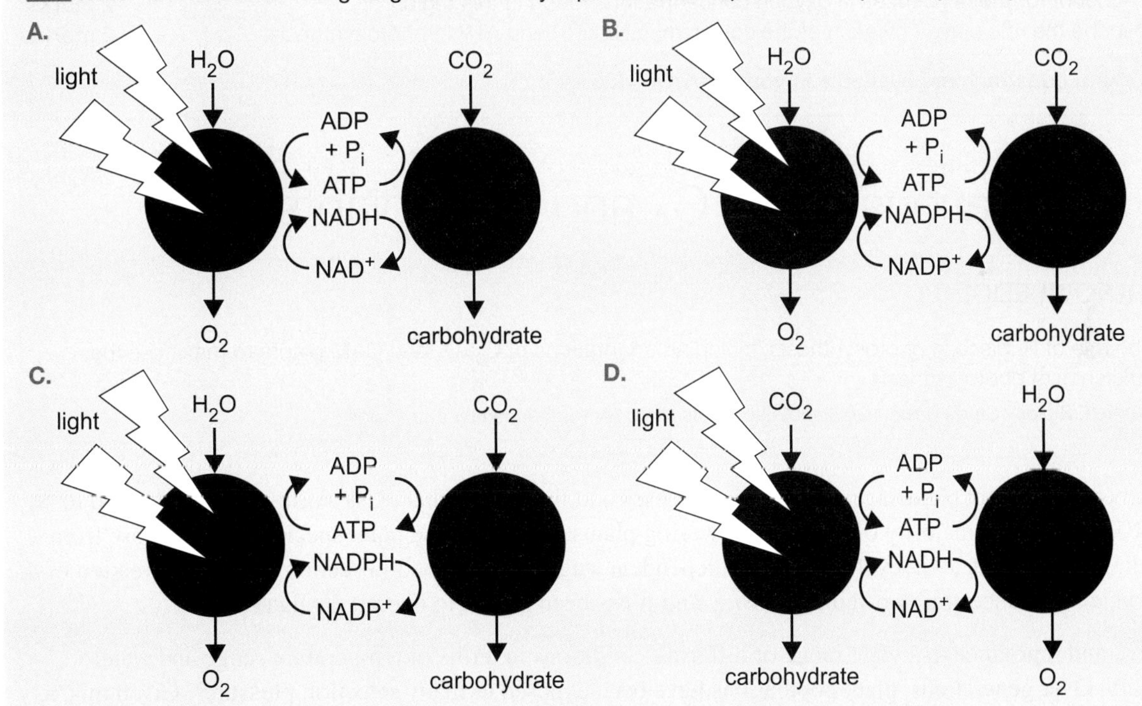

Question 5 (5 marks)

Source: *VCAA 2019 Biology Exam, Section B, Q2*

a. A chloroplast is surrounded by a double membrane.

  i. Name **two** molecules, as inputs for photosynthesis, that would need to diffuse from the cytosol of the plant cell across the chloroplast membranes and into the chloroplast. **1 mark**

  ii. Under high magnification, the internal structure of a chloroplast is visible. The diagram below shows part of this structure.

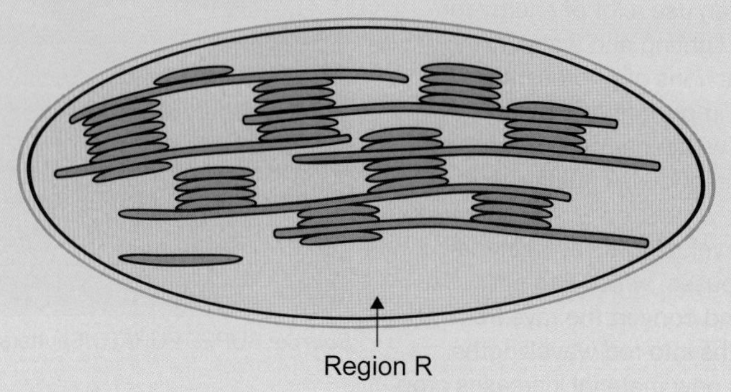

Source: Kazakova Maryia/Shutterstock.com

A higher concentration of oxygen is found in Region R when a plant is photosynthesising compared to when it is not photosynthesising.

Account for the differences in oxygen concentrations found in this region. **2 marks**

b. Describe the role played by each of the coenzymes NADPH and ATP in photosynthesis. **2 marks**

More exam questions are available in your learnON title.

4.3 Adaptations in C_3, C_4 and CAM plants

KEY KNOWLEDGE

- The role of Rubisco in photosynthesis, including adaptations of C_3, C_4 and CAM plants to maximise the efficiency of photosynthesis

The number of different plant species, including mosses and liverworts, that exist worldwide is estimated to be about 410 000, and the majority of these are flowering plants. Most of these plant species make glucose from carbon dioxide via the Calvin cycle (or light-independent stage) as described in section 4.2.5. This version of the Calvin cycle is the ancestral or original pathway and it has been shown to operate in algae.

Plants live and reproduce in a wide range of differing conditions in terms of temperature range and water availability. Over generations, plant populations have been exposed to many selection pressures. Given this, it is perhaps not surprising that some variations related to the Calvin cycle have evolved in some plants.

According to the process by which they fix carbon into glucose, plants can be organised into three groups: C_3 plants that carry out the original Calvin cycle, and C_4 plants and CAM plants that have each evolved a different variation of how the Calvin cycle operates.

4.3.1 C_3, C_4 and CAM plants

C_3 plants — use the original Calvin cycle

C_3 plants comprise about 85 per cent of terrestrial plants worldwide.

They include:

- the major crop plants, wheat (*Triticum* spp.) and rice (*Oryza sativa*), which are food staples for much of the world's population
- barley (*Hordeum vulgare*)
- rye (*Secale cereale*)
- oats (*Avena sativa*)
- soybean (*Glycine max*)
- sugar beet (*Beta vulgaris*)
- potato (*Solanum tuberosum*).

FIGURE 4.15 Different C_3 plants: a. soybean b. potato c. sugar beet

The C_3 label for this group of plants comes from the fact that the immediate organic product in their Calvin cycle is a three-carbon molecule of phosphoglyceric acid (PGA).

C_3 plants grow best in cool to temperate moist conditions. They use the Rubisco enzyme to fix inorganic carbon dioxide from the air and it enters the Calvin cycle joined to a carrier molecule (RuBP). The entire pathway of the Calvin cycle — from carbon dioxide to glucose — takes place in the stroma of the leaf mesophyll cells (see figure 4.16). The issues surrounding this will be explored further in section 4.3.2.

FIGURE 4.16 A highly simplified and stylised representation of the carbon dioxide to glucose pathway in C_3 plants

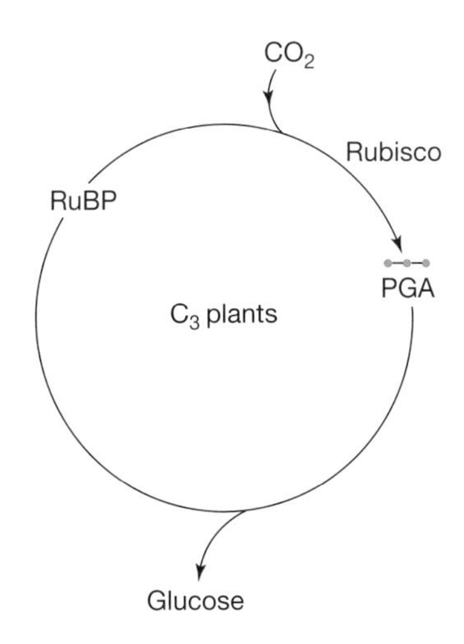

C_3 plants plants that carry out the original Calvin cycle using Rubisco and are prone to photorespiration

C_4 plants — different leaf anatomy

C_4 plants thrive in warm temperate regions and in tropical regions. They comprise about 3 per cent of land plant species. Among the C_4 plants are members of the grass family, including important crops such as:

C_4 plants plants that carry out an adapted Calvin cycle, in which carbon fixation and glucose production occur in different cells

- maize/corn (*Zea mays*)
- sugar cane (*Saccharum officinarum*)
- millet (*Panicum miliaceum*)
- sorghum (*Sorghum bicolor*)
- Mitchell grasses (*Astrebla* spp.). These grasses are the dominant vegetation on the Mitchell Downs, an extensive area in central Queensland that extends into the Northern Territory.

FIGURE 4.17 Different C_4 plants: a. sugar cane b. sorghum c. tussocks of Mitchell grass dominating the open grasslands of the Mitchell Grass Downs in central Queensland

These plants have the C_4 label because when they fix carbon dioxide, the immediate product is an organic acid with four C atoms, namely oxaloacetic acid (OAA).

In C_4 plants the anatomy of the leaves is different from that of C_3 and CAM plants (see figure 4.18). The leaves of C_4 plants have:

- bundle sheath cells, each with many chloroplasts enclose the vascular tissue in leaves
- mesophyll cells that are arranged in a close association around the bundle sheath cells.

Glucose production in C_4 plants is split into two stages, with carbon fixation taking place in mesophyll cells and glucose production via the Calvin cycle in bundle sheath cells. This will be further explored in section 4.3.3.

FIGURE 4.18 Diagram showing the 3D representation of the leaf anatomy of a. C_3 plants and b. C_4 plants

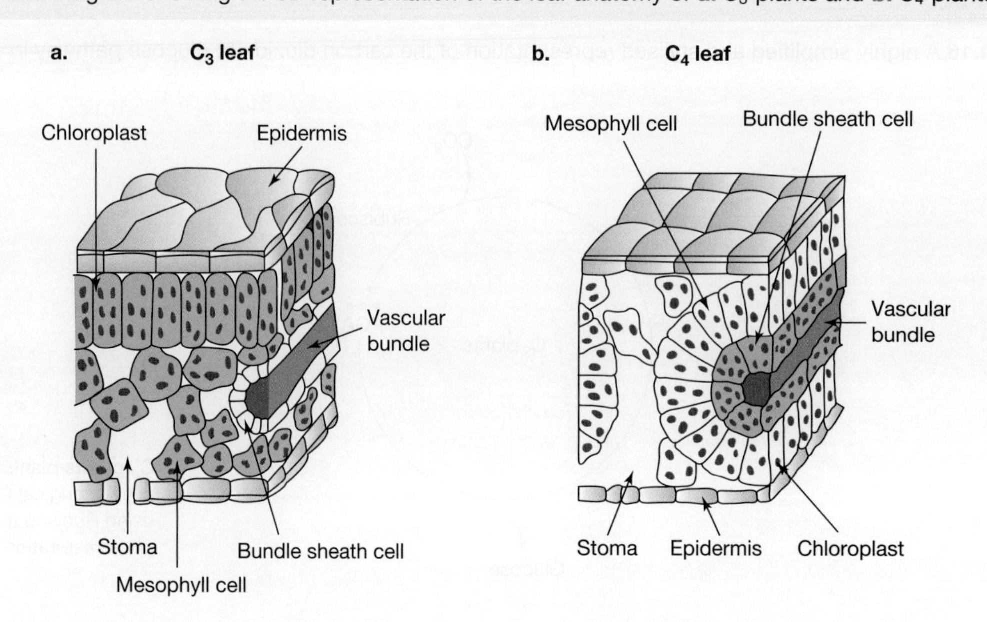

CAM plants — operate by day and night

CAM plants thrive in hot and arid environments and in regions exposed to drought. These plants constitute about 8 per cent of land plants and include:

- cacti
- many succulents, such as jade plants (*Crassula ovata*)
- orchids
- pineapples (*Ananas comosus*).

FIGURE 4.19 Different CAM plants: a. cacti b. moulded wax agave (*Echevaeria agavoides*) c. pineapple

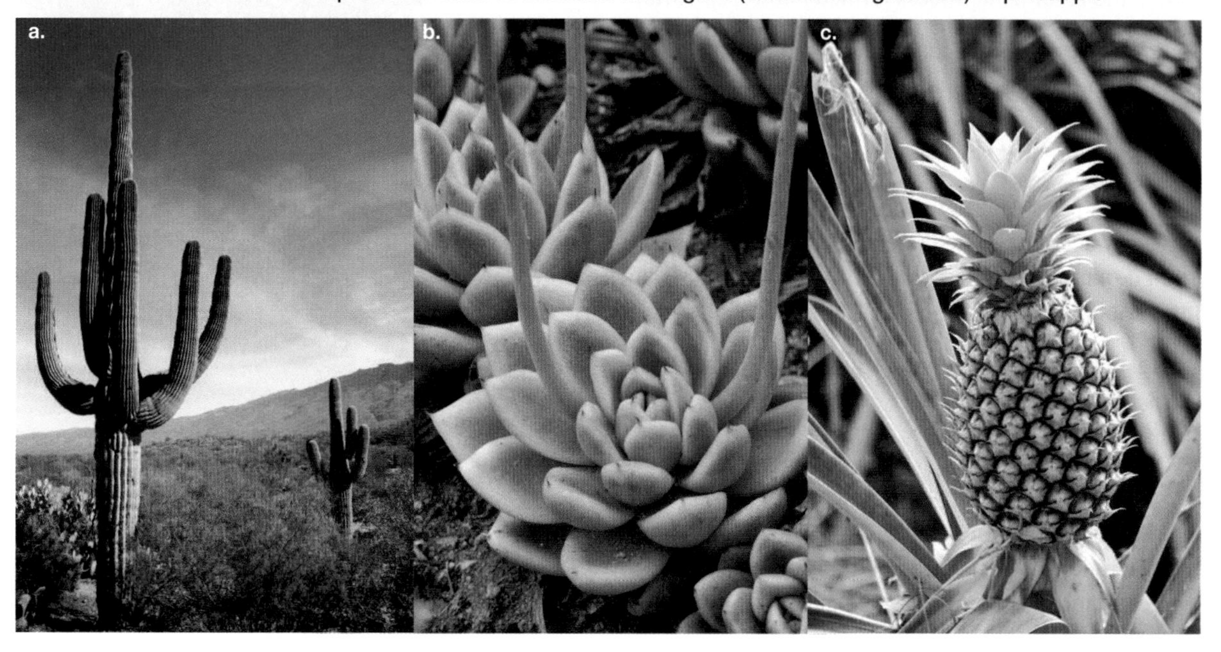

CAM means crassulacean acid metabolism, and CAM plants have this label because their particular variation of carbon fixation was first discovered in plants belonging to the family *Crassulaceae*.

Comparing the three plant types

In all three plant types, it is important that C_3, C_4 and CAM plants are able to maximise photosynthesis in changing conditions, making this process as efficient as possible. The adaptations of each type of plant to allow for this differs, which will be investigated in subtopics 4.3.2, 4.3.3 and 4.3.4.

Why have variations evolved in C_4 and in CAM plants? It's all because of a problem with the Rubisco enzyme and a wasteful process that may occur known as **photorespiration**.

CAM plants plants that thrive in arid conditions and have their two stages of the Calvin cycle occurring at different times

photorespiration a process in which plants take up oxygen rather then carbon dioxide in the light, resulting in photosynthesis being less efficient

4.3.2 Photorespiration in C_3 plants: it's all about Rubisco

Rubisco is the critical enzyme in C_3 plants that brings carbon dioxide from the air into the Calvin cycle where the glucose is made.

However, photosynthesis in C_3 plants is not always a totally efficient process because, instead of binding with carbon dioxide, its normal substrate in the Calvin cycle, Rubisco can bind with oxygen instead. If Rubisco binds oxygen rather than carbon dioxide, the result is a process termed *photorespiration*.

FIGURE 4.20 The molecular model of the enzyme Rubisco (ribulose bisphosphate carboxylase oxygenase)

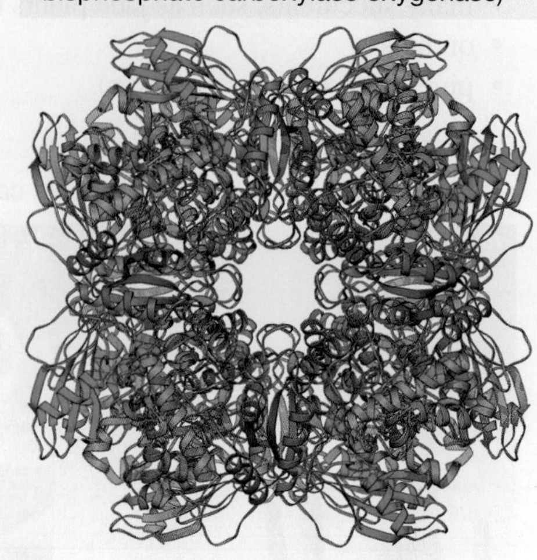

When photorespiration occurs, an estimated 20 to 40 per cent of the energy produced in photosynthesis in C_3 plants does not go to glucose production. This energy is 'lost' when, instead of binding with carbon dioxide, some Rubisco enzymes bind oxygen.

This can happen because the active site of the Rubisco enzyme can readily accommodate the oxygen molecules as well as the carbon dioxide molecules, so the two different molecules are in competition for the active site of Rubisco.

Photorespiration can arise in C_3 plants in two situations:

1. as temperatures increase
2. as conditions dry out.

When temperatures increase

In the cool to temperate conditions in which C_3 plants thrive, photorespiration is not a problem, as Rubisco will preferentially bind carbon dioxide. However, as leaves are exposed to higher temperatures, the rate of photorespiration increases faster than the rate of photosynthesis.

At low temperatures, Rubisco preferentially binds carbon dioxide. But, as temperatures increase, the ability of the Rubisco enzyme to distinguish between carbon dioxide and oxygen decreases and, as a result, Rubisco will increasingly bind oxygen. In addition, at higher temperatures, the solubility of carbon dioxide in the cytosol drops more rapidly than that of oxygen, so that there is more oxygen available in the mesophyll cells that at lower temperatures.

When conditions dry out

When conditions become dry and water availability declines, C_3 plants close their stomata to prevent water loss (see figure 4.21). This closure blocks the entry of carbon dioxide needed as input to the Calvin cycle and limits the exit of oxygen produced in the light-dependent stage of photosynthesis.

FIGURE 4.21 An open stoma (at left) and a closed stoma (at right)

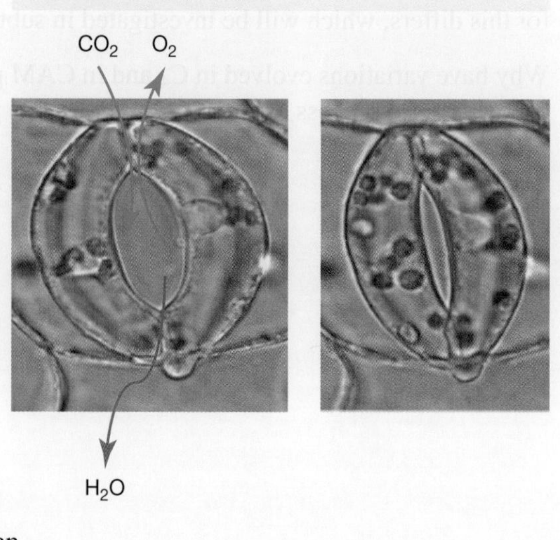

This creates a high oxygen and low carbon dioxide environment in leaf mesophyll cells, in which the Rubisco enzyme will increasingly bind oxygen rather than bind carbon dioxide in the process of glucose production via the Calvin cycle. Under these conditions, photorespiration rates increase.

Open stomata on leaves allow carbon dioxide to enter and the oxygen, generated in the light-dependent stage of photosynthesis, to exit. These processes are accompanied by a major loss of water vapour. Closed stomata prevent water loss but restrict the entry of carbon dioxide and the exit of oxygen.

Rubisco, photosynthesis and photorespiration

Rubisco works most efficiently when:
- carbon dioxide levels in leaves are high
- oxygen levels are low (as happens when water is freely available)
- when temperatures are moderate.

Under these conditions, the Rubisco enzyme will preferentially bind carbon dioxide. However, when the temperature rises, and when the concentration of carbon dioxide is low relative to that of oxygen, as happens when C_3 plants are water stressed and close their stomata, the Rubisco enzyme will preferentially bind oxygen, resulting in photorespiration. When this happens, instead of photosynthesis producing glucose from carbon dioxide, photorespiration produces carbon dioxide.

Summary of photorespiration

- Occurs when the Rubisco enzymes capture oxygen, instead of carbon dioxide
- Lowers the efficiency of photosynthesis
- Takes place when the $CO_2 : O_2$ ratio is low — that is, low CO_2 and high O_2
- Increases with increasing temperature
- Occurs more frequently on hot, dry days when C_3 plants close their stomata to prevent water loss
- Produces no glucose
- Produces carbon dioxide.

Figure 4.22 shows a highly simplified representation of photorespiration in C_3 plants compared to the usual Calvin cycle. The toxic by-product of PG (phosphoglycolic acid) is removed through a complex series of reactions. These reactions use energy and release carbon dioxide in a process that converts PG molecules to useful PGA molecules, which can bind with more PG.

FIGURE 4.22 A highly simplified diagram comparing photorespiration and Calvin cycle of photosynthesis. The input in photorespiration is oxygen and the output is carbon dioxide, rather than glucose. The problem molecules produced are the PGs (phosphoglycolic acid) that are shunted through a complex energy-requiring process to salvage them as PGA (phosphoglyceric acid).

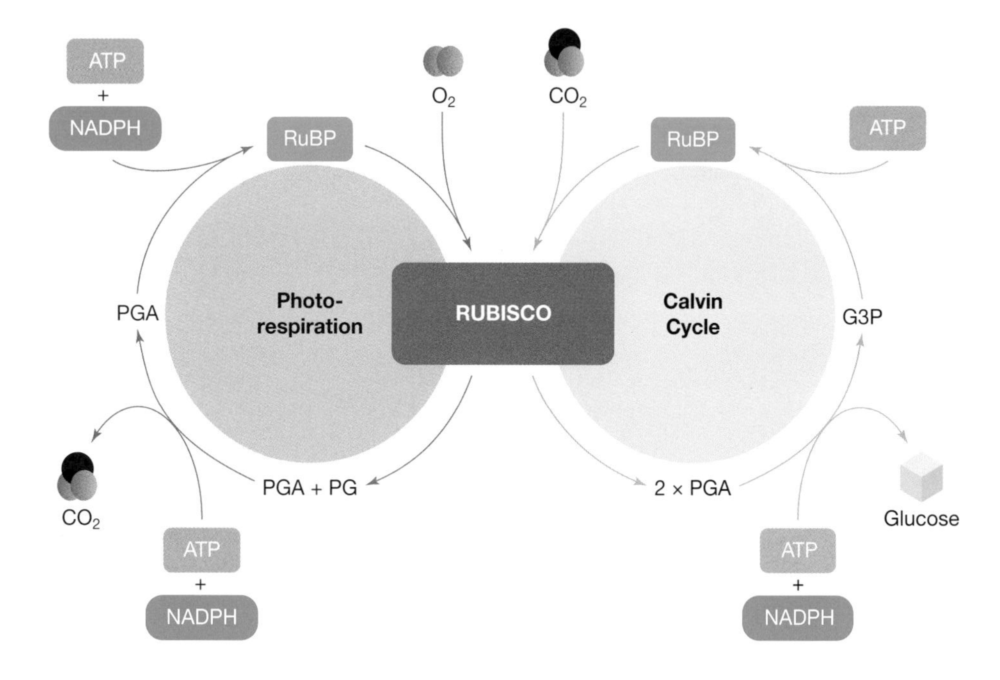

Photorespiration is a problem in C_3 plants, but not in C_4 or CAM plants. C_4 and CAM plants have evolved mechanisms to minimise or prevent photorespiration. These involve separating the process of carbon dioxide fixation from the process of glucose production by the Calvin cycle. As you will see in the following sections, C_4 plants do this by carrying out these processes in different cell types, and CAM plants do this by carrying out these processes at separate times (by night and by day).

4.3.3 Mechanisms to prevent photorespiration in C_4 plants

C_4 plants have methods to minimise photorespiration. These focus on differences in the location of carbon dioxide fixation and the Calvin cycle during the light-independent stage.

Minimising photorespiration in C_4 plants

- The pathway from carbon dioxide to glucose occurs in two stages that take place in two different cell types.
- The first stage of this pathway — carbon dioxide to malic acid — takes place in leaf mesophyll cells.
- In this stage, C_4 plants do not use the Rubisco enzyme. Instead they use the PEP carboxylase enzyme to catalyse the binding of carbon dioxide to an acceptor molecule. PEP carboxylase can only bind carbon dioxide molecules, so photorespiration cannot occur.
- The second stage — glucose production via the Calvin cycle — occurs in bundle sheath cells.
- In this stage, C_4 plants produce a steady supply of carbon dioxide from the breakdown of malic acid that raises the carbon dioxide concentration in their leaves. As a result, the Rubisco enzyme preferentially binds carbon dioxide, not oxygen, and brings it into the Calvin cycle.

Figure 4.23 shows a highly simplified representation of stages in C_4 plants during the production of glucose.

FIGURE 4.23 Diagram showing a simplified version of the light-independent stage of photosynthesis on C_4 plants that minimises or prevents photorespiration. The key mechanism is the physical separation into different cell types of the carbon-fixation process and the glucose-making process.

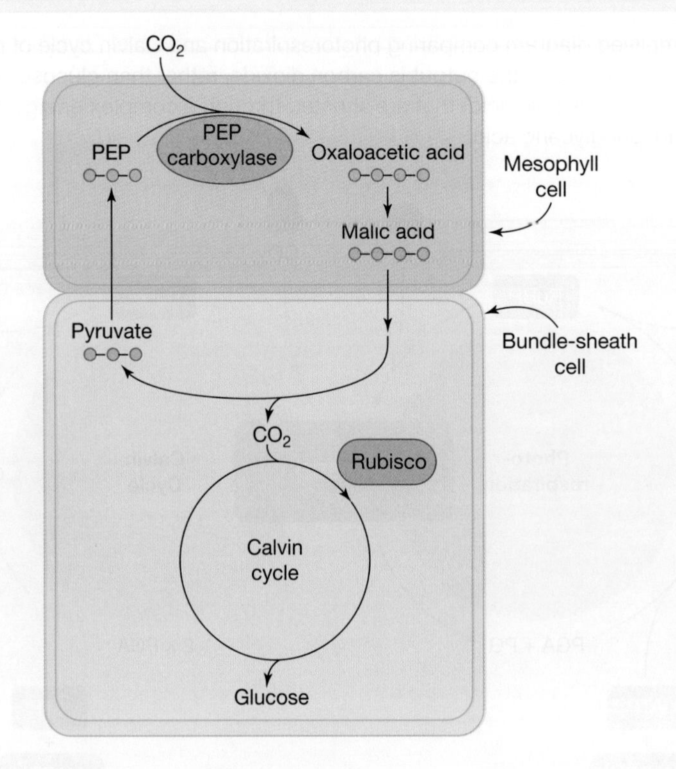

Why photorespiration is not a problem in C_4 plants

In C_4 plants, the pathway from carbon dioxide to glucose in C_4 plants occurs in two stages:

1. **A carbon fixation stage in mesophyll cells: from carbon dioxide to malic acid**
   - Occurs in the mesophyll cells that surround the bundle sheath cells
   - Uses the enzyme PEP carboxylase to join carbon dioxide to a carrier molecule phosphenol pyruvate (PEP), forming an organic acid
   - The PEP carboxylase enzyme can only bind carbon dioxide at its active site. Unlike Rubisco, it is not capable of binding oxygen.
   - The end product of this stage is malic acid.

   The use of the PEP carboxylase enzyme by C_4 plants to fix carbon dioxide into an organic acid eliminates the major problem of photorespiration.
2. **Calvin cycle in bundle sheath cells: from malic acid to glucose**
   - Occurs in the bundle sheath cells of C_4 plants
   - Involves the transport of malic acid (MA) from mesophyll cells into bundle sheath cells
   - In the bundle sheath cells, malic acid is continuously converted to pyruvate and carbon dioxide:

   $$\text{malic acid } (C_4H_6O_5) \rightarrow \text{pyruvate } (C_3H_4O_3) + \text{carbon dioxide } (CO_2)$$

   - The released carbon dioxide creates a high-concentration CO_2 environment in the bundle cells.
   - As in the usual Calvin cycle, the Rubisco enzyme joins carbon dioxide to an organic acceptor molecule (RuBP) that enters the Calvin cycle for glucose production.

The steady production of carbon dioxide into the bundle sheath cells means that the Rubisco enzyme will preferentially bind carbon dioxide, not oxygen.

4.3.4 Mechanisms to prevent photorespiration in CAM plants

In CAM plants the complete pathway from carbon dioxide to glucose occurs in two stages in CAM plants at different times.

Minimising photorespiration in CAM plants

- The carbon fixation stage takes place only at night when stomata are open.
- The Calvin cycle that produces glucose occurs only during the day when stomata are closed.
- Both stages take place in mesophyll cells.

Other mechanisms that prevent photorespiration in CAM plants are:

- *carbon fixation at night.* Inorganic carbon dioxide from the air is fixed by the PEP carboxylase enzyme — the same as in C_4 plants. PEP carboxylase can only bind carbon dioxide — there is no chance of binding oxygen and causing photorespiration! The products of this reaction are four-carbon organic acids, such as malic acid, that, as they are formed, are stored in vacuoles in the mesophyll cells.
- *Calvin cycle by day.* The organic acids are released from storage and broken down to releases carbon dioxide. This creates a high concentration of carbon

FIGURE 4.24 The stomata are only open at night.

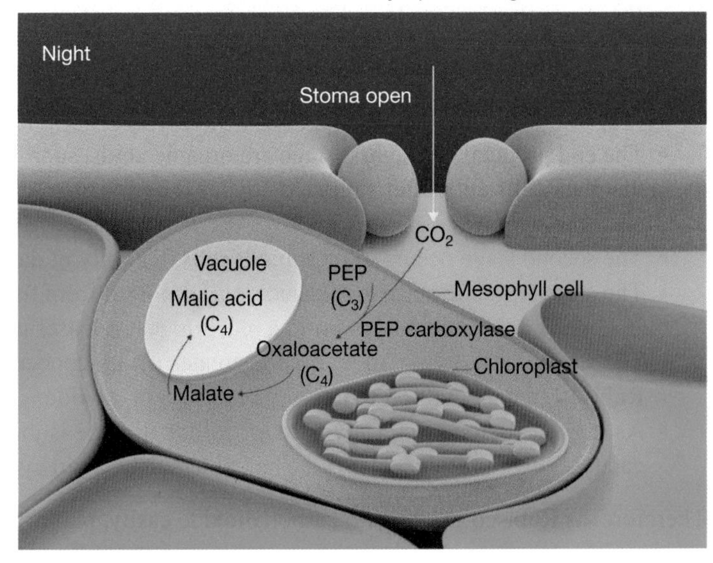

dioxide in the mesophyll cells producing an environment in which the Rubisco enzyme preferentially binds carbon dioxide for entry to the Calvin cycle.

Figure 4.25 shows the two stages of the CAM pathway of glucose production.

FIGURE 4.25 Diagram showing a simplified representation of the CAM plant pathways that are separated in time, with carbon fixation occurring at night and glucose production occurring during the day.

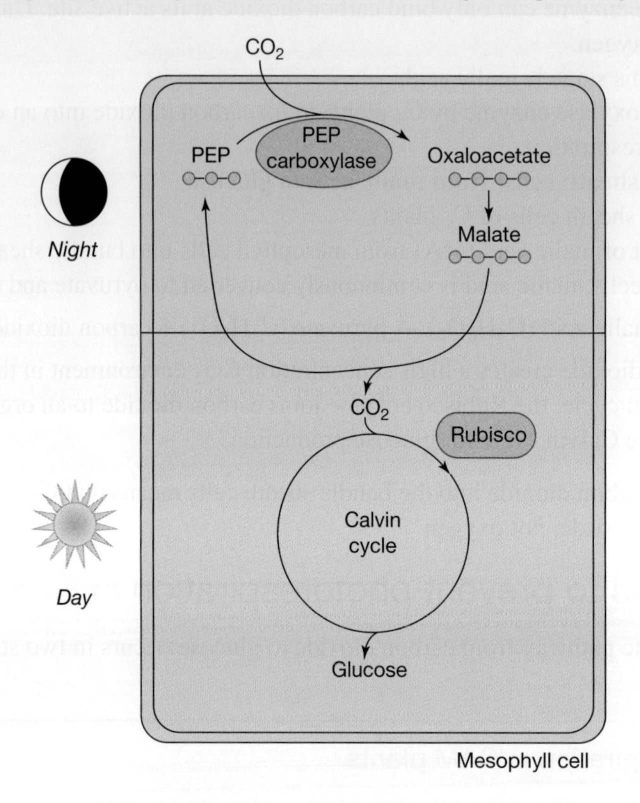

Why photorespiration is not a problem in CAM plants

1. The first stage

- The first stage occurs in mesophyll cells, but only at night.
- It occurs when these plants have their stomata open, allowing the free passage of carbon dioxide into the leaves.
- Carbon dioxide is fixed into organic acids in a reaction catalysed by the enzyme PEP carboxylase that can only bind carbon dioxide.
- The end products of the first stage are organic acids, such as malic acid, and this is stored in vacuoles in the plant cells until after sunrise on the next day.

2. The second stage

- The second stage occurs in mesophyll cells, but only in daylight when stomata are closed.
- The stored organic acid (malic acid) is transported from the vacuoles into the stroma of chloroplasts.
- The malic acid is broken down in a reaction that releases carbon dioxide.
- The steady release of carbon dioxide creates a high concentration of CO_2, an environment in which the Rubisco enzyme will preferentially bind carbon dioxide.
- As happens in the usual Calvin cycle, carbon dioxide is joined to an organic acceptor in a reaction catalysed by the Rubisco enzyme and enters the Calvin cycle.

Therefore, as Rubisco can bind to carbon dioxide easily, photorespiration is reduced.

4.3.5 Comparing photosynthesis in C_3, C_4 and CAM plants

The variations in the light-independent stage of photosynthesis in C_3, C_4 and CAM plants are shown in a simplified representation in figure 4.26.

FIGURE 4.26 Diagram showing the simplified pathways of the light-independent stage of photosynthesis in C_3, C_4 and CAM plants. Note that C_4 plants separate carbon fixation from the Calvin cycle into two different cell types. In CAM plants, both these processes occur within the same cell type (mesophyll cells), but achieve this separation by carrying out carbon fixation at night and the Calvin cycle by day.

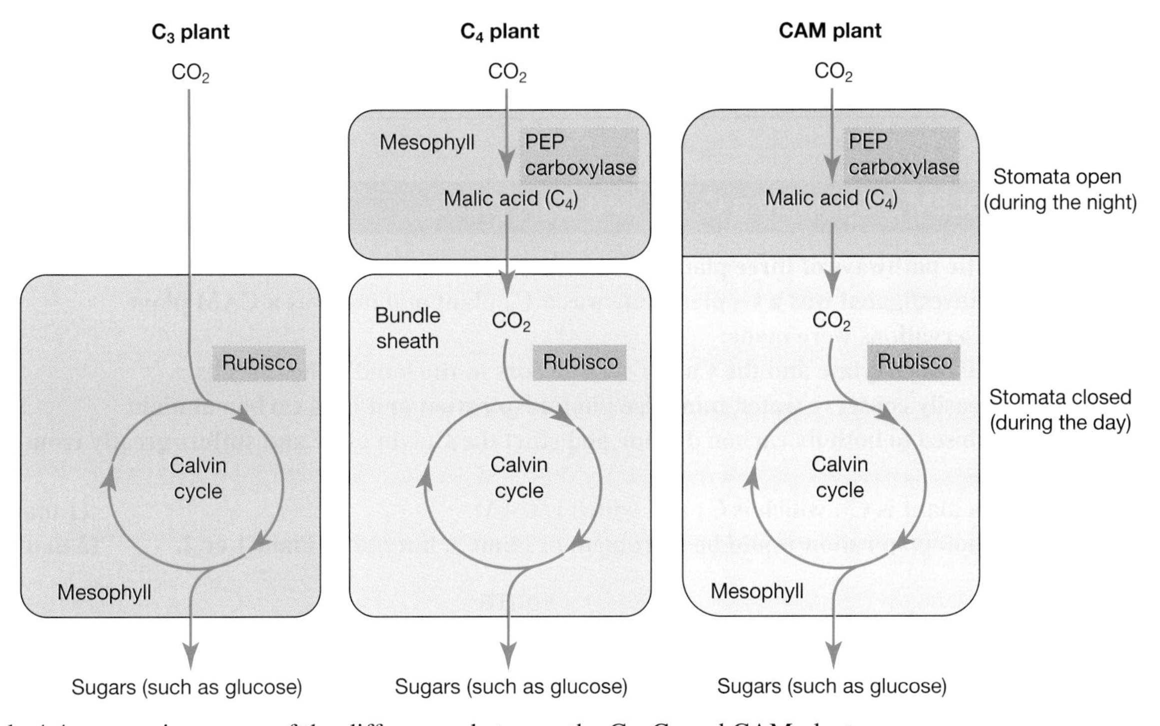

Table 4.4 summarises some of the differences between the C_3, C_4 and CAM plants.

TABLE 4.4 Some of the differences in the light-independent stage (Calvin cycle) of photosynthesis. The original pathway, as seen in C_3 plants, varies in C_4 plants and in CAM plants. One variation is the location of processes and the other is the timing of the processes

	C_3 plants	C_4 plants	CAM plants
Enzyme to fix carbon dioxide from air	Rubisco	PEP carboxylase	PEP carboxylase
Acceptor molecule of CO_2 from air	Ribulose bisphosphate (RuBP)	Phosphoenol pyruvate (PEP)	Phosphoenol pyruvate (PEP)
First product of carbon fixation	Phosphoglyceric acid (PGA), a 3C molecule	Oxaloacetic acid (OAA), a 4C molecule	• at night: OAA • by day: PGA
Location and number of carbon fixation events	One, in mesophyll cells	Two, in different cell types • first in mesophyll cells • second in bundle sheath cells	Two, both in mesophyll cells • first by night • second by day
Location of Calvin cycle	Mesophyll cells	Bundle sheath cells	Mesophyll cells
Enzyme to start Calvin cycle	Rubisco	Rubisco	Rubisco

(continued)

TABLE 4.4 Some of the differences in the light-independent stage (Calvin cycle) of photosynthesis. The original pathway, as seen in C_3 plants, varies in C_4 plants and in CAM plants. One variation is the location of processes and the other is the timing of the processes *(continued)*

	C_3 plants	C_4 plants	CAM plants
Presence of chloroplasts in bundle sheath cells	No	Yes	No
Open stomata required for efficient photosynthesis	Yes	No	Yes, at night only
Photorespiration in high temps and low CO_2 concentrations	High	Low to zero	Low to zero
Optimal temperature range	15–25 °C	30–40 °C	>40 °C

tlvd-1426

SAMPLE PROBLEM 2 Comparing C_3, C_4 and CAM plants

The photosynthetic pathways of three plants were explored.
One plant being investigated was a C_3 plant, one was a C_4 plant and one was a CAM plant.
The following observations were made:
Plant 1: Uses PEP carboxylase and the Calvin cycle occurs in the bundle sheath cells
Plant 2: Able to easily conserve water, minimise photorespiration and fixes carbon at night
Plant 3: Uses Rubisco to both fix carbon dioxide and start the Calvin cycle and suffers greatly from photorespiration

a. **Identify which plant is C_3, which is C_4 and which is CAM.** **(1 mark)**
b. **Justify why photorespiration would be a problem in Plant 3, but not in Plant 1 or 2.** **(3 marks)**

THINK	WRITE
a. Review the features of each plant type.	
C_3: affected by photorespiration and uses Rubisco for various stages	Plant 3 is the C_3 plant.
C_4: uses both PEP carboxylase and Rubisco, but uses two different cell types (mesophyll and bundle sheath cells)	Plant 1 is the C_4 plant.
CAM: differs between day and night, and only fixes carbon at night, allowing for minimal water loss	Plant 2 is the CAM plant. (1 mark for all three correct)
b. 1. Examine what the question is asking you to do. It is a *justify* question so requires a detailed explanation. As it is worth three marks, it requires three aspects: • why photorespiration is a problem in Plant 3 • why photorespiration is not a problem in Plant 1 • why photorespiration is not a problem in Plant 2.	In Plant 3 (the C_3 plant), photorespiration occurs when Rubisco can bind to oxygen rather that carbon dioxide, leading to a loss of energy and lower efficiency of photosynthesis (1 mark). This is different to Plant 1 (the C_4 plant) as a different enzyme, PEP carboxylase, fixes carbon dioxide, eliminating the cause of photorespiration (in which Rubisco binds to oxygen). Carbon dioxide moves into the bundle sheath so Rubisco will bind to this preferentially (1 mark).
2. Explain each component of the question, linking to both photorespiration and Rubisco.	In Plant 2 (the CAM plant), photorespiration is also not an issue because of the use of PEP carboxylase and extra steps in the process, allowing for Rubisco to more easily bind to CO_2 (1 mark).

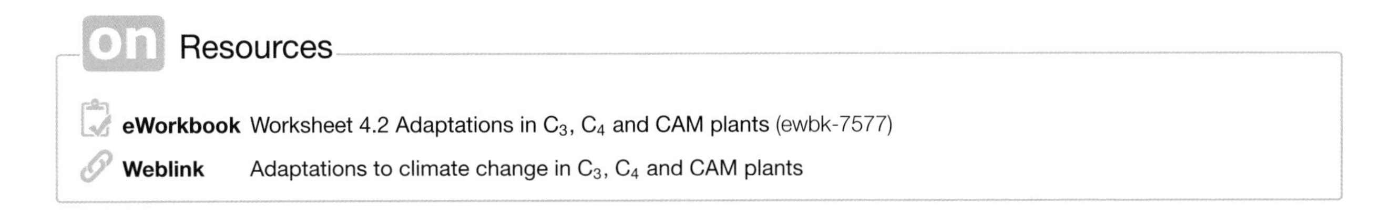

KEY IDEAS

- C_3 plants (such as wheat) carry out the original carbon cycle. C_4 plants (such as corn) and CAM plants (such as succulents) have each evolved a different variation in how the Calvin cycle operates.
- C_3 plants produce a three-carbon molecule (PGA) during the Calvin cycle. These plants used the Rubisco enzyme to fix inorganic carbon dioxide from air.
- C_4 plants fix carbon dioxide in a four-carbon molecule (OAA) using a molecule PEP carboxylase. Carbon fixation occurs in mesophyll cells and glucose production occurs in bundle sheath cells.
- CAM plants thrive in hot and arid environments and their two stages occur at different times. Carbon fixation only takes place at night and glucose production only takes place during the day.
- In C_3 plants, photosynthesis is not totally efficient as Rubisco can bind to oxygen instead, leading to photorespiration, in which much of the energy does not go into glucose production and is lost.
- Photorespiration is of particular concern in higher temperatures and drier conditions.
- Photorespiration is not a problem in C_4 or CAM plants, as they have developed mechanisms which enable them to minimise photorespiration by separating the process of carbon dioxide fixation from the process of glucose production by the Calvin cycle.

4.3 Activities

learnon

To answer questions online and to receive **immediate feedback** and **sample responses** for every question, go to your learnON title at **www.jacplus.com.au**. A **downloadable solutions** file is also available in the resources tab.

4.3 Quick quiz	4.3 Exercise	4.3 Exam questions

4.3 Exercise

1. Describe three distinct differences between C_3 and C_4 plants.
2. What conditions are best suited to C_3, C_4 and CAM plants?
3. *Astrebla lappacea* is the most common type of Mitchell grass, a C_4 plant.
   a. Why is *A. lappacra* better at surviving in conditions of drought compared to a C_3 plant such as wheat (*Triticum aestivum*)?
   b. What two cells of the leaf are important in photosynthesis of *A. lappacra*? How is each cell type used?
   c. Outline the role of Rubisco and PEP carboxylase in the function of *A. lappacra.*
4. Photorespiration is a process that is of particular concern to C_3 plants.
   a. Describe the process of photorespiration.
   b. Outline the role Rubisco plays in this process.
   c. In which environmental conditions are the effects of photorespiration more significant?
5. Plants such as cacti are known as CAM plants, and are able to survive in very arid conditions.
   a. Describe the differences between CAM plants during the daytime as opposed to nighttime.
   b. How does this allow CAM plants to minimise photorespiration?

4.3 Exam questions

Question 1 (2 marks)

Source: *VCAA 2011 Biology Exam 1, Section B, Q7e*

Rubisco is an enzyme found in chloroplasts. Its normal function is to catalyse the reaction in which carbon dioxide is a substrate. In certain plants, when the level of carbon dioxide is low in the leaf, Rubisco uses oxygen as the substrate and releases hydrogen peroxide and ammonia.

Explain why it is beneficial for a plant to have a high level of carbon dioxide in its leaves.

Question 2 (1 mark)

MC CAM plants thrive in hot and arid environments and have various adaptations to maximise the efficiency of photosynthesis. CAM plants are

A. more common globally than C_3 plants.
B. prone to photorespiration.
C. able to undergo carbon fixation at night and the Calvin cycle during the day.
D. able to use bundle sheath cells for carbon fixation during the day.

Question 3 (1 mark)

MC *Cyperus papyrus* is an endangered plant native around the Nile River. This plant has a long history of use. One of the main uses was in the production of papyrus paper, made from the pith of the plant. This paper was most commonly used in Egypt, where *C. papyrus* was once abundant.

C. papyrus is most suited to grow in temperatures around 30 °C in tropical conditions. Unlike many other plants, it does not require open stomata to stay open for the Calvin cycle to occur. Instead, malic acid produced in the mesophyll cells can be broken down to produce carbon dioxide, which can be taken to bundle sheath cells.

Based on this information, it can be assumed that *C. papyrus* is

A. a C_3 plant.
B. a C_4 plant.
C. a CAM plant.
D. free of chloroplasts.

Question 4 (9 marks)

a. Complete the following table to compare C_3, C_4 and CAM plants. **6 marks**

	Enzyme used to fix carbon dioxide from air	Location of carbon fixation	Location of Calvin cycle	Time of day carbon fixation occurs
C_3				
C_4				
CAM				

b. Draw a clear diagram showing carbon fixation and the Calvin cycle in C_3, C_4 and CAM plants. **3 marks**

Question 5 (7 marks)

A farmer in a temperate moist environment in northern Australia is growing two crops in adjacent fields: corn and wheat. While both plants are suited to grow in this these conditions, they undergo photosynthesis in different ways. Wheat is a C_3 plant, whereas corn is a C_4 plant.

a. With reference to Rubisco, identify two key differences between photosynthesis in wheat and corn. **2 marks**

The farmer is having a discussion with another farmer who lives in a drought-ridden region in central Victoria. He has been trying to grow both wheat and corn. This farmer has noticed that the photosynthetic efficiency of the wheat is much less than that of the corn.

b. Identify the process that is reducing the efficiency of photosynthesis in the wheat. **1 mark**

c. Explain why the process you identified in question b. in more of an issue in wheat than it is in corn. **4 marks**

More exam questions are available in your learnON title.

4.4 Factors that affect the rate of photosynthesis

KEY KNOWLEDGE

- The factors that affect the rate of photosynthesis: light availability, water availability, temperature and carbon dioxide concentration

Source: VCE Biology Study Design (2022–2026) extracts © VCAA; reproduced by permission.

The faster the rate of photosynthesis, the more glucose (and oxygen) that is produced in a shorter time frame. Many factors influence the rate in which the light-independent and light-dependent stages occur.

The rate at which plants photosynthesise depends on:

- the amount of light reaching their leaves
- the temperature of the environment
- the availability of water
- the concentration of carbon dioxide.

Other factors that can affect the rate of photosynthesis are the amount of chlorophyll, the availability of nutrients, such as Mg^{2+} that is needed for chlorophyll synthesis, and phosphates and nitrates. Under certain circumstances, any one of these factors — light availability, carbon dioxide or temperature — may become a **limiting factor** that stops any further increase in the rate of photosynthesis.

4.4.1 Light availability

It is a warm bright sunny day and a canola crop growing in a field bathed in sunlight can carry out photosynthesis at a faster rate than on the previous equally warm, but overcast, day. Light intensity or illuminance is the factor producing the difference in the rates of photosynthesis by the canola plants on these two days.

The photosynthetic rate increases as light intensity increases, until it reaches a maximal point.

FIGURE 4.27 Simple graph showing the change in rate of photosynthesis with increasing light intensity or illuminance (when both temperature and CO_2 levels are constant)

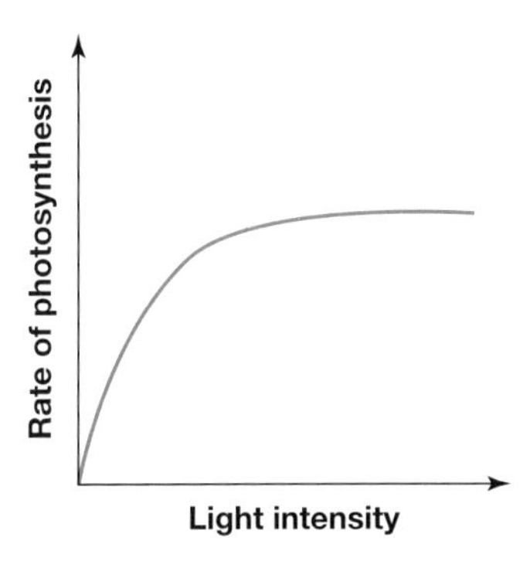

The following can be observed about changing light intensity:

- At low light intensities, the photosynthesis rate is slow or absent.
- With increasing illuminance, the photosynthesis rate increases, as seen by the steep upward sloping line of the graph. The optimal light intensity is the one at which the rate of photosynthesis is the greatest.
- Beyond the optimal light intensity, further increases in light intensity have no effect and the rate of photosynthesis stays constant. This is called the **light saturation point** and it is marked by the flattening or plateauing of the graph. At this point, some other factor is limiting the rate of photosynthesis. We will explore other limiting factors later in this subtopic.

The low light intensity in winter months and shorter day lengths are limiting factors on the rate of photosynthesis. This creates a problem for managing the grass surfaces in sports stadiums. In order to increase the rate of photosynthesis to levels that maintain healthy growth of the grass, banks of lights are used to illuminate the grass during the night. Figure 4.28 shows banks of lights over the grass surface at a soccer oval.

limiting factor environmental condition that restricts the rate of biochemical reactions in an organism

light saturation point the point in which increasing the light intensity no longer increases the rate of photosynthesis

On a bright sunny day, the illuminance or incident light on a surface reaches a maximum of about 110 000 lux (lumens per square metre). On an overcast day, that value falls markedly to about 1000 to 2000 lux. At sunrise and at sunset, the light intensity falls to about 400 lux, while moonlight (at full moon) is about 1 lux.

FIGURE 4.28 Artificial lighting is used to illuminate the grass surface of a soccer field at night.

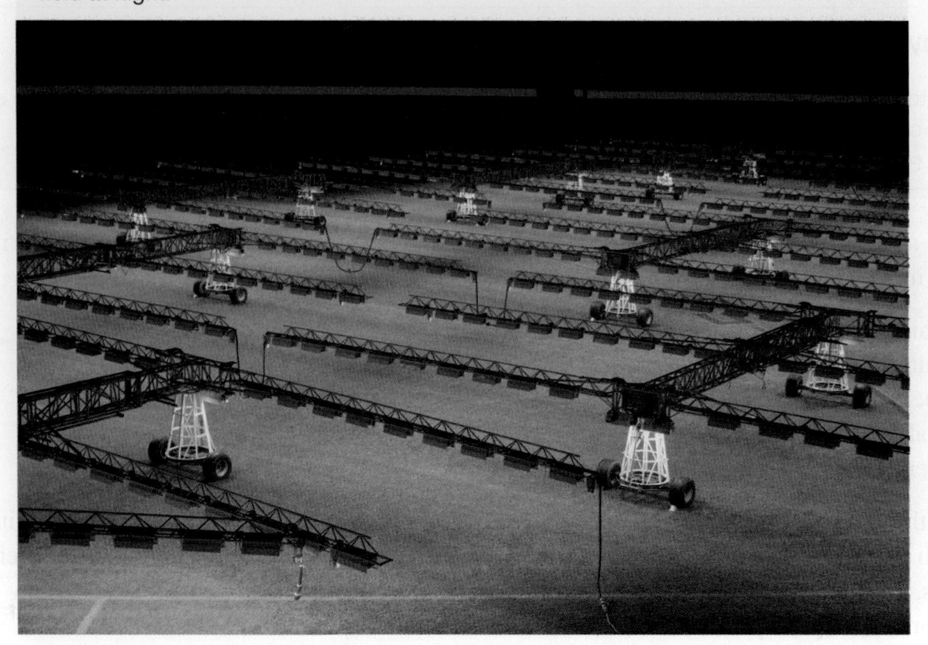

CASE STUDY: Wavelengths of light and the effect of different pigments

Photosynthetic organisms have a range of pigments, including chlorophylls, that trap the sunlight necessary for photosynthesis.

The most useful wavelengths of light for photosynthesis were first identified in an experiment carried out by a German botanist, T.W. Engelmann (1843–1909), in 1881. A thin strand of the green alga *Spirogyra* sp. floating in water was exposed to visible light covering a range of wavelengths. Engelmann used a prism to produce these different wavelengths.

Large numbers of a kind of bacteria that gather in areas with a high concentration of oxygen were also added to the water. These bacteria are termed aerophilic. The location of the bacteria shows where this oxygen is being released and where the rate of photosynthesis is highest. The experimental result supports the conclusion that the biologically useful wavelengths for this alga are those of red and violet light.

FIGURE 4.29 Engelmann's experiment. Only certain wavelengths of visible light are trapped by producer organisms. What are the useable wavelengths?

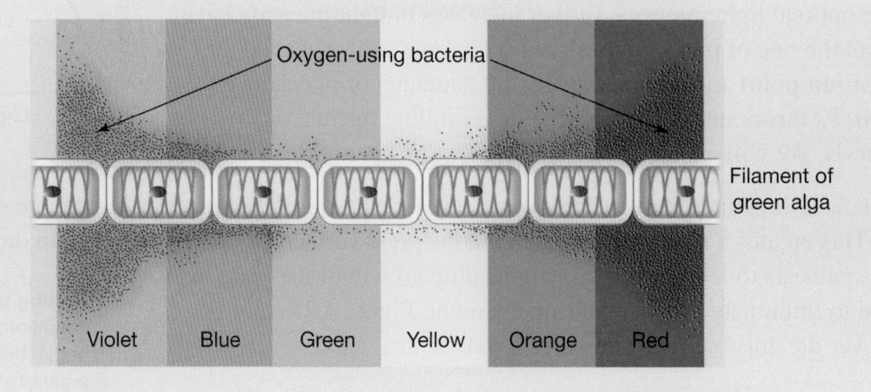

This experiment identified the wavelengths of visible light that were trapped by the photosynthetic green alga.

Different pigments absorb coloured light differently. Some pigments, such as chlorophyll, best absorb violet, blue and red light, so in plants with photosynthesis, the rate is highest in these coloured lights. Plants that contain phycocanin (such as cyanobacteria), however, are better at absorbing yellow light.

FIGURE 4.30 Absorption of light of various wavelengths for different plant pigments

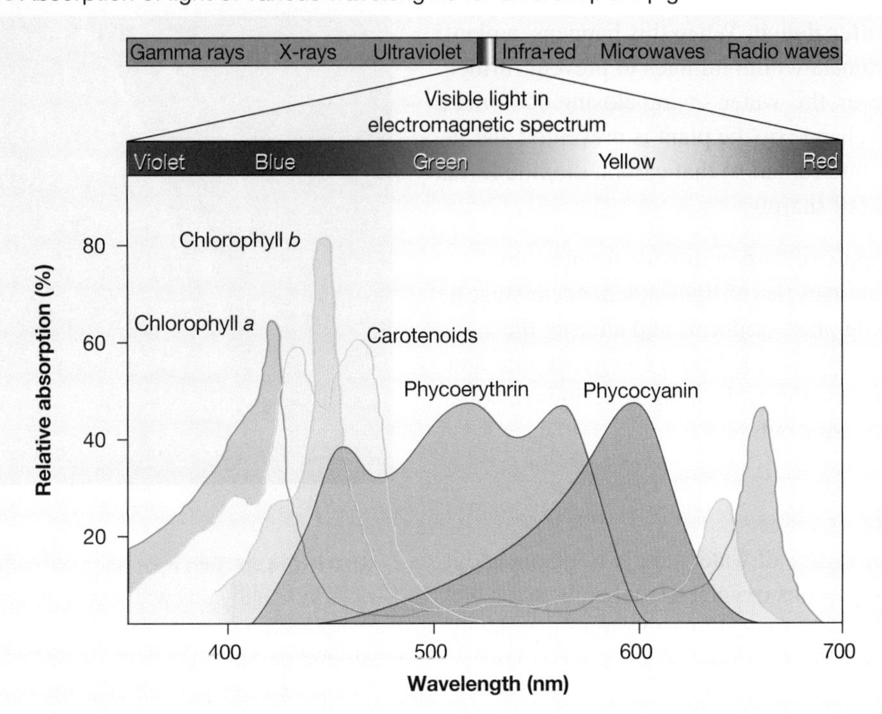

4.4.2 Water availability

An adequate water supply to photosynthetic cells of a terrestrial plant is essential for the normal operation of the photosynthesis pathway. The water supply to a plant can affect the rate of photosynthesis — too little water, and the rate of photosynthesis slows and stops (especially as water is a vital input in the light-dependent stage of photosynthesis). Too much water can have a similar effect.

Note the following:

- The source of water for terrestrial plants is the water content of the soil in which they are growing.
- The uptake of water and dissolved mineral nutrients by terrestrial plants depends on the root system of a plant.
- Water is taken up from the soil through root hair cells and transported to photosynthetic cells via the xylem transport system.
- Water is lost from plants as vapour during transpiration via the leaf stomata.
- When water loss from leaves by transpiration exceeds uptake of water by the root hair cells, a situation of water deficit exists.

The amount of water available to a terrestrial plant affects its rate of photosynthesis in the following ways:

- If soils dry out and the water supply becomes too little, the rate of photosynthesis declines and then stops because closed stomata prevent the uptake of carbon dioxide needed for the Calvin cycle.
- If the water supply increases too much causing waterlogging of the soil, the rate of photosynthesis will also decline and stop because the lack of oxygen for cellular respiration in root cells stops water uptake.

Water deficit

Studies of individual plants show that the photosynthetic rate of plants falls as their water content decreases in response to limited availability of water.

As water becomes even less available, plants are exposed to the stress of **water deficit**. When this happens, a plant can close their stomata within minutes to prevent further water loss. However, this water-saving closing of stomata means that gas exchange by the plant is prevented and its uptake of CO_2 is interrupted so that carbon dioxide levels within the leaves fall sharply.

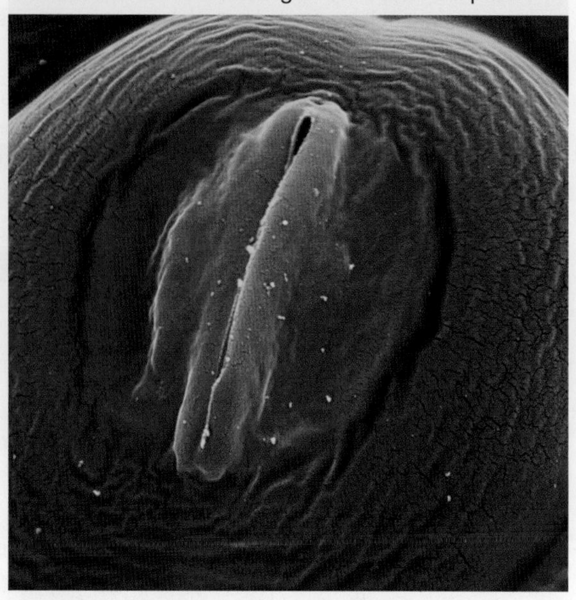

FIGURE 4.31 The closing of stomata on a plant

Preventing the supply of carbon dioxide to mesophyll cells stops the Calvin cycle (or light-independent stage), effectively stopping photosynthesis and altering the balance that normally exists between energy capture and energy use in photosynthesis. Prolonged periods of water deficit can cause plant death.

Waterlogging

Healthy soil is the source of water supply to plants and it also the source of gaseous oxygen to root cells to enable ATP production through the cellular respiration pathway.

Figure 4.32 is a diagram showing root hairs surrounded by healthy soil in which soil particles are surrounded by films of water and spaces between soil particles are filled with air.

FIGURE 4.32 Two root hairs extending from root cells in healthy soil with many air spaces. Soil particles are shown cross-hatched, light areas around particles are films of water, and the larger round clear areas in the soil are air spaces.

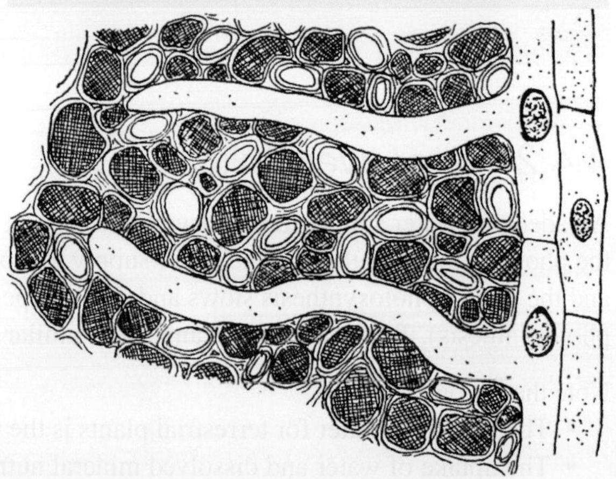

Soils become **waterlogged** when more rain falls than the soil can absorb or the atmosphere can evaporate. When this happens, the root zone of a plant becomes saturated with excess liquid water, which causes:

- the air spaces that normally exist in soil to become filled with water
- the oxygen content of the soil to become depleted
- as a result, plant root cells are not able to respire in this anoxic soil condition.

As we will see later in this topic, oxygen is an essential input to the life-supporting process of cellular respiration. In waterlogged soils, there is not enough oxygen to enable plant root cells to respire adequately and gain the energy for living. If waterlogging continues over an extended period, roots become permanently damaged, water uptake is stopped and the plant eventually dies. No water supply to mesophyll cells of the leaves means that an essential input to the light-dependent stage of photosynthesis is not available and the rate of photosynthesis falls to zero.

water deficit when there is a limited amount of water

waterlogged when excess water has reached a plant

4.4.3 Temperature

Temperature is a factor that affects the rate of photosynthesis and hence plant growth and, ultimately, crop yields. The biochemical reactions in the light-independent Calvin cycle are catalysed by several different enzymes, including Rubisco. The effect of temperature on enzymes was introduced in section 3.5.1.

The progress of chemical reactions depends on collisions between reactants and enzymes:

- At low temperatures, low collision rates produce a low rate of photosynthesis.
- As the temperature rises, the rate of photosynthesis initially increases as the rate of molecular collisions increases.

However, the increase in the rate of photosynthesis with increasing temperature does not continue indefinitely. Once the optimal temperatures of the enzymes involved are exceeded, the rate of photosynthesis decreases rapidly as heat denaturation of enzymes begins.

FIGURE 4.33 Graph showing the changes in the rate of photosynthesis with increasing temperature

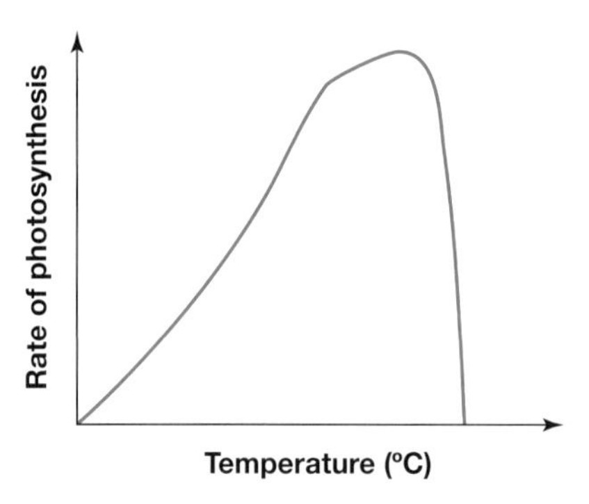

Very soon, the rate drops to zero and photosynthesis stops because the enzymes are denatured. Their altered shapes means that they can no longer function as enzymes. As these enzymes are vital in both the light-dependent and light-independent stages of photosynthesis, the rate drops to zero.

Because of the relationship between temperature and photosynthesis rates, when crop plants are cultivated in glasshouses, care is taken to ensure that temperature extremes are avoided by installing thermostats and greenhouse heaters and cooling devices.

As the ambient temperature is increased, the rate of photosynthesis also increases due to an increase in collisions between the reactants and the enzymes involved in photosynthesis. Eventually, as the heat passes a certain threshold, the enzymes start to denature, in which the tertiary structure of an enzyme is lost. This causes the rate to again decrease.

4.4.4 Carbon dioxide concentration

Carbon dioxide is one of the inputs to the light-independent stage of photosynthesis. The rate of photosynthesis is affected by the concentration of carbon dioxide.

As the concentration of carbon dioxide is progressively increased, the rate of photosynthesis will increase until it levels off due to limiting factors.

FIGURE 4.34 Graph showing the changes in the rate of photosynthesis with increasing carbon dioxide

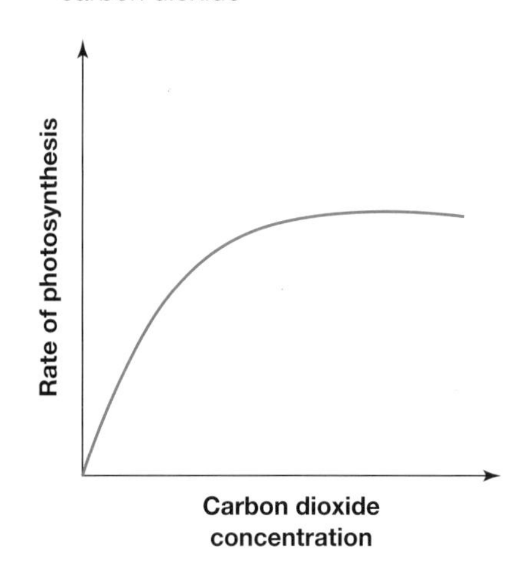

Note the initial linear relationship between the increase in carbon dioxide concentration and the rate of photosynthesis, but only as long as carbon dioxide molecules continue to be built into glucose molecules. A point is reached when the rate of photosynthesis begins to level out, as shown by the flattening of the graph.

One reason for this may be that the enzymes involved in carbon fixation are working at their maximum rate so that no further increase in rate is possible under the prevailing conditions. Another possible reason is that the availability of essential coenzymes, such as NADPH, may have become a limiting factor.

elog-0264

INVESTIGATION 4.2

online only

The effect of carbon dioxide on photosynthesis

Aim

To investigate how the rate of photosynthesis is affected by different concentrations of carbon dioxide

4.4.5 Interrelationship between various factors

During photosynthesis, various factors are constantly influencing the rate of photosynthesis; they are not operating in isolation.

An example of this can be seen in the impact of factors such as light and carbon dioxide. At low light intensities, light is the limiting factor. However, as the light intensity increases, carbon dioxide concentration becomes the limiting factor. Compare the pink line (carbon dioxide concentration of 0.03 per cent) with the blue line (carbon dioxide concentration of 0.13 per cent) in figure 4.35a. The amount of carbon dioxide is limiting the maximum rate of photosynthesis at higher light intensities.

Figure 4.35b shows that temperature is also a limiting factor on the rate of photosynthesis. Increasing the temperature from 15 °C to 25 °C produces a marked increase in the rate of photosynthesis as the plant's enzymes are close to their optimum working temperature. This shows the interrelationship between three factors: light intensity, CO_2 and temperature.

FIGURE 4.35 a. Plot of the rates of photosynthesis against light intensity at two different concentrations of carbon dioxide (with temperature held constant) **b.** Plot of the rates of photosynthesis at two different temperatures: 15 °C and 25 °C (at a constant carbon dioxide concentration)

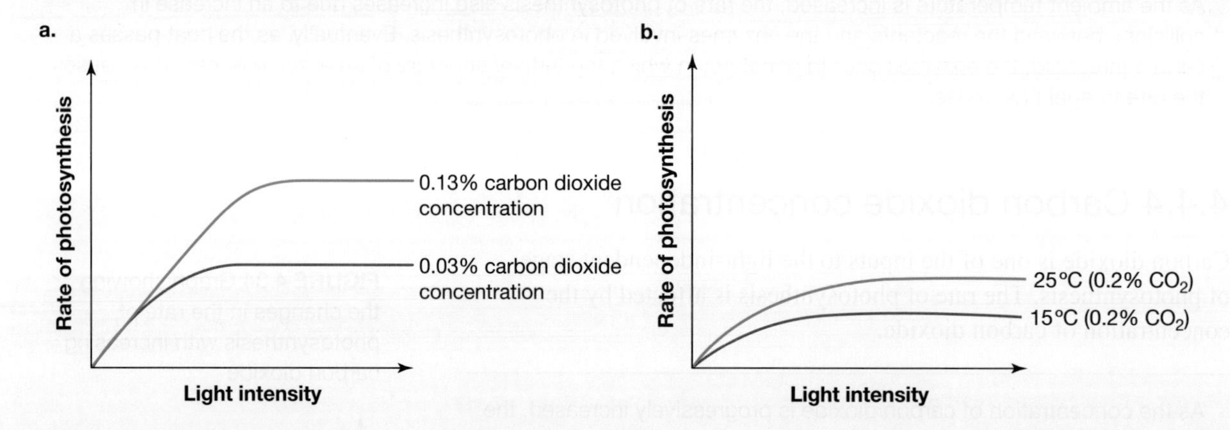

tlvd-1428

SAMPLE PROBLEM 3 Analysing factors affecting photosynthesis

Provide an explanation for each of the following observations:

a. **The amount of oxygen produced is highest in a plant at 35 °C. However, at both 25 °C and 45 °C, the rate of oxygen production is reduced.** **(2 marks)**

b. **Increasing the input of carbon dioxide increases the rate of photosynthesis until this rate suddenly levels off.** **(2 marks)**

c. **Plants that have excess water availability do not photosynthesise as well as a plant with slightly less water.** **(2 marks)**

THINK	WRITE
a. As this question is worth two marks, you should address both the increased temperature and the decreased temperature. TIP: You may write your explanation as one paragraph, or separate your key ideas into dot points as shown. This question asks for an explanation, so you must elaborate on your ideas.	• 35 °C is likely to be the optimal temperature. • When the temperature is lower than this, there are fewer collisions between substrates and enzymes involved in photosynthesis, so the rate of photosynthesis is slower (1 mark). • When the temperature is too high, the enzymes involved in photosynthesis begin to denature, so are much less effective at catalysing reactions (1 mark).
b. As this question is worth two marks, you should consider why carbon dioxide increases the rate for the first mark and why the rate levels off for the second mark.	• Carbon dioxide is an input for the light-independent stage. Therefore, as carbon dioxide increases, the rate in which glucose can be produced increases, showing an increase in the rate of photosynthesis (1 mark). • This levels off eventually as other factors may be limited, such as water, light or enzyme availability (1 mark).
c. As this question is worth two marks, you should consider the importance of water in photosynthesis and why waterlogging may prevent photosynthesis despite water being an input for the process.	• Water is vital for photosynthesis as it is an input for the light-dependent stage (1 mark). • When there is too much water, a plant becomes waterlogged, leading to a depletion of oxygen content in the soil. This means the plant cannot respire and, therefore, roots are damaged and can no longer take up water (1 mark).

Resources

eWorkbook Worksheet 4.3 Reviewing factors impacting photosynthesis (ewbk-7579)

KEY IDEAS

- Light availability, carbon dioxide concentration, temperature and water availability can all affect the rate of photosynthesis.
- As light intensity increases, the rate of photosynthesis increases. Eventually this rate reaches a maximal point as other factors limit the rate of photosynthesis.
- As temperature increases, the rate of photosynthesis increases. This is because collisions between the enzymes involved in photosynthesis and substrates increase due to molecules having higher energy. However, eventually, the rate decreases as enzymes begin to denature.
- As carbon dioxide increases, the rate of photosynthesis increases. Eventually this rate reaches a maximal point as other factors limit the rate of photosynthesis.
- As water availability increases, the rate of photosynthesis increases. However, if there is too much water, this can also harm the plant and lead to a decrease in the photosynthetic rate.
- A factor is referred to as limiting, if, in short supply, it restricts the rate of photosynthesis.

4.4 Activities

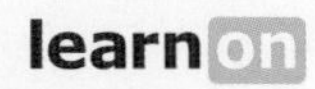

To answer questions online and to receive **immediate feedback** and **sample responses** for every question, go to your learnON title at **www.jacplus.com.au**. A **downloadable solutions** file is also available in the resources tab.

4.4 Quick quiz	4.4 Exercise	4.4 Exam questions

4.4 Exercise

1. Briefly explain why the rate of photosynthesis does not continue to increase as the air temperature continues to increase.
2. Why does the rate of photosynthesis not increase indefinitely as light intensity increases?
3. What is meant by the 'plateauing' of a trend line?
4. Describe three quantitative ways that you can measure the rate of photosynthesis.
5. The graph of the rate of photosynthesis of two plants are shown below.

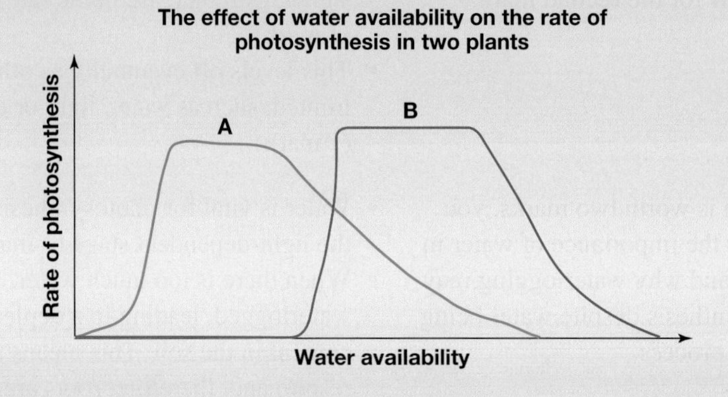

   a. Describe the patterns you see in the two graphs. What implications does this have for each plant?
   b. One plant is a cactus and the other plant is from an orchid. Which graph would you expect to belong to which plant? Justify your response.
6. Draw a graph comparing the photosynthetic rate when under different temperature conditions. Plot two lines on your graph: one at 0.30 per cent carbon dioxide and another at 0.70 per cent carbon dioxide.
7. A student wishes to conduct an experiment on the effect of light intensity on photosynthetic rate. Write a clear hypothesis and methodology for this investigation and outline expected results.

4.4 Exam questions

Question 1 (1 mark)

Source: *VCAA 2014 Biology Exam, Section A, Q8*

MC An increase in the atmospheric CO_2 level increases the rate of photosynthesis.

The rate of photosynthesis increases because

A. the rate of the light-independent reactions on the thylakoid membranes of the chloroplasts increases.
B. water loss from the leaf decreases, resulting in the availability of water for photosynthesis increasing.
C. the increased CO_2 level lowers the pH inside the chloroplasts and increases the rate of enzyme-catalysed reactions.
D. the rate of the light-independent reactions in the stroma increases with the increase in CO_2 level.

Question 2 (1 mark)

Source: *VCAA 2019 Biology Exam, Section A, Q15*

MC An experiment was carried out at a constant temperature and with a constant carbon dioxide concentration in order to determine the effect of changing light intensity on the photosynthetic rate. The following is a graph of the results.

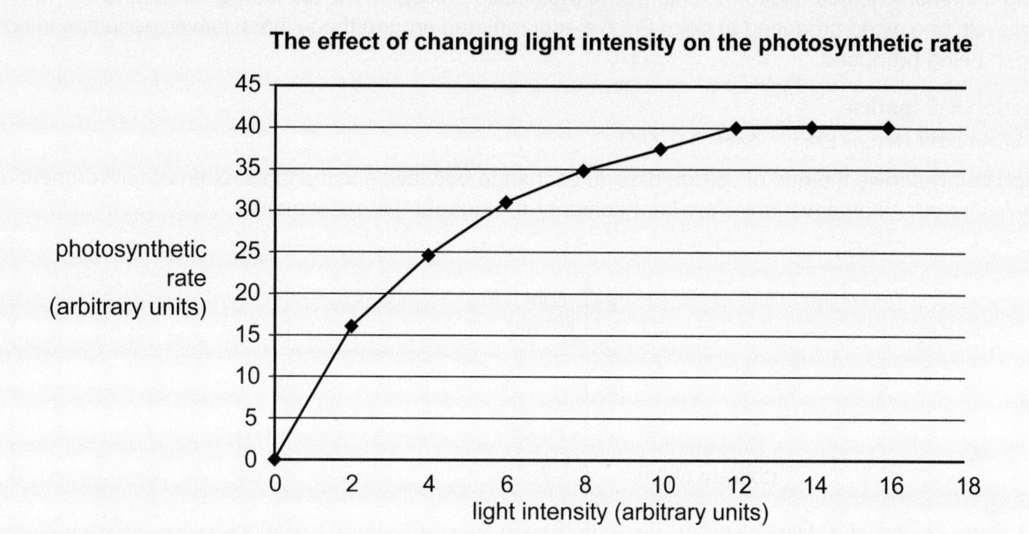

Based on your knowledge and the information in the graph, which one of the following conclusions can be reached?

A. Photosynthesis ceases to occur at a light intensity of 14 arbitrary units.
B. Plants do not undergo photosynthesis at a light intensity of 1 arbitrary unit.
C. Light intensity is a limiting factor when the photosynthetic rate is less than 40 arbitrary units.
D. Increasing the amount of carbon dioxide at a light intensity of 16 arbitrary units would lead to a decrease in the photosynthetic rate.

Use the following information to answer Questions 3 and 4.

The graph below shows the net output of oxygen in spinach leaves as light intensity is increased. Temperature is kept constant during the experiment.

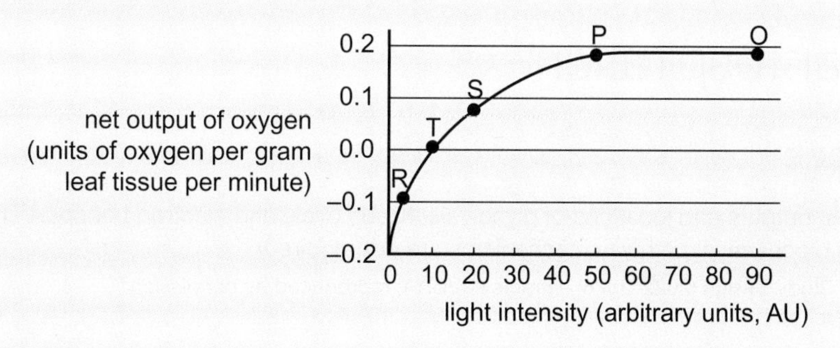

Question 3 (1 mark)

Source: *VCAA 2017 Biology Exam, Section A, Q13*

MC Which one of the following conclusions can be made based on the graph?

A. At point T photosynthesis is no longer occurring.
B. The optimal level of light intensity for photosynthesis is 40 AU.
C. At point S the amount of oxygen output is a third of that at point P
D. Below 10 AU of light intensity the aerobic respiration rate is greater than the photosynthesis rate.

Question 4 (1 mark)

Source: *VCAA 2017 Biology Exam, Section A, Q14*

MC The rate of oxygen output remains constant between points P and O because

A. heat has denatured the enzymes involved in the photosynthesis reactions.

B. the concentration of available carbon dioxide limits the rate of photosynthesis.

C. the light intensity has damaged the chlorophyll molecules present in the spinach chloroplasts.

D. high levels of oxygen produced at point P have accumulated around the spinach leaves, resulting in no more oxygen being produced.

Question 5 (2 marks)

Source: *VCAA 2011 Biology Exam 1, Section B, Q7d*

The graph below shows the rate of carbon dioxide exchange between a leaf and its external environment as light intensity is altered. All other variables are kept constant throughout the experiment.

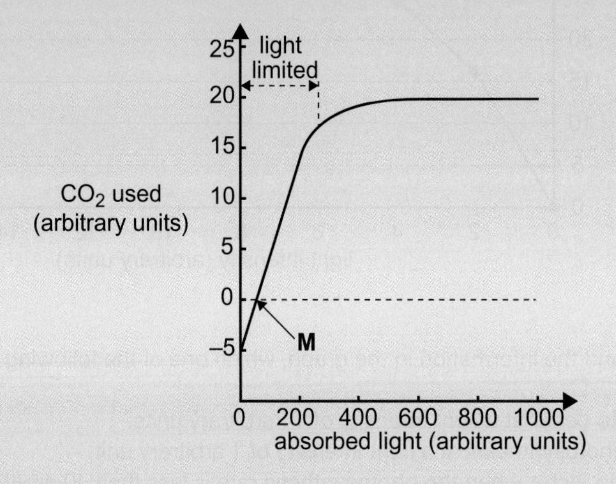

i. Outline what is occurring at point M in terms of chemical reactions. **1 mark**

ii. Explain why the graph line becomes nearly horizontal from about 600 units of absorbed light. **1 mark**

More exam questions are available in your learnON title.

4.5 Cellular respiration

KEY KNOWLEDGE

- The main inputs, outputs and locations of glycolysis, Krebs cycle and electron transport chain including ATP yield (details of biochemical pathway mechanisms are not required)

Cellular respiration is vital for our survival, providing us essential energy to grow, reproduce and function. Cellular respiration may be:

- aerobic — oxygen-requiring
- anaerobic — non-oxygen requiring.

Both these types of cellular respiration are used in humans. For humans, aerobic respiration is the most vital.

cellular respiration process of converting chemical energy into a useable form by cells, typically ATP

4.5.1 Role of cellular respiration

All living things require a constant source of chemical energy in the form of ATP to perform the basic functions of life. Cellular respiration is the biochemical process in cells that produces this ATP. There is no holiday period for cellular respiration — the use of ATP for living is continuous and, as a result, cellular respiration occurs all the time in every living cell.

In the simplest terms, cellular respiration is the process in all living organisms that produces ATP for use by cells.

Various processes in the body require this ATP to allow cells and the organism to stay alive.

In the human body, these include:
- the homeostatic mechanisms that keep internal conditions within cells within narrow limits
- the excretory processes that remove metabolic wastes
- the production of antibodies and the other activities of the immune system that defend against infection
- the transcription and translation processes involved in protein synthesis
- the transmission of nerve impulses along neurons
- the active transport of molecules across cell membranes
- the contraction of skeletal and cardiac muscle cells.

In almost all animals and plants, and in the cells of most tissues, cellular respiration to produce ATP can occur only if oxygen is available — this is termed **aerobic cellular respiration**.

However, some bacteria and archeans live in permanent oxygen-poor or even oxygen-free environments. These microbes carry out **anaerobic cellular respiration**. Both aerobic and anaerobic cellular respiration use energy released from the transfer of electrons to form ATP.

Other organisms such as some bacterial species and yeast (unicellular fungi) carry out **fermentation** to produce their ATP. This will be covered in subtopic 4.6.

Whether oxygen is present or not, all organisms have pathways to produce ATP for their energy needs. For this reason, ATP is called the universal energy currency of all living cells.

aerobic cellular respiration oxygen-requiring process that converts chemical energy into ATP
anaerobic cellular respiration a process that converts chemical energy into ATP in the absence of oxygen
fermentation a metabolic process that produces some ATP in the absence of oxygen

EXTENSION: Why make ATP? Why not just use glucose?

Glucose has a far greater chemical energy content than that of ATP. So why not use the chemical energy of glucose directly as the energy source for cells?

TABLE 4.5 Comparison of glucose and ATP as an energy source for cells

Organic molecule	Total energy content per molecule	Number of steps needed to release full energy content	Can molecule be stored?
Glucose	2800 kJ	20+ steps Glucose + $6O_2 \rightarrow 6CO_2 + 6H_2O$	Yes: as glycogen in animals and as starch in plants
ATP	30 kJ	1 step ATP $\rightarrow$ ADP +Pi	No, but it is rapidly regenerated

- ATP is the more useable form of energy for cells because its energy can be quickly released in a single step, making energy *instantly* available for use by cells. In contrast, the release of energy from glucose involves a complex multistep pathway that is about 100 times slower than that for ATP energy release.
- The direct use of glucose leads to the production of excessive heat and waste.

4.5.2 Biochemical pathway of aerobic cellular respiration

We can show, in simple form, the production of ATP from glucose in several ways:

- As a word equation:

glucose + oxygen → carbon dioxide + water + energy

- As a balanced equation (ATP could alternatively be placed as 30–32 ATP on the right side of the equation).

$$C_6H_{12}O_6 + 6O_2 \xrightarrow{ADP + Pi \rightarrow ATP} 6CO_2 + 6H_2O$$

- As an equation that highlights the total ATP production from cellular respiration:

glucose + oxygen → carbon dioxide + water + 30–32 ATP

- As a figure with a focus on inputs and outputs as seen in figure 4.36:

FIGURE 4.36 Inputs and outputs of cellular respiration

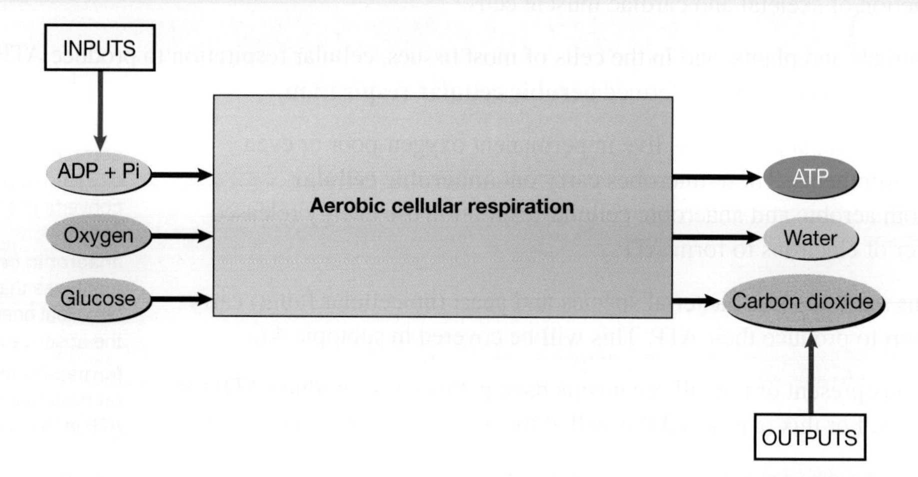

These simple representations just show inputs and the final outputs of the cellular respiration pathway. They do not reveal the essential role of oxygen and the role of NADH and $FADH_2$ coenzymes.

The ATP produced by cellular respiration does not accumulate in cells. Instead, after being generated, ATP transfers its energy to drive the many energy-requiring reactions that are proceeding in living cells and it is converted to ADP. The resulting ADP is rapidly recycled to generate more ATP. In total, aerobic cellular respiration produces about 30 moles of ATP from one mole of glucose in eukaryotic cells (32 ATP in prokaryotes).

The total yield of ATP is often debated. The value of 36–38 moles of ATP is the yield obtained in ideal conditions. However, in most cells, ideal conditions are not reached, resulting in a lower actual yield of 30–32 moles of ATP. This debate on the yield will be explored in further detail in section 4.5.7.

TIP: In exams and assessments you should list the actual ATP produced in cells (real-life efficiency). This is the accepted yield by VCAA (updated in the recent Study Design). This value is a range of 30–32 ATP.

BACKGROUND KNOWLEDGE: Where do the inputs to cellular respiration come from?

In subtopic 4.2, we investigated the process of photosynthesis. Photosynthesis and aerobic cellular respiration can be seen as complementary processes. The inputs of cellular respiration (oxygen and glucose) are the outputs of photosynthesis. The outputs of cellular respiration (water and carbon dioxide) are in turn the inputs of photosynthesis.

To access more information about the sources of glucose and oxygen used in cellular respiration, please download the digital document.

FIGURE 4.37 A rainbow lorikeet (*Trichoglossus moluccanus*) obtains much of its glucose from the sugar-rich nectar of flowering plants.

on Resources

Digital document The relationship between photosynthesis and cellular respiration (doc-35829)

elog-0266

INVESTIGATION 4.3

Photosynthesis and respiration — a balance

Aim

To investigate the processes of cellular respiration and photosynthesis in different light conditions in *Elodea* plants

The stages of aerobic cellular respiration

Aerobic cellular respiration is a complex biochemical pathway that involves three distinct interconnected stages as follows:

- Glycolysis: splits 6 carbon (6C) glucose molecules into two 3 carbon (3C) molecules
- Krebs cycle: makes a supply of energy-rich loaded coenzymes
- Electron transport chain: transfers energy from electrons supplied by loaded coenzymes to make ATP.

FIGURE 4.38 The three stages involved in aerobic cellular respiration

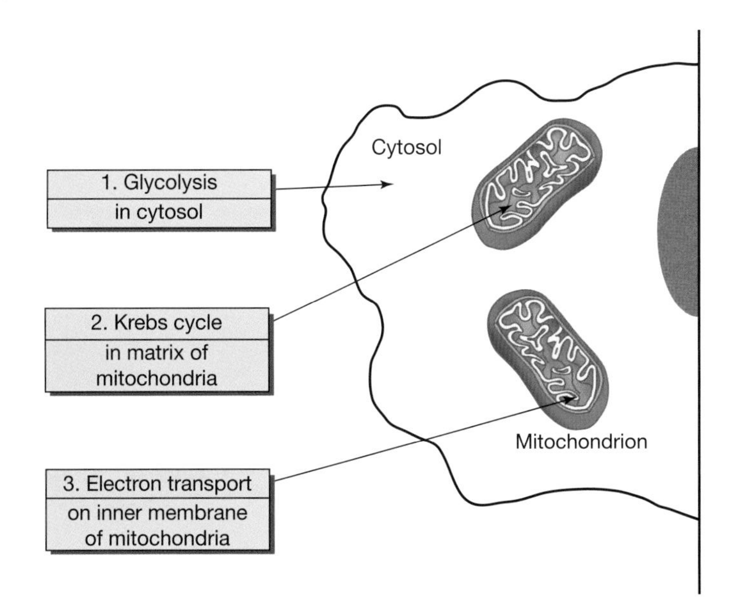

TIP: A good way to remember the order of the stages of aerobic cellular respiration is to count the number of words in each stage. The first stage, glycolysis, has one word. The second stage, the Krebs cycle, has two words. Finally, the third stage, electron transport chain, has three words.

elog-0268

INVESTIGATION 4.4 online only

Respiration involving oxygen — aerobic respiration

Aim

To observe aerobic respiration in germinating seeds

4.5.3 Glycolysis — the first stage of aerobic cellular respiration

Glycolysis is a fundamental pathway in almost all organisms. Glycolysis:

- is the first stage of aerobic cellular respiration
- begins with the input of glucose, a six-carbon sugar molecule, and ends with two three-carbon molecules known as **pyruvate** (or pyruvic acid)
- involves a 10-step pathway, with each step being catalysed by a specific enzyme
- occurs in the cytosol of cells where the required enzymes are present
- does not require oxygen at any step
- results in the net yield of two ATP molecules (four ATP are produced but two ATP are required and used up)
- produces two loaded NADH molecules for every molecule of glucose that enters this pathway.

glycolysis the first stage of cellular respiration, in which glucose is broken down into pyruvate

pyruvate a three-carbon molecule produced during glycolysis

(*Note:* You will see later that glycolysis is also the start of the anaerobic fermentation pathways in bacteria and yeasts and, in special circumstances, in mammalian skeletal muscle cells.)

Inputs and outputs of glycolysis

Glycolysis is the process in which one glucose molecule is split into two pyruvate molecules.

Inputs:
- Glucose
- ADP + Pi (x 2)
- NAD^+(x 2)

Outputs:
- Pyruvate (x 2)
- ATP (x 2)
- NADH (x 2)

Figure 4.39 shows a highly simplified representation of glycolysis.

FIGURE 4.39 Diagram showing a representation of glycolysis

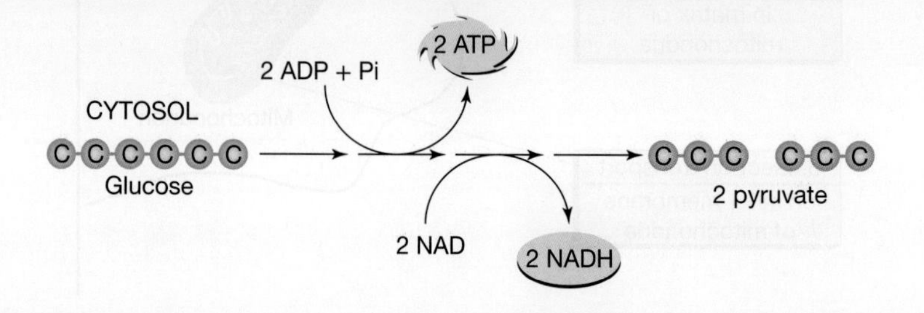

Similar information is captured in the following glycolysis equation. This shows that, as glucose is broken down to form two pyruvates, it loses four hydrogens along with their electrons. These hydrogens and their associated electrons are accepted by the coenzyme NAD^+ to form NADH. Some excess H^+ also remains.

$$\underset{\textbf{glucose}}{C_6H_{12}O_6} + 2\,NAD^+ + 2\,ADP + 2\,Pi \rightarrow \underset{\textbf{pyruvate}}{2\,C_3H_4O_3} + 2\,NADH + 2\,H^+ + 2\,ATP$$

The glycolysis stage of cellular respiration releases only a small percentage — about 6 per cent — of the chemical energy in glucose to form ATP. Clearly, pyruvate molecules still contain a lot of chemical energy (and this is converted to ATP in the next stages of cellular respiration).

4.5.4 The mitochondria

While the glycolysis step of cellular respiration takes place in the cytosol, the remaining stages of the cellular respiration pathway — the Krebs cycle and the electron transport chain — occur within cell organelles called **mitochondria** in the cytosol.

A representation of a mitochondrion showing the various membranes that form its structure is shown in figure 4.40a. Mitochondria cannot be resolved using light microscopy, but their structure has been revealed using electron microscopy (see figure 4.40b).

Note that mitochondria:
- are cell organelles in the cytosol of almost every cell of eukaryotes
- have an internal structure that can only be revealed using electron microscopy techniques
- are enclosed within a double membrane — an outer smooth one and an inner one that is folded into **cristae** (singular: crista); between them is the inter-membrane space.
- within their inner membrane is a fluid-filled space called the **matrix**.

mitochondria organelles in eukaryotic cells that are the major site of ATP production; singular: mitochondrion

cristae folds in the inner membrane of the mitochondria where the electron transport chain occurs

matrix the gel-like solution within the mitochondria

The second and third stages of cellular respiration occur in different locations within the mitochondria.
- The enzymes of the Krebs cycle are in solution in the matrix.
- The enzymes of the electron transport chain are embedded in the inner membrane and its cristae.

FIGURE 4.40 a. A simple 3D representation of a mitochondrion showing its main structural features b. False-coloured transmission electron micrograph of a mitochondrion surrounded by cytosol and other organelles

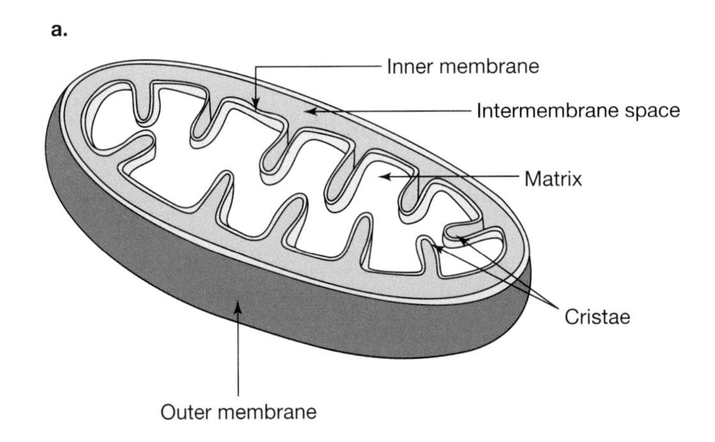

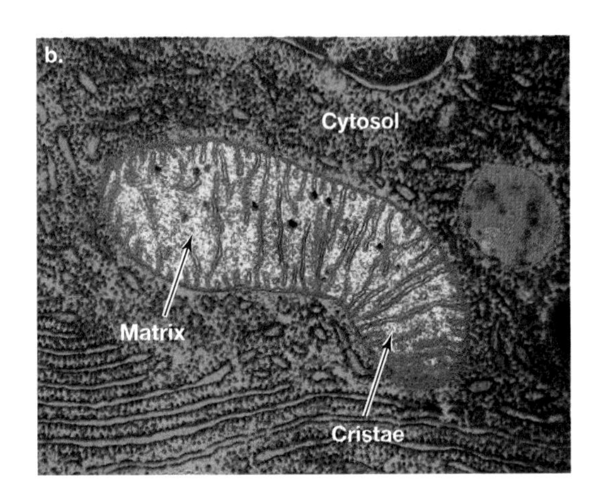

The number of mitochondria in plant and animal cells can vary depending on factors, a key one being the level of metabolic activity in the cell.

- Mammalian cells typically have from about 80 to about 2000 mitochondria per cell, with liver cells and muscle cells at the upper end of this range.
- Plant mesophyll cells have about 200 to about 600 mitochondria per cell.

Figure 4.41 shows two living mammalian cells labelled with fluorescent probes, including a red probe to identify individual mitochondria. The number of mitochondria present indicates that these cells are metabolically very active.

FIGURE 4.41 Living cells labelled with fluorescent probes that bind to specific cell organelles or macromolecules. Mitochondria are shown by the red fluorescence, the Golgi apparatus by green and the DNA by blue.

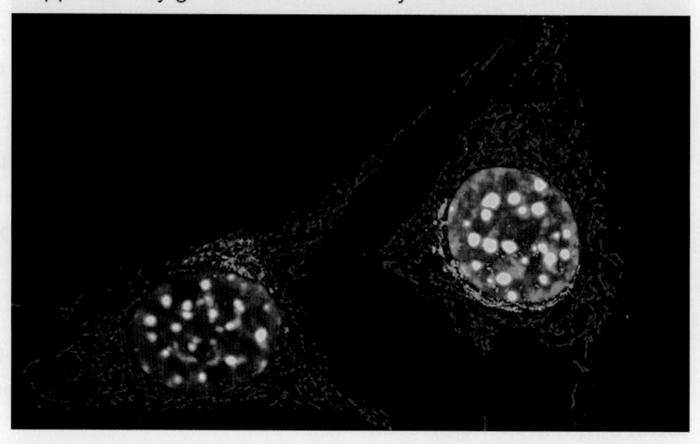

4.5.5 Krebs cycle — the second stage of aerobic cellular respiration

Glycolysis ends with the production of two molecules of pyruvate, but pyruvate molecules *cannot* directly enter the **Krebs cycle**. Before entering the Krebs cycle, a short linking process known as pyruvate oxidation occurs.

Pyruvate oxidation

Pyruvate molecules are transported from the cytosol across the mitochondrial membranes into the matrix. Here, in an energy-releasing reaction, each pyruvate loses a C and an H atom, forming a 2C acetyl group that is delivered to the Krebs cycle by coenzyme A (as acetyl coenzyme A or acetyl-CoA). This pyruvate oxidate step can be shown most simply as:

FIGURE 4.42 The breakdown of pyruvate in pyruvate oxidation, showing carbon atoms only

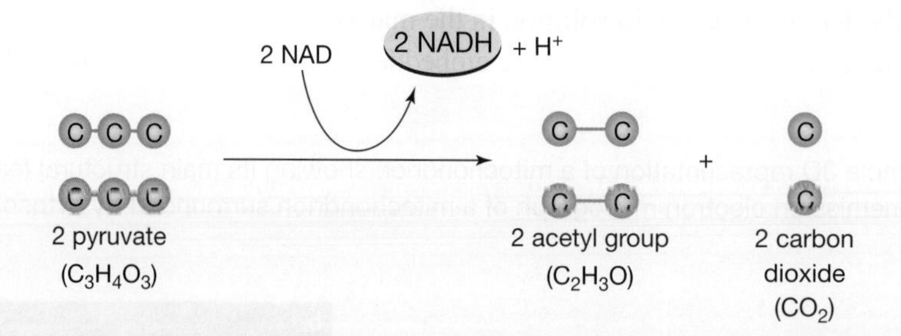

The Krebs cycle

The Krebs cycle also occurs in the mitochondrial matrix. Some of the key features of the Krebs cycle:

- It is the second stage of the cellular respiration pathway.
- It comprises an eight-step cyclic pathway with each step being catalysed by a specific enzyme.
- Oxygen is not directly involved in this cycle.
- Acetyl groups derived from pyruvate are the input to this cycle, transported by coenzyme A (CoA).
- Only two ATP molecule are produced directly.
- The main outputs of the cycle are the energy-rich loaded NADH and $FADH_2$ coenzymes.

The inputs and outputs of this stage can be summarised as follows:

- *Inputs:* 2 acetyl CoA (acetyl groups), 6 NAD^+, 2 FAD and 2 ADP + Pi
- *Outputs:* 4 CO_2, 6 NADH, 2 $FADH_2$ and 2 ATP.

Krebs cycle second stage of aerobic respiration in which coenzymes are loaded and carbon dioxide is produced

Combining pyruvate oxidation and the Krebs cycle

While they are distinct, often when we refer to the Krebs cycle, we discuss the pyruvate oxidation step and the Krebs cycle step together. It is fine to include pyruvate oxidation as part of the Krebs cycle when discussing the overall inputs and outputs. The joining of these as one step has been accepted in previous VCAA exams.

Figure 4.43 shows the combined inputs and outputs of the Krebs cycle (and pyruvate oxidation) that is the second stage of the aerobic cellular respiration pathway. This all occurs in the mitochondrial matrix. The energy of the loaded NADH and $FADH_2$ coenzymes produced are used in the next stage of cellular respiration when most of the ATP is produced.

FIGURE 4.43 Inputs and outputs of the Krebs cycle (and pyruvate oxidation). Note that during this process, pyruvate is broken down into acetyl groups, which are transferred with the aid of coenzyme A.

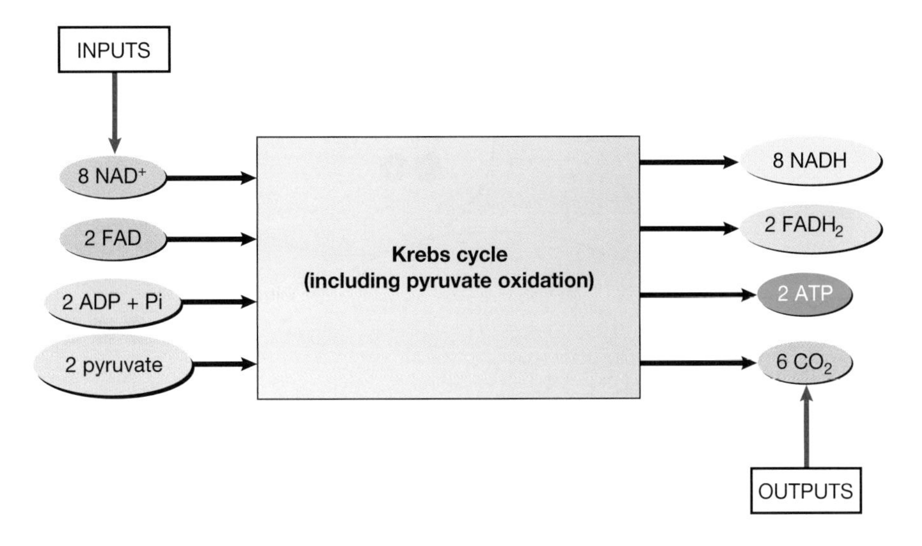

Inputs and outputs of the Krebs cycle (and pyruvate oxidation)

The Krebs cycle and pyruvate oxidation allows for the further breakdown of pyruvate, and leads to the formation of NADH and $FADH_2$ coenzymes.

It can be split into two stages: pyruvate oxidation and the Krebs cycle (or citric acid cycle). The inputs and outputs of these combined aspects (as shown in figure 4.43) are:

Inputs:
- Pyruvate (× 2)
- ADP + Pi (× 2)
- NAD^+ (× 8)
- FAD (× 2)

Outputs:
- CO_2 (× 6) — 2 during pyruvate oxidation and 4 during Krebs cycle component
- ATP (× 2)
- NADH (× 8) — 2 during pyruvate oxidation and 6 during Krebs cycle component
- $FADH_2$ (× 2)

EXTENSION: The mechanisms of the Krebs cycle

The output of glycolysis is two 3C (or 3 carbon) pyruvates. Figure 4.44 is a simplified representation of the Krebs cycle that starts with the pyruvate oxidation that forms an acetyl group. As well as the inputs and outputs, this representation shows the numbers of carbon atoms (as purple circles) in each reactant

This figure shows the outputs from one pyruvate that produces one acetyl group that enters the Krebs cycle. You can clearly see all outputs of the Krebs cycle in this diagram, remembering that there are **TWO** pyruvates that go into the cycle.

FIGURE 4.44 A simplified representation of the Krebs cycle for a single pyruvate entering the mitochondrial matrix. The acetyl group derived from pyruvate is brought into the Krebs cycle when it joins to a 4C acceptor molecule.

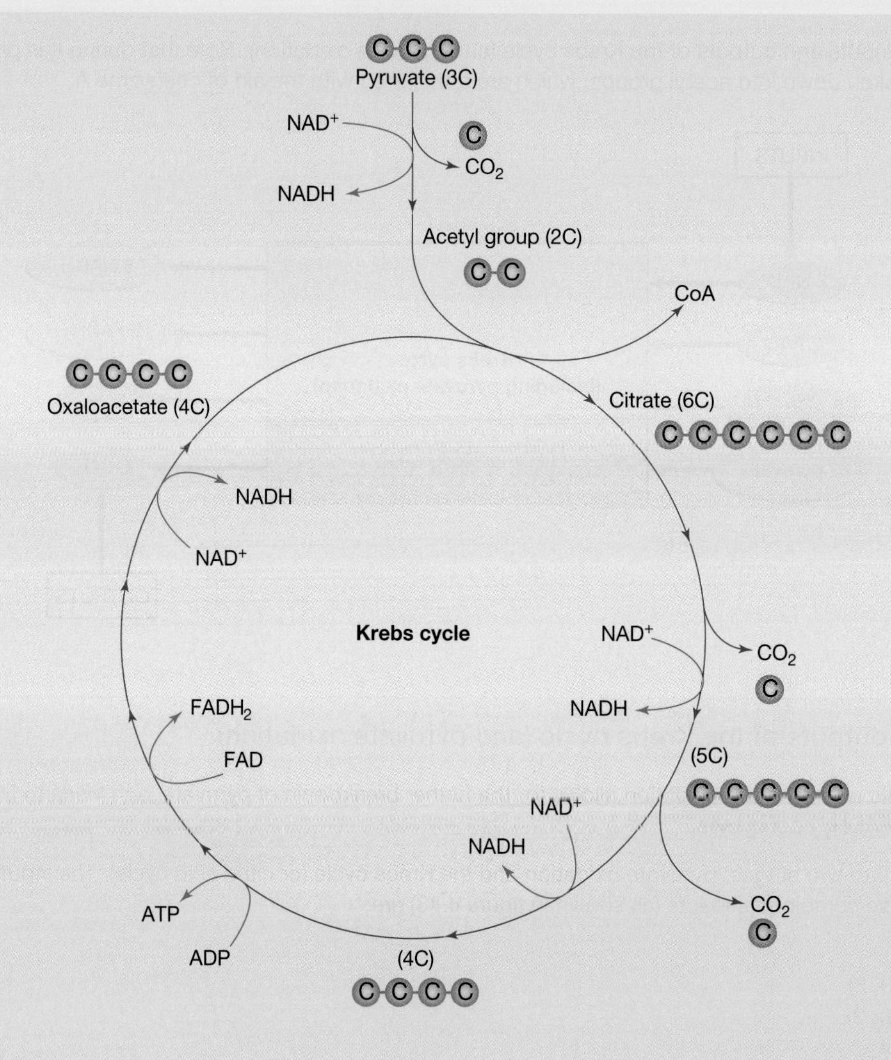

4.5.6 Electron transport chain — the third stage of aerobic cellular respiration

The **electron transport chain** stage of cellular respiration consists of a series of electron transfers that release the energy that has already been stored in the high-energy loaded coenzymes, NADH and $FADH_2$, mainly during the Krebs cycle. The energy released is ultimately transferred to drive production of ATP.

electron transport chain third stage of aerobic respiration in which there is a high yield of ATP

The electron transport chain (ETC):

- consists of a series of enzyme complexes that are embedded in the inner membrane and cristae of mitochondria.
- releases the energy from high-energy molecules created in the earlier stages (the NADH and $FADH_2$).
- can only operate if a supply of oxygen is available. If this supply is reduced or preventing (for example by carbon monoxide, which competes with oxygen on haemoglobin), cellular respiration will stop and death will likely result.

The following processes occur in the ETC:

- Electrons are transferred along these enzyme complexes, moving from electron donors to acceptors through a series of reactions.
- The first input of high-energy electrons to the ETC comes from loaded NADH coenzymes.
- $FADH_2$ also donates its high-energy electrons to an acceptor, but further down the chain.
- As electrons transfer from one enzyme complex to the next, the energy released is ultimately used to power the production of ATP from ADP and Pi.
- The final electron acceptor at the end of the ETC is oxygen.
- Oxygen accepts electrons and hydrogen ions, forming water — $\frac{1}{2}O_2 + 2e^- + 2H^+ \rightarrow H_2O$ — that is the final output, a reaction catalysed by cytochrome oxidase.

Inputs and outputs of the electron transport chain

The electron transport chain involves the production of large numbers of ATP by using high energy electrons from NADH and $FADH_2$. Note this shows the actual ATP produced in real-life conditions. This value is variable between eukaryotes and prokaryotes and may also vary between tissues.

Inputs:

- O_2 (× 6)
- ADP + Pi (× 26–28)
- NADH (× 10)
- $FADH_2$ (× 2)

Outputs:

- H_2O (× 6)
- ATP (× 26–28)
- NAD^+ (× 10)
- FAD (× 2)

EXTENSION: Exploring the electron transport chain in detail

Figure 4.45 shows a highly simplified representation of the electron transport chain with the path of electrons shown in pink.

- NADH is the initial donor of electrons to enzyme complex I, the first acceptor in the chain.
- Electrons then pass via complexes II and III to complex IV.
- Complex IV transfers the electrons to oxygen, the final electron acceptor.
- In accepting the electrons, oxygen reacts with them and with H^+ ions to form water.

The energy released from electron transport along the ETC is ultimately used to produce ATP from ADP and Pi. However, this energy is not used directly to phosphorylate ADP to form ATP. What happens is that the energy is used to push hydrogen ions (protons) across the inner mitochondrial membrane into the inter-membrane space.

The accumulation of protons in the inter-membrane space creates an electrochemical gradient, with a higher concentration of protons in the inter-membrane space relative to the concentration in the matrix. After the gradient is established, protons diffuse down the electrochemical gradient. The path back to the matrix takes the protons through an ion channel in a protein complex called ATP synthase. ATP synthase traps the kinetic energy of the passage of the protons through the ion channel to generate ATP from ADP.

FIGURE 4.45 Diagram showing the passage (red line) of electrons down the electron transport chain. NADH and $FADH_2$ coenzymes are electron donors and oxygen is the final electron acceptor.

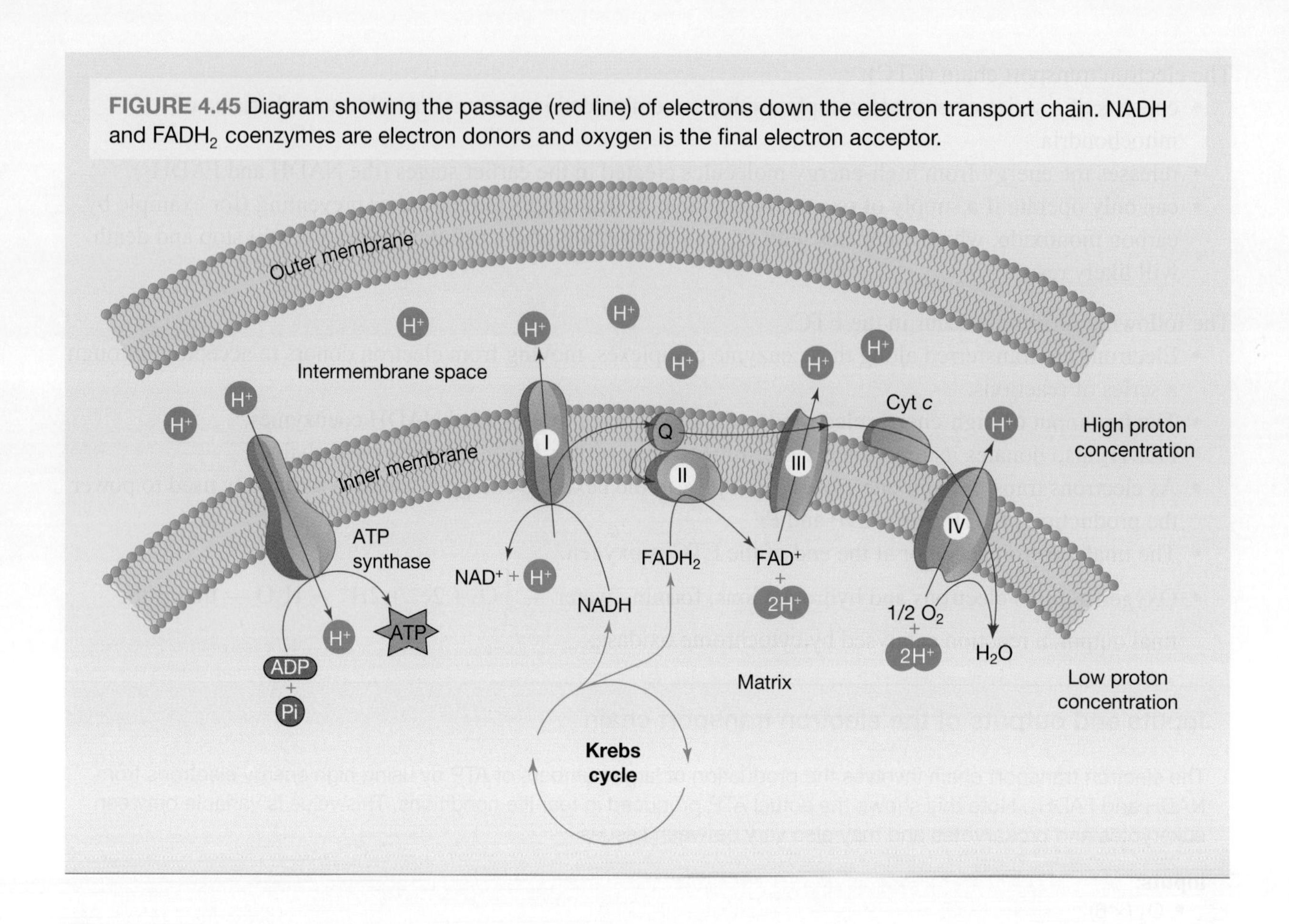

4.5.7 Putting the three stages together

Figure 4.46 is a stylised representation of the aerobic cellular respiration pathway, starting with the input of one molecule of glucose, that is split into two pyruvates that is then converted to six carbon dioxides.

FIGURE 4.46 The stages of aerobic cellular respiration (NAD^+ and FAD have been omitted)

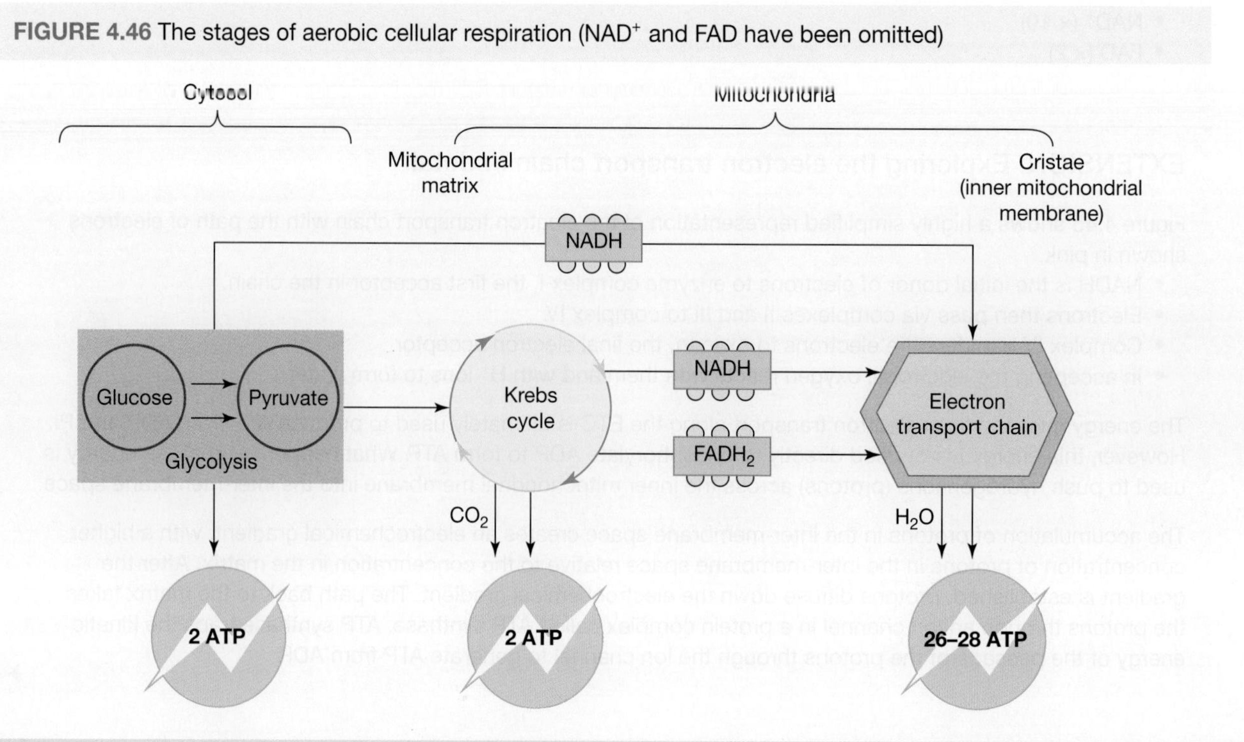

TABLE 4.6 Summary of stages in aerobic cellular respiration

	Location	Inputs	Outputs	ATP yield
Glycolysis	Cytosol	Glucose NAD^+ ADP + Pi	Pyruvate NADH ATP	2
Krebs cycle (including pyruvate oxidation)	Matrix of mitochondria	Pyruvate NAD^+ FAD ADP + Pi	CO_2 NADH $FADH_2$ ATP	2
Electron transport chain	Cristae	O_2 NADH $FADH_2$ ADP + Pi	H_2O NAD^+ FAD ATP	26–28

SAMPLE PROBLEM 4 Analysing aerobic cellular respiration

tlvd-1430

Three samples were being explored for their cellular respiration rate:

1. **A cell with mitochondria**
2. **A cell without mitochondria**
3. **A mitochondria in culture**

   a. **All samples were provided with a supply of oxygen, glucose, ADP, NAD^+ and FAD. How much ATP would be produced in each sample? Justify your responses.** **(3 marks)**

   b. **Samples were placed in a solution containing pyruvate, NADH and $FADH_2$. Would the amount of ATP produced in any samples change? Explain your response.** **(2 marks)**

   c. **Three stages occur in aerobic respiration. What is name of the final stage and where does this occur?** **(2 marks)**

   d. **List all the inputs and outputs for this the stage identified in part c.** **(2 marks)**

THINK	WRITE
a. 1. Consider how much ATP is produced in each stage: Glycolysis: 2 ATP Krebs cycle: 2 ATP Electron transport chain: 26–28 ATP	In sample A, all stages of cellular respiration can occur; therefore, 30–32 molecules of ATP can be produced (1 mark). In sample B, as there are no mitochondria, only glycolysis can occur, so only 2 molecules of ATP can be produced (1 mark).
2. Determine what stages would occur in each sample and add the total ATP produced.	In sample C, though mitochondria is present, there is no cytosol for glycolysis. As the products of glycolysis are required for both the Krebs cycle and electron transport chain, no ATP can be produced (1 mark).
b. 1. Explore each sample and how pyruvate, NADH and $FADH_2$ would affect results.	
2. In sample A and B, these are all produced already by glycolysis, so it wouldn't make any difference to the ATP produced.	The amount of ATP produced would only change in sample C (1 mark).
However, in sample C, the inputs for both the Krebs cycle and electron transport chain are present.	As the inputs for both the Krebs cycle and electron transport chain are provided, they would be able to proceed. Therefore 28–30 ATP would be able to be produced (1 mark).

c. Review the stages and locations, ensuring you address both parts of the question in your response.

The final stage is the electron transport chain (1 mark).
This occurs in the cristae (1 mark).

d. Ensure you list all inputs and outputs. It can often be helpful to match inputs and outputs (for example, if ADP + Pi goes in ATP must be an output).

Inputs:
- Oxygen
- NADH
- $FADH_2$
- 26–28 ADP + Pi (1 mark).

Outputs:
- Water
- NAD^+
- FAD
- 26–28 ATP (1 mark).

ATP yield and the debate on ATP

During aerobic cellular respiration, each glucose is oxidised to carbon dioxide. The electron transport chain is the stage when the energy held in loaded coenzymes (NADH and $FADH_2$) is released and used in ATP production.

There are many different sources that state different yields for the ATP produced during aerobic respiration (an actual yield of 30–32 ATP compared to a theoretical yield of 36–38 ATP). The amount of ATP that can be produced varies under different conditions, in different tissues and between different organisms.

Where do these values come from?

There is no disagreement about the numbers of NADH and $FADH_2$ produced during aerobic cellular respiration. The debate surrounds how much energy is transferred by these molecules in the electron transport chain.

In the ideal conditions (temperature of 298 kelvin, pressure of 1 atmosphere, pH of 7.0, and initial concentrations of reactants and products being equal), the theoretical maximum yield is 36–38 ATP. The conversion rates within the mitochondria in these ideal conditions are:

- 1 NADH yields 3 ATP
- 1 $FADH_2$ yields 2 ATP.

However, the conditions in cells are not necessarily standard, so this is rarely achieved, due to inefficiencies in the process. Therefore, it is generally accepted that, in most situations, the actual yield of ATP is 30–32 moles of ATP per mole of glucose. These values reflect modern research and real-life efficiencies.

In these actual conditions, the conversion rates within the mitochondria are usually:

- 1 NADH yields 2.5 ATP
- 1 $FADH_2$ yields 1.5 ATP.

Which yield should be used?

Due to the slight variation in the amount of ATP that can be produced, ranges are acceptable when answering questions on aerobic cellular respiration. When answering exam questions, you should use the actual, real-life efficiencies. In the most recent Study Design, the range of 30–32 ATP is accepted by VCAA and should be used in your answers (this was updated from previous Study Designs).

Table 4.7 shows this actual energy yield for each glucose that is oxidised to carbon dioxide in the aerobic cellular respiration pathway, using the conversions listed. Each loaded NADH formed in the cytoplasm requires the energy of one ATP to move across the mitochondrial membranes in eukaryotes (so the two NADH made in glycolysis require a total of two ATP to move across the mitochondrial membrane). Therefore, prokaryotes produce two more ATP molecules than eukaryotes.

TABLE 4.7 Energy yield (in the form of ATP) in eukaryotic cells for each molecule of glucose that is totally oxidised via the aerobic cellular respiration pathway. ATPs (with red arrows) are generated by the electron transport chain passing high energy electrons from NADH and $FADH_2$.

Stage and location	Energy-rich products formed	Final yield as ATP (with real life efficiencies)
Glycolysis (in cytosol)	2 ATP 2 NADH	2 ATP
Krebs cycle (in matrix of mitochondria)	2 ATP 8 NADH 2 $FADH_2$	2 ATP
Electron transport chain cristae of mitochondria		→ 2 × 2.5 = 5 ATP (or 3*) → 8 × 2.5 = 20 ATP → 2 × 1.5 = 3 ATP
	TOTAL	30 or 32* ATP

* In eukaryotes, NADH produced in the cytosol has an energy cost of 2 ATP to be taken across the mitochondrial membrane into the mitochondria, so that the total is 30 ATP, not 32 ATP as in prokaryotes.

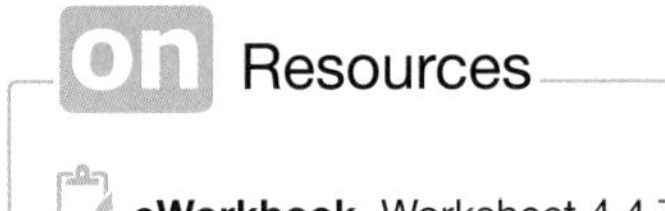

Resources

eWorkbook Worksheet 4.4 The stages of aerobic respiration (ewbk-7581)
Worksheet 4.5 Case study: The impact of carbon monoxide on cellular respiration (ewbk-7583)

Interactivity Photosynthesis or respiration (int-3039)

KEY IDEAS

- Cellular respiration is a process in which the chemical energy stored in organic molecules, such as glucose, is transferred into the chemical energy of ATP for use by cells for staying alive.
- The overall equation for cellular respiration is:

$$6O_2 + C_6H_{12}O_6 \rightarrow 6CO_2 + 6H_2O + 30\text{–}32 \text{ ATP}$$

- Aerobic cellular respiration is the main form of cellular respiration in which oxygen is required, in which large amounts of ATP are formed.
- 'Loaded' coenzymes (NADH and $FADH_2$) act as hydrogen and electron acceptors.
- The three stages of aerobic respiration are glycolysis, the Krebs cycle and the electron transport chain.
- Glycolysis occurs in the cytosol of cells, and is involved in the breakdown of glucose into two molecules of pyruvate, the formation of two molecules of ATP and two molecules of NADH (from an input of ADP + Pi and NAD^+).
- The Krebs cycle occurs in the mitochondrial matrix, and involves inputs of pyruvate, NAD^+, FAD and ADP + Pi and an output of CO_2, NADH, $FADH_2$ and two molecules of ATP.
- A short step before the Krebs cycle, known as pyruvate oxidation, allows for pyruvate to be used by converting it to acetyl molecules, which are transported to the Krebs cycle by Coenzyme A (CoA).
- The electron transport chain occurs in the cristae of the mitochondria and is where a majority of ATP is formed. The inputs are O_2, ADP + Pi and the NADH and $FADH_2$ made during glycolysis and the Krebs cycle. The outputs of this stage are H_2O, ATP, NAD^+ and FAD.

4.5 Activities

To answer questions online and to receive **immediate feedback** and **sample responses** for every question, go to your learnON title at **www.jacplus.com.au**. A **downloadable solutions** file is also available in the resources tab.

4.5 Quick quiz	4.5 Exercise	4.5 Exam questions

4.5 Exercise

1. Identify the stage(s) of aerobic respiration that occur in:
   a. the cytosol
   b. the mitochondrial matrix.
2. Which molecules or components that are part of aerobic respiration:
   a. are coenzymes loaded with electrons and hydrogen ions
   b. the end product of glycolysis
   c. the final acceptor of the electron transport chain?
3. Identify the following statements as true or false and justify your response.
   a. Electron transport precedes the Krebs cycle.
   b. Glucose transfers its chemical energy to ATP through the aerobic respiration pathway.
   c. The stage of aerobic respiration that releases the greatest amount of energy for ATP production is the electron transport stage.
4. A researcher stated that in ideal conditions each molecule of NADH leads to the production of 3 molecules of ATP and each molecule of $FADH_2$ leads to 2 molecules of ATP, leading to a total yield of 36–38 ATP. Why might the actual yield be lower than this?
5. Briefly explain how red blood cells can survive without mitochondria.
6. Describe three cell types that would likely have large numbers of mitochondria. Justify your response.
7. Oligomycin A is a molecule that inhibits ATP synthase, which is the final enzyme involved in the electron transport chain.
   a. Oligomycin A is commonly used as an anti-fungal agent. Suggest how it might be effective in acting as an anti-fungal.
   b. A side effect of taking oligomycin is increased lactic acid in the urine and the blood, which is usually a by-product of anaerobic respiration. Why might this be the case?
   c. Why would other anti-fungal agents be recommended instead of oligomycin?

4.5 Exam questions

Question 1 (1 mark)

Source: *VCAA 2018 Biology Exam, Section A, Q9*

MC Which of the following gives the inputs and outputs of the electron transport chain in an animal cell?

	Inputs	Outputs
A.	NADH, ADP, oxygen, Pi	ATP, NAD^+, water
B.	NADH, ADP, water, Pi	ATP, NAD^+, oxygen
C.	NAD^+, ADP, oxygen, Pi	NADH, ATP, water
D.	NADPH, ADP, water, Pi	$NADP^+$, ATP, oxygen

Use the following information to answer Questions 2 and 3.

Shown below is a simplified diagram summarising a series of biochemical process in a plant cell.

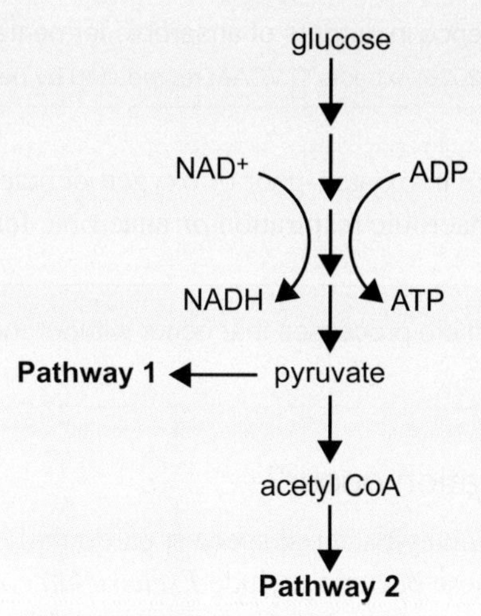

Source: Adapted from MG Stovell et al., 'Assessing metabolism and injury in acute human traumatic brain injury with magnetic resonance spectroscopy: Current and future applications', Frontiers in Neurolgy, 12 September 2017, <https://doi.org/10.3389/fneur.2017.00426>

Question 2 (1 mark)

Source: *VCAA 2020 Biology Exam, Section A, Q5*

MC Which one of the following is a correct statement?

A. Pathway 2 releases oxygen as a by-product.
B. Pathway 1 requires carbon dioxide as an input.
C. ATP is produced in Pathway 1 and is used by the cell as an energy source.
D. NADH created in Pathway 2 carries electrons into the electron transport chain.

Question 3 (1 mark)

Source: *VCAA 2020 Biology Exam, Section A, Q6*

MC The final products of Pathway 1 are produced in the

A. cristae.
B. cytosol.
C. mitochondrial matrix.
D. chloroplast membranes.

Question 4 (1 mark)

Source: *VCAA 2019 Biology Exam, Section A, Q14*

MC In glycolysis, the ATP yield per molecule of glucose is

A. 4 ATP produced and 2 ATP used for a net gain of 2 ATP.
B. 2 ATP produced and 4 ATP used for a net loss of 2 ATP.
C. 36 to 38 ATP produced for a net gain of 2 ATP.
D. 36 to 38 ATP used for a net loss of 2 ATP.

Question 5 (3 marks)

Source: *VCAA 2011 Biology Exam 1, Section B, Q7a and b*

a. Write the word or chemical equation for aerobic cellular respiration. **1 mark**

b. Cyanide inactivates metabolic reactions at the cristae of mitochondria. Cyanide poisoning often results in death. Explain why. **2 marks**

More exam questions are available in your learnON title.

4.6 Anaerobic fermentation

KEY KNOWLEDGE

- The location, inputs and the difference in outputs of anaerobic fermentation in animals and yeasts

Source: VCE Biology Study Design (2022–2026) extracts © VCAA; reproduced by permission.

Some organisms that live and reproduce in oxygen-poor or oxygen-deficient environments generate the energy needed for living (ATP) using *either* anaerobic respiration *or* anaerobic fermentation.

Anaerobic respiration and fermentation are processes that occur without the presence of oxygen, producing a net of 2 ATP molecules.

When does anaerobic fermentation occur?

Anaerobic fermentation is common in many bacterial species, particularly those living in an environment where oxygen availability can be variable. These bacteria include *Escherichia coli* (resident in your gut), *Salmonella enterica* (a common cause of food poisoning) and *Yersinia pestis* (the black death bacteria). Some bacteria use aerobic cellular respiration when oxygen is plentiful, but switch to anaerobic fermentation when oxygen is scarce — the best of both worlds! These types of bacteria are referred to facultative anaerobes.

Anaerobic fermentation also occurs in different species and strains of yeast, a unicellular fungus.

Human skeletal muscle cells also carry out anaerobic fermentation. When a person's skeletal muscle cells work very strenuously in bursts of power or speed, these cells switch to anaerobic fermentation to supply their ATP needs. This leads to the production of lactic acid as a by-product.

Anaerobic fermentation occurs in human skeletal muscle cells when the supply of oxygen to the cells by aerobic cellular respiration cannot keep up with their demand for ATP.

This situation can arise, for example, during short intense bursts of skeletal muscle activity such as Olympic athletes sprinting 100 metres, cyclists sprinting at the end of a stage in a bike road race (see figure 4.47), a weightlifter lifting a barbell in one movement from the floor to overhead. Because anaerobic fermentation does not require an input of oxygen, athletes may not even take a breath when performing at this extreme level. Depending on the intensity level of muscle activity, fermentation can last for no more than one or a few minutes before lactic acid builds up in the muscle cells and muscle fatigue hits.

In bacteria and in yeast cells, anaerobic fermentation can continue without interruption as long as needed. However, in mammalian skeletal cells, anaerobic fermentation can supply ATP for a very limited time.

FIGURE 4.47 In skeletal muscle cells, fermentation can provide ATP for a short period to power intense muscle activity, such as seen in this short intensive sprint to the finishing line.

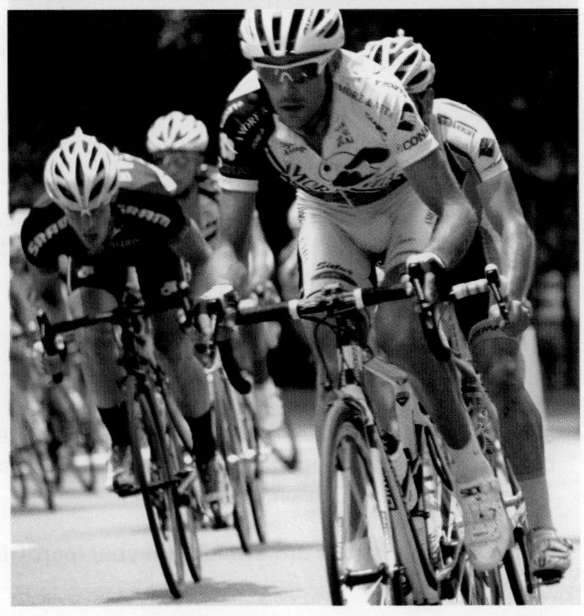

4.6.1 Role of anaerobic fermentation

Fermentation is an anaerobic process that produces ATP for the energy requirements of cells through a pathway that:

- operates under anaerobic conditions
- takes place totally within the cytosol of cells
- produces far less ATP per glucose molecule metabolised than that produced by aerobic cellular respiration
- produces ATP at a rate about 100 times faster than that for aerobic cellular respiration
- does not involve an electron transport chain.

Anaerobic fermentation, which occurs in the cytosol, produces less ATP per molecule of glucose than aerobic cellular respiration. However, its faster rate of production enables more ATP to be produced per unit of time.

4.6.2 Biochemical pathway of anaerobic fermentation

Two common anaerobic fermentation pathways are:

- lactic acid fermentation
- alcohol (ethanol) fermentation.

FIGURE 4.48 Comparing the types of anaerobic fermentation pathways. Both pathways rely on glycolysis to produce pyruvate.

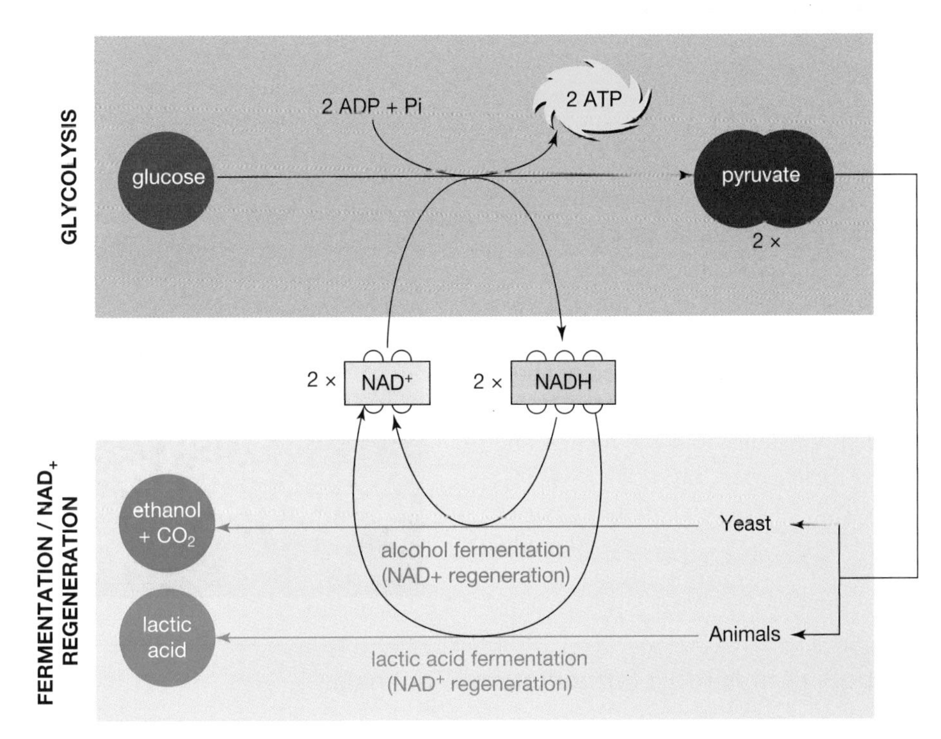

Both anaerobic fermentation pathways consist two stages, with both occurring in the cytosol:

1. Glycolysis:
   - starts with the input of a glucose molecule
   - ends with the output of two pyruvate molecules
   - is the only stage of anaerobic fermentation that is energy-releasing, generating two molecules of ATP and two loaded NADH coenzyme molecules
   - is identical to the glycolysis stage of aerobic cellular respiration pathway.

FIGURE 4.49 Reviewing glycolysis

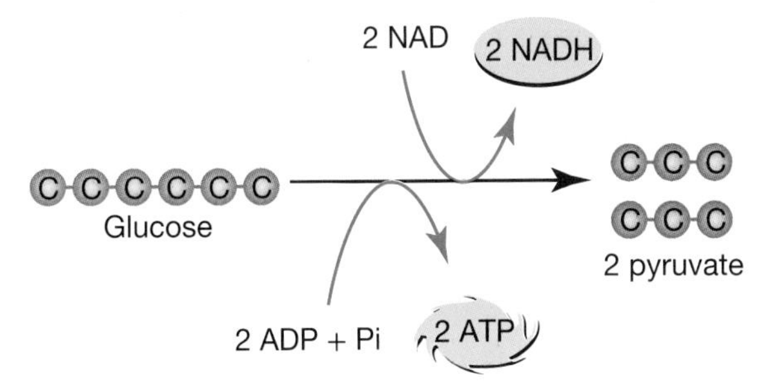

2. A short add-on stage (sometimes referred to as fermentation or NAD+ regeneration):
   - starts with pyruvate molecules
   - does not generate any energy-rich products
   - ends with different outputs, depending on the enzymes present in the cells involved
   - enables unloaded NAD^+ to be formed, allowing for them to be recycled and loaded again in glycolysis.

For **lactic acid fermentation**, the add-on stage is a one-step reaction, catalysed by the enzyme lactate dehydrogenase that produces lactic acid:

pyruvate	→	**lactic acid**
CH_4O_3		$C_3H_6O_3$

lactic acid fermentation a process that occurs without oxygen to produce two molecules of ATP and lactic acid

alcohol fermentation a process that occurs without oxygen to produce two molecules of ATP and ethanol

For **alcohol fermentation**, the end product is the alcohol, ethanol (C_2H_5OH), that is produced in a two-step reaction, each catalysed by a specific enzyme:

pyruvate	→	**acetaldehyde**	→	**ethanol**
$C_3H_4O_3$		C_2H_4O		C_2H_5OH

Note that these reactions have been simplified (and have not been balanced). Carbon dioxide is also produced in the process, during the conversion of pyruvate to acetaldehyde.

4.6.3 Anaerobic fermentation in animals

You have seen that during intensive muscle activity, human skeletal muscle cells switch to anaerobic fermentation because of a shortage of oxygen.

A cheetah (see figure 4.50) that is quietly stalking her prey uses aerobic cellular respiration to supply ATP to her skeletal muscles. However, once she races from cover in an explosive burst of speed in pursuit of a young gazelle, the cheetah will rely on anaerobic fermentation for her ATP supply. She will reach a top speed of about 100 km per hour in her pursuit. Unless the cheetah reaches her intended prey within about 100 metres, she will be forced to give up the chase. Just as happens in people, the accumulation of lactic acid produces muscle fatigue. The accumulation of lactic acid, the end product of anaerobic fermentation, may save the prey.

FIGURE 4.50 At the speeds attained by the cheetah, anaerobic lactic acid fermentation will supply ATP to her skeletal muscles for less than one minute. After that, the cheetah must increase breathing to deal with the oxygen deficit.

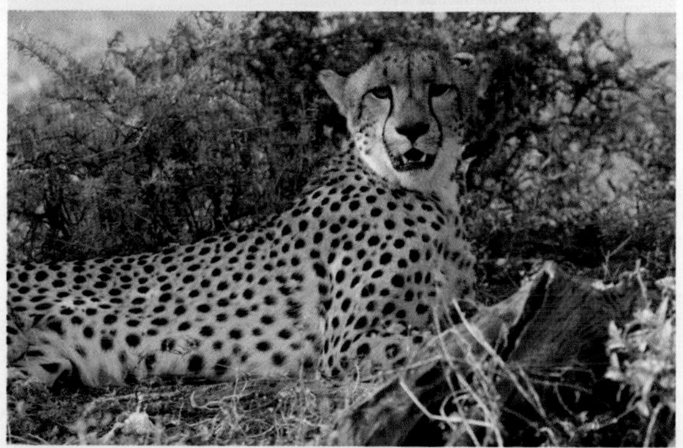

Inputs and outputs of anaerobic fermentation in animals

Lactic acid fermentation is an anaerobic process that leads to the production of small amounts of ATP in the absence of oxygen.

Inputs:
- Glucose
- NAD^+ (× 2)
- ADP + Pi (× 2)

Outputs:
- Lactic acid
- NAD^+ (× 2)
- ATP (× 2)

The NAD^+ is continually recycled between the glycolysis and NAD^+ regeneration stages, loaded to form NADH during glycolysis, and unloaded during NAD^+ regeneration.

Figure 4.51 shows a simple representation of the biochemical pathway of lactic acid fermentation. The add-on reaction produces lactic acid but no output of energy. Why bother adding this step? The add-on stage is important because it regenerates the NAD^+ that was used up in the glycolysis stage. Without the add-on, the fermentation pathway would stop running because of the lack of NAD^+ needed for the glycolysis stage.

FIGURE 4.51 Diagram showing a simplified version of the anaerobic fermentation pathway that produces lactic acid and regenerates unloaded NAD^+, enabling the pathway to continue

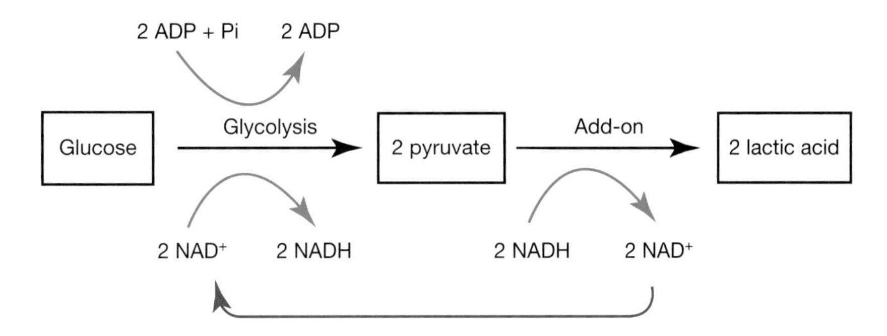

4.6.4 Anaerobic fermentation in yeasts

Yeasts are unicellular fungi that typically reproduce by budding. Different yeast species exist and, within one species, different strains or variants are recognised. A widely-studied yeast is *Saccharomyces cerevisiae* that has several strains: baker's yeast that is used in bread making and 'brewer's yeast, which plays an essential role in brewing beers and in wine making.

Note that:

- the two strains of *S. cerevisiae*, baker's yeast and brewer's yeast, use anaerobic fermentation to break down glucose to generate the ATP needed for their cells.
- this process is called alcohol fermentation
- the end products of alcohol fermentation process are ethanol (C_2H_5OH), an alcohol, and carbon dioxide (CO_2)
- baker's yeast and brewer's yeast differ in the relative proportion of alcohol and carbon dioxide that they produce — you can't make bread with brewer's yeast and vice versa.
- yet another strain of *S. cerevisiae* produces a different end product, lactic acid.

Inputs and outputs of anaerobic fermentation in most yeast species

Alcohol fermentation is an anaerobic process that leads to the production of small amounts of ATP in the absence of oxygen. This occurs in most species of yeast.

Inputs:

- Glucose
- NAD^+ (× 2)
- ADP + Pi (× 2)

Outputs:

- Ethanol
- CO_2
- NAD^+ (× 2)
- ATP (× 2)

The NAD^+ is continually recycled between the glycolysis and NAD^+ regeneration stages, loaded to form NADH during glycolysis, and unloaded during NAD^+ regeneration.

elog-0270

INVESTIGATION 4.5

online only

Anaerobic fermentation in yeast

Aim

To observe the rate of fermentation in yeast under different glucose concentrations

CASE STUDY: Applications of anaerobic fermentation in yeast

The end products of the anaerobic fermentation of glucose of several species and strains of yeast are shown in table 4.8. Several different strains of *S. cerevisiae* all produce the same end products (ethanol and CO_2). However, these various yeast strains differ in their optimal temperatures, in the relative proportions of end products they generate, in the ancillary flavours that they produce, and the concentration of alcohol they can survive — for example, beer yeasts tolerate up to about 5 per cent alcohol and then the cells stop fermenting; in contrast, most wine yeasts tolerate up to about 12 per cent alcohol.

As such, the outputs of anerobic fermentation in yeast are used for a variety of applications, including in bread making and the production of wine.

To access more information on case studies relating to the applications of anaerobic fermentation in yeast, please download the digital document.

FIGURE 4.52 The holes in the bread are due to the accumulation of carbon dioxide, one of the outputs of alcohol fermentation.

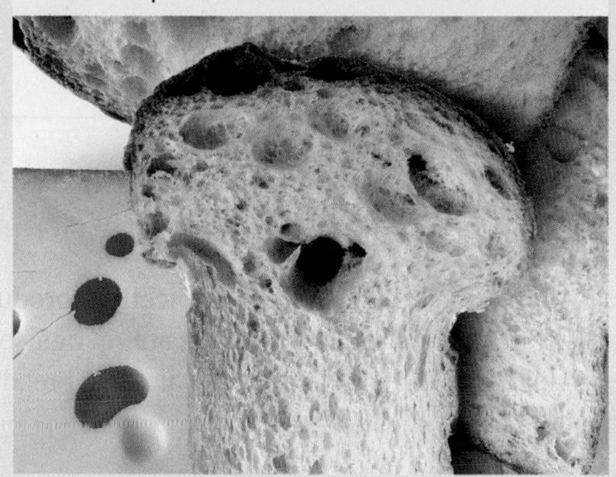

TABLE 4.8 Outputs (end products) of anaerobic fermentation in several yeast species and some uses in various industries

Yeasts	End products of fermentation	Commercial uses
Saccharomyces pasteurianus	Ethanol and CO_2	Lager style beers
Saccharomyces cerevisiae	Ethanol and CO_2	Ale style beers
Saccharomyces cerevisiae	Ethanol and CO_2	Wines
Saccharomyces cerevisiae	Ethanol and CO_2	Bread and pastries
Saccharomyces cerevisiae	Lactic acid	Pickles, sauerkraut
Pichia kudriavzevii	Ethanol and CO_2	Chocolate

Resources

Digital document Case study: Applications of anaerobic fermentation in yeast (doc-35828)

EXTENSION: Anaerobic fermentation in bacteria

Anaerobic fermentation is not just used by animals and yeast. Bacteria also use anaerobic fermentation to produce ATP.

To access more information on this extension, please download the digital document.

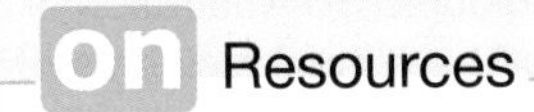

Resources

Digital document Extension: Anaerobic fermentation in bacteria (doc-35830)

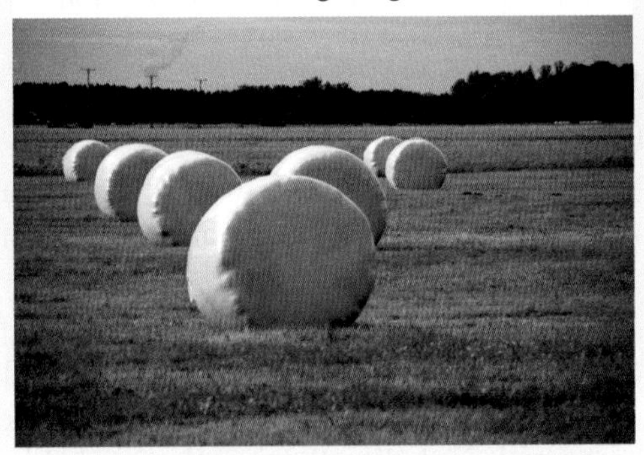

FIGURE 4.53 Bales of tightly-packed cut grass or other green crop, wrapped to exclude air for making silage

4.6.5 Comparing cellular respiration and fermentation

Aerobic cellular respiration is the means of ATP production by animals, plants, fungi and microbes living in environments where oxygen is available either from the air or in solution in water. Organisms living in oxygen-deficient or oxygen-free environments cannot meet their energy needs from aerobic cellular respiration and must use other processes. One of these alternative processes is fermentation. Table 4.9 shows a comparison of these two processes.

TABLE 4.9 Comparison of aerobic cellular respiration and fermentation

	Aerobic cellular respiration	Anaerobic fermentation
Requirement for oxygen	Yes	No
Occurrence	Mainly plants and animals, fungi, protozoa and many bacteria	Fungi (yeasts) and bacteria; skeletal muscle (limited period)
Number of stages	Three: glycolysis, Krebs cycle and electron transport chain	Two: glycolysis plus short add-on stage to regenerate NAD^+
Presence of electron transport chain	Yes, with oxygen as the terminal electron acceptor	No
Inputs	Glucose and oxygen	Glucose
End products	Carbon dioxide and water	Various, including ethanol, lactic acid and carbon dioxide
Energy yield per molecule of glucose	High, approximately 30–32 ATP	Low, 2 ATP
Rate of ATP production	Slower	Faster
Cellular location	Cytosol and mitochondria	Within cytosol of cell

There are many key features of each of these processes.

Aerobic cellular respiration

This process:

- is the ATP-producing pathway that occur in plants and animals and many microbes, including many pathogenic bacterial species
- requires an electron transport chain for the release of energy from electron transport between donors and acceptor in the chain
- can proceed only in the presence of oxygen that comes from an external source

- uses oxygen to act as the terminal electron acceptor in the electron transport chain, forming water
- produces most of its ATP (around 70 per cent) in the electron transport chain.
- has a high energy yield of about 30–32 ATP molecules in ideal conditions for each molecule of glucose that undergoes aerobic cellular respiration.

Anaerobic fermentation

This process:

- is an anaerobic process for making ATP using energy from the breakdown of a variety of organic molecules, such as glucose
- does *not* require oxygen
- occurs in fungi, such as yeasts, in skeletal muscle during strenuous exercise and in several kinds of bacteria including *Salmonella, Staphylococcus* and *E. coli* (figure 4.54), and archaeans that live in environments where oxygen is in very short supply, such as a cow's rumen, or where oxygen is absent, such as in anoxic deep-sea sediments
- has a very low total energy yield — as low as 2 ATP molecules — from each molecule of glucose that is fermented
- produces ATP without any involvement of an electron transport chain.

FIGURE 4.54 False-coloured 3D illustration of *Escherichia coli* (*E. coli*) — normal residents of the human gut. These bacteria use fermentation when living in the anaerobic conditions of the gut, but when in an oxygen-rich environment these bacteria can switch to aerobic cellular respiration.

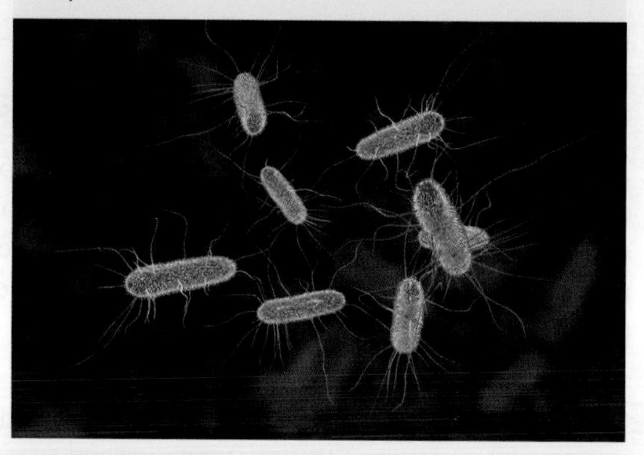

EXTENSION: Anaerobic respiration

Not all microbes living in anoxic environments use fermentation to supply their ATP energy needs. Some species of bacteria and archaea live in anoxic environments, such as marshes, bogs, deep ocean sediments and regions of 'dead' water with no dissolved oxygen, and they gain their energy from anaerobic respiration.

They have electron transport chains in their membranes, but the terminal electron acceptor is not oxygen and is varied. The terminal electron acceptor may be ions, such as nitrate (NO_3^-) or sulfate (SO_4^{2-}) or ferric iron (Fe^{3+}) from their environments. Depending on the terminal electron acceptor used, the energy yield from the anaerobic respiration of glucose can vary from just under 30 to about 4 ATP.

Resources

eWorkbook Worksheet 4.6 Anaerobic fermentation (ewbk-7585)

Video eLesson Fermentation experiment (eles-4336)

Interactivity Anaerobic fermentation (int-8323)

KEY IDEAS

- Anaerobic fermentation occurs in the absence of oxygen.
- This process occurs in the cytosol, and produces two molecules of ATP.
- Examples include lactic acid fermentation and alcohol fermentation.
- Two main stages occur in anaerobic respiration: glycolysis and NAD^+ regeneration.
- In animals, lactic acid is formed in fermentation.
- In yeasts, usually ethanol and carbon dioxide are formed in fermentation.

4.6 Activities

learnON

To answer questions online and to receive **immediate feedback** and **sample responses** for every question, go to your learnON title at **www.jacplus.com.au**. A **downloadable solutions** file is also available in the resources tab.

4.6 Quick quiz on	4.6 Exercise	4.6 Exam questions

4.6 Exercise

1. Where and when might you expect to find:
   a. lactic acid being produced in the human body
   b. an anaerobe producing carbon dioxide gas?
2. Briefly explain the following observations:
   a. In a liquor store, it is not possible to find a bottle of wine with an alcohol content of 20 per cent.
   b. Production of ATP by aerobic respiration is much slower than by anaerobic fermentation.
   c. Leavened bread (which contains baker's yeast) is raised and fluffy, and unleavened bread is flat and hard.
3. Outline how anaerobic fermentation generates ATP.
4. Bacteria are part of the microflora of the digestive system, in particular in the human intestines. Like all other cells, these bacteria must obtain ATP from glucose. Do you think intestinal bacteria use aerobic respiration or anaerobic fermentation for this purpose? Explain your answer.
5. Identify the following statements as true or false and justify your response.
   a. If deprived of oxygen, a person could survive for several hours using anaerobic fermentation.
   b. When competing, marathon runners gain their energy using aerobic respiration.
6. A student was conducting an investigation exploring anaerobic fermentation in yeast. They placed warm water in two conical flasks. In one flask, they added a packet of yeast and 10 grams of sugar. In the other flask they added a packet of yeast. Each conical flask was sealed in a balloon.
   a. What differences would you expect to see in the balloon between the two flasks? Why?
   b. The yeast used was baker's yeast, a strain of *Saccharomyces cerevisiae.* How would you know if lactic acid or ethanol was produced?
   c. Another student set up a third conical flask with 13 grams of sugar. What effect would this have on the results of the investigation?

4.6 Exam questions

Question 1 (1 mark)

Source: *VCAA 2019 Biology Exam, Section A, Q12*

MC During which process would the production of lactic acid be observed?

A. aerobic cellular respiration
B. fermentation in animals
C. fermentation in yeasts
D. photosynthesis

Question 2 (1 mark)

***Source:** VCAA 2009 Biology Exam 1, Section A, Q10*

MC Fermentation in yeast

A. produces ethanol.
B. requires lactic acid.
C. involves the Kreb's cycle.
D. requires the presence of oxygen.

Question 3 (2 marks)

***Source:** VCAA 2016 Biology Exam, Section B, Q2a and b*

Plant materials containing cellulose and other polysaccharides are reacted with acids to break them down to produce glucose. This glucose is then used by yeast cells for fermentation.

a. Why is fermentation important for yeast cells? **1 mark**
b. What are the products of fermentation in yeast cells? **1 mark**

Question 4 (3 marks)

a. Identify the 3-carbon intermediate product that is produced during anaerobic fermentation. **1 mark**
b. Although the starting points of anaerobic fermentation in human muscle tissue and yeast cells are the same, they produce different end products. Briefly explain. **2 marks**

Question 5 (1 mark)

MC A certain organism lives successfully deep in mud at the bottom of a swamp, an environment that is permanently free of oxygen. It is reasonable to state that this organism must

A. metabolise glucose to form lactate.
B. carry out anaerobic fermentation.
C. obtain its oxygen through photosynthesis.
D. have no requirement for ATP.

More exam questions are available in your learnON title.

4.7 Factors that affect the rate of cellular respiration

KEY KNOWLEDGE

- The factors that affect the rate of cellular respiration: temperature, glucose availability and oxygen concentration

Aerobic cellular respiration is commonly shown by the equation:

$$\textbf{glucose} + \textbf{oxygen} \rightarrow \textbf{carbon dioxide} + \textbf{water}$$

In theory, the rate of a reaction can be determined by measuring the rate at which the substrates disappear or the products appear. However, in cellular respiration, water is both a reactant and a product, so that is ruled out.

In practice, convenient measures of aerobic cellular respiration rates are based on:

- the rate of production of carbon dioxide
- the rate of uptake of oxygen
- the rate of uptake of glucose.

Using any one of these methods, it is possible to explore factors that affect the rate of cellular respiration, such as temperature, availability of glucose and oxygen concentration. Experiments on the effect of various factors on cellular respiration may involve whole organisms or cells in culture or even mitochondria.

4.7.1 Temperature

In sections 3.5.1 and 4.4.3 the effect of temperature on enzymes and photosynthesis was investigated. Temperature can also affect the rate at which cellular respiration occurs. Each reaction in the cellular respiration pathways is catalysed by a specific enzyme. Each enzyme has an optimal temperature for its catalytic action. The rate of cellular respiration is maximal at this optimal temperature, and decreases with deviations from this point.

At low temperatures, collisions between substrate and enzyme molecules are less frequent, resulting in a low rate of respiration. As the temperature increases further, the respiration rate also increases until the optimal temperature for an organism's cells is reached. At temperatures above the optimal, heat denaturation of the enzymes begins and a steep decline in respiration rate occurs.

FIGURE 4.55 Graph showing the effect of temperature on the rate of cellular respiration

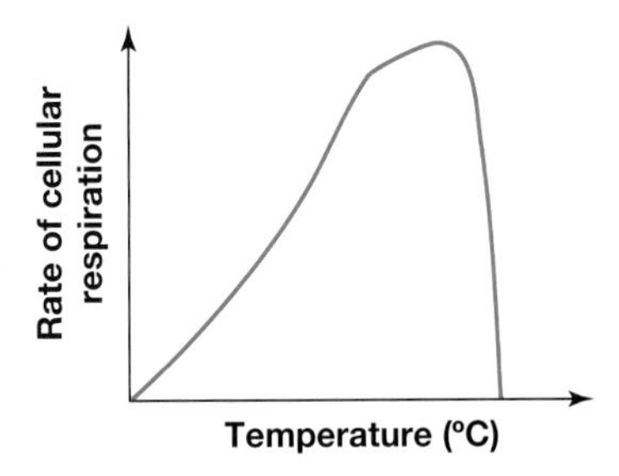

CASE STUDY: The effect of temperature of the cellular respiration of different species

Mammals and birds maintain their internal temperature within narrow limits through the operation of homeostatic mechanisms. This is in contrast to other organisms whose internal environment will vary with the temperature of the external environment. So, to explore the effect of temperature on cellular respiration in whole organisms, experiments are often conducted in plants or animals that do not maintain a constant body temperature.

Figure 4.56 shows the resting rate of oxygen consumption of a whole animal, namely a species of killifish (*Fundulus heteroclitus*), plotted against water temperature. The rate of oxygen consumption when the fish is resting quietly in a respirometer is a good indicator of the resting rate of cellular respiration. Note that the graph begins to flatten out as the temperature exceeds 30 °C. Beyond this temperature, the critical maximum temperature of the fish would be reached and the oxygen consumption would rapidly decrease, eventually reaching 0.

FIGURE 4.56 a. A fish in a respirometer chamber that is having its oxygen consumption measured **b.** Plot of respiration rate of a killifish over a range of temperatures

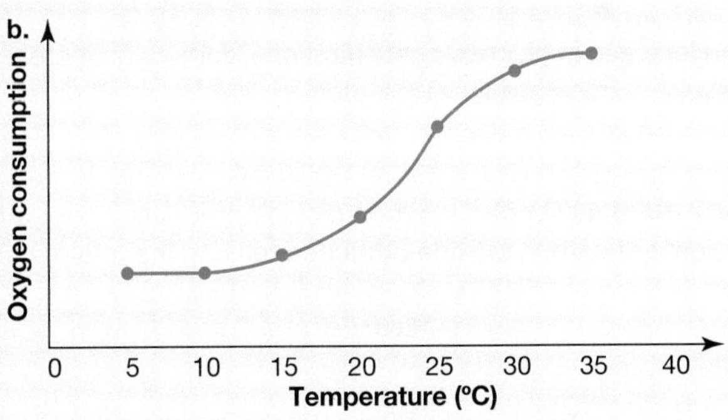

Figure 4.57 shows the total oxygen consumption over time for germinating corn kernels at two ambient temperatures.

At 22 °C, the rate of oxygen consumption of the corn kernels is 1.6 mL per 20 minutes, which converts to 0.08 mL/min, while the rate at 12 °C is 0.8 mL per 20 min, which converts to 0.04 mL/min. Comparing these two particular temperatures, the rate of cellular respiration is higher at the higher temperature. However, there is an upper limit to this because enzymes are involved. If the temperature goes too high, the enzymes responsible for cellular respiration will denature and respiration will slow and then stop.

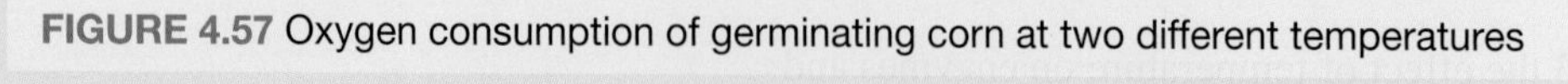

FIGURE 4.57 Oxygen consumption of germinating corn at two different temperatures

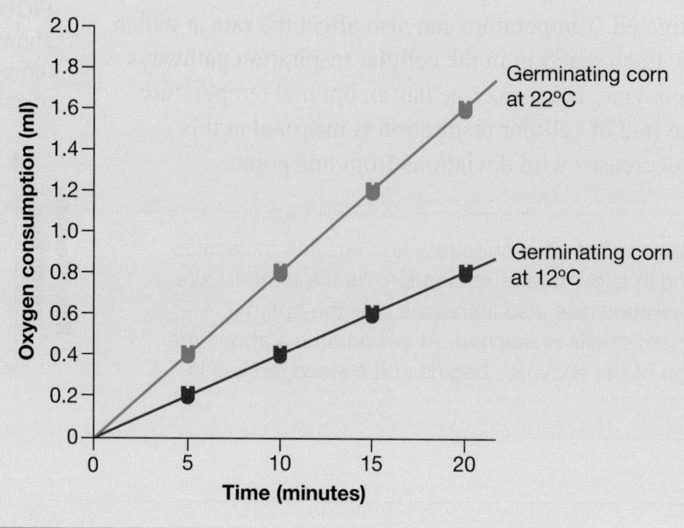

4.7.2 Glucose concentration

Glucose is the key input to aerobic cellular respiration. Figure 4.58 shows the graph of the volume of carbon dioxide produced over a fixed period by equivalent quantities of yeast cells suspended in glucose solutions of increasing concentrations. In this case, the rate of carbon dioxide produced over a fixed time interval can be used to measure the rate of cellular respiration.

As the glucose concentration increases, the carbon dioxide production over a fixed period of time also increases, initially sharply — there are plenty of available active sites on the rate-determining enzyme of the cellular respiration pathway. But, with further increases in glucose concentration, the rate of carbon dioxide production and cellular respiration slows and eventually flattens. At this point, the active sites of the first rate-determining enzyme in the cellular respiration pathway are saturated with substrates, so that no further increase in the amount of carbon dioxide can occur.

An increase in glucose concentrations leads to an increase in the rate of cellular respiration. This rate eventually levels off from rate-limiting factors.

FIGURE 4.58 Graph of carbon dioxide production over a fixed period against the concentration of glucose. The volume of carbon dioxide produced reflects the rate of cellular respiration.

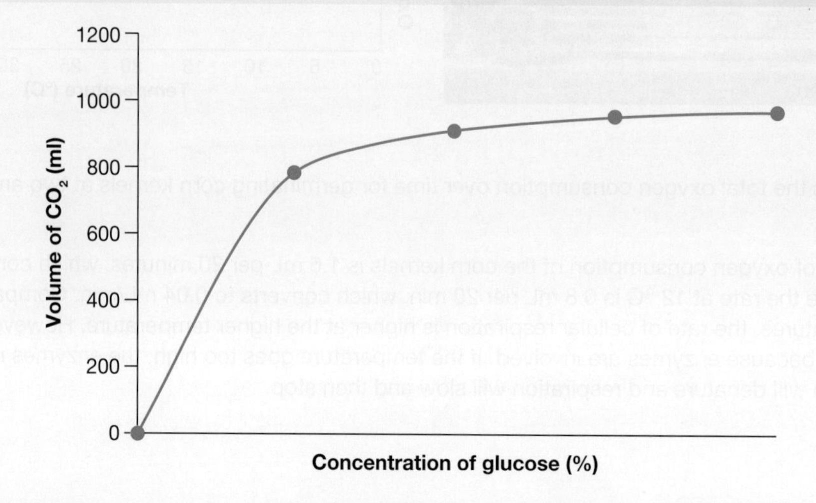

on Resources

Digital document Case study: Visualising glucose uptake in body cells (doc-36167)

4.7.3 Oxygen concentration

Oxygen is consumed in aerobic cellular respiration. Oxygen is the final acceptor of electrons that have passed down the electron transport chain. In this process, water is produced:

$$\frac{1}{2}O_2 + 2e^- + 2H^+ \rightarrow H_2O$$

An ongoing supply of oxygen is essential for cellular respiration to proceed. The lethal effects of carbon monoxide that blocks the delivery of oxygen to cells is evidence of this.

Figure 4.59 shows a graph of the rate of cellular respiration against the oxygen concentration. Note that, at the lowest oxygen concentration, the rate of aerobic cellular respiration is effectively zero.

As the oxygen concentration increases, the rate of cellular respiration (and carbon dioxide production) also increases up to a point. The graph eventually flattens out and any further increase in oxygen concentration is not matched by any further increase in carbon dioxide concentration. At this point, the rate of supply of electrons via the electron transport chain has reached its maximum so that no further increase in respiration rate can occur.

FIGURE 4.59 Graph showing the effect on cellular respiration of whole pears exposed to different oxygen concentrations

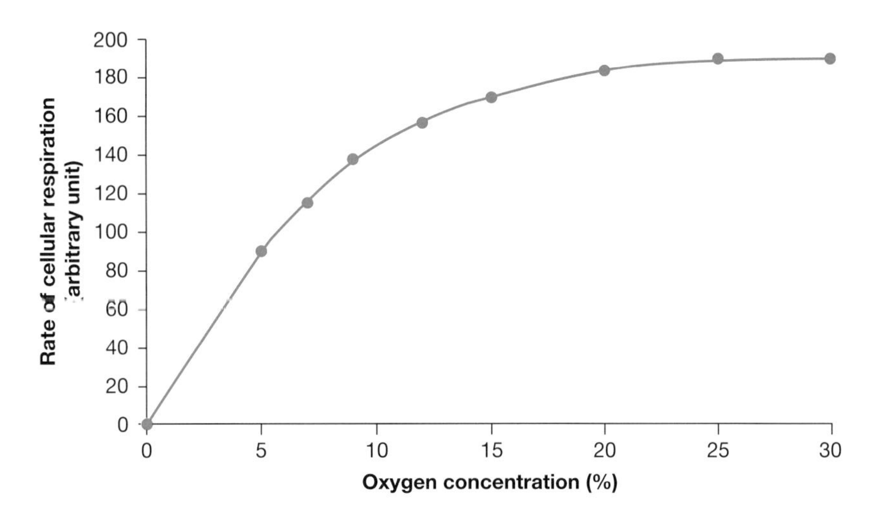

As oxygen concentration increases, so does the rate of cellular respiration. Like with glucose concentration, this eventually levels off due to other factors limiting the rate.

tlvd-1432

SAMPLE PROBLEM 5 Exploring factors affecting the rate of cellular respiration

a. A scientist wishes to measure the rate of cellular respiration. Identify two ways in which they could easily measure this. (2 marks)

b. One factor that affects the rate of cellular respiration is oxygen. Explain how changing levels of oxygen leads to a change in the rate of cellular respiration. (2 marks)

c. Why is ATP still produced when there is no oxygen present? (1 mark)

d. Starch is a polymer made of many molecules of glucose. How would an increase in starch affect the production of carbon dioxide? (2 marks)

THINK	WRITE
a. Processes can be measured by exploring either the inputs or the outputs. Consider two of these that would be easy to measure. (For example, ATP production would be hard to measure compared to CO_2 production.)	Two ways that cellular respiration may be measured: • Carbon dioxide concentration (how much is being produced) (1 mark) • Oxygen concentration (how much is being used) (1 mark)
b. Consider all factors that need to be discussed. Ensure two key points are discussed as the question is worth two marks.	• As oxygen levels increase, the rate of cellular respiration also increases, as there is more oxygen available for use in the electron transport chain (1 mark). • Eventually, the oxygen hits a level where the maximal rate of cellular respiration has been reached (due to other limiting factors), and the rate of cellular respiration levels off (1 mark).
c. Consider when ATP is made: glycolysis, Krebs cycle and ETC.	2 ATP are produced in glycolysis, which is an anaerobic process that does not require oxygen. (1 mark).
d. Examine the idea that starch is made of glucose. Link this to how the rate of cellular respiration is affected by glucose production.	As starch contains glucose, it may be broken down into these monomers (1 mark). Therefore, this glucose can be used to increase the rate of cellular respiration as there is an increase in the input for the process. (1 mark)

Resources

eWorkbook Worksheet 4.7 Reviewing factors impacting cellular respiration (ewbk-7587)

KEY IDEAS

- Temperature, glucose concentration and oxygen concentration can affect the rate of cellular respiration.
- As temperature increases, the rate of cellular respiration increases. This is because collisions between the enzymes involved in photosynthesis and substrates increase due to molecules having higher energy. However, eventually, the rate decreases as enzymes begin to denature.
- As oxygen levels increase, the rate of photosynthesis increases. Eventually this rate reaches a maximal point as other factors (such as the amount of glucose) limit the rate of cellular respiration.
- As glucose levels increase, the rate of photosynthesis increases. Eventually this rate reaches a maximal point as other factors (such as the amount of oxygen) limit the rate of cellular respiration.

4.7 Activities

To answer questions online and to receive **immediate feedback** and **sample responses** for every question, go to your learnON title at **www.jacplus.com.au**. A **downloadable solutions** file is also available in the resources tab.

4.7 Quick quiz on	4.7 Exercise	4.7 Exam questions

4.7 Exercise

1. **MC** The aerobic cellular respiration may be summarised by the word equation:

glucose + oxygen + ADP + Pi → carbon dioxide + water + ATP

The final stage of aerobic cellular respiration is the electron transport chain.
In a human muscle cell, the rate of the electron transport increases if there is
A. an increase in water in the mitochondria.
B. a temperature decrease to below 15 °C.
C. an increased concentration of oxygen in the capillaries.
D. an exposure to an incredibly acidic pH.

2. The optimal temperature for cellular respiration in humans is 37 °C. What would happen to the rate of cellular respiration at:
   a. 52 °C
   b. 20 °C?
   For each response, provide reasons that these changes in rate would occur.
3. Describe three quantitative ways that you can measure the rate of cellular respiration.
4. Why is measuring the amount of carbon dioxide produced by yeast not the best way to determine the rate of aerobic respiration?
5. A student was examining the aerobic respiration rate in four samples of yeast. They determined the rate of respiration by measuring oxygen consumption.
   Each yeast sample was placed in different conditions:
   - Sample 1: 35 °C with 20 per cent glucose
   - Sample 2: 14 °C with 20 per cent glucose
   - Sample 3: 35 °C with 30 per cent glucose
   - Sample 4: 35 °C with 40 per cent glucose.

   a. Would Sample 1 or Sample 2 be expected to have the higher oxygen consumption? Justify your response.
   b. The student was surprised when they found the exact same results for Sample 3 and Sample 4. Explain why this likely occurred.
   c. The student did not use a control group in their investigation. What should have been used as a control group?
6. Draw a clear graph showing how a change in oxygen concentration would affect the rate of cellular respiration.
7. A student wished to explore how increasing levels of sucrose affected the rate of cellular respiration in plant cells. Sucrose is a disaccharide composed of glucose and fructose.
   a. Outline a scientific method that would allow the student to investigate this.
   b. What conclusions would likely be drawn from this experiment?

4.7 Exam questions

Question 1 (1 mark)

Source: *VCAA 2019 Biology Exam, Section A, Q13*

MC The rate of aerobic cellular respiration in a human cell may increase if the
A. temperature of the cell is lowered from 37 °C to 35 °C.
B. oxygen concentration available to the mitochondria increases.
C. carbon dioxide concentration in the cytosol of the cell increases.
D. rate of facilitated diffusion of glucose into the cytosol of the cell decreases.

Question 2 (8 marks)

Source: *VCAA 2006 Biology Exam 1, Section B, Q4*

2,4-dinitrophenol is a chemical that is toxic to mitochondria. When added to mitochondria this chemical allows electron transport to occur but prevents the phosphorylation of ADP to ATP. The chemical achieves this by breaking the essential link between electron transport and ATP synthesis. This toxin causes mitochondria to produce heat instead of ATP. The greater the amount of toxin added, the quicker is its action.

a. If mitochondria are poisoned with 2,4-dinitrophenol by what process could a plant cell produce more ATP? **1 mark**

b. Where in the mitochondria does electron transport and ATP production occur? **1 mark**

A researcher wanted to study cellular respiration in insect cells. She cultured some muscle cells from the common field cricket, *Teleogryllus oceanicus*, then studied the effects of adding 2,4-dinitrophenol to these cells. An agricultural company may want to fund this research.

c. Give one reason why an agricultural company might want to fund research on the effects of this toxin on field crickets. **1 mark**

The experiment is summarised in the table below. Temperature observations in each trial were made at equal time intervals.

Observations made at equal time intervals	**Temperature °C**		
	control (no 2,4-dinitrophenol)	trial 1 (2,4-dinitrophenol added)	trial 2
1 (start)	28	28	28
2	27	28	29
3	28	29	
4	29	31	
5	28	36	
6	28	23	
7	27	21	

d. In terms of energy production, why did the temperature go up in trial 1 and not in the control? **1 mark**

e. Explain why the temperature went down after the fifth observation in trial 1 **1 mark**

f. Trial 2 had twice the concentration of 2,4-dinitrophenol added. Complete the table by writing in temperatures in the spaces provided to predict the trend. **2 marks**

Another researcher suggested adding pyruvate to the cells to cancel out the effects of this toxin.

g. Explain what effect adding pyruvate would have on cancelling out the effect of this toxin. **1 mark**

Question 3 (3 marks)

Explain, at the molecular level, why an increase in concentration of oxygen or glucose will increase the rate of cellular respiration.

Use the following information to answer Questions 4 and 5.

Hexokinase is an enzyme that catalyses the first step of glucose breakdown in the reactions of glycolysis. A scientist performed a simple experiment to test the action of hexokinase.

Identical volumes of glucose solutions, each with a differing concentration, were placed in eight separate tubes. An equal amount of hexokinase was added to each tube. The tubes were incubated at 37 °C for 30 minutes. The amount of glucose consumed in each tube was recorded. The rate of glucose consumption was calculated for each tube and this is shown on the graph.

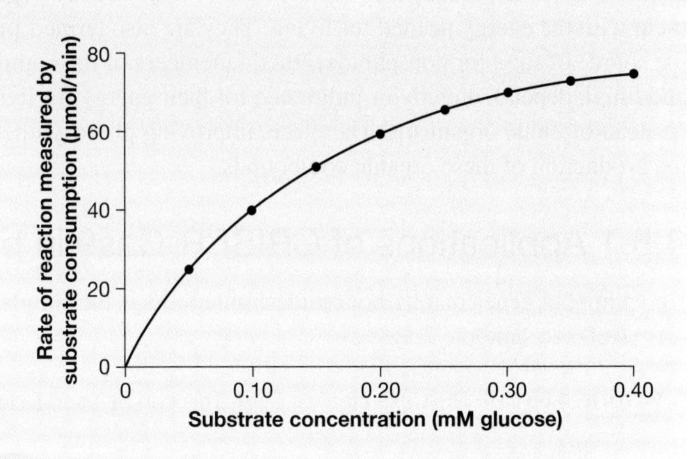

Question 4 (1 mark)

MC It can be concluded from the information provided that

A. increasing glucose concentration decreases the rate of glycolysis.
B. the rate of glycolysis is limited by the amount of hexokinase in the tubes.
C. greater amounts of oxygen were consumed at higher substrate concentrations.
D. the maximum rate was achieved when all available mitochondria were engaged in the reaction.

Question 5 (1 mark)

MC The test tubes were incubated at 37 °C because:

A. room temperature is approximately 37 °C.
B. the optimum temperature for hexokinase is 37 °C.
C. glucose is denatured at temperatures above or below 37 °C.
D. temperature was the independent variable in this experiment.

More exam questions are available in your learnON title.

4.8 Improving photosynthetic efficiencies and crop yields using CRISPR-Cas9

KEY KNOWLEDGE

- Potential uses and applications of CRISPR-Cas9 technologies to improve photosynthetic efficiencies and crop yields

The world's population is expected to increase to from its current population of 7.6 billion to 9.7 billion by 2050 (and over 11 billion by 2100). With this rapidly increasing population, it is important that we have the resources to feed a growing population. One way we can do this is by improving photosynthetic efficiencies and crop yields.

Biotic and abiotic stresses pose a risk by potentially decreasing crop yields. Development is needed in new technologies to improve yields and meet the demands of our growing population. Gene editing technologies, such as **CRISPR-Cas9**, have emerged as potential solutions to addressing these challenges in agriculture. CRISPR-Cas9 can precisely modify an organism's genomic sequence in order to achieve the desired trait that works against these biotic and abiotic stresses.

CRISPR-Cas9 a tool for precise and targeted genome editing that uses specific RNA sequences to guide an endonuclease, Cas9, to cut DNA at the required positions

Many organisms can undergo photosynthesis. Organisms with this ability are described as being photoautotrophic because they use sunlight to manufacture **organic compounds**, such as sugars, that provide them with the energy needed for living. They are also termed **producers** because, in an ecosystem, they are the source of food for non-photosynthetic members of the community. All other organisms, such as animals and fungi, depend, directly or indirectly, for their energy for living on the organic compounds produced by photoautotrophic organisms. Therefore, improving photosynthetic efficiency is incredibly important to maximise the production of these organic compounds.

4.8.1 Applications of CRISPR-Cas9 in plant breeding

Knocking out genes that harbour undesirable traits is the simplest and most common application of CRISPR-Cas9 (refer to subtopic 2.3).

FIGURE 4.60 CRISPR-Cas9 cleaves DNA. This can be used to knock out a gene.

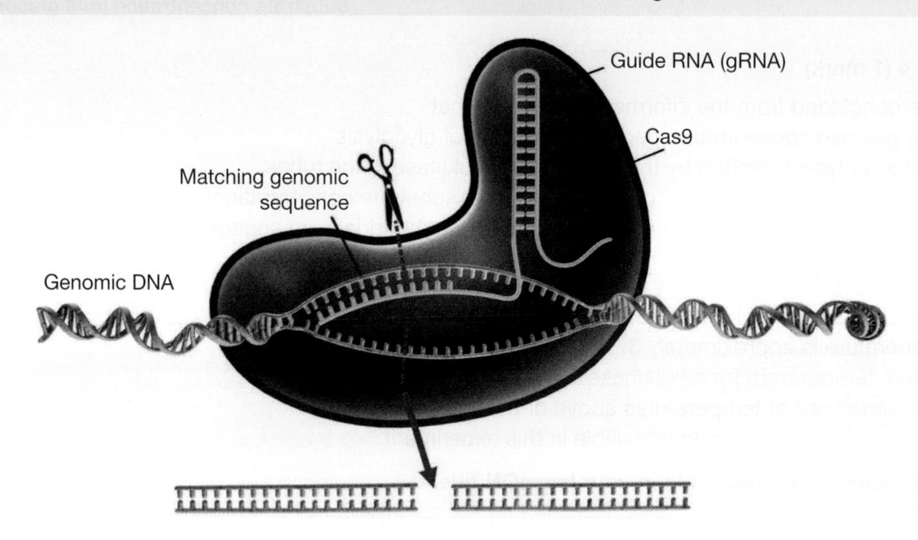

Traits and techniques that can be improved through the application of CRISPR-Cas9 include:

- photosynthetic efficiency
- crop yield
- crop quality
- biotic and abiotic stress resistance
- hybrid-breeding (two genetically different parent lines are produced).

organic compound any carbon and hydrogen containing compound

producer organism that can build organic matter from simple inorganic substances; also known as autotrophs

Improving photosynthetic efficiency

The improvement of photosynthetic efficiencies allows for the maximisation of the production of organic compounds (such as glucose), which are vital for sustaining life in autotrophic organisms, including humans.

As explored earlier in this topic, one of the major limitations of photosynthesis in C_3 plants is due to Rubisco and the impact of photorespiration on reducing photosynthetic efficiency. Rubisco is an important enzyme involved in photosynthesis. However, Rubisco cannot distinguish well between oxygen and carbon dioxide. This binding to oxygen rather than carbon dioxide reduces the efficiency of photosynthesis. Rubisco is produced through the transcription and translation of related genes, and thus these genes present a target site for improving the efficiency of Rubisco through genome editing. This could involve knocking out genes, or sections of genes, that result in Rubisco binding to O_2 or inserting genes that would improve efficiency.

CRISPR-Cas9 could potentially be used to improve the efficiency of Rubisco in carbon dioxide fixation by reducing its ability to bind to oxygen and undergo photorespiration, and instead improve its ability to associate with carbon dioxide. More effective forms of Rubisco could greatly improve the outcomes and speed of photosynthetic pathways in many of the crops that form a vital part of diets worldwide (including wheat, rice and barley). By reducing photorespiration and improving photosynthetic efficiency, crop yield from C_3 plants could be dramatically increased. This could account for billions of dollars worth of usable food being produced globally.

Increasing crop yields

The primary aim for editing a crop's genome is to protect and secure crop yields for a growing worldwide population. However, increasing total yields depends on many factors.

Traditional gene-editing techniques, such as knocking out genes, has had a negative effect on crop yields. Not only was it found that the crop yield (number of grains) decreased, but also the size and mass of the grains decreased. CRISPR-Cas9 could eliminate these negative side effects and improved these yield-related traits. CRISPR-Cas9 can be used to target certain genes that impact crop yield; such as by inserting genes to improve yield or knocking out genes that have a negative effect.

Improving crop quality

The use of CRISPR-Cas9 can greatly enhance not just the quantity, but the quality of the crops produced. Genome editing via CRISPR-Cas9 has been used to alter the gluten and nutrient content, storage quality and visual appearance of crops.

Examples of improved crop quality include the following:

- Rice has been altered to improve its eating and cooking qualities by reducing the grain's overall sugar content. Scientists were able to use CRISPR-Cas9 to knock out the Waxy gene.
- The nutrient quality of rice was improved by increasing the amount of amylose and reducing the starch in the rice. This was achieved by altering the genes that produce the proteins that produce starch. Rice high in amylose can help reduce the effects of many environmentally influenced health conditions, such as type 2 diabetes.
- CRISPR-Cas9 has been used to alter the α-gliadin gene. This gene is responsible for the gluten proteins found in cereal crops. Gluten is a known trigger for sufferers of coeliac disease.
- The shelf life of tomatoes and their appearance has been improved using CRISPR-Cas9 to enhance the levels of lycopene. Lycopene is the bright red carotenoid that gives the fruit their colour.
- The oleic acid oil content of seeds such as *Camelina sativa* and *Brassica napus* (as seen in figure 4.61) have also been improved through the use of CRISPR-Cas9.

FIGURE 4.61 The oleic acid oil content of seeds of **a.** *Camelina sativa* and **b.** *Brassica napus* can be enhanced using CRISPR-Cas9.

Biotic and abiotic stress resistance

Biotic (living) and abiotic (non-living) stresses are the main factors affect crop yield and quality. CRISPR-Cas9 has been used to increase crop resistance to biotic factors, such as bacterial, fungal and viral diseases, and pests. This ensures the crop can be used with greater efficiency, particularly those staples that are vital for world food supply.

Rice is one such food source that most of the world's population depends on. Rice has a small genome. Studies conducted have already produced positive results for CRISPR-Cas9 editing to improve resistance to biotic and abiotic stresses. Overall the rice crop yield has been improved.

Other crops treated with CRISPR-Cas9 to develop biotic and abiotic resistance include:

- wheat
- maize (corn)
- banana plants
- cotton
- soybean
- tomato
- potatoes
- citrus
- grapes.

CASE STUDY: Abiotic and biotic stress resistant crops

Table 4.10 provides a summary of stress/traits that have been targeted using CRISPR-Cas9. In recent years, there has been a significant rise in the use of CRISPR-Cas9 for various purposes to improve the yield of crops. Articles that explore these modifications are also listed.

TABLE 4.10 Summary of crops modified to be resistant to biotic and abiotic stress

Crop	Type	Target gene	Stress/trait	Reference
A.thaliana/ N. benthamiana	Biotic	dsDNA of virus (A7, B7 and C3 regions)	Beet severe curly top virus resistance	Ji et al., 2015
A. thaliana	Biotic	eIF(iso)4E	Turnip mosaic virus (TuMV) resistance	Pyott et al., 2016
N. benthamiana	Biotic	BeYDV	Bean yellow dwarf virus (BeYDV) resistance	Baltes et al., 2015
N. benthamiana	Biotic	ORFs and the IR sequence sDNA of virus	Tomato yellow leaf curl virus (TYLCV) and Merremia mosaic virus (MeMV)	Ali et al., 2015
Rice	Biotic	OsERF922 (ethylene responsive factor)	Blast resistance	Wang F. et al., 2016
Rice (IR24)	Biotic	OsSWEET13	Bacterial blight disease resistance	Zhou et al., 2015
Bread wheat	Biotic	TaMLO-A1, TaMLO-B1, and TaMLOD1	Powdery mildew resistance	Wang et al., 2014
Cucumber	Biotic	eIF4E (eukaryotic translation initiation factor 4E)	Cucumber vein yellowing virus (CVW), Zucchini yellow mosaic virus (ZYMV) and Papaya ring spot mosaic virus type-W (PRSV-W)	Chandrasekaran et al., 2016
Maize	Abiotic	ARGOS8	Increased grain yield under drought stress	Shi et al., 2017

(continued)

TABLE 4.10 Summary of crops modified to be resistant to biotic and abiotic stress *(continued)*

Crop	Type	Target gene	Stress/trait	Reference
Tomato	Abiotic	SlMAPK3	Drought tolerance	Wang et al., 2017
A. thaliana	Abiotic	UGT79B2, UGT79B3	Susceptibility to cold, salt, and drought stresses	
A. thaliana	Abiotic	MIR169a	Drought tolerance	Zhao et al., 2016
A. thaliana	Abiotic	OST2 (OPEN STOMATA 2) (AHAI)	Increased stomatal closure in response to abscisic acid (ABA)	Osakabe et al., 2016
Rice	Abiotic	OsPDS, OsMPK2, OsBADH2	Involved in various abiotic stress tolerance	Shan et al., 2013
Rice	Abiotic	OSMPK5	Various abiotic stress tolerance and disease resistance	Xie and Yang, 2013
Rice	Abiotic	OsMPK2, OsDEP1	Yield under stress	Shan et al., 2014
Rice	Abiotic	OsDERF1, OsPMS3, OsEPSPS, OsMSH1, OsMYB5	Drought tolerance	Zhang et al., 2014
Rice	Abiotic	OsAOX1a, OsAOX1b,OsAOX1c, OsBEL	Various abiotic stress tolerance	Xu et al., 2015
Rice	Abiotic	OsHAK- 1	Low cesium accumulation	Cordones et al., 2017
Rice	Abiotic	OsPRX2	Potassium deficiency tolerance	Mao et al., 2018
Rice	Nutritional	25604 gRNA for 12802 genes	Creating genome-wide mutant library	Meng et al., 2017
Maize	Nutritional	Zm1PKIA Zm1PK and ZmMRP4	Phytic acid synthesis	Liang et al., 2014
Wheat	Nutritional	TaVIT2	Iron content	Connorton et al., 2017
Soybean	Nutritional	GmPDS11 and GmPDS18	Carotenoid biosynthesis	Du et al., 2016
Tomato	Nutritional	Rin	Fruit ripening	Ito et al., 2015
Potato	Nutritional	ALS1	Herbicide resistance	Butler et al., 2016
Cassava	Nutritional	MePDS	Carotenoid biosynthesis	Odipio et al., 2017

Speeding up hybrid breeding

Hybrid breeding is a powerful approach as it enables the creation of outsprings with the desired characteristics from two different breeds. Gene editing using technologies such as CRISPR-Cas9 have proven to be an effective and fast way to enhance desired traits. CRISPR-Cas9 and hybrid breeding has allowed scientists to shorten the growth time for a plant, enabling it to reach maturity earlier, increasing crop yields in the long term.

FIGURE 4.62 Speed breeding accelerates generation time of major crop plants for research.

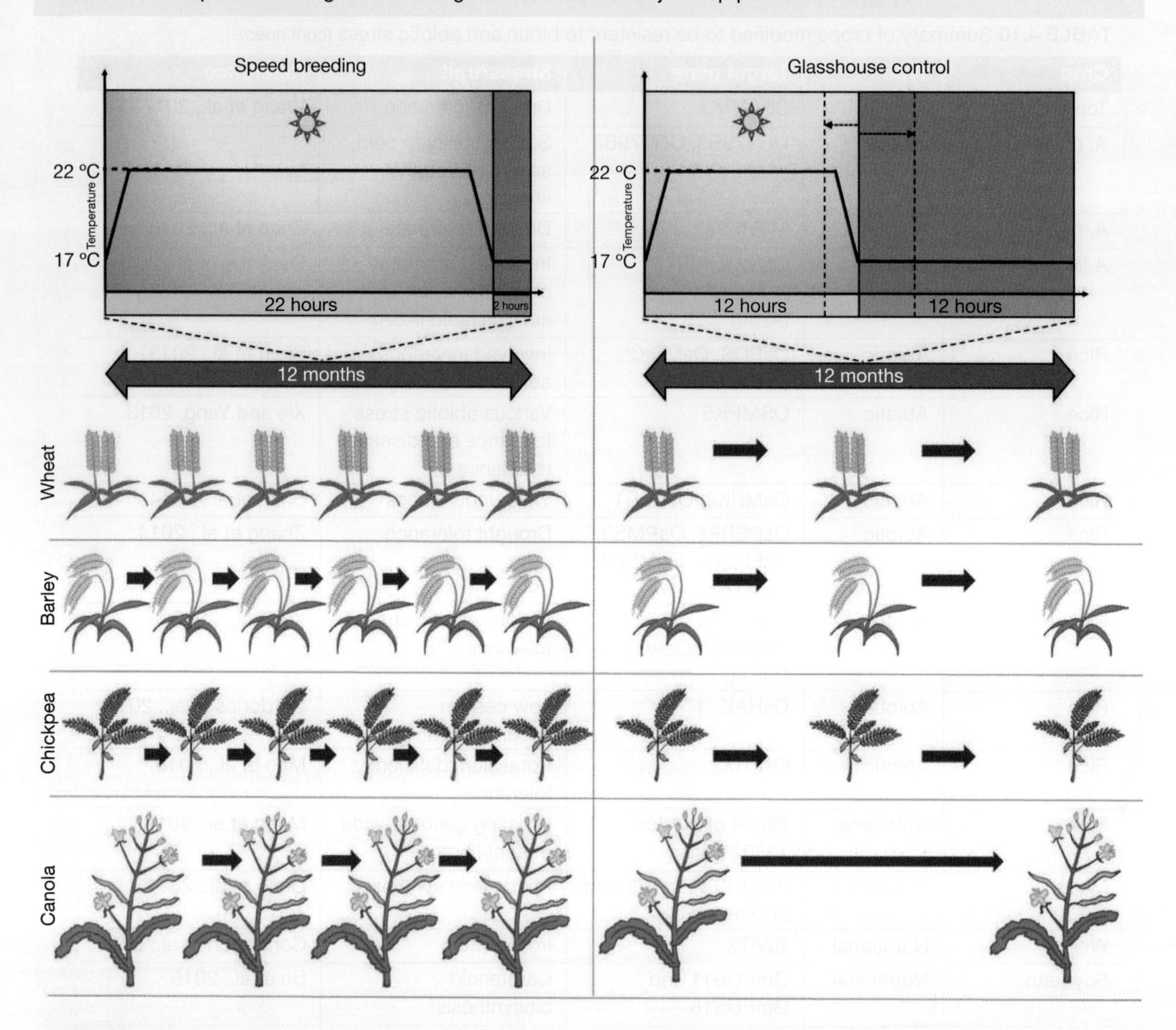

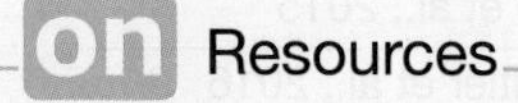

Resources

eWorkbook Worksheet 4.8 CRISPR-Cas9 and photosynthesis (ewbk-7589)

Weblinks CRISPR for crop improvement: an update review

The power of CRISPR-Cas9-induced genome editing to speed up plant breeding

CRISPR genome editing to address food security and climate change

KEY IDEAS

- CRISPR-Cas9 technologies involve the editing of genomes. This may include adding genes, replacing genes or deleting genes.
- Knocking out genes that harbour undesirable traits is the simplest and most common application of CRISPR-Cas9.
- Many improvements to photosynthetic efficiencies can be made using CRISPR-Cas9, enhancing the amount of sugar such as glucose produced.
- Traits and techniques that can be improved using CRISPR-Cas9 include crop yield and quality, biotic and abiotic stress resistance and hybrid breeding.

4.8 Activities

To answer questions online and to receive **immediate feedback** and **sample responses** for every question, go to your learnON title at **www.jacplus.com.au**. A **downloadable solutions** file is also available in the resources tab.

4.8 Quick quiz on	4.8 Exercise	4.8 Exam questions

4.8 Exercise

1. MC What is Cas9 and what is its role in plant genomes?
   A. An RNA molecule that binds to target DNA via complementary base pairing
   B. A DNA sequence that binds the Cas9 protein
   C. A viral protein that disrupts bacterial membranes
   D. An enzyme that cuts both strands of DNA at sites specified by an RNA guide
2. What is the main advantage of using CRISPR-Cas9 for genome editing in agriculture over traditional methods?
3. Identify and describe two different scenarios in which CRISPR-Cas9 has been shown to be beneficial in agriculture.
4. How does CRISPR differ from previous gene editing techniques?
5. Gene editing technology such as CRISPR-Cas9 remains unregulated due to government policies. Propose two specific measures to help regulate and keep pace with CRISPR-Cas9 technology. In explaining these measures, include details of the CRISPR-Cas9 technology and its applications.

4.8 Exam questions

Question 1 (1 mark)

MC The main gene that CRISPR-Cas9 targets to improve photosynthetic efficiencies in C_3 plants is

A. PEP carboxylase.
B. glucose-6-P dehydrogenase.
C. Rubisco.
D. alcohol dehydrogenase.

Question 2 (1 mark)

MC Potatoes (*Solanum tubersum*) are a major crop globally. Potatoes are particularly prone to infections by microorganisms and viruses. One microorganism, *Pytophthora infestans*, causes a serious disease in potatoes known as 'potato blight'. This disease has historically led to famine. The most well-known famine occurred in Ireland, between 1845 and 1852. In this famine, the population of Ireland dropped by approximately 20 per cent, due to both death and emigration.

Which of the following would not be a beneficial use of CRISPR-Cas9 to confer resistance to *Pytophthora infestans*?

A. Removing a gene that causes the *Solanum tubersum* to be susceptible to the *Pytophthora infestans*
B. Adding a new gene that enhances the resistance to *Pytophthora infestans*
C. Replacing an allele that causes susceptibility with one that enables resistance
D. Adding a region that upregulates the transcription of a gene that allows *Pytophthora infestans* to infect the plant

Question 3 (1 mark)

MC Which of the following is not true regarding CRISPR-Cas9?

A. CRISPR Cas-9 requires guide RNA to help target a specific gene.
B. It can improve photosynthetic efficiencies by altering genes responsible for the process of photorespiration.
C. It cannot be used to change the structure of enzymes.
D. It can upregulate genes to allow plants to reach maturity earlier.

Question 4 (7 marks)

Rice (*Oryza sativa*) is a food staple for billions of individuals globally. Rice naturally contains small amounts of cadmium, which is usually taken up by the plant through soil. While usually only a small amount of cadmium remains in the body, long-term consumption can lead to negative effects, such as organ damage and kidney disease.

A certain gene known as *OsNramp5* is found in rice. This produces a protein that enables the uptake of metals such as cadmium for soil. CRISPR-Cas9 technologies has been used to target this gene in a subspecies of rice, referred to as *indica* rice.

a. Explain how CRISPR-Cas9 may be used to knock out the *OsNramp5* gene. **3 marks**
b. Outline the benefits that this may have for individuals. **2 marks**
c. Suggest two possible concerns related to the use of CRISPR-Cas9 in modifying *indica* rice. **2 marks**

Question 5 (4 marks)

a. Identify two examples of how CRISPR-Cas9 can be used to improve crop yields. **2 marks**
b. Why is it important that CRISPR-Cas9 is able to target a specific location, rather than randomly integrating DNA into the genome of plants? **2 marks**

More exam questions are available in your learnON title.

4.9 Uses and applications of the anaerobic fermentation of biomass

KEY KNOWLEDGE

- Uses and applications of anaerobic fermentation of biomass for biofuel production

Source: VCE Biology Study Design (2022–2026) extracts © VCAA; reproduced by permission.

4.9.1 Reviewing the applications of anaerobic fermentation

As previously explored, in most eukaryotic organisms, and in the cells of most tissues, cellular respiration can occur only if oxygen is available; this type of cellular respiration is termed aerobic respiration.

However, some organisms that live in oxygen-free (anoxic) environments, including some microbes (such as certain bacteria and yeast) must produce energy (ATP) in the absence of oxygen through a process known as anaerobic fermentation.

Anaerobic fermentation is another pathway for breaking down glucose. This process has many industrial applications, including yeast used in the manufacturing of alcoholic beverages such as wines and beers. Energy obtained through fermentation can only occur through glycolysis.

4.9.2 What is biomass?

Microbes are present in the air that we breathe, the soil and water around us, decaying organic matter, the food we eat and the water we drink. Because they are not visible to an unaided eye, we are far less aware of the presence of microbes than the presence of macroscopic organisms. The number of microbes is much greater than the number of plants and animals on Earth. These microbes are vital in the breakdown and fermentation of **biomass**, the organic matter from living things such as plants and animals.

biomass the organic material from plants and animals; it is a renewable source of energy.

Biomass contains stored energy from the Sun and energy stored by plants through the process of photosynthesis. Some sources are shown in figure 4.63. When biomass is burned, heat is released, allowing for the production of electricity. This biomass may also be converted through fermentation by microbes to liquid **biofuels** or biogas that can be burned as fuels.

FIGURE 4.63 Types of biomass that can be used

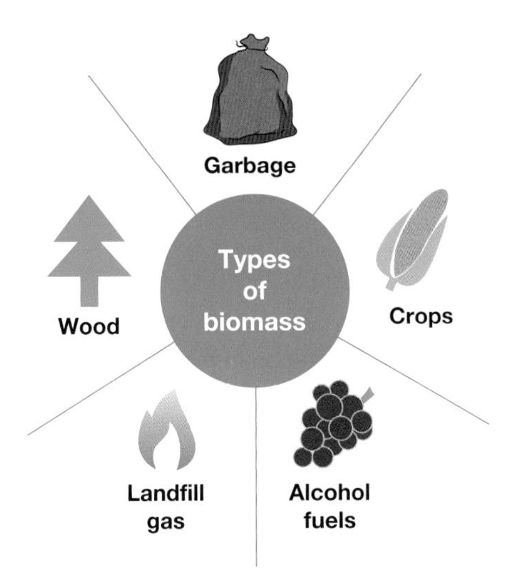

4.9.3 The anaerobic fermentation of biomass and applications

Anaerobic fermentation is a biological process that converts solid or liquid biomass to a gas (biogas) in the absence of oxygen. Biogas is predominantly a mixture of methane (60–65 per cent), carbon dioxide (30–35 per cent), water vapour (4–5 per cent) and small proportions of other gases (e.g. hydrogen sulphide).

Any solid or liquid waste residues can be used as compost and fertilisers. Anaerobic fermentation is a very effective method of treating high moisture content organic wastes, and many implementations of anaerobic fermentation are driven by waste management needs, with biogas as a valuable by-product. Feedstocks suitable for anaerobic fermentation include sewage sludge, agricultural and industrial organic wastes, and animal by-products.

Fermentation is an anaerobic process whereby glucose is broken down through one of the following pathways:

$$\text{glucose} \rightarrow \text{pyruvate} \rightarrow \text{acetaldehyde} \rightarrow \text{ethanol}$$

$$\text{glucose} \rightarrow \text{pyruvate} \rightarrow \text{lactic acid}$$

When yeast or bacteria are added to the biomass material, these organisms usually break down the sugars to produce ethanol and carbon dioxide. The ethanol produced is distilled in order to obtain a higher concentration of alcohol to achieve the required purity for the use as biofuel. The by-products left over from the fermentation process can be used as animal feed.

Fermentation of forest and industrial residues

Raw materials containing sugars can be used as fermentation substrates. The fermentable raw materials can be grouped as directly fermentable sugary materials, starchy, lignocellulosic (plant dry matter — biomass) materials and industrial wastes. Materials containing a sugar backbone require the least costly pre-treatment, whereas materials constructed from starchy, lignocellulosic materials and household wastes require costly pre-treatment in order to convert them into fermentable substrates that can be broken down into biofuels.

biofuel any fuel source derived from biomass

Fermentation of agricultural wastes

Unfortunately, crops such as corn and sugarcane are unable to meet the global demand for bioethanol production, as these crops' primary purpose is to produce food and animal feed. Therefore, agricultural wastes are an attractive option for bioethanol production. Agricultural wastes are cost effective, renewable and abundant. Bioethanol from agricultural waste is a promising technology, although the process has several challenges and limitations, such as biomass transport and handling, and efficient pre-treatment methods in order to breakdown the plant matter.

The goals of an effective pre-treatment process are:

1. to form sugars directly or subsequently by hydrolysis
2. to avoid loss and/or degradation of sugars formed
3. to limit formation of inhibitory products
4. to reduce energy demands
5. to minimise costs.

FIGURE 4.64 Outlining the cycle of biomass fermentation in the production of energy

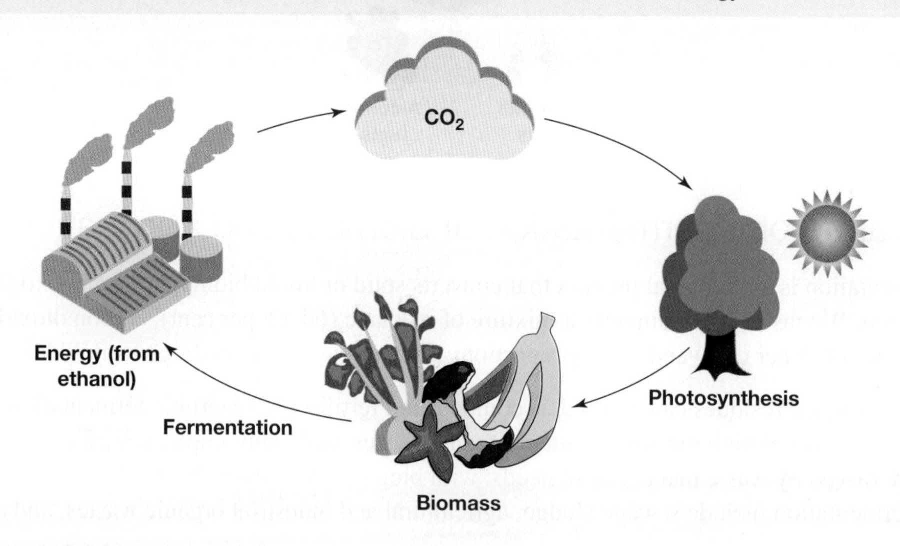

4.9.4 Applications and uses of biomass

There are many sources of biomass and ways it can be applied and used for energy.

Wood and agricultural products

Most biomass currently used comes from grown energy. Wooden products, such as logs, chips, bark and sawdust, account for a large amount of biomass. Other biomass sources can include agricultural waste products, such as fruit pits and corn cobs. Wood and wood waste are used to generate electricity. Paper mills and sawmills use much of their waste products in order to generate steam and electricity for their use. However, since they use so much energy, they need to buy additional electricity from fossil fuels.

FIGURE 4.65 A biomass energy factory in Lithuania using wood biomass as an energy source

Garbage waste

Burning garbage is a way to turn household waste into a usable form of energy. One thousand kilograms of garbage contains about as much heat energy as

220 kilograms of coal. Garbage is not all biomass; perhaps half of its energy content comes from plastics, which are made from petroleum and natural gas. Power plants that burn garbage for energy are called *waste-to-energy plants*. These plants generate electricity much as coal plants do, except that combustible garbage — not coal — is the fuel used to light their steam boilers. Garbage waste can also go into landfill, where it undergoes decomposition by bacteria and fungi.

Landfill gas and biogas

Bacteria and fungi decompose dead plants and animals, causing them to rot or decay. When waste is disposed in landfills, all the oxygen is quickly used up, and fermentation by bacteria and fungi occurs. This involves the decomposition of the biomass, producing methane as a by-product. New regulations require landfill to collect methane gas for safety and environmental reasons. Methane gas is highly explosive when ignited. Landfill facilities are able to collect the methane gas, purify it and use it as fuel. Methane can also be produced using energy from agricultural and human wastes. Biogas digesters (also known as a *biogas plant*) are airtight containers, where waste put into the containers is fermented without oxygen to produce a methane-rich gas. This gas can be used to produce electricity, or for cooking and lighting.

FIGURE 4.66 An example of part of a biogas plant

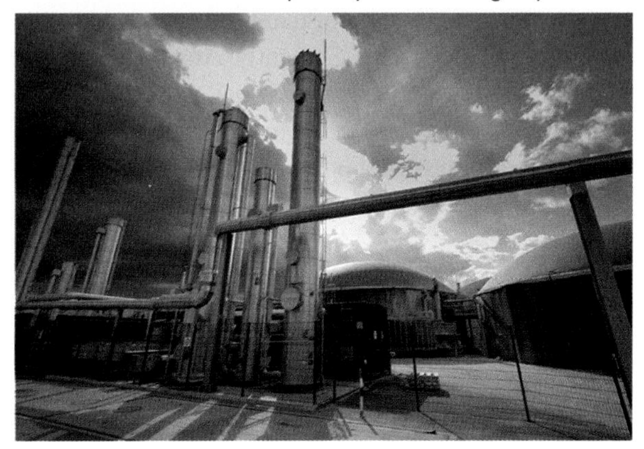

Ethanol

Ethanol-based fuels are manufactured from fermented sugars (starches and glucose) found in photosynthetic plants. These sugars are a product of photosynthesis and are then able to be fermented by bacteria such as *Zymomonas mobilis* to produce ethanol as a by-product. In Australia, bioethanol is produced from starch waste of sorghum. This is a type of grain used in the manufacturing of breakfast cereals and molasses. There have been some investigations into using cellulose to produce ethanol.

FIGURE 4.67 E10 is the most commonly available ethanol-blended fuel in Australia.

Two types of ethanol-blended fuels are produced in Australia: Ethanol-10 (E10) and Ethanol 85 (E85). E10 fuel contains 10 per cent ethanol and is the most commonly available ethanol fuel. E85 contains 85 per cent ethanol and has been designed for specialised and modified vehicles, and is available at selected petrols stations around Australia.

Biodiesel

Biodiesel fuels in Australia are made from animal fats (primarily made up of triglycerides), recycled cooking oil and a variety of vegetable oils. There has been a sharp decline in the use of biodiesel in Australia since 2015 — only an estimated 40 million litres of biodiesel was produced in Australia between 2017 and 2018. The sharp decline in biodiesel manufacturing is attributed to several factors, including a decline in the international oil price and insufficient tax relief for producers. Australia currently produces two biodiesel blends: B5 (5 per cent blend of biodiesel) and B20 (20 per cent blend).

A company in north-west Victoria is paving the way in biodiesel production. Barnawartha Biodiesel has the capacity to produce 50 million litres of biodiesel each year. There are currently four companies that produce biodiesel in Australia. Combined they have the capacity to produce 170 million litres of biodiesel annually.

Biofuel and biomass in Australia

Australia is fortunate that the production of biofuel does not create any conflicts between land requirements and crop growth, unlike some other countries. Australia's biofuel is produced from wastes and by-products, which has prevented conflict between the agricultural and fuel industries.

In Australia, it is the aim of many states to enhance their use of renewable energy such as biomass for the generation of electricity. Currently, Australia is behind many other countries with its use of bioenergy. In Victoria, there was a target of 25 per cent of all electricity being produced by biomass by 2020, and 40 per cent by 2025. Most biomass in Victoria comes from wood and wood waste. The Australian territory that is leading the way in biomass fuel production is the ACT. Currently in Australia the main use of biomass is in heating.

FIGURE 4.68 Bioenergy production sites in Victoria

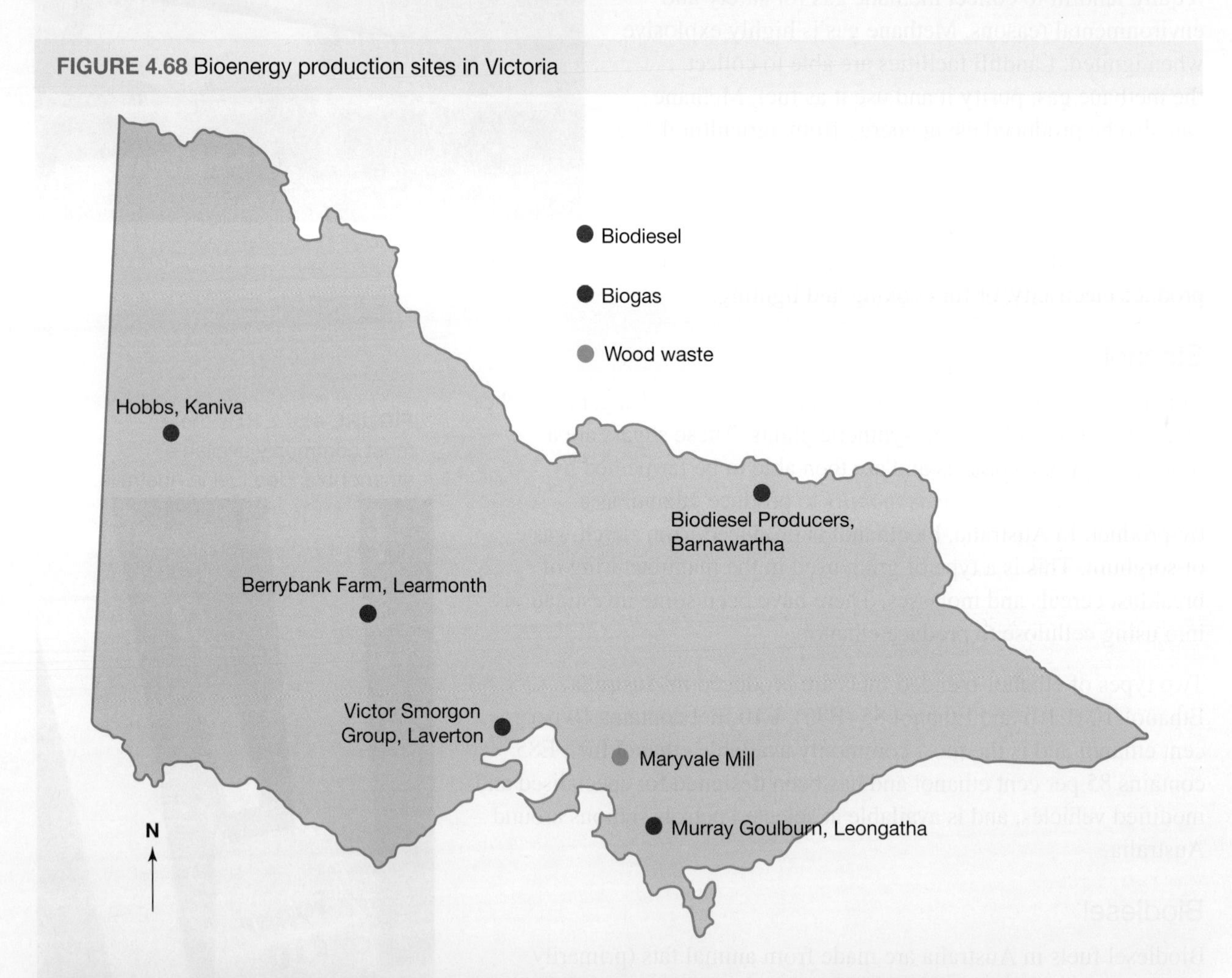

elog-0844

INVESTIGATION 4.6

online only

Fermentation of biomass

Aim

To investigate how biomass can be used for fermentation to produce biofuels

4.9.5 Sustainability and biofuel production

According to the International Energy Agency, Australia's biofuel growth will emerge from the production of advanced biofuels (plant-based wastes from agricultural industries). Along with financial and technological challenges, sustainability concerns need to be addressed, such as:

- *food versus fuel*. With the world's population expected to increase, there needs to be a balance between increased production of biofuels and food availability. Biofuels will need to be made using waste products from plants, such as corn leaves, stalks and cobs. Further technological developments are required to strike this balance.
- *land requirements*. To meet the 2060 target for biofuel production it is estimated an additional 100 million hectares of land will be required. Currently the biofuel industry accounts for 30 million hectares of arable land globally. This increase in demand for land for fuel is expected to occur at the same time as the global population increases.
- *energy efficiency*. Creating biofuels from feedstocks is currently an inefficient process. The energy required to produce and then collect the raw materials required is far greater than the fuel produced. For biofuels to be sustainable, biofuel production needs to become more energy efficient.

Resources

eWorkbook Worksheet 4.9 Biomass and anaerobic fermentation (ewbk-7591)

Weblinks Geoscience Australia — Bioenergy
Biomass opportunities in Victoria
Environment, Land, Water and Planning — Bioenergy

KEY IDEAS

- Anaerobic fermentation is an anoxic process that occurs in various organisms in which organic molecules (such as glucose) are broken down to produce ATP and other by-products such as carbon dioxide and ethanol.
- Biomass is organic material from plants and animals, and it is a renewable source of energy.
- The anaerobic fermentation of biomass can be used in biofuel production.
- Examples of sources of biomass include garbage waste, wood products, biogas, ethanol and biodiesel.
- By using fermentation, we can greatly enhance how biomass is used for bioenergy production.
- Bioenergy is a renewable energy source, and can be applied not only to energy efficiency, but also to other applications, such as land requirements and food.

4.9 Activities

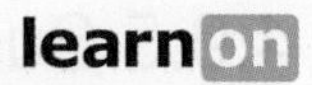

To answer questions online and to receive **immediate feedback** and **sample responses** for every question, go to your learnON title at **www.jacplus.com.au**. A **downloadable solutions** file is also available in the resources tab.

4.9 Quick quiz	4.9 Exercise	4.9 Exam questions

4.9 Exercise

1. Describe what is meant by the term *biomass* and provide two examples of biomass.
2. How can the fermentation of biomass be used to produce biofuel?
3. Outline two clear advantages of using biomass and two disadvantages.
4. Some scientists and researchers have said that biomass is the future of energy in Australia. Do you agree with this statement? Justify your response.
5. Describe two applications each for ethanol and carbon dioxide that is produced through the fermentation of biomass.

4.9 Exam questions

Question 1 (1 mark)

MC Biomass is

A. organic material from plants and animals.
B. a non-renewable source of energy.
C. only produced as a by-product of photosynthesis.
D. only produced in the presence of oxygen.

Question 2 (1 mark)

MC Which of the following is not a source of biomass?

A. Ethanol
B. Biogas and landfill gas
C. Wood products
D. Carbon dioxide

Question 3 (1 mark)

MC Bacteria and fungi are able to decompose dead plants and animals, often converting complex sugars to simpler forms. As organisms decay, they produce an odourless and colourless gas that is highly explosive if ignited. This gas can be collected and used as fuel in biogas plants. The likely identity of this gas is

A. carbon dioxide.
B. oxygen.
C. methane.
D. propane.

Question 4 (4 marks)

a. While more sustainable than other sources, there are some concerns about the use of biomass and its sustainability. Outline two considerations that need to be made in regards to the sustainability of the use of biomass. **2 marks**
b. Will we ever run out of biomass? Justify your response. **2 marks**

Question 5 (6 marks)

a. Ethanol is an example of a biofuel. What is meant by the term biofuel? **1 mark**
b. In Australia, a specific grain known as sorghum is used in the production of biomass. Sorghum provides starch waste for the process of fermentation. Starch is made of monomers of glucose. Outline the process in which ethanol is produced from sorghum waste. **3 marks**
c. Explain two advantages of using ethanol rather than petrol. **2 marks**

More exam questions are available in your learnON title.

4.10 Review

4.10.1 Topic summary

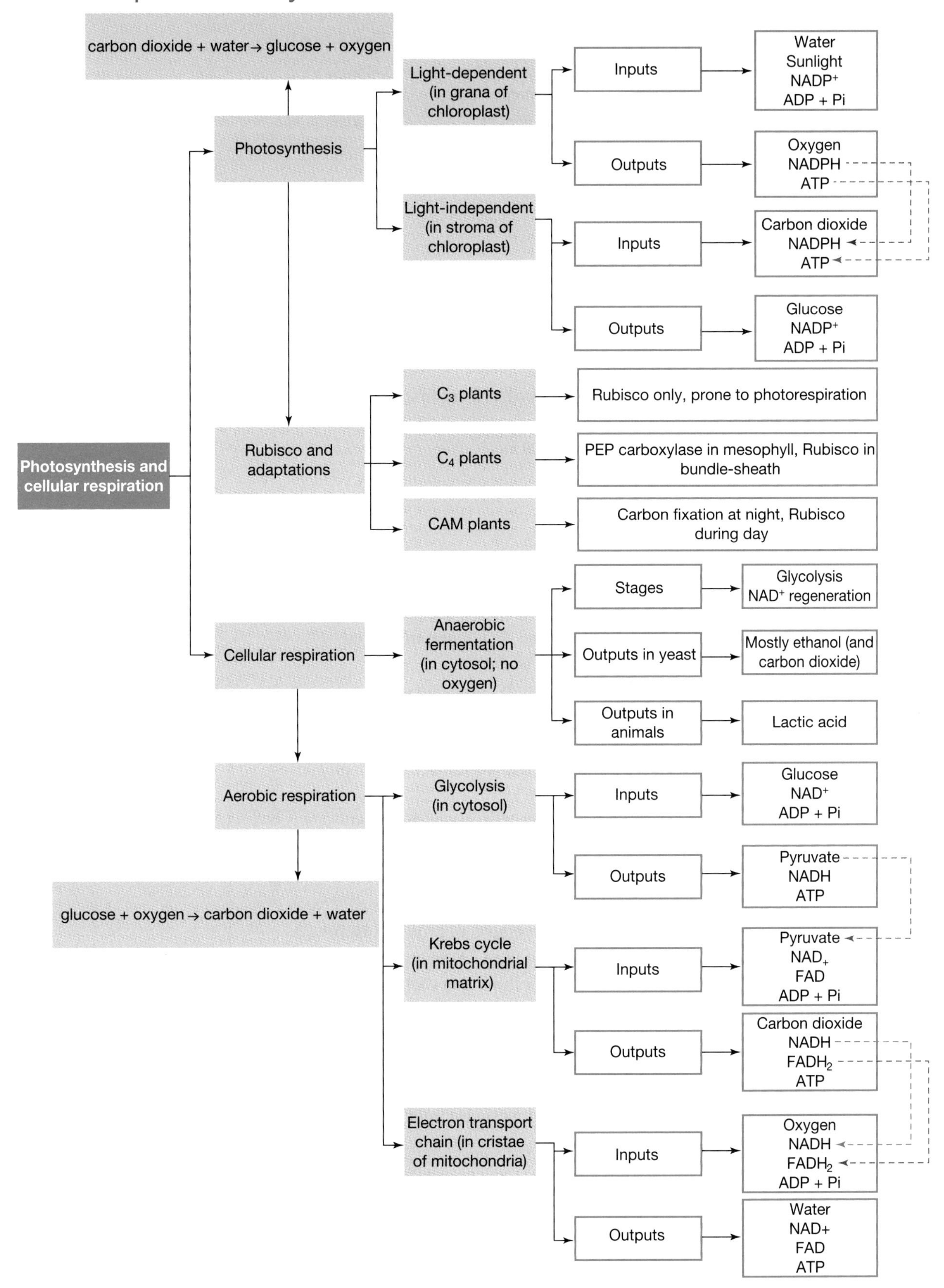

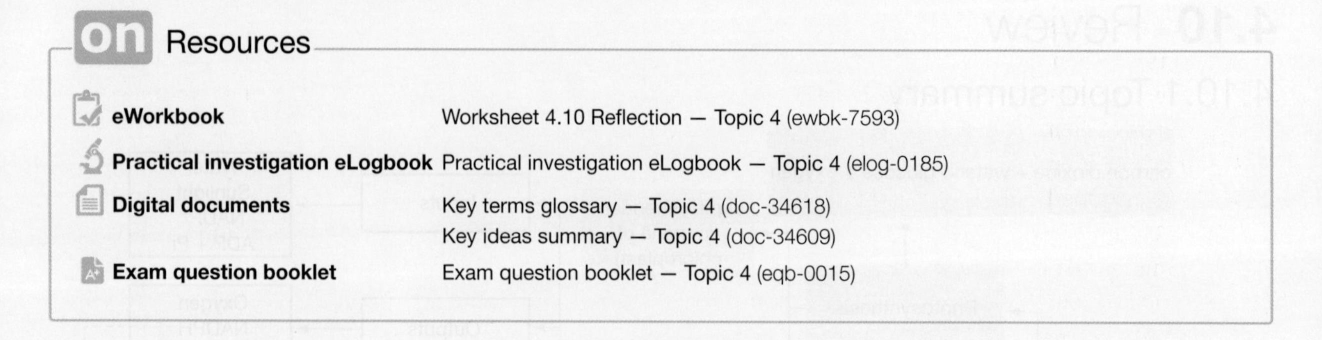

4.10 Exercises

learnon

To answer questions online and to receive **immediate feedback** and **sample responses** for every question, go to your learnON title at **www.jacplus.com.au**. A **downloadable solutions** file is also available in the resources tab.

4.10 Exercise 1: Review questions

1. Put the following events in photosynthesis in order, from first to last:
   A. Generation of ATP
   B. Entry of carbon dioxide into leaf via stomata
   C. Trapping of sunlight energy
   D. Fixation of carbon
   E. Production of glucose
   F. Splitting of water molecules
   G. Release of 3-C molecules from the Calvin cycle
2. Live yeast was mixed with flour and water to make a dough. The dough was kept at a constant temperature of 30 °C and its volume measured every five minutes. The graph of these results is shown.
   a. What was the increase in the volume of dough at the end of 30 minutes?
   b. What was the percentage increase in the volume of dough during the 10-minute interval between the 30-minute and 40-minute stages of the experiment?
   c. What kind of cellular respiration would be occurring in the yeast cells?
   d. Explain the cause of the increase in volume of the dough.

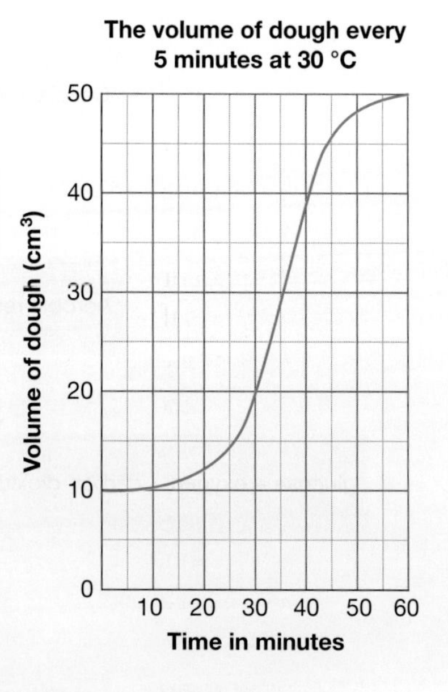

3. Student P said that the process of carbon fixation is 'the synthesis of ATP through the actions of the enzyme ATP synthase to produce useful energy.' Student Q said, 'No. It's the conversion of CO_2 to an organic compound'. Student R said, 'That's not right. I think that carbon fixation is the input to the Calvin cycle.'
   a. Which student(s) most accurately identified carbon fixation?
   b. Where you considered a student's description to be inaccurate, identify the inaccuracy involved.
   c. In which stage of photosynthesis does carbon fixation occur?
   d. Is the process of carbon fixation powered directly by light energy or indirectly by light energy? Explain your decision.
4. Identify the locations where the following processes or events occur in the chloroplast of a leaf cell:
   a. carbon fixation
   b. light trapping
   c. release of oxygen
   d. input of carbon dioxide
   e. production of glucose in C_4 plants
   f. splitting of water.

5. Formulate a possible valid biological explanation for the following observations:
   a. Farmers managing crops growing in glasshouses increased their yields by extending the period over which photosynthesis takes place. The farmers who chose to provide this light by burning paraffin oil lamps in their glasshouses obtained better results than the farmers who used standard electric-powered lighting. (Paraffin oil produces CO_2 when burnt.)
   b. The enzyme Rubisco is regarded as the most abundant enzyme on this planet.
   c. A tomato grower increased the yield of his glasshouse-grown hydroponic tomatoes by using artificial lighting after daylight hours.
   d. Aquatic plants growing in an aquarium under bright lights were observed to produce bubbles.
   e. A cheetah chasing a gazelle has to stop its pursuit after a distance of about 100 metres.
   f. Carbon monoxide (CO) competes with oxygen for active sites on haemoglobin. High levels of CO can lead to death.
6. Examine the graph shown.
   a. In which stage of photosynthesis is carbon dioxide used?
   b. What might have caused the rate of photosynthesis to plateau?
   c. Suggest why the rate of photosynthesis in the plant cells reaches higher values when illuminated by high-intensity light compared with that achieved under low-intensity lighting.

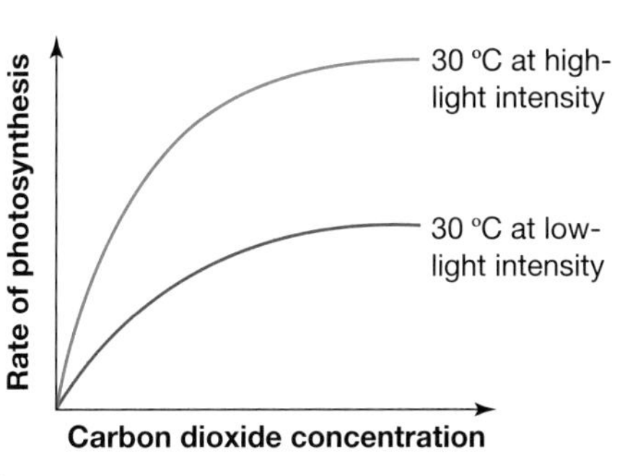

7. Consider a molecule of glucose ($C_6H_{12}O_6$) produced through photosynthesis in a leaf cell:
   a. What is the original source of the carbon atoms (C) in glucose?
   b. What is the original source of the hydrogen (H) atoms in glucose?
   c. How and where was this hydrogen produced?
   d. Which loaded carrier molecule transports hydrogen into the reactions involving production of glucose?
   e. Oxygen is present in carbon dioxide (CO_2) and oxygen is present in water (H_2O). From which source does the oxygen in glucose come? Explain.
8. Consider a fish tank containing several fish and some aquatic plants. Make a diagram showing one fish and one aquatic plant in a water-filled tank, outlining the inputs and outputs related to cellular respiration and photosynthesis and any cycling that might occur:
   a. when the tank is brightly illuminated
   b. when the tank is in darkness.
9. In the mitochondria of liver cells, the folds of the cristae increase the surface area of the inner membrane to about five times that of the outer membrane. What benefit, if any, might this increase in surface area bring?
10. Identify the following statements as true or false. Where you judge a statement to be false, rewrite it so that it would be judged to be true.
   a. Cellular respiration occurs all the time in animal cells but only in the dark in plant cells.
   b. Anaerobic fermentation in mammals yields half the amount of ATP compared to that produced in aerobic respiration.
   c. Carbon fixation is the breakdown of glucose to carbon dioxide.
   d. The easiest way of measuring the rate of aerobic respiration in an animal would be to measure the rate of disappearance of glucose.
11. Cellular respiration is a vital process in humans.
   a. Define the process of aerobic respiration.
   b. Create a table outlining the inputs and outputs of the various stages of aerobic respiration.

   Cyanide is a well-known toxin that inhibits a primary mitochondrial enzyme known cytochrome *c* oxidase. This enzyme is fundamental in the electron transport chain.
   c. What location in the mitochondria would you expect cytochrome *c* oxidase to be located?
   d. Explain why cyanide poisoning can lead to death if untreated.

12. There are many technologies that can be used to enhance the efficiency of photosynthesis. One such example of this is through CRISPR-Cas9.
    a. What is CRISPR-Cas9?
    b. How might CRISPR-Cas9 be used to improve photosynthetic efficiencies?
    c. CRISPR-Cas9 can also improve crop yields. Outline two examples of this.
13. Write a summary outlining how the anaerobic fermentation of biomass can be used for biofuel production. In your response, you should include at least two examples and outline the pros and cons of this process.

4.10 Exercise 2: Exam questions

on Resources

Teacher-led videos Teacher-led videos for every exam question

Section A — Multiple choice questions

All correct answers are worth 1 mark each; an incorrect answer is worth 0.

Question 1

Source: *VCAA 2008 Biology Exam 1, Section A, Q6*

The following graph shows the relationship between light intensity and net oxygen uptake or output by a particular green plant.

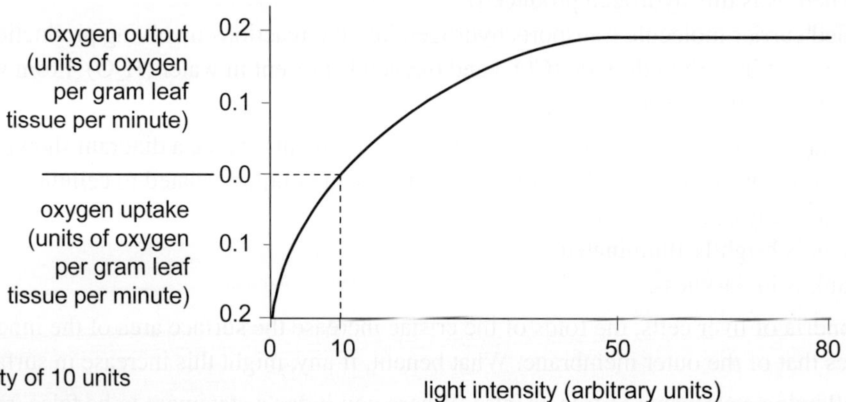

At a light intensity of 10 units

A. the rate of photosynthesis is zero.
B. the rate of aerobic respiration is zero.
C. oxygen produced by photosynthesis is equal to the oxygen used by aerobic respiration.
D. oxygen produced by photosynthesis is equal to twice the oxygen used by aerobic respiration.

Question 2

Source: *VCAA 2016 Biology Exam, Section A, Q11*

Which one of the following statements about photosynthesis in chloroplasts is correct?

A. The grana are the site of the light-independent stage.
B. Chlorophyll found in the stroma traps light for use during the light-dependent stage.
C. The light-dependent stage produces ATP for use during the light-independent stage.
D. The light-independent stage captures carbon dioxide for use during the light-dependent stage to produce glucose.

Question 3

Source: *VCAA 2017 Biology Exam, Section A, Q15*

A variegated leaf from a plant is shown below. Cells from sections M and K were examined and simple sketches were produced.

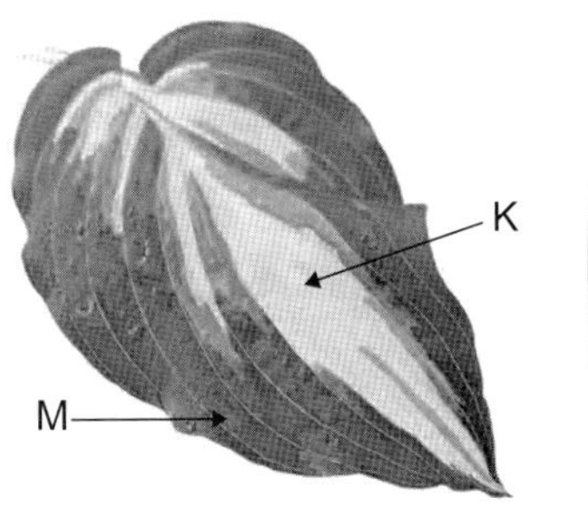

A typical cell from section M

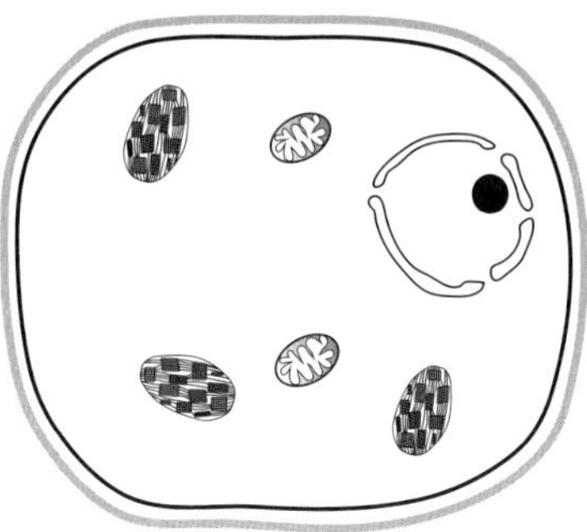

A typical cell from section K

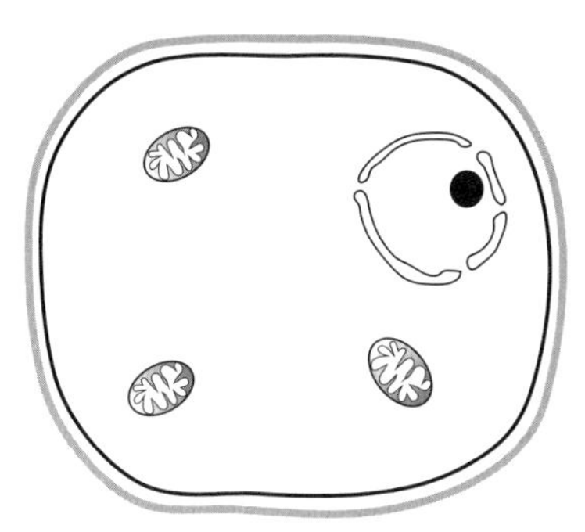

Source: Le Do/Shutterstock.com

From this information, it can be concluded that

A. cells in section K would be unable to carry out aerobic respiration.
B. light-independent reactions of photosynthesis can occur in cells from section K.
C. there is chlorophyll present in cells from section K but not in cells from section M.
D. glucose would be manufactured from carbon dioxide and water in cells from section M but not in cells from section K.

Question 4

Source: *VCAA 2018 Biology Exam, Section A, Q8*

The diagram below shows a section through a part of a mitochondrion.

R outside mitochondrion
inside mitochondrion
S
U
T

The sites of the pathways in aerobic respiration are

A. R – glycolysis, S – Krebs cycle, T – electron transport chain.
B. U – glycolysis, T – Krebs cycle, R – electron transport chain.
C. R – glycolysis, U – Krebs cycle, T – electron transport chain.
D. T – glycolysis, R – Krebs cycle, S – electron transport chain.

Question 5

Source: Adapted from VCAA 2007 Biology Exam 1, Section A, Q7

A student was asked to identify differences between the overall processes of photosynthesis and aerobic respiration in eukaryotic cells. The student prepared the table below to outline the differences.

The only correct comparison listed by the student is

	Photosynthesis	Aerobie respiration
A.	Exergonic	Endergonic
B.	All stages occur within chloroplasts	All stages occur within mitochondria
C.	Electron transport not involved	Electron transport involved
D.	Uses water as a reactant in the first stage	Forms water as a product in the final stage

Question 6

Source: VCAA 2011 Biology Exam 1, Section A, Q20

The reaction ADP + Pi $\longrightarrow$ ATP

A. is irreversible.
B. occurs without the presence of enzymes.
C. occurs in yeast cells during fermentation.
D. only occurs in cells containing mitochondria.

Question 7

Oxygen is produced

A. during the light-independent reaction.
B. during the light-dependent reaction.
C. during the Krebs cycle.
D. in the stroma of the chloroplast.

Question 8

Rubisco is most important in

A. C_3 plants.
B. C_4 plants.
C. CAM plants.
D. all of the above.

Question 9

Mammalian cells in tissue E are observed to have much higher numbers of mitochondria than cells in tissue F. It is reasonable to predict that, over the same time period, tissue E will

A. produce more glucose than tissue F.
B. have a higher demand for oxygen than tissue F.
C. produce less ATP than tissue F.
D. take up less carbon dioxide than tissue F.

Question 10

During bread making, yeast is used is order to allow the bread to rise. This occurs because

A. yeast undergoes aerobic respiration and is producing water.
B. yeast undergoes anaerobic fermentation and is producing carbon dioxide.
C. yeast undergoes glycolysis and is producing pyruvate.
D. yeast undergoes photosynthesis and is producing oxygen.

Section B — Short answer questions

Question 11 (7 marks)

Source: *Adapted from VCAA 2013 Biology Section B, Q1*

Yeast is a single-celled, microscopic fungus that uses sucrose as a food source. An experiment was carried out to investigate cellular respiration by a particular species of yeast.

Yeast cells were placed in a container and a sucrose solution was added. An airtight lid was placed on the container. The percentages of oxygen and ethanol in the container were recorded over a one-hour period. The experiment was carried out at room temperature. The results are shown in the following table.

	Percentage of oxygen	Percentage of ethanol
At the start of the experiment	21	0
at the end of the experiment	18	4

a. Explain any changes that have been observed in oxygen and ethanol levels within the airtight container. **2 marks**

Levels of carbon dioxide were also monitored during the experiment.

b. Predict whether the carbon dioxide concentration inside the airtight container would increase, stay the same or decrease within the time the experiment was carried out. Explain the reasoning behind your prediction. **2 marks**

Scientists are looking at ways to increase the efficiency of photosynthesis in plants, including the way in which carbon dioxide is captured.

c. i. Name the stage of photosynthesis in which carbon dioxide is captured. **1 mark**

ii. The stage of photosynthesis in which carbon dioxide is captured requires other inputs. Name two other inputs and describe the role played by each in this stage of photosynthesis. **2 marks**

Question 12 (5 marks)

Source: *VCAA 2010 Biology Exam 1, Section B, Q3*

Elysia chlorotica is a bright green sea slug, with a soft leaf-shaped body. It has a life span of 9 to 10 months. This sea slug is unique among sea slugs as it is able to survive on solar power.

E. chlorotica acquires chloroplasts from the algae it eats, and stores them in the cells that line its digestive tract.

Young *E. chlorotica* fed with algae for two weeks can survive for the rest of their lives without eating.

a. What is the product of photosynthesis that provides the energy that enables *E. chlorotica* to survive for so long without eating? **1 mark**

b. The product of photosynthesis must undergo a three-stage process for the slug to access the energy in the product. Name and give a brief description of each of these stages. **3 marks**

c. A watery environment can have a low concentration of dissolved gases. Explain how having chloroplasts allows *E. chlorotica* to overcome this disadvantage. **1 mark**

Question 13 (8 marks)

Source: *VCAA 2012 Biology Exam 1, Section B, Q8*

Climate change has been linked to an excess of carbon dioxide in the atmosphere. The burning of coal is a major contributor to this excess of carbon dioxide.

Microalgae such as *Chlorella* can use greater amounts of carbon dioxide than land plants and they do not require prime soil, reliable rainfall and a particular climate. *Chlorella* can be grown cheaply in existing or engineered ponds which are supplied with carbon dioxide from a coal-burning power station nearby.

The following diagram represents a summary of the processes (labelled M, N, O, P) occurring in a *Chlorella* cell.

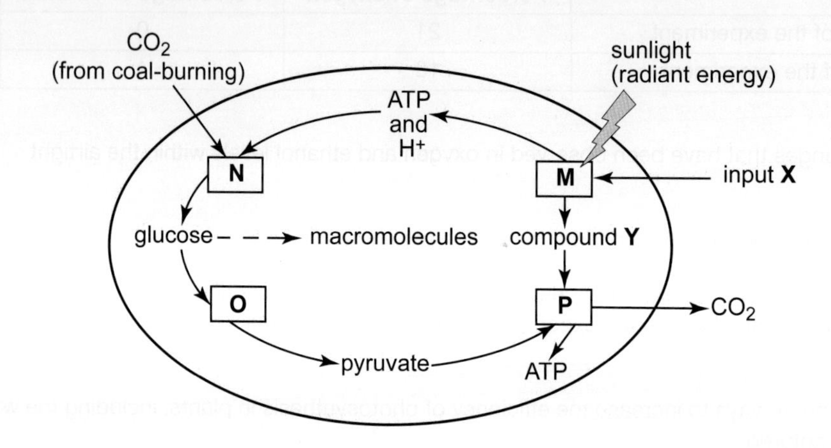

a. Name

i. input X. **1 mark**

ii. compound Y. **1 mark**

b. With reference to the diagram above, complete the following table. **3 marks**

Process	Name of process	Site of process
M		grana of chloroplast
O	glycolysis	
P	Stages of cellular respiration	

Chlorella pond farms could reduce 50% of the carbon dioxide that is produced by coal-burning power stations. Consider the summary of processes occurring in a *Chlorella* cell.

c. Given that carbon dioxide is an output of process P, explain how *Chlorella* farming could prevent 50% of the carbon dioxide emitted by coalburning power stations from entering the atmosphere. **2 marks**

d. What are two conditions, other than carbon dioxide supply, that an engineer or biologist maintaining a *Chlorella* pond farm would need to control to keep the growing conditions at an optimum level? **1 mark**

Question 14 (7 marks)

***Source:** VCAA 2008 Biology Exam 1, Section B, Q3*

The following diagrams show

Graph one The rate of photosynthesis in a green plant at different wavelengths of light

Graph two The estimated absorption of the different wavelengths of light by the different plant pigments

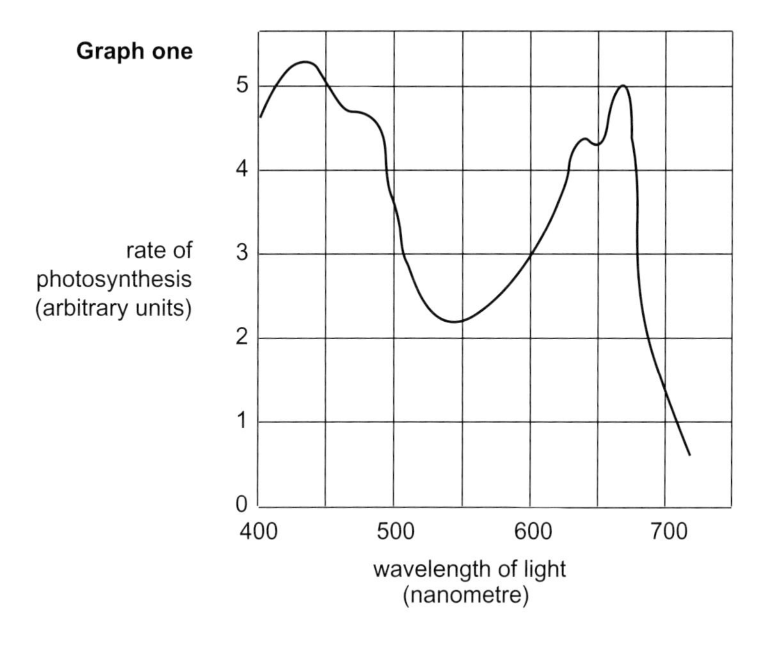

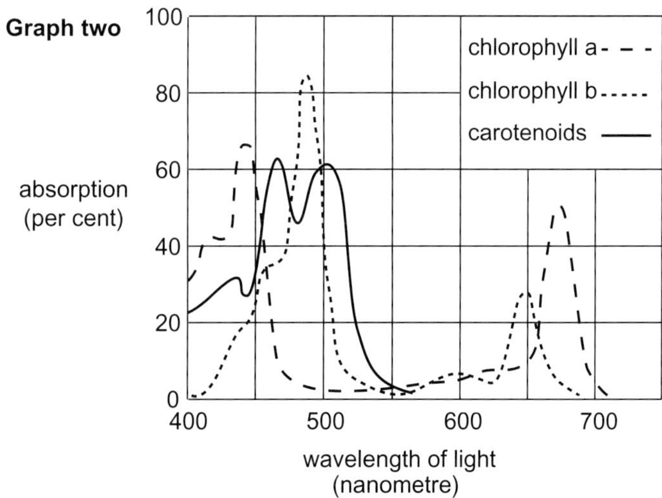

a. Explain why the graph showing the rate of photosynthesis has approximately the same shape as the absorption graphs of the plant pigments. **1 mark**

The following diagram shows a simplified representation of the first stage of photosynthesis

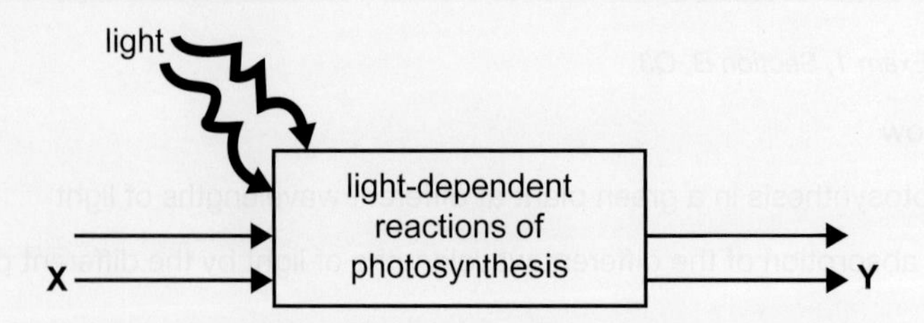

b. **i.** Name one input item that **X** could represent. **1 mark**

ii. Name one output item that **Y** could represent. **1 mark**

c. The breakdown of glucose in aerobic respiration can be represented by the simplified equation

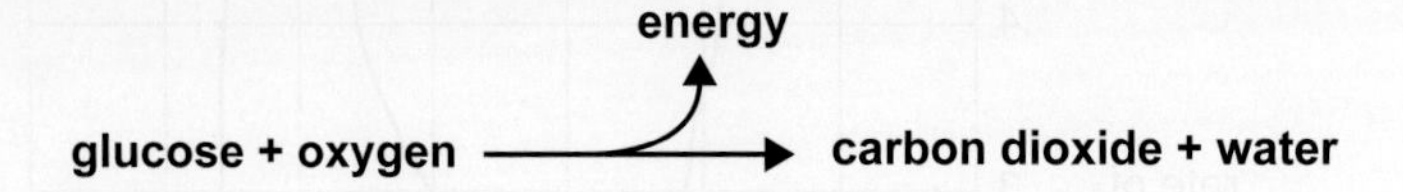

What is the energy yield per molecule of glucose as a result of aerobic respiration? **1 mark**

The breakdown of glucose in aerobic respiration can also be represented as occurring in three particular stages as indicated below.

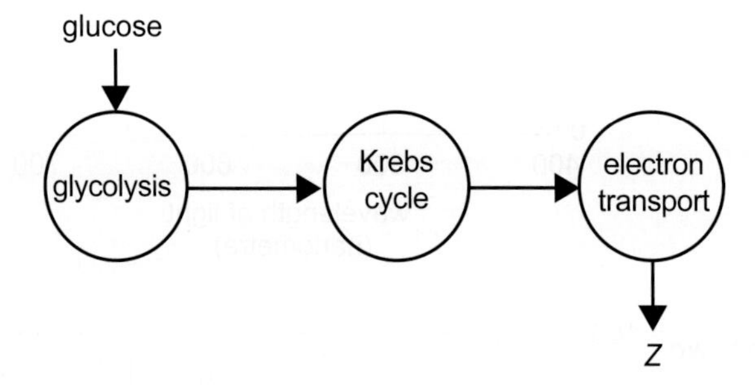

d. **i.** Within a cell, where does the electron transport stage of aerobic respiration occur? **1 mark**

ii. Describe what happens during the electron transport stage. In your answer include the name of product **Z**. **2 marks**

Question 15 (9 marks)

One important step in the process of bread making is the addition of yeast into the dough and allowing it to undergo anaerobic fementation.

a. Identify the gas produced that causes the bread to rise. **1 mark**

b. How do the products of anaerobic fermentation in yeast differ from anaerobic fermentation in animals? **2 marks**

c. Why are animal cells unable to use the same type of fermentation as yeast? **1 mark**

d. Explain how yeast is able to meet its energy requirements without using aerobic respiration. **1 mark**

e. Design an experiment to test the effect of temperature on the rate of anaerobic fermentation in yeast. **4 marks**

4.10 Exercise 3: Biochallenge online only

on Resources

eWorkbook Biochallenge — Topic 4 (ewbk-8085)

Solutions Solutions — Topic 4 (sol-0660)

Past VCAA examinations

Sit past VCAA examinations and receive immediate feedback, marking guides and examiner's report notes.
Access Course Content and select 'Past VCAA examinations' to sit the examinations online or offline.

Test maker
Create unique tests and exams from our extensive range of questions, including past VCAA questions.
Access the assignments section in learnON to begin creating and assigning assessments to students.

Online Resources

Below is a full list of **rich resources** available online for this topic. These resources are designed to bring ideas to life, to promote deep and lasting learning and to support the different learning needs of each individual.

eWorkbook

- 4.1 eWorkbook — Topic 4 (ewbk-1883) ☐
- 4.2 Worksheet 4.1 The stages of photosynthesis (ewbk-7575) ☐
- 4.3 Worksheet 4.2 Adaptations in C_3, C_4 and CAM plants (ewbk-7577) ☐
- 4.4 Worksheet 4.3 Reviewing factors impacting photosynthesis (ewbk-7579) ☐
- 4.5 Worksheet 4.4 The stages of aerobic respiration (ewbk-7581) ☐
- Worksheet 4.5 Case study: The impact of carbon monoxide on cellular respiration (ewbk-7583) ☐
- 4.6 Worksheet 4.6 Anaerobic fermentation (ewbk-7585) ☐
- 4.7 Worksheet 4.7 Reviewing factors impacting cellular respiration (ewbk-7587) ☐
- 4.8 Worksheet 4.8 CRISPR-Cas9 and photosynthesis (ewbk-7589) ☐
- 4.9 Worksheet 4.9 Biomass and anaerobic fermentation (ewbk-7591) ☐
- 4.10 Worksheet 4.10 Reflection — Topic 4 (ewbk-7593) ☐
- Biochallenge — Topic 4 (ewbk-8085) ☐

Solutions

- 4.10 Solutions — Topic 4 (sol-0660) ☐

Practical investigation eLogbook

- 4.1 Practical investigation eLogbook — Topic 4 (elog-0185) ☐
- 4.2 Investigation 4.1 Finding out about photosynthesis (elog-0262) ☐
- 4.4 Investigation 4.2 The effect of carbon dioxide on photosynthesis (elog-0264) ☐
- 4.5 Investigation 4.3 Photosynthesis and respiration — a balance (elog-0266) ☐
- Investigation 4.4 Respiration involving oxygen — aerobic respiration (elog-0268) ☐
- 4.6 Investigation 4.5 Anaerobic fermentation in yeast (elog-0270) ☐
- 4.9 Investigation 4.6 Fermentation of biomass (elog-0844) ☐

Digital documents

- 4.1 Key science skills — VCE Biology Units 1–4 (doc-34326) ☐
- Key terms glossary — Topic 4 (doc-34618) ☐
- Key ideas summary — Topic 4 (doc-34609) ☐
- 4.2 Extension of photosynthesis (doc-36166) ☐
- 4.5 The relationship between photosynthesis and cellular respiration (doc-35829) ☐
- 4.6 Case study: Applications of anaerobic fermentation in yeast (doc-35828) ☐
- Extension: Anaerobic fermentation in bacteria (doc-35830) ☐
- 4.7 Case study: Visualising glucose uptake in body cells (doc-36167) ☐

Teacher-led videos

- Exam questions — Topic 4 ☐
- 4.2 Sample problem 1 Exploring the stages of photosynthesis (tlvd-1424) ☐
- 4.3 Sample problem 2 Comparing C_3, C_4 and CAM plants (tlvd-1426) ☐
- 4.4 Sample problem 3 Analysing factors affecting photosynthesis (tlvd-1428) ☐
- 4.5 Sample problem 4 Analysing aerobic cellular respiration (tlvd-1430) ☐
- 4.7 Sample problem 5 Exploring factors affecting the rate of cellular respiration (tlvd-1432) ☐

Video eLessons

- 4.2 Photosynthesis (eles-4332) ☐
- 4.6 Fermentation experiment (eles-4336) ☐

Interactivities

- 4.2 Photosynthesis summary equation (int-0107) ☐
- 4.5 Photosynthesis or respiration (int-3039) ☐
- 4.6 Anaerobic fermentation (int-8323) ☐

Weblinks

- 4.3 Adaptations to climate change in C_3, C_4 and CAM plants ☐
- 4.8 CRISPR for crop improvement: an update review ☐
- The power of CRISPR-Cas9-induced genome editing to speed up plant breeding ☐
- CRISPR genome editing to address food security and climate change ☐
- 4.9 Geoscience Australia — Bioenergy ☐
- Biomass opportunities in Victoria ☐
- Environment, Land, Water and Planning — Bioenergy ☐

Exam question booklet

- 4.1 Exam question booklet — Topic 4 (eqb-0015) ☐

Teacher resources

There are many resources available exclusively for teachers online

To access these online resources, log on to **www.jacplus.com.au**.

AREA OF STUDY 2 How are biochemical pathways regulated?

OUTCOME 2

Analyse the structure and regulation of biochemical pathways in photosynthesis and cellular respiration, and evaluate how biotechnology can be used to solve problems related to the regulation of biochemical pathways.

PRACTICE EXAMINATION

STRUCTURE OF PRACTICAL EXAMINATION		
Section	**Number of questions**	**Number of marks**
A	20	20
B	8	30
	Total	**50**

Duration: 50 minutes

Information:

- This practice examination consists of two parts. You must answer all question sections.
- Pens, pencils, highlighters, erasers and rulers are permitted.

SECTION A — Multiple choice questions

All correct answers are worth 1 mark each; an incorrect answer is worth 0.

1. All living cells require a supply of energy to carry out their functions. The direct source of energy for most cell functions is
   A. oxygen.
   B. glucose.
   C. nutrient-rich food.
   D. adenosine triphosphate.

2. Energy released from the breakdown of organic compounds in a cell is used to make ATP. The released energy joins two molecules together, forming a bond. The joining reaction is represented in the diagram. What are the two molecules?

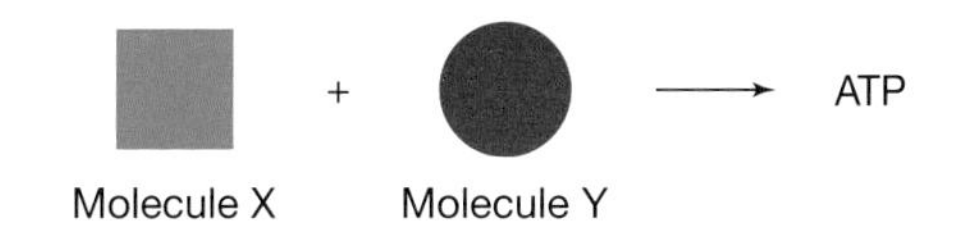

   A. ADP and ADP
   B. ADP and inorganic phosphate (Pi)
   C. ATP and inorganic phosphate (Pi)
   D. (Pi) and (Pi)

3. Coenzymes are organic compounds that act with an enzyme to increase the rate of a reaction. Which of the following substances is a coenzyme?
   A. Water
   B. Oxygen
   C. Sodium ions (Na^+)
   D. $NADP^+$

4. In which of the following processes is the coenzyme ATP **not** produced?
   A. Glycolysis
   B. The light-dependent stage
   C. The light-independent stage
   D. The Krebs cycle

5. NAD^+, $NADP^+$ and FAD are all coenzymes. They act as carriers of
   A. electrons and protons.
   B. electrons and ATP.
   C. electrons only.
   D. protons only.

6. A light-trapping pigment, chlorophyll, is embedded in
   A. the thylakoid membrane.
   B. the stroma.
   C. the stomata.
   D. the outer membrane.

7. One major energy transformation process occurring in the living world is photosynthesis. The main purpose of photosynthesis is to
   A. change glucose and carbon dioxide into oxygen and water.
   B. release the chemical energy trapped in starch.
   C. use radiant energy to produce chemical energy.
   D. convert chlorophyll to glucose and oxygen.

8. Chloroplasts in green plants are the means by which plants
   A. synthesise the proteins essential for living.
   B. carry out aerobic respiration in sunlight.
   C. replicate the genetic material of the plant.
   D. capture the radiant energy of sunlight.

9. The initial reactions of the C_4 pathway occur in the
   A. stomata plasma membrane.
   B. mesophyll cells.
   C. chloroplast stroma.
   D. bundle sheath cell.

10. Select the term that completes the following statement.
    Two interdependent biochemical pathways which comprise photosynthesis are:
    (1) the splitting of water and the formation of ______________________________
    (2) the formation of carbohydrates from CO_2.
    A. ADP
    B. NADH
    C. NAD^+
    D. ATP

11. The reactions that allow plants such as cacti and orchids to take up carbon at night are referred to as
   A. the C_4 pathway.
   B. the Calvin cycle.
   C. photorespiration.
   D. crassulacean acid metabolism (CAM).
12. Lactic acid is an output of which of the following reactions?
   A. Fermentation in animals
   B. Fermentation in yeast
   C. Photosynthesis
   D. Aerobic cellular respiration
13. Mammalian muscle cells are observed to have much higher numbers of mitochondria than skin cells. It is reasonable to predict that, over the same time period, muscle cells
   A. show a lower rate of heat production than skin cells.
   B. have a higher demand for ATP than skin cells.
   C. produce more oxygen than skin cells.
   D. take up less carbon dioxide than skin cells.
14. Cellular respiration releases the chemical energy held in glucose molecules. The energy released from cellular respiration appears in cells as
   A. heat energy only.
   B. the chemical energy of ATP only.
   C. both heat energy and the chemical energy of ATP.
   D. the chemical energy of both oxygen and ATP.
15. Which of these is a difference between aerobic respiration and anaerobic fermentation in humans?
   A. Aerobic respiration releases less energy than anaerobic fermentation.
   B. Glucose is completely broken down to maximise ATP yield in aerobic respiration but not in anaerobic fermentation.
   C. Carbon dioxide is produced in anaerobic fermentation but not in aerobic respiration.
   D. Lactic acid is produced in aerobic respiration but not in anaerobic fermentation.
16. What products are made during anaerobic fermentation in a majority of yeasts?
   A. Carbon dioxide and water
   B. Ethanol and water
   C. Ethanol and carbon dioxide
   D. Lactic acid and water
17. Mitochondria
   A. are located within the cell nucleus.
   B. are the site of glycolysis.
   C. are the major site of cellular ATP production.
   D. have a structure comprising a single membrane.
18. The stage of respiration that generates most ATP is
   A. glycolysis in cytoplasm.
   B. the Krebs cycle in the matrix.
   C. the electron transport chain in cristae.
   D. glycolysis in the matrix.
19. In what type of organisms can CRISPR be found naturally?
   A. Prokaryotes
   B. Fungi
   C. Viruses
   D. Protists

20. Genome editing technology using CRISPR has become a powerful tool for research in C_3 and C_4 plants for developing improved plant varieties with addition of important traits and removal of undesirable traits. Which of the following components is not involved in CRISPR?

A. Cas9 enzyme
B. DNA
C. sgRNA
D. *Taq* polymerase

SECTION B — Short answer questions

Question 21 (5 marks)

A chloroplast is surrounded by a double membrane. The diagram below shows the internal structure of the chloroplast.

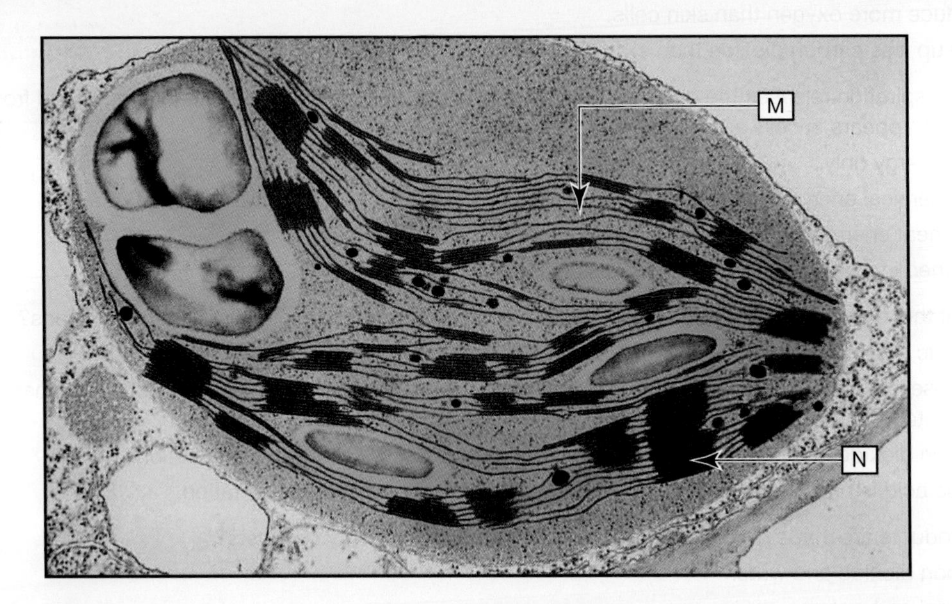

a. Identify structures M and N. **1 mark**
b. State which stages of photosynthesis take place in structures M and N. **1 mark**
c. Name two inputs for each of the two stages of photosynthesis. **2 marks**
d. A parasitic plant (species P) relies for its survival on the sugars that it absorbs from the vascular system of a host plant (species Q). Identify the key difference in how the two plant species obtain the energy needed for living. **1 mark**

Question 22 (3 marks)

Cytochrome *c* oxidase is an enzyme that acts on oxygen in the final step of aerobic respiration. Cyanide ions (CN^-) inhibit cytochrome *c* oxidase. The cyanide ions bind to cytochrome *c* oxidase at a point away from the active site. This action changes the shape of the active site of cytochrome *c* oxidase.

a. Define the active site. **1 mark**
b. The action of cyanide on cytochrome *c* oxidase is irreversible and thus it stays bound to cytochrome *c* oxidase, leading to eventual death. Draw a labelled diagram to represent the action of cyanide on cytochrome *c* oxidase. **1 mark**
c. Explain why is the inhibition of cytochrome *c* oxidase by cyanide is classifed as non-competitive inhibition. **1 mark**

Question 23 (6 marks)

The purpose of cellular respiration is to transfer the chemical energy stored in glucose into a useable form of chemical energy. The diagram below shows the internal structure of a mitochondrion.

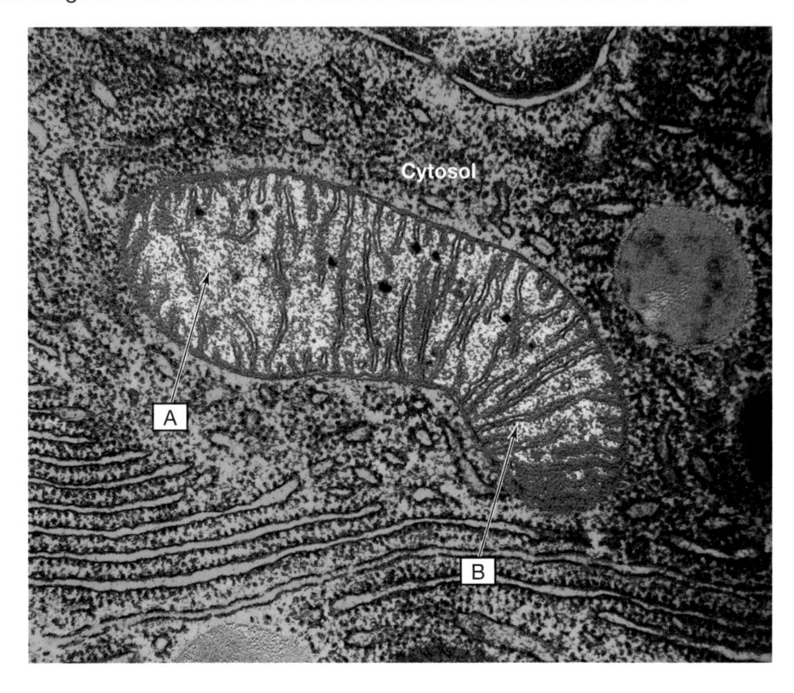

a. The first main stage of aerobic respiration is glycolysis. Identify the location where this stage occurs. **1 mark**

b. Identify structures A and B. **1 mark**

c. Give the balanced equation of aerobic cellular respiration. **1 mark**

d. Name the stage of aerobic respiration that takes place in B. List the inputs and outputs of this stage. **3 marks**

Question 24 (6 marks)

A scientist performed an experiment to test the action of hexokinase, an enzyme that catalyses the first step of glucose breakdown in the reactions of glycolysis. Identical volumes of glucose solutions, each with a differing concentration, were placed in eight separate tubes. An equal amount of hexokinase was added to each tube. The tubes were incubated at 37 °C for thirty minutes. The amount of glucose consumed in each tube was recorded. The rate of glucose consumption was calculated for each tube, and this is shown on the graph.

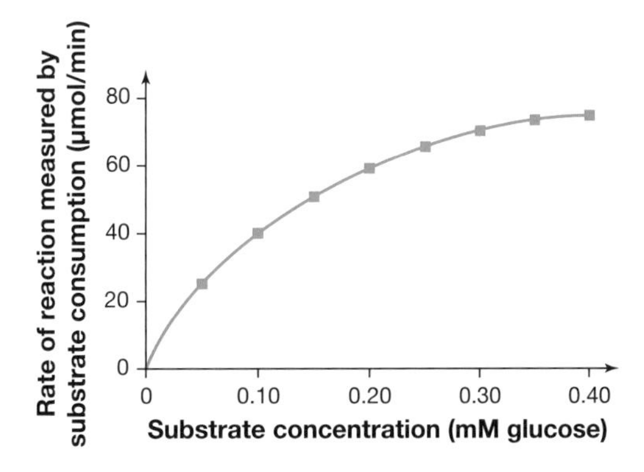

a. Identify the dependent variable, the independent variable and a control variable from the experiment. **3 marks**

b. Describe the relationship between the independent and dependent variables in the graph. **1 mark**

c. Explain, using data, why the graph levels off beyond 0.40 mM substrate concentration. **2 marks**

Question 25 (6 marks)

CRISPR-Cas9 is a rapidly developing genome editing technology that has been successfully applied in many organisms, including model and crop plants.

a. Name two major components of the natural CRISPR system. **2 marks**

b. Identify one specific use of CRISPR-Cas9 technologies for plants. **1 mark**

c. In the diagram below, label structures A, B and C. **3 marks**

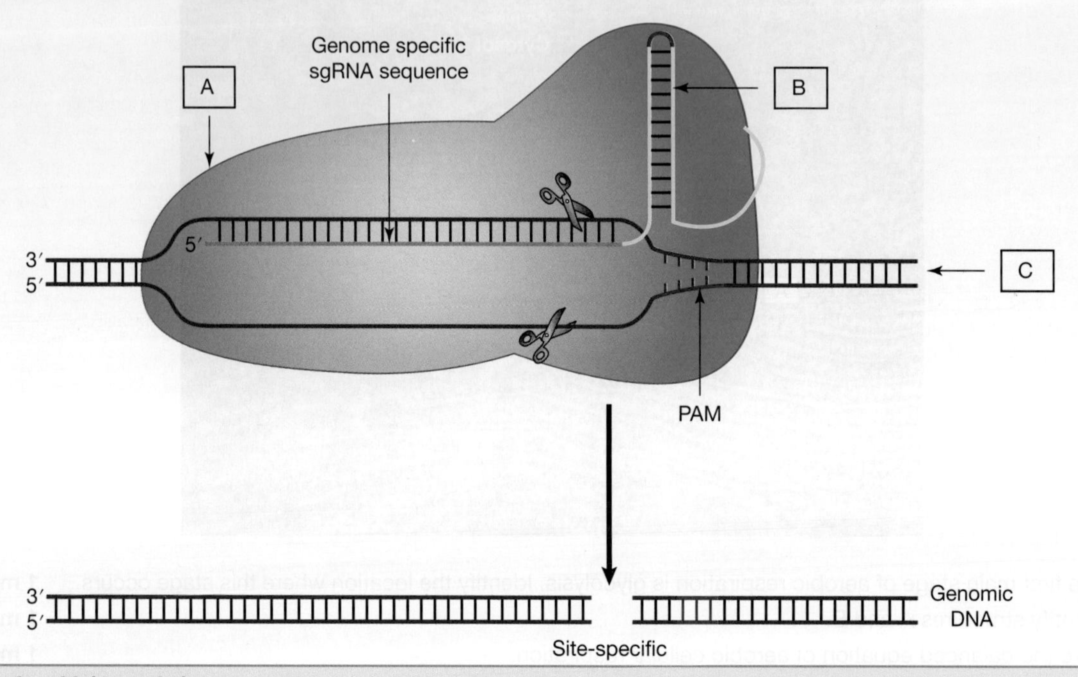

Question 26 (4 marks)

The rate of photosynthesis in plants and algae is affected by several environmental factors.

a. State two factors other than light intensity that affect the rate of photosynthesis. **1 mark**

b. In the graph below, identify the limiting factor. **1 mark**

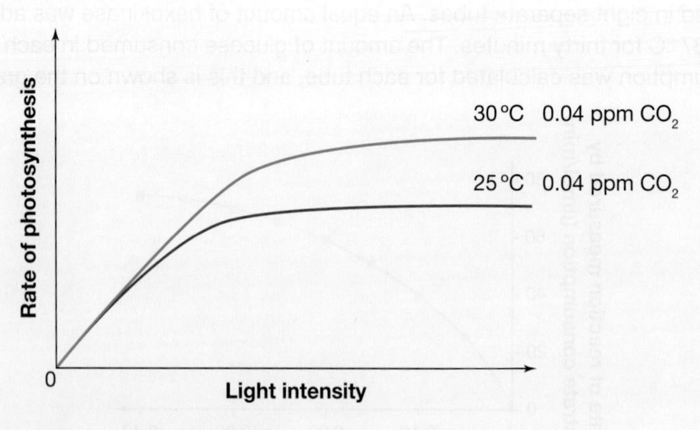

c. Describe the trend in the graph shown in part **b.** **2 marks**

END OF EXAMINATION

UNIT 3 | AREA OF STUDY 2

PRACTICE SCHOOL-ASSESSED COURSEWORK

ASSESSMENT TASK — Comparison and evaluation of biological concepts, methodologies and methods, and findings from three student practical activities

In this task you will analyse the structure and regulation of biochemical pathways in photosynthesis and cellular respiration, and evaluate how biotechnology can be used to solve problems related to the regulation of biochemical pathways.

- Students are permitted to use pens, pencils, highlighters, erasers and rulers.
- You need to make sure all results from your investigations are recorded in your logbook. You should use your data collected in your logbook to help write your report.

Total time: 2–3 lessons to conduct investigations, 70 minutes to write the report
Total marks: 40 marks

TASK

Enzymes are biological catalysts — integral in the cell's ability to undergo cellular respiration and photosynthesis. These processes are interrelated, as the outputs of cellular respiration provide the inputs of photosynthesis, and vice versa. Enzymes lower the activation energy required for both reactions to proceed. Without enzymes, chemical reactions would occur much too slowly to sustain life.

In this task you will be required to complete three separate experiments in groups of 2–4.

Following this you will undertake a comparison and evaluation of biological concepts, methodologies and findings, creating an individual scientific report.

Investigation 1: Investigating the effect of wavelength on photosynthesis

Introduction

The rate of photosynthesis is affected by limiting factors such as temperature, the availability of carbon dioxide and water, and the availability and wavelength of light.

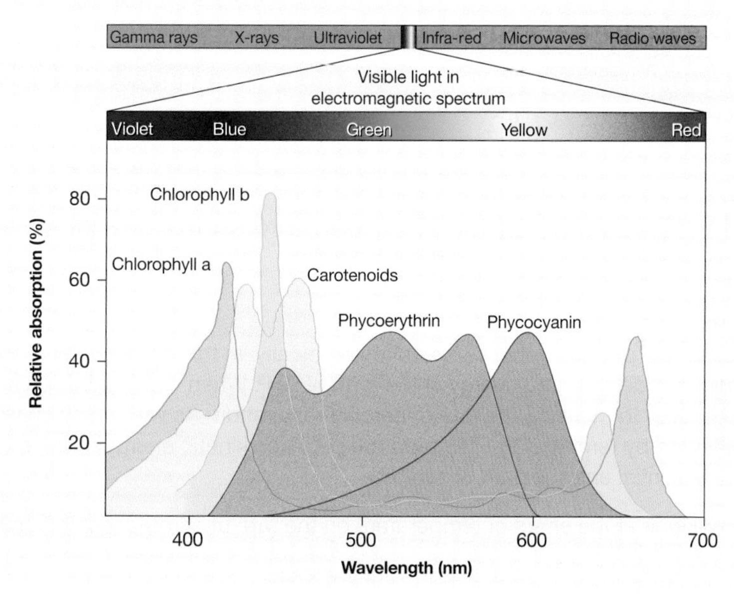

Chlorophyll will absorb different wavelengths of light at different rates. As light energy is used to break the bonds between hydrogen and oxygen in the light-dependent reaction of photosynthesis, this affects the rate of photosynthesis.

This practical investigation will investigate the wavelength of light required for photosynthesis to occur at its optimum rate.

Before you begin, write a clear aim and hypothesis in your logbook.

Materials

- 3 sheets of cellophane (1 red, 1 blue, 1 green)
- Heat lamp
- 5 × 50 mL beakers
- 1 × 50 mL measuring cylinder
- 4 rubber bands or clear tape
- 12 fresh spinach leaves
- Hole punch
- 1 × 10 mL syringe
- 1% × 30 mL bicarbonate solution ($NaHCO_3$)
- 3 mL liquid detergent

Method

1. Using the measuring cylinder, measure 30 mL of bicarbonate solution and add it to a 50 mL beaker. Add 3 drops of liquid detergent.
2. Using the hole punch, obtain 10 leaf discs from the fresh spinach.
3. Remove the plunger and load the discs into the syringe. Replace the plunger and push down so there is no air. Be careful not to damage the discs.
4. Fill the syringe with approximately 5 mL of the 1% bicarbonate solution (with detergent) by gently pulling back on the plunger.
5. Hold your thumb over the bottom of the syringe and pull back on the plunger to create a vacuum. Hold for 10 seconds. Repeat until all the discs have sunk to the bottom
6. Pour the discs and the solution into a 50 mL beaker. Top up the beaker to 20 mL using the bicarbonate solution.
7. Cover the beaker with red cellophane, using a rubber band or tape to secure.
8. Repeat steps 4–7 for blue cellophane, green cellophane and no cellophane.
9. Place the beakers under a lamp for 25 minutes, recording the number of discs that float each minute.

Results

Copy and complete the following table in your logbook. You should ensure you extend this so you are recording results for every minute up to 25 minutes. You may also add more data by collating and including class results, or conducting multiple trials to improve accuracy.

TABLE 1 Number of floating leaf discs each minute under different coloured light

Minutes	Red light	Blue light	Green light	White light
1				
2				

Investigation 2: Investigating the effect of temperature on the rate of aerobic respiration in yeast

Introduction

In the presence of glucose and oxygen, yeast can respire aerobically. In this process the glucose is broken down to produce water, carbon dioxide gas, and energy in the form of ATP. When carbon dioxide is produced, bubbles can be seen. This can be used to measure the rate of aerobic respiration. As with many chemical reactions, aerobic respiration is affected by temperature. Temperature provides kinetic energy to the molecules, but if the temperature is too high, essential enzymes are denatured.

In this experiment you will look at the effect of temperature on the rate of aerobic respiration in yeast.

Before you begin, write a clear aim and hypothesis in your logbook.

Materials

- 50 mL of activated yeast solution
- 10 test tubes with stoppers
- 5 delivery tubes
- Test tube rack
- 5% × 100 mL sucrose solution
- Water baths (0 °C, 20 °C, 40 °C, 60 °C and 80 °C)
- 5 thermometers
- 2 × 50 mL measuring cylinders
- Timer

If creating your own water baths

- Kettle
- 5 × 500 mL beakers

Method

1. Set up water baths at the five different temperatures listed (the 0 °C water bath will likely be an ice bath).
   Note: You may use commercially available water baths, or set up your own using 200 mL of water at the set temperature in a beaker as shown in the diagram (water may be boiled in a kettle and cooled as required).
2. Measure 10 mL of yeast and 20 mL of 5% sucrose solution using separate test tubes. Pour both into a test tube.
3. Suspend the test tube in the 0 °C water bath, cover it with the stopper and connect the delivery tube as shown.
4. At the end of the delivery tube, fill a test tube three-quarters full of tap water, where carbon dioxide bubbles can be observed as a product. Use the test tube rack to hold the test tube.
5. Wait 2 minutes to allow cellular respiration to begin.
6. Record the number of bubbles seen in minute 1 and minute 2.
7. Repeat steps 2–6 for each temperature.

Results

Copy and complete the following table in your logbook. You may also add more data by collating and including class results, including more time frames or conducting multiple trials to improve accuracy.

TABLE 2 Number of bubbles each minute at different temperatures

Temperature (°C)	Number of bubbles after 1 minute	Number of bubbles after 2 minutes	Total number of bubbles
0			
20			
40			
60			
80			

Investigation 3: Investigating the effect of pH on the enzyme catalase

Introduction

Enzymes are important in regulating the rates of biochemical pathways, allowing for reactions to occur at rates at which life can be sustained. There are many enzymes involved in the processes of cellular respiration and photosynthesis, increasing the speed of the multitude of reactions involved in these biochemical pathways.

Catalase is an enzyme that is not directly involved in photosynthesis or cellular respiration, but is found within the peroxisome of nearly all living organisms that are exposed to oxygen and capable of aerobic respiration. Potatoes, in particular, contain high amounts of catalase. Catalase can break down hydrogen peroxide, which is a byproduct of metabolic processes such as cellular respiration and causes damage to cells. Catalase breaks hydrogen peroxide into oxygen and water as seen in the following equation:

$$2H_2O_2 \rightarrow 2H_2O + O_2$$

Before you begin, write a clear aim and hypothesis in your logbook.

Materials

- Timer
- Potato
- Corer
- Buffer solutions at pH values 2, 4, 7, 8 and 10 (2 mL at each pH)
- Forceps
- Ruler
- 3% × 6 mL fresh hydrogen peroxide solution
- 11 wide-mouthed test tubes and stoppers
- 2 test tube racks

Methodology

1. Label the tubes 2, 2E, 4, 4E, 7, 7E, 8, 8E, 10, 10E and H_2O.
2. To the test tubes labelled 2, 4, 7, 8, 10 and H_2O, add 0.5 mL of hydrogen peroxide.
3. To the test tubes labelled 2, 4, 7, 8 and 10, add 1 mL of the corresponding buffer solution and fix the stopper.
4. Hold a ruler next to each test tube (2, 4, 7, 8, 10 and H_2O) and measure the height of the bubbles (if any). The height of the bubbles represents the rate of the reaction (the rate in which O_2 is produced from the breakdown of H_2O_2).
5. Use the corer to collect six pieces of potato. Cut them into 2 cm lengths.
6. In the last test tube (H_2O) containing hydrogen peroxide, add 1 mL of distilled water and 1 piece of 2 cm cored potato. Put on a stopper and measure the height of the bubbles with a ruler.
7. To the test tubes 2E, 4E, 7E, 8E and 10E, add 1 mL of the corresponding buffer solution.
8. Add 1 piece of 2 cm cored potato to each of these test tubes and cover with a stopper.
9. Leave for 3 minutes and observe any changes.
10. Add 0.5 mL of hydrogen peroxide to each tube and fix the stopper.
11. Hold a ruler next to each test tube and measure the height of the bubbles.

Results

Copy and complete the following tables in your logbook. You may also wish to record further observations or add more data by collating and including class results, or conducting multiple trials to improve accuracy.

TABLE 3 Controls (measurements before potato was added)

Test tube (pH buffer and H_2O_2)	Bubble height (cm)
2	
4	
7	
8	
10	
Water + H_2O_2	

TABLE 4 Effect of pH on catalase activity

Test tube (potato + pH buffer and H_2O_2)	Bubble height (cm)
2	
4	
7	
8	
10	
Water + H_2O_2	

Assessed task

Your formal SAC questions and your report will be completed under examination conditions.

You must include the following:

- An introduction, including information about the aims of your investigations, relevant background information and connections between cellular respiration, photosynthesis and regulation of biochemical pathways **4 marks**
- A results section, with graphical representations of the experimental data in all of your investigations; each graph should include a clear title and labelled axes with units (*x*-axis independent variable, *y*-axis dependent variable) **3 marks**
- A four-paragraph discussion (as broken down below) **30 marks**
- A conclusion. **3 marks**

Discussion breakdown

You should set your discussion into four distinct paragraphs. Within these paragraphs, you should address the following points:

Interpretation and evaluation of results

- Identification of the dependent and independent variable in each investigation **3 marks**
- Identification of two controlled variables per experiment **3 marks**
- A clear summary and analysis of your results for each investigation, including any patterns and trends **3 marks**
- Identification of the control group in each experiment and an outline of the purpose of control groups **2 marks**

Link to relevant biological concepts

- An explanation of how competitive inhibition of enzymes would affect cellular respiration and photosynthesis **3 marks**
- An outline of limiting factors for photosynthesis and cellular respiration **3 marks**
- A summary of how photosynthesis, cellular respiration and enzymes interrelate to each other **4 marks**

Limitations in data and methods with suggested improvement

- A critical evaluation of the scientific methodology and methods used, including reference to accuracy, precision and validity **3 marks**
- Suggestions to improve accuracy **2 marks**

Implications and further investigations

- Suggestions for further experimentation **2 marks**
- Suggestions for how the findings in your investigations would be used in a biotechnological setting to help solve problems (such as those related to the production of biomass or improving the yield in cellular respiration and photosynthesis). **2 marks**

Resources

Digital document Unit 3 AOS2 School-assessed coursework (doc-34864)

UNIT 4 How does life change and respond to challenges?

AREA OF STUDY 1

How do organisms respond to pathogens?

OUTCOME 1

Analyse the immune response to specific antigens, compare the different ways that immunity may be acquired, and evaluate challenges and strategies in the treatment of disease.

AREA OF STUDY 2

How are species related over time?

OUTCOME 2

Analyse the evidence for genetic changes in populations and changes in species over time, analyse the evidence for relatedness between species and evaluate the evidence for human change over time.

AREA OF STUDY 3

How is scientific inquiry used to investigate cellular processes and/or biological change?

OUTCOME 3

Design and conduct a scientific investigation related to cellular processes and/or how life changes and responds to challenges, and present an aim, methodology and method, results, discussion and a conclusion in a scientific poster.

5 Responding to antigens and acquiring immunity

KEY KNOWLEDGE

In this topic, you will investigate:

Responding to antigens and acquiring immunity

- physical, chemical and microbiota barriers as preventative mechanisms of pathogenic infection in animals and plants
- the innate immune response including the steps in an inflammatory response and the characteristics and roles of macrophages, neutrophils, dendritic cells, eosinophils, natural killer cells, mast cells, complement proteins and interferons
- initiation of an immune response, including antigen presentation, the distinction between self-antigens and non-self antigens, cellular and non-cellular pathogens and allergens

Acquiring immunity

- the role of the lymphatic system in the immune response as a transport network and the role of lymph nodes as sites for antigen recognition by T and B lymphocytes
- the characteristics and roles of the components of the adaptive immune response against both extracellular and intracellular threats, including the actions of B lymphocytes and their antibodies, helper T and cytotoxic T cells
- the difference between natural and artificial immunity and active and passive strategies for acquiring immunity

PRACTICAL WORK AND INVESTIGATIONS

Practical work is a central component of learning and assessment. Experiments, supported by a **practical investigation eLogbook** and **teacher-led videos**, are included in this topic to provide opportunities to undertake investigations and communicate findings.

5.1 Overview

Numerous **videos** and **interactivities** are available just where you need them, at the point of learning, in your digital formats, learnON and eBookPLUS at **www.jacplus.com.au**.

5.1.1 Introduction

All organisms show complex relationships for survival. From the simplest organisms, such as bacterial cells, through to the most complex, such as human beings, all organisms have some kind of effective immune system to protect them. Environments contain many pathogens, which are agents, usually microorganisms, that cause diseases in their hosts.

FIGURE 5.1 A false-coloured scanning electron micrograph showing dendritic cells (in grey) engulfing a yeast spore (in yellow)

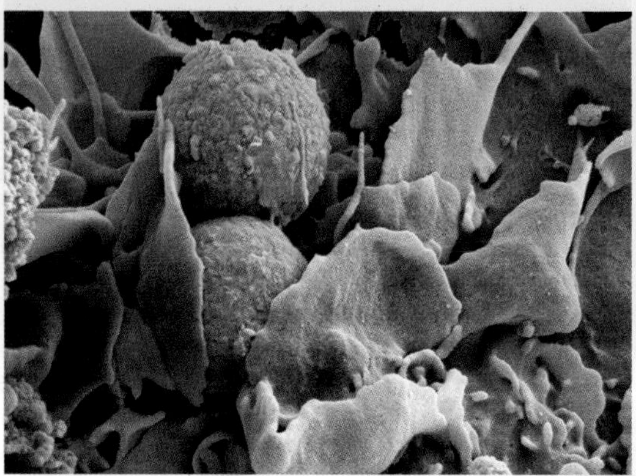

The immune system consists of many biological structures and processes within an organism that protect against disease. To function properly, an immune system must detect a wide variety of pathogens and distinguish them from the organism's own healthy tissue. We are constantly exposed to pathogens in food and water, on surfaces, and in the air. Features of the immune system such as pathogen identification, innate and specific responses, amplification, retreat, and remembrance are essential for protection against pathogens and will be explored in this topic.

LEARNING SEQUENCE

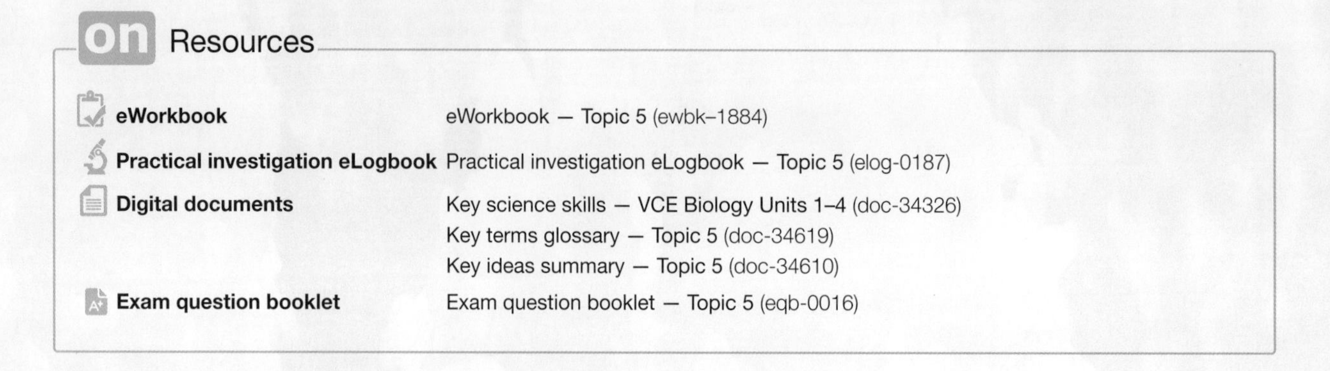

on Resources

eWorkbook — eWorkbook — Topic 5 (ewbk-1884)

Practical investigation eLogbook — Practical investigation eLogbook — Topic 5 (elog-0187)

Digital documents — Key science skills — VCE Biology Units 1–4 (doc-34326)
Key terms glossary — Topic 5 (doc-34619)
Key ideas summary — Topic 5 (doc-34610)

Exam question booklet — Exam question booklet — Topic 5 (eqb-0016)

5.2 Antigens, pathogens and allergens

KEY KNOWLEDGE

- The distinction between self-antigens and non-self antigens, cellular and non-cellular pathogens and allergens

Source: Adapted from VCE Biology Study Design (2022–2026) extracts © VCAA; reproduced by permission.

The ability of our **immune system** to protect us from disease and infection has been fundamental to our survival and our evolution. Without an immune system, we would struggle to fight against even the simplest of pathogenic agents, leading to disease and death.

Our immune system has incredible mechanisms to assist us in fighting **pathogens**.

Defence against infection depends on the ability of the body's immune system to:
- identify cells or molecules that are foreign or 'non-self', and react to and eliminate them
- recognise the body's own cells and the compounds they produce as 'self' and not react against them.

Understanding different antigens and pathogens are vital to understand how our immune system functions.

5.2.1 Antigens

Antigens are molecules or parts of a molecule that stimulate immune response through the adaptive immune response. In jawed vertebrates, they are 'antibody generators', leading to the production of antibodies.

Antigens, as well as stimulating the immune response through the production of **antibodies** or **immunoglobulins**, also may lead to **inflammation** and a cell-mediated immune response through **cytotoxic T cells**. These processes will be explored further in this topic.

Antigens can be classed as '**self-antigens**' or '**non-self antigens**'. This is particularly important in our adaptive immune response.

immune system the body system that helps resist infection and disease through specialised cells and proteins

pathogens agents that cause diseases in their hosts

antigens molecules or parts of molecules that stimulate an immune response

antibodies proteins produced by plasma cells in response to antigens and which react specifically with the antigen that induced their formation; also called immunoglobulins

immunoglobulins antigen-binding proteins produced by B cells and released in blood and lymph

inflammation an innate reaction by the immune response to foreign particles or injury resulting in redness and swelling

cytotoxic T cells T cells that are activated by cytokines to bind to antigen–MHC-I complexes on infected host cells and kill infected body cells

self-antigens antigens on cells that are recognised by self-receptors as being part of the same body

non-self antigens antigens that do not belong to the body's own cells

FIGURE 5.2 A simplified diagram showing the various receptors on immune cells

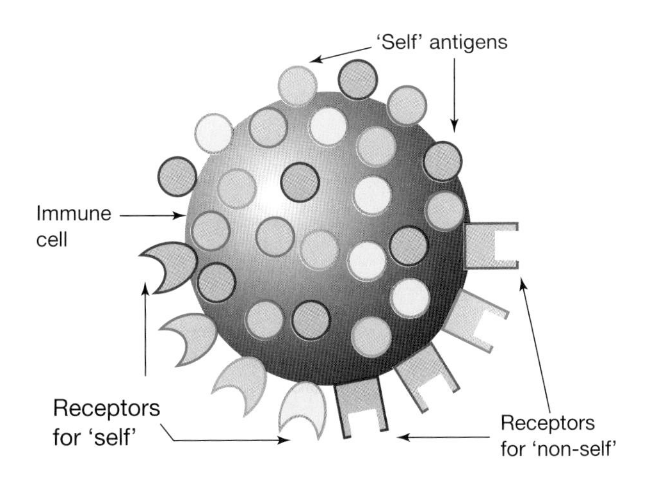

Self-antigens are those that are not foreign and are thus usually tolerated by the immune system. Non-self antigens can be identified as invaders and can be attacked by the immune system.

Distinguishing self from non-self

Figure 5.2 shows a stylised diagram of an immune cell that checks for non-self invaders. Note that the plasma membrane of this immune cells carries:

- self-antigens that identify this immune cell as 'self'
- **cell surface receptors** for self-antigens so that this cell can identify and not attack other body cells
- cell surface receptors for foreign antigens so that the immune cell can identify foreign material and signal other immune cells to eliminate it.

Antigen recognition depends on the detection of antigens by the receptors. The **major histocompatibility complex** (MHC) is a group of receptor proteins present on the surface of body cells. It helps the immune system distinguish the body's own proteins from proteins made by foreign invaders such as viruses and bacteria. When the immune system fails to recognise self-cells, **autoimmune disease** can result, where the immune system targets its own body cells.

In humans, MHC markers are also known as **human leukocyte antigens** (HLA). When discussing these markers on humans, either term is acceptable.

cell surface receptors regions of a trans-membrane molecule exposed at the surface of a cell that act in cell signalling by receiving and binding to extracellular molecules

major histocompatibility complex receptor proteins on the surface of cells that identify the cells as 'self'

autoimmune diseases diseases in which the immune system fails to identify 'self' material and makes antibodies against the body's own tissues

human leukocyte antigens antigens present on human cell surfaces that determine the 'self' status of a person's cells

MHC-I a type of major histocompatibility complex found on nucleated cells

MHC-II a type of major histocompatibility complex found on specific white blood cells involved in the adaptive immune response

antigen-presenting cells cells of the immune system that use MHC-II markers on their surface to present antigens to helper T cells to elicit an immune response

macrophages phagocytic antigen-presenting cells derived from monocytes that may be found in various tissues throughout the body and can engulf foreign material

dendritic cells a type of antigen-presenting cell and phagocyte that can activate T lymphocytes in the adaptive immune response

Classes of MHC markers

There are different classes of MHC markers, including HLA I and HLA II markers:

- **MHC-I** is present on all nucleated cells of the body. It contains a site that forms the structural binding site for an antigen. These are what allows cells to be recognised as 'self' so they will not be attacked by cells such as natural killer cells or cytotoxic T cells.
- **MHC-II** is presented mainly on specific white blood cells, including **antigen-presenting cells** (such as **macrophages** and **dendritic cells**) and helper T cells. Macrophages and dendritic cells present antigens on their surface on these MHC-II markers, which bind to helper T cells to help activate the adaptive immune response.

The role of these cells in the immune response will be explored further in subtopics 5.5 and 5.11.

FIGURE 5.3 MHC-I and MHC-II markers on cells. **a.** All nucleated cells have MHC-I markers **b.** Specific immune cells such as macrophages, dendritic cells and T cells have both MHC-I and MHC-II markers.

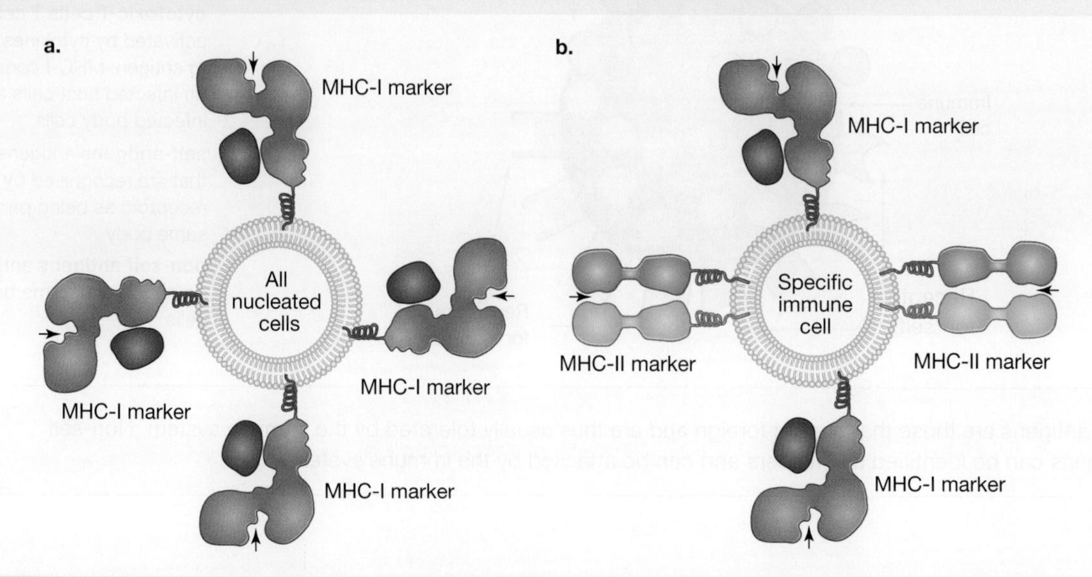

What substances are antigens?

Antigens may be lipoproteins, polysaccharides and lipopolysaccharides, lipids, nucleic acids, and even some metals, such as mercury. An antigen may be a molecule or part of a molecule. Some antigens originate from the external environment, such as proteins in or on bacteria and viruses, or even a short segment of these proteins. Other possible antigen sources include chemicals in snake venom, drugs, dust, proteins in food and protein markers on cells from an unrelated person. Many of these are **cell surface markers**.

cell surface markers proteins (or glycoproteins) present on the plasma membrane that distinguish various cell types and discriminate between self and non-self antigens

CASE STUDY: Discovery of surface antigens and their link to transplants

In 1958, the French researcher Jean Dausset (1916–2005) discovered that the immune system uses surface antigens on human cells to identify the cells that are self and the cells that are foreign or non-self. Dausset was the first to recognise that survival rates of transplanted organs were greatest when the donor and the recipient were identical twins, as compared to the situation when the donor and the recipient were unrelated. This observation pointed to the importance of matching the key antigens of the donor organ with the transplant recipient.

To access more information on the discovery of surface antigens and case studies on transplant and tissue typing, and to complete an analysis task relating to these, please download the worksheet.

FIGURE 5.4 Light microscope image of a section through a transplanted kidney that shows signs of acute rejection. The numerous purple dots are the nuclei of cells of the immune system. These cells have migrated to the transplanted kidney and are mounting an immune attack on the kidney cells. (Image courtesy of the National Toxicology Program, US Department of Health and Human Services.)

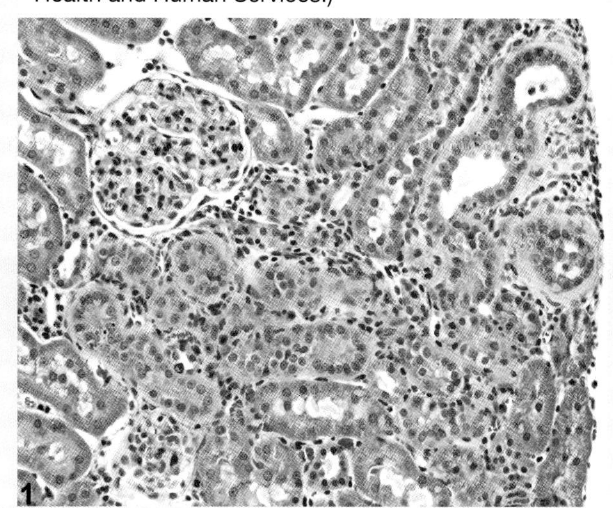

on Resources

eWorkbook Worksheet 5.1 Case study: Transplants and tissue typing (ewbk-5242)

on Resources

Video eLesson Antigens and blood type (eles-4194)

Interactivity Antigens (int-0045)

5.2.2 Pathogens — sources of non-self antigens

A pathogen or infectious agent is a biological agent that causes disease or illness to its host. Pathogens can be cellular or non-cellular.

Cellular pathogens are classified as living organisms. They can reproduce independently.
Non-cellular pathogens are non-living. They cannot reproduce without a host. They hijack the host's processes in order to replicate.

TABLE 5.1 Pathogens can be cellular (living) or non-cellular (non-living).

Cellular				Non-cellular	
Bacteria	Worms (parasites)	Fungi	Protozoa	Viruses	Prions

Pathogens can cause a variety of different illnesses and disease, some of which are outlined in table 5.2.

Note: You do not need to memorise the names of these pathogens, but you should recognise that there is a variety of both cellular and non-cellular pathogens that can cause disease.

TABLE 5.2 Examples of different cellular and non-cellular pathogens

Causative agent	Disease	Example of how it spreads
CELLULAR		
Bacteria		
Streptococcus pneumoniae	Pneumonia	From person to person through inhalation of contaminated droplets, such as from the sneeze of an infected person
Corynebacterium diphtheria	Diphtheria	From person to person through close respiratory contact
Mycobacterium tuberculosis	Tuberculosis	From person to person through inhalation of contaminated droplets that reach the lung alveoli
Bacillus anthracis	Anthrax	Via inhalation or ingestion of spores or via a break in skin that enables entry of spores
Yersinia pestis	Black Death (bubonic plague)	Via bites from infected fleas, from an infected rodent
Vibrio cholerae	Cholera	Via ingestion of faecal-contaminated water or food
Streptococcus pyogenes	Toxic shock syndrome, strep throat and others	Via upper respiratory tract or skin lesion
Clostridium tetani	Tetanus	Via deep puncture wound
Treponema pallidum	Syphilis	From person to person through direct sexual contact with syphilis sore of an affected person
Helicobacter pylori	Stomach ulcers	From person to person through saliva of infected person
Salmonella typhi	Typhoid fever	Via ingestion of faecal-contaminated water or food; can also be transmitted from person to person

Clostridium botulinum	Botulism, a form of food poisoning	Via ingestion of improperly canned food that exposes a person to a bacterial exotoxin that has not been heat-inactivated
Salmonella enterica	Salmonellosis, a form of food poisoning	Via ingestion of faecal-contaminated foods such as raw eggs, and improperly cooked chicken
Fungi		
Candida albicans	Thrush (oral candidiasis)	Via mouth-to-mouth contact with an infected person
Trichophyton spp.	Athlete's foot (tinea)	Via contact with an infected person or with contaminated surfaces
Protozoa		
Giardia intestinalis	Giardiasis	Via ingestion of water or food contaminated with *Giardia* cysts
Plasmodium falciparum	Malaria	Via bite of an infected mosquito
NON-CELLULAR		
Viruses		
Influenza viruses	Influenza	Via inhalation of airborne droplets from sneeze or cough of an infected person
Varicella zoster virus (VZV)	Chickenpox (varicella)	Via nose or mouth from direct contact with an infected person
Rubella virus	Rubella (German measles)	Via nasopharynx from airborne droplets in sneeze/cough of an infected person
Hepatitis B virus	Viral hepatitis type B	Via contact with body fluids of an infected person
Human immuno-deficiency virus (HIV)	Can progress to Acquired immuno-deficiency disease (AIDS)	Via sexual contact with an infected person or from use of contaminated needles
Poliovirus	Poliomyelitis (polio)	Via mouth from faeces or sneeze droplets of infected person
Rhinovirus	Common cold	Via inhalation of droplets in air from an infected person or via hand-to-hand contact with an infected person
SARS coronavirus (SARS CoV)	Severe acute respiratory syndrome	Via inhalation of infected droplets or by touching contaminated surfaces and then touching one's mouth or nose
SARS coronavirus 2 (SARS-CoV-2)	Coronavirus disease (COVID-19)	Via inhalation of infected droplets or by touching contaminated surfaces and then touching one's mouth or nose
Ebolavirus	Ebola virus disease	Via direct contact with body fluids of an infected person or contact with contaminated needles or syringes.
Prions		
	Creutzfeldt–Jakob disease (CJD)	Via tissue grafts from infected cadavers, including corneas or brain membrane, or via growth hormone isolated from infected pituitary glands of cadavers
	Variant Creutzfeldt–Jakob disease (vCJD)	Via ingestion of beef from cattle infected with bovine spongiform encephalopathy (BSE), commonly known as 'mad cow disease'
	Kuru	Via past practice of ceremonial ingestion of brain tissue of a deceased infected person

FIGURE 5.5 Different diseases are caused by pathogens: a. *Yersinia pestis*, which causes the Black Death; b. *Candida albicans* and oral thrush; c. *Ebolavirus* and warning signs about the disease; d. *Giardia intestinalis* and *Plasmodium falciparum*.

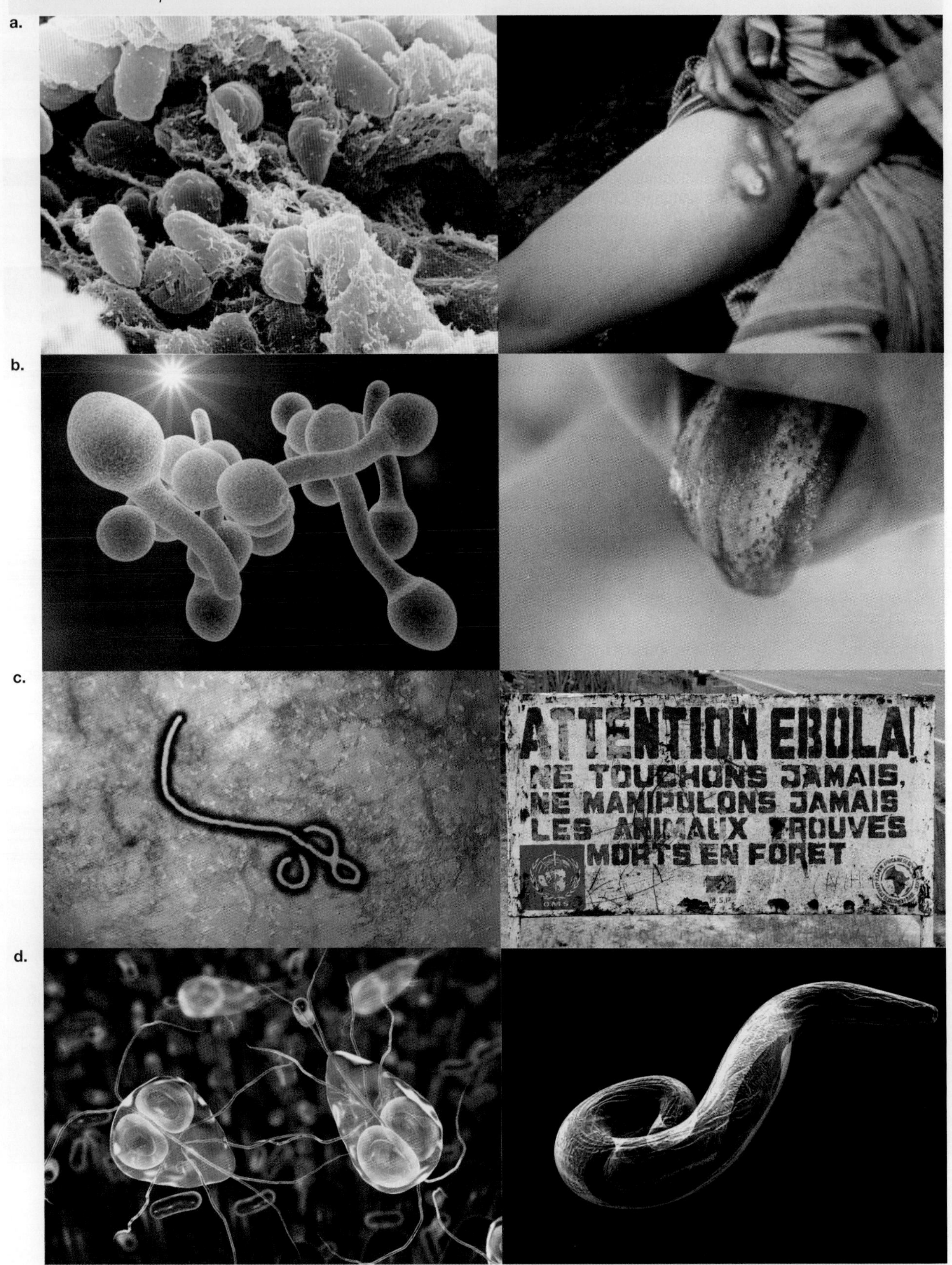

5.2.3 Cellular pathogens

Cellular pathogens include bacteria, fungi, protozoa, worms and arthropods. Cellular pathogens differ from non-cellular pathogens in that they are capable of reproducing independently without a host cell. Many of these are often referred to as microorganisms (or microbes). Some cellular pathogens are simpler prokaryotes, whereas others are eukaryotes.

Bacteria

Bacteria are prokaryotic microbes, and their genetic material is double-stranded DNA. Only a very small percentage of bacteria are human pathogens, but they are responsible for a number of diseases. Under favourable conditions, bacteria can, on average, reproduce every 20 to 30 minutes. Over a period of 12 hours, enormous numbers of bacterial cells can potentially be produced through binary fission. If pathogenic bacteria can cross the body's defences, such an infection has the potential to develop rapidly into a disease.

Bacterial pathogens can be either **intracellular** or **extracellular**:

- Some bacterial pathogens can live and reproduce *only* on the external surfaces of human cells or in body fluids, such as plasma, **lymph** or interstitial fluid. For example, *Vibrio cholera*, which causes the uncontrolled diarrhoea of cholera, is only extracellular.
- Other bacteria can live and reproduce either inside or outside the cells of their host, for example *Neisseria meningitidis*, the cause of bacterial meningitis.
- Some other bacterial pathogens can only survive and reproduce within host cells. One example is *Chlamydia trachomatis*, the cause of a sexually transmitted infection that, if left untreated, can develop into pelvic inflammatory disease and infertility.

cellular pathogens any disease-causing agent made up of cells that can reproduce independently without relying on the host machinery

intracellular anything that is within a cell

extracellular locations within the body that are outside cells, such as blood plasma and extracellular fluid

lymph interstial fluid surrounding the tissues that is filtered through the tiny holes between the capillaries into the lymphatic system

FIGURE 5.6 Bacterial species come in various different shapes and contain many structures that allows it to be pathogenic

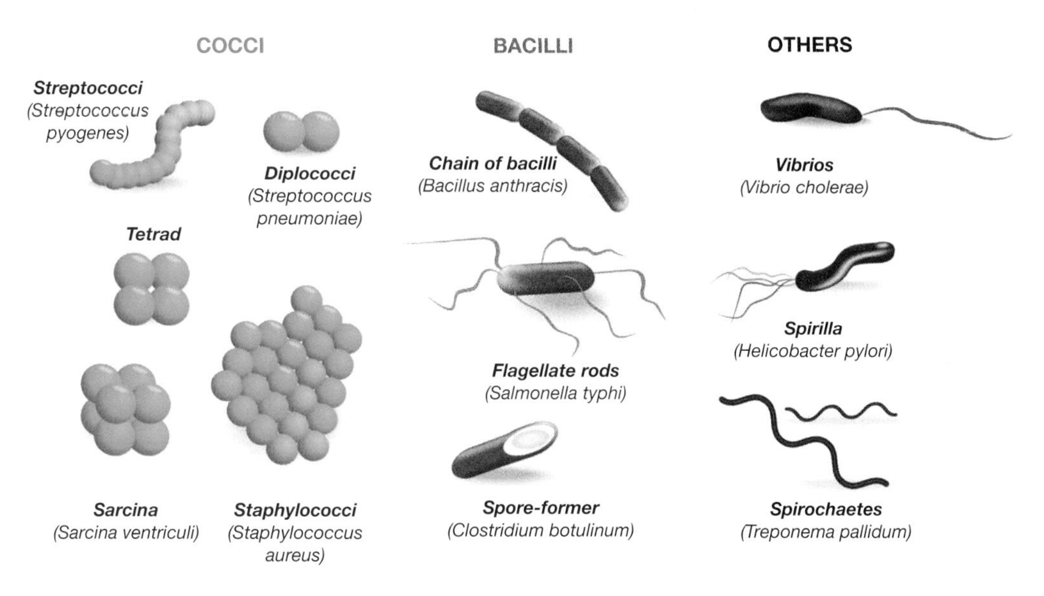

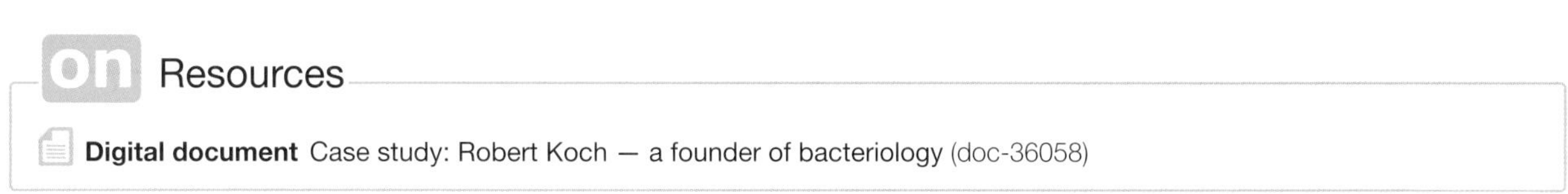

Some bacteria produce toxins

A major factor that increases the virulence of some bacteria is their ability to produce specific toxins. Bacterial toxins are substances produced by some bacteria that damage particular tissues of their host organism and cause disease. Bacterial toxins are of two types:

- **exotoxins** — secreted toxins
- **endotoxins** — parts of the outer membrane that are released when bacteria die.

TABLE 5.3 Summary of differences between bacterial exotoxins and endotoxins

Exotoxins	Endotoxins
• Soluble proteins • Produced by some Gram-positive and Gram-negative bacteria • Released by living bacterial cells • Many different types, each producing specific damage • Typically destroyed by heat	• Lipopolysaccharides • Produced only by Gram-negative bacteria • Part of the outer membrane of bacterial cell walls; released when bacteria die • One type only, producing a diverse range of effects • Heat stable

These toxins are recognised as foreign and thus can also stimulate an immune response. However, often these toxins have already caused harm to cells.

exotoxins toxins that are secreted into the surrounding medium by a microorganism as it grows

endotoxins toxic parts of the outer membrane of some Gram-negative bacteria that are released when the bacteria die

Exotoxins and endotoxins are both recognised by the immune system. While you do not need to memorise the difference between them for the Study Design, it provides an example of how molecules can act as antigens.

EXTENSION: Examining endotoxins and exotoxins

There are various ways in which pathogens can cause disease. Some of these are listed in Table 5.4.

TABLE 5.4 Factors involved in disease production by bacteria

Disease	Bacterium responsible	Factor involved in disease	Mode of action
Botulism (one form of food poisoning)	*Clostridium botulinum*	Exotoxins that are neurotoxins	Paralysis
Salmonellosis (most common form of food poisoning)	Several species of *Salmonella*	Invasive properties	Invades tissue lining the intestine
Cholera (severe gastroenteritis)	*Vibrio cholerae*	Exotoxin	Alters intestinal permeability
Diphtheria (sore throat and fever)	*Corynebacterium diphtheriae*	Exotoxin	Prevents protein synthesis
Scarlet fever (fever and rash)	*Streptococcus pyogenes*	Exotoxin	Invades tissue and damage blood vessels
Tuberculosis (lesions in lung and other tissues)	*Mycobacterium tuberculosis*	Invasive properties	Invades tissue

What are exotoxins?

Exotoxins are highly toxic soluble proteins that are produced by living bacterial pathogens as part of their metabolism and are released into their surroundings. Exotoxins can spread throughout the body and cause system-wide damage, but can also be recognised by the immune system, stimulating an immune response.

Several bacteria, mainly Gram-positive bacteria, produce exotoxins that can damage or kill cells of all kinds. Other bacteria produce exotoxins that can damage cells of particular kinds, such as nerve cells. Exotoxins act in different ways on host tissues; some exotoxins damage plasma membranes and disrupt transport of compounds across these membranes, some inhibit protein synthesis, and some block normal nerve function.

Exotoxin-producing bacteria include the following:

- *Clostridium tetani* releases a neurotoxin that blocks muscle relaxation, resulting in tetanus, a disease that is characterised by painful muscle spasms and lockjaw and that can result in respiratory failure.
- *Vibrio cholerae* releases a toxin that damages the cells of the gut lining, leading to the uncontrolled production of watery diarrhoea that is seen in the disease cholera.
- *Streptococcus pyogenes* releases an exotoxin that kills cells and can lead to major organ failure as seen in streptococcal toxic shock syndrome.

One of the most lethal exotoxins comes from the bacterium *Clostridium botulium.* Less than 70 micrograms for a 70-kg individual can lead to death. This botulinum toxin can be inactivated by heating at 85 °C or higher for 5 minutes. This is the botox used for cosmetic purposes.

FIGURE 5.7 Exotoxins are released via exocytosis.

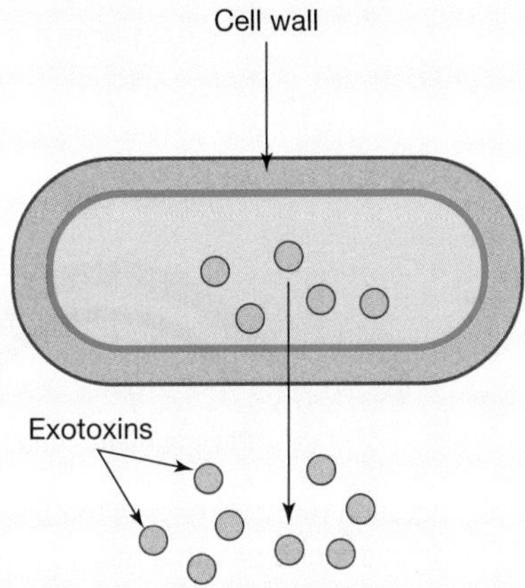

What are endotoxins?

Endotoxins are toxins derived from the lipid portion of the lipopolysaccharide (LPS) of the outer membrane of Gram-negative bacteria. These endotoxins are produced by some Gram-negative pathogenic bacteria. Endotoxins are released only after these bacteria die and their outer membrane breaks down. Examples of endotoxin-producing bacteria include *Salmonella typhi,* the cause of typhoid fever, and *Neisseria meningitidis*, the cause of meningitis.

FIGURE 5.8 Endotoxins are part of the membrane of some bacteria.

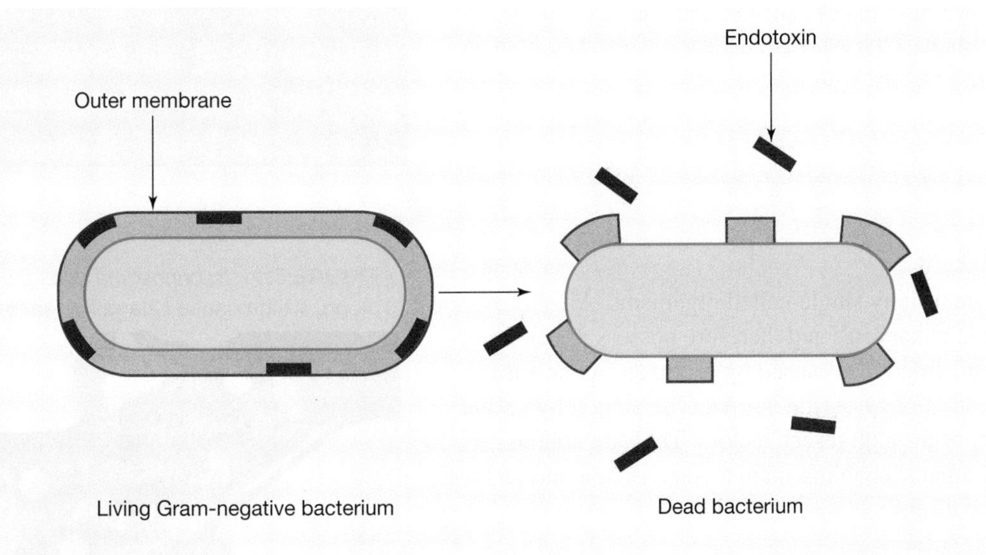

Resources

Digital document Case study: The story of Matthew Ames and exotoxins (doc-36168)

Weblink The story of Matthew Ames

EXTENSION: Other features of bacteria

Gram-positive or Gram-negative?

Bacteria can be classed as Gram-positive or Gram-negative. This classification is linked with the structure of their cell wall and their colour upon staining.

FIGURE 5.9 Diagram comparing the structure of Gram-positive and Gram-negative bacteria

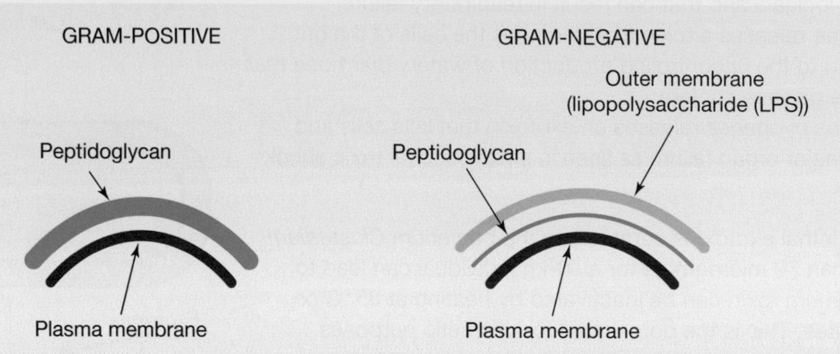

Gram-positive bacteria have a thick level of peptidoglycan. Gram-negative bacteria have an additional membrane of lipopolysacchirdides that allows them to be more resistant to antibiotics.

Capsule or no capsule?

Many pathogenic bacteria, both Gram-positive and Gram-negative, have a gelatinous layer of polysaccharide that lies outside their cell walls. This outermost layer, known as a capsule, protects these bacteria from one of the defence mechanisms of the human body — the defence cells of the body known as macrophages. As a result, the presence of a capsule increases the virulence of pathogenic bacteria; that is, it increases the likelihood of disease resulting from an infection by these bacteria.

FIGURE 5.10 Bacterial cells surrounded by a capsule in a micrograph

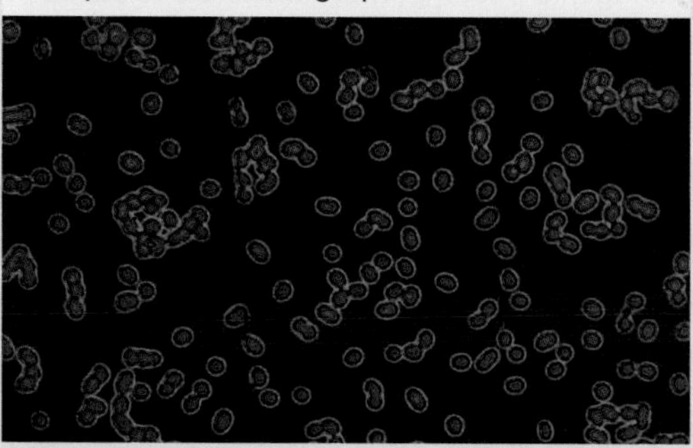

Protozoans

Protozoans are usually single-celled organisms. All protozoans are eukaryotes and therefore possess a membrane-bound nucleus. Infections caused by protozoa can be spread through ingestion of cysts (the dormant life stage), by sexual transmission, or through insect vectors. They are able to multiply in humans, enabling them to survive in a human host while causing disease.

FIGURE 5.11 *Trypanosoma cruzi* parasites in blood, which cause Chagas disease

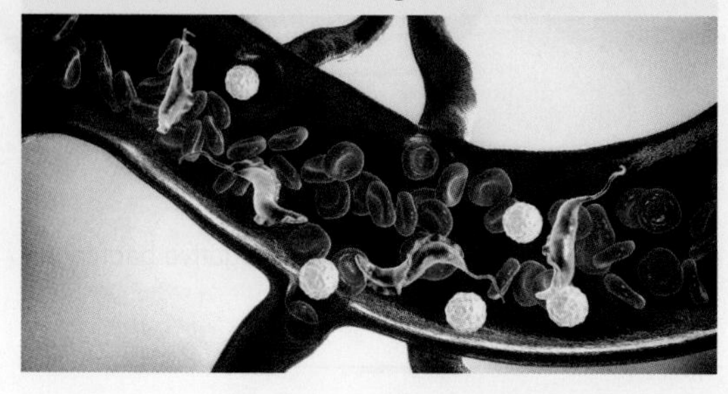

Common infectious diseases caused by protozoans include malaria, giardiasis and toxoplasmosis, which affect the blood, gut and **lymphatic system** respectively. As outlined in table 5.2, *Giardia intestinalisis* is spread through ingestion of contaminated food and *Plasmodium falciparum*, which causes malaria, is spread via the bite of infected mosquitoes (see figure 5.5d).

lymphatic system a network of tissues and organs that plays a key role in the immune response of the mammals

Fungi

Fungi are eukaryotic organisms. They lack chlorophyll and are distinguished from all other living organisms by their principal modes of vegetative growth and nutrient intake. Fungi grow from the tips of filaments (hyphae) that make up the bodies of the organisms (mycelia), and they digest organic matter externally before absorbing it into their mycelia.

Fungi are opportunistic pathogens. In animals, fungal infections are usually external or cutaneous infections, such as fungal nail infections, ringworm or thrush. However, in some cases (especially among those who are immunocompromised), fungal infections can breach outer defences and be far more invasive. For example, infection by the fungus *Aspergillus* can lead to death in some individuals.

Most parasitic fungi cause diseases in plants, and some can lead to widespread damage. An example of this is shown in figure 5.13.

FIGURE 5.12 Different types of fungal nail infection

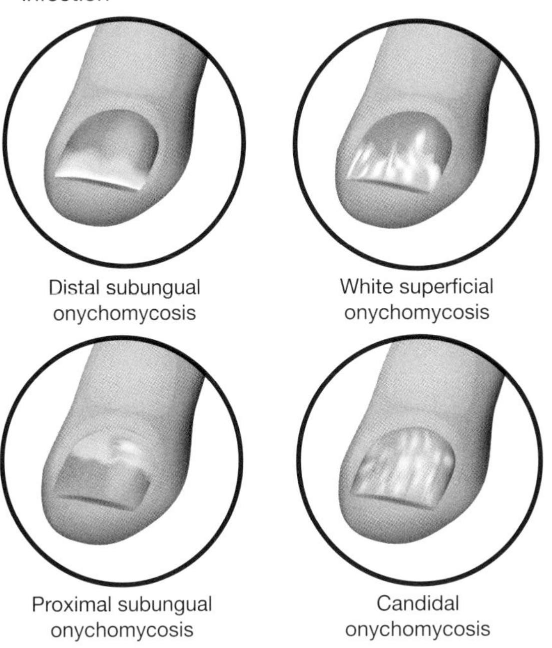

FIGURE 5.13 Fungal infection known as powdery mildew on the leaf of a plant

Parasites

Another group of disease-causing organisms is parasitic helminth worms, including roundworms, hookworms and whipworms. Helminth infections are transmitted via soil contaminated with human faeces that contains eggs of these worms. Many helminths do not fall into the category of microbes as they are visible to the unaided eye.

FIGURE 5.14 Hookworm

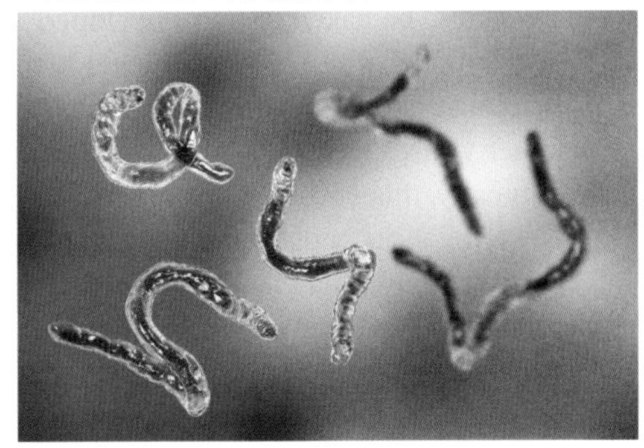

elog-0673

INVESTIGATION 5.1 online only

Microorganisms — where are they found?

Aim

To recognise that microorganisms live in a range of habitats

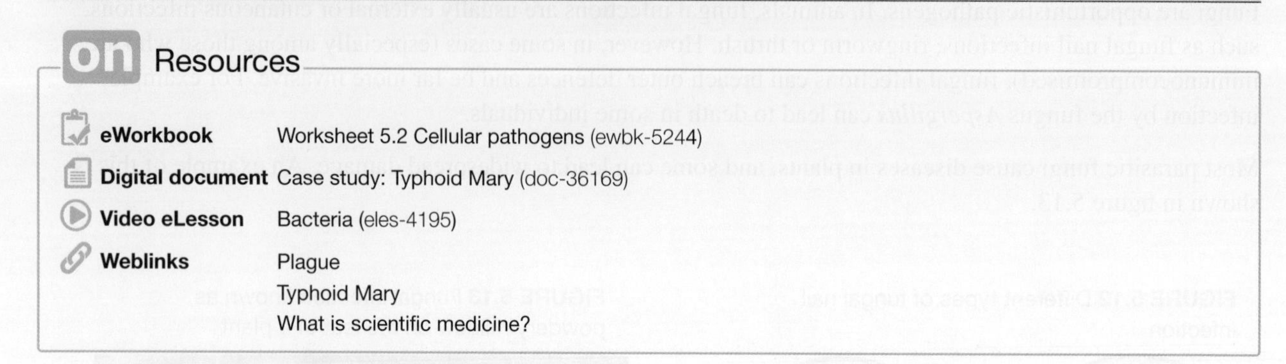

Resources

eWorkbook	Worksheet 5.2 Cellular pathogens (ewbk-5244)
Digital document	Case study: Typhoid Mary (doc-36169)
Video eLesson	Bacteria (eles-4195)
Weblinks	Plague Typhoid Mary What is scientific medicine?

5.2.4 Non-cellular pathogens

Viruses, **prions** and **viroids** are not alive, but they have the ability to cause diseases. They are unable to reproduce independently of host cells and do not have metabolic activity.

Viruses

Viruses may be defined as non-cellular agents consisting of genetic material, either DNA or RNA, that can replicate only within a host cell. They are obligate intracellular parasites and force the host cell to produce more viral particles.

Outside a host cell, viruses exist in a form called a **virion** or viral particle. Virions are the means by which a virus can transfer from one host cell to another.

Viruses are very much smaller than bacteria. Their extremely small size means that they can only be seen using electron microscopy.

All viral particles have a simple non-cellular structure that includes:

- genetic material, either DNA or RNA, organised as a single molecule or as several molecules
- a protein shell, known as a **capsid**, that surrounds the genetic material.

The combination of genetic material plus its surrounding protein coat or capsid is the nucleocapsid.

- For some viruses, the nucleocapsid comprises their total structure; these viruses are said to be naked or **non-enveloped viruses**.
  - Examples of naked viruses include those that cause polio, hepatitis A, and human papillomavirus (HPV).
- Other viruses have an additional outer envelope that surrounds the nucleocapsid. Typically, this envelope is a segment of the plasma membrane that is 'captured' when the virus buds from the infected host cell. The budding process does not destroy the plasma membrane, so that the host cell is not killed. As a result, some virus-infected cells may continue to shed viruses for some time, so that the viral infection can persist until the cell eventually dies. Viruses with an outer envelope are more sensitive to heat, drying, acid, and detergent treatment than naked viruses. As a result, enveloped viruses can be destroyed more easily by sterilisation.
  - Examples of these **enveloped viruses** include those that cause chickenpox, influenza and HIV.

viruses non-cellular pathogens that use the host cell in order to replicate their genetic material

prions infectious particles made of protein that lack nucleic acids

viroids simple forms of viruses that lack a capsid

virion the extracellular form of a virus that can transfer between hosts

capsid protein shell enclosing the genetic material of a virus

non-enveloped viruses viruses that lack an outer membrane; also referred to as naked

enveloped viruses viruses with an outer envelope composed of part of the plasma membrane of the host cell when the viral particles bud from the cell

FIGURE 5.15 Comparing a. enveloped and b. naked viruses

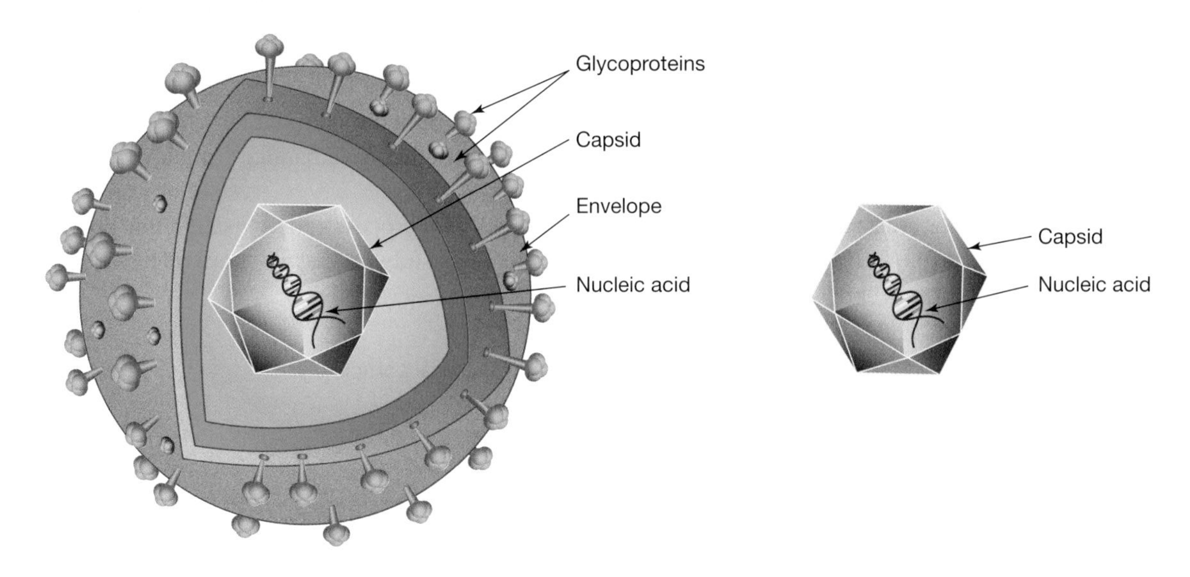

EXTENSION: Viral genetic material

The genetic material of all eukaryotic organisms (plants, animals and fungi) and all prokaryotic organisms (bacteria and archaea) is double-stranded DNA (dsDNA).

In contrast, the genetic material of viruses is much more diverse. The genomes of different viruses are composed of either DNA or RNA, and this DNA and RNA may be either double-stranded (ds) or single-stranded (ss). For some viruses, their genetic material is present as a single nucleic acid molecule. In other viruses, the genetic material consists of several segments of nucleic acid. For example, the genome of the influenza A virus consists of eight segments of ssRNA, the HIV genome consists of two segments of ssRNA, the polio virus genome is a single segment of ssRNA, and the genome of the smallpox virus is a single segment of dsDNA.

Retroviruses are a specific type of RNA virus that produce DNA from their RNA genome (using reverse transcription) and incorporate this DNA into the genome of the host. Retroviruses also contain enzymes that allow for this — reverse transcriptase to produce DNA and integrase to incorporate their DNA.

TABLE 5.5 Comparing different viruses

Virus	Disease	Structure	Genetic material
Varicella zoster virus	Chickenpox (varicella)	Enveloped	dsDNA
Smallpox virus	Smallpox	Enveloped	dsDNA
Parvovirus B19	Erythema infectiosum	Naked	ssDNA
Rotavirus	Gastroenteritis	Naked	dsRNA
Ebola virus	Ebolavirus disease	Enveloped	ssRNA (negative)
Influenza viruses	Influenza	Enveloped	ssRNA (negative)
Measles virus	Measles	Enveloped	ssRNA (negative)
Rabies virus	Rabies	Enveloped	ssRNA (negative)
SARS coronavirus 2	COVID-19	Enveloped	ssRNA (positive)
Poliovirus	Poliomyelitis	Naked	ssRNA (positive)
Rhinovirus	Common cold	Naked	ssRNA (positive)
HIV	AIDS	Enveloped	ssRNA-retro

ss = single-stranded; ds = double-stranded

The rate of mutation of RNA viruses is much higher than that of DNA viruses. As a result, RNA viruses, such as coronaviruses and the influenza A virus, are constantly changing and can produce new strains. For this reason, **vaccines** against influenza that are effective in one year can be ineffective if a new strain of influenza emerges.

vaccines soluble antigens derived from the causative agents of diseases that are administered to individuals, providing them with protection

Virus release

An infected host cell becomes a factory for the production of multiple copies of the virus. The mode of release of viral particles from an infected cell may be by budding or by cell **lysis**. This allows it to spread and infect other cells.

- Enveloped viruses are released from an infected cell by a process of budding, in which virions are released until the infected cell finally dies. Figure 5.16 shows viral particles that have budded from a virus-infected eukaryotic cell.
- Naked virus particles are commonly released from the host cell in a process called lysis. In this process, the infected host cell 'explodes' as its plasma membrane disintegrates and viral particles are released into the extracellular fluid from where they can infect other cells.

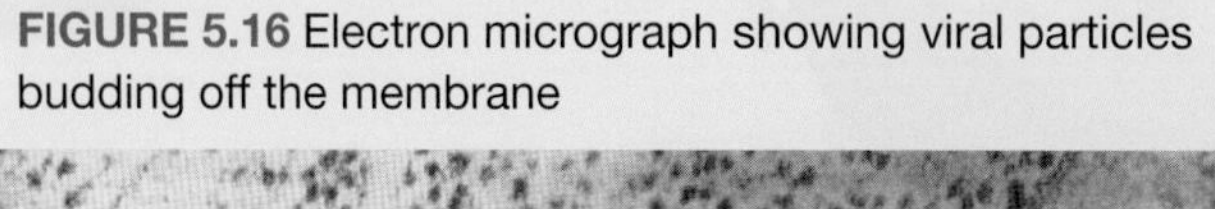

FIGURE 5.16 Electron micrograph showing viral particles budding off the membrane

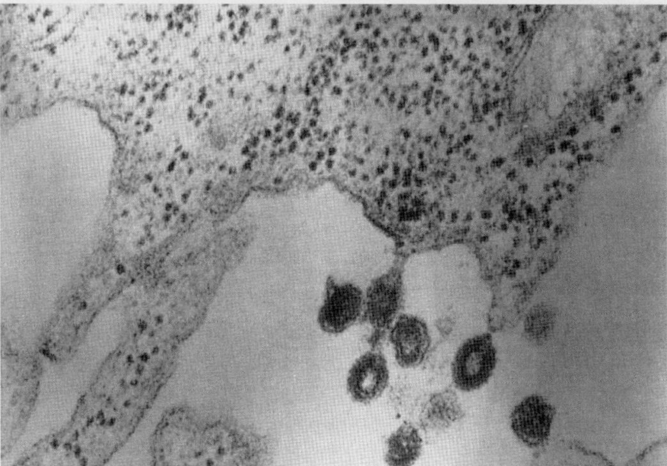

lysis destruction of cells by rupturing the membrane of the cell

FIGURE 5.17 Enveloped viruses using part of the plasma membrane as they are released

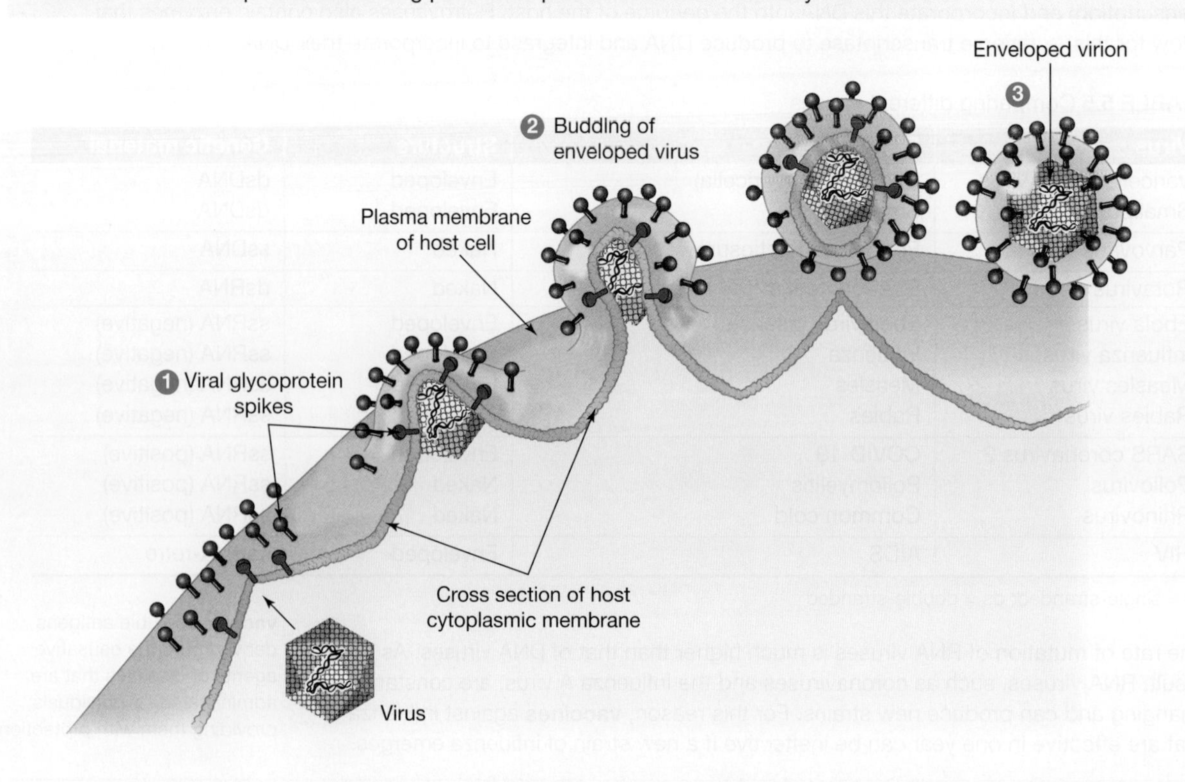

Viroids

Viroids are simpler forms of viruses that lack a capsid. They are the smallest known pathogens and usually only infect plants. Viroids are made of single-stranded RNA molecules that range between 100 and 500 nucleotides. Like viruses, they need to infect a host cell in order to replicate.

An example of a viroid is the coconut cadang-cadang viroid, which infects coconut palms and is lethal, killing million of coconut palms annually.

Prions

Prions are infectious particles made of protein and lacking nucleic acids.

This infectious protein triggers normal proteins in the brain to fold into an abnormal structure. It can affect both humans and animals. The most common form of prion disease that affects humans is Creutzfeldt–Jakob disease (CJD). Prions are extremely small, smaller than viruses, and even through an electron microscope only aggregations (clusters) can be seen, not individual prions.

Prion proteins are most abundant in nerve cells. The prion protein can exist in two forms:
- a normal harmless cellular form of the protein, denoted PrP^{C}
- a harmful infectious prion form of the protein, denoted PrP^{Sc}.

(*Note:* Pr = prion; P = protein; C = cellular; Sc = scrapie.)

The amino acid sequence, that is, the primary structure of the two forms, is identical. The difference between the harmless and the disease-causing form is the secondary structure of the protein involved. This makes them experts at evading the immune system, as they are not often recognised as foreign.

EXTENSION: How do prions 'reproduce'?

The normal form of the prion protein is found mainly in nerve cells. The normal PrP^{C} protein can be transformed to the harmful disease-causing PrP^{Sc} prion by contact with the harmful prion. This contact causes the PrP^{C} protein to unfold and then re-fold abnormally so that its secondary structure is converted to that of the harmful PrP^{Sc} prion.

FIGURE 5.18 Secondary structure of **a.** a normal protein compared to **b.** an infectious prion (image courtesy of Dr Fred Cohen, University of California, San Franscisco)

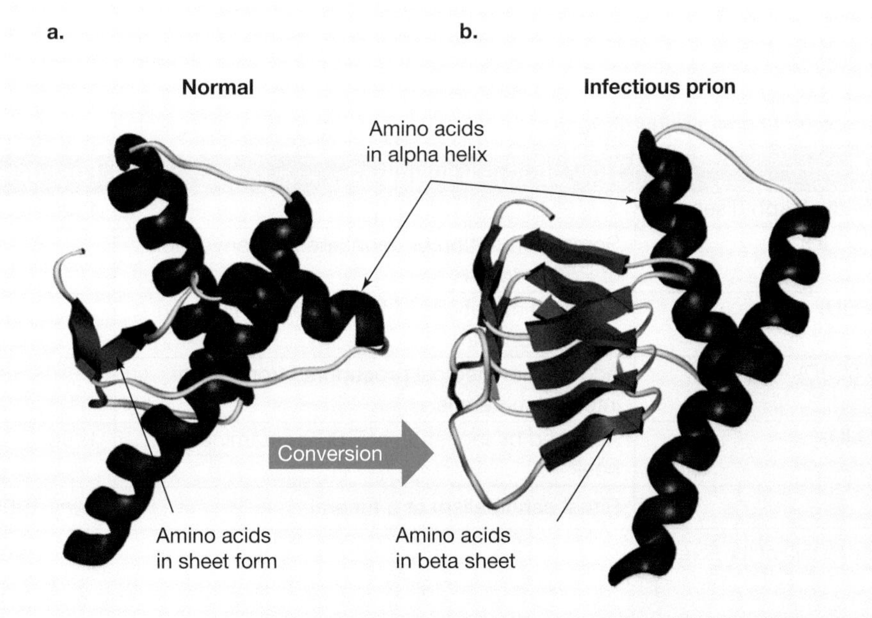

As each new harmful prion is formed, it too can convert other normal protein into harmful prions, and so on. This sets up a chain reaction that rapidly multiplies the numbers of harmful prions (refer to figure 5.19).

FIGURE 5.19 Harmful PrP^{Sc} prions do not remain as monomers but aggregate into rods that form plaques in nerve cells. The accumulation of PrP^{Sc} prions in the brain causes progressive nerve cell death that is most visible as 'holes' or lesions in the brain.

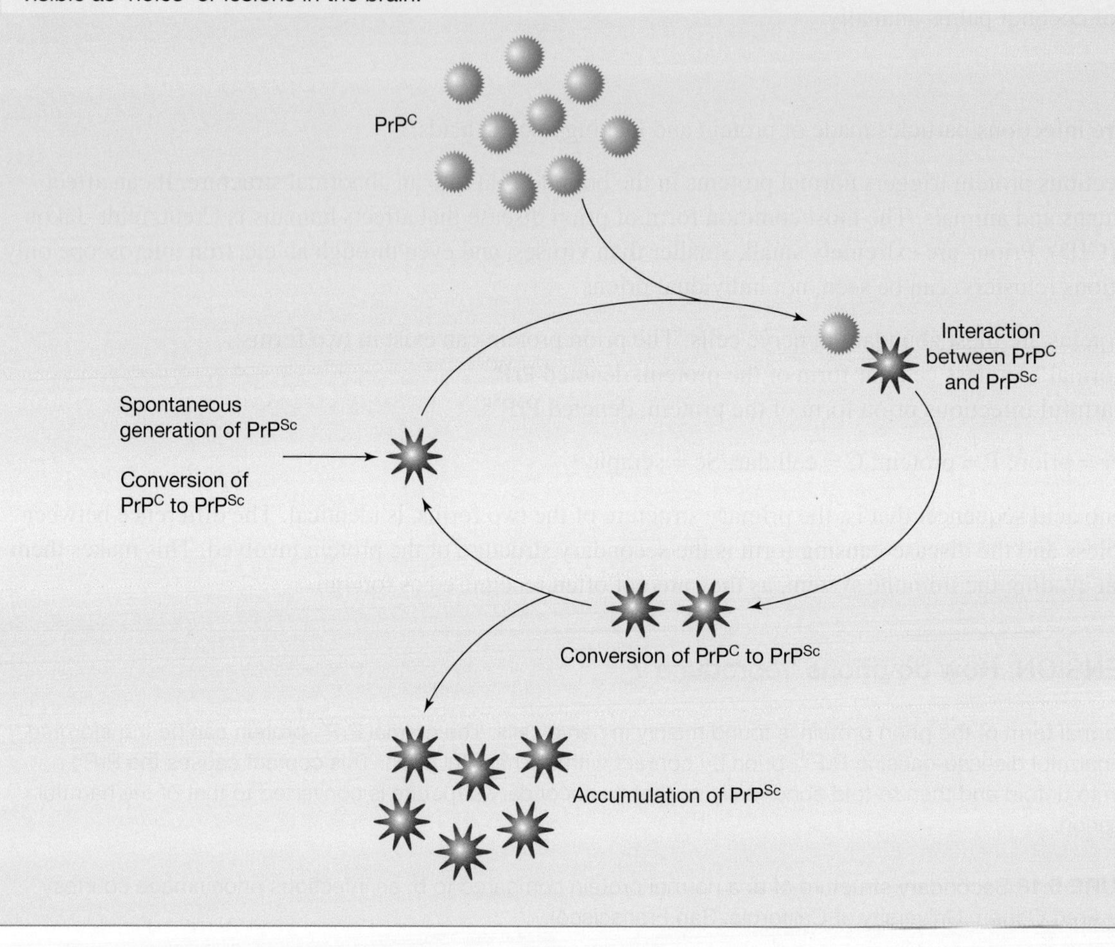

Human prion diseases

TABLE 5.6 Examples of human prion diseases

Human prion disease	How prion is transmitted	First described case
Classic Creutzfeldt–Jakob disease		1920
Sporadic CJD (sCJD)	Somatic mutation or spontaneous conversion of PrP^C into PrP^{Sc}	
Familial CJD (fCJD)	Inherited mutations of PNPR gene on human chromosome 20	
Iatrogenic CJD (iCJD)	Acquired in medical procedures from prion-infected materials	
Variant CJD (vCJD)	Acquired by ingestion of BSE-contaminated beef	1996
Kuru	Ritual cannibalism at a funeral	Early 1900s#

#Date based on oral history of the Fore people.

Each prion disease is distinct, but all of these diseases show the following features:

- a long **incubation period**, sometimes measured in years
- a progressive deterioration of brain function with an inevitable fatal outcome
- distinctive changes to the brain including loss of neurons and development of lesions ('holes') in specific regions of the brain that produce a spongiform appearance that is visible on microscopic examination.

No treatment is currently available for people with prion diseases. The main issue with prions is that they do not stimulate an immune response. As they are misfolded forms of an individual's own protein, they are not recognised as foreign. This makes them very different from other pathogens.

incubation period the time period between infection and the first appearance of the symptoms of a disease

Prion diseases can't be cured, but certain medicines may help slow their progress. It is important that individuals with prion diseases, such as CJD, do not donate organs or tissue.

FIGURE 5.20 An MRI of a brain infected with CJD

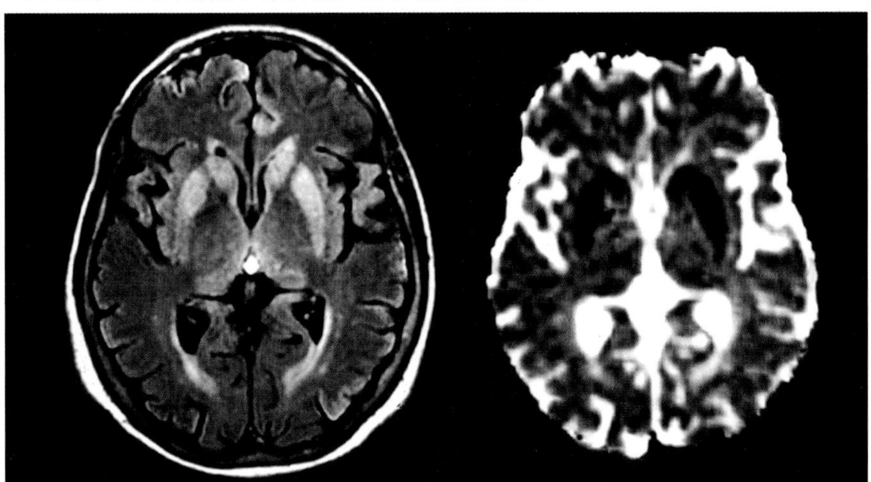

CASE STUDY: Kuru

In Papua New Guinea, a prion disease known as kuru was common among the Fore people until the 1960s. Kuru, like all other prion diseases, causes neurodegenerative symptoms leading to spongiform encephalopathy.

One of the symptons of kuru is random spurts of laughter as the prions slowly convert normal proteins into the pathogenic form.

Why was kuru common in this tribe when it is so rare elsewhere? One of the practices conducted by individuals of the Fore people was ritual cannibalism. Part of the Fore belief system was that eating the bodies of the dead was important in freeing their spirit. To pass on the wisdom of decreased ancestors, the women and children often ate the brain.

Kuru was found mostly in women and children due to this practice of eating brains infected with the prion. Although the Fore people no longer practice this cannibalism, the prion can have an incubation of over 50 years, meaning that there may still be individuals infected with kuru.

tlvd-1776

SAMPLE PROBLEM 1 Cellular and non-cellular pathogens

At a local doctor's surgery, three patients reported as having experienced vomiting, diarrhoea and abdominal pain. The doctor was concerned that they may all be experiencing gastroenteritis as a result of food poisoning. Stool samples were taken from all three patients to determine the cause of the infection. Information about each pathogen from each individual is as follows:

- **Individual 1: The pathogen was found to be Gram-negative and rod-shaped, and was able to reproduce on an agar plate by binary fission.**
- **Individual 2: The pathogen was a single-celled organism. Various cysts were located in the stool containing the *Giardia* parasite.**
- **Individual 3: The pathogen was found to be naked and not contain a protein envelope. Upon exploring the pathogen further, it was found that it was unable to replicate without a host cell and was much smaller than the other pathogens.**

a. **Based on these results, identify which individuals were infected with cellular pathogens and which were infected by non-cellular pathogens. Explain your response. (3 marks)**

b. **All individuals mounted an immune response against the pathogen due to the presence of non-self antigens. Explain what non-self antigens are. (1 mark)**

THINK	WRITE
a. Consider differences between non-cellular and cellular pathogens. • Cellular pathogens are organisms that are able to reproduce independently. • Non-cellular pathogens require a host to reproduce. Link this with the pathogens found in each of the individuals.	Individual 1 was infected with a *cellular* pathogen (a bacterium) as it could reproduce without a host (1 mark). Individual 2 was infected with a *cellular* pathogen (a protozoan) as it is specifically described as an organism (1 mark). Individual 3 was infected with a *non-cellular* pathogen (a virus) as it could not replicate without a host (1 mark).
b. This is a one-mark explain question. You therefore need to provide one detailed statement about non-self antigens	Non-self antigens are any antigens that are recognised as foreign and do not belong to the body's own cells (1 mark).

Resources

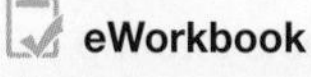

eWorkbook Worksheet 5.3 Non-cellular pathogens (ewbk-5246)

Video eLesson Pathogenic agents: prions (eles-4162)

5.2.5 Allergens

An **allergen** is any antigen that causes the immune system to produce an abnormal and inappropriate overreaction when a person is exposed to it. For most indiviuduals, these antigens are recognised as harmless, but in some sensitised individuals, an immune response is mounted.

allergen an antigen that elicits an allergic response

allergy an abnormal immune response to a substance that is harmless for most people

allergic response rapid immune response to normally harmless antigens such as dust or pollen; involves production of IgE antibodies by B lymphocytes and release of histamines by mast cells

An **allergy** is present when a person's immune system reacts abnormally to substances in the environment that are harmless to most people. This reaction is an **allergic response**, or a hypersensitive reaction. Allergic responses involve cells of both the innate and adaptive immune systems.

Many allergens are small, highly soluble proteins present on the surfaces of dry particles, such as pollen grains and dust. Table 5.7 lists some common allergens.

TABLE 5.7 Common sources of allergens

Category	Sources of allergens
Foods	Peanuts, shellfish, eggs, soy, milk and dairy products, wheat, tree nuts
Pharmaceuticals	Penicillin, aspirin
Plants and plant products	Latex, pollens, sap of rhus tree
Animal products	Shed pet skin cells (dander), excreta of dust mites, insect stings (bees and wasps)
Other	Fungal spores

A sensitised person may make direct contact with a particular allergen by:

- inhaling the allergen (for example grass pollen or dust mite excreta)
- ingesting it (for example peanuts or prawns)
- touching it (for example latex gloves or leaves of the rhus tree)
- when it is injected (for example bee stings or drugs).

Most allergic reactions, although very uncomfortable, are mild to moderate, such as redness and swelling from a mosquito bite, or the sneezing, runny nose and watery eyes of hayfever. These allergic reactions do not cause major problems. They typically affect a specific area of the body, such as the eyes (red, itchy and watery), the nose and sinuses (runny) and the skin (rash and hives). The effects of allergy are often linked to a chemical known as histamine, which is released by immune cells known as **mast cells**. Histamine causes an **inflammatory response**.

Type I hypersensitivity is an allergic reaction. Examples of type I hypersensitivity include:

- allergic asthma
- allergic conjunctivitis
- allergic rhinitis ('hayfever')
- **anaphylaxis**
- atopic dermatitis (eczema)
- urticaria (hives).

The two key players in the immune response are mast cells, which release histamine (which causes inflammation), and antibodies (known as IgE). The immune system's response to these allergens will be further examined in the case study in section 5.5.9, where the role of mast cells is further explored.

mast cells immune cells containing histamine, which is involved in allergic responses and inflammation

inflammatory response a reaction to an infection, typically associated with the reddening of the skin owing to an increased blood supply to that region

anaphylaxis acute and potentially lethal allergic reaction to an allergen to which a person has become hypersensitive

FIGURE 5.21 Hives and hayfever are both examples of allergic responses.

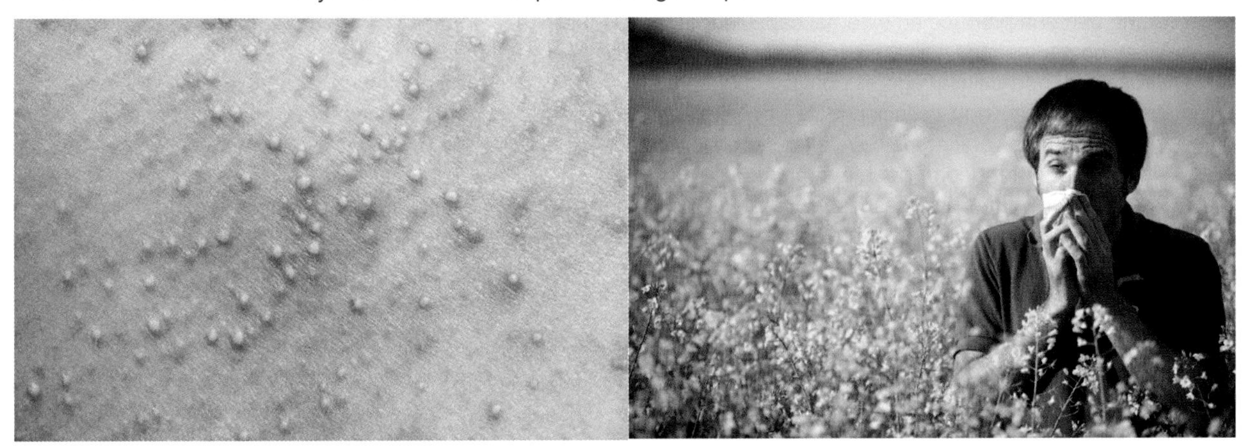

EXTENSION: Anaphylaxis and diagnosing allergies

Many people have various allergies that they are tested for. Anaphylaxis, which occurs in rare cases, is a very severe and life-threatening allergic reaction with a rapid onset that involves many parts of the body. This requires immediate medical attention such as the self-administration of epinephrine (or adrenaline) using an autoinjector (such as an EpiPen®), as epinephrine counteracts histamine.

To access more information on anaphylaxis and testing for allergies, please download the digital document.

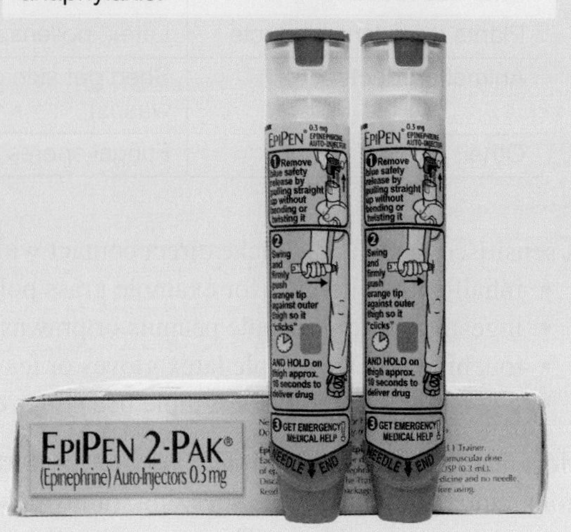

FIGURE 5.22 EpiPens are vital in treating anaphylaxis.

Resources

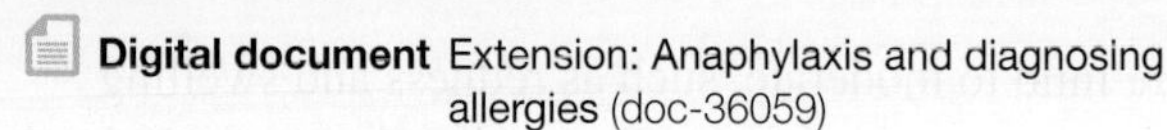

Digital document Extension: Anaphylaxis and diagnosing allergies (doc-36059)

Resources

eWorkbook Worksheet 5.4 The effect of allergens (ewbk-5248)

Weblink Allergy and Anaphylaxis Australia

KEY IDEAS

- The human immune system can identify cells and cell products as self or non-self.
- Antigens are any molecules or parts of a molecule that initiate an immune response.
- MHC markers on the surfaces of all nucleated cells are self-antigens, which are normally tolerated by a person's immune system.
- Cells with MHC markers that differ from a person's own 'self' markers are identified by that person's immune system as non-self and come under immune attack.
- In humans, MHC markers are also known as HLA markers.
- Diseases may be caused by a variety of cellular pathogens, including bacteria, fungi and protozoa, and by non-cellular agents such as viruses and prions.
- Some bacteria produce an external capsule that makes them more virulent.
- Many bacteria cause disease because of the toxins that they produce.
- Viruses are non-cellular agents consisting of genetic material, enclosed within a protein capsid.
- Viruses and other non-cellular agents can only reproduce within the living target cells of their particular hosts.
- Virus structures vary; some have an outer envelope and others are naked. The genetic material may also be either single-stranded or double-stranded DNA or RNA.
- Prions are infectious particles made of protein and lacking nucleic acids that are found mainly in nerve cells.
- Allergens are substances that are harmless but produce an inappropriate immune reaction in susceptible individuals upon exposure. This is termed an allergic response.
- Allergies involve the actions of mast cells, IgE antibodies and histamine.

5.2 Activities

To answer questions online and to receive **immediate feedback** and **sample responses** for every question, go to your learnON title at **www.jacplus.com.au**. A **downloadable solutions** file is also available in the resources tab.

5.2 Quick quiz on	5.2 Exercise	5.2 Exam questions

5.2 Exercise

1. **MC** In early 2018, there was a national outbreak of listeria related to rockmelons. Listeria infection is particularly dangerous for pregnant women, unborn babies and elderly people, and is caused by the prokaryote *Listeria monocytogenes*. This pathogen is able to divide rapidly through binary fission.
   It is reasonable to say that *Listeria monocytogenes* is
   A. a virus.
   B. a protist.
   C. a fungus.
   D. a bacterium.
2. What is the difference between a cellular and a non-cellular pathogen?
3. Identify the following statements as true or false and justify your response.
   a. All pathogens require a host cell in order to replicate.
   b. 'Self' is determined by the MHC markers on a person's nucleated cells.
   c. An example of a non-self cell is a bacterial cell that enters the body through a cut in the skin.
4. Self-tolerance refers to the ability of the immune system to not mount a response against its own cells. Describe the consequences if self-tolerance did not occur.
5. Compare and contrast MHC-I and MHC-II markers.
6. In late 2019, a novel strain of coronavirus was detected in the city of Wuhan. Identify this virus as cellular or non-cellular and outline its mechanism of action.
7. An individual has an allergic reaction against pollen. Outline why antihistamines may be an appropriate prevention for this allergy.
8. Describe why the immune system is unable to recognise and mount an immune response against prions.
9. On Monday 14 August 1922, a group of people were enjoying a picnic near Loch Maree in Scotland. Among the refreshments were sandwiches containing duck paste that had been prepared at the hotel where the picnickers were staying. That evening, two of the picnickers became ill; they died the following morning. Other picnickers became ill, and two died on Wednesday. In all, within a week, eight people were dead, all having eaten the duck paste sandwiches. The inquest determined that the deaths were a result of food poisoning caused by the bacterium *Clostridium botulinum.* Clostridium is an anaerobic, exotoxin-producing bacterium. Explain how the bacterium lead to the death of all eight individuals.

5.2 Exam questions

Question 1 (1 mark)

Source: *VCAA 2020 Biology Exam, Section A, Q21*

MC The property of the immune system that enables it to fight infections and destroy cancer cells is the
A. ability to kill all invading organisms.
B. ability to adapt to donor tissue, facilitating transplants.
C. ability to distinguish self from non-self biological molecules.
D. generation of complement proteins and other chemical barriers.

Question 2 (1 mark)

Source: *VCAA 2014 Biology Exam, Section A, Q14*

MC An example of 'self' material in an adult human female is
A. pollen inhaled from flowers in her garden.
B. sperm cells present in her reproductive tract.
C. cells lining her nose and trachea.
D. malarial parasites inside her red blood cells.

Question 3 (1 mark)

Source: *VCAA 2012 Biology Exam 1, Section A, Q9*

MC Major Histocompatibility Complex (MHC) class 1 molecules

A. release complement proteins.
B. are found only on B and T cells.
C. present foreign antigens to B and T cells.
D. produce antibodies that are specific to each antigen.

Question 4 (4 marks)

Source: *Adapted from VCAA 2008 Biology Exam 1, Section B, Q5a, c, d*

Normally in mammals, if tissue from another individual enters the body, the foreign cells are recognised as 'non-self' by the immune system. The tissue is then rejected unless special drugs are used.

a. i. Which cells of the immune system are initially responsible for recognising **non-self** cells introduced by an organ transplant? **1 mark**

ii. How do the cells you have named in part i. distinguish between **self** and **non-self** cells. **1 mark**

The drawing provided, made in 1886, shows the Tasmanian devil, *Sarcophilus harrasii.*

The Tasmanian devil is the largest surviving carnivorous marsupial in Australia. It is officially in danger of extinction due to the deadly Devil Facial Tumour Disease (DFTD), a type of cancer.

DFTD is an unusual type of cancer because it can be passed from one individual to another when deep wounds occur as they fight over food or as they mate. Tumour cells in the mouth or cheek of an infected animal break off and enter a deep wound on an uninfected animal. The tumour cells multiply in the body of the newly infected devil, eventually forming new tumours that kill the animal.

Recent research has shown that the immune system of an unaffected Tasmanian devil responds in the usual way to tissue from other mammalian species. However, a devil accepts tumour cells from another devil as if they are 'self' cells. The tumour cells are ignored, no immune response develops against them, and so the cancerous cells multiply.

b. Suggest why DFTD tumour cells are accepted as **self** cells by previously uninfected Tasmanian devils. **1 mark**

c. Would you consider tumour cells which have entered the body of an unaffected devil to be pathogens? Support your answer **1 mark**

Question 5 (1 mark)

Source: *VCAA 2012 Biology Exam 1, Section B, Q7a*

Yellow fever is caused by a virus transmitted through the bite of a particular species of mosquito.

Would you describe a virus as a cellular or non-cellular pathogen? Justify your answer.

More exam questions are available in your learnON title.

5.3 Subdivisions of immunity

KEY KNOWLEDGE

- The innate immune response
- Characteristics and roles of the components of the adaptive immune response

Source: Adapted from VCE Biology Study Design (2022–2026) extracts © VCAA; reproduced by permission.

Immunity is resistance to infectious disease. The term 'immunity' is derived from the Latin *immunitas* = exemption from duty. However, in a biological sense, immunity means 'exemption from disease'.

The cells and tissues involved in resistance to infection are part of the body's immune system. The human immune system is the collection of organs, tissues, cells and molecules that protects us from various damaging agents around us, including pathogens, toxins and other foreign molecules. The essence of immune defence is the ability of the immune system to distinguish between the body's own cells and molecules, and the foreign cells and molecules that carry distinctive 'non-self' antigens.

Divisions of immunity

Immunity has two major subdivisions whose combined operations protect the body from infectious diseases:

1. innate immunity, also known as non-specific immunity
2. adaptive immunity, also known as specific or acquired immunity.

FIGURE 5.23 The divisions of the immune system

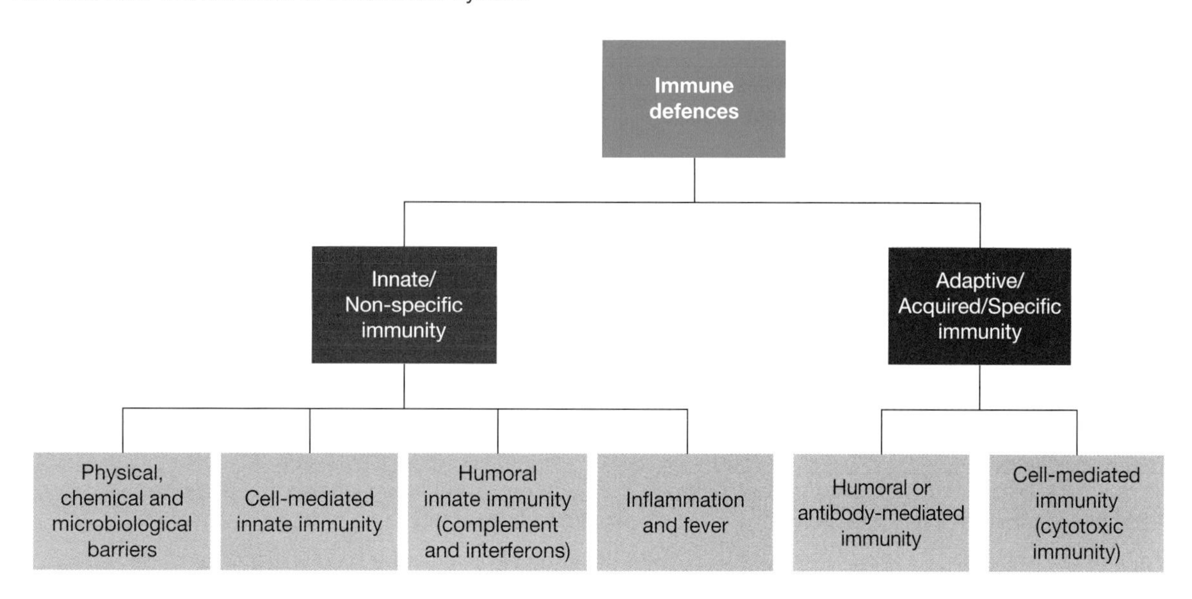

5.3.1 Innate immunity

Innate immunity refers to non-specific defence mechanisms that come into play immediately (or within a very short time frame) of an antigen's appearance in the body. However, it uses the exact same response regardless of the pathogen (hence it is referred to as non-specific). The innate immune system is the body's pre-existing defence against invasions by pathogens. These mechanisms include barriers such as intact skin, chemical secretions, dissolved proteins in the blood, and immune cells (white blood cells) that attack particles, cells and antigens in the body. These are explored in further detail in subtopics 5.4, 5.5 and 5.6. The term innate refers to abilities that come naturally from birth, which is the case in innate immunity.

innate immunity the type of immunity that is present from birth, is fast acting but not long lasting, and produces non-specific (generic) responses against classes of pathogens

adaptive immunity an immune response that is specific to a particular antigen and develops through contact with an antigen

immunological memory ability of the adaptive immune response to remember antigens after primary exposure

memory cells long-lived cells specific to an antigen that are retained in lymph nodes and can respond to future reinfection

5.3.2 Adaptive immunity

Adaptive immunity refers to antigen-specific immune response. The adaptive immune system comes into operation only after the innate immune defences are evaded or overwhelmed, and it takes days to develop a full response. Adaptive immunity provides defences that are specifically tailored for attack against each particular pathogen. The formation of **immunological memory** (through **memory cells**) is also a key feature of adaptive immunity and makes future responses against a specific antigen more efficient. Adaptive immunity is only present in vertebrates.

TABLE 5.8 Summary of differences between innate (non-specific) immunity and adaptive (specific) immunity

Innate immunity	Adaptive immunity
Also known as non-specific immunity	Also known as specific or acquired immunity
Response is not antigen specific	Response is antigen specific
Produces non-specific (generic) responses against classes of pathogens, not against specific pathogens	Produces specific responses, with tailor-made antibodies against each particular pathogen
Immune responses occur mainly at sites of infection	Immune responses occur mainly in secondary lymphoid organs, such as lymph nodes
Only reacts against pathogens	Reacts against both pathogens and foreign molecules such as toxins
Present from birth	Develops only after infection or after immunisation
Activity is always present so that maximum response is rapid and immediate	Normally inactive or silent so that maximum response is slower — within days or weeks
Defence provided is not long-lasting	Defence provided after an infection is long-lasting, in some cases even a lifetime
No 'memory' of prior infections, so that an identical response occurs with every infection	'Memory' of prior infections so that response is faster and stronger if the same microbe re-infects
Major responses are: 1. Cellular attack on bacteria and virus-infected cells 2. Attack by soluble proteins. These two responses (1 and 2) combine to produce inflammation.	Major responses are: 1. Cellular responses attack infected cells (that is, intracellular pathogens). 2. Antibody responses target extracellular pathogens and non-self antigens.

tlvd-1777

SAMPLE PROBLEM 2 Comparing and contrasting types of immunity

Compare and contrast innate and adaptive immunity. **(2 marks)**

THINK	WRITE
1. Identify what the question is asking. The question is asking you to *list* the similarities and differences between innate and adaptive immunity.	
2. Compare the two types of immune response by writing down the similarities. • Both defend against pathogens and foreign antigens. • Both can destroy intracellular and extracellular pathogens.	Both innate and adaptive immune responses provide protection against foreign antigens and pathogens (1 mark).
3. Contrast the two types of immune response by giving the differences. Ensure that you address both the innate response and the adaptive response in your answer. • Innate immunity is non-specific to a particular antigen, but adaptive immunity is specific. • Innate immunity forms no memory, but adaptive immunity forms a memory for future responses. • Innate immunity is quick but short-lasting, whereas adaptive immunity is slower but long-lasting.	However, innate immune response is non-specific, fast and forms no memory, whereas adaptive immune response is specific, slow and forms memory cells for a particular antigen, allowing for an amplified response on future exposure (1 mark).

5.3.3 Cell-mediated and humoral responses

Both subdivisions of immunity protect against infection within the body through the actions of:

1. **cell-mediated immunity**
2. **humoral immunity**.

Traditionally, these terms are often linked to the adaptive immune response. However, in the innate response, there are also cell-mediated and humoral responses, involving immune cells or dissolved proteins or molecules respectively.

cell-mediated immunity immune response that is mediated by immune cells

humoral immunity immune response mediated by soluble molecules in the blood, lymph and interstitial fluid that disable pathogens

FIGURE 5.24 Comparing cell-mediated and humoral responses

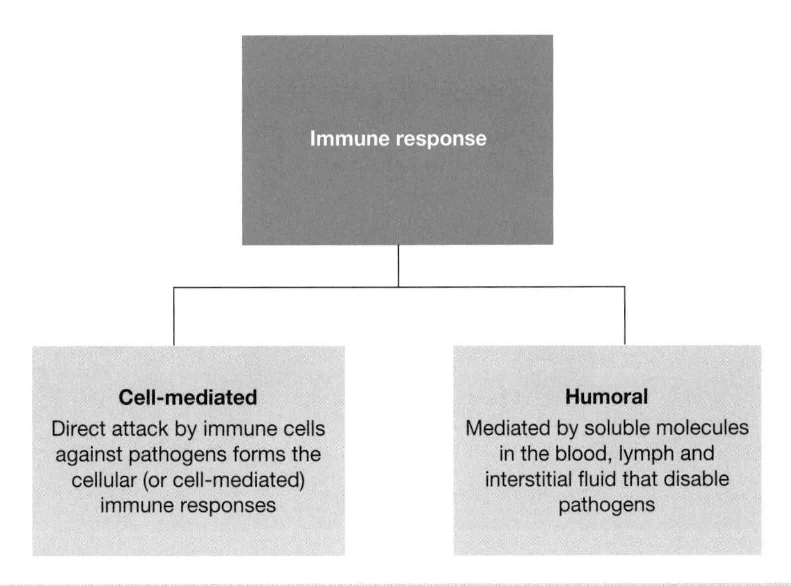

Cells of the immune system

All the cells of the immune system are white blood cells derived from multipotent stem cells in the **bone marrow**. One subgroup of immune cells is the **lymphocytes**, which include **B lymphocytes** and **T lymphocytes** found in large numbers in **lymph nodes**. Other immune cells, such as **neutrophils**, circulate in the bloodstream; others, for example macrophages and dendritic cells, are located close to points where pathogens can gain entry to the body, such as in membranes of the throat, airways and gut. These different cell types will be investigated in subtopics 5.5, 5.10 and 5.11.

FIGURE 5.25 Blood cells are derived from bone marrow.

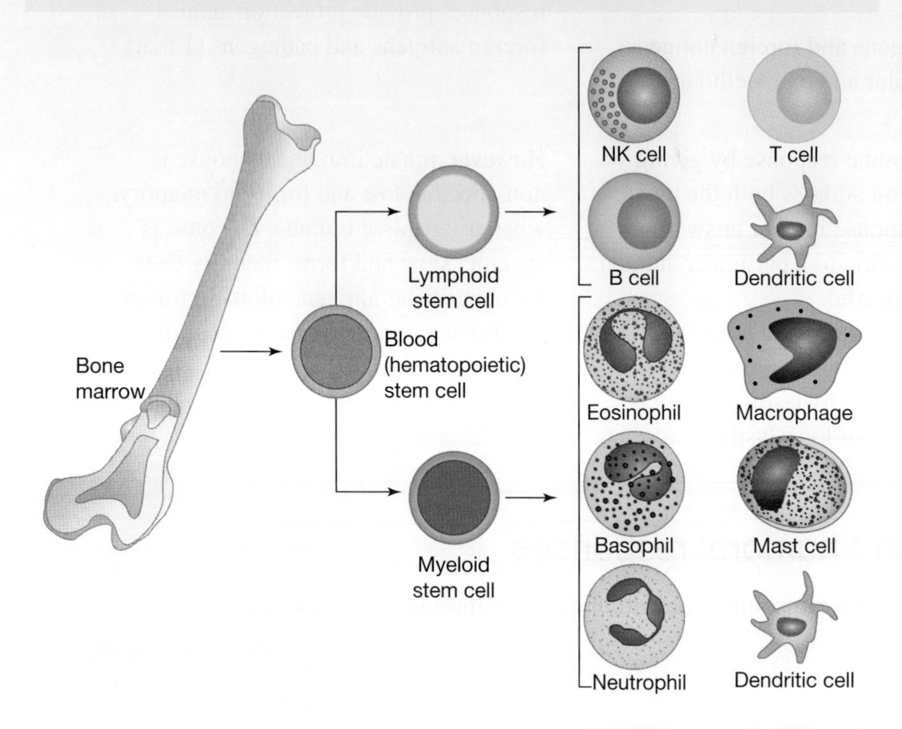

bone marrow fatty substance in the internal cavity of bones; the site of blood cell formation

lymphocytes class of white blood cells found in all tissues including blood, lymph nodes and spleen, and which play a role in specific immunity

B lymphocytes also called B cells; white blood cells that recognise antigens or pathogens and produce a large number of antibodies specific to an antigen

T lymphocytes also called T cells; white blood cells that mature in the thymus and participate in the adaptive immune response

lymph nodes organs of the lymphatic system where B cells and T cells are activated and adaptive immune responses occur

neutrophils the most common type of white blood cell; one kind of phagocyte

FIGURE 5.26 Photomicrographs of human blood smears: **a.** neutrophil (upper right) and lymphocyte (lower left), **b.** basophil, **c.** eosinophil, **d.** monocyte (left) and lymphocyte (right)

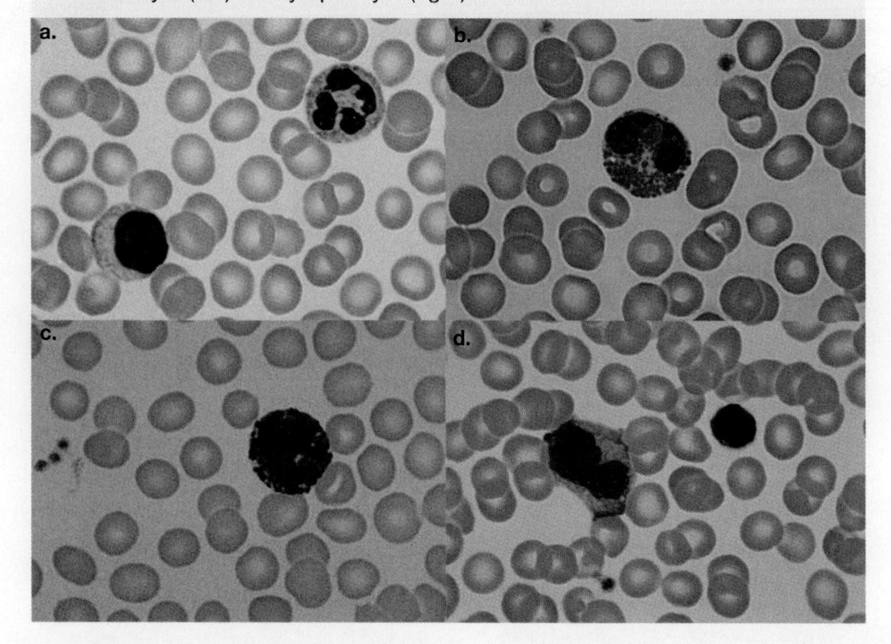

5.3.4 Lines of defence

Immunity operates through three lines of defence against pathogens:

1. The **first line of defence** consists of physical, chemical and microbiological barriers to prevent pathogens from gaining entry to the body. This includes barriers such as intact skin and mucous membranes. (refer to subtopic 5.4)
2. The **second line of defence** consists of the actions of immune cells and soluble proteins mounting a rapid but non-specific attack against pathogens that gain entry to the body. This includes inflammation and the action of phagocytes. (refer to subtopics 5.5 and 5.4)
3. The **third line of defence** consists of the actions of immune cells and antibodies tailored specifically to attack each invading pathogen. The third line of defence involves the recognition of specific antigens by lymphocytes. This includes the action of B cells and their antibodies, and the actions of cytotoxic T cells (refer to subtopics 5.8, 5.9, 5.10 and 5.11).

first line of defence part of the defence against pathogens provided by barriers of the innate immune system that prevent entry of pathogens into the body

second line of defence part of the defence provided by the immune cells and soluble proteins of the innate immune system against attacking pathogens that gain entry to the body

third line of defence part of the defence provided by the immune cells of the adaptive immune system through the various actions of T cells and B cells

The first and second lines of defence involve operations of the innate immune system.
The third line of defence comes into operation only if the second line of defence fails. This line of defence involves the operations of the adaptive immune system.

Resources

eWorkbook Worksheet 5.5 Comparing types of immunity (ewbk-5250)

Weblink The immune system interactivity
Examining the components of the immune system

KEY IDEAS

- The immune system comprises various organs, cells, and molecules that act to protect the body from infectious microbes.
- Immunity is resistance to infectious diseases.
- Immunity depends on the ability to distinguish between 'self' and 'non-self' markers (antigens) on cells.
- The immune system has two major subdivisions: innate (non-specific) immunity and adaptive (specific) immunity.
- Immunity provides three lines of defence against infectious disease.
- The first and second lines of defence involve operations of the innate immune system.
- The third line of defence involves operations of the adaptive immune system.
- Innate and adaptive immunity both operate through cellular responses and the actions of soluble proteins (humoral responses).
- Innate immunity refers to non-specific defence mechanisms that respond to the presence of foreign pathogens but are not specific to the pathogen or antigen type. It does not involve the development of immunological memory.
- Adaptive immunity refers to antigen-specific immune response. The adaptive immune system comes into operation only after the innate immune defences are evaded or overwhelmed, and this takes days to develop fully.
- Adaptive immune response in vertebrates can be classified as humoral or cell-mediated.

5.3 Activities

learnON

To answer questions online and to receive **immediate feedback** and **sample responses** for every question, go to your learnON title at **www.jacplus.com.au**. A **downloadable solutions** file is also available in the resources tab.

5.3 Quick quiz on	5.3 Exercise	5.3 Exam questions

5.3 Exercise

1. Name two divisions of the immune system.
2. Identify the subdivision of the immune system that
   a. is the first to respond to a pathogen.
   b. has a memory of previous infections.
   c. responds to each pathogen in a specific manner.
   d. includes physical barriers to the entry of pathogens to the body.
   e. provides long-lasting defence against pathogens that have caused disease.
3. Compare and contrast humoral and cell-mediated immunity.
4. Describe two advantages and two disadvantages of an innate immune response.

5.3 Exam questions

Question 1 (1 mark)

Source: *Adapted from VCAA 2011 Biology Exam 1, Section A, Q8*

MC Nonspecific defences of the immune system that act against bacteria include

A. antibodies.
B. phagocytes.
C. T lymphocytes.
D. plasma cells.

Question 2 (2 marks)

The adaptive branch of immunity is a vital part of the immune system. Severe combined immunodeficiency (SCID) is a disease in which an individual fails to develop a functional immune system. Suggest two possible consequences of this.

Question 3 (1 mark)

MC Factors that do not play a role in the nonspecific immune response include

A. complement proteins.
B. interferon proteins.
C. intact skin.
D. antibodies.

Question 4 (1 mark)

MC In the third line of defence against infection

A. lymphocytes respond to specific antigens.
B. pathogens are prevented from entering the body.
C. the body secretes bactericidal agents.
D. no immunological memory is formed.

Question 5 (4 marks)

a. Define the term immunity. **1 mark**
b. A branch of immunity has differing responses to different antigens and can form memory, providing lifelong immunity. State the name of this branch of immunity. **1 mark**
c. Outline two key differences between humoral and cell-mediated immunity. **2 marks**

More exam questions are available in your learnON title.

5.4 Physical, chemical and microbiota barriers

KEY KNOWLEDGE

- Physical, chemical and microbiota barriers as preventative mechanisms of pathogenic infection in animals and plants

Source: VCE Biology Study Design (2022–2026) extracts © VCAA; reproduced by permission.

The first and external line of defence is achieved by barriers that prevent entry of pathogens into the body. This first line of defence involves the immediate strategy: '*Keep pathogens out*.' The best protection against infection is preventing pathogens from crossing the body surfaces.

There are different types of barriers that prevent the pathogens from entering into the body:

- **physical barriers**
- **chemical barriers**
- **microbiological barriers**

Examples of some of these in humans can be seen in figure 5.27.

physical barriers innate barriers that act to prevent the entry of pathogens into the body

chemical barriers innate barriers that use enzymes to kill pathogens and prevent invasion into a host

microbiological barriers innate barriers involving normal flora in the body

FIGURE 5.27 There are various aspects involved in keeping pathogens out. These may be physical (pink), chemical (blue) or microbiological (purple). In many locations, there are multiple barriers to block pathogens. Microbiological barriers are the natural floral present in that location.

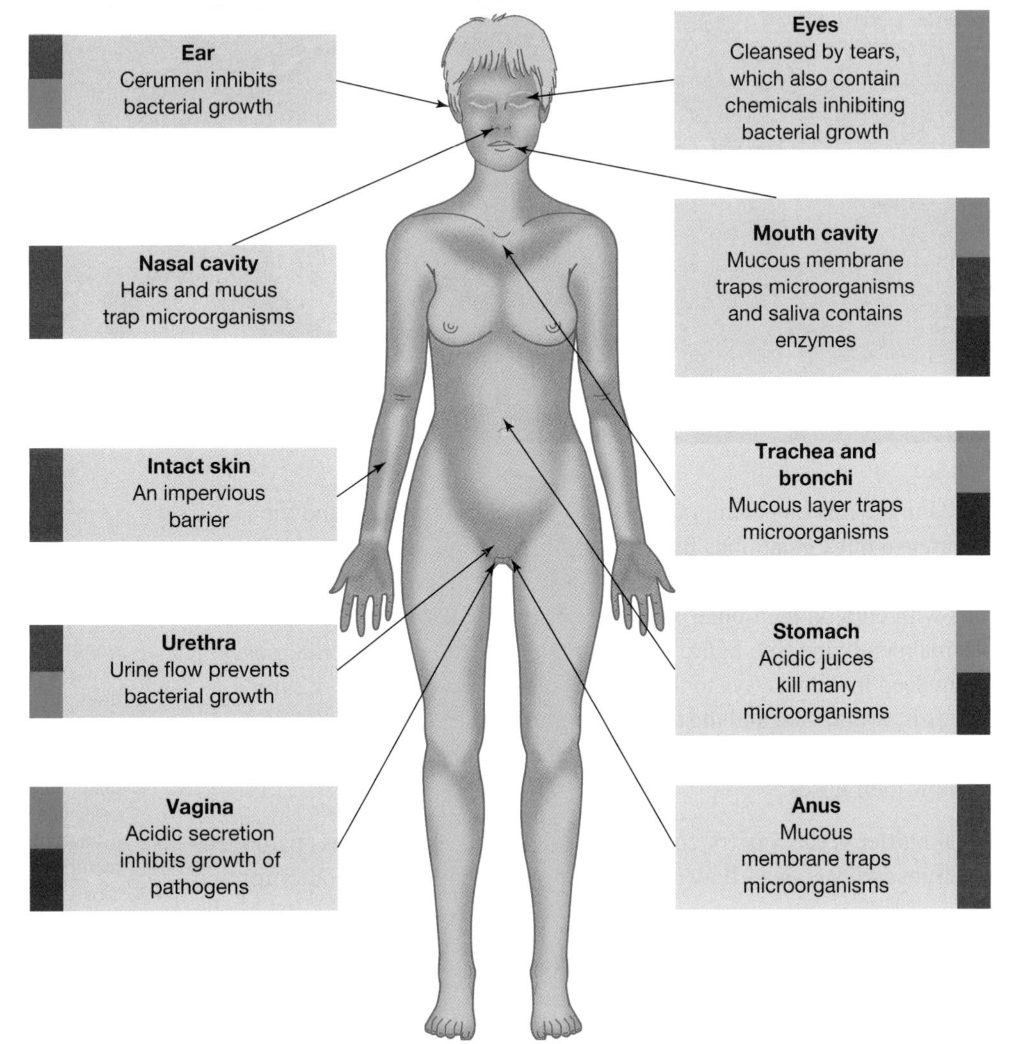

5.4.1 Physical barriers in animals

In vertebrates, epithelial tissue prevents the pathogens from crossing the body surfaces. Epithelial tissues include the external layers of the skin and the internal linings or **mucous membranes**. These epithelial tissues form a *physical barrier* to the entry of microbes and other pathogens because their cells are densely packed, leaving no intercellular space between them.

The major barriers that prevent entry of pathogens are intact skin, mucous membranes and their secretions.

- **Intact skin:** The intact skin constitutes an important physical and chemical barrier to microbial infection. The epidermis of the skin is composed of many layers of cells (keratinocytes) with the outermost layer consisting of flattened dead cells. The constant shedding of these dead surface cells is an effective barrier against entry of pathogens.

TIP: It is important that you refer to INTACT skin and not just skin when discussing this as a physical barrier to infection. If the skin is not intact, it is not able to properly prevent the invasion of pathogens

- **Mucous membranes**: The inner spaces of the airways, the gut and the urogenital tract are lined by mucous membranes consisting of epithelial cells.

FIGURE 5.28 a. Coloured scanning electron micrograph of mucous membrane, showing the epithelium **b.** Stained longitudinal section through the skin. The upper layers are being shed constantly.

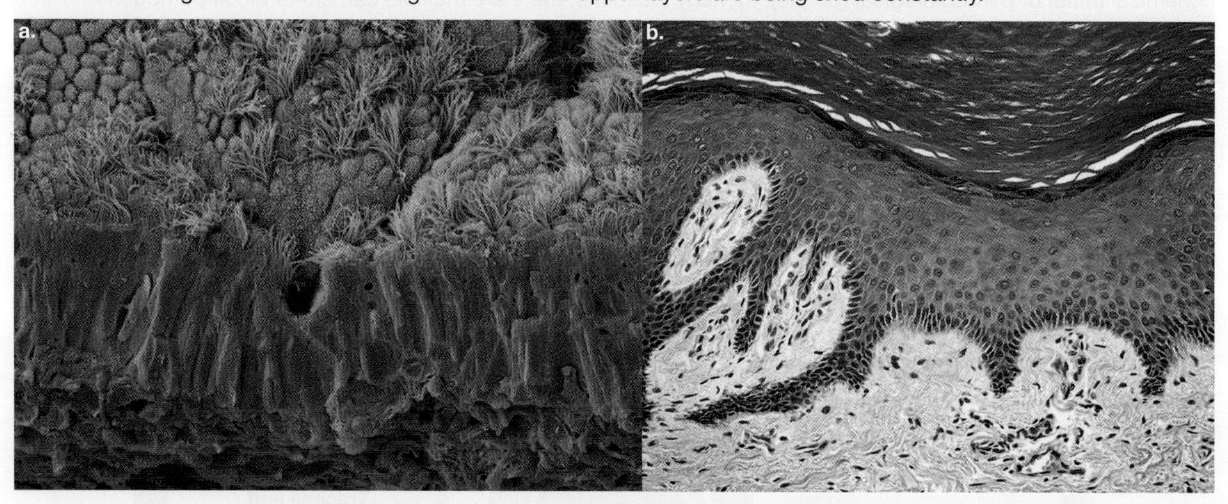

Special cells in the mucous membranes of the gut, the genital tract and the airways secrete a thick gelatinous fluid called **mucus** that can trap particulate matter, including pathogens. In the airways, the mucous membranes include many cells with **cilia** on their outer surfaces. Cilia are fine hair-like outfoldings of the plasma membrane (see figure 5.29a). Regular beating of the cilia moves mucus from deep in the airways to the back of the throat (see figure 5.29b). Some mucus is swallowed, and the acid secretions of the stomach destroy any bacteria. Other mucus is expelled from the back of the throat when people cough or blow their noses.

Another physical barrier is ear wax (or cerumen), which reduces the access pathogens have to the ear drum and ear canal. It also protects the ear from dust and other foreign particles.

mucous membranes cellular linings of the inner spaces within the airways, the gut and the urogenital tract

mucus a gelatinous fluid secreted by cells of the mucous membranes

cilia (singular = cilium) in eukaryote cells, fine hair-like outfoldings formed by extensions of the plasma membrane involved in synchronised movement

FIGURE 5.29 The mucous membrane: a. photograph and b. a schematic representation

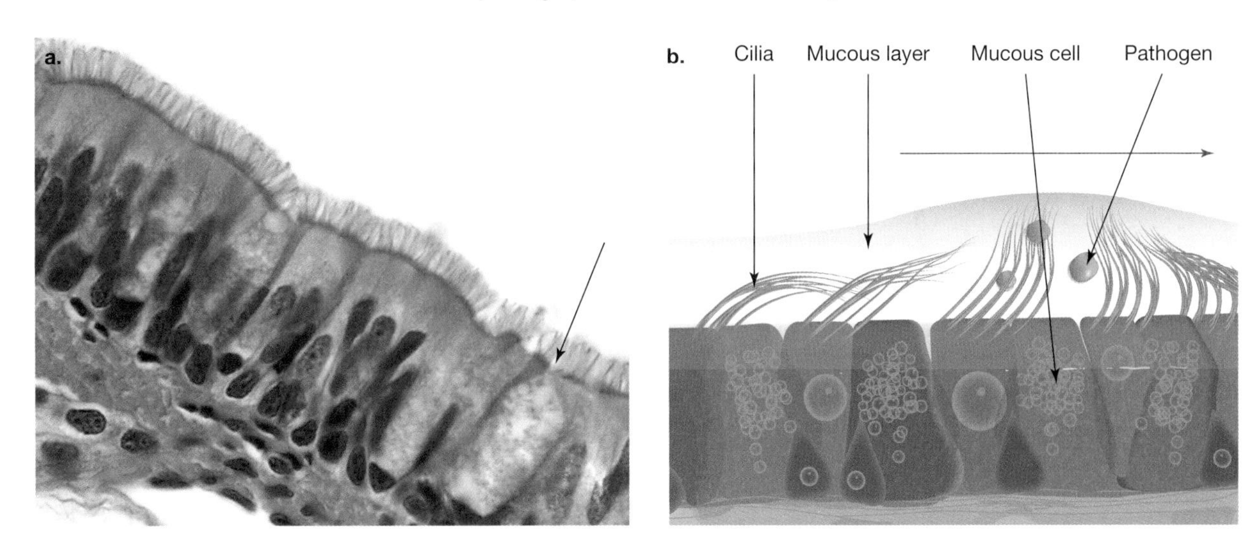

elog-0675

INVESTIGATION 5.2 online only

The importance of intact skin

Aim

To observe the importance of intact skin as a physical barrier against pathogens

5.4.2 Chemical barriers in animals

Chemical barriers destroy pathogens on the outer body surface, at body openings, and on inner body linings. Sweat, mucus, tears, and saliva all contain an enzyme called **lysozyme** that kills pathogens.

Sebaceous glands in the skin produce a secretion called **sebum**, which provides a protective and anti-microbial film on the skin. Sweat secreted onto the skin contains an antimicrobial protein called **dermicidin**, which acts against a wide range of pathogens — particularly fungal and bacterial agents. In addition, stomach acid and digestive enzymes kills pathogens that enter the digestive tract in food or water.

FIGURE 5.30 The structure of lysozyme

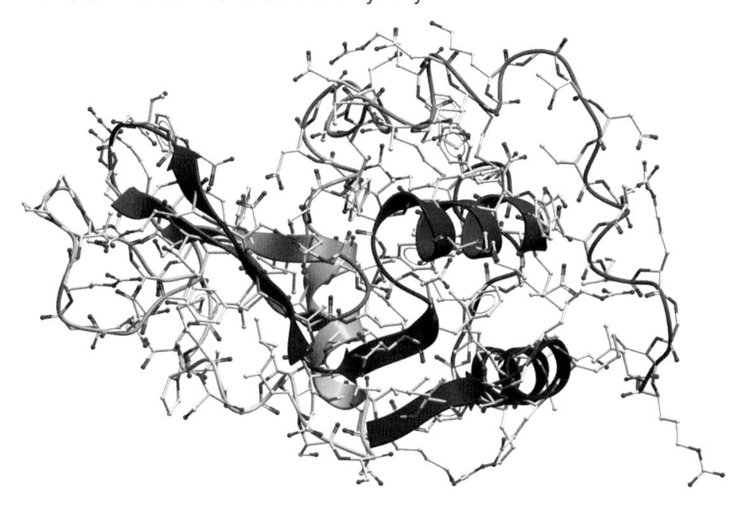

lysozyme an enzyme present in body secretions such as saliva and tears that helps in the first line of defence

sebum the oily secretion produced by sebaceous glands of the skin

dermicidin an antimicrobial protein that acts as a chemical barrier is animals, which is found is secreted sweat

FIGURE 5.31 Sebaceous glands are found close to hair follicles. The sebaceous gland secretes sebum from special cells called sebocytes.

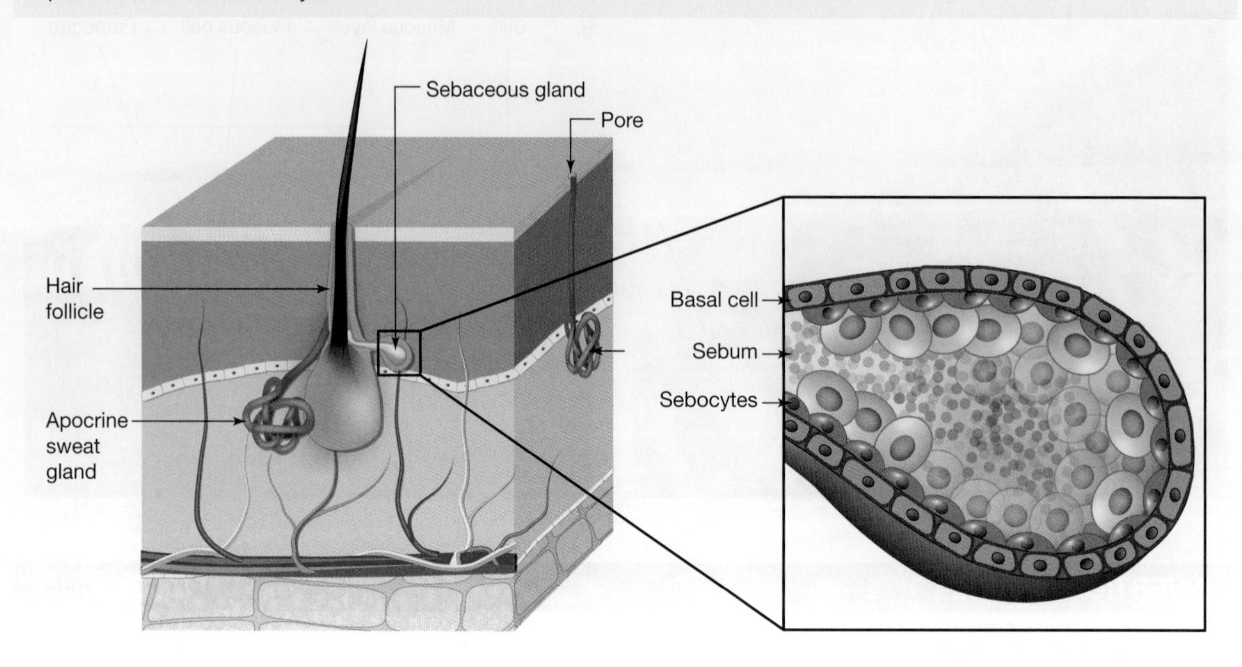

5.4.3 Microbiological barriers in animals

A microbiological barrier to the entry of pathogens is the presence of normal flora. The term 'normal flora' refers to the non-pathogenic bacteria that are the normal residents in particular regions of the body, including the gut, the mouth and throat, and the genital tract. The presence of these harmless bacteria inhibits the growth of pathogenic microbes. For example, the normal flora of the vagina is *Lactobacillus acidophilus*. These bacteria produce lactic acid, which prevents the establishment of pathogenic microbes.

FIGURE 5.32 Examples of normal flora present in the body

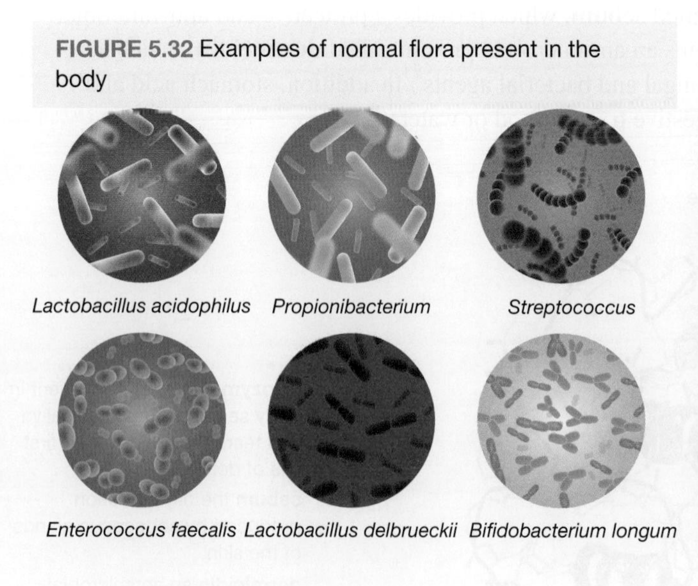

FIGURE 5.33 Lactobacillus organisms and vaginal squamous epithelial cell

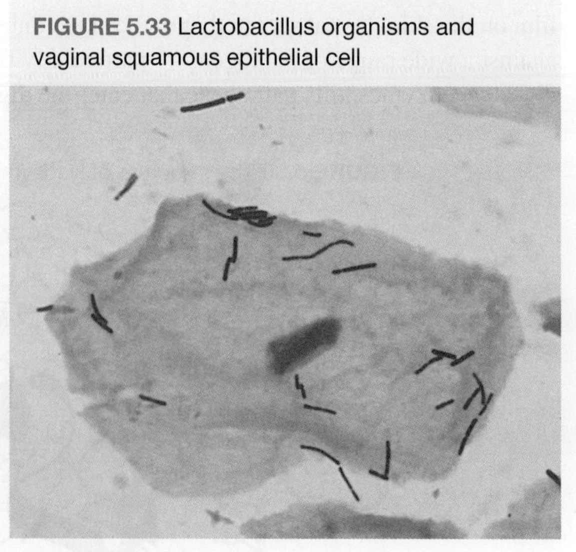

TABLE 5.9 Examples of some normal flora and their locations

Microbiological factor	Location
Lactobacillus spp. (including *L. acidophilus* and *L. delbrueckii*)	Gastrointestinal tract, mouth and vaginal tract
Bifidobacterium longum	Gastrointestinal tract
Propionibacterium spp.	Skin, sweat and sebaceous glands
Streptococcus sanguinis	Mouth
Enterococcus faecalis	Gastrointestinal tract
Staphylococcus aureus	Nose and skin

 Resources

 **Weblink** Microbiological barriers

5.4.4 Barriers in plants

Physical barriers in plants

Plants do not have an immune system comparable to animals, but they have developed an array of structural, chemical and protein-based defences to detect pathogens and stop them before they are able to cause extensive damage. They can remove infected parts or grow gall tissue around an infection to prevent further spread.

FIGURE 5.34 A cross-section of a leaf, showing some of the various physical barriers that protect plants from pathogens

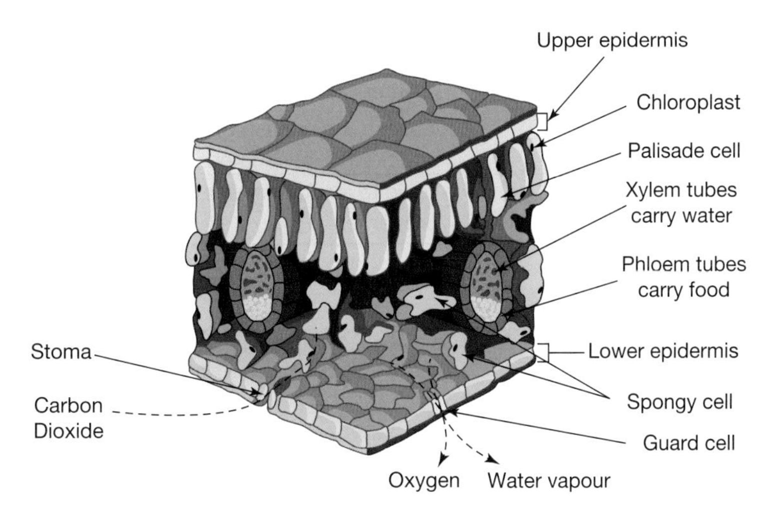

Physical barriers include:

- the cuticle — a waxy covering on leaves that reduces water accumulation and helps prevent cells becoming infected
- thick bark — an external layer that acts similar to intact skin
- stomata — stomata can be closed to prevent pathogens entering; they may also be sunken and sit lower down, to further protect the plant (refer to figure 5.35)
- cell wall — the cellulose cell wall can help protect cells from infection from viruses
- leaf orientation — vertical leaves make it harder for pathogens to attack as water is unable to accumulate on the leaf surface (refer to figure 5.36)
- thorns and spikes — some plants have thorns and spikes (modified leaves and branches) that protect the plants from grazing animals.

FIGURE 5.35 Stomata of the surface of a leaf

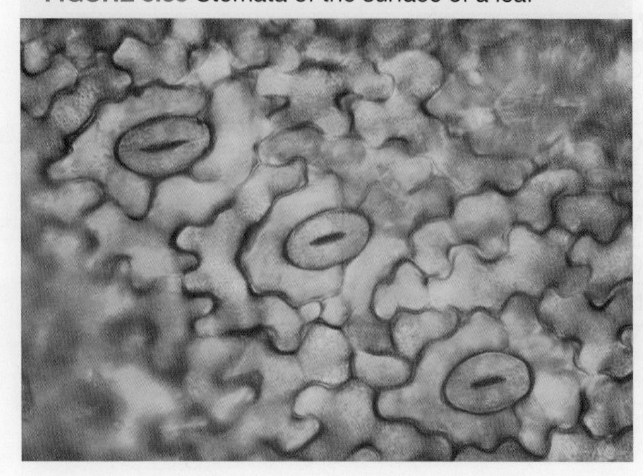

FIGURE 5.36 The *Mimosa pudica* has leaves that fold inwards when touched or shaken for protection.

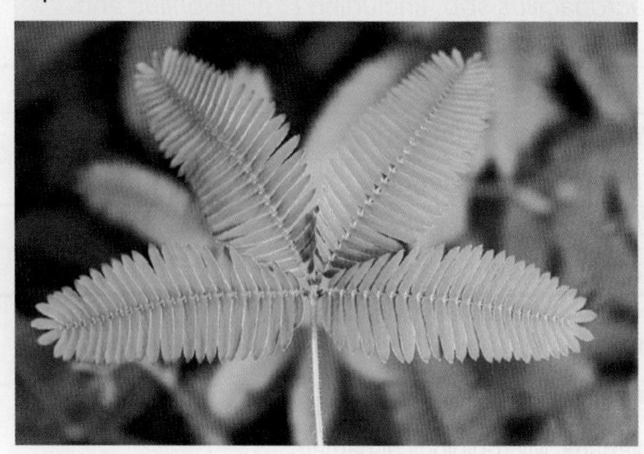

Chemical barriers in plants

Plants also have a variety of chemical defences against pathogens. Plants produce antimicrobial chemicals, antimicrobial proteins, and antimicrobial enzymes that prevent pathogens from entering the plant.

TABLE 5.10 Some chemicals produced by the plants to fight the pathogens

Chemical	Function
Saponin	Plant proteins that disrupt bacterial and fungal cell membranes
Defensins	Chemicals in many plant cell membranes that interfere with fungal cell membranes
Caffeine	Toxic to fungi and insects
Tannins	A bitter-tasting substance that may deter predators
Cyanogenic glycosides	Causes the release of hydrogen cyanide when plant is attacked or consumed
Cinnamon (monoterpenoid)	Toxic to insects and have antibacterial and antifungal properties

elog-0677

INVESTIGATION 5.3

online only

Innate barriers in plants

Aim

To explore different chemical and physical barriers in plants

tlvd-1778

SAMPLE PROBLEM 3 Physical and chemical barriers in plants

The immune system is a host defence system comprising many physical and chemical barriers and processes within an organism that protect against disease. Name and outline the actions of two physical and two chemical barriers that protect plants from infections. (4 marks)

THINK	WRITE
1. List two physical and chemical barriers in plants.	Physical barriers: thick bark and waxy cuticles Chemical barriers: defensins and caffeine
2. Provide a general description of the defence mechanisms mentioned. As this is a four-mark question, 1 mark is awarded for each of the barriers with their appropriate outline.	Thick bark is an external layer of dead cells preventing the entry of pathogens (1 mark). The waxy cuticles on leaves prevent the formation of moisture (1 mark). This stops the pathogen invasion. Caffeine is toxic to many insects and fungi (1 mark), and defensins have antimicrobial action (1 mark).

Resources

eWorkbook Worksheet 5.6 Innate barriers (ewbk-5252)

KEY IDEAS

- Barriers that provide innate resistance include physical, chemical and biological barriers in both plants and animals.
- Physical barriers in animals include intact skin and mucous membranes.
- Chemical barriers include various secretions. Secretions that contribute to preventing entry include mucus, tears, saliva and urine.
- The so-called 'normal flora' that are the non-pathogenic bacterial residents of regions of the body act as a microbiological barrier to pathogenic bacteria.
- Plants have physical and chemical barriers to stop pathogenic invasion. They do not have a second or third line of defence, so they rely on preventing pathogenic entry.

5.4 Activities

learnon

To answer questions online and to receive **immediate feedback** and **sample responses** for every question, go to your learnON title at **www.jacplus.com.au**. A **downloadable solutions** file is also available in the resources tab.

5.4 Quick quiz 	5.4 Exercise	5.4 Exam questions

5.4 Exercise

1. Name the three types of barriers that form the first line of defence against pathogens.
2. Give an example of
   a. an epithelial tissue.
   b. a mucous membrane.
3. Where would you expect to find a ciliated mucous membrane?
4. How do the 'normal flora' of the body contribute to the first line of defence?
5. The immune system is a host defence system comprising many chemical and microbiological barriers and processes within an organism that protect against disease. Name and outline the actions of two physical and chemical barriers that protect humans from infections.

6. Briefly outline how intact skin defends against entry of pathogens to the body.
7. While taking a course of antibiotics, it is often recommended to take probiotics, which contain live bacteria. Explain the purpose of this.
8. Individuals with cystic fibrosis produce thickened mucus (that is difficult to expel) due to defects in a plasma membrane protein. Suggest why individuals with cystic fibrosis are more prone to lung infections, despite producing mucus.

5.4 Exam questions

Question 1 (4 marks)

Source: *VCAA 2018 Biology Exam, Section B, Q3a and b*

Plants are a rich source of nutrients for many organisms, including bacteria, fungi and viruses. Although plants lack an immune system that is comparable to animals, plants have evolved chemical barriers to stop invading pathogens from causing significant damage.

a. Describe **two** chemical barriers that could be present in a plant that is protecting itself from an invading pathogen. **2 marks**

b. Humans have a sophisticated immune response to invading pathogens.
State **two** ways that pathogens are prevented from entering the internal environment of the human body. **2 marks**

Question 2 (1 mark)

Source: *VCAA 2015 Biology Exam, Section A, Q17*

MC Which one of the following is an example of a plant defence against a pathogen?

A. production of antibodies
B. an active immune system
C. extensive cell death throughout the plant
D. waxy leaf surfaces acting as physical barriers

Question 3 (1 mark)

Source: *VCAA 2017 Biology Exam, Section B, Q2a*

Although plants have no immune systems, they do have many chemical and physical methods of defence against pathogens and attacks by insects.

Describe one example of a physical method of defence in a plant.

Question 4 (3 marks)

Source: *VCAA 2006 Biology Exam 1, Section B, Q7a and b*

Plants and animals are both susceptible to infection.

a. i. Name one feature of plants that inhibits the entry of infective organisms. **1 mark**
ii. Name one feature of humans that inhibits the entry of infective organisms. **1 mark**

Assume that infection has occurred in a plant.

b. Outline one way in which a plant responds to minimise damage to its tissues. **1 mark**

Question 5 (1 mark)

Source: *VCAA 2009 Biology Exam 1, Section A, Q19*

MC First-line defence mechanisms in humans include

A. development of fever.
B. action of phagocytes.
C. use of antibiotics.
D. presence of cilia.

More exam questions are available in your learnON title.

5.5 The components of innate immunity

KEY KNOWLEDGE

- The innate immune response and the characteristics and roles of macrophages, neutrophils, dendritic cells, eosinophils, natural killer cells, mast cells, complement proteins and interferons

Source: Adapted from VCE Biology Study Design (2022–2026) extracts © VCAA; reproduced by permission.

5.5.1 What is the innate immune response?

Microorganisms do occasionally breach the first line of defence. It is then up to the innate and adaptive immune systems, which were introduced in subtopic 5.3, to recognise and destroy them without harming the host.

FIGURE 5.37 When the first line is breached, the second line of defence comes into play

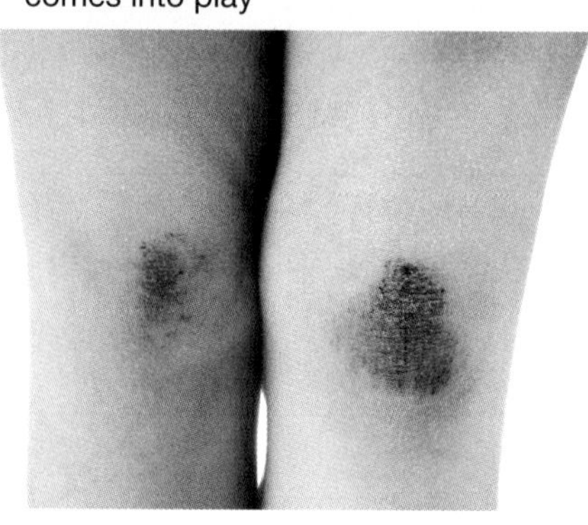

Consider the following situation. You fall over and graze your knee so that the epidemis of the skin is eroded and the dermis is exposed (see figure 5.37). After brushing the dirt away, you notice you are bleeding. As your skin is no longer intact, pathogens from the dirt have gained access to the internal environment, placing you at risk of infection.

The second line of defence comes into action when pathogens have entered the tissues or the bloodstream. It is an innate system — an inborn system that lacks specificity and memory. For example, after you graze your knees, the area becomes red, inflamed and swollen, and pus eventually forms. This is all part of the second line of defence.

The innate immune system must be able to distinguish self from non-self to ensure it is only acting against foreign pathogens. The innate immune system relies on the recognition of particular types of molecules that are common to many pathogens but are absent in the host. These pathogen-associated molecules (called **pathogen-associated molecular patterns**) stimulate two types of innate immune responses — inflammation and **phagocytosis** by cells such as dendritic cells, neutrophils and macrophages. Both of these responses can occur quickly, even if the host has never been previously exposed to a particular pathogen.

FIGURE 5.38 Components involved in the second line of defence

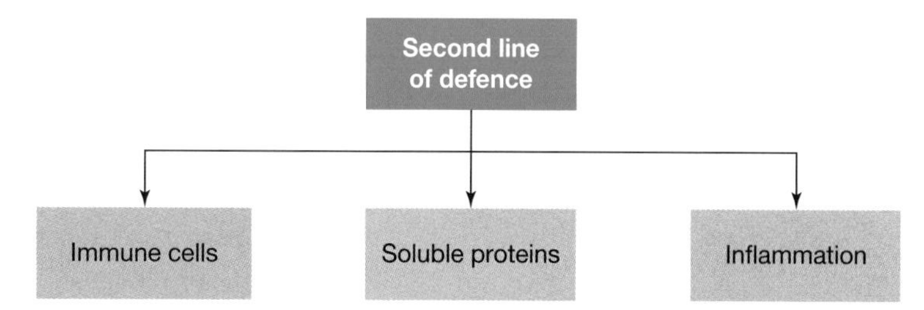

The second line of defence has three key aspects: immune cells, inflammation and soluble proteins. These aspects are explored further in the following sections.

5.5.2 Cells involved in innate immunity

There are many types of white blood cells (immune cells) or **leukocytes** that work to protect the body. They are found in blood, lymph and other tissues as outlined in table 5.11. Each of these cell types will be explored further in upcoming subtopics.

The innate leukocytes include natural killer cells (NK cells), mast cells, eosinophils and phagocytic cells (including macrophages, neutrophils and dendritic cells).

pathogen-associated molecular patterns molecules that are found in pathogens but are not found in a host, allowing them to be recognised as foreign

phagocytosis bulk movement of solid material into cells where the cell engulfs a particle to form a phagosome

leukocytes white blood cells that are involved in protecting the body from infectious disease

TABLE 5.11 Summarising the cells of the innate immune system

Cell type	Stylised image	Description & characteristics	Major role in innate immunity
		In blood	
• Monocytes: 2–12% of circulating white blood cells		Circulating white blood cells (2–12%); can leave bloodstream and move into tissues, where they differentiate into macrophages	Precursors of macrophages
• Neutrophils: ***'first into the fray'*** 30–80% of circulating white blood cells		Most abundant circulating white blood cells (30–80%) with distinctive multi-lobed nuclei; short-lived; first cells to arrive at infection site in response to signals from other cells; one of the major cells involved in phagocytosis, which kills engulfed microbes by toxic chemicals	Identify and mount phagocytic attack on microbes
• Eosinophils: up to 7% of circulating white blood cells		Circulating white blood cells; bilobed nucleus; phagocytic; release toxic chemicals from granules; major defence against parasites that are too large to be attacked by phagocytosis, e.g. parasitic worms	Defence against larger parasites; attack by using toxic chemicals released from cytoplasmic granules
• Basophils: up to 2% of circulating white blood cells*		Rare circulating white blood cells; phagocytic; contain cytoplasmic granules that are rich in histamine and heparin	Release of histamine and other molecules as part of inflammatory response; play a role in allergic reactions
• Natural killer (NK) cells: ***'the assassins'*** 1–6% of circulating white blood cells		NK cells contain granules filled with potent chemicals; recognise and attack cells lacking 'self' markers; degranulation releases proteases and also perforin proteins, which insert holes in plasma membrane of foreign cells; induce programmed cell death (apoptosis)	Elimination of virus-infected cells and cancer cells by degranulation
		In tissues	
• Macrophages: ***'the vacuum cleaners'***		Develop from monocytes that migrate into tissues; phagocytic; present in almost every tissue of body; eliminate microbes and cell debris; initiate acute inflammatory responses by secretion of various cytokines (also play a role as antigen-presenting cells)	Identify and eliminate pathogens by phagocytosis; also remove dead cells and cell debris
• Dendritic cells: ***'sentinels on patrol'*** in tissues		Resident in tissues; phagocytic; characteristic star shape; present in skin epidermis, and on surface of linings of airways and gut; migrate via lymphatic vessels to lymph glands where they act as antigen-presenting cells	Phagocytosis of pathogens and presentation of antigens to initiate adaptive immune response
• Mast cells: ***'border guards'***		Contain granules rich in histamine and heparin; found in tissues close to the external environment; involved in early recognition of pathogens; release chemical signals that attract other immune cells to infection site; involved in acute inflammatory response	Release of histamine and other active molecules during acute inflammation; play role in allergies

* Basophils are not required to be known for the Study Design, but are another example of an immune cell.

Basophils, eosinophils, neutrophils and mast cells are often referred to as granulocytes. This is due to the presence of granules within their cytoplasm. These granules often contain antimicrobial agents or enzymes that assist in the destruction of microbes and other non-cellular pathogens.

5.5.3 Phagocytes and phagocytosis

One of the most important mechanisms conducted by immune cells is the process of phagocytosis, which involves the engulfment and destruction of a pathogen or cellular debris. Cells that engulf and destroy foreign material are known as **phagocytes** or phagocytic cells.

Neutrophils, macrophages and dendritic cells are types of phagocytes. These are able to engulf pathogens. Macrophages and dendritic cells are also able to activate the immune system and are often referred to as antigen-presenting cells.

Steps in phagocytosis

1. The pathogen is identified by a **pattern recognition receptor (PRR)** and engulfed by outfoldings of the plasma membrane of the phagocyte.
2. The pathogen is engulfed in a vesicle called a **phagosome.**
3. Lysosomes fuse with the phagosome (forming a phagolysosome).
4. Toxic chemicals from the lysosome (include free radicals, lysozymes and proteases) digest and destroy the pathogen.
5. Indigestible material is discharged from the phagocytic cell by a process of exocytosis.

phagocytes types of white blood cell, including neutrophils and macrophages, that can engulf and destroy foreign material

pattern recognition receptor (PRR) protein receptors present on phagocytic cells of the innate immune system that enable these cells to recognise and bind to pathogens, with recognition being at a generic level

phagosome a membrane-bound vesicle formed within a phagocytic cell that encloses the engulfed pathogen

In dendritic cells and macrophages, antigens or sub-units from the digested pathogen are presented on the MHC-II markers to activate the adaptive immune system. This initiation is explored in subtopic 5.8.

FIGURE 5.39 The process of phagocytosis

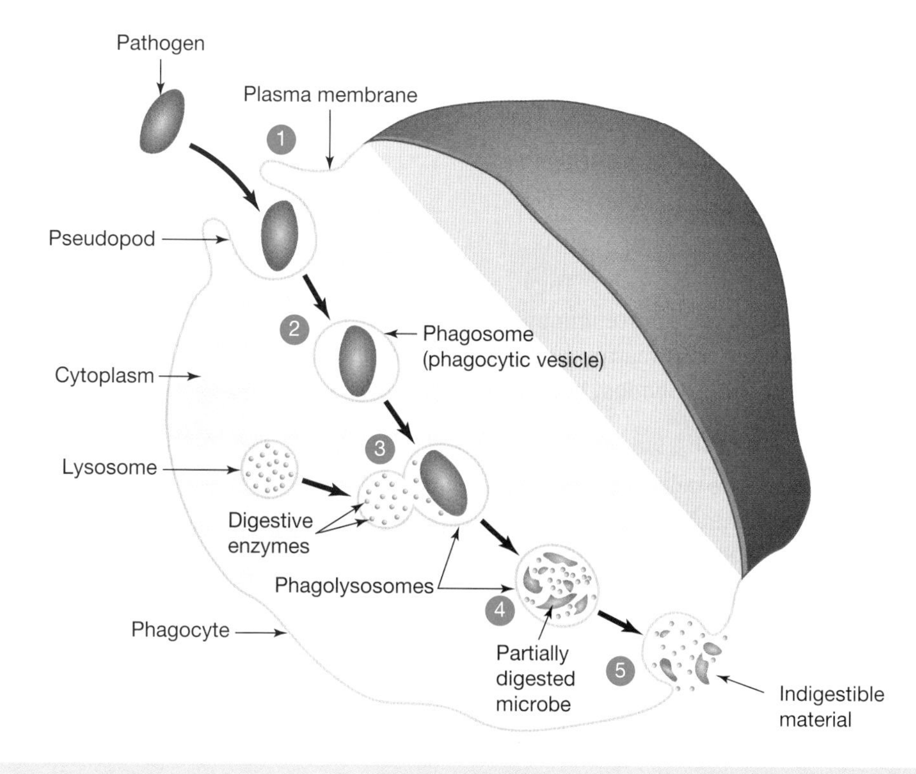

EXTENSION: How phagocytes recognise foreign material

In order to defend against pathogens, cells of the innate immune system must first identify pathogens as 'foreign' or non-self before they can eliminate them.

Before any immune mechanism can go into action, there must be a recognition that something exists for it to act against. Normally this means foreign material such as a virus, bacterium or other infectious organism. This recognition is carried out by a series of recognition molecules or receptors. Some of these circulate freely in blood or body fluids; others are fixed to the membranes of various cells or reside inside the cell cytoplasm. In every case, some constituent of the foreign material must interact with the recognition molecule, like a key fitting into the right lock.

FIGURE 5.40 PRRs on the surface of a macrophage recognise molecular patterns present on the pathogen.

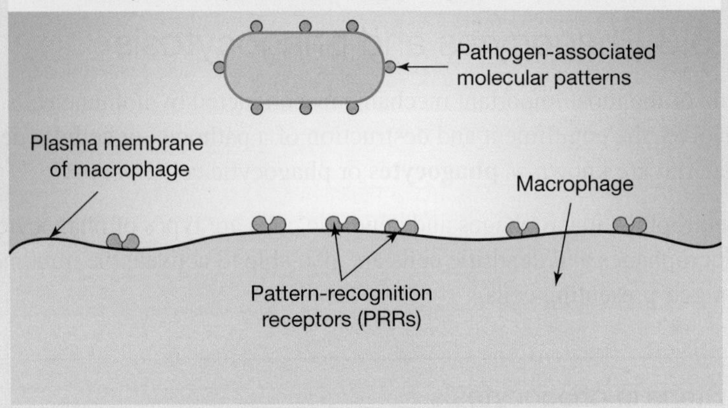

Phagocytic cells of the innate immune system, such as macrophages and neutrophils, have many receptors known as pattern recognition receptors (PRRs). These identify pathogens by recognising and binding non-specific pathogen-associated molecular patterns found on groups and patterns. This level of identification is a bit like being able to identify cats and dogs at a generic level, as, for example, 'That's a cat!'. In an immune example, molecular patterns signal that a pathogen is a bacterium but do not provide information about the specific type of bacteria (e.g. if it is *Yersinia pestis* or *Salmonella typhi*). Thus, these are different from the antigens outlined in section 5.2.1, which lead specifically to the production of antibodies.

Example of these molecular patterns include:
- glycans
- endotoxins in cell membranes
- lipopolysaccharides in some bacteria
- flagellin
- nucleic acid variants (such as double-stranded RNA, which is not found in animals).

 Resources

 **Video eLesson** Phagocytosis (eles-4343)

5.5.4 Macrophages

Macrophages are mature forms of monocytes. They are considered important actors in both innate and adaptive immunity. Monocytes are produced by stem cells in the bone marrow and circulate through the blood. They then undergo differentiation, becoming macrophages and settling in body tissues.

Macrophages are large phagocytic cells. They are found in most tissues, but they are most concentrated in tissues close to epithelial layers, working to destroy pathogens and cellular debris by phagocytosis.

FIGURE 5.41 A false-coloured scanning electron microscope (SEM) image of a macrophage

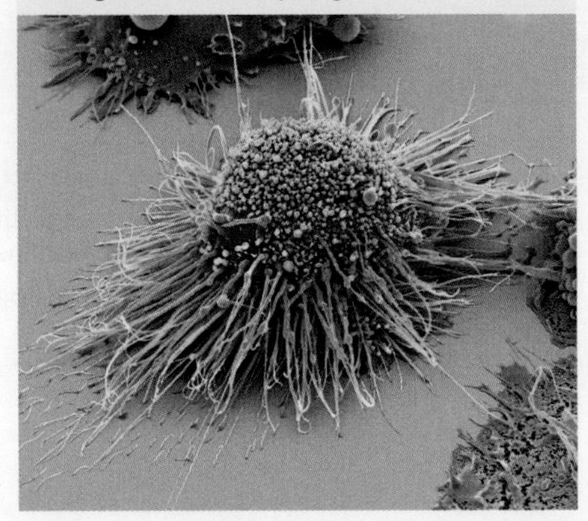

Macrophages are phagocytic cells found in tissues, which identify eliminate and engulf pathogens by phagocytosis and clear dead cells and debris. They are also antigen-presenting cells that can activate the adaptive immune response (a feature they share with dendritic cells).

5.5.5 Neutrophils

Neutrophils are the most abundant white blood cells of the immune system. Like macrophages, they are phagocytes. They can ingest and destroy viruses and microorganisms, especially bacteria.

Neutrophils have a life span of only a few days and are continuously produced from stem cells in the bone marrow. They enter the bloodstream and circulate for a few hours, after which they leave circulation and die. Neutrophils are mobile and are attracted to foreign materials by chemical signals, some of which are produced by the invading microorganisms themselves, others by damaged tissues, and still others by the interaction between microbes and proteins in the blood plasma.

FIGURE 5.42 Neutrophils circulate in the blood stream.

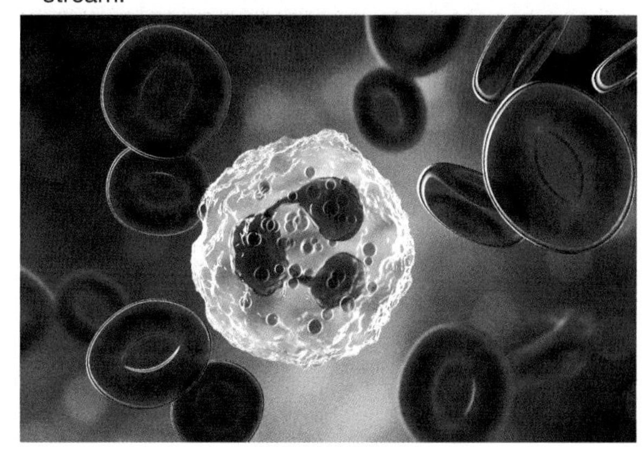

Neutrophils are abundant immune cells that are usually the first to the site of an infection, engulfing and destroying pathogens.

5.5.6 Dendritic cells

Dendritic cells can engulf pathogens and present their antigens on their surface. Dendritic cells are considered to be professional antigen-presenting cells. Like macrophages, they act to activate the adaptive immune response. They reside in and patrol the skin and mucosal surfaces, thus playing an important role in the innate immune system. They can then migrate to lymph nodes, leading to the subsequent activation of T cell responses to provide a cell-mediated immunity against microbial pathogens.

Dendritic cells have many branched projections, providing them with a very large surface area to volume ratio. This improves their phagocytic and antigen-presenting properties.

FIGURE 5.43 A dendritic cell (in blue) interacting with a lymphocyte (image courtesy of Olivier Schwartz and Institut Pasteur)

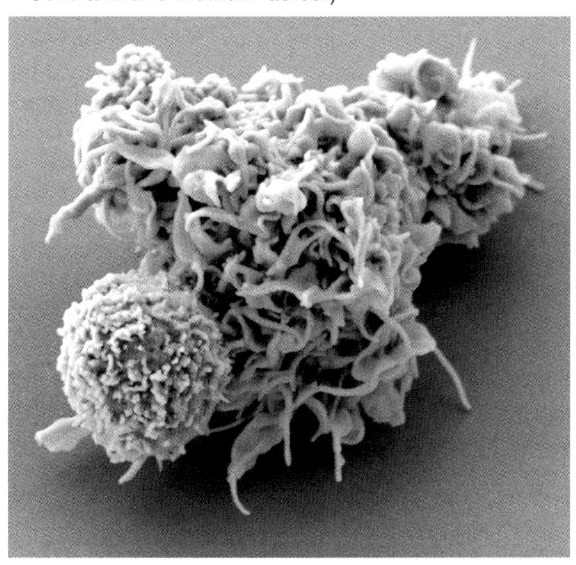

Dendritic cells are professional antigen-presenting cells, engulfing pathogens and presenting their antigens to T cells to activate the adaptive immune response.

The presentation of antigens by dendritic cells and macrophages will be explored in subtopic 5.8.

5.5.7 Eosinophils

Eosinophils are granulocytes that are present in the respiratory, gastrointestinal, and urinary tracts. They assist in defending against larger parasitic agents that are too large to be engulfed by phagocytosis (such as parasitic worms). They contain granules with toxic chemicals and histamine. Their effector function is mediated by **degranulation** and release of histamine.

They also contain some granules containing ribonucleases (RNases), which are enzymes that degrade RNA into smaller components. This helps them to fight viral infections.

The main function of eosinophils is to destroy microbial pathogens, mainly parasites. However, they also play an important role in the allergic processes together with mast cells, particularly around hayfever and asthma.

FIGURE 5.44 Eosinophils contain granules and a distinctively shaped nucleus.

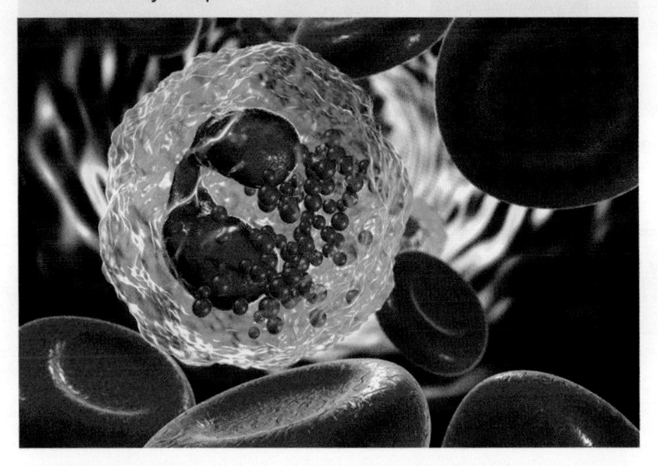

FIGURE 5.45 Eosinophil granulocytes in inflammation

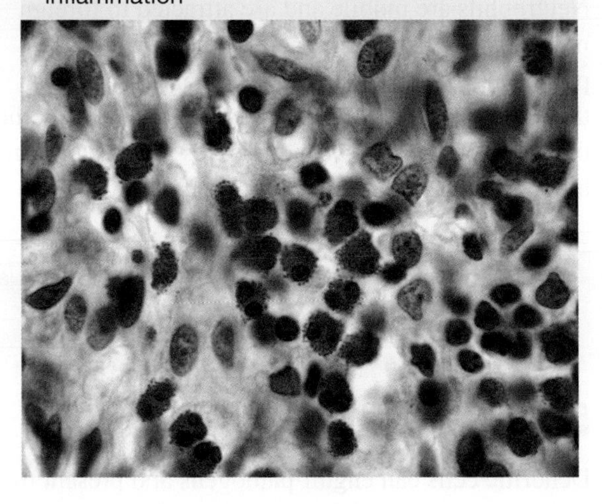

5.5.8 Natural killer (NK) cells

Once pathogens have gained entry to body cells and become intracellular, they cannot be directly attacked by innate immune cells. Instead, they are eliminated by cells known as **natural killer cells**. Natural killer (NK) cells are lymphocytes, which are specific types of white blood cells found in lymph.

Natural killer cells are lymphocytes that kill virus-infected cells through apoptosis, destroying both the cell and any viruses it contains. This prevents the virus infecting other cells.

NK cells do not affect extracellular pathogens; instead, they destroy virus-infected cells. Although this involves the apoptosis of self-cells, it also leads to the death of the virus, preventing its replication and spread. They also have a role in destroying cancerous cells. Both virus-infected and cancerous cells often have abnormal or missing MHC markers that signal their destruction by NK cells.

Killing virus-infected cells by apoptosis, rather than by lysis, is important. *Why?* Lysis of a virus-infected cell simply explodes the cell, releasing the virus particles into the extracellular fluid so that the virus particles can infect other cells. In contrast, apoptosis destroys both the cell and any viruses it contains in a systematic manner, preventing the further spread of the virus.

eosinophils a type of white blood cell that contain granules, enabling them to kill larger parasitic agents

degranulation the process by which immune cells release various chemicals (such as histamine and antimicrobials) stored within secretory vesicles known as granules

natural killer cells special white blood cells involved in the innate immune response that kill virus-infected cells

FIGURE 5.46 a. Transmission electron microscope of natural killer cells with several granules (image courtesy of Professor Colin Brooks and Dr George Sale) b. False-coloured SEM image of a natural killer cell (blue) attacking a cancer cell (pink) (image courtesy of Dr Donna Stolz and Mr Marc Rubin of the Centre for Biological Imaging at the University of Pittsuburg)

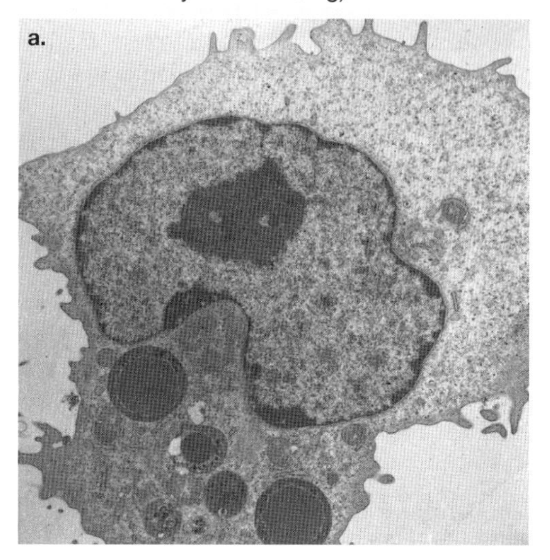

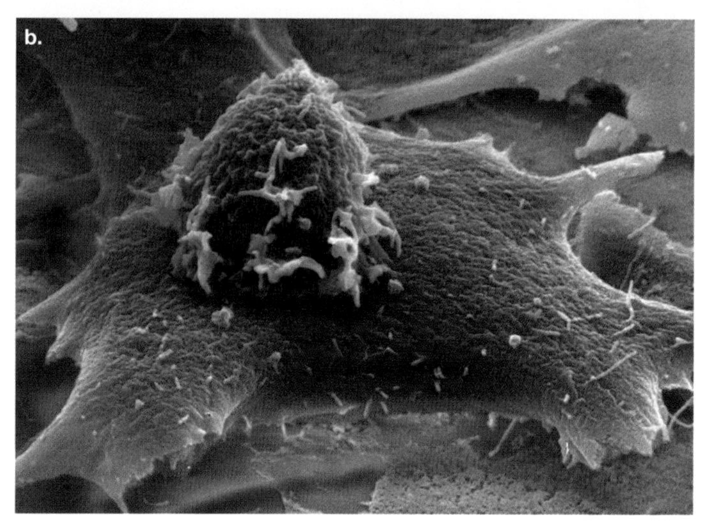

How do natural killer cells destroy virus-infected cells?

Natural killer cells undergo a process known as degranulation to destroy virus-infected cells. Degranulation is the release of anti-microbial and toxic molecules from membrane-bound granules stored in the cytoplasm of some innate immune cells. The granules in NK cells contain active protease enzymes, known as **granzymes**, and a pore-forming protein called **perforin**.

- Perforin molecules form a ring structure that punches a hole in the plasma membrane of the target cells, enabling entry of the proteases (known as granzymes) into the target cell.
- Once inside the cytoplasm of an infected target cell, the proteases (usually granzyme B) induce apoptosis.
- Elimination of an infected cell or a cancer cell by degranulation of an NK cell is completed within hours.

FIGURE 5.47 Diagram showing the release of perforin and granzymes by NK cells. Perforin forms a pore in the membrane of the target cell, allowing granzymes (proteases) to enter the virus-infected cell.

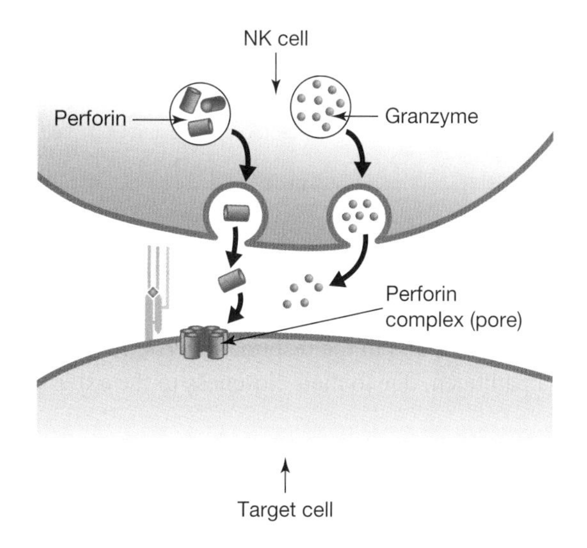

granzymes active protease enzymes present in granules that form part of the immune defences of NK cells and cytotoxic T cells

perforin a protein, released by some immune cells, that produces a pore in the membrane of cells undergoing an immune attack

EXTENSION: How do NK cells decide to kill cells?

Body cells infected with viruses and cancer cells are part of a person's body. How do natural killer (NK) cells recognise these infected and abnormal cells as targets for elimination?

NK cells identify targets for elimination even though the targets are body cells. The decision '*kill or don't kill*' by an NK cell is believed to be determined by receptors on the plasma membrane of the NK cell: one receptor is for 'kill' and another is an inhibitory receptor that blocks the kill signal.

- All body cells carry **ligands** that can bind to 'kill' receptors on NK cells.
- This binding activates the 'kill' receptors that signal an NK cell to eliminate any cell bound to it.
- However, this kill signal is inhibited by normal HLA markers on healthy body cells. The activation of the inhibitory receptors sends a signal that blocks the kill signal. As a result, an NK cell does not identify healthy body cells as targets for elimination.

Virus infections of body cells suppress the expression of class I HLA markers. Likewise, cancer cells produce abnormal HLA markers. These missing or abnormal HLA markers make these cells susceptible to being killed by NK cells because they cannot inhibit the 'kill' signal. When contact is made with an infected cell that is missing or has abnormal HLA markers, the NK cell responds to the 'active kill' signal and degranulates, spilling the lethal contents of its granules onto the target cell.

FIGURE 5.48 Identification of cells to kill by NK cells **a.** The NK cell recognises the MHC marker so does not destroy the cell. **b.** The infected cell lacks the correct self-marker, so it is targeted by the NK cell.

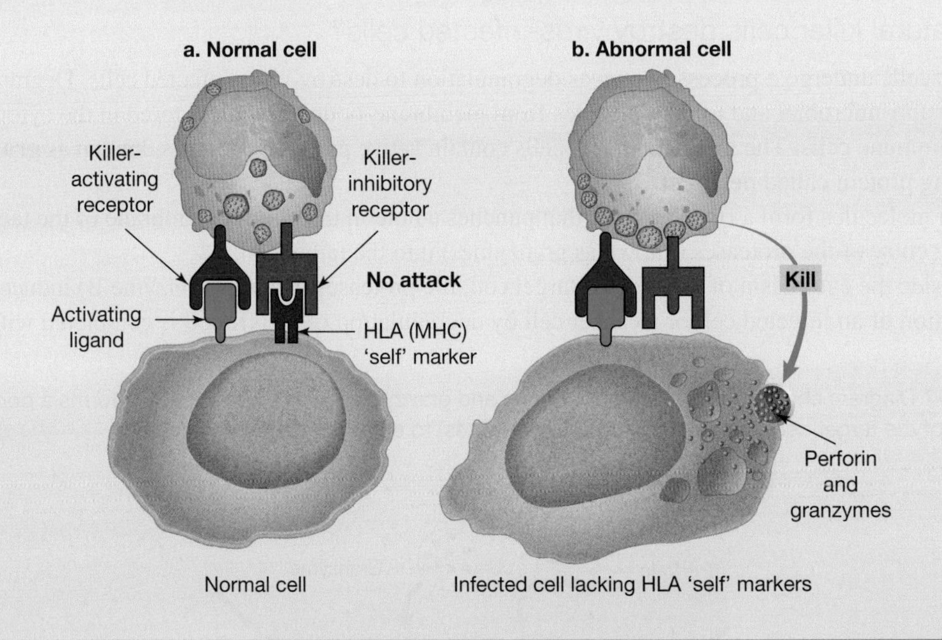

5.5.9 Mast cells

Mast cells mediate inflammatory responses. They are found in tissue, especially beneath the surface of the skin, near vessels and in the respiratory system. They act very early in the infection by a pathogen, due to their closeness to the external environment.

Mast cells contain various chemicals in granules, including cytokines, histamine and heparin. Upon stimulation, mast cells release the contents of their granules via degranulation. This causes **histamine** and heparin to be released in tissues.

ligands any molecule that binds to a specific target to form an active complex

histamine a substance involved in inflammation and allergic reactions that causes blood vessels to dilate and become more permeable to immune cells

Mast cells are vital in the inflammatory response. As well as releasing histamines that increase vascular permeability, they also release cytokines that attract other immune cells to help destroy the pathogen.

Histamine leads to the increased permeability of blood vessels and causes smooth muscles to contract. The release of cytokines attracts of other immune cells to the site of the infection. Heparin is also linked to inflammation, initiating the production of a hormone known as bradykinin, which contributes to the inflammation associated with mast cells.

The inappropriate activation of mast cells and release of histamine leads to allergic reactions and hypersensitivity. This is linked with allergens, which were introduced in section 5.2.5.

FIGURE 5.49 Mast cells with granules containing histamine

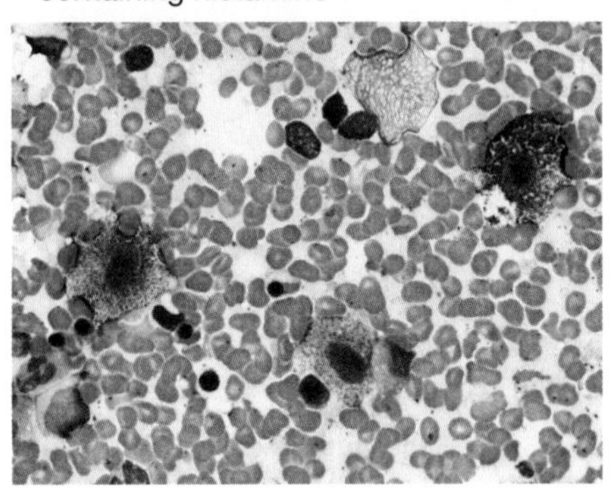

CASE STUDY: Mast cells and the allergic response

In section 5.2.5 we were introduced to allergens. Allergens are harmless substances that initiate an immune response in a sensitised individual. This reaction is controlled by both mast cells and antibodies.

The key players in an allergic reaction include:

- mast cells
- antibodies, specifically immunoglobulin E (IgE)

There are several stages in an allergic reaction, such as the hayfever caused by exposure to pollen. First, a person must be sensitised to a potential allergen; otherwise, an allergic reaction will not occur. The allergic response is shown in figure 5.50 outlined in the listed steps.

Sensitisation to an allergen: first exposure

1. Potential allergens, such as an antigen on airborne pollen cells, are inhaled, consumed or contacted.
2. Cells of the immune system identify this antigen as non-self, and an immune response is activated.
3. The production of specific antibodies against the particular allergen occurs. These are **immunoglobulin E (IgE) antibodies**, which are produced by B cells in the tissue around the site of entry of the allergen. (The mechanism by which B cells release antibodies will be explored in subtopic 5.10.)
4. The IgE antibodies attach to the surface receptors of mast cells that are located in the linings throughout the body, including the airways. Mast cells that are coated with IgE antibodies are said to be primed or sensitised.
5. The presence of primed mast cells means that the person is now sensitised to that particular allergen so that the next exposure will cause an allergic response.

Allergic reaction: second exposure

1. When the next exposure to the particular allergen occurs, the IgE antibodies on the primed mast cells recognise the allergen and bind to it.
2. The binding of the allergen to the IgE antibodies on the mast cells activates them and they degranulate, releasing their contents of chemical mediators, including histamine.
3. The release of histamine results in the effects of inflammation: an increased blood flow to the region (causing redness), increased permeability of blood vessels (causing swelling and pain), sneezing, coughing, nasal congestion, itchy and watery eyes, and runny nose.
4. Migration of more immune cells continues the inflammatory response (refer to subtopic 5.6 for further detail on the inflammatory response). Mast cells produce chemical messengers that signal various circulating immune cells to migrate from the bloodstream into the affected tissue to join the immune attack on the allergen, leading to sustained inflammation.

immunoglobulin E (IgE) antibodies a type of antibody produced in response to exposure to a particular allergen; involved in allergic reactions

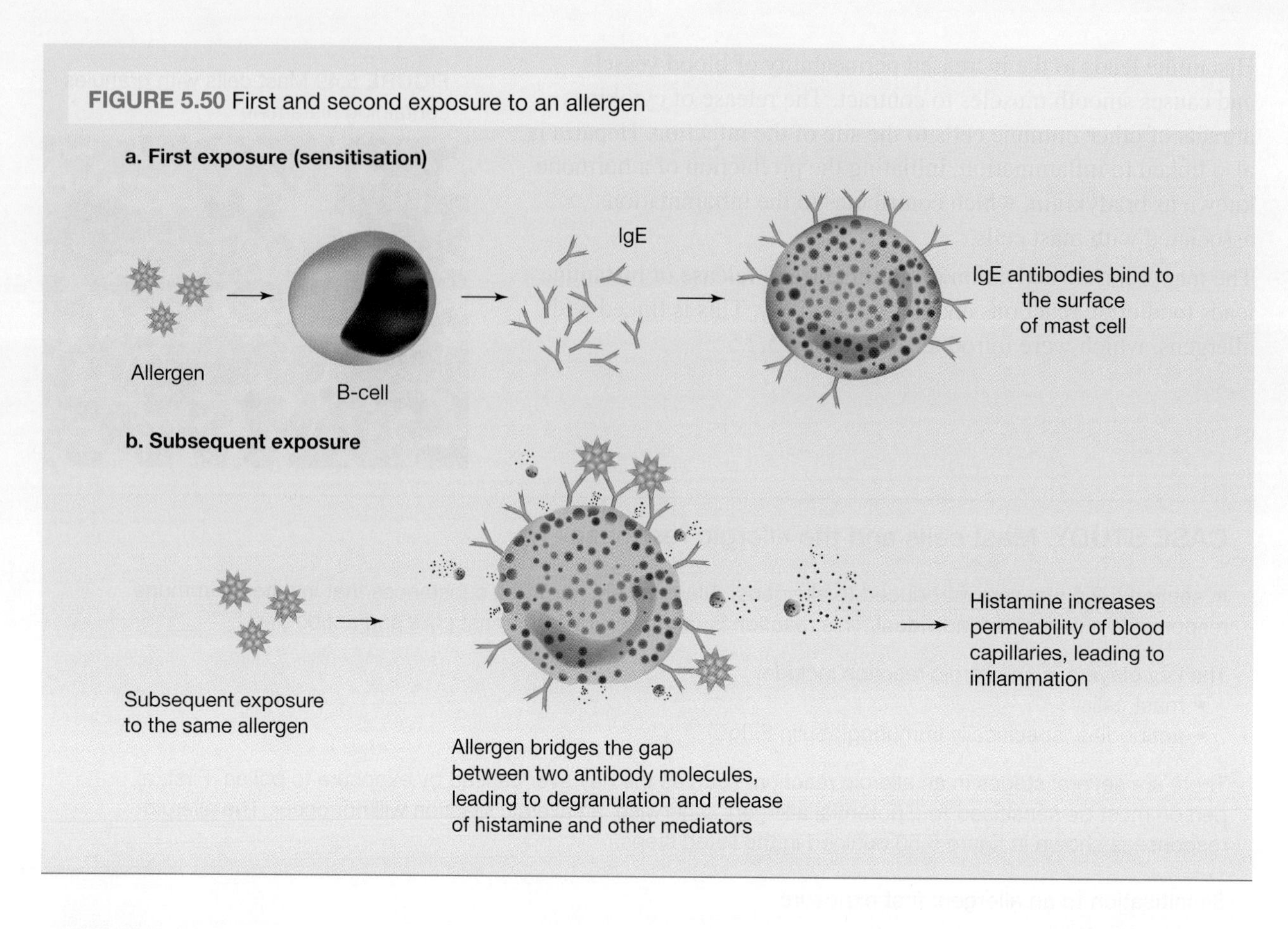

FIGURE 5.50 First and second exposure to an allergen

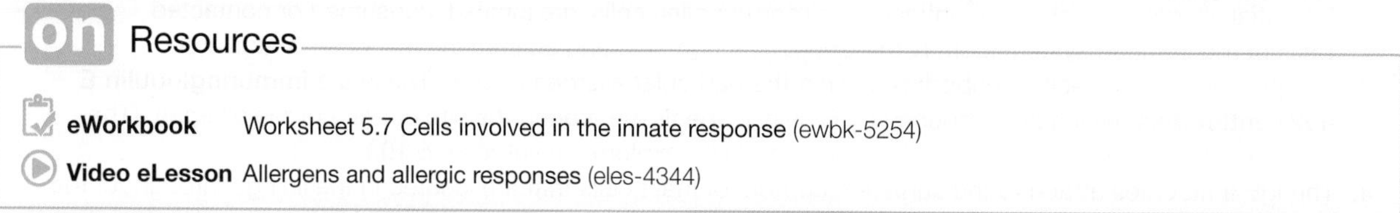

5.5.10 Humoral innate immunity

In the previous sections, we explored the cells of the innate immune response. The other important part of the innate immune system is the humoral component, which involves the action of soluble proteins and their derivatives in extracellular body fluids such as lymph and blood. ('Humor' is a medieval term for body fluid.)

Humoral responses involve the actions of soluble active molecules. Key soluble molecules involved in innate immunity include proteins known as **complement proteins** and **cytokines**.

complement proteins proteins that assist other innate immune cells and can destroy bacterial cells by lysis

cytokines signalling molecules of the immune system

interferons proteins secreted by some cells, in response to a virus infection, that helps uninfected cells resist infection by that virus

Cytokines are the signalling molecules of the immune system. Some of the most important cytokines involved in innate immunity are **interferons**. Other cytokines include interleukins (which mediate communication between immune cells), tumour necrosis factors (TNF, which regulate immune cells) and chemokines (which mediate the attraction of cells).

5.5.11 Complement proteins

Dissolved in the plasma of your blood are proteins that form what is called complement. In total, the complement system comprises a large group of protease proteins (more than 25 proteins). These plasma proteins are called 'complement' because they complement or add to the function of immune cells.

There are various types of complement proteins that contribute to the innate immune system in various ways.

The role of complement

Complement proteins contribute to the immune system by:
- opsonising pathogens, making them more susceptible to phagocytosis
- recruiting immune cells involved in an inflammatory response (chemotaxis)
- destroying bacterial pathogens by lysis through the initiation of the membrane-attack complex.

Opsonisation

Some complement proteins cover the surface of a pathogen and form a complex with the surface antigens on the pathogen. This process is termed **opsonisation**. Opsonisation makes pathogens more susceptible to elimination by phagocytosis. This is because phagocytes, such as macrophages, have receptors for complement proteins on their plasma membranes, and these bind to the opsonised microbes. The main complement protein involved in opsonisation is known as complement component 3b (C3b).

FIGURE 5.51 Complement allows phagocytes to more easily bind to the pathogen, increasing the likelihood of phagocytosis occuring.

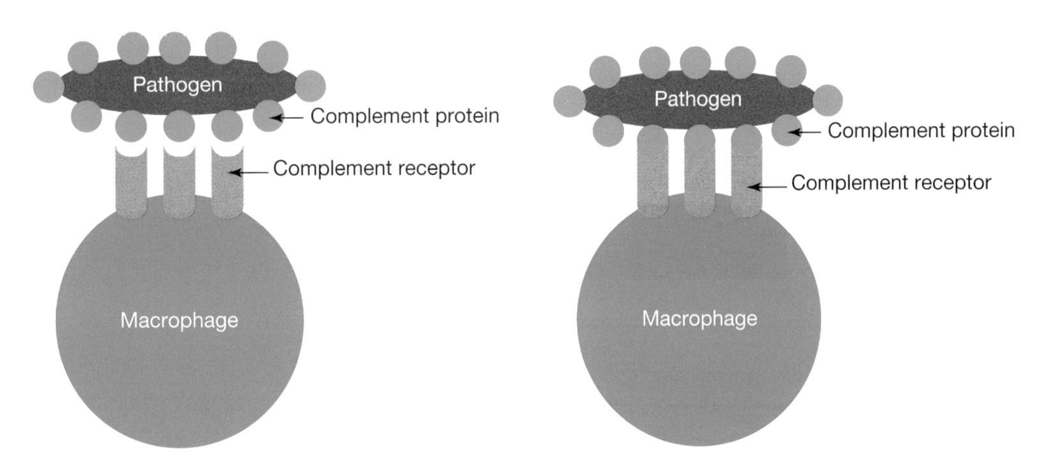

Chemotaxis

Chemotaxis refers to the movement of cells in response to a chemical stimulus. Small complement peptides that diffuse from the pathogen surface act as chemical signals, attracting immune cells involved in the inflammatory response to the site of the infection.

Lysis of pathogens

Some complement proteins are involved in the direct 'explosive' killing of extracellular pathogens. This occurs when a **membrane-attack complex** (MAC) forms on the plasma membrane of the pathogen due to the interaction of various complement proteins. The MAC inserts into the plasma membrane of the pathogen and produces a pore that allows fluid to enter, causing the pathogen cell to swell and burst — explosive death by osmotic shock!

opsonisation the coating of the surface of pathogen cells by complement proteins, making the pathogens more susceptible to phagocytosis

chemotaxis movement of a cell or organism in response to a chemical substance (such as complement or cytokines)

membrane-attack complex (MAC) one of the defence mechanisms resulting from activation of complement proteins that destroys pathogen cells by osmotic shock

FIGURE 5.52 A membrane attack complex forms on the membrane of a pathogen.

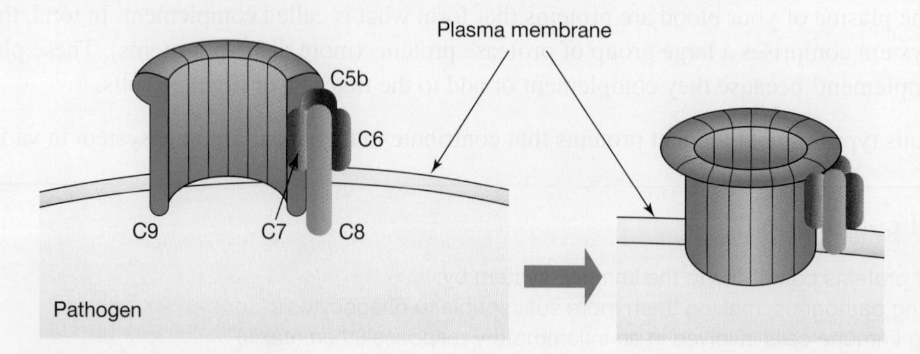

How is complement activated?

The complement proteins circulating in the bloodstream are inactive enzymes. Complement proteins are activated when they make direct contact with molecules on the surface of a pathogen. The process involved in complement activation is as follows:

1. The activation of an initial complement protein (known as C3) starts a sequence of reactions that take place on the surface of a pathogen. The first protein in the series enzymatically alters the next protein in the series.
2. The product of the first reaction then activates the next enzyme in the series, which, in turn, activates the next protein, and so on. The activation of a complement protein occurs when the protein is cut (cleaved) into two fragments — a larger activated protein and a smaller peptide fragment. This sequence of reactions starts a **cascade** that can neither be stopped nor reversed.

cascade a multi-step process in which each step must occur in a set order, with each step triggering the next in the sequence

FIGURE 5.53 Complement cascade — various complement proteins are activated during an immune response.

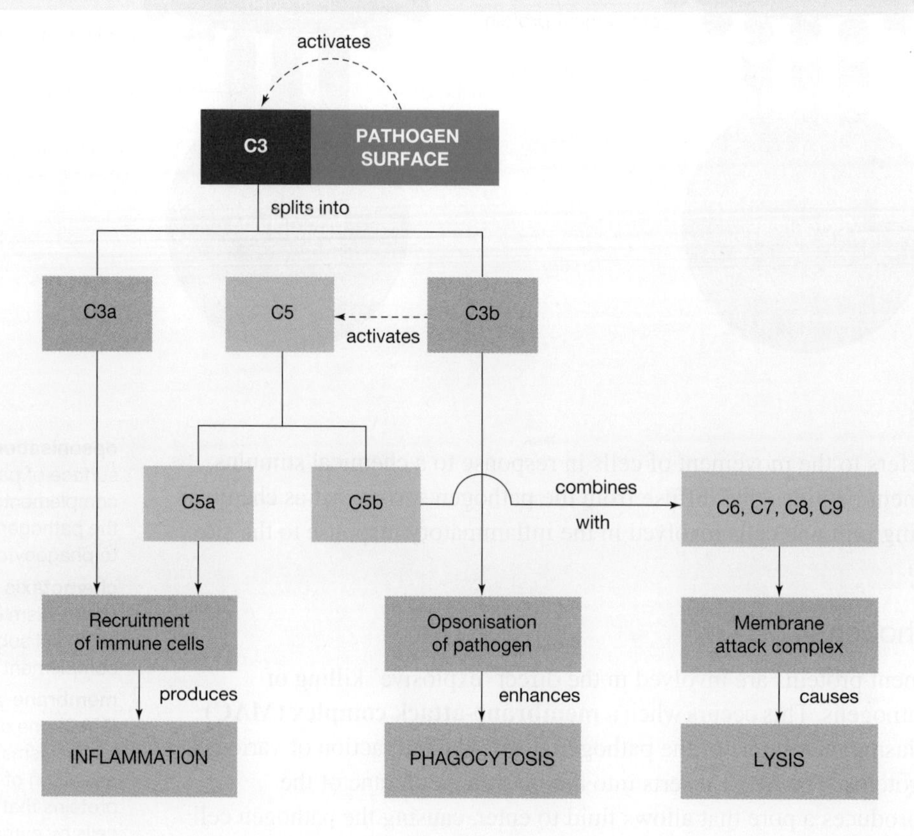

TIP: You do not need to know the names of the specific complement proteins (such as C5), but rather understand that complement proteins have a variety of functions and act in a cascade, with some complement proteins activating others.

5.5.12 Interferons

Interferons are a group of cytokines (signalling proteins) made and released by host cells in response to the presence of several viruses. The name of these proteins reflects the fact that they 'interfere' with viral replication. Once a body cell is infected with viruses, the cell secretes interferons. That cell is doomed, but the interferons that it secretes into its surroundings act as warning signals to nearby cells so that they can prepare in advance for a possible virus infection.

Interferons act in can act in either an autocrine (target the cell they were released from) or paracrine fashion (target nearby cells). They activate the infected cells to produce enzymes that break down viral proteins.

The role of interferons

Interferons bind to receptors on neighbouring cells, producing signals that:

1. induce transcription of a number of specific genes that encode production of *inactive* forms of antiviral enzymes, inhibiting protein synthesis and destroying RNA. These enzymes are activated only if a virus succeeds in infecting the cell and, if activated, they block the synthesis of viral proteins.
2. make the plasma membrane less fluid, making its fusion with viral particles more difficult and so reducing the chance of viral infection in these cells
3. cause virus-infected cells to undergo apoptosis
4. activate immune cells, such as natural killer (NK) cells, that eliminate virus-infected cells by apoptosis.

FIGURE 5.54 The various actions on nearby cells of interferons released from a virus-infected cell

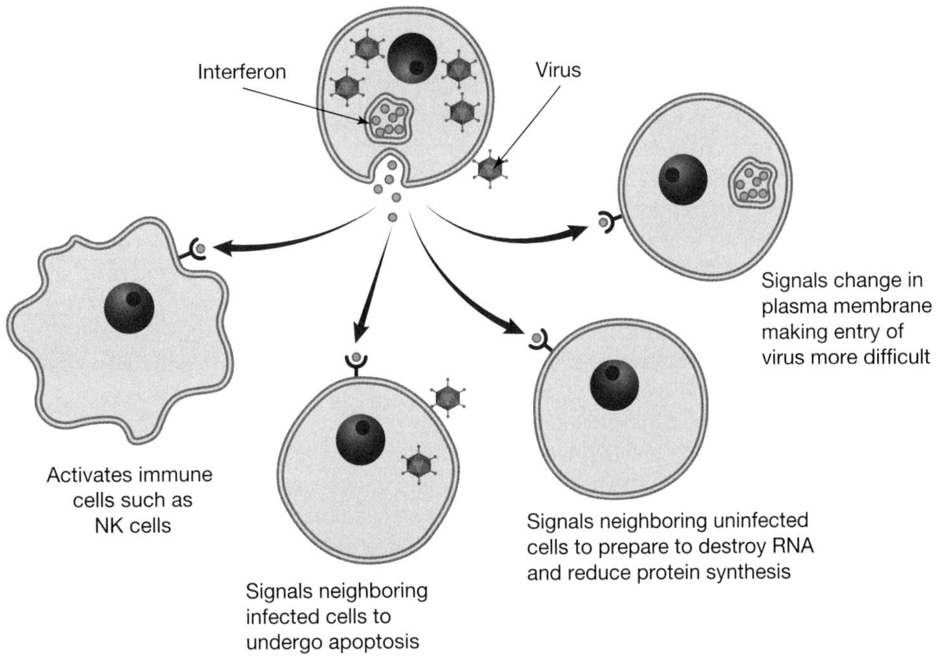

Interferons are particularly important if a virus is able to infect a nearby cell. If a virus does not have to travel to reach its target cell, it can cause infection quickly. In the early stages of infection, interferon is a vital part of the immune defence against viruses. When the infection develops into disease (such as the cold or flu), interferons have failed and the adaptive immune system must come into play.

tlvd-1779

SAMPLE PROBLEM 4 Identifying and describing soluble proteins

Cells infected with viruses secrete soluble substances that travel to adjacent cells and activate them to make antiviral proteins that are involved in innate immune response.

a. **Name the substance secreted by virus-infected cells.** **(1 mark)**

b. **Outline two actions of the substance mentioned in part** a. **(2 marks)**

THINK	WRITE
a. State the name of the protein secreted from virus-infected cells. You must be specific (cytokines alone is not an acceptable answer).	Interferons (1 mark)
b. Write down a brief description of the ways interferons stop viral infection.	Infected cells secrete interferons, which bind to the receptors on the neighbouring cells and signal uninfected cells to destroy viral RNA and reduce protein synthesis (1 mark). Interferons also activate other immune cells such as NK (natural killer) cells to destroy virus-infected cells (1 mark).

Resources

eWorkbook Worksheet 5.8 Soluble molecules in the innate response (ewbk-5256)

Video eLesson The interferon mechanism against viruses (eles-4347)

KEY IDEAS

- The second line of defence of the innate immune system comes into immediate operation if pathogens overcome the barriers of the first line of defence.
- The second line of defence involves the operation of innate immune cells and humoral factors.
- The innate immune system relies on the recognition of particular types of molecules that are common to many pathogens (called pathogen-associated molecular patterns) but are absent in the host.
- The innate leukocytes include natural killer cells (NK), mast cells, eosinophils and phagocytic cells, include macrophages, neutrophils and dendritic cells.
- Macrophages, dendritic cells and neutrophils directly attack and eliminate extracellular pathogens through the process of phagocytosis. Macrophages and dendritic cells are also antigen-presenting cells and can activate the adaptive immune system.
- Pattern recognition receptors on phagocytic cells generically identify pathogens for elimination.
- Natural killer (NK) cells attack pathogens indirectly by identifying and eliminating pathogen-infected body cells and cancer cells through the process of degranulation.
- Defensive molecules of innate immunity include complement proteins and cytokines
- Complement proteins are approximately 20 types of soluble proteins that destroy extracellular pathogens by binding to their surfaces. These proteins serve as markers to indicate the presence of a pathogen to phagocytic cells and to enhance engulfment; this process is called opsonisation. Certain complement proteins can combine to form attack complexes that open pores in bacterial cell membranes.
- Interferons secreted by a virus-infected cell signal nearby cells to make various advance preparations to reduce the chance of a viral infection. They signal virus-infected cells to undergo apoptosis.

5.5 Activities

To answer questions online and to receive **immediate feedback** and **sample responses** for every question, go to your learnON title at **www.jacplus.com.au**. A **downloadable solutions** file is also available in the resources tab.

5.5 Quick quiz on	5.5 Exercise	5.5 Exam questions

5.5 Exercise

1. Identify two methods by which innate immune cells eliminate pathogens.
2. Complete the following passage using the listed terms: *engulfed, MHC-II markers, pattern recognition receptors, lysosome, destroy, exocytosis, phagolysosome, phagosome, dendritic cells*
   During phagocytosis, a pathogen is identified by through ____________. The pathogen is ____________ in a vesicle, forming a ______________. This fuses with a ____________, forming a ____________. Toxic chemicals and enzymes from the lysosome digest and __________ the pathogen. The indigestible material is discharged by the process of __________. In macrophages and _____________, some of this material is presented on __________.
3. Give an example of each the following that is part of innate immunity.
   a. The most common circulating white blood cell
   b. The 'assassins'
   c. A phagocytic cell
   d. A cell that eliminates intracellular pathogens
4. Identify the following as true or false and justify your response.
   a. The level of identification of pathogens by the cells of innate immunity is more specific than that of the cells of adaptive immunity.
   b. Natural killer (NK) cells eliminate pathogens directly by degranulation.
   c. Macrophages are tissue-based cells of the innate immune system that eliminate pathogens by phagocytosis.
   d. Complement proteins are involved in the elimination of intracellular pathogens.
   e. Interferons released by a virus-infected cell can signal nearby cells to produce antiviral proteins.
5. A bacterial cell is 'opsonised'. Briefly explain what has happened to this bacterial cell.
6. Compare and contrast dendritic cells and macrophages.
7. Draw a flowchart showing how NK cells kill virus-infected cells.
8. Explain why apoptosis of virus-infected cells is preferred over lysis.
9. In some immunodeficiency diseases (where the immune system doesn't function properly), there is a deficiency in the complement system. Explain how complement deficiencies can lead to disease.
10. Interferons can be used to treat autoimmune diseases such as multiple sclerosis, in which cells of the immune system target the myelin sheath around the axons in nerve cells. Suggest two reasons why this treatment may be used.

5.5 Exam questions

Use the following information to answer Questions 1 and 2.

The diagram below shows an immune cell responding to a substance. This process occurs during certain types of allergic reactions.

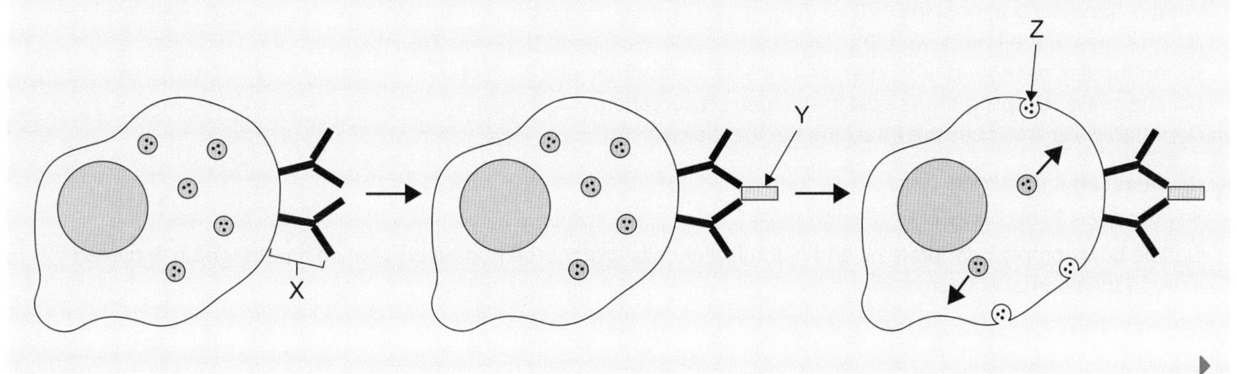

Question 1 (1 mark)

Source: *VCAA 2020 Biology Exam, Section A, Q22*

MC Which type of immune cell is featured in the diagram above?

A. Mast cell
B. Neutrophil
C. Macrophage
D. Dendritic cell

Question 2 (1 mark)

Source: *VCAA 2020 Biology Exam, Section A, Q23*

MC What do the structures X and Y in the diagram above represent?

	X	Y
A.	Antibody	Allergen
B.	receptor	Antibody
C.	glycolipid	G protein
D.	antigen	allergen

Question 3 (1 mark)

Source: *VCAA 2019 Biology Exam, Section A, Q20*

MC The diagram below shows the process of phagocytosis. This process is vital for immunity against extracellular infections.

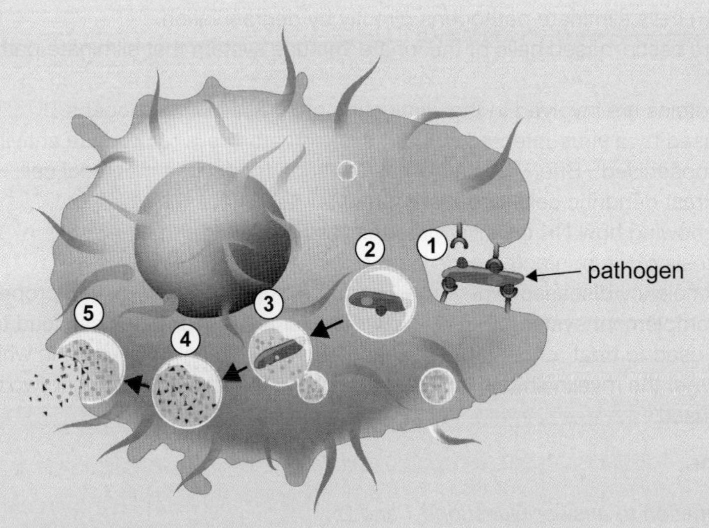

Source: Designua/Shutterstock.com

What is happening at position 3?

A. Enzymes that break down the microorganism are released into the vesicle.
B. Antibodies are added to the vesicle to kill the microorganism.
C. The cell is sampling the vesicle for antigen presentation.
D. Intracellular microbes are attacking the microorganism

Question 4 (2 marks)

Source: *VCAA 2018 Biology Exam, Section B, Q3c*

Outline how complement proteins and natural killer cells protect the human body once a pathogen has gained entry to the internal environment.

Question 5 (1 mark)

Source: *VCAA 2017 Biology Exam, Section A, Q22*

MC Which of the following matches a cell correctly with its role in an immune response?

	Cell	Role
A.	macrophage	Stimulates inflammation by secreting interferon
B.	dendritic cell	Presents fragment of antigens to T helper cells
C.	mast cell	Engulfs bacteria and debris
D.	neutrophil	Secretes antibodies

More exam questions are available in your learnON title.

5.6 The inflammatory response in innate immunity

KEY KNOWLEDGE

- The innate immune response including the steps in an inflammatory response

Source: Adapted from VCE Biology Study Design (2022–2026) extracts © VCAA; reproduced by permission.

5.6.1 What is inflammation ?

Inflammation is a reaction to an infection, injury or damaged tissue that results in heat, pain, swelling and redness due to accumulation of fluids and proteins and an increased blood supply to the infected region.

Inflammation is the immune system's response to harmful stimuli, such as pathogens. It acts by removing the stimulus and initiating the healing process. Inflammation is an early protective response of the innate immune system and is localised around the site of infection. Inflammation does not only occur in response to pathogens; it may also occur when cells are damaged by other factors, as for example, thermal burns to the skin, corrosive chemical spills, frostbite and sunburn.

The main symptoms of inflammation are:

- redness — due to vasodilation of blood vessels, leading to red blood cells released into tissue
- pain — due to the systemic response (such as fever), the stimulation of nerve endings through the release of bradykinin and histamine, and swelling putting pressure on pain receptors
- heat — due to increased blood flow
- swelling — due to the movement of fluid into tissues after vasodilation
- pus — due to dead phagocytes and cell debris.

FIGURE 5.55 Pathogens can enter when the skin is cut, leading to inflammation.

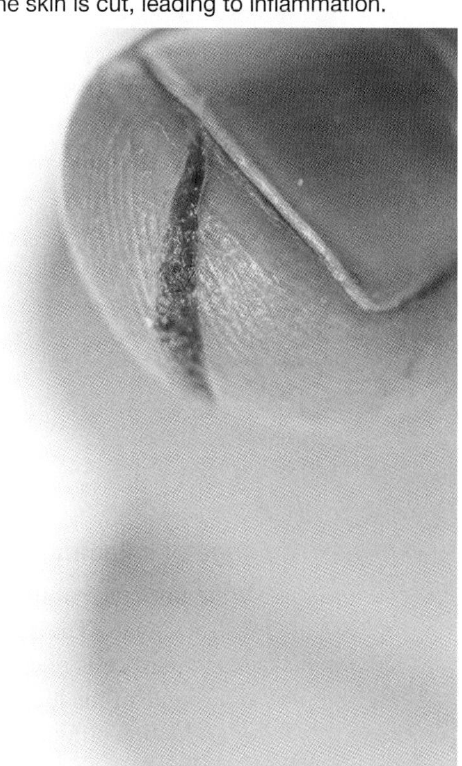

5.6.2 Stages of inflammation

A number of steps are involved in an inflammatory response.

The stages of inflammation

The inflammatory response can be divided into three main stages:
- the vascular stage, which involves the actions of blood vessels
- the cellular stage, which involves the actions of immune cells
- the resolution stage, when inflammation is stopped and the normal state is restored.

There are various players in the inflammatory response, some of which are highlighted in figure 5.56.

FIGURE 5.56 Some of the main events in inflammation

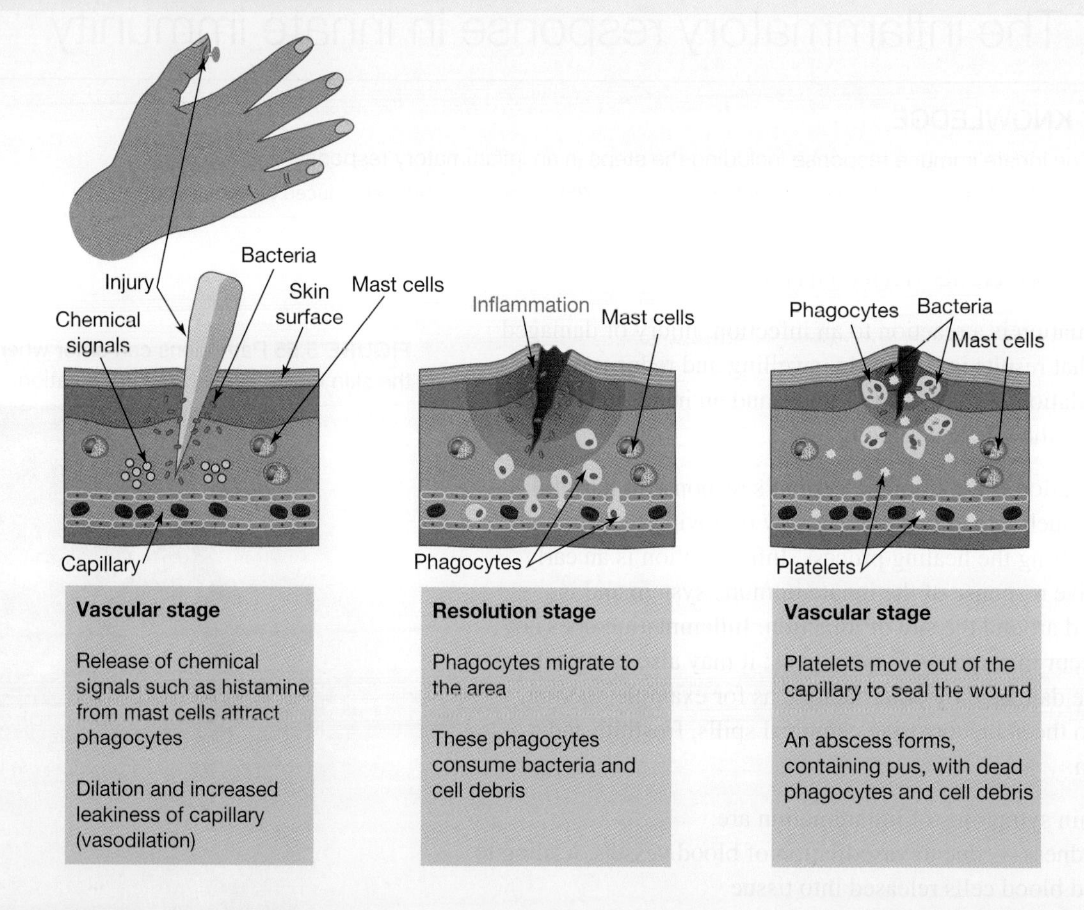

1. The vascular stage of inflammation

The vascular stage is the first stage of inflammation. The following events occur during this stage:
- Release of chemical signals: The damaged cells release cytokines and prostaglandins that attract neutrophils, and the mast cells release histamines, which triggers the dilation of blood vessels.
- The arteriole and venule of the local capillary bed dilate. This allows the blood flow to the damaged area to increase. This vasodilation produces redness and the increased blood flow produces heat.
- Increased permeability of local capillaries: The capillaries in the area around the cut become more 'leaky' so that protein-rich fluid (exudate) escapes from the capillaries into the infected region. The exudate causes the swelling (edema) that is typical of inflammation. The swelling causes pressure on the surrounding tissue, which stimulates pain receptors and contributes to the pain that is characteristic of inflammation.
- Clotting agents in the exudate assist in clot formation, which isolates the infection.

FIGURE 5.57 The vascular stage of inflammation involves vasodilation due to histamine. a. The release of histamine leading to b. vasodilation of blood vessels.

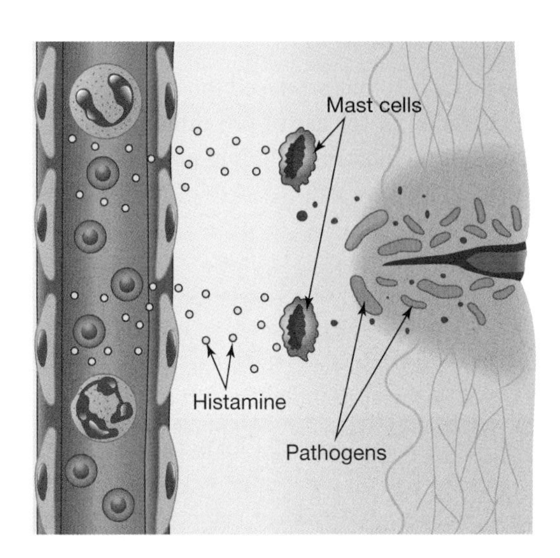

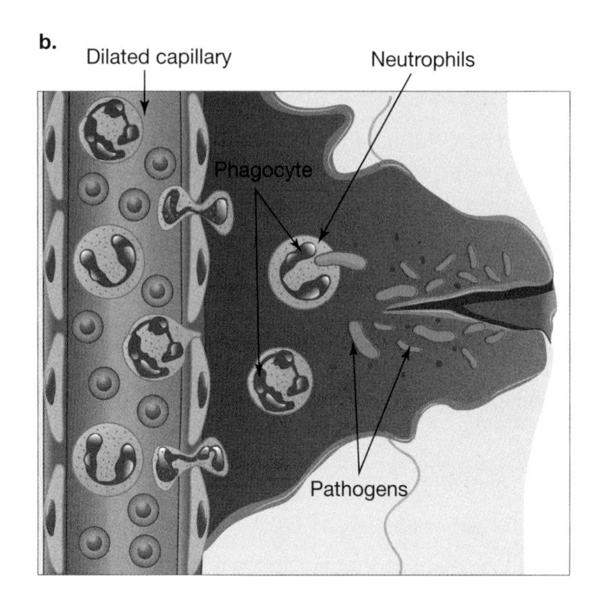

2. The cellular stage of inflammation

The cellular stage involves further intervention from immune cells, particularly phagocytic cells. The following events occur during this stage:

- Escape of immune cells from capillaries: The expansion of the capillary bed enables neutrophils to squeeze between the endothelial cells that line the capillaries (see figure 5.58).
- Migration of neutrophils to the infection site: Neutrophils are the first immune cells to arrive at the infection site, attracted by cytokines released by damaged cells. Other immune cells, including macrophages from nearby tissues, follow. These cells release signals, such as histamine and more cytokines, that attract more phagocytic cells to the infection site.
- Phagocyte attack on bacteria: Both neutrophils and macrophages attack bacterial pathogens by engulfing them in a process called phagocytosis (refer back to section 5.5.3). Phagocytic cells also remove cell debris from the infection site by a similar process. Pus consists mainly of dead phagocytic cells and other immune cells, and also contains living cells and cell debris.

FIGURE 5.58 Neutrophils and other phagocytes can squeeze between endothelial cells after the vascular stage occurs.

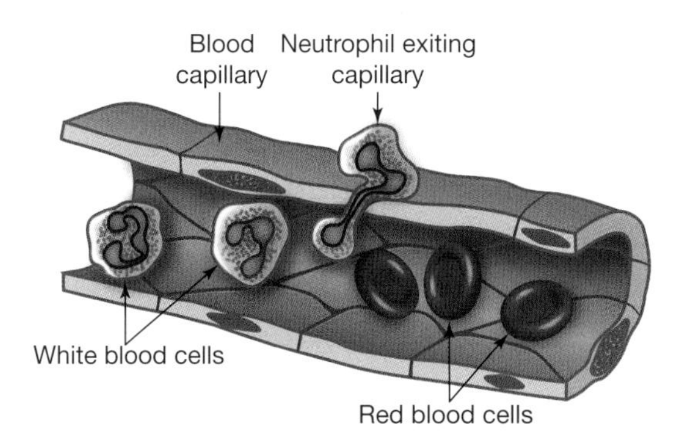

FIGURE 5.59 Pus is a symptom of infection, and contains dead phagocytes and cell debris.

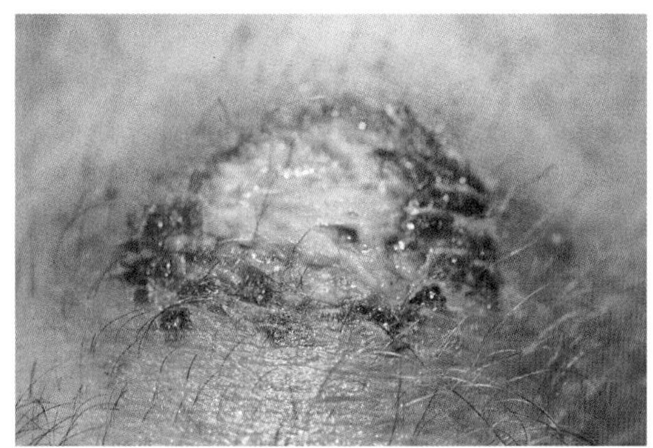

3. The resolution stage of inflammation

While the infection is present, pro-inflammatory cytokines are released to maintain the inflammatory response. This response continues until the pathogen has been eliminated. However, once the infection is under control and tissue repair underway, it is important that the normal state is restored. This is termed the **resolution stage**.

resolution stage the final stage of inflammation, in which the normal state in restored

Resolution is a complex process that includes a reversal of all the processes that produced the acute inflammation, for example, the reversal of capillary dilation and cessation of release of pro-inflammatory cytokines. Resolution involves the release of many active molecules or mediators, including anti-inflammatory cytokines.

CASE STUDY: Chronic inflammation

If normality is not established during the resolution stage of inflammation, chronic inflammation can result. Many diseases are caused by chronic inflammation, as the immune system does not shut off in the absence of infection.

Many of the diseases that are linked with chronic inflammation are autoimmune diseases. These include:

- ulcerative colitis: chronic inflammation of the lining of the colon (as shown in figure 5.60)
- rheumatoid arthritis: chronic inflammation of the joints
- multiple sclerosis: chronic inflammation and autoimmunity of myelin sheath on neurons
- systematic lupus: system-wide chronic inflammation and autoimmunity.

To access more information on various case studies on chronic inflammation and complete an analysis task relating to these, please download the worksheet.

FIGURE 5.60 Observing the inflammation of the colon of an indivudal affected with UC

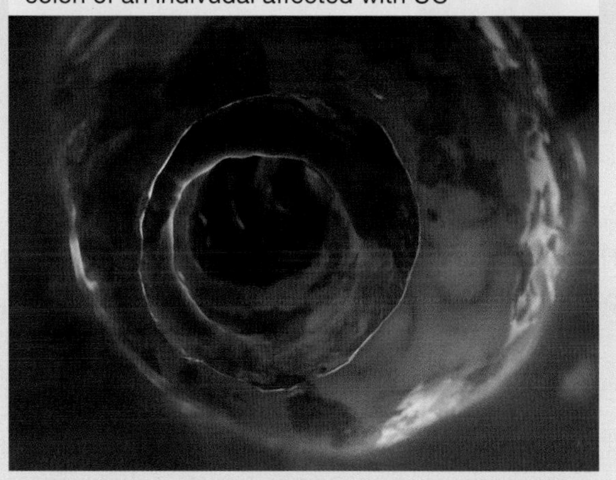

on Resources

 **eWorkbook** Worksheet 5.9 Inflammation and case studies of chronic inflammation (ewbk-5258)

 **Video eLesson** Inflammatory response (eles-4345)

KEY IDEAS

- Inflammation is a reaction to an infection, injury or damaged tissue that results in redness, swelling, heat and pain due to accumulation of fluids and proteins and an increased blood supply to the infected region.
- The three stages of an inflammatory response are the vascular stage, the cellular stage and the resolution stage.
- The signs of inflammation are redness, swelling, heat and pain.
- The vascular stage of inflammation involves changes in the capillaries supplying the site of infection.
- Actions by cells in the cellular stage of inflammation include various cell movements, as well as phagocytic activity at the infection site.
- If the inflammatory response is not turned 'off' when an infection is contained, a damaging situation of chronic inflammation can arise.

5.6 Activities

To answer questions online and to receive **immediate feedback** and **sample responses** for every question, go to your learnON title at **www.jacplus.com.au**. A **downloadable solutions** file is also available in the resources tab.

5.6 Quick quiz	5.6 Exercise	5.6 Exam questions

5.6 Exercise

1. MC Inflammation is the immune system's early protective response to harmful stimuli, such as pathogens, damaged cells, toxic compounds, or irradiation. It acts by removing the stimulus and initiating the healing process. This response
   A. is specific to an antigen.
   B. involves phagocytes migrating to the site of infection.
   C. involves the production of antibodies.
   D. is a part of adaptive immunity.
2. Why are mast cells important in the process of inflammation?
3. Briefly explain why the area around an infection becomes swollen.
4. Identify two key changes in capillary vessels that occur during the vascular stage of inflammation.
5. Identify the following statements as true or false and justify your response.
   a. The main response of the innate immune system to invading pathogens is acute inflammation.
   b. The stages in an inflammatory response in order are the cellular stage, the resolution stage and the vascular stage.
   c. Resolution of an inflammatory response requires the production of more pro-inflammatory molecules.
   d. Cytokines released by cells at the infection site attract more immune cells to the site.
   e. The redness around an area of inflammation is the result of the constriction of blood vessels.
6. Draw a diagram showing the process of inflammation. Label all the cells involved in the process and the molecules produced by them.
7. Cytokines are signalling molecules of the immune system. One role of cytokines is in the activation and attraction of NK cells, macrophages and dendritic cells at the site of the infection. In some cases, an event called a cytokine storm may occur, in which excessive cytokines are released, which may lead to increased heart rate and possible death. Linking this to inflammation, explain why this event may lead to possible death.

5.6 Exam questions

Question 1 (1 mark)

Source: *VCAA 2015 Biology Exam, Section A, Q18*

MC A girl is carrying a piece of wood. A small piece breaks off and becomes embedded in her finger. The next day, she notices an inflammatory response occurring in her finger.

In the region around the small piece of wood embedded in her finger

A. mast cells would release antibodies.
B. the skin tissue would become pale and cold.
C. the capillaries would become more permeable.
D. red blood cells would leave the blood vessels and engulf foreign material.

Question 2 (1 mark)

Source: *VCAA 2010 Biology Exam 1, Section A, Q5*

MC An inflammation reaction involves the

A. release of histamines.
B. agglutination of bacteria.
C. production of immunoglobulin.
D. vasoconstriction of blood vessels.

Question 3 (1 mark)

Source: *VCAA 2012 Biology Exam 1, Section A, Q10*

MC Consider the following diagram showing a bacterial infection within a human. Cells moving from the blood vessel towards the bacteria

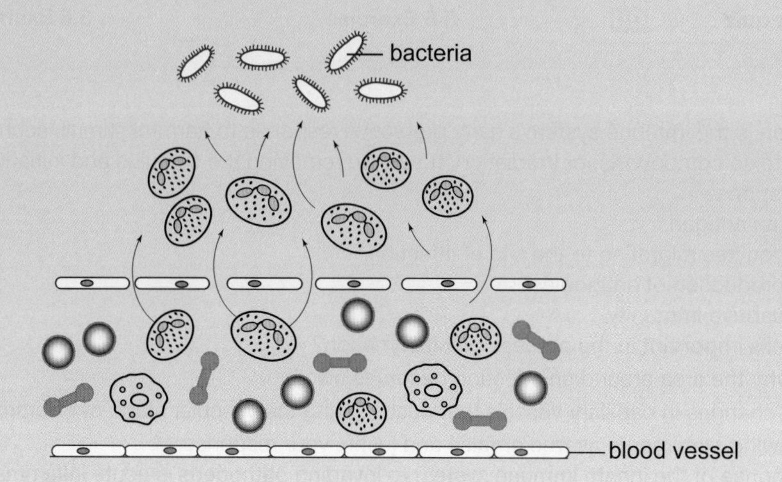

A. are natural killer cells.
B. would act as phagocytes.
C. cause vasodilation of the blood vessel.
D. release histamine in response to tissue damage.

Question 4 (1 mark)

Source: *VCAA 2016 Biology Exam, Section A, Q20*

MC The inflammatory response is a defence mechanism that evolved in higher organisms to protect them from infection and injury.

This response
A. includes phagocyte migration to the site of the injury.
B. is part of the adaptive immune system.
C. is specific to the type of foreign body.
D. involves the production of lymphocytes.

Question 5 (2 marks)

Source: *Adapted from VCAA 2006 Biology Exam 1, Section B, Q7c*

Plants and animals are both susceptible to infection.

Assume that infection has occurred in a human.

Outline two general features of inflammation that minimise the impact of the infection.

More exam questions are available in your learnON title.

5.7 The role of the lymphatic system

KEY KNOWLEDGE

- The role of the lymphatic system in the immune response as a transport network and the role of lymph nodes as sites for antigen recognition by T and B lymphocytes

5.7.1 The function of the lymphatic system

The lymphatic system plays a key role in the immune response of mammals. It acts as a transport network and works very closely with the circulatory systems, carefully monitoring the body for signs of infection.

The lymphatic system has several functions:

- production and maturation of immune cells
- the removal of excess fluids from body tissues
- absorption and transportation of fatty acids to the digestive system
- allowing for the process of antigen recognition by T and B lymphocytes.

5.7.2 Lymphatic system structure

The lymphatic system is a network of tissues and organs. The tissue of the lymphatic system is called lymphoid tissue, because it contains large numbers of lymphocytes.

The lymphatic system consists of:

- lymph
- lymphatic vessels
- **primary lymphoid organs**
- **secondary lymphoid organs** (including lymph nodes).

primary lymphoid organs part of the lymphatic system that comprises the bone marrow and thymus

secondary lymphoid organs part of the lymphatic system that comprises the lymph nodes and the spleen

FIGURE 5.61 A simplified representation of the components of the lymphatic system

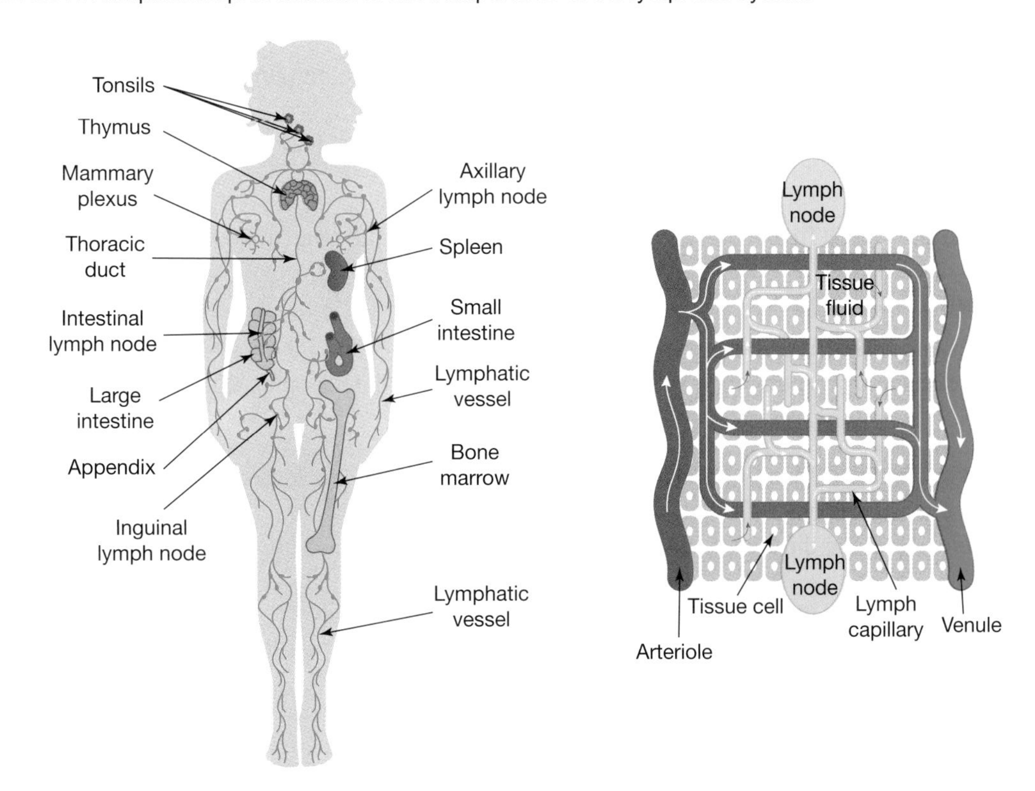

Lymph

Lymph is the fluid in the lymphatic system that gets squeezed out of blood vessels. Interstial fluid surrounding the tissues gets filtered through the tiny holes between the capillaries into the lymphatic system. Due to the presence of valves in lymphatic vessels, the lymph can only move one direction.

FIGURE 5.62 The interaction between lymph nodes and capillaries

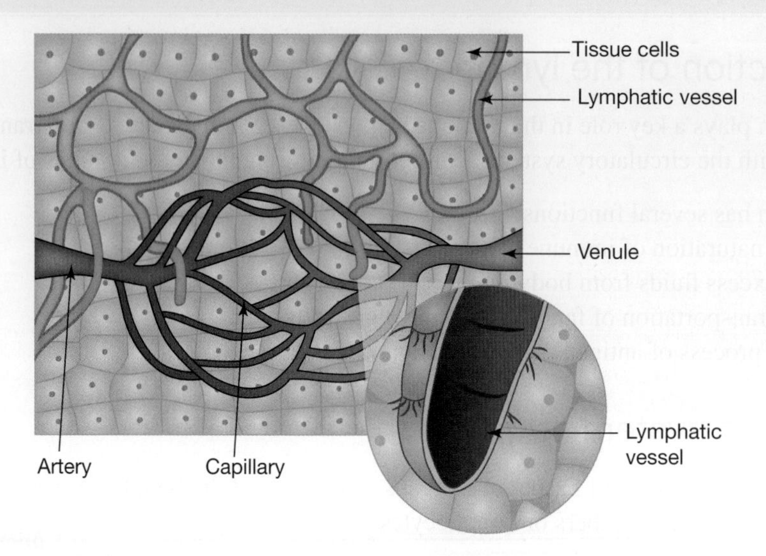

Primary lymphoid organs

The bone marrow and the thymus are called primary lymphoid organs because they are the sites where mature lymphocytes (B cells and T cells) develop from precursor cells. These are also the sites where tolerance of lymphocytes develops and any B and T cells that target self-cells are eliminated. This is referred to as **self-tolerance**.

Bone marrow

Bone marrow:

- is the source of pluripotent stem cells, from which all the cells of the immune system (and other blood cells, such as red blood cells and platelets) originate
- is the site of maturation of B cells (unlike T cells, which mature elsewhere).

Thymus

The thymus is the site where T cells mature after being released from the bone marrow.

Secondary lymphoid organs

The lymph nodes and spleen are secondary lymphoid organs. These organs are the sites where mature B cells and T cells are activated by meeting their complementary antigens and developing into effector cells.

Spleen

The spleen is an organ that is found in nearly all vertebrates. In humans, it is located behind the stomach in the upper left abdomen. The spleen:

- filters the blood passing through it, clearing the blood of bacteria and viruses as well as worn-out red blood cells
- contains T cells and B cells that detect and respond to infectious agents in the blood
- contains other immune cells, including macrophages and dendritic cells.

self-tolerance inability of an adaptive immune system to respond to the body's own self-antigens

Lymph nodes

Lymph nodes are main sites of the adaptive immune response. The adaptive immune response will be explored further in the upcoming subtopics.

Lymph nodes are the sites in which antigen recognition occurs. This is where antigen-presenting cells display their antigens to their specific T and B lymphocytes. This leads to the expansion of the appropriate lymphocytes involved in adaptive immunity.

FIGURE 5.63 Light microscope image of a section of the cortex of a lymph node showing the dense population of lymphocytes, mainly B cells

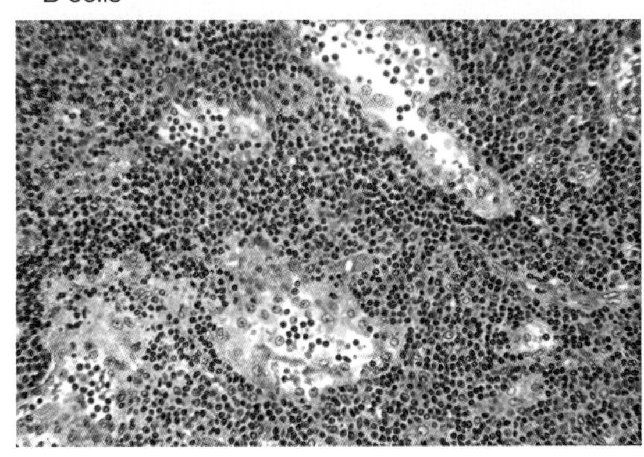

FIGURE 5.64 A swollen lymph node during infection

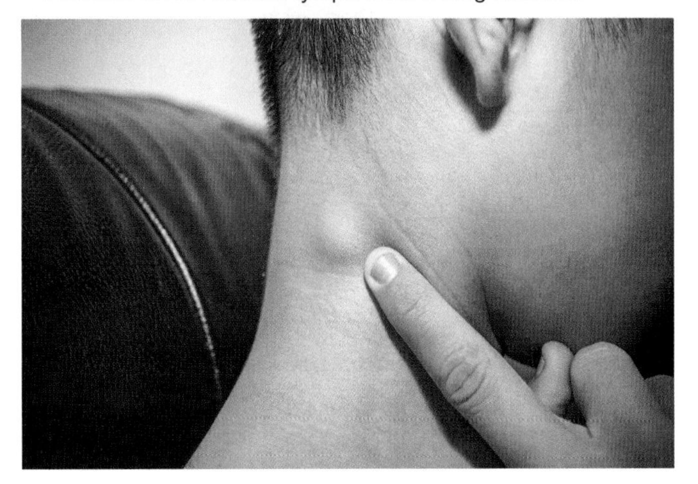

Lymph nodes:

- are small bean-shaped structures located along lymphatic vessels and present in high numbers in strategic locations within the body — in the armpits, groin, neck and abdomen
- are located along blood vessels and lymphatic vessels, which enable B cells and T cells to enter and exit the lymph nodes. Most enter the lymph nodes via incoming arteries, but about 10 per cent enter via the incoming lymphatic vessel
- are the sites where immune responses occur and any 'new' foreign antigen meets and activates appropriate B cells and T cells (as part of clonal selection and expansion)
- swell when infections occur, because the numbers of B and T cells in the lymph nodes increase — this produces the so-called 'swollen glands'
- consist of an internal structure comprising an outer cortex, an inner cortex and a central medulla:
  - The outer cortex contains follicles with large numbers of B cells that divide and diversify; antigen-presenting cells, such as dendritic cells, are also found here.
  - The inner cortex mainly contains T cells and other immune cells such as dendritic cells.
  - The medulla contains fewer cells, but has some macrophages and B cells, including special antibody-producing B cells called **plasma cells**.
- trap cancer cells travelling in lymph vessels — for this reason, people who have a cancer surgically removed, such as a breast cancer, typically will also have the lymph nodes draining the affected organ removed.
- are protected by valves in the incoming and outgoing lymphatic vessels so lymph can only travel one direction through the lymph nodes. These valves are present in lymphatic vessels throughout the lymphatic system. Imagine this like airport security — once you have passed the security checkpoint, you cannot go back out the entrance. If you try to reenter, you will need to be screened again.

plasma cells B cells that are short-lived and secrete soluble antibodies against the specific antigen

FIGURE 5.65 a. A diagram of a lymph node b. A light micrograph of a lymph node from a rabbit. The follicles in the outer cortex contain inactivated B cells, which when activated, develop into a germinal centre. T cells are mainly concentrated in the paracortex (inner cortex).

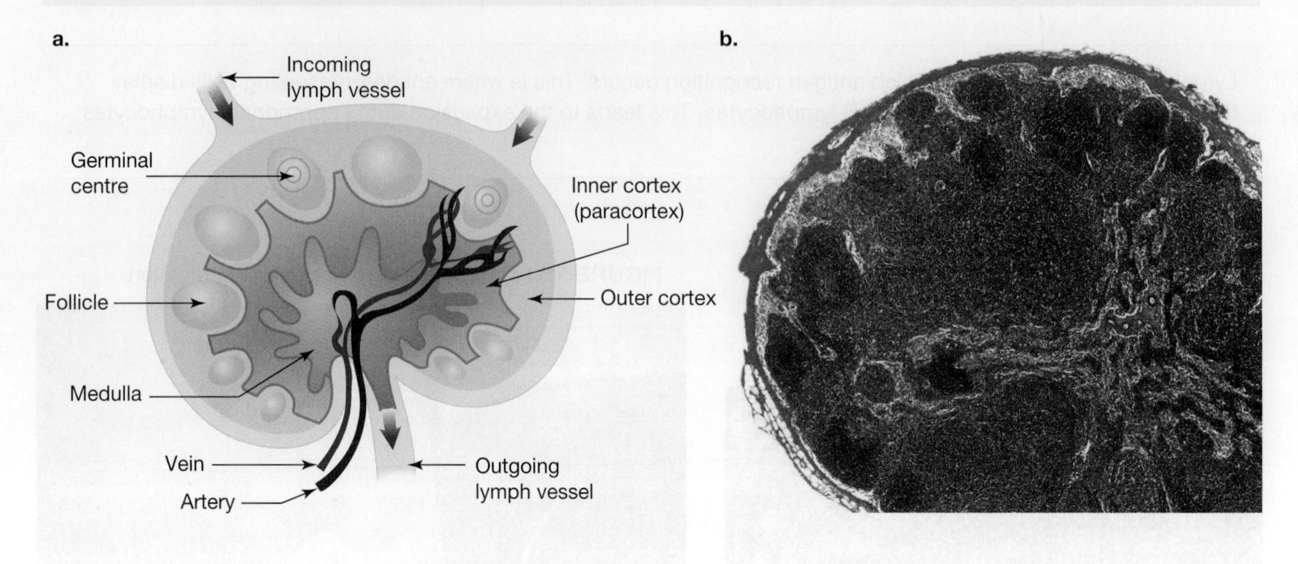

elog-0679

INVESTIGATION 5.4

online only

Modelling the swelling of lymph nodes

Aim

To model how clonal expansion leads to swelling in lymph nodes

Resources

eWorkbook Worksheet 5.10 The role of the lymphatic system (ewbk-5260)

KEY IDEAS

- The lymphatic system produces lymphocytes and transports them to the lymph nodes to elicit the immune response.
- Some lymphocytes live in lymphoid organs and others circulate in the blood and lymphatic system.
- All the cells of the immune system originate as stem cells in the bone marrow.
- Primary lymphoid organs are sites where immune cells are produced and mature. They include bone marrow and the thymus.
- Secondary lymphoid organs are the sites where immune cells are activated by meeting antigens and where immune responses occur. They include lymph nodes and the spleen.
- Lymph nodes are the site of antigen recognition, in which T and B lymphocytes come into contact with their specific antigens. This results in clonal selection and expansion.

5.7 Activities

To answer questions online and to receive **immediate feedback** and **sample responses** for every question, go to your learnON title at **www.jacplus.com.au**. A **downloadable solutions** file is also available in the resources tab.

5.7 Quick quiz	5.7 Exercise	5.7 Exam questions

5.7 Exercise

1. Give an example of each of the following.
   a. A primary lymphoid organ
   b. A secondary lymphoid organ
   c. A lymphocyte
2. Identify the component of the human immune system that:
   a. is the source of all immune cells
   b. is the site of maturation of T cells
   c. filters the blood that passes through it
   d. traps cancer cells that escape into the lymphatic vessels
   e. is the site where 'new' foreign antigens are shown to T cells.
3. Label parts A and B in the diagram shown. Explain the structure and function of each of these.

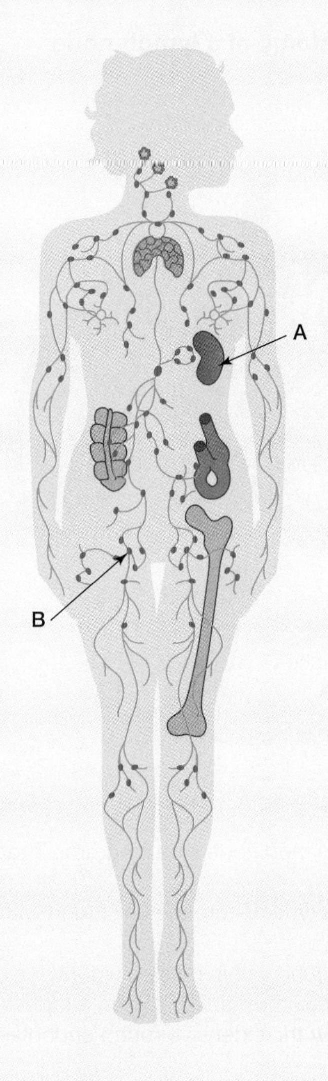

4. Complete the following passage using the listed terms:
   lymph node, thymus, T cells, B cells, primary, secondary, lymph, antigen recognition
   Pathogens can enter the bloodstream. A fluid known as _________ is squeezed out of the blood vessels and moves into the lymphatic system, moving one way towards the _____________. This is the site of __________, where clonal selection and expansion occurs as lymphocytes are exposed to their particular antigen. Lymphocytes are produced in the bone marrow. _________ move to the _________ to mature, whereas ________ remain in the bone marrow to mature. Bone marrow and the thymus are ____________ lymphatic organs, unlike the lymph nodes, which are ____________ lymphatic organs.
5. During illness, a doctor will often check your lymph nodes for swelling. Explain why swollen lymph nodes are an indication of infection.
6. Due to trauma and damage to the spleen, an individual may need to undergo a splenectomy, which involves the removal of the spleen. Although individuals can survive without a spleen, there are many post-surgery risks associated with living without a spleen. Describe what risks would be associated with this and suggest possible precautions individuals without a spleen should take.

5.7 Exam questions

Question 1 (5 marks)

Source: *VCAA 2015 Biology Exam, Section B, Q5*

Consider the following diagram of a lymph node.

Anatomy of a lymph node

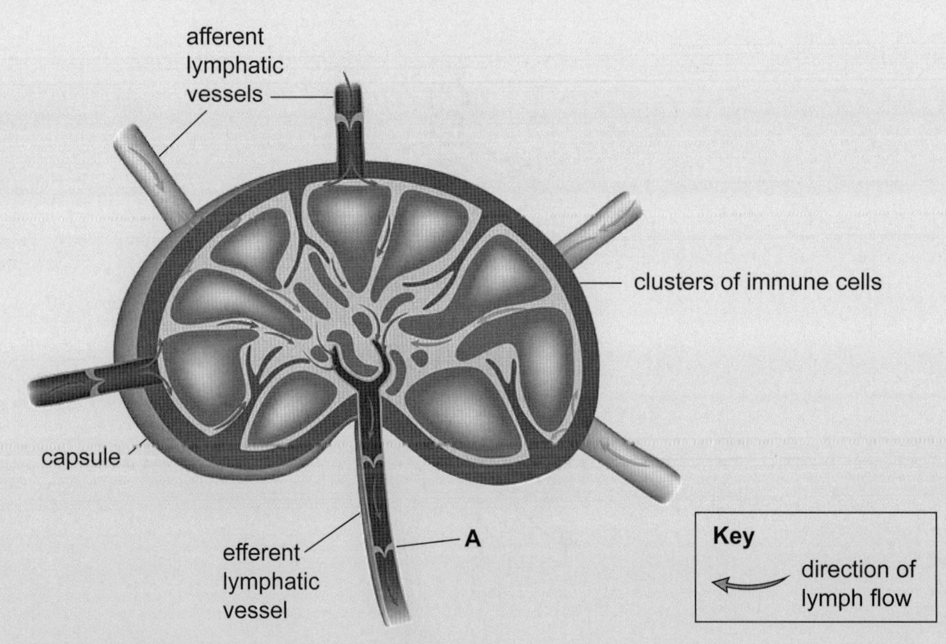

Source: Alila Medical Media/Shutterstock.com

a. Describe the role of the structure labelled A, found within the efferent lymphatic vessel. **1 mark**

Immune cells are clustered within the lymph node (see diagram). There is more than one type of immune cell within each of these clusters.

b. Name and describe the role of **one** type of immune cell found within these clusters that plays a role in the innate immune response. **2 marks**
c. Another of the immune cell types found within these clusters has a large nucleus and extensive rough endoplasmic reticulum, and plays an important role in an adaptive immune response.
   Name this cell type and explain how the extensive rough endoplasmic reticulum assists this cell to perform its function. **2 marks**

Question 2 (1 mark)

Source: *Adapted from VCAA 2017 Biology Exam, Section A, Q23*

MC The lymphatic system includes the lymph nodes, spleen and tonsils.

In these particular organs

A. clotting factors are inactivated to help seal a wound.
B. clonal selection and proliferation of B cells occurs.
C. non-self antigens are identified by red blood cells.
D. the initial response to an allergen is triggered.

Question 3 (1 mark)

Source: *VCAA 2011 Biology Exam 1, Section A, Q7*

MC The lymphatic system contains

A. B cells only.
B. T cells only.
C. B cells and T cells.
D. neither B cells nor T cells.

Question 4 (1 mark)

Source: *VCAA 2013 Biology Section A, Q18*

MC DiGeorge syndrome is a rare, congenital disease that can disrupt the normal development of the thymus gland.

This disease would result in the

A. swelling of lymph nodes.
B. overproduction of B cells.
C. reduced production of T cells.
D. release of histamines from mast cells.

Question 5 (2 marks)

Source: *VCAA 2011 Biology Exam 1, Section B, Q2a*

The thymus is an important organ in the immune system. As humans grow older, there is a change in the weight of the thymus and an increase in the proportion of fat it contains.

Examine the following table.

Age	At birth	10 years	20 years	30 years	60 years
Average weight of thymus (gram)	20	35	20	15	5

Explain the likely consequence of this change in thymus weight in an individual.

More exam questions are available in your learnON title.

5.8 Initiation of an adaptive immune response

KEY KNOWLEDGE

- Initiation of an immune response, including antigen presentation
- The actions of helper T cells

Source: Adapted from VCE Biology Study Design (2022–2026) extracts © VCAA; reproduced by permission.

5.8.1 Initiating the adaptive immune system

As discussed in previous subtopics and above, many aspects of the innate immune response are initiated by the presence of pathogenic agents, the use of complement proteins, and the release of chemicals such as cytokines and histamines that allow for chemotaxis and the attraction of other immune cells.

The innate immune system is vital in initiating a response by the adaptive immune system. This is achieved through the use of antigen-presenting cells, which include dendritic cells and macrophages.

Pathogens can enter the human body by ingestion, injection or inhalation. Typically, these pathogens are met close to the entry point by phagocytic cells that engulf and eliminate them. In particular, dendritic cells are active in this role because they are located on and in epithelial tissues of the skin and the linings of the airways and the gut — all potential points of entry of pathogens to the body.

The initiation of an immune response can be split into two components — phagocytosis and antigen presentation.

FIGURE 5.66 Phagocytosis and antigen-presentation by dendritic cells to helper T cells. **a.** The microbe is engulfed by the dendritic cell. **b.** The microbe is destroyed and fragments are presented on MHC-II molecules. **c.** Dendritic cells present the antigen to specific helper T cells.

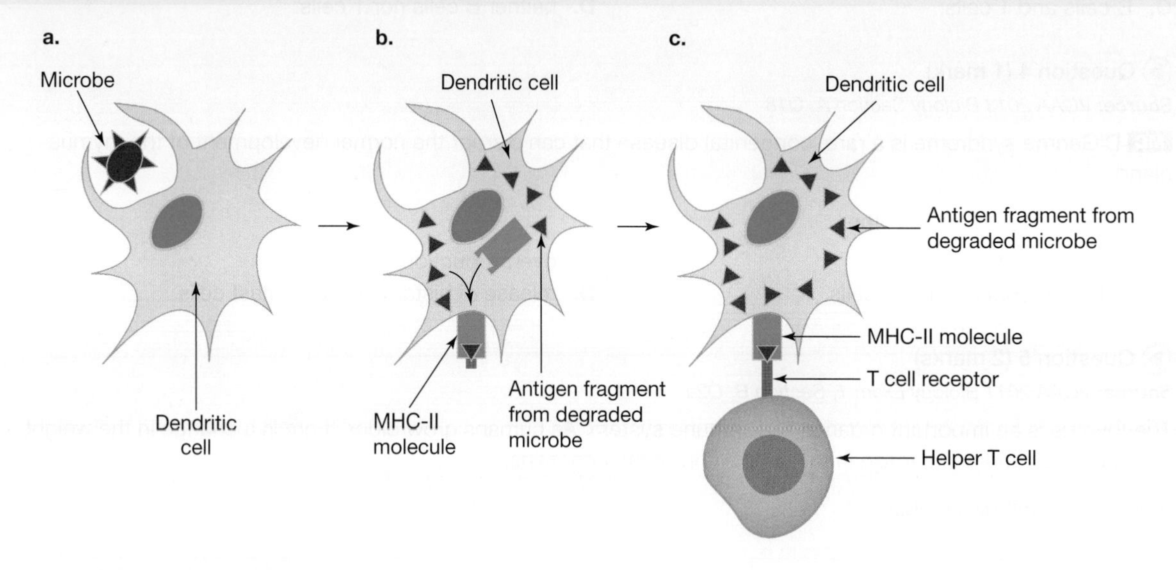

Steps involved in antigen presentation and initiation of an immune response

1. The pathogen is engulfed into a phagosome, which fuses with a lysosome (refer to information on phagocytosis in section 5.5.3).
2. The pathogen is digested by enzymes in the lysosome.
3. Some of the fragments (digested parts of an antigen) are released by exocytosis.
4. Some of the digested fragments of antigen (or peptides) are displayed on the MHC-II receptors on the surface of the macrophage or dendritic cell.
5. These dendritic cells and macrophages move to the lymph node and present them to naïve **helper T cells** that carry specific receptors for that antigen. This process is referred to as antigen presentation (as seen in figure 5.66).
6. These helper T cells undergo expansion, and **clones** are produced.
7. Helper T cells assist other immune cells by releasing specific cytokines (named interleukins). These can initiate either the humoral adaptive immune system (through stimulating the **clonal selection** and **clonal expansion** of B cells) or the cell-mediated immune system (through stimulating the clonal selection and expansion of cytotoxic T cells). These pathways will be investigated in subtopics 5.9, 5.10 and 5.11.

helper T cells a class of T cell that bind to antigen–MHC-II complexes on antigen-presenting cells and activate B cells and cytotoxic T cells

clones groups of cells, organisms or genes with identical genetic make-up

clonal selection an event occurring in lymph nodes in which those lymphocytes with receptors that can recognise a new antigen come into contact with that particular antigen

clonal expansion a process of multiple cycles of cell division of a lymphocyte specific to a particular antigen, resulting in the production of large numbers of identical lymphocytes

FIGURE 5.67 A dendritic cell presenting an antigen to a helper T cell

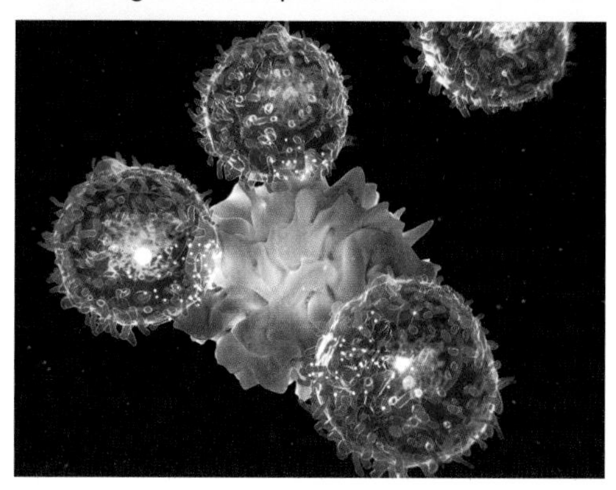

5.8.2 Helper T cells

Helper T cells do not directly kill the pathogens; rather, they help with immune responses. They express CD4 receptors and help with the activation of cytotoxic T cells, B cells and other immune cells. Due to the presence of CD4 receptors (as shown in figure 5.68), they are referred to as CD4+ T cells.

FIGURE 5.68 a. Helper T cells have important receptors that enable them to function. b. The binding of a helper T cell to the presented antigen of an MHC-II marker

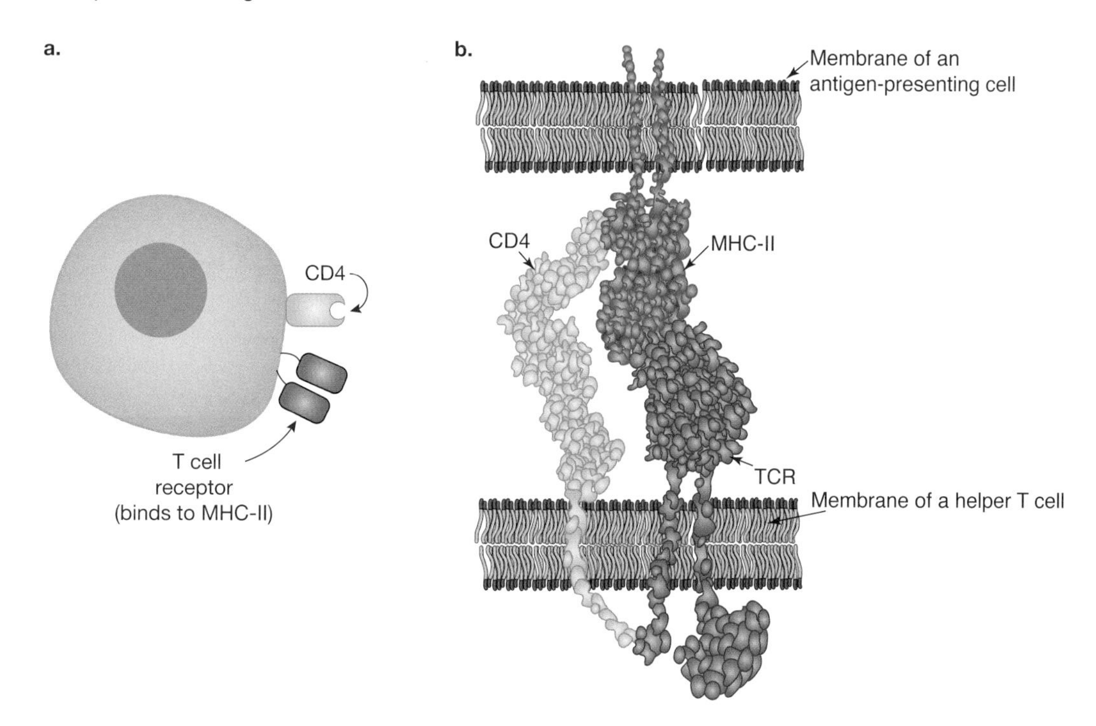

Helper T cells are activated when they bind to the antigens presented by the dendritic cells. This binding mainly involves **T cell receptors** aided by CD4 receptors.

Activated helper T cells proliferate, and the effector helper T cells stimulate immune responses in the secondary lymphoid organs, where adaptive immune cells are concentrated.

T cell receptor a molecule found on the surface of T cells that is responsible for recognising fragments of antigen as peptides bound to major histocompatibility complex (MHC) molecules

The function of helper T cells

Activated effector helper T cells secrete various cytokines, such as interleukin 2, that:

- help activate cytotoxic T cells to seek and destroy cells infected by viruses with a specific antigen
- help activate B cells into becoming antibody-producing plasma cells that will produce and release antibodies to defend against pathogens carrying that antigen
- help activate macrophages to remove antibody-coated pathogens by phagocytosis.

Effector helper T cells stimulate the activities of both the cell-mediated arm of adaptive immunity and the humoral arm of adaptive immunity.

FIGURE 5.69 Helper T cells are important in activating other aspects of the immune system.

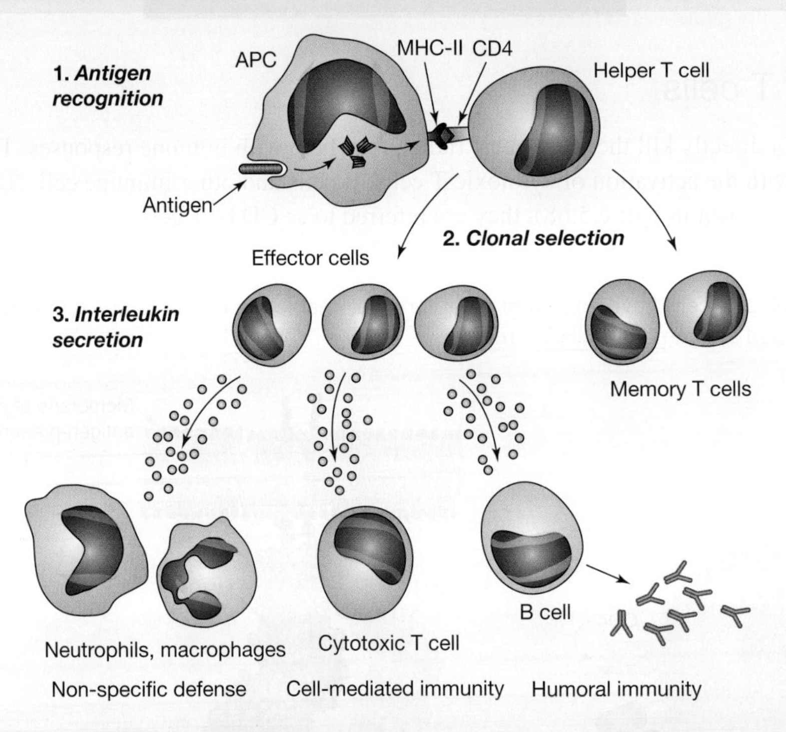

tlvd-1780

SAMPLE PROBLEM 5 The impact of HIV on helper T cells

Human immunodeficiency virus (HIV), a member of the retrovirus family, is the causative agent of acquired immunodeficiency syndrome (AIDS). HIV infects vital cells in the human immune system. The main cells that HIV targets are helper T cells.
Name the immune responses that are likely to be impacted by HIV. (3 marks)

THINK

Think about the type of immune responses that involve helper T cells. You only need to name the response. Because this is a 3-mark question, you need to ensure that you give 3 responses.

WRITE

- Innate immune response (1 mark)
- Humoral adaptive immune response (1 mark)
- Cell-mediated adaptive immune response (1 mark)

CASE STUDY: Helper T cells and HIV

HIV is a retrovirus that carries its genetic information in the form of two fragments of single-stranded RNA. Figure 5.70a shows a diagram of an HIV particle.

FIGURE 5.70 a. The strucutre of HIV **b.** The attachment of HIV to T helper cells

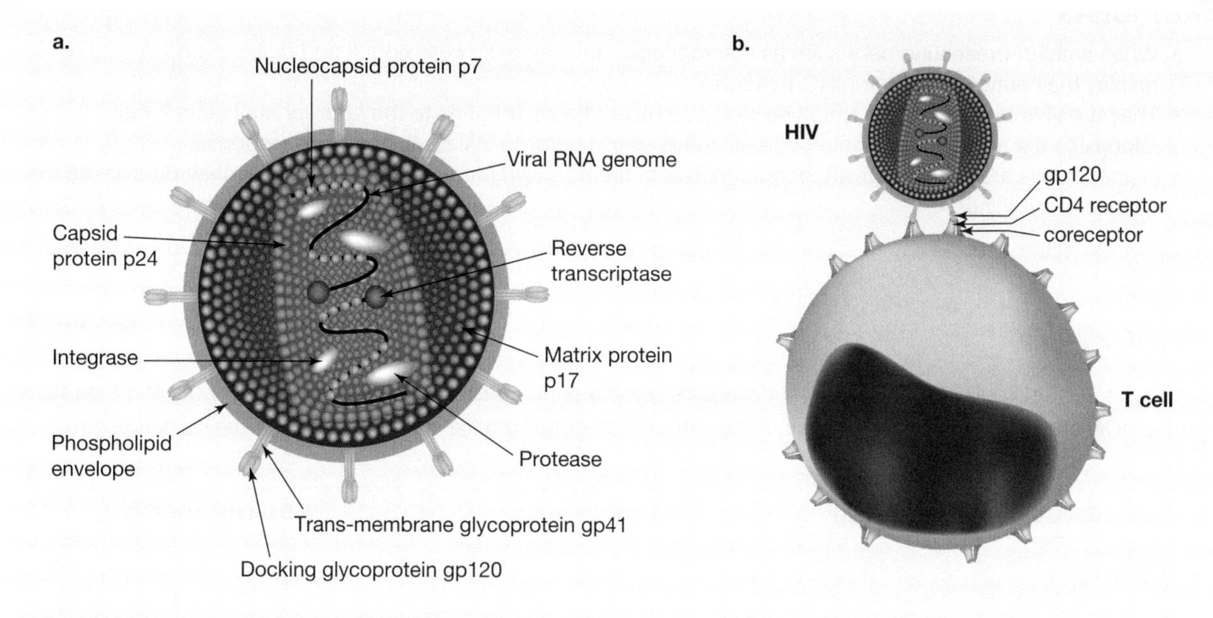

HIV targets helper T cells

Like all viruses, HIV can only replicate in a specific host cell. The host cells for HIV are the helper T cells that carry CD4 receptors on their outer surface. To gain entry to these cells, the docking glycoproteins of the virus (gp120) must first bind to CD4 receptors (marker proteins) on the surface of the helper T cells (refer to figure 5.70a).

FIGURE 5.71 Electron micrograph of HIV exiting a helper T cell

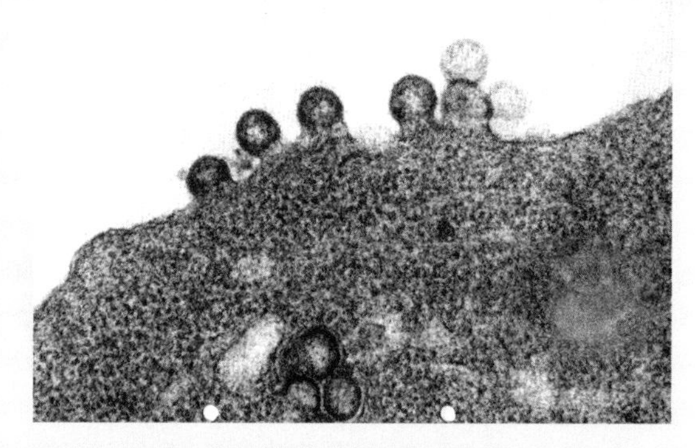

Once the viral RNA and proteins gain entry to a helper T cell, the virus takes control of the host cell. Rather than performing its normal functions, the helper T cell becomes a virus factory, producing multiple copies of the viral genetic material (RNA) and all the viral proteins, both enzymes and structural proteins, needed to make new viruses. Some HIV RNA integrates into the host genome and can remain dormant for years.

In a healthy adult, the helper T cell count in the blood ranges from 500 to 1200 cells per microlitre (μL). In a person with an HIV infection, the helper T cell count may drop to less than 200 cells per μL. When the $CD4^+$ cell count of a person falls below 200 cells per μL, that person is identified as having a stage III HIV infection, or AIDS. If untreated, the CD4 cell count will continue to fall.

The gradual depletion of the helper T cell population means that the immune system ceases to protect against pathogens. The immune systems of people with untreated AIDS are suppressed, and these people are at risk of multiple infections. These infections are termed 'opportunistic' because the pathogens take advantage of the weakened immune system. Opportunistic infections can affect many organs, and include fungal diseases such as cryptococcal meningitis, bacterial infections such as tuberculosis, and viral diseases such as Kaposi's sarcoma. Death can occur from these opportunistic infections.

Resources

eWorkbook Worksheet 5.11 How the adapive immune response is initiated (ewbk-5262)

Digital document Case study: The mechanism of action of HIV (doc-36057)

KEY IDEAS

- When antigen-presenting cells such as macrophages and dendritic cells engulf and digest pathogens, they display their antigens to their MHC-II markers.
- They are able to move to the lymph nodes and present these antigens to the appropriate helper T cell (alongside the appropriate cytotoxic T cell and B cell) in order to initiate the immune response.
- Helper T cells are vital to activate both branches of the adaptive immune response, which they do through the release of cytokines.

5.8 Activities

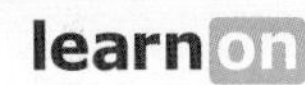

To answer questions online and to receive **immediate feedback** and **sample responses** for every question, go to your learnON title at **www.jacplus.com.au**. A **downloadable solutions** file is also available in the resources tab.

5.8 Quick quiz | **5.8 Exercise** | **5.8 Exam questions**

5.8 Exercise

1. Identify the following statements as true or false and justify your response.
   a. Dendritic cells acquire antigens by phagocytosis.
   b. The only target cells for antigen presentation are antibody-producing B cells.
   c. The loss of helper T cells would be expected to create a life-threatening situation.
   d. The antigens presented by dendritic cells are peptide fragments of degraded pathogen proteins.
2. Why is it important to begin treating HIV as soon as an individual is infected?
3. Draw a clearly labelled diagram showing the process of the initiation of an immune response via antigen-presenting cells.
4. Explain what would be expected to happen if a dendritic cell came across a helper T cell that did not have the appropriate receptor for the specific antigen being presented.
5. **a.** Complete this diagram by labelling the structures numbered 1, 2, and 3.

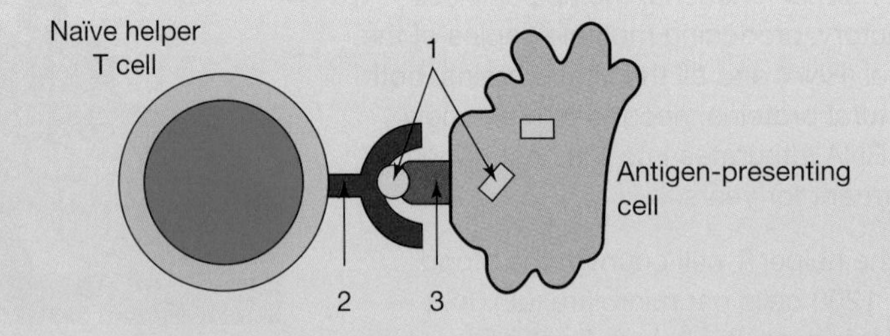

   b. What is the likely identity of the antigen-presenting cell?
   c. What is an alternative name for a helper T cell: CD4+ cell or CD8+ cell?
   d. Compete the following sentences by choosing from the alternatives supplied:
      i. Binding of a receptor on a helper T cell to a presented antigen (*kills/activates/opsonises*) the T cell.
      ii. Another cell that can act as an antigen-presenting cell is a (*mast cell/macrophage/neutrophil*).
      iii. For presentation, antigens must be linked to (*an MHC molecule/a complement protein/an antibody*).

5.8 Exam questions

Question 1 (1 mark)

Source: *VCAA 2020 Biology Exam, Section A, Q19*

MC The role of T helper cells is to

A. destroy cells that are infected with bacteria.
B. control the adaptive immune response.
C. generate antibodies.
D. engulf parasites.

Question 2 (2 marks)

Source: *Adapted from VCAA 2010 Biology Exam 1, Section B, Q7c*

Once a macrophage has destroyed a gluten fragment, it displays a piece of the fragment on its membrane using a special major histocompatibility complex (MHC) marker. A T-helper cell then attaches to the MHC marker-antigen complex.

The macrophage T-helper cell complex is shown.

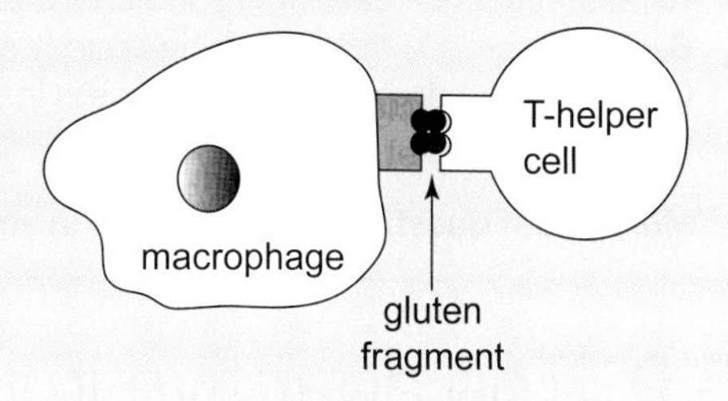

The macrophage T-helper cell complex stimulates other cells and chemicals to target and damage epithelial cells that line the intestine.

Name one cell or chemical that would be stimulated by the macrophage T-helper cell complex and state its function.

Question 3 (3 marks)

Source: *Adapted from VCAA 2006 Biology Exam 1, Section B, Q8b, c and d*

HIV is a virus which targets helper T cells.

A person infected with the HIV virus was monitored for several years for the level of T cells and HIV particles. The results are summarised in the following graph.

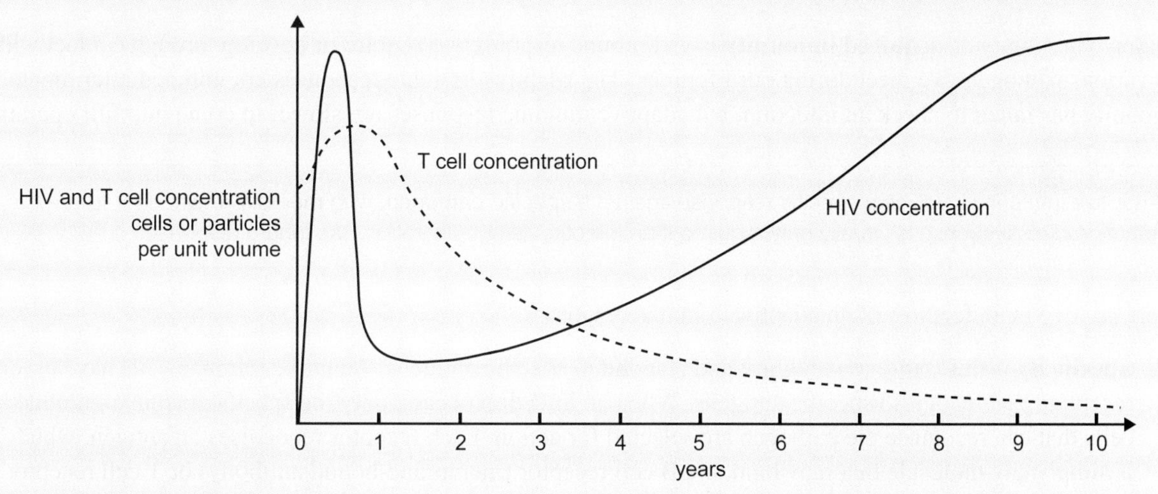

a. Explain what is happening in the first year after infection with the HIV virus. **1 mark**
b. In the second to fifth years (inclusive) after infection the patient has swollen lymph nodes. Explain. **1 mark**
c. Nine years after infection, describe what has happened to the immune system of the patient. **1 mark**

Question 4 (1 mark)

MC Antigen-presenting cells can display antigens on their surface and display these antigens to other immune cells. These cells recognise these antigens as being non-self and release cytokines, causing the other lymphocytes to become activated and increase in number.

Which of the following lymphocytes release cytokines to activate other lymphocytes?

A. Helper T cells
B. Macrophages
C. Plasma cells
D. Neutrophils

Question 5 (1 mark)

Source: *Adapted from VCAA 2017 Biology Exam, Section A, Q24*

Multiple sclerosis (MS) is an autoimmune disease, in which autoantibodies are produced by an individual which target an individual's self-cells. In sufferers of MS, the myelin coating of nerve cell axons is damaged. This damage results in poor transmission of nerve messages between the brain, the spinal cord and the rest of the body. One aspect of MS diagnosis is imaging the brain to detect visible areas of demyelination, called plaques.

Researchers investigating MS have analysed various tissue samples from patients.

MC In these samples they would expect to find

A. an abundance of allergens in nerve cells.
B. cancer cells in MS plaques in brain tissue.
C. increased numbers of T helper cells in spinal fluid.
D. an absence of T cytotoxic cells in the spinal cord and brain.

More exam questions are available in your learnON title.

5.9 The adaptive immune response

KEY KNOWLEDGE

- The characteristics and roles of the components of the adaptive immune response against both extracellular and intracellular threats

Source: Adapted from VCE Biology Study Design (2022–2026) extracts © VCAA; reproduced by permission.

5.9.1 Features of adaptive immune response

Adaptive immunity or **acquired immunity** is an immune response we acquire or develop through contact with the various pathogens we meet during our lifetimes. The adaptive immune responses are initiated after innate immunity has failed to check an infection, but adaptive immune responses are slower to come into full operation.

Adaptive immunity involves a specific response against a specific pathogen, with memory retained for future infection. This response is usually only required if an infection is not cleared by the innate response.

There are two main features of an adaptive immune response:

1. **Specificity** is the ability to recognise and respond to specific antigens. Adaptive immune cells have unique receptors that recognise specific antigens. When an infection occurs, only the specific adaptive immune cells that can recognise the pathogen are selected for action. Each receptor recognises an antigen, which is simply any molecule that may bind to a B cell receptor (membrane-bound antibody) or T cell receptor (TCR).
2. Immunological memory is the ability of adaptive immune cells to remember antigens after primary exposure. It enables more rapid and stronger response in the case of future infections by the same pathogen. Most of the cells involved in the adaptive immune response to infection are removed by apoptosis after a particular pathogen has been eliminated. However, a small number of memory cells remain in the body. Memory cells are already primed to produce a more rapid and stronger adaptive immune response if a pathogen re-infects. The pathogen is often eliminated before any symptoms appear. The action of memory B cells can be seen in the faster and greater production of antibodies on the second exposure to an antigen as compared to the initial exposure. Memory T cells are also produced, leading to a faster activation of the cell-mediated immune response.

acquired immunity an immunity that develops during a person's lifetime

specificity the ability to recognise and respond to a specific antigen

FIGURE 5.72 Due to the formation of memory, the secondary response in adaptive immunity is always stronger than the primary response.

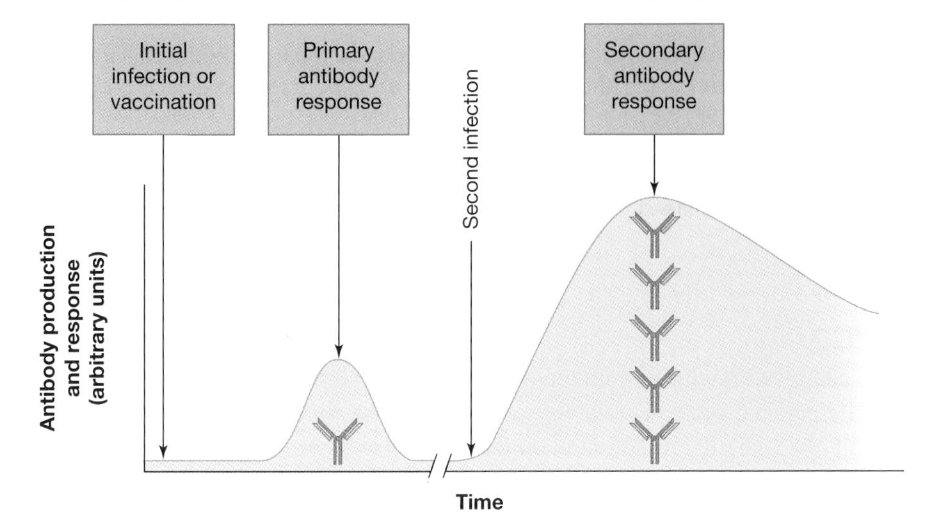

An initial infection (or vaccination) produces a primary response about 10 days after infection. This response is shown by the appearance of antibodies specific to antigens of the pathogen concerned. In the secondary response to the same pathogen, memory immune cells are activated more rapidly, and antibodies are produced more quickly and in greater quantities than in the primary response.

TABLE 5.12 Differences between primary and secondary immune response

	Primary immune response	Secondary immune response
Definition	Immune response against initial antigen exposure	Immune response against subsequent exposure to the same antigen
Response	Low and short-lived	Prompt, powerful and prolonged
Antibody-producing cells	Naïve B cells	Memory B cells
Antibody levels	Antibody levels peak in the primary response at about day 14 and then begin to drop off as the plasma cells begin to die.	Because there are many more memory cells than there were naïve B cells for the primary response, more plasma cells are generated in the secondary response, and antibody levels are consequently 100- to 1000-fold higher.

5.9.2 Components of adaptive immune response

The key components of adaptive immunity include:

- special white blood cells (lymphocytes) — T cells and B cells
- special antigen-binding proteins — antibodies, also known as immunoglobulins
- lymph nodes — secondary lymphoid organs where B cells and T cells meet foreign antigens and are activated, and where adaptive immune responses occur.

The two branches of adaptive immunity

Adaptive immunity operates in two different ways:

1. *Humoral immunity* involves the actions of antibodies that identify and bind to extracellular pathogens, to toxins and to other extracellular foreign antigens. Antibodies are products of special B cells termed plasma cells.
2. *Cell-mediated immunity* involves various actions of T cells. Cytotoxic T cells eliminate body cells that are infected by pathogens or have abnormal or missing self markers. Note that cell-mediated responses do not directly attack pathogens. Instead, they remove infected cells and, by this means, eliminate intracellular pathogens.

Cells of adaptive immunity

The main cells of the adaptive immune system are members of the class of white blood cells known as lymphocytes. The lymphocytes that play the key roles in adaptive immunity are T cells and B cells. Lymphocytes are small cells (about 7 to 10 μm in diameter, similar in size to red blood cells). They are found in the blood and the lymph, and are highly concentrated in lymph nodes. T cells and B cells develop in the primary lymphoid organs — T cells in the thymus and B cells in the bone marrow.

TIP: The word cells and lymphocytes can be used interchangeably when describing T and B cells. A good way to differentiate between T and B cells in that **T** cells mature in the **t**hymus and **B** cells mature in the **b**one marrow.

From the primary lymphoid organs, these cells migrate to secondary lymphoid organs, such as the lymph nodes.

This is where the **naïve** B and T cells are activated by exposure to their complementary antigens and undergo cycles of cell division and differentiate into effector cells. Lymph nodes are the locations where the various adaptive immune responses to pathogens take place.

naïve refers to an immune cell that has not yet been activated

FIGURE 5.73 False-coloured scanning electron microscope image of harvested whole blood showing two lymphocytes (yellow), a single red blood cell (red) and a cluster of activated platelets (orange) (image courtesy of Jonathan M. Franks and Dr Donna Stolz, University of Pittsburgh Center for Biologic Imaging)

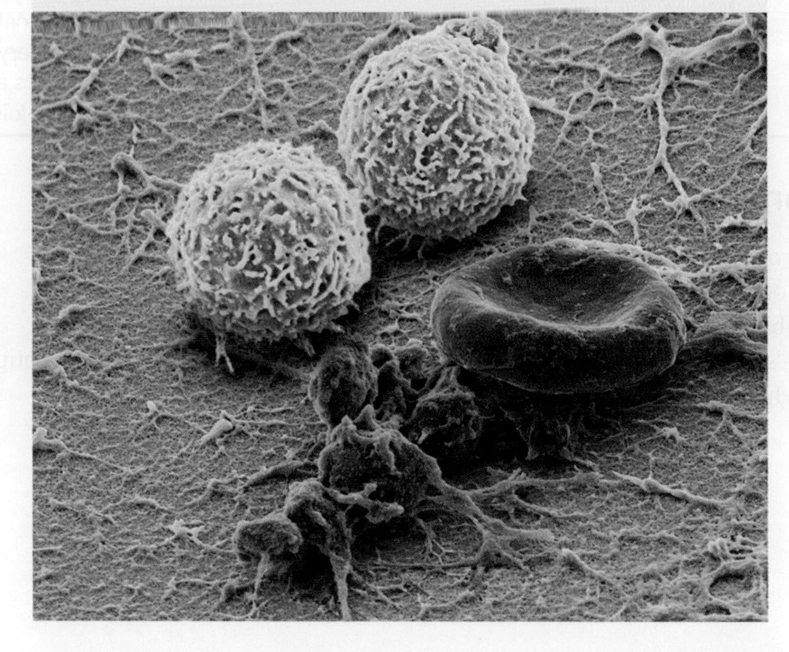

The roles of T and B cells

- T cells deliver the cell-mediated immune defences that include the direct elimination of pathogen-infected cells and other abnormal cells, such as cancer cells.
- B cells deliver the humoral immune defences by secreting antibodies that bind to surface antigens on pathogens and label them for elimination; antibodies also bind to soluble toxins.

TABLE 5.13 Comparing the features and function of B and T lymphocytes

Lymphocyte	Features	Function
B lymphocytes	• Mature in bone marrow • Each B cell has many surface receptors that recognise one specific antigen. • Activated by direct exposure to raw antigens • Develop into plasma cells that produce antibodies against specific antigens.	• Activated B (plasma) cells produce antibodies against extracellular antigens. • Memory B cells retain memory of antigens met previously.
T lymphocytes	• Mature in thymus and then migrate to lymph nodes • Each T cell has many surface receptors that recognise one specific antigen. • Activated by exposure to antigens presented to them on the surface of other cells (APCs) • Differentiate into various types of T cells including helper T cells and cytotoxic T cells	• Activated cytotoxic T cells eliminate infected and abnormal cells. • Activated helper T cells send signals that stimulate B cells to secrete specific antibodies. • Memory T cells retain memory of antigens met previously.

on Resources

 **eWorkbook** Worksheet 5.12 The branches of adaptive immunity (ewbk-5264)

 **Weblink** The immune system in action

KEY IDEAS

- Adaptive or acquired immunity forms the third line of defence against infection and is specific to particular pathogens.
- The adaptive immune response involves B and T lymphocytes.
- B and T cells are lymphocytes (white blood cells) that are able to recognise antigens that distinguish 'self' from 'other' in the body.
- Adaptive immunity has a humoral component and a cell-mediated component.
- Humoral adaptive immunity targets extracellular antigens and is achieved through the action of antibodies produced by B lymphocytes.
- Cell-mediated immunity attacks infected or abnormal cells and is brought about by T cells.
- The adaptive immune response is slower to develop than the innate immune response, but it can act much more powerfully and quickly than the innate immune response against pathogens that it has seen before.
- Once B and T lymphocytes mature, the majority of them enter the lymphatic system, where they are stored in lymph nodes.

5.9 Activities

To answer questions online and to receive **immediate feedback** and **sample responses** for every question, go to your learnON title at **www.jacplus.com.au**. A **downloadable solutions** file is also available in the resources tab.

5.9 Quick quiz	5.9 Exercise	5.9 Exam questions

5.9 Exercise

1. Complete the following passage using the listed terms:
   Adaptive, B cell, T cells, memory, antibodies, intracellular, extracellular, primary, secondary
   At birth, a baby has no ________ immune defences. Over time this develops with exposure to antigens. Adaptive immunity differs from innate immunity in that it is specific to particular antigens and leads to ________ cells being formed. This leads to the ________ response being stronger and faster compared to the ________ response.
   There are two branches of the adaptive immune system. Cell-mediated immunity involves the formation of ________ and is used for ________ pathogens. Humoral immunity involves ________, which produce ________, leading to the destruction of ________ pathogens.
2. Describe two features that make adaptive immunity different from innate immunity.
3. Draw a Venn diagram to compare and contrast cell-mediated and humoral adaptive immunity.
4. Name the kind of protein that circulates throughout the blood in humoral adaptive immunity.
5. One disease that affects the adaptive immune system is known as severe combined immunodeficiency (SCID). This disease is often colloquially referred to as bubble baby disease. In this very rare genetic disorder, functional T cells and B cells are not produced. Describe the symptoms you would expect to see in an individual with SCID and justify the reason for your response.

5.9 Exam questions

Question 1 (1 mark)

Source: *VCAA 2020 Biology Exam, Section A, Q20*

MC Which one of the following describe a feature common to both T cells and B cells?

A. having immunological memory
B. rapidly responding to pathogens on first exposure
C. providing a physical barrier to the entry of pathogens
D. being able to attach to both microorganisms and viruses.

Question 2 (1 mark)

Source: *VCAA 2007 Biology Exam 1, Section A, Q18*

MC A specific immune response involves

A. helper T lymphocytes releasing cytotoxins.
B. T lymphocytes recognising antigens presented by macrophages.
C. suppression of the response by memory B cells after the infection.
D. B lymphocytes, each with a number of different surface antibodies.

Question 3 (1 mark)

Source: *VCAA 2015 Biology Exam, Section B, Q4b*

Scientists developing new vaccines for EVD are conducting trials in animal subjects. To evaluate the effectiveness of a new vaccine, both humoral and cell-mediated responses are measured in the animal subjects.

Explain how these two immune responses are different. Give **two** differences in your answer.

Question 4 (4 marks)

Source: *VCAA 2016 Biology Exam, Section B, Q5c*

Yellow fever is a potentially fatal, mosquito-borne, viral disease that occurs in many countries in Africa, the Caribbean, and Central and South America. An effective and safe vaccine has been available since 1938.

Recent research shows that the vaccine gives lifelong immunity.

Name two different cell types that would be important in providing lifelong immunity and explain the role of each in providing lifelong immunity.

Question 5 (7 marks)

Source: *VCAA 2012 Biology Exam 1, Section B, Q4*

Australia is currently experiencing an epidemic of pertussis. Pertussis is a highly contagious respiratory infection caused by the bacteria *Bordetella pertussis* (whooping cough). Pertussis vaccine is offered as part of an immunisation program for children at two months, four months, six months, four years and in Year 10 of secondary school.

a. **i.** Name the cells that are responsible for the production of antibodies. **1 mark**

ii. Two children have been immunised according to the schedule. One is two months old and the other is four months old.

What difference would there be in the children's levels of antibodies against *Bordetella pertussis*? **1 mark**

Consider a Year 10 student. Memory cells will have been produced during the periods of immunisation when the student was younger.

b. What are **two** advantages of having these memory cells when the student receives their immunisation in Year 10? **2 marks**

In Victoria in the past two years the number of cases of pertussis has increased dramatically. In 2010 there were over 6500 reported cases of pertussis; 66 per cent of these cases were adults and most of these adults had been immunised in childhood.

c. **i.** Outline a likely reason for the high percentage of adults with pertussis in 2010. **1 mark**

ii. Describe one process that could be introduced by the Department of Health, Victoria, to reduce the number of adults being infected with pertussis. **1 mark**

d. The human immune response to antigens of *Bordetella pertussis* can be measured by the level of antibodies in the blood.

Is this test a measure of cell-mediated immunity? Explain your answer. **1 mark**

More exam questions are available in your learnON title.

5.10 Humoral adaptive immunity and B lymphocytes

KEY KNOWLEDGE

- The characteristics and roles of the components of the adaptive immune response against intracellular threats, including the actions of B lymphocytes and their antibodies

Source: Adapted from VCE Biology Study Design (2022–2026) extracts © VCAA; reproduced by permission.

5.10.1 What is the humoral adaptive immune response?

The humoral adaptive response, also called the antibody-mediated response, involves B cells (or B lymphocytes) that recognise antigens or pathogens and produce a large number of antibodies specific to an antigen. These antibodies defend only against extracellular antigens, such as antigens on the surface of a pathogen in the blood, or soluble antigens, such as toxins or venoms, in extracellular body fluids such as lymph or blood. Therefore, each B cell has a large number of specific identical B cell receptors.

B lymphocytes

B lymphocytes play important role in the adaptive humoral immune system and are responsible for mediating the production of antigen-specific immunoglobulin (typically known as antibodies) directed against extracellular pathogens. They are formed in the bone marrow and continue to mature in the bone marrow.

FIGURE 5.74 TEM of an activated plasma cell, with extension endoplasmic reticulum

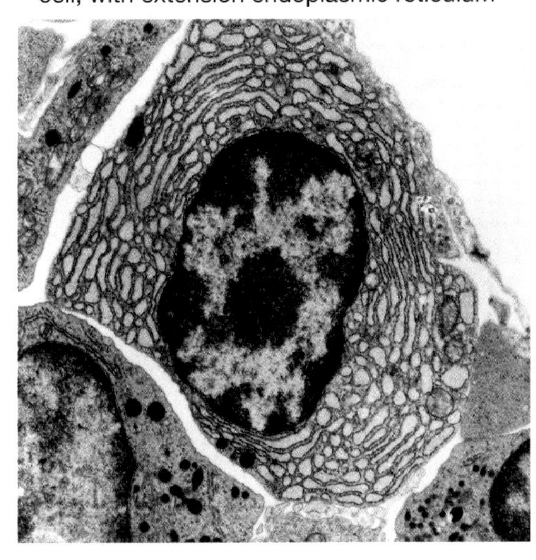

There are two types of B lymphocytes:

- Plasma cells secrete antibodies specific to the infection. Each plasma cell can only produce one type of antibody.
- Memory B cells remain in the lymphatic tissue to remember the antigen for future infections.

5.10.2 Activating the humoral adaptive response

In order for antibodies specific to an antigen to be produced, a specific chain of events occurs. These events involve the actions of antigen-presenting cells, helper T cells and B cells. The chain of events that occurs in the activation and action of this response is as follows:

The steps in the activation of humoral adaptive immunity

1. A 'new' antigen that gains entry to the body reaches the lymph nodes (usually via antigen-presenting cells), and it comes into contact with many naïve B cells that do not recognise it. Eventually, however, the antigen meets a B cell that can recognise and bind to it. This is called clonal selection.
2. Helper T cells that have also bound to the antigen release cytokines to activate the correct B cells.
3. The action of helper T cells and the binding of the antigen to its 'selected' B cell activates the B cell to differentiate and proliferate (divide) into two types of daughter cells — plasma cells and memory B cells. The result is the production of a large clone of B cells, all having identical antigen-binding receptors — this is called clonal expansion.
4. Most of these cells differentiate into short-lived plasma cells that secrete soluble antibodies against the specific antigen.
5. Memory B cells remain in the lymphoid tissue for after the infection has resolved. They initiate immune responses more rapidly and strongly upon re-exposure to the same antigen, producing large amounts of the specific antibody.

FIGURE 5.75 Upon exposure to the appropriate antigen, B cells undergo clonal selection and expansion after stimulation by helper T cells. Most of these B cells are plasma cells that produce and release large numbers of antibodies. Some of the B cells are memory B cells.

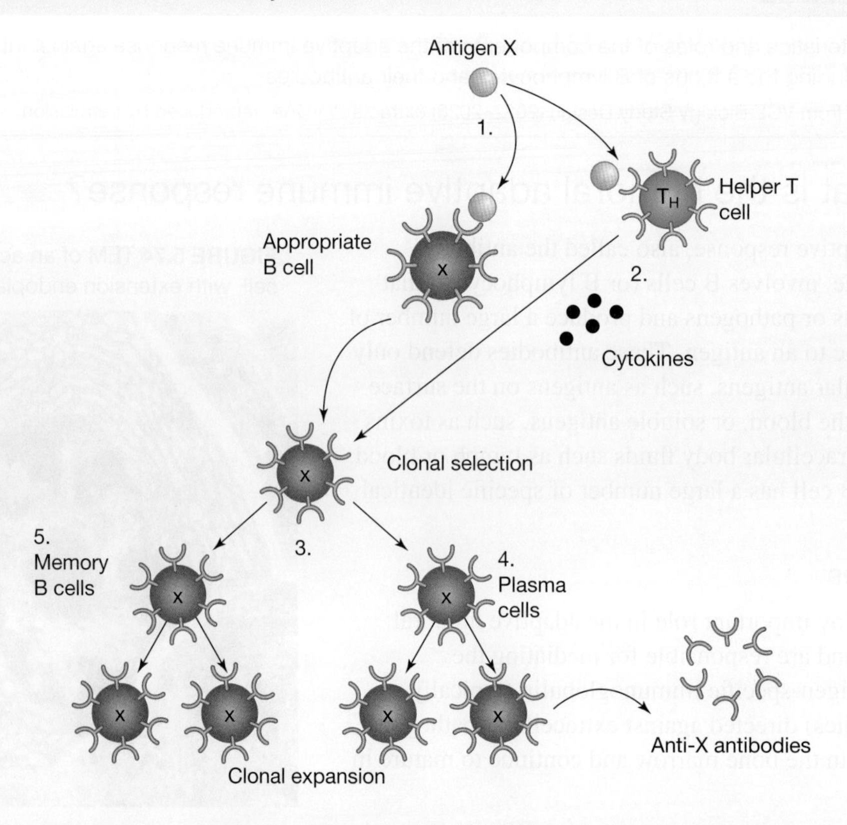

5.10.3 Antibody structure and function

Antibodies, also known as immunoglobulins, are antigen-binding proteins produced by B cells and released in blood and lymph. There are five main classes of immunoglobulins — IgA, IgD, IgE, IgG and IgM. Each of these have slightly different functions (for example, IgE is specific in the allergic response).

Antibody structure

All antibody molecules have the same basic structure made up of four polypeptide chains — two heavy chains and two light chains joined to form a Y-shaped molecule. This means that antibodies are proteins with a quaternary structure.

constant region the section of an antibody that does not vary between antibodies of the same class

variable regions the region that is unique to each antibody, where a specific antigen is able to bind

antigen-binding sites regions of an antibody molecule to which an antigen binds; also called variable regions

In terms of their amino acid sequences, both the heavy chains and the light chains have two distinct regions:

- a **constant region** that does not vary between antibodies of the same class
- a **variable region** that differs between antibodies.

The amino acid sequence varies greatly in the variable regions, located at the end of each arm of the Y-shaped molecule. These regions are called the **antigen-binding sites**, where the antibody binds to its specific antigen. This variable region gives the antibody its specificity for binding antigen.

FIGURE 5.76 Diagram of an antibody

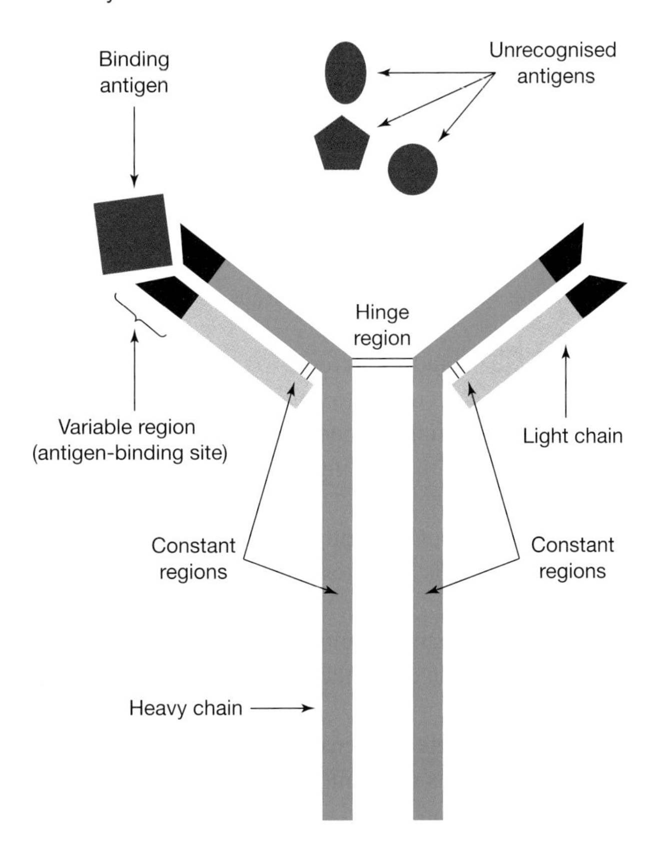

The antigen-binding site must be identical on both sides on the antibody. Remember that each antibody can only bind one specific antigen.

FIGURE 5.77 a. An antigen bound to an antibody. Note that the antigen includes a shape that can fit the antigen-binding site of the antibody. b. Diagram showing stylised antibodies with differently shaped antigen-binding sites

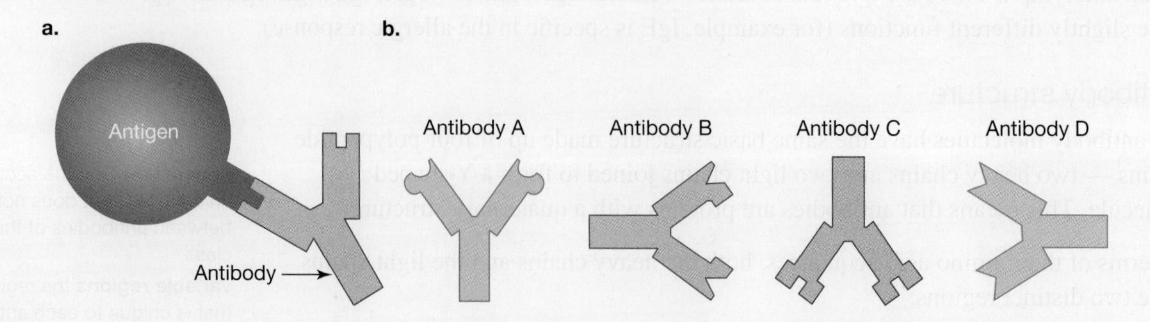

elog-0683

INVESTIGATION 5.5 online only

Modelling antibodies and the diversity of the variable region

Aim

To model the structure of antibodies and the cause of their diversity

Antibody function

Antibodies directly identify and bind to extracellular foreign antigens, either neutralising them or tagging them for destruction. Many antibodies are free molecules in solution in the blood or lymph, or in secondary lymphoid organs. Other antibodies are present as surface receptors on B lymphocytes.

Antibodies do not directly destroy pathogens but defend against infectious pathogens and toxins in the following ways (refer to figure 5.78):

- **Neutralisation** of pathogens — antibodies bind to surface antigens on pathogens and form a coating that neutralises pathogens by blocking their receptors so that the pathogens cannot attach to healthy body cells and infect them.
- Neutralisation of toxins — antibodies bind to bacterial toxins, animal toxins and venoms. The antibodies bind to and neutralise the harmful effects of the toxin or venom. For example, a person bitten by a venomous snake is given antivenom by intravenous injection. Antivenom is a solution of specific antibodies that combine with the venom molecules, preventing them from binding to cell surface receptors — this renders the venom harmless.
- Opsonisation — antibodies bind to the surface antigens on pathogens to form antigen–antibody complexes and tag the pathogen for destruction. This activates phagocytes and complement proteins, leading to the destruction of the pathogen.
- Agglutination — antibodies bind to the surface antigens on pathogens to form antigen–antibody complexes. By doing this to various pathogens, it can cause them to clump together and be more visible to the immune system.
- Precipitation — antibodies bind to soluble antigens, making them insoluble. This causes them to precipitate out of the solution, creating a solid that is much more visible to the immune system. This occurs due to the cross-linking between antigens and antibodies.
- Inflammation — antibodies can trigger the release of histamine, causing inflammation. They also activate a complement cascade.

neutralisation the process of binding of an antibody to toxins or antigens on the surface of the pathogens that inhibits their action

TIP: A good way to remember the effects of antibodies is to use the acronym PIANO:
Precipitation
Inflammation
Agglutination
Neutralisation (of pathogens and toxins)
Opsonisation.

FIGURE 5.78 Mechanisms of antibody action

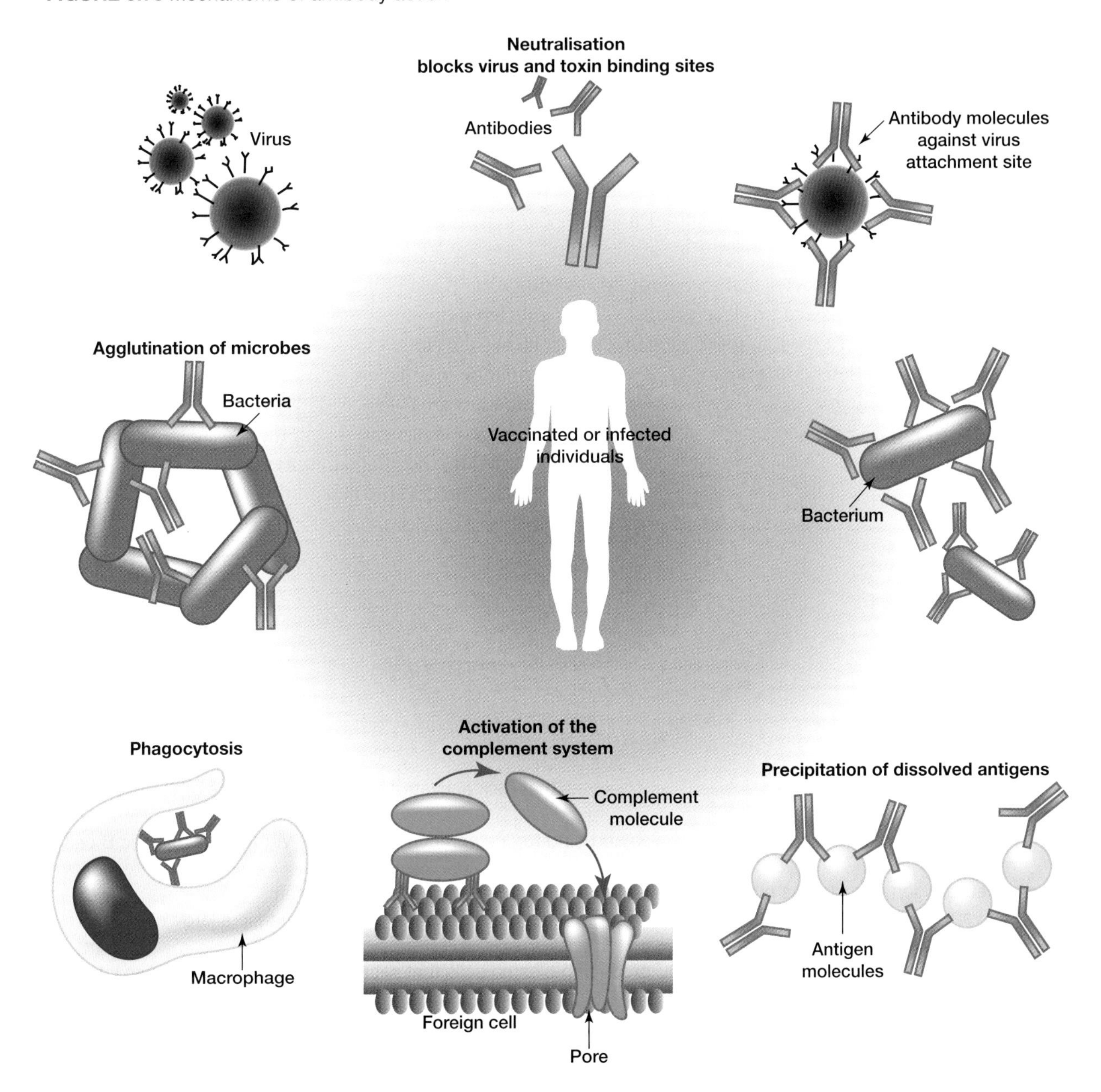

tlvd-1781

SAMPLE PROBLEM 6 Understanding the function of B cells

Bruton type agammaglobulinemia is a genetic condition that causes a failure of B cell development and an inability to make immunoglobulins in an individual. A person suffering from this condition has increased tendency to develop infections.

a. **Name the type of immune response absent in a person suffering from this condition.** **(1 mark)**

b. **Identify the type of pathogens an individual with this condition would struggle to remove from their body. Justify your response.** **(2 marks)**

c. **Explain why bacterial infections are common in people suffering from such condition.** **(2 marks)**

THINK	WRITE
a. Think about an immune response that involves immunoglobulins.	Humoral adaptive response (1 mark)
b. Identify a link between humoral immune response, immunoglobulins (antibodies) and the type of pathogens removed by it. The question is asking you to both identify the pathogen and justify your reasoning.	An individual would struggle to remove extracellular pathogens from the body (1 mark). Immunoglobulins and the humoral response only can target pathogens that are external to cells (1 mark).
c. Think about B lymphocytes that recognise antigens or pathogens and produce a large number of antibodies specific to an antigen. These antibodies defend only against *extracellular* antigens.	Since there is a failure of B cell development in affected individuals, immunoglobulin-secreting plasma cells are also absent, so there is no antibody formation in affected individuals (1 mark). Since antibodies are absent to bind to extracellular foreign antigens, either neutralising them or tagging them for destruction, the affected individuals have an increased tendency to develop bacterial infection (1 mark).

Resources

eWorkbook Worksheet 5.13 Antibodies and the humoral adaptive response (ewbk-5266)

KEY IDEAS

- Humoral adaptive response involves B lymphocytes that gets activated by specific antigens or cytokines released by T helper cells.
- Each B cell is pre-programmed to recognise and bind to one particular antigen by virtue of its specific B cell receptor.
- A new antigen 'selects' the matching naïve B cells, binds to them and activates them.
- Activated B cells proliferate, producing several generations of daughter cells, resulting in a large clone of B cells.
- Activated B cells differentiate in antibody-producing plasma cells that secrete free-floating antibodies into blood and other body fluids.
- Antibodies (B cell receptors) are present on the surface of B cells.
- Antibodies are secreted in the blood by plasma B cells.
- Antibodies do not directly destroy pathogens but defend against infectious pathogens and toxins in the following ways: opsonisation, neutralisation and complement activation.

5.10 Activities

To answer questions online and to receive **immediate feedback** and **sample responses** for every question, go to your learnON title at **www.jacplus.com.au**. A **downloadable solutions** file is also available in the resources tab.

5.10 Quick quiz	5.10 Exercise	5.10 Exam questions

5.10 Exercise

1. **MC** A bacterium that can cause infection is shown.

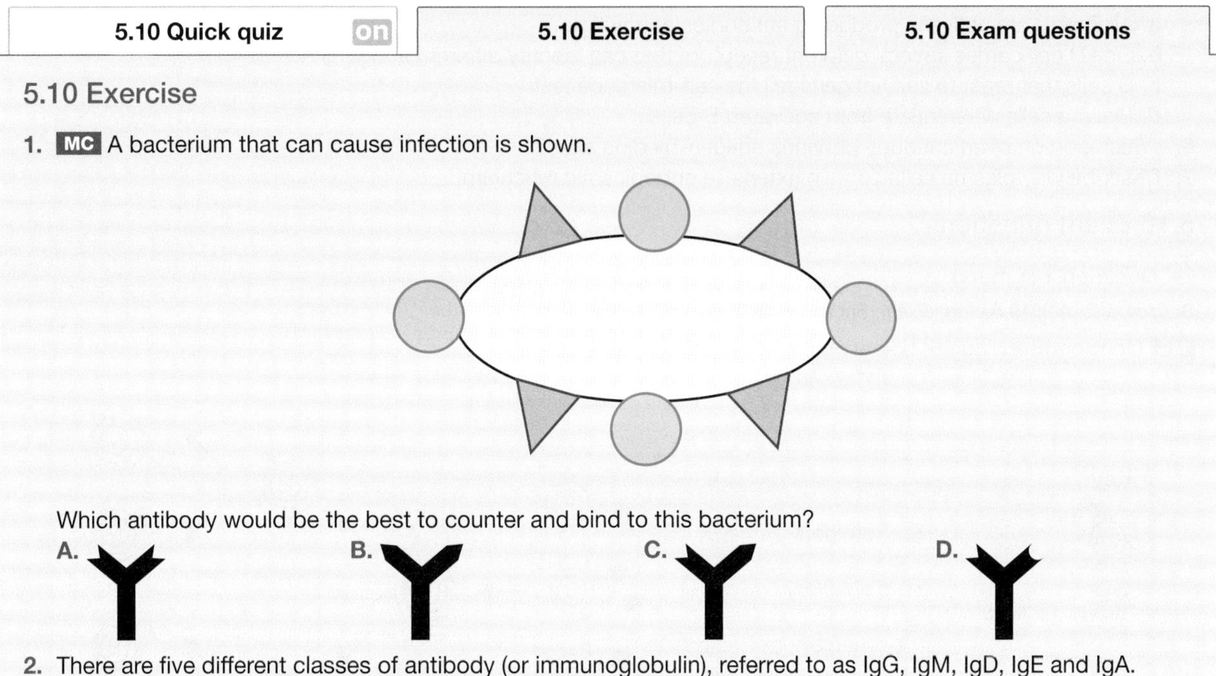

Which antibody would be the best to counter and bind to this bacterium?

A. B. C. D.

2. There are five different classes of antibody (or immunoglobulin), referred to as IgG, IgM, IgD, IgE and IgA. Some of these have multiple Y-shaped molecules combined, as shown in the images below.

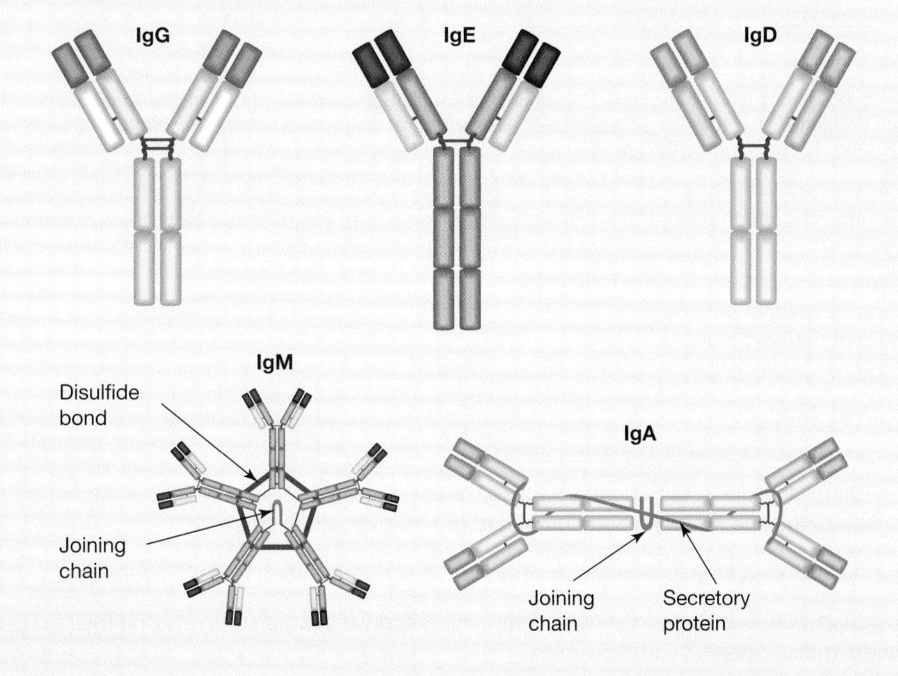

a. Explain why IgG is more likely to be passed to babies through the placenta than IgM.
b. Which of the antibodies would be most useful for agglutination of microbes? Justify your response.

3. Where would you expect to locate the following?
a. B cells undergoing clonal expansion
b. B cells undergoing testing for self-tolerance
c. B cells differentiating into antibody-producing plasma cells

4. Consider the clonal selection theory.
   a. What is meant by the term 'clone'?
   b. What is 'selected'?
   c. Summarise the process that allows the clonal selection of B cells to occur.
5. Identify the following statements as true or false, and justify your response.
   a. All B cells are active in producing antibodies at any time.
   b. One B cell carries several different receptors that can identify different antigens.
   c. B cells that react to self antigens fail the self-tolerance test.
   d. Plasma cells differentiate from activated B cells
6. Which portion of an antibody provides antigen-binding sites?
7. Describe why plasma cells contain extensive endoplasmic reticulum.

5.10 Exam questions

Question 1 (1 mark)

Source: *VCAA 2019 Biology Exam, Section A, Q21*

MC Consider the diagram below of clonal selection in B cells.

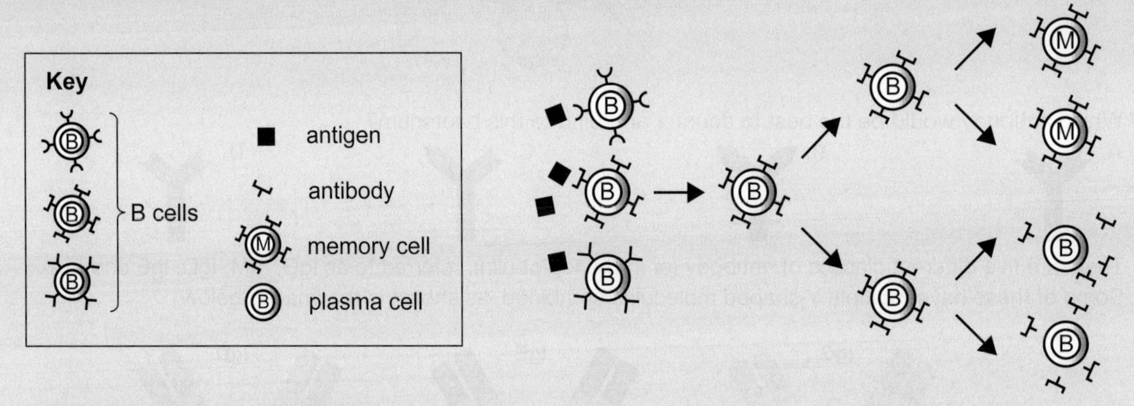

Source: adapted from Tang Zheng et al., 'An Improved Clonal Selection Algorithm and its Application to Traveling Salesman Problems', in *IEICE Transactions on Fundamentals of Electronics, Communications and Computer Sciences*, vol. E90–A, no. 12, December 2007, p. 2931

In adaptive immunity, which part of this process allows long-term (sometimes lifetime) protection against pathogens?

A. recognition of one antigen by one B cell clone
B. production of specific antibodies
C. generation of memory cells
D. production of plasma cells

Question 2 (4 marks)

Source: *Adapted from VCAA 2013 Biology Section B, Q4b*

In 1995, the Australian Bureau of Statistics released a report showing that only 53 per cent of children aged between three months and six years had completed the immunisation schedule.

After a major advertising campaign by the government, the immunisation rate increased to 92 per cent.

Create a flowchart with four steps to show the immune response in a person receiving their first vaccination. The following words must appear in your flow chart.

plasma cells; memory cells; B cells; antibodies; cytokines

The first step in your flow chart should be: *antigens detected by T-helper cells*.

Question 3 (1 mark)

***Source:** VCAA 2015 Biology Exam, Section A, Q16*

Consider the following diagram of four pathogens and three antibodies.

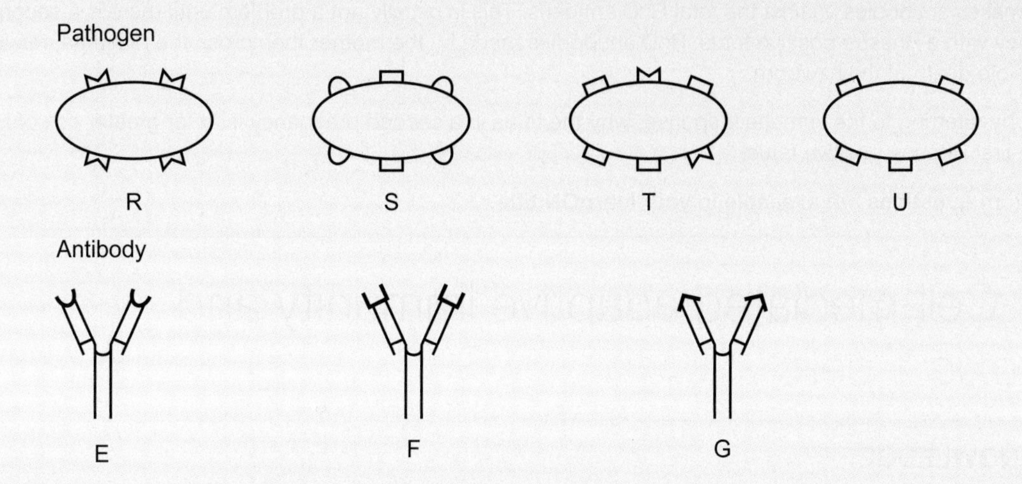

Which one of the following statements is correct?

A. Antibody E would be effective against both pathogen S and pathogen R.
B. Antibody F is effective against three of the pathogens.
C. There are no antibodies effective against pathogen U.
D. Antibody G is only effective against pathogen R.

Question 4 (4 marks)

***Source:** VCAA 2010 Biology Exam 1, Section B, Q4*

EB12 is a receptor on the B cells of mice that helps determine if a cell becomes a plasma or a memory cell. Scientists used three different strains of mice to investigate B cell immunity. None of the strains had been exposed to the influenza virus. The strains were as follows.

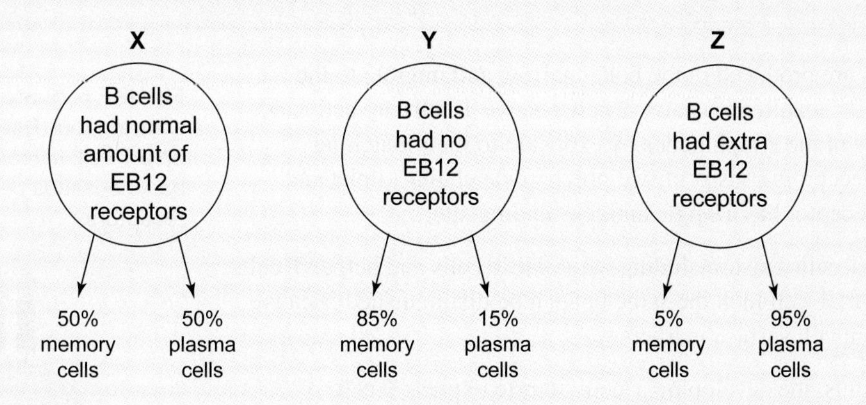

The three mice strains were infected with the influenza virus.

a. Explain which strain, X, Y or Z, would be best at destroying the fast-acting influenza virus. **2 marks**
b. Explain how blocking the action of EB12 receptors could result in the production of a more efficient vaccine. **2 marks**

Question 5 (2 marks)

Source: *VCAA 2014 Biology Exam, Section B, Q5b*

Hemolytic disease of the newborn (HDN) can occur if a Rhesus-negative mother is pregnant with a Rhesus-positive fetus. During pregnancy and birth, some fetal blood cells may enter the mother's bloodstream. The mother makes antibodies against the fetal RhD antigens. This is usually not a problem until there is a second pregnancy with a Rhesus-positive fetus. RhD antibodies made by the mother then cross the placenta, resulting in the possible death of the newborn.

Explain, by referring to the immune response, why the fetus in a second pregnancy is at far greater risk of HDN than the first Rhesus-positive fetus.

More exam questions are available in your learnON title.

5.11 Cell-mediated adaptive immunity and cytotoxic T cells

KEY KNOWLEDGE

- The characteristics and roles of the components of the adaptive immune response against intracellular threats, including the actions of helper T and cytotoxic T cells.

Source: Adapted from VCE Biology Study Design (2022–2026) extracts © VCAA; reproduced by permission.

5.11.1 What is cell-mediated immunity?

cell-mediated adaptive response a specific response in which cytotoxic T cells destroy virus-infected cells using perforin and granzyme B

The **cell-mediated adaptive response** eliminates infected cells and involves T lymphocytes (or T cells).

'Cell-mediated' refers to the fact that the response is carried out by cytotoxic cells that eliminate intracellular pathogens.

Precursor T cells are produced in the bone marrow and migrate to the thymus, where they mature into naïve T cells. Naïve T cells are activated to effector T cells in the lymph nodes by exposure to their matching antigens. T cell receptors consist of two polypeptide chains (alpha and beta), and each receptor has a single antigen-binding site.

FIGURE 5.79 Stylised diagram showing four T cell receptors on a T cell. In reality, a T cell has thousands of these surface receptors.

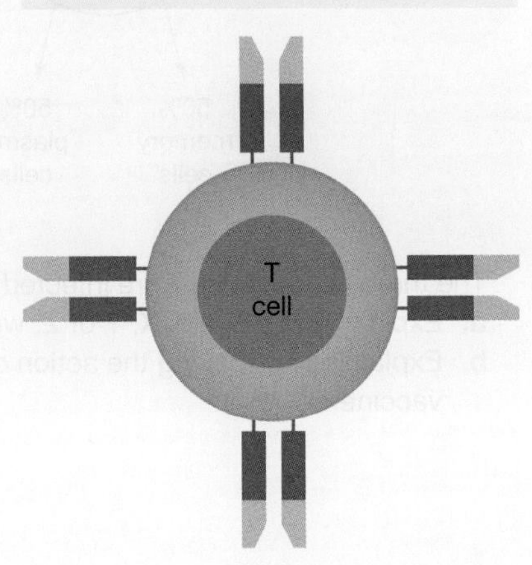

Several kinds of T cells exist, including cytotoxic T cells and helper T cells. These two kinds of T cell look the same but can be distinguished because they have different cell surface markers.

While in the thymus, the developing T cells start to express T cell receptors (TCRs) and other receptors called CD4 and CD8 receptors. All T cells express T cell receptors in addition to either CD4 or CD8 That is, some T cells will express CD4 and others will express CD8. CD4 receptors are expressed by helper T cells, and CD8 receptors are expressed by cytotoxic T cells.

The cell-mediated immune system includes several cell types:

- *dendritic cells*, which present antigens to T cells. Dendritic cells (and macrophages) are innate, and they present antigens without any specificity or recognition of the antigen type.
- *helper T cells*, which participate in antigen recognition and activation of both cytotoxic T cells and B cells

- *cytotoxic T cells*, which can kill virus-infected cells with or without antibodies. Macrophages and helper T cells produce cytokines that activate helper and cytotoxic T cells, leading to the killing of the pathogen or tumour cell.

5.11.2 Cytotoxic T cells (T_C)

The major role of cytotoxic T cells (T_C cells) is to monitor body cells for the presence, on their surface, of foreign antigens that have been generated within the body cell by intracellular pathogens. Such antigens include viral proteins produced during the multiplication of a virus within a cell. Infected body cells advertise their infected state on their surfaces by displaying pathogen antigens that are linked to MHC-I molecules. These infected cells can also be taken up by antigen-presenting cells. The antigen-presenting cells may also present these antigens to T cells.

The monitoring of body cells by cytotoxic T cells requires that contact be established between the T cell and the cell to be monitored. If any cells are identified as being infected (or abnormal), cytotoxic T cells set about destroying them. By destroying infected cells, cytotoxic T cells stop infections from spreading to nearby body cells.

Cytotoxic T cells recognise the foreign antigens that are linked to MHC-I molecules on the surface of an infected body cell. Figure 5.80 shows a diagram of a cytotoxic T cell making contact with an infected body cell. Note the foreign antigen (shown in red) that is linked to an MHC-I marker (shown in green). The T cell receptor (TCR) recognises this MHC-linked antigen as a signal that this body cell is infected.

FIGURE 5.80 A cytotoxic T cell recognising and binding to a foreign antigen on an MHC-I marker on the surface of an infected cell

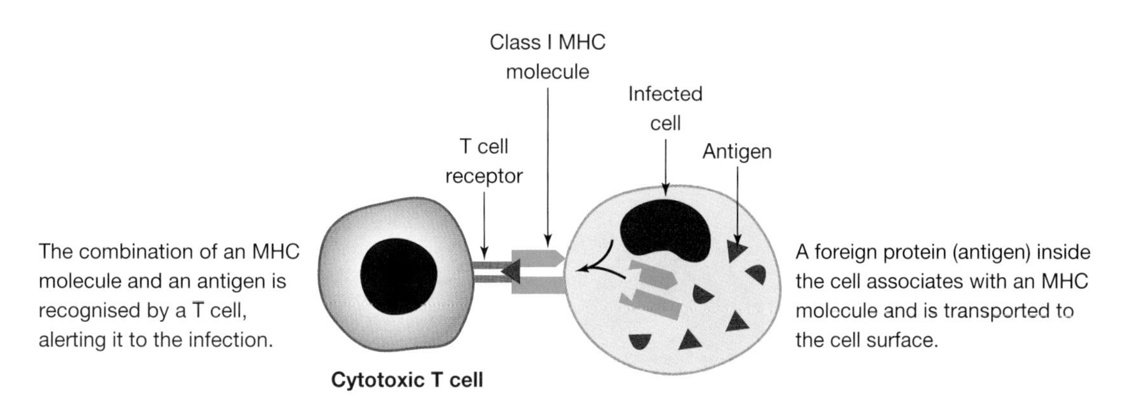

As well as recognising virus-infected cells, cytotoxic T cells can also recognise abnormal proteins such as those resulting from DNA mutations in cancer cells. This often results in abnormal MHC-I markers. Through this recognition, cytotoxic T cells can target and eliminate cancer cells.

The combination of signals from helper T cells and the binding of this antigen to the T cell receptor activates these cytotoxic T cells. These activated cells undergo cycles of cell division and form effector cytotoxic T cells. These cell divisions bring about a several-thousand-fold increase in the number of effector cytotoxic T cells that can recognise the particular antigen. Although this clonal expansion may take a week to occur, the adaptive immune system is now well equipped to destroy infected cells.

5.11.3 Activating the cell-mediated adaptive response

The cell-mediated response mostly involves T cells. The T cells respond to any cell that displays abnormal MHC markers, including cells invaded by pathogens, tumour cells, or transplanted cells. A specific chain of events is required for this response to be activated.

The steps in cell-mediated adaptive immunity

1. Antigen-presenting cells (usually dendritic cells) displaying foreign antigens on their MHC-II markers bind to their specific helper T cells (usually within lymph nodes).
2. Helper T cells undergo clonal selection and expansion, forming effector helper T cells and memory T cells.
3. Interleukins (a type of cytokine) are secreted by the helper T cell to stimulate immature T cells.
4. Immature cytotoxic T cells also bind to the self-cell with abnormal foreign MHC-I markers or APC displaying foreign antigens. Note that APC will use MHC-II markers, but other infected cells use MHC-I markers.
5. Upon stimulation by cytokines released from helper T cells, cytotoxic T cells proliferate, producing activated cytotoxic T cells and memory T cells through clonal selection and expansion.
6. Cytotoxic T cells destroy the cells through apoptosis. These steps are shown in figure 5.81.

FIGURE 5.81 The steps involved in cell-mediated immunity. The MHC-I markers are often abnormal or display foreign antigens in infected cells. Antigen-presenting cells can also present foreign antigens to the cells (mainly helper T cells) on their MHC-II markers to initiate clonal selection and expansion.

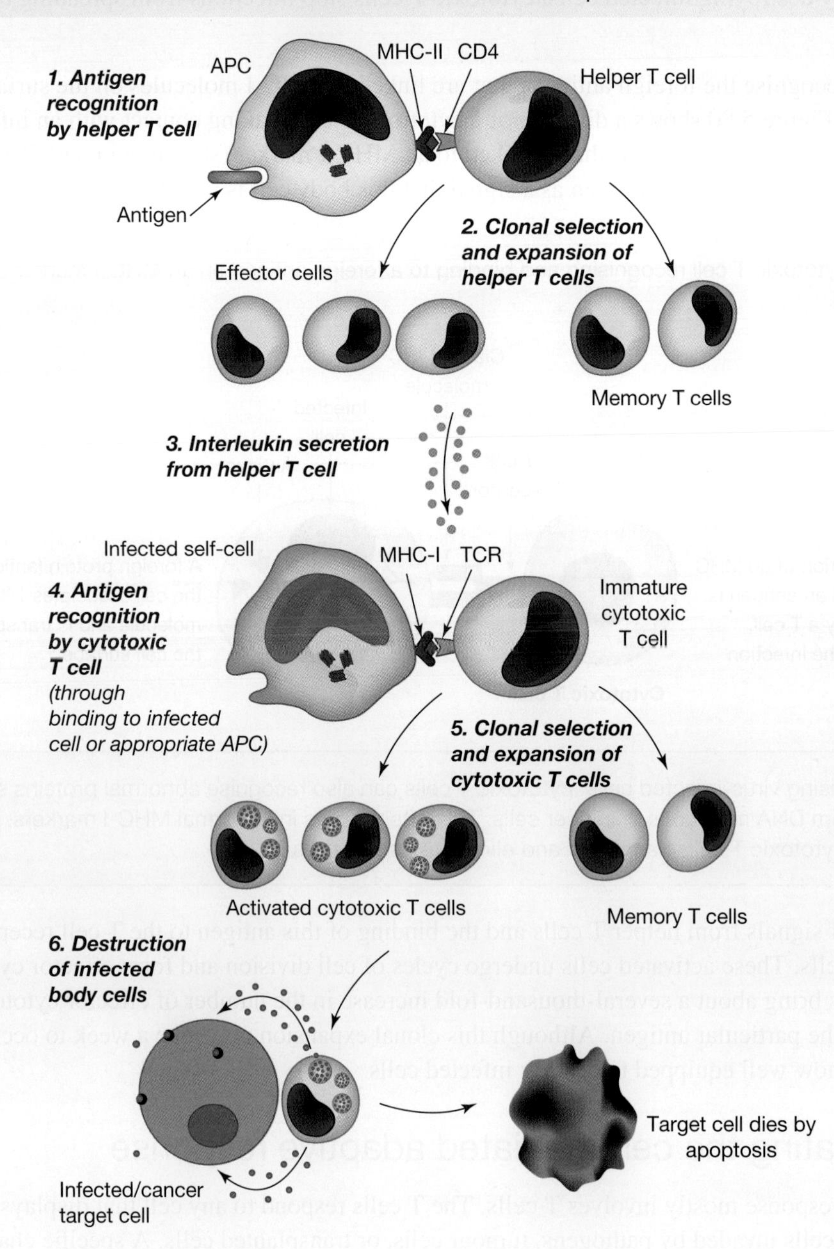

5.11.4 Apoptosis and cytotoxic T cells

Effector cytotoxic T cells attack infected body cells as follows:

1. Cytotoxic T cells release perforin, a protein that inserts into the plasma membranes of target cells and creates pores.
2. Cytotoxic T cells also release destructive granules, called granzyme B, that enter the infected cell via the pore and initiate cell death by apoptosis (see figure 5.82).
3. The cytotoxic T cells are then free to attack other infected cells that display the same foreign antigen.

FIGURE 5.82 A cytotoxic T cell initiating apoptosis in an infected cell. This involves the action of perforin, which forms pores in the infected cells, and granzyme B, which induces apoptosis, leading to the controlled death of a cell.

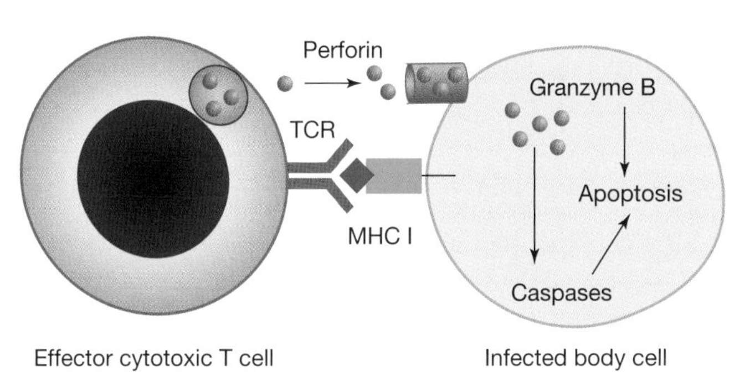

FIGURE 5.83 A cytotoxic T cell with the CD8 markers on its plasma membrane in purple, microtubules in green, and its collection of lysosomes containing destructive granules in orange/pink (image courtesy of Alex Ritter and Gillian M. Griffiths, Cambridge Institute for Medical Research)

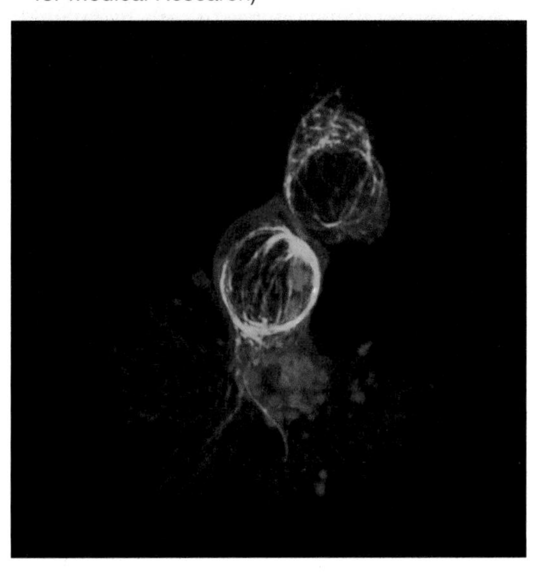

Activated cytotoxic T cells kill infected body cells through apoptosis to ensure the virus cannot spread. Figure 5.83 shows a fluorescence image of a cytotoxic T cell that has recognised and will destroy a pathogen-infected cell.

It is important to remember that cytotoxic T cells initiate apoptosis in infected body cells. Without cytotoxic T cells, necrosis may instead occur, which will damage nearby cells. The mechanism of action of cytotoxic T cells is quite similar to natural killer cells. However, cytotoxic T cells are specific for certain antigens and form memory cells.

tlvd-1817

SAMPLE PROBLEM 7 Comparing and contrasting natural killer and cytotoxic T cells

Both natural killer cells and cytotoxic T cells are immune cells that are capable of killing foreign pathogens and also cancer cells. However, they are categorised differently. Compare and contrast natural killer cells and cytotoxic T cells. (2 marks)

THINK	WRITE
1. Identify what the question is asking. The question is asking you to *list* the similarities and differences between natural killer cells and cytotoxic T cells.	
2. Compare the two types of cells by writing down the similarities. • Both kill virus-infected cells. • Both release perforin and granzyme from granules to initiate apoptosis.	Both cytotoxic T cells and natural killer cells trigger immune response and destroy virus-infected cells through apoptosis using perforin and granzyme B (1 mark).
3. Contrast the two types of cells by giving the differences. • Cytotoxic T cells are antigen-specific, are involved in adaptive immunity and have memory for a specific antigen. • Natural killer cells are non-specific, are involved in innate immunity and do not form memory cells.	Cytotoxic T cells are specific to particular pathogens and retain the memory for a particular antigen. Natural killer cells are non-specific and trigger the innate immune response. They do not have memory for a particular antigen (1 mark).

Resources

eWorkbook Worksheet 5.14 Cell-mediated adaptive immunity (ewbk-5268)

KEY IDEAS

- T cells deliver the cell-mediated components of adaptive immunity.
- Each T cell carries multiple copies of one specific T cell receptor that can recognise just one antigen.
- T cells include cytotoxic T cells and helper T cells.
- T cells can only recognise antigens when they are located on the surface of another cell and are linked to MHC molecules.
- Cytotoxic T cells attack pathogen-infected cells (and also cancer cells) using chemical compounds, resulting in the death of the target cell by apoptosis.

5.11 Activities

To answer questions online and to receive **immediate feedback** and **sample responses** for every question, go to your learnON title at **www.jacplus.com.au**. A **downloadable solutions** file is also available in the resources tab.

5.11 Quick quiz on	**5.11 Exercise**	**5.11 Exam questions**

5.11 Exercise

1. **MC** Which of the following statements is correct?
   A. Cytotoxic T cells produce antibodies to kill pathogens.
   B. Cytotoxic T cells cannot produce memory cells.
   C. A single T cell has multiple copies of one kind of T cell receptor.
   D. Cytotoxic T cells can destroy both extracellular and intracellular pathogens.
2. Identify each of the following.
   a. The molecule that can create holes in the plasma membrane of an infected cell
   b. The location where T cells are activated
3. What is the role of MHC markers in the cell-mediated response?
4. Outline the steps of how cytotoxic T cells lead to the apoptosis of infected cells.
5. In type 1 diabetes, beta cells in the pancreas are destroyed. Explain how this may occur. Ensure you reference T lymphocytes in your response.

5.11 Exam questions

Question 1 (1 mark)

Source: *VCAA 2013 Biology Section A, Q19*

MC Cytotoxic T cells are

A. antibodies.
B. able to kill virus-infected cells.
C. part of the humoral response.
D. part of the second line of the immune defence.

Question 2 (1 mark)

Source: *VCAA 2019 Biology Exam, Section A, Q19*

MC In adaptive immunity, which cells directly destroy virally infected cells?

A. B cells
B. plasma cells
C. T helper cells
D. T cytotoxic cells

Question 3 (1 mark)

Source: *VCAA 2014 Biology Exam, Section A, Q19*

MC Australian scientists have grown a miniature mammalian kidney from stem cells taken from adult skin. In the future, scientists aim to grow full-size kidneys for transplants in patients with kidney disease using the patient's own skin cells. This would overcome the problem of rejection of the transplanted kidney by the immune system.

Rejection of transplanted organs results mainly from an attack on the
A. donor organ by the patient's memory B cells.
B. donor organ by the patient's cytotoxic T cells.
C. patient's immune cells by lymphocytes in the donated organ.
D. patient's immune system by immunosuppressant drug treatment.

Question 4 (2 marks)

Source: *VCAA 2008 Biology Exam 1, Section B, Q5b*

Describe the process of tissue rejection after recognition of the non-self cells has occurred.

Question 5 (2 marks)

Cytokines are signalling molecules and are important components of the immune system. IL-2 is a cytokine produced by T cells. IL-2 acts on other T cells to make them divide rapidly (clonal expansion).
a. What event would trigger the production of IL-2? **1 mark**
b. What is the advantage of the T cells' response to IL-2? **1 mark**

More exam questions are available in your learnON title.

5.12 Natural and artificial immunity

KEY KNOWLEDGE

- The difference between natural and artificial immunity, and active and passive strategies for acquiring immunity

5.12.1 Types of acquired immunity

Antibodies are our most powerful protection against infectious disease — each antibody is specifically tailored to counter one particular pathogen. When we have the specific antibodies against a particular pathogen, we are said to 'be immune to' or 'have immunity against' the disease caused by that pathogen. This can be also referred to as acquiring immunity. The antibodies will, in fact, be against specific antigens of the pathogen concerned or antigens of its toxin. So, we can gain immunity to infection by acquiring the specific antibodies against the antigens of the pathogen responsible. Immunity based on specific antibodies can be termed specific immunity.

Immunity is active or passive and natural or artificial depending on the type of immune response.

FIGURE 5.84 The different types of specific immunity

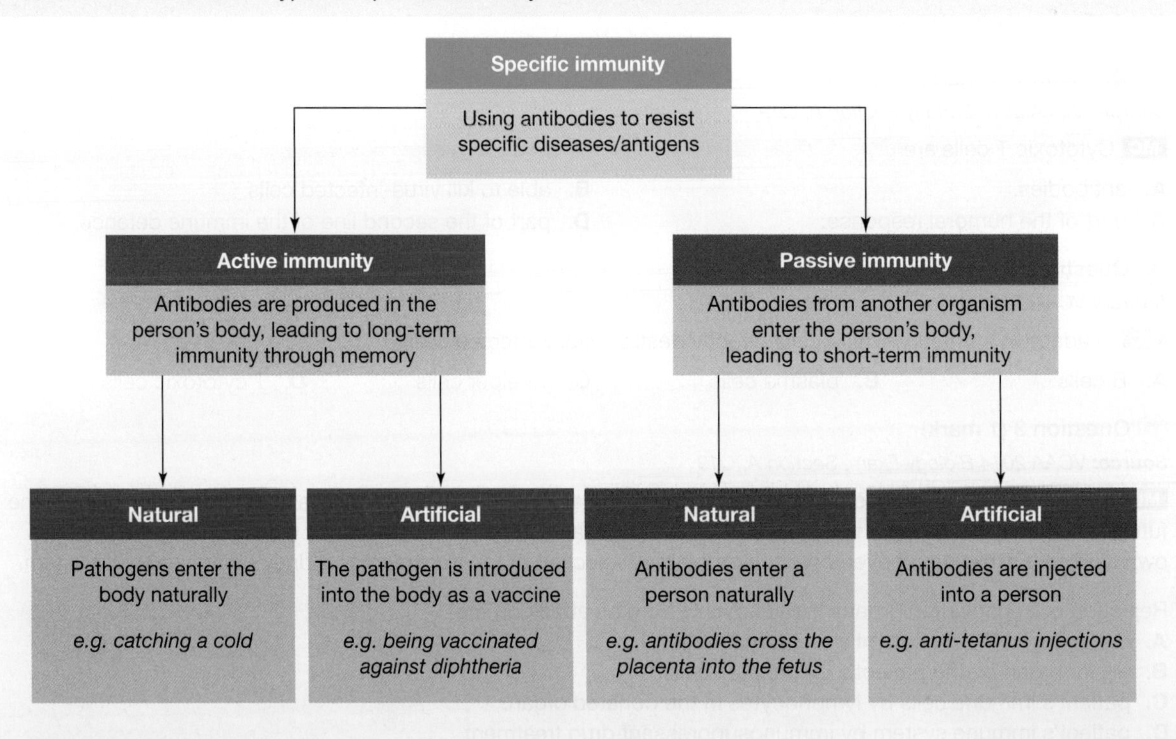

Specific immunity can be achieved in various ways, depending on:

1. the source of the antibodies:
   - **active immunity** antibodies are produced by a person's own adaptive immune system
   - **passive immunity** antibodies are acquired by a person from an external source
2. the means of gaining immunity:
   - **natural immunity** occurs naturally, without deliberate intervention
   - **artificial immunity** is induced through a deliberate intervention.

active immunity the production of antibodies by a person in response to exposure to a particular antigen

passive immunity short-term immunity acquired from an external source of antibodies

natural immunity a form of specific immunity in which antibodies are produced or obtained through natural means

artificial immunity immunity that is formed through deliberate exposure and intervention

TABLE 5.14 Comparing active and passive immunity

Active immunity	Passive immunity
Produced by body's own immune system in response to antigen	Received passively by host
Secondary response is more enhanced due to memory	No immunological memory
Long-lasting	Short-lasting
More effective	Less effective
Immunity develops over weeks	Immunity is immediate

5.12.2 Active immunity

Active immunity involves the production of antibodies by a person in response to exposure to a particular antigen. In addition, B memory cells and T memory cells are produced and react quickly if another encounter occurs with the same organism.

Active immunity can be acquired in two ways: naturally or artificially (induced).

natural active immunity a type of immunity in which the body produces antibodies in response to a normal infection by a pathogen

artificial active immunity the deliberate administration of disabled antigens to elicit the production of antibodies

vaccination an artificially active process in which an individual is injected with either antigens or weakened pathogens in order to produce their own antibodies and memory cells

Natural active immunity

Natural active immunity, also called acquired immunity, develops from natural infection and results in immunological memory.

This type of immunity is *natural* because it is acquired naturally after an infection without any artificial intervention; it is *active* because the antibodies are made by the immune system of the infected person.

When a person comes into contact with a particular disease-causing agent for the first time, no antibodies against the pathogen (or the antigen) are present. It takes several days for the appropriate plasma cells and antibodies to form, and during that time the person begins to show symptoms of the disease.

Within several days, antibodies have formed with identical sites that bind to the antigens. As the number of antibodies increases, the infecting pathogens begin to be destroyed and the person starts to recover. If sufficient antibodies are made to destroy all the infecting pathogens, the person recovers completely.

An important aspect in active immunity is in the formation of memory cells, which are retained after the infection is resolved. This results in immunity against further infection by the same pathogen.

FIGURE 5.85 Natural active immunity occurs through natural exposure to a pathogen.

Artificial active immunity

Artificial active immunity involves the deliberate and artificial introduction of a disabled pathogen or its toxin to the body. Most commonly, this is done by the injection of a particular vaccine that causes the adaptive immune system to produce antibodies to the introduced antigen.

This artificial way of stimulating adaptive immune response to produce active immunity is called **vaccination** (or immunisation).

Vaccines are used to activate the immune system to produce antibodies against specific disease-causing organisms without actually causing the disease.

FIGURE 5.86 Artificial active immunity occurs through vaccination.

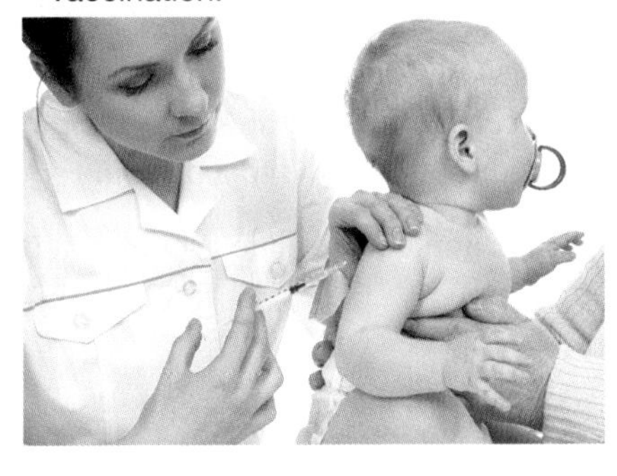

Vaccines are of different types. Vaccines may contain:

- living **attenuated pathogens** that do not cause disease but can still reproduce. This type of vaccine stimulates a much higher and longer-lasting level of antibodies compared to vaccines that use inactivated or killed pathogens.
- killed bacteria or inactivated viruses (you can't kill viruses because they are non-living). The treatment to kill bacteria or to inactivate viruses commonly involves the use of heat or chemicals.
- one or more sub-units of a pathogen that act as antigens stimulating antibody formation.
- bacterial toxins that have been treated to form harmless **toxoids**. This type of vaccine is used for those pathogens that cause disease by secreting exotoxins. Just like toxins, toxoids act as antigens but they are unable to cause disease.

When a vaccine is first injected into a person, the immune system shows a **primary antibody response**. A second injection of vaccine produces a **secondary antibody response**. Often, multiple injections are used to amplify the antibodies. These are known as boosters. The antibody is specific to the treated pathogen used in the vaccine, so if the person comes into contact with the pathogen at some future date, memory cells and antibodies will be ready to act and the person is immune to infection.

FIGURE 5.87 Antibody levels in response to vaccination. The point when an individual is given the vaccination is shown by the pink dots.

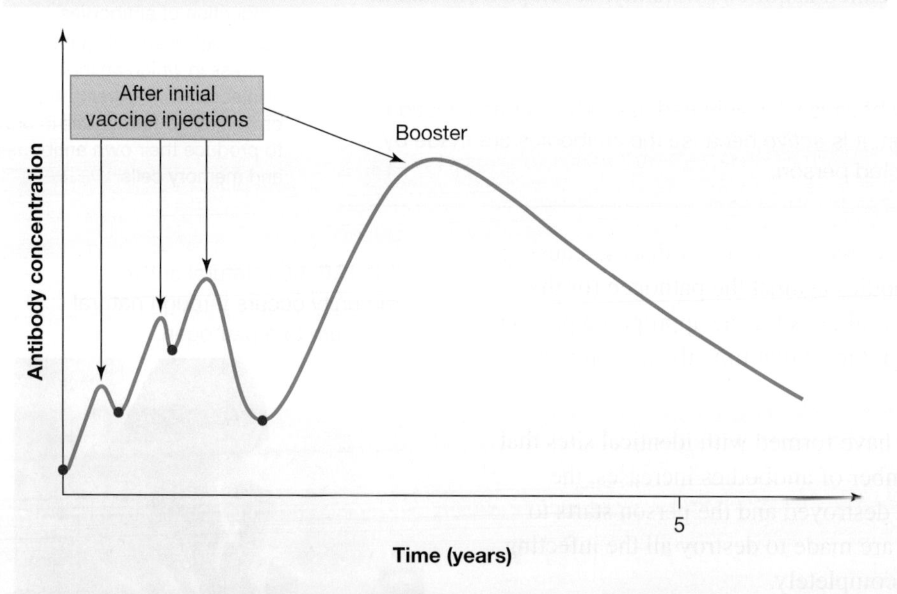

attenuated pathogen a pathogen that, while still living, has been treated so that it is no longer able to cause disease

toxoids inactivated toxins used for active immunisation

primary antibody response production of antibodies induced in an individual by the first exposure to an antigen

secondary antibody response the rapid production of high levels of specific antibodies to a foreign antigen that occurs in a person who was previously exposed to the same antigen

Vaccination programs are explored further in Topic 6, in subtopic 6.4.

5.12.3 Passive immunity

Passive immunity involves the transfer of antibodies in a person from an external source.

The advantage of passive immunity is that it gives *immediate* protection to the person receiving the antibodies. However, introduced antibodies decline relatively quickly and do not provide long-lasting immunity to the receiver. Figure 5.88 shows the rapid decline in the concentration of antibodies introduced to a person.

Passive immunity can be acquired in two ways: naturally or artificially (induced).

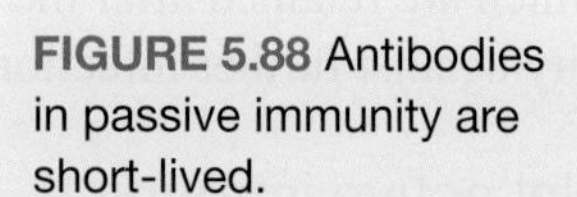

FIGURE 5.88 Antibodies in passive immunity are short-lived.

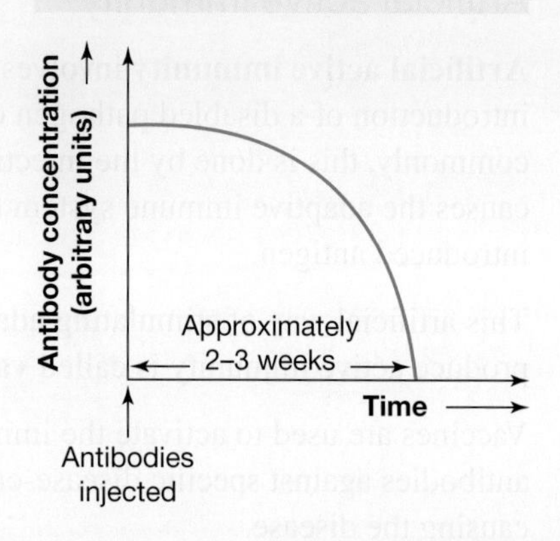

Natural passive immunity

Natural passive immunity occues when an individual receives antibodies from a natural source, such as through breastfeeding or via the placenta.

A developing foetus receives maternal antibodies across the placenta from its mother. These antibodies (immunoglobulin G [IgG]) provide important protection for the foetus and the newborn baby because the baby's immune system does not mature fully until after birth.

A baby continues to acquire antibodies through the mother's milk. Colostrum is the thick yellowish milk the mother produces for the first two or three days after a baby's birth, and this is particularly rich in antibodies. One of the reasons breastfeeding is recommended is because of the protection against infection that the mother's antibodies give the baby (see figure 5.89).

FIGURE 5.89 Natural passive immunity includes through breastfeeding and during pregnancy as IgG moves across the placenta. Note that the time before 0 months marks gestation, with the assumption that a childbirth occurs at 9 months gestation.

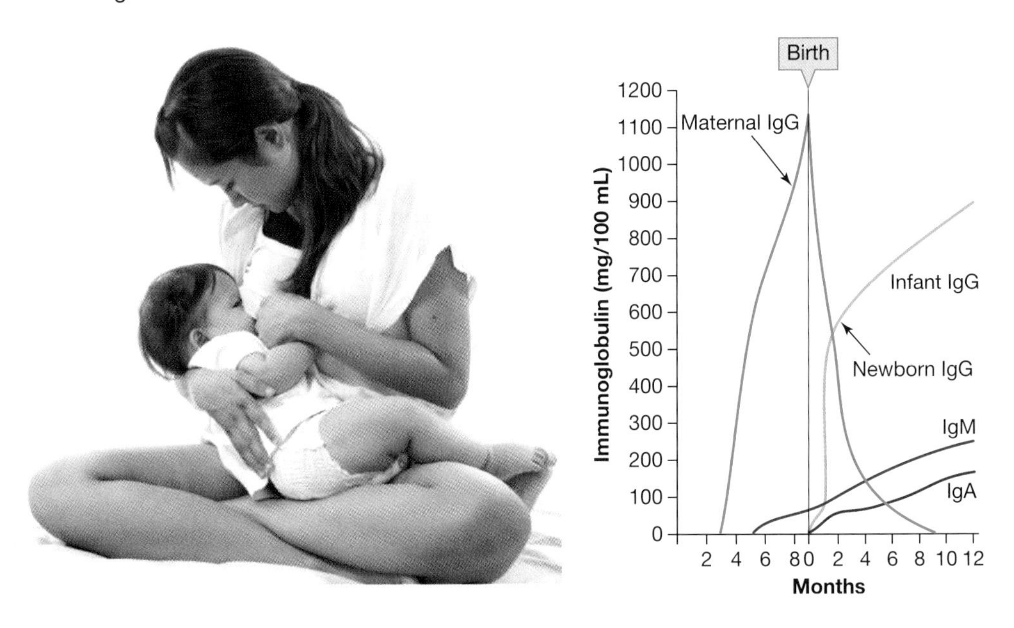

Artificial passive immunity

Artificial passive immunity is an immediate but short-term immunity provided by the injection of antibodies. Specific antibodies known as antivenoms (or antivenins) and antitoxins may also be used for this short-term immunity.

These antibodies come from an external source. Many are produced in animals for human use. However, due to the chance of allergic reaction, antibodies from other humans are preferable. Artificial passive immunity is important when treatments are time-dependent.

There are many examples outlining artificial passive immunity.

Snakebite and antivenom

A person receives antibodies artificially in the treatment of snakebite. When a person receives antivenom because they have been bitten by a venomous snake, they are receiving antibodies that act specifically against the snake venom, so they are given immediate protection. The antivenom combines with antigens of the venom so that the venom is no longer free to damage body cells.

natural passive immunity a form of immunity in which an individual receives antibodies from a natural means, such as through breastfeeding

artificial passive immunity the administration of antibodies to provide an immediate, specific immune response

Stepping on a rusty nail and tetanus shots

Clostridium tetani is a bacterium that enters the body through wounds (often puncture wounds). This bacterium releases toxins (exotoxins) that causes the destruction of tissue. Tetanus is a potentially fatal illness that causes severe muscle spasms. If you step on a rusty nail, it is recommended that you immediately get a tetanus shot. As this is time-critical, the injection would contain antitoxins against the antigens on the surface of this toxin.

FIGURE 5.90 Tetanus shots are required after stepping on rusty nails.

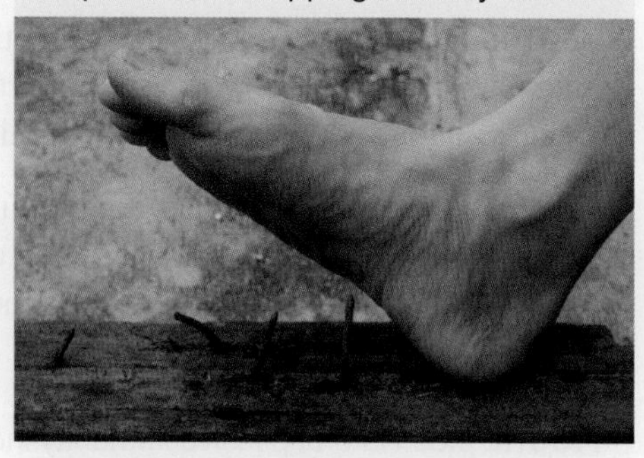

Animal bites and rabies shots

When bitten by an animal, particularly a rabid animal, it is recommended to get a rabies shot. Rabies is caused by a virus and is usually spread through the infected saliva of animals. Although vaccines can be effective if given as soon as a bite occurs, in time-critical cases, antibodies against rabies are instead given. As rabies leads to over 10 000 deaths worldwide each year, it is vital that treatment occurs as soon as possible.

FIGURE 5.91 Rabid dogs can transmit rabies.

Hepatitis A infection

If a member of your family develops infectious hepatitis (hepatitis A), you are at risk of being infected. If you immediately receive an injection of antibodies specific to hepatitis A, the infection may be avoided. The antibodies used in such injections are obtained from blood collected from voluntary donors by the Australian Red Cross Society. The antibodies, or immunoglobulins, are extracted from the blood plasma collected from people known to have had hepatitis A. The plasma contains many different antibodies, including those against hepatitis A. Note, however, that the immune defence provided by 'donated' antibodies is only temporary and does not provide long-time protection against hepatitis, but can provide protection at critical times.

CASE STUDY: Making antivenom

One way snake antivenom (or antivenin) is produced is in Percheron horses that receive injections of venom. Initially, only a very small dose of venom is injected into the horse. The dose is non-lethal so the horse is not harmed and produces antibodies that are specific to the venom. A slightly higher dose of venom is then injected into the horse, which responds by producing higher levels of antibodies. At an appropriate time, blood samples are taken from the horse. The antibodies are extracted from the serum, purified and used in vaccines for people and other animals that are bitten by snakes. Different antivenoms are produced for various species of snake. Other animals, such as rabbits, may also be used to produce antivenom.

Due to ethical concerns surrounding animal testing and the high cost of production, other methods continue to be explored for the production of antivenom, such as through bacterial transformation.

tlvd-1818

SAMPLE PROBLEM 8 Identifying and describing types of immunity

Some marsupials are born with no adaptive immune system because their primary immune tissue (in the bone marrow and thymus) does not mature until 30 days after birth and their humoral immunity does not function effectively until 90 days after birth. These young marsupial animals rely heavily on maternal milk for protection against infections.

a. **List the type of immunity involved.** **(1 mark)**
b. **Explain how this type of immunity help marsupials to fight off infections.** **(2 marks)**

THINK	EXPLAIN
a. Think about the type of substances present in the maternal milk that help to fight off infections.	Antibodies present in maternal milk — natural passive immunity (1 mark).
b. Explain how antibodies helps the marsupials. As this question is worth 2 marks, you should make two key points: you need to discuss how they are unable to produce their own antibodies and how these are provided from the mother.	Maternal milk provides antibodies to fight against infections. The immune system is underdeveloped and cannot produce antibodies quickly enough on its own or has no memory B cells (1 mark), so the antibodies provide ready-made antibodies from the mother to a marsupial baby (1 mark).

Resources

eWorkbook Worksheet 5.15 Comparing the types of acquired immunity (ewbk-5270)

KEY IDEAS

- Active immunity involves the production of antibodies and memory cells by an individual upon exposure to an antigen.
- Passive immunity involves receiving antibodies from an external source.
- Immunity can be naturally acquired (for example through normal exposure to an antigen or through breastfeeding) or artificially acquired (for example through immunisations).
- Vaccinations are artificially acquired active immunity. They involve the introduction of weakened or dead forms of the pathogen or subsets into an individual, so they will not gain the disease but will produce antibodies and memory cells against the foreign agent to product against future exposure.

5.12 Activities

learnON

To answer questions online and to receive **immediate feedback** and **sample responses** for every question, go to your learnON title at **www.jacplus.com.au.** A **downloadable solutions** file is also available in the resources tab.

5.12 Quick quiz (on)	5.12 Exercise	5.12 Exam questions

5.12 Exercise

1. **MC** An example of passive immunity is
   - **A.** when an infant receives antibodies from their mother through the placenta or breast milk
   - **B.** when a person receives an injection of attenuated virus
   - **C.** when a person recovers from flu infection
   - **D.** when a person gets inflammation
2. What is the difference between
   - **a.** active immunity and passive immunity
   - **b.** natural immunity and artificial (induced) immunity?
3. Identify the type of immunity each of the following is an example of.
   - **a.** A person receives antivenom after suffering a snake bite.
   - **b.** A foetus receives maternal IgG across the placenta.
   - **c.** A baby girl receives the MMR vaccine.
   - **d.** A young boy is recovering from the flu.
4. Which gives longer lasting protection: antibodies that have been actively acquired or antibodies that have been passively acquired? Explain your answer.
5. Before travelling to some countries, it is recommended to get a rabies immunisation. How does this differ from a rabies injection after getting bitten by a rabid dog?
6. A family of five was exposed to chicken pox. The six-year-old child and the father both contracted chicken pox. The nine-year-old child, the three-month-old baby and the mother did not contract the disease. The mother had already had chicken pox when she was 12 years old. The nine-year-old child and the three-month-old baby had never had the chicken pox.
   - **a.** Why might the six-year-old have been the only sibling to contract the disease?
   - **b.** What would you expect if the family were exposed to the disease again in two years?
   - **c.** What type of immunity did the mother have against chicken pox?
   - **d.** The three-month-old baby had never been exposed to chicken pox before. What is the most likely reason they did not contract the disease?
   - **e.** Explain the differences between the immunity of the mother and the nine-year-old child.

5.12 Exam questions

Question 1 (1 mark)

Source: *VCAA 2018 Biology Exam, Section A, Q23*

MC Rabies is a viral disease spread to people by infected animals. A person bitten by an infected animal should be given an injection of specific antibodies.

Following the injection, this person should have

- **A.** natural active immunity.
- **B.** artificial active immunity.
- **C.** natural passive immunity.
- **D.** artificial passive immunity.

Question 2 (4 marks)

Source: *VCAA 2013 Biology Section B, Q4a and c*

In 1995, the Australian Bureau of Statistics released a report showing that only 53 per cent of children aged between three months and six years had completed the immunisation schedule.

- **a.** Is childhood immunisation an active or passive form of protection against a disease? Justify your answer. **2 marks**
- **b.** Many immunisation schedules include regular booster injections. Explain how these work and why they are necessary. **2 marks**

Question 3 (1 mark)

Source: *VCAA 2015 Biology Exam, Section A, Q14*

Chickenpox (varicella) is a highly contagious viral disease caused by the varicella-zoster virus (VZV). A technician measured the concentration of antibodies to VZV in a person's blood over a 120-day period. An event occurred on day 30 that significantly altered the concentration of antibodies.

The concentration of antibodies over the 120 days is displayed in the graph below.

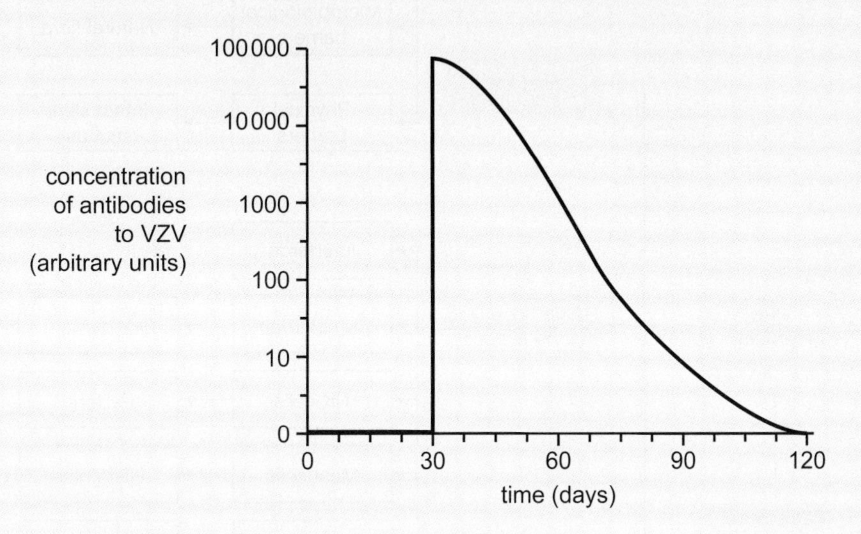

MC Which one of the following events could have occurred on day 30?

A. an exposure of the person to VZV
B. a booster vaccination against VZV for the person
C. an injection of antibodies to VZV into the person
D. an oral dose of antibiotics was given to the person

Question 4 (2 marks)

Source: *VCAA 2014 Biology Exam, Section B, Q5c*

Hemolytic disease of the newborn (HDN) can occur if a Rhesus-negative mother is pregnant with a Rhesus-positive fetus. During pregnancy and birth, some fetal blood cells may enter the mother's bloodstream. The mother makes antibodies against the fetal RhD antigens. This is usually not a problem until there is a second pregnancy with a Rhesus-positive fetus. RhD antibodies made by the mother then cross the placenta, resulting in the possible death of the newborn.

A treatment called immunoprophylaxis has reduced the incidence of newborn death due to HDN. In this treatment, the Rhesus-negative mother receives injections of RhD antibodies with her first Rhesus-positive pregnancy and again at the birth.

Would you consider this treatment to be an active or a passive form of immunotherapy? Explain your response.

Question 5 (1 mark)

Source: *VCAA 2008 Biology Exam 1, Section A, Q14*

MC Diphtheria is a disease caused by the bacterium *Corynebacterium diphtheriae.*

A six-month old baby boy, whose mother and father both had diphtheria as children, will develop active immunity against diphtheria if he

A. is being breast-fed by the mother.
B. receives a blood transfusion from the father.
C. receives an injection of dead diphtheria bacteria.
D. receives injections of gamma globulin from the mother.

More exam questions are available in your learnON title.

5.13 Review

5.13.1 Topic summary

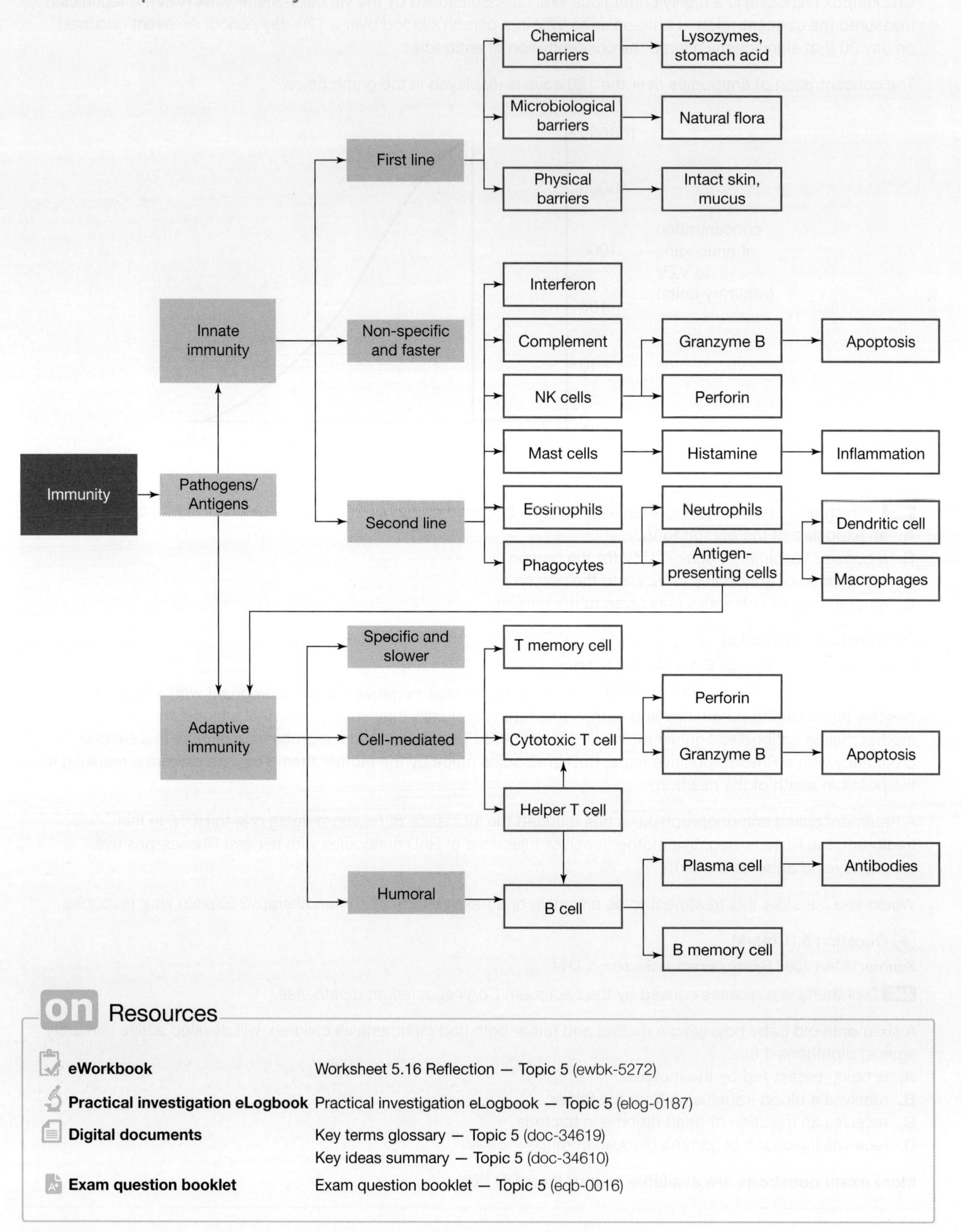

Resources

eWorkbook	Worksheet 5.16 Reflection — Topic 5 (ewbk-5272)
Practical investigation eLogbook	Practical investigation eLogbook — Topic 5 (elog-0187)
Digital documents	Key terms glossary — Topic 5 (doc-34619) Key ideas summary — Topic 5 (doc-34610)
Exam question booklet	Exam question booklet — Topic 5 (eqb-0016)

5.13 Exercises

To answer questions online and to receive **immediate feedback** and **sample responses** for every question go to your learnON title at **www.jacplus.com.au**.

5.13 Exercise 1: Review questions

1. Identify the immune events that have occurred in the following people.
   a. A person who has been sensitised to the pollen of a particular plant
   b. A person who has been protected from the bite of a redback spider by administration of an antivenom
   c. A person who has received the Gardasil vaccine against the human papillomavirus (HPV).
2. State one difference and one similarity between the members of the following pairs.
   a. Active immunity and passive immunity
   b. Natural immunity and artificial immunity
   c. Dead pathogen and attenuated pathogen
   d. Allergy and allergen
3. Antibodies are produced against a variety of pathogens.
   a. Draw a labelled diagram of an antibody.
   b. Name the cells that produce large amount of antibodies in our immune system.
   c. Where are the cells named in part **b** made and matured?
4. Complete the following table.

Type of immunity	How is it acquired?	Short term or long term?	Involves immune response	Involves memory cells
Natural active immunity				Yes
	Vaccination			
Natural passive immunity				
				Yes

5. List three different ways an antibody can inhibit pathogens.
6. Allergic reactions involve the abnormal response of an immune system to substances that are harmless to most people. Ojas is allergic to dust and smoke. She develops watery and itchy eyes when exposed to these substances. Name and explain the process that happens to Ojas's eyes when exposed to dust and smoke.
7. Name two types of pathogens. Give an example of each type.
8. Compare and contrast the roles of humoral and cell-mediated adaptive immunity with regards to the types of pathogens destroyed by each of them and the process of pathogen destruction by each of them.
9. Draw a diagram to outline all the three lines of defences that a pathogen comes into contact.
10. The Mantoux test is designed to determine whether a person has antibodies against the bacteria responsible for tuberculosis. When the test is carried out, a small amount of 'test' solution is injected under the skin on the arm. If the test is positive, a red area appears around the site of injection.
    a. What material must be in the test solution to react with antibodies in the blood?
    b. What do you think causes the redness around the injection site?
    c. Explain the various ways in which a person might have obtained antibodies against tuberculosis.

11. Examine the following graph showing the time course of three events that form part of one particular defensive response of the immune system.

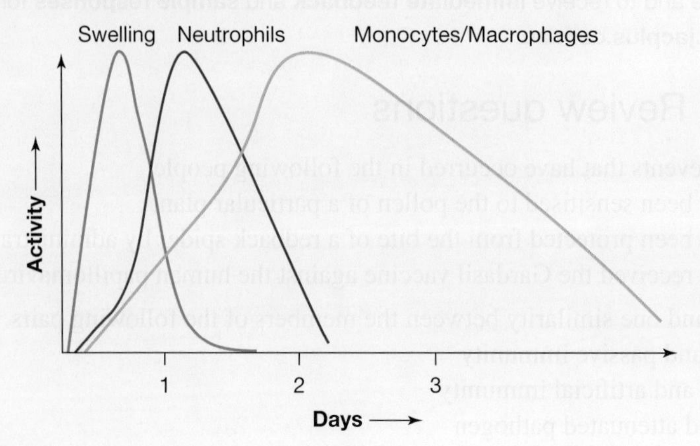

a. Briefly outline the various events depicted in this graph.
b. What immune response is shown in this graph?
c. On what evidence did you make your decision?

12. A pathogen has crossed the mucous membrane of the airways and an immune cell (immune cell A) can only identify this pathogen as follows: 'This is a bacterium; it's not a virus.'
a. Is this an example of a specific or a generic identification?
b. To which line of defence does this immune cell belong — first, second or third?
c. What is a possible identity of this immune cell?
d. Identify a cellular response to this pathogen that might then be expected to occur.

Another immune cell recognises this pathogen as a particular species of Gram-negative bacteria, as follows: 'This is *Vibrio cholera.*'
e. To which line of defence does this immune cell belong — first, second or third?
f. Identify a humoral response to this pathogen that might be expected to occur.
g. What kind of immune cell can produce this response?

13. Suggest a possible explanation in biological terms for each of the following observations.
a. Babies born with an inherited defect in the protein perforin that makes this protein non-functional are at risk of death from viral infection or tumours early in their lives.
b. A history of recurrent virus infections in a patient suggests a defect in the immune defences carried out by T cells.
c. Photomicrographs of blood smears taken from healthy people are expected to show a majority of red blood cells.
d. A white blood cell in a photomicrograph of a blood smear taken from a healthy person is more likely to be a neutrophil than a basophil.
e. A person feeling unwell often has swollen glands.
f. A person who suffers and recovers from an infectious disease such as measles has lifelong immunity to that disease.
g. Pus may appear during the process of the healing of a skin wound.
h. At birth, a baby has some level of adaptive immunity.

14. Consider each of the following immune responses and identify each as part of either innate immunity or adaptive immunity or both.
a. Inflammatory response to infection
b. Activation of the complement cascade to opsonise bacteria
c. Movement of mucus from airways by action of cilia
d. Recognition and elimination of a virus-infected cell
e. Presentation of antigens to naïve immune cells
f. Physical barrier to infectious agents
g. Secretion of cytokines

15. a. Rejection of a transplanted organ may be *hyperacute*, occurring in the first 24 hours after transplantation. This is caused by pre-existing antibodies in the recipient that bind to antigens on the transplant and results in the activation of the complement system and migration of neutrophils to the transplant site. (Hyperacute rejection can be prevented by pre-screening for these anti-graft antibodies.)
Is hyperacute rejection an immune response of the innate immune system or the adaptive immune system? Explain your choice.
    b. A second form of rejection is termed *acute* rejection and it usually begins after the first week of transplantation. This is caused by a mismatch of MHC antigens and involves the action of T cells. (Acute rejection is minimised by the use of immunosuppressive therapy following transplantation.)
        i. Is acute rejection an immune response of the innate immune system or the adaptive immune system? Explain your choice.
        ii. Explain why the acute rejection does not appear until after the first week of transplantation.
16. Pathogens are agents of disease. These infectious microorganisms, such as fungi, bacteria, and nematodes, can cause several diseases in plants and damage tissues. Plants have developed a variety of strategies to fight against these pathogens.
    a. Describe two physical barriers that detect pathogens and stop them before they are able to cause extensive damage.
    b. Humans have a complex and very effective immune system. Describe two chemical barriers in humans that prevent pathogens from entering the body.
    c. Outline the process by which natural killer cells provide protection once a pathogen has gained entry inside the human body.
17. Polio is a highly infectious disease caused by a virus. It invades the nervous system and can cause total paralysis in a matter of hours. There is no cure for polio; it can only be prevented. The polio vaccine, given multiple times, can protect a child for life.
    a. Define vaccine.
    b. Explain how a vaccine provides lifelong immunity in the body.
    c. Vaccines provide protection against many diseases by giving immunity to the body.
        i. Name 2 cells involved in providing lifelong immunity.
        ii. Name the part of the body where these cells mature.
        iii. Describe the role of each of the cells named in part (i).

5.13 Exercise 2: Exam practice questions

on Resources

Teacher-led videos Teacher-led videos for every exam question

Section A — Multiple choice questions

All correct answers are worth 1 mark each; an incorrect answer is worth 0.

Question 1

Source: *VCAA 2010 Biology Exam 1, Section A, Q3*

One of the similarities between the defence mechanisms of a plant and an animal includes the

A. production of memory cells.
B. release of immune cells through a circulatory system.
C. use of an epidermal layer to inhibit the invasion of pathogens.
D. production of salicylic acid to warn cells of an invading pathogen.

Question 2

Source: *VCAA 2013 Biology Section A, Q14*

As part of the first line of defence in the human immune system, naturally occurring barriers to invading pathogens include

A. lysozymes in tears.
B. the production of antibodies.
C. the engulfing of pathogens by phagocytes.
D. inflammation at the site of infection.

Question 3

Source: *VCAA 2013 Biology Section A, Q15*

Defence mechanisms against bacterial pathogens include

A. the production of interferon.
B. neutralisation by histamines.
C. destruction by complement proteins.
D. agglutination by cytotoxic T cells.

Question 4

Source: *VCAA 2016 Biology Exam, Section A, Q24*

In the search for a malaria vaccine, scientists have focused on a protein called circumsporozoite protein (CSP). CSP is secreted by the malaria parasite and is present on its surface.

For the vaccination to work, the scientists want CSP to act as

A. an antigen.
B. an allergen.
C. an antibody.
D. a complement protein

Question 5

Source: *VCAA 2009 Biology Exam 1, Section A, Q21*

After an individual is exposed to a microbial infection, the immune system increases its activities.

The following graph summarises the timeline of the level of those activities.

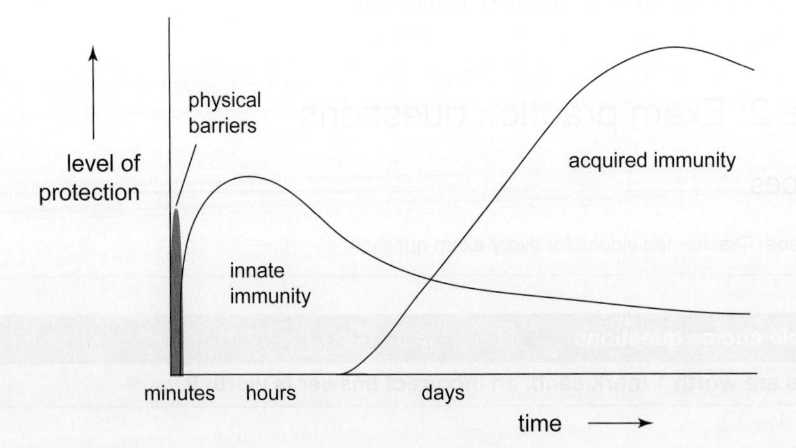

It is reasonable to conclude that

A. physical barriers involve macrophages.
B. innate immunity lacks involvement of living cells.
C. lymph nodes are involved in the acquired immunity phase.
D. the protection developed against the disease ceases at the end of the infection.

Question 6

Source: *VCAA 2013 Biology Section A, Q20*

In the lymphatic system

A. clonal selection occurs.

B. mast cells are produced.

C. vessels have thick, muscular walls.

D. lymph is pumped by the heart.

Question 7

Source: *VCAA 2016 Biology Exam, Section A, Q23*

A park ranger was injected with an antivenom serum to treat a snakebite. The treating doctor explained that the injection would not protect him against future snakebites.

This is because antivenom serum is used to achieve

A. active and natural immunity.

B. passive and natural immunity.

C. active and induced (artificial) immunity.

D. passive and induced (artificial) immunity.

Question 8

Source: *Adapted from VCAA 2017 Biology Exam, Section A, Q26*

A daily blood sample was obtained from an individual who received a single vaccination against a particular strain of the influenza virus. The individual had no prior exposure to this strain of influenza. The graph below shows the concentration of antibodies present in the individual's blood for this strain of influenza over a period of 65 days.

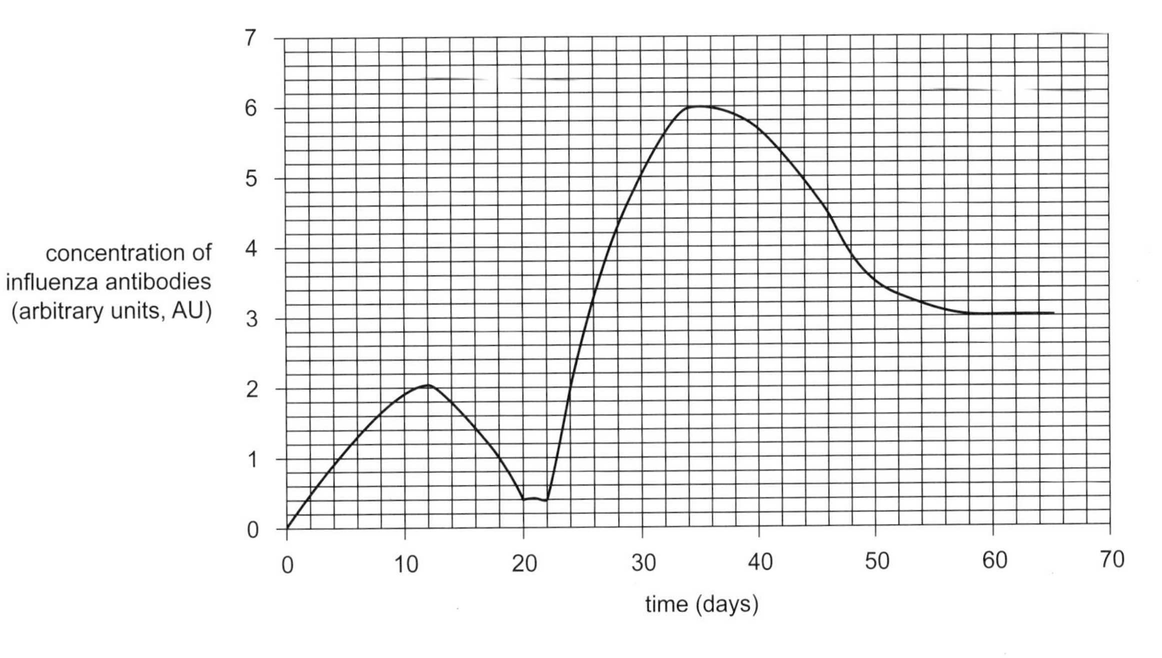

Which one of the following conclusions can be made using this data?

A. Memory B cells were activated by exposure to the same strain of the influenza virus on day 22.

B. B plasma cells specific to this strain of influenza were most numerous on day 12.

C. Herd immunity to this particular strain of influenza was achieved by day 55.

D. The vaccination containing weakened influenza antigens occurred on day 10.

Question 9

Source: *VCAA 2009 Biology Exam 1, Section A, Q24*

If the red blood cells of a blood donor clot when they enter a recipient patient, the patient will die. The blood of both recipient and donor has to be tested and typed. Humans are divided into four different groups with regard to the ABO blood-grouping system. The characteristics of these groups are summarised in the following table.

	group A	group B	group AB	group O
Protein on red blood cells	A	B	A and B	Neither A nor b
Antibodies in plasma	Antibodies to B protein	Antibodies to A protein	No antibodies to either A or B protein	Antibodies to both A and B proteins

A transfusion would be safe if a

A. group A person received blood from a group B person.
B. group B person received blood from a group O person.
C. group O person received blood from a group A person.
D. group A person received blood from a group AB person.

Question 10

Source: *VCAA 2019 Biology Exam, Section A, Q23*

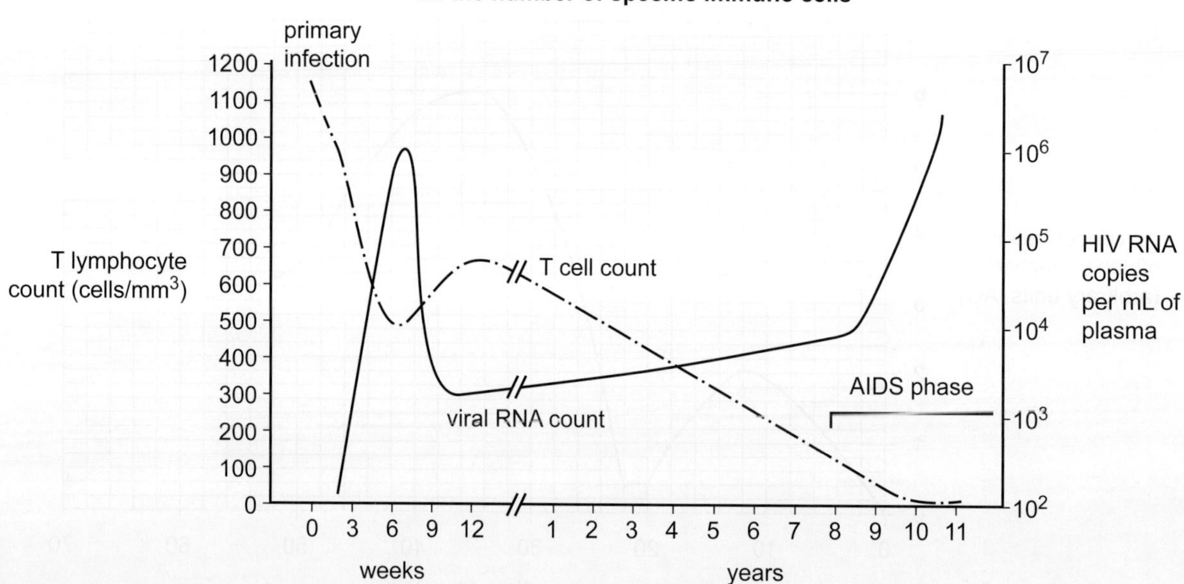

Source: adapted from Paola Paci et al., 'A Discrete/Continuous Model of Anti-HIV Response and Therapy', conference paper, Tenth International Conference on Computer Modeling and Simulation, May 2008, p. 484

The graph above shows the HIV RNA count in the blood and the effect of the virus on the number of specific immune cells in an untreated patient. Based on your knowledge and the information in the graph above, what is the effect of the change in the number of T cells over time?

A. a decrease in the viral RNA count during the AIDS phase
B. loss of effective function of the adaptive immune system
C. failure of macrophages to engulf HIV
D. failure of complement proteins

Section B — Short answer questions

Question 11 (7 marks)

Source: *VCAA 2020 Biology Exam, Section B, Q5*

a. The human immune system uses several different type of cells to eliminate virally infected cells. Name **one** of these cells and outline how it eliminates virally infected cells. **2 marks**

b. i. State the role played by the lymphatic system in an immune response. **1 mark**

ii. Describe the sequence of events that occurs in the secondary lymphoid tissue that results in the production of antibodies. **4 marks**

Question 12 (6 marks)

Source: *VCAA 2019 Biology Exam, Section B, Q4*

The diagram below illustrates how 'thunderstorm asthma' occurs. Thunderstorm asthma is an allergic condition that can be very serious. In 2017 it led to a number of deaths in Victoria. The condition is a combination of hay fever and asthma.

People who suffer from either can be susceptible to thunderstorm asthma.

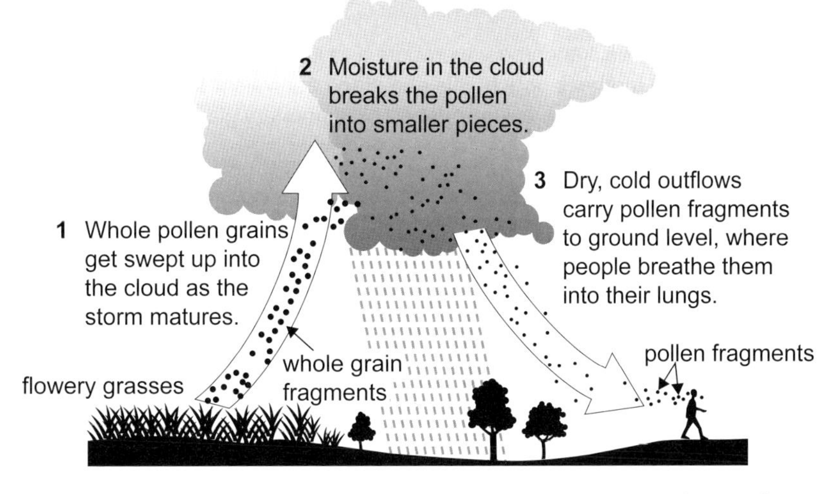

Source: adapted from Asthma Australia, < www.asthmaaustralia.org.au/national/home >

a. Pollen fragments cause allergic reactions when they enter the body by interacting with a specific type of cell and a protein of the immune system.
Name the cell and the protein involved. **2 marks**

b. List **two** strategies that a person could take to reduce their risk of developing thunderstorm asthma. **2 marks**

c. Explain if it could be possible to develop a vaccine against thunderstorm asthma. **2 marks**

Question 13 (7 marks)

Source: *VCAA 2019 Biology Exam, Section B, Q3*

The human immune system consists of a series of defensive barriers that protect the body from infection. When bacteria come into contact with the body, they immediately encounter these defences and must bypass each barrier if they are to survive and infect the body.

a. When bacteria come into contact with the body, they must gain access to the living tissues to become pathogens.
List **two** possible routes the bacteria could use to access the living tissues of the body. **2 marks**

b. Once bacteria are within or have access to the living tissues of the body, but before cells are aware of their presence, the bacteria will encounter chemical barriers.
List **one** of these chemical barriers and explain its function. **2 marks**

c. When an inflammatory response starts, the first cellular responders will be cells from the innate immune system. One of these cells releases histamine.
How does histamine contribute to the inflammatory response? **1 mark**

d. If bacteria are not destroyed by innate immune responses, adaptive immune responses become involved.
Describe how an adaptive immune response is initiated during a bacterial infection. **2 marks**

Question 14 (7 marks)

Source: *Adapted from VCAA 2017 Biology Exam, Section B, Q4b–d*

Australian marsupials, such as wallabies, kangaroos, wombats and koalas, give birth to very undeveloped young called joeys. When a joey enters the mother's pouch, it is at a stage equivalent to a seven-week-old human fetus. It spends many weeks in the pouch feeding on milk produced by mammary glands. Although the pouch provides protection from predators, it is neither sealed nor sterile.

The joey's primary immune tissue (in the bone marrow and thymus) does not mature until 30 days after birth and its humoral immunity does not function effectively until 90 days after birth. Biologists have analysed milk samples from several marsupial species and found that they contain various antibodies. Some of the antibodies in the mother's milk remain in the joey's gut, while others cross the gut wall and enter the joey's bloodstream.

a. Describe at a molecular level how antibodies perform their function. **2 marks**

b. Name the type of immunity that the joey obtains from the antibodies in the milk and explain how this form of immunity is beneficial to the joey. **3 marks**

Scientists have found that the milk of the tammar wallaby (*Macropus eugenii*) contains high levels of peptides with antibiotic properties, as well as lysozyme, complement proteins, cytokines and venom inhibitors.

Scientists tested the tammar wallaby milk peptides and found them to be 10 times more effective than antibiotics such as tetracycline and ampicillin, which are commonly used to fight human diseases. The scientists are keen to find a pharmaceutical company that will support further testing and development of these peptides with antibiotic properties.

c. i. Name the part of the immune system to which these peptides and the other listed chemicals belong. **1 mark**

ii. Select one of the chemicals below, found in tammar wallaby milk, and describe its role in protecting the joey against pathogens.

lysozyme   complement proteins   cytokines   venom inhibitors **1 mark**

Question 15 (7 marks)

Source: *Adapted from VCAA 2014 Biology Exam, Section B, Q4*

a. Define the term 'pathogen'. **1 mark**

The diagram below shows a generalised pathogen with antigens on its surface. The immune system responds to antigens by making antibodies.

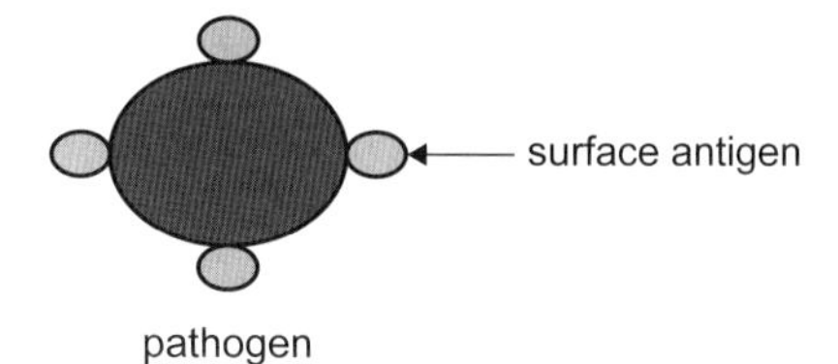

b. Draw an antibody that would be effective against this pathogen. Label the different parts of the antibody. **2 marks**

One way that antibodies work is by forming antigen–antibody complexes.

The diagram below shows four pathogens.

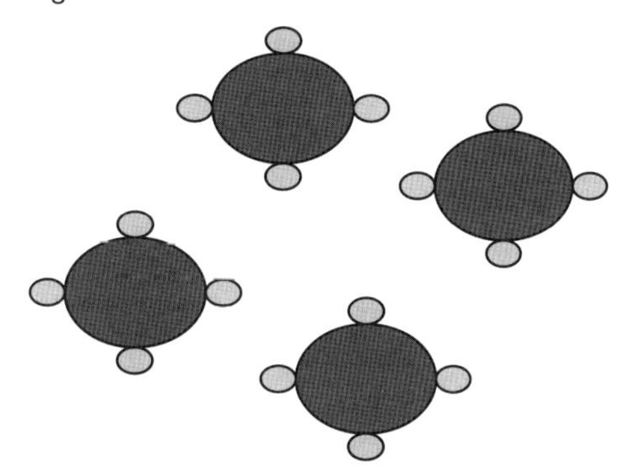

c i. Illustrate on the diagram how the antigen–antibody complex forms, using at least **four** antibodies in your drawing. **2 marks**

ii. Explain how an antigen–antibody complex provides protection against these pathogens. **2 marks**

5.13 Exercise 3: Biochallenge online only

Resources

eWorkbook Biochallenge — Topic 5 (ewbk-8086)

Solutions Solutions — Topic 5 (sol-0661)

Past VCAA examinations

Sit past VCAA examinations and receive immediate feedback, marking guides and examiner's report notes.
Access Course Content and select 'Past VCAA examinations' to sit the examinations online or offline.

teachon

Test maker
Create unique tests and exams from our extensive range of questions, including past VCAA questions.
Access the assignments section in learnON to begin creating and assigning assessments to students.

Online Resources

Below is a full list of **rich resources** available online for this topic. These resources are designed to bring ideas to life, to promote deep and lasting learning and to support the different learning needs of each individual.

eWorkbook

- 5.1 eWorkbook — Topic 5 (ewbk-1884) ☐
- 5.2 Worksheet 5.1 Case study: Transplants and tissue typing (ewbk-5242) ☐
- Worksheet 5.2 Cellular pathogens (ewbk-5244) ☐
- Worksheet 5.3 Non-cellular pathogens (ewbk-5246) ☐
- Worksheet 5.4 The effect of allergens (ewbk-5248) ☐
- 5.3 Worksheet 5.5 Comparing types of immunity (ewbk-5250) ☐
- 5.4 Worksheet 5.6 Innate barriers (ewbk-5252) ☐
- 5.5 Worksheet 5.7 Cells involved in the innate response (ewbk-5254) ☐
- Worksheet 5.8 Soluble molecules in the innate response (ewbk-5256) ☐
- 5.6 Worksheet 5.9 Inflammation and case studies of chronic inflammation (ewbk-5258) ☐
- 5.7 Worksheet 5.10 The role of the lymphatic system (ewbk-5260) ☐
- 5.8 Worksheet 5.11 How the adaptive immune response is initiated (ewbk-5262) ☐
- 5.9 Worksheet 5.12 The branches of adaptive immunity (ewbk-5264) ☐
- 5.10 Worksheet 5.13 Antibodies and the humoral adaptive response (ewbk-5266) ☐
- 5.11 Worksheet 5.14 Cell-mediated adaptive immunity (ewbk-5268) ☐
- 5.12 Worksheet 5.15 Comparing the types of acquired immunity (ewbk-5270) ☐
- 5.13 Worksheet 5.16 Reflection — Topic 5 (ewbk-5272) ☐
- Biochallenge — Topic 5 (ewbk-8086) ☐

Solution

- 5.13 Solutions — Topic 5 (sol-0661) ☐

Practical investigation eLogbook

- 5.1 Practical investigation elogbook — Topic 5 (elog-0187) ☐
- 5.2 Investigation 5.1 Microorganisms — where are they found? (elog-0673) ☐
- 5.4 Investigation 5.2 The importance of intact skin (elog-0675) ☐
- Investigation 5.3 Innate barriers in plants (elog-0677) ☐
- 5.7 Investigation 5.4 Modelling the swelling of lymph nodes (elog-0679) ☐
- 5.10 Investigation 5.5 Modelling antibodies and the diversity of the variable region (elog-0683) ☐

Digital documents

- 5.1 Key science skills — VCE Biology Units 1–4 (doc-34326) ☐
- Key terms glossary — Topic 5 (doc-34619) ☐
- Key ideas summary — Topic 5 (doc-34610) ☐
- 5.2 Case study: Robert Koch — a founder of bacteriology (doc-36058) ☐
- Case study: The story of Matthew Ames and exotoxins (doc-36168) ☐
- Case study: Typhoid Mary (doc-36169) ☐
- Extension: Anaphylaxis and diagnosing allergies (doc-36059) ☐
- 5.8 Case study: The mechanism of action of HIV (doc-36057) ☐

Teacher-led videos

- Exam questions — Topic 5 ☐
- 5.2 Sample problem 1 Cellular and non-cellular pathogens (tlvd-1776) ☐
- 5.3 Sample problem 2 Comparing and contrasting types of immunity (tlvd-1777) ☐
- 5.4 Sample problem 3 Physical and chemical barriers in plants (tlvd-1778) ☐
- 5.5 Sample problem 4 Identifying and describing soluble proteins (tlvd-1779) ☐
- 5.8 Sample problem 5 The impact of HIV on helper T cells (tlvd-1780) ☐
- 5.10 Sample problem 6 Understanding the function of B cells (tlvd-1781) ☐
- 5.11 Sample problem 7 Comparing and contrasting natural killer and cytotoxic T cells (tlvd-1817) ☐
- 5.12 Sample problem 8 Identifying and describing types of immunity (tlvd-1818) ☐

Video eLessons

- 5.2 Antigens and blood type (eles-4194) ☐
- Bacteria (eles-4195) ☐
- Pathogenic agents: prions (eles-4162) ☐
- 5.5 Phagocytosis (eles-4343) ☐
- Allergens and allergic responses (eles-4344) ☐
- The interferon mechanism against viruses (eles-4347) ☐
- 5.6 Inflammatory response (eles-4345) ☐
- 5.9 Antibody activity (eles-4346) ☐

Interactivities

- 5.2 Antigens (int-0045) ☐
- 5.9 The body's response to antigens (int-5770) ☐

Weblinks

- 5.2 The story of Matthew Ames ☐
- Plague ☐
- Typhoid Mary ☐
- What is scientific medicine? ☐
- Allergy and Anaphylaxis Australia ☐
- 5.3 The immune system interactivity ☐
- Examining the components of the immune system ☐
- 5.4 Microbiological barriers ☐
- 5.9 The immune system in action ☐

Exam question booklet

- 5.1 Exam question booklet — Topic 5 (eqb-0016) ☐

Teacher resources

There are many resources available exclusively for teachers online.

To access these online resources, log on to **www.jacplus.com.au**

6 Disease challenges and strategies

KEY KNOWLEDGE

In this topic, you will investigate:

Disease challenges and strategies

- the emergence of new pathogens and re-emergence of known pathogens in a globally connected world, including the impact of European arrival on Aboriginal and Torres Strait Islander peoples
- scientific and social strategies employed to identify and control the spread of pathogens, including identification of the pathogen and host, modes of transmission and measures to control transmission
- vaccination programs and their role in maintaining herd immunity for a specific disease in a human population
- the development of immunotherapy strategies, including the use of monoclonal antibodies for the treatment of autoimmune diseases and cancer.

PRACTICAL WORK AND INVESTIGATIONS

Practical work is a central component of learning and assessment. Experiments and investigations, supported by a **practical investigation eLogbook** and **teacher-led videos**, are included in this topic to provide opportunities to undertake investigations and communicate findings.

6.1 Overview

Numerous **videos** and **interactivities** are available just where you need them, at the point of learning, in your digital formats, learnON and eBookPLUS and at **www.jacplus.com.au**.

6.1.1 Introduction

The world's population is constantly increasing, with greater numbers of people moving greater distances. More interactions are happening between humans and pathogens, resulting in a continuous struggle between the two groups. Population movement is having an impact on both humans and our health systems globally. This can be seen in increases in outbreaks of diseases such as HIV/AIDS; the emergence of new diseases such as severe acute respiratory syndrome (SARS) and coronavirus disease (COVID-19); and the re-emergence of ancient diseases such as tuberculosis (TB).

FIGURE 6.1 Airport screening for the 2020 COVID-19 coronavirus outbreak. Thermal imagery is being used to check for signs of fever.

Diseases have existed since before the first humans walked the Earth. However, it was not until the work of scientists, such as Robert Koch and Louis Pasteur, during the nineteenth century that there was greater understanding about disease transmission, impacts and control measures. In the twenty-first century, increased interaction between humans, and between humans and other animals and microorganisms, has meant greater risk of exposure to new pathogens and diseases. It has become a global challenge to control the spread of infectious diseases. However, new treatments are constantly being developed, from new antibiotics and antiviral drugs to immunotherapy strategies, such as the use of monoclonal antibodies like herceptin (which can be seen in the topic opener image).

LEARNING SEQUENCE

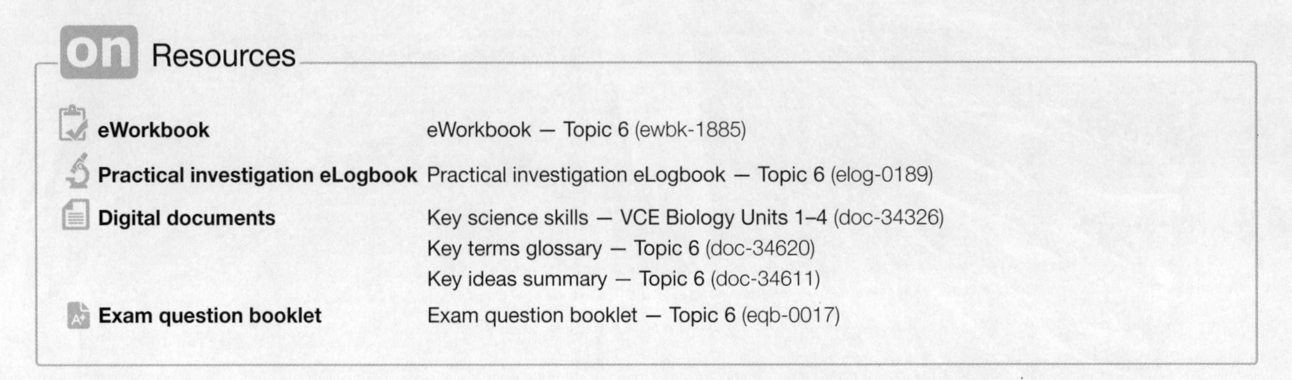

on Resources

eWorkbook	eWorkbook — Topic 6 (ewbk-1885)
Practical investigation eLogbook	Practical investigation eLogbook — Topic 6 (elog-0189)
Digital documents	Key science skills — VCE Biology Units 1–4 (doc-34326) Key terms glossary — Topic 6 (doc-34620) Key ideas summary — Topic 6 (doc-34611)
Exam question booklet	Exam question booklet — Topic 6 (eqb-0017)

6.2 The emergence and re-emergence of pathogens

KEY KNOWLEDGE

- The emergence of new pathogens and re-emergence of known pathogens in a globally connected world, including the impact of European arrival on Aboriginal and Torres Strait Islander peoples

6.2.1 Different types of diseases

Disease is a condition in a living animal or plant body that impairs the normal functioning of an organ, part, structure or system. Broadly, we can split diseases into two main types: **non-infectious** and **infectious**.

Non-infectious, or non-communicable, diseases cannot spread from affected people to healthy people via the environment. Examples include environmental, nutritional and inherited diseases.

Infectious or communicable diseases can be transmitted from one individual to another. These diseases are caused by pathogenic agents.

Non-infectious diseases

Non-infectious diseases include:
- genetic diseases
- degenerative diseases
- nutritional diseases
- social diseases
- nutritional diseases
- cancers
- physiological malfunctions
- cardiovascular diseases.

Infectious diseases

In the early twentieth century, infectious diseases, such as pneumonia, cholera, diphtheria, tuberculosis (TB) and influenza, were some of the diseases that were determined to be the main causes of death in human beings.

However, by the end of the twentieth century, scientists had developed vaccines against some **pathogens**, which resulted in the near eradication of diseases such as mumps, measles, rubella, whooping cough, diphtheria and polio — at least in the developed world. Worldwide, only one disease affecting humans has been completely eradicated, which is smallpox. Therefore, many infectious diseases continue to be a threat to public health.

Every year, around five new diseases emerge in human beings due to factors such as environmental change, population growth and urbanisation.

Emerging diseases are defined as:
- a disease caused by a newly identified or previously unknown agent
- a disease that has existed in other species but whose incidence in humans has increased in the past two decades, either locally or internationally.

A **re-emerging disease** is a disease which reappears after a significant decline in its incidence. Re-emerging diseases were once controlled but have increased to a level that causes significant health issues.

disease a condition in a living animal or plant body that impairs the normal functioning of an organ, part, structure or system

non-infectious diseases that cannot spread from affected people to healthy people via the environment

infectious diseases that can be transmitted between individuals and are caused by pathogens

pathogens agents that cause diseases in their hosts

emerging disease a disease caused by a newly identified or previously unknown agent

re-emerging disease reappearance of a known disease after a significant decline in incidence

TABLE 6.1 Examples of emerging and re-emerging diseases

Category	Disease	Agent	Year identified
Newly identified or unknown variant of a known disease (emerging)	AIDS: Acquired Immune deficiency syndrome	HIV: Human immunodeficiency virus	1981
	Variant Creutzfeldt–Jakob disease (vCJD or mad cow)	Prion BSE	1996
	Severe Acute Respiratory Syndrome (SARS)	SARS associated coronavirus	2003
	Middle East Respiratory Syndrome	MERS coronavirus	2012
	Zika	Zika virus	2015
	COVID-19 (coronavirus)	SARS-CoV-2	2019
Increase in incidence (re-emerging)	Ebola haemorrhagic fever	Ebola virus	1976
	Dengue	Dengue virus	1943
	Cholera	Vibrio cholera	1854

FIGURE 6.2 A map showing global examples of emerging and re-emerging diseases as of 2020

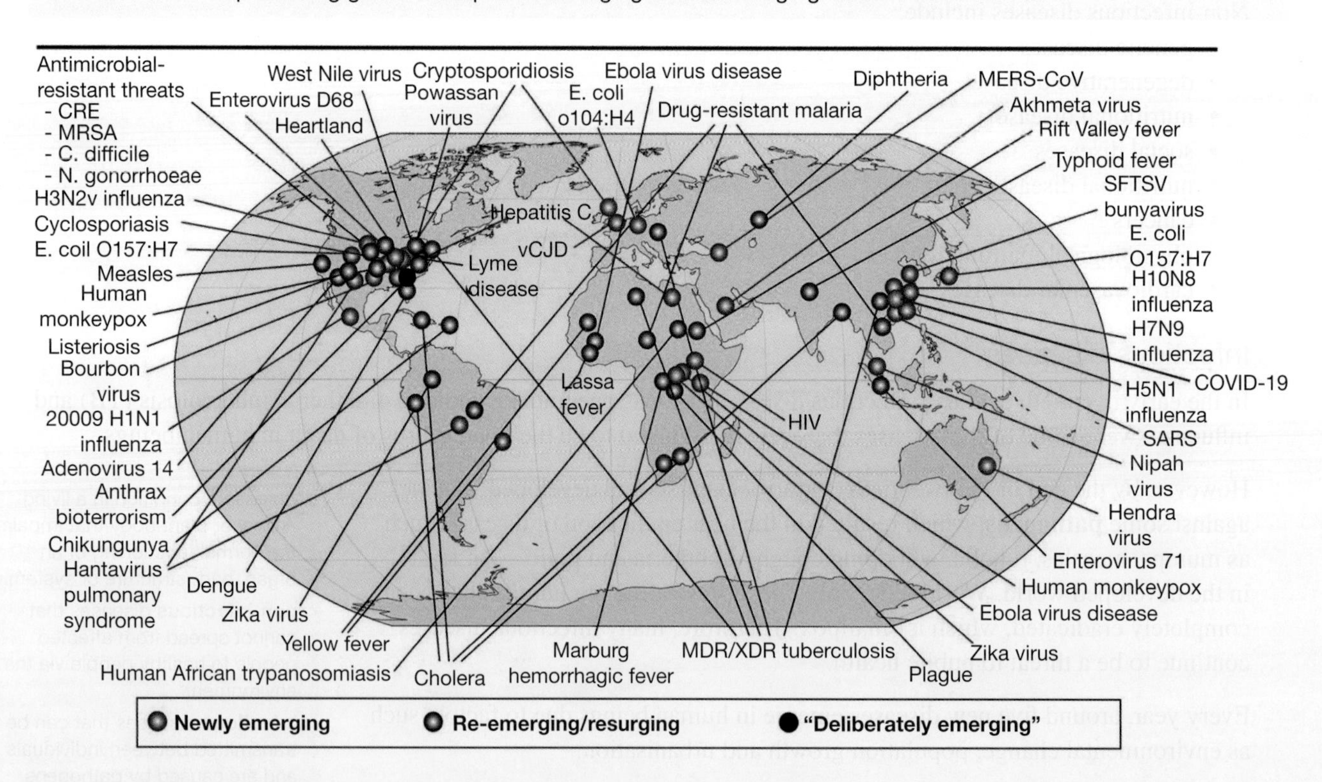

Some other examples of the emergence or re-emergence of disease include:

- unknown microbes or new strains or variants causing new diseases — for example, SARS and Zika virus
- known agents causing new diseases — for example, the Hanta virus in the United States in 1993 caused respiratory disease instead of kidney disease
- microbes of animal origin causing diseases in humans — for example, the influenza virus in pigs causing swine flu in humans.

In many cases, the disruption of ecosystems by humans and increased travel has increased our exposure to new or re-emerging disease-causing agents.

6.2.2 Pathogens, pandemics and epidemics

In 1918, the spread of the influenza virus (the 'Spanish flu') was facilitated by the deployment of large numbers of soldiers from different countries to the battlefields of Europe, from where it spread to Russia and even Greenland. In 1919, the spread of the virus was further facilitated by the demobilisation of soldiers returning to their home countries.

The world is more connected than ever before. The increased availability of air travel has allowed more individuals to move easily between countries. Technology and communication has improved so that people can connect across the world, which can be beneficial for providing health information and warnings, but it can also lead to miscommunication and the spread of incorrect information.

The greater mobility of people in the twenty-first century, particularly by air travel, has created the opportunity for pathogens to be exported from one country by infected passengers to another country in a day or less.

As the world is now so much more connected, pathogenic agents that in the past may have been isolated to only one region can more easily spread across multiple countries and regions. These can then develop into epidemics and/or pandemics. The terms **pandemic** and **epidemic** both relate to the uncontrolled spread of infectious diseases, but they differ in geographic spread. A pandemic affects a much larger geographical area compared with an epidemic. It affects multiple world regions (figure 6.3). In an interconnected world, an epidemic can develop into a pandemic (but not vice versa).

pandemic a situation when, over a relatively short time, many people worldwide contract a specific disease as it spreads from a region of origin

epidemic the widespread occurrence of an infectious disease in a community or in a restricted geographic area at a particular time

FIGURE 6.3 The six different World Health Organization (WHO) regions and the location of the WHO offices in each region

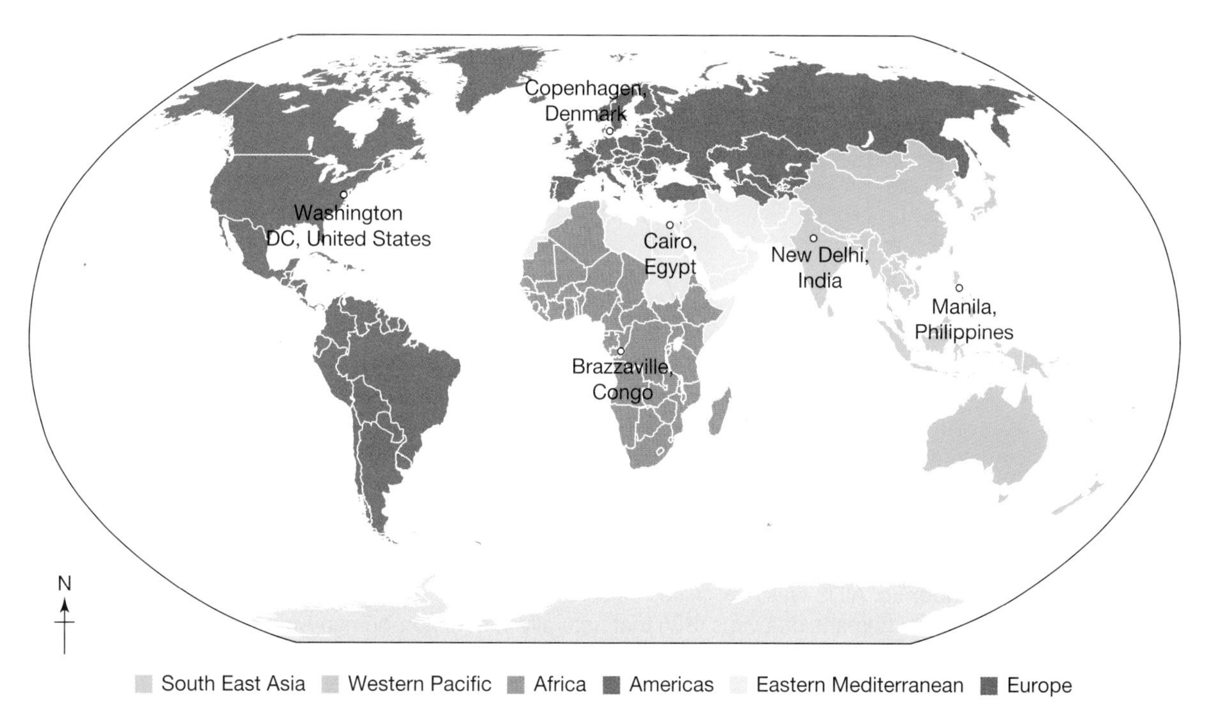

What is a pandemic?

A pandemic (from the Greek: *pan* all; *demos* people) refers to the global outbreak of a disease. Other definitions include:

- an outbreak of a disease that occurs over a wide geographic area and affects an exceptionally high proportion of the population (Merriam-Webster)
- the worldwide spread of a new disease (WHO).

For the WHO to declare an event as a pandemic, the infection must spread easily and sustainably among human populations in at least three countries in at least two different WHO regions.

Note that the term 'pandemic' refers to the spread of a disease, not to the severity of the disease — a pandemic could involve the global spread of a pathogen that causes a mild disease. However, the pandemics that attract greatest public interest and media coverage are those in which the disease involved is severe and has a high death rate.

CASE STUDY: Examples of pandemics

Examples of pandemics since the start of the twentieth century have included:

- **The influenza pandemics:** these include the 1918 (Spanish flu), 1951 (Asian flu), 1968 (Hong Kong flu) and the 2009 pandemics (11 June 2009–10 August 2010). Each of these pandemics resulted from a major change in the influenza A virus, known as a 'shift', that created a new or novel influenza A virus to which the population had no immunity.

FIGURE 6.4 Death rates for five-year age groups for the 1919 influenza pandemic in New South Wales. Note that the second wave of the pandemic was more deadly than the first wave. (Reproduced by permission, NSW Ministry of Health, 2016 — Influenza Report 1920.)

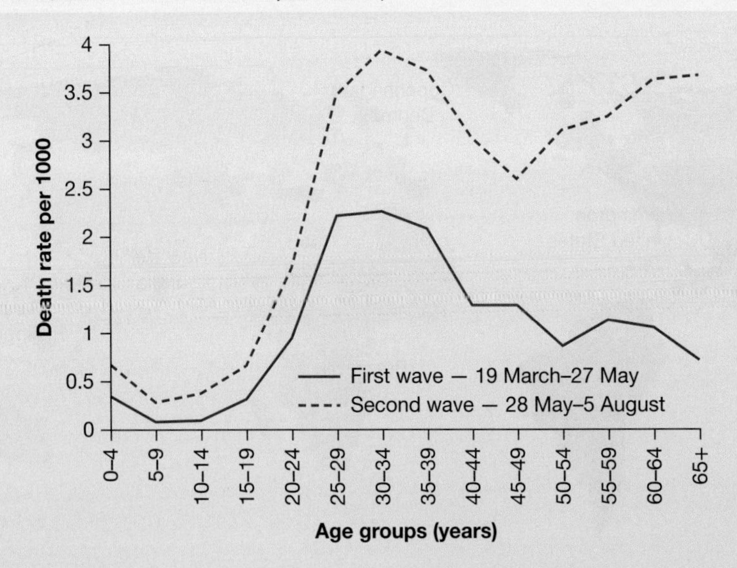

- **The HIV/AIDS pandemic:** AIDS was recognised as a new disease in 1981. Its origin traces back to a cross-species transmission to humans of a virus (SIV) that naturally infects non-human African primates. Since the beginning of the pandemic, almost 71 million people have been infected with the HIV virus and about 34 million people have died from HIV.
- **Cholera:** Cholera has killed tens of millions of people throughout history. While pandemics of cholera are less common now, it still poses a severe threat in developing countries, with outbreaks occurring often. Over the past 200 years, there have been seven main cholera pandemics. The most recent pandemic occurred in the 1970s.
- **COVID-19 (SARS-CoV-2):** On 11 March 2020, COVID-19 caused by the virus SARS-CoV-2 (often referred to as coronavirus) was declared a pandemic by the WHO.

CASE STUDY: COVID-19

In recent times, the biggest health pandemic seen was that of COVID-19 (coronavirus), a disease caused by the virus SARS-CoV-2. Differences between COVID-19, the common cold and influenza can be seen in table 6.2. While the first infections were reported in 2019, the pandemic had its greatest impact in 2020. This affected the lives of individuals globally, not just through widespread infections and fatalities but on a vast economic scale. As a result of coronavirus, many countries locked down and implemented regulations around quarantine, the use of masks and social distancing. The spread of coronavirus worldwide reflected the hazards of a globally connected world.

Many documented cases exist of travellers who spent time in countries where COVID-19 was present and who inadvertently brought the virus back to their home countries. An outline of some of the key events in this pandemic are as follows:

- In December 2019, several individuals in the city of Wuhan, China, were reported to have unusual cases of pneumonia.
- In early January 2020, these unusual cases of pneumonia were determined to be caused by a novel coronavirus.
- On 9 January 2020, the first death in China from coronavirus was reported.
- By 13 January 2020, the first case of coronavirus was officially reported outside of China. A woman, who had travelled from Wuhan, was diagnosed in Thailand.
- By late January 2020, Wuhan was placed under quarantine.
- The first case of coronavirus was diagnosed in Australia on 25 January 2020.
- On 30 January 2020, the coronavirus was declared a global emergency by the WHO. On March 11, the WHO officially declared coronavirus a pandemic.
- At the start of March 2020, around 90 000 global cases of coronavirus had been reported. By the start of May, this had risen to over 4 million. In January 2021, almost a year after the first Australian case was documented, 100 million cases had been recorded globally, with 2 million deaths.
- Almost a quarter of these cases were documented in the United States. Other countries that had high case numbers included India, Brazil, France, Russia, Spain and the United Kingdom.
- Australia, at the start of March 2020, had 30 diagnosed cases. By mid May, there were around 7000 diagnosed cases, with approximately 100 deaths. By late November, Australian cases had reached around 28 000 (with 900 deaths). However, the number of new cases in Australia had dramatically dropped, apart from some small outbreaks linked to hotel quarantine.
- As numbers continued to skyrocket globally, international travel was prevented and regulated. These regulations eventually reduced travel across state borders within Australia.
- People in Victoria faced some of the toughest restrictions in Australia, with individuals only permitted to leave their homes for exercise, medical services, purchasing essential goods and to attend work (if essential). Most schools adjusted to home learning as Victoria (and the rest of Australia) worked to get the virus under control.
- In late 2020, numerous vaccines (such as the Pzifer mRNA vaccine and AstroZeneca viral vector vaccine) were approved for use to try to combat the spread of COVID-19 and see the end of the pandemic.

TABLE 6.2 Comparisons of influenza, the common cold and COVID-19

	Influenza	Common cold	COVID-19
Cough	Common	Common	Common
Difficulty breathing	N/A	N/A	Common
Headaches	Common	Common	Occasional
Runny nose	Occasional	Occasional	Occasional
Fatigue	Common	N/A	Occasional
Fever	Common	Rare	Common
Aches and pains	Common	N/A	Occasional
Sneezing	N/A	Common	N/A

FIGURE 6.5 Data showing total number of cases of COVID-19 between January and November 2020 (data taken at the end of each month) **a.** globally **b.** in Australia

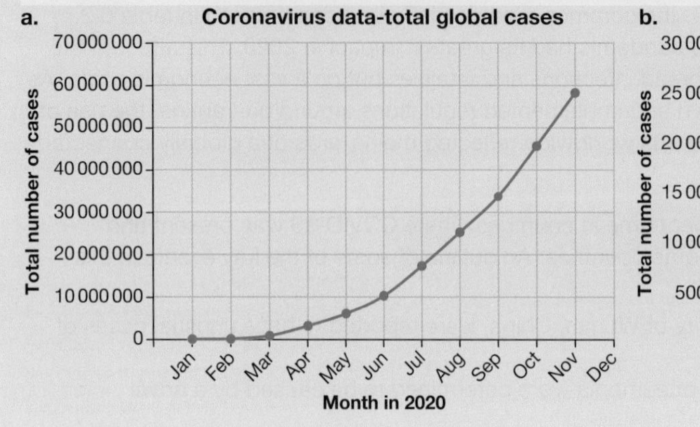

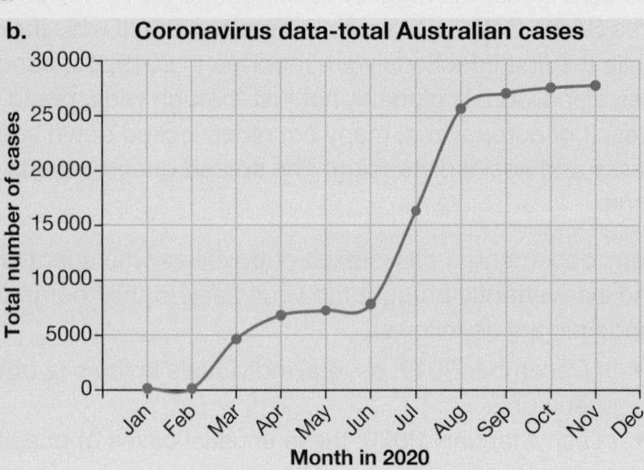

Resources

Weblinks Timeline of COVID-19
Data related to COVID-19

How do pandemics occur?

The conditions that favour the emergence of a disease and the start of a pandemic include the following:

1. A new pathogen or a novel strain of an existing pathogen suddenly appears in geographic areas where the human populations have not previously come into contact with the pathogen.
   The suddenness of appearance means that:
   - most people will have little or no immunity to the pathogen
   - a vaccine is unlikely to exist or is available only in limited amounts that are not sufficient to prevent the spread of the pathogen.
2. The pathogen is the cause of an illness, often serious, in people. In many cases, a disease that can develop into a pandemic is caused by a pathogen that infects not only people but also other hosts, such as birds, pigs, bats and monkeys. These various non-human hosts act as reservoirs of the pathogen, moving to human hosts via a suitable vector.
3. The pathogen can be transmitted easily from person to person. This spread may occur:
   - by airborne particles coughed or sneezed by an infected person (as in influenza virus)
   - by contact with blood or other body fluids of an infected person (as in AIDS and Ebola virus fever)
   - via vectors, such as mosquitoes (as in Zika virus) or fleas (as in the plague), that transmit the pathogen from an infected person to an uninfected person.
4. Uncontrolled spread of the pathogen occurs across a wide geographic area. This spread may be facilitated by the movement of infected individuals from their home range and/or by the migration of infected vectors that transmit the pathogen from infected to non-infected individuals.

What is an epidemic?

An epidemic (from the Greek: *epi* upon; *demos* people) refers to the widespread occurrence of an infectious disease in a community or in a restricted geographic area at a particular time. With any epidemic, a concern is that the disease may spread more widely and become a pandemic.

CASE STUDY: Examples of epidemics

- **Severe acute respiratory syndrome (SARS):** The SARS epidemic in China (2002–2003) was caused by a new coronavirus that jumped from another species to humans; transmission of SARS occurred by respiratory droplets from coughs or sneezes of infected individuals.
- **Cholera (Haiti 2010–present):** this epidemic was caused by the bacterial species *Vibrio cholera*, which is transmitted through faecal-contaminated water or food; these bacteria produce a toxin that binds to cells of the intestinal wall and interferes with the normal flow of water, sodium and chloride ions.
- **Ebola (West Africa 2013–2016):** the Ebola virus is transmitted by direct contact of broken skin or mucous membranes with blood or other body fluids. This epidemic led to around 10 000 deaths from 2013 to March 2016.
- **Yellow fever epidemics:** Yellow fever virus is an RNA virus that belongs to the genus *Flavivirus*, the same genus as the Zika virus. Yellow fever disease is transmitted by the bite of infected female mosquitoes that acquire the virus when they feed on infected people or infected monkeys. One yellow fever epidemic occurred in Angola in 2016. By October 2016 the number of suspected cases was 4347, including 884 confirmed cases with 377 deaths, throughout all provinces of the country. In December 2016, Angola declared the end of this outbreak. Another epidemic of yellow fever occurred in 2016–2017 in Brazil. Due to recurring epidemics of this disease in South America, vaccinations for yellow fever are often recommended.

To access more information about the Ebola and yellow fever epidemics, please download the digital document.

Resources

Digital document Case study: The Ebola, Zika and yellow fever epidemics (doc-36102)

SAMPLE PROBLEM 1 Comparing pandemics and epidemics

In 2009, there was an influenza pandemic that caused concern around the world as it was similar to Spanish flu virus of 1918. Each of these pandemics resulted from a major change in the influenza virus.

What is the difference between a pandemic and an epidemic? (2 marks)

THINK	WRITE
The question is asking you to provide differences between an epidemic and a pandemic. Provide clear differences between the two, considering the main difference of geographical area.	An epidemic is an outbreak of a disease in a localised area or population (1 mark). If an epidemic spreads widely, it can become pandemic. A pandemic is an outbreak of a disease that occurs over a wide geographic area, affecting at least three countries in two different regions (1 mark).

Resources

eWorkbook Worksheet 6.1 Comparing epidemics and pandemics (ewbk-7849)

Weblink Managing epidemics — WHO

6.2.3 The impact of European arrival on Aboriginal and Torres Strait Islander peoples

In 1770, James Cook, during a Pacific voyage, proclaimed part of Australia's eastern coast as Crown land for the British government. In 1788, the First Fleet arrived from Britain and invaded the land that was already owned by the pre-existing Aboriginal peoples. British colonisation in Australia led to the dramatic decline in populations of Australian Aboriginal peoples (as shown in figure 6.6). One of the main reasons for this significant decrease was the introduction of new diseases to these populations to which they had no prior exposure.

The Torres Strait Islander peoples traded with the Aboriginal peoples of Cape York and the peoples of Papua New Guinea before the Europeans arrived. The initial European contact with Torres Strait Islander peoples was made in 1606, when Luiz Vaez de Torres sailed through the Torres Strait. After 1770, many British ships used the Torres Strait as a passage.

Despite only a small number of individuals being in the First Fleet (around 1500 in total), they brought in various new diseases and pathogens. Prior to this invasion and the passage through the Torres Strait, there was no evidence of these diseases in Aboriginal and Torres Strait Islander populations.

The major epidemic diseases brought to Australia during the early contact stage were smallpox, chickenpox syphilis, tuberculosis, influenza and measles. These diseases devastated Aboriginal and Torres Strait Islander peoples, who had no immunity against them. Because these peoples had never been exposed to most of the new infectious microbes, there were large numbers of susceptible individuals and low herd immunity, resulting in the spread of new infectious diseases and significant fatalities.

FIGURE 6.6 The changing population size of Australian Aboriginal peoples before and after colonisation in 1788. The First Fleet, which invaded Australia in 1788, included around 1500 individuals. Note that the figures for the Aboriginal peoples are estimates that were determined at a later date (backcast).

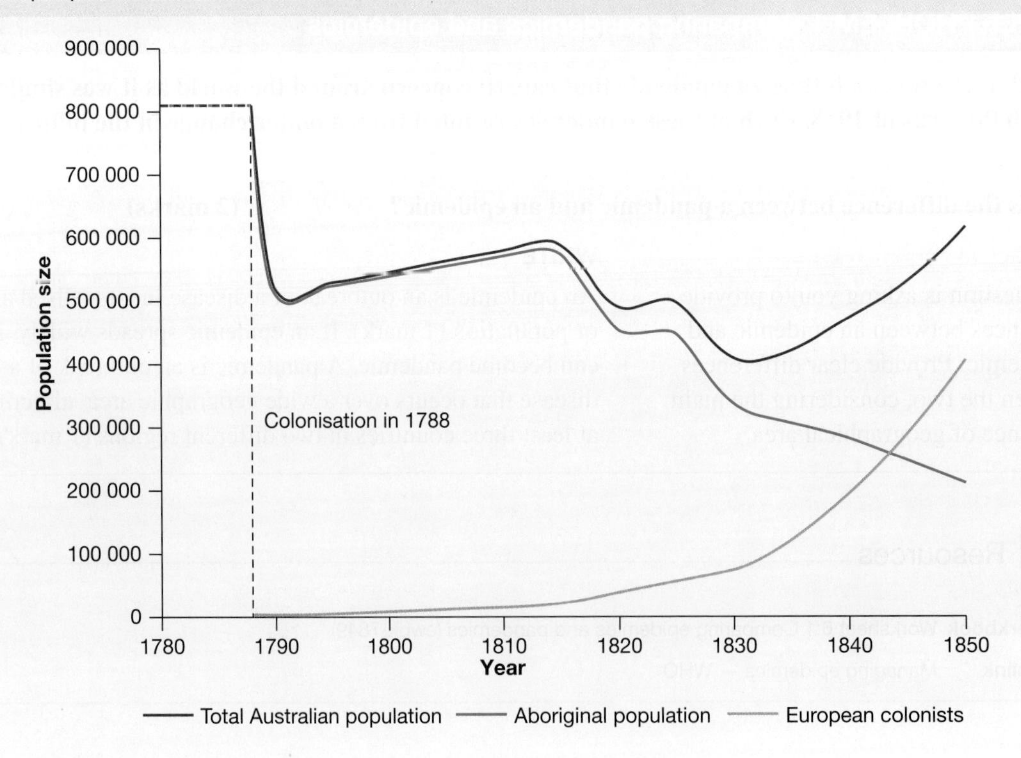

There are several factors that contributed to the devastating effects of introduced diseases on Aboriginal and Torres Strait Islander peoples:

- previous exposure to pathogens
- route of transmission of pathogens

- population density of populations
- health and nutritional status of Aboriginal and Torres Strait Islander peoples
- intergroup social relationships.

Table 6.3 shows a list of diseases that were found in Aboriginal and Torres Strait islander peoples before and after European colonisation. The effect of these diseases and pathogens caused irreversible damage to Aboriginal and Torres Strait Islander peoples.

TABLE 6.3 Comparing diseases present in populations of Aboriginal and Torres Strait Islander peoples before and after colonisation

Before colonisation	After colonisation
Botulism	Chicken pox
Ross River Fever	Influenza
Query (Q fever)	Measles
Hepatitis B	Mumps
Mononucleosis	Rubella
Tetanus	Smallpox
Salmonellosis	Cholera
Scabies	Malaria
Roundworm	Typhus
Streptococcal diseases	Syphilis

CASE STUDY: The introduction of smallpox

Smallpox was one of the most virulent and deadly diseases that infected and killed millions of people. In 1789, a major smallpox epidemic broke out in Australia. The outbreak of this disease did not affect the British colonists. However, the Aboriginal peoples had no previous exposure to the smallpox virus and therefore had no resistance to the disease. Since there was no immunological protection or herd immunity to the disease, the epidemic resulted in a significant number of deaths in Australian Aboriginal peoples (estimated as 70 per cent of the population in areas that were settled). Questions still remain about the nature of this exposure. Was it a deliberate action by colonists or the accidental spread of a devastating disease?

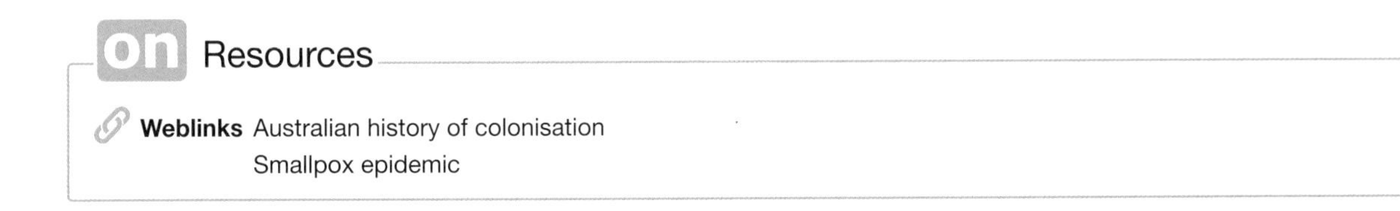

KEY IDEAS

- Human demographic changes, such as increasing population and urbanisation results in overcrowding, which speeds up the chances of spreading diseases.
- Increased international travel, especially without taking appropriate vaccines and other protective measures, leads to increased infection in travellers who then bring the infection home with them.
- Without proper control measures, disease outbreaks can develop into an epidemic (a large spread in a restricted geographical area). Further spreading to countries other regions may lead to a pandemic.
- The arrival of Europeans in Australia and the Torres Strait exposed the the original inhabitants to many new diseases to which they had no resistance, leading to widespread fatalities.

6.2 Activities

learnON

To answer questions online and to receive **immediate feedback** and **sample responses** for every question, go to your learnON title at **www.jacplus.com.au**. A **downloadable solutions** file is also available in the resources tab.

6.2 Quick quiz on	6.2 Exercise	6.2 Exam questions

6.2 Exercise

1. MC A new disease spreads through the United States, China and Japan. This is known as
   A. an epidemic. B. a pandemic. C. an endemic disease. D. an outbreak.
2. List two conditions that would be needed for a pandemic to develop.
3. Give three examples of emerging diseases.
4. Using an example, explain what is meant by a re-emerging disease.
5. Identify the following statements as true or false. Justify your response.
   a. People exposed to new or novel pathogens would be expected to have little or no immunity to those pathogens.
   b. Some pathogens have multiple hosts.
   c. A pandemic could develop into an epidemic.
   d. Deaths from the Spanish flu were higher in fit young adults than in older people.
6. Explain the impact of European arrival on the Aboriginal and Torres Strait Islander peoples.
7. While smallpox is now eradicated, the effect of novel and new diseases affects many individuals worldwide.
   a. Explain what would happen if Australians colonised an area that had previously been isolated from the rest of the world.
   b. What preventative measures could be taken to protect native inhabitants?

6.2 Exam questions

Question 1 (1 mark)

MC Pandemics are often caused by viruses. The reason that pandemics are often caused by viruses, rather than other pathogens, is that viruses

A. are more infectious than other pathogens.
B. are carried overseas by wind and ocean currents.
C. often evolve to become resistant to drug treatments.
D. mutate into new strains to which humans have no immunity.

Question 2 (1 mark)

MC Which of the following is not a condition which would favour the start of a pandemic?

A. Exposed individuals have not come in contact with the pathogen previously
B. The illness only is able to infect humans.
C. The pathogen can spread across a wide geographical range.
D. The pathogen can be easily transmitted from person to person.

Question 3 (2 marks)

There are several recently emerged diseases that have threatened to become pandemics, such as SARS (severe acute respiratory syndrome), Avian flu and Zika virus.

Explain how a disease that starts in a small population, such as a family or a small village, can spread around the world within weeks or a few months.

Question 4 (5 marks)

In recent years, outbreaks of Ebola have occurred in Africa. Ebola is caused by the Ebola virus which leads to a haemorrhagic fever. The chance of dying after contracting Ebola is much higher compared to other diseases.

There have been many outbreaks of Ebola documented since it was first identified in 1976. Over the past ten years, there have been three separate outbreaks of Ebola, all in Africa:

- Western Africa: primarily in Guinea, Liberia and Sierra Leone between 2013 and 2016, leading to over 10 000 deaths

- Kivu: primarily in Democratic Republic of the Congo and Uganda between 2018 and 2020, leading to over 2000 deaths
- Democratic Republic of Congo: smaller outbreak in the Democratic Republic of Congo (in a different province).

a. Ebola is classified as a re-emerging disease. Define the term re-emerging disease and highlight how this differs to an emerging disease. **2 marks**

b. Would the Western African outbreak be classified as a pandemic or epidemic? Justify your response. **1 mark**

c. Outline the evidence that suggests that Ebola is an infectious disease rather than a non-infectious disease. **2 marks**

Question 5 (4 marks)

In 1788, the First Fleet invaded and colonised Australia, land that was already owned and inhabited by Aboriginal peoples. The arrival of Europeans in Australia led to the introduction of many pathogenic agents that were not found on Australia previously.

a. Diseases such as smallpox, syphilis, measles and tuberculosis devastated the Aboriginal peoples, leading to a dramatic drop in population. However, it did not significantly affect the European colonists. Explain these observations. **2 marks**

b. The initial European contact with Torres Strait Islander peoples was made in 1606. Would the same observations about the effect of European diseases on Aboriginal peoples have been seen in Torres Strait Islander peoples? Explain your response. **2 marks**

More exam questions are available in your learnON title.

6.3 Identifying and controlling the spread of pathogens

KEY KNOWLEDGE

- Scientific and social strategies employed to identify and control the spread of pathogens, including identification of the pathogen and host, modes of transmission and measures to control transmission

Source: VCE Biology Study Design (2022–2026) extracts © VCAA; reproduced by permission.

6.3.1 Bringing outbreaks under control

The emergence of new diseases and the re-emergence of older diseases is a continual threat to public health. Outbreaks, epidemics and pandemics all have the potential to lead to a large loss of human life and long-term health consequences.

As well as this, epidemics and pandemics can have a massive effect on the economy, not just in the cost of preventing the spread of the disease, but because of the secondary consequences that quarantining individuals may have, such as the closure of businesses, growth in unemployment and decline in stockmarkets. Therefore, it is vital that pathogens be identified and controlled as quickly as possible, through both the rapid identification of the pathogen and **host** and a thorough understanding of the mode of pathogenic transmission.

When an outbreak of a serious disease occurs and begins to spread rapidly through person-to-person contact, there is a risk of an epidemic. An epidemic, if not brought under control, has the potential to develop into a pandemic. In this situation, the 'boots-on-the-ground disease detectives' of the Epidemic Intelligence Service (EIS) are quick to move into the affected region to work with local officials. The EIS is part of the US Centers for Disease Control and Prevention (CDC). Other groups, including the World Health Organization (WHO) also work to rapidly obtain answers and bring outbreaks under control. In Australia, the Australian Health Protection Principal Committee (AHPPC) coordinates the response to possible health crises.

host an organism that can get a disease

Personnel such as those from the EIS, WHO and AHPPC seek rapid answers to key questions such as:

- What is the cause of the sickness?
- If a pathogen is identified as the cause, how can it be treated and how can the disease be prevented from spreading?
- What measures are needed to prevent another outbreak?

FIGURE 6.7 Photograph taken in Liberia during the 2014 West African Ebola outbreak. A CDC staff member is being assisted by a Médecins Sans Frontières (MSF) staff member in an essential decontamination process before exiting an Ebola treatment unit. A hypochlorite spray is known to destroy Ebola virus.

Source: Centers for Disease Control

How do people get infected?

The epidemiological triad model (figure 6.8) is used to explain the three main components of disease causation. The model consists of:

- an agent or pathogen: the disease-causing organism
- a host: the target of a disease (i.e. an organism that carries and/or gets infected by an agent)
- the environment: conditions that allow the disease to be transmitted.

FIGURE 6.8 The epidemiological triad model

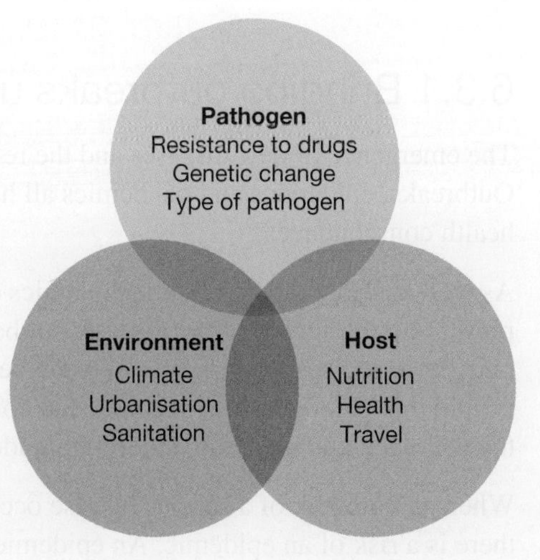

6.3.2 Identifying pathogens

When an infectious disease breaks out, speedy identification of the **causal pathogen** is critical. Knowing the identity of the pathogen guides decisions about treatment of the disease and preventing it from spreading.

Identifying viruses

Viruses, as non-cellular pathogens, were previously introduced in section 5.2.4 in topic 5. Unlike most bacteria, viruses cannot be replicated in standard microbiological broths or on agar plates. Instead, because they are obligate intracellular parasites, viruses require living cells in order to replicate. Viruses must be cultured inside suitable host cells.

causal pathogen a pathogen that is causing a specific disease under investigation

viruses non-cellular pathogens that use the host cell in order to replicate their genetic material

Finding a suitable host cell is often a difficult part of viral identification. The virus needs to be isolated from a patient sample. The sooner this can be done during a disease outbreak the better, as more time can be used to create vaccinations and treatments before the spread of the virus is out of control.

Once a host cell is found suitable, it is then grown and multiplied in the laboratory to produce a cell line. The cell line is then used to grow the virus. Usually, cell lines to grow the virus need to be able to easily replicate. Some cells lines that are used are kidney cells, fibroblasts from lungs, chicken eggs and carcinoma cells (such as the Hep G2 cell line).

Once a sample of the virus is available, identification can proceed.

A range of techniques can be used to identify viruses, and they include physical methods, immunological methods and molecular methods. Viruses must be grown in cell culture for these techniques to be used.

Physical methods

Physical methods can assist in identifying viruses based on size and shape.

These methods include:

- x-ray crystallography, which has determined the structure of many viruses
- electron microscopy, which has given us images that distinguish various kinds of virus. Figure 6.9 shows the contrasting shapes of Ebola virus and influenza A virus as revealed by transmission electron microscopy.

FIGURE 6.9 Transmission electron micrographs of viruses with differing shapes.
a. False-coloured Ebola virus b. Negative stained influenza A virus particle or virion

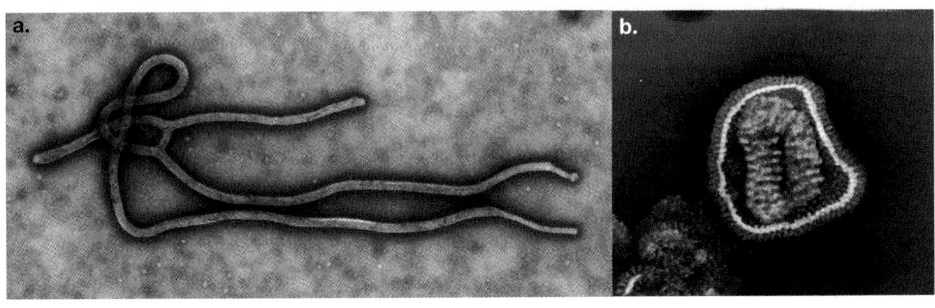

Source: Centers for Disease Control

Immunological methods

Immunological methods detect specific viral antigens or antibodies. One of the main techniques used is the enzyme-linked immunosorbent assay (**ELISA**) technique which allows for the diagnosis of diseases (including viral diseases).

There are many different types of ELISA technique. The main types, highlighted in figure 6.10 are:

- **direct ELISA**
- **indirect ELISA**
- **sandwich ELISA**.

physical methods methods of identifying pathogens based on size and shape

immunological methods methods used to diagnose pathogens based on the presence of specific antigens or antibodies

ELISA a technique known as enzyme linked immunosorbent assay, which can detect specific antigens or antibodies

direct ELISA a form of ELISA where a primary antibody with an enzyme indicator directly binds to an antigen attached to a surface

indirect ELISA a form of ELISA where a primary antibody attached to a secondary antibody with an enzyme indicator binds to an antigen attached to a surface

sandwich ELISA a form of ELISA where an antigen attached to an antibody with an enzyme indicator binds to an antibody attached to a surface

Direct ELISA

Direct ELISA is the simplest form of ELISA. In direct ELISA, a viral antigen is placed on a surface. Matching primary antibodies bind to this antigen. These primary antigens have an enzyme indicator directly attached to them (as shown in figure 6.10a). The steps of direct ELISA are as follows:

1. A plate (called microtiter plate) with wells is coated with an antigen (or toxins) specific to the disease being tested.
2. An antibody specific to a particular antigen is added to each well. This antibody also has an associated enzyme indicator.
3. During incubation, antibodies present in the sample bind to the antigen in the well.
4. The wells are then washed using a detergent solution to remove any unbound antibodies.
5. The substrate for the enzyme is added, leading to a colour change if an antigen–antibody complex formed. This indicates a positive test.

Indirect ELISA

Indirect ELISA is similar to direct ELISA. However, the primary antibody does not have an enzyme indicator. Instead, the enzyme indicator is attached to a secondary antibody. This secondary antibody is attached to the enzyme indicator. This can help amplify the signal, as multiply secondary antibodies can be used (as shown in figure 6.10b).

Sandwich ELISA

Another type of ELISA, known as sandwich ELISA, involves antibodies bound to the surface (as shown in figure 6.10c). This differs to direct and indirect ELISA, which use bound antigens. The steps in this process are as follows:

1. A 'capture' antibody is used to identify the presence of a specific viral antigen through an antigen–antibody reaction.
2. All unbound material is washed away with a detergent solution.
3. A second antibody with an enzyme indicator is then added. This binds to the antigen. As such, the antigen is 'sandwiched' between two antibodies.
4. The substrate for the enzyme indicator is added. If colour appears, the specific viral antigen is present.

Often, rather than having the enzyme indicator on the antibody that binds to the antigen, an additional antibody is added with this indicator. This binds to the antibody (in a similar way to indirect ELISA).

FIGURE 6.10 The ELISA technique can be used to detect either viral antigens or antibodies for the virus. **a.** Direct ELISA uses an antigen captured on a support and only one antibody with an enzyme indicator. **b.** Indirect ELISA also uses an captured antigen, but involves a primary and secondary antibody. **c.** Sandwich ELISA involves an antibody captured on the surface to detect antigens.

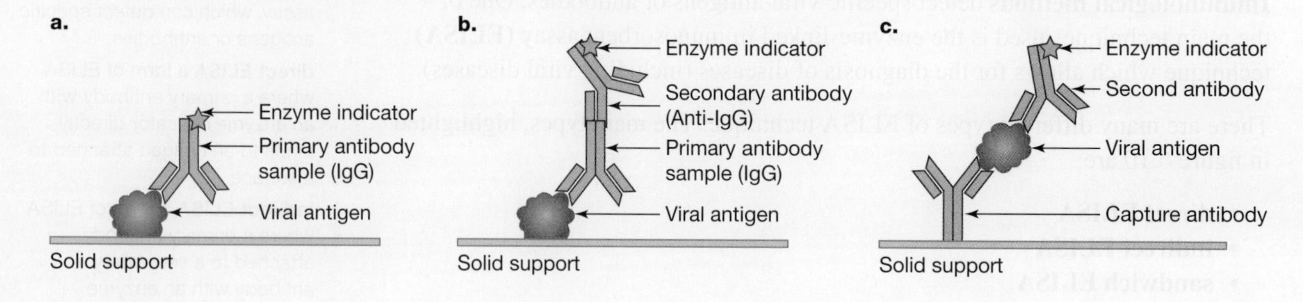

CASE STUDY: Using ELISA to diagnose HIV

The ELISA test can detect the presence of antibodies to specific viral disease in cells or body fluids (see figure 6.11). This is exemplified in the screening test used to detect the presence of antibodies to human immunodeficiency virus (HIV) in a person's serum.

In this ELISA screening test, a serum sample from the person to be tested is diluted 400-fold and applied to a plate on which HIV antigens are immobilised. If antibodies to HIV are present in the serum, they can bind to the HIV antigens. All unbound material is washed away. A second antibody linked to an enzyme is then added. Because of its high sensitivity and ease of application, ELISA is a powerful tool in the identification of viruses and is used routinely to test and diagnose many viral diseases, including HIV.

FIGURE 6.11 Results from an ELISA test. Positive results are shown in blue.

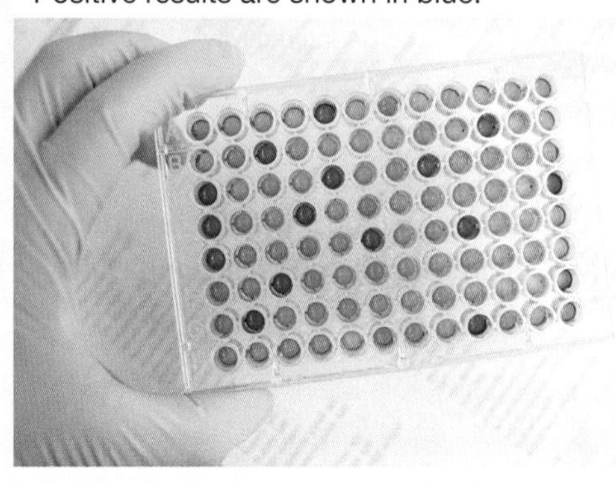

Molecular methods

Molecular techniques can be used to identify viruses.

Viruses are diverse. One way in which they are diverse is in their nucleic acids. Viruses can have either DNA or RNA, and this can be single-stranded or double-stranded (refer to table 5.5 in subtopic 5.2). The specific sequences of certain viruses are known and can be used to confirm their identities.

Molecular techniques include the use of in situ hybridisation with **probes** to detect and locate specific genetic sequences that are diagnostic of particular viruses. This involves using a short radioactively labelled strand of nucleic acid (a probe) to bind to a specific sequence in tissue (in situ) through complementary base pairing. If the specific sequence is present, this can be easily located through the radioactive label on the probe.

FIGURE 6.12 Areas of virus assembly within the cell are pink. Host and viral DNA (deoxyribonucleic acid) is blue.

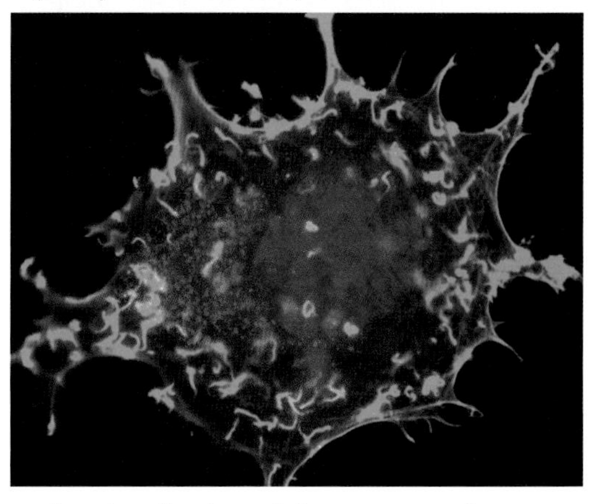

Reverse techniques can be used to identify RNA viruses. This allows for visualisation of the presence of different viruses and the type of viruses present.

DNA sequencing can also be used to assist in the identification of viruses and their different strains. DNA sequencing involves determining the sequence of DNA using a specialised machine (up to 400 to 600 million bases over 10 hours). This involves adding nucleotides with coloured dyes (a different dye for each nucleotide is used). These attach to the DNA sequence. Computers can then analyse and construct the DNA sequence from this data, as shown in figure 6.13.

FIGURE 6.13 A sample of a section of sequenced DNA

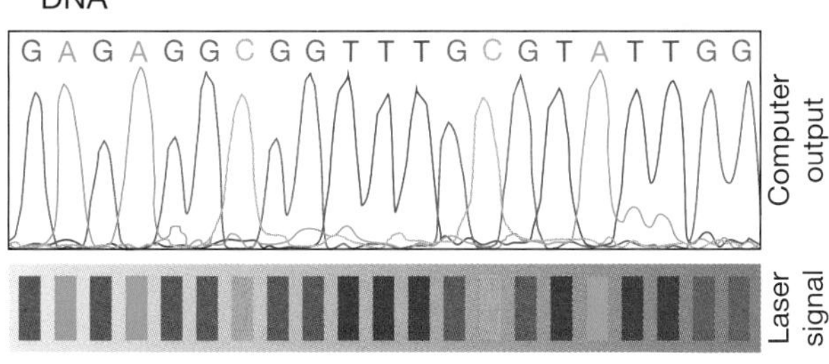

molecular techniques methods using DNA or RNA to identify a pathogen

probe single-stranded segment of DNA (or RNA) with a radioactive label that locates and binds to a target sequence

on Resources

- **Video eLesson** Sandwich ELISA (eles-4364)
- **Weblinks** Types of ELISA
  ELISA interactivity

Identifying bacteria

Identification of bacterial species is important for several reasons, such as deciding on antibacterial therapy for patients or identifying the clinical significance of bacterial infections.

> Various techniques can be used to identify bacteria and they fall into three categories: phenotypic, immunological and genotypic methods.

Phenotypic methods

Phenotypic methods use techniques that involve identifying particular traits or features in bacteria. They include:

- use of microscopy to differentiate bacteria on the basis of differences in cell shape, size and response to Gram stain, and physical features such as the presence or absence of a capsule. (Many of these differences were introduced in section 5.2.3 in topic 5. Refer to the upcoming case study box to learn more about Gram staining).
- use of a range of biochemical tests eliciting different bacterial responses
- use of different media to differentiate bacteria on the basis of variation in growth patterns. This may be done by observing growth on non-selective, selective and differential media.
  - Non-selective media or agar can be used to detect and count the number of bacteria in the sample.
  - Selective media contain compounds, such as antibiotics or growth nutrients, that selectively inhibit or enhance the growth of specific bacteria. For example, selective media can be used to distinguish aerobic bacteria from facultative anaerobes (which can survive with or without oxygen) and obligate anaerobes (which can only survive in anaerobic conditions).
  - Differential media contain a substrate that, under the action of an enzyme, produces a coloured or fluorescent product. This can identify bacteria according to various chemical reactions that are carried out during growth.

phenotypic methods techniques to differentiate bacteria on the basis of microscopy, different media and biochemical tests

FIGURE 6.14 Numerous biochemical tests being conducted at once to identify *Listeria monocytogenes*

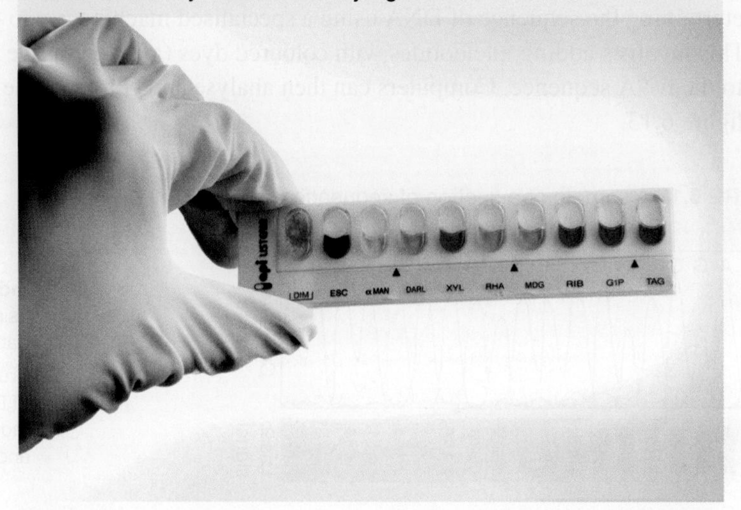

An example of the process used to identify bacteria based on different properties is shown in figures 6.15 and 6.16.

FIGURE 6.15 Diagram showing the identification of four species of Gram-negative bacteria based on their reactions to several biochemical tests

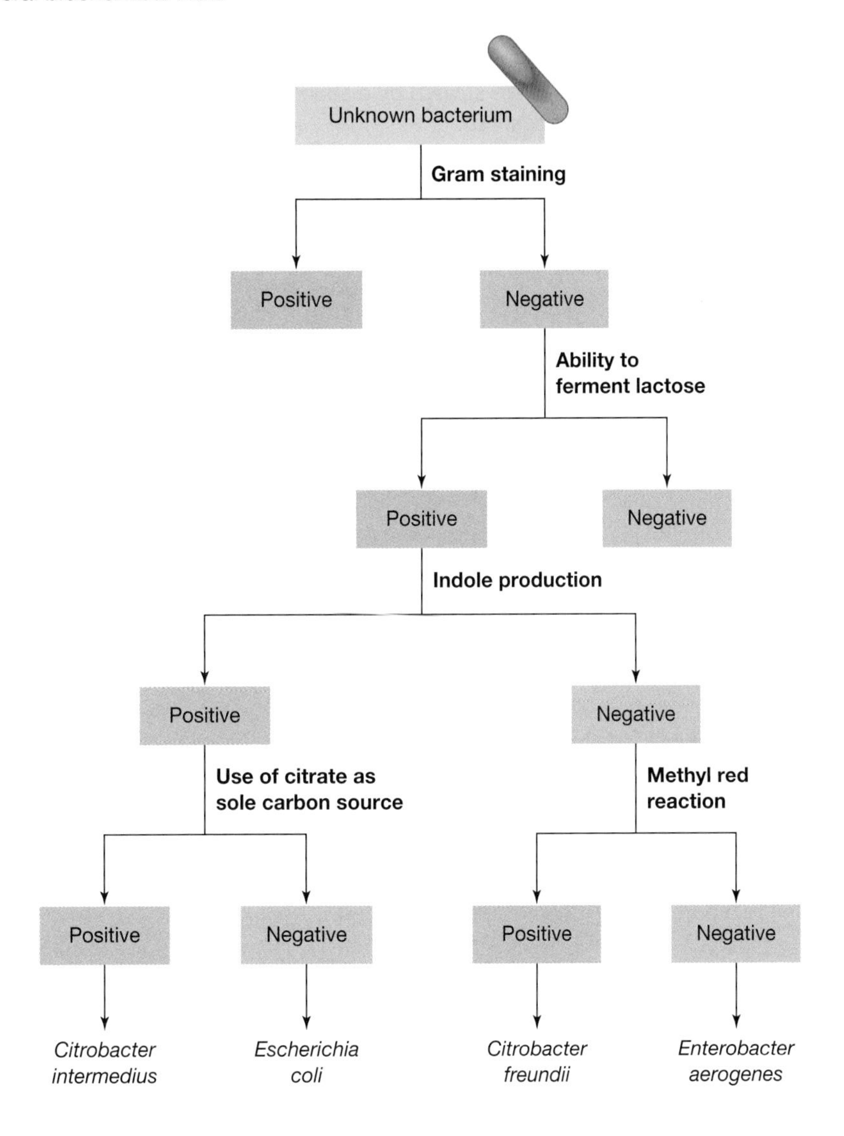

FIGURE 6.16 a. Positive indole test b. Methyl red test (red is positive) c. Citrate test (blue is positive)

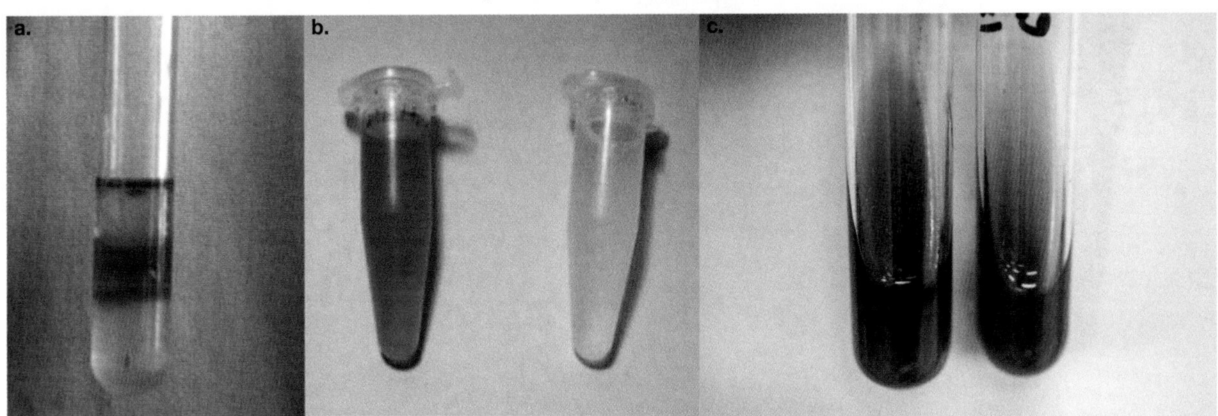

CASE STUDY: What is Gram staining?

A special stain, known as Gram stain, is commonly used for the general identification of bacteria. This stain separates bacteria into two main groups, Gram-positive and Gram-negative, depending on the structure of their cell wall.

This staining technique involves the use of two dyes as outlined in figure 6.17. Some bacteria retain the first dye (crystal violet) and appear dark blue — these bacteria are said to be Gram-positive. In contrast, other bacteria do not retain the crystal violet dye and instead are stained by a second dye (safranin) so that they appear red or pink — these bacteria are said to be Gram-negative.

FIGURE 6.17 Steps in the Gram staining technique

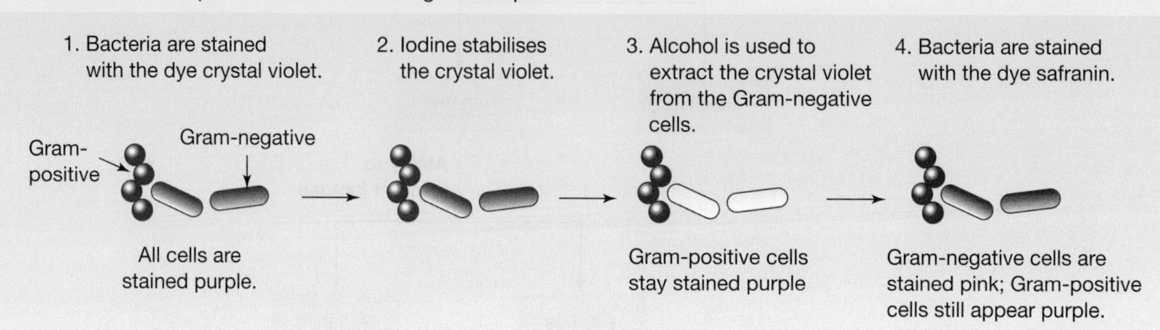

The contrasting responses of bacteria to Gram staining are due to differences in the structure of the external cell wall that lies outside the plasma membrane of bacterial cells. The cell wall of Gram-positive bacteria contains a thick layer of **peptidoglycan** (as shown in figure 6.18b), which retains the crystal violet dye used in the Gram staining technique.

Gram-negative bacteria have an outer membrane made of lipopolysaccharide (LPS) (see figure 6.18b). This LPS outer membrane enables these bacteria to expel or exclude certain drugs and antibiotics.

Due to the presence of this LPS outer membrane, Gram-negative bacteria are generally far more resistant to antibiotic treatment than are Gram-positive bacteria. Gram-positive bacteria are normally susceptible to antibiotics, such as penicillin and the sulfonamide drugs, but these antibiotics are not effective against Gram-negative bacteria. So, the result of Gram staining can give an indication of the type of antibiotic treatment that could usefully be given to a person with a bacterial disease.

FIGURE 6.18 a. Light microscope (LM) image showing the results of Gram staining of two kinds of bacteria. The majority of the bacteria are Gram-positive cocci stained purple, while the other bacteria are Gram-negative and are stained pink. **b.** Differences in Gram-positive and negative bacteria

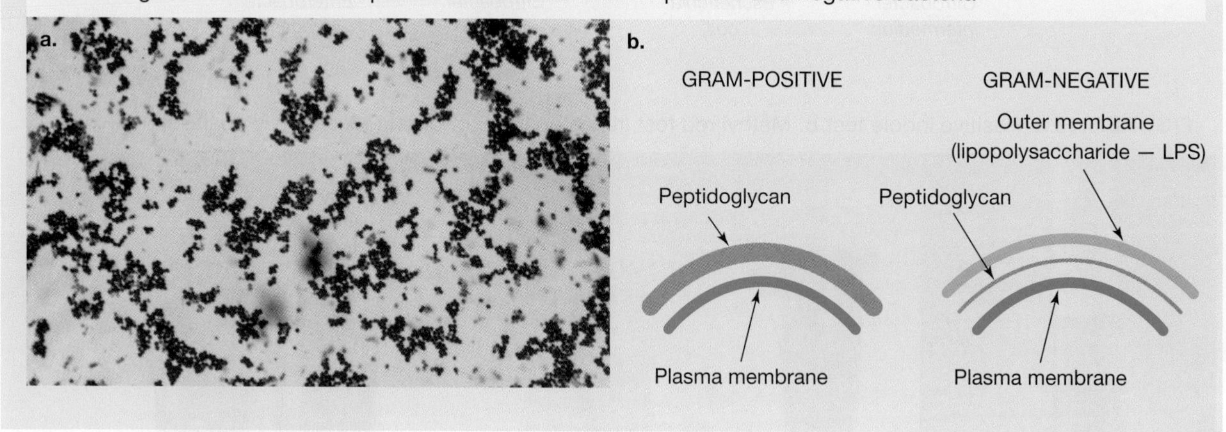

peptidoglycan a polymer consisting of sugars and amino acids that forms a major part of the cell wall of Gram-positive bacteria

If a bacterial infection is suspected, Gram staining is carried out on either body fluids or cell samples. This procedure is a quicker method of identifying the presence of bacteria than that of culturing them (growing bacteria on a medium such as nutrient agar).

TABLE 6.4 Examples of Gram-positive and Gram-negative bacteria

Gram-positive bacteria	Gram-negative bacteria
Streptococcus pyogenes	*Escherichia coli*
Staphylococcus pneumonia	*Vibrio cholerae*
Staphylococcus aureus	*Yersinia pestis*
Clostridium botulinum	*Neisseria meningitidis*
Clostridium tetani	*Helicobacter pylori*
Bacillus anthracis	*Legionella pneumophila*
Corynebacterium diphtheria	*Treponema pallidum*

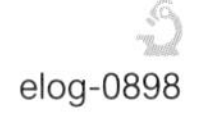
elog-0898

INVESTIGATION 6.1 online only

Gram staining and biochemical testing

Aim

To identify different types of bacteria using various phenotypic methods

Genotypic and molecular methods

Genotypic and molecular methods involve the examination of the genetic material of bacteria and use techniques such as gene probes, sequence analyses and plasmid fingerprinting to identify bacteria. Because of their speed and accuracy, genotypic techniques are increasingly important in the identification of bacteria and identification of the various **serotypes** of a single bacterial species. Figure 6.19 shows one example of a labaratory test to detect different serotypes. These serotypes of *Streptococcus* are tested on cards. Solutions containing antibodies that target specific parts of each serotype are added. If a sample is a certain serotype, a positive result is caused by the binding of the antibody with the specific antigen. The three main techniques used for serotyping are as follows:

- Gene probes are specifically designed, radioactively labelled sequences of nucleic acids that bind to specific genes. This may be a gene that is found in a certain type of bacterium or a specific serotype.
- Sequence analyses involve determining the order of sequences of the nucleotides in the bacterial DNA.
- Plasmid fingerprinting involves using DNA profiling techniques (similar to those explored in topic 2) to identify the genetic profiles of specific plasmids and thus determine a bacterial species and strain. This can allow for the bacterial species and strain to be determined.

FIGURE 6.19 Identifying *Streptococcus* serotypes

genotypic and molecular methods methods of identifying bacteria by examining its genetic material

serotypes variants within a species of bacterium or virus that are distinguished by their surface antigens

Immunological methods

Immunological methods use techniques including monoclonal antibodies, ELISA and immunofluorescence to identify bacteria. These immunolobical methods are used in the following ways:

- Monoclonal antibodies are antibodies that are designed to have a specific antigen-binding site. These can be formed to target and bind to particular bacterial antigens. Monoclonal antibodies will be further explored in subtopic 6.5.
- ELISA was introduced in the subsection on identifying viruses, and works in a similar way for bacteria (refer back to figure 6.10). ELISA for bacteria can be used not only to detect antigens or antibodies, but also to detect toxins specific to a certain bacterium. Direct, indirect and sandwich ELISA may all be used for this. Another type of ELISA, known as reverse ELISA, may also be used to identify particular strains of bacteria. This technique does not use traditional wells, but leaves the antigens suspended in fluid.
- Immunofluorescence, in a similar way, uses an antibody with a fluorescent marker to bind to and detect specific antigens or antibodies in serum.

tlvd-1783

SAMPLE PROBLEM 2 Using ELISA and immunological methods

ELISA is a technique designed for detecting substances such as peptides, proteins, antibodies and hormones using antibodies and colour changes. ELISA is a common medical and research lab technique.

a. **Why is ELISA able to be used for both the detection of viruses and of bacteria?** **(2 marks)**

b. **Describe how ELISA can be used in to detect substances.** **(4 marks)**

c. **Other than ELISA, phenotypic techniques can be used in the detection of bacteria. Provide two examples of phenotypic techniques.** **(2 marks)**

THINK	WRITE
a. 1. Consider factors that both viruses and bacteria have in common that would be useful for ELISA. Outline the selected factor.	Both viruses and bacteria have antigens that stimulate an immune response and the production of antibodies (1 mark).
2. Describe how this would be linked to ELISA.	Therefore, ELISA works for both of these, as the specific antigens and antibodies can be used to allow a positive or negative result to be observed (1 mark).
b. 1. The question is asking you to describe and is worth four marks. Therefore, you need to consider four key factors in how ELISA is used. As this question does not mention direct, indirect or sandwich ELISA, you may select which you describe. 2. Consider the factors involved in ELISA. You should ensure you mention both the antigens and antibodies, the antibody-antigen complex and the use of a substrate to signal a positive test. *Note:* While these have been shown in dot points for clarity, you may write a clear paragraph with four distinct points, with clear interlinking of ideas.	• A plate with wells is coated with an antigen (bacterial toxin) specific to the disease being tested (1 mark). • An antibody specific to a particular antigen is added to each well. During incubation, antibodies present in the sample bind to the antigen in the well (1 mark). • Next, enzyme- linked secondary antibodies are added to the well. If the antigen is present, then the secondary antibodies bind to the antigen (1 mark). • A substrate is added which reacts with enzymes to produce a colour change, indicating a positive test (1 mark).

c. 1. This question asks only for examples and not an explanation. You should ensure you only follow the instructions of the question and do not provide extra information that may be incorrect.
2. Consider aspects of phenotypic testing: biochemical tests, Gram staining, growth in different media and microscopy.

- Testing the growth of bacteria in different media
- Biochemical tests (such as indole production)

Other pathogenic agents

While bacteria and viruses are some of the more common causes of diseases that are easily transmissible, they are not the only source of infections. In similar ways, other pathogens such as fungi, protozoa, prions and parasites need to be quickly identified to prevent the spread of disease. This can be done by identifying particular physical features, by genetic and molecular means or immunological techniques.

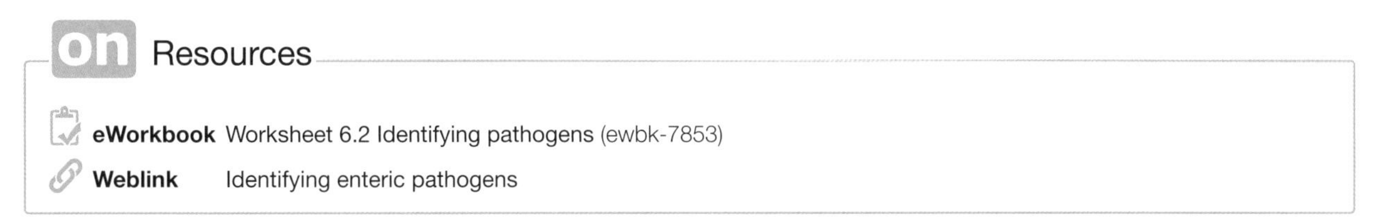

on Resources

eWorkbook Worksheet 6.2 Identifying pathogens (ewbk-7853)

Weblink Identifying enteric pathogens

6.3.3 Identifying the host

Reservoirs and hosts

Transmission occurs when a pathogen-agent leaves its **reservoir** or host through a point of exit, is transmitted and enters through a point of entry to infect a susceptible host.

It is important to be able to identify both the reservoir and/or the host when exploring infectious diseases. This can assist with **quarantine** (isolating people who are infectious) or through other preventative measures (such as the use of insect repellent in areas where mosquito-borne infections are prevalent).

transmission passing a pathogen on to another individual

reservoir the habitat in which a pathogen lives, grows and multiplies

quarantine the act of isolating infected individuals to prevent the spread of a disease

FIGURE 6.20 The chain of infection from reservoir to hosts

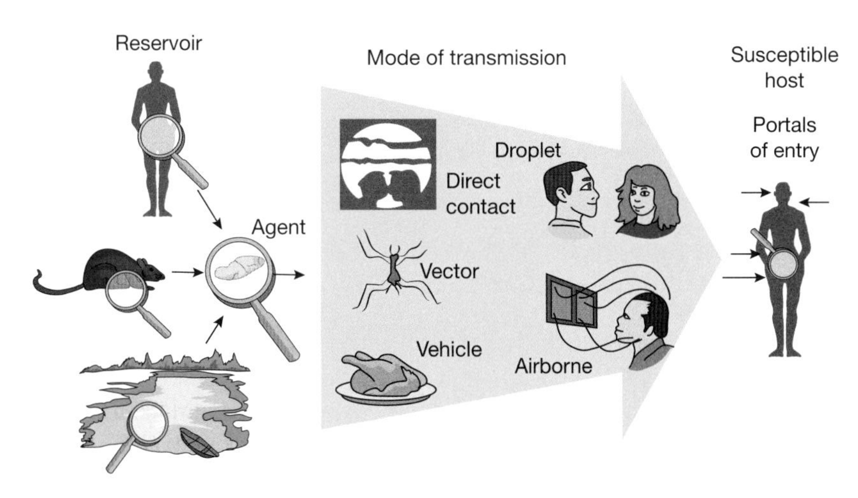

The reservoir of a pathogen is the habitat in which it lives, grows and multiplies. Reservoirs may or may not be the source from which the pathogen is transferred to its host. Reservoirs do not experience symptoms of the disease.

Reservoirs can include:
- humans
- animals: **zoonotic diseases** are infectious diseases that are transmitted from animals to humans. Some examples of zoonotic diseases include plague (rodents), anthrax (sheep) and rabies (dogs and other mammals).
- the environment: plants, soil and water are also reservoirs for some infectious agents.

A host refers to the organism who can get the disease (such as humans). Many infectious agents can infect more than one host (for example, swine flu was able to infect both humans and pigs).

Susceptibility of a host depends on many factors, including:
- genetic factors: an individual's genetic condition affects their susceptibility to infection. For example, an individual suffering from sickle cell anaemia is less likely to get malaria.
- specific immunity: initial exposure to a pathogenic agent provides protection to the host in the form of specific antibodies. Such antibodies develop during secondary exposure to the pathogen and provide protection to the host.
- sex
- age
- nutrition.

Index cases

During an outbreak, epidemic or pandemic, it is vital to be able to determine the host and reservoir of a disease.

Often, this involves finding 'patient zero'. Patient zero or the **index case**, is the first individual to have a case of an infectious disease. This allows for not only the spread of an infectious disease to be better tracked, but also allows for the pathogenic agent (and other possible reservoirs) to be detected as early as possible.

It can be quite difficult to determine patient zero, particularly if a disease has a long incubation period. Often, many people are diagnosed around a similar time, such as in the case of COVID-19. In the case of Ebola, patient zero was more easily traced. This was also the case of an outbreak of typhoid fever, where patient zero was determined to be Mary Mallon (read about this in the case study box in 6.3.4).

While finding patient zero can help, it is more important to identify how a pathogen can be spread, and what organisms are able to spread a pathogenic agent, in order to improve preventative measures. This is often done through epidemiological studies.

6.3.4 Modes of pathogen transmission

If a pathogen can gain entry to the human body and reach the target cells, it may multiply rapidly and produce an infection. If the body's immune system cannot overcome an infection, the infection will develop into a disease. So, an infection is not equivalent to a disease, but infection is a necessary pre-condition for an infectious disease.

Before an infection can develop into a disease, several events must occur. The pathogen, whether bacterial, fungal or viral must first:
- gain entry to the body and reach target site(s) in the body; typically, the portals of entry are the skin and the mucous membranes of the respiratory, uro-genital, and digestive systems of the body
- overcome the defence mechanisms of the body

zoonotic diseases diseases that have been transferred from other animals
index case the first individual known to have a case of an infectious disease

- become established at one or more sites
- multiply rapidly, causing harm to the host and producing the symptoms of the disease.

Modes of transmission

Infectious diseases may spread from infected people to healthy people by various means:

- by **direct transmission**, such as by person-to-person contact, through kissing or sexual contact (e.g. chickenpox, chlamydia and conjunctivitis)
- by **indirect transmission**, such as:
  - by airborne droplets or particles, such as an uncovered sneeze or cough (e.g. influenza, rotavirus and coronavirus)
  - by contact with contaminated objects, such as bedding, cups or medical instruments (e.g. glandular fever or tetanus)
  - by ingestion of contaminated food or water (e.g. salmonella, cholera and gastroenteritis)
  - by biological vehicles, such as contaminated blood, sputum, or faeces (e.g. HIV and hepatitis B and C)
  - by **vectors** that carry pathogenic agents and spread them to people through bites from infected ticks, mites, fleas or mosquitoes, or through contaminated particles that they leave on material, such as fly droppings on food. (e.g. bubonic plague, psittacosis, anthrax, West Nile virus disease and rabies). Many of these vectors transmit pathogens from their natural host (e.g. birds, bats or rodents) to people. These diseases are referred to as zoonotic diseases or zoonoses.

Many diseases have multiple forms of transmission. For example, glandular fever can be spread both through indirect means by being exposed to infected saliva (such as contaminated objects like drink bottles) or through direct contact such as kissing.

FIGURE 6.21 Modes of transmission

When can pathogens be transmitted?

Often when individuals feel unwell, they are recommended to self-isolate to prevent the spread of pathogenic agents. This idea of self-isolation was a common occurrence during the COVID-19 pandemic, when unwell individuals were urged to get tested and stay home. However, the issue with the spread of diseases is that individuals can be infectious without showing symptoms.

direct transmission mechanism of transmission of pathogenic agents that involves direct person-to-person contact, such as by kissing or sexual contact

indirect transmission mechanism of transmission of pathogenic agents that does not involve direct person-to-person contact, such as by airborne droplets or by ingestion of contaminated food

vectors organisms that carry pathogenic agents and spread them to other organisms

Some pathogens can be transmitted from an infected person to a healthy person only after the infected person shows visible symptoms of the disease. These pathogens include those responsible for diseases such as Ebola virus disease, typhoid and pertussis. In contrast, other pathogens, including those causing diphtheria, rubella, influenza, measles and tuberculosis, can be transmitted, not only by people showing obvious disease symptoms, but also by infected people during the incubation period when they show no symptoms of the disease.

Incubation period

The period after infection and before the first symptoms of a disease appear is called the **incubation period**. An incubation period is the interval between a person's exposure to a pathogen and the onset of disease symptoms in that person. During the incubation period, the disease-causing agent multiplies to concentrations that are sufficient to produce the symptoms of the disease.

TABLE 6.5 Incubation periods for several diseases

Disease	Pathogen	Incubation period
Salmonellosis	Bacteria	6 to 72 hours
Influenza	Virus	1 to 3 days
Common cold	Virus	1 to 3 days
Coronavirus (COVID-19)	Virus	2 to 14 days
Typhoid	Bacteria	8 to 14 days
Measles	Virus	10 to 21 days
Tuberculosis	Bacteria	2 to 12 weeks
HIV	Virus	Weeks to months
Kuru	Prion	10 to 13 years

Asymptomatic carriers

In addition, some people can be infected by a pathogen but be in good health and never show any signs or symptoms of the disease concerned. Such people are said to be **asymptomatic carriers** of the pathogen concerned and they can be a source of infection of people with whom they come in contact. Perhaps the most famous asymptomatic carrier of an infectious disease was Mary Mallon (1869–1938), who was infected with the *Salmonella typhi* bacteria, which cause typhoid fever.

incubation period the period between infection and the first appearance of the symptoms of a disease

asymptomatic carrier person with an infectious disease showing no symptoms but able to infect others

CASE STUDY: Typhoid Mary

Mary Mallon (1869–1938) worked as a cook for several wealthy families in New York City in the early 1900s. In each household where she was employed, family members came down with typhoid fever. Health authorities recognised the connection between Mary and the outbreak of typhoid fever in the family members in that household. This was the first time that an asymptomatic carrier of a pathogen had been identified. Mary is presumed to have transmitted the bacteria to about forty-seven people, three of whom died from typhoid fever.

To learn more about Typhoid Mary, please download the digital document.

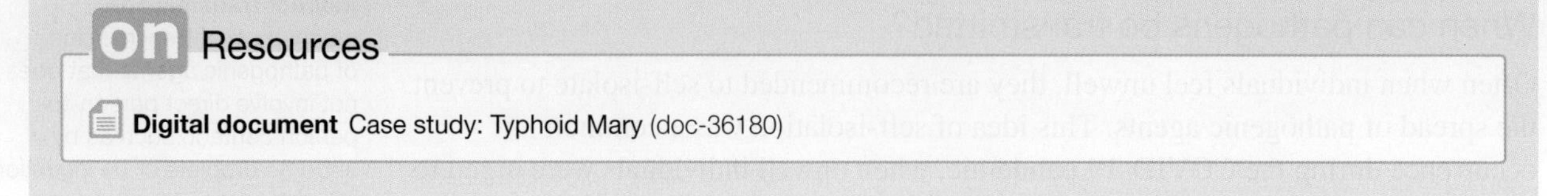

on Resources

Digital document Case study: Typhoid Mary (doc-36180)

6.3.5 Measuring the spread of a pathogen

Monitoring the spread

Part of controlling the spread of disease is through careful surveillance to measure and determine the extent of the spread of a pathogen. As part of this, the WHO often categorises the disease into different alert phases. These phases allow for the appropriate social and scientific courses of action to be undertaken. The earlier the intervention is made, the quicker a disease can be identified and controlled, reducing the chance of a global pandemic. An example of these phases in influenza is shown in the case study box. The WHO (and other health organisations) aim to intervene as soon as possible to avoid pandemic alert periods and pandemic periods.

CASE STUDY: Influenza and pandemic alert phases

TABLE 6.6 The WHO pandemic alert phases for influenza

WHO phases	
INTER-PANDEMIC PERIOD	
1	No new influenza virus subtypes have been detected in humans. An influenza virus subtype that has caused human infection may be present in animals. If present in animals, the risk of human disease is considered to be low.
2	No new influenza virus subtypes have been detected in humans. However, a circulating animal influenza virus subtype poses a substantial risk of human disease.
PANDEMIC ALERT PERIOD	
3	Human infection(s) detected with a new subtype, but no human-to-human spread, or at most rare instances of spread to a close contact.
4	Small cluster(s) detected with limited human-to-human transmission but the spread is highly localised.
5	Larger cluster(s), but human-to-human spread is still localised, suggesting that the virus is becoming increasingly better adapted to humans, but may not yet be fully transmissible (substantial pandemic risk).
PANDEMIC PERIOD	
6	There is increased and sustained transmission in the general population.

Quantifying the spread using R_0 values

The spread of a pathogen is often quantified using the **R_0 value** (the basic reproduction number). This value shows the expected number of individuals that are infected by one individual.

TABLE 6.7 Example R_0 values of different pathogens

Infectious disease	R_0
COVID-19 (Coronavirus)	3.5
SARS	4
Measles	18
HIV	4
Ebola	2

R_0 value the basic reproduction number that identifies the expected number of individuals a person with a certain disease will infect

FIGURE 6.22 Comparing the spread of diseases with different R_0 values

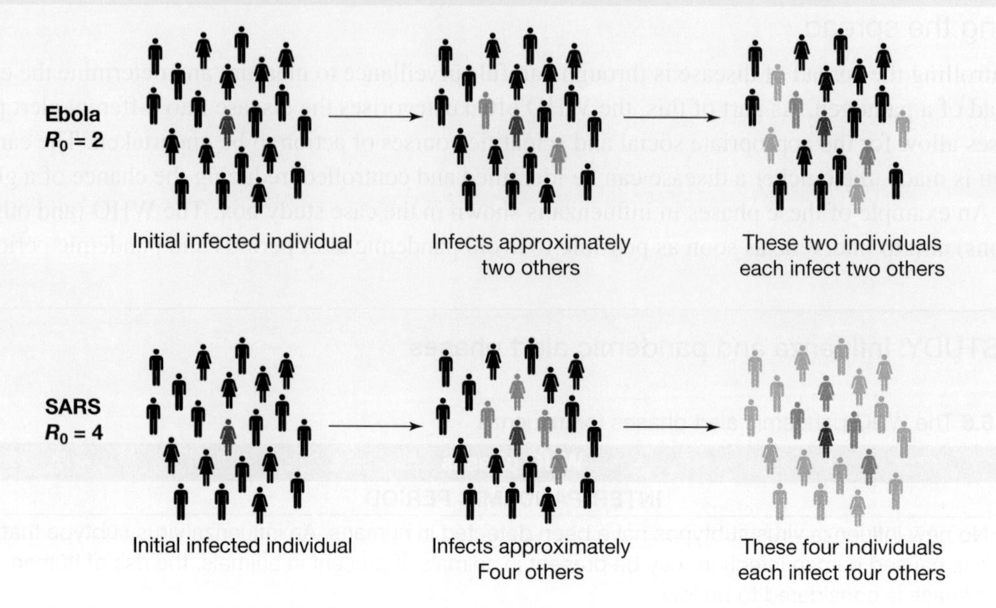

6.3.6 Controlling the spread of pathogens

Controlling the spread of pathogens includes various measures that prevent and contain the spread of infectious disease.

There are many factors that contribute to the spread of disease, including the climate, diet and availability of medical aid. By implementing the change in behaviour and human intervention, the rate of disease transmission can be controlled or inhibited.

Methods of disease control

- *Prevention*. The transmission of disease can be prevented by changing behaviours. Examples include practising personal hygiene such as washing hands, using condoms to prevent the spread of sexually transmitted diseases and using insect repellent to prevent the spread of a particular disease by vectors. The access to improved sanitation and clean drinking water is also a vital measure to prevent against diarrhoeal and parasitic diseases, such as cholera.
- *Vaccination.* Vaccination is a way of providing long-term protection against infectious diseases. Vaccines help in preventing and, in some cases, eradicating diseases. For example, child immunisation schedules have resulted in the dramatic decline of diseases such as measles and polio.
- *Medication.* Antibiotics for bacterial infections are just one of many medications now used to manage infectious diseases.
- *Surveillance.* The global monitoring of disease outbreaks is another tool used to control the spread of several diseases.
- *Modification of the environment*. The environment can be made less suitable for the microbes to grow and be transmitted. Examples may include vector control like in case of malaria.
- *Infection control standards.* These help in preventing the spread of infectious diseases, and include sterilisation, isolation (quarantine) and proper hygiene.

FIGURE 6.23 Mechanisms of controlling the spread of disease

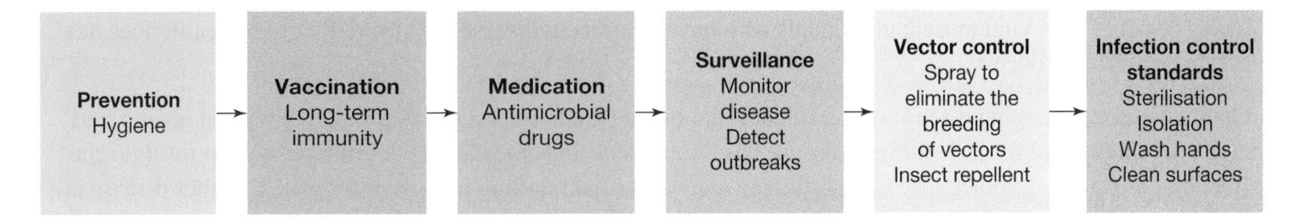

FIGURE 6.24 Preventative measures such as mosquito nets and the use of quarantine can help prevent the spread of infection.

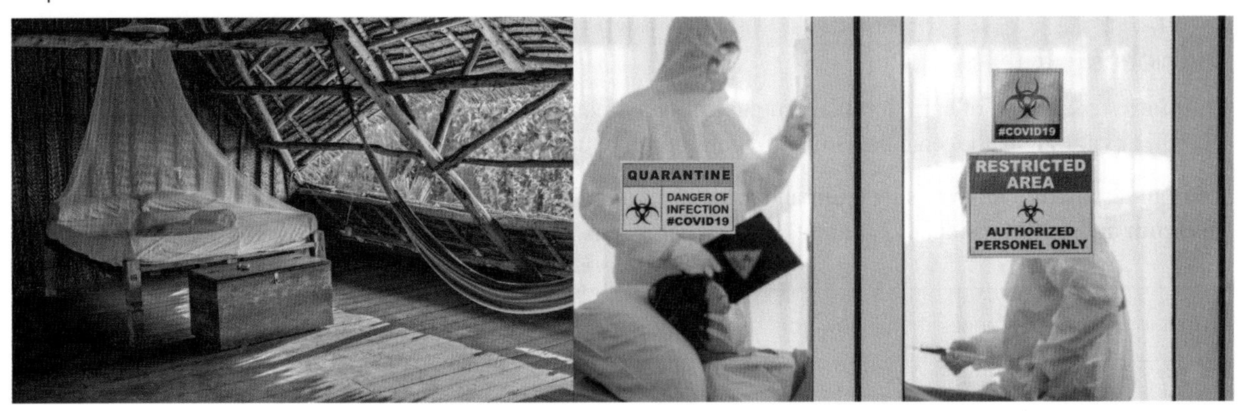

tlvd-1785

SAMPLE PROBLEM 3 Analysing methods to prevent the spread of disease

In recent times, COVID-19, a respiratory disease caused by a coronavirus (COVID-19) resulted in many countries, including China, Italy, the UK, Australia and Spain being placed into lockdown. Numerous workplaces and schools were closed and large events (including the Australian Grand Prix, AFL and NRL seasons and theatre programs) were postponed or cancelled.
Outline why the lockdown of countries, including a ban on all travel and closing of major events and busy locations, would be used to help prevent the spread of disease. (2 marks)

THINK	WRITE
1. Carefully examine the question. The first aspect you need to address is why travel bans would prevent disease spread. Ensure you specifically link your answer back to COVID-19.	Isolating individuals by banning travel reduces the chance that COVID-19 can spread to other individuals. Some people may leave a location, not knowing they are infected, and spread it globally. By stopping travel, the chance of coronavirus travelling further is reduced (1 mark).
2. The second aspect you need to address is why closing major events and locations would be helpful to control the disease. Again, ensure you specifically link your answer back to COVID-19.	Closing major locations and events also reduces the chance that an infected individual will pass the coronavirus causing COVID-19 to other individuals, particularly in a location with a large number of people such as a sportsevent. Usually, an infected individual will spread COVID-19 to at least two other people. This number would be much higher if there were large numbers of individuals in close proximity (1 mark).

Defence against infection

While prevention is always the preferred method to combat the spread of infection, in many cases this is not always possible. It is vital to treat individuals who have contracted diseases to ensure that their health does not decline or mortality occurs.

When a novel or emerging disease is identified, it is often a race to not only find a vaccination, but also to find an effective treatment for it. Not every disease has a treatment. In some cases, the immune system resolves the infection on its own. In other cases, the pathogen can be dormant, hiding in cells and causing further disease at a later date (such as with HIV and herpes). Sometimes, the only option is to treat the symptoms rather than the disease, to ensure an individual is in the best health possible for their immune system to fight the infection.

Treatments for bacterial and viral infections are different: antibiotics are used to treat bacterial infections, and antiviral agents are used for viral infections.

Antibiotics

Antibiotics are a class of antimicrobial drug used in the treatment and prevention of bacterial infections. They act either by killing pathogenic bacteria or by inhibiting their growth.

Some antibiotics are **narrow spectrum** and act against a limited variety of microorganisms; others are **broad spectrum** and act against many different kinds of pathogens.

Antibiotics are substances that, in low concentrations, inhibit the growth of or kill microorganisms.

Antibiotics can be:
- naturally produced by other microorganisms (such as penicillin produced by *Penicillium* moulds)
- semi-synthetic and produced partially by chemical synthesis (such as ampicillin, which is derived from penicillin)
- synthetic and produced wholly by chemical synthesis (such as sulphonamides).

Actions of antibacterial drugs

The general action of an antibiotic is either by directly killing microorganisms (bactericidal) or by inhibiting their growth (bacteriostatic).

Different antibiotics act on different bacterial targets (refer to figure 6.25). For example:
- penicillin inhibits cell wall synthesis in bacteria and so targets actively reproducing bacteria. Since human cells do not have peptidogly can cell walls, penicillin does not attack human cells and so has low toxicity.
- chloramphenicol, erythromycin, tetracyclines and streptomycin inhibit protein synthesis by acting on the ribosomes (70S) of prokaryotic cells.
- sulfanilamide acts as an antimetabolite by competitively inhibiting enzyme activity in bacteria.
- rifampin and quinolones inhibit nucleic acid synthesis.

Using sensitivity tests to determine the best antibiotic

One of the biggest crises we face is antibiotic resistance in many bacteria. Many strains are becoming resistant to certain antibiotics, which creates challenges for their treatment and management (this resistance will be further explored in subtopic 7.7). To control the spread of an infection, it is vital to use an antibiotic that will kill the particular bacterium.

antibiotics a class of antimicrobial drug used in the treatment and prevention of bacterial infections that act either by killing pathogenic bacteria or by inhibiting their growth

narrow spectrum antibiotics that act against a limited variety of microorganisms

broad spectrum antibiotics that act against many different kinds of pathogens

FIGURE 6.25 Various targets of different antibiotics on pathogenic bacteria

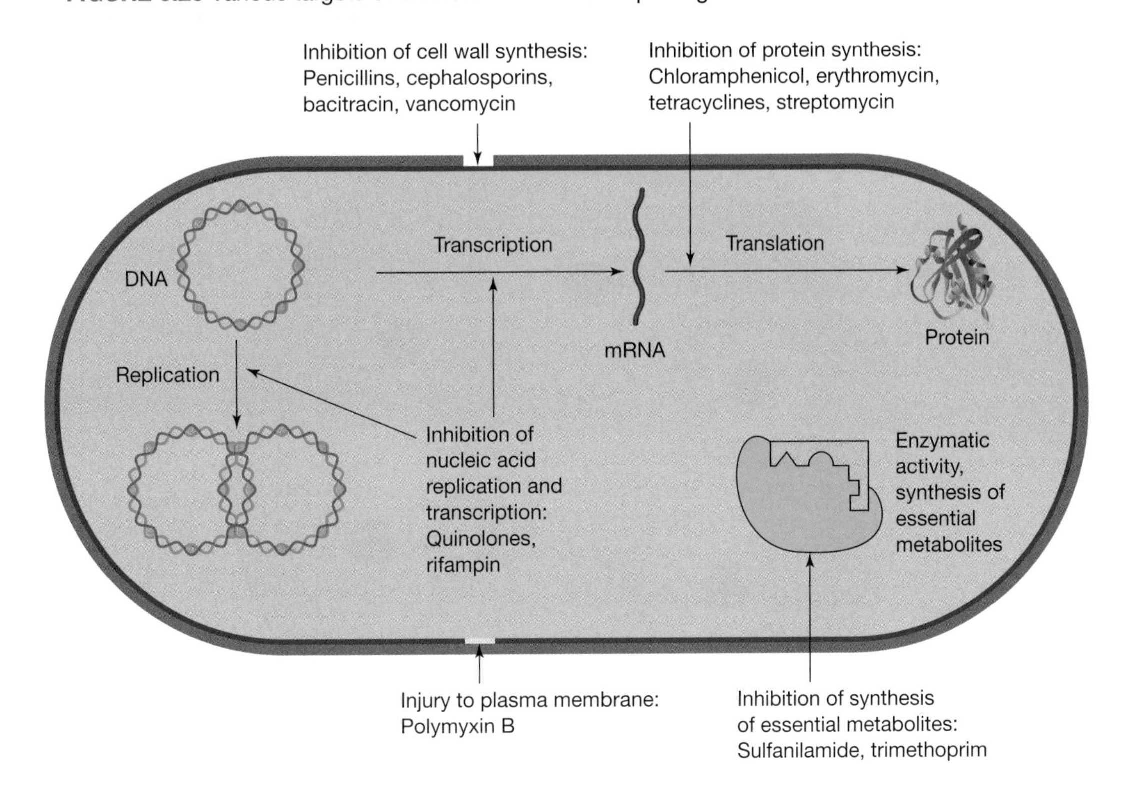

TABLE 6.8 Different antibiotics and their modes of action

Drug	Spectrum	Mode of action	Possible side effects
Erythromycin	Narrow (Gram +)	Inhibits protein synthesis so bacteria are unable to make essential compounds	Gastrointestinal upset Liver damage
Penicillin	Narrow (Gram +)	Inhibits cell wall synthesis so bacteria cannot reproduce	Allergic responses
Sulfonamides	Broad (Gram + and –)	Compete against and inhibit the bacteria	Allergic responses Kidney and liver damage
Tetracyclines	Broad (Gram + and –)	Inhibit protein synthesis so bacteria are unable to make essential compounds	Gastrointestinal upset Teeth discolouration Sometimes kidney and liver damage

Sensitivity tests are carried out to determine the antibiotic that would be most effective to treat a bacterial infection. In one kind of sensitivity test, bacteria are spread across the surface of a solid nutrient plate and small discs with known concentrations of different antibiotics are put on the surface of the plate (see figure 6.26). If the bacteria are sensitive to a particular antibiotic, they will be killed by that antibiotic and so will not grow around the disc containing that drug. The lack of growth is shown by a clear zone around the disc of antibiotic. The larger the clear zone around a disc, the more effective the antibiotic in that disc.

sensitivity test test to establish the most effective drug to use for treatment against a particular bacterial infection

FIGURE 6.26 Testing drug sensitivity of *Staphylococcus aureus*. From 12 o'clock, the drugs are fusidic acid (FD10), pencillin G (P10), ciprofloxacin (CIP5), rifampicin (RD5), gentamicin (CN10) and vancomycin (VA30). This bacterium is very sensitive to FD10, CIP5 and RD2. It is resistant to P10. (Image courtesy of Marjory Martin.)

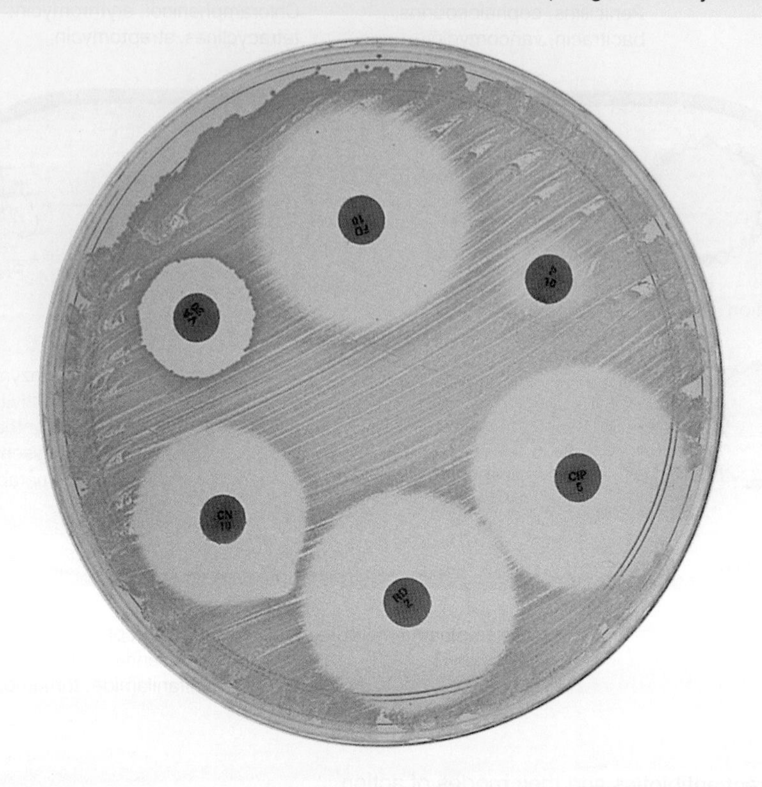

elog-0900

INVESTIGATION 6.2 online only

Optimal antibiotic concentrations

Aim

To determine optimal concentration of ampicillin to inhibit the growth of *E. coli*.

Antiviral drugs

Antiviral drugs are a type of medication that is used specifically for treating viral infections. Typically, these drugs are effective only when the viruses are located within cells and are undergoing replication. Antibiotics are ineffective against viral infections.

Most of the antiviral drugs presently available are designed to address the viruses that cause influenza A and B, hepatitis B and C (which can cause liver cancer), HIV (which causes AIDS), and various herpes viruses that cause many diseases ranging from cold sores to genital herpes infections. Advances in both our understanding of the details of the viral replication cycle in host cells and of the 3D structure of viral proteins has contributed to the development of many highly specific and effective antiviral drugs.

The possible modes of action of antiviral agents include:

- prevent viral attachment and/or entry
- prevent replication of the viral genome
- prevent synthesis of specific viral protein(s)
- prevent assembly or release of new infectious virions.

FIGURE 6.27 a. Diagram showing the replication cycle of HIV in human CD4 T-helper cells and antiviral targets, b. Antiviral targets across various viruses

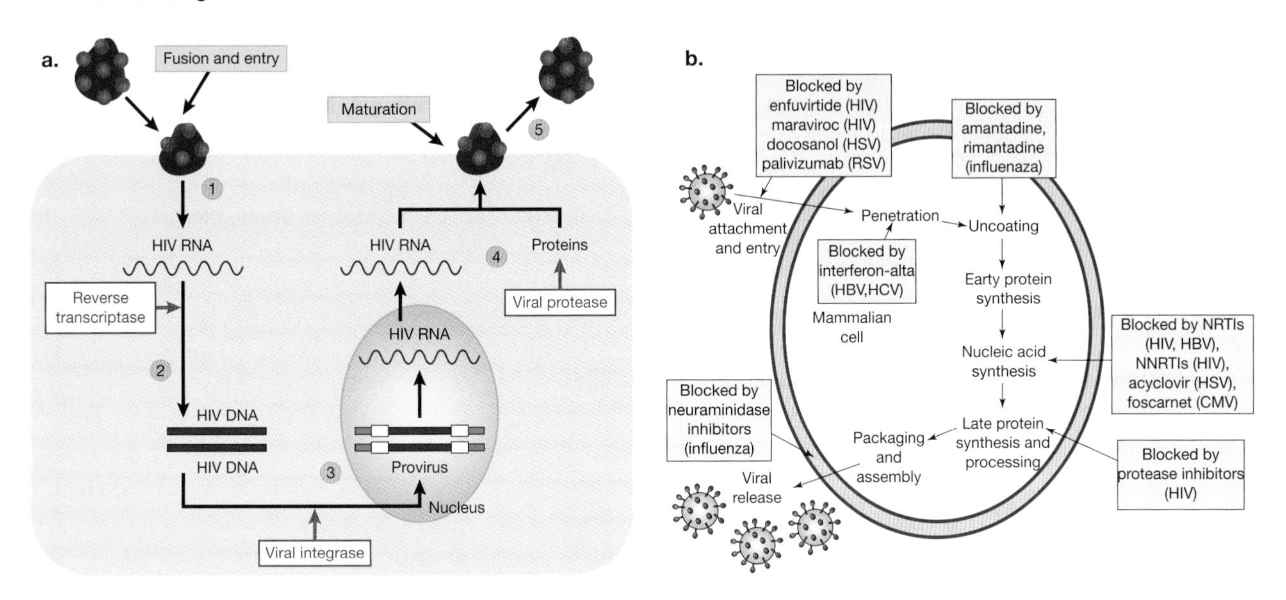

TABLE 6.9 Mode of action of various antiviral drugs and the viruses they target

Mode of action of antiviral drug	Example of drug	Target of drug
Block entry of virus to host cell by binding to surface receptors on host cell	Maraviroc	HIV/AIDS
Block fusion of virus with host cell by binding to surface protein (gp41) on virus	Enfuvirtide	HIV/AIDS
Block uncoating of virus by inhibiting action of viral membrane protein M2	Amantadine rimantadine	Influenza A
Block DNA replication through use of nucleoside analogues • interfering with reverse transcriptase • interfering with viral DNA polymerase.	Zidovudine (AZT) Rilpivirine Lamivudine Acyclovir	HIV/AIDS HIV/AIDS Chronic hepatitis B Herpes simplex virus
Block integration of viral DNA into host genome through use of integrase inhibitor	Raltegravir	HIV/AIDS
Block viral protease enzymes by use of protease inhibitors	Saquinavir Boceprevir	HIV/AIDS Hepatitis C
Interfere with assembly of new infectious virions	Mitoxantrone Daclatasvir	Vaccinia virus Hepatitis C
Block release of new virions from host cells by inhibiting action of neuraminidase	46 seltamiv (Relenza) 46 seltamivir (Tamiflu)	Influenza A and B
Stimulate immune system	Interferon alpha	Multiple viral infections, including hepatitis B

TIP: While you don't need to remember the names or actions of each antibiotic and antiviral drug, it is important to consider that they have different mechanisms of action. These actions differ greatly between antibiotics and antivirals and are often specific to certain pathogens or types of pathogens.

EXTENSION: Rational drug design

Many pathogens are constantly changing into new subtypes, making them harder and harder to treat. The ability of viruses and bacteria to do this will be further investigated in topic 7.

One concept that is useful in the treatment of changing pathogenic agents is **rational drug design**. The process involves finding a target enzyme on a pathogen that causes infection and designing a drug that blocks its active site.

To access more information about rational drug design and examples of different designer drugs, please download the digital document.

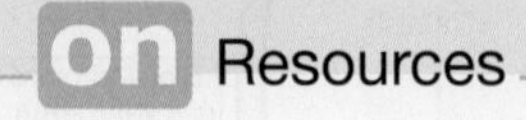

Resources

Digital document Extension: Rational drug design (doc-36104)

Other chemical agents to control pathogens

Other ways of controlling the spread of pathogens are:

- *sterilisation.* This is the removal or killing of the microbes from surfaces. Sterilisation is done by heat through such methods as autoclaving. This is a very effective method and relies on pressurised steam at a high temperature.
- *chemical agents.* Antiseptics and disinfectants are also used to control the spread of pathogens. The inhibition or killing of pathogenic organisms on non-living surfaces, such as taps and door handles, is called disinfection. Antiseptics are used for inhibiting the growth of pathogens on living surfaces such as the skin.

rational drug design construction of a drug to fit the active site of a molecule so that the natural action of the molecule cannot occur

elog-0902

INVESTIGATION 6.3

online only

Sensitivity testing using spices and different household disinfectants

Aim

To explore the bacteriostatic and bactericidal actions of different household substances using sensitivity testing

Resources

eWorkbook Worksheet 6.3 Controlling the spread of pathogens (ewbk-7855)

KEY IDEAS

- Identification of pathogens is important for both treatment and prevention.
- Methods to identify viruses include physical, immunological and molecular methods.
- Methods to identify bacteria include phenotypic, immunological and genotypic methods.
- It is also important to identify the host and/or the reservoir of a pathogenic agent to help prevent the spread of disease.
- The incubation period of a disease is the interval between a person's exposure to a pathogen and the onset of disease symptoms in that person.
- Asymptomatic carriers of a disease show no symptoms of a disease but can spread it.
- Infectious diseases may be spread directly (by person-to-person contact) or they may be spread indirectly.

- Methods to control the spread of disease includes prevention, vaccination, surveillance, modification of environment, infection control measures and medications.
- Antibiotics are agents that kill or inhibit bacteria using different bacterial targets.
- Antiviral drugs have been developed that target key viral enzymes involved in the viral replication cycle.

6.3 Activities

learnON

To answer questions online and to receive **immediate feedback** and **sample responses** for every question, go to your learnON title at **www.jacplus.com.au**. A **downloadable solutions** file is also available in the resources tab.

6.3 Quick quiz on	**6.3 Exercise**	**6.3 Exam questions**

6.3 Exercise

1. **MC** Commencing in 2018, an outbreak of Ebola occurred in the region of Kivu in Democratic Republic of the Congo.
   Which of the following would not be recommended to contain the outbreak?
   A. Quarantining infected individuals
   B. Testing individuals before they are permitted to travel out of the affected region
   C. Allowing families to partake in traditional funeral and burial methods of individuals who died from the disease
   D. Providing healthcare workers with personal protective equipment while treating affected individuals
2. Identify the following statements as true or false. Justify your responses.
   a. Viruses are not destroyed by antibiotics.
   b. The presence of a particular viral protein in a person's serum could be identified using ELISA.
   c. Different viruses cannot be distinguished on the basis of their shape.
   d. Inhibiting the viral enzyme integrase is one strategy for controlling HIV.
3. Hand sanitisers are commonly used to kill 99.99 per cent of germs without water. These sanitisers act as antiseptics. What are antiseptics and how do they differ from antibiotics?
4. What is the term for mosquitoes transmitting a disease such as Zika virus?
5. Influenza, also known as the flu, is a highly contagious respiratory illness caused by influenza viruses.
   a. An individual has claimed that 'preventing influenza through annual vaccinations is better than treating the disease using antivirals'. State whether this statement is true or false, and justify your response.
   b. How does flu get transmitted?
6. A certain bacteria underwent testing. It was found to be Gram-negative and tested positive for indole production and fermenting lactose, but was unable to use citrate as its sole carbon source.
   a. Using figure 6.15, identify the likely identity of this pathogen.
   b. Was this type of test phenotypic or genotypic? Justify your response.
7. **a.** A certain disease has a long incubation period. Describe the consequences this might have for the spread of the disease.
   b. The same disease have an R_0 value of 5.0. Outline what this represents and explain how it might affect the number of individuals infected.
8. Bacteria and viruses are both pathogens capable of causing diseases. However, there are many differences between the two.
   a. Explain why there are differences between techniques used to identify bacteria and techniques used to identify viruses.
   b. Describe the differences between antiviral drugs and antibiotics in both their use and mechanisms of action.
9. Explain the difference between direct, indirect and sandwich ELISA. Use a diagram to support your response.
10. During the COVID-19 pandemic, two recommendations were washing hands with soap (or sanitiser) for at least 20 seconds and wearing face masks. Describe how each of these would be useful in preventing the spread of disease.

6.3 Exam questions

Question 1 (1 mark)

Source: *VCAA 2020 Biology Section A, Q34*

MC Bovine spongiform encephalopathy (BSE) is a prion disease of cattle. It is sometimes called mad cow disease. It is caused by feeding cattle food that contains prions from other infected animals. The time between infection and symptoms appearing can be up to five years. There are concerns that variant Creutzfeldt-Jakob disease (vCJD) in humans could be caused by eating infected cattle meat.

Yellow fever is a viral disease that affects humans. The yellow fever virus can cause symptoms three to six days after infection. The virus is carried by a mosquito vector.

Which combination of approaches would be most effective at controlling the risk of outbreaks of both vCJD and yellow fever?

	vCJD	Yellow fever
A.	Prevent all cattle that show symptoms of mad cow disease from reproducing.	Remove breeding grounds for mosquitoes.
B.	Test all cattle for the presence of the prions.	Ensure that all healthcare professionals wear gloves when working with infected patients.
C.	Destroy all cattle that have been fed infected food containing the prions.	Ensure that people take measures to reduce their chances of being bitten by mosquitoes.
D.	Stop selling cattle meat.	Instruct people who are infected with yellow fever to wear masks in public places.

Use the following information to answer Questions 2 and 3

A diagnostic test for HIV infection includes the following steps.

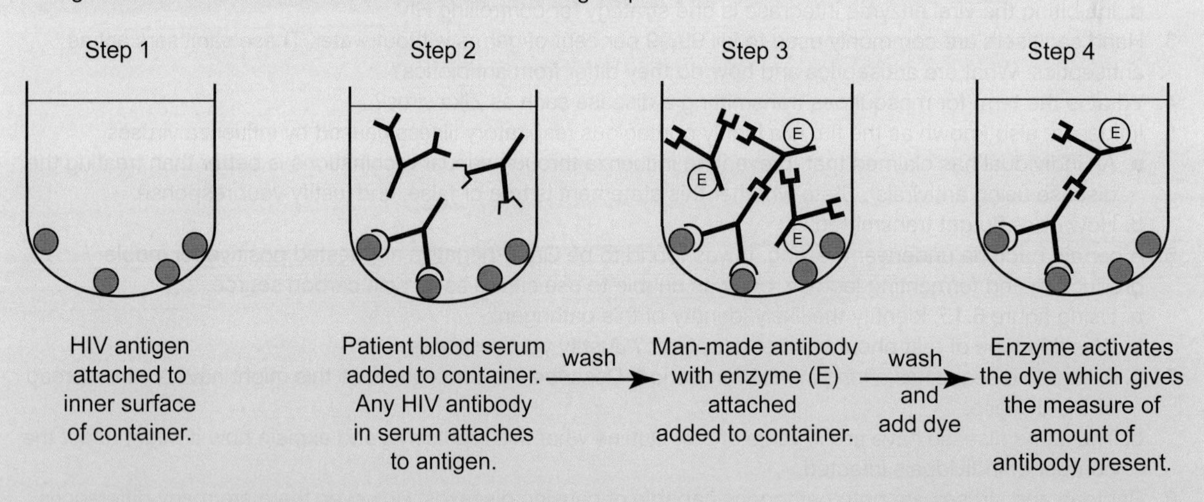

Question 2 (1 mark)

Source: *VCAA 2012 Biology Exam 1, Section A, Q24*

MC This test for HIV is reliable because the

A. dye reacts with the patient's blood serum.
B. enzyme has an active site for the HIV antigen.
C. man-made antibody has the same shape as the HIV antigen.
D. HIV antigen has a complementary shape specific to the HIV antibody.

Question 3 (1 mark)

Source: *VCAA 2012 Biology Exam 1, Section A, Q25*

MC A diagnostic test for HIV infection includes the following steps.

The results of the tests of three patients are given in the following table.

Positive control	Negative control	Patient R	Patient S	Patient T
1.689	0.153	0.155	0.675	1.999

Numbers are expressed as optical density at 450 nm. The more intense the dye is, the higher the optical density.

The cut-off value indicating a positive result is 0.500. Values below 0.300 are considered to be negative.

The results of these tests suggest that

A. patient T has not been exposed to HIV.
B. patient R has been exposed to the HIV antigen.
C. patient S has responded to exposure to HIV by developing antibodies.
D. the positive control contained fewer HIV antibodies than the negative control.

Question 4 (1 mark)

Source: *VCAA 2013 Biology Section A, Q17*

MC Ross River fever is caused by a virus that lives in kangaroos and wallabies. When a female mosquito bites an infected animal, it picks up viral particles. When the mosquito bites a human, the virus enters the bloodstream. The virus then reproduces in blood cells, resulting in fever, rashes and joint pain.

The most effective way to reduce the incidence of Ross River fever in Australia would be to

A. prevent humans from living near the Ross River.
B. use an attenuated form of the virus to create a human vaccine.
C. increase spending on anti-inflammatory drugs to treat the symptoms.
D. isolate kangaroos and wallabies in nature reserves near the Ross River.

Question 5 (1 mark)

Source: *VCAA 2012 Biology Exam 1, Section B, Q7c*

As a requirement for re-entry, travellers returning to Australia from Africa and South America must have proof of vaccination against yellow fever.

Explain why this precaution is taken and what course of action Australian authorities may take for an unvaccinated person wanting to re-enter Australia.

More exam questions are available in your learnON title.

6.4 Vaccination programs and herd immunity

KEY KNOWLEDGE

- Vaccination programs and their role in maintaining herd immunity for a specific disease in a human population

Source: VCE Biology Study Design (2022–2026) extracts © VCAA; reproduced by permission.

6.4.1 Vaccination programs

The aim of vaccination programs is to reduce the impact of vaccine-preventable infectious diseases through achieving high rates of immunisation in the community. **Vaccination programs** are one of the best ways to protect the community against certain diseases.

vaccination programs mandated programs that set a schedule in which vaccinations against specific diseases should be administered

When a large percentage of the population are immunised against some specific diseases, it becomes harder for those diseases to spread.

BACKGROUND KNOWLEDGE: Reviewing how vaccinations lead to immunity

Vaccines contain a pathogen in a weakened, live or killed state, or proteins or toxins from the organism to trigger the immune response. This is known as artificial adaptive immune response. This was first introduced in topic 5 as an example of artificial active immunity in section 5.12.2.

Various means are used to produce vaccines with no disease-causing capability. All these various types of vaccines are able to provoke the adaptive immune system to produce antibodies specific to the antigens of the pathogens or their toxins.

In a vaccination:

1. the vaccine (containing either live attenuated pathogens that cannot cause disease, inactivated or killed pathogens, inactivated toxins or subunits of pathogens) is injected into a person
2. the immune system produces antibodies and memory cells against the pathogen
3. additional injections lead to an amplified production of antibodies
4. the antibody is specific to the treated pathogen used in the vaccine so, if the person comes into contact with the live organisms at some future date, memory cells and antibodies will be ready to act and the person is immune to infection.

TABLE 6.10 Various types of vaccine

Type of vaccine	Examples
Live attenuated vaccines	Measles, mumps, rubella (MMR) Yellow fever Polio (Sabin vaccine)
Inactivated or killed vaccines	Polio (Salk vaccine) (IPV) Rabies
Inactivated toxin of bacteria (called a toxoid)	Diphtheria, tetanus (part of DTPa vaccine)
Sub-units of bacteria or viruses	Hepatitis B (hepB) Haemophilus influenza type b (Hib) Human papillomavirus (Gardasil)

Why do we need boosters?

Vaccines are sometimes given as two (or three) injections at shorter intervals followed by a booster after a longer interval.

In general, killed or inactivated vaccines produce a weaker immune response compared to the response from using live attenuated vaccines, and the immunity lasts for a shorter period. As a result, killed or inactivated vaccines have to be administered more than once. This is vital in order to maintain immunity against a disease. This is shown in figure 6.28.

FIGURE 6.28 Booster shots increase the immunity against a disease.

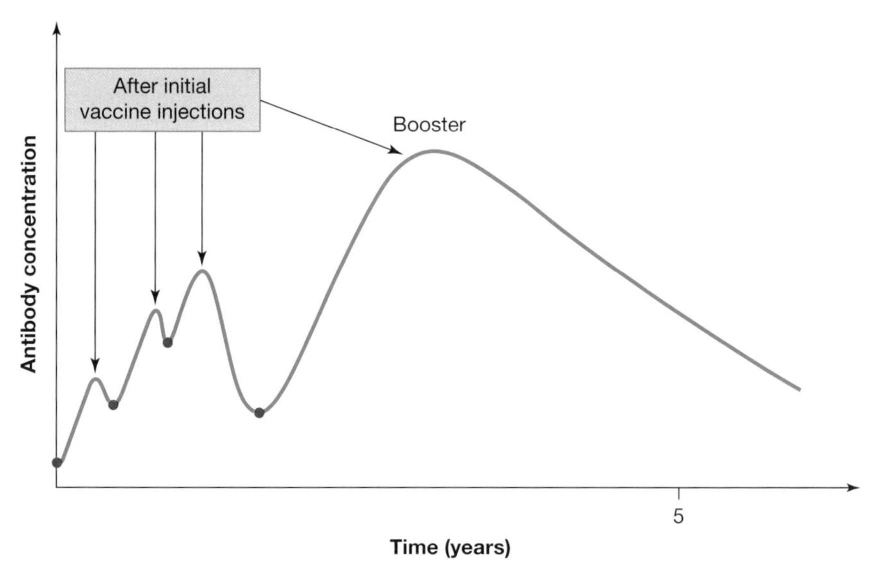

on Resources

Video eLesson Memory cells and vaccination (eles-4366)

Vaccination programs in Australia

The National Immunisation Program Schedule in Australia, effective from 1 April 2019, shows the vaccinations required. Some minor variations exist between Australian states. The specific schedule in Victoria is shown in the provided weblink.

TABLE 6.11 National Immunisation Program Schedule for babies, children and adolescents in Australia, effective from 1 April 2019. Those diseases in one dot point are administered as one single vaccine that is multivalent.

Age	Diseases to be immunised against
Birth	• Hepatitis B (preferably within 24 hours of birth)
2 months	• Diphtheria, tetanus, pertussis (whooping cough), hepatitis B, *Haemophilus influenzae* type b and polio • Pneumococcal • Rotavirus
4 months	• Diphtheria, tetanus, pertussis (whooping cough), hepatitis B, *Haemophilus influenzae* type b and polio • Pneumococcal • Rotavirus
6 months	• Diphtheria, tetanus, pertussis (whooping cough), hepatitis B, *Haemophilus influenzae* type b and polio • Pneumococcal (for at-risk only)

(continued)

TABLE 6.11 National Immunisation Program Schedule for babies, children and adolescents in Australia, effective from 1 April 2019. Those diseases in one dot point are administered as one single vaccine that is multivalent. *(continued)*

Age	Diseases to be immunised against
12 months	• Meningococcal ACWY • Measles, mumps and rubella • Pneumococcal • Hepatitis A (for at-risk only)
18 months	• *Haemophilus influenzae* type b • Measles, mumps, rubella and varicella/chickenpox • Diphtheria, tetanus, pertussis (whooping cough) • Hepatitis A (for at-risk only)
4 years	• Diphtheria, tetanus, pertussis (whooping cough), polio • Pneumococcal (for at-risk only, different vaccine to those used before 1 year old)
12–13 years	• Human papillomavirus (HPV) • Diphtheria, tetanus, pertussis (whooping cough)
14–16 years	• Meningococcal ACWY

The most recent changes made in 2019 include the introduction of the meningococcal ACWY vaccine for 14- to 16-year-olds.

Many of these vaccinations are intramuscular (injected into the muscle). For younger babies, this is usually in the outer thigh. For individuals older than a year, this is more commonly in the upper arm (into the deltoid). The only orally taken vaccination is for rotovirus, which causes diarrhoeal diseases in babies.

Several other programs are delivered as part of the National Immunisation Program.

Examples include:

- Influenza vaccinations for children between 6 months and 5 years, at-risk individuals, elderly and pregnant women
- Pertussis vaccination (whooping cough) in pregnant women
- Shingles in individuals aged between 70 and 79 years.

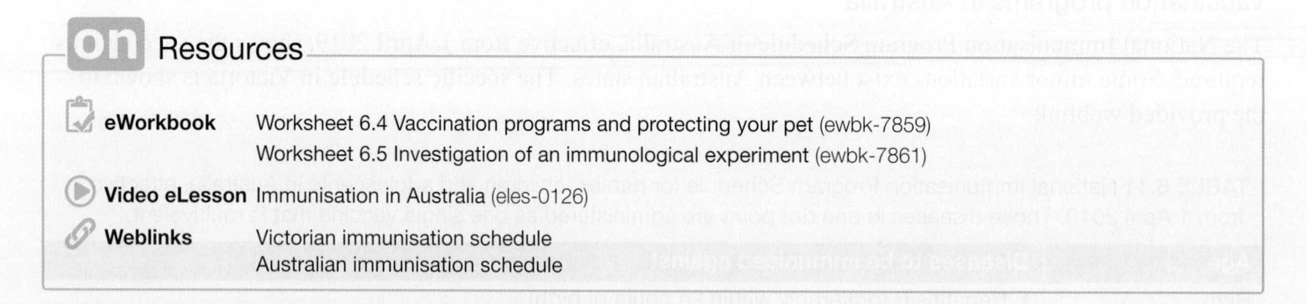

on Resources

eWorkbook	Worksheet 6.4 Vaccination programs and protecting your pet (ewbk-7859) Worksheet 6.5 Investigation of an immunological experiment (ewbk-7861)
Video eLesson	Immunisation in Australia (eles-0126)
Weblinks	Victorian immunisation schedule Australian immunisation schedule

6.4.2 The importance of mass vaccination

Mass vaccinations have eliminated diseases

Presently, only two diseases have been eradicated worldwide: smallpox and rinderplast. Only one of these, smallpox, affects humans.

CASE STUDY: Eradication of smallpox

Smallpox is a devastating disease. Before the 1980s, smallpox, a disease caused by the variola virus, was widespread in many countries. Australians travelling overseas were required to carry a vaccination booklet (see figure 6.29 showing that they had been vaccinated against various diseases, including smallpox.

The vaccination for smallpox was discovered due to its similarity to a more minor pox disease known as cowpox. It had been found that milkmaids who had previously had cowpox were immune to smallpox. Edward Jenner used this idea to develop a vaccine against smallpox. This vaccine was refined over time.

The World Health Organization (WHO) began an extensive vaccination program against smallpox in 1959 and by December 1979, the WHO announced that the disease had been eradicated worldwide. Because of the widespread vaccination against smallpox, the infecting virus could no longer find appropriate hosts who were not immune. Since 1977, no naturally transmitted cases of smallpox have been recorded anywhere in the world.

FIGURE 6.29 **a.** Smallpox was a devastating disease that has been eradicated. **b.** Prior to the 1980s, Australians travelling abroad had to carry vaccination booklets to prove they had been immunised (Image courtesy of Marjory Martin.)

a.

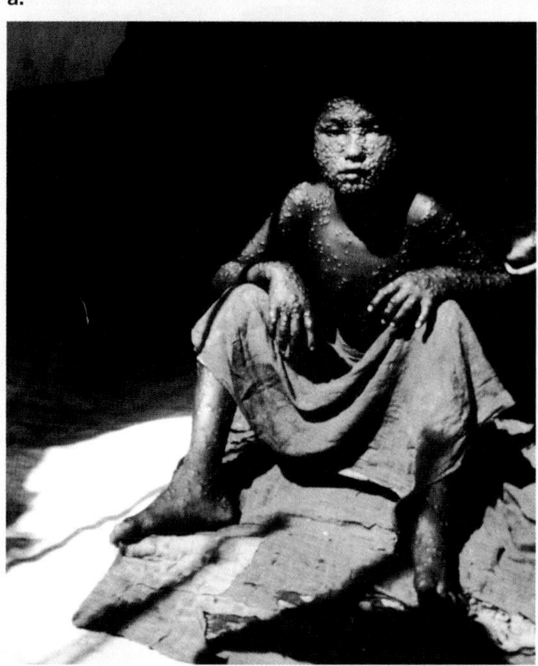

b.

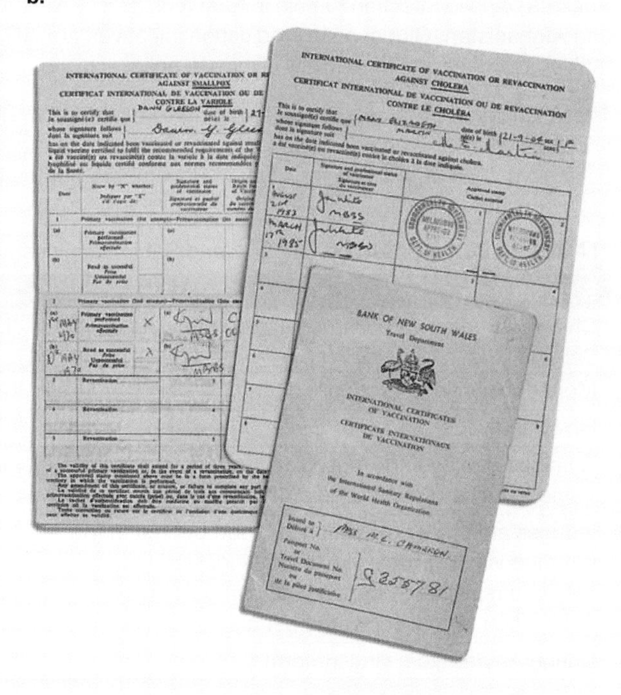

INTERNATIONAL CERTIFICATE OF VACCINATION OR REVACCINATION AGAINST SMALLPOX

CERTIFICAT INTERNATIONAL DE VACCINATION OU DE REVACCINATION CONTRE LA VARIOLE

INTERNATIONAL CERTIFICATE OF VACCINATION OR REVACCINATION AGAINST CHOLERA

CERTIFICAT INTERNATIONAL DE VACCINATION OU DE REVACCINATION CONTRE LE CHOLÉRA

BANK OF NEW SOUTH WALES

Travel Department

INTERNATIONAL CERTIFICATES OF VACCINATION

CERTIFICATS INTERNATIONAUX DE VACCINATION

Two laboratories in the world hold stocks of smallpox virus: the US Centers for Disease Control and Prevention (CDC) and the Russian State Research Center of Virology and Biotechnology (VECTOR); they are held in highly secure laboratories. All other samples of smallpox have been destroyed.

Some groups argue that these smallpox stocks should also be destroyed in order to prevent any possibility of their accidental release. Other groups argue that scientific opportunities will be lost if all smallpox stocks are destroyed and that they are needed for research into the development of antiviral drugs. As of 2020, the question still remains whether they should be destroyed or not.

CASE STUDY: On the road to eradicate poliomyelitis

Poliomyelitis or polio is caused by the pathogen poliovirus. Poliovirus is spread by person-to-person contact through the exchange of nasal or oral secretions, or by contact with faecal-contaminated material.

The virus enters the mouth and replicates in the gut. In most cases, a polio infection results in a mild illness, but in a very small number of infected people the polio virus migrates from the gut, travels through the bloodstream and attacks nerve cells, resulting in paralysis.

Polio has been eradicated from Australia and almost all other countries through vaccination programs that have given people immunity to polio, with the result that the virus cannot find hosts for multiplication.

- The last case of locally transmitted poliovirus in Australia was that of a 22-year-old man in Victoria in 1986.
- The last confirmed case of polio in Australia was that of a 22-year-old student who returned from Pakistan in 2007 with the disease.

FIGURE 6.30 Polio can cause paralysis.

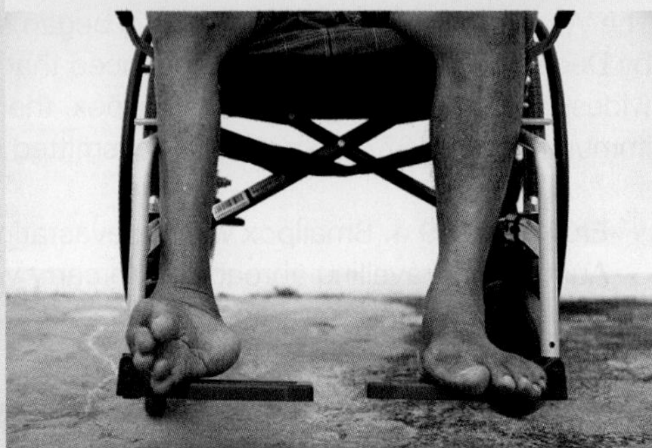

Pockets of polio infection remain in rural regions of Pakistan and Afghanistan. Global polio eradication initiatives are still making major efforts to officially eliminate polio.

The vaccination programs that are leading towards the worldwide elimination of polio are based on two different vaccines: the Salk vaccine and the Sabin vaccine (see table 6.12 below). In Australia, the oral Sabin vaccine was used from 1966 until it was replaced in November 2005 by the Salk vaccine.

TABLE 6.12 Comparison of the Salk and Sabin vaccines against poliovirus

Salk vaccine	Sabin vaccine
Chemically inactivated virus developed by Jonas Salk	'Live' attenuated virus developed by Albert Sabin
More expensive to produce	Cheaper to produce
First used widely in 1955	First available in 1962
Administered by intra-muscular injection — requires trained staff and sterile equipment	Administered by mouth (orally) — very easy to deliver
Induces antibody production in blood but not in the gut mucosa; if virulent viruses are ingested later they can replicate in the gut and can be a source of infection for unvaccinated people.	Induces antibody production in both the gut mucosa and the blood; if virulent polioviruses are ingested later, they will be eliminated by the mucosal antibodies.
Cannot mutate to a virulent form	Can mutate and revert to a virulent form

Reducing the incidence of other diseases

The incidence of a number of other diseases — the bacterial diseases diphtheria (caused by *Corynebacterium diphtheriae*), tetanus (*Clostridium tetani*) and pertussis (also known as whooping cough and caused by *Bordetella pertussis*), and the viral diseases measles, mumps and rubella (also called German measles) — has also been significantly decreased by the use of vaccines, in particular through the immunisation programs for babies and children and susceptible groups.

In Australia, contagious diseases have been contained and almost eliminated through mass vaccinations. Figure 6.31 shows the incidence of meningococcal disease in Australia in the period 1995 to 2019. Note the steady decline in the number of cases since the introduction of the pneumococcal vaccine in 2002. However, given the growth in international travel, the reintroduction of some diseases to Australia from offshore where these diseases are still endemic is always a possibility.

FIGURE 6.31 Incidence of invasive meningococcal disease in Australia in the period 1996–2019

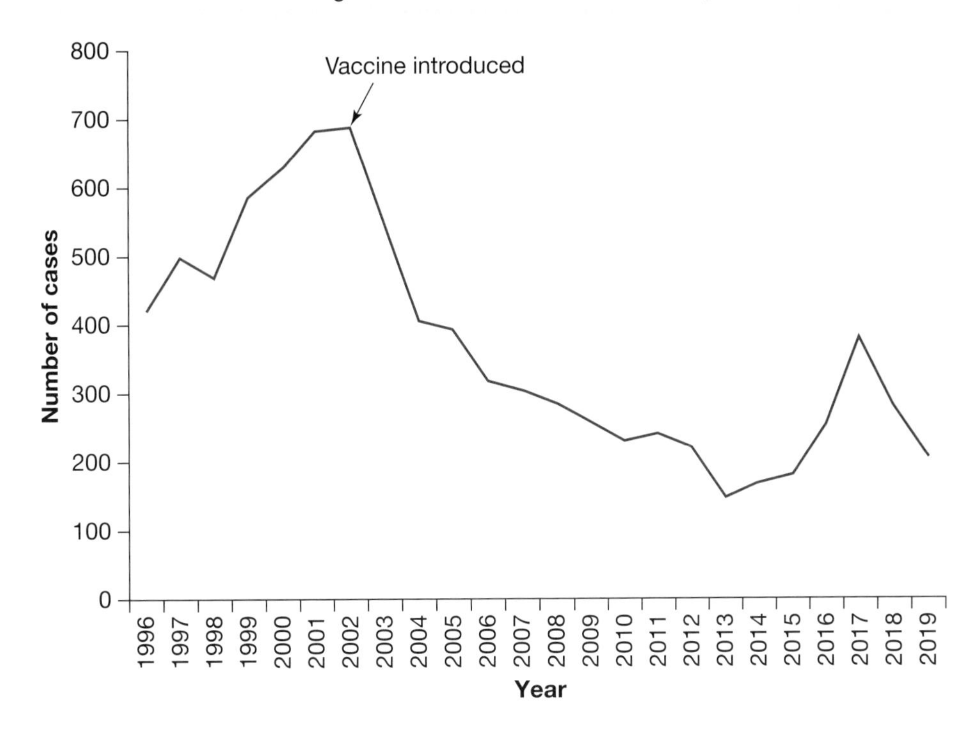

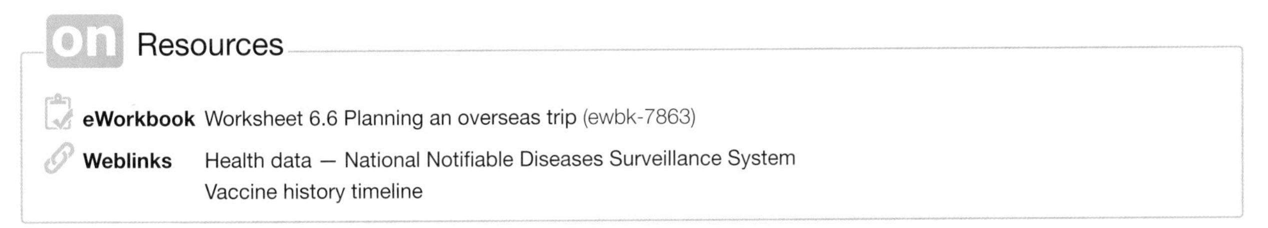

6.4.3 Herd immunity

As individuals, we can be protected against an infectious disease by being immunised against it — this is an example of *direct* protection, which we gain by developing antibodies against the disease in response to the injection of an antigen of the pathogen concerned.

In addition, *indirect* protection from infectious disease can exist, but only at the level of a population. This indirect protection is termed **herd immunity**, also known as community immunity.

Herd immunity is the indirect protection of populations from infection where that protection is created by the presence of immune individuals in the population and the protection is received by unvaccinated individuals.

The immune people in a population are those who acquired their immunity either artificially through vaccination or naturally by having recovered from the disease.

There are a number of factors that affect herd immunity:

1. Herd immunity operates only when a high proportion of the population has immunity to a particular disease.
2. Herd immunity applies only to those infectious diseases that are contagious — that is, diseases that are spread directly from person to person, such as influenza, measles, mumps, pertussis (whooping cough) and meningococcal disease.

herd immunity indirect protection, at the population level, against an infectious disease; the protection is created by the presence in the population of a high proportion of individuals who are vaccinated against the particular disease

Herd immunity is important for vulnerable members of a community who either cannot be vaccinated or for whom vaccination is ineffective. Such individuals include newborn babies, people whose immune systems are defective because of an inherited disorder (such as SCID) or an acquired condition (such as acquired immunodeficiency syndrome — AIDS), elderly people and people undergoing chemotherapy for cancer.

In a population, the higher the proportion of immune individuals to an infectious disease, the lower the chance that an unimmunised person will come into contact with an infectious individual. If enough people are immune, an infectious disease will not get the chance to become established and spread. If the pathogen cannot find a susceptible host, the infectious disease that it causes will, over time, gradually disappear from a population.

FIGURE 6.32 Herd immunity protects those who are not immunised.

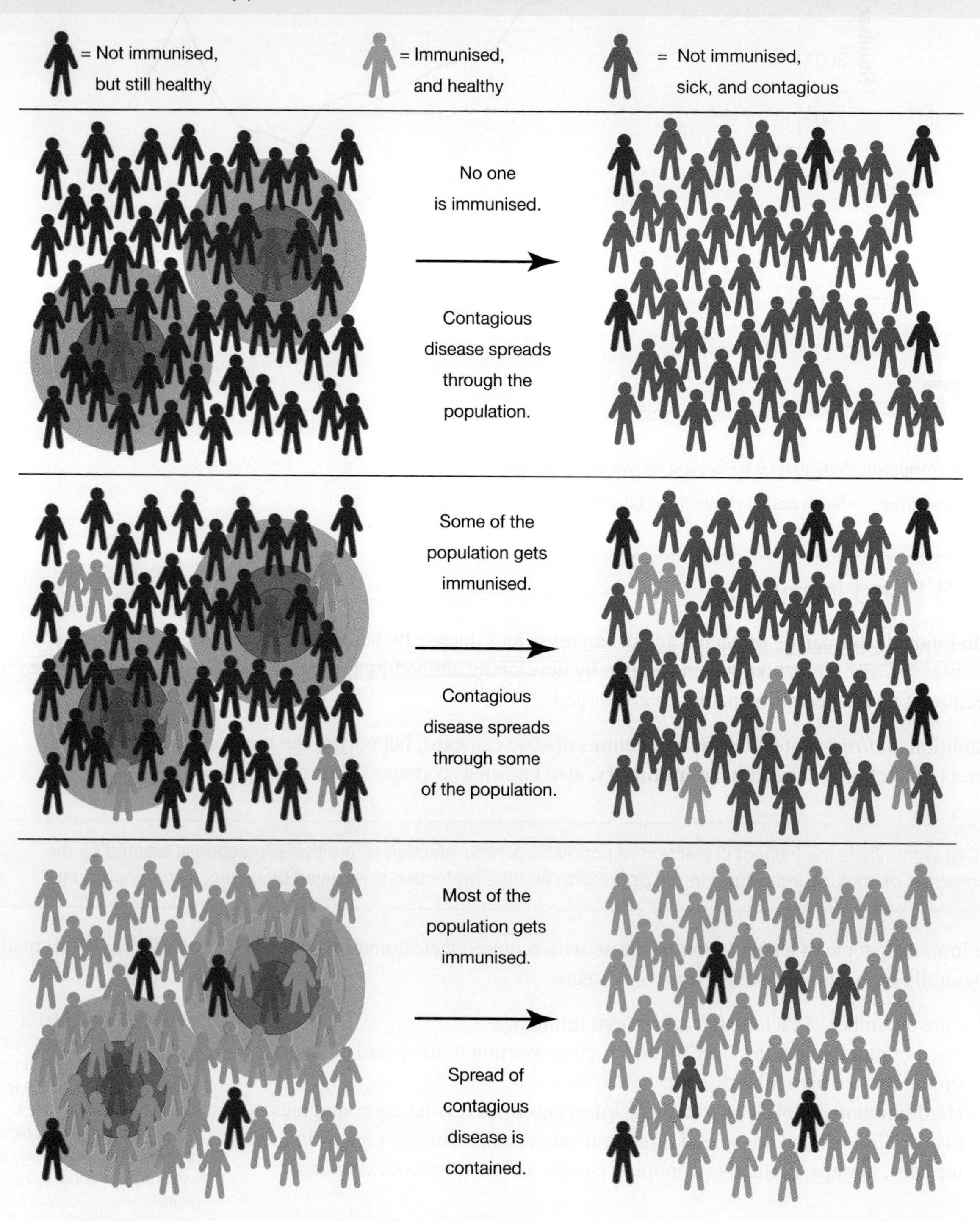

High levels of immunised individuals in a population have been achieved through mass immunisation programs, such as through the National Immunisation Program (NIP), which was implemented in Australia in 1997 (see section 6.4.1). Mass vaccination programs have generated herd immunity and have successfully prevented the spread of infectious diseases. In several cases, mass vaccinations have led to the elimination of diseases, either worldwide or from a region, as for example, smallpox and polio. Recommendations for herd immunity to be effective is 95 per cent of individuals being vaccinated.

elog-0904

INVESTIGATION 6.4 online only

Modelling herd immunity

Aim

To model how herd immunity works and determine how the number of immunised individuals impacts its effectiveness

Vaccination rates

While vaccination rates, particularly in children, have increased over the last 10 years, the 95 per cent aspirational target for vaccinations has not been reached in many populations, as seen in figure 6.33. Opposition to vaccination poses a challenge to herd immunity by allowing preventable diseases to persist in, or reappear in, communities because of a decline in vaccination rates.

FIGURE 6.33 Immunisation coverage rates of all children in Australia and immunisation coverage rates of Aboriginal and Torres Strait Islander children

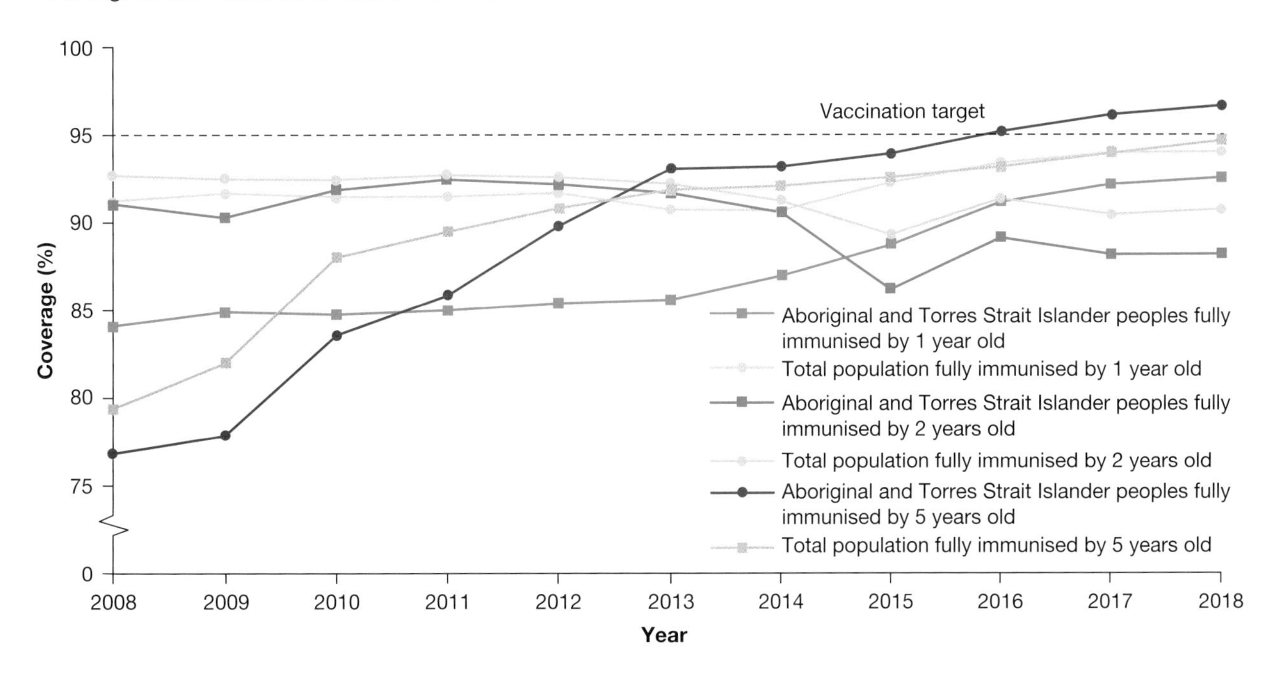

tlvd-1786

SAMPLE PROBLEM 4 Understanding the immunisation schedule and herd immunity

Refer to table 6.11.

The immunisation for diphtheria, tetanus, pertussis (whooping cough), hepatitis B, *Haemophilus influenzae* type b and polio (known as Infanrix hexa) is given at two months, four months and six months.

a. **Given that a baby has been already immunised at two months with this vaccine, explain why booster shots are given.** **(2 marks)**

b. **Why is the polio vaccination given when there have been no cases in Australia for over 10 years?** **(1 mark)**

c. **Explain the benefit for giving six immunisations at once. How would this be associated with herd immunity?** **(2 marks)**

THINK	WRITE
a. The question is asking to give a reason for giving booster doses to babies. As it is is worth 2 marks, you need to give two distinct points that clearly link back to Infanrix hexa. Consider the reasoning for boosters: producing an antibody response at a higher and faster rate.	Killed or inactivated vaccines produce weaker responses and immunity lasts for a shorter period of time (1 mark). By giving a booster shot of Infanrix hexa, the level of antibodies and memory cells is increased and the level remains high for a longer period (1 mark).
b. Consider the key idea as to why we still vaccinate against polio. It is because it is not yet eradicated. Formulate your response using this as a basis.	Polio is yet to be eradicated worldwide. Therefore, there is still a chance for it to come into Australia and become prevalent. By having babies immunised, it reduces the chance that it can become endemic in Australia (1 mark).
c. Carefully read what the question is asking you to do. You are required to focus on two aspects: why six immunisations are given and the association with herd immunity.	Giving six immunisations at once (in one injection) results in less harm to the baby, as they don't need multiple injections and cases of inflammation occurring (1 mark). It also means they are more easily covered against numerous diseases. This allows a greater chance of herd immunity to be reached for a greater number of diseases, providing a greater opportunity for 95 per cent of the population to be immune to six different diseases (1 mark).

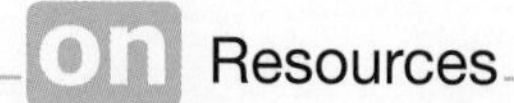

Resources

eWorkbook Worksheet 6.7 Herd immunity (ewbk-7865)

Video eLesson Herd immunity (eles-4367)

KEY IDEAS

- Vaccination entails the exposure of a person to an attenuated, dead or inactivated pathogen that induces active antibody formation in the person receiving the vaccine.
- Herd immunity is an indirect form of protection against infectious contagious diseases that exists in populations containing a high proportion of immune people.
- The protection created by herd immunity applies to unimmunised members of the population.
- Some people within a population who depend on herd immunity are those who cannot be vaccinated because of age or those with malfunctioning immune systems and for whom vaccinations would be ineffective.
- Herd immunity is put at risk when immunisation rates fall because of opposition to vaccination.

6.4 Activities

learnon

To answer questions online and to receive **immediate feedback** and **sample responses** for every question, go to your learnON title at **www.jacplus.com.au**. A **downloadable solutions** file is also available in the resources tab.

6.4 Quick quiz on	**6.4 Exercise**	**6.4 Exam questions**

6.4 Exercise

1. Refer to the National Immunisation Schedule for babies, children and adolescents (table 6.11).
   a. What is the first vaccine given?
   b. What vaccine is administered at 18 months?
2. Does herd immunity exist in all populations? Explain.
3. Identify two ways in which a person can become immune to a disease.
4. Give an example of:
   a. a vulnerable member of the community who cannot benefit from immunisation
   b. a disease that has been eliminated worldwide by mass vaccination.
5. A person stated: 'Meningococcal disease has almost been eliminated. There is no need to get vaccinated against this disease'. Is this a valid statement? Explain.

6.4 Exam questions

Question 1 (2 marks)

Source: *VCAA 2018 Biology Exam, Section B, Q5b*

Controlling the number of measles cases in a population relies on herd immunity.

What is herd immunity and how does it help control the number of cases of this disease?

Question 2 (6 marks)

Vaccination leads to immunity in the recipient to a particular disease. Medical authorities in remote areas of developing countries can rarely ensure delivery of vaccination to all those at risk of a particular disease so alternative methods are being researched.

a. Identify two difficulties that prevent easy access to vaccination in remote areas. **2 marks**
b. One investigation on delivery of vaccination in remote areas of developing countries involves genetic material (DNA) and plant tissue. DNA from a pathogen is injected into embryonic plant DNA and the plant is allowed to develop and grow. The adult plants are dried and eaten by humans. Explain why such a strategy may be successful for the development of immunity. **2 marks**
c. When plants are used to produce proteins the same as those in particular pathogens, for use as a vaccine, certain precautions must be taken to ensure that the vaccine is effective. Explain two precautions. **2 marks**

Question 3 (1 mark)

MC Measles is a serious disease and the complications of measles can be fatal. Before funded vaccination programs against measles were introduced for Australian babies in 1970, serious measles epidemics happened every two to three years. By 1989, 85 per cent of babies had been immunised against measles. In 1993, a measles vaccination program was also introduced in schools for 10–14 year olds. The rate of measles dropped from 27 cases per 100 000 people in 1993 to just 7 cases per 100 000 in 1995. In 1998, primary schools introduced another additional measles vaccination program to 5 year olds. Ninety-four per cent of school children were immunised in that year. Only 2 cases of measles per 100 000 people were reported.

In February 2016, several cases of measles were reported in two inner-northern suburbs of Melbourne. This drew attention that in these suburbs, the proportion of fully-immunised children was only 83 per cent.

From this information it can be inferred that

A. vaccination is more effective for 5-year-olds than for babies 10- to 14-year-olds.
B. measles has become a less dangerous disease since vaccinations were introduced.
C. booster vaccination programs increase the chance of achieving herd immunity.
D. herd immunity for measles can be achieved successfully with an immunisation rate of 83 per cent.

Question 4 (1 mark)

MC Achieving herd immunity against a particular disease is an advantage because it

A. increases the chance of infection.
B. provides protection for individuals who are not immune.
C. allows more children to attend school and daycare centres.
D. reduces the cost of vaccination programs.

Question 5 (3 marks)

Source: *VCAA 2016 Biology Exam, Section B, Q5a and b*

Yellow fever is a potentially fatal, mosquito-borne, viral disease that occurs in many countries in Africa, the Caribbean, and Central and South America. An effective and safe vaccine has been available since 1938.

a. What is a vaccine? **1 mark**
b. For the vaccine to be effective, it is recommended that travellers to these regions have the vaccination approximately two to four weeks before travelling. Why is this time frame recommended? **2 marks**

More exam questions are available in your learnON title.

6.5 Development of immunotherapy strategies

KEY KNOWLEDGE

- The development of immunotherapy strategies, including the use of monoclonal antibodies for the treatment of autoimmune diseases and cancer

Source: VCE Biology Study Design (2022–2026) extracts © VCAA; reproduced by permission.

6.5.1 What is immunotherapy?

Immunotherapy strategies can be used to treat varying diseases. Immunotherapy involves altering the immune response to fight diseases such as **cancer** and **autoimmune diseases**.

This differs from treatments such as chemotherapy and radiotherapy, which both act to stop cells from growing (for example, chemotherapy, when used for cancer, destroys cells that can replicate quickly, which can also include body cells).

immunotherapy a type of treatment that alters the immune response in an individual to combat diseases

cancer a disease in which cells divide in an uncontrolled manner, forming an abnormal mass of cells called a tumour

autoimmune diseases diseases in which the immune system fails to identify 'self' material and makes antibodies against the body's own tissues

There are several types of strategies that we can consider to be immunotherapy, many of which can be used in cancer treatment and prevention:

- Vaccination — human papilloma virus (HPV) can cause many cancers, particularly cervical cancers in females. There is a vaccination against HPV which produces active immunity, stimulating the immune system to produce antibodies and memory cells against HPV.
- CAR T Cell therapy — Special T cells (chimeric antigen receptor T cells or CAR T cells) are extracted from a patient and reprogrammed to recognise cancer cells. These are then replaced into the patient's blood so they can destroy cancer cells.
- Monoclonal antibodies — these are antibodies designed to target specific cells and cause an immune response.
- Immune inhibitors — these allow T cells to be more active and target immune cells by inhibiting the blocking of T cells that act to control the immune response.
- Cytokine therapy (with specific interferons and interleukins) — activates the immune system to better destroy cancer cells.

FIGURE 6.34 a. The process of CAR T cell therapy b. How this allows for the recognition of cancer cells

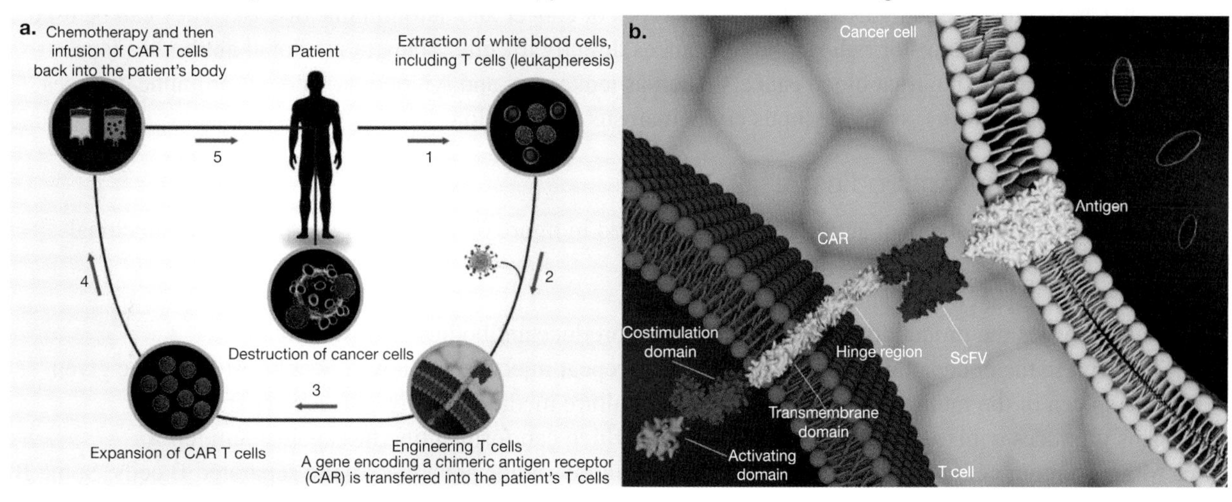

Now many treatments, such as monoclonal antibodies, are being designed and tailored for specific individuals, as cancer and autoimmune diseases differ between individuals. Thus, immunotherapy is designed to target specific cells and allow the immune system to target the specific cells or components of the body that are causing disease.

6.5.2 What are monoclonal antibodies?

Monoclonal antibodies (MAbs) are a relatively new class of drug that can be used in the treatment of cancers.

MAbs are specially designed sets of antibodies, with every antibody in the set binding to the same antigen (protein marker).

Monoclonal antibodies (MAbs) are:

- artificially produced antibodies that bind to one specific type of antigen
- produced in the laboratory by stimulating the production of B lymphocytes in mice injected with a specific type of antigen.

monoclonal antibodies specifically designed antibodies used in the treatment of some diseases such as cancer and autoimmune disease

FIGURE 6.35 Different monoclonal antibodies (IgG1 and IgG2a respectively)

Monoclonal antibodies are being used to treat some types of cancer and autoimmune disease. They can be used alone or to carry drugs, toxins or radioactive substances directly to cells. Many monoclonal antibodies (such as rituximab) can be used to treat blood cancers (such as leukaemia and lymphoma) and autoimmune diseases (such as rheumatoid arthritis, systemic lupus erythematosus and multiple sclerosis).

Making monoclonal antibodies

The process to make monoclonal antibodies is summarised in figure 6.36. The steps to produce monoclonal antibodies are:

1. A mouse is injected with antigen X.
2. This activates the production of its B cells, which produce antibodies against antigen X.
3. To increase the concentration of these antibodies, repeat injections followed by a booster may be given.
4. The spleen of the mouse is removed, placed in a culture medium and its cells are separated.
5. This produces a mixture of B cells, only some of which can form antibodies against antigen X.
6. Mouse tumour cells (myeloma cells) that can constantly divide are added to the separated B cells. Some B cells fuse with tumour cells to form new cells called hybridomas.
7. The unfused cells die, leaving only hybridoma cells.
8. Individual hybridoma cells are cultured in a new medium — one cell per well — and allowed to divide repeatedly. This is the cloning step during which multiple identical copies of each individual hybridoma cell are produced.
9. Each individual clone is screened for the presence of the required antibody so that clones of cells that produce antibodies against antigen X are identified.
10. The selected clones can be grown indefinitely in mass culture, and the required antibodies against antigen X can be harvested as required from the culture medium.

This amazing process uses immortal tumour cells to help create antibodies against other cancer cells. Usually, B cells cannot divide, but by fusing a B cell with a tumour cell that can divide indefinitely, a non-stop factory with an antibody production line is created. In 1984, Georges Kohler and César Milstein won a Nobel Prize for their work developing this technique.

The term *monoclonal antibodies* comes from their function, structure and source. They are 'monoclonal' because the antibodies come from clones of one parent cell and, importantly, they are antibodies specific for one known antigen.

FIGURE 6.36 The process to produce monoclonal antibodies

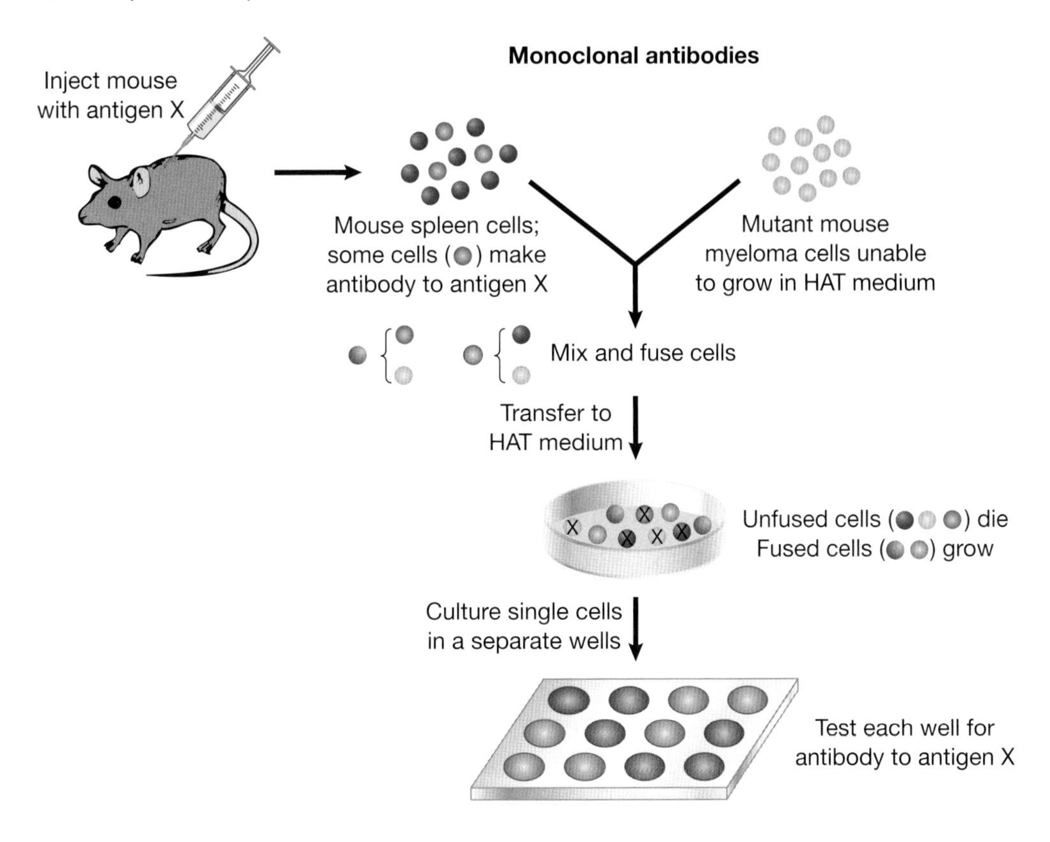

6.5.3 Use of monoclonal antibodies to treat cancer

Researchers can design antibodies that target particular antigens on cancer cells, and they can make multiple copies of these antibodies in the laboratory. Table 6.13 shows some of the monoclonal antibodies that have been designed for use in the treatment of various cancers. Monoclonal antibodies are most commonly not used alone — they are used in combination with chemotherapy and/or radiotherapy. In some cases, the use of a particular MAb may be useful only for a particular subset of cancer patients, as, for example, the use of Herceptin for those breast cancer patients who are HER2-positive.

TABLE 6.13 Some monoclonal antibodies used in the treatment of particular cancers. The two names of each MAb are its generic name and, in brackets, its trade name. All generic names all have a –mab suffix.

Monoclonal antibody	Examples of type(s) of cancer treated with MAbs	Mode of action
Bevacizumab (Avastin)	Various, including advanced colorectal cancers; lung, brain and breast cancers	Blocks growth of new blood vessels to cancer
Alemtuzumab (Campath)	Chronic lymphocytic leukaemia (CLL)	Attaches to a surface protein on cancer cells and signals immune cells to eliminate cancer cells
Trastuzumab (Herceptin)	Some breast and stomach cancers	Blocks signals for cancer cells to divide
Brentuximab (Adcetris)	Hodgkin lymphoma	Carries an attached chemotherapy drug to cancer
Ibritumomab (Zevalin)	Non-Hodgkin lymphoma	Carries an attached radioisotope (Y-90) to cancer
Rituximab (Rituxan®)	Non-Hodgkin lymphoma and chronic lymphocytic leukaemia (CLL)	Block growth signals of cancer cells

Modes of action of MAbs in cancer treatments

As highlighted in table 6.13, there are many different monoclonal antibodies that can be used in the treatment of cancer. These can act in different ways.

Modes of action

The four main modes of action of MAbs are to:

- stop the growth of new blood vessels
- signal immune cells to attack
- block growth factors
- deliver anticancer or radioisotopes to cancer cells.

Monoclonal antibodies used in cancer treatments can be divided into naked MAbs and conjugated MAbs. Naked MAbs do not have any other molecules joined to them, while conjugated MAbs have an additional group attached. Both types of MAbs are used in the treatment of cancers.

Use of naked monoclonal antibodies in treating cancers

1. **Stopping the growth of new blood vessels to cancers**: A solid cancer, such as a malignant tumour, cannot grow beyond a size of about two millimetres in diameter without being supplied with new blood vessels.These tumours release growth factors, such as VEGF (vascular endothelial growth factor), that diffuse to nearby blood vessels and signal them to sprout new blood vessels (see figure 6.37). The new blood vessels grow into the tumour, transporting oxygen and nutrients to the cancer cells, enabling further growth of the tumour.
   Monoclonal antibodies, such as bevacizumab (Avastin), block the growth of new blood vessels to malignant tumours. They do this by binding to a growth factor (VEGF) released by cancer cells. This binding stops the communication between the tumour and nearby blood vessels. Because the blood vessels do not receive the signal, they do not sprout new vessels. Without an increased blood supply to provide oxygen and nutrients to its cells and remove wastes from its cells, a malignant tumour cannot continue growing.

FIGURE 6.37 Blood vessels sprouting towards a solid cancer in response to growth factor signals from cancer cells.

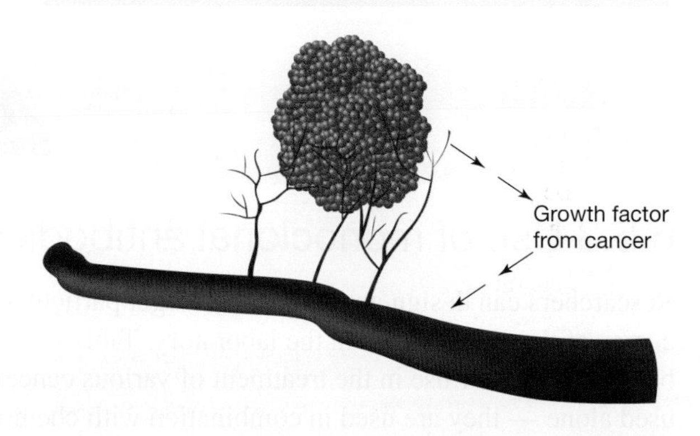

2. **Signalling immune cells to attack cancers:** Some cancers are not very ‘visible’ to immune cells. Some MAbs bind to antigens on cancer cells (see figure 6.38) and act as markers that attract immune cells to attack the cancer cells.
   One monoclonal antibody used for this purpose is alemtuzmab (Campath). Campath may be used in the treatment of chronic leukaemia, a cancer of one class of white blood cells. Campath binds to the CD52 antigen on certain white blood cells. The attached antibody acts as a signal that attracts immune cells to the cancer, and the immune cells begin eliminating these white blood cells, including the leukaemic cells.

FIGURE 6.38 Monoclonal antibodies bind to one specific protein (antigen) on a cancer cell.

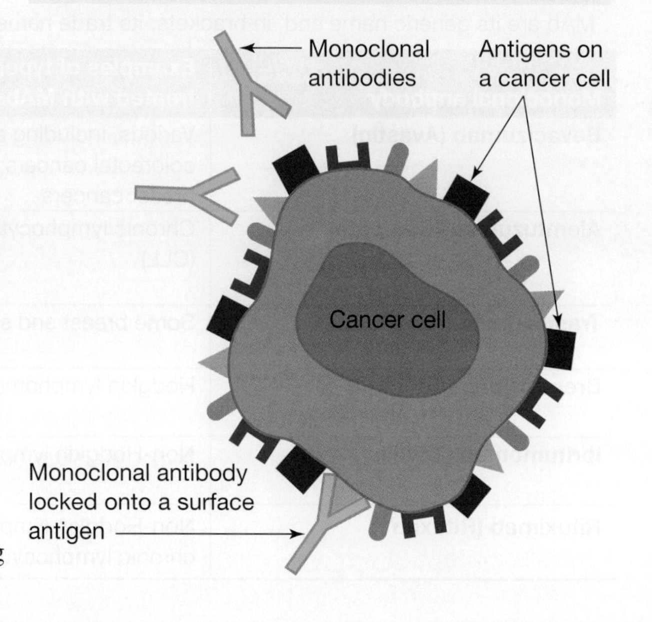

3. **Blocking signals for cell division:** When growth factors bind to receptors on cancer cells, this produces a signal for the cancer cells to divide. In healthy breast tissue, surface receptors receive signals from growth factors for normal cell replacement. These receptors are known as HER2 (human epidermal growth factor receptor 2). About 20 per cent of breast cancers are classified as HER2-positive breast cancers. In these cancer cells, the receptor is over-expressed, resulting in increased signalling that directs the cancer cells to divide uncontrollably. The monoclonal antibody trastuzumab (Herceptin) binds to these receptors, blocking them from receiving signals from growth factors (see figure 6.39). This slows or prevents the growth of the cancer. (A polarised light micrograph of Herceptin crystals was shown on the topic opener page.)

FIGURE 6.39 Herceptin binds to HER2 receptors on breast cancer cells, blocking them from growth factors.

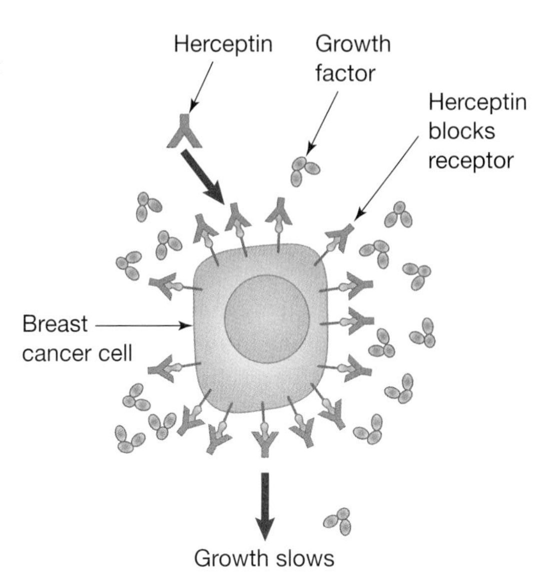

Conjugated monoclonal antibodies

Conjugated monoclonal antibodies are MAbs joined to a second molecule, such as a chemotherapy drug or a radioisotope particle.

1. **Delivering anticancer drugs to cancer cells:** Some monoclonal antibodies, such as brentuximab (Adcetris), can be joined to molecules of a chemotherapy drug and then deliver the cancer-killing drug directly to the target antigen on particular cancer cells. In the case of Adcetris, its target antigen is the CD30 protein on lymphoma cells, and it can be used to deliver a chemotherapy drug in some cases of Hodgkin lymphoma.
2. **Delivering radioisotopes to cancer cells:** Some monoclonal antibodies can be joined to a particle of a radioisotope and deliver it to cancer cells. Once the antibody links to its target antigen on the cancer cell, the radioisotope emits radiation that can destroy the cancer cell. One such MAb is britumomab (Zevalin), which has as its target antigen the CD20 protein on the surface of B lymphocytes. Zevalin can be used in the treatment of some types of non-Hodgkin lymphoma.

FIGURE 6.40 A monoclonal antibody joined to a radioisotope or chemotherapy drug, which can bind to a cancer cell with the right target antigen

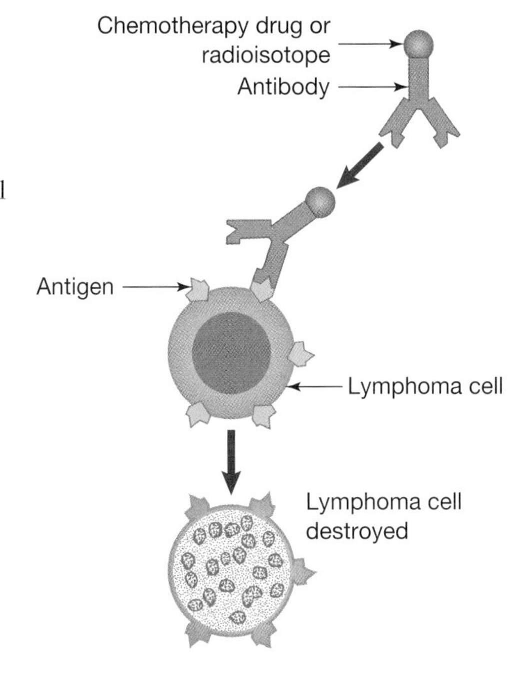

 Resources

 **Weblink** Lymphoma Australia — Monoclonal antibodies

6.5.4 Use of monoclonal antibodies to treat autoimmune diseases

A key feature of the immune system is its ability to distinguish between 'self' and 'non-self'. Normally, a person's immune system does not attack tissues whose cells carry that person's own self HLA markers. Sometimes, however, this 'self' recognition fails, and a person's immune system attacks and destroys their own body cells, tissues or organs, resulting in autoimmune diseases.

In autoimmune diseases, the body produces autoantibodies — that is, antibodies that attack the body's own cells. Instead of defending the body against foreign pathogens, the immune system attacks the person's own body cells. Autoimmunity occurs when T cells and/or B cells are inappropriately activated, resulting in autoimmune disease.

When the immune system produces autoantibodies, these antibodies can lock onto self antigens on the person's own body cells, resulting in an immune attack on those cells. What causes the body to mistakenly produce antibodies against its own cells is not known, but several factors, including viral infections, genetic factors, hormones and drugs, have been suggested as possible triggers for this malfunction.

In topic 5 (in subtopic 5.6), chronic inflammation was discussed. Many diseases that are linked to chronic inflammation are due to autoimmune diseases. Examples of these are shown in the following case study box.

CASE STUDY: Examples of autoimmune disease

More than 80 different autoimmune diseases have been identified. It is estimated that 1 in 20 people in Australia have an autoimmune disease. Some autoimmune diseases affect one type of cell or one organ of the body, while other autoimmune diseases affect several body systems.

Some examples of autoimmune diseases include:

- Multiple sclerosis (MS): Autoantibodies target myelin around nerve fibres. This results in numbness, slurred speech, fatigue and a lack of coordination.
- Type 1 diabetes: Autoantibodies target beta islet cells in the pancreas. This results in the inability to produce the hormone insulin, affecting the ability to regulate blood glucose.
- Systematic lupus erythematosus (SLE): Autoantibodies targets many tissues through the body (multisystem). This leads to symptoms such as light sensitivity, rashes, fever and fatigue.

To access more information about different autoimmune diseases, please download the digital document.

FIGURE 6.41 In MS, the myelin sheath is destroyed by autoantibodies.

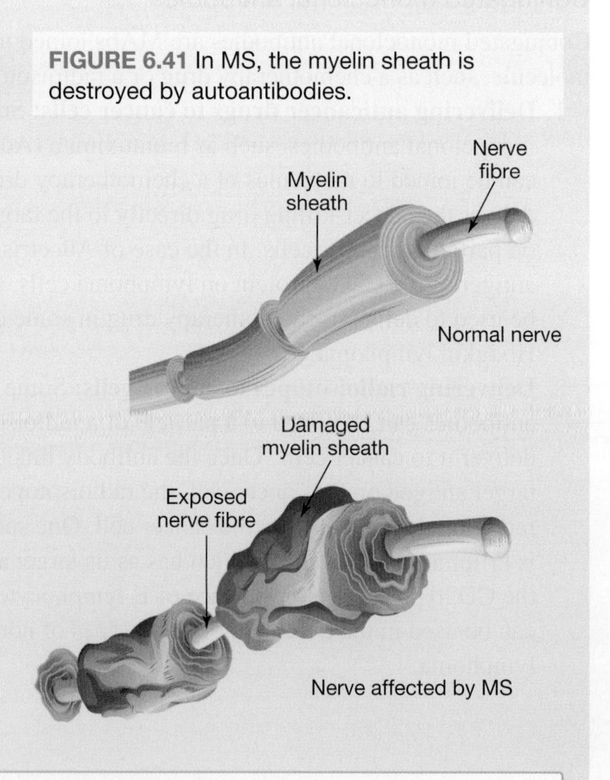

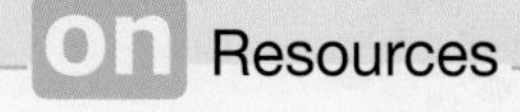 Resources

Digital document Case study: Examples of autoimmune diseases (doc-36103)

Strategies for treating autoimmune disease with monoclonal antibodies

While monoclonal antibodies used in cancer are often used to increase the immune response, they can also be used to decrease the immune response. As autoimmune disease is caused by the immune system attacking 'self' cells, monoclonal antibodies can be designed to act against specific cells of the immune system which cause autoimmune disease.

There are currently no reliable and safe strategies to cure autoimmune diseases such as systemic lupus or multiple sclerosis. Severe cases require cytotoxic drugs, which frequently cause serious side effects. The development of monoclonal antibodies has led to new therapeutic strategies through which the treatment can be focused more directly towards specific cells. Research into immunotherapy and the use of monoclonal antibodies for autoimmune disease continues. The success of this research could help millions of individuals with autoimmune diseases globally.

Certain antibodies are used as immunosuppressants and can thereby help in organ transplantation.

The most important advantage of monoclonal antibodies is the ability to produce pure antibody without a pure antigen. For example, a suspension of lymphocytes can produce monoclonal antibodies, each of which react with one specific antigen. This can be useful in treating different autoimmune diseases in individuals with unique self-antigens.

TABLE 6.14 Different monoclonal antibodies used for treating autoimmune disease

Monoclonal antibody	Examples of autoimmune diseases treated with MAbs	Mode of action
Infliximab (Inflecta®, Remicade)	Various, including Crohn's disease, ulcerative colitis and rheumatoid arthritis	Binds to cytokines (TNF) to prevent binding to receptors
Adalimumab (Humira®)	Various, including Crohn's disease, ulcerative colitis and rheumatoid arthritis	Inactivates the cytokine TNF-α
Basiliximab (Simulect®)	Organ rejection after transplantation and chronic inflammation of skin	Acts against a cytokine receptor (IL-2) of T cells to stop signalling
Natalizumab (Tysabri®)	MS and Crohn's disease	Prevents ability of immune cells to attach to other cells and move into the central nervous system
Omalizumab (Xolair®)	Bullous pemphigoid and allergic asthma	Binds to free IgE that often causes allergic responses
Rituximab (Rituxan®)	Rheumatoid arthritis and MS	Targets proteins on surface of B cells

Monoclonal antibodies can be used to alter the course of an autoimmune disease by directing the antibodies against major histocompatibility antigens to prevent them triggering an autoimmune response. The strategy was based on the fact that many autoimmune diseases are linked with HLA or MHC antigens. These antigens, also called MHC-II antigens, play important role in promoting the immune response.

Both humoral and cell-mediated immune responses are initiated when antigens are presented by dendritic cells (APCs) in association with MHC-II markers to helper T cells. Because MHC-II are required to promote immune response and certain MHC-II are associated with autoimmunity, it might be possible to stop autoimmunity by using monoclonal antibodies to block certain MHC-II antigens. It is important that only specific markers are targeted to ensure that immunity against non-self material is still functional.

FIGURE 6.42 Blocking MHC-II markers can prevent autoimmunity.

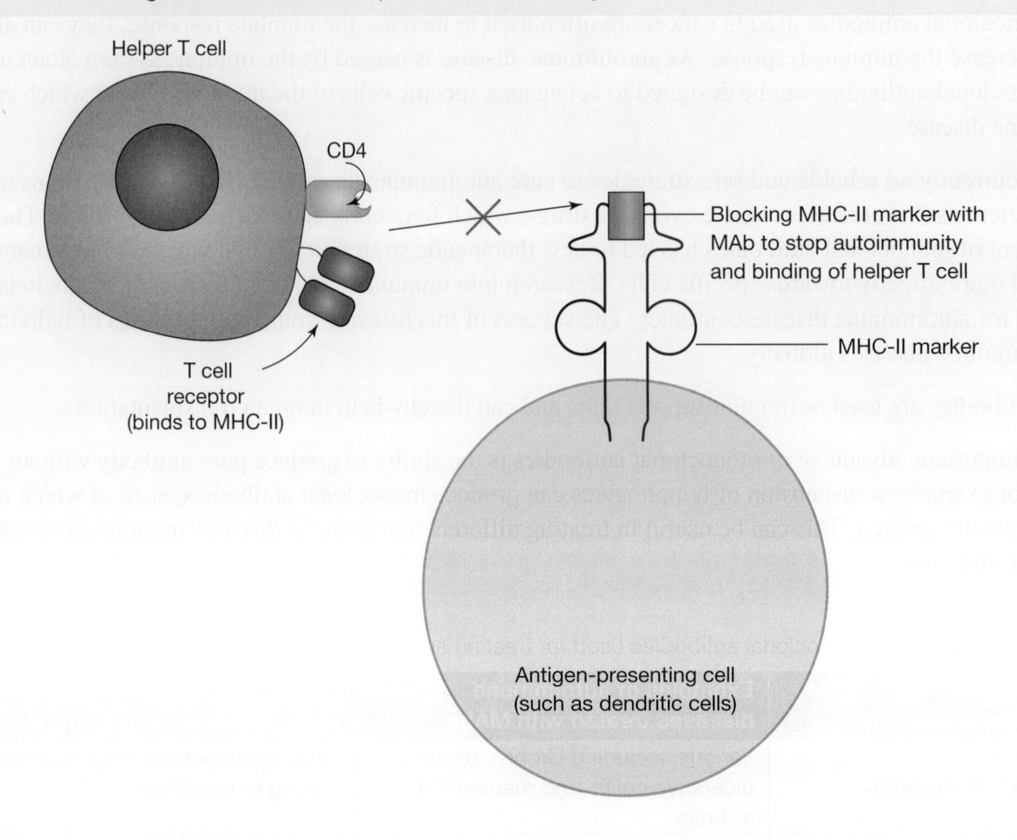

tlvd-1787

SAMPLE PROBLEM 5 Comparing monoclonal antibodies

Monoclonal antibodies can be used in the treatment of both cancer and autoimmune diseases.

a. Compare and contrast the use of monoclonal antibodies in cancer and autoimmune diseases. **(2 marks)**

b. One target for monoclonal antibodies is the CD40 molecule, which is expressed on a range of tumour cells. Why might targeting CD40 with monoclonal antibodies conjugated with chemotherapy drugs be advantageous over using chemotherapy drugs alone? **(2 marks)**

THINK	WRITE
a. This question is asking you to both compare and contrast. Therefore, you need to identify differences (contrast) and similarities (compare).	Both cancer and autoimmune disease can be treated using monoclonal antibodies that are made to target a **specific** protein (or antigen) that causes disease (1 mark). However, in cancer treatment, monoclonal antibodies usually lead to the activation of the immune response (by making cancer cells more visible). In autoimmune diseases, monoclonal antibodies are used to suppress the immune response against self-cells (1 mark).

b. Consider the side effects of chemotherapy alone, and which of these side effects are less likely when monoclonal antibodies are used. This is a two-mark question, so you need to be able to include:
- how this occurs
- why it is advantageous.

Chemotherapy targets all fast-growing and replicating cells which, while including cancer cells, often include many other cell types in the body. By targeting only CD40, it is mostly only tumour cells being targeted (1 mark).
Therefore, using monoclonal antibodies is advantageous as it will result in fewer of the side effects of chemotherapy, which usually target healthy cells (1 mark).

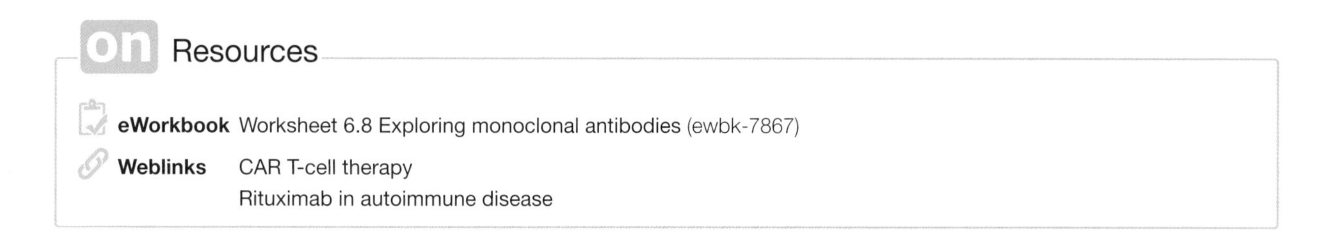

KEY IDEAS

- Monoclonal antibodies (MAbs) are produced by a single clone of a cell and consist of identical antibody molecules that bind to the same antigen.
- Monoclonal antibodies are being used in the treatment of various cancers.
- Various monoclonal antibodies block the division of cancer cells through different modes of action.
- Some monoclonal antibodies are attached to other chemotherapy drugs and radioisotopes which are also used for targeted delivery to cancer cells.
- Monoclonal antibodies may also be used in the treatment of autoimmune diseases and target immune cells to decrease the immune response against self cells.

6.5 Activities

learnon

To answer questions online and to receive **immediate feedback** and **sample responses** for every question, go to your learnON title at **www.jacplus.com.au**. A **downloadable solutions** file is also available in the resources tab.

6.5 Quick quiz on	**6.5 Exercise**	**6.5 Exam questions**

6.5 Exercise

1. Give an example of a MAb that can block the growth of new blood vessels into a solid mass of cancer cells.
2. What is the expected outcome of preventing the growth of new blood vessels into a solid mass of cancer cells?
3. Why is it useful to combine monoclonal antibodies with chemotherapy drugs or radioisotopes?
4. a. Briefly describe how Herceptin can slow or stop the growth of HER2-positive breast cancer cells.
   b. Could another MAb, such as Zevalin be used for the same purpose? Briefly explain.
5. a. How do autoimmune diseases occur?
   b. Why are monoclonal antibodies useful in the treatment of autoimmune diseases?
6. Summarise the steps used to produce monoclonal antibodies.

6.5 Exam questions

Question 1 (1 mark)

Source: *VCAA 2020 Biology Exam, Section A, Q35*

MC Monoclonal antibodies attaching to antigens on a cancer cell are shown in the diagram below.

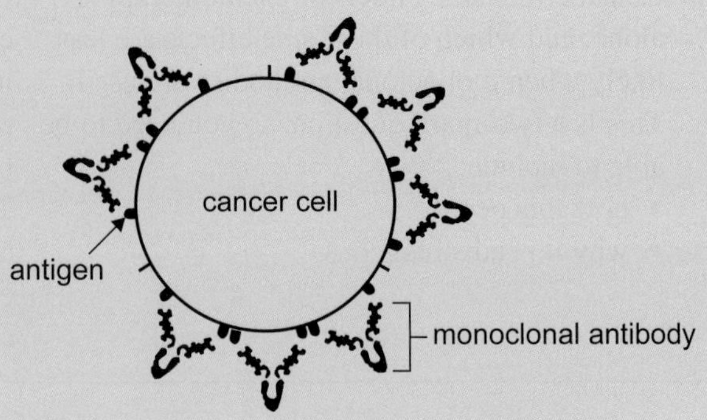

Source: adapted from N. Halim, 'Monoclonal antibodies: A-25 roller coaster ride'. *The Scientist*, 20 February 2000, <www.the-scientist.com>

Monoclonal antibodies

A. are used to suppress B cells acting on cancer cells.
B. make it easier for cells of the immune system to detect cancer cells.
C. can bind to dendritic cells to stimulate them to destroy cancer cells.
D. can attach to many structurally different proteins found on the surface of cancer cells.

Question 2 (1 mark)

Source: *VCAA 2018 Biology Exam, Section A, Q24*

MC Monoclonal antibodies can be produced and used to treat different types of cancers.

Which one of the following is a correct statement about monoclonal antibodies?

A. Monoclonal antibodies are carbohydrate molecules.
B. Monoclonal antibodies produced from the same clone of a cell are specific to the same antigen.
C. Monoclonal antibodies pass through the plasma membrane of a cancer cell and attach to an antigen within the cell.
D. Monoclonal antibodies produced to treat stomach cancer will be identical to monoclonal antibodies produced to treat breast cancer.

Question 3 (1 mark)

MC Monoclonal antibodies are cultured from a special 'fused' cell.

A specific, antibody-producing B-cell is obtained from a mouse that was previously treated with the desired antigen. The specific B-cell is fused with a hybridoma (cancer) cell. The resulting 'fused' cell divides indefinitely (like a cancer cell).

The main advantage of the 'fused cell' technique is that

A. the antibody produced by the fused cells will target only hybridoma cells.
B. laboratory mice will not develop cancer through this procedure.
C. the fused cells provide a long-lasting supply of specific antibody for harvest in the laboratory.
D. the antigen confers life-long immunity in the mice for a range of human diseases.

Question 4 (1 mark)

Monoclonal antibodies can be made to differ in their structure according to their intended purpose.

Explain the structure of a type of monoclonal antibody designed to target specific cancer cells.

Question 5 (1 mark)

MC Monoclonal antibodies

A. occur naturally in mice.
B. occur naturally in humans.
C. are artificially produced in humans.
D. are artificially produced in a laboratory.

More exam questions are available in your learnON title.

6.6 Review

6.6.1 Topic summary

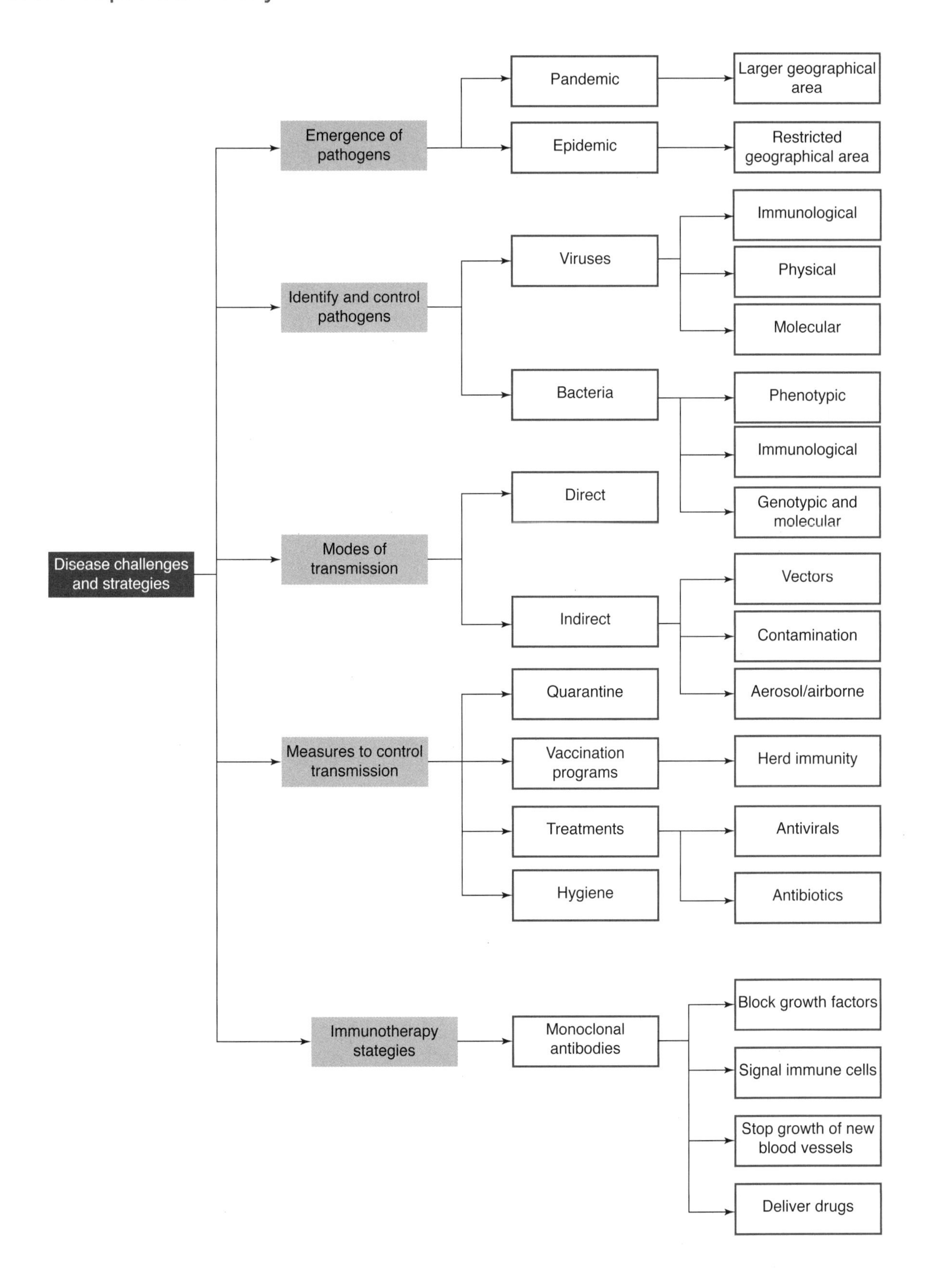

on Resources

eWorkbook	Worksheet 6.9 Reflection — Topic 6 (ewbk-7871)
Practical investigation eLogbook	Practical investigation eLogbook — Topic 6 (elog-0189)
Digital document	Key terms glossary — Topic 6 (doc-34620) Key ideas summary — Topic 6 (doc-34611)
Exam question booklet	Exam question booklet — Topic 6 (eqb-0017)

6.6 Exercises

learnon

To answer questions online and to receive **immediate feedback** and **sample responses** for every question, go to your learnON title at **www.jacplus.com.au**. A **downloadable solutions** file is also available in the resources tab.

6.6 Exercise 1: Review questions

1. In 2020, the Australian Department of Health advised people returning from China to remain isolated inside their homes for at least 14 days due to an outbreak of coronavirus. Explain the reasoning behind this process.
2. The following graph shows the death rates from influenza for males in various age groups in New South Wales during two influenza pandemics:

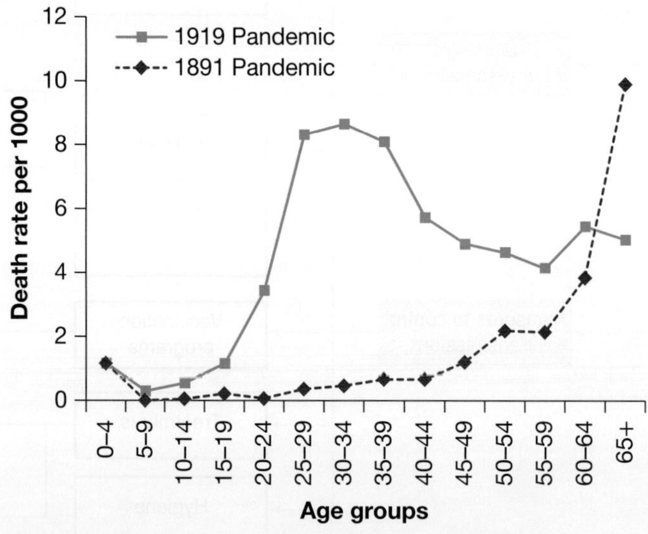

Source: Reproduced by permission, NSW Ministry of Health, 2016.

   a. Explain why the 1891 and 1919 influenza outbreaks were classified as pandemics, rather than epidemics.
   b. What pattern in the death rates is apparent in the 1919 influenza pandemic?
   c. In 1919, compare the death rate for males in the 25 to 29 age group to males in the 65+ age group?
   d. What pattern in the death rates is apparent in the 1891 influenza pandemic?
   e. In 1891, compare the death rate for males in the 25 to 29 age group to males in the 65+ age group.
   f. Does this data indicate that the same strain of influenza was responsible for these two pandemics? Explain.
   g. Overall in Australia, the death rate from influenza from the 1919 pandemic was about three deaths per 1000 people. However, death rates differed in various countries. In the Pacific country of Western Samoa during the 1918–1919 pandemic, there were 8500 deaths in a total population of only 38 000, equating to approximately 224 deaths per 1000 people. Suggest two reasons these death rates differed.

3. Formulate a biologically valid explanation for each of the following observations:
   a. New influenza vaccines are typically developed on an annual basis.
   b. Epidemics tend to begin without warning.
   c. Antibiotics are not effective against virus infections.
   d. Monoclonal antibodies can reduce the side effects of chemotherapy and radiotherapy drugs in cancer treatment.
4. The following diagram shows the life cycle of *Plasmodium*, a malarial parasite, which involves two hosts.

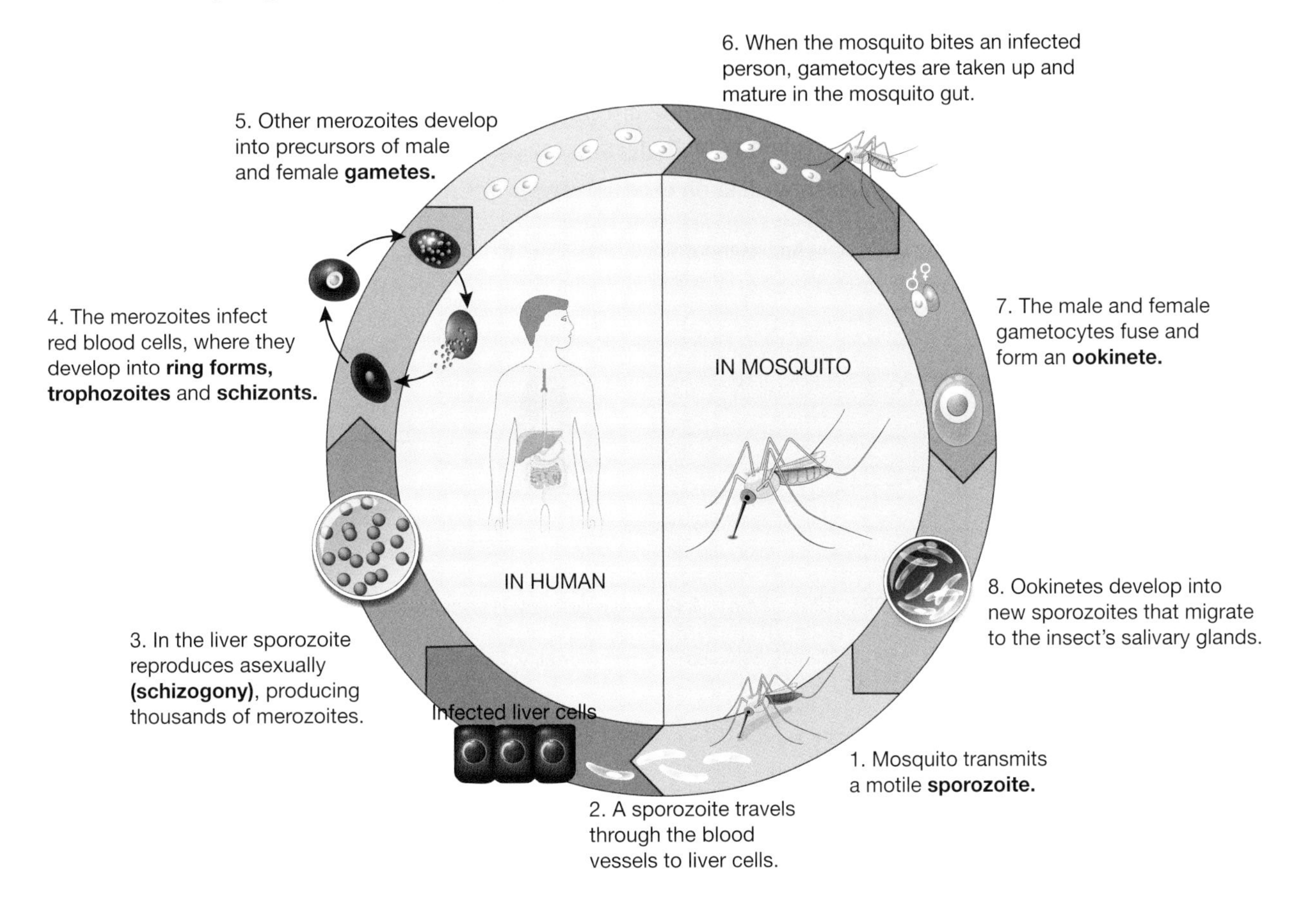

   a. Define the role of mosquito in transmission of malaria.
   b. Suggest two measures to prevent the spread of malaria.
   c. An antigen known as apical membrane antigen 1 (AMA1) appears on the surface of the merozoite during the blood-stage of *Plasmodium* parasite and is a necessary component for the invasion of red blood cells. If a vaccine is created to protect the people from malaria, what information would be required?
5. Identify one key difference between the members of the following pairs:
   a. infection and disease
   b. Gram-negative and Gram-positive bacteria.
6. Various kinds of bacteria can be grown in a special medium known as thioglycolate broth, which contains a chemical that absorbs oxygen. However, oxygen diffuses into the broth from the air so that the highest concentration of oxygen is present in the broth at the top of the tube but the concentration decreases with depth, with a zero concentration of oxygen at the bottom of the tube.
   This broth is used to identify different kinds of bacteria in terms of their oxygen requirements.

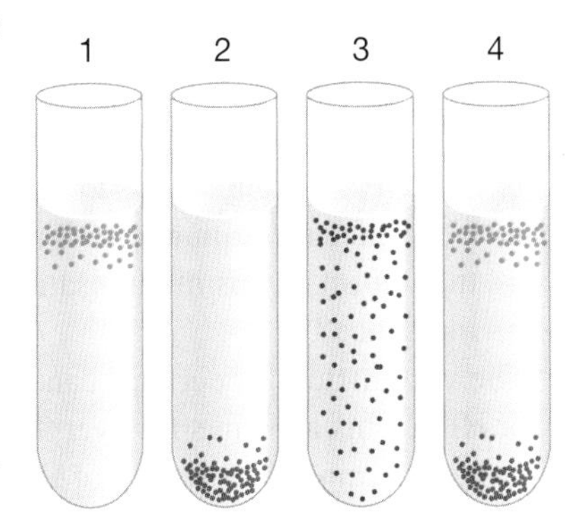

Consider the following kinds of bacteria in terms of their oxygen requirements:

i. obligate anaerobic bacteria
ii. obligate aerobic bacteria
iii. a mixture of obligate anaerobes and obligate aerobes
iv. facultative aerobic bacteria.

a. Match each group of bacteria (i to iii) to its predicted pattern of distribution in one of the tubes of thioglycolate broth (1 to 4) in the provided figure.
b. Other bacteria are termed aerotolerant; these bacteria do not require oxygen for their energy production, but they are not poisoned by oxygen as is the case for obligate anaerobes. Draw a tube showing a likely distribution of aerotolerant bacteria in a tube of thioglycolate broth.
c. A sample of bacteria from a suspected *Clostridium* infection was placed in thiogylcolate broth. What pattern would be expected? Explain your choice.

7. Examine the following graphs, which show data on cervical cancer in Australia.

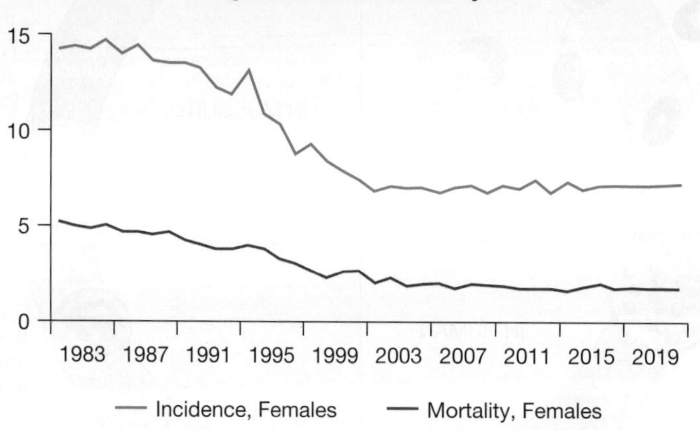

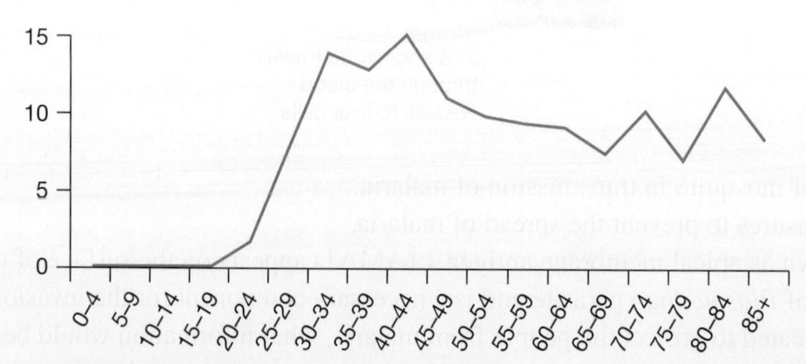

Source: Australian Institute of Health and Welfare.

a. What general trend is apparent in the incidence of cervical cancer in the period from 1983 to 2019?
b. What general trend is apparent in the mortality rate for cervical cancer in the period from 1983 to 2019?
c. Which age group shows the most marked 'jump' in incidence compared to the preceding age group?
d. Cervical cancer immunisation was introduced in Australia in 2007 for 13-year-old females. Would the impact of this immunisation program be expected to have an immediate impact on cervical cancer incidence?

The vaccination for cervical cancer is actually a vaccination against a virus known as human papilloma virus or HPV. This accounts for over 90 per cent of cervical cancer cases. The current HPV vaccine Gardasil 9 protects against nine strains of HPV.

e. In 2013, the HPV vaccine was extended to males. Why might males need the HPV vaccination when they are not at risk of cervical cancer?

8. *Listeria monocytogenes* causes the bacterial disease listeriosis, which is usually not serious in most healthy people, but can cause miscarriages and may be fatal in elderly people and in people whose immune systems are weakened. *L. monocytogenes* infection most commonly occurs via consumption of contaminated food. Unusually, *L. monocytogenes* can grow at temperatures as low as 0 °C, although relatively slowly. The generation time (doubling time) of *L. monocytogenes* growing in dairy products was found to be about 30 hours at 4 °C, about 2 hours at 21 °C and about 0.8 hours at 35 °C.
   a. Starting with 50 *Listeria* cells in contaminated milk, about how many bacterial cells would be expected to be present after a period of 16 hours:
      i. when the milk is stored at 21 °C
      ii. when the milk is left in the summer sun at 35 °C
      iii. when the milk is stored in the refrigerator?
   b. Which of these situations is most similar to what can happen if *Listeria* bacteria gained entry to the human body?
   c. Why is correct food storage an important public health consideration?

6.6 Exercise 2: Exam practice questions

on Resources

Teacher-led videos: Teacher-led videos for every exam question

Section A — Multiple choice questions

All correct answers are worth 1 mark each; an incorrect answer is worth 0.

Use the following information to answer Questions 1 and 2

Source: *VCAA 2018 Biology Exam, Section A, Q32 and 33*

The table below compares how eight diseases spread and the number of people likely to be infected by one other infected person

Disease	measles	whooping cough	rubella	polio	smallpox	mumps	severe acute respiratory syndrome (SARS)	Ebola
How it spreads	airborne droplets	airborne droplets	airborne droplets	fecal–oral route	airborne droplets	airborne droplets	airborne droplets	bodily fluids
Number of people infected from one other person	12 to 18	12 to 17	6 to 7	5 to 7	5 to 7	4 to 7	2 to 4	1 to 4

Data: © 2018 Thomson Reuters

Question 1

Source: *VCAA 2018 Biology Exam, Section A, Q32*

What would be the most effective method of preventing the spread of measles during an outbreak?

A. Wash hands thoroughly after going to the toilet.
B. Establish a 'clean needle' exchange program.
C. Vaccinate all infected people.
D. Isolate all infected people.

Question 2

Source: VCAA 2018 Biology Exam, Section A, Q33

Based on the information in the table, which one of the following statements is correct?

A. Ebola is the most contagious disease.
B. Polio and smallpox have a similar infection rate.
C. More people would die from measles than any other disease shown.
D. The fecal–oral route is the most effective means of spreading pathogens.

Question 3

Source: VCAA 2018 Biology Exam, Section A, Q36

Chagas' disease, a potentially life-threatening illness in humans, is caused by the protozoan parasite *Trypanosoma cruzi*. The disease is found mainly in Latin American countries, where the parasite is spread by an insect called a triatomine.

In this example, the role of the triatomine is to

A. introduce a virus into the protozoan parasite.
B. infect the insect with Chagas' disease.
C. introduce a gene into a bacterium.
D. enable *T. cruzi* to enter the host.

Question 4

Source: VCAA 2017 Biology Exam, Section A, Q37

Yellow fever is a viral disease that is transmitted primarily by mosquitoes.

An outbreak of yellow fever was reported to have occurred in an area of Brazil in January 2017. This outbreak was reported to be spreading to other areas within Brazil.

Which one of the following is a correct statement about this outbreak of yellow fever?

A. This outbreak of yellow fever is considered to be a pandemic.
B. Infected individuals who travel to other areas of Brazil will not increase the spread of the disease.
C. This outbreak of yellow fever is occurring in populations with high vaccination rates for yellow fever.
D. Elimination of mosquito breeding sites in areas with yellow fever will reduce the number of individuals affected.

Question 5

Source: VCAA 2018 Biology Exam, Section A, Q16

Lupus is a condition that results in the increased secretion of antibodies that attach themselves to healthy cells in a patient's body. The accumulation of these antibodies causes inflammation, joint pain, rash, fatigue and fever.

Lupus is an example of

A. an allergic reaction.
B. an autoimmune disease.
C. an immune deficiency disease.
D. a complement protein response.

Question 6

Malaria is caused by a parasite known as Plasmodium, which normally spreads to humans by mosquito bites. When a mosquito bites a human, the protist enters the bloodstream.

Using the information given above, it can be concluded that

A. protists must be non-cellular pathogens.
B. the mosquito is a vector.
C. the mosquito must also show symptoms of malaria.
D. malaria is an incurable disease.

Question 7

Source: *VCAA 2018 Biology Exam, Section A, Q29*

The graph below shows the death rates from acquired immune deficiency syndrome (AIDS) and also the number of people infected with the human immunodeficiency virus (HIV). Before 1995 many people infected with the virus went on to develop AIDS, which led to their deaths.

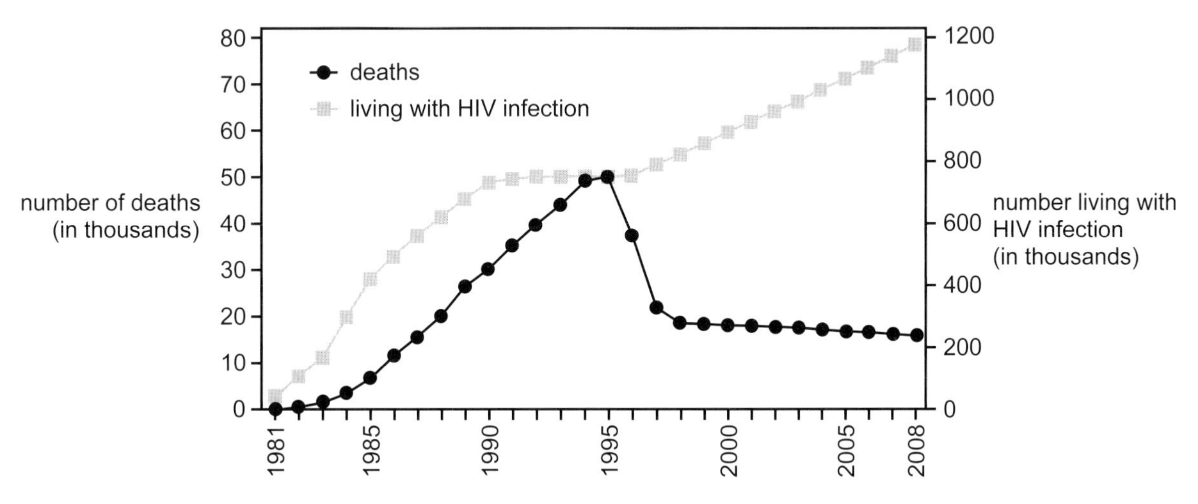

Source: National Institute on Drug Abuse, 'How does drug abuse affect the HIV epidemic?', research report, (www.drugabuse.gov)

Based on the information in the graph, what is the most likely reason for the change in death rates, even though infection rates continued to climb after 1995?

A. People had access to new antiviral drugs.
B. People were educated about what caused HIV infection.
C. People infected with HIV were isolated from the rest of the public.
D. A widespread vaccination program for HIV was introduced within a targeted population.

Question 8

Antibiotics are chemicals that are

A. effective against animal cells.
B. synthesised naturally by protozoans.
C. have bactericidal or bacteriostatic actions.
D. cannot bind to viruses because of a lack of genetic information.

Question 9

Which of the following is not a way to identify a viral pathogen?

A. Physical methods such as x-ray crystallography
B. Molecular methods to find specific nucleic acid sequences
C. Immunological methods using capture antibodies
D. Phenotypic methods by growing the viruses in on different agar plates

Question 10

Monoclonal antibodies currently used clinically

A. can protect against a wide variety of viruses and bacteria.
B. can reduce the inflammation associated with rheumatoid arthritis.
C. are derived from the plasma of individuals already immune to these organisms.
D. have broad specificity for many antigens.

Section B — Short answer questions

Question 11 (11 marks)

SARS or severe acute respiratory syndrome was first reported in China in November 2002. It took just a few months for this highly contagious and sometimes fatal respiratory illness to spread through 26 countries in Asia, North and South America and Europe. This outbreak ceased in 2004. The causative agent was a coronavirus (a different strain as to the coronavirus that caused COVID-19).

a. Does this suggest that SARS was an existing or a new virus? Explain. **2 marks**
b. Identify a factor that might have contributed to the speed of this disease's spread. **1 mark**
c. Would this event be identified as a SARS epidemic or a SARS pandemic? Explain. **2 marks**

In Singapore, there were 238 cases of SARS with 33 deaths. Across all countries affected by SARS, there were 8096 cases with 774 deaths.

d. Compare the mortality rates between Singapore and other countries. **1 mark**
e. Which, if any, of the following agents might have been tried in the treatment of SARS: penicillin or monoclonal antibodies? Explain. **2 marks**

Read the following:

When SARS was first reported, its cause was not known. A key task was to find the causative agent and the top suspect was a coronavirus, now known as SARS coronavirus. Scientists in the Netherlands demonstrated that the SARS coronavirus was indeed the causative agent of SARS.

f. Why is identifying the causative agent of a new disease important? **1 mark**

After SARS, changes to public health practices and procedures were introduced in the affected countries in order to minimise or prevent epidemics of SARS in the future. One change in Taiwan was to replace handshakes (for greeting people) with courtesy bows.

g. Would this change be expected to reduce the risk of transmission of SARS? **1 mark**
h. Suggest three other possible changes that might have been introduced. **1 mark**

Question 12 (10 marks)

Source: *VCAA 2020 Biology Exam, Section A, Q10*

Measles in Samoa

Measles is one of the most contagious viruses affecting humans. Measles spreads when an infected person coughs or sneezes and the virus is breathed in by another person, or by direct contact with bodily fluids. In a susceptible population — people who have neither been vaccinated nor had measles previously — one person with measles could infect 12 to 18 other people.

The Pacific island nation of Samoa had a significant measles outbreak in 2019. This started when a person who had measles arrived in Samoa by plane in August 2019. In the following months over 5000 measles cases were recorded and more than 70 people died.

A measles outbreak was declared by the Samoan Government in October 2019. On 15 November the Samoan Government declared a 30-day state of emergency as the number of measles cases continued to rise and more people died. Ninety per cent of the deaths were among children less than five years old. More than one in five Samoan babies aged six to 11 months contracted measles during this outbreak and more than one in 150 babies in this age group died. Fewer deaths occurred in babies who were less than six months old.

Prior to the measles outbreak in Samoa, the vaccination rate for measles for five-year-old children in the country had fallen to 31% in 2018. One of the responses of the government to the outbreak was a mandatory vaccination program for all people. By early December 2019 over 90% of the population had been vaccinated.

In Australia a measles-containing vaccine (MMR vaccine) is recommended for children aged 12 months of age or older. A single dose of the measles vaccine provides protection for between 95% and 98% of recipients, while two doses protects 99% of vaccinated people. In 2018 in Australia, 95% of five-year-old Australian children were fully vaccinated.

Reference: K Gibney, 'Measles in Samoa: How a small island nation found itself in the grips of an outbreak disaster', The Conversation, 12 December 2019, https://theconversation.com/au

a. Is the measles outbreak discussed in the article above best described as an epidemic or a pandemic? Give your reasoning. **2 marks**

b. Consider the Samoan children who were less than five years old during the measles outbreak. Of this group, what age were the children who were least likely to die from measles? Explain why children of this age would be less likely to die. **2 marks**

People who are vaccinated are unlikely to be affected by the measles virus.

c. **i.** What is the percentage difference between vaccinated five-year-old children in Samoa and Australia in 2018? **1 mark**

ii. The MMR vaccine contains antigens for measles, mumps and rubella. What form of immunity is given when a person is vaccinated with the MMR vaccine? **1 mark**

iii. Some children, for example those undergoing chemotherapy, cannot be vaccinated. Explain how high vaccination rates can also protect unvaccinated individuals. **2 marks**

d. Describe **two** strategies, other than vaccination, that could reduce the transmission of measles. **2 marks**

Question 13 (9 marks)

Source: *VCAA 2019 Biology Exam, Section B, Q9*

Zika fever is a rapidly emerging viral disease. It is most commonly transferred from one person to another by the Aedes species of mosquito.

Zika fever in people was discovered in Uganda in 1947. It was thought that a bite from a mosquito had transferred the virus from monkeys to humans.

The symptoms of Zika fever are usually mild and 80% of infected humans do not show symptoms. Infection of pregnant women, however, can cause severe defects in their babies.

a. One way that diseases, such as Zika fever, are thought to occur is when a pathogen infects humans from an animal host. Identify **one** social or economic factor that could lead to this transfer between hosts. **1 mark**

b. When scientists attempt to identify a particular disease, they can look for specific antibodies in infected humans. Scientists trying to identify Zika fever infections found that testing for the antibodies produced against the Zika virus often gave them incorrect results. This was because the antibody tests that had been developed could not always identify the difference between the antibodies produced against the Zika virus and the antibodies produced against other viruses. Explain why making a correct identification of a viral pathogen is important in the control of a disease. **3 marks**

c. Explain why the antibody tests could not identify the difference between the antibodies produced against the Zika virus and the antibodies produced against other viruses. In your response, refer to the structure of the antibody. **2 marks**

d. Aedes mosquitoes are not found on every continent. They cannot fly great distances. Vaccines are currently being trialled for the Zika virus. Describe **three** different approaches, other than vaccination, that government health officials could use to reduce the spread of the Zika virus. **3 marks**

Question 14 (3 marks)

***Source:** VCAA 2010 Biology Exam, Section B, Q8a and b*

Measles is a highly contagious, serious disease caused by an RNA virus. There were regular epidemics of the disease until the introduction of mass vaccination. The following graph indicates the incidence of measles in Victoria from 1962 to 1979.

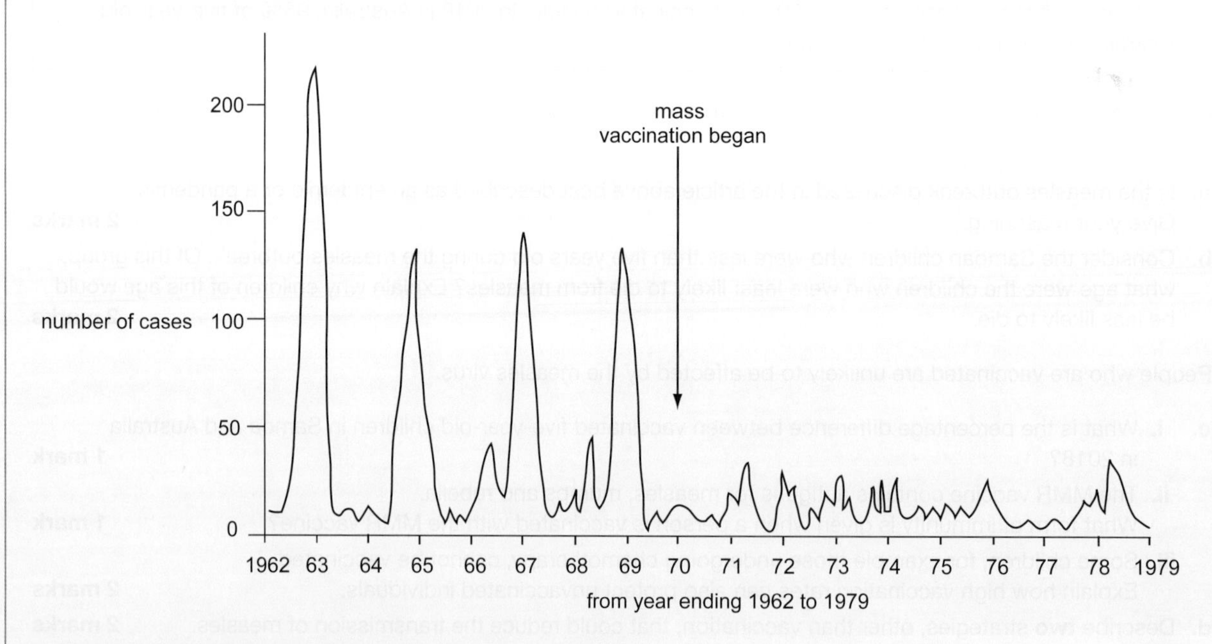

a. What was the time period between successive epidemics? **1 mark**

b. An unaffected person without immunity has a 90 per cent chance of catching measles if they live in the same house as a person with the disease. If a child is suspected of having measles, a serum sample is taken and tested for measles-specific IgM and IgG antibodies.
What conclusion could be made if high levels of these antibodies were found and what action would be taken? **2 marks**

Question 15 (5 marks)

A traditional treatment for cancer is chemotherapy, which involves using strong chemical drugs to kill cancer cells. While the treatment can be effective in killing the cancer cells, the drugs cause many side effects in patients.

Monoclonal antibodies are a relatively new kind of treatment used for treating cancers.

a. Explain how monoclonal antibodies work in killing the cancer cells. **2 marks**

b. Name the type of macromolecule monoclonal antibodies are made of. **1 mark**

c. Some monoclonal antibodies, such as brentuximab, are used to deliver a chemotherapy drug to treat cancers like Hodgkin lymphoma. Suggest a name for these type of monoclonal antibodies and explain how they work. **2 marks**

6.6 Exercise 3: Biochallenge online only

on Resources

eWorkbook Biochallenge — Topic 6 (ewbk-8087)

Solutions Solutions — Topic 6 (sol-0662)

Past VCAA examinations

Sit past VCAA examinations and receive immediate feedback, marking guides and examiner's report notes. Access Course Content and select 'Past VCAA examinations' to sit the examinations online or offline.

teachon

Test maker

Create unique tests and exams from our extensive range of questions, including past VCAA questions. Access the assignments section in learnON to begin creating and assigning assessments to students.

Online Resources

Below is a full list of **rich resources** available online for this topic. These resources are designed to bring ideas to life, to promote deep and lasting learning and to support the different learning needs of each individual.

eWorkbook

- 6.1 eWorkbook — Topic 6 (ewbk-1885) ☐
- 6.2 Worksheet 6.1 Comparing epidemics and pandemics (ewbk-7849) ☐
- 6.3 Worksheet 6.2 Identifying pathogens (ewbk-7853) ☐
- Worksheet 6.3 Controlling the spread of pathogens (ewbk-7855) ☐
- 6.4 Worksheet 6.4 Vaccination programs and protecting your pet (ewbk-7859) ☐
- Worksheet 6.5 Investigation of an immunological experiment (ewbk-7861) ☐
- Worksheet 6.6 Planning an overseas trip (ewbk-7863) ☐
- Worksheet 6.7 Herd immunity (ewbk-7865) ☐
- 6.5 Worksheet 6.8 Exploring monoclonal antibodies (ewbk-7867) ☐
- 6.6 Worksheet 6.9 Reflection — Topic 6 (ewbk-7871) ☐
- Biochallenge — Topic 6 (ewbk-8087) ☐

Solutions

- 6.6 Solutions — Topic 6 (sol-0662) ☐

Practical investigation eLogbook

- 6.1 Practical investigation eLogbook — Topic 6 (elog-0189) ☐
- 6.3 Investigation 6.1 Gram staining and biochemical testing (elog-0898) ☐
- Investigation 6.2 Optimal antibiotic concentrations (elog-0900) ☐
- Investigation 6.3 Sensitivity testing using spices and different household disinfectants (elog-0902) ☐
- 6.4 Investigation 6.4 Modelling herd immunity (elog-0904) ☐

Digital documents

- 6.1 Key science skills — VCE Units 1-4 (doc-34326) ☐
- Key terms glossary — Topic 6 (doc-34620) ☐
- Key ideas summary — Topic 6 (doc-34611) ☐
- 6.2 Case study: The Ebola, Zika and yellow fever epidemics (doc-36102) ☐
- 6.3 Extension: Rational drug design (doc-36104) ☐
- Case study: Typhoid Mary (doc-36180) ☐
- 6.5 Case study: Examples of autoimmune diseases (doc-36103) ☐

Teacher-led videos

- Exam questions — Topic 6 ☐
- 6.2 Sample problem 1 Comparing pandemics and epidemics (tlvd-1782) ☐
- 6.3 Sample problem 2 Using ELISA and immunological methods (tlvd-1783) ☐
- Sample problem 3 Analysing methods to prevent the spread of disease (tlvd-1785) ☐
- 6.4 Sample problem 4 Understanding the immunisation schedule and herd immunity (tlvd-1786) ☐
- 6.5 Sample problem 5 Comparing monoclonal antibodies (tlvd-1787) ☐

Video eLessons

- 6.3 Sandwich ELISA (eles-4364) ☐
- 6.4 Memory cells and vaccination (eles-4366) ☐
- Immunisation in Australia (eles-0126) ☐
- Herd immunity (eles-4367) ☐

Weblinks

- 6.2 Timeline of COVID-19 ☐
- Data related to COVID-19 ☐
- Managing epidemics— WHO ☐
- Australian history of colonisation ☐
- Smallpox epidemic ☐
- 6.3 Types of ELISA ☐
- ELISA interactivity ☐
- Identifying enteric pathogens ☐
- 6.4 Victorian immunisation schedule ☐
- Australian immunisation schedule ☐
- Health data— National Notifiable Diseases Surveillance System ☐
- Vaccine history timeline ☐
- 6.5 Lymphoma Australia — Monoclonal antibodies ☐
- CAR T-cell therapy ☐
- Rituximab in autoimmune disease ☐

Exam question booklet

- 6.1 Exam question booklet — Topic 6 (eqb-0017) ☐

Teacher resources

There are many resources available exclusively for teachers online.

To access these online resources, log on to **www.jacplus.com.au**

AREA OF STUDY 1 How do organisms respond to pathogens?

OUTCOME 1

Analyse the immune response to specific antigens, compare the different ways that immunity may be acquired and evaluate challenges and strategies in the treatment of disease.

PRACTICE EXAMINATION

STRUCTURE OF PRACTICAL EXAMINATION		
Section	**Number of questions**	**Number of marks**
A	20	20
B	7	30
	Total	**50**

Duration: 50 minutes

Information:

- This practice examination consists of two parts. You must answer all question sections.
- Pens, pencils, highlighters, erasers and rulers are permitted.

SECTION A — Multiple choice questions

All correct answers are worth 1 mark each; an incorrect answer is worth 0.

1. Which of the following is **not** an innate form of immunity?

A. Interferon
B. Intact skin
C. Macrophage
D. Plasma cell

2. HIV is a virus that targets T helper cells, leading to their eventual death. A consequence of the loss of T helper cells would be

A. the loss of functionality of the innate immune system.
B. the loss of functionality of both the adaptive and innate immune systems.
C. the loss of functionality of the adaptive immune system.
D. the loss of functionality of the humoral response but not the cell-mediated response.

3. A young boy accidentally receives a small cut on his arm. Foreign material such as dust and bacteria enter the open wound. One of many immune system responses to foreign material entering the body is the activation of mast cells to release histamine in tissue fluid near the site of the cut. In response to the histamine, cells lining the nearby blood vessels increase their permeability. The advantage of increased permeability of blood vessels is that

A. blood flows out of the body more quickly.
B. phagocytes can exit blood vessels more easily.
C. foreign material can enter blood vessels more rapidly.
D. nothing will be able to enter or exit the blood vessels.

4. A person receives a liver transplant from an unrelated person. It is reasonable to claim that
   A. the antigens in the recipient liver are identical to those in the donor liver.
   B. the antigens in the transplanted liver are made of proteins or polysaccharides.
   C. the transplanted liver will produce antibodies against the recipient.
   D. the recipient will produce antigens against the transplanted liver.

5. Cellular pathogens include
   A. viruses.
   B. prions.
   C. retroviruses.
   D. bacteria.

6. Virus-infected cells are destroyed directly by
   A. cytotoxic T cells.
   B. helper T cells.
   C. plasma cells.
   D. B cells.

7. Non-specific barriers that inhibit entry of pathogens into the human body include
   A. antibodies.
   B. antigens.
   C. non-pathogenic bacteria on the skin.
   D. memory cells.

8. When tissue damage occurs, a series of events in blood leads to the development of insoluble fibres at the site of the damage as shown in the provided diagram. The function of these fibres is most likely to inhibit

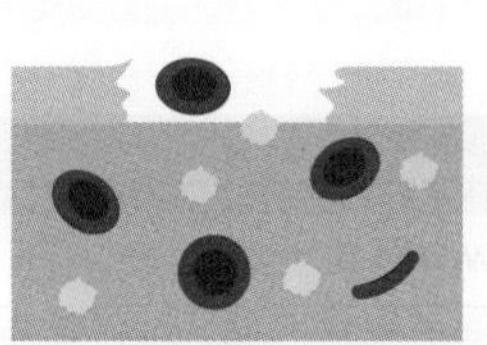

Source: gritsalak karalak/Shutterstock

   A. the spread of a pathogen.
   B. the action of white blood cells.
   C. the action of red blood cells.
   D. the clumping of platelets.

9. Which statement is correct regarding phagocytes?
   A. Antibodies are released by phagocytes.
   B. Phagocytes are responsible for blood clotting.
   C. Phagocytes mature in the thymus.
   D. Foreign material is engulfed and digested by phagocytes.

10. One of the functions of complement proteins is to promote the inflammatory response. Which of the following would be expected in a patient when complement proteins are activated as a result of infection?
   A. Reduced temperature
   B. More phagocyte activity
   C. Decreased blood flow to an infected area
   D. Decrease in capillary permeability

11. Interferons

A. stimulate virus reproduction.
B. are proteins produced by virus-infected cells.
C. initiate viral infection of some cells.
D. are lipids that damage one kind of virus.

12. The production of antibodies is a fundamental part of the adaptive immune response. Which cells are primarily involved in the production of antibodies?

A. Helper T cells
B. Cytotoxic T cells
C. Memory cells
D. Plasma cells

13. After suffering a disease caused by a bacterium, a person may have life-long immunity against further infection. The development of life-long immunity requires

A. the presence of fever.
B. the production of memory cells.
C. inflammation of tissues.
D. damage to red blood cells.

14. The shapes of the antigen-binding sites on antibodies differ across different antibodies. Many antibodies have two sites for antigens to bind. The shapes of antigen-binding sites

A. are due to differences in the constant region.
B. are identical in one antibody.
C. are created by heavy regions.
D. are complementary to at least four different antigens.

15. Herd immunity is

A. passed from one individual to another in a population.
B. can be transmitted from one generation to the next.
C. only shown in populations of herding animals, such as sheep and cows.
D. achieved when a large proportion of the population is vaccinated.

16. Antigen-presenting cells can move antigens to their surface and display these antigens to other immune cells. Which of the following cells are activated by antigen-presenting cells?

A. Helper T cells
B. Plasma cells
C. Dendritic cells
D. Neutrophils

17. Rabies is a disease caused by a deadly virus spread to humans through an infected animal bite. A person bitten by an infected animal is given an injection of antibodies to prevent the disease causing death. This is an example of

A. natural active immunity.
B. artificial active immunity.
C. natural passive immunity.
D. artificial passive immunity.

18. Some autoimmune diseases are caused by B cells that are unable to identify 'self' tissue. It has been suggested that destruction of an affected person's B cells may be a cure. If such a procedure was carried out, the difficulties that could arise would include

A. an excess of plasma cells in the individual.
B. increased susceptibility to other diseases.
C. an oversupply of memory B cells.
D. lack of phagocytic activity.

19. Monoclonal antibodies bind to only one specific
    A. enzyme.
    B. antigen.
    C. antibody.
    D. signalling molecule.

20. In 1928, the first natural antibiotic, penicillin was discovered and began being used extensively in the 1940s. Penicillin acts by inhibiting cell wall synthesis in bacteria. It would be reasonable to conclude that penicillin
    A. would also attack human cells as they also have cell walls.
    B. is narrow spectrum so is useful against all types of bacteria.
    C. can also be used to destroy viral pathogens.
    D. is less effective today than when it was first discovered, as some bacteria have developed resistance.

SECTION B — Short answer questions

Question 21 (3 marks)

Although plants have no immune system, they have many physical and chemical barriers to protect them against pathogens.

a. Describe one example of a physical barrier in plants. **1 mark**

b. Describe one example of a chemical barrier in plants. **1 mark**

c. The physical and chemical barriers are non-specific ways of protecting an individual against pathogens. Give one major difference between specific and non-specific immune responses. **1 mark**

Question 22 (4 marks)

The diagram below shows a lymph node.

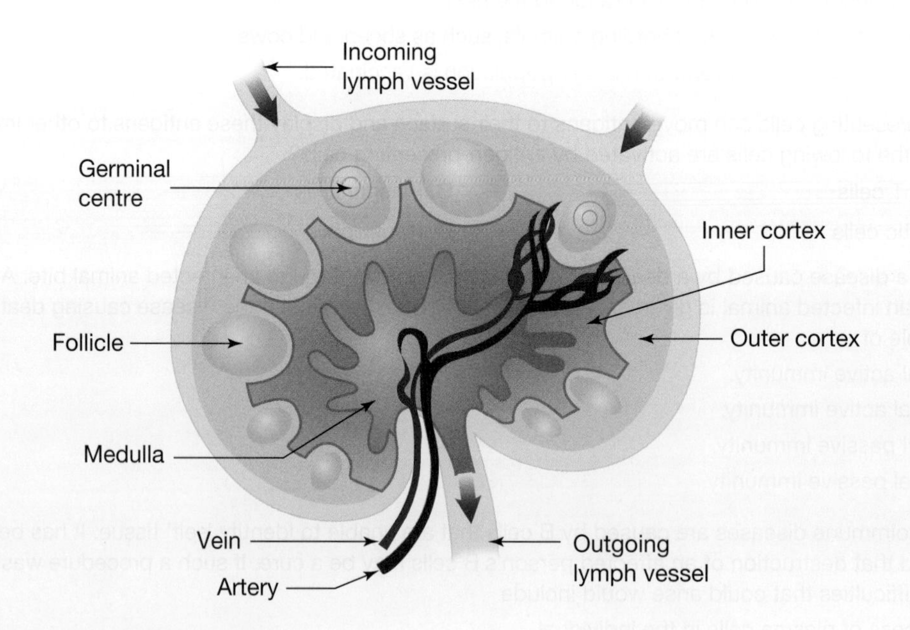

a. Describe the function of lymph nodes. **1 mark**

b. Identify structures A and B in the diagram. **1 mark**

c. A number of immune cells are found in the lymph nodes. Name and describe the role of one of the immune cells of innate immunity found in lymph nodes. **2 marks**

Question 23 (3 marks)

When an organism is infected by a pathogen, it responds by producing large numbers of antibodies.

a. Outline the process that leads to the production of these antibodies. **2 marks**

b. In what way is the production of memory cells related to the production of plasma cells? **1 mark**

Question 24 (5 marks)

Indirect protection from infectious diseases can exist, but only at the level of a population. This indirect protection is termed herd immunity.

a. Define herd immunity. **1 mark**

b. State two advantages of herd immunity. **2 marks**

c. State two conditions under which herd immunity is effective. **2 marks**

Question 25 (5 marks)

Pathogens contain surface antigens that are recognised by the immune system.

a. How do antigens on a pathogen differ from antigens on the surface of body cells? **1 mark**

b. Draw a labelled diagram of an antibody that would act against a pathogen. **2 marks**

c. Outline two ways in which the antigen–antibody complexes can act against a pathogen. **2 marks**

Question 26 (5 marks)

Pandemics have occurred many times in human history and have been responsible for deaths of thousands of people.

a. Define pandemic. **1 mark**

b. There are several recently-emerged diseases that have threatened to become pandemics, such as SARS (severe acute respiratory syndrome), Avian flu and Zika virus. Explain how a disease that starts in a small population, such as a family or a small village, can spread around the world within weeks or a few months. **2 marks**

c. List two strategies that may prevent an epidemic from becoming a pandemic. **2 marks**

Question 27 (5 marks)

Myeloid leukemia is a type of cancer. Monoclonal antibodies are used in treating it.

a. Suggest two ways in which monoclonal antibodies may act against cancer. **2 marks**

b. Outline the steps involved in the production of monoclonal antibodies. **3 marks**

END OF EXAMINATION

PRACTICE SCHOOL-ASSESSED COURSEWORK

ASSESSMENT TASK — Analysis and evaluation of a selected biological case study

In this task you will analyse the immune response to specific antigens, compare the different ways that immunity may be acquired, and evaluate challenges and strategies in the treatment of disease.

- Students are permitted to use pens, pencils, highlighters, erasers and rulers.
- Students can take in a blank copy of the stimulus material.
- The stimulus material is freely available in the public domain and can be accessed at any time in the lead up to the task.

Total time: 70 minutes (10 minutes reading, 60 minutes writing)
Total marks: 50 marks (to obtain score out of 40, divide raw score by 1.25)

THE 1918 INFLUENZA PANDEMIC

Preparing for the task

- Access the stimulus material via the weblink provided: 'The 1918 influenza pandemic: insights for the 21st century'.
  - Print out or save a copy for annotation and mark up.
  - The reference for this article is:
    David M Morens & Anthony S Fauci 2007, 'The 1918 influenza pandemic: insights for the 21st century', *The Journal of Infectious Diseases*, volume 195, issue 7, 1 April 2007, pp. 1018–28, https://doi.org/10.1086/511989
- Read the article.
- Mark up the article as follows:
  - green highlighter for text/sections that relate to key knowledge — responding to antigens
  - yellow highlighter for text/sections that relate to key knowledge — acquiring immunity
  - blue highlighter for text/sections that relate to key knowledge — disease challenges and strategies.
- Annotate the article:
  - For each highlighted section, write a short dot point list that connects the content from the study design to the information in the article.
- Want to challenge yourself or prepare further?
  - Convert your annotations into a mind map that creates a visual representation of the connections between the article and the key knowledge from the study design.

ASSESSED TASK

- You may only have **a blank copy** of the article for reference.
- Once writing time begins, you may write on your article, but this writing will not be assessed.
- Make sure that all your answers and examples are given using the article as the context.
- You should write your responses on loose leaf paper to prepare for submission. Ensure your name is on each page.

CASE STUDY

The 1918 influenza pandemic was caused by the influenza A virus subtype H1N1. Influenza A viruses are RNA viruses. There are several subtypes of influenza virus. The subtypes are labelled according to the various antigens (hemagglutinin and neuraminidase) found on their surface. Each subtype has an H number (for the type of hemagglutinin proteins) and an N number (for the type of neuraminidase enzyme). There are 18 different known H antigens (H1 to H18) and 11 different known N antigens (N1 to N11).

Each viral subtype evolves into a variety of strains with different capability for causing disease. Some may be pathogenic to one species but not to another, and some are pathogenic to multiple species.

1. H1N1 is considered a pathogen. Define 'pathogen.' **1 mark**
2. Name one chemical barrier and one physical barrier that may prevent viral infection by H1N1 in humans. **2 mark**
3. Although the H1N1 virus is pathogenic to humans, some virus types can also affect plants. Name one barrier to infection employed by plants. **1 mark**
4. Determine whether H1N1 is considered a cellular or non-cellular pathogen. Justify your answer. **2 marks**
5. Based on the description above, draw an H1N1 viral particle. Include the following labels: RNA, protein coat, hemagglutinin and neuraminidase. **4 marks**
6. Outline the process of antigen presentation, from infection to activation of the adaptive immune response, that would be followed in a person infected by H1N1. **4 marks**
7. Explain why the lymph nodes are referred to as the link between the innate and adaptive immune responses. **3 marks**

During its course, the 1918 influenza pandemic killed approximately 25% of the world's population at the time, making it one of the deadliest pandemics in human history. It is hypothesised that the deadly nature of the infection was not due as much to the influenza infection but due to secondary infections from bacteria, such as *Streptococcus pneumoniae,* leading to bacterial pneumonia.

8. State two differences between a virus and a bacterium. **2 marks**
9. Outline the response by the adaptive immune system to an infection by the H1N1 virus. Include references to specific cell types in your answer. **4 marks**
10. By referring to the article, compare the response by the adaptive immune system to an infection by H1N1 and an infection by *Streptococcus pneumoniae.* Include references to specific cell types in your answer. **4 marks**

Use the graph provided to answer Questions 11 and 12 (this is figure 2B in the article).

FIGURE 2B Deaths from influenza per 1000 persons by age and time period

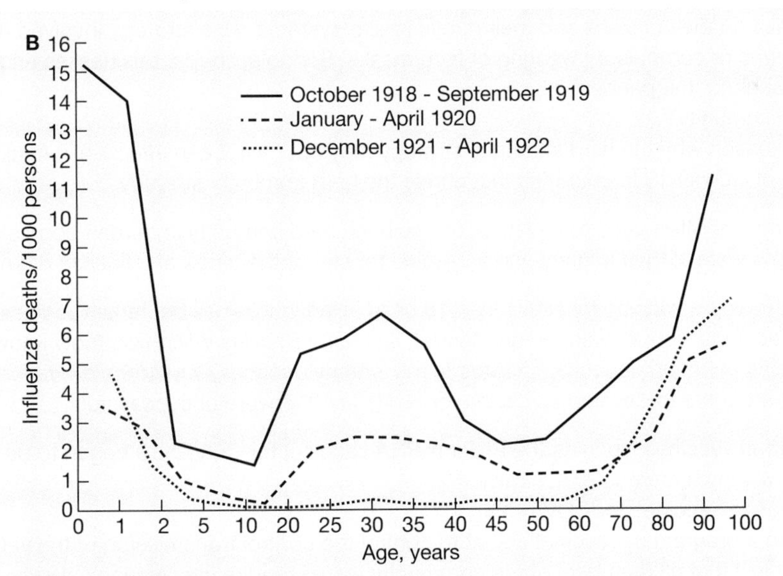

Source: David M. Morens & Anthony S. Fauci, 2007, 'The 1918 influenza pandemic: insights for the 21st century', *The Journal of Infectious Diseases*, vol. 195, issue 7, 1 April 2007, pp. 1018–28

11. With reference to the graph above, summarise the changes in deaths per 1000 persons for 20–40-year-olds from October 1918 to April 1922. **3 marks**
12. Using your knowledge of acquired immunity, explain the reasons for the trends in death rate for 20–40-year-olds observed in figure 2B and summarised in your answer to question 11. **4 marks**

The 1918 influenza pandemic provides the scientific and medical communities with an interesting point of difference when it is compared with other influenza pandemics.

In general, death rates for influenza are highest in the very young, the very old and those with underlying medical conditions that affect the immune response. The 1918 influenza pandemic affected these groups of people but also saw an unusual spike in the death rate for healthy young adults aged 20–40 years old and a lower than expected death rate in the elderly.

Regarding young adults, one hypothesis that aims to explain this unexpected death rate is that the H1N1 infection caused a 'cytokine storm' that triggered the stronger immune systems of young adults to produce a strong inflammatory response.

Symptoms of a cytokine storm include fever, fatigue, loss of appetite, muscle and joint pain, nausea, vomiting, diarrhea, rashes, fast breathing, rapid heartbeat and low blood pressure.

13. Identify two steps of the inflammatory response that may produce the symptoms of a 'cytokine storm' that were observed in H1N1 patients. Explain each of these steps. **4 marks**
14. Interferon is a type of cytokine that is released by cells infected by H1N1. Describe two actions of interferon which act to limit the spread of H1N1 within an individual. **2 marks**

Even though the influenza pandemic of 1918 took place over a century ago, it is still regarded as an important point of reference for governments and clinicians in planning for health policy, future pandemics and disease control efforts.

In recent years, the world experienced the COVID-19 pandemic. COVID-19 is a disease caused by a type of novel coronavirus known as 'severe acute respiratory coronavirus 2' (SARS-CoV2). The term 'novel' is used to describe SARS-CoV2 because it had not been seen in human populations before and was considered 'new'.

The article referenced in this case study was published in 2007, over a decade before the COVID-19 pandemic.

15. Suggest how the scientific and social strategies used to identify and control the spread of H1N1 in the influenza pandemic of 1918 may have differed from those used in the COVID-19 pandemic. **2 marks**

In response to the COVID-19 pandemic, some countries adopted a 'herd immunity' strategy to manage the impact of the disease on their citizens and their public health systems. This strategy involved not imposing mandatory lock downs or compulsory wearing of face masks. Schools, shops, businesses and restaurants all remained open throughout the pandemic.

16. Define 'herd immunity'. **2 marks**
17. Outline one reason why the herd immunity strategy may not have been effective in minimising the impacts of COVID-19 in countries that chose the herd immunity strategy. **1 mark**

Vaccination programs are often used to develop and maintain herd immunity in human populations. In 1918, vaccination programs were in their infancy and a vaccine for any of the types of influenza did not yet exist.

Initial attempts to develop a vaccine for H1N1 in 1918 used killed whole-cell bacterial vaccines. Vaccines for various species of bacteria were developed and tested, including *Bacillus influenzae* (now known as *Haemophilus influenzae*) and strains of pneumococcus, streptococcus, staphylococcus, and *Moraxella catarrhalis* bacteria.

18. Explain why attempts to develop a vaccine for H1N1 in 1918 were unsuccessful. **1 mark**

In late 2020, numerous vaccines (such as the Pfizer mRNA vaccine) for COVID-19 began to be mass produced. In 2021, this vaccine was implemented in Australia.

19. With reference to vaccines for COVID-19:
    a. how would a program of vaccination act to control the control transmission of the virus? **2 marks**
    b. how likely is it that all members of the community would require vaccination? Justify your answer. **2 marks**

Digital document Unit 4 AOS 1 School-assessed coursework (doc-34868)

Weblink The 1918 influenza pandemic: insights for the 21st century

7 Genetic changes in a population over time

KEY KNOWLEDGE

In this topic, you will investigate:

Genetic changes in a population over time

- causes of changing allele frequencies in a population's gene pool, including environmental selection pressure, genetic drift and gene flow, and mutations as the source of new alleles
- biological consequences of changing allele frequencies in terms of increased and decreased genetic diversity
- manipulations of gene pools through selective breeding programs
- consequences of bacterial resistance and viral antigenic drift and shift in terms of ongoing challenges for treatment strategies and vaccination against pathogens.

Source: VCE Biology Study Design (2022–2026) extracts © VCAA; reproduced by permission.

PRACTICAL WORK AND INVESTIGATIONS

Practical work is a central component of learning and assessment. Experiments and investigations, supported by a **practical investigation eLogbook** and **teacher-led videos**, are included in this topic to provide opportunities to undertake investigations and communicate findings.

7.1 Overview

Numerous **videos** and **interactivities** are available just where you need them, at the point of learning, in your digital formats, learnON and eBookPLUS and at **www.jacplus.com.au**.

7.1.1 Introduction

The word evolution means to change and, in terms of biological evolution, this refers to the change in a population's genetic material over time. The theory of evolution is not just an idea about how organisms' genetic material has changed over time. Evolution is a scientific theory, just as gravity, cell theory and atomic theory are. It is a natural aspect of the world which is supported by a range of observable facts and evidence that have been repeatedly tested and confirmed. Charles Darwin and Alfred Russel Wallace were the first to accurately hypothesise on how species were changing in the mid-nineteenth century using available evidence. Since then many scientists have continued to investigate evolution and add to the observable and testable evidence that support the theory.

FIGURE 7.1 The genetic variation shown in the colouration of Bennetts wallabies

This topic will investigate the causes and mechanisms of how different populations' genetic material changes over time and how this affects their genetic diversity. It will also explore how we have manipulated the gene pools of domesticated species, and how we respond to rapid changes in the genetic material of bacteria and viruses.

LEARNING SEQUENCE

on Resources

eWorkbook	eWorkbook — Topic 7 (ewbk-1886)
Practical investigation eLogbook	Practical investigation eLogbook — Topic 7 (elog-0191)
Digital documents	Key science skills — VCE Biology Units 1–4 (doc-34326) Key terms glossary — Topic 7 (doc-34621) Key ideas summary — Topic 7 (doc-34612)
Exam question booklet	Exam question booklet — Topic 7 (eqb-0018)

7.2 Gene pools and allele frequencies

KEY KNOWLEDGE

- Causes of changing allele frequencies in a population's gene pool

Source: Adapted from VCE Biology Study Design (2022–2026) extracts © VCAA; reproduced by permission

7.2.1 What are gene pools?

A **population** refers to an interbreeding group of organisms of the same species living in the same region at the same time. So, we can talk about the present population of lions (*Panthera leo*) on the Serengeti Plains shown in figure 7.2 or the human population of Milan during the period of the Black Death. A population may be as small as a group of one species on a remote island or isolated on a mountain top, or it may be as large as the **gene pool** of one species on a continent.

FIGURE 7.2 A population of lions can interbreed.

Each individual in a population has two copies of each **gene** in the form of **alleles**. The total set of this genetic information in an individual is referred to as the **genotype**.

For example, in sheep, the gene that controls wool colour has the alleles white (*W*) and black (*w*), with white being dominant to black. Sheep genotypes at this gene locus are expressed as a combination of two alleles: for example, either *WW* or *Ww* for a white sheep, and *ww* for a black sheep (see figure 7.3).

FIGURE 7.3 The genotype of an individual sheep

FIGURE 7.4 The gene pool of a population

population members of one species living in one region at a particular time

gene pool sum total of genetic information present in a population

gene a section of DNA that codes for a protein

alleles alternate forms of a gene

genotype the genetic makeup of an organism; also refers to the combination of alleles at a particular gene locus

BACKGROUND KNOWLEDGE: Reviewing genes and alleles

A gene is an inherited instruction carried on a chromosome which can carry an instruction to produce a specific protein. Alleles are different forms of a specific gene. Therefore, there can be multiple alleles for each gene.

Each population has a gene pool that consists of all the alleles present for each gene within the population. The greater the **variation** and number of alleles within a gene pool, the greater the **genetic diversity**.

7.2.2 Allele frequencies in a population

Gene pools are described by the frequencies (proportions) of the alleles of each gene present. The frequency of an allele within the gene pool is referred to as the **allele frequency**.

To understand allele frequencies, look at figure 7.4 and work through the following information, which will give us the frequencies of the alleles for the gene for wool colour in sheep:

1. There are 10 sheep, each with 2 alleles, so the total number of alleles is $10 \times 2 = 20$
2. Using the genotypes of the sheep as shown, we can count the number of each allele:
   Number of *w* alleles = 6 and number of *W* alleles = 14
3. So,
   The frequency of the *w* allele $= \frac{6}{20} = \frac{3}{10} = 0.3$

   The frequency of the *W* allele $= \frac{14}{20} = \frac{7}{10} = 0.7$

The frequency of the allele that determines the dominant phenotype is typically denoted by the letter '*p*' and that of the allele that determines the recessive phenotype is denoted by the letter '*q*'.

So, for this limited gene pool, we can write: freq $(W) = p = 0.7$ and freq $(w) = q = 0.3$.

variation differences exhibited among members of a population owing to the action of genes

genetic diversity the amount of genetic variation there is within a population's gene pool

allele frequency the proportion of a specific allele in a population

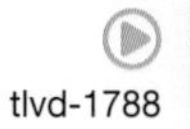
tlvd-1788

SAMPLE PROBLEM 1 Calculating allele frequencies

The students in your Year 12 cohort were all gathered together. Out of 50 students, it was found that 18 students had blue eyes (*bb*) and 32 students had brown eyes. Of the brown eyed students, 21 were heterozygous (*Bb*). The rest were homozygous (*BB*).

a. **Calculate the frequency of *b* in the population of students.** **(1 mark)**
b. **Calculate the frequency of *B* in the population of students.** **(1 mark)**
c. **Which allele has the greater frequency?** **(1 mark)**

THINK	WRITE
a. 1. Determine the total number of alleles in the population.	Total number of alleles $= 2 \times 50$ $= 100$
2. Determine the total number of *b* alleles, ensuring you remember to count those in the heterozygous individuals and count each of the alleles in the blue-eyed individuals (who possess two *b* alleles).	Number of *b* alleles $= 18 + 18 + 21$ $= 57$
3. Calculate the frequency.	Frequency $(b) = \frac{57}{100}$ $= 0.57$ (1 mark)
b. 1. Determine the total number of alleles in the population.	Total number of alleles $= 2 \times 50$ $= 100$
2. Calculate the number of individuals who are homozygous (*BB*).	Number of homozygous individuals $= 32 - 21$ $= 11$

3. Determine the total number of B alleles, ensuring you remember to count those in the heterozygous individuals and both alleles in the homozygous individuals.

$$\text{Number of } B \text{ alleles} = 11 + 11 + 21 = 43$$

4. Calculate the frequency. *Note:* As there are only two alleles in this case, you may have also subtracted your answer in part a. from 1 to determine the frequency.

$$\text{Frequency } (B) = \frac{43}{100} = 0.43 \text{ (1 mark)}$$

c. Identify the allele which has the greater frequency. The b allele with a frequency of 0.57, has a greater frequency than the B allele, with a frequency of 0.43 (1 mark).

elog-0034

INVESTIGATION 7.1

online only

Variations — no two the same

Aim

To measure and record variation in a population

CASE STUDY: Can you drink milk?

Allele frequencies are continually changing within gene pools, due to factors such as mutation, natural selection and chance events, which will be explored further in this topic.

One gene in which gene pools have had changing allele frequencies is the *LCT* gene on the number-2 chromosome which codes for the enzyme allowing individuals to break down lactose.

Human infants up to about three to four years of age can digest lactose. However, after that time, the majority (about 70 per cent) of the world's population start to lose their lactase enzyme activity and, by about seven or eight years of age, cannot digest lactose. Such people are said to be lactase nonpersistent (sometimes termed lactose intolerant).

FIGURE 7.5 Goats are one example of a herd animal that can provide milk.

About 10 000 years ago, a mutation occurred in a hunter-gatherer population that resulted in a new allele *C*. This allele allowed for the persistence of lactase enzyme activity into adulthood. The spread of dairy farming across Europe is associated with the spread of the *C* allele for lactase persistence and changing allele frequencies in the gene pools of populations.

This case study provides an excellent summary of changing allele frequencies and selective advantage. To access more information about this case study on lactose persistence and complete an analysis task, please download the worksheet.

Resources

eWorkbook Worksheet 7.1 Case study: Can you drink milk? (ewbk-8234)

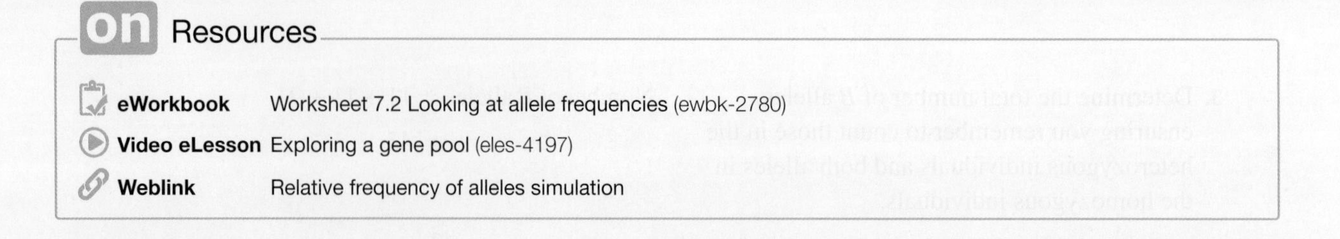

KEY IDEAS

- A population is a group of interbreeding individuals of the same species that live in a given area.
- A gene pool consists of all the alleles present for each gene within a population.
- The proportion of each allele of a gene within a gene pool is its allele frequency.
- Populations with more variation in their alleles have gene pools with greater genetic diversity.

7.2 Activities

learnon

To answer questions online and to receive **immediate feedback** and **sample responses** for every question, go to your learnON title at **www.jacplus.com.au**. A **downloadable solutions** file is also available in the resources tab.

7.2 Quick quiz on	7.2 Exercise	7.2 Exam questions

7.2 Exercise

1. State the difference between a genotype and a gene pool.
2. Describe how allele frequencies can be calculated within a gene pool.
3. Calculate the allele frequencies in each of the following situations:
   a. In a population of 50 individuals, 22 individuals have attached earlobes (*aa*) and 28 individuals have unattached earlobes. Eleven of these individuals with unattached earlobes are homozygous (*AA*). The rest are heterozygous.
   b. In a population of 187 flies, 54 have white eyes (*ww*). The rest have red eyes. Of these, 119 are homozygous (*WW*).
4. A gene known as *ALDH2* leads to the production of aldehyde hydrogenase, an enzyme which is important in the breakdown of alcohol into non-toxic byproducts. A mutation in this gene produces a faulty enzyme leading to an inability to breakdown alcohol upon consumption.
   This trait is dominant, meaning individuals that are either homozygous or heterozygous for the faulty allele cannot produce fully functioning aldehyde hydrogenase.
   a. Would the allele frequency of this faulty allele be the same in all populations? Explain your response.
   b. The percentage of individuals in China, Japan and Korea that are deficient in aldehyde dehydrogenase is approximately 40 per cent. Does this mean the frequency of the faulty allele is 0.40? Justify your response.

7.2 Exam questions

Question 1 (1 mark)

Source: *VCAA 2006 Biology Exam 2, Section A, Q10*

MC All the alleles in a population are referred to as the

A. phenotypic family.
B. proteome.
C. gene pool.
D. genotype.

Question 2 (1 mark)

Source: *VCAA 2019 Biology Exam, Section A, Q26*

MC Consider the diagram below showing the gene pool of a population over 20 generations.

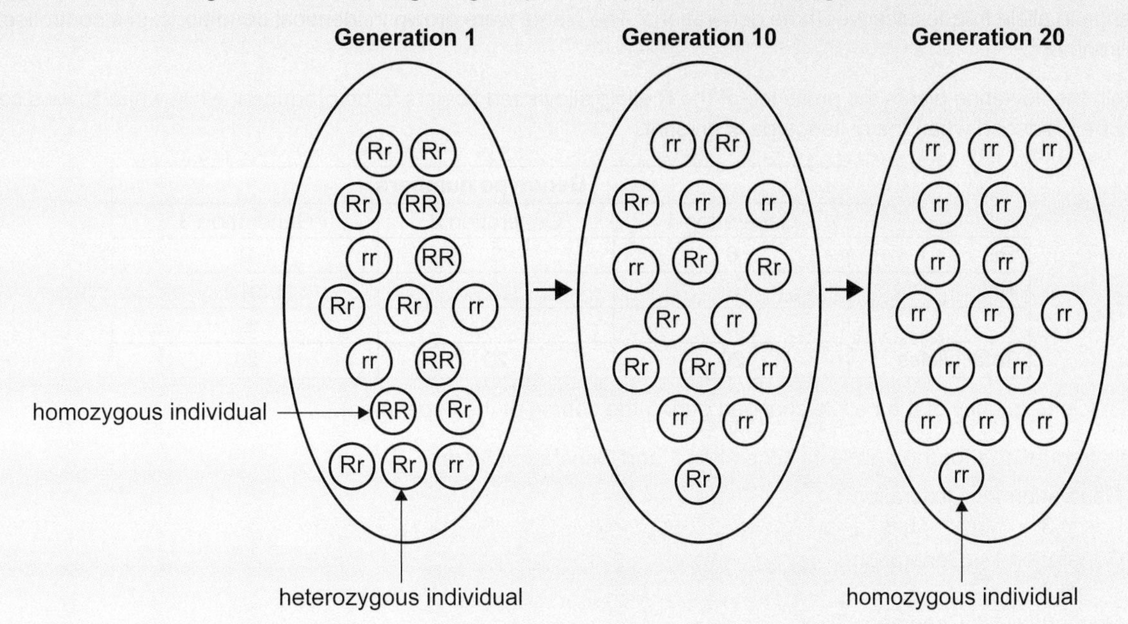

It would be correct to conclude that, over the 20 generations

A. genetic diversity is increasing in this population.
B. individuals with the genotype RR had a selective advantage in this population.
C. the frequency of each allele is equal in Generation 1 but not in other generations.
D. new advantageous alleles for this gene were introduced as individuals joined this population.

Question 3 (1 mark)

Source: *Adapted from VCAA 2009 Biology Exam 2, Section B, Q3d*

The endangered pygmy possum (*Burramys parvus*) lives in three restricted alpine areas, Mt Buller, Bogong High Plains and Mt Kosciusko.

About 2000 individuals remain in the wild. Studies show that there is a lot of genetic diversity between the three populations. Due to the isolation of these populations, scientists think that each population has a separate gene pool.

Explain what is meant by gene pool.

Question 4 (1 mark)

Source: *VCAA 2007 Biology Exam 2, Section A, Q17*

MC Lactase is an enzyme in humans which breaks down lactose, one of the sugars in milk. Milk is a safe and nutritious food which is readily available year-round. Although most adults around the world lose the ability to produce lactase as they mature, more than 90% of Europeans have a lactase-producing allele which remains active into adulthood.

Scientists analysed DNA in bone samples from a number of Neolithic Europeans (dated between 5840 BC and 5000 BC) and found that none of them had the adult lactase allele.

The most likely explanation for this data is that

A. the adult lactase-producing allele which remains active into adulthood arose millions of years ago in ancestors of modern humans.
B. possession of the adult lactase-producing allele which remains active into adulthood confers a significant evolutionary advantage.
C. the adult lactase-producing allele which remains active into adulthood did not arise in Europe.
D. modern Europeans are not descended from the Neolithic Europeans tested.

Question 5 (1 mark)

Source: *VCAA 2009 Biology Exam 2, Section A, Q24*

MC A scientist took a small population of 10 flowering plants and conducted an experiment to examine the change in allele frequenciesover three generations. The plants were grown in identical conditions, in a controlled environment.

Within the flowering plants the presence of the **R** allele allows red flowers to be produced, while white flowers can only be produced when the **rr** genotype is present.

	Genotype numbers		
	Generation 1	Generation 2	Generation 3
RR	6	3	0
Rr	2	3	4
rr	2	4	6
Total alleles	**20**	**20**	**20**

The allele frequency of the **r** allele changed during the course of the experiment.

Allele frequencies for the **r** allele in Generation 1 and Generation 3 are

A. Generation 1 Generation 3
0.3 0.8

B. Generation 1 Generation 3
0.8 0.3

C. Generation 1 Generation 3
0.7 0.3

D. Generation 1 Generation 3
0.3 0.7

More exam questions are available in your learnON title.

7.3 Environmental selection pressures, genetic drift and gene flow

KEY KNOWLEDGE

- Causes of changing allele frequencies in a population's gene pool, including environmental selection pressures, genetic drift and gene flow

Source: Adapted from VCE Biology Study Design (2022–2026) extracts © VCAA; reproduced by permission

The allele frequencies in most populations tend to remain stable generation after generation. However, some populations do show changes in allele frequencies in their gene pools over time.

The agents that can cause allele frequencies to change over time include:
- changes in environmental selection pressures leading to natural selection
- genetic drift
- gene flow

Each of these agents will be explained within this subtopic.

7.3.1 Environmental selection pressures and natural selection

The environment in which a population lives exposes the organisms to a wide range of selection pressures. **Environmental selection pressures** are external agents which influence the ability for an individual to survive.

Selection pressures can fall under three main categories: physical, biological or chemical. Some categories of agents that can act as environmental selection pressures are shown in figure 7.6.

FIGURE 7.6 Examples of selective pressures

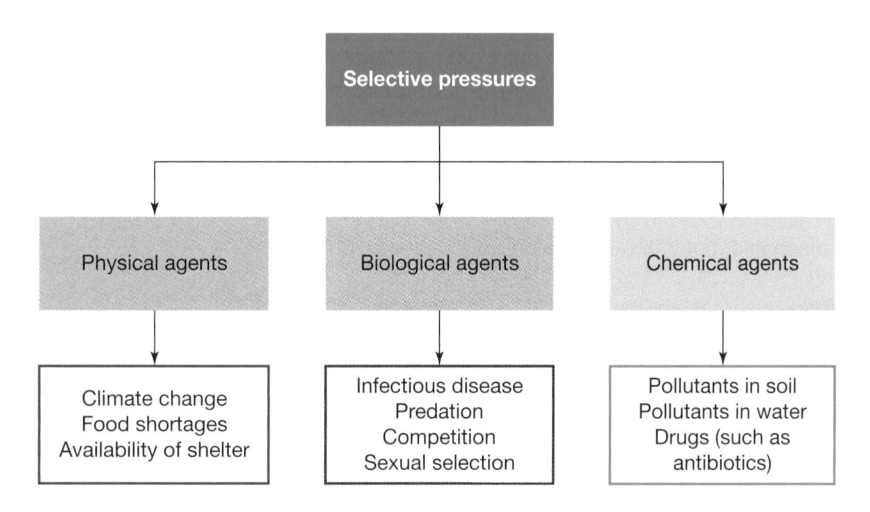

In the wild, members of a population must compete with each other for access to living space, energy supplies and mating partners in their habitat. Members of a population are also exposed to competition from other species, to predation, to parasites and to disease-causing microorganisms. Individual members within a population react differentially to these pressures, with some being more successful than others. These individuals within a population's gene pool can have phenotypic traits which allow them to survive in higher numbers and consequently reproduce more offspring. These individuals therefore contribute more of their alleles to the next generation and this phenotype is likely to increase in the gene pool. This is referred to as biological fitness. This is opposed to physical fitness (linked with health and athletic ability), which cannot necessarily be passed on to the next generation or enhance reproductive success.

environmental selection pressures external agents which influence the ability of an individual to survive in their environment

Natural selection occurs when any selective pressure acts on a population in the wild and produces differences in the survival and reproduction rates of variants in that population.

Members of a polymorphic population living *under a particular set of environmental conditions* are exposed to a range of selecting agents and the selection pressures that they produce. Under these circumstances, the various phenotypes (variants) may show different survival and reproductive rates. This can be observed in figure 7.7.

FIGURE 7.7 These moths have different chances of survival against predators due to variations in their appearance.

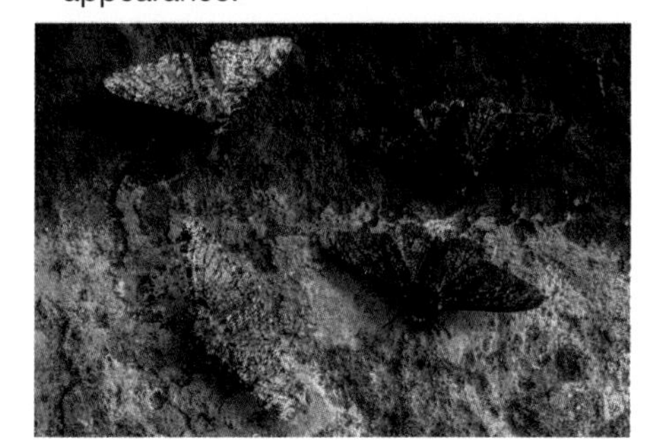

A phenotype that makes the greater contribution to the gene pool in the next generation has a higher **fitness** value and is said to be at a '**selective advantage**'. Phenotypes that make lesser contributions to the gene pool of the next generation are 'less fit' or are said to be 'selected against'.

An observed fitness value for a phenotype is not fixed, but applies under a particular set of environmental conditions. For example, in figure 7.7, the fitness of the moth depends on the environmental conditions — in this case, the colour of the tree.

Natural selection

Natural selection occurs when the allele frequencies in a population's gene pool change due to environmental selection pressures, creating a selective advantage for particular phenotypes (refer to figure 7.8). Individuals that have phenotypes with a selective advantage are more likely to survive and pass on the alleles to their offspring, which can alter the allele frequencies within gene pools.

FIGURE 7.8 The mechanism of natural selection on a beetle population

The mechanism of natural selection

1. There is variation within the population's gene pool. This is created by new alleles arising from mutations which are inheritable (some variation also comes from sexual reproduction).
2. There is a struggle for all individuals in the gene pool to survive. Environmental selection pressures act upon the population.
3. Individuals that are better adapted to their environment are more likely to survive and reproduce, passing their alleles on to the next generation.
4. The alleles that allow for survival will be inherited by subsequent generations and they can increase in frequency in the gene pool over time.

Remember that there needs to be some initial variation in a population for natural selection to occur. Some variants need to have a selective advantage within environmental selective pressures.

When there are alleles which have differing selective advantages within a population, the allele frequencies within the gene pool can change over time. The alleles with a selective advantage will increase in frequency and those alleles which do not give an advantage will decrease in frequency. Over many generations natural selection can cause the allele to become common within a population.

This can often lead to a decrease in genetic diversity, as those alleles that lead to a beneficial trait may become fixed, while others that confer a selective disadvantage may be lost. Table 7.1 outlines some examples of the mechanism of natural selection acting on populations.

fitness the ability to survive and pass genetic material on to the next generation

selective advantage relative higher genetic fitness of a phenotype compared with other phenotypes controlled by the same gene

natural selection process in which organisms better adapted for an environment are more likely to pass on their genes to the next generation

TABLE 7.1 The action of natural selection on populations

Variation present in population's gene pool	Environmental selection pressure acting on population	Action of natural selection
Beetles with green colouration and red colouration	Predation from birds	• The beetles live on green leaves and therefore the green beetles (selective advantage) are better camouflaged than the red beetles (refer to figure 7.8). • The red beetles are more likely to be eaten by birds before they reproduce. • The green beetles are more likely to survive and reproduce, passing these alleles on to future generations.
Plants with faster growth rate and plants with a slower growth rate	Limited space and access to sunlight	• The plants with faster growth rate (selective advantage) can access more sunlight to photosynthesise than the slower growing plants. • The plants with the slower growth rate are more likely to die before they reproduce. • The plants with the faster growth rate are more likely to survive and reproduce, passing these alleles on to future generations.
Viral-resistant rabbits and non-resistant rabbits	Presence of virus which kills rabbits	• The viral-resistant rabbits (selective advantage) have an increased chance of survival. • Non-resistant rabbits are more likely to die before they reproduce. • The viral-resistant rabbits are more likely to survive and reproduce, passing these alleles on to future generations.

CASE STUDY: Malaria and sickle cell anaemia — how genetic fitness can vary in different environments

In human history, important selecting agents affecting populations have included widespread infectious diseases that are major causes of death. Malaria is one such disease. For people living in areas of the world affected by the malarial parasites, their fitness depends on whether they are able to produce haemoglobin S. The parasite cannot survive in red blood cells containing haemoglobin S, which is due to the H^S allele. However, the malarial parasite thrives in normal red blood cells, which contain haemoglobin A, produced through the action of the H^A allele, so this group is at risk of death from malaria. On the other hand, when red blood cells contain only haemoglobin S, this causes the disease sickle-cell anaemia.

FIGURE 7.9 a. Coloured scanning electron micrograph (SEM) image of red blood cells. The lower cells are infected with malarial parasites as indicated by the yellowish swollen lumps on the cells, which eventually rupture and spread the infection. b. A healthy red blood cell (left) versus a sickled red blood cell (right).

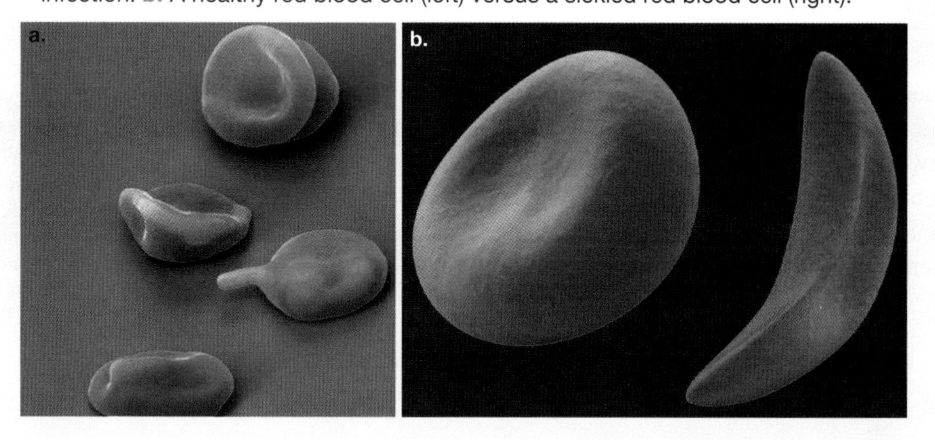

Table 7.2 shows how genetic fitness can vary in different environments. Note that the H^AH^S individual is at a selective advantage relative to the others in a malaria-affected environment, but loses that advantage in a malaria-free environment.

TABLE 7.2 The fitness of a phenotype depends on the environment.

Genotype	Phenotype	Relative fitness
In malaria-affected environments:		
H^AH^A	Usual red blood cells with haemoglobin A only	Can die from malaria *Reduced fitness*
H^AH^S	Cells capable of sickling with haemoglobins A and S	Minor effects of sickling but resistant to malaria *Most fit*
H^SH^S	Severe sickle-cell anaemia with haemoglobin S only	Resistant to malaria but can die from sickle-cell anaemia *Least fit*
In malaria-free environments:		
H^AH^A	Usual red blood cells	No effects of sickling *Most fit*
H^AH^S	Cells capable of sickling	Minor effects of sickling *Reduced fitness*
H^SH^S	Severe sickle-cell anaemia	Can die from sickle-cell anaemia *Least fit*

elog-0035

INVESTIGATION 7.2

online only

Modelling natural selection

Aim

To model natural selection and its effect on allele frequency

on Resources

Video eLessons Selection and tick populations (eles-4198)
Change agents and natural selection (eles-4199)

7.3.2 Genetic drift

Chance factors can cause allele frequencies in a population to randomly change over time rather than being influenced by selection (so it does not matter if an allele causes harm or leads to benefit). This is usually due to **sampling error** in populations that occur between generations. This means that alleles passed on or retained may not be representative of the initial population. When chance operates on the allele frequencies in a population, the direction of the change is unpredictable and can vary from one generation to the next. The resulting pattern of random change is known as **genetic drift**.

sampling error differences that occur when a sample or subset is not representative of an initial population

genetic drift random changes, unpredictable in direction, in allele frequencies from one generation to the next owing to the action of chance events

The change in allele frequency from one generation to the next in genetic drift is random and this is in marked contrast to the directional change that occurs in natural selection.

Unlike the action of natural selection, genetic drift does *not* favour one allele over another; both are equally subject to being affected by genetic drift — it is completely random.

FIGURE 7.10 Changes in allele frequency in two replicate populations due to chance over generations. Despite having the same initial frequency of the *p* allele in their populations, genetic drift has led to this being different over time.

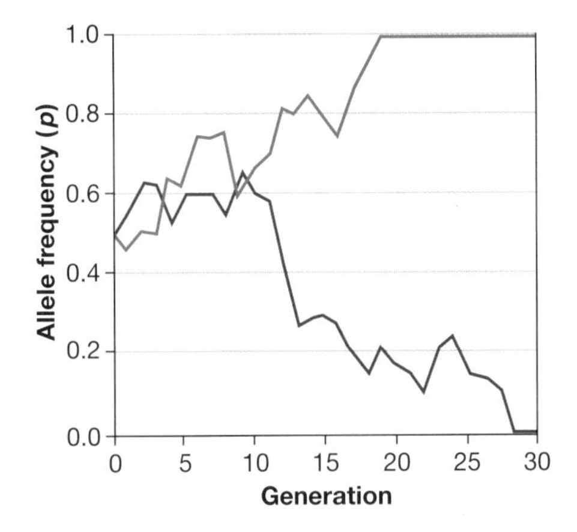

Let's examine an example of this. Consider a population of 100 individuals. Forty of these individual are carriers for cystic fibrosis but are not aware of this. By chance, only 56 people in that population have children. It may be that of the 56 people, 40 were the carriers of the cystic fibrosis allele and 16 were not — a complete chance event. This would be expected to cause the allele frequencies to change drastically over the next generation, all due to a chance event. Over time, it might lead to the fixation of one allele.

The smaller the population size, the greater the potential impact of genetic drift. In a very small population, genetic drift can lead to the decrease, and eventual loss, of favourable alleles from the gene pool. For this reason, when a species is reduced to one or a few small populations, the species is at great risk of extinction.

Two specific examples of genetic drift are through a bottleneck effect or a founder effect.

TIP: Genetic drift is the terminology used in the Study Design. When answering questions about chance events that lead to changing allele frequencies, you should describe the process as genetic drift. From here, you may then specify the type of genetic drift that is shown if appropriate (bottleneck effect or founder effect).

Bottleneck effect can change gene pools

Bottleneck effects come into operation when the size of a population is drastically reduced for at least one generation. The few survivors that reproduce to give the next generation may by chance be an unrepresentative sample of the gene pool of the original population.

This reduction may be the result of:
- a natural disaster, such as a bushfire or flood
- a new disease to which the population has not previously been exposed
- human activity, such as destruction of habitat or large scale poaching.

bottleneck effect chance effects on allele frequencies in a population as a result of a major reduction in population size

All of these may result in a rapid decrease in the size of a population. The original large population may have included a diverse set of alleles in its gene pool. The small post-disaster population is likely, by chance, to have a much less diverse set of alleles in its gene pool (see figure 7.11).

FIGURE 7.11 The bottleneck effect caused by a catastrophic event. Only a few individuals survive to breed. Due to chance, this new population may not be representative of the initial population.

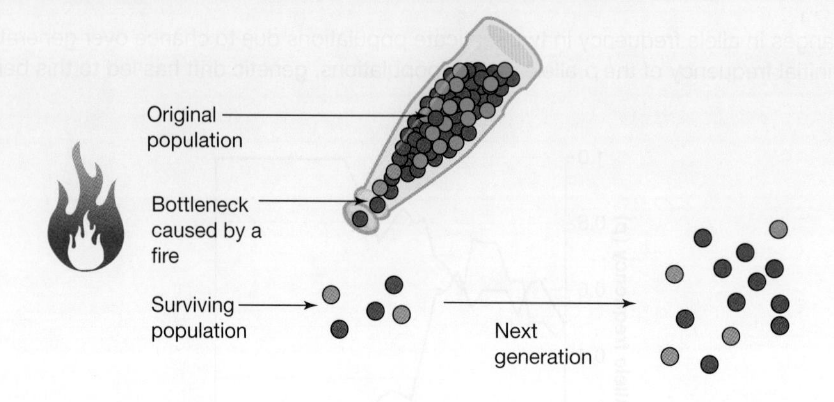

After the disaster has ceased, the population will rebuild. However, the allele frequencies can be very different from those in the original (pre-disaster) population, and the genetic diversity that is the the population's insurance policy may be greatly reduced.

CASE STUDY: Koalas and population bottlenecks

Koalas have been the victims of population bottlenecks, which has greatly affected their genetic diversity over time, presenting lots of challenges for individuals working to prevent their extinction.

FIGURE 7.12 A koala being rescued in Australia after the devastation of the bushfires

Koalas have faced many events that have led to a decrease in their genetic diversity including:

- destruction of habitat (eucalyptus trees)
- hunting and poaching
- drought
- bushfire
- introduced diseases (particularly with *Chlamydia* infections).

As a result, koalas have limited genetic diversity due to various genetic bottlenecks, placing them at a high risk of extinction. Some of the main aims of koala conservation are trying to increase diversity in the gene pool and working to save koalas in times of natural disaster, such as during the 2018–2019 and 2019–2020 bushfires.

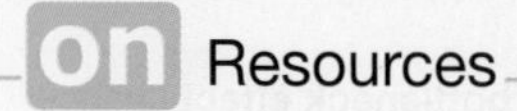

Resources

Weblink Koalas and bushfires: saving their genetic material

Founder effect can change gene pools

A **founder effect** occurs when a new colony is started by a few members from a larger population. This small group that forms a new colony is referred to as the founder population. This **founder population** can be as small as a mating pair or an individual inseminated female. This small population size means that the gene pool of the new colony is highly likely to:

- have reduced genetic variation
- be a non-random or **unrepresentative sample** of the original population (see figure 7.13).

FIGURE 7.13 The founding members of a new population can be less genetically diverse that the original population.

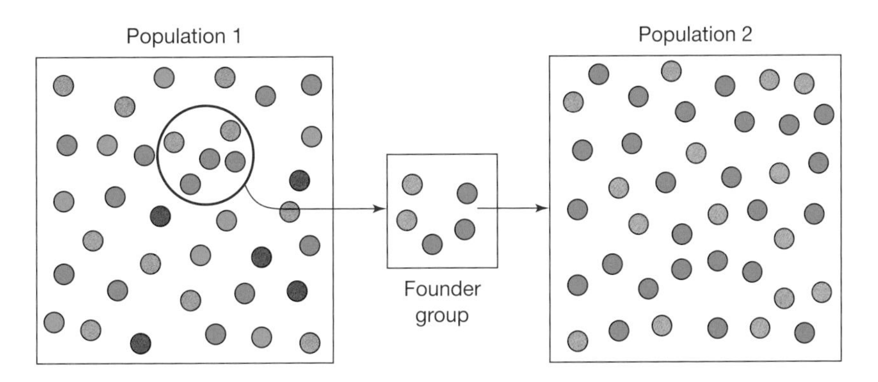

To understand the influence of sample size, think about netting samples of fish from a large tank that contains 20 orange fish and 20 blue fish. If a sample of only four fish is netted, this sample might comprise fish of one variety only. If the sample size is larger, say 14, the sample will almost certainly include fish of both varieties and be representative of the tank population. *With a sample of 14, the chance that the fish will all be blue is very small (around 1 in a half million).*

founder effect chance effects on allele frequencies in a population that is formed from a small unrepresentative sample of a larger population

founder population a small group of organisms that starts a new population

unrepresentative sample a sample or subset that is not representative of the allele frequencies in an initial population

FIGURE 7.14 The sample size of the new colony likely will affect its diversity. **a.** A small sample size is likely to have less variation. **b.** A larger sample size has more variation in the fish in the new colony.

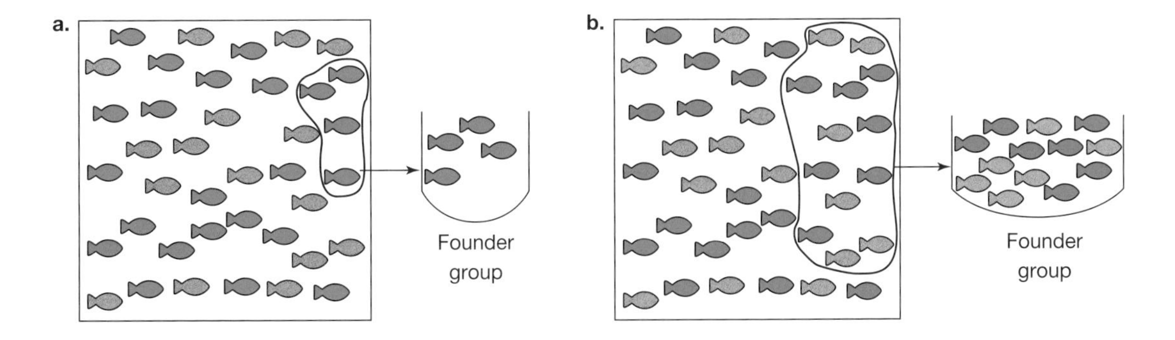

CASE STUDY: Founder effect and Macaroni penguins

FIGURE 7.15 Macaroni penguins

Populations of macaroni penguins (*Eudyptes chrysolophus*) live on subantarctic islands and on the Antarctic Peninsula (see figure 7.15). Most have black faces but a small proportion have white faces. On Macquarie Island, however, the population is composed almost entirely of the white-faced variety. *How did this occur?* It may be by chance that the small founder population of Macaroni penguins that first occupied Macquarie Island consisted of the white-faced variety only. So, when a small unrepresentative or non-random sample of a population leaves to colonise a new region, this is known as the founder effect.

CASE STUDY: Founder effect and the prevalence of Huntington's disease

Examples of founder effect are also seen in humans. Founder effects may be the cause of unusually high incidences of particular inherited diseases in a particular geographic area. One person may come into a region and reproduce there, introducing into the local gene pool an allele that determines a particular disease.

A medical historian has shown that virtually all the families in Tasmania, and in the south-eastern states of Victoria and South Australia, who are affected by Huntington's disease (HD) can trace their family trees back to an English woman who migrated to Tasmania in 1842 with her husband and seven children.

This woman, 'Mary' was born in 1806 in Somerset, England, and she inherited the *HD* allele from her father. Mary and her husband had a further seven children who were born in Tasmania. Nine of Mary's fourteen children developed HD, and all of these individuals had children.

elog-0036

INVESTIGATION 7.3 online only

Modelling genetic drift

Aim

To model the process of genetic drift and explore how this can lead to the fixation of alleles

Resources

Video eLesson Genetic drift (eles-4200)

7.3.3 Gene flow

Gene flow is an important cause of changing alleles in a population.

Gene flow is the movement of alleles between interbreeding populations. When individual organisms move or migrate between populations, they are transferring their alleles from one gene pool to another if they interbreed.

This gene flow can be through the interbreeding of individuals in different populations or through **immigration** (the movement into a population) or **emigration** (the movement out of a population). This movement of alleles can increase the genetic diversity of a population when a new allele is introduced (refer to figure 7.16). When there is no gene flow between populations, they become isolated and any new alleles that arise will remain in the one population.

gene flow the movement of individuals and their genetic material between populations

immigration the movement of individuals and their alleles into a population, and thus into a gene pool

emigration the movement of individuals and their alleles out of a population, and thus out of a gene pool

FIGURE 7.16 The gene flow of individuals and their alleles between populations. The allele frequencies have changed with the movement of individuals.

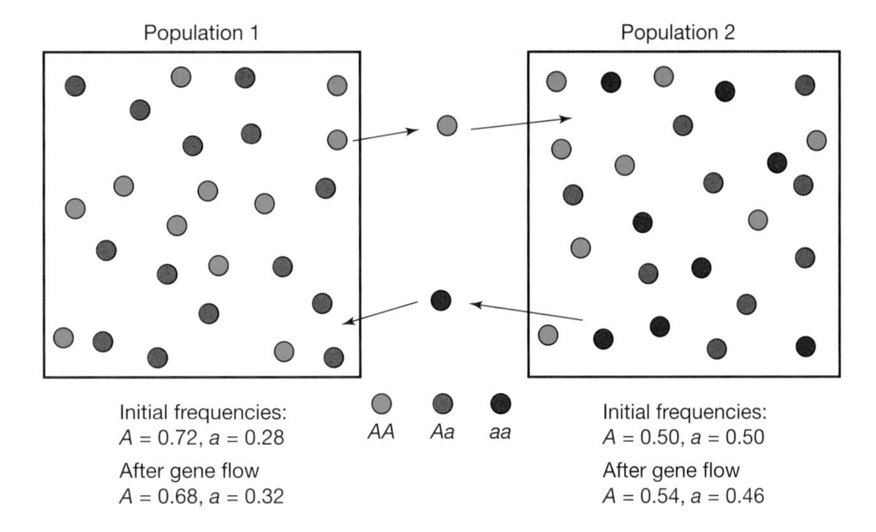

Regional human populations have differences between their gene pools that reflect the long-term effects of natural selection in the prevailing environmental conditions, as well as chance events and gene mutations. The gene pool of western Europeans has a higher frequency of the allele for cystic fibrosis compared to the gene pool of southern Europeans. The gene pool of that population has a relatively higher frequency of the allele for beta thalassaemia. The immigration of groups of people from one region to a new population in a different region can introduce new alleles into the gene pool of the new population or can alter its existing allele frequencies (as shown in figure 7.17).

Emigration can also change allele frequencies if the emigrant group is not a representative sample of the original population. Imagine a small hypothetical population that comprises mainly green-furred organisms, both homozygous *BB* and heterozygous *Bb*, and a few homozygous *bb* red-furred organisms. If a group emigrates from that population, the gene pool of that population may change. If, for example, most the red-furred organisms emigrate, the frequency of the *b* allele will decrease (as shown in figure 7.18).

FIGURE 7.17 Immigration is the movement of individuals into a population.

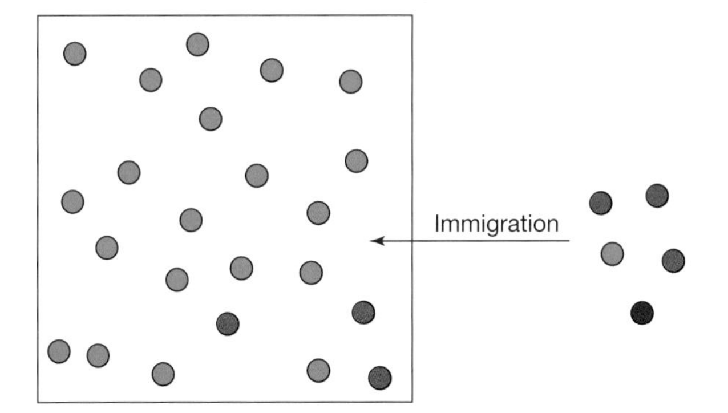

FIGURE 7.18 Emigration is the movement of individuals out of a population.

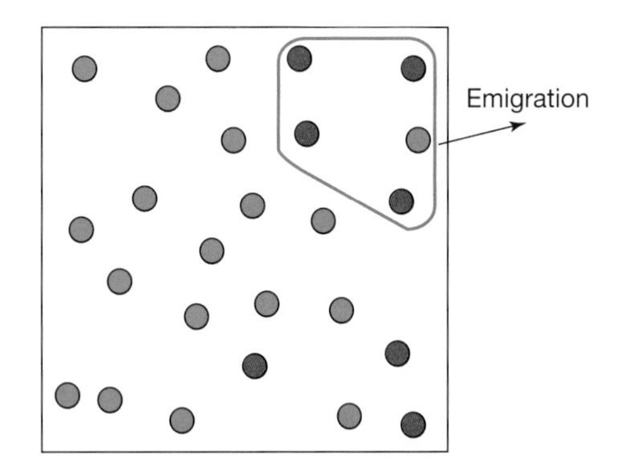

Population size and allele frequency are affected by many factors, including emigration, immigration, births and deaths. Birth and immigration add new alleles to a population, whereas death and emigration lead to the loss of alleles.

change in population = births - deaths + immigration - emigration

tlvd-1789

SAMPLE PROBLEM 2 Explaining allele frequencies in different populations

The common wombat, *Vombatus ursinus*, is a nocturnal Australian marsupial that is herbivorous and lives in burrows. Its colouration can vary between brown, black and sandy. This allows the common wombat to camouflage successfully into its environment. Common wombats are distributed into three populations, as shown in the figure.

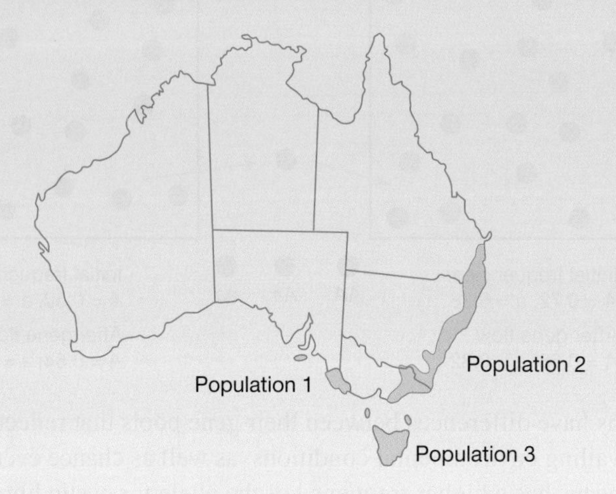

a. Common wombats cannot migrate naturally between the three populations. Describe how this will affect the allele frequencies between the populations. (2 marks)

b. A small colony of isolated common wombats in southern Victoria is composed of ash-white and albino wombats. Explain how this colour change may have occurred. (2 marks)

THINK	WRITE
a. 1. In a *describe* question, you give a detailed account in sentence (or bullet) form.	
2. Consider that no migration equals no gene flow between the populations. Link this back to the allele frequencies of the three populations.	The three populations are isolated and therefore will have no gene flow to allow for the movement of alleles (1 mark). Therefore, the allele frequencies may change due to environmental selection pressures, but not due to gene flow (1 mark).
b. 1. In an *explain* question, you need to account for the reason of why or how something occurs using your scientific knowledge.	
2. Consider if an increase in ash-white or albino wombats would be due to natural selection or genetic drift. It is likely genetic drift (bottleneck event). Natural selection is unlikely as the lighter wombats would likely be at a disadvantage.	The population of ash-white and albino wombats are isolated from other populations. Within their population a bottleneck event may have occurred where due to chance only individuals with this phenotype survived (1 mark). As the population grew in size all individuals were light coloured (1 mark).

CASE STUDY: The introduction of cystic fibrosis and beta thalassaemia alleles

The introduction of new alleles to a gene pool can be advantageous in many situations, increasing variation and genetic diversity and bringing in alleles that may allow for a selective advantage. But in some cases, new alleles coming into a gene pool can be deleterious, leading to genetic disease.

Australia is now a multicultural population. Each of the groups that migrated to Australia has brought with it specific alleles that are common in the home regions of each group but had previously not been present in the gene pool of Australians.

The high incidence of cystic fibrosis in Australia is due to the immigration of people from western Europe, where the allele that causes cystic fibrosis is common. This allele had not been seen in Australia before this immigration occurred. Now, approximately 1 in 25 Caucasian Australians are carriers for cystic fibrosis, showing a drastic change in the gene pool.

Since World War II, migration of people from southern Europe, the Middle East, India and South-East Asia introduced the beta thalassaemia (*t*) allele into the gene pool of the Australian population. Beta thalassaemia affects the haemoglobin in red blood cells, reducing their ability to transport oxygen. The allele for beta-thalassaemia is on chromosome 11 on a gene known as the *HBB* gene.

FIGURE 7.19 Photograph showing Ron Cooke, his wife and 10 children on Southampton Dock prior to boarding a ship in early 1969 to emigrate to Australia

FIGURE 7.20 Beta thalassaemia affects red blood cells as shown in a blood smear from a. an unaffected individual b. an individual with thalassaemia

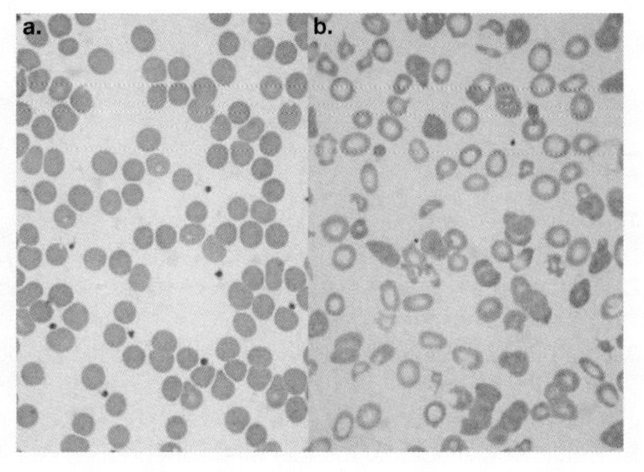

Resources

eWorkbook Worksheet 7.3 Natural selection, gene flow, genetic drift and evolution (ewbk-2781)

KEY IDEAS

- Populations are exposed to environmental selection pressures in the form of physical, biological and chemical agents.
- Various phenotypes in a population may differ in their fitness value according to their environment.
- Environmental selection pressures will act on populations and change allele frequencies.
- The mechanism of natural selection will select for phenotypes in a population with higher fitness values.
- Genetic drift can alter the allele frequency in a gene pool by chance events and does not select phenotypes based on their fitness value.
- Genetic drift can be categorised as a bottleneck effect or a founder effect.
- Bottlenecks usually occur after a catastrophic event and may result in a decrease in the diversity of the allele frequency in the gene pool.
- Founder effect occurs when a small sample of one population becomes the foundation of a new population. Their new gene pool may not be representative of all the alleles of the original gene pool.
- Gene flow, such as the migration of a group into an existing population, may result in an increase in the diversity of the receiving population.
- Gene flow between populations can introduce new alleles and/or change allele frequencies.

7.3 Activities

learnON

To answer questions online and to receive **immediate feedback** and **sample responses** for every question, go to your learnON title at **www.jacplus.com.au**. A **downloadable solutions** file is also available in the resources tab.

7.3 Quick quiz (on)	7.3 Exercise	7.3 Exam questions

7.3 Exercise

1. a. What is does the term *biological fitness* mean?
   b. Contrast between the terms *biological fitness* and *physical fitness*
2. Describe how environmental selection pressures effect allele frequencies of a population through the process of natural selection.
3. In a plant population, some phenotypes survive periods of drought better than other phenotypes. Explain what would occur to this population if it were exposed to drought conditions for several generations.
4. Compare how natural selection and genetic drift change allele frequencies in a population's gene pool.
5. Bottleneck effect and founder effect are both chance factors that can act on the gene pool of a population.
   a. Identify one way in which they are similar.
   b. Identify one way in which they are different.
6. a. Complete the following table comparing the bottleneck and founder effects.

	Description	Diagram	Effect on the gene pool
Bottleneck effect			
Founder effect			

   b. Tristan da Cunha is a group of remote volcanic islands in the Atlantic Ocean. In 1814, a small group of British colonists established a settlement on this island. One of these early colonists was a carrier for a rare disorder known as retinitis pigmentosa. This condition leads to degenerative vision loss.
   The incidence of this disease globally is around 1 in 4000. By the 1960s, the incidence of retinitis pigmentosa on Tristan da Cunha was around 1 in 60. Around 1 in 25 individuals were also carriers for this recessive condition.
   Explain why there is a significantly high incidence of retinitis pigmentosa on Tristan da Cunha.

7.3 Exam questions

Question 1 (1 mark)

Source: *VCAA 2020 Biology Exam, Section A, Q28*

MC The graph below shows the frequency of five alleles in a population over 10 generations.

The changes in allele frequency is represented as a percentage of the frequency of the five alleles.

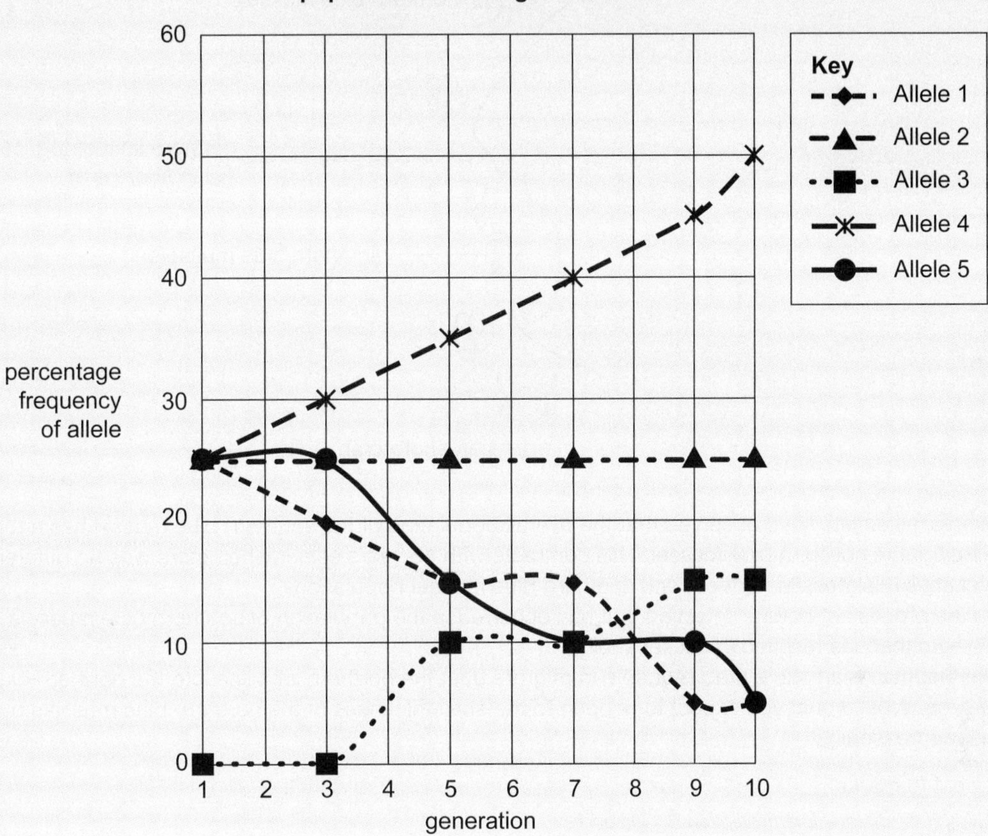

Based on the information in the graph, which one of the following is most likely?

A. The founder effect can explain the change in allele frequency in the first three generations.
B. Genetic diversity within the population remains unchanged over the 10 generations.
C. The appearance of Allele 3 after Generation 3 may be explained by gene flow.
D. A bottleneck event occurred between Generation 7 and Generation 9.

Question 2 (1 mark)

Source: *VCAA 2006 Biology Exam 2, Section A, Q24*

MC The following statements (not in correct order) summarise the steps in natural selection.

1. Some individuals are better suited to a particular environment.
2. Over time there is an increase in particular characteristics in the population.
3. There is variation within a population, some of which is genetic.
4. Individuals better suited to the environment are more successful at survival and reproduction.

The order of statements which best describe natural selection are

A. 1, 3, 2, 4
B. 3, 1, 4, 2
C. 2, 3, 1, 4
D. 1, 2, 4, 3

Question 3 (5 marks)

Source: *VCAA 2010 Biology Exam 2, Section B, Q4a and b*

The blue mussel, *Mytilus edulis*, lives along the northeastern coastline of the USA. A species of Asian shore crab, *Hemigrapsus sanguineus*, was accidentally introduced into the area about 15 years ago. As shown below, the Asian shore crab has only migrated to the southern half of the total area inhabited by the blue mussel.

The Asian shore crab feeds off the blue mussels. The thinner the mussel shell, the easier it is for the crab to crush and eat the mussel.

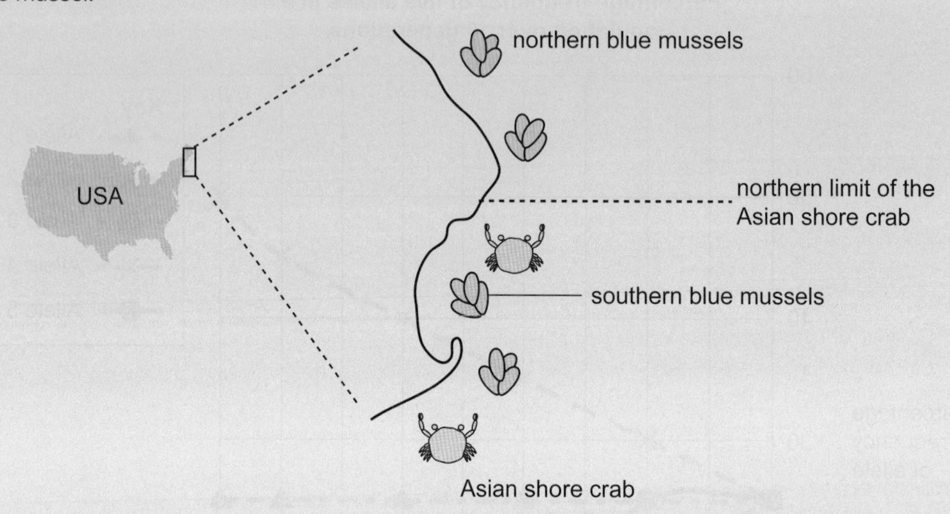

In recent times, scientists have observed that the overall population of the southern blue mussel has a thicker shell than that of the northern blue mussel. This contrasts with 15 years earlier when there was no difference in the range of shell thickness in northern and southern blue mussel populations.

a. Explain the process of natural selection that has occurred in the population of southern blue mussels over the last 15 years that has resulted in thicker shells. **3 marks**

b. Assume that the Asian shore crab is unable to migrate past the northern limit line into the northern blue mussel area. What would you expect to happen to the shell thickness of the northern blue mussels over time? Explain your reasoning. **2 marks**

Question 4 (5 marks)

Source: *VCAA 2017 Biology Exam, Section B, Q5a and b*

The rufous bristlebird (*Dasyornis broadbenti*) is a ground-dwelling songbird. The rufous bristlebird is found in gardens near thick, natural vegetation and builds nests in shrubs close to the ground. The rufous bristlebird feeds on ground-dwelling invertebrates. It is a weak flyer and is slow to go back to areas from which it has been previously eliminated. Two distinct populations of rufous bristlebird exist in Victoria. The distribution of each population is shown on the map of Victoria. The distance between Population A and Population B is over 200 km.

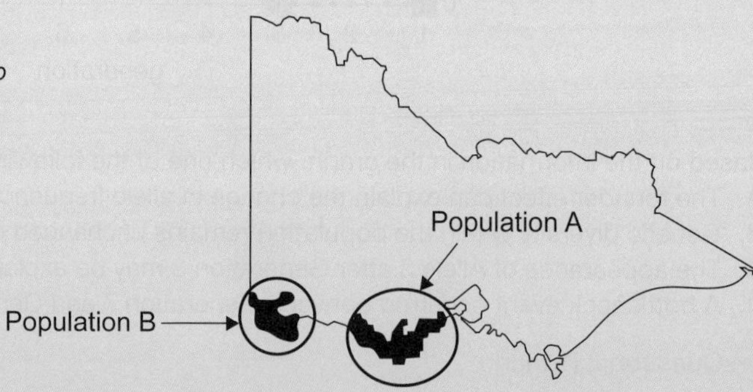

Source: *Flora & Fauna Guarantee Action Statement*, 1993, no. 49; © The State of Victoria, Department of Sustainability and Environment, 2003

a. Define the term 'gene flow' and explain whether gene flow is likely to occur between these two populations. **3 marks**

b. Both of the rufous bristlebird populations in Victoria are small. Referring to the theory of natural selection, explain why the rufous bristlebird is at risk of extinction. **3 marks**

Question 5 (1 mark)

Source: *VCAA 2012 Biology Exam 2, Section A, Q21*

MC Retinitis pigmentosa (RP) is an autosomal recessive trait that results in progressive blindness in humans.

On the island of Tristan da Cunha in the Atlantic Ocean, the frequency of the allele causing RP is four times greater in its population of a few hundred individuals than in the original British population from which it was colonised in the early 1800s. No natural disasters have occurred on the island since it was colonised.

The process that is most likely responsible for this observation related to allele frequency is
A. genetic drift.
B. founder effect.
C. bottleneck effect.
D. natural selection.

More exam questions are available in your learnON title.

7.4 Mutations as the source of new alleles

KEY KNOWLEDGE

- Causes of changing allele frequencies in a population's gene pool, including mutations as the source of new alleles

Source: Adapted from VCE Biology Study Design (2022–2026) extracts © VCAA; reproduced by permission

7.4.1 Mutations as a source of new alleles

Generally, the genetic material of an organism is stable, both in its base sequence and in its chromosomal location, and is passed unchanged from generation to generation.

However, the genetic material can change. This change is very commonly a change in the base sequence of a gene. When this happens, a new form of a gene, called an allele, is created. Such a change is known as a **mutation** and it can alter the instructions that are encoded in the DNA.

This may be seen in the unexpected birth of a baby with achondroplasia, a form of dwarfism, to two parents of average height (see figure 7.21). This event is the visible expression of one change in the base sequence of a gene in a gamete (either egg or sperm) of one of the parents. Other gene mutations result from a change in the location of a gene, and this move alters its expression.

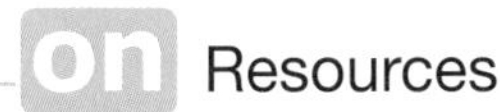 Resources

Weblinks Outcome of mutations
How often do mutations happen?

FIGURE 7.21 Laila El Garaa of Morocco is a short-statured woman who competed in the Paralympics.

mutation a permanent alteration in the DNA sequence of a gene that leads to the formation of new alleles

7.4.2 Causes of mutation

Mutations may be spontaneous or may be induced by known exposure to **mutagens** (or mutagenic agents), such as irradiation (for example, x-rays and gamma rays), some chemicals (for example, benzene, dioxane and mustard gas) and some viruses.

FIGURE 7.22 Causes of mutation

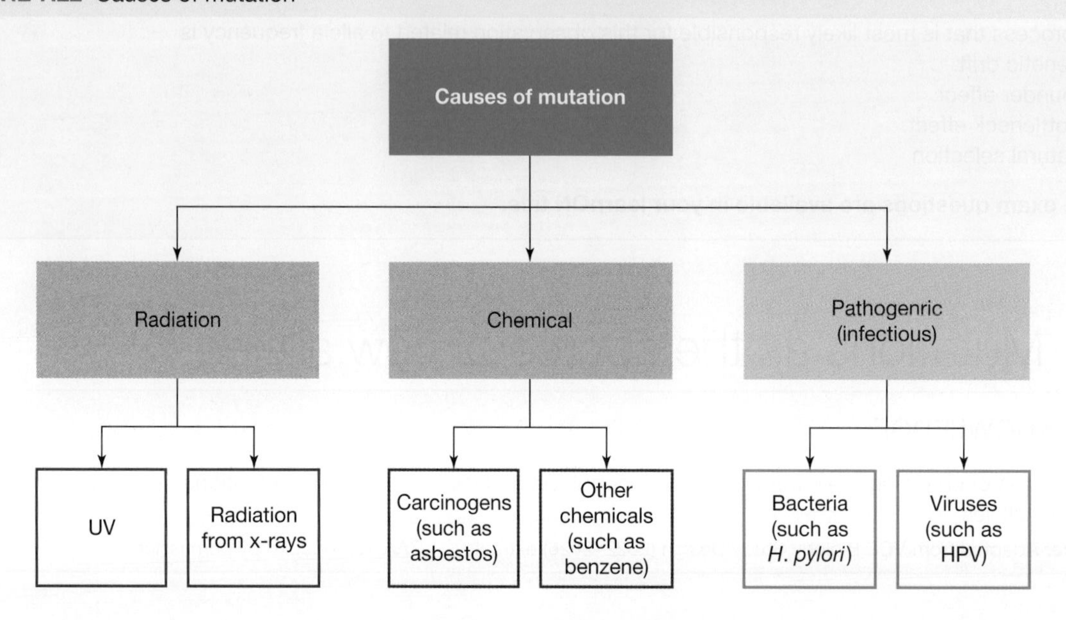

Protection against mutagenic agents is necessary to protect the DNA in both **somatic cells** and **germline cells**. Have you had an x-ray during a dental check-up? The use of a lead apron during this procedure is to protect body tissues from exposure to radiation (see figure 7.23). Doses received by people working with radiation are monitored by wearing badges that contain radiation-sensitive devices.

FIGURE 7.23 A lead apron is worn to protect the body's tissues from radiation.

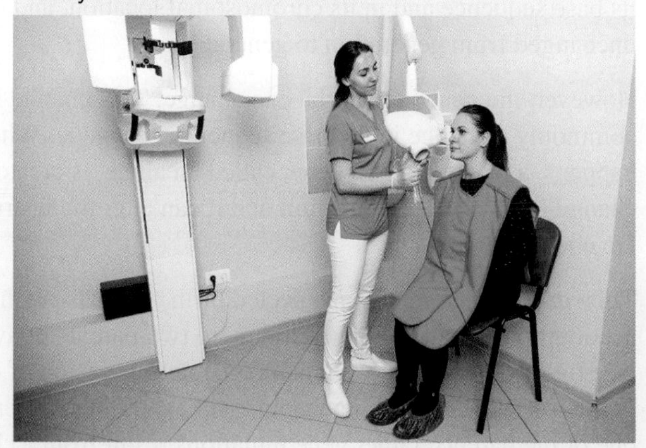

Mutation rates are not equal for all genes and the spontaneous mutation rate for achondroplasia, one form of inherited dwarfism, is much higher than that for some other inherited disorders (see table 7.3). This is in accord with the fact that about 90 per cent of the cases of achondroplasia are sporadic (or spontaneous), and are due to mutation. On average, more sporadic cases occur in association with increased paternal age at the time of conception.

mutagen chemical or physical agent that can cause mutation in DNA

somatic cells all body cells other than the reproductive cells (sperm and egg)

germline cells sex cells (sperm and egg) that are also known as gametes, that may pass on genetic information to the next generation

TABLE 7.3 Spontaneous mutation rates at different gene loci in various human population samples

Mutation	Mutations per 100 000 gametes	Locus
Retinoblastoma		
United Kingdom	1.2	Number-13 chromosome
United States	1.8	
Achondroplasia		
Northern Ireland	14.3	Number-4 chromosome
Japan	12.2	
Huntington's disease		
United States	0.5	Number-4 chromosome

CASE STUDY: Mutations and Chernobyl

The world has been recently reminded of the Chernobyl disaster with the release of an Emmy–award winning series recounting the horrors of the nuclear disaster at the time it occurred. However, the effects of the Chernobyl disaster lasted long after the nuclear accident.

The Chernobyl disaster occurred in 1986 when a nuclear reactor exploded during routine safety inspections. This released large amounts of iodinising radiation, which is a known mutagen.

One of the long-term consequences of the Chernobyl disaster has been an increased incidence of thyroid cancer due to mutations that occurred as a result of the radiation. Many reports also suggest an increase in mutations and congenital defects after the Chernobyl incident due to the inheritance of mutations by the next generation. This was found not only in humans, but also in the flora and fauna of the area.

FIGURE 7.24 An area in the city of Pripyat, near the Chernobyl site, which was abandoned after the disaster

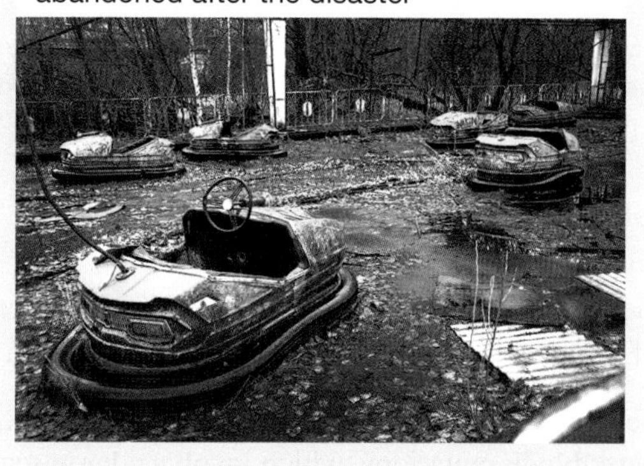

The area around Chernobyl was made an exclusion zone. Even today, very few people live there due to higher levels of radiation.

Persistence of mutations within a gene pool

In order for a gene mutation to be sustained in a population, it is first necessary that the mutation occur in the germline of an organism so that it may be passed to the next generation.

If the germline mutation produces an allele that controls a dominant trait, its phenotypic expression will appear in the next generation. If, however, the new allele controls a recessive trait, it may be several generations before the trait is expressed phenotypically in the population.

After a mutation is expressed in a phenotype, it will be the actions of change agents, in particular natural selection, that determine whether the allele is lost from the population or whether it becomes established in the gene pool of a population. Some gene mutations disappear after one or a few generations, because of strong selection pressure acting against them.

In addition, in a small population, even advantageous mutations may be lost through the action of genetic drift. As genetic drift involves chance events, all traits and associated alleles have a chance to be retained or lost, regardless of whether it is an advantage or not. The influence of natural selection and genetic drift on alleles can be seen in figure 7.25.

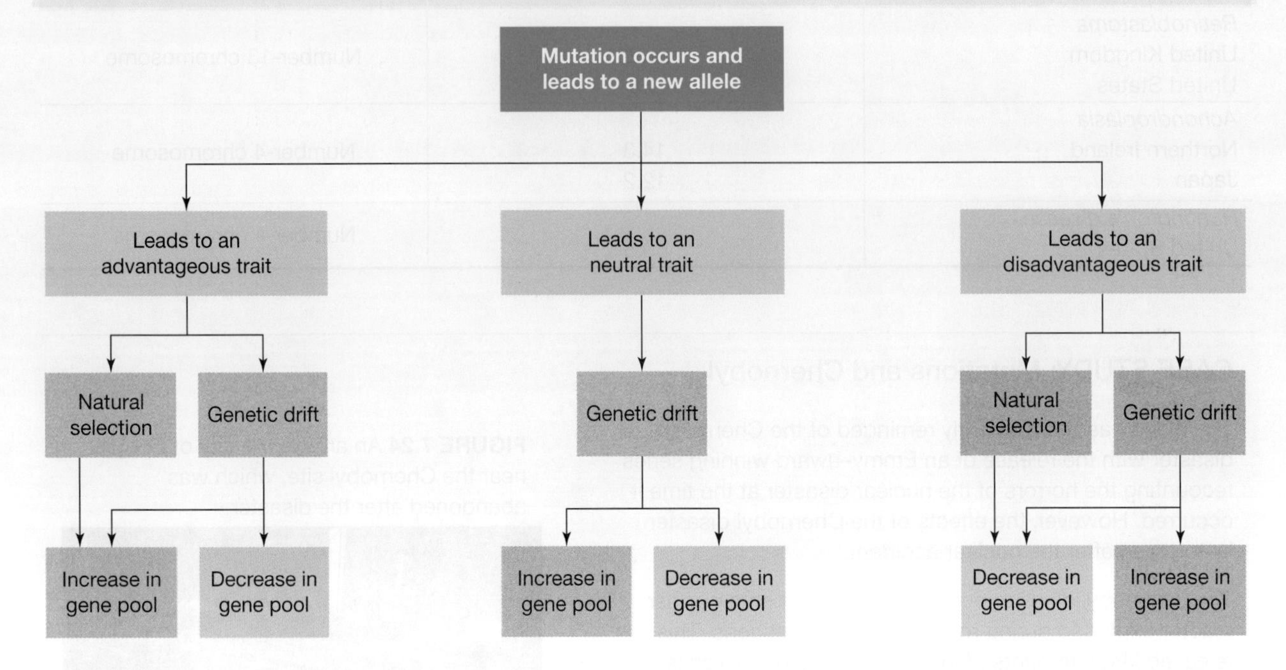

FIGURE 7.25 Natural selection and genetic drift can affect whether a mutation and associated allele is lost or retained in a population.

7.4.3 Types of mutation

Gene mutations are changes in the genetic material or DNA. DNA provides the genetic code for the production of proteins through transcription and translation (as explored in topic 1).

Several kinds of change in DNA can be identified, such as point mutations, which involve a change of a single base; block mutations, which involve a large section of a chromosome; and chromosomal mutations, which involve entire chromosomes.

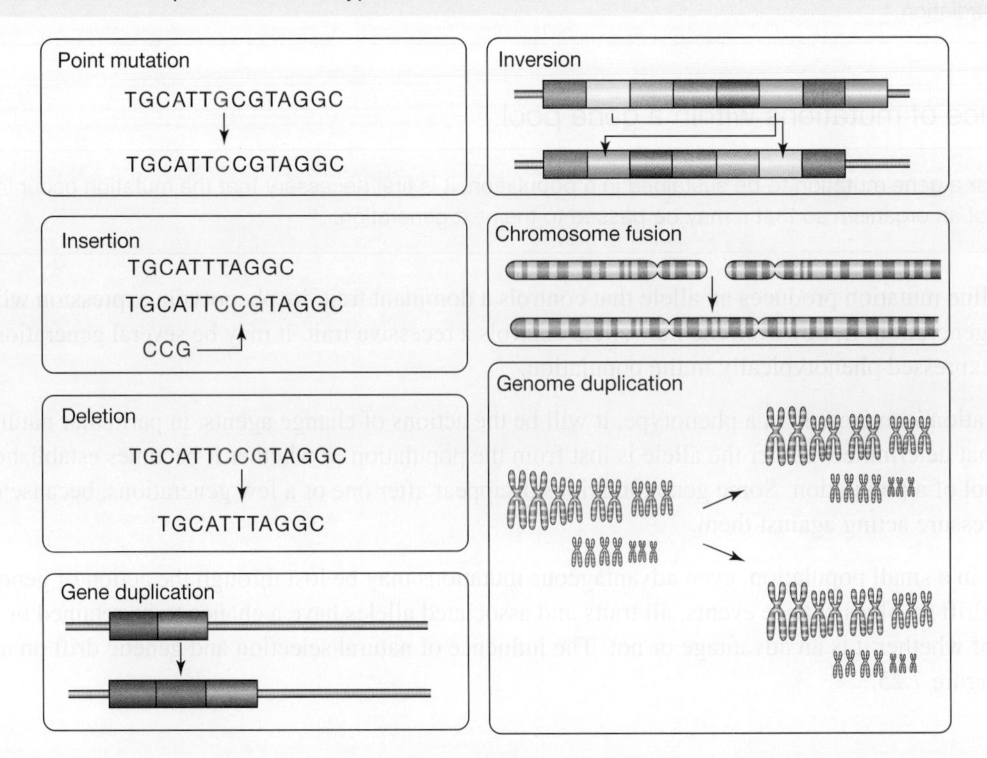

FIGURE 7.26 Some examples of different types of mutation

7.4.4 Point mutations

The sequence of bases in DNA and genes provides the code for proteins. In subtopic 1.4, protein synthesis was introduced in which DNA is transcribed into mRNA and mRNA is translated to produce proteins. Therefore, any changes in DNA through mutation may affect the amino acids added in translation, leading to a new allele.

Point mutations affect a single nucleotide in DNA, either by **substitution**, insertion or deletion.

Point mutations include:

1. Substitution mutations (change of a single nitrogenous base to a different nitrogenous base), including:
   - silent mutations
   - nonsense mutations
   - missense mutations (conservative or non-conservative)
2. Frameshift mutations (change that results in alteration to the reading frame), including:
   - insertions
   - deletions.

These mutations commonly affect the nucleotide sequence of a structural gene. However, these mutations may also affect a segment of DNA that regulates the expression of a structural gene. These mutations may result in the creation of a protein which is faulty or non-functional.

7.4.5 Substitution mutations

In a substitution, one base in a DNA triplet is replaced by another.

Three kinds of base substitutions are recognised: silent, nonsense and missense.

Silent mutations

Silent mutations are mutations that do not result in any change in phenotype, as the mutation leads to the same amino acid being encoded for. This is because the genetic code is degenerate (refer back to section 1.4.1) and many amino acids are encoded by several DNA triplets.

A silent mutation is a nucleotide substitution in DNA that does not result in a change in the amino acid sequence of polypeptide chain encoded by this gene.

For example, the amino acid valine (*val*) is encoded by four DNA triplets on the coding strand: GTT, GTC, GTA and GTG (the complementary triplets appear on the template strand). Therefore, a mutation resulting in a change to the last nucleotide will not change the amino acid in the polypeptide chain.

FIGURE 7.27 A example of a silent mutation

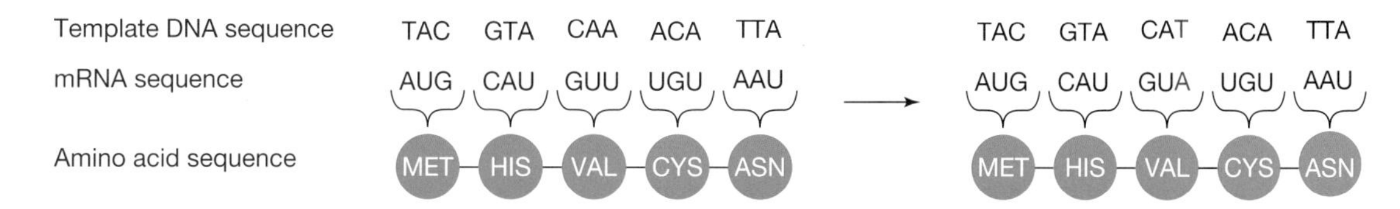

This DNA substitution mutation will not affect the expression of the gene because the mutated DNA triplet still codes for the same amino acid (*val*), hence the label 'silent'. Silent mutations most commonly occur in the third nucleotide of a DNA triplet.

While these mutations are 'silent' in terms of their polypeptide chain, they are of course visible when the DNA sequence is examined.

substitution a gene mutation where one nucleotide is replaced with another

silent mutation a type of mutation in which a single base change does not change the amino acid and final protein expressed

Nonsense mutations

Nonsense mutations cause a STOP codon to be coded for earlier than intended, resulting in a shortened polypeptide to be formed. Instead of continuing adding amino acids as is intended, the translation of the protein is stopped and thus the protein formed is incomplete. The STOP signals in amino acids are coded by the codons UAA, UAG and UGA.

> A nonsense mutation in DNA is a nucleotide substitution that changes a DNA triplet so that, instead of coding for an amino acid, it codes for a STOP signal, leading to a shortened protein product.

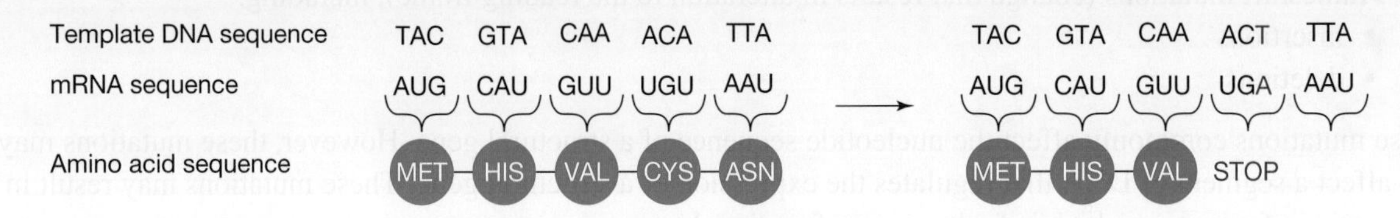

FIGURE 7.28 A example of a nonsense mutation

Such a mutation will likely have a serious effect on the polypeptide product of the gene because translation will stop at the mRNA codon transcribed by the mutated triplet DNA sequence. In turn, this will shorten the polypeptide chain encoded by the gene. The degree of shortening of the polypeptide gene product will depend on the position of the point mutation relative to the start of the coding sequence — the earlier in the DNA sequence, the greater the shortening of the polypeptide.

Missense mutations

A **missense mutation** is a nucleotide substitution in DNA that results in a single amino acid alteration in the polypeptide product of the gene. Missense mutations may be conservative or non-conservative.

> A missense mutation involves a change in one amino acid to another amino acid.

Conservative mutation

When a missense mutation occurs and a different amino acid is incorporated into a polypeptide chain, the outcome depends on whether the mutated polypeptide can carry out its normal function. This is likely to be the case in a **conservative missense mutation** and such a mutation can be tolerated. A conservative missense mutation of DNA results in the substitution of one amino acid by a different amino acid with similar chemical properties (see the example in figure 7.29). The chemical properties of the polypeptide are said to be 'conserved' (refer to the appendix for the chemical properties of amino acids).

nonsense mutation a type of mutation in which a single base change leads to a STOP signal being received, resulting in a truncated protein

missense mutation a type of mutation in which a single base change leads to the change in the amino acid translated in the protein chain

conservative missense mutation a type of missense mutation in which the substituted amino acid is similar in properties to the initial amino acid

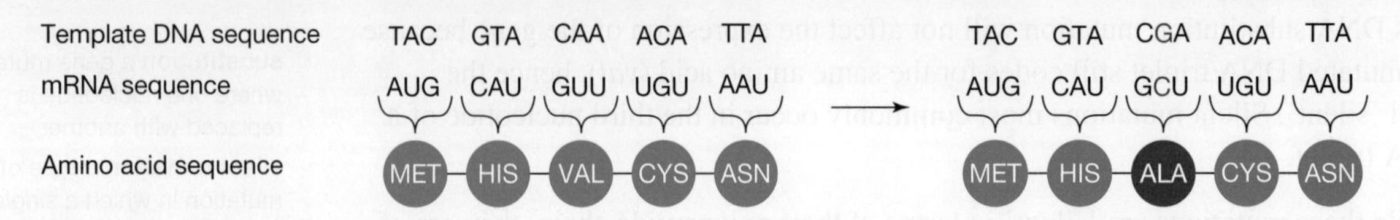

FIGURE 7.29 A example of a conservative missense mutation

This mutation results in the replacement of the amino acid valine (*val*) with alanine (*ala*). Both of these amino acids have hydrophobic side chains and thus similar properties, so may still result in a functional protein.

Non-conservative mutation

In other cases, the missense mutation results in the substitution of an amino acid by one that has different chemical properties (for example from a hydrophillic amino acid to a hydrophobic amino acid). Such a mutation is said to be a **non-conservative missense mutation**, as the chemical properties are not conserved. In this case, the resulting polypeptide is unlikely to be able to carry out its normal function and such a mutation will have severe clinical effects.

Consider the replacement of histidine (*his*), which contains a charged side chain, with tyrosine (*tyr*), which contains a hydrophobic chain. This will affect the formation of secondary structures and the folding of the polypeptide chain into the three-dimensional tertiary structure.

FIGURE 7.30 A example of a non-conservative missense mutation

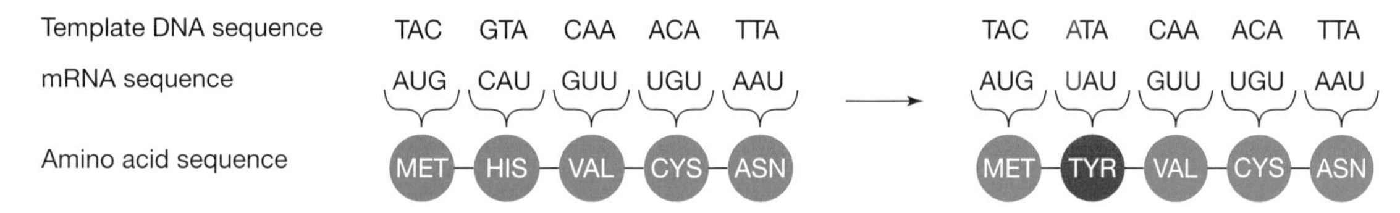

Examples of non-conservative missense mutations are:

- haemoglobin M disease, which results from a missense mutation of a substitution of histidine (*his*) with tyrosine (*tyr*) in the alpha chain of haemoglobin. Haemoglobin M can carry oxygen but cannot release it to body cells where it is needed.
- achondroplasia, which is the result of the replacement of the amino acid glycine (*gly*) with arginine (*arg*) at amino acid number 380 in the receptor protein for a growth factor. We can write this replacement as *Gly*380*Arg*. This replacement is caused by a single base substitution in the DNA triplet that encodes the 380th amino acid in this protein that mutates as CCA→GCA.

7.4.6 Frameshift mutations

Single nucleotide insertions have a major effect on the genes involved because they alter the DNA triplet at the point of the insertion and also affect every DNA triplet following that point. Likewise, single base deletions have the same effects. These single base insertions or deletions are known as **frameshift mutations**.

Frameshift mutations result in the 'reading frame' being altered, affecting all codons from the point of the mutation. Remember that the genetic information in DNA is organised into groups of three bases or triplets that are then read in order, with one triplet coding for one amino acid. In addition, both insertion and deletion may form an early stop codon where one should not exist. This causes translation to stop early, leading to the production of an incomplete polypeptide.

This idea of changing the reading frame can be seen in the sentence shown in table 7.4.

The initial sentence is THE FAT CAT SAT. If a letter is added or removed and the letters are rearranged back in groups of three, the sentence becomes nonsensical.

non-conservative missense mutation a type of missense mutation in which the substituted amino acid is very different in properties to the initial amino acid

frameshift mutation type of mutation in which, as a result of insertion or deletion of a base, all codons from that point are affected

TABLE 7.4 Adding or deleting letters changing the reading frame

Initial	THE	FAT	CAT	SAT	
Insertion (of letter X)	THX	EFA	TCA	TSA	T
Deletion (of letter H)	TEF	ATC	ATS	AT	

Frameshift mutations alter the base sequence of a gene so that the message it encodes no longer makes sense. This is referred to as changing the reading frame.

It is important to note that frameshift mutations can also occur with insertions and deletions of more than one nucleotide. For example, the insertion or deletion of two nucleotides will change the reading frame. Insertions or deletions of nucleotides in multiples of three, however, will not cause a frameshift, as the reading frame after the mutation will not be affected.

Insertion mutations

An insertion mutation occurs when one nucleotide base is added to a DNA strand.This type of mutation is often more damaging than a substitution mutation because it causes a frameshift, with all the bases from the point of insertion being moved down by one position to make room for the extra nucleotide.

FIGURE 7.31 A example of an insertion resulting in a frameshift

C
Template DNA sequence TAC GTA CAA ACA TTA → TAC GCT ACA AAC ATT A
mRNA sequence AUG CAU GUU UGU AAU → AUG CGA UGU UUG UAA U
Amino acid sequence MET HIS VAL CYS ASN → MET TYR VAL CYS STOP

Deletion mutations

A nucleotide deletion also causes a frameshift with all bases from the point of the deletion being moved back by one position to compensate for the deletion.

FIGURE 7.32 A example of a deletion resulting in a frameshift

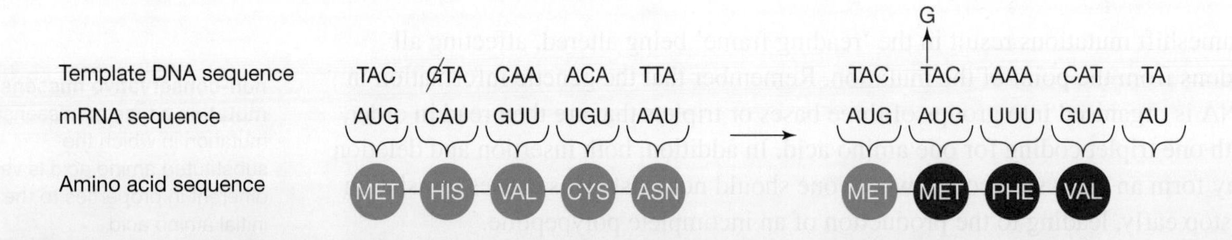

elog-0037

INVESTIGATION 7.4 online only

Investigating mutations

Aim

To understand different types of mutations and how they impact polypeptide formation

tlvd-1790

SAMPLE PROBLEM 3 Identifying mutations and explaining their impacts

Examine the mRNA nucleotide sequence below:

5′	AUG	GGG	CGU	AGC	UAC	AGG	CUU	3′

Codon Table:

		SECOND BASE							
		U		C		A		G	
FIRST BASE	U	UUU	*Phe*	UCU	*Ser*	UAU	*Tyr*	UGU	*Cys*
		UUC	*Phe*	UCC	*Ser*	UAC	*Tyr*	UGC	*Cys*
		UUA	*Leu*	UCA	*Ser*	UAA	*STOP*	UGA	*STOP*
		UUG	*Leu*	UCG	*Ser*	UAG	*STOP*	UGG	*Trp*
	C	CUU	*Leu*	CCU	*Pro*	CAU	*His*	CGU	*Arg*
		CUC	*Leu*	CCC	*Pro*	CAC	*His*	CGC	*Arg*
		CUA	*Leu*	CCA	*Pro*	CAA	*Gln*	CGA	*Arg*
		CUG	*Leu*	CCG	*Pro*	CAG	*Gln*	CGG	*Arg*
	A	AUU	*Ile*	ACU	*Thr*	AAU	*Asn*	AGU	*Ser*
		AUC	*Ile*	ACC	*Thr*	AAC	*Asn*	AGC	*Ser*
		AUA	*Ile*	ACA	*Thr*	AAA	*Lys*	AGA	*Arg*
		AUG	*Met*	ACG	*Thr*	AAG	*Lys*	AGG	*Arg*
	G	GUU	*Val*	GCU	*Ala*	GAU	*Asp*	GGU	*Gly*
		GUC	*Val*	GCC	*Ala*	GAC	*Asp*	GGC	*Gly*
		GUA	*Val*	GCA	*Ala*	GAA	*Glu*	GGA	*Gly*
		GUG	*Val*	GCG	*Ala*	GAG	*Glu*	GGG	*Gly*

a. **Using the codon table provided, translate the mRNA nucleotide sequence.** **(1 mark)**

b. **A guanine nucleotide is inserted before the thirteenth nucleotide of the mRNA sequence.**

   i. **Identify which type of mutation this is.** **(1 mark)**

   ii. **Explain how this type of mutation will affect the structure and function of the protein produced.** **(2 marks)**

THINK

a. 1. Identify what the question is asking you to do. The question is asking you to *translate*. Translate means using the mRNA nucleotide sequence to determine the amino acid sequence.

2. Consider the amino acid table provided. It is a codon table. Therefore, the mRNA nucleotide sequence is used to identify the amino acid sequence. A common mistake is creating a tRNA sequence — this only needs to occur if the amino acid chart is an anticodon table.

WRITE

List the amino acids in order:
Met-Gly-Arg-Ser-Tyr-Arg-Leu (1 mark)

b. i.	**1.** Identify what the question is asking you to do. The first part of the question asks you to *identify*. In an identify question, you can list answers with one word from the options given.	
	2. The key word to identify in question is 'inserted'. When a nucleotide is inserted (added) or deleted and changes the reading frame, then it is a frameshift mutation.	Frameshift mutation (1 mark)
c. ii.	**1.** Identify what the question is asking you to do. The question asks you to explain. In an *explain* question, you need to account for the reason why or how something occurs using your scientific knowledge.	
	2. First consider how an inserted nucleotide will change the codon sequence and therefore the amino acid sequence. Use this to explain how a changed amino acid will change the secondary and tertiary folding of a protein (structure).	The amino acid sequence after the insertion will be changed (the last three amino acids will be *Val-Glu-Ala* instead of *Tyr-Arg-Leu*). Therefore, the polypeptide chain will be folded differently (1 mark).
	3. Consider how a different structure will affect the function of a protein.	This will result in a protein that is non-functional (1 mark).

7.4.7 Other types of mutations

The point mutations discussed involve single base changes. However, some mutations are much larger than just a single base. Many of these mutations lead to changes in genes and thus can be the sources of new alleles.

- Some mutations involve more than one nucleotide being inserted or deleted. Adding nucleotides in groups of three is less damaging than single nucleotides as it does not affect the reading frame (known as non-frameshift).
- Some mutations, known as trinucleotide repeats involve mutations that repeat a large number of nucleotides.
- Some mutations, known as block mutations, involve large sections of DNA being translocated, inverted, deleted or duplicated (such as in Cri-du-chat syndrome)
- Some mutations are chromosomal and involve the deletion or addition of an entire chromosome (such as in trisomy 21 or Down syndrome).

Block and chromosomal mutations

Block mutations are chromosomal changes affecting large segments of a chromosome. These block mutations most commonly arise as a result of spontaneous errors in crossing over during meiosis, or they may be induced by mutagenic agents, such as x-rays.

These block mutations include:

- deletion of part of a chromosome
- duplication or gain of part of a chromosome
- translocation or reciprocal exchange between non-homologous chromosomes
- inversion when a segment of a chromosome rotates through 180 degrees.

Figure 7.33 shows some of these block mutations. In each case, chromosome breakages occur followed by rearrangement and rejoining.

FIGURE 7.33 a. Normal chromosome and the same chromosome showing a duplication b. Normal chromosome and the same chromosome showing a deletion c. Normal chromosome and the same chromosome showing an inversion d. An example of a 14/21 translocation

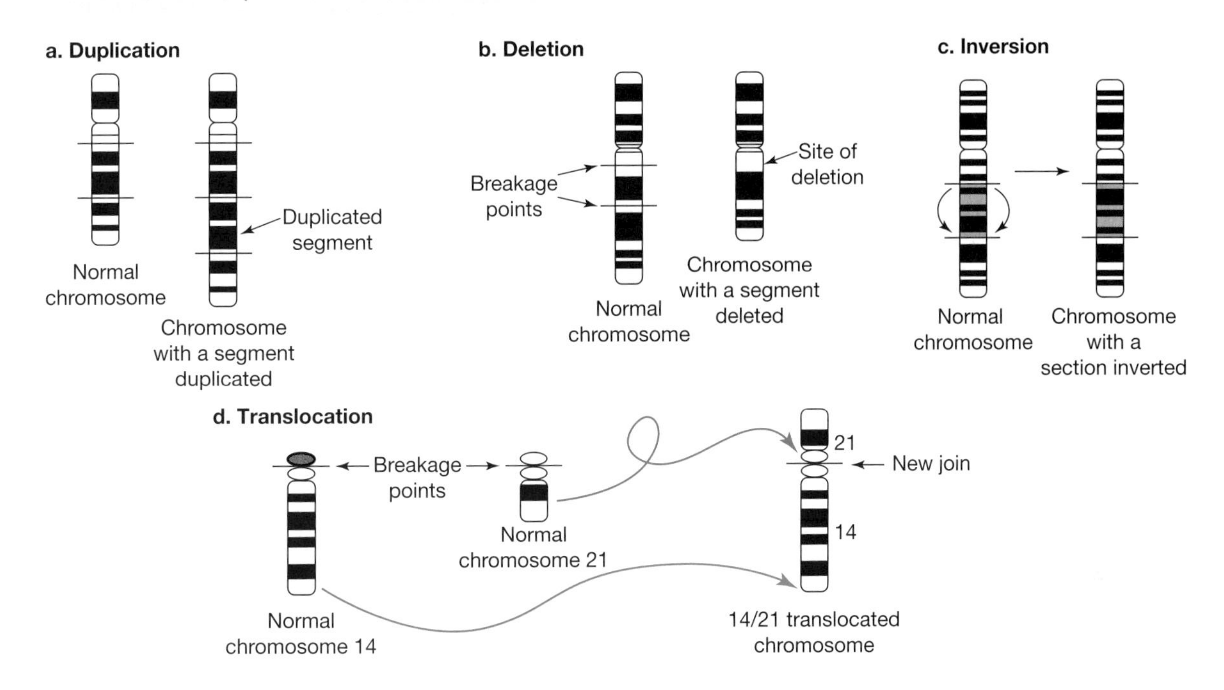

None of these chromosomal block mutations involves a loss of genetic material, but they do alter the location of some genes, relocating them in a new position on a chromosome. Research has shown that relocating can alter the activity or the expression of a gene — such effects are termed position effects. In some cases, a gene at a breakpoint may be silenced or its expression may become unstable or, in the case of some reciprocal translocations, abnormalities such as cancers occur.

Figure 7.34 shows the details of a reciprocal translocation that results in chronic myeloid leukaemia. In a reciprocal translocation, sections from two chromosomes change location. The relocation of the ABL and the BCR genes in close proximity on the Philadelphia chromosome has been shown to be the cause of chronic myeloid leukaemia.

FIGURE 7.34 Diagram showing the reciprocal exchange that occurs between chromosome 9 and chromosome 22 in chronic myeloid leukaemia

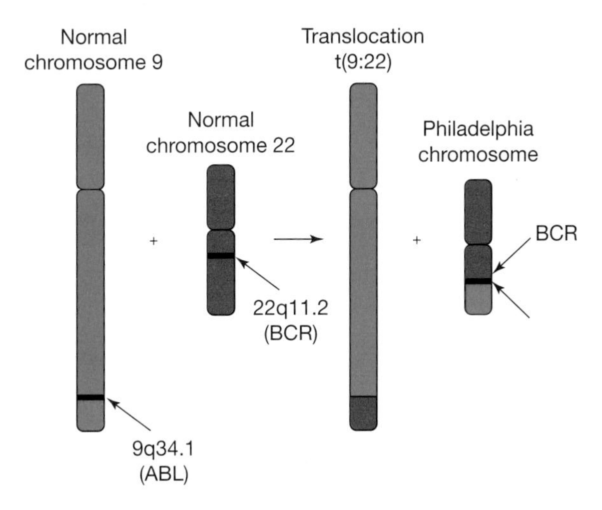

Figure 7.35 shows the unstable expression of a gene for eye colour in the compound eye of a fruit fly (*Drosophila melanogaster*) after relocation of the gene involved to a new chromosomal location by an inversion. Note that some eye facets are dark where the gene is active but others are pale where the gene is not expressed. The variegation results from the unstable expression of the gene determining red eye colour because of a chromosomal inversion that has relocated it.

FIGURE 7.35 Images of the compound eye of the fruit fly *Drosophila melanogaster*, showing the normal red eye (left), the normal white eye (middle) and the variegated eye (at right). (Images courtesy of A.M. Bauer and S.C.R. Elgin, Washington University, St Louis.)

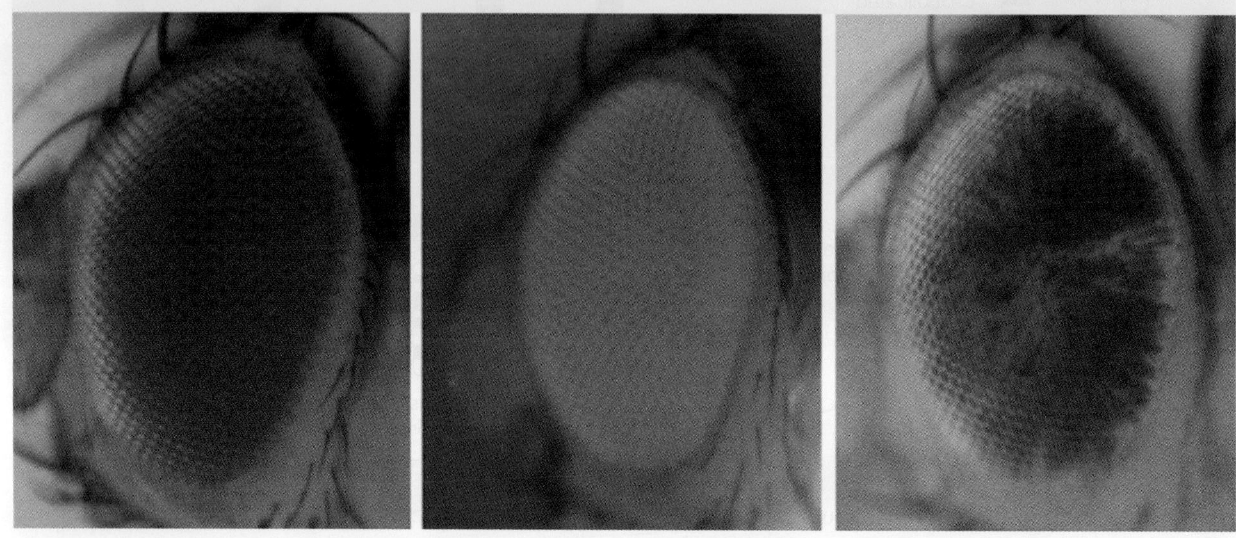

CASE STUDY: Trinucleotide repeat expansion mutations

One class of gene mutations involves the addition or deletion of a large number of short sequences of bases.

Several normal human genes contain multiple copies of a three-base (trinucleotide) sequence, such as CCG and CAG. One kind of mutation, known as a trinucleotide repeat expansion (TRE), involves additional repeats of these sequences beyond the normal range. These TRE mutations are the cause of two inherited disorders: fragile X syndrome (FMR1) and Huntington's disease (HD) (see table 7.5).

TABLE 7.5 Examples of trinucleotide repeat expansion mutations

Gene	Mutant condition	Trinucleotide repeats
FMR1	Fragile X syndrome	CCG: 200 to 1000+ repeats
HD	Huntington's disease	CAG: 36 to 120 repeats

To access more information on trinucleotide repeat expansion mutations, please download the digital document.

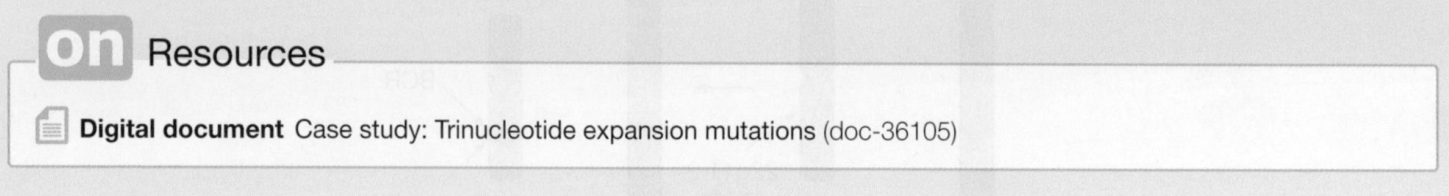
on Resources

Digital document Case study: Trinucleotide expansion mutations (doc-36105)

Polyploidy and aneuploidy

As well as block mutations, changes to entire chromosomes can occur, such as:

- polyploidy — the entire genome is replicated (so instead of 2 of each chromosomes there may be 3 of every chromosome). This does not occur in humans.
- aneuploidy — the loss or gain of a single chromosome (the loss of a chromosome is termed monosomy and the gain is termed trisomy).

TABLE 7.6 Some examples of chromosome changes and approximate incidence rates. Which syndrome is an example of a trisomy? A monosomy? Note that the XYY condition does not have a clinical name.

Chromosome changes	Resulting syndrome	Approximate incidence rates
Addition: whole chromosome		
extra number-21 (47, +21)	Down syndrome	1/700 live births
extra number-18 (47, +18)	Edwards syndrome	1/3000 live births
extra number-13 (47, +13)	Patau syndrome	1/5000 live births
extra sex chromosome (47, XXY)	Klinefelter syndrome	1/1000 male births
extra Y chromosome (47, XYY)	n/a	1/1000 male births
Deletion: whole chromosome		
missing sex chromosome (46, XO)	Turner syndrome	1/5000 female births
Deletion: part chromosome		
missing part of number-4	Wolf-Hirschhorn syndrome	1/50 000 live births
missing part of number-5	Cri-du-chat syndrome	1/25 000 live births

on Resources

eWorkbook Worksheet 7.4 Types of mutation (ewbk-2782)

Video eLessons Changes in chromosomes (eles-4201)
Types of mutations (eles-4214)

Interactivities Mutations (int-1453)
Chromosomal changes (int-0181)

Weblinks Mutations interactive

KEY IDEAS

- Mutations are changes to genetic materials which can results in new alleles.
- Mutations may occur spontaneously or be induced by exposure to a mutagen.
- Mutations which occur in germline cells can continue in a population's gene pool.
- Mutations can result in faulty or non-functional proteins.
- Point mutations can be classified as either substitution mutations or frameshift mutations.
- Substitution mutations can affect the polypeptide chain differently and result in either silent, nonsense or missense mutations.
- Silent mutations do not change the amino acid in a protein.
- Missense mutations lead to the change in an amino acid in a protein. This can be conservative or non-conservative, which depends on the properties of the amino acid.
- Nonsense mutations introduce a stop codon, leading to a truncated protein.
- Frameshift mutations change the sequence of amino acids more significantly than most substitution mutations. This is because they alter the reading frame.
- Other mutations can involve large sections of chromosomes (block mutations), multiple nucleotides or even entire chromosomes.

7.4 Activities

learnon

To answer questions online and to receive **immediate feedback** and **sample responses** for every question, go to your learnON title at **www.jacplus.com.au**. A **downloadable solutions** file is also available in the resources tab.

7.4 Quick quiz on	7.4 Exercise	7.4 Exam questions

7.4 Exercise

1. Identify three causes of a mutation.
2. Outline the difference in the impact that somatic mutations and germline mutations can have on a population's gene pool.
3. Compare substitution mutations and frameshift mutations by identifying a similarity and a difference between them.
4. A certain triplet in the DNA coding has the base sequence CAA.
   a. Describe the effect the following mutations will have on the amino acid sequence:
      i. The first nucleotide is changed to guanine.
      ii. The second nucleotide in changed to guanine.
      iii. The third nucleotide in changed to guanine.
   b. Which of the three nucleotide positions has the greatest impact on the protein? Explain your response and why this is the case.
5. A frameshift mutation will always have a greater consequence on a polypeptide chain than a substitution mutation. Is this statement correct? Justify your response.
6. Describe how each of the following leads to the formation of a new allele:
   a. Missense mutation
   b. Trinucleotide repeat
   c. Block mutation: inversion
   d. Frameshift mutation: deletion.

7.4 Exam questions

Question 1 (1 mark)

Source: *Adapted from VCAA 2018 Biology Exam, Section A, Q21*

MC Use an amino acid chart for this question.

Consider the following sequence of six amino acids that make up part of a polypeptide.

---- phe ---- leu ---- pro ---- val ---- tyr ---- ala ----

A mutation within the gene coding for this sequence of six amino acids resulted in the following six amino acids in the same position.

---- phe ---- leu ---- ala ---- val ---- tyr ---- ala ----

Another mutation within the original gene resulted in a shortened protein. The protein was truncated (cut short) after the amino acid valine, as shown below.

---- phe ---- leu ---- pro ---- val

This truncated protein resulted from the codon

A. UAU changing to UAA.
B. UAC changing to UGC.
C. GUA changing to UGA.
D. GUG changing to UGG.

Question 2 (1 mark)

Source: *VCAA 2007 Biology Exam 2, Section A, Q14*

MC New alleles arise in a sexually reproducing population by

A. mutations in DNA sequences prior to meiosis.
B. random fertilisation of gametes during reproduction.
C. random assortment of homologous chromosomes during meiosis.
D. exchange of chromatin between homologous chromosomes during meiosis.

Question 3 (1 mark)

Source: *VCAA 2017 Biology Exam, Section A, Q29*

MC A newborn baby was diagnosed with Patau syndrome. Her karyotype showed three copies of chromosome 13.

This is an example of

A. frameshift mutation.
B. block mutation.
C. aneuploidy.
D. polyploidy.

Question 4 (1 mark)

Source: *VCAA 2011 Biology Exam 2, Section B, Q5ei*

A genetic condition called HERDA affects certain quarter horses. It is an autosomal recessive trait.

A DNA sequencing test is available that reveals the HERDA status of a horse. The sequence for the different genes is shown below.

sequence for a normal gene is	AAG	AAG	AAG	GGG	CCT	AAA
The sequence for the HERDA gene is	AAG	AAG	AAG	AGG	CCT	AAA

What is the name given to this type of mutation?

Question 5 (1 mark)

Source: *VCAA 2015 Biology Exam, Section B, Q8aii*

Consider the template strand of a hypothetical gene, shown below. The exons are in bold type.

3' **TAC AAA** CCG GCC **TTT GCC AAA** CCC AAC CTA **AAT ATG AAA ATT** 5'

Note:

1. The DNA triplet **TAC** indicates START and codes for the amino acid methionine that remains in the polypeptide.
2. The DNA triplets **ATC**, **ATT** and **ACT** code for a STOP instruction.

An allele for this gene codes for a polypeptide with only five amino acids. This is caused by a mutation in one of the exons. This mutation is a result of one nucleotide change.

By referring to the original sequence above, identify the nucleotide change that must have occurred to bring about this shorter polypeptide.

More exam questions are available in your learnON title.

7.5 Biological consequences of changing allele frequencies

KEY KNOWLEDGE

- Biological consequences of changing allele frequencies in terms of increased and decreased genetic diversity

Source: VCE Biology Study Design (2022–2026) extracts © VCAA; reproduced by permission

7.5.1 Genetic diversity within populations

The genetic diversity of a population is the level of genetic variation within a gene pool. A population with high genetic diversity has a gene pool with a large number of alleles for each gene. This diversity can allow a population to be resilient to changes in the environment. If there is a sudden change in the environment, there is a higher chance that some individuals will be able to adapt to the changed conditions and survive.

FIGURE 7.36 Trees have a higher genetic diversity compared to grasses.

Populations with high genetic diversity tend to be large in size and have gene flow with other populations' gene pools within its species. Woody plants or trees are an example of a group of species with high genetic diversity. Their genetic diversity is higher than other groups of plants such as grasses. This higher genetic diversity allows woody plants to adapt to changing environmental conditions.

Different events can greatly affect the genetic diversity of a population. An increased genetic diversity brings more variation into a population, helping to reduce the chance of extinction. However, it may lead to other biological consequences, such as the addition of new alleles that lead to genetic disorders (such as the introduction of the cystic fibrosis allele into Australian populations).

Decreased genetic diversity leads to a limited variation in a population, which may make organisms more vulnerable if there are changes in selective pressures, increasing the chance of extinction.

As highlighted in subtopics 7.3 and 7.4, various mechanisms affect allele frequencies. Their effects on genetic diversity are summarised in table 7.7.

TABLE 7.7 Summary of changing allele frequencies and the impact on genetic diversity

	What usually happens to genetic diversity	Reason
Natural selection	Decreased	During natural selection, a particular phenotype has a selective advantage, so particular alleles become more common, while others are removed.
Gene flow	Increased	Usually, the movement of alleles results in new alleles coming into a population.
Genetic drift	Decreased	Random chance events may lead to a loss of alleles in a population.
Bottleneck	Decreased	An event leading to the death of many members of a population may lead to a loss of alleles in a population.
Founder effect	Decreased	A population is descended from individuals with a limited diversity in their genetic material.
Mutations	Increased	Mutations are the source of new alleles, increasing diversity in a population.

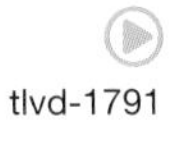
tlvd-1791

SAMPLE PROBLEM 4 Identifying genetic drift and the impact on genetic diversity

In the 1890s the northern elephant seal was hunted to the point of near extinction, leading to fewer than 100 individuals surviving (speculated to be as low as 20). Since then, the species was protected and now there are over 120 000 individuals living in the Pacific Ocean, mating along beaches on the west coast of North America. The southern elephant seal is a close relative and was not hunted to such low numbers.

a. **Identify the type of genetic drift that occurred in the northern elephant seal population. (1 mark)**
b. **Explain why the northern elephant seal population has a lower genetic diversity than the southern elephant seal population. (2 marks)**

THINK	WRITE
a. 1. The first part of the question is asking you to *identify*. In an identify question, you can list or name an answer with one word or a short sentence.	
2. Consider what the two types of genetic drift are: bottleneck effect and founder effect. The seals went from a very small to a large population without a location change.	Bottleneck effect (1 mark)
b. 1. This question asks you to *explain*. In an explain question, you need to account for the reasons why or how something occurs using your scientific knowledge. As this question is worth two marks, you should cover two points comparing the seal populations.	
2. Consider that the survivors of a bottleneck effect are due to chance and not because they have a selective advantage. Also consider that with 100 individuals the population would need to inbreed to reproduce. What effect would these two factors have on genetic diversity?	The genetic diversity is lower in the northern elephant seal because their population dropped to fewer than 100 individuals. As this was from a bottleneck, the chance of different alleles being retained is random (1 mark). As the population began to increase, individuals would have had to inbreed and this would have further decreased the population's genetic diversity (1 mark).

7.5.2 Biological consequences of low genetic diversity

Populations that are small and do not have gene flow with other populations generally have a lower genetic diversity. The smaller population size can also increase the rate of inbreeding and this can further reduce the genetic diversity. This can limit the population's ability to adapt to changing environmental conditions, and can make them vulnerable to becoming susceptible to disease and even to becoming extinct. Decreased genetic diversity can occur in the wild through the genetic drift mechanisms of the bottleneck effect and founder effect and in domesticated settings due to selective breeding.

The bottleneck effect and the founder effect play a significant role in the lowering of genetic diversity because alleles are selected by chance and not because they allow individuals within a population to survive. This can lead to alleles becoming lost or fixed within a gene pool. An allele is lost if it no longer present within the gene pool and fixed when there is only one allele present for a particular gene (as shown in figure 7.37). Both of these changes will reduce the genetic diversity of a population's gene pool.

FIGURE 7.37 Allele frequencies may be lost or fixed over time, resulting in decreased diversity. While this shows a gradual change, the loss or fixation of an allele through genetic drift can be quite sudden.

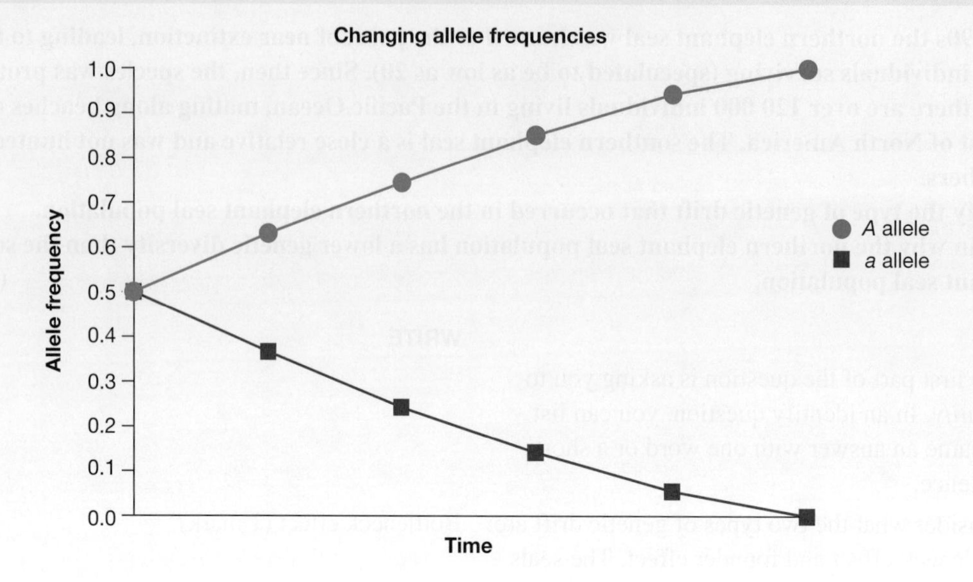

Consequences of low genetic diversity from bottleneck and founder effect

Some examples of species that have undergone a bottleneck effect include:

- *the Australian humpback dolphin*. This species of dolphin (figure 7.38a) experienced a bottleneck effect between 1250 and 3750 years ago due to an El Niño Southern Oscillation climatic event. A warming of the dolphin's habitat lead to a sudden decrease in its population size. This species is currently listed as vulnerable due to having low genetic diversity and limited gene flow between its populations.
- *the Cheetah*. Cheetahs (figure 7.38b) experienced a bottleneck approximately 10 000 years ago which reduced their numbers dramatically. This resulted in the remaining small population becoming inbred. Species with good genetic diversity can have around 20 per cent variation in their alleles; however, the cheetah population has only 1 per cent variation. Because of this limited variation, cheetahs have accumulated a number of harmful mutations associated with reproduction. This has resulted in reproductive challenges for male cheetahs and lethal genetic disorders.

FIGURE 7.38 a. An Australian humpback dolphin b. A cheetah with her five cubs

Founder effect can lead to a specific allele becoming fixed in frequency, leading to an allele to be fixed in a population. This may be anything from particular traits through to the fixation of genetic diseases in populations.

CASE STUDY: The founder effect and Alzheimer's disease in Colombia

There is a high prevalence of a form of early-onset Alzheimer's disease in the town of Yarumal and surrounding district in Colombia, South America. About 5000 people presently carry a particular mutation that causes this disease, which typically shows its effects when people are in their early forties. This early-onset form of Alzheimer's disease is a dominant trait. Researchers analysed the sequence of the DNA segment involved in the affected people and found that all these people carried the same mutation. It has been estimated that this particular mutation was introduced to the population about 375 years ago. Further, it has been concluded that this allele was brought into the district by a Spanish conquistador in the early sixteenth century. This conquistador did not remain in the district but, as a result of his sexual activities, his genetic material remained through at least one child whom he fathered with a local woman. This child survived to pass this DNA segment to the next generation. As mentioned, about 5000 people are now known to carry this DNA and will develop the destructive symptoms of early-onset Alzheimer's disease.

7.5.3 Protecting genetic diversity

EXTENSION: Saving genetic variation

As wild populations are destroyed through land clearance and as larger areas are devoted to the cultivation of smaller numbers of commercial crop varieties, the safeguarding of genetic variation in wild populations is critical. In the case of plant varieties, some contribution to this is being achieved through the establishment of seed banks.

To access more information on saving genetic diversity and seed plants, please download the digital document.

on Resources

Digital document Extension: Saving genetic variation (doc-36106)

FIGURE 7.39 The Norwegian island of Spitsbergen, north of the Arctic Circle, is the location of the Svalbard Global Seed Vault

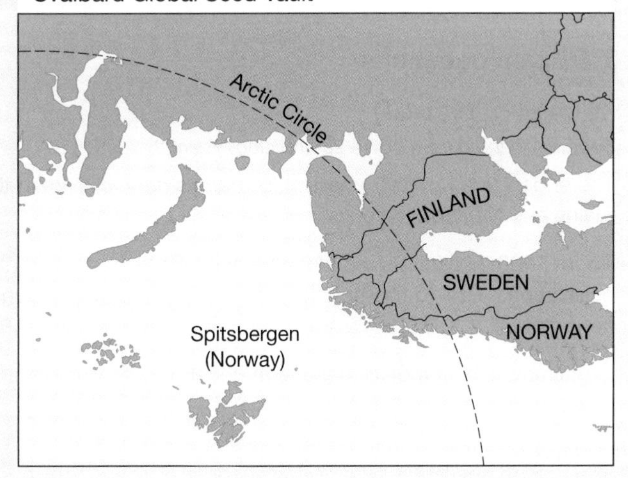

on Resources

eWorkbook Worksheet 7.5 Biological consequences and genetic diversity (ewbk-2783)

Weblink Svalbard Global Seed Vault

KEY IDEAS

- Genetic diversity is the amount of genetic variation in a population's gene pool.
- Mutations and gene flow often increase genetic diversity.
- Genetic drift and natural selection often decrease genetic diversity.
- Populations with high genetic diversity have a greater chance of successfully adapting to changing environments.
- Low genetic diversity can occur in populations due to a bottleneck effect, a founder effect and limited gene flow.
- Populations with low genetic diversity are less likely to be able to adapt to changing environments.
- Populations with low genetic diversity are more likely to become vulnerable to extinction.

7.5 Activities

learnon

To answer questions online and to receive **immediate feedback** and **sample responses** for every question, go to your learnON title at **www.jacplus.com.au**. A **downloadable solutions** file is also available in the resources tab.

7.5 Quick quiz on	7.5 Exercise	7.5 Exam questions

7.5 Exercise

1. Identify a factor that can increase and decrease the genetic diversity of population.
2. Describe how a population's size can affect the genetic diversity of a gene pool.
3. The frequency of a particular allele in the gene pool shows a steady decline over each successive generation. Is this change the result of genetic drift? Justify your response.
4. Outline the impact of alleles becoming lost or fixed will have on the gene pool.
5. Explain the consequence of inbreeding within a population.

7.5 Exam questions

Question 1 (1 mark)

Source: *Adapted from VCAA 2009 Biology Exam 2, Section B, Q3e*

The endangered pygmy possum (*Burramys parvus*) lives in three restricted alpine areas, Mt Buller, Bogong High Plains and Mt Kosciusko.

About 2000 individuals remain in the wild. Studies show that there is a lot of genetic diversity between the three populations. Due to the isolation of these populations, scientists think that each population has a separate gene pool.

Explain how exchange of genetic material may be beneficial in the survival of endangered species like the pygmy possum.

Question 2 (1 mark)

MC Which of the following is the most likely to lead to an increase in genetic diversity?

A. Genetic drift
B. Natural selection
C. Gene flow
D. Mutation

Question 3 (6 marks)

Source: *VCAA 2011 Biology Exam 2, Section B, Q6b and c*

The prairie chicken (*Tympanuchus cupido pinnatus*) is a grassland bird native to North America. A prairie chicken spends its entire life within several kilometres of its birthplace. Prior to European settlement, prairie chickens numbered in the millions across the Midwest of the United States of America. As a result of the grasslands being replaced by plant food crops, the distribution of prairie chickens has diminished, as shown below.

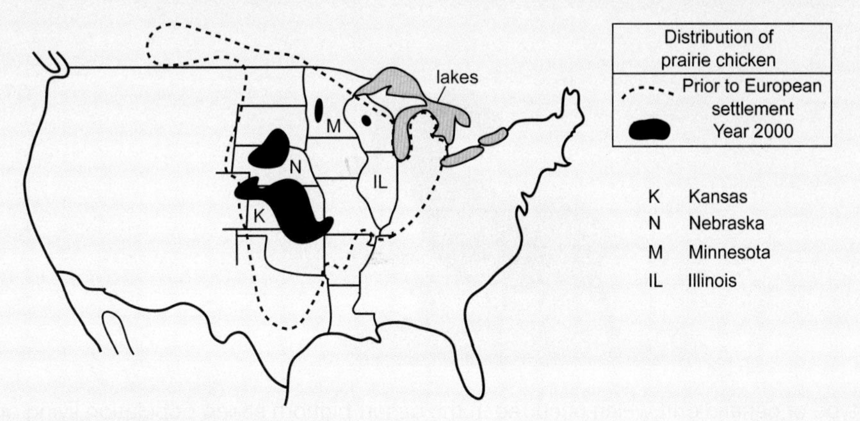

By 1994, Kansas, Nebraska and Minnesota still supported large and widespread populations; however, in the state of Illinois, the number of prairie chickens fell to less than fifty individuals in two separate geographical areas.

Illinois – prairie chicken distribution

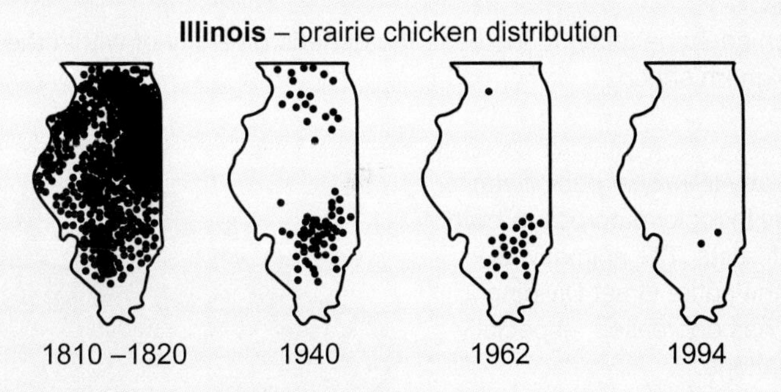

Representative samples of prairie chickens from the four states were selected for testing. Each prairie chicken had six gene loci tested. The average number of alleles at each gene locus for each prairie chicken group is shown in the graph below.

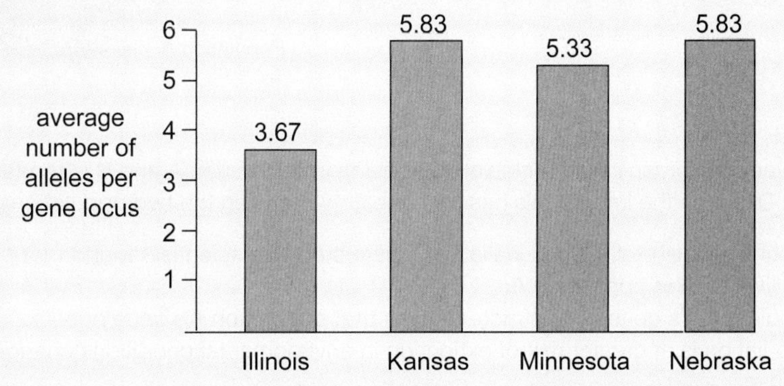

a. i. Explain the significance of the results for the Illinois birds compared to the results of the birds from the three other states. **2 marks**

ii. Explain why the results for the Kansan birds and Nebraskan birds are similar to each other. **1 mark**

b. Measures were taken in the 1990s to prevent the Illinois prairie chicken from dying out completely.

i. Explain why low genetic diversity in a population threatens the survival of the population. **2 marks**

ii. Describe one measure that could be used to prevent the Illinois prairie chicken from dying out. **1 mark**

Question 4 (5 marks)

A group of 20 desert bighorn sheep were introduced to Tiburon Island which is off the coast of Mexico in 1975 from a herd in Arizona in the United States. No other sheep were introduced to the island and by 1999 there were 650 desert bighorn sheep living there.

a. Identify the type of genetic drift which occurred in the desert bighorn sheep population living on Tiburon Island. **1 mark**
b. Explain why the Tiburon Island desert bighorn sheep has a lower genetic diversity than the Arizona desert bighorn sheep. **2 marks**
c. Explain an action which could be taken to increase the genetic diversity of within the Tiburon Island population of desert bighorn sheep. **2 marks**

Question 5 (6 marks)

The brush-tailed rock wallaby (*Petrogal penicillata*) is a small wallaby which lives in rocky outcrops in both rainforest and temperate forests. The animal was once widespread across the entire Great Dividing Range but its distribution has decreased, especially within the Victorian populations.

The distribution of the brush-tailed rock wallaby is shown on the map. These are sections that contain rocky outcrops.

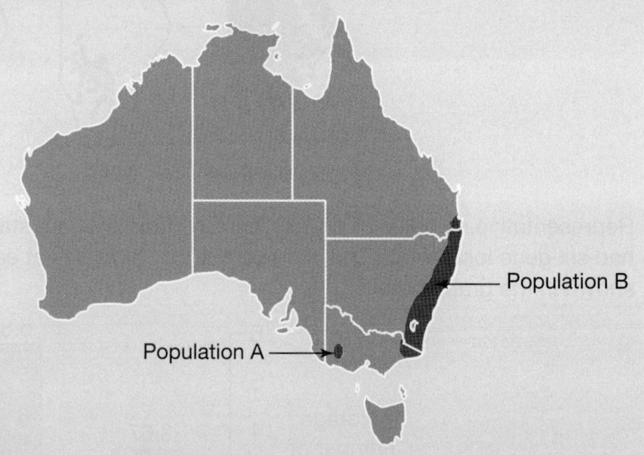

a. The distance between the brush-tailed rock wallabies living in population A and B is too far for them to travel between. Describe the impact this has on gene flow between the two populations. **2 marks**

The genetic diversity of the brush-tailed rock wallabies was tested and scientists found that population B in Victoria was more diverse than population A.

b. Describe why population B's gene pool is more diverse that population A's gene pool. **2 marks**
c. Identify which population is at greater risk of extinction in environmental conditions change rapidly. Justify your response. **2 marks**

More exam questions are available in your learnON title.

7.6 Selective breeding programs

KEY KNOWLEDGE

- Manipulations of gene pools through selective breeding programs

Source: VCE Biology Study Design (2022–2026) extracts © VCAA; reproduced by permission

7.6.1 What is selective breeding?

From the time when humans went from a hunter-gather society to an agricultural society, humans have been changing the gene pools of domesticated species. This is known as **artificial selection**. The deliberate manipulation of domesticated species' gene pools has occurred through **selective breeding** programs.

Selective breeding is an example of artificial selection. In this procedure, only those organisms that display a desirable trait in their phenotype or are known carriers of the trait are chosen to reproduce. Selective breeding is an example of how humans can intervene in natural evolutionary processes by artificially manipulating the gene pool of a population.

The mechanism of selective breeding is comparable to that of natural selection, but there is one key difference. Instead of there being a struggle for all individuals in the gene pool to survive and for selection pressure to 'select' individuals, humans intervene, and we select the trait or phenotype which is most desirable for our purposes. Individuals with this trait are selected to breed and they pass on this trait to their offspring. This will occur over many generations and the number of individuals with the trait will greatly increase. This may lead to the allele causing the trait to become fixed in frequency. What makes a trait desirable can vary between species and regions around the world.

The mechanism of artificial selection

1. There is variation within the population's gene pool.
2. Humans select individuals with a desirable trait.
3. These individuals breed (reproduce) and pass their alleles on to the next generation.
4. The alleles that lead to the desired phenotype will be inherited by subsequent generations and they can increase in frequency in the gene pool over time.

Some examples of species which are bred for desirable traits are outlined in table 7.8.

TABLE 7.8 Desirable traits selectively bred in different species

Species	Trait
Cattle	• Higher milk yields • Quality of meat • Absence of A1 milk protein • Pest resistance
Dogs	• Temperament • Appearance ('cuteness') • Hunting ability • Type of coat
Apples	• Size • Sweetness • Crispness • Pest resistance

artificial selection the process by which humans breed animals or plants in such a way to increase the proportion of chosen phenotypic traits

selective breeding a process of mating that is not random, but uses parents chosen by the breeder on the basis of particular phenotypic characteristics that they display

In plants, often artificial pollination is used in order to select desirable traits in cultivated plants.

TIP: When discussing artificial selection and selective breeding, ensure you mention human intervention in the process, and remember that the phenotype being chosen is not necessarily one that gives an advantage for an organism, but is instead desired by humans.

CASE STUDY: Selective breeding in dogs

Selective breeding of animals is not restricted to agricultural stock, such as cattle and sheep. Selective breeding has been used by animal fanciers and hobbyists to produce the great variety of breeds that may be seen in various pets, such as dogs. This selective breeding continues today.

The Australian National Kennel Council recognises 200 different breeds of dog, from Affenpinschers to Yorkshire Terriers. Figure 7.40 shows four dog breeds that show striking variation in terms of their coats — from almost zero hair in the Chinese Crested Dog to the Puli with its profuse corded coat.

FIGURE 7.40 a. The Chinese Crested Dog b. Shih Tzu c. Shar-Pei d. Puli

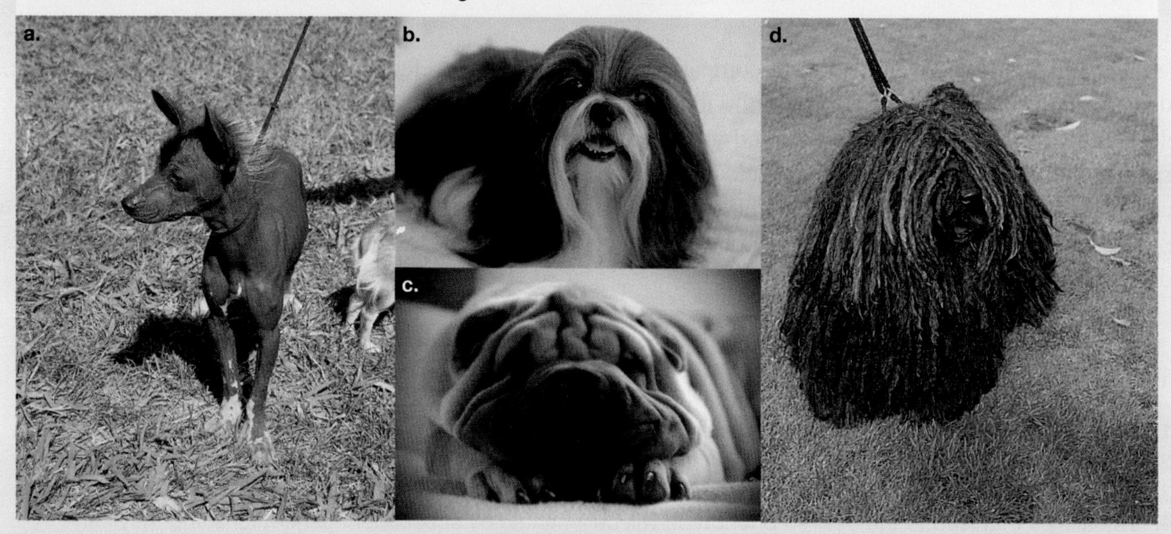

There are many other examples of selective breeding in other species, including cattle and plants. To access more information on other case studies of selective breeding and complete an analysis task, please download the worksheet.

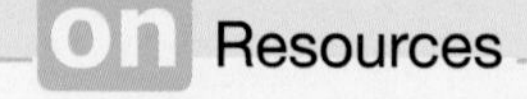

Resources

eWorkbook Worksheet 7.6 Case study: Investigating other examples of selective breeding (ewbk-2784)

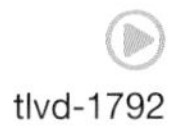
tlvd-1792

SAMPLE PROBLEM 5 Outlining the processes in selective breeding

Dairy cows produce milk with the variants of A1 and A2 beta-casein proteins within it. These proteins only differ from each by one amino acid. The dairy cows can produce milk with only one of these proteins or both. It has been suggested that drinking milk with only the A2 protein allows for easier digestion for people who have lactose intolerance. In 1993 a genetic test was developed in New Zealand to identify cows which only produced milk with the A2 milk. This allowed some dairy farmers to produce milk that did not contain the A1 protein.

Outline how farmers could selectively breed a herd of cows that only produce A2 milk. (4 marks)

THINK

1. Identify what the question is asking you to do. In an *outline* question, you give a step-by-step account of what is occurring.
2. Consider the steps that occur in selective breeding and which desired trait is being selected for. Apply your knowledge of selective breeding to the A2 milk scenario.

WRITE

1. There is variation in the milk proteins cows produce: A1 and A2 (1 mark).
2. Having milk with only the A2 protein is the desired trait (1 mark).
3. Farmers use a genetic test to identify cattle with only the A2 alleles (1 mark).
4. Only these cattle are bred and used on the farm (1 mark).

7.6.2 The effect of selective breeding on gene pools

Selective breeding or artificial selection occurs when breeders, fanciers and farmers prefer particular inherited features in their show animals and livestock because of their economic value or aesthetic appeal, and use selective breeding to enhance those features and increase their frequency in the breeding stock. In doing this, breeders manipulate the gene pool of their breeding stock.

Artificial selection can maintain features in a population that are economically important or aesthetically appealing but which are disadvantageous in terms of survival and reproduction. In natural populations in the wild, these features would come under negative selection pressure so that individuals with these features would not be maintained, and the alleles responsible would tend to be lost from the gene pool.

Artificial selection in domesticated species, particularly in so-called show varieties, can favour features that are disadvantageous for survival and reproduction and would be selected against in the wild. Traits such as these would likely be lost through natural selection. However they are retained from artificial selection, selective breeding and human manipulation of gene pools.

Examples of features maintained only by artificial selection can be seen in domesticated animals such as:

- hairless Sphynx cats (see figure 7.41a) and Chinese Crested dogs (see figure 7.40a). The lack of an insulating fur coat in these animals means that, in cold environments, the homeostatic mechanisms that regulate core body temperature are at risk of failing.
- English bulldogs with greatly shortened muzzles that result in breathing problems (see figure 7.41b).
- Shar-Pei dogs (see figure 7.40c) have various health problems related to their rounded snout and wrinkles. They often do not make it past the age of 10. Common problems in Shar-Pei dogs include eye problems (which may lead to blindness) and ear infections due to yeast hiding within the wrinkled structure.

FIGURE 7.41 a. Hairless Sphynx cats b. English bulldogs

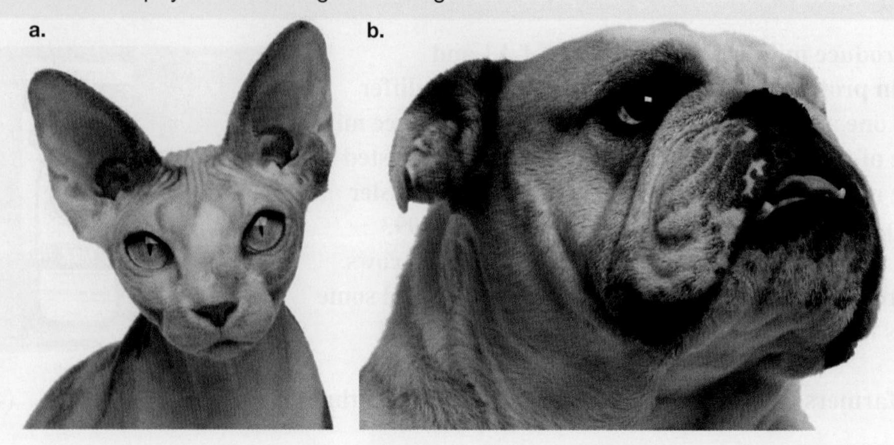

Other examples of disadvantageous features maintained thorough selective breeding can be seen in the short-legged dog breeds, such as dachshunds. These dogs are at a greatly increased risk of a disc herniation, in which one of the discs between the spine vertebrae is displaced. This creates pressure on the spinal cord, resulting in pain and, in some cases, temporary or permanent nerve damage.

Disadvantageous features such as these are maintained in the gene pools of these populations only through human intervention and selective breeding, which raises many ethical questions.

EXTENSION: What technologies are used in selective breeding?

In many situations, selective breeding occurs when individuals choose organisms with desirable traits to breed in order to increase the likelihood of a specific phenotype appearing.

In commercial herds and flocks, new reproductive technologies resulting in selective breeding are outlined in table 7.9.

TABLE 7.9 Technologies used in selective breeding

Technique	Process
Artificial insemination	Semen is collected from a selected stud animal and then introduced by artificial means into the reproductive tract of females of the same species.
Sex selection	Semen is collected from a stud bull and sperm is examined (using fluorescent dye) to determine if it carries the X or the Y chromosomes. The desired sperm is then used.
Multiple ovulation and embryo transfer (MOET)	A female (with a desirable trait) receives injections of the follicle-stimulating hormone (FSH), which stimulates her to super-ovulate, or produce multiple eggs. An injection of gonadotrophin-releasing hormone (GnRH) is also given to make all the eggs mature at the same time. Embryos at six to seven days of development are removed from the reproductive tract of a female and transplanted into the tracts of other females of the same species. These females act as surrogate mothers and carry the embryos to term and give birth.
Oestrus synchronisation	Breeding cycles are manipulated so that all members of a group of sexually receptive and ovulating at the same time

on Resources

eWorkbook Worksheet 7.7 Making the most of new variants — Artificial selection (ewbk-2785)

Digital document Extension: Technologies in selective breeding (doc-36184)

Weblink Dogs that changed the world — selective breeding problems

KEY IDEAS

- Selective breeding is one form of artificial selection used in plant and animal breeding.
- Selective breeding will select for desired traits for human benefit rather than traits which increase the survival chances of the population.
- Selective breeding can decrease the genetic diversity of gene pools.
- Loss of genetic variation in a population may result in failure of the population to survive environmental change.

7.6 Activities

learnon

To answer questions online and to receive **immediate feedback** and **sample responses** for every question, go to your learnON title at **www.jacplus.com.au**. A **downloadable solutions** file is also available in the resources tab.

7.6 Quick quiz on	7.6 Exercise	7.6 Exam questions

7.6 Exercise

1. Define the term *selective breeding.*
2. Describe two traits for which a cattle farmer may select.
3. Compare the similarities and differences of artificial selection with natural selection.
4. The English bulldog has been bred for the trait of having a shortened muzzle. Suggest why this selective breeding may be problematic.
5. Explain how selective breeding has changed the allele frequencies within domesticated species' gene pools.

7.6 Exam questions

Question 1 (1 mark)

Source: *VCAA 2020 Biology Exam, Section A, Q27*

MC Tasmanian devils (*Sarcophilus harrisii*) were originally broadly distributed across Australia. When sea levels rose 12 000 years ago, an island, now referred to as Tasmania, was formed. The small number of Tasmanian devils on Tasmania was cut off from the Auatralian mainland populations. The population in Tasmania showed less genetic variation than the mainland populations. Mainland populations became extinct approximately 3000 years ago.

Over the last 20 years, the total Tasmanian devil population on Tasmania has halved. Many of the deaths have been the result of Tasmanian devil facial tumour disease (DFTD). Scientists have taken some Tasmanian devils that do not have DFTD to mainland Australia to set up a conservation program. The scientists have shown that greater genetic diversity among offspring in this program is observed when the Tasmanian devils are kept in isolated male-female pairs rather than in larger groups.

The conservation program for Tasmanian devils is an example of

A. a population bottleneck.
B. allopatric speciation.
C. selective breeding.
D. natural selection.

Question 2 (1 mark)

Source: *VCAA 2010 Biology Exam 2, Section A, Q20*

MC The wild sunflower plant has been cultivated by humans over several generations. During that time, selection for or against particular sunflower traits has been carried out.

A comparison of some of the traits in wild and cultivated sunflowers is given below.

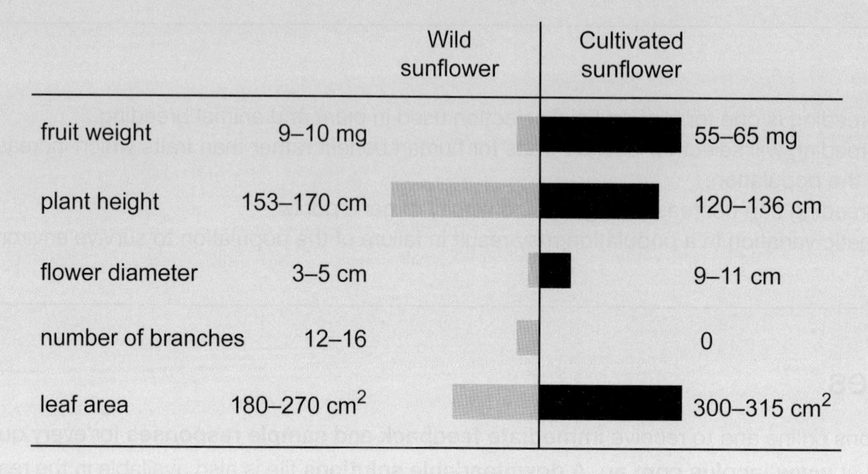

From the information above, we can assume that humans have selected against large

A. leaf area.
B. plant height.
C. fruit weight.
D. flower diameter.

Question 3 (3 marks)

Source: *VCAA 2007 Biology Exam 2, Section B, Q6b and c*

Australian agricultural biologists are currently researching weeping rice grass (*Microlaena stipoides*), a deep-rooted native relative of rice. Their aim is to produce drought-tolerant grain crops, pasture grass for livestock and domestic lawns.

M. stipoides thrives in a variety of soil types from coastal to mountain habitats. It does not spread in an uncontrolled way as many introduced grasses do. It requires less fertiliser and liming of soil than introduced crop species.

At present *M. stipoides*' seed is only half the size of domestic rice.

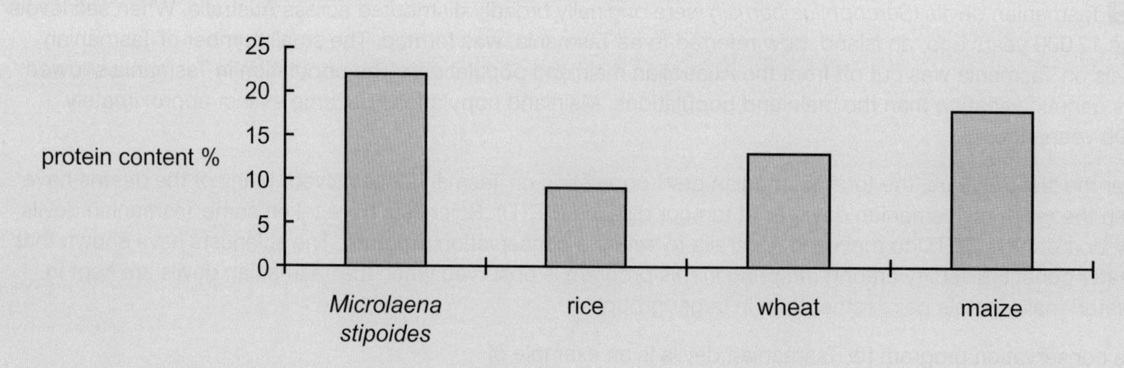

A researcher stated that weeping rice grass has not undergone selection or breeding for larger seed size.

a. What might be the advantage of breeding this species for larger seed size? **1 mark**

b. If you were a farmer involved in a systematic breeding program, outline the steps you should take to develop a variety of weeping rice grass with large seeds. **2 marks**

Question 4 (1 mark)

Source: *VCAA 2017 Biology Exam, Section A, Q30*

MC In the 18th century, farmer Robert Bakewell separated large, fine-boned sheep with long, shiny wool from his native stock to interbreed for future sheep flocks.

This is an example of

A. genetic fitness.
B. natural selection.
C. selective breeding.
D. allopatric speciation.

Question 5 (3 marks)

Source: *VCAA 2006 Biology Exam 2, Section B, Q6e, f and g*

a. Selective breeding has been used to improve the milk yield of cattle herds in Australia. Identify a key difference between selective breeding and random mating in a herd of cattle. **1 mark**

b. Selective breeding has been used to improve the milk yield of cattle herds in Australia. What is the impact of selective breeding on genetic variability in a herd of cattle? **1 mark**

c. The quality and yield of milk in cattle has been improved by artificial insemination in which semen from a selected bull is used. Explain how the use of artificial insemination may intervene in the evolutionary process. **1 mark**

More exam questions are available in your learnON title.

7.7 The consequences of rapid genetic change in pathogens

KEY KNOWLEDGE

- Consequences of bacterial resistance and viral antigenic drift and shift in terms of ongoing challenges for treatment strategies and vaccination against pathogens

The time it takes for species to change and evolve can be measured over thousands or even millions of years. A mutation that creates a new allele which is advantageous to a species can take many generations for it to increase its frequency within a gene pool. The generation time of a species will determine how long it takes for any inherited changes to affect the gene pool of a species. A shorter generation time means that there is a shorter time frame between successive generations, and thus the inherited changes will affect the gene pool faster. For example, *Escherichia coli* (*E. coli*), which has a generation time of approximately 20 minutes, will have a much more significant change in its gene pool compared with the human species, which have a generation time of approximately 26 years. Bacteria reproduce and viruses replicate at a much faster rate. This increases their rate of evolution and, consequently, how we treat these pathogens. These pathogens were introduced in topic 5.

7.7.1 Consequences of bacterial resistance

How do bacteria become resistant?

Bacteria are single-celled organisms which reproduce through binary fission. This asexual form of reproduction occurs when a single bacterium divides into two identical cells. This results in bacteria being one on the fastest reproducing organisms, with species such as *E. coli* dividing every 20 minutes and *Staphylococcus aureus* every 30 minutes. This rapid reproduction allows one bacterium to create a colony of more than 1 million in less than 10 hours. Like all organisms, bacteria will have random mutations occurring in their nucleotide sequence which produce new cells that are not identical. While the majority of these mutations will have a negative or neutral effect on the bacteria, some will give a bacterium a selective advantage to survive.

As introduced in section 6.3.6, **antibiotics** are a group of drugs that either kill bacteria directly or prevent them from reproducing by disrupting their cell structure or processes. Antibiotics are prescribed when a bacterial infection is diagnosed. Usually the antibiotic will destroy all of the bacteria inside a person. What has been increasingly occurring, however, is that some bacteria have become resistant to antibiotics. This occurs through the process of natural selection where the presence of antibiotics is acting as the environmental selection pressure.

The mechanism of natural selection works on bacteria as it does in other organisms, as shown in figure 7.42. An example of antibiotic resistance in *S. aureus* is seen in the sensitivity test shown in figure 7.43.

FIGURE 7.42 The process of natural selection selecting for antibiotic resistant bacteria

This is a population of dividing bacteria.

During division, one of the bacterium undergoes a mutation in its DNA that results in antibiotic resistance.

When an antibiotic is added, all of the sensitive bacteria are killed.

The antibiotic resistant bacterium, however, is unaffected by the antibiotic.

The resistant bacterium can continue to divide, forming a population of antibiotic resistant bacteria.

FIGURE 7.43 Testing the antibiotic sensitivity of *Staphylococcus aureus*. When there is bacterial growth around an antibiotic disc (i.e. no zone of inhibition), the bacteria is resistant to the antibiotic.

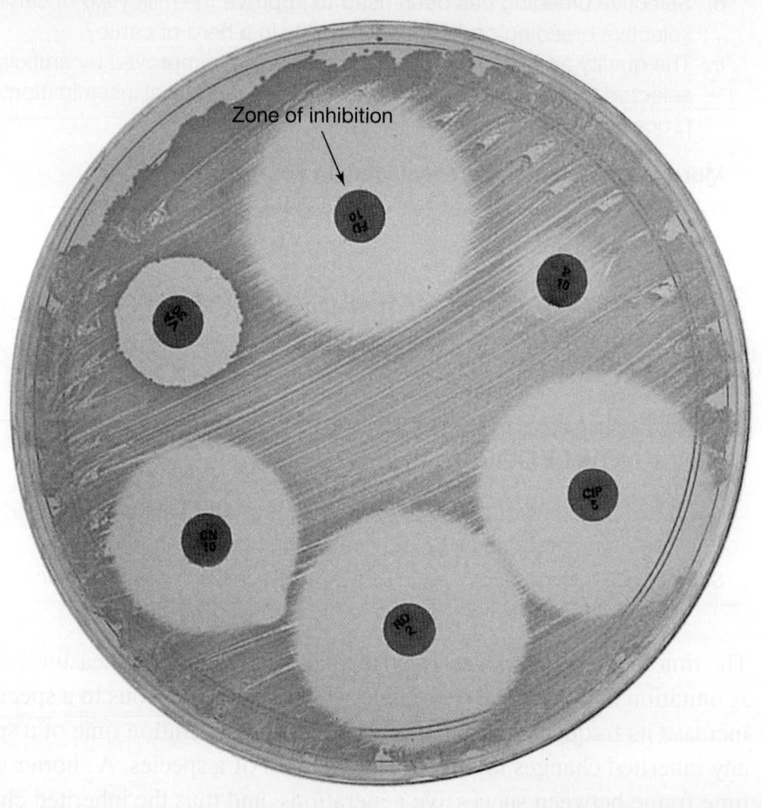

There are many factors that increase the chance that bacterial resistance may occur. Some of these are highlighted in table 7.10.

antibiotics a class of antimicrobial drug used in the treatment and prevention of bacterial infections that act either by killing pathogenic bacteria or by inhibiting their growth

TABLE 7.10 Causes of the increase in bacterial resistance

Cause	Why this causes resistance
Doctors over-prescribing antibiotics	This leads to there being more opportunities for bacteria to mutate and consequently evolve to become resistant.
Patients not finishing a course of antibiotics	Full course of antibiotics will kill the infectious bacteria and the beneficial bacteria flora. But if the full course is not taken, some infectious bacteria may survive and be more likely to reproduce and evolve.
Increased use of antibiotics in livestock farming	A common practice among farmers was to give livestock, such as cattle, antibiotics in order to promote faster growth and to reduce the infection rate. However, the resistant bacteria developed in the livestock can be ingested and transferred to humans.
Poor hygiene and sanitation	This increases the spread and transmission of bacteria.
A lack of infection control in medical centres	This increases the likelihood of resistant bacteria being transferred between patients.

EXTENSION: Other mechanisms of acquiring resistance

Not all bacteria need to inherit a trait of antibiotic resistance. They can also acquire this trait through the process of horizontal gene transfer (refer to figure 7.44). This process can occur in a variety of different ways.

a. *Bacterial transformation.*
An antibiotic-resistance gene is released from a dying bacterium and can picked up by another bacterium and inserted into their nucleic material.

b. *Bacterial transduction.*
A bacteriophage transfers an antibiotic-resistance gene from one bacterium to another.

c. *Bacterial conjugation.*
Bacterial cells can undergo conjugation when there is cell to cell contact. A plasmid containing an antibiotic-resistance gene can be transferred.

FIGURE 7.44 The transfer of antibiotic resistance genes via horizontal gene transfer

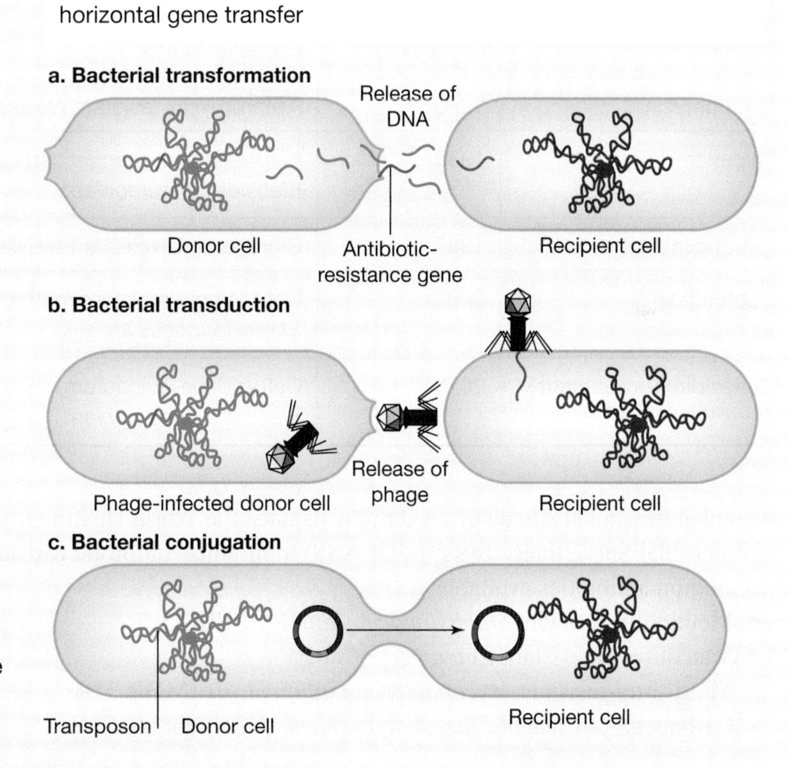

What consequences does this resistance have?

With the number of bacteria with resistance to antibiotics rising globally, it is becoming more challenging to treat common infectious diseases (refer to table 7.11).

Diseases which could once be treated with a dose of antibiotics can now be life-threatening to individuals with an antibiotic-resistant strain of infection because the bacteria can continue to reproduce in the body.

Some examples of bacterial infections that are becoming increasingly difficult to treat with antibiotics are:

- pneumonia
- tuberculosis
- gonorrhoea
- salmonella.

Some recent cases of gonorrhoea (nicknamed 'super gonorrhoea'), have been found to be resistant to at least six different main categories of antibiotic, which are nearly all the antibiotics available to treat the infection.

TABLE 7.11 A timeline of bacterial resistance to antibiotics

Antibiotic approved or released	Year released	Resistant strain identified	Year identified
Penicillin	1943	Penicillin-resistant *Streptococcus pneumoniae*	1967
		Penicillinase-producing *Neisseria gonorrhoeae*	1976
Vancomycin	1958	Plasmid-mediated vancomycin-resistant *Enterococcus faecium*	1988
		Vancomycin-resistant *Staphylococcus aureus*	2002
Amphotericin	1959	Amphotericin B-resistant *Candida auris*	2016
Methicillin	1960	Methicillin-resistant *Staphylococcus aureus*	1960
Extended-spectrum cephalosporins	1980 (Cefotaxime)	Extended-spectrum beta-lactamase-producing *Escherichia coli*	1983
Azithromycin	1980	Azithromycin-resistant *Neisseria gonorrhoeae*	2011
Imipenem	1985	*Klebsiella penumoniae* carbapenemase (kpc)-producing *Klebsiella pneumoniae*	1996
Ciprofloxacin	1987	Ciprofloxacin-resistant *Neisseria gonorrhoeae*	2007
Daptomycin	2003	Daptomycin-resistant methicillin-resistant *Staphylococcus aureus*	2004
Ceftazidime-avibactam	2015	Ceftazidime-avibactam-resistant KPC-producing *Klebsiella pneumoniae*	2015

To combat the global spread of bacterial resistance the World Health Organization (WHO) has set up the Global Antimicrobial Surveillance System (GLASS) to monitor antibiotic resistance. These bacteria are able to resist the common antibiotics available:

- Methicillin-resistant *Staphylococcus aureus* (MRSA)
- Vancomycin-resistant *Enterococci* (VRE)
- Multi-drug-resistant *Mycobacterium tuberculosis* (MDR-TB)
- Carbapenem-resistant *Enterobacteriaceae* (CRE) gut bacteria.

This presents incredible challenges for treatment for specific bacterial strains. Research is being continually undertaken to address the antibiotic resistance that has evolved in bacterial species. New antibiotics (to which the bacteria are not resistant) or new treatment options (such as draining areas infected with MRSA or the time length of treatments being altered) need to be explored to help treat these resistant bacteria.

elog-0038

INVESTIGATION 7.5

online only

Investigating bacterial resistances

Aim

To investigate the resistance of bacteria to a variety of different antibiotics

Resources

Weblinks Resistance map
WHO: bacterial resistance
The antibiotic resistance crisis

7.7.2 Consequences of antigenic shift and drift in viruses

Viruses are non-cellular pathogens which take over a host's cells protein synthesis processes in order to replicate and spread. Viruses, like other pathogens, have antigens on their surface (such as in influenza A in figure 7.45) which allow our immune system to recognise them as non-self and initiate both the innate and adaptive immune responses. Once these processes have occurred, the body will create B memory cells to be able to initiate a rapid secondary response if the virus is encountered again.

FIGURE 7.45 **a.** Model showing a 3D representation of a typical influenza A virus. **b.** Transmission electron micrograph of influenza A virus particles (virions). Note the rough outer margin that consists of the surface proteins, hemagglutinin (HA) and neuraminidase (NA).

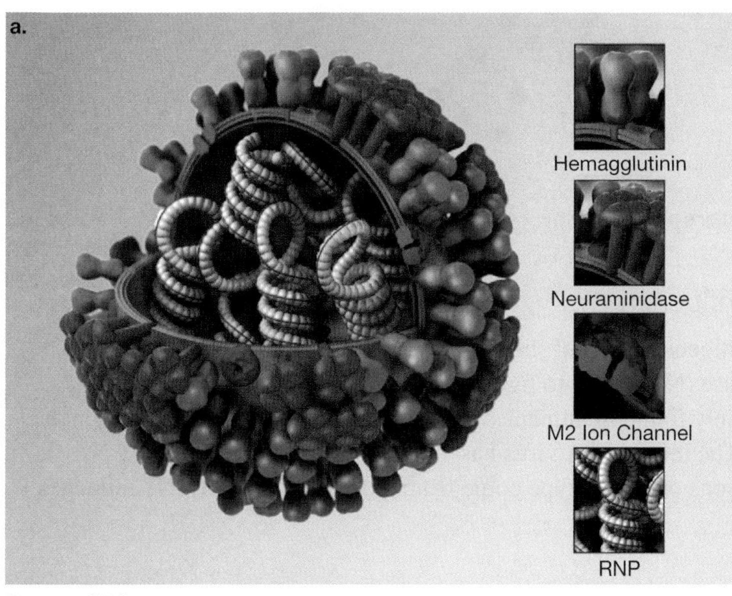

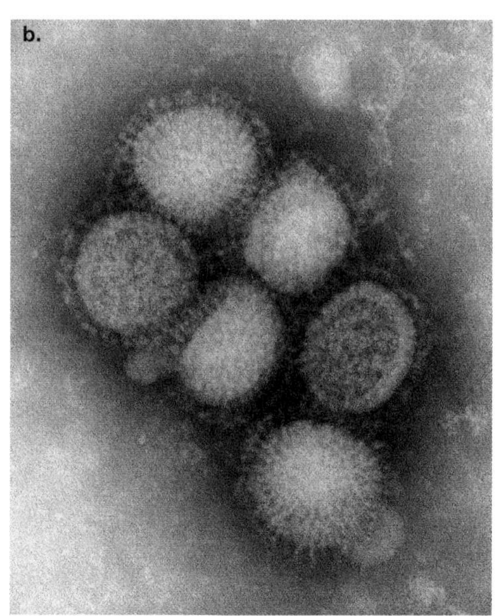

Source: CDC

Antigenic drift and shift

If this is the case, how can we be infected with viral diseases such as influenza more than once?

The reason this occurs is because viruses mutate at a high rate during replication and these antigens can change. This is especially true for viruses which have RNA as their nucleic material, as RNA does not have proof reading mechanisms for mutations as DNA does. Viruses which have RNA as their nucleic material include influenza, SARS-CoV-2, HIV, Ebola and hepatitis.

When certain viruses mutate during replication, the antigens on its surface can be altered. The degree to which the antigens are altered will result in either **antigenic drift** or **antigenic shift** (refer to figure 7.46).

antigenic drift when a point mutation alters a virus's nucleic material, resulting in small changes so that it continues to be recognised and reacted to by the body's immune system

antigenic shift when two or more strains of a virus combine to form a new strain of the virus so that it is no longer recognised and reacted to by the body's immune system

Antigenic drift is when a point mutation alters a virus's nucleic material, resulting in small changes to its antigens.

Antigenic shift occurs when two or more strains of a virus combine to form a new strain of the virus with antigens from each of the original strains.

FIGURE 7.46 The change in a virus's antigen through antigenic drift and antigenic shift

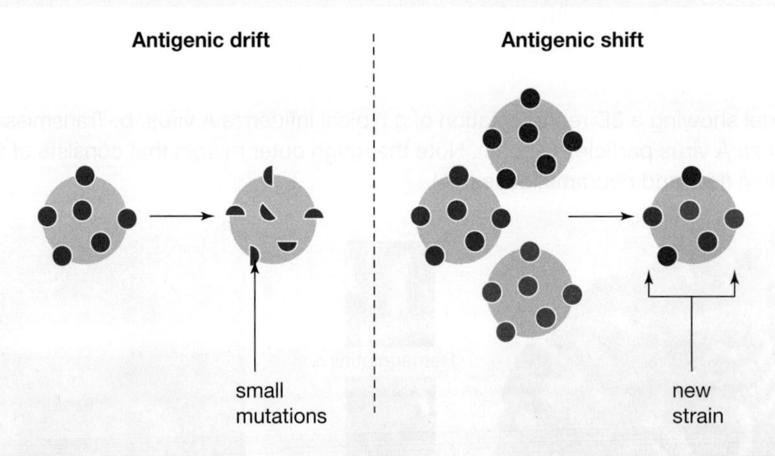

Applying antigenic drift and shift to influenza

Influenza is a virus with the highest rates of antigenic drift and shift. The two main antigens on an influenza virus are hemagglutinin (HA) and neuraminidase (NA) (refer to figure 7.45). There are three main types of influenza: A, B and C. Influenza B and C are only found in humans, while influenza A can also infect animals, such as birds (avian flu) and pigs (swine flu). The influenza A virus has 17 different HA antigens and 9 NA antigens and this is where the naming conventions of the subtype come from (for example the H1N1 influenza strain).

FIGURE 7.47 The naming of influenza. Two different examples are shown.

a.

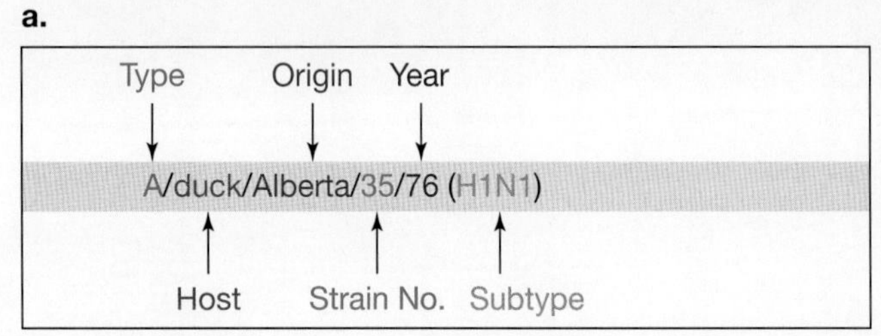

b.

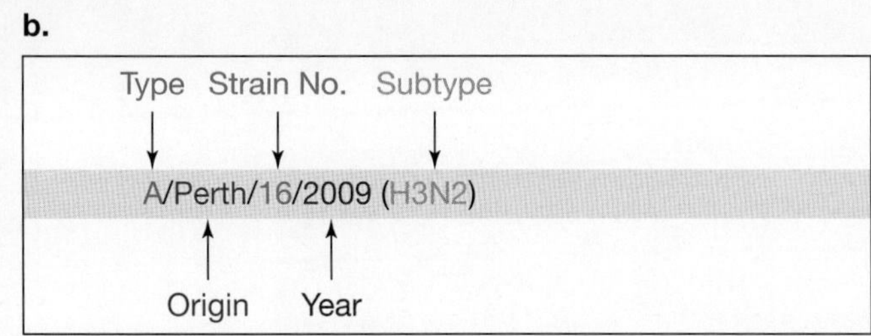

Antigenic drift in influenza

The RNA of the genetic material of the influenza virus continually undergoes frequent mutations as it replicates. Each newly replicated virus can have a nucleotide change or point mutation in one of its genes. These small changes continually occurring in the influenza virus are termed antigenic drift. All influenza virus types A, B and C are subject to the gradual change of antigenic drift.

Over time, however, the accumulation of these small changes means that antigenic properties of the mutated viruses have been altered to a major degree, and a new subtype is identified. When this happens, the virus will no longer be recognised by immune system memory cells, but will be treated as a new pathogen.

Antigenic shift in influenza

Influenza A viruses can also undergo an occasional but major sudden change in their two surface proteins, haemagglutinin (HA) and neuraminidase (NA). These proteins are embedded in the outer envelope of the influenza A virus and give the virus its antigenic identity.

This is known as antigenic shift. When one host is infected with two different kinds of influenza A virus, a new combination of genetic material can be produced by re-assortment. This may produce a novel influenza subtype (one that is yet to be seen). Thus, no-one is likely to have immunity to this subtype and will develop influenza. This is one of the main causes of pandemics and epidemics.

FIGURE 7.48 Antigenic shift in influenza A

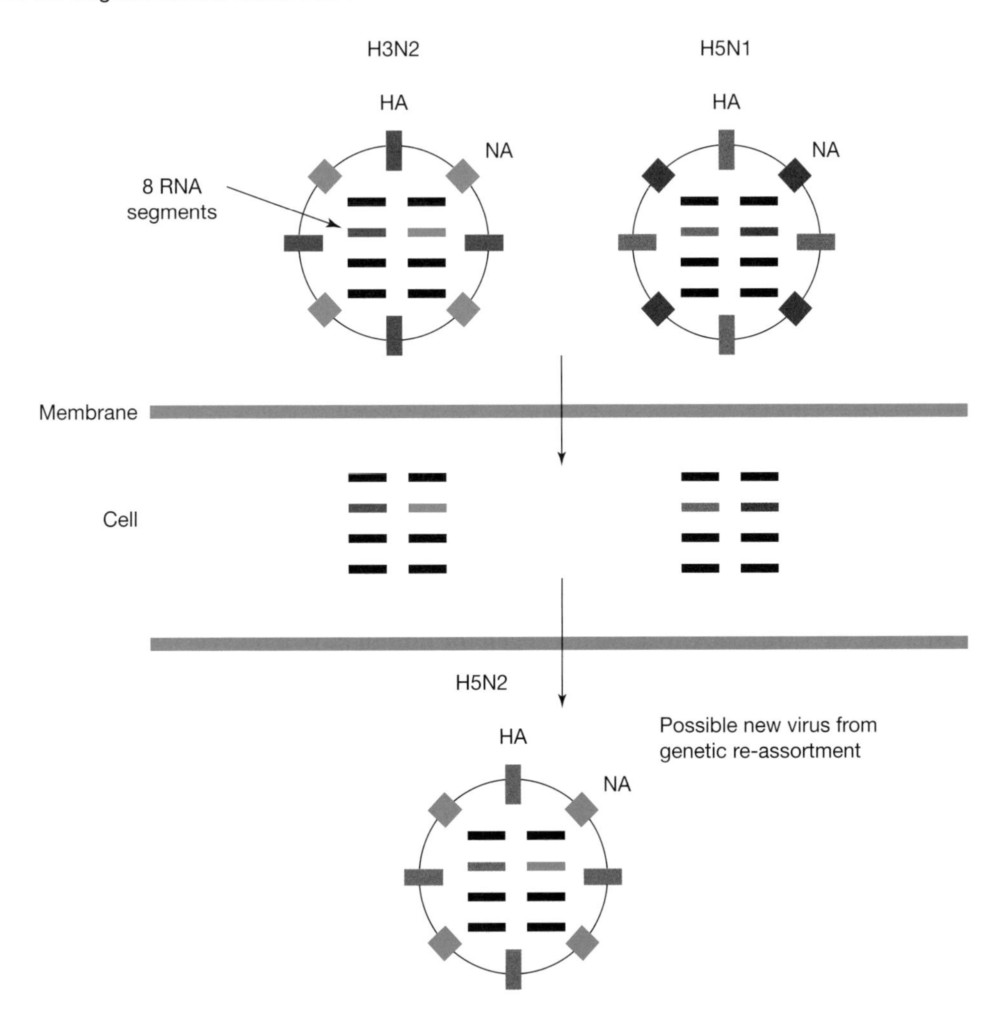

FIGURE 7.49 a. A livestock and poultry market provides an opportunity for cross infection of a host by influenza viruses from different species. b. The genetic re-assortment produces an antigenic shift that creates a new subtype of influenza A virus that can infect people.

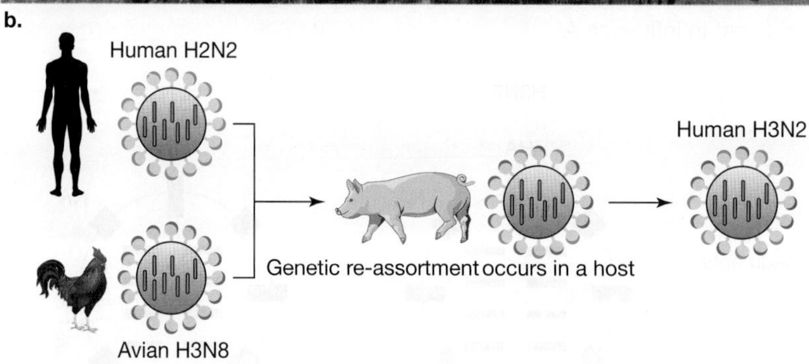

TIP: A good way to remember what occurs in antigenic shift is that antigenic **s**hift leads to **s**ignificant changes that change the **s**ubtype of the virus due to altering **s**urface proteins (shift, significant, subtype and surface proteins all start with s).

tlvd-1793

SAMPLE PROBLEM 6 Explaining and justifying challenges relating to vaccinations

A student had the flu last winter and his body created an immune response to overcome all symptoms of the virus after ten days. His parents have arranged for him to have a flu vaccination this year but he argues that he does not need one as his body has already built an immunity to the virus.

Does the student need the receive the flu vaccination? Justify your choice. (3 marks)

THINK	WRITE
Consider the flu virus. Do its antigens remain constant or change from year to year? If the virus's antigens are changing, will the body's immune system recognise them? Your answer should respond to the question for one mark. The other two marks are awarded by providing two clear points to justify your choice.	Yes (1 mark). The flu virus mutates at a fast rate, which leads to antigenic drift and/or shift. This results in different antigens on this year's flu viruses compared with last year's (1 mark). Therefore, the body's immune system will not recognise the virus and the student would be susceptible to being infected (1 mark).

Consequences of antigenic drift and shift

The consequences for antigenic drift and shift within the immune system differ from each other. If antigenic drift occurs and a slight variation of the influenza strain is produced, then a person who had already been exposed to that strain may have a partial immune response. This occurs because the B memory cells are able to initiate plasma cells to make antibodies which are still able to make some antibody-antigen complexes. However, if an antigenic shift occurs in influenza, then the body's immune system will identify the virus as new and a new immune response will need to be initiated.

FIGURE 7.50 The immune response to influenza after prior exposure

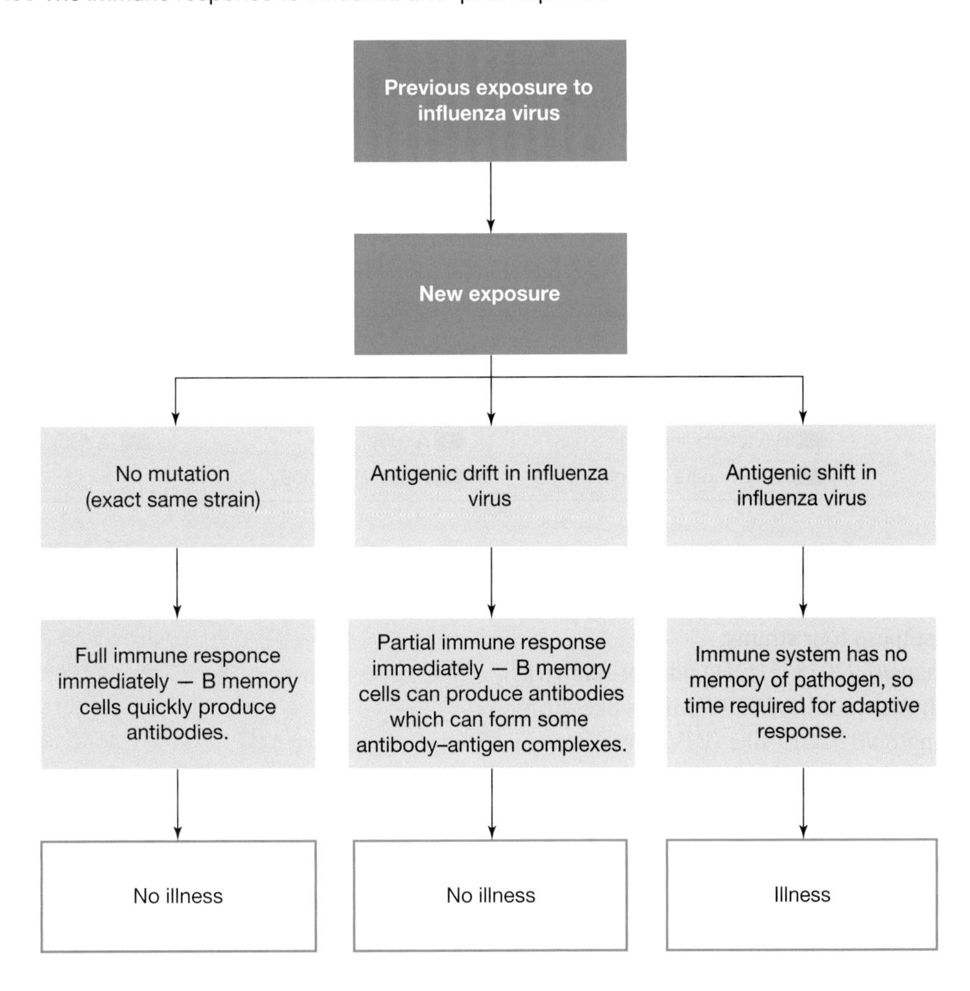

The consequence of influenza constantly mutating is that a single vaccination cannot be made and administered to provide long-term protection.

For individuals to protect themselves from influenza, a vaccination should be administered every year. The World Health Organization (WHO) works with global governments to track the strains of influenza present in their populations throughout the year. The strains that are most common in a region are referred to as seasonal strains.

As seen in figure 7.51 there were eight strains of influenza tracked throughout 2019 in the southern hemisphere.

FIGURE 7.51 The tracking of different influenza strains in the southern hemisphere

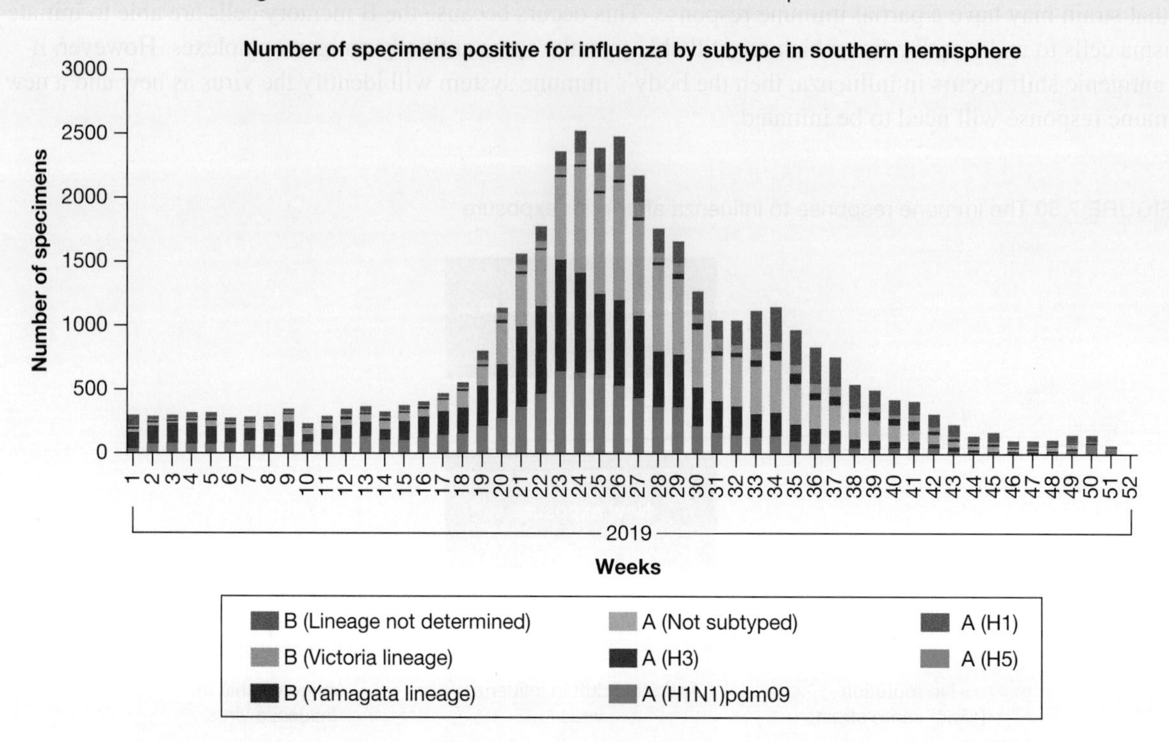

The WHO recommends the Australian government make a vaccine covering the most common seasonal strains of influenza for each year. The influenza vaccine for Australia in 2019 was a quadrivalent vaccine because it protected against these four strains:

- an A/Brisbane/02/2018 (H1N1) pdm09-like virus
- an A/South Australia/34/2019 (H3N2)-like virus
- a B/Washington/02/2019-like (B/Victoria lineage) virus
- a B/Phuket/3073/2013-like (B/Yamagata lineage) virus.

This differs from the influenza vaccine of 2021, which contained protection against two additional strains:

- an A/Victoria/2570/2019 (H1N1) pdm09-like virus
- an A/Hong Kong/2671/2019 (H3N2)-like virus

These two strains replaced the two influenza A strains in the 2019 vaccine. The other two components remained the same.

The influenza strains in future vaccines will be different based on the antigenic drift and shift of the virus. To produce yearly vaccines that are the most effective, constant surveillance of the world's influenza strains is required. The seasonal influenza strains in the northern hemisphere will likely travel to the southern hemisphere and vice versa (travel restrictions from COVID-19 have reduced the risk of this).

The changing nature of viruses was a challenge for the development of vaccines against SARS-Cov-2 with the appearance of variant strains (such as the UK and South African strains). Continual research is being undertaken in relation to the vaccines that have been approved or are upcoming to try to protect against variant forms.

on Resources

eWorkbook Worksheet 7.8 Comparing antigenic shift and drift (ewbk-2786)

Weblinks Influenza vaccine: outmanoeuvring antigenic shift and drift
Modelling antigenic shift and drift
WHO Influenza updates
The natural evolution of SARS-CoV-2

KEY IDEAS

- Bacteria reproduce rapidly and this provides more opportunities for mutations to occur.
- Bacterial resistance occurs through the process of natural selection, with resistant bacteria having a selective advantage.
- Bacterial resistance genes can be acquired through horizontal gene transfer.
- Some diseases are becoming increasingly difficult to treat with antibiotics.
- The reduction and careful prescribing of antibiotics is required to reduce the rate of bacterial resistance.
- Changes in viral antigens results from antigenic drift and antigenic shift.
- Antigenic shift produces new strains of a virus which require a new immune response.
- Viruses that have a rapid rate of antigenic drift and/or shift require the strains to be tracked and novel vaccinations produced.

7.7 Activities

learnon

To answer questions online and to receive **immediate feedback** and **sample responses** for every question, go to your learnON title at **www.jacplus.com.au**. A **downloadable solutions** file is also available in the resources tab.

7.7 Quick quiz on	7.7 Exercise	7.7 Exam questions

7.7 Exercise

1. Identify the environmental selection pressure which directs bacterial resistance.
2. Outline how the process of natural selection can select for bacterial resistance.
3. Compare the similarities and differences between antigenic drift and antigenic shift.
4. Explain the impact antigenic shift in an influenza strain will have on a body's immune response.
5. Explain why a new vaccine for the influenza virus is required every year.

7.7 Exam questions

Question 1 (1 mark)

Source: *VCAA 2013 Biology Section A, Q39*

MC Health professionals are concerned about the overprescription of antibiotics. Many antibiotics have become ineffective against certain species of bacteria.

Any rise in incidence of antibiotic-resistant bacteria is due to

A. these bacteria having acquired immunity to antibiotics.
B. the overuse of antibiotics causing mutations in bacteria.
C. the introduction of selectively bred, antibiotic-resistant bacteria.
D. antibiotic-resistant phenotypes being favoured through natural selection.

Question 2 (1 mark)

Source: *VCAA 2019 Biology Exam, Section A, Q31*

MC Relenza was developed by researchers in Australia in 1999 as a drug that could be used against influenza viruses. The researchers found that one of the influenza virus's surface proteins was resistant to change. Relenza was developed to interact with this surface protein.

In 2009, a report stated that influenza viruses resistant to Relenza had been identified. Resistance occurred because of a mutation in the viral DNA. This mutation caused glutamine to replace lysine in a protein.

For Relenza to no longer be effective, this mutation must have caused a change in the shape and charge of

A. the enzyme needed for ongoing virus replication.
B. the ribosomes of the influenza virus.
C. the capsule of the influenza virus.
D. Relenza.

Question 3 (1 mark)

Source: *VCAA 2018 Biology Exam, Section A, Q35*

MC The emergence of antibiotic-resistant diseases in humans means that

A. antibiotics are causing resistance mutations in bacteria.
B. some bacteria are less sensitive to antibiotics.
C. viruses are becoming resistant to antibiotics.
D. humans are less sensitive to antibiotics.

Question 4 (1 mark)

Source: *VCAA 2019 Biology Exam, Section A, Q35*

MC There is an increasing trend of infectious bacteria showing resistance to antibiotics.

The implication of this resistance in the treatment of bacterial infections is that

A. antibiotic treatments will be replaced with antiviral drugs.
B. simple bacterial infections may become life-threatening for patients.
C. hospital stays for patients with resistant bacteria will become shorter.
D. a person's immune system will adapt to overcome antibiotic resistance.

Question 5 (8 marks)

Influenza continues to be one of the leading causes of death worldwide. One of the most notable cases of influenza was the Spanish flu pandemic of 1918, which is estimated to have killed around 50 million individuals worldwide. The cause of this form of influenza was from the H1N1 subtype. H1N1 also caused numerous influenza cases in Australia in 2019.

a. Differences between the H1N1 strain of 2019 and the H1N1 strain during the Spanish Flu were found despite being the same subtype. What would account for these differences? **1 mark**

In 2019, many individuals were affected with influenza type A. In 2019, two types of influenza type A were found to be most prevalent: H1N1 and H3N2. Differences in these subtypes of influenza are because the surface proteins, hemagglutinin (H) and neuraminidase (N) are different. Both H1N1 and H3N2 can also affect pigs. A third type of influenza, H1N2, can also be found in pigs, which is caused through antigenic shift.

b. Explain how antigenic shift would lead to this subtype, using a diagram to show this process. **2 marks**

Vicki, a 62-year old female, received a vaccination for influenza in 2017. However, in 2019 she developed influenza.

c. Explain why Vicki developed influenza despite having the flu vaccination two years earlier. **2 marks**
d. What recommendations would you make for the process of creating vaccines against influenza? **3 marks**

More exam questions are available in your learnON title.

7.8 Review

7.8.1 Topic summary

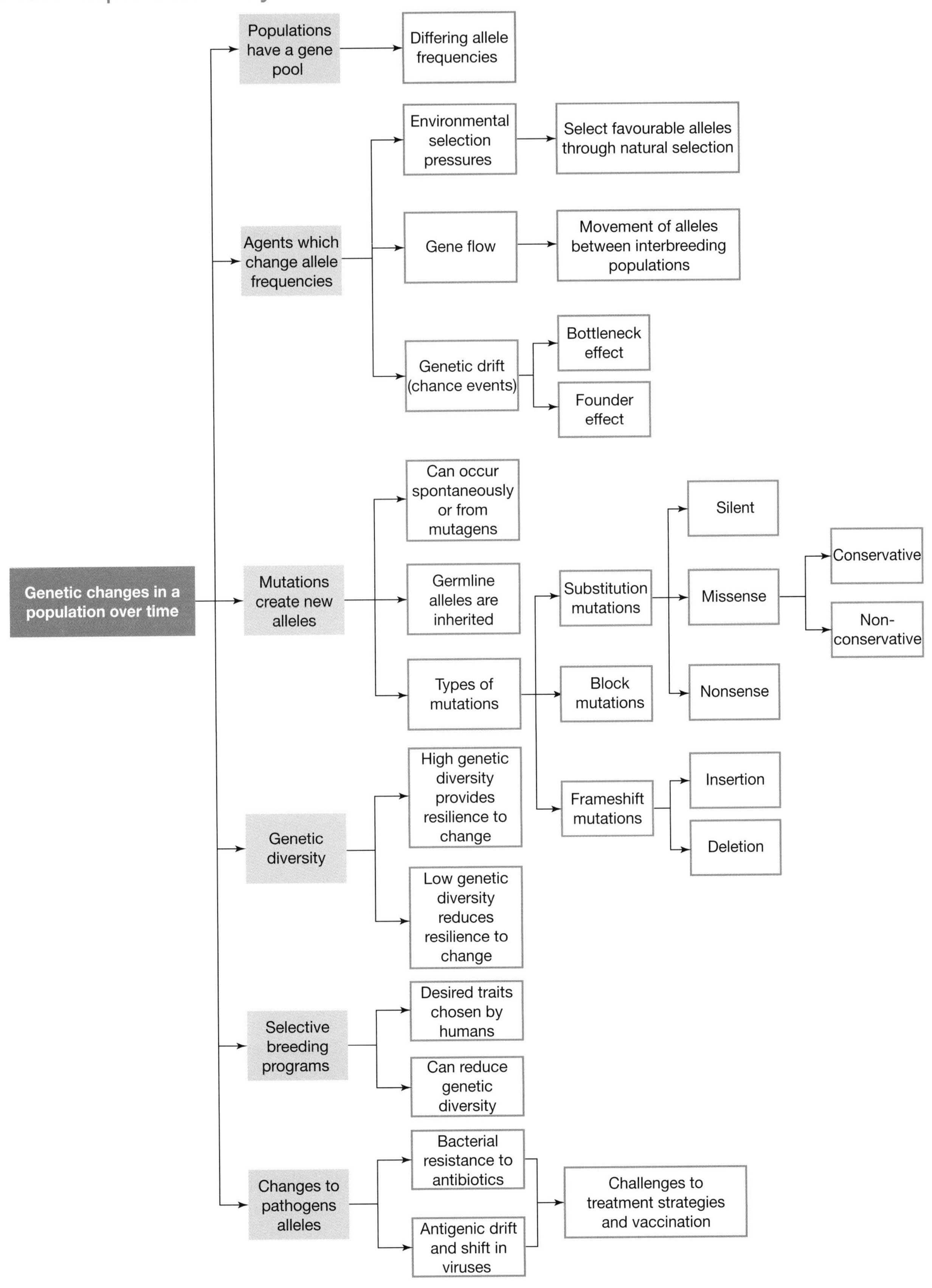

on Resources

eWorkbook	Worksheet 7.9 Reflection — Topic 7 (ewbk-2787)
Practical investigation eLogbook	Practical investigation eLogbook — Topic 7 (elog-0191)
Digital document	Key terms glossary — Topic 7 (doc-34621) Key ideas summary — Topic 7 (doc-34612)
Exam question booklet	Exam question booklet — Topic 7 (eqb-0018)

7.8 Exercises

learnon

To answer questions online and to receive **immediate feedback** and **sample responses** for every question, go to your learnON title at **www.jacplus.com.au**. A **downloadable solutions** file is also available in the resources tab.

7.8 Exercise 1: Review questions

1. Identify the difference between the members of the following pairs:
   a. gene flow and gene pool
   b. natural selection and artificial selection
   c. germline mutation and somatic mutation.
2. In an animal species, body colour is inherited, with yellow being dominant over purple. Consider two populations of this species: Population A consists of 9 purple and 1 yellow individual; population B consists of 900 purple and 100 yellow individuals.
   a. In which population would genetic drift be more likely to lead to loss of the allele controlling the yellow colour? Justify your response.
   b. All individuals in population A are homozygous. What is the frequency of each allele, where *Y* is the allele for a yellow body and *y* is the allele for a purple body?
3. Consider the following mutations in the coding strand of DNA and identify the kind of mutation that they represent. Assume that the code will be read starting from the first base shown.
   a. ... T T T T C T A G G G T C → ... T T T T G T A G G G T C
   b. ... T T T T C T A G G G T C → ... T T T T C T A A G G G T C
   c. ... T T T T C T A G G G T C → ... T T T C T A G G G T C
   d. ... T T T A T T G T C C C T → ... T T T A T C G T C C C T
4. The sheep strike blowfly, *Lucilia cuprina*, is a major pest for sheep farmers. In 1955, the insecticide dieldrin was first introduced to control this pest. The introduction was initially highly successful, but, within two years, this pesticide began to be less effective against these blowflies.
   a. Explain what was happening in this two-year period.
   b. Outline what would be expected to happen if dieldrin continued to be used.
5. Read the following abstract of a research report by Iannuzzi, M.C. *et al.* published in the *American Journal of Human Genetics* vol. 48 m, page 227 (1991):

   > Cystic fibrosis (CF) is a recessive disease caused by mutations in the CF transmembrane conductance regulator (*CFTR*) gene. We have identified in exon 7 two frameshift mutations, one caused by a two-nucleotide insertion and the other caused by a one-nucleotide deletion; these mutations, CF1154insTC and CF1213delT, respectively, are predicted to shift the reading frame of the protein and to introduce termination codons at residues 369 and 368.

   a. What is a frameshift mutation?
   b. Two frameshift mutations are identified in this case of cystic fibrosis. Note that CF means cystic fibrosis and the following number identifies a base position in the DNA of that gene. What do you think is meant by the shorthand CF1154insTC and CF1213delT?
   c. What effect did these frameshifts have on the protein encoded by this mutated gene?

6. This figure shows examples of inherited variation in dogs.
   Identify, giving an explanation for your decision, the most likely genetic basis for:
   a. the observed variation in body size between the two dogs
   b. the observed variation (black/yellow) in the base colour of the two dogs (ignore the white spotting).

   Consider the proposal that these size differences emerged in wild dog populations by natural selection.
   c. Is this assumption reasonable or not? Explain your choice.
   d. If you decided that the assumption is not reasonable, suggest an alternative explanation.

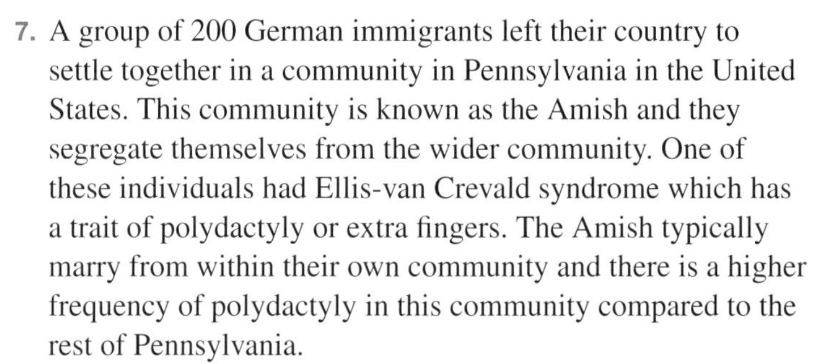

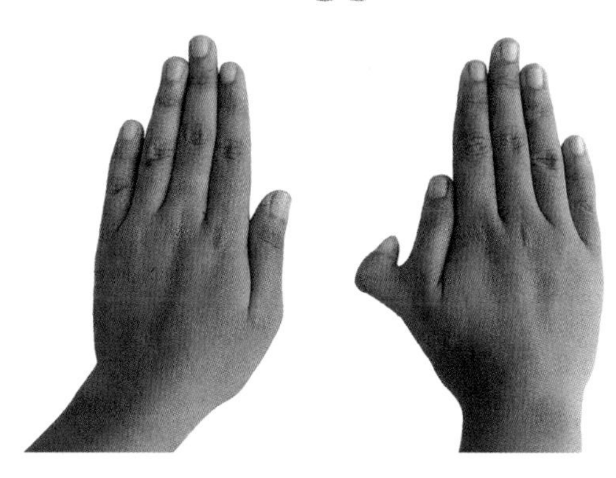

7. A group of 200 German immigrants left their country to settle together in a community in Pennsylvania in the United States. This community is known as the Amish and they segregate themselves from the wider community. One of these individuals had Ellis-van Crevald syndrome which has a trait of polydactyly or extra fingers. The Amish typically marry from within their own community and there is a higher frequency of polydactyly in this community compared to the rest of Pennsylvania.
   a. Identify the type of genetic drift which occurred in the Amish population.
   b. Explain why the Amish population has a higher frequency of polydactyly compared to other communities.
8. The platypus, *Ornithorhynchus anatinus*, is an Australian monotreme that lives in burrows besides fresh water systems. Recent studies have shown its numbers have been decreasing due to habitat interference and predation from introduced animals such as the fox. Their distribution is shown in the diagram below.

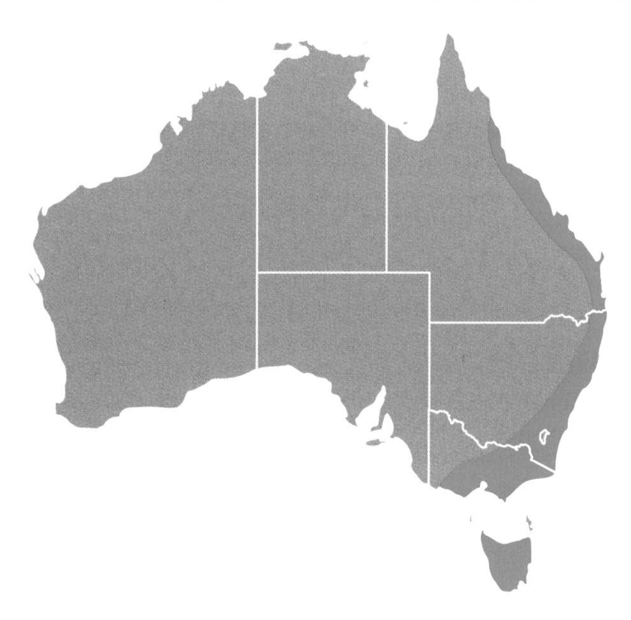

   a. The mainland platypus and the Tasmanian platypus are isolated from each other. Describe how this can affect the allele frequencies between the populations.
   b. The Tasmanian platypus will walk across open land to move between water sources but the mainland platypus does not. Explain why this difference in behaviour may have occurred.

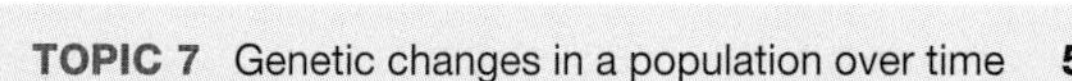

9. Some cattle are bred to be dairy livestock and some to be beef livestock. The breed called the Belgium blue was originally bred as a dairy cow. However, a mutation in one cow's myostatin gene caused 'double muscling'. The Belgium blues have 40 per cent more muscle mass than other breeds and were subsequently bred for their meat. Outline how farmers could selectively breed a herd of Belgium blue cattle which have 'double muscling'.

10. A student went to her doctor with a sore throat and it was determined that the infection was viral. She wants to get better as fast as possible and asked the doctor for some antibiotics. Should the doctor write her a prescription for antibiotics? Justify your choice.

7.8 Exercise 2: Exam questions

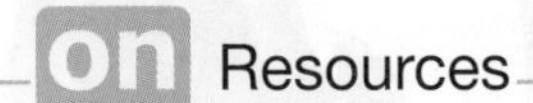

Resources

Teacher-led videos Teacher-led videos for every exam question

Section A — Multiple choice questions

All correct answers are worth 1 mark each; an incorrect answer is worth 0.

Question 1

Source: *VCAA 2009 Biology Exam 2, Section A, Q11*

A mutation is

A. a product of natural selection.
B. caused by immigration and emigration.
C. a change in an allele due to a change in DNA.
D. a random change in gene frequencies from one generation to the next.

Question 2

Source: *VCAA 2010 Biology Exam 2, Section A, Q13*

In populations

A. genetic drift will have less effect in a large population compared to a small population.
B. bottlenecks enable a population to become better equipped for future changes in the environment.
C. some organisms develop mutations in order to better suit them to their environment compared to other members of the population.
D. allele frequencies remain constant if the number of individuals leaving the population equals the number of individuals entering it.

Question 3

Source: *VCAA 2008 Biology Exam 2, Section A, Q23*

Lucilia cuprina, the sheep blowfly, lays its eggs in wounds and the wet fleece of sheep. The larvae hatch and burrow into the sheep's skin, causing distress, reduced wool production and sometimes death. Particular chemicals were used in the past to control the *L. cuprina* but these became less effective as sheep blowfly developed a resistance to the chemicals.

The cause of the increased resistance to the chemicals was most likely due to

A. farmers successively reducing the levels of insecticide applied to sheep.
B. the insecticide producing a change in a gene which enhanced the survival of the blowfly.
C. a chance mutation in a blowfly gene conferring a survival advantage in the chemical environment.
D. the insecticide producing a change in phenotype which enhanced reproduction of the blowfly.

Question 4

Source: *VCAA 2015 Biology Exam, Section A, Q40*

Northern elephant seals, *Mirounga angustirostris*, were nearly hunted to extinction in the 1890s, with only about 20 individuals left at the end of the century. The population has now grown to more than 120 000. In the 1890s, southern elephant seals, *Mirounga leonina*, were not as severely hunted and currently there are estimated to be 600 000 southern elephant seals.

Based on this information, it is true to say that

A. northern elephant seals have evolved as a result of the 'founder effect'.
B. northern elephant seals would show less genetic variation than southern elephant seals.
C. southern elephant seals would have experienced greater genetic drift than northern elephant seals.
D. the mutation rate in northern elephant seals would have been greater than in southern elephant seals.

Question 5

Source: *VCAA 2014 Biology Exam, Section A, Q33*

Consider the diagram below that models changes in allele frequencies for one trait in a population over two generations. The original population is shown on the left.

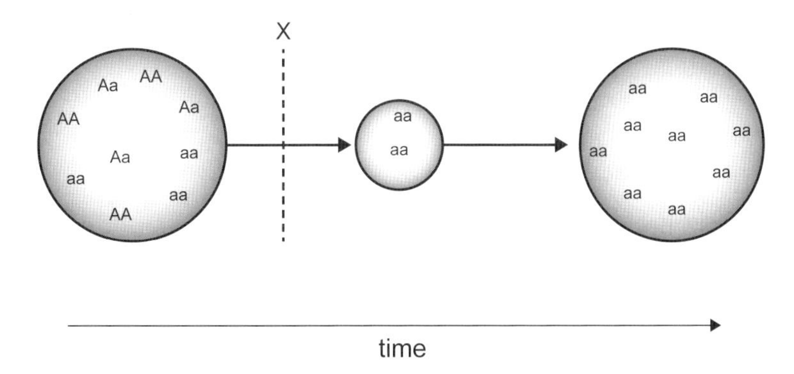

If the diagram above models the founder effect, then event X is

A. migration.
B. a natural disaster.
C. random mating.
D. a random assortment of alleles.

Question 6

Source: *VCAA 2016 Biology Exam, Section A, Q37*

Tiburon is an isolated island off the coast of Mexico. Desert bighorn sheep became extinct on this island hundreds of years ago. In 1975, 20 desert bighorn sheep were taken from a population in the American state of Arizona (shown on the map) and were reintroduced to Tiburon Island. By 1999, the population of desert bighorn sheep on Tiburon Island had risen to 650.

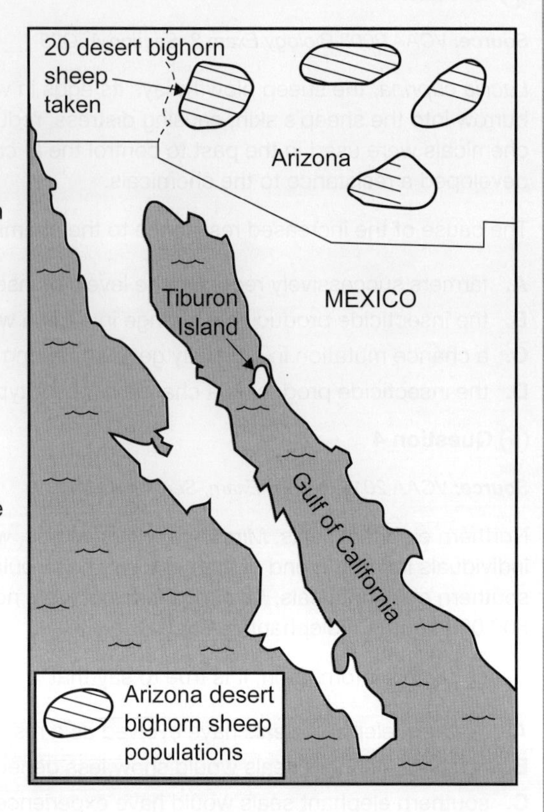

Which one of the following statements about this 1999 population of desert bighorn sheep on Tiburon Island is correct?

A. The gene pool of this population will be identical to the gene pool of the Arizona populations.
B. This population has less genetic variation than the Arizona populations and is an example of the founder effect.
C. This population will have become a new species because the mutation rate on Tiburon Island will be much higher than in Arizona.
D. Having been through a population bottleneck, the current population will now show increased genetic variation compared to the Arizona populations.

Question 7

Source: *VCAA 2011 Biology Exam 2, Section A, Q22*

Selective breeding over many generations has produced gradual changes in farm animals.

It is reasonable to state that such gradual changes in a herd of cattle

A. result from random mating in the herd in each generation.
B. are due to the occurrence of gene mutations in each generation.
C. will improve traits that enhance survival of the animals in the wild.
D. result from the restriction of breeding to chosen animals in the herd.

Question 8

Source: *VCAA 2012 Biology Exam 2, Section A, Q18*

Anomalocaris fossils have been found at Emu Bay in South Australia. *Anomalocaris* was a predatory, shrimp-like invertebrate measuring 60 cm in length. It had long, spiny, frontal appendages and a powerful, disc-shaped mouth made of overlapping, hard plates. The Emu Ba fossils were found in layers of shale and dated back to about 520 million years ago. *Anomalocaris* fossils that have been found around the world suggest that this genus existed for at least 50 million years.

A likely explanation for the extinction of *Anomalocaris* is that

A. its populations had high genetic diversity.
B. its disc-shaped mouth was unsuitable for ingesting prey.
C. it produced offspring that were suited to their environment.
D. selection pressures changed dramatically due to rapid climate change.

Question 9

Koalas have adapted over time to be able to consume eucalyptus leaves. Eucalyptus leaves are toxic to most other organisms. Koalas have specific genes that allow them to better digest chemicals from eucalyptus in leaves.

This ability to digest the toxins in eucalyptus was inherited by offspring and has been retained in the population.

Which of the following is correct in regard to this case study?

A. This shows artificial selection and has led to an increase in genetic diversity.
B. This shows artificial selection and has led to a decrease in genetic diversity.
C. This shows natural selection and has led to an increase in genetic diversity.
D. This shows natural selection and has led to a decrease in genetic diversity.

Question 10

Haemochromatosis is a genetically inherited condition in which the body absorbs too much iron. When untreated, this accumulation of iron can lead to organ damage.

A common cause of haemochromatosis is a mutation in the *HFE* gene located on chromosome 6.

The initial sequence is shown above the mutated sequence:

AAC	TTA	TGT	AGG	GGA
Thr	*Leu*	*Cys*	*Arg*	*Gly*
AAC	TTA	TAT	AGG	GGA
Thr	*Leu*	*Tyr*	*Arg*	*Gly*

It can be reasonably assumed that

A. this type of mutation is known as a missense mutation.
B. this type of mutation would lead to a shortened polypeptide.
C. this type of mutation is caused by a duplication.
D. this type of mutation is known as a frameshift mutation.

Section B — Short answer questions

Question 11 (8 marks)

Source: *VCAA 2020 Biology Exam, Section B, Q7*

Only 35% of the world's adult human population can digest lactose, which is found in milk. These people continue to produce the enzyme lactase throughout their lives. Most people who can digest lactose have European ancestry. There is evidence that people kept animals for milk in Europe 10 500 years ago.

About 7500 years ago in central Europe, a gene mutation occurred in the lactase gene, where cytosine was replaced by thymine. The allele produced by this mutation allows individuals to produce lactase and to digest lactose throughout their lives.

Researchers have estimated that populations in Europe with this mutation produced more offspring than populations who did not have this mutation.

a. Name the type of mutation that occurred 7500 years ago in central Europe. **1 mark**

b. i. The increase in frequency of the allele for lactase persistence happened relatively quickly in some populations.
Explain why the frequency of this allele increased relatively quickly. **3 marks**

ii. In some present-day populations there are no individuals with the mutation.
Give **two** reasons for the absence of the mutation in these populations. **2 marks**

c. Cows are the main source of milk in Europe. Modern dairy cow breeds can produce 25 L of milk each day — much more than their wild ancestors.
Describe how an increase in production of cow's milk could be achieved by farmers over many generations of cows. **2 marks**

Question 12 (7 marks)

Source: *VCAA 2018 Biology Exam, Section B, Q7*

Populations of the lizard species *Anolis sagrei* are found on the many islands of the Bahamas. There is natural variation between the phenotypes of individuals within each population.

a. Explain how natural variation can exist between individuals within a lizard population. **3 marks**

In 2004 a hurricane killed all populations of *A. sagrei* lizards on seven of the smaller islands. Scientists randomly chose seven males and seven females from a remaining population on a large island. They introduced one male and one female to each of the seven smaller islands. Over the next three years, the scientists noted that the size of the populations increased on each of the seven smaller islands. The scientists measured the genetic diversity within each of the populations and found there was lower genetic diversity in each new population compared with the population on the large island.

b. Explain the reasons for the lower genetic diversity of the new populations on the smaller islands compared with the population on the large island. **2 marks**

c. The scientists noted that after three years there was a significant decrease in the average length of the hind legs of the lizards living on the smaller islands compared with those on the large island.
Explain what may have happened on the smaller islands to produce this decrease in the average length of hind legs. **2 marks**

Question 13 (7 marks)

Source: *Adapted from VCAA 2013 Biology Exam, Section B, Q9*

The southern brush-tailed rock wallaby (*Petrogale penicillata*) is threatened with extinction in Victoria.

Since European settlement, it has suffered from hunting for its fur, clearing of habitat and predation by foxes. Early in 2012, the population of brush-tailed rock wallabies in the Grampians National Park numbered only four individuals.

Source: Richard Lydekker, *A Hand-book to the Maruspialia and Monotremata*, R Bowdler Sharpe (ed.), WH Allen & Co. Limited, London, 1894

a. What is meant by extinction? **1 mark**

b. Eighteen brush-tailed rock wallabies were released into the Grampians in late 2012. The wallabies had been bred in captivity in zoos and nature reserves in Victoria, South Australia, New South Wales and the Australian Capital Territory. Care was taken to ensure that the gene pool of the released wallabies was as diverse as possible.

i. What is a gene pool? **1 mark**

ii. Using your knowledge of natural selection, explain why it is an advantage to have a diverse gene pool among the released wallabies. **2 marks**

Wildlife officers are hoping that the Grampians population will increase to at least 50 wallabies in five years and that it will be maintained over time. A small population of wallabies could be affected by genetic drift and, possibly, a genetic bottleneck effect.

c. Explain the meaning of the terms genetic drift and bottleneck effect with reference to allele freqencies. **2 marks**

d. The Grampians wallabies will be closely monitored by surveillance cameras and radio collars after their release. Suggest one further measure that wildlife officers should carry out to help maintain the population over time. **1 mark**

Question 14 (10 marks)

Sarcophilus harrisii, better known as the Tasmanian devil, is a carnivorous Australian species. It was once found across all of mainland Australia, until eventually being isolated to just Tasmania.

The Tasmanian devil is a highly endangered species. Signficant conservation efforts are underway to bring their numbers back up.

Tasmanian devils' genetic diversity can be linked to an initial founder effect in their introduction to Tasmania.

a. Describe founder effect, and outline the effect this would likely have had on the genetic diversity of Tasmanian devils. **2 marks**

One of the biggest impacts on the population of Tasmanian devils was a disease known as devil facial tumour disease (DFTD).

A number of Tasmanian devils who are immune to DFTD have been found in the population. While there were only a very small number of individuals immune to the disease in the early 2000s, the number of immune individuals in the wild has been found to have increased.

b. Identify the process that has allowed for this increase to occur and clearly outline how this may have occurred. **3 marks**

c. How might selective breeding programs be used in the conservation of the Tasmanian devil? **2 marks**

In 2020, a small breeding population of Tasmanian devils was reintroduced into New South Wales. This is the first time they have been on the mainland in over 3000 years.

d. Identify one difference in the selective pressures faced between the Tasmanian and NSW populations. **1 mark**

e. Would you expect the gene pools and allele frequencies to remain the same in the two populations? Justify your response. **2 marks**

Question 15 (6 marks)

Source: *VCAA 2016 Biology Exam, Section B, Q8*

Two species of *Cryptasterina* sea stars are found in coastal Queensland. *Cryptasterina pentagona* is found in warmer water further north, while *Cryptasterina hystera* is found further south in cooler water.

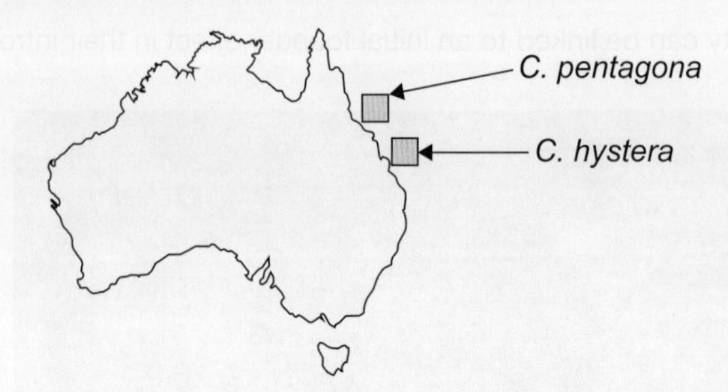

Researchers have concluded that these two species arose from a recent common ancestor via natural selection. They believe that, over thousands of years, the sea environment has changed, with the boundary line between cold water and warm water moving further north. They have found that water temperature and predation of sea star larvae by cold-water predators are important selection pressures for these sea stars.

a. Using the information **above**, explain how natural selection can lead to differences in phenotypes between these two sea star species. **4 marks**

b. One of the phenotypic differences between these two species of sea stars is their method of reproduction. *C. pentagona* reproduces sexually and its sperm and eggs are free-floating in the ocean. *C. hystera* self-fertilises and its fertilised eggs are kept within the sea star until maturity.
The researchers found that one species of *Cryptasterina* has a significantly higher diversity of alleles in its gene pool than the other species.
Using this information about reproduction strategies, which species of *Cryptasterina* would you expect to have the highest diversity of alleles? Explain your answer. **2 marks**

7.8 Exercise 3: Biochallenge online only

on Resources

eWorkbook Biochallenge — Topic 7 (ewbk-8088)

Solutions Solutions — Topic 7 (sol-0663)

Past VCAA examinations

Sit past VCAA examinations and receive immediate feedback, marking guides and examiner's report notes.
Access Course Content and select 'Past VCAA examinations' to sit the examinations online or offline.

teachon

Test maker

Create unique tests and exams from our extensive range of questions, including past VCAA questions.
Access the assignments section in learnON to begin creating and assigning assessments to students.

Online Resources

Below is a full list of **rich resources** available online for this topic. These resources are designed to bring ideas to life, to promote deep and lasting learning and to support the different learning needs of each individual.

eWorkbook

- 7.1 eWorkbook — Topic 7 (ewbk-1886) ☐
- 7.2 Worksheet 7.1 Case study: Can you drink milk? (ewbk-8234) ☐
- Worksheet 7.2 Looking at allele frequencies (ewbk-2780) ☐
- 7.3 Worksheet 7.3 Natural selection, gene flow, genetic drift and evolution (ewbk-2781) ☐
- 7.4 Worksheet 7.4 Types of mutation (ewbk-2782) ☐
- 7.5 Worksheet 7.5 Biological consequences and genetic diversity (ewbk-2783) ☐
- 7.6 Worksheet 7.6 Case study: Investigating other examples of selective breeding (ewbk-2784) ☐
- Worksheet 7.7 Making the most of new variants — Artificial selection (ewbk-2785) ☐
- 7.7 Worksheet 7.8 Comparing antigenic shift and drift (ewbk-2786) ☐
- 7.8 Worksheet 7.9 Reflection — Topic 7 (ewbk-2787) ☐
- Biochallenge — Topic 7 (ewbk-8088) ☐

Solutions

- 7.8 Solutions — Topic 7 (sol-0663) ☐

Practical investigation eLogbook

- 7.1 Practical investigation eLogbook — Topic 7 (elog-0191) ☐
- 7.2 Investigation 7.1 Variations — no two the same (elog-0034) ☐
- 7.3 Investigation 7.2 Modelling natural selection (elog-0035) ☐
- Investigation 7.3 Modelling genetic drift (elog-0036) ☐
- 7.4 Investigation 7.4 Investigating mutations (elog-0037) ☐
- 7.7 Investigation 7.5 Investigating bacterial resistances (elog-0038) ☐

Digital documents

- 7.1 Key science skills — VCE Biology Units 1–4 (doc-34326) ☐
- Key terms glossary — Topic 7 (doc-34621) ☐
- Key ideas summary — Topic 7 (doc-34612) ☐
- 7.4 Case study: Trinucleotide expansion mutations (doc-36105) ☐
- 7.5 Extension: Saving genetic variation (doc-36106) ☐
- 7.6 Extension: Technologies in selective breeding (doc-36184) ☐

Teacher-led videos

- Exam questions — Topic 7 ☐
- 7.2 Sample problem 1 Calculating allele frequencies (tlvd-1788) ☐
- 7.3 Sample problem 2 Explaining allele frequencies in different populations (tlvd-1789) ☐
- 7.4 Sample problem 3 Identifying mutations and explaining their impacts (tlvd-1790) ☐
- 7.5 Sample problem 4 Identifying genetic drift and the impact on genetic diversity (tlvd-1791) ☐
- 7.6 Sample problem 5 Outlining the processes in selective breeding (tlvd-1792) ☐
- 7.7 Sample problem 6 Explaining and justifying challenges relating to vaccinations (tlvd-1793) ☐

Video eLessons

- 7.2 Exploring a gene pool (eles-4197) ☐
- 7.3 Selection and tick populations (eles-4198) ☐
- Change agents and natural selection (eles-4199) ☐
- Genetic drift (eles-4200) ☐
- 7.4 Changes in chromosomes (eles-4201) ☐
- Types of mutations (eles-4214) ☐

Interactivities

- 7.4 Mutations (int-1453) ☐
- Chromosomal changes (int-0181) ☐

Weblinks

- 7.2 Relative frequency of alleles simulation ☐
- 7.3 Koalas and bushfires: saving their genetic material ☐
- 7.4 Outcome of mutations ☐
- How often do mutations happen? ☐
- Mutations interactive ☐
- 7.5 Svalbard Global Seed Vault ☐
- 7.6 Dogs that changed the world — selective breeding problems ☐
- 7.7 Resistance map ☐
- WHO: bacterial resistance ☐
- The antibiotic resistance crisis ☐
- Influenza vaccine: outmanoeuvring antigenic shift and drift ☐
- Modelling antigenic shift and drift ☐
- WHO Influenza updates ☐
- The natural evolution of SARS-CoV-2 ☐

Exam question booklet

- 7.1 Exam question booklet — Topic 7 (eqb-0018) ☐

Teacher resources

There are many resources available exclusively for teachers online.

To access these online resources, log on to **www.jacplus.com.au**

8 Changes in species over time

KEY KNOWLEDGE

In this topic, you will investigate:

Changes in species over time

- changes in species over geological time as evidenced from the fossil record: faunal (fossil) succession, index and transitional fossils, relative and absolute dating of fossils
- evidence of speciation as a consequence of isolation and genetic divergence, including Galápagos finches as an example of allopatric speciation and *Howea* palms on Lord Howe Island as an example of sympatric speciation.

PRACTICAL WORK AND INVESTIGATIONS

Practical work is a central component of learning and assessment. Experiments and investigations, supported by a **practical investigation eLogbook** and **teacher-led videos**, are included in this topic to provide opportunities to undertake investigations and communicate findings.

8.1 Overview

Numerous **videos** and **interactivities** are available just where you need them, at the point of learning, in your digital formats, learnON and eBookPLUS and at **www.jacplus.com.au**.

8.1.1 Introduction

When the Earth formed 4.5 billion years ago, there were no living organisms. Today the number of known species is in the millions and the number of extinct species is even greater. How did these species change over time and what evidence can we use to track these changes? Fossils are one kind of the evidence we can use to understand how species change. Fossilised organisms give us a view back in time to see not only how an organism appeared but also its environmental conditions. How did the Earth's biodiversity look 10 000, 1 million and 1 billion years ago? The answer is of course vastly different. Species are not fixed in time and space; they evolve and create new species through the process of speciation.

FIGURE 8.1 Paleontologists excavating fossilised bones of a plant-eating ceratopsian dinosaur (*Centrosaurus* sp.)

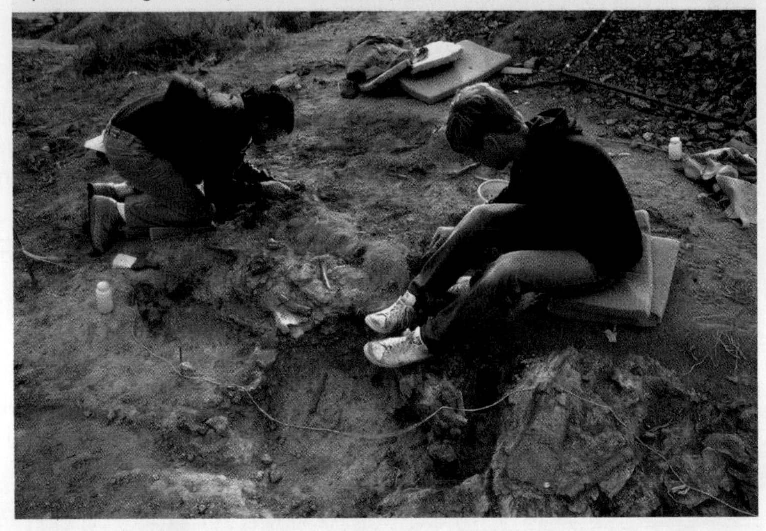

This topic investigates how species have changed over time and the evidence which is used to illustrate this. The fossil record is examined to determine how fossils are made and which evidence the different types of fossils provide. The topic also explores how new species are constantly being formed through the processes of allopatric and sympatric speciation.

LEARNING SEQUENCE

on Resources

eWorkbook	eWorkbook — Topic 8 (ewbk–1887)
Practical investigation eLogbook	Practical investigation eLogbook — Topic 8 (elog-0193)
Digital documents	Key science skills — VCE Biology Units 1–4 (doc-34326) Key terms glossary — Topic 8 (doc-34622) Key ideas summary — Topic 8 (doc-34613)
Exam question booklet	Exam question booklet — Topic 8 (eqb-0019)

8.2 Changes in species over geological time

KEY KNOWLEDGE

- Changes in species over geological time as evidenced from the fossil record

8.2.1 Geological time

The age of the Earth and more particularly its rocks is measured in **geological time**.

Our understanding of geological time is recent. The change in thinking about Earth from being just a few thousand years old to the concept of a very ancient Earth thousands of millions of years old was initially due to the work of two Scottish thinkers: James Hutton, the farmer and amateur geologist, and Charles Lyell, the pre-eminent geologist of his time.

Based on modern techniques of radiometric dating of lunar rocks and meteorites, we now know that the Earth is about 4500 million years old (or 4.5 billion years) — an interval that provides sufficient time for evolutionary and geological processes to occur.

Geological ages almost defy comprehension. We can gain some understanding by representing the age of Earth as a 100-metre track, with the starting line being the time of formation of Earth more than 4500 million years ago (Mya) and the finishing line being the present time.

- A human life of 72 years on this scale would be an undetectable 0.002 millimetres near the end of the race.
- Just the tip of a fingernail (one millimetre) from the finish line would take us back 45 000 years.
- The 200 000-year history of our species, *Homo sapiens*, would be placed less than 5 millimetres from the finish line.
- The entire history of humanity (as represented by fossil evidence of members of the genus *Homo*) would be just 6 centimetres from the finish line.
- Two steps (two metres) back from the finish line would take us to about 90 million years ago, to the time when Australia was part of the Gondwana supercontinent and dinosaurs dominated life on Earth.
- Multicellular organisms only appear in the last 14 metres of the race.
- The first 10 metres of the race would have no signs of life. The first signs of life (thought to be between 3500 and 4000 million years old) would appear between 10 and 20 metres from the starting line.

This time scale can be seen in figure 8.2, where the track is shown as a circular image, showing the time scale of life compared to the formation of the Earth.

Perhaps the most remarkable fact about this geological time scale is that the first indirect evidence of life on Earth appeared about 3850 million years ago, but it was not until about 620 million years ago, in the Ediacaran period, that the first multicellular animals appeared in the fossil record.

The first members of the genus *Homo* appeared only 2.8 million years ago, and the first modern humans (*Homo sapiens*, our species) appeared about 200 000 years ago.

geological time a system of chronological dating that relates geological strata to time

FIGURE 8.2 The geological time scale can be represented as a circle. The appearance of the first hominins take up a very small part of geological time.

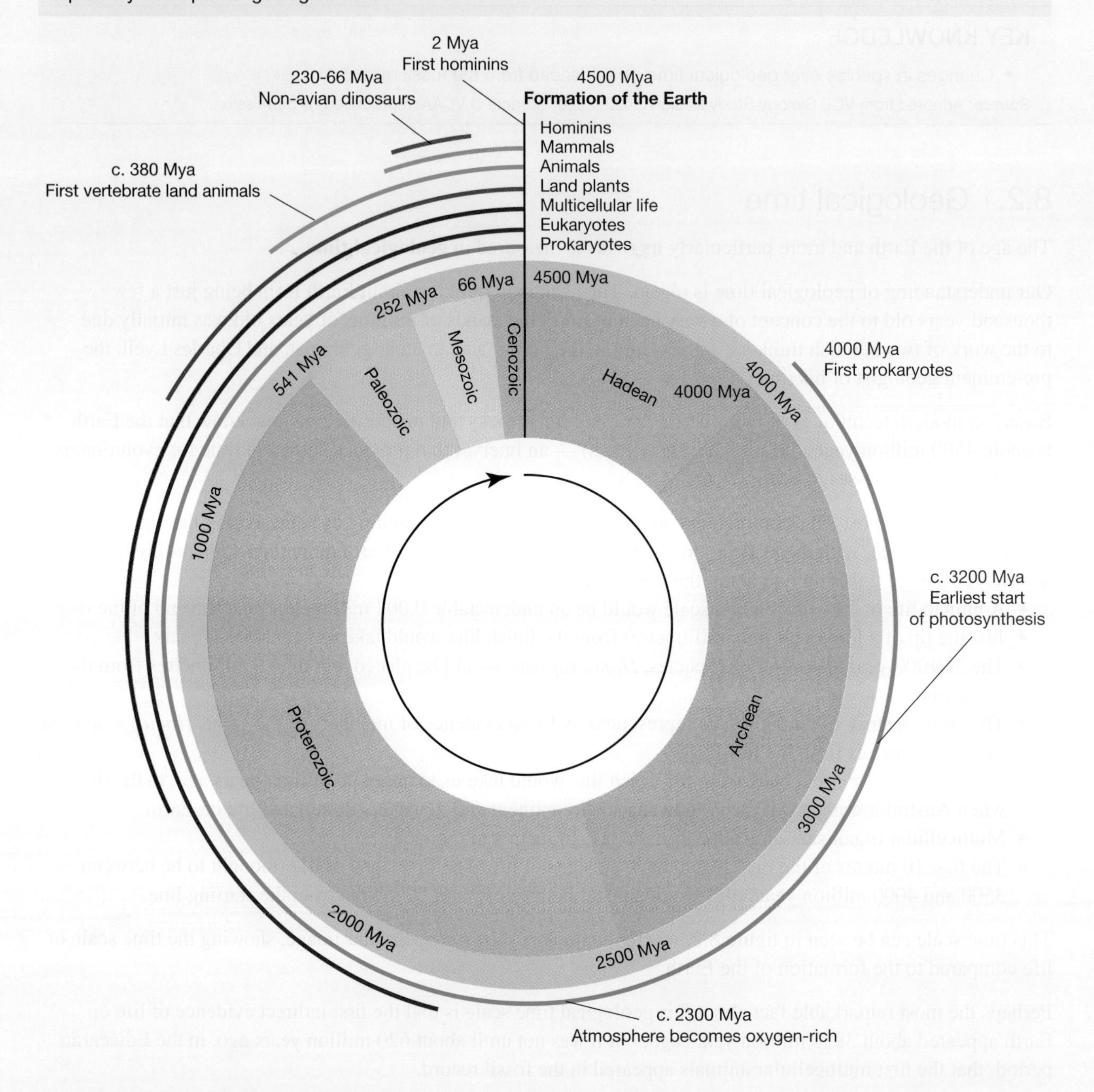

Time intervals in geological time

The geological history of the Earth is divided into various time intervals as a hierarchy that includes eons, eras and periods (see figure 8.3). From largest to smallest, these are:

eons → eras → periods → epochs → ages.

This geological time scale was developed in the eighteenth and nineteenth centuries when scientists organised sedimentary rock strata in the same region into groups of similar **relative ages** and also recognised similarities in rock strata in different regions because they contained identical fossils. The major divisions of the time scale based on relative ages were established by about 1840, but the **absolute ages** were not identified until the twentieth century.

relative age the age of an object expressed in relative terms so that it is identified as younger or older than another object

absolute age the age of an object, such as a rock or fossil, expressed in actual years

FIGURE 8.3 Geological time scale, as published by the American Geological Institute. Dates are in accord with the International Commission of Stratigraphy. Mya refers to millions of years ago. Note that some eras, periods and epochs have been omitted.

	Eon	Era	Period	Epoch	Mya
	Phanerozoic	Cenozoic	Neogene	Holocene	0.01
				Pleistocene	2.6
				Pliocene	5.3
				Miocene	23.0
			Paleogene	Oligocene	33.9
				Eocene	56.0
				Paleocene	66.0
		Mesozoic	Cretaceous		145.0
			Jurassic		201.3
			Triassic		252.2
		Paleozoic	Permian		299.0
			Carboniferous		358.9
			Devonian		419.2
			Silurian		443.8
			Ordovician		485.4
			Cambrian		541.0
Precambrian	Proterozoic				2500
	Archean				4000
	Hadean				

TIP: Although you do not have to memorise these dates, you should understand that geological time is broken up into smaller groupings that often differ in their fossil records.

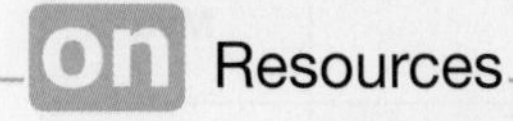
Resources

Weblink Understanding scales in geological time

8.2.2 Changing life forms over time evidenced with fossils

Since life first appeared, the Earth has been a continually changing planet. The diversity of species has increased as time has moved on, with many new species appearing over geological history. Three thousand Mya, the only living organisms were prokaryotes, but in the modern era, plants, animals, fungi and protists are in great variety.

FIGURE 8.4 A representation of life forms throughout history

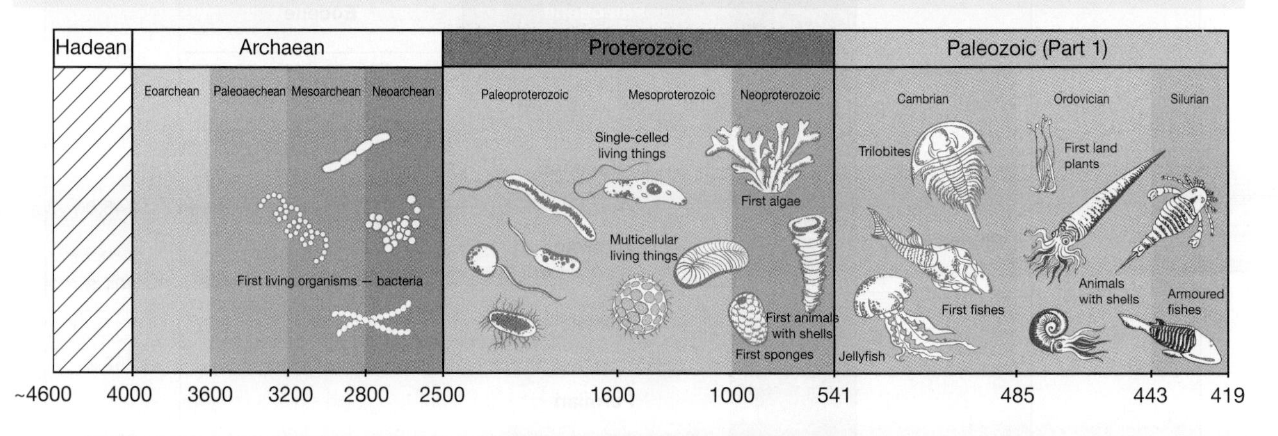

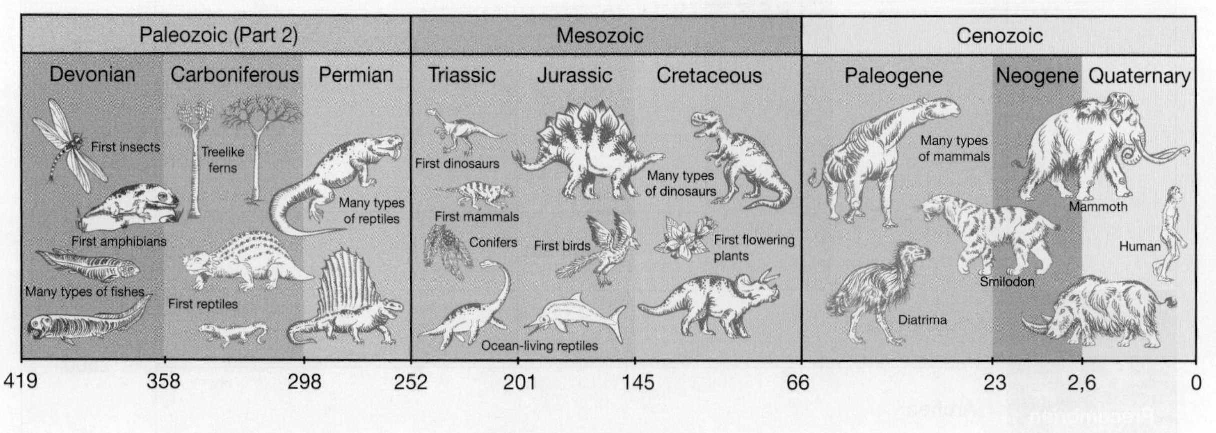

CASE STUDY: Changing life forms over time evidenced with fossils

Imagine that we could travel back in time on planet Earth down the geological time scale. The further back in time we travel, the less familiar the scene around us will be. These changes can also be seen in the fossil record.

Some examples of different species we would see across geological time are:

- the Pleistocene epoch: around 60 000 years ago, when large mammals (megafauna) were found
- the Cretaceous period: around 110 million years ago, when dinosaurs roamed
- the Ordovician period: around 460 million years ago, with primitive plants and invertebrates
- the Archaean eon: around 3000 million years ago, when only prokaryotes could survive in the oxygen-free atmosphere.

FIGURE 8.5 Reconstructed views of **a.** the Pleistocene epoch and **b.** the Cretaceous period

To access more information about these different times, please download the digital document.

Resources

Digital document Case study: Changing life forms over time evidenced with fossils (doc-36107)

It is clear that there have been significant changes in the Earth's species over time. Table 8.1 shows the approximate times of some major landmarks in these changes. Until about 400 Mya, plants and animals were almost all limited to life in the oceans. Since plants and animals became established on land, Earth's life forms have become more diverse.

TIP: You do not need to memorise the dates at which different organisms were formed or fossils were found, but you should have a general idea about the order in which they appeared (for example, amphibians and reptiles came into existence before mammals).

TABLE 8.1 Approximate times of appearance of various life forms on Earth. Note that for most of Earth's history, the only life forms were unicellular organisms — first, prokaryotic microbes, which later were joined by unicellular eukaryotic organisms.

Approximate time	Emergence of life forms
~ 4500 Mya	The lifeless (abiotic) Earth
~ 3800 Mya	First prokaryotic cells
~ 3500 Mya	First photosynthetic bacteria
~ 2500 Mya	First aerobic (oxygen-using) microbes
~ 2100 Mya	First unicellular eukaryotes
~ 600 Mya	First multicellular eukaryotes
~ 500 Mya	First vertebrates (jawless fish)
~ 479 Mya	First insects
~ 450 Mya	First land plants
~ 365 Mya	First amphibians
~ 360 Mya	First ferns
~ 330 Mya	Earliest reptiles
~ 250 Mya	First conifers
~ 160 Mya	First mammals
~ 130 Mya	First flowering plants

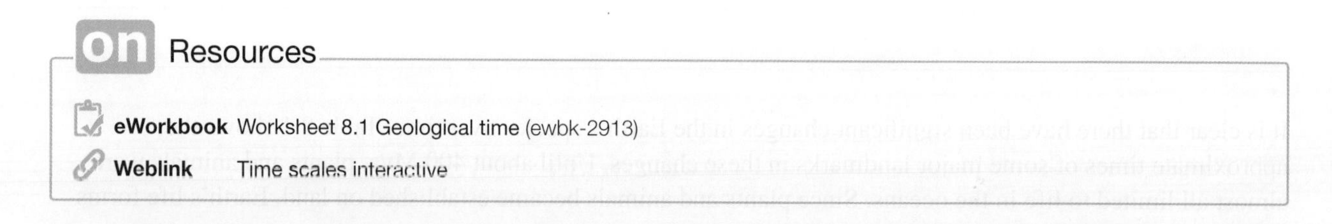

elog-0076

INVESTIGATION 8.1

online only

Geological time

Aim

To model geological time and show when different species appeared over this time

CASE STUDY: Australian megafauna

In the early 1830s, Major Thomas Mitchell (1792–1855), an explorer and surveyor, collected some fossil bones in a cave near Wellington, New South Wales. These bones were sent to England to Sir Richard Owen (1804–1892), the leading vertebrate fossil expert at that time (and an outspoken opponent of Darwin's theory of evolution).

From these bones, Owen reconstructed the skeleton of an extinct marsupial mammal, now known as *Diprotodon* (see figure 8.6). This animal was a member of a diverse group of animals known as the Australian megafauna ('giant animals'). The term 'megafauna' refers to large extinct vertebrates that lived during the Pleistocene epoch and which grew to sizes larger than their close relatives.

To access more information about other species part of Australia's megafauna, please download the digital document.

FIGURE 8.6 a. Skeleton of *Diprotodon optatum*, an extinct marsupial and one of the Australian megafauna. (Image courtesy of Judith Kinnear.) **b.** Reconstruction of *Diprotodon optatum*

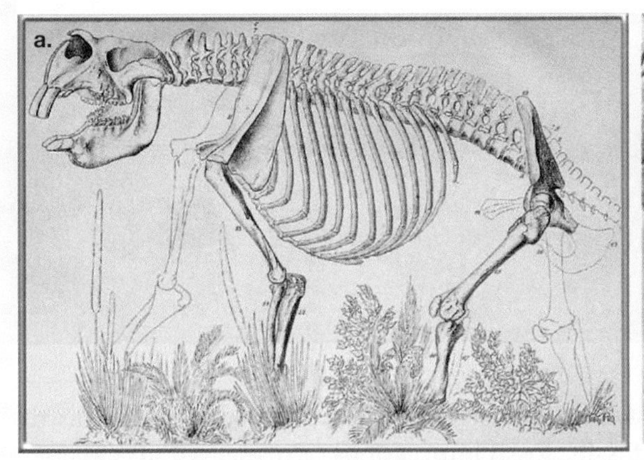

on Resources

Digital document Case study: Australian megafauna (doc-36108)

on Resources

eWorkbook Worksheet 8.2 Changing species — exploring mass extinctions over time (ewbk-2914)

Digital document Extension: Mass extinction (doc-36110)

Weblink Megafauna

8.2.3 Types of fossils

Our window to life on Earth in the geological past is through fossils. Fossils provide evidence of the kinds of organisms that once lived on Earth, and the fossil record illustrates how the life forms of this planet have changed over geological time.

The study of fossils is known as palaeontology, and a person who specialises in that field is known as a palaeontologist.

Fossils can be categorised as:
- physical fossils — remains of structures
- trace fossils — evidence of activities
- biosignatures — inferred evidence.

FIGURE 8.7 Physical fossils: **a.** mineralised fossil of an ammonite shell (*Dactylioceras commune*); **b.** impression fossil of the hard exoskeleton of a trilobite (*Elrathia kingie*)

a.

b.

1. **Physical fossils**
   Physical fossils are the remains of all or part of the structures of organisms, such as bones, teeth or leaves. These fossils are commonly formed as **mineralised fossils**, **fossil impressions** or **preserved organisms**:
   - **Mineralised fossils:** the organic material of the structure is replaced by minerals. Figure 8.7a shows an ammonite fossil; the original shell has been replaced, molecule by molecule, by pyrite, which is an iron sulphide material.
   - **Fossil impressions:** the organic matter has disappeared, but the organism has left an impression of its structure. Figure 8.7b shows a trilobite fossil; the original material of the exoskeleton has disappeared, but the impression was preserved in the fine-grained mudstone. Two examples of fossil impressions are **moulds** and **casts** (see figure 8.14c).
   - **Preserved organisms:** Sometimes, rather than being mineralised, an organism (or part of an organism) is completely preserved in a substance (see figure 8.13). Substances that allow for this type of preservation include amber, ice and tar. Amber, for example, is fossilised tree resin. Organisms can become caught in the resin before it hardens. As the resin fossilises, the trapped organism can become fully preserved.
2. **Trace fossils**
   Trace fossils are preserved evidence of the activities of organisms, such as footprints (figure 8.8a), tooth marks, tracks, burrows and coprolites (fossilised dung, shown in figure 8.8b).
   Among the best-known trace fossils are sets of dinosaur footprints, called dinosaur trackways. Figure 8.8a shows part of the dinosaur trackway at Lark Quarry. The tracks tell the story of a Cretaceous day, 95 million years ago, when a mixed group of small dinosaurs were on the edge of a lake. On the approach of a large carnivorous dinosaur, the herd of small dinosaurs scattered, leaving their tracks in the mudflats beside the lake. This mud solidified, the lake level rose and the footprints were progressively covered by more and more sediment.

physical fossil the remains of all or part of the structures of organisms, such as bones, teeth or leaves

mineralised fossil a physical fossil formed when the organic material of the structure is replaced by minerals

fossil impression a physical fossil formed when the organic matter has disappeared but the organism has left an impression of its structure

preserved organism a physical fossil formed when an organism is completely preserved in a substance such as amber, ice or tar

mould type of fossil that results when a buried organism decays, leaving an impression of the original organism

cast type of fossil formed when a buried structure decays leaving a cavity, which is later filled by different material that forms a model of the original structure

trace fossils preserved evidence of the activities of organisms, such as footprints, toothmarks, tracks, burrows and coprolites

FIGURE 8.8 a. A small part of the dinosaur trackway at Lark Quarry, near Winton in central Queensland. (Image courtesy of Judith Kinnear.) b. Coprolite fossil

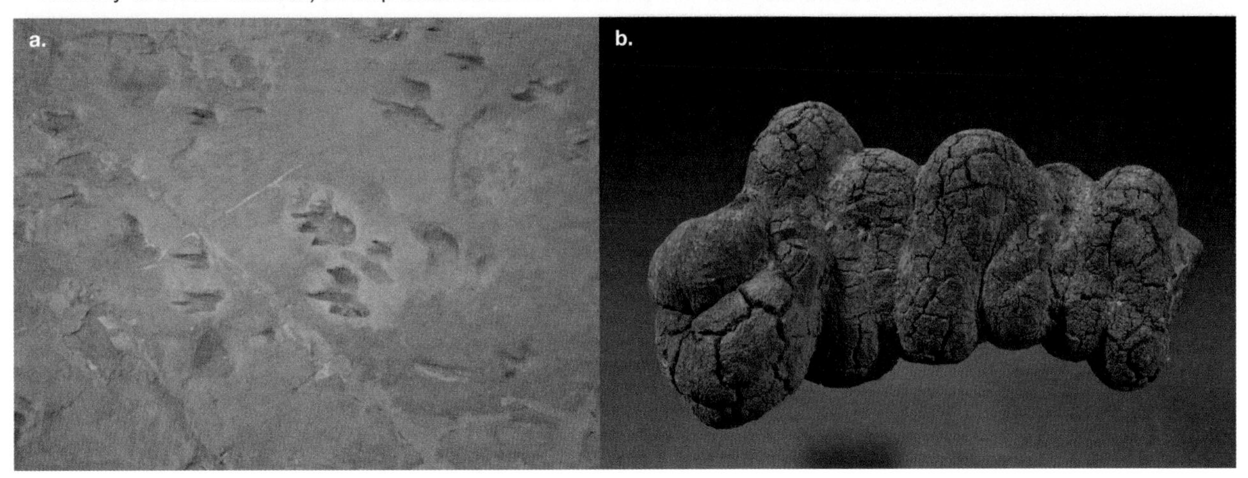

3. **Biosignatures**

Biosignatures or biomarkers provide evidence from which past life may be inferred to exist.

Fossils, including trace fossils, are definite evidence of past life. Another clue to past life comes from biosignatures. A biosignature is a physical or chemical sign preserved in minerals, rocks or sediments that *can be inferred* to be cellular or to have resulted from the metabolic activities of an organism. Biosignatures are important in the study of rocks from the Archaean period, when the only life forms were microbes (as microbes are unlikely to leave fossils).

Examples of biosignatures include:

- corrosion pits in rocks caused by chemosynthetic microbes
- detection in rocks of isotopic ratios of bio-essential elements, such as carbon and sulphur, that are similar to the ratios produced by life processes in microbes
- stable molecules, such as steranes (compounds derived from plasma membranes), that can persist long after all other signs of the cells from which they came have disappeared.

FIGURE 8.9 Fossil stromatolites from the Bitter Springs formation east of Alice Springs. (Specimen from the Houston Museum of Natural Science.)

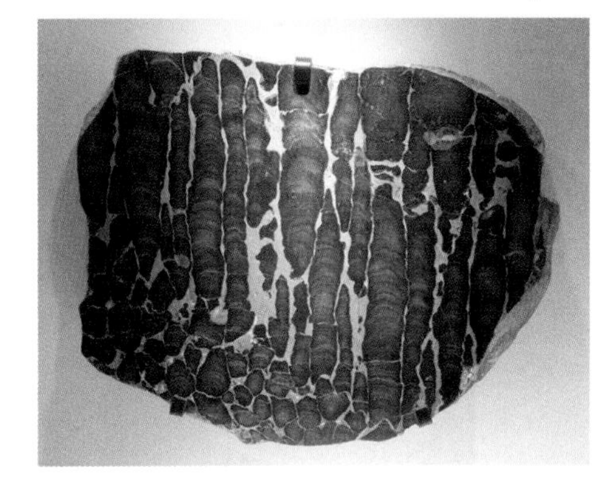

Interpretation of a potential biosignature as a sign of past life must be made with caution. In some cases, its origin might also be from a non-living (abiotic) source, and in other cases, it may be the result of later contamination. The older the microbial signatures, the more difficult it is to prove that they originated from the action of microbes.

However, one unambiguous physical biosignature is the stromatolites from the Archaean eon (see figure 8.9).

Fossil stromatolites are accepted as having a biological origin and signify the existence of microbial life *at least* 3.5 thousand million years ago. The microbes that formed these structures cannot be detected, but the pattern of accumulated sedimentary grains in these ancient structures cannot be explained by any abiotic processes.

biosignatures chemical or physical traces that can be inferred to have resulted from the action of life forms

elog-0077

INVESTIGATION 8.2 online only

Examining casts, moulds and impressions

Aim

To produce casts, moulds and fossil impressions, and compare the structure of these

Macroscopic and microscopic fossils

Many fossils are **macroscopic**, meaning they can be seen and studied with an unaided eye or very low magnification. Some fossils, however, are microscopic. These **microfossils** require the use of either light or electron microscopy to visualise them and to identify the details of their shape and other morphological features.

The earliest life forms on Earth were prokaryotic microbes. As a result, evidence of these early life forms often involves searching for rocks of a relevant age that might have been formed under conditions suitable for early life, then making thin slices of these rocks and searching for microfossils.

Another important group of microfossils is those formed by pollen and spores. Because of their resistant outer coats, both pollen and spores are very durable and can be preserved in sedimentary rocks. Equally important is the fact that the outer coats of pollen and spores have distinctive patterns on their surfaces by which they can be identified. These fossils are one tool that can assist in identifying the climate that prevailed in a region in the geological past.

macroscopic of a size that is visible to the unaided eye

microfossils very small fossils of a size that can only be studied using a microscope

EXTENSION: Bringing modern technology to the study of fossils

Today, high-resolution x-ray CT scanners are used to create detailed 3D images of fossils. These industrial CT scanners are similar to medical CT scanners, but they produce more intense x-rays that penetrate the fossil. The resulting images enable palaeontologists to view far more detail of a fossil, both external and internal, and from different angles.

Figure 8.10 shows the external and internal detail of a fossil as revealed by high-resolution x-ray CT technology.

FIGURE 8.10 False-coloured three-dimensional digital reconstruction of the skull of *Rooneyia viejaensis*, a primate from the Eocene of West Texas, derived from high-resolution x-ray computed tomography data. (Scanned at the University of Texas High-Resolution X-ray CT Facility, and courtesy of the Texas Memorial Museum.)

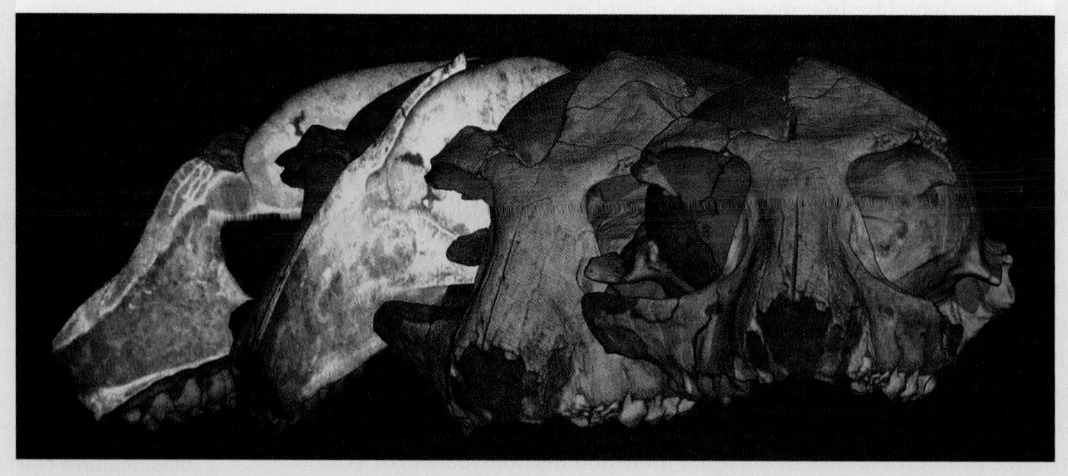

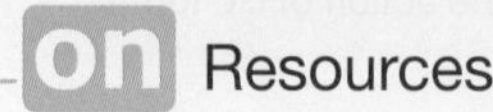

Resources

Weblinks Understanding *Rooneyia viejaensis*
Types of fossils

8.2.4 Formation of fossils

When organisms die, microbial action causes their decomposition; after a period, no trace remains. Very rarely, the remains of organisms are preserved long after death. This process of preservation is known as **fossilisation**.

Great excitement accompanies the discovery of a near complete skeleton, such as the 1.6-million-year-old skeleton of the Turkana boy, now called Nariokotome Boy (*Homo erectus*), found in 1984; the discovery of Penny the plesiosaur in Queensland in 1989 (see figure 8.11); and the 15 partial skeletons of *Homo naledi* found in the Rising Star Cave system in South Africa in 2013. Often, only smaller components of a fossil are found, such as the skull of *Australopithecus anamensis*, which was uncovered in Ethiopia in 2016.

FIGURE 8.11 Fossil bones of the *Kronosaurus queenslandicus*, known as 'Penny the plesiosaur', on display at the Kronosaurus Korner in Richmond, central Queensland. (Image courtesy of Judith Kinnear.)

The chance of fossilisation of a dead organism is very low. As a result, the preservation of a significant proportion of an entire skeleton of a large vertebrate in one location is an unusual event. Scavengers would be expected to consume tissues and, in so doing, disperse the remains.

The following conditions are required to allow the process of fossilisation and discovery to occur (see figure 8.12).

1. The most common condition is that the remains are rapidly buried in sediments after death. These sediments include:
   - silt deposits on a lake bed or on the flood plain of a river
   - sediments on the sea floor
   - ash from a volcanic eruption
   - windborne particles in a hot, dry desert.

   Some other conditions also allow fossilisation, such as:
   - burial in conditions that are alkaline, oxygen-depleted or even anoxic (oxygen-free), for example the mud at the bottom of a stagnant pool or a still body of water, a marsh, a peat bog or an asphalt seep
   - occurrence of death in a very cold environment where the remains are preserved by freezing.
2. After a dead organism is buried, continued deposits of sediments bury it more and more deeply. Gradually, the weight of the overlying sediments compresses the original sediment layer so that it becomes rock. Mud deposits become mudstone or shale, sand deposits become sandstone, and volcanic ash becomes tuff.
3. Over time the rock is eroded, uplifted through the movement of tectonic plates or excavated by palaeontologists so that the fossil is exposed.

fossilisation process of preserving parts of organisms that lived in the geological past

FIGURE 8.12 An example of how an organism may become fossilised

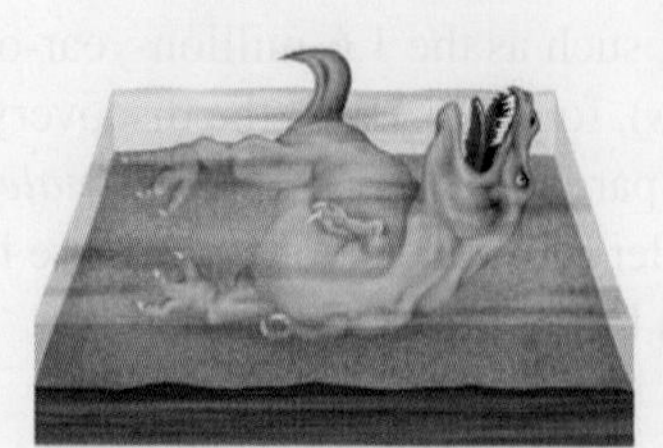

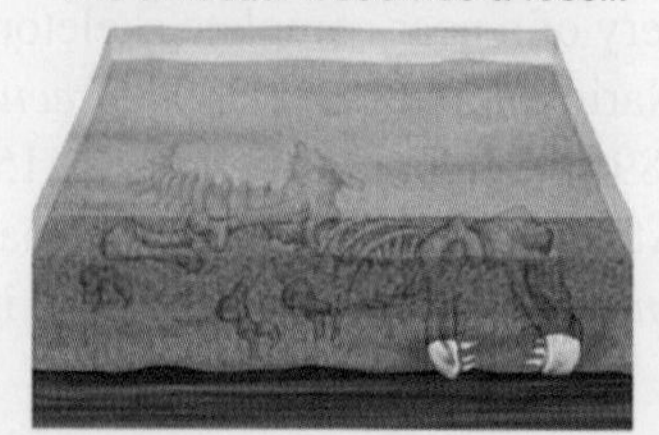

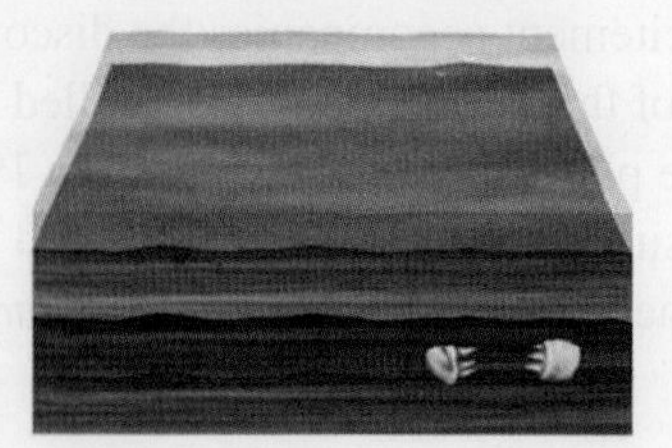

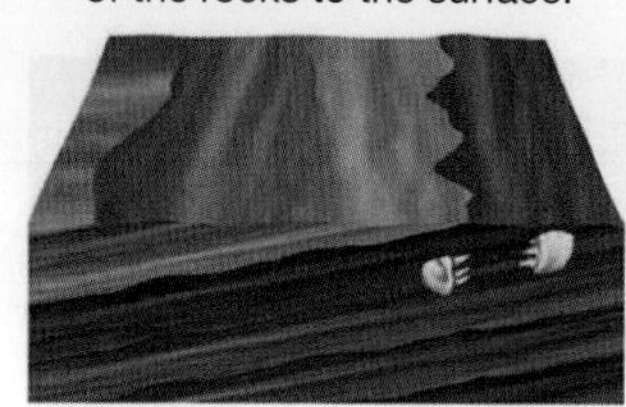

It is important to consider that the fossil record does not give us a complete picture of the evolution of species due to the specific conditions required for fossilisation to occur. Many species that may be extinct may have no fossil record.

elog-0078

INVESTIGATION 8.3 online only

Exploring the process of fossilisation

Aim

To model the process of fossilisation and understand the conditions required for fossilisation to occur

How much of an organism is fossilised?

When organisms are fossilised, most often only their hard tissues are preserved. In rare cases, however, the entire organism is preserved as a fossil. This can occur, for example, where insects become trapped in tree sap that becomes amber (see figure 8.13). Another situation is where organisms are buried and become frozen in the Arctic permafrost, such as has occurred with woolly mammoths. In these cases, DNA and other organic material can be extracted from these fossils.

FIGURE 8.13 A winged termite fly preserved in amber

In other cases, organisms, or parts of organisms are fossilised in an altered form. This occurs when:

- organisms are compressed under layers of sediment and their tissues are replaced by a thin carbon film showing their outlines (see figure 8.14a).
- the organic material of the dead organism is replaced by minerals, as can be seen, for example, in petrified wood and in opalised snail and clam shells (see figure 8.14b)
- fossils are found preserved within opal, such as the dinosaur *Fostoria dhimbangunmal*. These fossils were found in New South Wales, in Lightning Ridge.

Sometimes all or parts of dead organisms become covered by sediments that later form sandstone or mudstone. The organisms decay, leaving a cavity known as a mould. When the cavity within a mould is later filled by other material, such as fine-grained sediment, a three-dimensional model of the organism, known as a cast, is formed (see figure 8.14d).

FIGURE 8.14 **a.** Carbonised fossil showing the outline of part of the leaf of a fern (courtesy of Judith Kinnear) **b.** Opalised clam and snails. The material of the shells has been replaced by opal. **c.** Cast and mould of a trilobite fossil (image courtesy of Judith Kinnear)

TIP: When answering a question about evidence that fossils provide for evolution or evidence from fossils that allow for an understanding of advantageous traits, ensure you only discuss features that actually would be present and visible after fossilisation. For example, the presence of body hair would be difficult to discuss as there is no evidence of this in most fossils.

tlvd-1727

SAMPLE PROBLEM 1 Describing and identifying fossils

In the Alberta region of Canada, a fossil ammonite (*Placenticeras meeki*) was fossilised with opal during the Cretaceous period. The ammonite was a carnivorous mollusc and lived in fresh waterways.

a. **Identify which type of fossil formed the ammonite.** **(1 mark)**
b. **Outline how this fossil could have been formed.** **(3 marks)**

THINK	WRITE
a. 1. Identify what the question is asking you to do. The first part of the question is asking you to *identify*. In an identify question, you can list or name an answer with one word or a short sentence.	
2. Consider the all the types of fossils: physical — mineralised or impression, trace or biosignature. Always consider your options first, because then you can select from a short list. Now consider that the ammonite structure was replaced with opal — this is mineral.	Mineralised fossil (1 mark)
b. 1. Identify what the question is asking you to do. The second part of the question asks you to *outline*. In an outline question, you need to provide a step-by-step account of the event.	
2. Consider what the steps of fossilisation are. Now you need to apply these steps to the ammonite example. Some important information in the stem was that the ammonite lived in fresh waterways and that opal replaced the ammonite's structure. Use numbered steps to order your answer. As the question is worth three marks, you should have three clear steps.	1. An ammonite died and was buried in sediment in a body of water (lake, river, etc.) before any scavengers could eat it (1 mark). 2. Over time more and more layers of sediment covered the ammonite and compressed the ammonite (1 mark). 3. The ammonite's shell structure was replaced with opal minerals over time (1 mark).

CASE STUDY: The La Brea Tar Pits

An unusual case of fossilisation occurs at the La Brea Tar Pits in Los Angeles, California. The tar pits are in fact asphalt seeps. This site is one of the world's most famous late Pleistocene fossil sites. Here, fossils are recovered, not from buried sediments, but from asphalt seeps. The sticky asphalt preserves the bones of any organism that is trapped in it (see figure 8.15).

More than 600 different species have been found in these asphalt seeps, with ages ranging from 50 000 to 10 000 years ago (late Pleistocene). Organisms found at La Brea include amphibians, reptiles, birds and mammals. Among the many predatory mammalian species found are dire wolves (*Canis dirus*), sabre-toothed cats (*Smilodon fatalis*) and the Pleistocene coyote (*Canis latrans*). To date, more than 2000 individual sabre-toothed cats have been recovered, and more than 4000 dire wolves. Other fossils recovered include ground sloths (*Paramylodon harlani*) and even mammoths (*Mammuthus columbi*).

FIGURE 8.15 a. An excavator, Christina Lutz, at work for the Page Museum clearing bones recovered from the La Brea Tar Pits. (Image courtesy of La Brea Tar Pits and Page Museum, USA.) **b.** An artist's impression of two dire wolves and a sabre-tooth tiger in the La Brea Tar Pits

 Resources

 **eWorkbook** Worksheet 8.3 Fossilisation and types of fossils (ewbk-2915)

 **Weblink** La Brea Tar Pits

KEY IDEAS

- The Earth is approximately 4500 million years old and its time is organised by the geological time scale.
- Life on Earth has changed over geological time.
- Fossils provide evidence of the forms of past life on Earth.
- Fossils as evidence of past life can be categorised as physical fossils, trace fossils or biosignatures.
- Physical fossils can be mineralised or form impressions.
- Fossilisation of a dead organism can occur in special conditions if it is rapidly buried and its environment lacks oxygen.

8.2 Activities

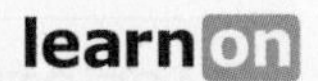

To answer questions online and to receive **immediate feedback** and **sample responses** for every question, go to your learnON title at **www.jacplus.com.au**. A **downloadable solutions** file is also available in the resources tab.

8.2 Quick quiz on	8.2 Exercise	8.2 Exam questions

8.2 Exercise

1. MC In examining the changes in life forms in Earth's geological history, it would be correct to state that
   A. animals on land appeared before prokaryotes.
   B. flowering plants first appeared after the first appearance of birds.
   C. mammals first appeared before amphibians.
   D. fish first appeared after reptiles.
2. Identify the geological ages in which the first prokaryotes, eukaryotes, insects and mammals evolved on Earth.
3. Compare the similarities and differences between
   a. mineralised fossils and fossil impressions
   b. a trace fossil and a physical fossil
   c. a cast and a mould.
4. Outline the steps required for the fossilisation of a dead organism.
5. Describe why an extremely small number of organisms become fossils.
6. Explain why marine organisms are more likely to be fossilised than terrestrial organisms.

8.2 Exam questions

Question 1 (1 mark)

Source: *VCAA 2018 Biology Exam, Section A, Q27*

MC The timeline below summarises the first appearance of some major groups of organisms in the evolution of life on Earth, as indicated by the fossil record. Three major groups are missing from the timeline and are represented by the letters P, Q and R.

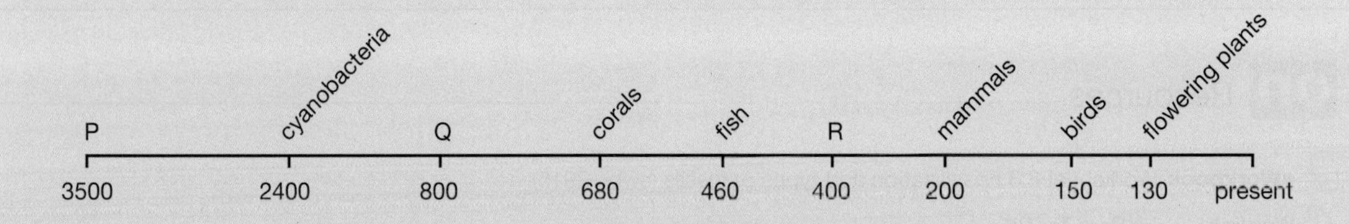

What are the correct groups of organisms labelled P, Q and R?

	P	Q	R
A.	bacteria	multicellular organisms	insects
B.	prokaryotes	non-flowering land plants	eukaryotes
C.	bacteria	reptiles	dinosaurs
D.	unicellular organisms	insects	non-flowering land plants

Question 2 (1 mark)

***Source:** Adapted from VCAA 2019 Biology Exam, Section A, Q28*

MC Consider the close-up image of a dinosaur footprint discovered by scientists.

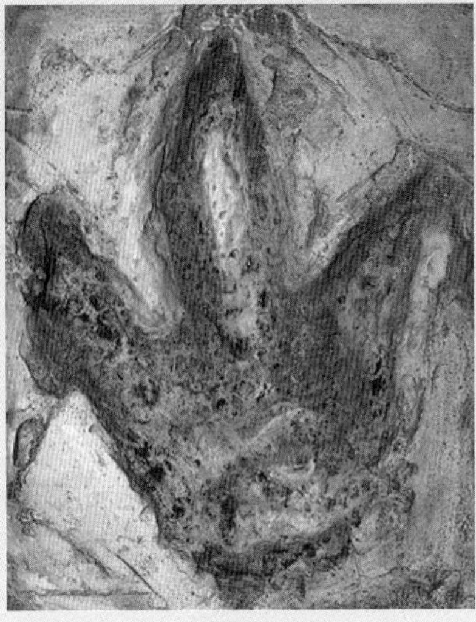

Source: Marcio Jose Bastos Silva/Shutterstock.com

This type of fossil is best described as

A. preserved remains.
B. a petrified fossil.
C. a trace fossil.
D. a cast.

Question 3 (2 marks)

***Source:** Adapted from VCAA 2016 Biology Exam, Section B, Q10a*

Over the past 20 years, a number of new hominin fossils have been discovered. *Homo erectus georgicus* was found near the banks of the Black Sea in Georgia and *Homo naledi* was found in a cave in South Africa.

Consider the conditions that may have led to the fossilisation of members of these species.

Identify one condition in the environment of each species that will have made fossilisation possible. The same answer cannot be used for both species.

a. *H. erectus georgicus* near the banks of the Black Sea **1 mark**
b. *H. naledi* in a cave in South Africa **1 mark**

Question 4 (2 marks)

***Source:** Adapted from VCAA 2011 Biology Exam 2, Section B, Q7d*

Two early hominins were fossilised in two different areas of Africa. One hominin was buried quickly during an avalanche while the other was buried slowly on a forest floor. The figure below represents these two different possibilities for fossilisation of the hominins.

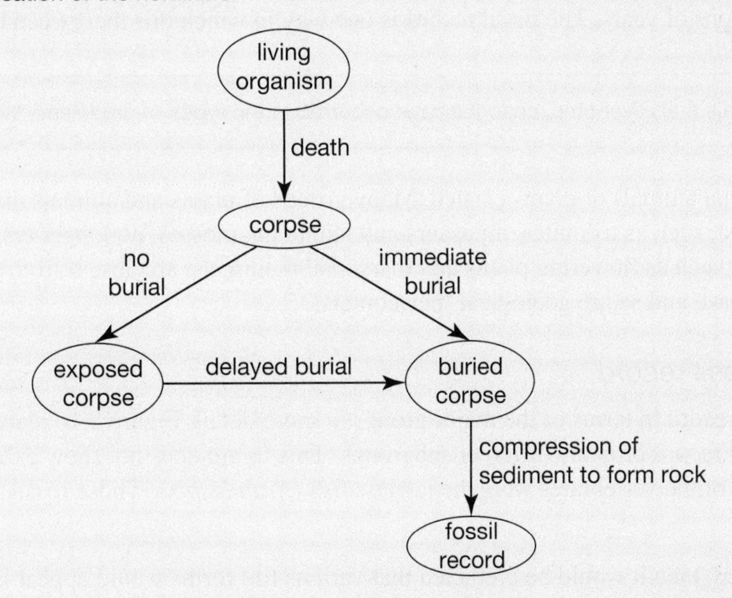

a. Outline one reason why the hominin in the forest would have a greater chance of being only partially fossilised. **1 mark**
b. State one factor not indicated on the diagram that would increase the chance of fossilisation of an organism. **1 mark**

Question 5 (1 mark)

Source: *VCAA 2008 Biology Exam 2, Section B, Q5c*

Three different kinds of plants, a cactus, euphorbia and milkweed, have similar adaptations for growing in desert environments.

They all have long, fleshy stems for water storage, protective spines and reduced leaves. Both the euphorbia and milkweed plants are believed to have evolved from leafy plants adapted to more temperate climates. They share a more recent ancestor than either one does to a cactus.

What is a possible explanation for there being so few cactus fossils?

More exam questions are available in your learnON title.

8.3 Evidence from fossils

KEY KNOWLEDGE

- Changes in species over geological time as evidenced from the fossil record: faunal (fossil) succession, index and transitional fossils, relative and absolute dating of fossils

8.3.1 Evidence from fossils

The Darwin–Wallace theory of evolution states that the various kinds of organisms living today evolved from ancestral kinds that lived in the geological past. According to this theory, the rich diversity of the animals, plants, fungi, bacteria and archaea living on Earth today did not appear simultaneously but is the product of evolution over thousands of millions of years. The fossil record is one way in which this theory can be tested.

The fossil record reveals that, over time, changes have occurred in the types of organisms living on this planet.

At one time, no terrestrial animals or plants existed. Many groups of plants and animals that were abundant at other times in the past, such as trilobites, dinosaurs and giant club mosses, no longer exist. Other kinds of organisms living today, such as flowering plants and marsupial mammals, are absent from the fossil record of the distant geological past, and so are geological 'newcomers'.

Changes in the fossil record

Let's look at the fossil record in terms of the major groups in more detail. Figure 8.16 identifies the time of first appearance in the fossil record of many different subgroups. This figure does not show extinctions. Note that the first representatives of the vertebrates were the jawless fish (*Arandaspis*), whose fossils first appeared in the Ordovician period.

If evolution has occurred, then it would be predicted that various life forms would appear in a predictable and consistent sequence in the fossil record, with ancestral species appearing in the fossil record before the species that evolved from them.

- Amphibians are believed to be ancestral to the reptiles and, consistent with this, the first amphibians appear in the fossil record before the first reptiles.
- The first reptiles appear in the fossil record before the first birds.

The fossil record provides evidence in support of the prediction that ancestral species will appear before the species that descend from them.

FIGURE 8.16 Simplified representation of the fossil record

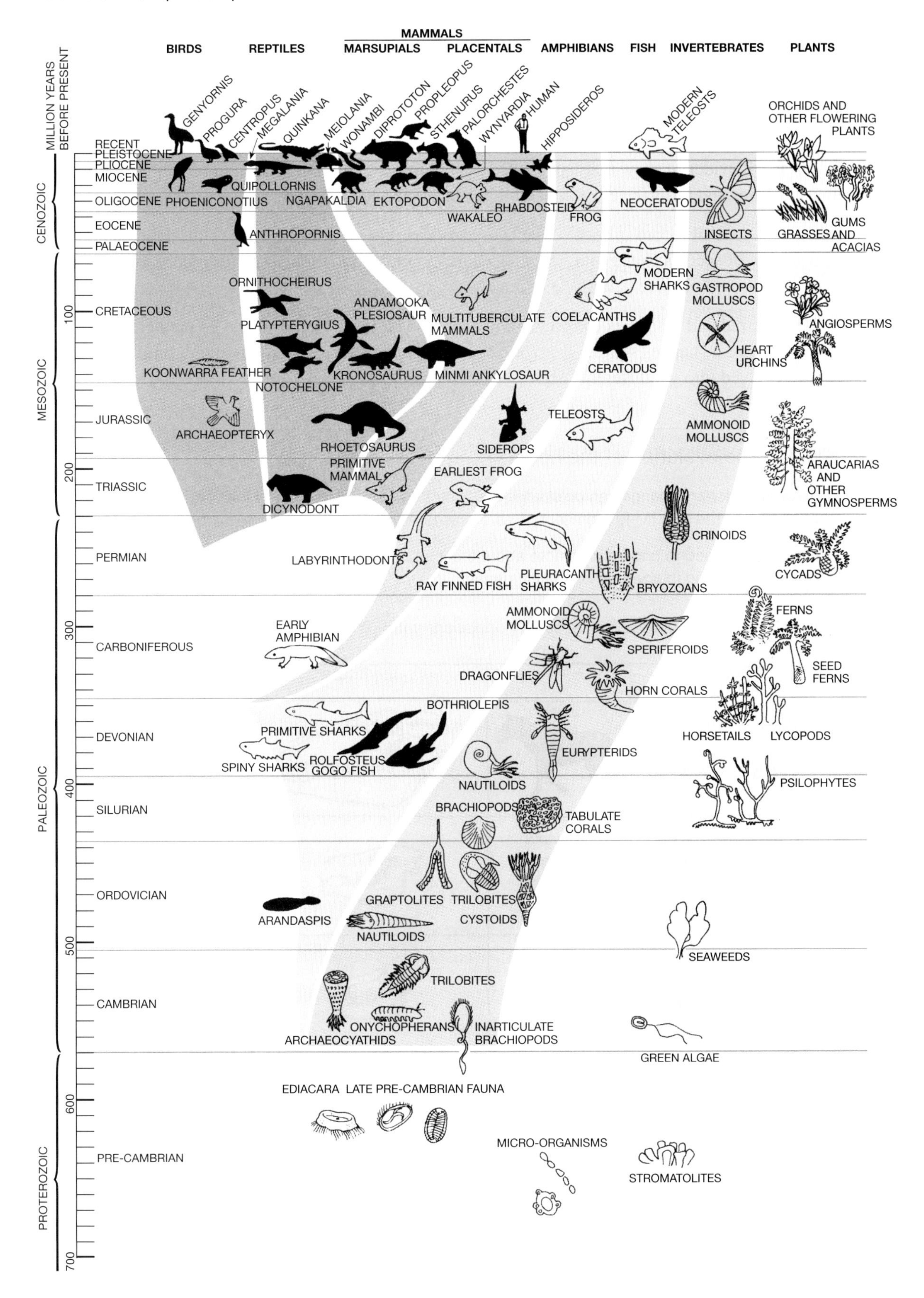

How can we interpret the fossil record?

Refer to the fossil record shown in figure 8.16. Some of the information that can be obtained is as follows:

- The time scale represents millions of years ago. The further down a fossil is located, the older the fossil is. Thus, assumptions can be made about the age of the organism.
- Looking at the birds (the darker blue), we can see that they first existed in the Permian period. One bird species, *Archaeopteryx*, was found to have existed in the Jurassic period, around 140 million years ago.
- Other analyses can be made in this fossil record. Ferns, for example, existed before flowering plants. Records of fossil evidence show that fish existed well before amphibians, which in turn appeared before mammals, reptiles and birds (refer back to table 8.1 in section 8.2.2).

The interpretation of a fossil record in such a way is a vital skill, and is fundamental to your understanding of the initial appearance of species over geological time. You do not need to memorise this record, but you should be able to analyse data and observe patterns and trends.

If evolution has occurred and if species can change over geological time, then it would be predicted that the fossil record would reveal changes starting from an ancestral species that evolved into one or more new species, and these in turn evolved into yet different species.

CASE STUDY: The evolution of the horse

One example of evolutionary change can be seen in the evolution of the horse family (see figure 8.17). Members of the genus *Equus* are the only living representatives of this family.

Fossil evidence can be used to understand more about the evolution of horses. For example, it provides evidence that the horse has increased in size with each new genus.

FIGURE 8.17 Natural selection acted on some populations within the horse family over millions of years.

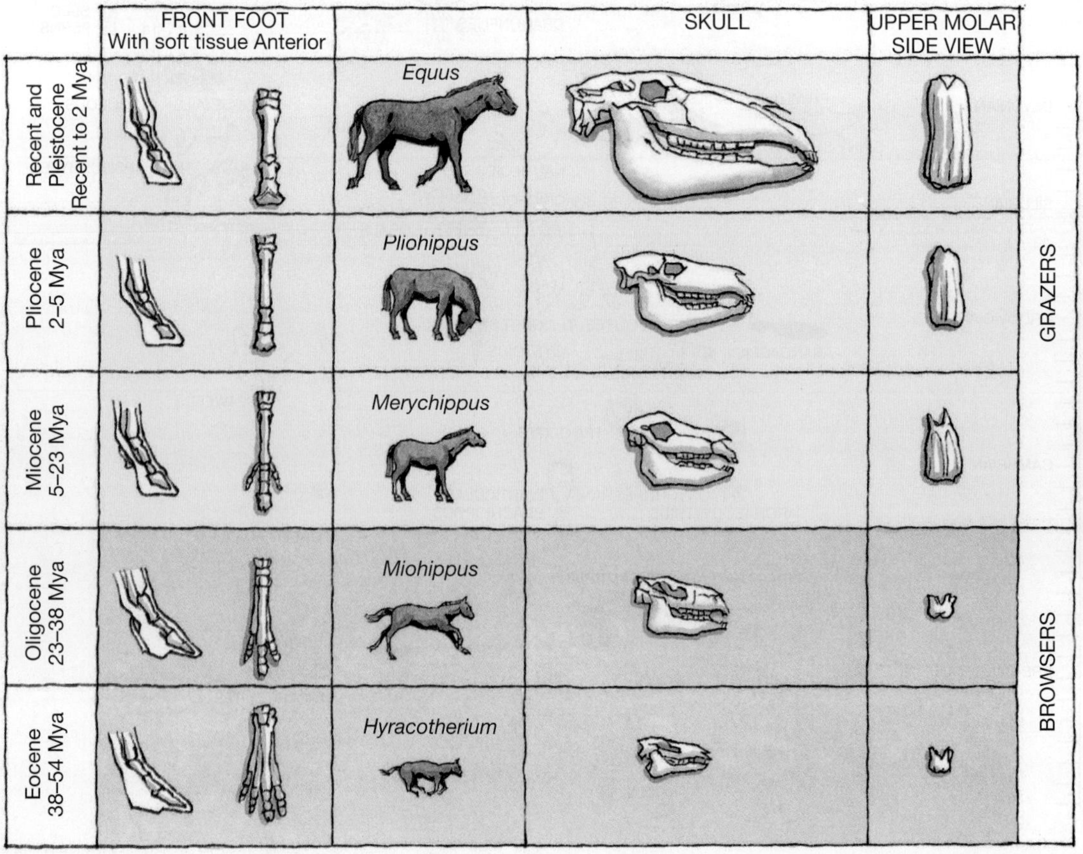

	FRONT FOOT With soft tissue Anterior		SKULL	UPPER MOLAR SIDE VIEW	
Recent and Pleistocene Recent to 2 Mya		*Equus*			GRAZERS
Pliocene 2–5 Mya		*Pliohippus*			GRAZERS
Miocene 5–23 Mya		*Merychippus*			GRAZERS
Oligocene 23–38 Mya		*Miohippus*			BROWSERS
Eocene 38–54 Mya		*Hyracotherium*			BROWSERS

To access more information about the evolution of horses, please download the digital document.

Resources

Digital document Case study: The evolution of horses (doc-36109)

Interactivity The evolution of the horse (int-5878)

Weblink Evolution of horses

8.3.2 Faunal succession

One method that is vital in understanding fossil records from different locations throughout the world is **faunal succession**. This method allows for the relative dates of the appearance of species to be determined.

The principle of faunal succession, otherwise known as fossil succession, is that sedimentary rock strata containing fossilised fauna and flora are arranged vertically in a specific manner over a large section of the Earth.

Because species change over time, the fossils that they leave are distinctive and can be used to identify rocks of the same age, even if they are in different elevations and in different locations.

faunal succession the principle that fossilised fauna and flora in sedimentary rock strata are arranged vertically in a specific order

FIGURE 8.18 Distribution of major groups over geological time

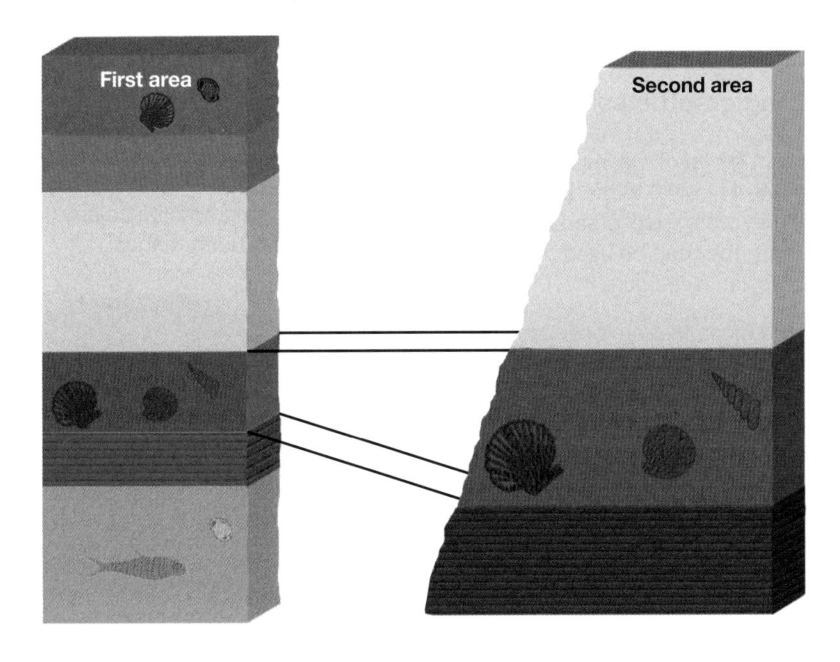

Different kinds of organisms do not occur randomly in the fossil record but are found only in rocks of particular ages and appear in a consistent order (as shown in figure 8.18). This is known as the principle of faunal succession. For example:

- the first prokaryotes (microbes) appear in the fossil record long before the first eukaryotes
- the first single-celled organisms appear before the first multicellular organisms
- the first simple soft-bodied animals appear before the first complex animals with exoskeletons or with shells
- the first amphibians appear in the fossil record before the first reptiles
- the first ferns appear in the fossil record before the first flowering plants occur, and so on.

Sequences such as these can be seen in the fossil record worldwide. As of the present day, no exceptions have been found.

FIGURE 8.19 Distribution of major groups over geological time

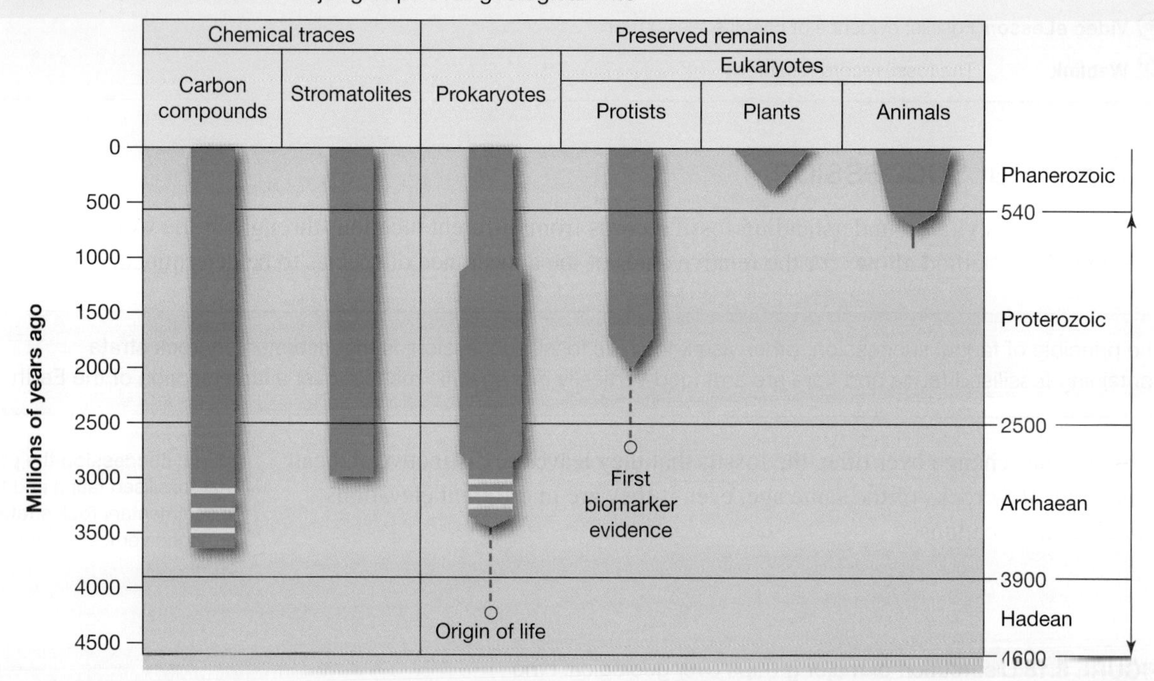

CASE STUDY: Using graptolites for faunal succession

Graptolites are an extinct group of marine animals that lived from the Ordovician (485–444 Mya) to the Silurian period (see figure 8.20). Because they were widespread and had distinctive changes in their anatomy, they can be used to date rocks and other species in faunal succession (see figure 8.21).

The fossilised fauna and fauna that are best used to determine faunal succession are index fossils. Index fossils are discussed in section 8.3.3.

FIGURE 8.20 *Tetragraptus fruticosus*, a graptolite colony (Image courtesy of Judith Kinnear)

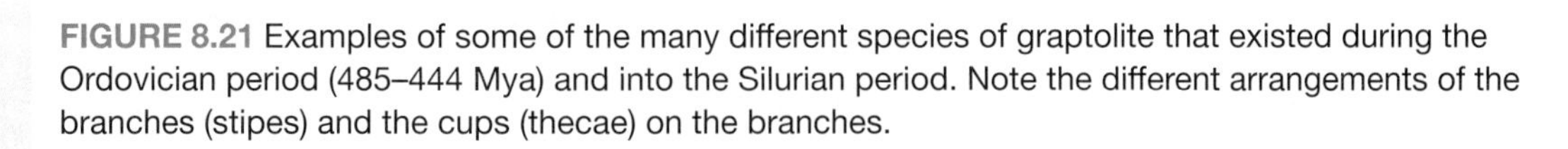

FIGURE 8.21 Examples of some of the many different species of graptolite that existed during the Ordovician period (485–444 Mya) and into the Silurian period. Note the different arrangements of the branches (stipes) and the cups (thecae) on the branches.

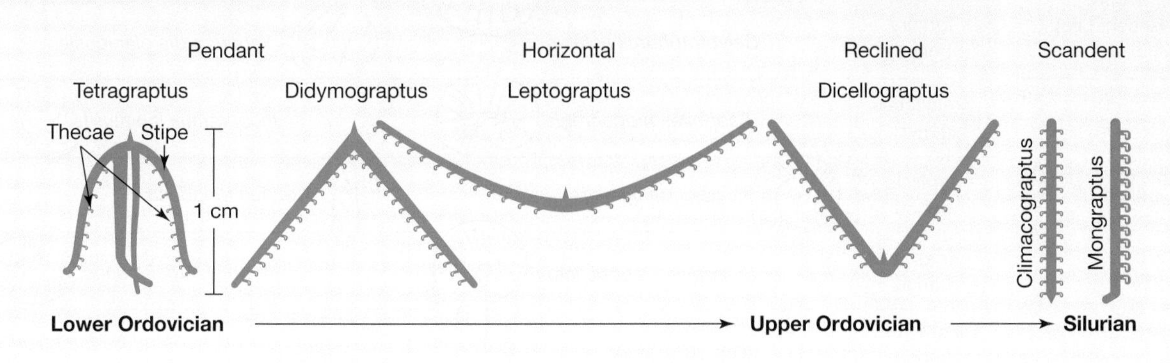

8.3.3 Index and transitional fossils

Index fossils

Index fossils are used to define a period of geological time. These are geologically short-lived species that have a limited occurrence in the fossil record and so are found only in a restricted depth of sedimentary rock strata, and are widely distributed.

The presence of index fossils in rock strata in widely separated regions of the world can identify these rocks as having the same age.

Before radiometric dating became available, correlating the fossils present in rocks was the best method available to date rock layers relative to one another.

Graptolites (examined in the earlier case study box) are an important group of index fossils for the following reasons:

1. Several hundred species of graptolites exist, each readily distinguishable by the number of branches (stipes) and the arrangement of the cups (thecae) on the branches (see figure 8.21).
2. They occur in the fossil record from the Cambrian to the Carboniferous periods.
3. Each different species existed for a limited time only in the fossil record; that is, graptolites underwent rapid evolution.
4. They are widely distributed globally because graptolites formed floating colonies that drifted on the ocean surface.

Other index fossils that are commonly used are shown in figure 8.22. All of these fossils were abundant during their lifetimes but were not able to survive changing environmental conditions. The majority of index fossils are marine specimens rather than terrestrial ones. This is because marine organisms are universal to the oceans and have a higher chance of being covered in the ever-moving sediment of the ocean floor.

Rocks of the same age are identified by the name of a geological period, such as Devonian or Jurassic. Likewise, fossils can be dated as the same age, or as younger or as older than a particular index fossil.

index fossils fossils of geologically short-lived species that are widely distributed but are found in a restricted depth of rock strata

FIGURE 8.22 A display of the Earth's most common index fossils and the time periods they belong to

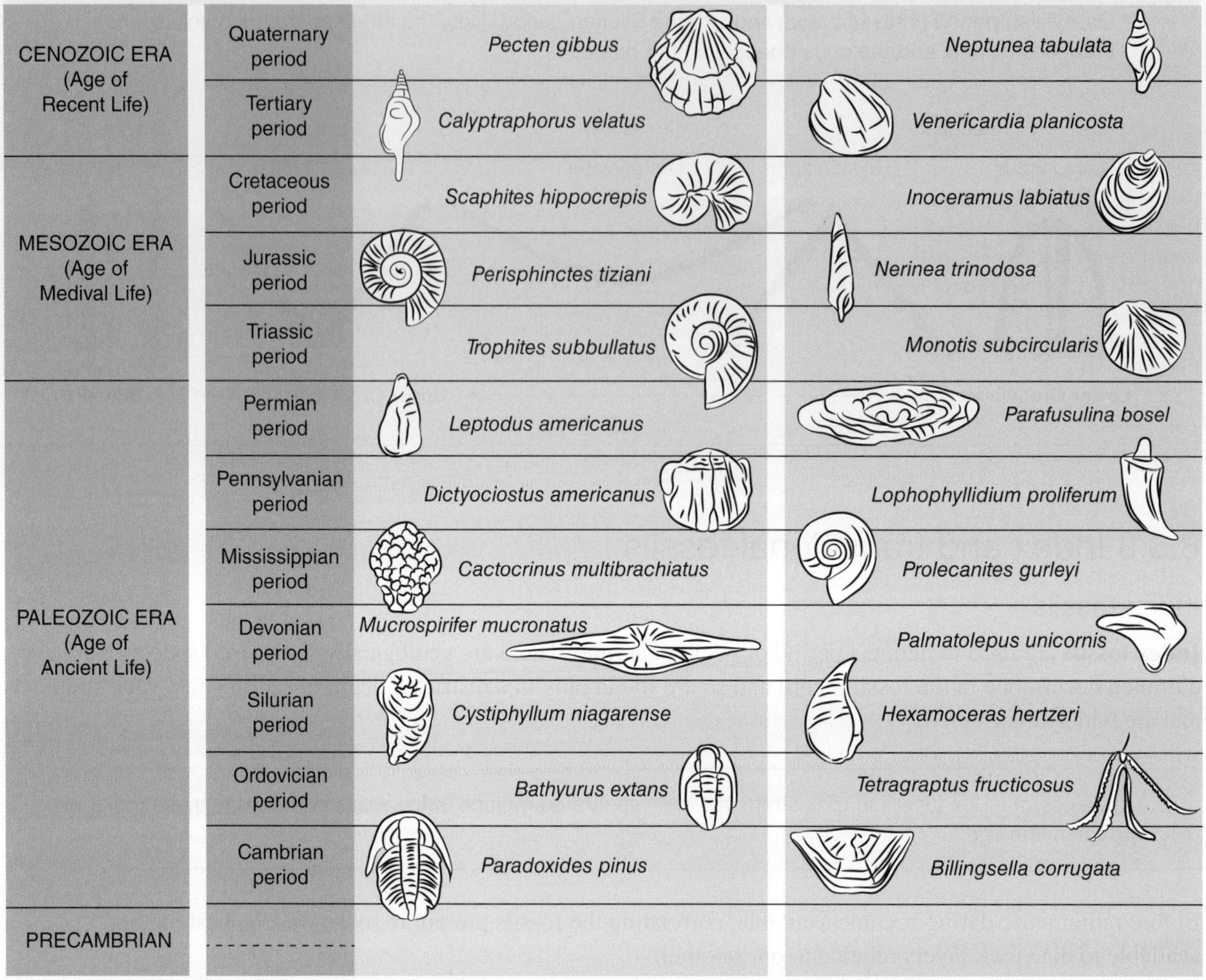

Era	Period	Index fossil	Index fossil
CENOZOIC ERA (Age of Recent Life)	Quaternary period	*Pecten gibbus*	*Neptunea tabulata*
	Tertiary period	*Calyptraphorus velatus*	*Venericardia planicosta*
MESOZOIC ERA (Age of Medival Life)	Cretaceous period	*Scaphites hippocrepis*	*Inoceramus labiatus*
	Jurassic period	*Perisphinctes tiziani*	*Nerinea trinodosa*
	Triassic period	*Trophites subbullatus*	*Monotis subcircularis*
PALEOZOIC ERA (Age of Ancient Life)	Permian period	*Leptodus americanus*	*Parafusulina bosel*
	Pennsylvanian period	*Dictyociostus americanus*	*Lophophyllidium proliferum*
	Mississippian period	*Cactocrinus multibrachiatus*	*Prolecanites gurleyi*
	Devonian period	*Mucrospirifer mucronatus*	*Palmatolepus unicornis*
	Silurian period	*Cystiphyllum niagarense*	*Hexamoceras hertzeri*
	Ordovician period	*Bathyurus extans*	*Tetragraptus fructicosus*
	Cambrian period	*Paradoxides pinus*	*Billingsella corrugata*
PRECAMBRIAN			

Transitional fossils

If new species arise by evolution from ancestral species, it would be predicted that the fossil record should reveal some fossils that are intermediate between forms. They are referred to as **transitional fossils**.

Transitional fossils show traits between an ancestral group and its descendants, acting as evidence of an intermediate form.

There are many examples of transition fossils or intermediate forms that relate an ancestral group with its descendants. These include:

- primitive amphibians that show a transitional stage between the simple pelvic (hip) girdle present in fish and the complex pelvic girdle in later more advanced amphibians, such as *Eryops megacephalus* (shown in figure 8.23).
- fossil mammal-like reptiles that show a transitional stage between reptiles with simple conical teeth and mammals with teeth differentiated into incisors, canines, pre-molars and molars.

transitional fossil the fossilised remains of a life form that exhibits traits common to both an ancestral group and its derived descendant group(s)

FIGURE 8.23 Skeleton of the prehistoric amphibian *Eryops megacephalus*. This amphibian shows several evolutionary adaptations marking the transition from life in water to life on land.

CASE STUDY: *Archaeopteryx lithographica*

Birds are believed to have evolved from a group of reptiles. If so, the fossil record might be expected to reveal organisms that show both the features of their reptilian ancestors and new features characteristic of birds.

The first unequivocal evidence of birds in the fossil record occurs in the late Jurassic period, about 150 Mya. A fossil skeleton of the earliest bird, *Archaeopteryx lithographica*, was found in a limestone quarry in Bavaria, Germany, in 1861. The fine-grained limestone preserved the faint impressions of feathers. In the absence of feather impressions, these organisms would have been classified as reptiles.

Like modern birds, *Archaeopteryx* showed the characteristic presence of feathers and a wishbone (furcula). However, it also showed some reptilian features now lost in modern birds: *Archaeopteryx* had teeth in its beak, claws on its wings, unfused (free) bones in its 'hand' and a long jointed bony tail (see figures 8.24 and 8.25). They were also not capable of the magnificent flight of modern birds.

FIGURE 8.24 a. *Archaeopteryx* and b. a modern flying bird. Features in black show distinctive reptilian features in a. and bird features in b.

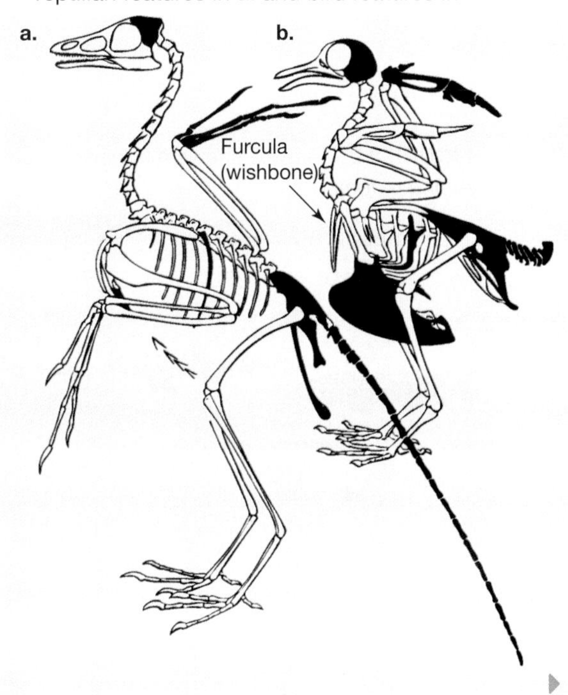

FIGURE 8.25 Fossil of *Archaeopteryx*, a flying reptile with feathers

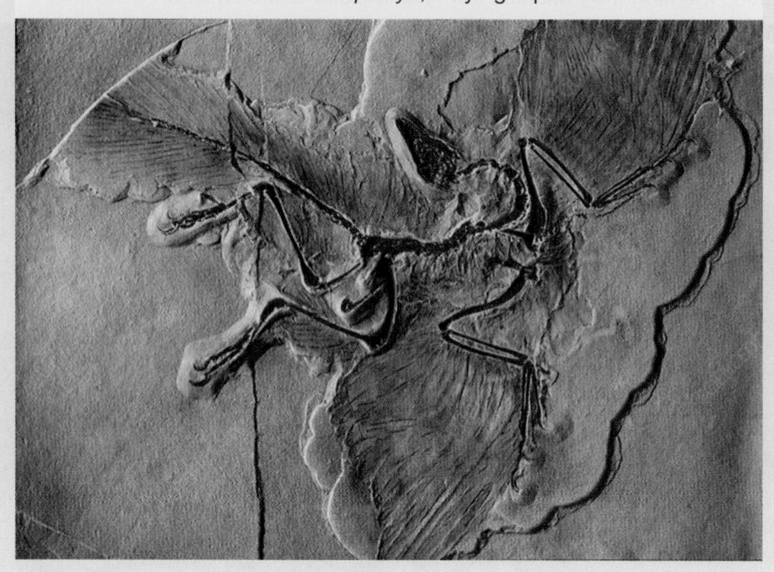

CASE STUDY: *Tiktaalik*

Tiktaalik roseae is another example of a transitional fossil. This creature lived 375 million years ago. It represents the transition of vertebrate life from aquatic environments to terrestrial environments (see figure 8.26). *Tiktaalik* has fish characteristics such as scales, gills and fins, but it also has some terrestrial features. The fins have the same bone structure as terrestrial vertebrates, which would have enabled *Tiktaalik* to lift its weight on land. The fossils also indicate a strong ribcage and a rotating neck, which help with walking on four limbs.

FIGURE 8.26 Computer artwork of *Tiktaalik roseae*

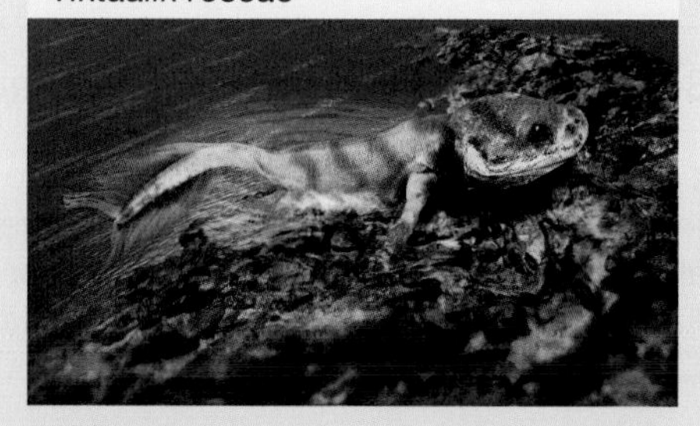

FIGURE 8.27 *Tiktaalik* shows a transition from aquatic species *Panderichthys* to the amphibian *Acanthostega*.

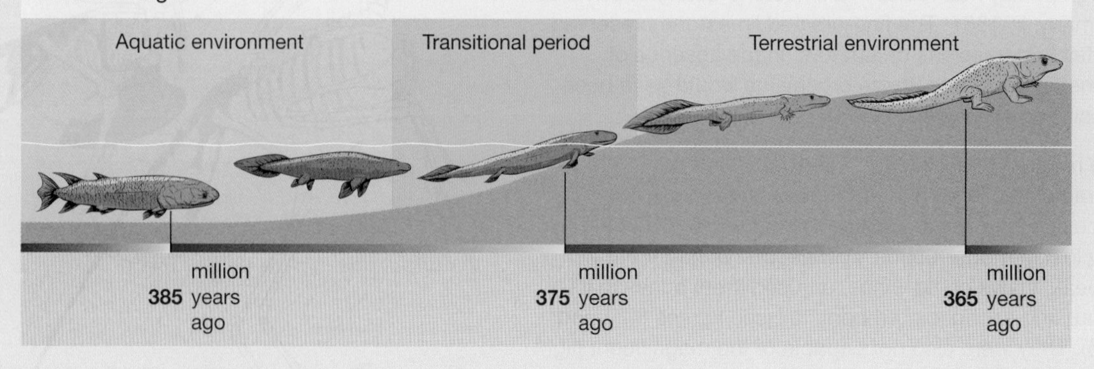

tlvd-1728

SAMPLE PROBLEM 2 Analysing fossils

The fossil shown is of *Archaeopteryx*. It has bird-like feathers but it also has the long bony tail of a reptile.

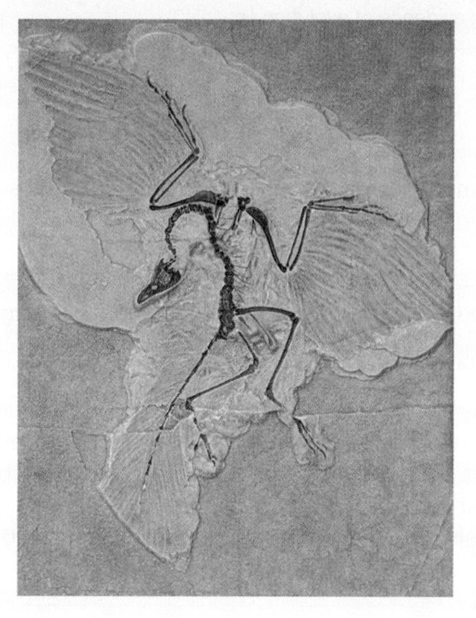

a. **State which type of fossil *Archaeopteryx* is.** **(1 mark)**
b. **Explain how this fossil provides evidence for evolution.** **(2 marks)**

THINK	WRITE
a. 1. Identify what the question is asking you to do. In a *state* question, you can list answers with one word or a short sentence.	
2. Consider that the fossil is described as having both bird and reptile features — a mix of two different groups.	Transitional fossil (1 mark)
b. 1. Identify what the question is asking you to do. In an *explain* question, you need to account for the reason of why or how something occurs using your scientific knowledge. As this question is for two marks, you should supply two reasons.	
2. Consider that evolution is the biological change in species over time. What would you expect to see if organisms were changing over time? You would expect to organisms with 'in between' traits of different species.	Transitional fossils provide evidence for evolution because they show fossils that have characteristics of one group of organisms (1 mark) — a reptile's long tail — and of another — a bird's feathers (1 mark). This provides evidence for how groups of species evolved their particular characteristics.

on Resources

eWorkbook Worksheet 8.4 Transitional and index fossils (ewbk-2916)
Video eLesson Transition fossils (eles-4370)
Interactivity Transition fossil: evidence of evolution (int-0240)

8.3.4 Dating fossils

To determine how old a fossil is and where it fits into the geological time scale, the rock or the fossil needs to be dated.

The age of anything can be stated in two ways — relative age or absolute age.

Relative age provides a comparative age, whereas absolute age provides a more precise numerical age.

For example:
- the statement 'Kym is older than Tran, who is younger than Shane' gives the relative ages of these people but does *not* give their actual ages. Relative age allows us to place the people in order of age: Kym, then Shane, then Tran, from oldest to youngest.
- the statement 'Kym is 34 years old, Shane is 18 years old and Tran is 16 years old' gives the absolute ages of these people.

Likewise, the ages of fossils or of geological structures, such as a deposit of mudstone or a layer of solidified volcanic ash (tuff), can be given in either relative or absolute terms.

Relative dating of fossils

Two methods that may be used to identify the relative ages of rocks and fossils are:
- the **stratigraphic method** (or the law of superposition) and the use of the principle of faunal succession
- the use of index fossils

stratigraphic method the method of obtaining the relative age of an object by its position within a given sequence of rock strata

Fossils are found in sedimentary rocks, such as mudstone, sandstone, shale and limestone. The stratigraphic method of dating sedimentary rocks gives the relative ages of the rock strata.

Relative dating and the law of superposition simply state that, for rock layers or strata, the oldest stratum is at the bottom and progressively younger layers lie above it, as shown in figure 8.28.

This method can only be used to identify the relative ages of rock strata in the same sequence. Based on position alone, the relative ages of strata from different sequences in widely separated regions of the world cannot be identified, due to differences in the formations of rock strata.

FIGURE 8.28 Fossils found in layer A will be older than fossils found in layer D.

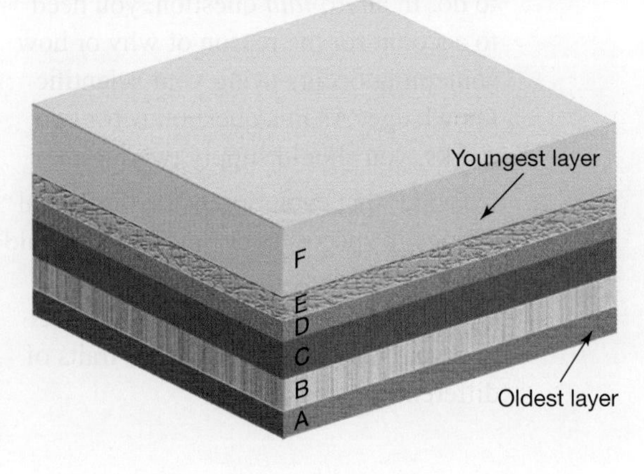

on Resources

Video eLesson Relative age of rocks (eles-4371)
Interactivity Relative age of rocks (int-0238)

Absolute dating of fossils

During the twentieth century, techniques for identifying the absolute age of rocks were developed. Absolute age give a more precise age than relative ages.

The most important method of estimating the absolute ages of rocks is the **radiometric dating technique**, which is based on the decay of certain radioactive elements. The radioactive elements concerned are those present in minerals in igneous rocks, and each element decays at a known rate.

The principle of radiometric dating depends on the fact that various elements exist in two or more forms known as **isotopes**, some of which are radioactive. This decay is shown in figure 8.29 and 8.30.

1. The radioactive isotopes, which can be called 'parents', spontaneously decay or break down over time to form stable 'daughter' products.
2. The rate of the decay is specific for each radioactive isotope. This is usually measured in terms of **half-life**, the time taken for half the original radioactive isotope to decay.

radiometric dating technique a technique for obtaining an absolute age that depends on the known rate of decay of a radioactive parent isotope to a stable daughter product

isotopes the different forms of an element which differ in the number of neutrons

half-life the time taken for an amount of radioactive isotope to decay to half of its initial amount

By using the half-life of a parent isotope and measuring the proportion of a daughter isotope compared to the radioactive parent isotope, the approximate age of a rock or fossil can be determined.

int-5876

FIGURE 8.29 After every half-life, 50% of the parent element has decayed into a daughter element. Note that this image should be read from left to right.

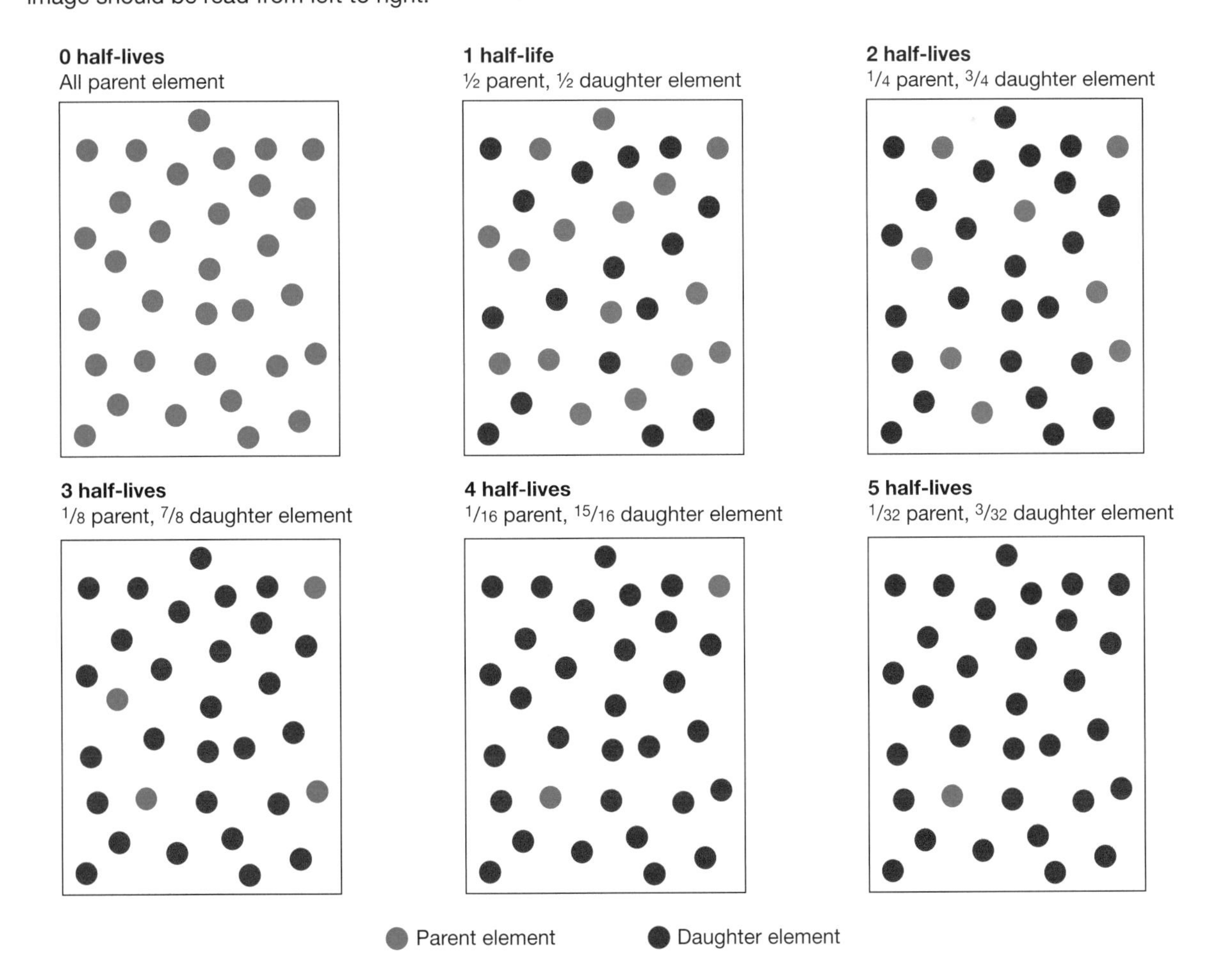

FIGURE 8.30 Decay of a radioactive parent element over a time period of 5 half-lives

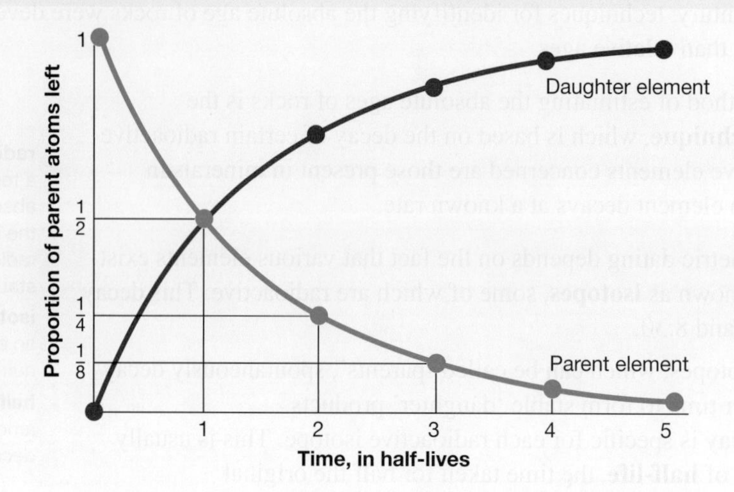

Table 8.2 shows some of the radioactive isotopes used for absolute dating and their stable daughter products. Each technique can be used over a particular age range that depends on the half-life of the parent radioactive isotope. For example, **carbon dating** is used for organic remains but is only useful for relatively young fossils or remains. Using the decay of radioactive elements, the age of the oldest rocks yet found on Earth has been estimated at 4300 million years. These rocks come from ancient bedrock in northern Canada.

carbon dating a form of absolute dating in which the amount of stable nitrogen-14 is compared to that of the radioactive parent isotope carbon-14

TABLE 8.2 Some radioactive parent isotopes and the daughter products of decay. The longer the half-life of a radioactive isotope, the older the material that can be dated using a particular radiometric method.

Radioactive parent isotope	Daughter product	Half-life (years)	Comments
Carbon-14	Nitrogen-14	5 730	Used for dating organic (carbon-based) remains; useful for ages up to about 60 000 years old. Often the amount of carbon-14 is compared to carbon-12, which does not decay.
Uranium-235	Lead-207	710 000 000	Used for dating igneous rocks containing uranium-based minerals; useful in the range from 10 million years and older
Potassium-40	Argon-40	1 300 000 000	Used for dating igneous rocks containing K-bearing minerals, such as feldspars; useful in the range from 0.5 million years and older
Rubidium-87	Strontium-87	47 000 000 000	Can be used to date the most ancient igneous rocks on Earth

TIP: Ensure you carefully check the age of a fossil (or other evidence) when answering questions about which radiometric dating technique would be used. A common exam mistake is stating that carbon dating would be used for absolute dating when a fossil is over 1 million years old. This is not possible to determine with carbon dating.

Radiometric dating cannot be applied to sedimentary rocks derived from the erosion of pre-existing rocks, because the minerals that they contain were formed before the sedimentary rocks themselves. In this situation, layers on either side of the sedimentary rock can be examined instead.

This can occur, for example, when fossils are embedded in a sedimentary rock layer that is located between layers of solidified volcanic ash (tuff). The age of a fossil can be inferred to lie within the range defined by the ages of the tuff layers above and below the fossil (see figure 8.31).

A particular radiometric technique can be used to date only material that contains a specific radioactive parent isotope. For example, carbon-14 dating *cannot* be used to estimate the age of zircon crystals ($ZrSiO_4$) in a rock, but it can be used to date linen fabric from an ancient Egyptian grave.

FIGURE 8.31 Fossils in a sedimentary layer located between layers of solidified volcanic ash. The age of the fossils can be inferred to lie in the range of 545 to 520 million years ago.

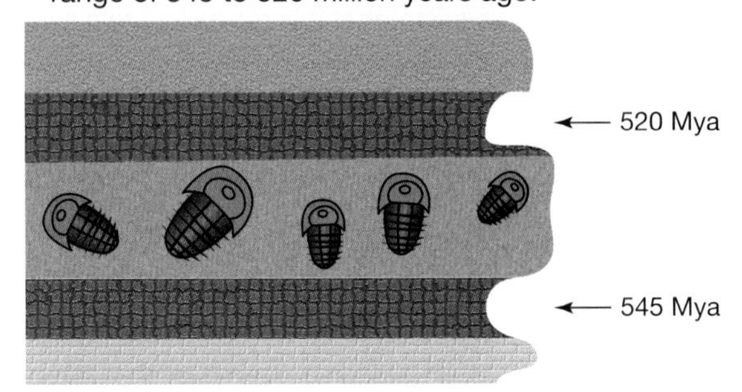

CASE STUDY: Exploring the different types of radiometric dating

Potassium–argon (K–Ar) dating

Absolute dating of rocks using K–Ar dating depends on the decay of the radioactive potassium-40 isotope. This isotope is present in minerals found in rocks that form when molten magma cools and solidifies, either at depth below the Earth's surface or on the surface. Rocks formed in this way are called igneous rocks (from Latin *ignis* = fire). Examples of these rocks are granite, which solidifies at depth below the Earth's surface, and basalt, which solidifies on the Earth's surface.

FIGURE 8.32 a. Granite and b. basalt are igneous rocks.

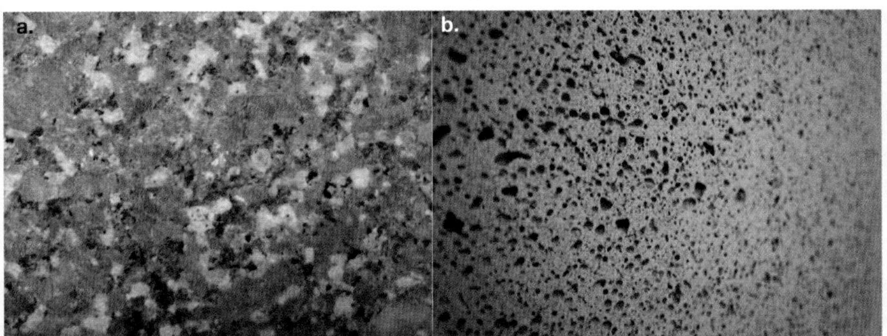

Potassium (K) is present in some of the minerals found in many igneous rocks. One common mineral is orthoclase, a potassium aluminium silicate ($KAlSi_3O_8$) that is found in granite (see figure 8.32a).

- A small fixed proportion (approximately 0.012%) of potassium is in the form of the radioactive isotope potassium-40; most is in the stable form of potassium-39.
- Potassium-40 decays at a known constant rate to form the inert gas argon-40, which accumulates within potassium-containing mineral crystals in rock.
- The older a rock, the longer the time available for potassium-40 to decay and for argon-40 to accumulate.
- Potassium-39 does not decay, so it provides a measure of the original amount of potassium-40 present.

Age estimates are done under carefully controlled conditions in which the sample of rock under test is heated in a vacuum to release argon-40 atoms, which are measured using a highly sensitive instrument. From this result and measurements of the amount of potassium-39, the absolute age of the rock can be calculated.

Carbon-14 dating

Carbon dating was developed in 1974 by Willard Libby.

All living organisms are built of carbon-containing organic matter, such as proteins (for example, the keratin of hair, nails, hooves and claws, and the collagen of bones) and structural carbohydrates (for example, the cellulose of plant tissues). When an organism is alive, the carbon in its organic matter is a mixture of two isotopes:

- the stable isotope carbon-12 (C-12)
- the radioactive isotope carbon-14 (C-14), which decays to nitrogen-14 (N-14).

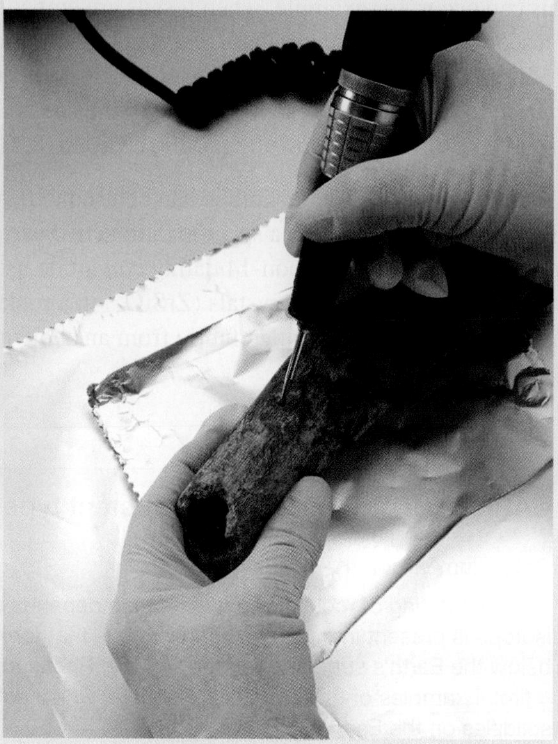

FIGURE 8.33 Sample being removed from bone for carbon dating

In life, the proportion of these two carbon isotopes is constant and matches that in the carbon dioxide in the atmosphere. This proportion remains constant during the life of an organism.

After the death of an organism, however, the proportion of carbon-14 decreases as this isotope decays and is not replaced. As the time after death increases, the ratio of C-14 to C-12 becomes progressively smaller. After one half-life of 5730 years, half the original amount of C-14 present in the organic material at the time of death will have disappeared, and so on for each successive half-life period. By measuring the ratio of C-14 to C-12 in a sample of organic material, an estimate of the time since the death of the organism that produced this material can be obtained. Alternatively, the proportion of C-14 to N-14 can be compared to determine the age of the sample. This is often more difficult, as most organisms have nitrogen within their bodies before their deaths, which can skew results.

C-14 dating can not only date fossils; it can also be used to date artefacts, such as the wooden handle of a tool, or a localised collection of ash and fragments of burnt wood found in a cave. C-14 dating can estimate the absolute age of this material provided the wood sample is not older than about 60 000 years. After this point, the amount of carbon-14 is too small to measure.

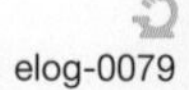
elog-0079

INVESTIGATION 8.4

online only

Modelling radioactive decay

Aim

To model radioactive decay and show how this is used in absolute dating to determine the age of fossils

tlvd-1729

SAMPLE PROBLEM 3 Using absolute age to determine the age of fossils

In Lake Mungo (a town in New South Wales), fossil remains of early *Homo sapiens* have been discovered.
Two such fossils, Mungo Woman (LM1 in red) and Mungo Man (LM3 in blue) were found at this site, as shown in the provided figure. Mungo Man is the oldest *Homo sapiens* remains in Australia.
The absolute age of LM1 was determined using absolute dating. In this case, the fossil was found to be around 23 000 years old.

a. **What type of absolute dating would have been used to determine the age of LM1? (1 mark)**
b. **How many half-lives of the parent isotope have occurred to determine the age of the fossil to be 23 000 years old? (2 marks)**
c. **What proportion of parent isotopes to the initial amount would be expected after this time? (1 mark)**

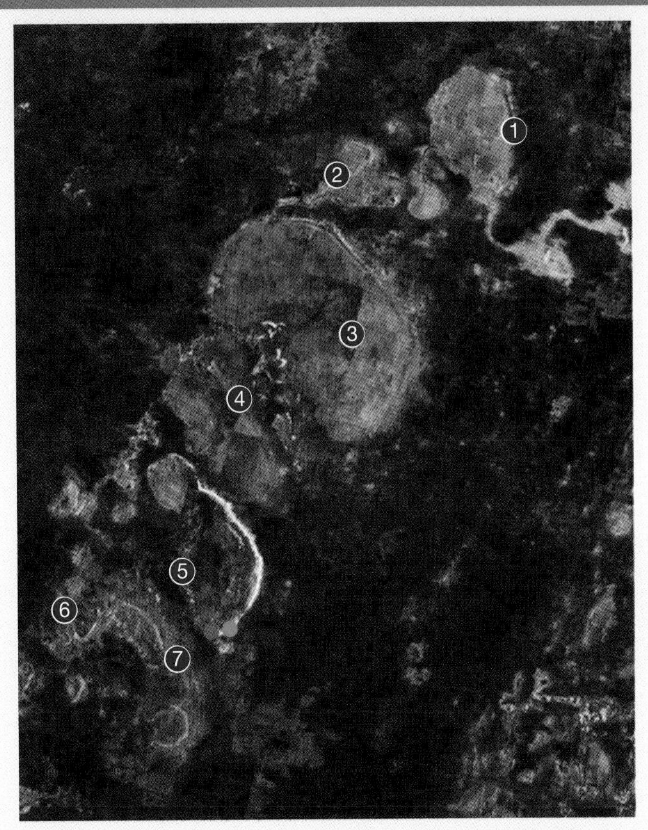

LM3 did not initially have its remains directly examined. Instead, relative dating was used, using the previously aged fossil of LM1.

d. **Describe the process of relative dating. (1 mark)**
e. **Where would you expect LM3 to be compared to LM1 in regards to layers in the Earth? Justify your response. (2 marks)**

THINK	WRITE
a. 1. You are being asked to *state* the type of absolute dating, so you should consider carbon, uranium–lead, potassium–argon and rubidium–strontium dating.	
2. Carefully read the information provided. The information states that the fossil was found to be *23 000 years old*. It is too young for uranium–lead, potassium–argon or rubidium–strontium dating.	Carbon dating (1 mark)
b. 1. This question requires you to complete a calculation. The half-life of carbon is 5730 years	Half-life of carbon = 5730 years (1 mark)
2. The fossil was determined to be 23 000 years old. Calculate the number of half-lives. **TIP:** You will not have a calculator in your Biology exam, so you can approximate to the nearest whole number.	$23\,000 \div 5730 \approx 4$ Therefore, approximately 4 half-lives of the carbon isotope have occurred (1 mark).

c. 1. This question requires you to *determine* the proportions. You should recall a half-life is the time for half of the initial parent isotope to decay.

2. You would expect the parent element to be:
- $\frac{1}{2}$ of its initial amount after 1 half-life
- $\frac{1}{4}$ of its initial amount after 2 half-lives
- $\frac{1}{8}$ of its initial amount after 3 half-lives
- $\frac{1}{16}$ of its initial amount after 4 half-lives

You would expect $\frac{1}{16}$ of the initial amount of the carbon-14 (the parent isotopes) to remain (1 mark).

d. 1. Identify what the question is asking you to do. You are being asked to *describe* relative dating. It is only a one-mark question, so you only need one key point.

Relative dating is the process of determining an estimated age of a fossil or geographical structure in comparison to another fossil or geographical structure, through using stratigraphy or index fossils (1 mark).

e. 1. The question is asking you both to state *where* the fossil would be located and to *justify* your response.

2. Consider where the fossil LM3 is located compared to LM1. The stem of the question specifically states that LM3 was the *oldest Homo sapiens* found. Thus, it must be older than LM1.

LM3 would be found in a lower layer than LM1 (1 mark).

3. Justify your response — provide evidence and reasoning for your statement.

LM3 is an older fossil than LM1, as it is stated to be the oldest *Homo sapiens* found on Australia. Therefore, it would have been buried earlier than LM1 and thus would be located in a lower layer, which is where older fossils are found (1 mark).

EXTENSION: Electron spin resonance

Carbon dating is not useful for organic material older than about 60 000 years. Another technique, **electron spin resonance** (ESR), is useful for determining the age of material between 50 000 years and 500 000 years old.

This dating technique depends on the fact that when objects are buried, they are bombarded by natural radiation from the soil, building up high-energy electrons (as seen in the tooth in figure 8.34). This absorbed radiation (and the rate of absorption) can be measured and used to calculate when the object was last outside the soil or exposed to heat, fire or sunlight.

To access more information about electron spin resonance, please download the digital document.

FIGURE 8.34 Radiation in soil bombarding a buried tooth

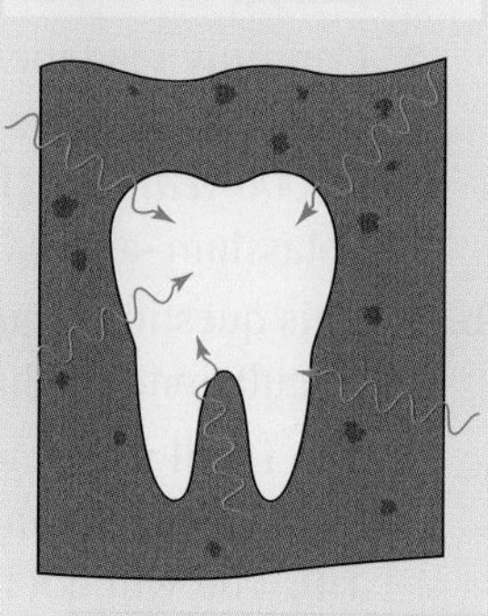

 Resources

 **Digital document** Extension: Electron spin resonance (doc-36111)

electron spin resonance a dating technique that determines the age of an object based on absorbed radiation and trapped high-energy electrons

Resources

eWorkbook	Worksheet 8.5 Absolute and relative ages (ewbk-2917) Worksheet 8.6 Reviewing fossils (ewbk-2918)
Video eLesson	Absolute age (eles-4203)
Interactivity	Absolute age (int-0239)

KEY IDEAS

- Evidence for biological change over time comes from the fossil record.
- Faunal succession is the principle that fossilied organisms are arranged vertically within sedimentary rock in a particularly manner.
- Index fossils are the fossils of widely distributed short-lived species that can be used to determine the age of other fossils.
- Transitional fossils have intermediate features of two species and provide evidence for how species have changed over time.
- The age of a fossil can be determined through relative or absolute dating.
- Relative dating uses stratigraphy and the principle of superposition to determine the age of a fossil.
- Relative dating relies on faunal succession and index fossils to provide the age of rocks and fossils.
- Radiometric techniques or absolute dating are based on the decay of naturally occurring radiometric isotopes and can be used to estimate the absolute ages of rocks.

8.3 Activities

learnON

To answer questions online and to receive **immediate feedback** and **sample responses** for every question, go to your learnON title at **www.jacplus.com.au**. A **downloadable solutions** file is also available in the resources tab.

8.3 Quick quiz	8.3 Exercise	8.3 Exam questions

8.3 Exercise

1. **MC** A new radioactive isotope was discovered. This isotope had a half-life of 15 000 years. Initially, a fossil was found to have 10 grams of the parent element. Approximately, how much of the parent element would be present after 50 000 years?
   A. 2.5 g **B.** 1.5 g **C.** 1 g **D.** 0.6 g
2. State the principle of superposition.
3. Identify which radiometric dating technique can be used with fossils that are
   a. 10 000 years old
   b. 2 million years old.
4. A fossil was dated using carbon dating. Carbon-14 has a half-life of 5730 years. Upon examining the parent and daughter elements, it was determined that three half-lives had occurred. Approximately how old is this fossil?
5. Compare the similarities and differences between
   a. relative dating and absolute dating of fossils
   b. radiometric carbon dating and electron spin resonance
   c. index fossils and transitional fossils.
6. Describe how fossils provide evidence of species changing over time.
7. A new fossil is discovered in a layer of sedimentary rock with an index fossil dated to 2.5 million years old. Explain how an absolute date of this fossil could be calculated.
8. Using the decay of radioactive elements, the age of the oldest rocks yet found on Earth has been estimated at 4300 million years. These rocks come from ancient bedrock in northern Canada. Rocks of this age are not widespread. Can you suggest why?

9. Three fossils of the same species were found in layers of rock close to the Earth's surface. These are labelled Fossil A, Fossil B and Fossil C on the diagram below.

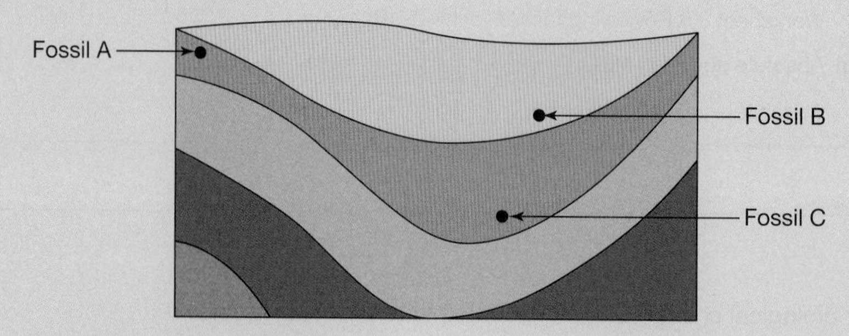

a. Fossil A was found closest to the Earth's surface. Is this the youngest fossil? Justify your response.
b. Identify the two fossils that are similar in age.
c. In the rock stratum below Fossil C, an index fossil was found. Describe how an index fossil can be used to date the age of another fossil.

8.3 Exam questions

Question 1 (1 mark)

Source: *VCAA 2018 Biology Exam, Section A, Q25*

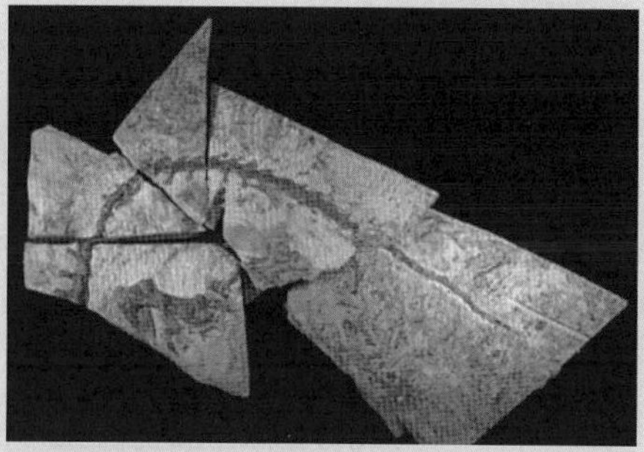

Sources: Stephen Poropat/Museums Victoria, <https://collections.museumvictoria.com.au/specimens/1572890> (left); MC Herne, AM Tait, V Weisbecker, M Hall, JP Nair, M Cleeland, SW Salisbury/Wikimedia Commons/CC-BY-4.0 (right)

MC The photograph shows a fossil of the ornithopod dinosaur *Diluvicursor pickeringi* found in 113-million-year-old rock in western Victoria. The fossil consists of a tail, a partial hind limb and some vertebrae. *D. pickeringi* grew to 2.3 m long. The diagram on the right is a reconstruction of the complete dinosaur. Evidence suggests that the dinosaur was fossilised in a log-filled hollow at the bottom of an ancient riverbed. Two stratigraphically younger fossils that had been found previously at a nearby site are closely related to *D. pickeringi*.

It is most probable that the two stratigraphically younger fossils would have been found in a layer of rock that

A. was closer to the present-day ground surface than the rock surrounding the *D. pickeringi* fossil.
B. contained a smaller quantity of carbon-14 than the rock surrounding the *D. pickeringi* fossil.
C. was located at a depth of 2.3 m below the ancient riverbed.
D. was formed from extremely hot, volcanic lava flow.

Question 2 (1 mark)

Source: *VCAA 2010 Biology Exam 2, Section A, Q18*

MC An index fossil may be used for identifying

A. the oldest rocks in a series of strata.
B. the age of rocks from the Jurassic period only.
C. the absolute age of a sedimentary rock stratum.
D. sedimentary rocks of the same age in different locations.

Question 3 (1 mark)

Source: *VCAA 2007 Biology Exam 2, Section A, Q5*

MC Biologists have suggested for a long time that reptiles evolved from fish-like ancestors. Recently, a 375 million-year-old fossil fish (*Tiktaalik roseae*) was found in Canada. This fossil had fins, scales and a lower jaw like those of a fish but had a crocodile-like skull, a mobile neck and forelimb bones resembling those of early reptiles.

Relative to fish and crocodiles, this rare *Tiktaalik* fossil represents an example of

A. a transitional form.
B. an index fossil.
C. convergent evolution.
D. coevolution.

Question 4 (4 marks)

Source: *VCAA 2012 Biology Exam 2, Section B, Q6*

One form of dating the age of a fossil is by radioactive carbon dating. The ratio of carbon-14 to nitrogen-14 ($C_{14} : N_{14}$) in the fossil is analysed and compared with the ratio of these elements in an organism living today.

The graph below shows the rate of decay for carbon-14.

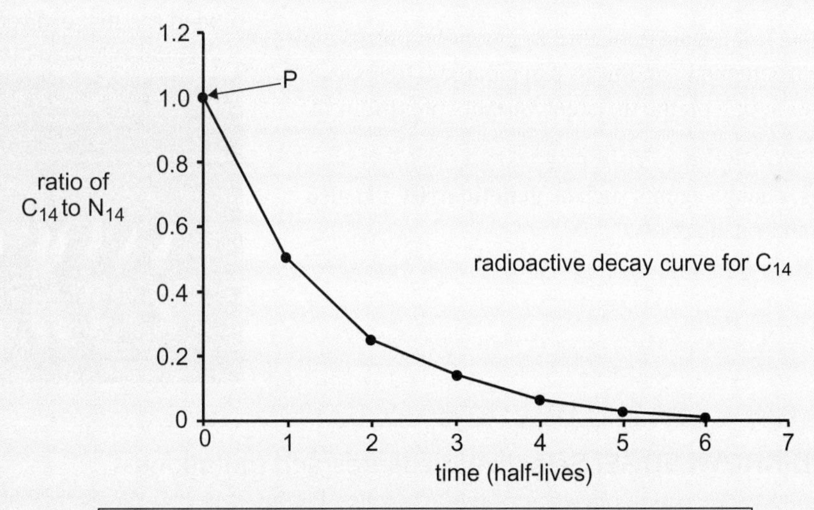

Key	Point P on the graph represents the ratio of C_{14} to N_{14} found in present-day organisms.

A fossil kangaroo skull was found in a limestone cave. The skull's $C_{14} : N_{14}$ ratio was analysed and found to contain one-quarter $\left(\frac{1}{4}\right)$ of the carbon-14 of a kangaroo that died in 2012.

a. **i.** Place an X on the curve to show the fossil's $C_{14} : N_{14}$ ratio. **1 mark**

ii. Given the half-life of carbon is approximately 6000 years, what is the approximate age, in years, of the kangaroo skull? **1 mark**

b. Carbon-dating analysis is not always possible and the age of the fossil can be estimated by dating the rock in which it is found.

i. Why is carbon-dating analysis not always possible? **1 mark**

ii. Name another absolute dating technique that can determine the age of the rock surrounding a fossil. **1 mark**

Question 5 (1 mark)

Source: *VCAA 2015 Biology Exam, Section A, Q35*

MC Potassium-40 has a half-life of 1.25 billion years. In igneous rocks closely associated with a fossil layer, the ratio of potassium-40 to its radioactive breakdown product, argon-40, is approximately 1:1.

The age of the fossils in the fossil layer will be close to

A. 125 million years.
B. 310 million years.
C. 1.25 billion years.
D. 2.5 billion years.

More exam questions are available in your learnON title.

8.4 Speciation

KEY KNOWLEDGE

- Evidence of speciation as a consequence of isolation and genetic divergence, including Galápagos finches as an example of allopatric speciation and *Howea* palms on Lord Howe Island as an example of sympatric speciation

8.4.1 What is speciation?

As discussed in Topic 7, the gene pools of large populations living under particular environmental conditions can change in response to selection pressures acting on the different variants present in the population.

In small populations, change agents such as **genetic drift** can also produce changes in their gene pools.

FIGURE 8.35 Different species of penguins that evolved after facing different selective pressures

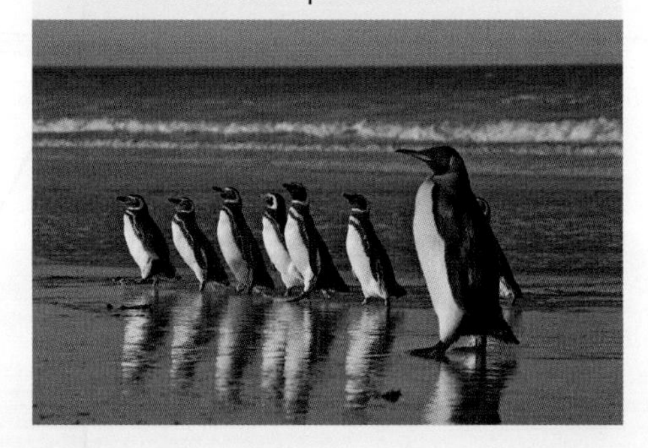

BACKGROUND KNOWLEDGE: Selective pressures and mutations

Populations of the same species living under different environmental conditions are likely to be subject to different selection pressures. This leads to natural selection, and the allele frequencies in their gene pools change, as different alleles are retained due to being advantageous. In addition, any new alleles added by mutation to these gene pools are likely to be dissimilar.

The theory of evolution by **natural selection** proposes that populations of the same species living in different locations under different conditions can evolve in different directions.

Over many generations, under different selection pressures, these populations can become increasingly different from each other in structure, physiology and behaviour. Eventually they can become so different that they form two distinct species.

The process of formation of new species is termed **speciation**.

genetic drift random changes, unpredictable in direction, in allele frequencies from one generation to the next owing to the action of chance events

natural selection the process in which organisms better adapted for an environment are more likely to pass on their traits to the next generation

speciation process of formation of new species

What are species?

For sexually reproducing organisms, species are recognised as different when they are not able to interbreed under natural conditions, or, if they interbreed, the offspring are either inviable or, if they survive, are sterile (not fertile).

When considering the speciation of extinct species, remember that it is difficult to determine if they are distinct species using the fossil record, as you are unable to determine if fertile offspring could have been produced between two organisms. In this case, comparison of DNA sequences and genomes may be used.

Isolating mechanisms

In order for speciation to occur, populations need to be separated and isolated in some way that prevents gene flow or interbreeding.

Several factors keep different species separated and prevent interbreeding or **gene flow.** These may come into operation before mating takes place — these are often called **pre-zygotic isolation mechanisms** (or pre-mating) — or they may operate after mating — these are **post-zygotic isolation mechanisms** (or post-mating). With penguins for example, a separation of population due to continental drift prevented gene flow. As penguins are unable to fly and therefore their separated populations are unable to interbreed, this is an example of pre-zygotic isolation.

gene flow the movement of alleles between interbreeding populations

pre-zygotic isolating mechanisms barriers that prevent an organism from finding and securing a mate

post-zygotic isolating mechanisms barriers that prevent a fertile offspring developing after mating

The methods of isolation

- Pre-zygotic isolation mechanisms include barriers to finding and securing a mate.
- Post-zygotic isolation mechanisms operate after the transfer of sperm from male to female and are due to chromosomal and chemical imbalances between the different species.

Different types of pre-zygotic and post-zygotic isolation mechanisms are outlined in table 8.3.

TABLE 8.3 Examples of pre-zygotic and post-zygotic isolation mechanisms

Pre-zygotic isolation mechanisms	Description
Temporal isolation	One species may be active by day and the other species is active at night, so they are unlikely to meet.
Geographic isolation	One species may live on mountain tops and the second in valleys — again, they are unlikely to meet.
Behavioural isolation	One species does not recognise the signs of sexual readiness in a second species or is unable to perform the correct courtship rituals — so they meet, but nothing comes of it.
Mechanical isolation	Mating may be attempted but, because of physical differences, it is not successful — so they meet, they try, but the parts don't fit.
Post-zygotic isolation mechanisms	**Description**
Incompatibility of gametes	Sperm cannot penetrate the outer coats surrounding the egg of the second species, so fertilisation does not occur.
Zygote mortality	Fertilisation occurs, but the zygote fails to develop.
Inviability of zygote	The zygote develops into an embryo but does not develop beyond that.
Sterility of hybrid	Hybrid offspring survive but are sterile.

BACKGROUND KNOWLEDGE: Hybrids

A hybrid is an offspring that has resulted from sexual reproduction of two different but closely related species. The offspring are usually sterile and are breed in captivity. An example of a hybrid is a zonkey (see figure 8.36), which is the result of a mating between a zebra and a donkey.

Other examples of hybrid animals include the following:

- Mules are crosses between a female horse and a male donkey.
- Hinnies are crosses between a female donkey and a male horse.
- Ligers are crosses between a lion and a tiger.
- Beefalos are cross between a cow and bison (these are unusual in that they are fertile).

FIGURE 8.36 A zonkey is a sterile offspring between a zebra and a donkey.

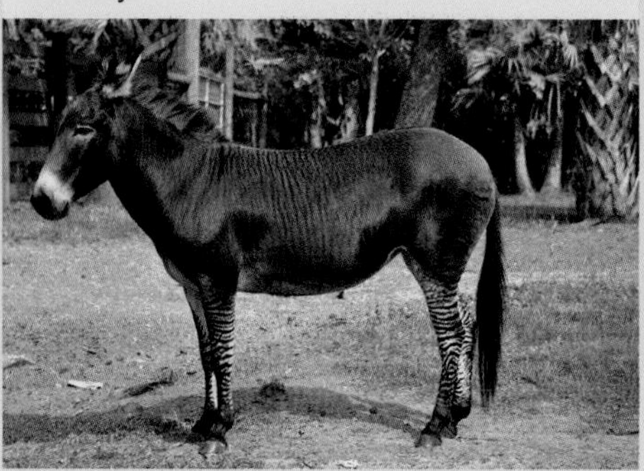

How does speciation occur?

The process of speciation relies on numerous processes in order to occur: **variation**, **isolation**, **selection** and **genetic divergence**. The general process for speciation is outlined in figure 8.37.

FIGURE 8.37 Outline of the process of speciation

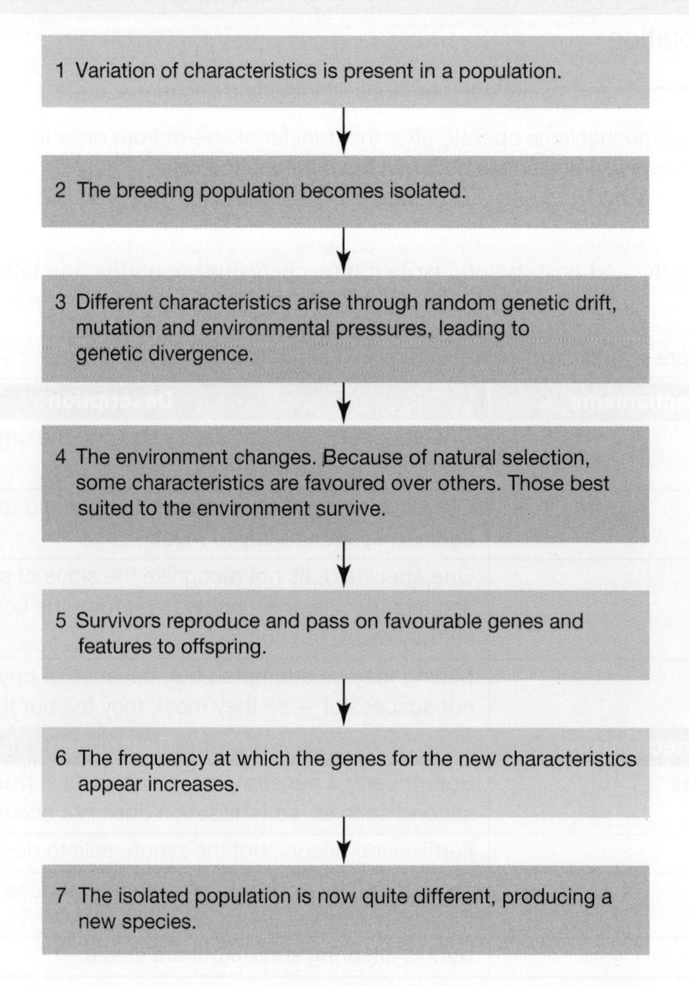

variation differences exhibited among members of a population owing to the action of genes

isolation the process in which populations are separated and gene flow is reduced or stopped

selection the process in which a variant is more advantageous under certain selective pressures, enhancing its chance of survival and reproduction

genetic divergence when two or more populations accumulate genetic changes, leading them to eventually be reproductively isolated

TIP: Remember that for speciation to occur, there needs to be some variation in a population, the isolation of populations, and different selective pressures applied to isolated populations. In any question requiring an explanation of the speciation process, you should ensure you include the terminology isolation and genetic divergence, which are specified in the Study Design.

BACKGROUND KNOWLEDGE: Subspecies

Before becoming different species, separated populations may be identified as distinct subspecies. At this stage, members of geographically separated populations are recognisably different from each other. However, if the populations come together again, members of the different subspecies groups can still interbreed and produce viable and fertile offspring.

Subspecies are denoted by the addition of a third part to the scientific name. Some of the different subspecies of the crimson rosella (*Platycercus elegans*) are shown in figure 8.38.

- A red-headed subspecies (*Playtycercus elegans elegans*) is found mainly on the south-east coastal regions of Queensland, New South Wales and Victoria.
- The orange-headed subspecies (*Platycercus elegans subadelaidae*) is found mainly in the Gulf region of South Australia.
- The yellow-headed subspecies (*Platycercus elegans flaveolus*) is found in the Murray–Darling Basin of New South Wales, Victoria and South Australia.

FIGURE 8.38 Different subspecies of rosella are still able to interbreed.

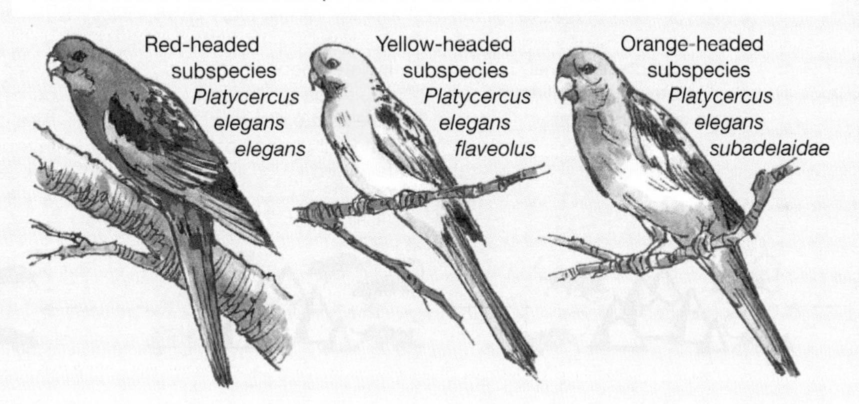

Types of speciation

When speciation occurrs, it can be categorised into two main forms: **allopatric speciation** and **sympatric speciation**. These types of speciation are divided based on the presence or absence of gene flow during the process.

allopatric speciation speciation that occurs between two populations that are geographically isolated

sympatric speciation speciation that occurs between two populations that have no geographical barrier between them

In allopatric speciation, populations are geographically separated.
In sympatric speciation, populations are not geographically separated.

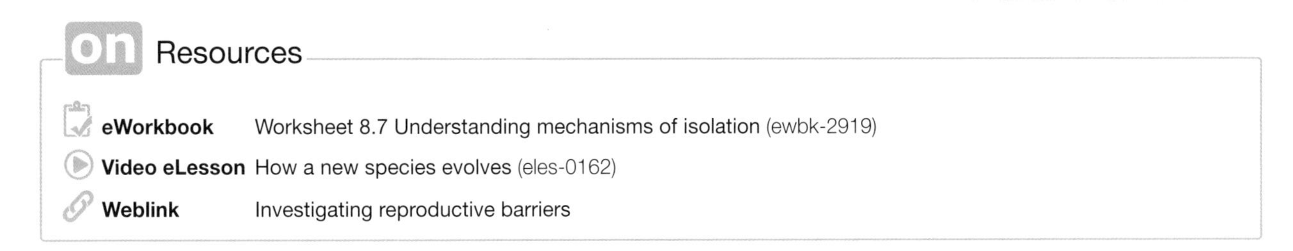

Resources

eWorkbook Worksheet 8.7 Understanding mechanisms of isolation (ewbk-2919)

Video eLesson How a new species evolves (eles-0162)

Weblink Investigating reproductive barriers

8.4.2 Allopatric speciation

When speciation occurs under circumstances where populations become geographically separated, it is termed allopatric speciation (allopatric = 'different homeland').

Factors that can split a population into geographically isolated groups may be:

- quick acting, such as habitat fragmentation owing to clearing or construction
- slow acting, such as the change of a river course
- even slower geological processes, such as the uplift of mountains or rising sea levels.

Steps in allopatric speciation

1. A single population (with initial variation) is divided by a physical barrier to become **geographically isolated**. There is no gene flow between the populations. The populations begin to experience genetic divergence (i.e. through mutation).
2. Over many generations, the isolated populations are subjected to different environmental selection pressures, which leads to different phenotypes being selected for by natural selection or genetic drift.
3. The isolated populations continue to experience genetic divergence over time. If they cannot produce fertile offspring when brought together, they are **reproductively isolated**. There are now two distinct gene pools and two different species (see figure 8.39).

FIGURE 8.39 Allopatric speciation involves separation of a population by a geographical barrier.

Single population — one species (with variation)

A geographical barrier separates the population into two populations.

Isolated populations subjected to different selection pressures; different phenotypes favoured

When populations come together again, they can no longer interbreed — two species.

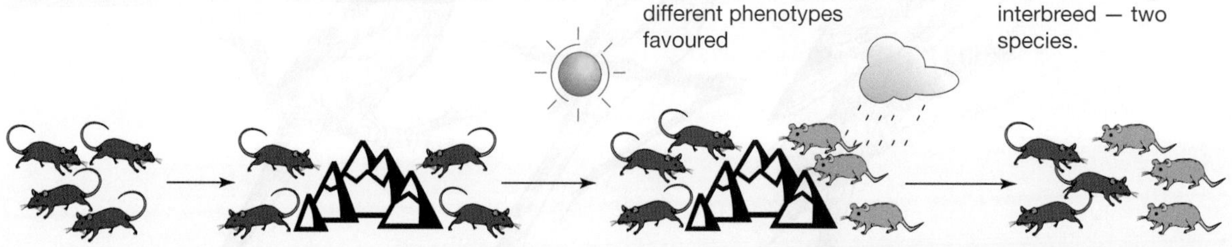

Example of allopatric speciation: Galápagos finches

The Galápagos Islands (shown in figure 8.40) and their finches have become synonymous with Charles Darwin and the study of evolution. During his voyage around the world in 1831 on the HMS *Beagle*, the ship stopped at the Galápagos Islands for an extended time. These islands are located to the west of Ecuador in South America. During this time Darwin noted that each of the islands had similar but distinctive plants and animals.

FIGURE 8.40 An aerial photograph of the Galápagos Islands

geographically isolated the isolation of two or more populations due to a geographical barrier

reproductively isolated the inability of species to breed and produce fertile offspring

FIGURE 8.41 Different species of finches on the Galápagos Islands have differences in their beak shapes. a. A tree finch b. A warbler finch

Darwin took particular notice of the finches that lived on each island and how they were similar and different not only to each other, but also to finches on the mainland of South America. His famous deduction was that each finch on the Galápagos Islands had evolved from a few ancestral finches that reached there from South America. These finches are an example of allopatric speciation occurring relatively quickly.

The different finch species (see figure 8.42) all have distinctively shaped beaks that assist them in eating different foods such as fruits, insects and seeds.

FIGURE 8.42 The diversity of beaks shown between species of Galápagos Island finches

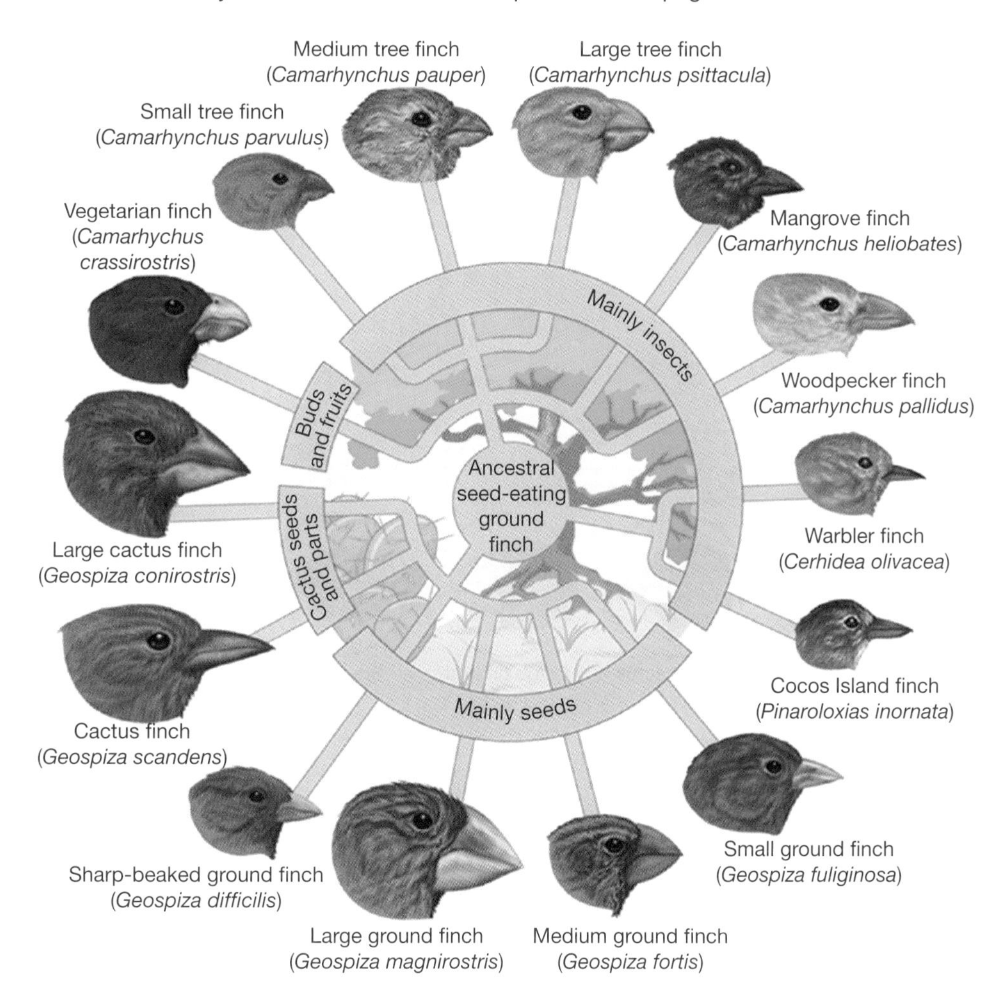

DNA evidence has found that all Galápagos Island finches are more closely related to each other than they are to the finches living on the mainland of South America. This indicates that all Galápagos Island finches have evolved from one common ancestor.

So why do these finches look different from each other?

A driving cause of these differences is the differing environmental selection pressures on each of the islands based on food availability.

- The larger islands of Isabela and Fernandina have higher elevations, which allows large trees to grow (see figure 8.43).
- The smaller islands of Española and Santa Maria have lower elevations and have vegetation of cacti and grasses.
- After the finches had arrived on an island, they became geographically isolated due to the surrounding ocean (they did not tend to fly between the islands). Over time, allopatric speciation occurred.

FIGURE 8.43 The distribution of finches across the Galápagos Islands and South America

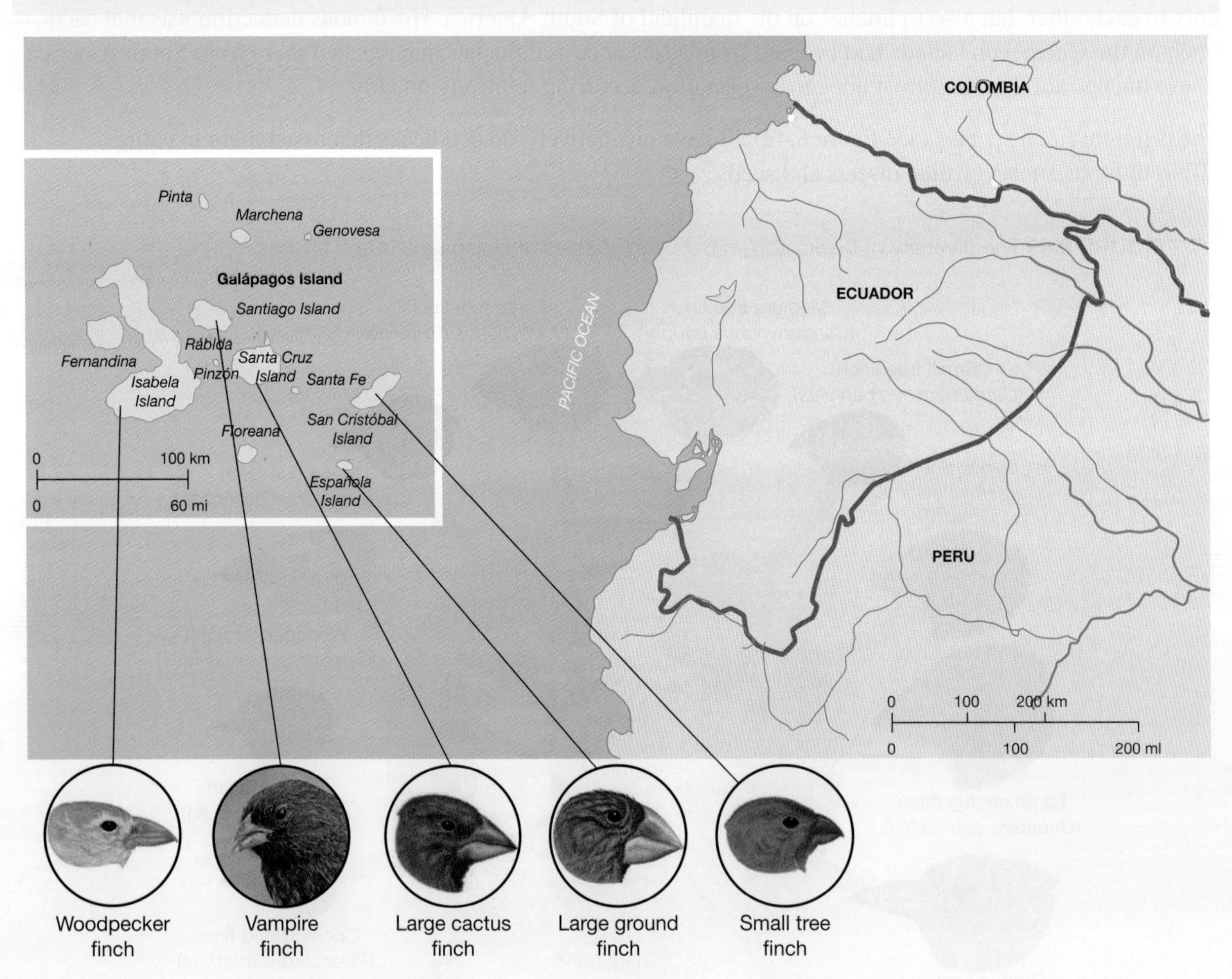

Some examples of how allopatric speciation occurred with the finches are outlined in table 8.4.

TABLE 8.4 Allopatric speciation of woodpecker and large cactus finches

	Woodpecker finch	**Large cactus finch**
Geographically isolated	Isabella Island	Española Island
Habitat of island	High elevations with a mixture of large trees, cacti and grasses	Low elevation, most plants are cacti
Mechanism of evolution	A large number of invertebrates live in crevices of trees on the island. Individuals with longer and narrow beaks had a selective advantage because they could eat more food to survive. They reproduced and passed on this trait.	The main food source is cactus, which provides pulp, flowers and fruit. Individuals with larger and stronger beaks had a selective advantage because they were able to tear the pulp of the cacti. They reproduced and passed on this trait.
Reproductively isolated	Over time the woodpecker finch became genetically different; now it cannot produce fertile offspring with any other finches.	Over time the large cactus finch became genetically different; now it cannot produce fertile offspring with any other finches.

Resources

 **Video eLesson** Allopatric speciation (eles-4372)

tlvd-1730

SAMPLE PROBLEM 4 Analysing allopatric speciation

The Galápagos Islands are home to a variety of different finch species that have evolved from one species of seed-eating ground finch. The finches show variation in their shape and in beak size, and thus have different diets.

a. **The finches live on different islands that vary in their vegetation and climate. Describe the impact this has on the finch populations. (2 marks)**

b. **Outline the process by which the different finches evolved to have separate beaks. (3 marks)**

THINK	WRITE
a. 1. Identify what the question is asking you to do. The first part of the question is asking you to *describe*. In a describe question, you give a detailed account in sentence form.	
2. Consider that the finches live on different islands — they are geographically isolated from each other. Then consider that the islands have different vegetation and climates — different environmental selection pressures.	The finches live on different islands that are geographically isolated from each other (1 mark). There are differing environmental selection pressures on the islands that give different beaks selective advantage (1 mark).
b. 1. Identify what the question is asking you to do. The first part of the question asks you to *outline*. In an outline question, you need to provide a step-by-step account of the event.	

2. Consider what the process is in this scenario — speciation. There is a geographical barrier between the species, so it is allopatric speciation. As this question is worth three marks, you should detail the three main steps of allopatric speciation, using the finches as the example.

TIP: You must specify finches as they are mentioned in the question. You will lose marks if you only have generic statements.

1. The ocean acts as a geographical barrier between the finches; therefore, no gene flow exists (1 mark).
2. Over many generations the finches with beaks best suited to eating food on their islands survive and pass on their genes (1 mark).
3. If enough genetic divergence occurs over time, the finches will be reproductively isolated (1 mark).

8.4.3 Sympatric speciation

Sympatric speciation (sympatric = 'same homeland') occurs when members of a population living in the same area diverge sufficiently to become two different species.

It was thought for some time that speciation could not occur under sympatric conditions because there is no physical barrier to prevent gene flow between the populations.

However, recent evidence indicated that the *Howea* palms on Lord Howe Island have evolved into two separate species through sympatric speciation. There is also evidence of sympatric speciation in other species.

Steps in sympatric speciation

1. A single population is divided by a pre-zygotic isolation barrier other than geographic isolation, such as temporal or behavioural isolation. There is some gene flow between the populations.
2. Over many generations, genetic divergence occurs. Eventually the isolated populations respond differently to environmental selection pressures, which leads to different phenotypes being selected for by natural selection or genetic drift.
3. The isolated populations change significantly over time. If until they can no longer reproduce fertile offspring when brought together, they are reproductively isolated. There are now two distinct gene pools and two different species.

FIGURE 8.44 Sympatric speciation involves separation of a population by a barrier that is not geographical.

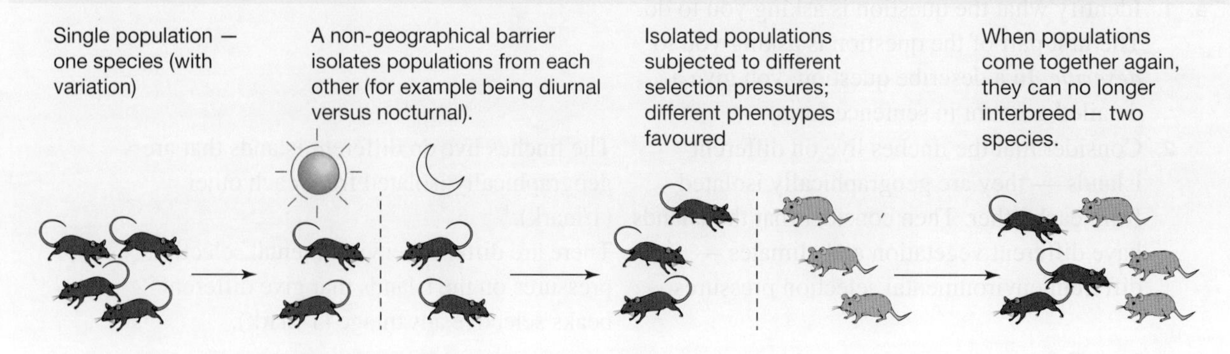

TIP: Ensure when discussing sympatric speciation that you do not mention geographical isolation. This only occurs with allopatric speciation.

Example of sympatric speciation: *Howea* palms on Lord Howe Island

The *Howea* palms exist on the small Lord Howe Island, which is approximately 15 km^2 in area and is found 600 km off the New South Wales coastline.

Although there are four different species of *Howea* palms on the island, it is the two species of *Howea belmoreana* and *Howea forsteriana* that have diverged by sympatric evolution.

There are many differences between *H. belmoreana* and *H. forsteriana,* as outlined in table 8.5.

TABLE 8.5 Comparison between *H. belmoreana* and *H. forseriana*

H. belmoreana	*H. forsteriana*
Soil specialist and will only grow in volcanic soil	Soil generalist and will grow in both volcanic and calcareous soils
Lives in a more specific habitat but out competes *H. forsteriana* in this habitat, thus having a greater population size (around 3 times)	Can live in a wider habitat, but *H. belmoreana* out competes it in volcanic soils
Flowers 6–7 weeks later	Flowers 6–7 weeks earlier

How do palm trees speciate from each other when there is no geographical barrier to gene flow?

Evidence collected from the *Howea* palms indicates that both pre-zygotic and post-zygotic isolation has occurred to stop gene flow between the species.

Lord Howe Island is a volcanic island with has two main soil types (see figure 8.45):

- volcanic soil — higher in nutrients required for plant growth
- calcareous soil — a sandy soil with lower levels of nutrients required for plant growth

FIGURE 8.45 a. The distribution of volcanic soil and calcareous soil on Lord Howe Island **b.** Different species of Howea palms: Hb — *H. belmoreana*; Hf — *H. forsteriana*

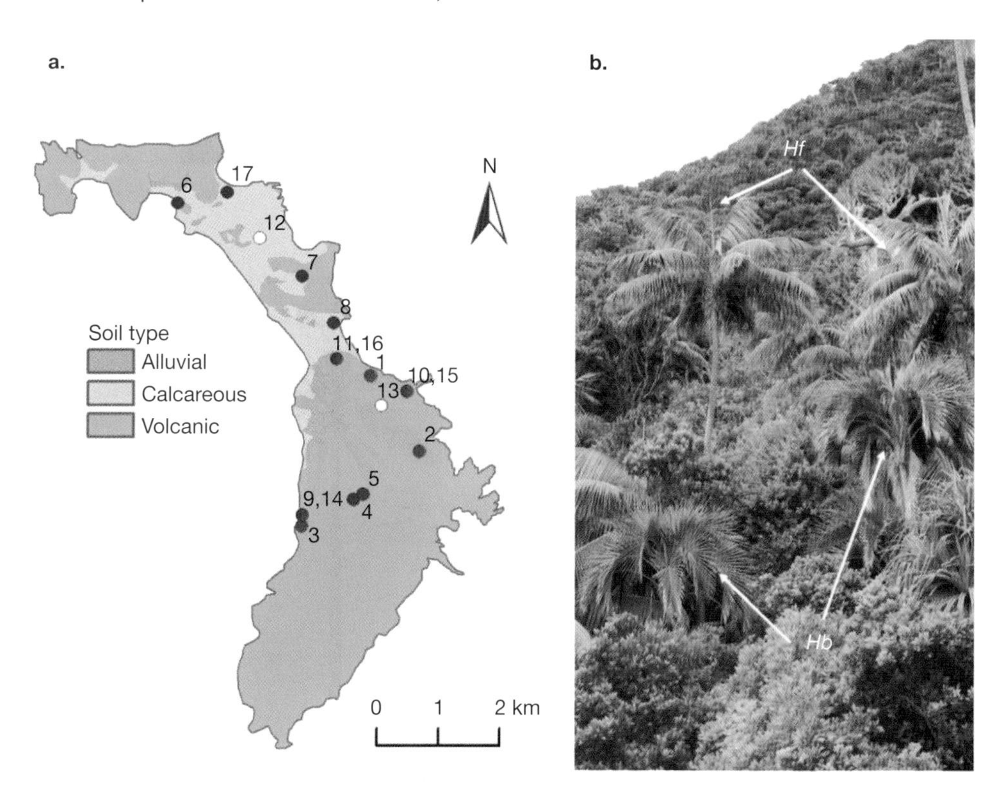

Evidence indicates that variations in soil preference have altered the two palms' flowering time so that they flower at different times of the year. Different flowering times create the pre-zygotic mechanism of temporal isolation.

Further evidence has found that any hybrids which do form have the post-zygotic mechanism of reduced hybrid fitness. These two types of isolating mechanisms have created reproductive isolation between the two species and therefore gene flow has ceased.

FIGURE 8.46 The effect of soil type on sympatric speciation of *Howea* palms

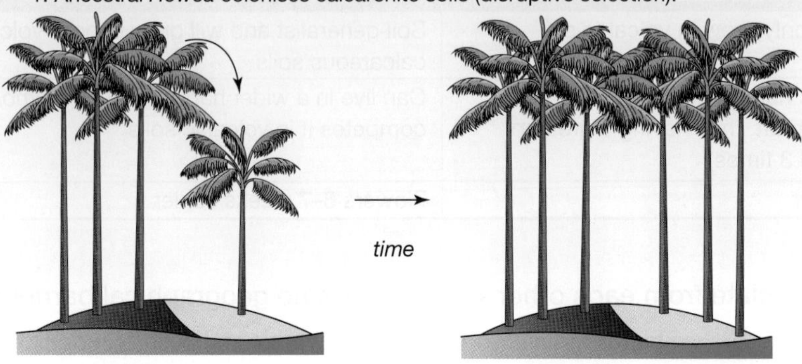

elog-0080

INVESTIGATION 8.5

online only

Modelling allopatric and sympatric speciation

Aim

To model and compare the processes of allopatric and sympatric speciation

EXTENSION: Polyploid speciation

Polyploidy refers to a condition in which an organism has more than two matched sets of chromosomes. When just the two matched sets of chromosomes are present, an organism (and its cells) are said to be diploid. The presence of one or more additional sets of chromosomes over the base diploid condition (2 sets) produces organisms that are termed triploid (3 sets), tetraploid (4 sets), hexaploid (6 sets) and even dodecaploid (12 sets).

Polyploidy is common in flowering plants and ferns because an organism with a different number of chromosomes can self-fertilise to make the next generation. The differing number of chromosomes results in an isolated gene pool and sympatric speciation may occur.

FIGURE 8.47 Different mechanisms of polyploidy

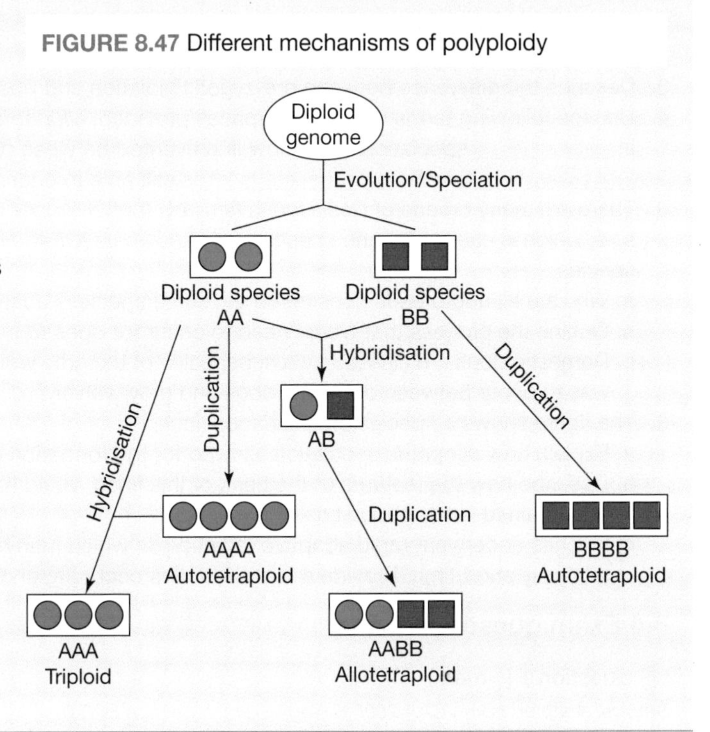

Resources

 **eWorkbook** Worksheet 8.8 Allopatric and sympatric speciation (ewbk-2920)

 **Weblinks** Types of speciation

Sympatric speciation in palms on an oceanic island

KEY IDEAS

- A species is a group of individuals that can produce fertile viable offspring when interbreeding.
- Speciation is the process of forming a new species, which requires isolation and genetic divergence.
- Species can be isolated from each other with the pre-zygotic isolating mechanisms; temporal, geographical, behavioural and mechanical isolation.
- Species can be isolated from each other with the post-zygotic isolating mechanisms; incompatible gametes, zygote mortality, inviable zygote and sterility of zygotes.
- Allopatric speciation occurs when there is geographical isolation to stop gene flow between populations. An example of allopatric speciation is seen in Galápagos finches.
- Sympatric speciation occurs without geographic isolation. Initially some gene flow can still occur and is eventually prevented by another isolation mechanism (such as temporal isolation). An example of sympatric speciation is seen in *Howea* plants on Lord Howe Island.

8.4 Activities

To answer questions online and to receive **immediate feedback** and **sample responses** for every question, go to your learnON title at **www.jacplus.com.au**. A **downloadable solutions** file is also available in the resources tab.

8.4 Quick quiz on	**8.4 Exercise**	**8.4 Exam questions**

8.4 Exercise

1. Define the term speciation.
2. Identify the isolating mechanism between *Howea* palms.
3. Describe the difference between pre-zygotic isolation and post-zygotic isolation.
4. Use the following terms to fill in the spaces provided: *allopatric, sympatric, geographical, gene flow.*
   In ____________ speciation, gene flow is prevented between organisms due to ____________ barriers. In ______________, some ________________ is still able to occur between populations, such as in *Howea* palms.
5. There are many breeds of *Canis lupus familaris* (better known as the domestic dog). Dogs vary greatly in size, which is caused by one specific gene known as *IGF-1*. However, all dogs are still considered the same species.
   a. Why are all dog breeds considered the same species? Justify your response.
   b. Outline the process that would need to occur for dogs to be classed as separate species.
   c. Domestic dogs are classed as a subspecies of the grey wolf. What expectations would you have if breeding was to occur between domestic dogs and grey wolves?
6. The figure shows a species of Galápagos finch.
   a. Explain how allopatric speciation allowed for the formation of this species.
   b. Describe how the features of the beak of this finch would have been selected for and retained in the population.
   c. Finches underwent rapid adaptive radiation, in which many species formed in a relatively short time. Provide a reason for this occurring and justify your response.

8.4 Exam questions

Question 1 (1 mark)

Source: *Adapted from VCAA 2008 Biology Exam 2, Section B, Q7f*

The relatively abundant Green-eyed Tree Frog (*Litoria genimaculata*) inhabits tropical rainforest in Far North Queensland. There are two main populations: a north population and a south population. The critically endangered Kuranda Tree Frog, *Litoria myola*, is found in very small populations in isolated fragments of wet tropical rainforest near Cairns in Queensland. The two species, *L. myola* and *L. genimaculata*, are very similar in appearance, although the male *L. genimaculata* is larger than the male *L. myola*. Biologists have also found that their mating calls differ markedly. There is some overlap between the ranges of these three frog groups.

In northern areas, where the distributions of the two species overlap, a few hybrid offspring (1.4% of the population) have been observed, resulting from matings between the two species. Biologists wish to establish whether or not the Green-eyed and Kuranda Tree Frogs are, in fact, distinct species.

How could biologists determine this experimentally?

Question 2 (1 mark)

MC A group of cichlid fish live in a small volcanic crater lake in Nicaragua. The lake is not connected to any other lakes by rivers or streams. Over time three species living in this small lake have genetically diversified and are now reproductively isolated. This speciation event is the result of

A. allopatric speciation with no gene flow.
B. allopatric speciation with gene flow.
C. sympatric speciation with no gene flow.
D. sympatric speciation with gene flow.

Question 3 (7 marks)

Source: *VCAA 2016 Biology Exam, Section B, Q9*

Galápagos tortoises (*Chelonoidis* spp.) can be found on many of the islands that make up the Galápagos Islands.

Originally, 14 different species were identified based on the islands on which they lived and on their morphology.

Santa Cruz, the second largest of the Galápagos Islands, has two isolated tortoise populations. Population A contains more than 2000 individuals covering an area of 156 square kilometres. Population B is a small population of 250 individuals covering an area of 40 square kilometres.

The position of the two populations on the island of Santa Cruz is shown below. The two populations are separated by a distance of 20 kilometres.

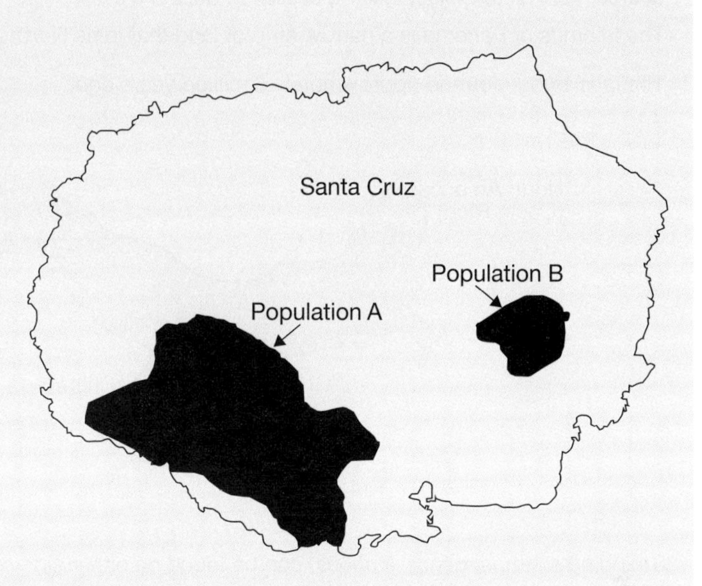

In 2015, scientists investigated whether the individuals of the two populations belong to the same species or whether they are two different species.

Average measurements of skull size were calculated for tortoises belonging to both populations A and B. The skulls were measured in six different places. The six measurements were also compared to average measurements taken from skulls of other Galápagos tortoise species. The results are shown in the table below. Comparisons have been made with three other Galápagos tortoise species.

Measurement position	Average skull measurement (mm)				
	Population B	Population A	*Chelonoidis vicina*	*Chelonoidis chathamensis*	*Chelonoidis ephippium*
1	118	98	86	80	74
2	40	37	28	27	25
3	21	18	16	14	12
4	26	23	21	18	17
5	10	9	8	7	6
6	19	17	16	14	13

Source (map and table): Adapted from N Poulakakis, DL Edwards, Y Chiari, RC Garrick, MA Russello, E Benavides et al., 'Description of a New Galápagos Giant Tortoise Species (*Chelonoidis*; Testudines: Testudinidae) from Cerro Fatal on Santa Cruz Island', *PLoS ONE*, 10(10): e0138779, doi:10.1371/journal.pone.0138779, 21 October 2015

a. Consider the data given.
Does the data support the hypothesis that individuals in Population A belong to a different species from individuals in Population B? Explain your answer. **2 marks**

b. Scientists have carried out genetic studies on the two populations.
Give an example of genetic evidence that may be produced by scientists to support the hypothesis that individuals of the two populations belong to different species. Explain your answer. **2 marks**

c. Some scientists thought that allopatric speciation may have occurred on the island of Santa Cruz.

i. Name a feature that scientists would look for in the island environment to support the occurrence of allopatric speciation. **1 mark**

ii. Explain how the feature named in **part c. i.** could contribute to allopatric speciation. **2 marks**

Question 4 (4 marks)

Source: *VCAA 2006 Biology Exam 2, Section B, Q8, b and c*

The Isthmus of Panama is a narrow strip of land that joins North and South America.

The land bridge formed approximately 3 million years ago.

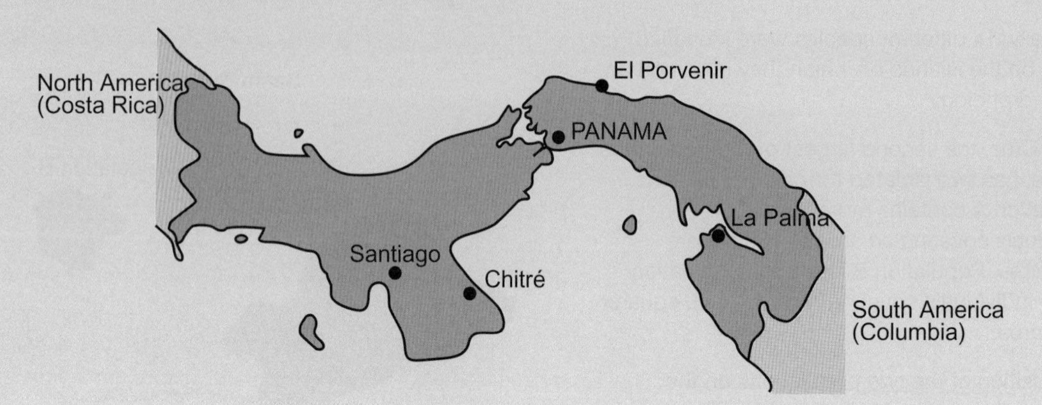

Snapping shrimps, genus *Alpheus*, can be found on either side of the land bridge. The two groups are phenotypically similar. However when the males and females from either side of the land bridge were brought together they snapped aggressively at each other and would not mate. They are now considered to be two different species.

a. Why is the inability to mate sufficient evidence to call the two groups different species? **1 mark**

b. What type of speciation has occurred in the snapping shrimp? **1 mark**

c. Explain how the differences between the shrimp on either side of the land bridge could have arisen. **2 marks**

Question 5 (3 marks)

Source: *VCAA 2007 Biology Exam 2, Section B, Q5d*

Eastern tiger snakes (*Notechi scutatus*) living on desolate islands off mainland Australia have longer jaws than the mainland populations of snakes. The diet of island snakes includes large prey, such as seagull chicks, while the diet of the mainland snakes consists of small prey, such as frogs and mice.

Researchers set up experiments using baby snakes from both locations. Snakes were fed either large or small mice over several months, until they reached maturity. The method and results are indicated in the table below.

	experiment 1		experiment 2	
	group A island snakes	**group B island snakes**	**group C mainland snakes**	**group D mainland snakes**
Length of eastern tiger snakes, jaws at birth	long	long	normal	normal
Type of prey given over several months	small mice	large mice	small mice	large mice
Length of eastern tiger snakes, jaws at maturity	normal	long	normal	normal

At present the island and mainland populations are both classified as the same species. It has been proposed that the two populations of snakes may eventually evolve into two separate species.

Outline the steps involved in the process of speciation, with particular reference to the snakes in the two populations. You may use a labelled diagram or flow chart to illustrate your answer.

More exam questions are available in your learnON title.

8.5 Review

8.5.1 Topic summary

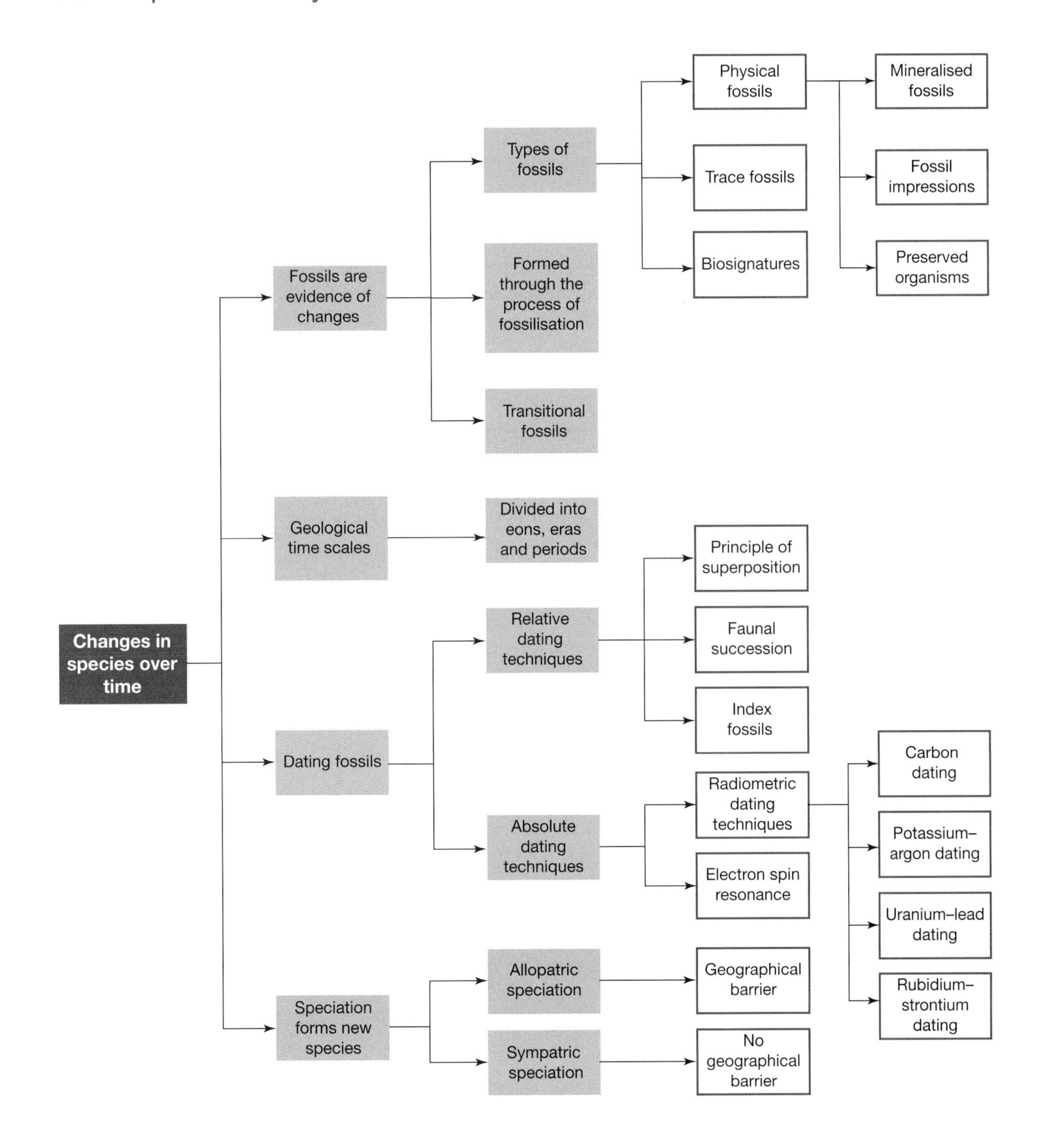

Resources

eWorkbook	Worksheet 8.9 Reflection — Topic 8 (ewbk-2922)
Practical investigation eLogbook	Practical investigation eLogbook — Topic 8 (elog-0193)
Digital documents	Key terms glossary — Topic 8 (doc-34622) Key ideas summary — Topic 8 (doc-34613)
Exam question booklet	Exam question booklet — Topic 8 (eqb-0019)

8.5 Exercises

learnON

To answer questions online and to receive **immediate feedback** and **sample responses** for every question, go to your learnON title at **www.jacplus.com.au**. A **downloadable solutions** file is also available in the resources tab.

8.5 Exercise 1: Review questions

1. Explain how absolute dating can be used to determine the age of fossils.
2. Theophrastos of Eresos (368–284 BC) wrote about fossil fish that were found in rocks in hills at a great distance from the sea. He called these fossils 'dug-up fishes', and he explained their occurrence as being the result of fish eggs being washed by flood waters to high land where the eggs hatched in moist soil. Briefly outline an alternative explanation for these 'dug-up fishes' that would be given by scientists today.
3. From the various radiometric techniques for dating objects, name the one that corresponds to each of the following.
   a. Can be used with a single mineral crystal from a rock
   b. Produces argon-40 as a daughter product
   c. Might be useful for dating the most ancient rocks on Earth
   d. Can be used only for dating rocks with uranium-based minerals
   e. Can be used for dating rocks that contain the mineral feldspar
   f. Could be used for dating a wooden spear thrower
   g. Might be useful to date blood on an ancient stone axe head
4. The Grand Canyon is one of the most famous gorges worldwide, located in Arizona in the United States. The canyon is estimated to have formed 5 to 6 million years ago, due to the Colorado River breaking down and cutting through the rock.
   Before the formation of the canyon, there was one squirrel species in the area. However, after the formation of the Grand Canyon, squirrel populations were divided by the river. Today, there are two distinct squirrel species — one species to the north of the canyon and one to the south of the canyon.
   a. Explain how the formation of the Grand Canyon led to the formation of two distinct species of squirrel. Draw a clear diagram to show this.
   b. Identify the type of speciation that this shows.
   c. What would be expected to be observed if a bridge was added and the squirrel populations were connected?
5. The figure shown is a reconstructed image of the prehistoric fish *Tiktaalik*. It is referred to as a transitional fossil.
   a. Describe why *Tiktaalik* is called a transitional fossil.
   b. Explain how transitional fossils provide vital evidence for how species change over time.

6. Populations of giraffes (*Giraffa camelopardalis*) from different regions on the African continent show distinctive differences in inherited traits, including colour and pattern. In all, up to nine subspecies of giraffe are recognised. The figure shows the geographic distributions of the various groups.

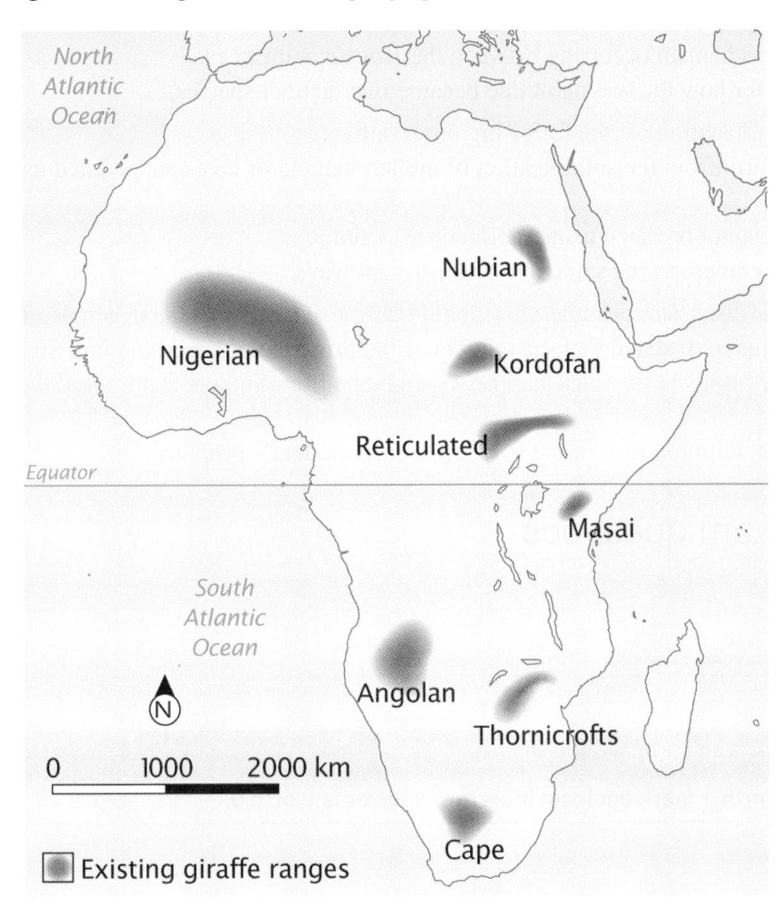

   a. What might be expected to happen over a very long period if the populations remain geographically separated?
   b. Suggest a likely origin for the different colours and patterns in the different subspecies.
   c. What test would tell if the giraffe populations are the same or different species?

7. In Siberia, which is part of the Arctic region of Russia, a female baby woolly mammoth was found. It was preserved in a glacier and has been there for 39 000 years.

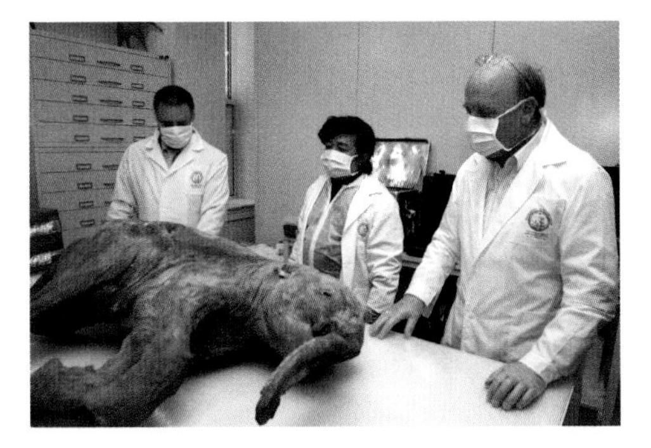

   a. Identify which type of fossil the woolly mammoth formed.
   b. Outline how this fossil could have been formed.
   c. Which radiometric dating technique was used to determine the woolly mammoth's age? Justify your response.

8. Lord Howe Island is home to the *Howea* palm trees species *Howea belmoreana* and *Howea forsteriana.* Although they live side by side, these two palms species have diverged from each other. The palm *H. forsteriana* flowers 6–7 weeks before *H. belmoreana* does.
   a. Identify the type of speciation that occurred between the *Howea* palm trees.
   b. State the isolating mechanism occurring between the *Howea* palm trees.
   c. Outline the process for how the two palm tree became two distinct species.
9. Suggest a reasonable explanation for the following observations:
   a. Igneous rocks that form from the solidification of molten magma or lava can be dated using radiometric techniques.
   b. Sedimentary rocks cannot be dated using radiometric techniques.
   c. Different people may interpret the same fossil in different ways.
10. A group of students was discussing whether they could find the age of an isolated sample of igneous rock that they were given. Student P stated that she would use the principle of superposition. Student Q argued that searching for index fossils in this rock sample would be helpful. Student S proposed that carbon-14 dating would be useful.
    Give your assessment, with reasons, of the validity of each student's proposal.

8.5 Exercise 2: Exam questions

Resources

Teacher-led videos Teacher-led videos for every exam question

Section A — Multiple choice questions

All correct answers are worth 1 mark each; an incorrect answer is worth 0.

Question 1

Source: *VCAA 2012 Biology Exam 2, Section A, Q16*

Anomalocaris fossils have been found at Emu Bay in South Australia. *Anomalocaris* was a predatory, shrimp-like invertebrate measuring 60 cm in length. It had long, spiny, frontal appendages and a powerful, disc-shaped mouth made of overlapping, hard plates. The Emu Bay fossils were found in layers of shale and dated back to about 520 million years ago. *Anomalocaris* fossils that have been found around the world suggest that this genus existed for at least 50 million years.

Factors that would have contributed to the fossilisation of this animal include

A. its predatory way of life.
B. the action of waves and currents.
C. the salt content of the water in which it lived.
D. its hard-plated mouth and spiny appendages.

Use the following information to answer Questions 2–5.

Trilobites existed from the Early Cambrian period (521 million years ago) until the end of the Permian period (250 million years ago). The chart below, based on fossil evidence, shows the phylogeny of some trilobite orders present in Earth's oceans over this time.

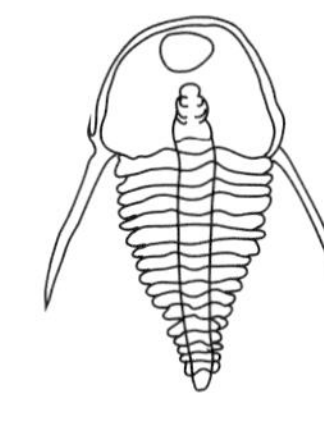

A trilobite fossil
order: Ptychopariida

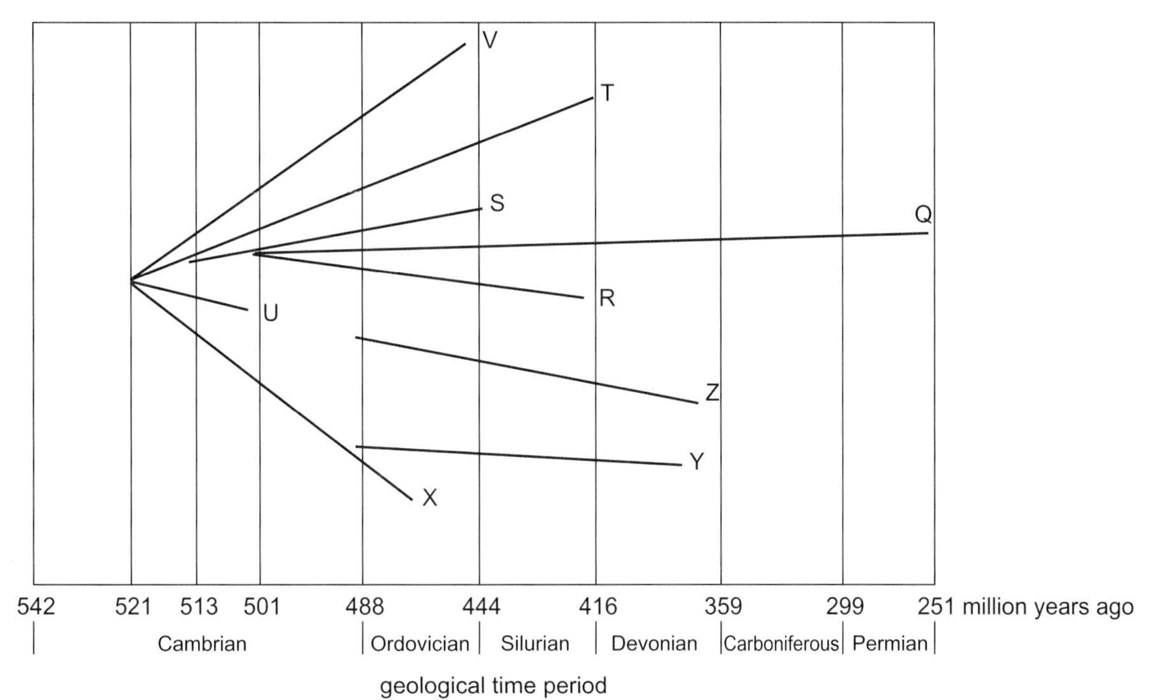

Question 2

Source: *VCAA 2014 Biology Exam, Section A, Q34*

Trilobite fossils in a particular layer of rock were used to date a fossil shell in the same layer. A palaeontologist dated the fossil shell to 328–359 million years old.

It is most likely that the trilobite fossils present were of the order

A. Y.
B. U.
C. Q.
D. T.

Question 3

Source: *VCAA 2014 Biology Exam, Section A, Q35*

The absolute age of the trilobite fossils was most likely determined by using

A. carbon dating.
B. transition fossils.
C. index fossils.
D. potassium–argon dating.

Question 4

Source: VCAA 2014 Biology Exam, Section A, Q36

The geological time periods shown on the chart differ in duration because the time periods reflect

A. the diversity of fossils and mass extinction events.
B. the absence of trilobite fossils in the Late Cambrian period.
C. different rates of radioactive decay.
D. different rates of fossilisation.

Question 5

Source: VCAA 2014 Biology Exam, Section A, Q37

The chance of a trilobite becoming fossilised is increased by

A. slow burial of its remains in dry sediment.
B. large temperature variations in the sediment containing its remains.
C. the presence of hard body parts.
D. the presence of scavengers at the time of its death.

Question 6

Source: VCAA 2009 Biology Exam 2, Section A, Q15

In 2006, two separate palaeontology laboratories were set up in the Sahara desert.

Laboratory 1 uncovered a burial site that contained human remains. Over 200 human remains were found and they were dated from 10 000 to 4500 years ago.

Laboratory 2 discovered a 110 million-year-old plant-eating dinosaur, *Nigersaurus*, in a nearby area in a different sedimentary layer.

To date the fossils accurately the two groups of palaeontologists would most likely have used

A. carbon-14 dating for both the human and dinosaur remains.
B. uranium-235 dating for both the human and dinosaur remains.
C. uranium-235 dating for the human remains and carbon-14 dating for the dinosaur remains.
D. carbon-14 dating for the human remains and uranium-235 dating for the dinosaur remains.

Question 7

A male songbird in Papua New Guinea is trying to mate with a female by singing a particular song. However, the female does not recognise the song and will not mate with the male. What is this an example of?

A. Temporal isolation
B. Geographical isolation
C. Behavioural isolation
D. Mechanical isolation

Question 8

Which of the following is an example of an isolating mechanism in allopatric speciation?

A. A river forming between two populations
B. Courtship behaviours not being recognised between groups
C. Differences in anatomical structures preventing breeding
D. One group being nocturnal and another group being diurnal

Question 9

Source: *VCAA 2016 Biology Exam, Section A, Q38*

In India, a group of scientists was studying fossils from a coal deposit formed during the Permian period (290–245 million years ago). They found three fossil species from the same genus in different levels (strata) of the coal. When radiocarbon dating on these fossils was performed, it showed exactly the same levels of carbon-14 in all three fossil species. The data is summarised in the table below.

Fossil species	Depth at which fossil was found in the coal deposit (m)	Proportion of carbon-14 (%)
Gangamopteris major	6.2	0.0001
Gangamopteris obliqua	8.1	0.0001
Gangamopteris clarkeana	4.7	0.0001

Which one of the following is the correct conclusion to draw from these findings?

A. There is no evolutionary relationship between these three fossil species.
B. *G. clarkeana* is the common evolutionary ancestor of *G. major* and *G. obliqua*.
C. As carbon dating is a more reliable dating technique than analysis of strata in coal deposits, the fossils of *G. major, G. oblique* and *G. clarkeana* are all of the same age.
D. An analysis of strata in coal deposits is a more reliable dating technique than carbon dating for Permian fossils; the fossil of *G. major* is younger than the fossil of *G. obliqua*.

Question 10

Index fossils are found in specific layers of rock around the world. The figure provided shows rock strata from three different regions.

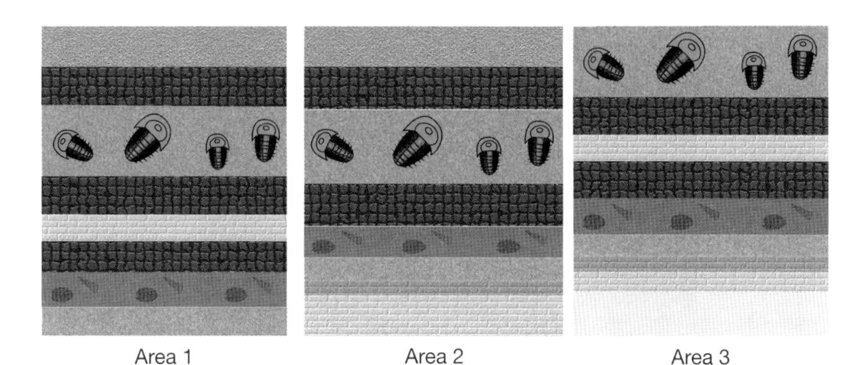

What do these rock strata indicate?

A. Area 1 has the oldest layer of stratum.
B. Area 2 has the oldest layer of stratum.
C. Area 3 has the oldest layer of stratum.
D. There is no difference between the three regions because they all have the same index fossils.

Section B — Short answer questions

Question 11 (4 marks)

Source: *Adapted from VCAA 2020 Biology Exam, Section B, Q8b and c*

The evolutionary relationship between seven species of cichlid fish was examined.

a. Fossils of species of fish are more likely to be found than fossils of land-dwelling animals. Explain why this is the case with reference to **two** conditions required for the fossilisation of an organism. **2 marks**

b. A group of scientists stated that a particular fossilised fish was 5000 years old. Outline a dating technique that could have been used by the scientists to determine the age of the fossil. **2 marks**

Question 12 (7 marks)

Source: *VCAA 2009 Biology Exam 2, Section B, Q4*

In 1877, German workers found a slab of stone containing the fossil of an ancient bird form. The fossil bird was called *Archaeopteryx*.

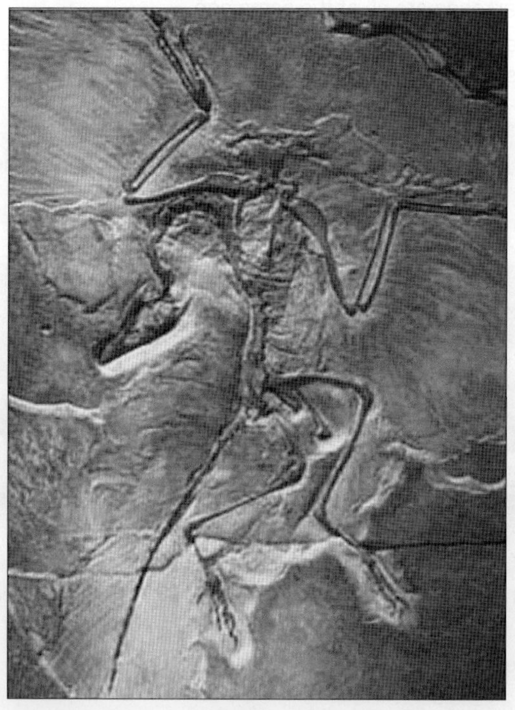

Fossil

a. **i.** Describe how this fossil could have been formed. **1 mark**

ii. Scientists use information gained from sedimentary rock to arrange animal and plant fossils into some kind of evolutionary sequence over time. Explain how such sequencing is possible. **2 marks**

b. **i.** Name one isolation barrier involved in allopatric speciation. **1 mark**

ii. Name one isolation barrier involved in allopatric speciation. Explain how isolation may result in speciation. **3 marks**

Question 13 (6 marks)

***Source:** VCAA 2010 Biology Exam 2, Section B, Q7a, b, cii*

The islands of Hawaii in the Pacific Ocean were formed as a result of volcanic action in which small land masses were thrown up by submarine volcanoes. The youngest of the islands lies to the east of the oldest. A similar pattern of deposition has been found across all islands, shown by the profile below.

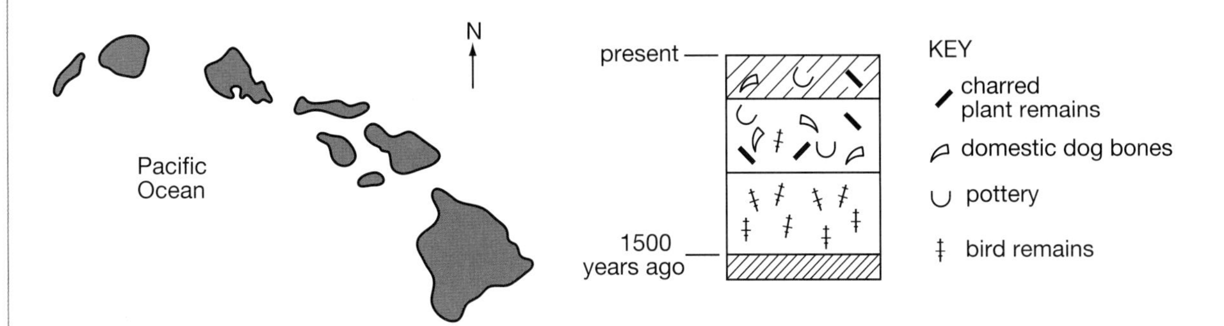

a. What assumption is made about the formation of strata when interpreting profiles such as this? **1 mark**

b. i. State a hypothesis to account for the disappearance of many of the bird species from the groups of islands. **1 mark**

ii. Provide evidence to support your hypothesis. **1 mark**

Biologists studied many species of the fruit fly, *Drosophila*, living on the Hawaiian islands. The species vary widely in appearance, behaviour and habitat. The diversity of *Drosophila* can be explained by the successive colonisation of newly formed islands by a small number of individuals 'island-hopping' from the neighbouring westerly island. This is represented in the diagram below.

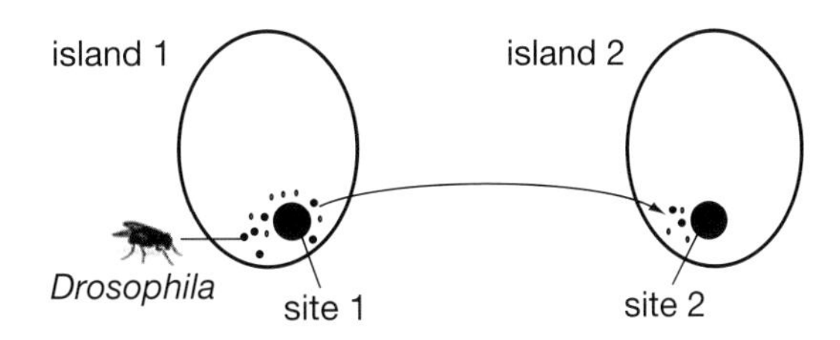

c. Explain how the new and old colonies became separate species. **3 marks**

Question 14 (6 marks)

Galápagos tortoises are distributed across most of the Galápagos Islands. The tortoises on different islands have differences in their shell and neck morphology.

Some of the tortoises have a saddle-backed shell; these shells are curved around the neck and allow the tortoise to lift their heads high off the ground. Other tortoises have a dome-shaped shell; these prevent the tortoises from lifting their heads up high. Pinta Island, Española Island and Isabela Island are all part of the Galápagos Islands, but they each have different climates and vegetation.

These three islands are all a significant distance from each other. The tortoises of each of the islands are different species as shown in the diagram.

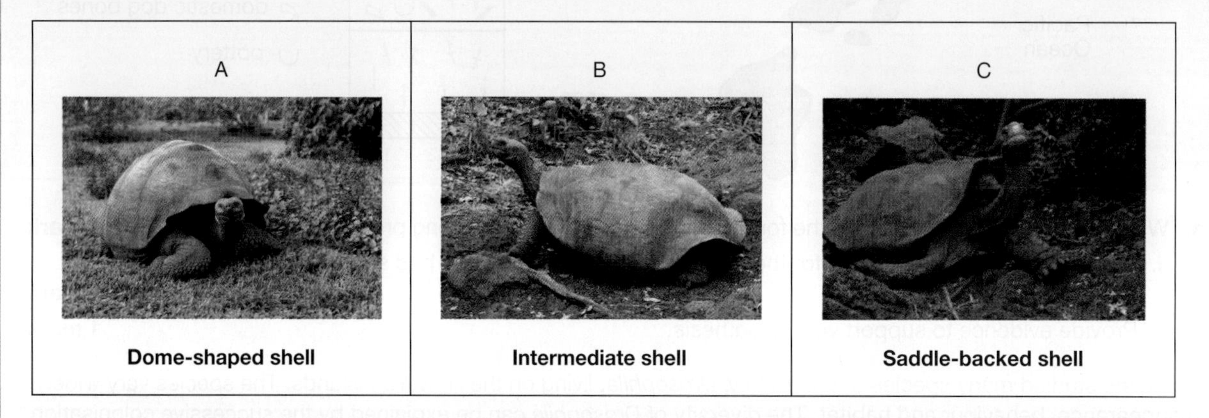

The correlation between the vegetation of the particular island and the morphology of the tortoises living on the island is outlined in the table.

Tortoise	Shell shape	Neck length	Vegetation on the island
Española Island tortoise	Saddle-shaped	Long	Arid environment with sparse ground vegetation and tall cacti
Isabela Island tortoise	Dome-shaped	Short	Abundant rainfall and ground vegetation
Pinta Island tortoise	Dome-shaped with a slight saddle	Medium	Ground vegetation and tall cacti.

a. The three tortoises are different species. Define the term species. **1 mark**

b. Scientists have identified with DNA evidence that all Galápagos tortoises have descended from one tortoise species that lives in South America, and that they became distinct species through the process of speciation.

- **i.** Identify the type of speciation that occurred. Justify your response. **2 marks**
- **ii.** Considering the data provided in the diagram and table, outline how the tortoises evolved different shell shapes. **3 marks**

Question 15 (5 marks)

Horses have been evolving for over 40 million years. Horse species evolved to become larger in size and for their toes to develop into a singular toe or hoof. This is unique to this family — most vertebrates have five toes on each digit. The diagram shows the evolution of different horse species in relation to their size and toe structures.

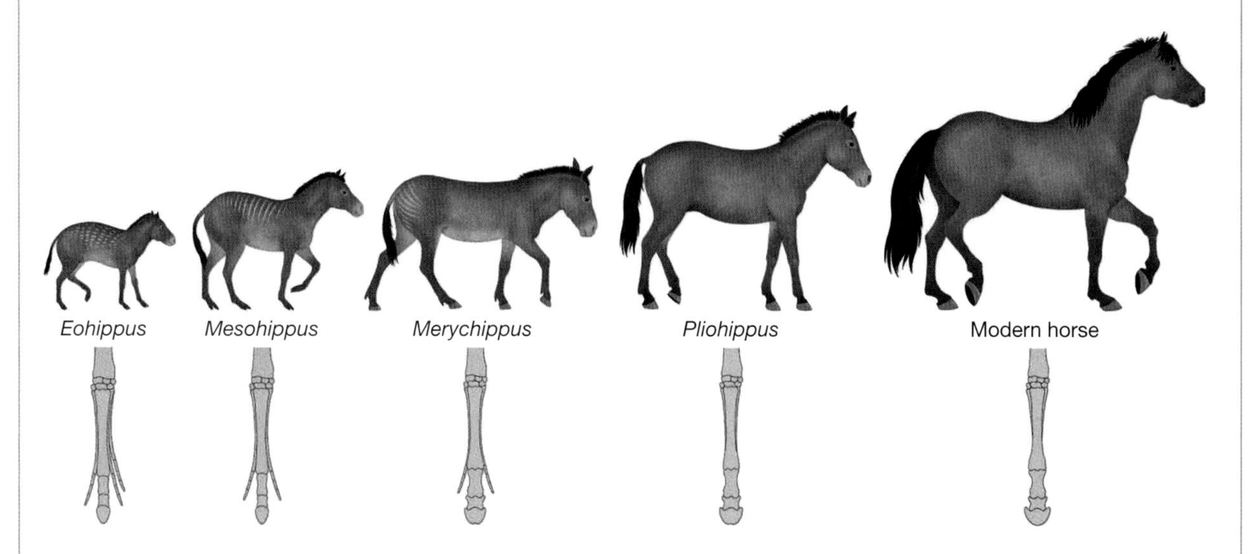

The drawings of the species *Euhippus, Mesohippus, Merychippus* and *Pliohippus* are based from fossils. *Mesohippus* and *Merychippus* are considered to be transitional species.

a. Identify why *Mesohippus* and *Merychippus* are transitional species. **1 mark**

b. Describe how the presence of transitional species provides evidence of evolution. **2 marks**

A group of palaeontologists found a new *Pliohippus* fossil specimen and calculated its absolute age to be 12 million years.

c. Explain which absolute age technique would have been used by the scientists to calculate the fossil's age. **2 marks**

8.5 Exercise 3: Biochallenge online only

eWorkbook Biochallenge — Topic 8 (ewbk-8089)

Solutions Solutions — Topic 8 (sol-0664)

Past VCAA examinations

Sit past VCAA examinations and receive immediate feedback, marking guides and examiner's report notes.
Access Course Content and select 'Past VCAA examinations' to sit the examinations online or offline.

teachon

Test maker

Create unique tests and exams from our extensive range of questions, including past VCAA questions.
Access the assignments section in learnON to begin creating and assigning assessments to students.

Online Resources

Below is a full list of **rich resources** available online for this topic. These resources are designed to bring ideas to life, to promote deep and lasting learning and to support the different learning needs of each individual.

eWorkbook

- 8.1 eWorkbook — Topic 8 (ewbk-1887) ☐
- 8.2 Worksheet 8.1 Geological time (ewbk-2913) ☐
- Worksheet 8.2 Changing species — exploring mass extinctions over time (ewbk-2914) ☐
- Worksheet 8.3 Fossilisation and types of fossils (ewbk-2915) ☐
- 8.3 Worksheet 8.4 Transitional and index fossils (ewbk-2916) ☐
- Worksheet 8.5 Absolute and relative ages (ewbk-2917) ☐
- Worksheet 8.6 Reviewing fossils (ewbk-2918) ☐
- 8.4 Worksheet 8.7 Understanding mechanisms of isolation (ewbk-2919) ☐
- Worksheet 8.8 Allopatric and sympatric speciation (ewbk-2920) ☐
- 8.5 Worksheet 8.9 Reflection — Topic 8 (ewbk-2922) ☐
- Biochallenge — Topic 8 (ewbk-8089) ☐

Solutions

- 8.5 Solutions — Topic 8 (sol-0664) ☐

Practical investigation eLogbook

- 8.1 Practical investigation eLogbook — Topic 8 (elog-0193) ☐
- 8.2 Investigation 8.1 Geological time (elog-0076) ☐
- Investigation 8.2 Examining casts, moulds and impressions (elog-0077) ☐
- Investigation 8.3 Exploring the process of fossilisation (elog-0078) ☐
- 8.3 Investigation 8.4 Modelling radioactive decay (elog-0079) ☐
- 8.4 Investigation 8.5 Modelling allopatric and sympatric speciation (elog-0080) ☐

Digital documents

- 8.1 Key science skills — VCE Biology Units 1–4 (doc-34326) ☐
- Key terms glossary — Topic 8 (doc-34622) ☐
- Key ideas summary — Topic 8 (doc-34613) ☐
- 8.2 Case study: Changing life forms over time evidenced with fossils (doc-36107) ☐
- Case study: Australian megafauna (doc-36108) ☐
- Extension: Mass extinction (doc-36110) ☐
- 8.3 Case study: The evolution of horses (doc-36109) ☐
- Extension: Electron spin resonance (doc-36111) ☐

Teacher-led videos

- Exam questions — Topic 8 ☐
- 8.2 Sample problem 1 Describing and identifying fossils (tlvd-1727) ☐
- 8.3 Sample problem 2 Analysing fossils (tlvd-1728) ☐
- Sample problem 3 Using absolute age to determine the age of fossils (tlvd-1729) ☐
- 8.4 Sample problem 4 Analysing allopatric speciation (tlvd-1730) ☐

Video eLessons

- 8.3 Fossils: evidence of past life (eles-4163) ☐
- Transition fossils (eles-4370) ☐
- Relative age of rocks (eles-4371) ☐
- Absolute age (eles-4203) ☐
- 8.4 How a new species evolves (eles-0162) ☐
- Allopatric speciation (eles-4372) ☐

Interactivities

- 8.3 The evolution of the horse (int-5878) ☐
- Transition fossil: evidence of evolution (int-0240) ☐
- Relative age of rocks (int-0238) ☐
- Forming daughter isotopes (int-5876) ☐
- Absolute age (int-0239) ☐

Weblinks

- 8.2 Understanding scales in geological time ☐
- Time scales interactive ☐
- Megafauna ☐
- Understanding *Rooneyia viejaensis* ☐
- Types of fossils ☐
- La Brea Tar Pits ☐
- 8.3 Evolution of horses ☐
- The fossil record resources ☐
- 8.4 Investigating reproductive barriers ☐
- Types of speciation ☐
- Sympatric speciation in palms on an oceanic island ☐

Exam question booklet

- 8.1 Exam question booklet — Topic 8 (eqb-0019) ☐

Teacher resources

There are many resources available exclusively for teachers online.

To access these online resources, log on to **www.jacplus.com.au**

9 Determining the relatedness over time

KEY KNOWLEDGE

In this topic, you will investigate:

Determining the relatedness over time

- Evidence of relatedness between species: structural morphology — homologous and vestigial structures; and molecular homology — DNA and amino acid sequences
- The use and interpretation of phylogenetic trees to show evidence of relatedness between species.

Source: VCE Biology Study Design (2022–2026) extracts © VCAA; reproduced by permission.

PRACTICAL WORK AND INVESTIGATIONS

Practical work is a central component of learning and assessment. Experiments and investigations, supported by a **practical investigation eLogbook** and **teacher-led videos**, are included in this topic to provide opportunities to undertake investigations and communicate findings.

9.1 Overview

Numerous **videos** and **interactivities** are available just where you need them, at the point of learning, in your digital formats, learnON and eBookPLUS and at **www.jacplus.com.au**.

9.1.1 Introduction

The millions of different species of plants, animals and microorganisms that live on Earth today are related by descent from common ancestors. What does it mean to be related? How do we decide which species are the most closely related? How do we decide which species branched off from which?

FIGURE 9.1 The evolutionary tree of life, showing diversification, branching and key characteristics of different species over time

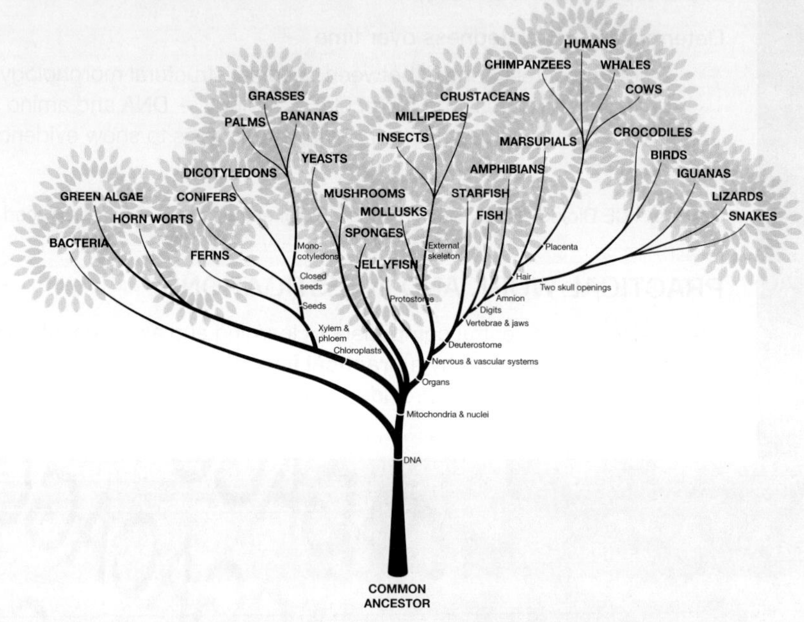

If species are related by evolution, they should show similarities in their structural morphology and molecular homology. Further, a greater degree of similarity should be evident in those species that are more closely related by descent and shared a common ancestor more recently than other species that are more distantly related.

This topic will investigate how homologous and vestigial structures of organisms can be used as evidence of relatedness between species. It will explore how comparisons of amino acid sequences and DNA between species can determine the extent of their relatedness. The interpretation and construction of phylogenetic trees to visualise relatedness will also be studied.

LEARNING SEQUENCE

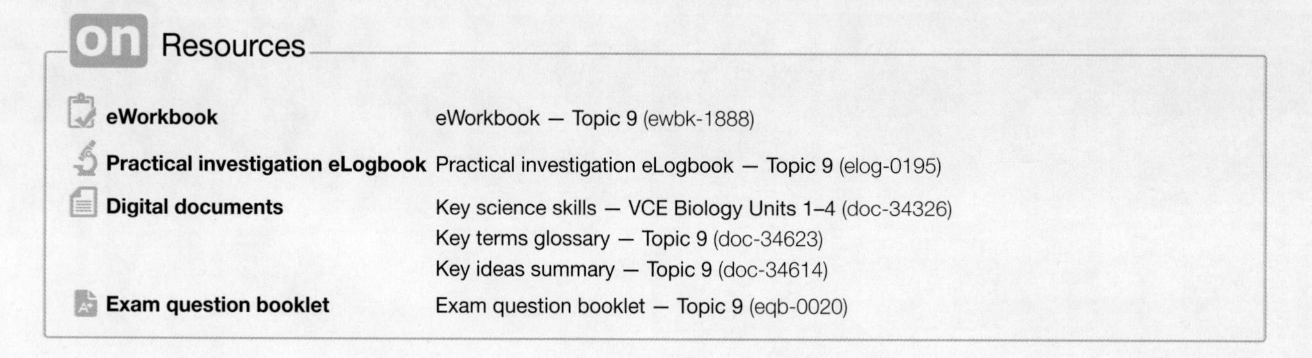

on Resources

eWorkbook	eWorkbook — Topic 9 (ewbk-1888)
Practical investigation eLogbook	Practical investigation eLogbook — Topic 9 (elog-0195)
Digital documents	Key science skills — VCE Biology Units 1–4 (doc-34326) Key terms glossary — Topic 9 (doc-34623) Key ideas summary — Topic 9 (doc-34614)
Exam question booklet	Exam question booklet — Topic 9 (eqb-0020)

9.2 Structural morphology

KEY KNOWLEDGE

- Evidence of relatedness between species: structural morphology — homologous and vestigial structures

Source: Adapted from VCE Biology Study Design (2022–2026) extracts © VCAA; reproduced by permission

9.2.1 What does it mean to be related?

In a biological sense, **relatedness** refers to how recently species split from a common ancestor. It is more than just familial relatedness, such as being related to aunties or cousins or siblings. Any species with a common ancestor is related. The question is not *are species related*? But instead *what degree of relatedness is there between organisms or species*?

So, for figure 9.2, we may ask the question: *Is species A more closely related to species B or C or D?* The answer cannot come by comparing the similarity of habitat or way of life (niche), or even similarity in appearance.

FIGURE 9.2 Species A is more closely related to species C than to other species.

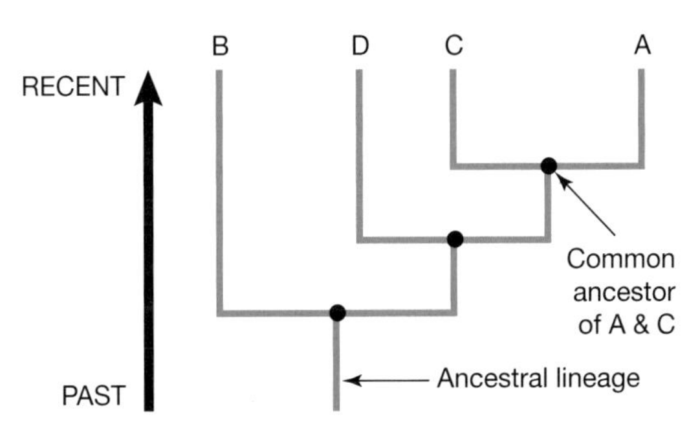

The answer comes from re-phrasing the question: *Does species A share a more recent common ancestor with species B, species C or species D?* The species (B, C or D) that shares this most recent common ancestor is the species that is most closely related to A. In figure 9.2, this is species C.

Many pieces of evidence are used to show relatedness between species. These include:

- **structural morphology:** comparing structures between individuals such as homologous and vestigial structures (this is often done through the fossil record)
- **molecular homology**: comparing DNA and amino acid sequences
- **developmental biology**: comparing the embryonic development of different organisms
- **biogeography**: exploring the locations of different species (both living and extinct species over geological time through fossils) and comparing this relative to continental drift.

relatedness how recently species split from a common ancestor

structural morphology the process of comparing similarities in body structures to infer relatedness

molecular homology the process involved in comparing similarities in molecular structures to infer relatedness, with a particular focus on DNA and amino acid sequences

developmental biology the process of comparing embryos of different species to infer relatedness

biogeography geographical distribution of species

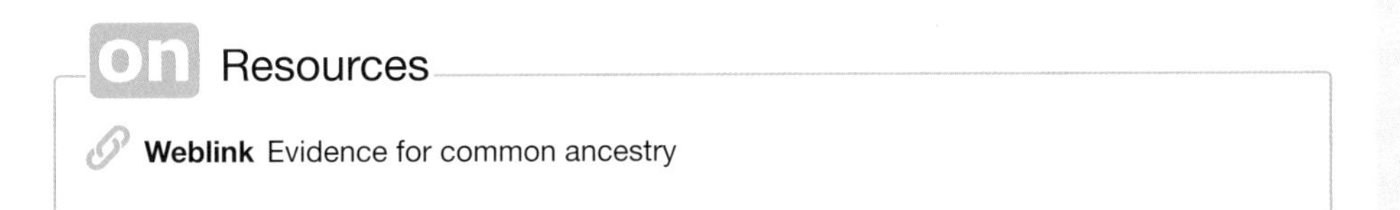

BACKGROUND KNOWLEDGE: Divergent evolution

Divergent evolution occurs when closely related species become more dissimilar over time, usually in response to different environmental conditions and different selection pressures. This can result in speciation, in which the species can no longer interbreed and produce fertile offspring.

Divergent evolution is the mechanism by which both homologous structures and vestigial structures arise and is due to the evolution of species from a common ancestor.

9.2.2 Homologous structures

The first mammals appeared about 200 million years ago and all are believed to have evolved from a reptilian ancestor. The various kinds of mammals that are alive today share a common ancestry.

If mammals are related by evolution from a common ancestor, it would be expected that they show similarities in their structural features, regardless of their way of life. This similarity of structure can be seen in bones of the forelimbs of various mammals (see figure 9.4).

Each of these mammalian limbs has a similar number of bones arranged in the same basic pattern. This similarity in basic structure exists, even though the limbs may serve different functions. The bones of the mammalian forelimbs are known as **homologous structures**.

> Homologous structures are those structures that have been derived from a common ancestor and thus show similarities in structure, even though they may have different functions.

FIGURE 9.3 Homologous structures are due to the process of divergent evolution, in which a structure is derived from a common ancestor.

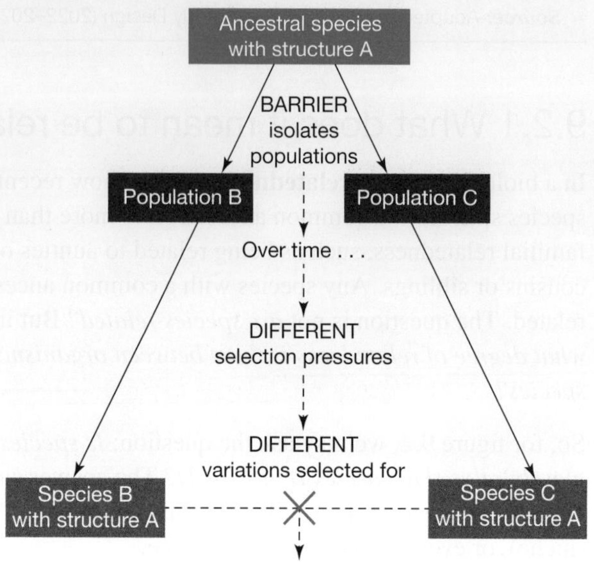

int-5879

FIGURE 9.4 The pentadactyl limb is a homologous structure derived from an ancestral vertebrate forelimb.

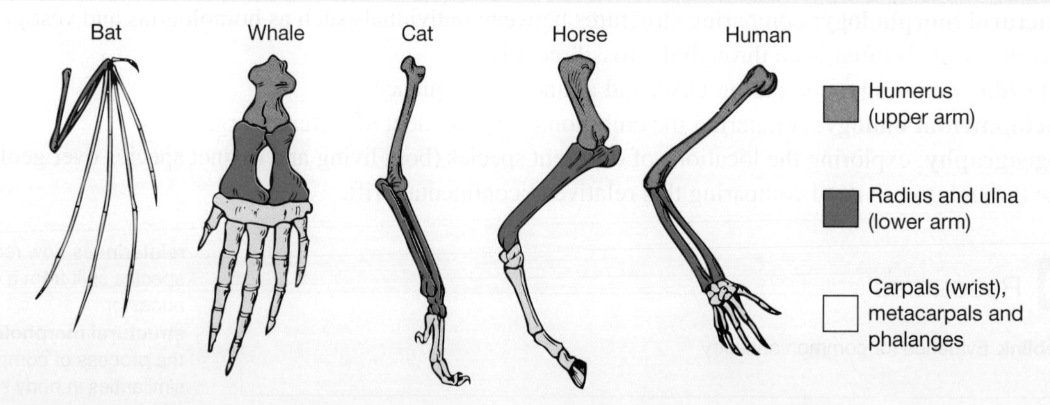

So, the bones of a bat's wing are homologous to those of a whale's flipper, even though one structure is specialised for flying and the other is specialised for swimming.

These morphological changes can be achieved through mutations that change the time or the place (when and where) genes are active. This includes through the use of master control genes, which control genes during embryonic development, and a subset of these known as homeotic genes, which control the anatomical structures during development.

Evolution commonly occurs by modification of pre-existing structures, not by the production of totally new structures. The homology seen in the pattern of bones in mammalian forelimbs illustrates how one basic pattern present in an ancestral species can, over time, be modified to produce a diversity of forelimbs adapted for flying, swimming, swinging and running.

homologous structures structures of organisms that have evolved from a common ancestor to have possible differences in form and function but retain similarities in structure

If we look at the limbs of terrestrial vertebrates, we will see that their bone structures are homologous. In the case of the evolution of the vertebrate limb, the pre-existing structures were the fins of lobe-finned fish. Fossils of these fish and also living lobe-finned fish have bones in their fins that are similar to the bones of modern terrestrial vertebrates.

elog-0073

INVESTIGATION 9.1

online only

Exploring homologous structures

Aim

To explore homologous structures that are derived from a common ancestor in different species

BACKGROUND KNOWLEDGE: Homologous or analogous?

Homologous structures are those that have a similar structure, but a different function. They are produced by the process of divergent evolution, by which new species have developed from a common ancestor. Homologous structures help determine relatedness between species.

This is in contrast with analogous structures. Analogous structures are those that have different structures but possess similar functions. Analogous structures often emerge when organisms are under similar selective pressures, so they develop and retain features that have similar functions and that are best suited to the environment. This is known as convergent evolution. The presence of analogous structures supports the theory of natural selection, in which advantageous traits under certain selective conditions are retained and passed on to the next generation.

One example of analogous structures are the fins of penguins, fish and dolphins. All of these structures enable these animals to better swim in an aquatic environment. However, they evolved independently due to similar selective pressures.

FIGURE 9.5 Fins, wings and flippers have similar functions in sharks, penguins and dolphins but have different structures and ancestral origins.

Another example of analogous structures is the development of wings in birds and insects. While these organisms do not have a common ancestor that possessed wings, they have all evolved the ability to fly, which has enhanced their survival. The wings of birds and insects, while providing the same function, have very different structures.

Analogous structures do not provide evidence of relatedness between species, as similar traits evolve independently.

FIGURE 9.6 Birds and insect wings are analogous rather than homologous.

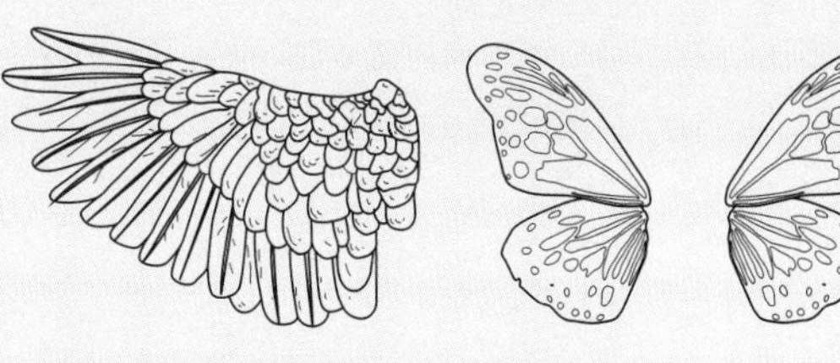

 Resources

Interactivity Comparative anatomy (int-0243)

9.2.3 Vestigial structures

Evidence of biological changes expressed in structural morphology may also be seen in the **vestigial structures** that are present in some species. If organisms are related by evolutionary descent, some species, because of changes in their way of life, will show the presence of functionless and reduced remnants of organs that were present and functional in their ancestor.

> Vestigial structures are structures that are non-functional remnants of structures that were functional in ancestral species.

A striking example of this change in structural morphology can be seen in whales. Whales are mammals that evolved from a terrestrial mammalian ancestor, but they are now specialised for life in the seas, although they still breathe air. The front limbs of whales have become flattened flippers, and no hind limbs are apparent.

FIGURE 9.7 The hip bone of a whale is an example of a vestigial structure.

However, their skeletons show the presence of a reduced pelvis and, in some cases, vestiges of the bones of the hind limbs (see figures 9.7 and 9.8).

The presence of vestigial hind limbs in whales reflects the fact that whales evolved from four-limbed terrestrial mammals. In this case, the modification of the pre-existing pattern involves the reduction and/or elimination of some bones. The modification that results in producing vestigial bones does not involve new gene mutations. Rather, it can be achieved by changing the regulation of existing genes, as, for example, by delaying or stopping the activation of specific genes. This is often evident in the control and expression of master genes, which control genes during development (refer to the extension box on master genes later in this subtopic).

FIGURE 9.8 A whale's skeleton showing the location of the vestigial hind limbs

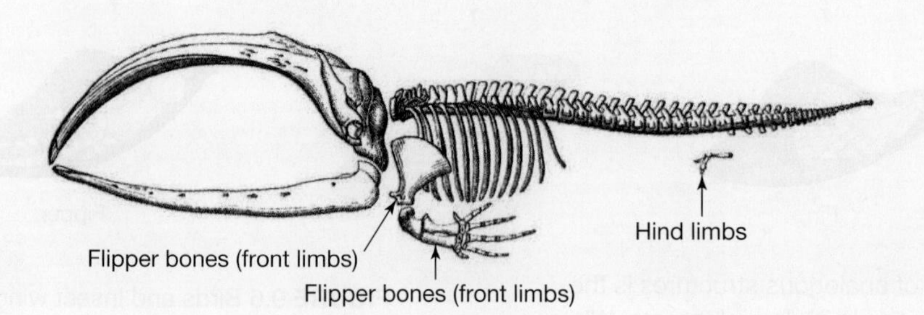

vestigial structures non-functional structures that are remnants of functional structures in ancestral species

Vestigial structures in the human body

Just like the whales, the human body has remnants of our ancestors present as vestigial structures. The following are some of the vestigial structures which can be present in humans (see figure 9.9):

FIGURE 9.9 Some of the vestigial structures present in humans

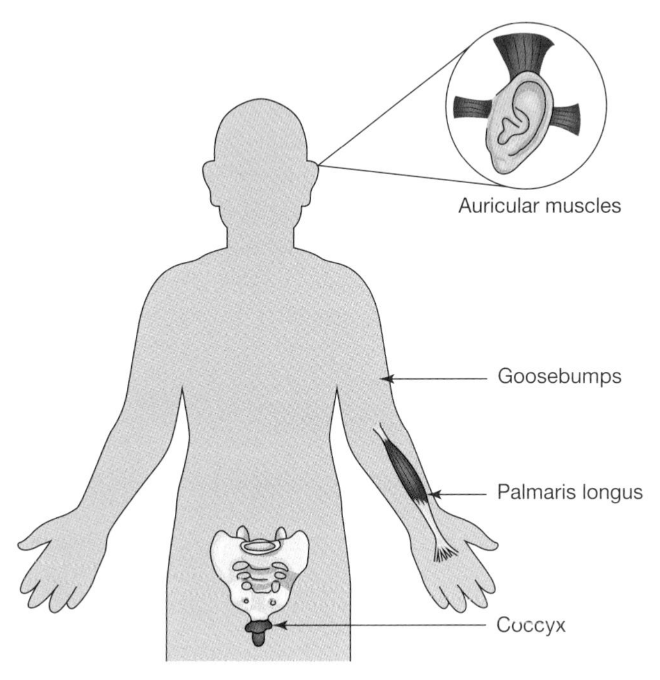

- **Goosebumps**. When you are cold or sometimes scared, your body can produce goosebumps. This is otherwise known as a pilomotor reflex. The pilomotor reflex is considered to be a vestigial structure in humans because it no longer carries out the same function it did for our ancestors. Our ancestors had more hair than modern humans do so, like cats and birds, when they had goosebumps it insulated the body by trapping warm air against the skin. This reflex also makes animals appear larger, which can deter predators from attacking them. As human hair is sparsely spread over the majority of the body, goosebumps no longer perform this function.

FIGURE 9.10 The formation of goosebumps does not lead to body insulation like it did in our ancestors.

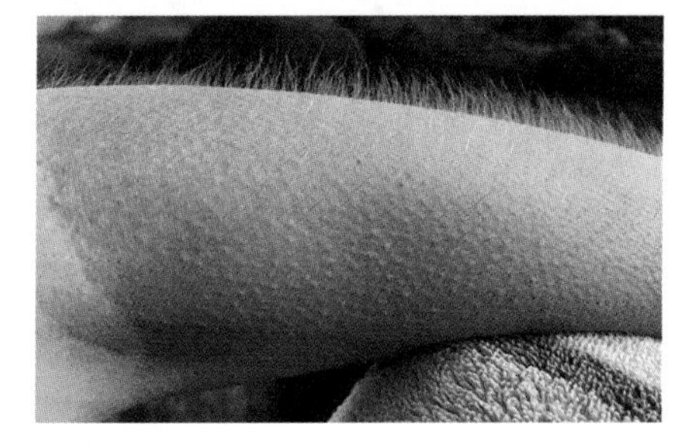

- **Forearm muscle**. There is an elongated and slender forearm muscle called the *palmaris longus* which can be present in both arms, one arm or neither in humans. This muscle is more developed in animals such as monkeys and lemurs which use their forearms to swing through trees. Approximately 10 per cent of people do not have the *palmaris longus* on either arm, but this does not have a negative impact on their grip strength. This vestigial structure provides evidence that human ancestors lived an arboreal life.
  Do you have *palmaris longus* muscles on either arm? Place your hands on a flat surface palms up and then bring your thumb and little finger together and stretch your wrist back slightly. If there is a thin band raised in the middle of your wrist you have the muscle (shown in figure 9.11).

FIGURE 9.11 The *palmaris longus* is still present in many individuals, but no longer serves a purpose.

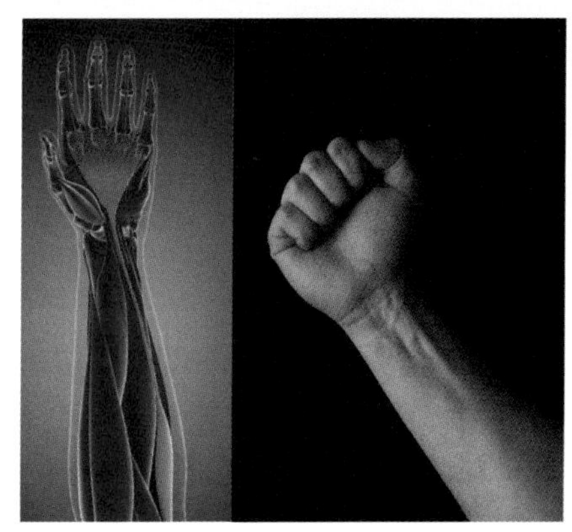

- **Ear muscles**. Can you wiggle your ears? For the majority of people, the answer will be no, but if you can you are using a vestigial structure. The ear has three external auricular muscles which most people do not use. However, these muscles are evidence that our ancestors were once able to move their ears, a trait that assists other mammals in locating the direction of sounds. This trait can be seen in monkeys, cats and dogs; when they detect a sound their whole ear can move in the direction of the sound.
- **Coccyx**. The coccyx bone is at the base of our vertebrae and is more commonly referred to as the tailbone. The presence of the coccyx is evidence that our human ancestors once had tails. For other primates such as monkeys and lemurs, a tail provides increased balance and mobility for living an arboreal life. However, the regulatory genes that control the development of the human tail 'switch off' after four weeks and the human tail does not continue to grow. There have been documented cases of babies being born with a 10- to 12-centimetre tail and these have been surgically removed.

FIGURE 9.12 The coccyx is a vestigial remnant of the tail.

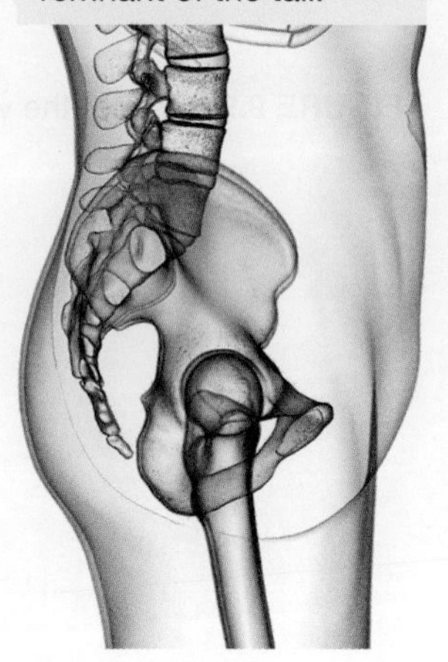

CASE STUDY: The changing ideas on the appendix

The appendix in humans was once considered a vestigial structure with no apparent function. However, recent research suggests that the appendix is not vestigial, having various functions in immunity and the protection of beneficial gut bacteria. The appendix, therefore, should not be listed as an example of a vestigial structure.

FIGURE 9.13 The appendix is no longer considered to be a vestigial structure.

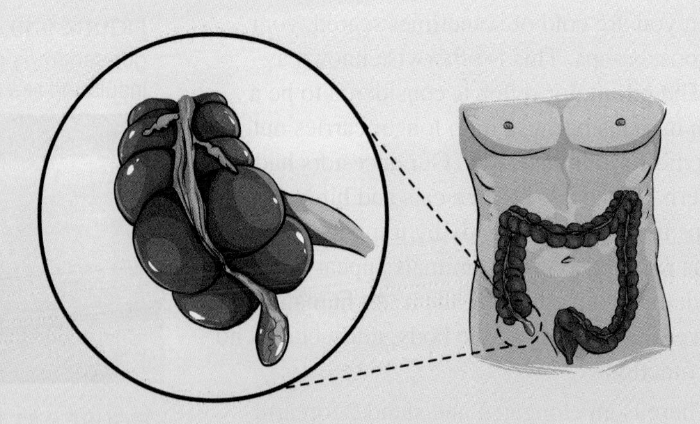

TIP: Be careful with using the term 'vestigial organs' rather than 'vestigial structures'. Not all structures discussed are technically organs (such as goosebumps), so it can lose you marks when used incorrectly. The Study Design specifies the term 'vestigial structures' so this should be used when you are answering questions.

elog-0074

INVESTIGATION 9.2

online only

Observing vestigial structures

Aim

To investigate vestigial structures in humans and compare this to class and population data

tlvd-1720

SAMPLE PROBLEM 1 Examining structural morphology and their evidence for evolution

Emus are large flightless birds that are native to Australia. They can reach 1.9 metres in height and are covered in feathers. They have wings measuring 20 centimetres on both sides of their body. The wings also have small claws at their tip. Emus can travel long distances and can run at speeds of up to 50 kilometres per hour.

a. **Identify the type of structure that emus' wings are.** **(1 mark)**

b. **Describe how this structure can be used to show the relatedness of emus to other birds that can fly.** **(2 marks)**

THINK	WRITE
a. 1. Identify what the question is asking you to do. The question is asking you to *identify*. In an identify question, you can list or name an answer with one word or a short sentence.	
2. Consider the structures that you need to compare in the Study Design: homologous structures and vestigial structures. An emu's wings are present but they are functionless. Which structure matches this?	Vestigial structure (1 mark)
b. 1. Identify what the question is asking you to do. The question is asking you to *describe*. In a describe question, you give a detailed account in sentence form.	
2. Consider how a vestigial structure can be used as evidence for relatedness. Why do emus have wings that are functionless and how does this determine relatedness? This question is worth two marks so needs two distinct points.	The emu's vestigial wings are now functionless but they would have been functional in their ancestors (2 marks). This provides evidence that emus share a common ancestor with other birds which have flight (2 marks).

Resources

eWorkbook Worksheet 9.1 Exploring homologous and vestigial structures (ewbk-2905)

Weblinks TED-Ed: Vestigial structures

7 vestigial features of the human body

EXTENSION: Master genes and evolutionary developmental biology (evo-devo)

The importance of master control and homeotic genes are thought to be incredibly significant in the evolution of species, particularly with the appearance of both vestigial and homologous structures and the differences that exist between them. This idea of evolutionary developmental biology (or evo-devo) compares the processes involved in the developmental processes of organisms and how these link to the evolution of species. Evo-devo allows us to understand not just only how structures form in organisms, but also how minor differences in structures exist between both homologous and vestigial structures in different species.

One example of a master gene involved in evolutionary developmental biology is *BMP4*, a master control gene expressed during embryonic development. The *BMP4* gene is expressed in a signalling molecule, the bone morphogenetic protein. A shift in the time of expression of the *BMP4* gene, the level of expression or where it is expressed can produce great levels of diversity in a homologous structure. For example, it affects the jaw structure and teeth in cichlid fish in African Lakes (shown in figure 9.14) and the formation of different beak structures in Galapagos finches.

FIGURE 9.14 Different jaw structures in cichlid fishes are affected by master genes that are expressed during development.

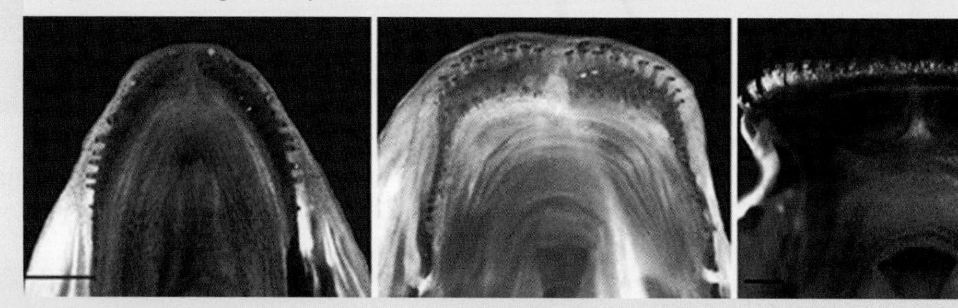

Thus, the regulation of genes, particularly those involved in development, is important in evolutionary biology. Further information about homeotic genes can be found in topic 1 in section 1.6.1.

EXTENSION: Developmental biology

Developmental biology is the study of the processes that result in the growth and development of multicellular organisms. These studies include comparing the development of different species in order (1) to identify their evolutionary relationships, and (2) to recognise how developmental processes have evolved.

Ancestral traits often appear and disappear at different stages of the embryological development of an organism. Because they share a common ancestry, all vertebrate embryos display some common features at some point during their development.

Regardless of whether or not they are present in their adult structure, all vertebrates display the following features during at least some period of embryonic development, as shown in figures 9.15 and 9.16:

- a tail, located posterior to the anus
- a cartilaginous notochord (in some jawless fish, it is retained, whereas in most vertebrates, it is replaced by the backbone or spine).
- pharyngeal arches (which develop into gills in aquatic vertebrates, but facial and jaw structures in terrestrial vertebrates).

FIGURE 9.15 Pharyngeal arches and tails are evident in the embryos of many organisms.

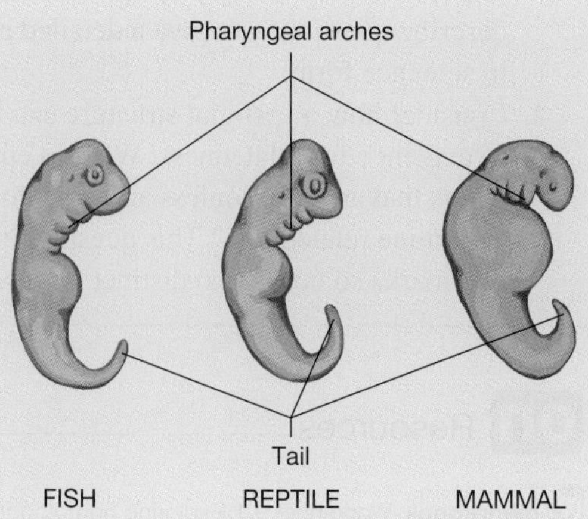

FIGURE 9.16 Comparing the embryonic development of various animals provides evidence of common ancestors and evolutionary patterns.

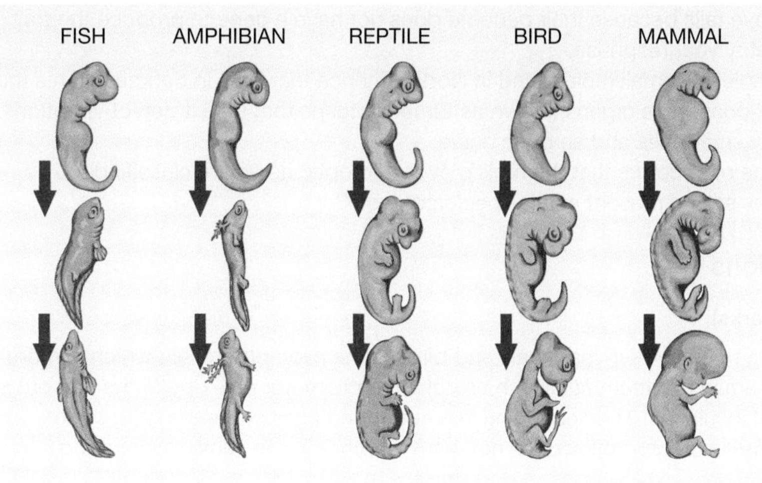

KEY IDEAS

- Evidence of relatedness between species can be shown with structural morphology as homologous structures and vestigial structures.
- Homologous structures show similarities in structures of different species due to a common ancestor.
- Homologous structures show how different species have adapted to different environmental selection pressures over time.
- Vestigial structures are features of an organism that have lost all or some of their original function.

9.2 Activities

learnon

To answer questions online and to receive **immediate feedback** and **sample responses** for every question, go to your learnON title at **www.jacplus.com.au**. A **downloadable solutions** file is also available in the resources tab.

9.2 Quick quiz on | **9.2 Exercise** | **9.2 Exam questions**

9.2 Exercise

1. Define the term *homologous structure*.
2. Using the figure shown, identify a similarity and a difference between the human and horse forelimb.

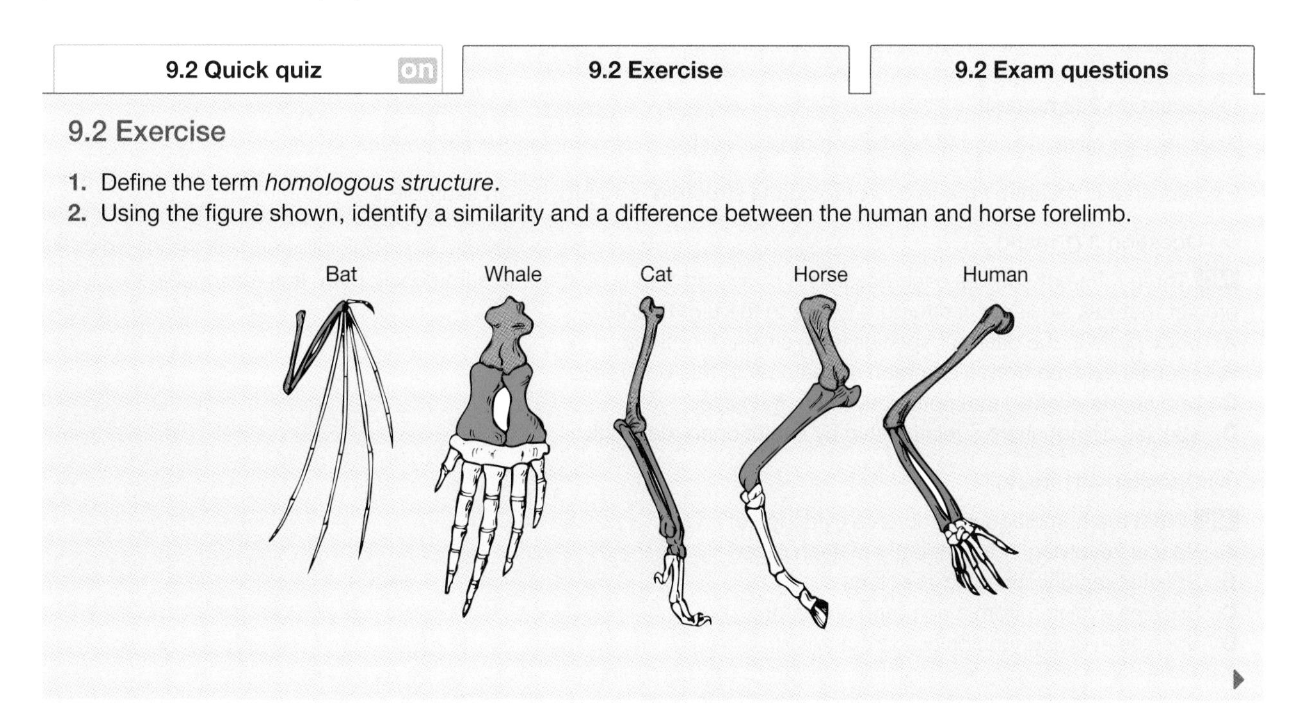

3. Explain why homologous structures provide evidence that all mammals have evolved from a common ancestor.
4. Explain why vestigial structures provide evidence of relatedness between species.
5. Human do not have tails because their genome does not have a gene to produce the trait. Is this statement true or false? Justify your response.
6. The star-nosed mole is a small mole found in North America that lives in darkness. This mole is completely blind and instead possesses organs known as Eimer's organs that help it detect vibrations. Despite this, the mole still has very small eyes and an optic nerve.
   a. Identify the type of structure that the mole's eyes and optic nerve are classified as.
   b. Why might eyes still be present in this species of mole?

9.2 Exam questions

Question 1 (3 marks)

A response to cold temperatures in mammals and birds is the pilomotor reflex, which is more commonly known as goosebumps in humans. In mammals such as cats, the pilomotor reflex puffs the hair out straight, creating an additional layer of insulation. The heat being lost is trapped against the skin and the cat can maintain warmth in cold conditions. The pilomotor reflex does not warm humans in the same manner and is considered to be a vestigial structure.

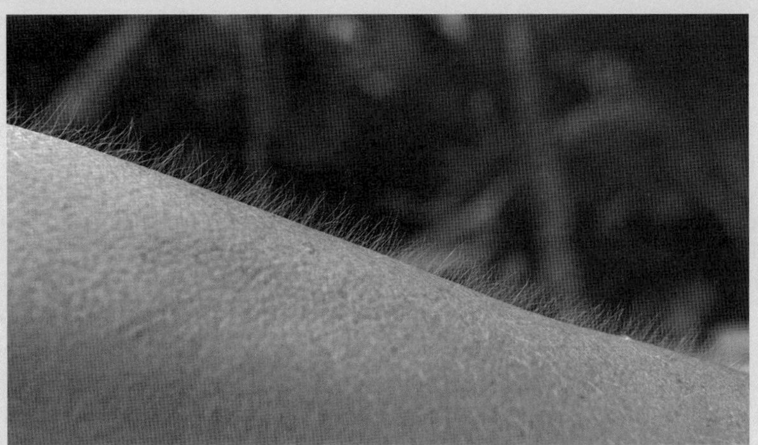

a. Define the term *vestigial structure*. **1 mark**
b. Describe how the pilomotor reflex can be used as evidence of relatedness between humans and cats. **2 marks**

Question 2 (3 marks)

Species that have descended from a common ancestor have homologous structures.

Explain what is meant by the term 'homologous structure' and provide an example.

Question 3 (1 mark)

MC Three different species each possess a body part that is built to the same basic plan. It is reasonable to predict that this similarity in structure is due to the fact that the

A. body parts in the different species have similar functions.
B. species evolved from a common ancestor with this body part.
C. body parts evolved independently in each species.
D. species do not share a relationship by evolutionary descent.

Question 4 (1 mark)

MC What are homologous structures evidence of?

A. Species evolving independently of each other
B. Species rapidly changing over time
C. Species evolving from a common ancestor
D. Species evolving funcionless structures

Question 5 (1 mark)

MC There are many remnants of structure in the human body that were functional in our ancestors but no longer serve a vital purpose for our species. One example is in a tendon known as the *palmaris longus*, which is still present in some members of *Homo sapiens* and missing from many others. This tendon was thought to mostly be used for climbing in our ancestors.

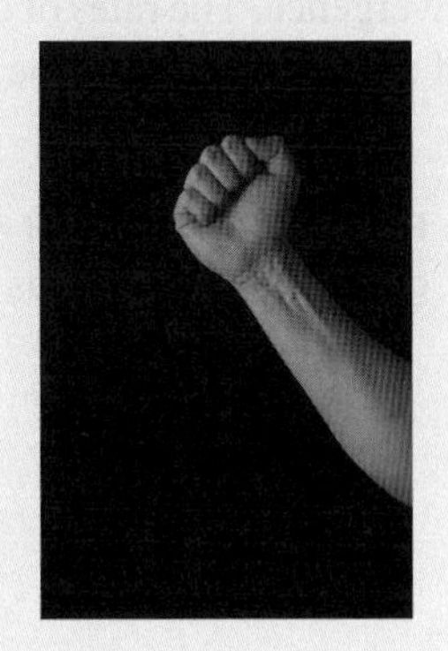

What is the most appropriate name for this structure?

A. A remnant structure
B. A homologous structure
C. A vestigial structure
D. A transitional structure

More exam questions are available in your learnON title.

9.3 Molecular homology

KEY KNOWLEDGE

- Evidence of relatedness between species: molecular homology — DNA and amino acid sequences

Source: Adapted from VCE Biology Study Design (2022–2026) extracts © VCAA; reproduced by permission

9.3.1 Comparing amino acid sequences

The proteins of all organisms — whether they are jellyfish, tomatoes, lobsters, ferns or people — are composed of the same set of 20 **amino acids**. Likewise, the genetic code that carries the information for making proteins is essentially the same in all organisms (refer back to subtopic 1.4 to review the genetic code). These observations are consistent with a common evolutionary ancestry for all living organisms.

Proteins from different species can be compared in terms of their amino acid sequences. Species that are more closely related are expected to have fewer differences in the amino acid sequences of their corresponding proteins than species that are more distantly related.

The number of protein differences between species can help determine relatedness. It is reasonable to conclude that, the more differences that are observed in the corresponding protein in two species, the further back their divergence in time. This is because there is more time available for changes to accumulate in DNA and, in turn, within proteins that are present in both species.

What proteins do we analyse?

When analysing proteins from different species, it is usually more helpful to compare proteins that have been conserved across species. This is the case for many proteins, particularly those involved in development and regulation; those involved in process such as protein synthesis and DNA storage (such as histones); and those involved in ATP production.

Some of these proteins (and the genes that code for them) are termed **orthologous**, which means they can be inferred as being inherited from a common ancestral species. This is very powerful in determining relatedness between species. Examples of orthologous proteins are cytochrome *c* and haemoglobin.

amino acids basic building blocks or sub-units of polypeptide chains and proteins

orthologous gene or proteins derived from a shared ancestor in which the primary function is conserved

It is also useful if the rate of change of a protein is reliable and constant, as it can better help estimate a time of divergence. The rates of change in both cytochrome *c* and haemoglobin are both relatively well known, with the rate of change in cytochrome *c* being relatively slow, and that of haemoglobin being faster, as shown in figure 9.17.

FIGURE 9.17 Rates of evolution for three different proteins. The different coloured bands represent different geological time periods and differences in the proteins between different species.

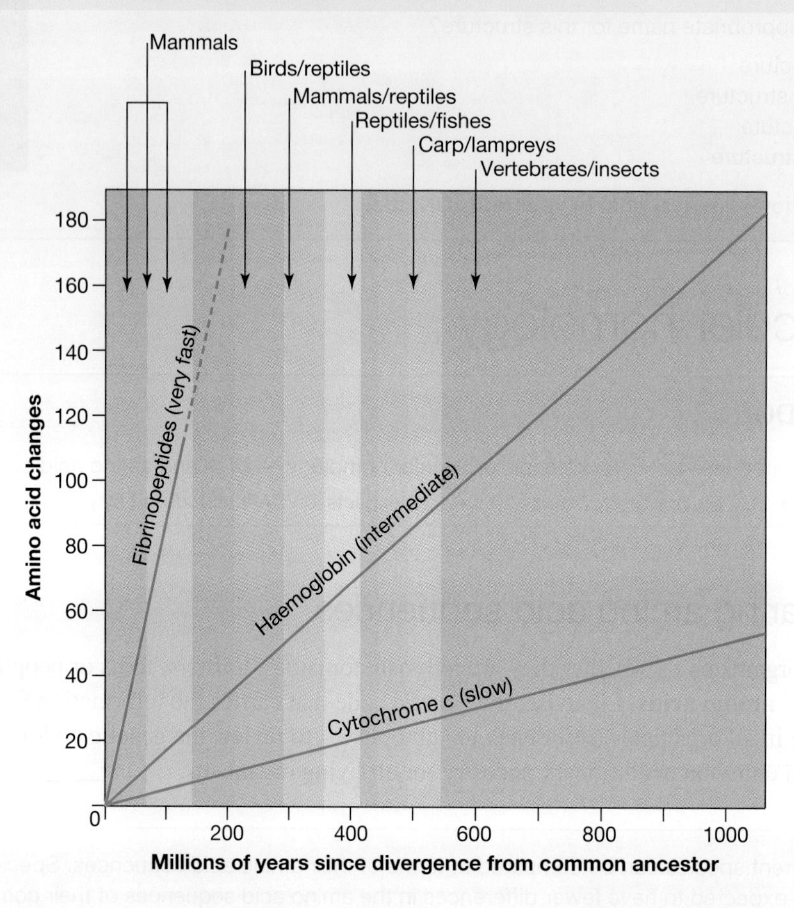

Comparing amino acids in haemoglobin

Haemoglobin is a vital protein that helps carry oxygen. In humans, it is mostly present in red blood cells. Haemoglobin is composed of up to four chains: two identical chains (alpha chains) consisting of 141 amino acids, and two other identical chains (beta chains) consisting of 146 amino acids.

The amino acid sequences of the haemoglobin chains have been identified for many mammals, birds, reptiles, amphibians, fish and invertebrates.

Table 9.1 shows the number of differences in the amino acid sequence of the beta chain of haemoglobin between humans and several species.

FIGURE 9.18 Haemoglobin is made up of four polypeptides.

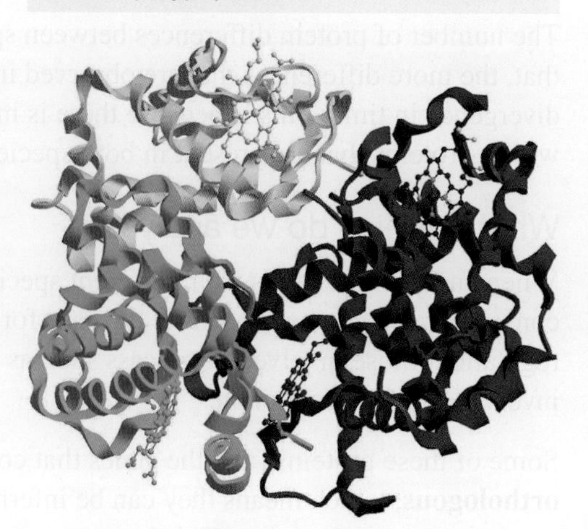

TABLE 9.1 Number of differences in the amino acid sequence of the beta chain of haemoglobin of various species compared with the human beta chain of haemoglobin

Vertebrate	Number of differences from human haemoglobin
Chimp (*Pan* spp.)	0
Gorilla (*Gorilla gorilla*)	1
Gibbon (*Hylobates* spp.)	2
Rhesus monkey (*Macaca mullata*)	8
Dog (*Canis familiaris*)	15
Mouse (*Mus musculus*)	27
Kangaroo (*Macropus* spp.)	38
Chicken (*Gallus gallus*)	45
Frog (*Rana* spp.)	67
Lamprey (*Petromyzon marinus*)	125
Mollusc (sea slug)	127

Based on this data, it is possible to estimate the relationships among the various species.

Data in this table support the conclusion that:

- gorillas are more closely related to humans than Rhesus monkeys or mice
- chickens are more closely to humans than frogs
- invertebrates (such as molluscs) are less related to humans than all other listed vertebrates.

Degrees of evolutionary relationships identified through this means are in agreement with relationships inferred from fossil evidence and from structural comparisons.

Comparing amino acids in cytochrome *c*

If other proteins are examined, a similar picture of relationships between organisms can be inferred.

An enzyme found in organisms from all of the five kingdoms is known as cytochrome *c*. Human cytochrome *c* has been compared with that from other organisms (see table 9.2).

FIGURE 9.19 The amino acid sequence of cytochrome *c*. Amino acids are coloured in accordance to particular properties.

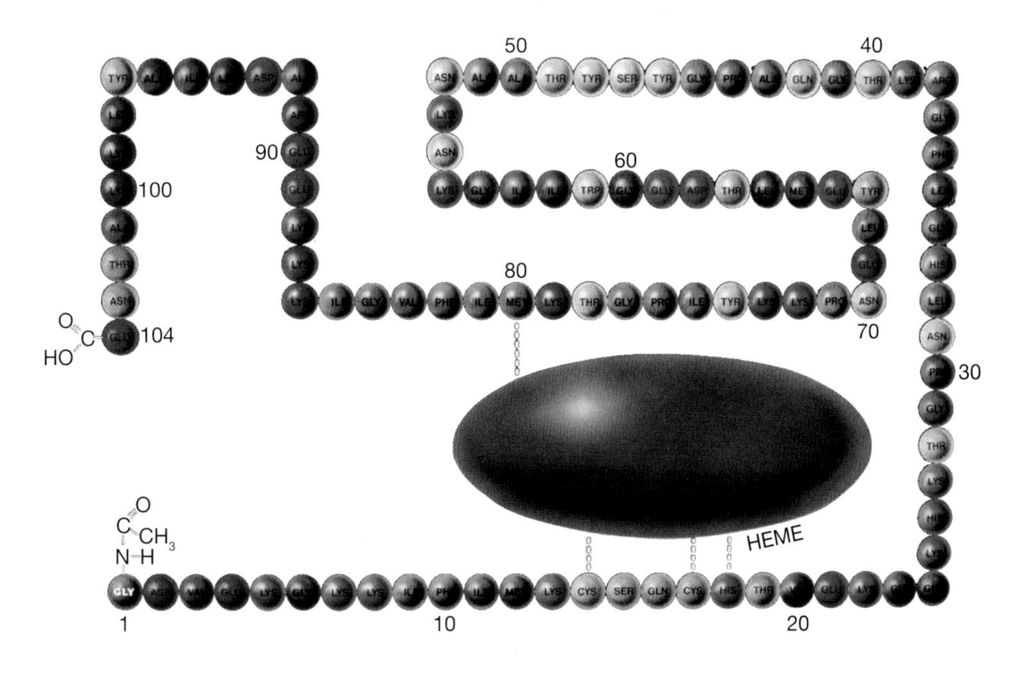

TABLE 9.2 Comparison of the sequence of amino acid sub-units in cytochrome *c* from various species. Cytochrome *c* also occurs in yeast cells. If compared with human cytochrome *c*, what would you predict?

Organism	Number of changes in amino acid sequence relative to the human cytochrome *c*
Human	—
Chimp	0
Rhesus monkey	1
Whale	7
Rabbit	9
Chicken	13
Fish (tuna)	22
Wheat	37

tlvd-1721

SAMPLE PROBLEM 2 Comparing amino acids sequences

Histones are a group of proteins that are found in the nucleus of eukaryotic cells. These proteins provide structural support to the chromosome by forming nucleosomes with chromatin. The table below shows the amino acid sequences (using the single letter codes) of a segment of Histone H1 of five different species. Amino acids that are conversed are highlighted.

Human	KKASKPKKAASKAPTKKPKATPVKKAKKKLAATPKKAKKPKTVKAKPVKASKPKKAKPVK
Chimp	KKASKPKKAASKAPTKKPKATPVKKAKKKLAATPKKAKKPKTVKAKPVKASKPKKAKPVK
Mouse	KKAAKPKKAASKAPSKKPKATPVKKAKKKPAATPKKAKKPKVVKVKPVKASKPKKAKTVK
Rat	KKAAKPKKAASKAPSKKPKATPVKKAKKKPAATPKKAKKPKIVKVKPVKASKPKKAKPVK
Cow	KKAAKPKKAASKAPSKKPKATPVKKAKKKPAATPKKAKKPKTVKAKPVKASKPKKAKPVK

By counting the number of differences between each organism and humans, and using the data for the molecular homology given for histone H1, identify the organisms which are:

a. most closely related to humans **(1 mark)**

b. most distantly related to humans. **(1 mark)**

THINK	WRITE
a. 1. Identify what the question is asking you to do. In an *identify* question, you can list or name an answer with one word or a short sentence.	
2. Compare the amino acid sequences for humans to the other four species. You may wish to make a table to help you to count and record the number of differences. **TIP** This question is only asking for a comparison to humans, so you don't need to calculate all differences between all species.	Chimps: 0 differences Mouse: 6 differences Rat: 5 differences Cow: 5 differences
3. If two species are closely related there are fewer differences between their amino acid sequences.	Chimp (1 mark)
b. 1. Use the same information as above, only this time identify the species with the most differences to the amino acid sequence.	Mouse (1 mark)

There are some challenges associated with amino acid sequencing. One important feature is due to the degeneracy of the genetic code, where multiple triplets can code for a particular amino acid. As such, some mutations, such as silent mutations, are not visible in amino acid sequencing and instead require DNA sequencing.

9.3.2 Comparing DNA

DNA sequences have been described as 'documents of evolutionary history'. Comparisons of DNA from different species may be made through:

1. direct comparison of DNA base sequences
2. comparison of whole genomes.

Comparing DNA base sequences

DNA molecules consist of a series of nucleotides containing nitrogenous bases that form a base sequence. The four nitrogenous bases found in DNA are adenine (A), cytosine (C), guanine (G) and thymine (T).

If evolution has occurred, we can predict that species closely related by evolutionary descent will show more similarities in the base sequences of their common genes.

Hence, direct comparisons of the DNA sequence of genes in different species can also be used to infer evolutionary relationships. This usually provides similar evidence to amino acid sequences (as the DNA base sequence provides the blueprint for the amino acid sequence).

For example, haemoglobin genes are present in all mammals. Sequences have been identified for the approximately 17 000 bases in this segment of DNA in human beings and other animals. The results show that these sequences are most similar between humans and chimpanzees.

Table 9.3 shows the DNA sequences from part of a haemoglobin gene.

TABLE 9.3 DNA sequences from a segment of a haemoglobin gene from four mammalian species — a human, orangutan, monkey (specifically a rhesus monkey) and rabbit. Differences to humans are highlighted in red. Note that dashes have been used to maintain the sequence alignment.

Species	DNA sequence of part of a haemoglobin gene
Human	TGACAAGAACA-GTTAGAG-TGTCCGAGGACCAACAGATGGGTACCTGGGTCCCAAGAAACTG
Orangutan	TCACGAGAACA-GTTAGAG-TGTCCGAGGACCAACAGATGGGTACCTGGGTCCCAAGAAACTG
Monkey	TGACGAGAACAAGTTAGAG-TGTCCGAGGACCAACAGATGGGTACCTGGGTCCCAAGAAACTG
Rabbit	TGGTGATAACAAGACAGAGATATCCGAGGACCAGCAGATAGGAACCTGGGTCCTAAGAAGCTA

Table 9.3 shows there are two differences between the DNA sequences of humans and the other primates: orangutans and rhesus monkeys. It also shows there are 13 differences between humans and rabbits. This indicates that, while all of the species are mammals, humans are more closely related to orangutans and rhesus monkeys than we are to rabbits.

EXTENSION: How is DNA sequenced?

The process of sequencing is automated and is done using instruments known as DNA sequencers (see figure 9.20). One automated system, known as the Sanger method, involves the use of four different coloured fluorescent dyes, each of which binds to a specific base (A, T, C or G) in DNA.

FIGURE 9.20 a. Applied Biosystems 3500 genetic analyser (Image courtesy of Applied Biosystems) b. Output from a DNA sequencer c. Scientist examining another sequence of bases in DNA

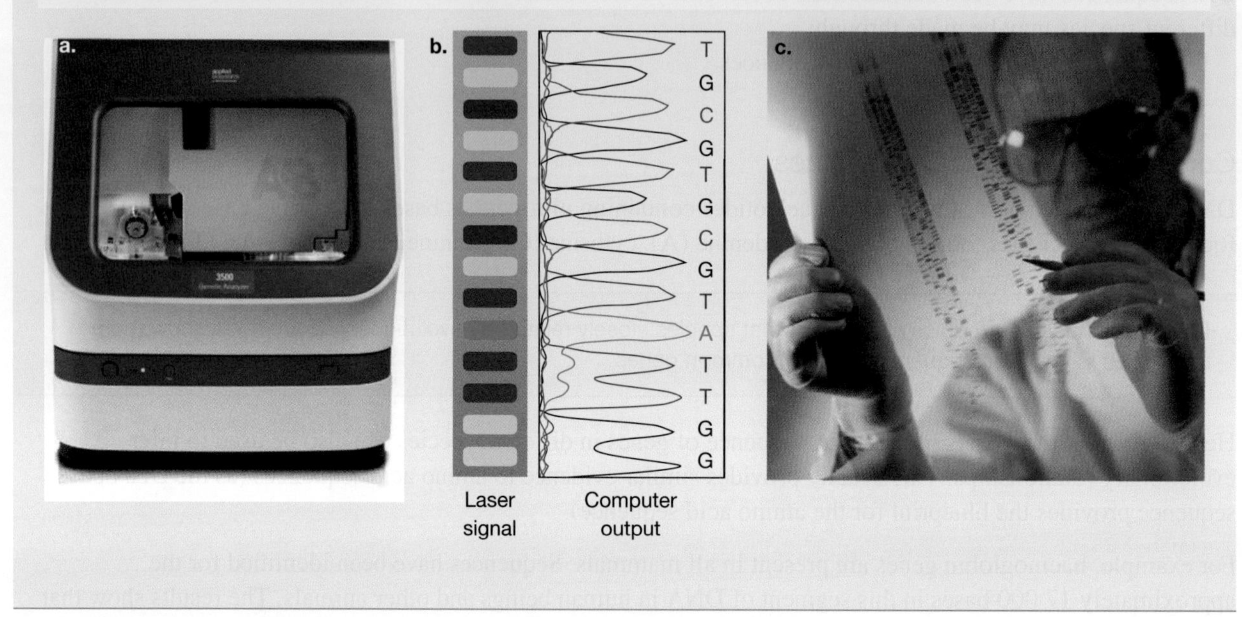

How can we use this to predict evolutionary relationships?

If evolution has occurred, we can predict that many of the genes present in an ancestral species will also be present in the species that evolved from it. If a particular gene or DNA sequence is present in members of a related group, that DNA sequence or that gene is said to be conserved among those species. The DNA sequences of conserved genes will, however, be very similar, such as the haemoglobin gene shown in table 9.3. It is reasonable to suggest that, when a DNA sequence (or a gene) is conserved, it has an important function.

Let's use a simple example to show how we can infer evolutionary relationships between species, based on differences in DNA sequences.

TABLE 9.4 Base sequences of a common segment of DNA of five related species

Species	DNA sequence
Sparrow	T C C A A C T C G T G C C T C G A T G A A G A C T A A G T T A T A C C A T A C A G A C G
Brown bear	T C G A G C T A G T G C A T C G A T G A C G A C T A A G T G A T A C C A T A A A G A C T
Cat	T C C A G C T C G T G T A T C G A T G A C G A C T A A G T G A T A C C A A A A A G A C T
Jaguar	T C C A G C T C G T G T A T C G A T G A C G A C T A A G T G A T A C C A A A A A G A C T
Elephant	T C C A G C T A G T G C A T C G A T G A C G A C T A C G T G A T A C C A T A A A G A C T

The number of differences between the species can be observed in table 9.5 (this in turn can be used to produce phylogenetic trees, which will be explored in subtopic 9.4).

TABLE 9.5 Number of base sequence differences (*Note:* Each comparison is only done once. Comparisons between individuals of the same species are not included.)

	Sparrow	Brown bear	Cat	Jaguar	Elephant
Sparrow		8	8	8	7
Brown bear			4	4	2
Cat				0	4
Jaguar					4
Elephant					

It can be assumed that the sparrow, which has many differences in its DNA sequence, is the least related to the other species shown. The cat and the jaguar are the most closely related, as no mutations appear to have occurred in the examined DNA sequence.

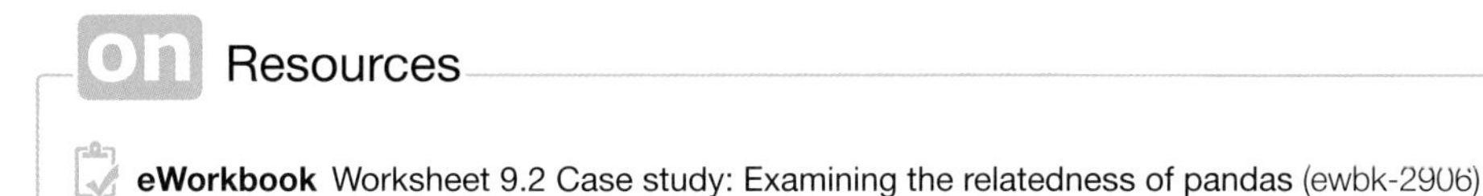

Resources

eWorkbook Worksheet 9.2 Case study: Examining the relatedness of pandas (ewbk-2906)

CASE STUDY: What DNA do we use — nuclear or mitochondrial?

When we consider DNA, we usually focus on nuclear DNA, as this is where a majority of genetic material is located. However, other organelles also contain DNA, such as the mitochondria. This is referred to as mitochondrial DNA (mtDNA).

Both nuclear DNA and mtDNA have their advantages when exploring evolutionary relationships. For example, nuclear DNA has a greater number of genes, including many that are conserved (in both coding and non-coding regions).

FIGURE 9.21 Comparing nuclear and mitochondrial DNA (mtDNA)

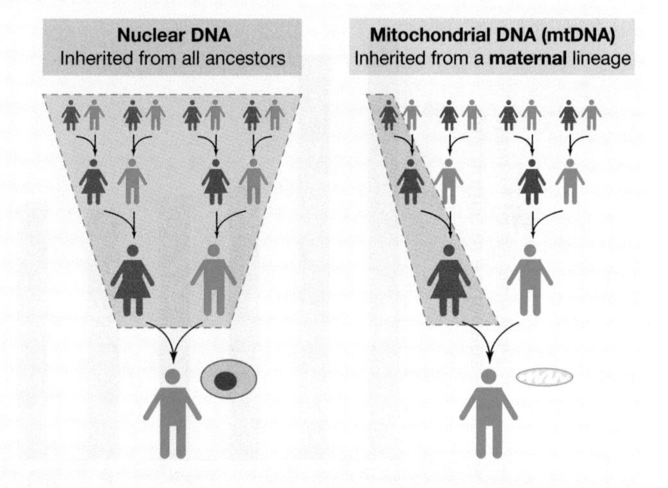

Mitochondrial DNA samples can be useful as:

- unlike nuclear/chromosomal DNA, it can be recovered from ancient skeletal remains such as teeth and bones (as there are many mitochondria per cell)
- it mainly passes from mother to children, generation after generation, barely altered by the recombination events that occur with chromosomal DNA in meiosis, so it is useful for tracking lineage
- it contains a non-coding region known as the D-loop. Within the D-loop are two regions of DNA that undergo mutational change at a much higher rate than the rest of the molecule.
- over the 200 000 years or so of the existence of modern humans *(Homo sapiens)*, many mutations have occurred in these hypervariable regions. As a result, mtDNA sequences differ between populations and between individuals.

The uses of mtDNA in determining evolutionary relationships and migration in humans will be further explored in topic 10. To access more information about case studies related to the use of mitochondrial DNA, please download the digital document.

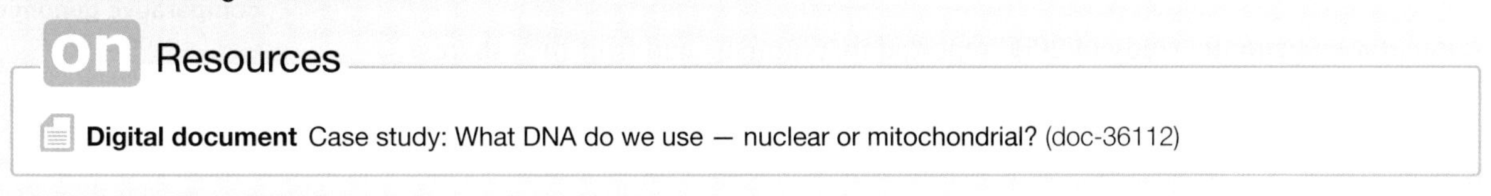

Resources

Digital document Case study: What DNA do we use — nuclear or mitochondrial? (doc-36112)

Comparing whole genomes

It is now possible to compare the genomes of different organisms — a field of study known as **comparative genomics**. The genome is the complete set of genes found in an organism. These comparisons can help to clarify the evolutionary history of species.

Because the amounts of data are so large (for example, the human genome contains 3000 million base pairs), computer technology is necessary for these studies. Information gained from comparative genomics has applications in medicine and industry.

Living species have evolved from common ancestors, and therefore the genomes of related species exhibit similarities. The more recent the divergence of two related species from a common ancestor, the greater the number of conserved DNA sequences and of their arrangement within the genome.

By comparing the genomes of living species, it is possible to:

- identify the degree of relationship between different species from the fraction of genes shared between their genomes
- make inferences about the phylogeny or evolutionary history of a species.

For example, human DNA was compared with DNA from 12 other vertebrate species (see figure 9.22). Note that DNA from the nonhuman species can be aligned or matched to the human DNA. Note also that the percentage of alignment varies, with the percentage being greatest between the most closely related species (as determined by other techniques). In this case, the greatest degree of similarity occurred in the human–chimp comparison.

FIGURE 9.22 Percentage of human DNA sequences that can be aligned in 12 vertebrate species. The different colours denote the DNA in different categories.

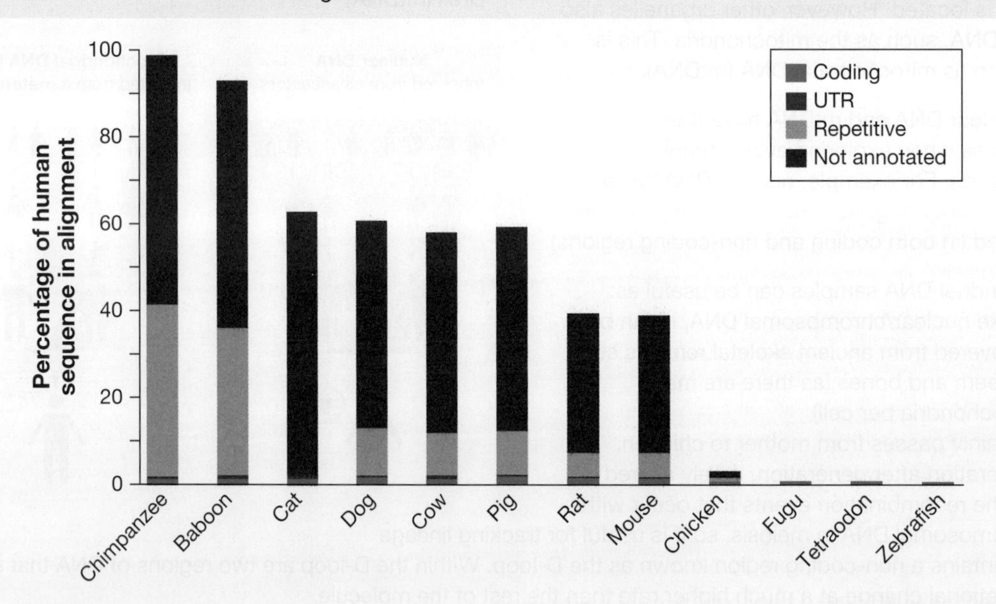

In figure 9.22, four different types of DNA are explored:

- coding DNA, which codes for proteins and is transcribed and translated
- UTR: untranslated regions which do not code for proteins
- repetitive: sequences that are repeated within a genome and not translated, such as satellite DNA
- not annotated: DNA whose purpose was either not specified or unknown.

A comparative genomic study in 2000 reported that humans and fruit flies share a basic set of genes, with about 60 per cent of the genes in humans and fruit flies being conserved.

comparative genomics the comparison of DNA or genes between different species to identify evolutionary relationships

EXTENSION: Comparing chromosomes

Chromosomes can also be compared between different species by examining chromosome banding patterns, or the translocation of different sections of chromosomes through chromosome painting.

The chromosomes of various species can be identified by a three-letter code that consists of the first letter of the name of the genus plus the first two letters of the specific name. So, the human *(Homo sapiens)* number-5 chromosome is denoted as HSA 5 and the common chimp *(Pan troglodytes)* number-7 chromosome is denoted as PTR 7.

FIGURE 9.23 Chromosome sets of humans, chimps, gorillas and orangutans. Chromosome 2 in chimps, gorillas and orangutans is actually split into two separate chromosomes showing that, over time, a chromosome fusion has likely occurred. (Image courtesy of Dr Mariano Rocchi.)

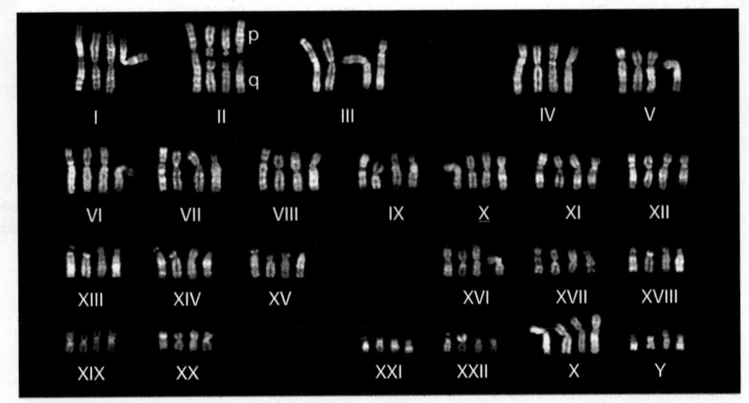

To access more information on comparing chromosomes and complete an analysis task relating to this, please download the worksheet.

Resources

eWorkbook Worksheet 9.3 Extension: Comparing chromosomes (ewbk-2907)

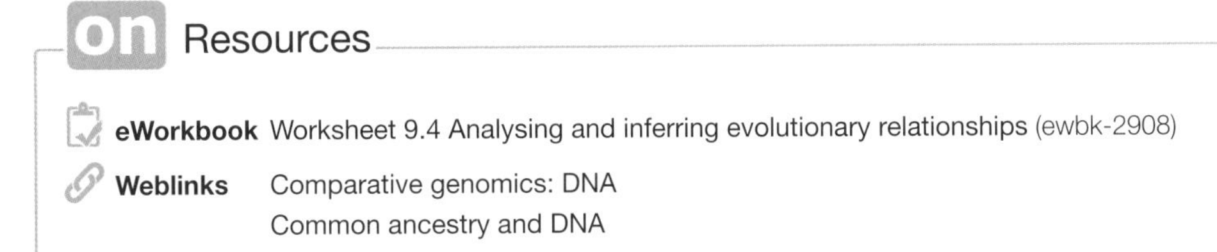

Resources

eWorkbook Worksheet 9.4 Analysing and inferring evolutionary relationships (ewbk-2908)

Weblinks Comparative genomics: DNA
Common ancestry and DNA

KEY IDEAS

- Species are related through evolution, with the degree of relatedness differing according to when two species last shared a common ancestor.
- Differences in the amino acid sequences of proteins can be used to assess degrees of relationship between species and infer their evolutionary history.
- Differences in the DNA nucleotide sequence in genes can be used to assess degrees of relationship between species and infer their evolutionary history.
- Fewer differences in DNA and amino acids sequences between species suggest a closer degree of relatedness between them.

9.3 Activities

learnON

To answer questions online and to receive **immediate feedback** and **sample responses** for every question, go to your learnON title at **www.jacplus.com.au**. A **downloadable solutions** file is also available in the resources tab.

9.3 Quick quiz on	9.3 Exercise	9.3 Exam questions

9.3 Exercise

1. The table below shows the number of differences in the amino acid sequence between humans and other vertebrates for the beta chain of haemoglobin. Using this information, order the vertebrates from most to least related to humans.

TABLE Differences in amino acids between four species

Vertebrate	Number of amino acid differences
Dog	15
Frog	67
Gorilla	1
Chicken	45

2. Describe why a comparison of the DNA nucleotide sequences of two species can show more evolutionary changes than a comparison of their amino acid sequences.
3. Examine these four DNA nucleotide sequences of four unknown species.

TABLE Differences in the nucleotide sequence between four species

Species	Nucleotide sequence of four closely related species
E	A A T G G C T A G A A C G T A C G A T A T T T A G C G T A
F	A A T C G C T A G A T C G T A C G G G A T T T A G C G T A
G	A A T G G C T A G A A C G T A C G A G A T T T A G C G T A
H	A A T G G C T A G A T C C T A C G G G A T T T A G C G T T

a. Identify the differences in these DNA sequences and use them to describe the relatedness between these species.
b. Assume these four species are a blue whale (a mammal), a common bottlenose dolphin (a mammal), a bat (a mammal) and a snake (a reptile). Outline which letter is likely to represent which species and justify your response.
c. If the sequence of an ancestral species was AATGGCTAGAACGTACGATATTTAGCGTA, determine the percentage sequence conservation of each species.

4. Explain why a conserved gene such as the one that codes for cytochrome *c* is useful when comparing DNA sequences between species.
5. Explain why molecular homology can be used to show relatedness between species.
6. The base sequence of a short section of DNA is shown below:
Human ATCGGC
Monkey ATCGGC
Gorilla ATCAGC
After further analysis of other aspects of the genome, it was found that humans are more closely related to gorillas than to monkeys. Outline one reason why this may not be evident in the sequence shown above.
7. Two students, Vicki and Chris, wished to investigate the relatedness between species, designated species A, B and C. They choose to investigate a gene that codes for a particular histone protein (involved in the packaging of DNA). Vicki found that there were three amino acid differences in the histone protein between species A and B. Chris found that there were four DNA base sequence changes between the species B and C. Upon investigation, it was found that species B and C were more closely related. Explain, linking to the results found by Vicki and Chris, why this might be the case.

9.3 Exam questions

Use the following information to answer Questions 1 and 2

Cytochrome c is a protein that consists of 104 amino acids. Many of these 104 sites on cytochrome c contain exactly the same amino acid across a large range of organisms. There are, however, some differences at certain sites. It is hypothesised that different organisms, all containing cytochrome c proteins, descended from a primitive microbe that lived over 2 billion years ago. The table below uses the three-letter codes for various amino acids found at specific sites for each organism.

Molecular homology of cytochrome c

Organism	Site 1	Site 4	Site 11	Site 15	Site 22
human	Gly	Glu	Ile	Ser	Lys
pig	Gly	Glu	Val	Ala	Lys
dogfish	Gly	Glu	Val	Ala	Asn
chicken	Gly	Glu	Val	Ser	Lys
Drosophila	Gly	Glu	Val	Ala	Ala
yeast	Gly	Lys	Val	Glu	Lys
wheat	Gly	Asp	Lys	Ala	Ala

Question 1 (1 mark)

Source: *VCAA 2016 Biology Exam, Section A, Q39*

MC Using only the data for the molecular homology of cytochrome c, which one of the following organisms is most closely related to the dogfish?

A. *Drosophila* **B.** chicken **C.** human **D.** yeast

Question 2 (1 mark)

Source: *VCAA 2016 Biology Exam, Section A, Q40*

MC Using only the data for the molecular homology of cytochrome c, which pair of organisms is most distantly related to wheat?

A. dogfish and *Drosophila*
B. *Drosophila* and yeast
C. *Drosophila* and pig
D. human and yeast

Question 3 (3 marks)

The coagulation factor VIII gene or *F8* produces an essential protein involved in blood clotting. A section of this gene was sequenced in three different species and the results are shown below.

Species	DNA sequence
X	T G G C G C G T A C G A - C A T T T A C G G T A
Y	T - G C G C G T A C G A - C A T T T A C A G G A
Z	T G T C G C G C A C G A - C A T T T A C G G T A

a. Which species share a more recent common ancestor? **1 mark**

b. Would the same number of differences be seen if the amino acids were sequenced? Justify your response. **2 marks**

Question 4 (1 mark)

Source: *VCAA 2008 Biology Exam 2, Section B, Q7d*

Scientists now believe that *L. myola* evolved from the *L. genimaculata* about 8000 years ago. It is proposed that a population of Green-eyed Tree Frogs became isolated due to changing climate conditions. What is one type of evidence that biologists could use to estimate an approximate date of divergence of these two species?

Question 5 (1 mark)

MC Examination of the genomes of five species (human, mouse, chicken, frog and fish) reveals that some matching DNA sequences are present in all five species. On the other hand, other DNA sequences are present only in the human and the mouse genomes.

It is reasonable to suggest that the DNA sequences conserved in all five species were examples of

A. genes that have a key function in the five species concerned.
B. so-called 'junk' DNA with no known function.
C. genes that are essential in mammals but play no role in the other species.
D. DNA acquired by gene duplication during evolution of the mammals.

More exam questions are available in your learnON title.

9.4 Phylogenetic trees

KEY KNOWLEDGE

- The use and interpretation of phylogenetic trees to show evidence of relatedness between species

9.4.1 What are phylogenetic trees?

Relationships between species (and larger taxonomic groups) may be shown visually by the use of **phylogenetic trees**.

phylogenetic tree a branching diagram showing inferred evolutionary relationships between life forms based on their observed physical and genetic similarities and differences

Figure 9.24 shows the phylogenetic tree of life that captures all groups of living organisms. This tree is a product of Darwin's evolutionary theory that illustrates the interconnectedness of all life forms through evolution.

FIGURE 9.24 The tree of life is a diagram that shows the degree of evolutionary relatedness of all the groups of life forms on planet Earth. Diagrams like this are called phylogenetic trees.

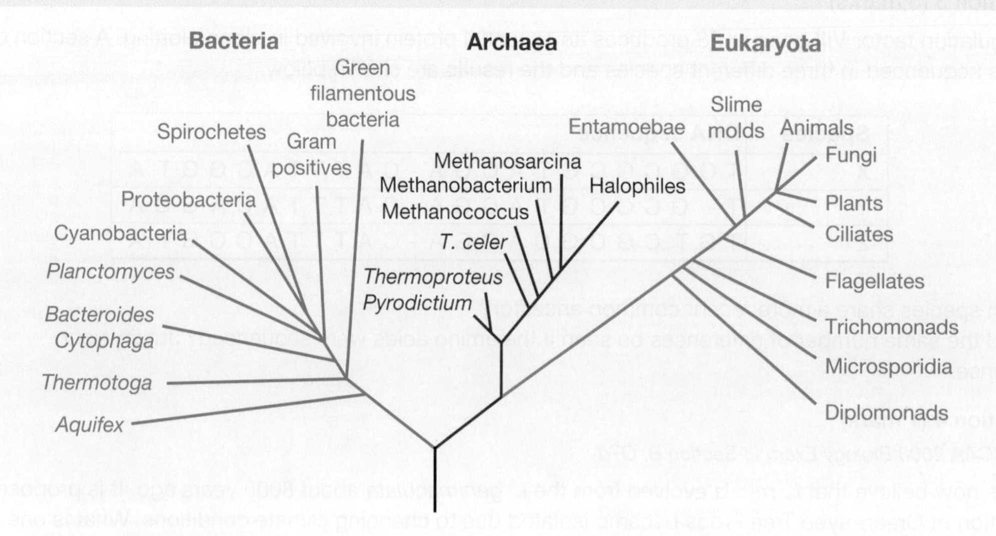

Phylogenetic trees are also called evolutionary trees. They are branching diagrams that show *inferred* evolutionary relationships or lines of evolutionary descent among biological groups or taxa (taxa are taxonomic groups of any ranks such as a species, family or class).

Phylogenetic trees illustrate evolutionary history as inferred from molecular data or other evidence. Molecular evidence includes amino acid sequences of proteins, RNA sequences and DNA sequences. In the case of DNA sequences, these may be nuclear or mitochondrial DNA and may be coding or non-coding DNA.

FIGURE 9.25 A phylogenetic tree examining different species of extinct and living penguins

Phylogenetic trees are not fixed, but are subject to change as new research and observations are made.

9.4.2 Components of phylogenetic trees

The many important features in phylogenetic trees are outlined below and shown in figure 9.26.

FIGURE 9.26 The components of phylogenetic trees

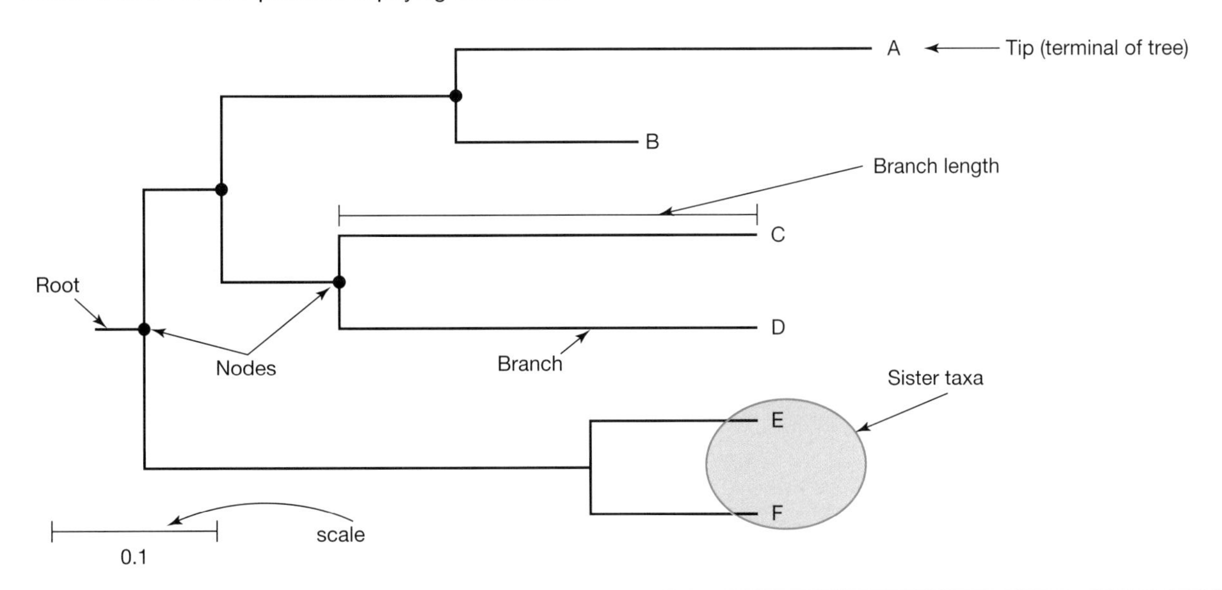

Features of phlyogenetic trees

There are various components of phylogenetic trees (as shown in figure 9.26), including:

- the **tips** or terminals of a 'tree', which are the descendant groups (taxa)
- **nodes**, which denote an ancestor of two (or more) descendants (referred to as the **recent common ancestor**)
- **branches**, which indicate a speciation event and show the relationship between an ancestor and a descendant of that ancestor
- the **root**, which is the common ancestor of all taxa shown in the tree
- **sister taxa**, which are two groups, such as two species, with a common ancestor that is not shared with other taxa.

The branch length also conveys key information in a phylogenetic tree:

- In some trees, a time scale is included that may be a geological time scale in millions of years. In this case, the branch length denotes a period of time. If a time scale is not given, it can be assumed that longer branches denote longer time periods.
- In other trees, a scale is included that indicates the amount of molecular change that has occurred between species (or other taxa). The amount of molecular change denoted by the branch length may be the number of differences in DNA base sequences between two species or the number of differences in the amino acid sequences of a protein.

tip the terminal point of a phylogenetic tree, representing a particular species

node shows an ancestor of two or more descendants on a phylogenetic tree

recent common ancestor an organism that two groups are descended from

branch a section of the phylogenetic tree which indicates a speciation event

root the common ancestor of all taxa on a phylogenetic tree

sister taxa two groups with a common ancestor that is not shared with another taxa

When a common ancestor occurs more recently in a phylogenetic tree, two species are more closely related.

The evolutionary distance between two species (or other taxa) is shown by the combined length of the two branches leading to each species.

9.4.3 Drawing phylogenetic trees

Phylogenetic trees can be drawn using diagonal, horizontal or vertical lines.

FIGURE 9.27 Simple phylogenetic trees of living insect groups drawn **a.** using diagonal branch lines **b.** using vertical branch lines **c.** using horizontal branch lines

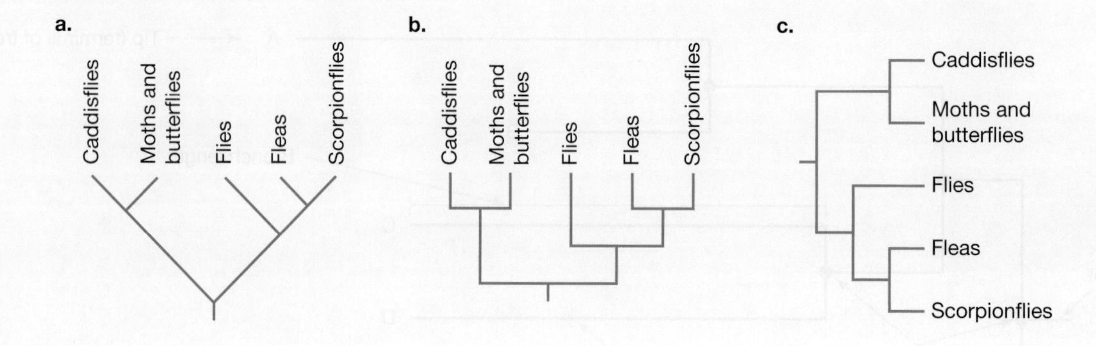

Steps to construct a phylogenetic tree

1. Determine the number of differences between the different species or groups provided.
2. Identify the sequences with the fewest differences.
3. Connect these groups with the fewest differences as sister taxa, joined by a node.
4. Determine the group with the fewest differences to the sister taxa and add this onto your tree as a branch.
5. Continue this until all groups are added onto the phylogenetic tree.
6. Determine the outgroup (if any), which is the most distantly related group (has the most differences to the other species). Place this on the phylogenetic tree.

Let's follow these steps to create a phylogenetic tree for five species. The number of differences between five species for a certain section of DNA can be observed in the table 9.6.

TABLE 9.6 Number of base sequence differences

	Lion	Hippopotamus	Jaguar	Meerkat	Nurse shark
Lion		4	1	2	8
Hippopotamus			5	6	4
Jaguar				3	9
Meerkat					10
Nurse shark					

This data can be used to construct a phylogenetic tree.

- The lion and the jaguar are most closely related, so are sister taxa. They will have a node that is more recent and come off this node as branches.
- The meerkat is the next most closely related, so is the next branch. The common ancestor (shown at the node) will be further back for the meerkat, lion and jaguar compared to the node for the lion and jaguar.
- This is followed by the hippotamus, which wll form another branch.
- The nurse shark is the least closely related, so it is the outgroup.

The locations of the branches may vary slightly. For example, the nurse shark could be at the top of the phylogenetic tree rather than the bottom. The location of the nodes and the links between the species are the most important part of the phlyogenetic tree.

FIGURE 9.28 Phylogenetic tree for species lion, hippopotamus, jaguar, meerkat and nurse shark. Note that only the common ancestor of all species is marked for simplicity.

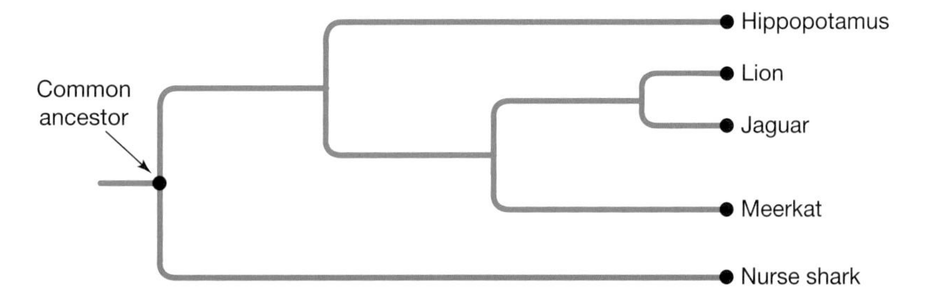

elog-0075

INVESTIGATION 9.3

online only

Using data to create phylogenetic trees

Aim

To use secondary data to create phylogenetic trees

on Resources

Weblink Making phylogenetic trees: PhytoT

tlvd-1722

SAMPLE PROBLEM 3 Making phylogenetic trees

All vertebrates use the haemoglobin protein to transport oxygen on red blood cells around their bodies. The amino acid sequences of four species of the alpha chain of the haemoglobin protein have been compared to the amino acid sequence of humans.

	Human	Crocodile	Starling	Shark	Chimp
Human	0	33	29	50	3
Crocodile		0	33	55	33
Starling			0	55	30
Shark				0	52
Chimp					0

Use the information provided to construct a phylogenetic tree to show the evolutionary relationships between the five animals. (5 marks)

THINK

1. Identify what the question is asking you to do. You are being asked to *construct* the tree, which in this case involves producing a diagram of a phylogenetic tree.
2. Review the steps of making a phylogenetic tree.
3. Compare the amino acid sequences for all five species.

   Which two species have the least number of differences? The answer is humans and chimps with three, so these two are placed in the top two boxes.
4. Consider which species has the fewest changes to humans and chimps: the starling with 29 and 30. Place it in the third box coming off as a new branch from the humans and chimps.
5. Of the remain two species, the crocodile has the fewest differences so is placed in the next box.
6. The final species left, the shark, has the most differences and is placed in the last box as the outgroup.

WRITE

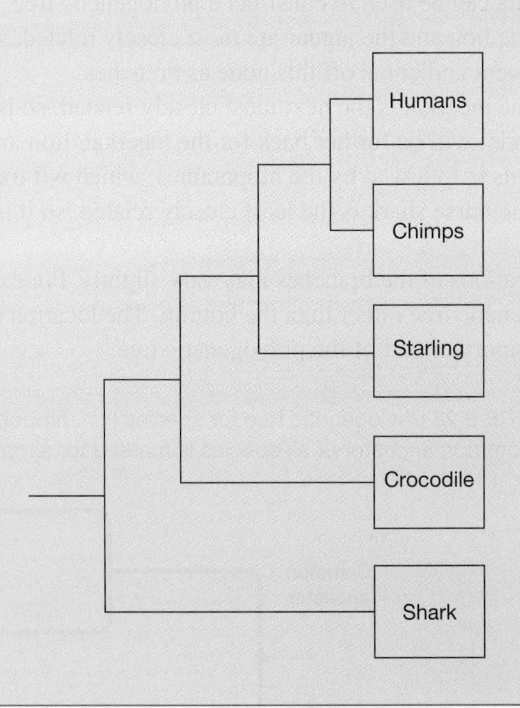

Interpreting phylogenetic trees

Figure 9.29 shows a phylogenetic tree of the taxonomic group called the vertebrates. Humans are included with the other placentals. This tree can be examined to infer various relationships. For example, of the mammals (that branched off 350 million years ago), humans (and placentals) are more closely related to marsupials than monotremes.

FIGURE 9.29 Phylogenetic tree of vertebrates

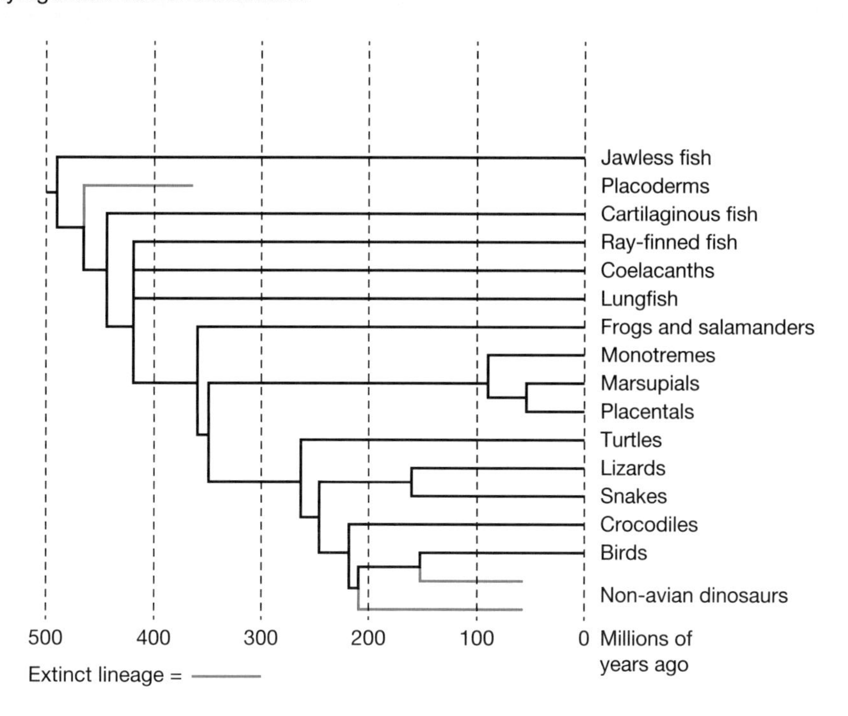

CASE STUDY: Analysis of a phylogenetic tree for salmon and trout

Let's look at a phylogenetic tree that shows the evolutionary relationship between species.

Pacific salmon and Pacific trout are members of the genus *Oncorhynchus*. Figure 9.30 shows mature males (lower row) and females (upper row) of Pacific salmon: from left, pink salmon (*O. gorbuscha*), sockeye salmon (*O. nerka*) and masu salmon (*O. masou*). Atlantic salmon are members of a different genus (*Salmo salva*).

FIGURE 9.30 The genus *Oncorhynchus* includes 17 species, commonly called salmon and trout. This image shows three mature male and female species of Pacific salmon in the mating season.

Figure 9.31 shows a phylogenetic tree of eight species in the genus *Oncorhynchus* that is the group of interest. The less closely related Atlantic salmon (*Salmo salva*) is included as an outgroup (i.e. a taxon outside the particular group of interest). An outgroup branches directly from the base of the phylogenetic tree.

FIGURE 9.31 Phylogenetic tree of eight species of salmon and trout of the genus *Oncorhynchus*, and the Atlantic salmon (*Salmo salva*) as the outgroup. Note that in this case, the branch length represents the number of nucleotide differences rather than a specific time period or the extinction of certain species.

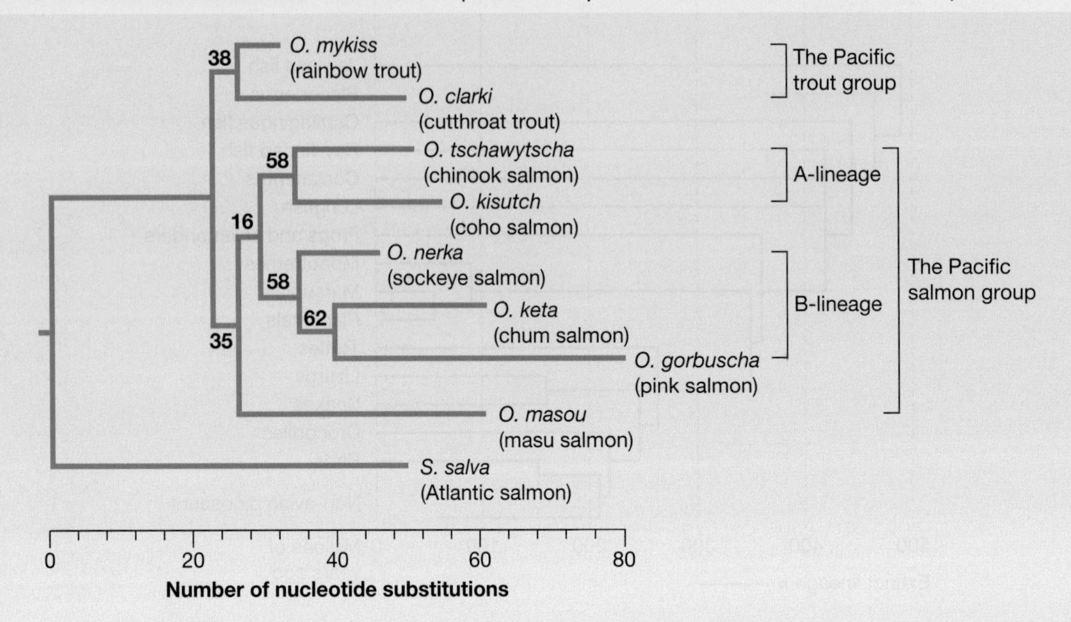

Many relationships can be inferred from this tree. For example, the trout species are more closely related to each other than they are to other salmon species.

Value of phylogenetic analysis

Phylogenetic studies have made valuable contributions to our understanding in many fields, including:

- identifying genetic variation within natural populations of a species and its subspecies
- reconstructing the evolutionary history of a species or a genus
- assisting our understanding of human prehistory, including early migrations of modern humans
- clarifying the relationship between modern humans, *Homo sapiens*, and the extinct *Homo* species
- identifying the origin of human viral diseases that have jumped from other species (zoonotic diseases).

cladogram diagram, based on cladistic study, that shows the inferred relationship between different group of organisms

CASE STUDY: Cladograms

Another type of diagram that can be used to illustrate evolutionary relatedness is a **cladogram**.

Like a phylogenetic tree, a cladogram is a diagram that shows the relationships between groups of organisms. However, rather than using molecular homology, cladograms focus on derived characters or features shared between species.

To access more information on cladograms, please download the digital document.

FIGURE 9.32 Cladogram of vertebrate groups

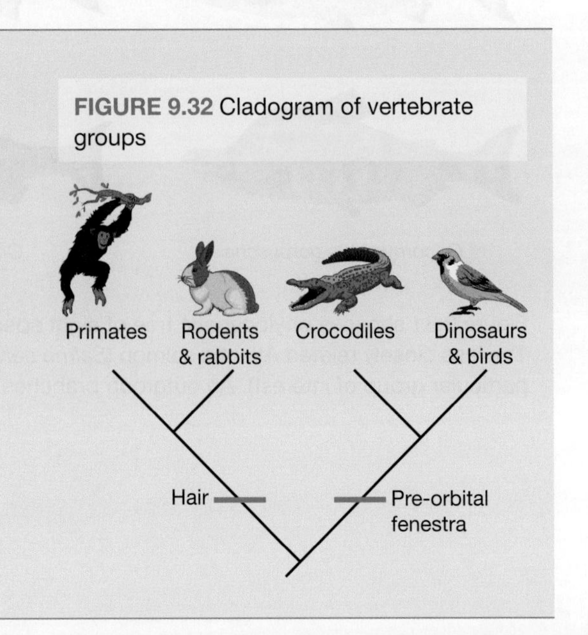

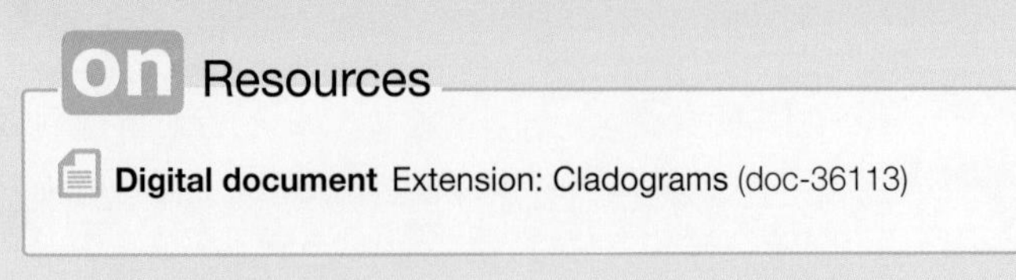

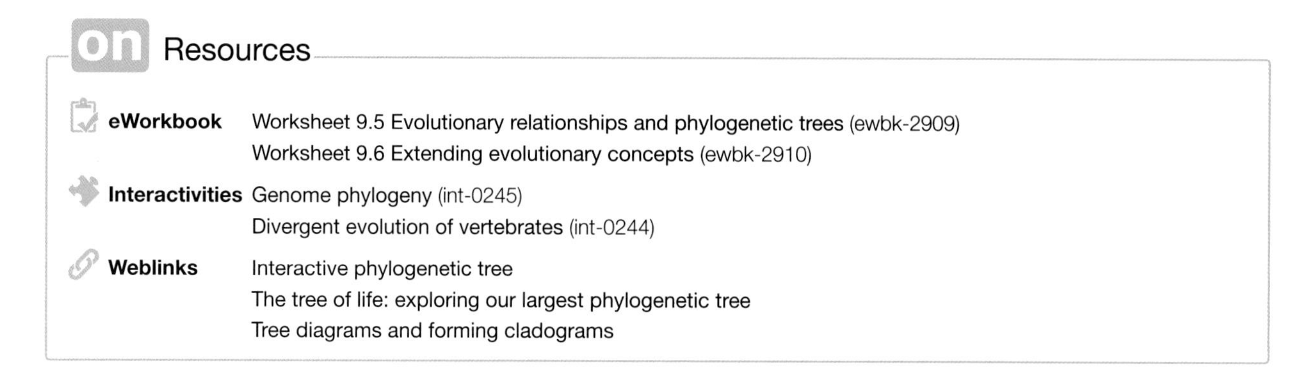

Resources

eWorkbook Worksheet 9.5 Evolutionary relationships and phylogenetic trees (ewbk-2909)
Worksheet 9.6 Extending evolutionary concepts (ewbk-2910)

Interactivities Genome phylogeny (int-0245)
Divergent evolution of vertebrates (int-0244)

Weblinks Interactive phylogenetic tree
The tree of life: exploring our largest phylogenetic tree
Tree diagrams and forming cladograms

KEY IDEAS

- Relationships between organisms can be shown in diagrams called phylogenetic trees.
- Phylogenetic trees enable inferences to be made about the evolutionary history of groups of organisms.
- Phylogenetic trees are typically based on molecular data, while cladograms use shared derived features (which can be molecular) that are inferred to have evolved from an original character present in an ancestral organism.
- Molecular data, including differences in DNA base sequences and amino acid sequences of proteins, are used as a basis for constructing phylogenetic trees.
- The greater the number of molecular differences between two taxa, the longer the period since they diverged.
- Inferences from phylogenetic trees and cladograms are subject to revision as new data become available.

9.4 Activities

learnon

To answer questions online and to receive **immediate feedback** and **sample responses** for every question, go to your learnON title at **www.jacplus.com.au**. A **downloadable solutions** file is also available in the resources tab.

9.4 Quick quiz on	**9.4 Exercise**	**9.4 Exam questions**

9.4 Exercise

1. State the purpose of a phylogenetic tree.
2. Identify one relevant difference between the members of the following pairs:
   a. a phylogenetic tree and a cladogram
   b. a node and a branch (in a phylogenetic tree).
3. Describe how an extinct species would be indicated on a phylogenetic tree.
4. Identify the type of evidence that is used to construct a phylogenetic tree.
5. Examine the phylogenetic tree provided that shows five species A, B, C, D and E. Which two species share the most recent common ancestor? Justify your response.

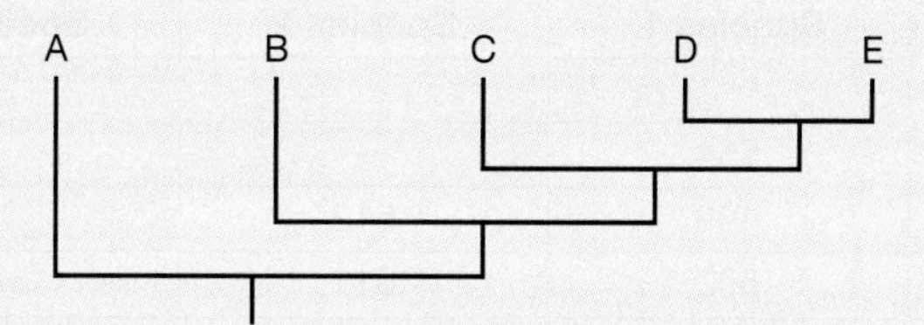

6. A short section of DNA from a conserved gene is shown in five species.
   Species F: AAT TAG GAC CAG
   Species G: CAT TAC CTC CGG
   Species H: AAT TAC GAC CAG
   Species I: CCT TAC CTC AGG
   Species J: AAT TAC CTC CAG
   Construct a phylogenetic tree based on this information.

9.4 Exam questions

Use the following information to answer Questions 1 and 2

Consider the following phylogenetic tree, which summarises the evolutionary relationships between certain fish species.

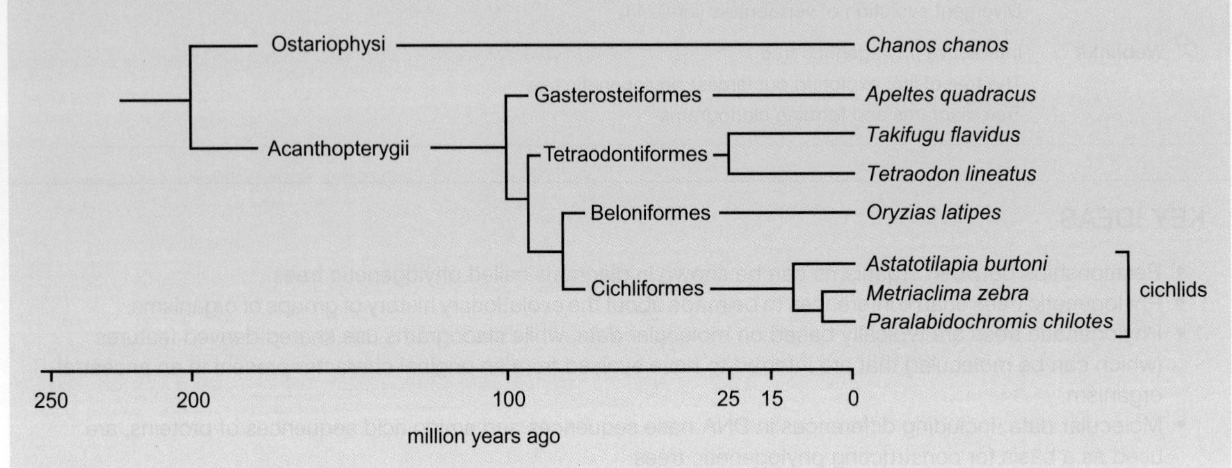

Question 1 (1 mark)

Source: *VCAA 2019 Biology Exam, Section A, Q24*

MC *O. latipes* is most closely related to

A. *A. quadracus.*
B. *T. lineatus.*
C. *P. chilotes.*
D. *T. flavidus.*

Question 2 (1 mark)

Source: *VCAA 2019 Biology Exam, Section A, Q25*

MC Which one of the following statements is correct?

A. Cichlids diverged to form three distinct species 100 million years ago.
B. *C. chanos* was the last species to diverge from the most distant common ancestor.
C. Gasterosteiformes, Beloniformes and Cichliformes do not share a common ancestor.
D. *T. flavidus* and *T. lineatus* diverged to form two distinct species 25 million years ago.

Question 3 (4 marks)

DNA in four different species were explored for a conserved gene and the number of DNA bases that are the **same** are shown.

	Species 1	Species 2	Species 3	Species 4
Species 1				
Species 2	554			
Species 3	499	512		
Species 4	592	611	542	

a. Identify which species are most closely related and which species are least closely related. **1 mark**
b. Construct a phylogenetic tree to show the relationship between these four species. **2 marks**
c. The four species are a snake, goldfish, human and chimpanzee. State the likely identity of each species (as species 1, 2, 3 or 4). **1 mark**

Question 4 (1 mark)

***Source:** VCAA 2011 Biology Exam 2, Section A, Q23*

MC Two possible phylogenetic relationships between eight groups of flowering plants are shown in the following diagrams.

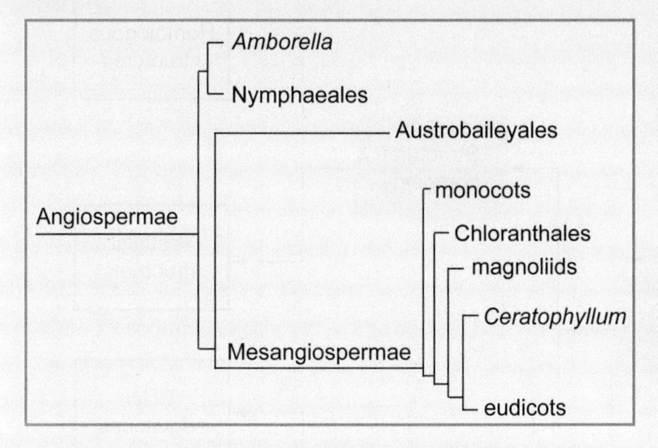

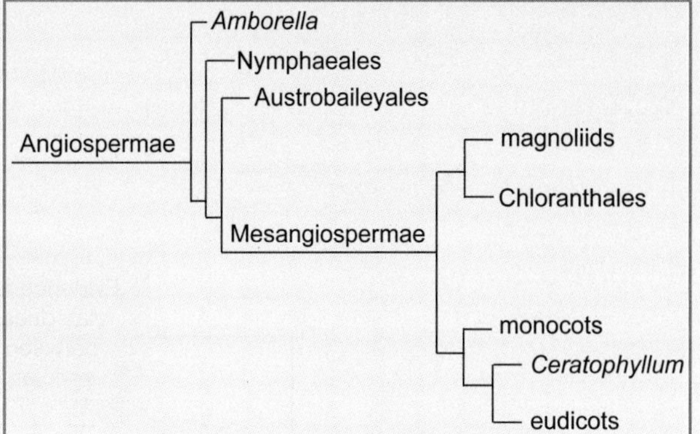

One similarity between the alternatives is

A. monocots diverged before Chloranthales.
B. *Ceratophyllum* and eudicots diverged from monocots.
C. *Amborella* and Nymphaeales diverged first from Angiospermae.
D. magnoliids were the first group to diverge from Mesangiospermae.

Question 5 (3 marks)

***Source:** VCAA 2020 Biology Exam, Section B, Q8a*

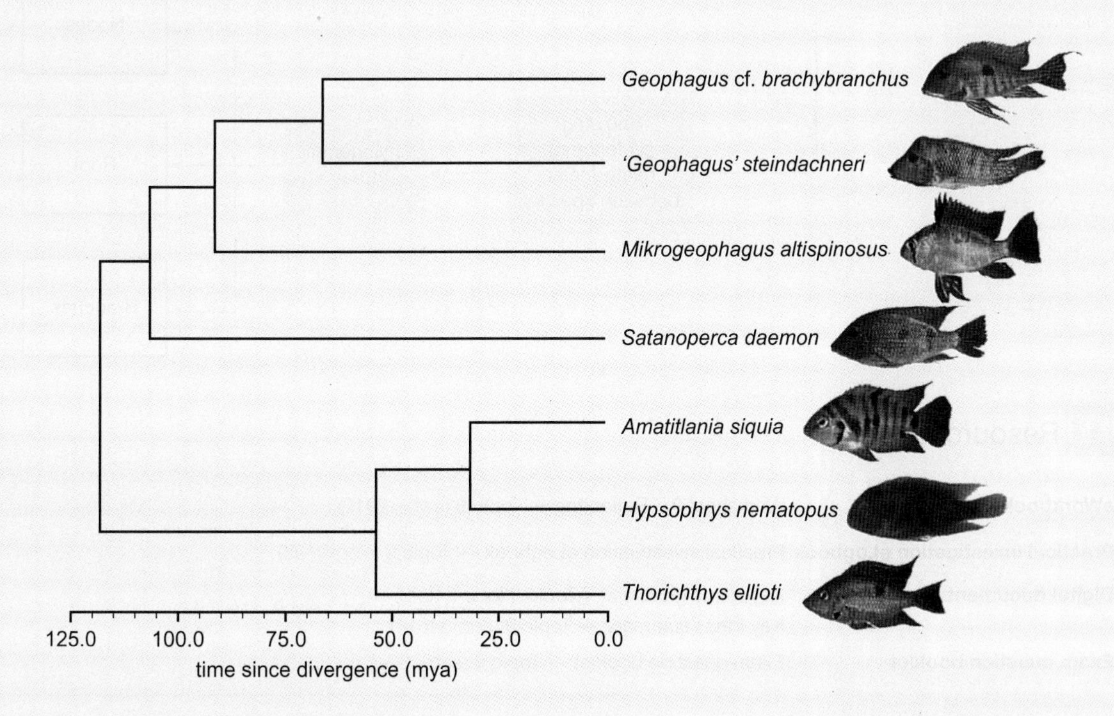

Molecular homology can be used to construct a phylogenetic tree.

Based on the information above, state which two species of cichlid fish would be expected to have the most similar amino acid sequences in their proteins. Justify your answer. **3 marks**

More exam questions are available in your learnON title.

9.5 Review

9.5.1 Topic summary

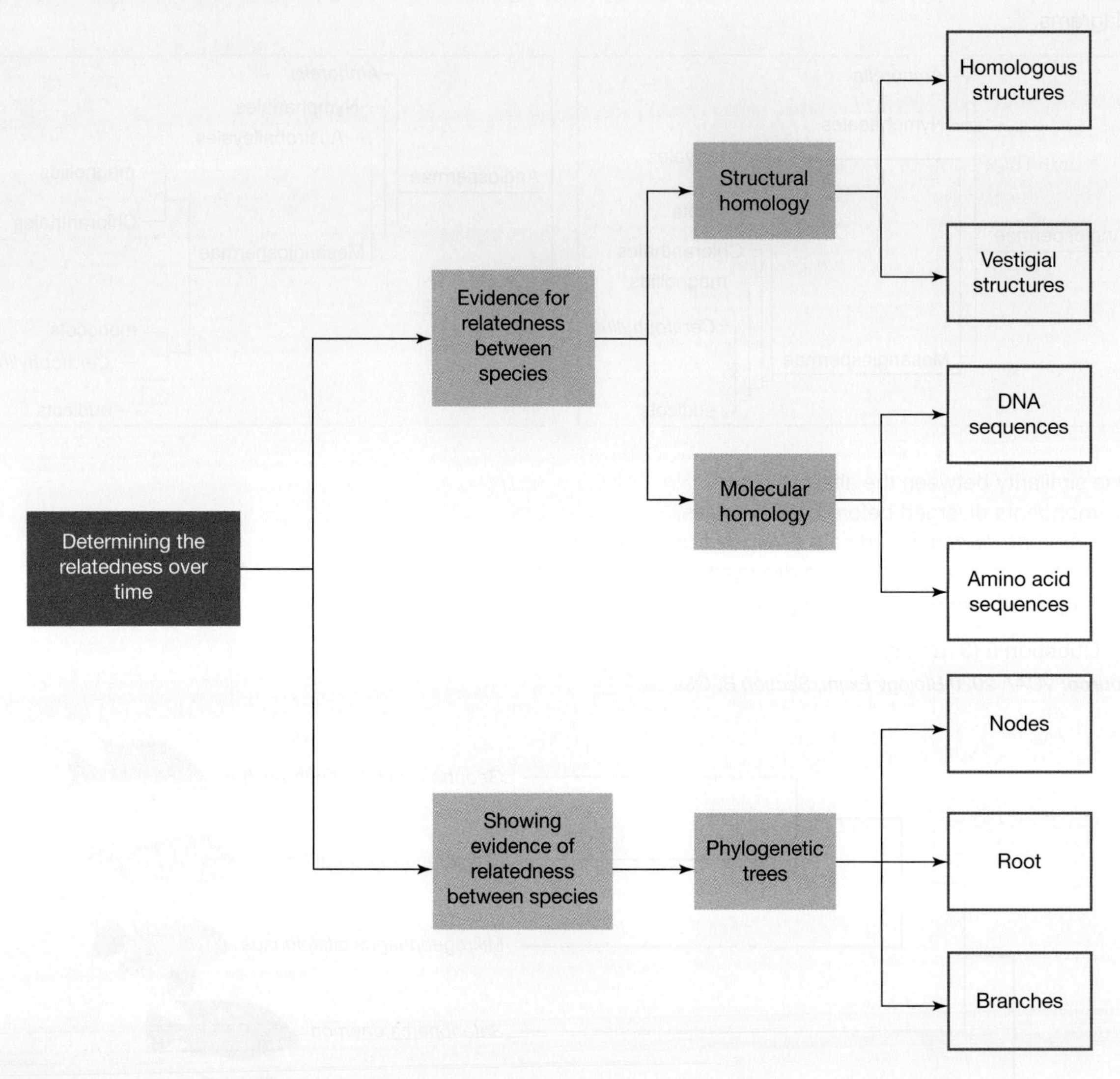

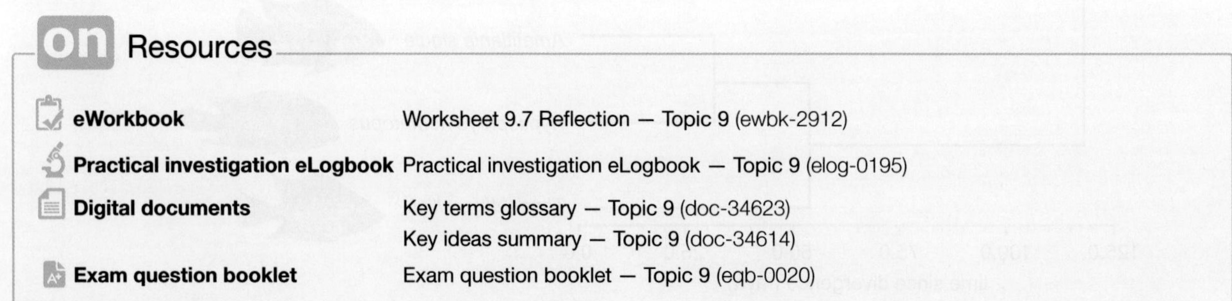

eWorkbook	Worksheet 9.7 Reflection — Topic 9 (ewbk-2912)
Practical investigation eLogbook	Practical investigation eLogbook — Topic 9 (elog-0195)
Digital documents	Key terms glossary — Topic 9 (doc-34623) Key ideas summary — Topic 9 (doc-34614)
Exam question booklet	Exam question booklet — Topic 9 (eqb-0020)

9.5 Exercises

learnon

To answer questions online and to receive **immediate feedback** and **sample responses** for every question, go to your learnON title at **www.jacplus.com.au**. A **downloadable solutions** file is also available in the resources tab.

9.5 Exercise 1: Review questions

1. Suggest possible logical explanations for the following observations:
   a. When the haemoglobin of a species of mammal is compared with that of other mammals, some are more similar than others.
   b. The percentage difference in the amino acid sequence of one protein from a cat and a dog is far less than that between a lizard and either mammal.
2. Refer to the figure shown, which shows a phylogenetic tree for the groups of animals called vertebrates.

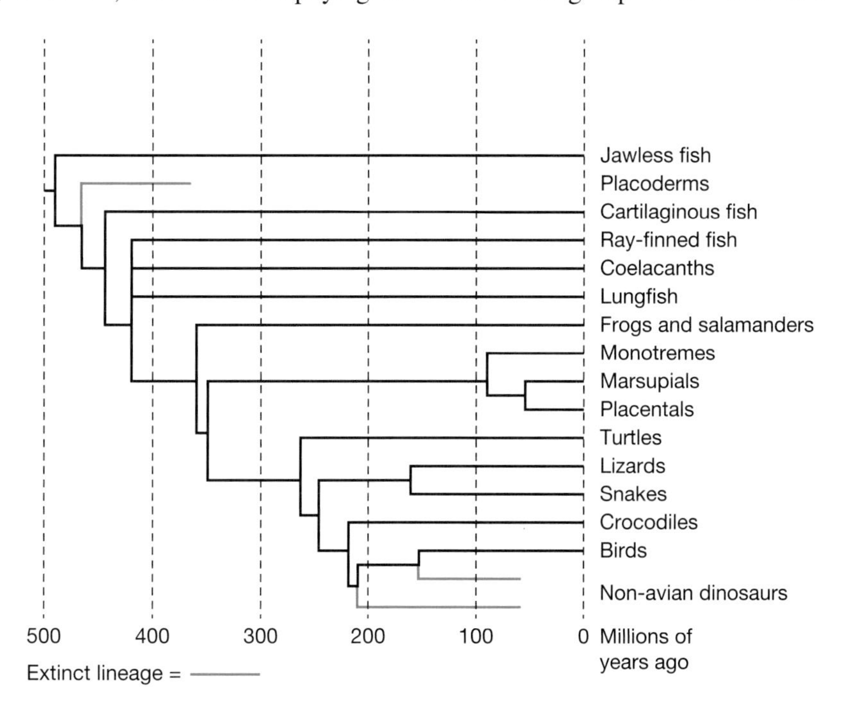

   a. What were the earliest vertebrates to appear?
   b. Are representatives of this group still living on Earth?
   c. Cartilaginous fish include sharks and rays, while ray-finned fish include the bony fish. Which group appeared earlier?
   d. Are marsupials, such as kangaroos, more closely related to birds or coelacanths? Justify your response.
3. Identify a key difference between the members of the following pairs:
   a. a homologous structure and a vestigial structure
   b. DNA and amino acid sequences.
4. a. Provide two examples of vestigial structures present in humans.
   b. What were these structures used for in our ancestors?
   c. What evidence do vestigial structures provide for evolution?
5. Homologous structures are used in structural morphology as evidence for evolution.
   a. What does the term *homologous structure* mean?
   b. The wings of a bird and a butterfly both enable them to fly. Why are the wings of birds and butterflies not considered homologous?

6. Refer to the figure and answer the following questions:
   a. How could evolutionary distance be measured in this phylogenetic tree?
   b. Based on this, approximately how many differences would there be between species A and B in regards to nucleotide differences?
   c. Complete the sentence from the choice below:
   Species pair E and F show (*a higher/a lower/the same*) degree of relatedness compared to species pair A and B.
   d. Explain your choice in (c) above.

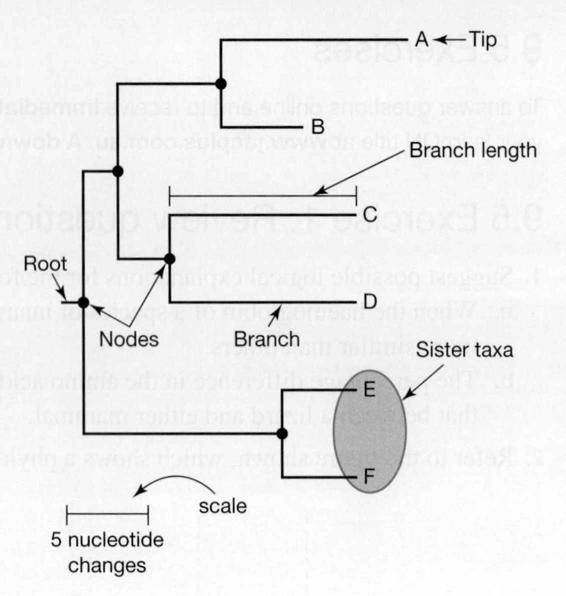

7. Examine the hypothetical evolutionary lineages shown in the figure. Examine this figure and answer the following questions:

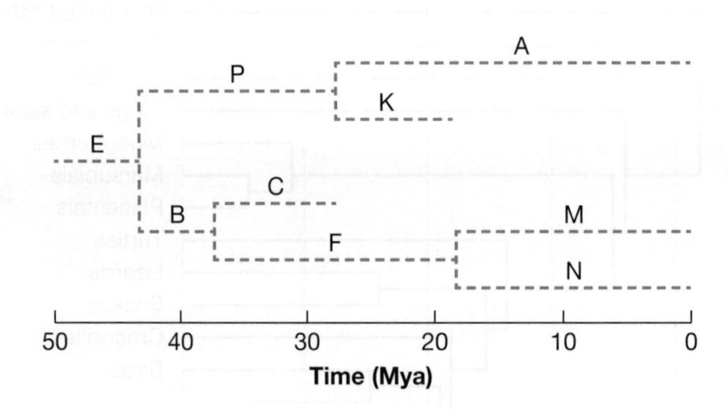

   a. Approximately when did species M and N diverge from a common ancestor?
   b. Is it reasonable to conclude that species M evolved from species N? Explain.
   c. Which species is the common ancestor of M and N?
   d. Did species C evolve from species F? Explain.
   e. Did species B and species K live during the same period?
   f. What is meant by the term *common ancestor*?
   g. Which species is the common ancestor of all the other species?
   h. Which species existed for the shortest period?

8. The data shown below are the percentage differences between the corresponding protein in four different species (A, B, C and D).
   Note: A and D only appear once in the table shown.

Species	B	C	D
A	4%	17%	36%
B		16%	32%
C			35%

   a. Use the data to outline the hypothetical evolutionary relationship of the four species.
   b. Assume that the four species are a canary (bird), cat, mouse and lion (mammals). Draw a clear phylogenetic tree to show this.

9. Carefully consider the phylogenetic trees shown below and answer the following questions:
   a. Do these three phylogenetic trees tell the same story or not about the evolutionary relationships of mosses, ferns, pines and roses? Explain your choice.
   b. Redraw one of these trees using horizontal lines rather than diagonal lines.

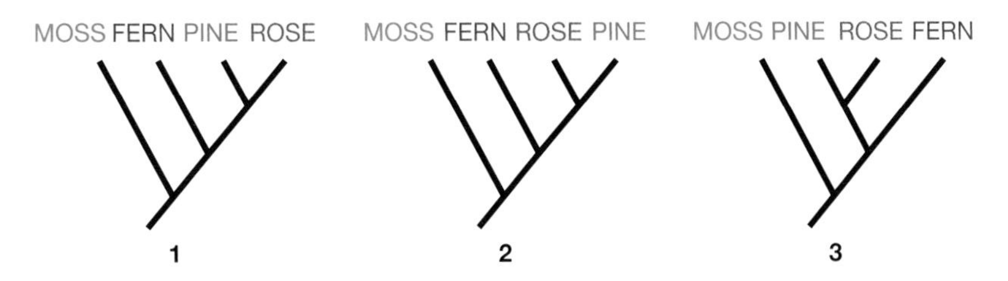

9.5 Exercise 2: Exam questions

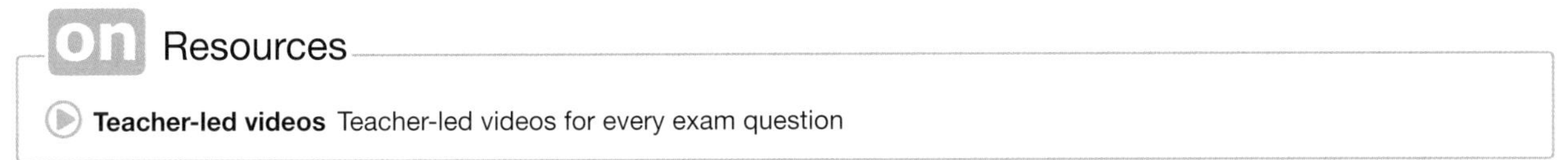

Section A — Multiple choice questions

All correct answers are worth 1 mark each; an incorrect answer is worth 0.

Question 1

Source: *VCAA 2016 Biology Exam, Section A, Q29*

Consider the following phylogenetic tree for different species of lice. The tree has been constructed based on molecular and morphological data.

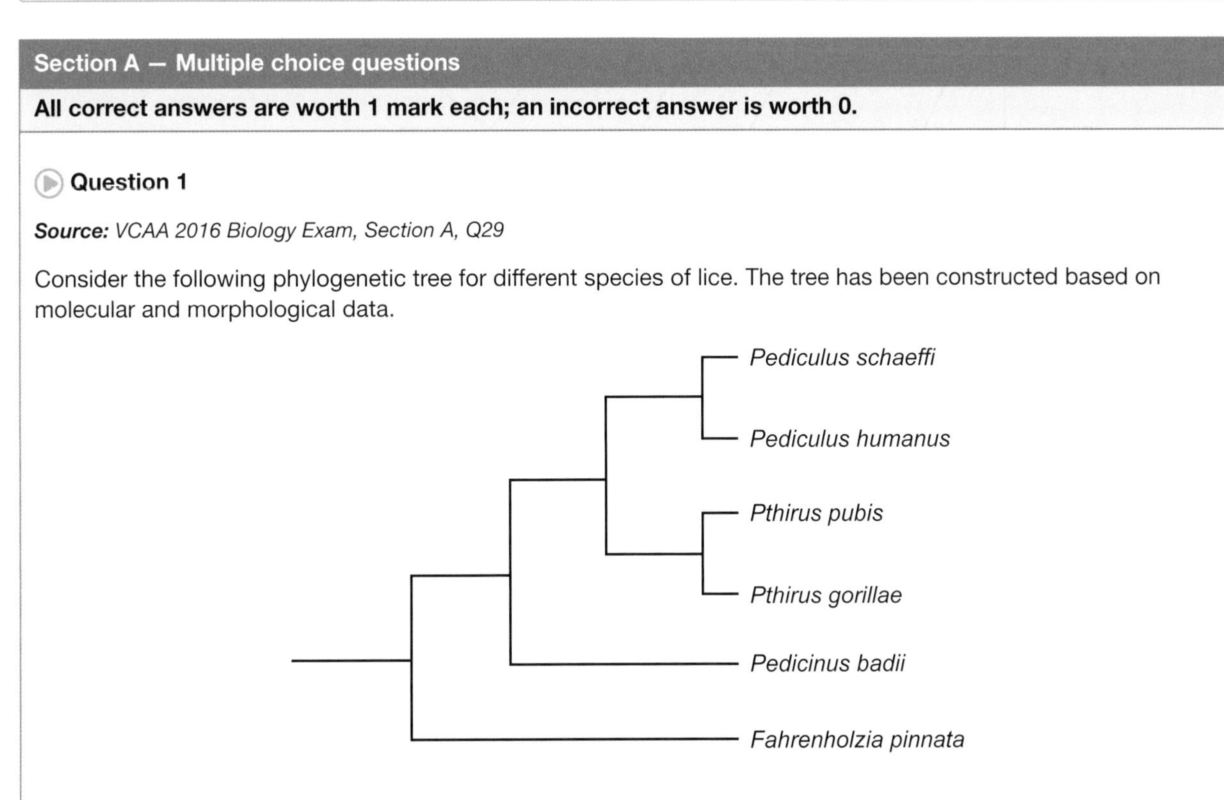

This information suggests that

A. *Pedicinus badii* shares a more recent common ancestor with *Pthirus gorillae* than with *Fahrenholzia pinnata*.
B. *Pediculus humanus* is more closely related to *Pedicinus badii* than it is to *Pthirus pubis*.
C. the six species of lice would have evolved by convergent evolution.
D. *Pediculus schaeffi* is the ancestor of *Pediculus humanus*.

Question 2

Source: *VCAA 2012 Biology Exam 2, Section A, Q2*

The genome of the woodland strawberry *Fragaria vesca* has been recently sequenced to show a relatively small genome of just 206 million base pairs. *F. vesca* is an ancestor of the garden strawberry and is a relative of apples and peaches.

The genome of *F. vesca*

A. is found only in the stem cells of the woodland strawberry.
B. includes all of the proteins made by *F. vesca*.
C. comprises all of the genes of *F. vesca*.
D. is the same as the genome of the apple.

Question 3

Source: *VCAA 2017 Biology Exam, Section A, Q32*

The phylogenetic tree below represents one model of the order and approximate time of appearance of the major groups of living organisms, and includes four groups represented by the letters R, S, T and U.

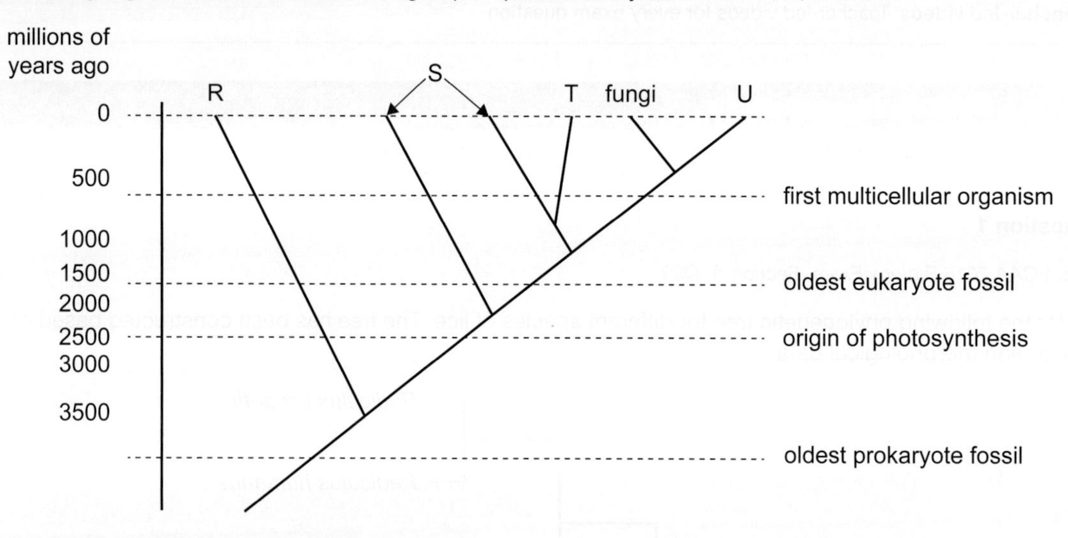

Which of the following shows the correct placement of the organisms on the phylogenetic tree?

A. R – animals, S – plants, T – bacteria, U – protists
B. R – bacteria, S – protists, T – plants, U – animals
C. R – protists, S – animals, T – bacteria, U – plants
D. R – plants, S – animals, T – bacteria, U – protists

Question 4

Gene conservation in species M and N would be expected to be greater than in species P and Q when

A. all four species are descended from a common ancestor.
B. species M and N shared a common ancestor more recently than did species P and Q.
C. species P and Q are more closely related by evolutionary descent.
D. species P and Q shared a common ancestor more recently than did species M and N.

Question 5

Source: *VCAA 2006 Biology Exam 2, Section A, Q25*

Comparisons of the amino acid sequences of the α-globin polypeptide have been made between humans and a number of other vertebrates. The number of differences is shown in the table below.

Organism	shark	kangaroo	carp	cow	newt
Amino acid differences in α-globin compared to human	79	27	68	17	62

An evolutionary relationship between these vertebrates was determined and illustrated in the figure below.

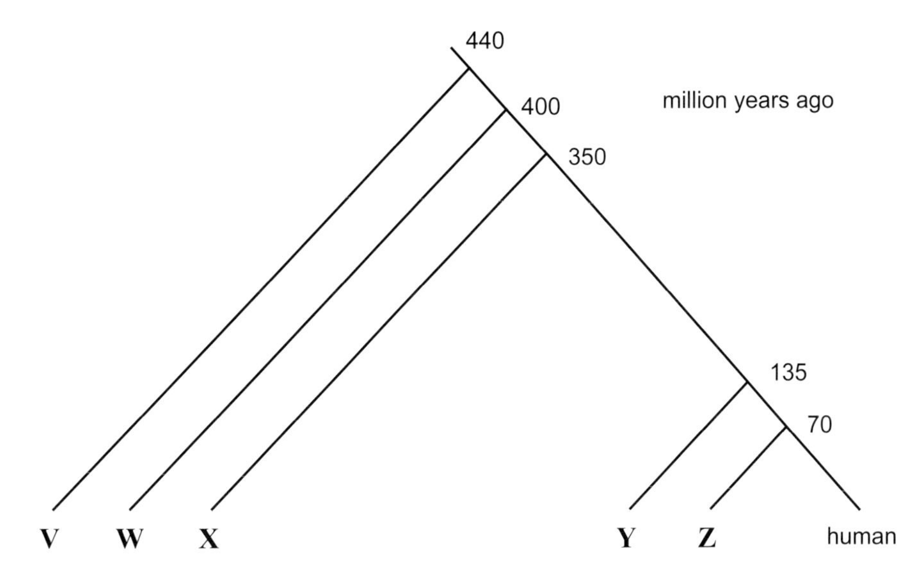

Based on the information provided, the correct placement of each animal on the figure to show the evolutionary relationship is

A. **V** = cow, **W** = kangaroo, **X** = newt, **Y** = carp, **Z** = shark
B. **V** = shark, **W** = carp, **X** = newt, **Y** = kangaroo, **Z** = cow
C. **V** = carp, **W** = shark, **X** = kangaroo, **Y** = newt, **Z** = cow
D. **V** = kangaroo, **W** = cow, **X** = newt, **Y** = shark, **Z** = carp

Question 6

How are phylogenetic trees most commonly constructed?

A. Comparison of DNA evidence
B. Comparison of the number of chromosomes in organisms
C. Derived characteristics
D. Homologous structures

Question 7

What are vestigial structures?

A. Structures that carry out a similar function but have evolved independently from each other.
B. Structures that are different in form and function but have evolved from a common ancestor.
C. Structures which are now functionless but were once functional in an ancestral species.
D. Structures which show a transition from two distinct species.

Question 8

All insects have a similar body pattern with three body segments, six legs and two pairs of wings. The following diagram shows three insects: a wasp, dragonfly and moth.

What are the differences between the wings of the three insects an example of?

A. A remnant structure
B. A homologous structure
C. A vestigial structure
D. A transitional structure

Question 9

The amino acid sequence of a small segment of a gene from three species was investigated and is shown in the table below:

Species	Amino acid sequence
Horse	*ser – arg – pro – lys – arg – arg – cys*
Cat	*ser – arg – pro – val – arg – arg – cys*
Lizard	*ser – arg – his – trp – arg – arg – val*

The amino acid comparison provides evidence that

A. the lizard and the horse share the most recent common ancestor.
B. the cat and the lizard share the most recent common ancestor.
C. the cat and the horse share the most recent common ancestor.
D. the cat and the horse share the most distant common ancestor.

Question 10

The amino acid sequences of six species were compared and used to construct the phylogenetic tree below.

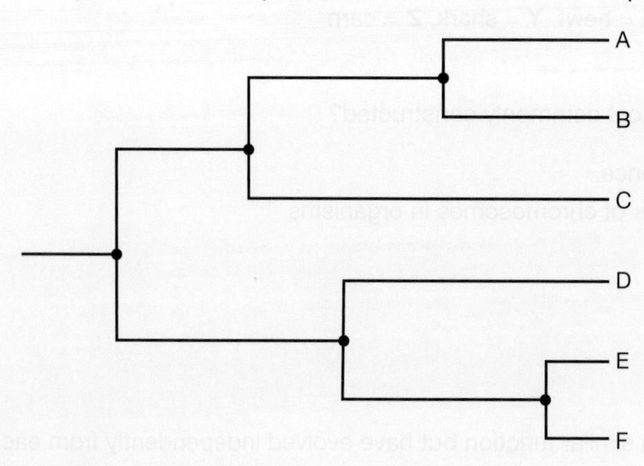

Using this inferred phylogeny, it is reasonable to conclude the percentage of matching amino acid sequences was

A. greater in species E and F than in species A and B.
B. greater in species C and D than in species A and C.
C. greater in species A and C than in species A and B.
D. greater in species B and C than in species D and E.

Section B — Short answer questions

Question 11 (6 marks)

Source: *Adapted from VCAA 2015 Biology Exam, Section B, Q9*

A fossil of an extinct species called *Indohyus major*, found in northern India, is thought to share a recent common ancestor with the group of living organisms called cetaceans. Cetaceans include dolphins and whales.

Indohyus major

Source: Nobu Tamura (http://spinops.blogspot.com)

a. Name the type of evolution that describes the relationship between *I. major* and cetaceans. **1 mark**

For a long time, scientists have believed that cetaceans are related to the group of terrestrial mammals classified as artiodactyls, which includes pigs and hippopotami. The name artiodactyl refers to the shape of the feet or hooves of these animals.

To work out the evolutionary relationships between *I. major* and living animals, scientists closely studied their bones and skeletal structures.

b. What name is given to the study of the similarities and differences between the bones and skeletal structures of animals, including fossils of extinct species? **1 mark**

The table below shows a summary of the scientists' findings.

Animal	Feet	Limb bones	Inner-ear bones
cetaceans (e.g. whales)	artiodactyl	thick	thick
suids (e.g. pigs)	artiodactyl	thin	thin
hippopotamids (e.g. hippopotami)	artiodactyl	thick	thin
I. major	artiodactyl	thick	thick

c. i. Using all the information provided, create a phylogenetic tree to show the evolutionary relationships between the following animals. Write the corresponding letter of the animal (**A.–D.**) in your phylogenetic tree. **2 marks**

A. cetaceans (e.g. whales and dolphins)
B. suids (e.g. pigs)
C. hippopotamids (e.g. hippopotami)
D. *I. major*

ii. Explain the reasoning behind your response to **part c.i.** **2 marks**

Question 12 (4 marks)

Source: *VCAA 2013 Biology Section B, Q10*

In humans, severe acute respiratory syndrome (SARS) is a serious form of pneumonia. SARS is caused by a coronavirus that was first identified in 2003.

Scientists suspected that the virus had been transmitted to humans from some other animal. Testing was completed on several animal species. Strains of the coronavirus similar to those found in humans were identified in different species of horseshoe bats (genus *Rhinolophus*) and palm civets (*Paguma larvata*).

Samples were taken from the different sources and the virus's RNA from each sample was sequenced.

a. What molecular information would the scientists obtain from sequencing RNA? **1 mark**

The molecular information enabled the scientists to draw an evolutionary tree for different strains of the coronavirus. The following evolutionary tree was drawn.

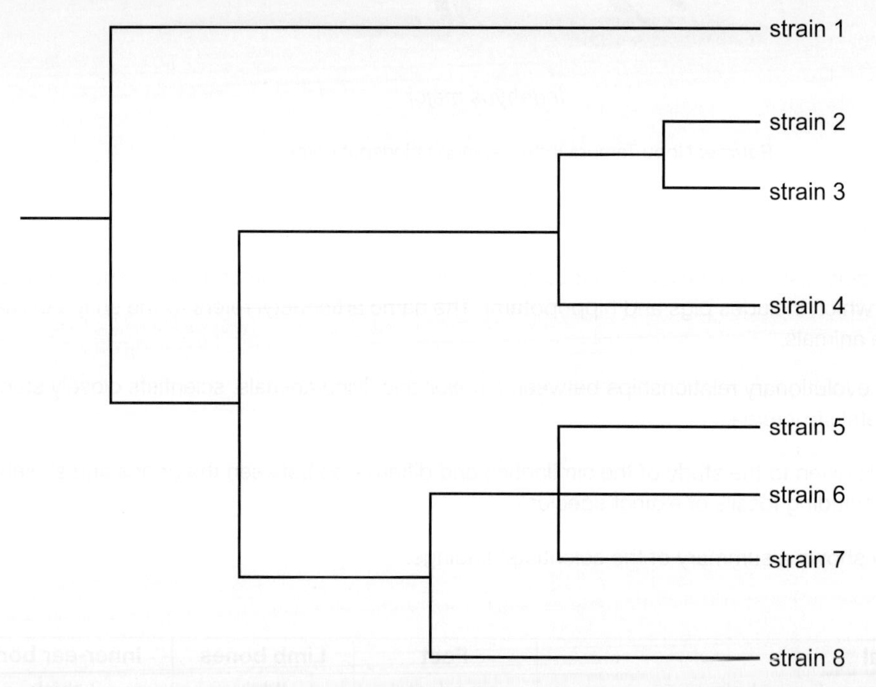

b. Coronavirus strains 2 and 3 are the most similar of the strains. Using your knowledge and the information given in the evolutionary tree, give **two** possible explanations as to why they are the most similar of the strains. **2 marks**

Strain 7 is found in palm civets, and strains 5 and 6 in humans. All other strains are found in different species of horseshoe bats.

c. What conclusion can be drawn about the origin of the strain of virus that causes SARS in humans? **1 mark**

Question 13 (1 mark)

Source: *VCAA 2008 Biology Exam 2, Section B, Q5a*

Three different kinds of plants, a cactus, euphorbia and milkweed, have similar adaptations for growing in desert environments. They all have long, fleshy stems for water storage, protective spines and reduced leaves. Both the euphorbia and milkweed plants are believed to have evolved from leafy plants adapted to more temperate climates. They share a more recent ancestor than either one does to a cactus.

Draw an evolutionary tree (also called phylogenetic tree or cladogram) to demonstrate this relationship.

Question 14 (6 marks)

Cytochrome *c* is a protein that consists of 104 amino acids and has the same amino acid sequence across a large range of organisms. The amino acid sequences of crytochrome *c* from a variety of species has been compared to human crytochrome *c*.

Organism	Number of differences in the amino acid sequence relative to human cytochrome *c*
Human	–
Chimp	0
Whale	7
Chicken	13
Tuna	22

a. Using the data for the molecular homology given for crytochrome *c*, identify the organism which is:

i. most closely related to humans. Justify your response.

ii. most distantly related to humans. Justify your response. **2 marks**

b. Using the data from the table only, create a phylogentic to show the relatedness between the five species. **2 marks**

c. There are no changes in the amino acid sequence of crytochrome *c* betweeen humans and chimps. Explain if this provide evidence that there are no changes in the gene for crytochrome *c* between the two species. **2 marks**

Question 15 (5 marks)

There is a rich fossil record of extinct Australian megafauna that lived over 40 000 years ago. These include the marsupial lion, *Thylacoleo carnifex* and the giant wombat, *Diprotodon*. The existence of two species have been identified through the fossil records and ancient Aboriginal artworks. These two species went extinct at a similar time.

The marsupial lion and the giant wombat have both been classified as belonging to the genus *Vombatiformes* by scientists through their structural characteristics. The trends seen in these traits are further supported by molecular evidence.

The table shows some of the characteristics of extinct and modern *Vombatiformes.*

Animal	Height	Teeth	Claws	Pouch
Marsupial lion: *Thylacoleo carnifex*	0.8 metres	Large slicing molars and incisors	Semi-opposable thumb on claws	Present
Diprotodon	2 metres	Chiselling incisors	Strong claws facing forwards	Backwards facing
Wombat	0.25–0.35 metres	Chiselling incisors	Strong claws facing forward	Backwards facing
Koala	0.6–0.8 metres	Incisors and molars	Opposable thumb on claws	Backwards facing

a. Use the information to create a phylogenetic tree to include the marsupial lion, *Diprotodon*, wombat and koala. **2 marks**

b. Justify your reasoning for response to 15a. **2 marks**

c. Outline why it would it be important to use molecular evidence and not just structural evidence to produce the phylogenetic tree. **1 mark**

9.5 Exercise 3: Biochallenge online only

on Resources

eWorkbook Biochallenge — Topic 9 (ewbk-8090)

Solutions Solutions — Topic 9 (sol-0665)

Past VCAA examinations

Sit past VCAA examinations and receive immediate feedback, marking guides and examiner's report notes.
Access Course Content and select 'Past VCAA examinations' to sit the examinations online or offline.

teachon

Test maker
Create unique tests and exams from our extensive range of questions, including past VCAA questions.
Access the assignments section in learnON to begin creating and assigning assessments to students.

Online Resources

Below is a full list of **rich resources** available online for this topic. These resources are designed to bring ideas to life, to promote deep and lasting learning and to support the different learning needs of each individual.

eWorkbook

- 9.1 eWorkbook — Topic 9 (ewbk-1888) ☐
- 9.2 Worksheet 9.1 Exploring homologous and vestigial structures (ewbk-2905) ☐
- 9.3 Worksheet 9.2 Case study: Examining the relatedness of pandas (ewbk-2906) ☐
  - Worksheet 9.3 Extension: Comparing chromosomes (ewbk-2907) ☐
  - Worksheet 9.4 Analysing and inferring evolutionary relationships (ewbk-2908) ☐
- 9.4 Worksheet 9.5 Evolutionary relationships and phylogenetic trees (ewbk-2909) ☐
  - Worksheet 9.6 Extending evolutionary concepts (ewbk-2910) ☐
  - Worksheet 9.7 Reflection — Topic 9 (ewbk-2912) ☐
- 9.5 Biochallenge — Topic 9 (ewbk-8090) ☐

Solutions

- 9.5 Solutions — Topic 9 (sol-0665) ☐

Practical investigation eLogbook

- 9.1 Practical investigation eLogbook — Topic 9 (elog-0195) ☐
- 9.2 Investigation 9.1 Exploring homologous structures (elog-0073) ☐
  - Investigation 9.2 Observing vestigial structures (elog-0074) ☐
- 9.4 Investigation 9.3 Using data to create phylogenetic trees (elog-0075) ☐

Digital documents

- 9.1 Key science skills — VCE Biology Units 1–4 (doc-34326) ☐
  - Key terms glossary — Topic 9 (doc-34623) ☐
  - Key ideas summary — Topic 9 (doc-34614) ☐
- 9.3 Case study: What DNA do we use — nuclear or mitochondrial (doc-36112) ☐
- 9.4 Extension: Cladograms (doc-36113) ☐

Teacher-led videos

- Exam questions — Topic 9 ☐
- 9.2 Sample problem 1 Examining structural morphology and their evidence for evolution (tlvd-1720) ☐
- 9.3 Sample problem 2 Comparing amino acids sequences (tlvd-1721) ☐
- 9.4 Sample problem 3 Making phylogenetic trees (tlvd-1722) ☐

Interactivities

- 9.2 Homologous structures (int-5879) ☐
  - Comparative anatomy (int-0243) ☐
- 9.4 Genome phylogeny (int-0245) ☐
  - Divergent evolution of vertebrates (int-0244) ☐

Weblinks

- 9.2 Evidence for common ancestry ☐
  - TED-Ed: Vestigial structures ☐
  - 7 vestigial features of the human body ☐
- 9.3 Comparative genomics: DNA ☐
  - Common ancestry and DNA ☐
- 9.4 Making phylogenetic trees: PhytoT ☐
- 9.5 Interactive phylogenetic tree ☐
  - The tree of life: exploring our largest phylogenetic tree ☐
  - Tree diagrams and forming cladograms ☐

Exam question booklet

- 9.1 Exam question booklet — Topic 9 (cqb-0020) ☐

Teacher resources

There are many resources available exclusively for teachers online.

To access these online resources, log on to **www.jacplus.com.au**

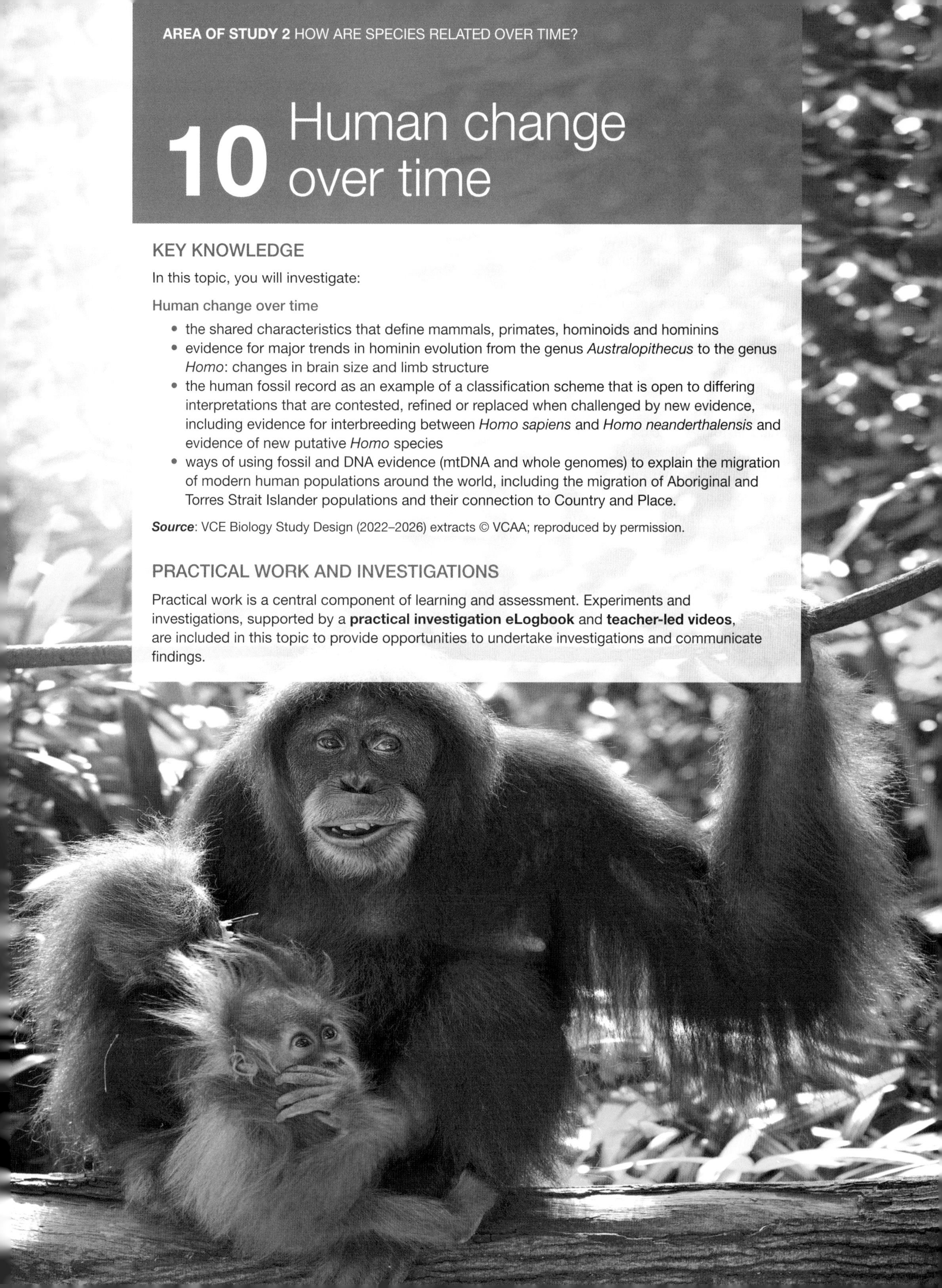

10 Human change over time

KEY KNOWLEDGE

In this topic, you will investigate:

Human change over time

- the shared characteristics that define mammals, primates, hominoids and hominins
- evidence for major trends in hominin evolution from the genus *Australopithecus* to the genus *Homo*: changes in brain size and limb structure
- the human fossil record as an example of a classification scheme that is open to differing interpretations that are contested, refined or replaced when challenged by new evidence, including evidence for interbreeding between *Homo sapiens* and *Homo neanderthalensis* and evidence of new putative *Homo* species
- ways of using fossil and DNA evidence (mtDNA and whole genomes) to explain the migration of modern human populations around the world, including the migration of Aboriginal and Torres Strait Islander populations and their connection to Country and Place.

PRACTICAL WORK AND INVESTIGATIONS

Practical work is a central component of learning and assessment. Experiments and investigations, supported by a **practical investigation eLogbook** and **teacher-led videos**, are included in this topic to provide opportunities to undertake investigations and communicate findings.

10.1 Overview

Numerous **videos** and **interactivities** are available just where you need them, at the point of learning, in your digital formats, learnON and eBookPLUS at **www.jacplus.com.au**.

10.1.1 Introduction

All organisms are classified through the Linnaean system based on their structural, physiological and chemical make-up. Each different kind of organism is a species. As *Homo sapiens*, we are classified as our own species within the genus *Homo*. This genus lies within the taxonomic tribe of Hominini (see figure 10.1). This classification reflects on our biological relationships with other species and our evolutionary history. While our closest living relative is the chimpanzee, it was not long ago that we shared Earth with other human species. How humans have changed over time and their migration across the world has been investigated through fossils and molecular evidence.

FIGURE 10.1 Models of members of the tribe Hominini

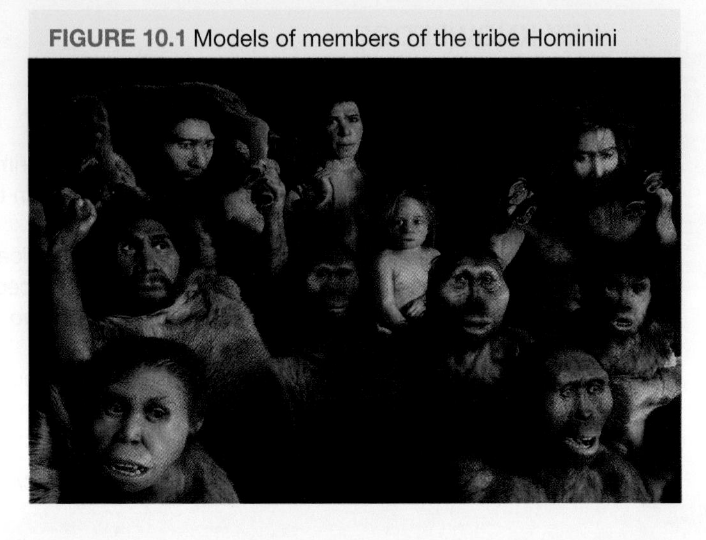

This topic will explore how humans are classified from mammals to hominins. It will investigate the trends in size and structure from the genus *Australopithecus* all the way to modern *Homo sapiens*; how new fossils can change the way species are classified; and how the movement of humans across Earth, including Aboriginal and Torres Strait Islander peoples, can be determined from fossil and genetic evidence.

LEARNING SEQUENCE

on Resources

eWorkbook	eWorkbook — Topic 10 (ewbk-1889)
Practical investigation eLogbook	Practical investigation eLogbook — Topic 10 (elog-0197)
Digital documents	Key science skills — VCE Units 1–4 (doc-34326) Key terms glossary — Topic 10 (doc-34624) Key ideas summary — Topic 10 (doc-34615)
Exam question booklet	Exam question booklet — Topic 10 (eqb-0021)

10.2 Shared characteristics of mammals, primates, hominoids and hominins

KEY KNOWLEDGE

- The shared characteristics that define mammals, primates, hominoids and hominins

Source: VCE Biology Study Design (2022–2026) extracts © VCAA; reproduced by permission.

Classification is an important way in which we group and organise organisms. It allows us to understand the relatedness between different species. The science of classification is referred to as taxonomy.

Through classification, biologists:

- put order into the living world
- show the relationship between different species
- gain insight into the evolution of a species.

Organisms with shared characteristics are separated into groups referred to as taxa. Each taxon falls within a taxonomic rank. This can be further divided into other taxa, as shown in the classification and taxonomic hierarchy for humans in figure 10.2.

This rank is usually simplified into a seven-level hierarchy of:

- kingdom
- phylum
- class
- order
- family
- **genus**
- **species**.

Additional ranks, such as subphylum and tribe, further split the taxa to enhance classification schemes and allow for organisms with similar traits to be grouped. Some of these are shown in figure 10.2.

The animal world is organised into large groups called phyla (singular phylum). Members of one phylum share a **common ancestor** and show similarities in their body structures, both external and internal.

About 35 different phyla are recognised. Modern humans are members of the phylum Chordata (animals with a notochord).

This phylum includes several classes, such as the class Mammalia, which includes humans.

FIGURE 10.2 The classification of humans

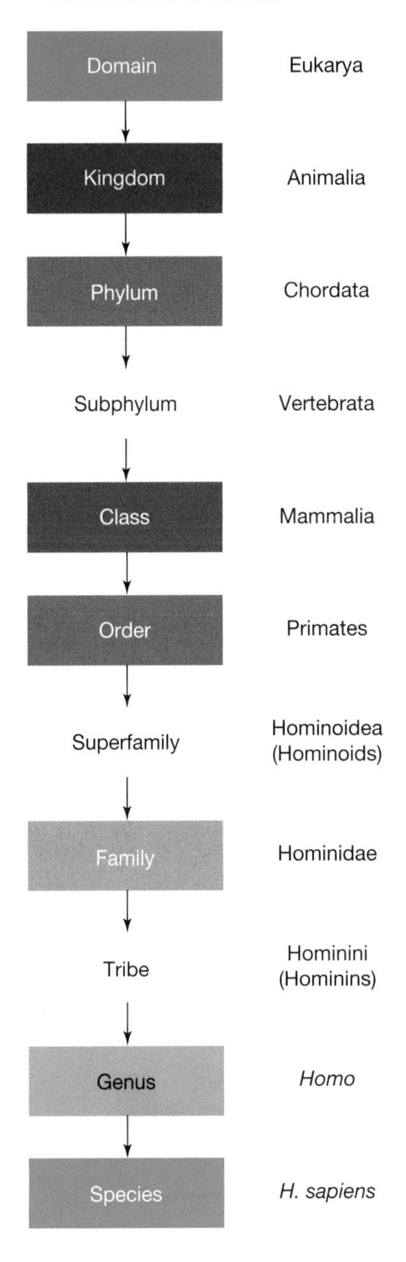

genus a taxonomic rank that is less specific than species

species organisms that are able to interbreed and produce fertile offspring

common ancestor the species that two or more descendants have diverged from

10.2.1 Shared characteristics of mammals

There are around 5400 species within the class Mammalia. **Mammals** are the familiar animals that include humans, common domestic pets such as cats and dogs, and farm animals such as sheep, cattle, pigs, horses and goats. Some other mammals include species in the wild such as whales, bats, walruses and bears.

Features of mammals

The characteristic features shared by all mammals are:

- fur or hair over their body surface
- milk-producing mammary glands
- teeth comprising of incisors, canines, premolars and molars
- a lower jaw made of a single bone
- three bones in the middle ear
- a diaphragm separating the chest cavity from the abdomen.

Many of these characteristics are so distinctive to mammals that just one of these features is sufficient to identify one. As a result, a **fossil** jawbone, or even a few teeth, can be used to identify whether an animal was a mammal. Some of these features are shown in figure 10.3. The teeth of mammals are different to other classes (such as Reptilia, seen in figure 10.4). In mammals, teeth are usually differentiated into incisors, canines, premolars and molars, that can cut, piece, grip, shear and grind.

mammals a group of animals that are characterised by the presence of fur or hair and milk-producing mammary glands

fossil evidence or remains of an organism that lived long ago

FIGURE 10.3 Different characteristics seen in mammals

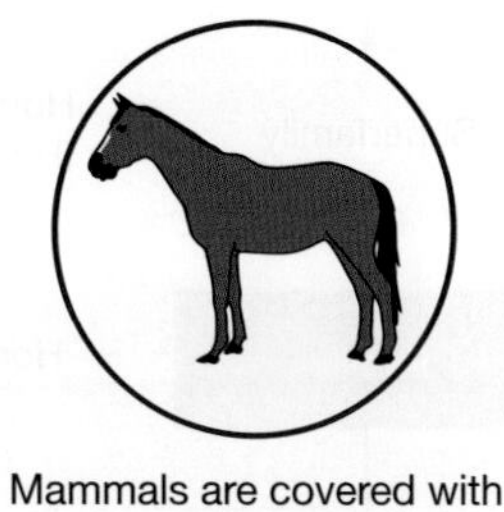

Mammals are covered with hair or fur.

Mammals breathe with lungs.

Most mammals give birth to live young.

Mammals are warm-blooded.

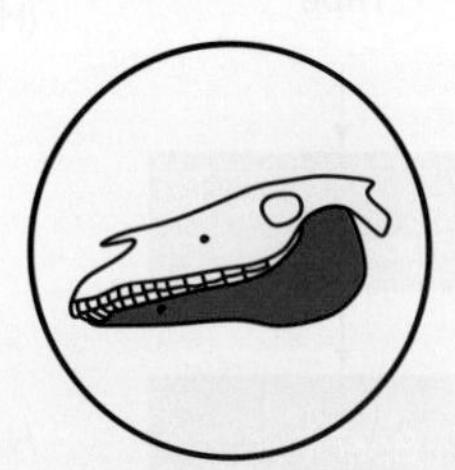

Mammals have single-boned lower jaws.

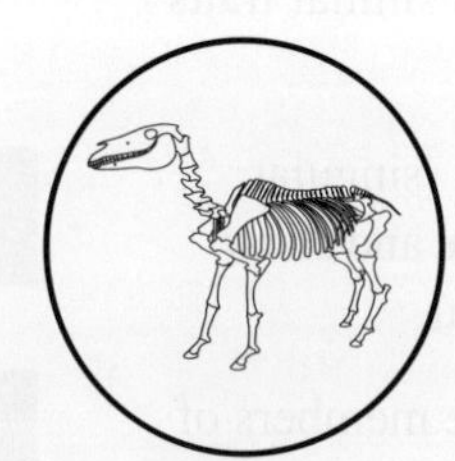

Mammals are vertebrates.

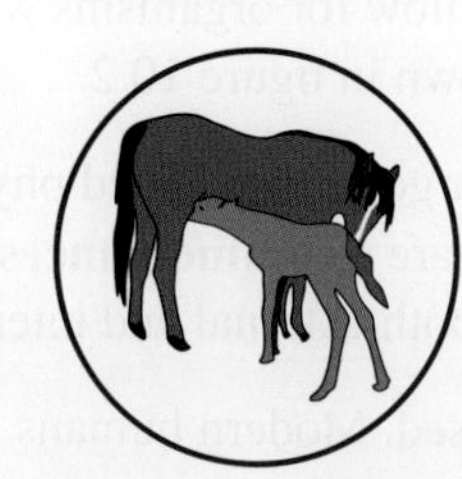

Mammals have mammary glands and feed milk to their babies.

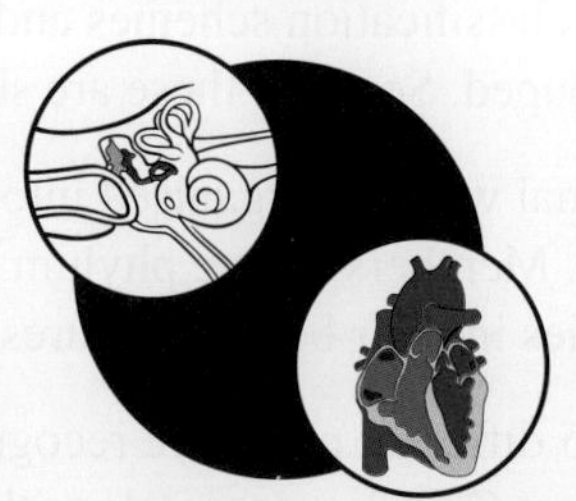

Mammals have three bones in the middle ear and a four-chambered heart.

FIGURE 10.4 Undifferentiated reptile teeth versus differentiated mammal teeth

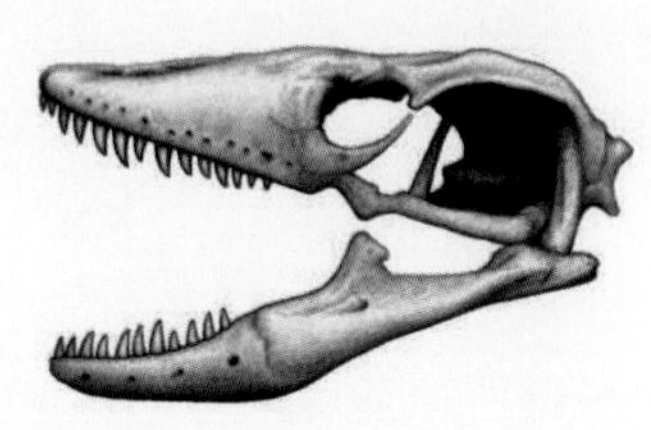

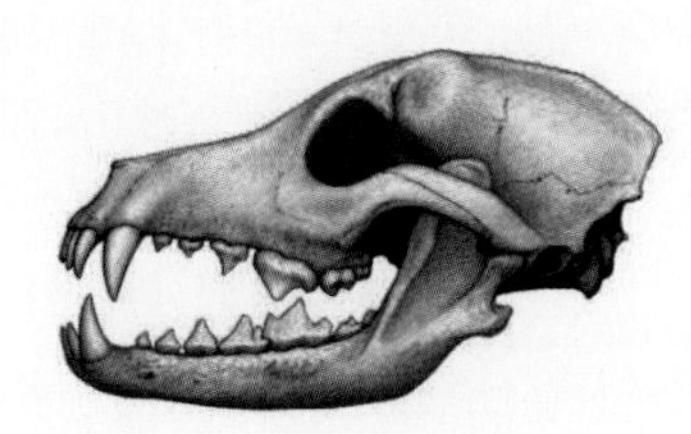

10.2.2 Shared characteristics of primates

Types of primates

The class Mammalia is further divided into 26 orders. **Apes** (including ***Homo sapiens***), monkeys and prosimians are all members of the order **Primates**.

There are approximately 400 different primate species living today on this planet. They range from the tiny mouse lemur (*Microcebus rufus*), which has an average mass of about 40 grams, to the western gorilla (*Gorilla gorilla*), with the adult male having a mass of about 175 kilograms.

Most primates are tree dwellers and are found in tropical forest and woodland habitats. As well as living species (summarised in table 10.1), primates also include extinct species of genus ***Homo***, such as ***Homo erectus*** and various species of the genus ***Australopithecus***.

ape a large primate that lacks a tail

Homo sapiens modern humans

primates a group of mammals with opposable thumbs, flat nails and binocular vision

Homo a genus of hominins that includes modern humans and closely related species with an increased brain size and a longer leg-to-arm ratio

Homo erectus an extinct human species appearing about two million years ago that was the first to migrate from Africa

Australopithecus a genus of hominins from which the *Homo* genus is considered to be descended; sometimes referred to as australopithecines

TABLE 10.1 Different examples of living primates

Group	Animals
Prosimians	• Lemurs (pictured) • Lorises • Tarsiers
New World monkeys (from South and Central America)	• Capuchin monkeys (*Cebus* spp.) (pictured) • Howler monkeys (*Alouatta* spp.) • Spider monkeys (*Ateles* spp.)
Old World monkeys (from Africa and Asia)	• Baboons (*Papio* spp.) (pictured) • Macaque monkeys (*Macaca* spp.)
Apes	• Gibbons (*Hylobates* spp.) of South-East Asia • Orang-utans (*Pongo* spp.) of Borneo and Sumatra (pictured) • Mandrills (*Mandrillus sphinx*) • Gorillas (*Gorilla* spp.) • Chimpanzees (*Pan* spp.) of Africa • Humans (*Homo sapiens*).

FIGURE 10.5 The different groups in the order of Primates

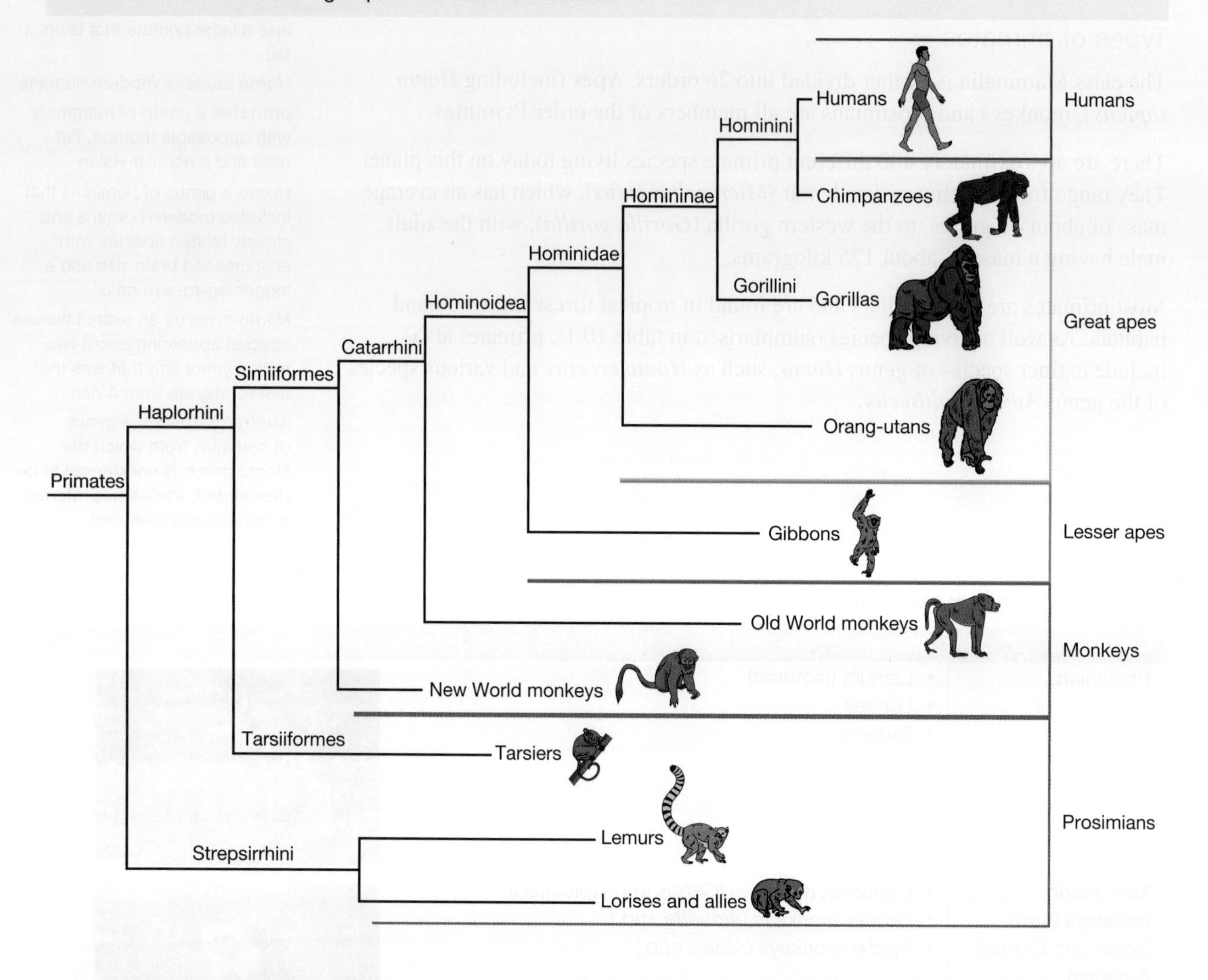

Characteristics of primates

No single characteristic is diagnostic of a primate but, taken together, the following features can help identify one.

Features of primates

The characteristics of primates include:
- opposable thumbs
- flat nails on digits
- binocular vision
- large brains relative to their body
- a long gestational period.

Exploring these features, the following can be observed.
- The hands and feet of a primate typically have:
  - five digits that can grasp or curl around objects due to possessing **opposable thumbs** and sometimes also an opposable large toe. These are able to be brought into contact with the other digits (see figures 10.6a and 10.6b).
  - flat nails on their digits.

opposable thumb a thumb that allows for grasping and can be rotated and moved to touch other fingers

FIGURE 10.6 a. Comparison of primate hands and feet b. The structure of these can allow for both the power grip and the precision grip.

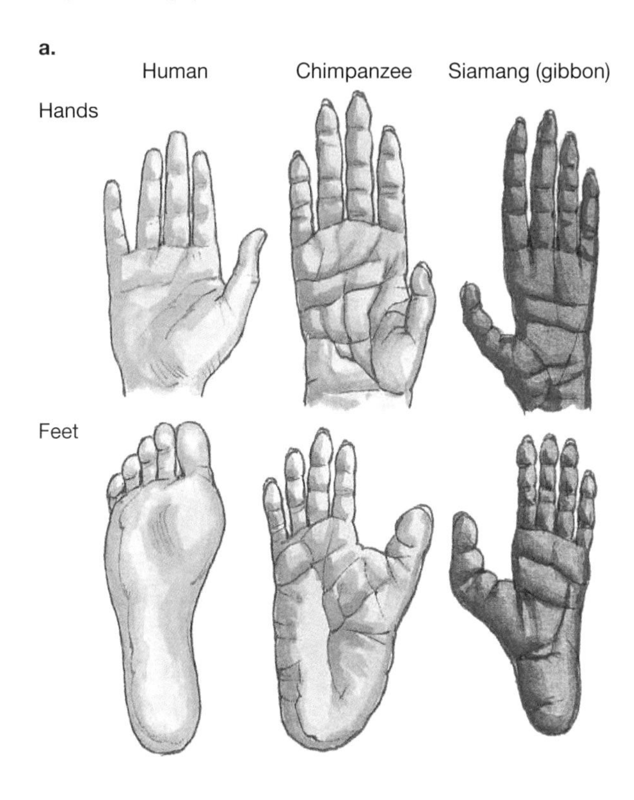

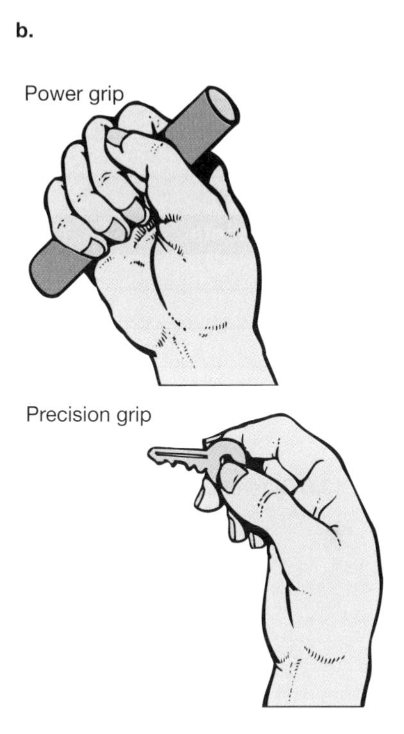

- Primates have an emphasis on vision. They have:
  - **binocular vision** due to possessing large, forward-facing eyes that give stereoscopic (3D) vision (see figure 10.7)
  - colour vision because they have three types of cone in the retinas of their eyes
  - a protective bone at the outer side of the eye socket.
- Primates have *fully rotating shoulder sockets* that are adapted to life swinging from branch to branch as primates move between trees.
- Compared to other mammals, primates have *large brains relative to their body sizes* — we primates are more brains than brawn! In addition, when a primate brain is compared to a brain of similar size from a non-primate mammal, the primate brain is seen to have several times as many neurons due to the increased folding of the cerebral cortex. Therefore, primate brains have a greater concentration of neurons per unit volume of brain tissue than is the case for non-primates. Table 10.2 and figure 10.8 show comparisons between the brains of primates and rodent species.

FIGURE 10.7 Stereoscopic vision results from forward-facing eyes.

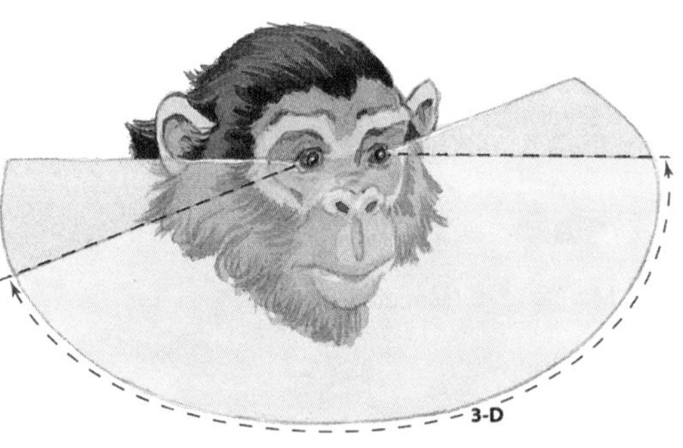

binocular vision a type of vision that allows for both peripheral vision and a perception of depth using two eyes

TABLE 10.2 Comparisons of the mass and the total number of neurons in brains from primates and from non-primates, in this case rodents (Data from Herculano-Houzel, S. *Frontiers in Human Neuroscience* 9 November 2009)

Species of primates	Brain mass (g)	Neurons (millions)
Owl monkey	15.7	1468
Squirrel monkey	30.2	3246
Capuchin monkey	53.2	3690
Macaque monkey	87.4	6376
Species of rodents	**Brain mass (g)**	**Neurons (millions)**
Rat	1.8	200
Guinea pig	3.8	240
Agouti	18.4	857
Capybara	76.0	1600

FIGURE 10.8 Differences in brain size and cerebral cortex folding in various species

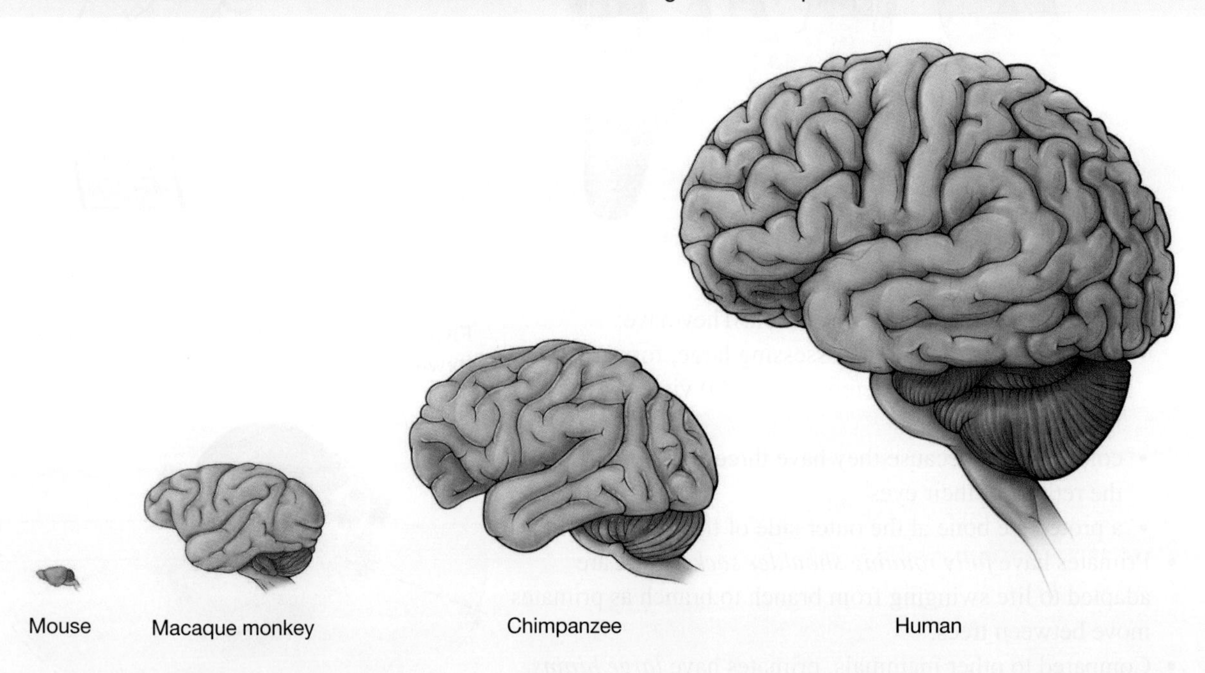

- Primates are *social mammals* and typically live in groups that, depending on the species, may be as large as a troop of several hundred animals or as small as a pair.
- Primates have a relatively *long gestation period* compared to non-primate mammals (see table 10.3). This longer gestation period allows for the more extensive foetal brain growth of primates. Primates typically produce a single young at each birth and provide parental care for an extended period — the longest period being in the case of human young.

FIGURE 10.9 Primates, including orang-utans, are highly social.

TABLE 10.3 Gestation periods for several mammals. Note that, for similar maternal body mass, a primate has a longer gestation period than a non-primate mammal.

Mammal	Average mass of adult female (kg)	Average gestation (days)
Primates		
Ring-tailed lemur (*Lemur catta*)	2.2	120
Owl monkey (*Aotus trivirgatus*)	0.7	133
White-cheeked gibbon (*Nomascus leucogenys*)	7.5	210
Chimpanzee (*Pan troglodytes*)	40	237
Orang-utan (*Pongo pygmaeus*)	36	251
Human (*Homo sapiens*)	60	265
Non-primates		
Mouse (*Mus musculus*)	0.3	21
Kangaroo (*Macropus giganteus*)	37	34
Cat (*Felis catus*)	5	60
Lion (*Panthera leo*)	130	110

TIP: You do not need to memorise every specific feature of each individual primate. You should be able to outline features that distinguish primates from other mammals and use provided data as evidence to support these trends.

Resources

Weblink Humans are primates

elog-0427

INVESTIGATION 10.1

Observing primates

Aim

To observe different primates and their shared characteristics

10.2.3 Shared characteristics of hominoids

Types of hominoids

Among the primates, humans are **hominoids** or members of the superfamily Hominoidea.

Figure 10.10 shows the classification of members of the order Primates to the level of superfamily Hominoidea.

Humans share membership of the superfamily Hominoidea with:
- the lesser apes (gibbons, siamangs)
- the great apes (chimpanzees, gorillas, orang-utans).

So, all the apes are hominoids.

This taxonomic group, superfamily Hominoidea, includes all those primates that fit the biological definition of 'ape'. This means that the hominoids include the following living species: gibbons (14 species), orang-utans (2 species), gorillas (2 species), chimpanzees (2 species) and humans.

hominoids a superfamily of primates that lack a prehensile tail, including apes and humans

FIGURE 10.10 Classification of primates to the level of superfamily Hominoidea

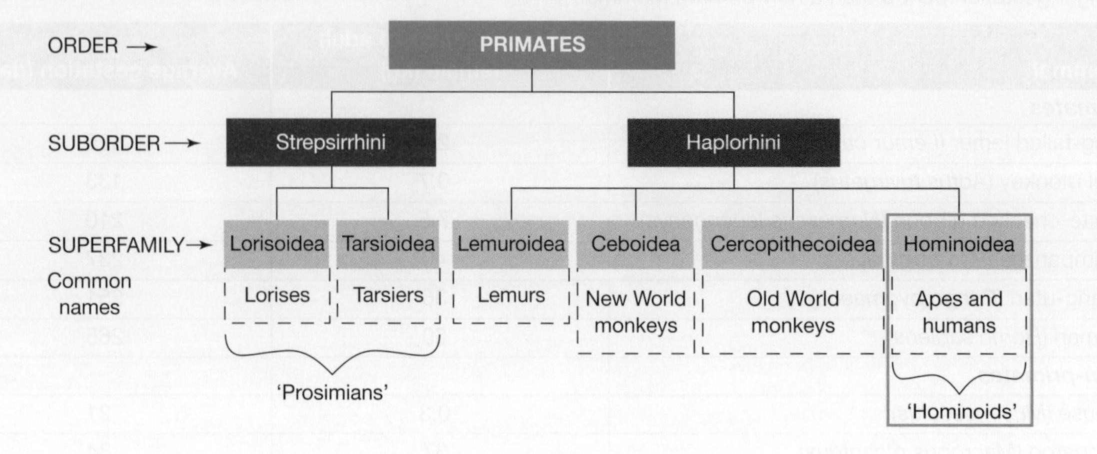

Characteristics of hominoids

The skeletons of some different hominoids are shown in figure 10.11. Many common features can be observed across all hominoids.

FIGURE 10.11 Skeletons of different hominoids

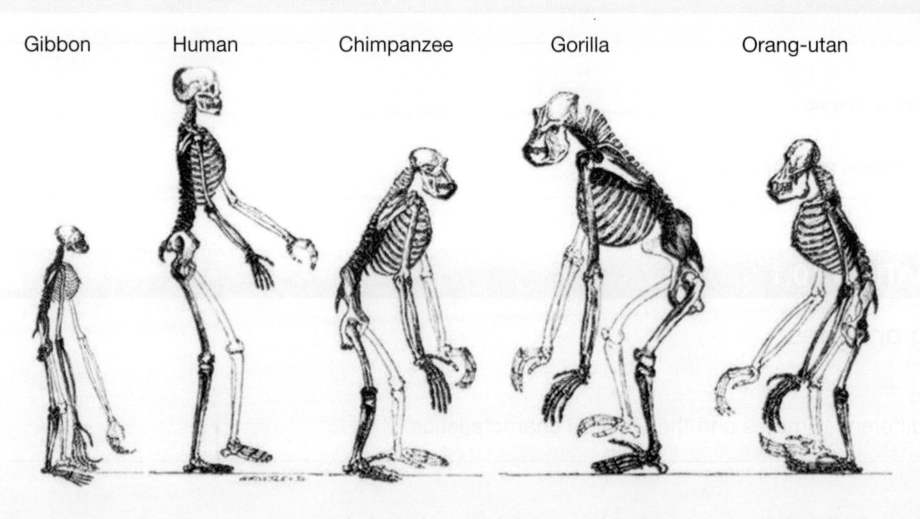

Features of hominoids

Shared features of hominoids (that distinguish them from other primates) include:

- the absence of a tail
- larger and more complex brains
- distinctive molar teeth in the lower jaw with five cusps
- relatively long upper limbs
- a wider chest
- shoulder joints that permit the arms to be rotated
- larger and more complex brains (that allow for greater intelligence, problem-solving and communication).

Each of these features will now be explored in further detail.

- Hominoid brains are larger and more complex than those of other primate groups, which allows for greater intelligence, problem-solving and communication.
- The axial skeleton (skull, spine and rib cage, as shown in figure 10.12) of hominoids is characterised by:
  - a reduced lumbar spine (shown in pink in figure 10.12b)
  - an expanded sacrum (shown in blue in figure 10.12b)
  - the absence of a tail (coccyx does not extend; shown in purple in figure 10.12b).

FIGURE 10.12 **a.** Apes can be distinguished from monkeys by their lack of a tail. **b.** The sacrum is longer in hominoids (such as humans), but the coccyx (tailbone) is much shorter.

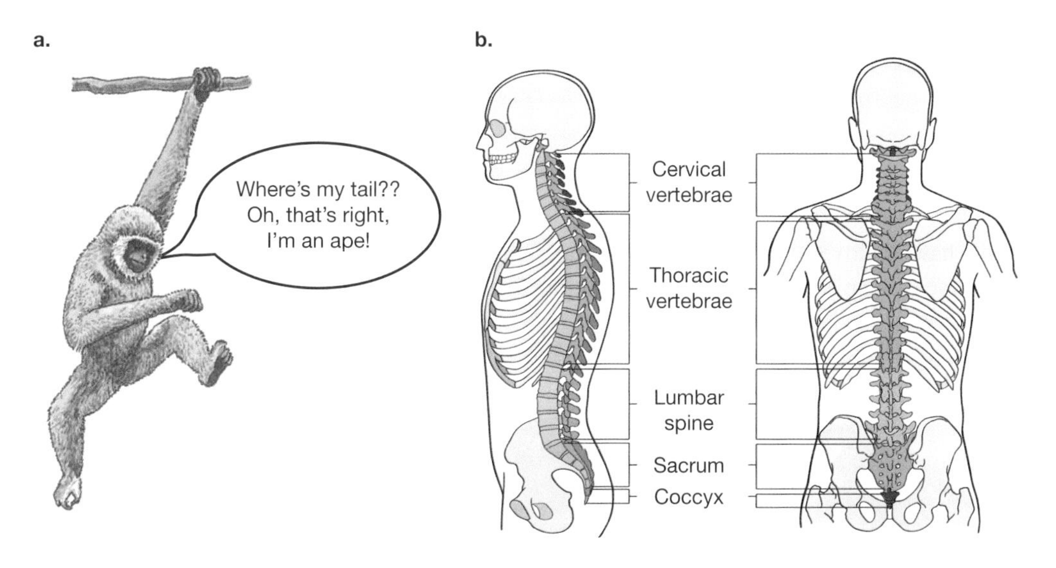

- The ribcage shape in hominoids creates a much wider chest, as shown in figure 10.13a. Other primates, such as monkeys (shown in figure 10.13b), do not show this wider chest.

FIGURE 10.13 **a.** The wide chest of a gorilla and **b.** the much narrower chest of a monkey

- Hominoids have distinctive molars in the lower jaw with five cusps (raised bumps) arranged in a 'Y5' pattern (see figure 10.14).

FIGURE 10.14 a. Diagram showing an example of the Y5 pattern of five cusps (raised bumps) on a hominoid molar **b.** A photograph of this Y5 pattern

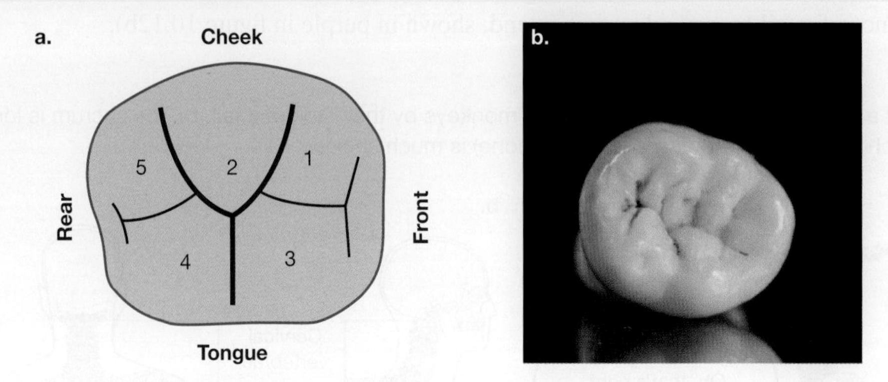

- Hominoids have relatively long upper limbs, and their shoulder joints permit the arms to be rotated around the shoulders, as in an overarm serve or bowling action.

FIGURE 10.15 a. The long limbs of hominoids as seen in an orang-utan and **b.** the rotating shoulder joint

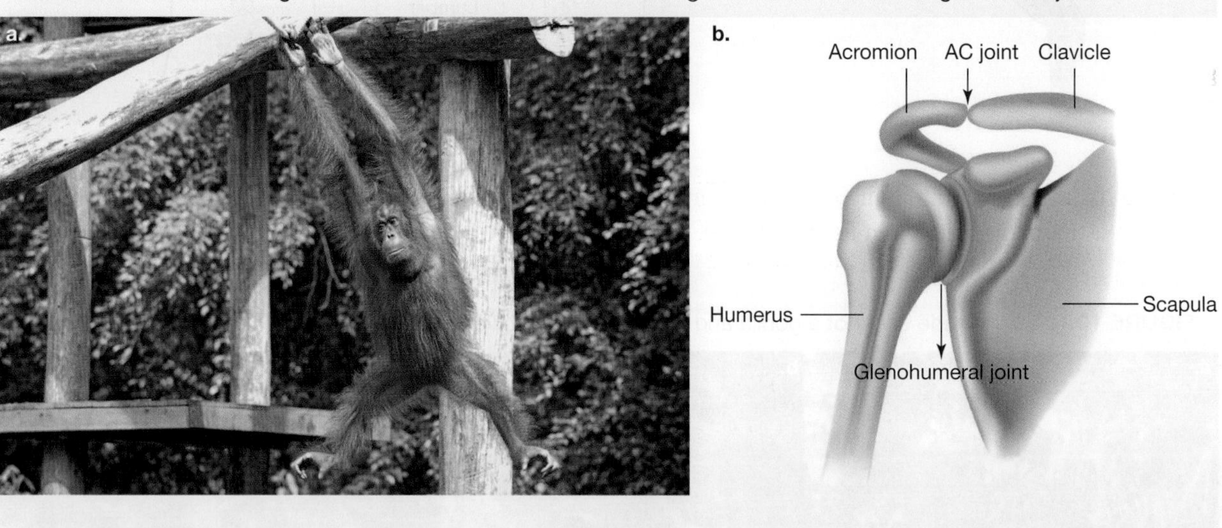

TIP: Be very careful with spelling hominoid in assessments and your exam, as many similar terms refer to different things. Hominids, for example, are the next level down (from the family Hominidae) and do not include the lesser apes. Therefore, misspelling hominoid as hominids will be marked as incorrect.

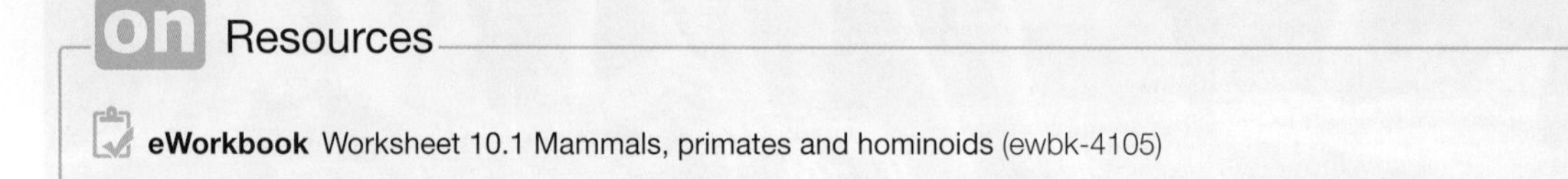

on Resources

eWorkbook Worksheet 10.1 Mammals, primates and hominoids (ewbk-4105)

10.2.4 Shared characteristics of hominins

Apes, such as gorillas and chimpanzees, and humans are all classified as hominoids; however, they are all classified into different genus groups (see figure 10.16).

FIGURE 10.16 Classification of members of superfamily Hominoidea (hominoids) to the level of genus. Hominins (Hominini) are highlighted.

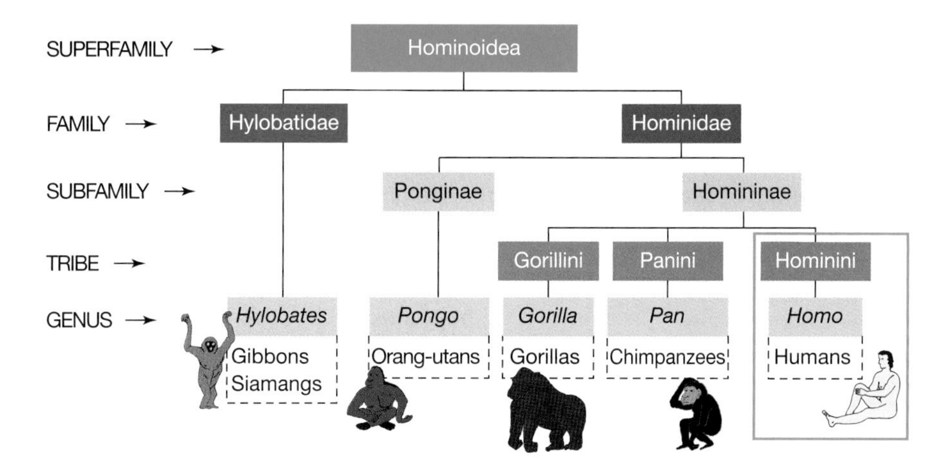

hominins modern human species and our extinct close relatives that could walk with a bipedal locomotion

bipedalism a form of locomotion involving routine movement on two feet

knuckle walk a style of walking on four limbs in which the forelimbs are in contact with the ground through the knuckles of flexed hands

Modern classification places humans in the taxonomic group known as tribe Hominini. Members of this tribe are known as **hominins**. It is only at this level of classification that humans are separated from all the other great apes. Note that our closest living relatives are members of the tribe Panini, which includes the common chimpanzee (*Pan troglodytes*) and the bonobo or pygmy chimpanzee (*Pan paniscus*).

TIP: Be careful when using old resources for information about hominins and hominids. In previous classification schemes (prior to 1980), the term *hominids* was used to refer to humans and their bipedal ancestors. This has been adjusted in modern times, in which the great apes are also referred to as hominids. This shows how classifications are not fixed, but are subject to review and revision that can occur when new information or interpretations provide a better explanation than those they replace.

Try to make sure you use up-to-date resources when researching the different classifications. *Hominin* is the correct term for humans and their erect-walking ancestors.

Characteristics of hominins

The term *hominin* refers to the modern human species and our extinct close relatives that could walk with bipedal locomotion.

Bipedalism (meaning 'two-footed') refers to when a species can walk erect on their hind legs in a sustained fashion. Bipedal locomotion is the key defining characteristic of hominins.

This is in contrast with other primates that walk on four limbs and are said to be quadrupedal (meaning 'four-footed'). The great apes, such as gorillas and chimpanzees, typically walk with their knuckles on the ground (see figure 10.17). This is known as **knuckle walking**.

FIGURE 10.17 Knuckle walking of gorillas

Bipedalism

Structures vital for bipedalism

Evidence of erect or bipedal walking can come from the:
- position of the hole (foramen magnum) in the base of the skull
- arrangement of the femur and tibia
- shape of the pelvis
- shape of the spine
- size of the heel bone and subsequent heel arch.

The position of the foramen magnum

The **foramen magnum** is the hole at the bottom of the **cranium** where the spinal cord leaves the brain and enters the vertebral column (backbone). Its position indicates how the skull sat on the vertebral column.

If the foramen magnum is towards the centre, the organism walks upright, whereas if it is at the back of the skull, the organism walks stooped on all fours. As shown in figure 10.18, hominins, such as *Homo sapiens* and *Australopithecus africanus*, have foramen magnums that are much more central on their skull compared to other primates.

FIGURE 10.18 a. The position of the foramen magnum in gorillas versus two hominins **b.** The position of the head on the vertebral column

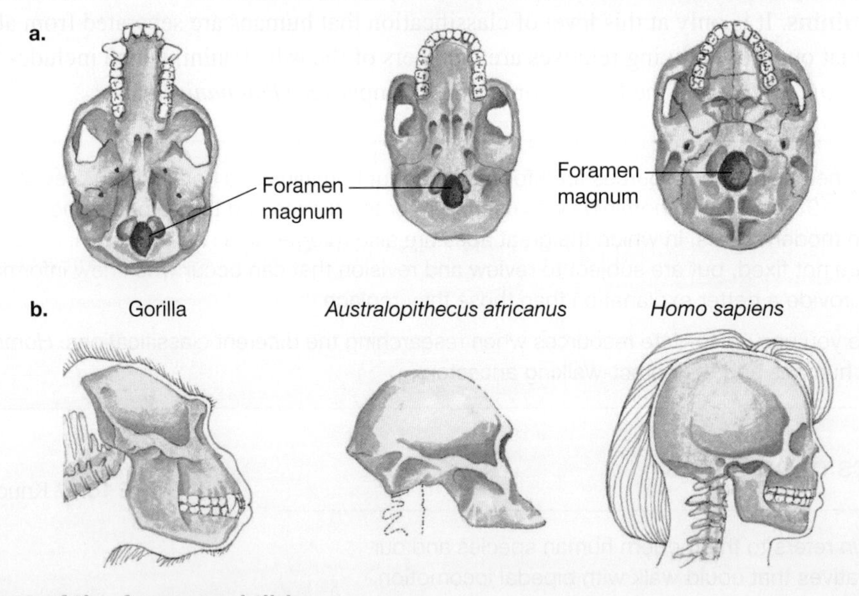

The arrangement of the femur and tibia

In primates that walk erect, such as humans, the femur (thighbone) is at an angle to the tibia (shinbone) (see figure 10.19a). As a result, our knees and feet lie below the centre of mass of the body, and the main weight of the body falls on the outside of the knee joints.

This arrangement, combined with a locking knee joint, enables us to stand erect with straight thighs and to walk erect, because we can support our body weight on one leg as the other leg steps forward.

The angle made by the vertical with the femur forms the **carrying angle** (see figure 10.19b). In modern humans, the carrying angle is in the range of 8 to 11 degrees. Chimpanzees (and the other great apes) cannot walk erect for a sustained period. Their femurs and tibias join in a straight line. The carrying angle in these prehuman great apes is about 1 degree.

foramen magnum a hole at the base of the skull in which the vertebral column attaches

cranium the vertebrate skull minus the lower jaw

carrying angle the angle formed between the femur and the vertical; also known as the bicondular angle

As a result, when erect, the knees and feet of the great apes lie outside the centre of mass of the body, and the main weight of the body falls on the inside of the knee joints. In attempting to walk erect, chimpanzees lean forward with widely spaced legs and feet, leading to knuckle walking.

FIGURE 10.19 a. The different angles of the femur and tibia in hominin species (*Homo sapiens*) compared to that of chimpanzees **b.** The carrying angle that forms between the femur and vertical

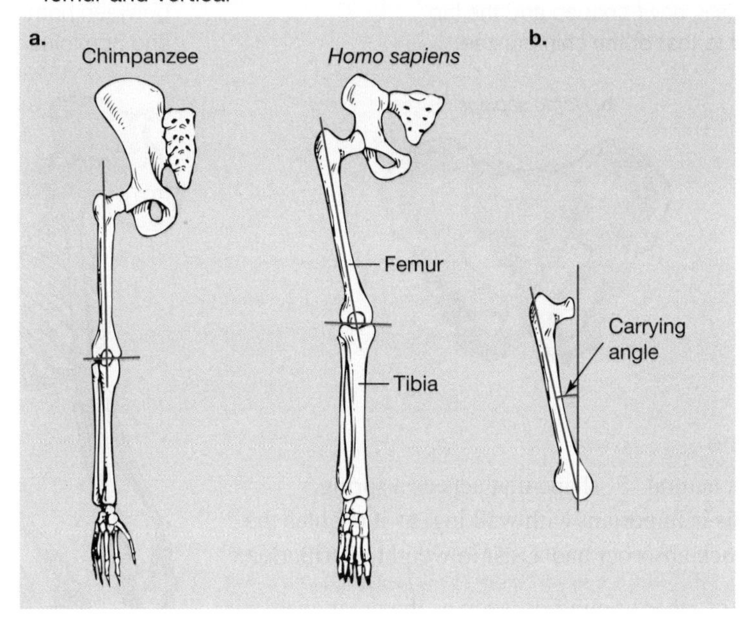

Leg length

The legs in hominins are also much longer. This leads to a greater leg : arm ratio compared to other hominoids (see figure 10.20). This assists with bipedalism and allowing for hominins to remain in an erect, upright posture.

FIGURE 10.20 A comparison of a gorilla with two hominin species showing the changes in the leg (in blue): arm (in pink) ratio

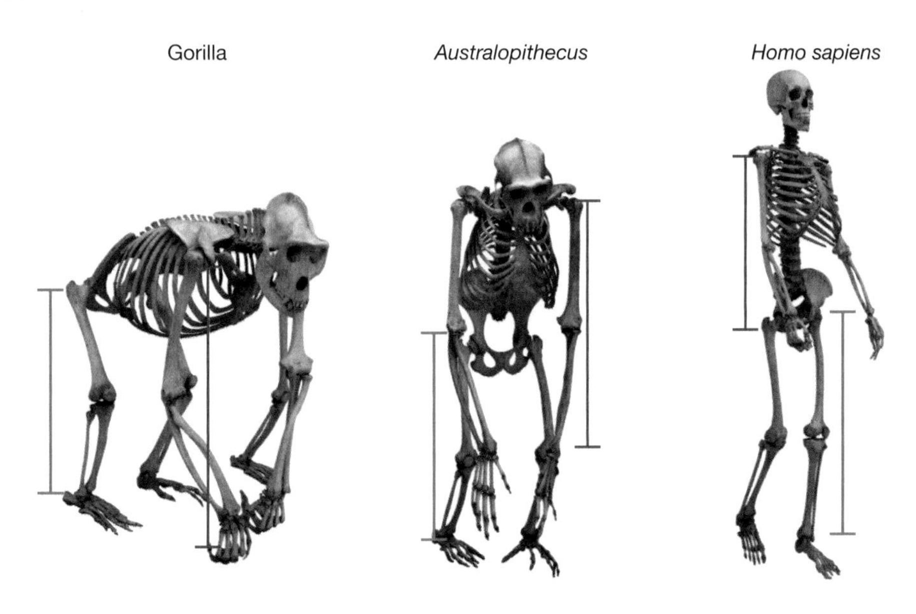

The shape of the pelvis

The structure of the pelvis is quite different between hominins and other hominoids, such as the great apes.

To assist with bipedalism in hominins, the pelvis is more bowl-shaped with a much shorter hip bone compared to the pelvis of a chimpanzee. This will be further explored in section 10.3.3 (refer to figure 10.37).

FIGURE 10.21 Structure of the pelvis of **a.** a chimpanzee and **b.** a modern human. In hominins, who can walk erect, the pelvis is bowl-shaped and the hip bone is short compared to that of the chimpanzee.

a. Chimpanzee

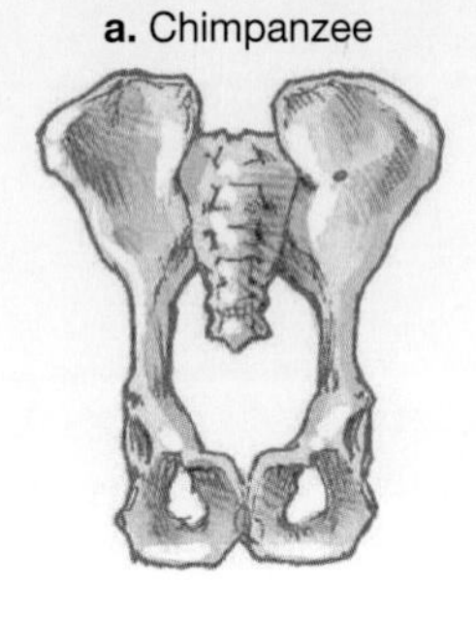

b. *Homo sapiens*

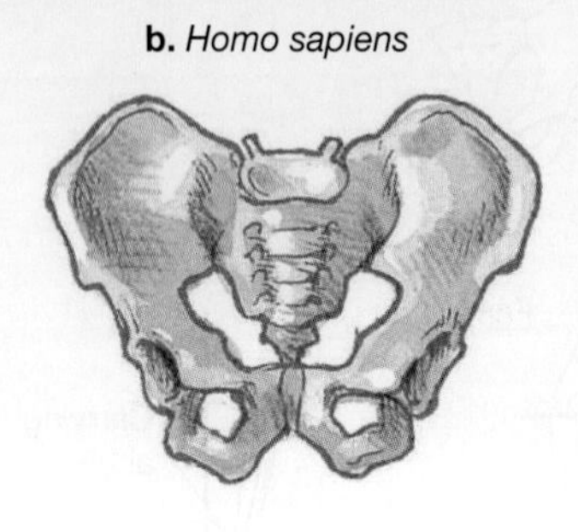

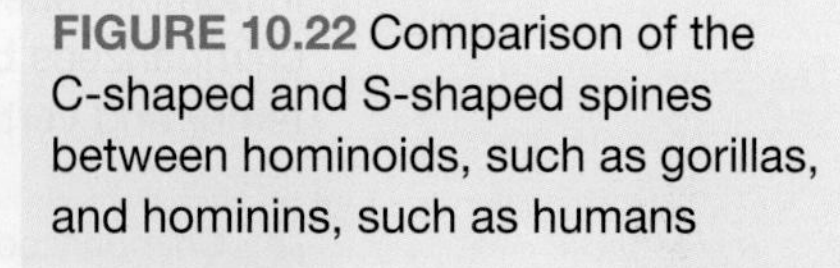

FIGURE 10.22 Comparison of the C-shaped and S-shaped spines between hominoids, such as gorillas, and hominins, such as humans

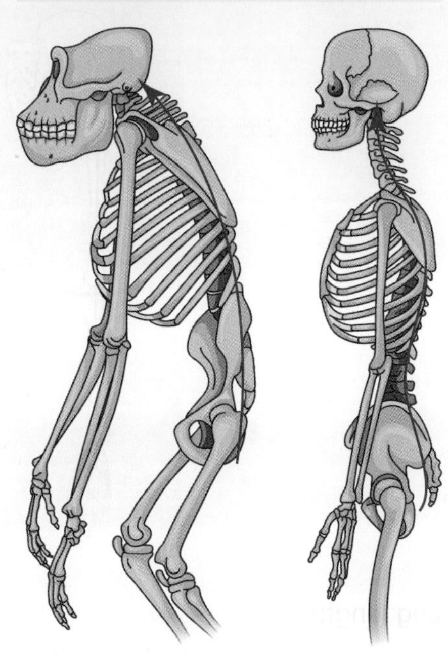

The shape of the spine

The hominin spine has a natural 'S' shape that acts as a spring, providing flexibility. This is important with walking, as it enables the spine to act as both a shock absorber and assist in weight distribution.

This differs to the spine of other hominoids, such as the great apes, who have a C-shaped spine, as shown in figure 10.22.

The size of the heel bone and subsequent arch

In hominins, the size of the heel bone is larger compared to other hominoids. The arch is also more significant. As well as this, the hallux (the big toe) is much more in line with the other toes, rather than being opposable.

These changes in foot structure allowed for bipedalism in hominins as the feet better support the weight of an individual, and also allow for better weight transference while walking. The differences in the feet between humans and chimpanzees can be seen in figure 10.23.

FIGURE 10.23 Comparison of feet between *Homo sapiens* and chimpanzees

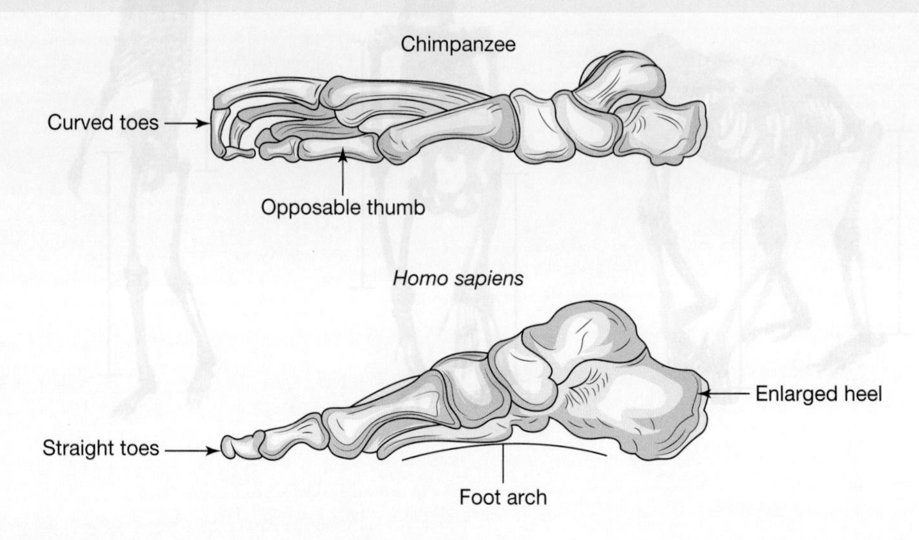

Other features of hominins

While this evidence of bipedal movement and an erect posture are their distinguishing characteristics, there are other features that are seen in hominins when compared with other hominoids, including:

- a larger and more complex brain
- a reduction in teeth size
- a more parabolic-shaped (or V-shaped) dental arch (shown in figure 10.24).

FIGURE 10.24 The upper jaws and teeth of a chimpanzee, *Australopithecus afarensis* and a modern human, showing the dental arch and size of the teeth. Note the changes seen in hominins compared to other hominoids — the dental arch has become more V-shaped, rather than the U-shaped arch seen in other hominoids.

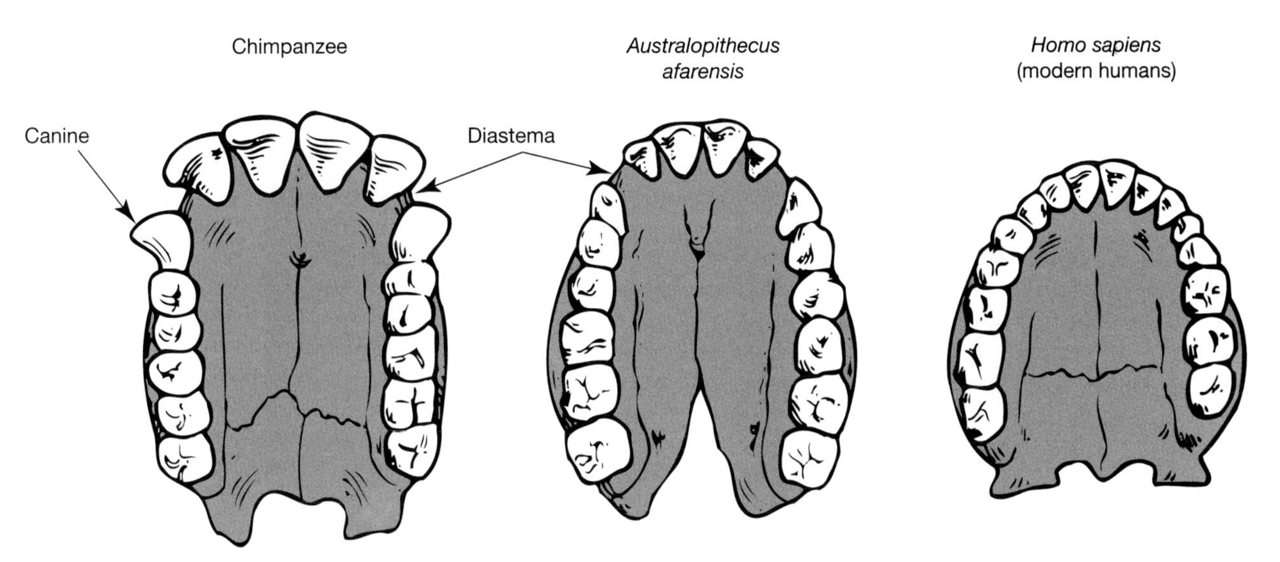

Summary of changes

A summary of the changes seen in some of the features of hominins is shown in table 10.4.

TABLE 10.4 Comparing features of other great apes (such as gorillas) to hominins

Feature	Other great apes	Hominins
Brain size	Smaller	Larger
Size of canine teeth	Larger	Smaller
Shape of dental arch	Box-shaped	More rounded
Gap (diastema) in tooth rows	Present	Absent
Position of foramen magnum	Towards the back of the skull	More central on skull
Spine	C-shaped	S-shaped
Arrangement of femur and tibia	Less angled (straight)	More angled
Size of heel bone	Smaller	Larger
Heel arch	Flat	Arched
Shape of the pelvis	Longer	Shorter and more bowl-shaped

CASE STUDY: Using footprints to explore bipedalism

Other evidence for the erect walking of hominin species comes from fossilised footprints.

A trail of fossilised footprints extending for more than 20 metres was discovered in Laetoli in northern Tanzania in 1978. The footprints were made about 3.6 million years ago by one or possibly two adult hominins and a juvenile when they walked across a surface covered in damp volcanic ash (see figure 10.25).

The Sun dried the wet ash, preserving the footprints. These impressions were later covered by more volcanic ash and the material became compacted and hard. The feet that made these prints were like our feet, with the big toe parallel to the other toes. In contrast, footprints of African apes would show an opposable big toe (refer back to figure 10.6a).

FIGURE 10.25 The Laetoli footprints. This fossil evidence captured in rock supports the conclusion that hominins walked upright about 3.6 Mya.

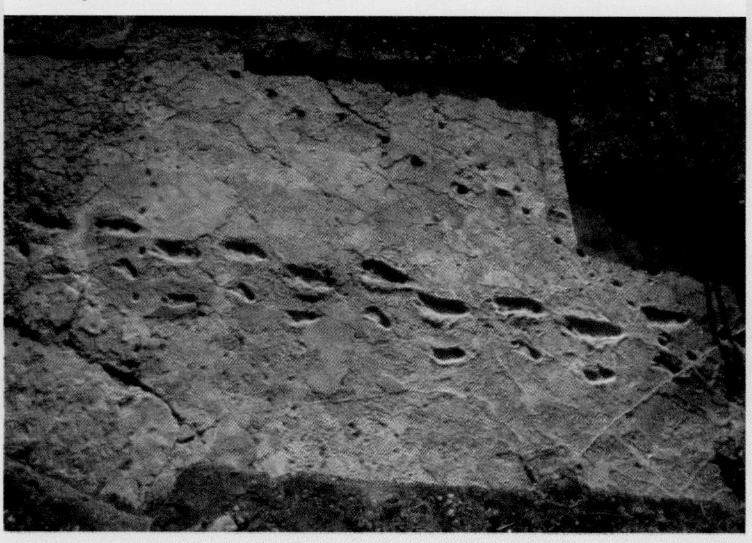

tlvd-1808

SAMPLE PROBLEM 1 Comparing skulls of hominins and hominoids

Examine the skulls of two different primates shown.

Skull A

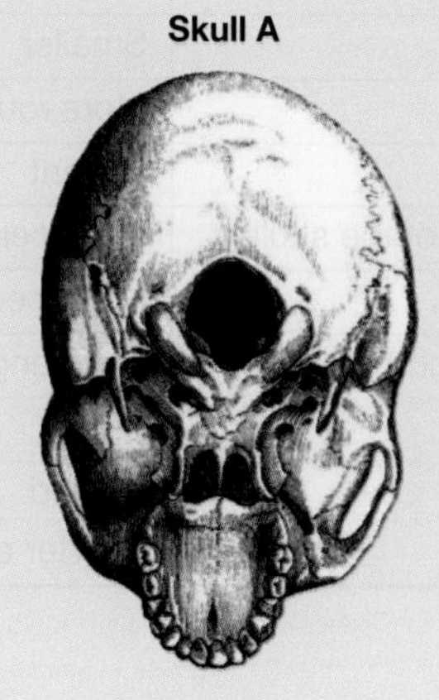

Skull B

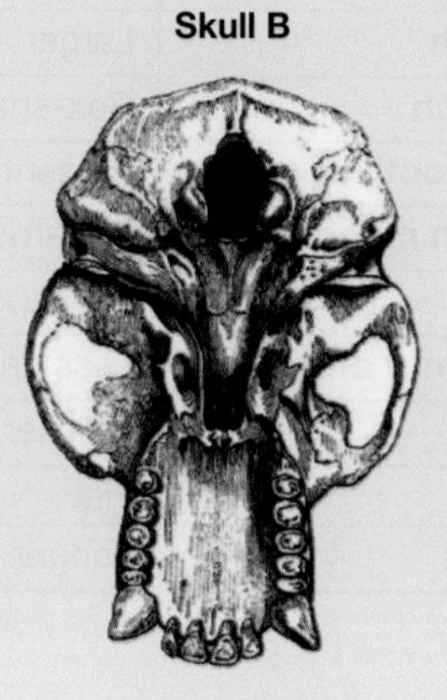

a. **Name three characteristics that the species represented by skulls A and B would have in common. (3 marks)**

b. **Construct a table to identify which primate has its highest classification level as hominoid and which has it as hominin. Justify your response. (4 marks)**

THINK

a. 1. Identify what the question is asking you to do. In a *name* question, you can list answers with one word or a short sentence.

2. Consider what both skulls A and B have in common — they are primates, as stated in the stem of the question. The question asks for three characteristics; these must be distinct from each other.

WRITE

Any three of the following:

- Opposable thumbs **OR** opposable toes
- Binocular vision **OR** stereoscopic vision
- Fully rotating shoulder joint
- Large brains relative to body size
- Relatively long gestation periods.

(1 mark for each correct answer)

b. 1. Identify what the question is asking you to do. You need to *identify* and *justify* using a table. You need to list an answer and explain your response with evidence.

2. Consider the differences between a hominoid and a hominin and then what evidence a skull can give you about this.

In these skulls, one feature that differs is the foramen magnum. In hominins, this is more central to allow for a bipedal locomotion (skull A). In other hominoids, this sits further back (skull B) as they walk with a quadrupedal locomotion.

3. Use this information to construct a table.

Skull	Classification level	Justification
A	Hominin (1 mark)	As hominins walk with a bipedal locomotion, the foramen magnum is located centrally on the skull due to the spine being vertical (1 mark).
B	Hominoid (1 mark)	As hominoids walk with a quadrupedal locomotion, the foramen magnum is located at the back of the skull due to the spine being horizontal (1 mark).

on Resources

eWorkbook Worksheet 10.2 Who's in the family (ewbk-4107)

Weblinks Hominid and hominin — what's the difference?
Australian Museum Human Evolution resources

KEY IDEAS

- Humans are members of the class Mammalia and share a distinctive set of features with all other mammals, such as possessing hair/fur and milk-producing mammary glands.
- Humans are members of the order Primates and share a distinctive set of features with all other primates, such as opposable thumbs and binocular vision.
- Humans and the other apes are members of the superfamily Hominoidea, and are called hominoids. Hominoids, unlike other primates, do not possess prehensile tails.
- Hominins, both living and extinct, are distinguished by the ability to walk upright.
- The term *hominin* includes modern humans and extinct species of the genus *Homo* and the genus *Australopithecus*.

10.2 Activities

learnon

To answer questions online and to receive **immediate feedback** and **sample responses** for every question, go to your learnON title at **www.jacplus.com.au**. A **downloadable solutions** file is also available in the resources tab.

10.2 Quick quiz on	10.2 Exercise	10.2 Exam questions

10.2 Exercise

1. MC *Homo sapiens* are classified as
   A. hominoids but not hominins.
   B. hominins but not hominoids.
   C. both hominoids and hominins.
   D. neither hominoids nor hominins.
2. MC Bipedalism is a feature specific to
   A. the genus *Homo*.
   B. hominins.
   C. hominoids.
   D. primates.
3. Identify the key characteristic that classifies a species as a hominoid.
4. Create a table that lists the key characteristics of humans that defines them as mammals, primates, hominoids and hominins.
5. Describe the differences between a hominoid and a hominin.
6. Describe how the walking style of humans differs to other great apes.
7. Consider the following statement: 'All species that have opposable thumbs are primates.' Is this statement correct or incorrect? Justify your response.
8. Examine the skull shown. Suggest the walking style of this organism. Justify your response.

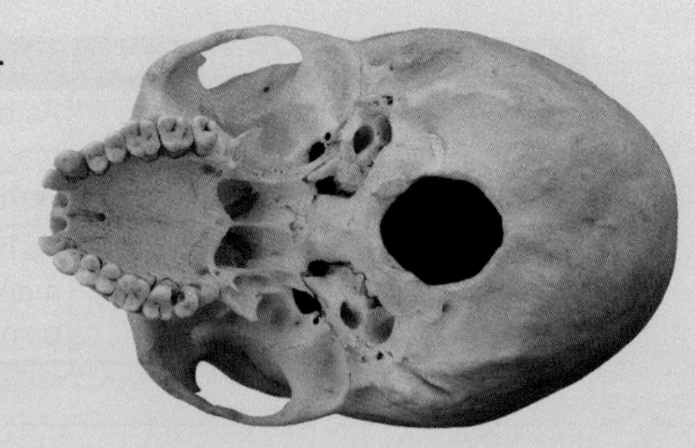

10.2 Exam questions

Question 1 (1 mark)

Source: *VCAA 2018 Biology Exam, Section A, Q38*

MC Consider the evolution of hominins.

Which one of the following statements about hominin evolution is correct?

A. *Homo sapiens* and *Homo neanderthalensis* are the only present-day hominin species.
B. Members of the *Australopithecus* genus are not classified as hominins.
C. *Homo erectus* was a bipedal primate.
D. All hominoids are also hominins.

Question 2 (1 mark)

Source: *VCAA 2018 Biology Exam, Section A, Q37*

MC Members of the order Primates are mammals.

Which combination of features is common to all primates and distinguishes them from other mammals?

	Feature 1	Feature 2	Feature 3
A.	Forward-facing eyes	Sloping forehead	Fur or hair
B.	Binocular vision	Opposable thumbs	Fully rotating shoulder joints
C.	Parabolic jaw	Tail	Nails instead of claws
D.	Even-sized teeth	Arms longer than legs	Bipedal stance

Question 3 (1 mark)

Source: *VCAA 2019 Biology Exam, Section A, Q40*

Source: Smallcreative/Shutterstock.com

MC What evidence in the image above enables the primate shown to be classified as a hominoid?

A. the presence of an opposable thumb
B. the absence of claws on the toes
C. the presence of hair
D. the absence of a tail

Question 4 (1 mark)

Source: *VCAA 2006 Biology Exam 2, Section A, Q22*

MC Consider the following diagrams of skulls.

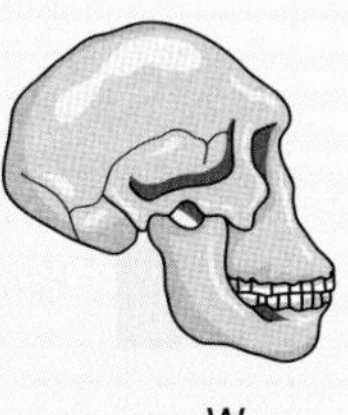

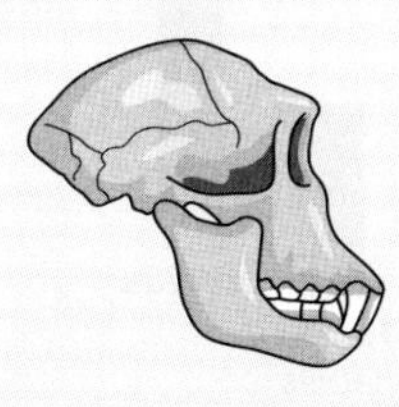

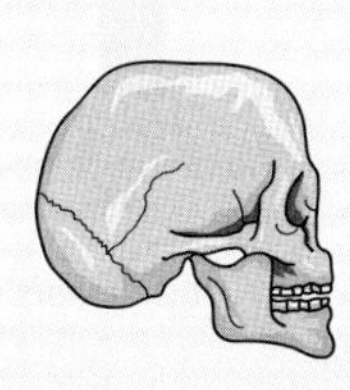

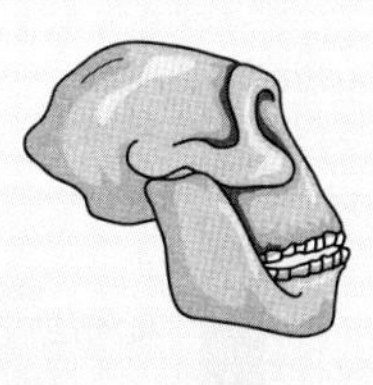

The skull most likely to be that of a chimpanzee is

A. W
B. X
C. Y
D. Z

Question 5 (4 marks)

Human are classified as primates.

a. Name three primates, other than humans. **1 mark**

b. List three distinguishing characteristics common to all primates. **1 mark**

c. Chimpanzees and humans belong to the same subfamily, Homininae, but different tribes. Describe two features that you might use to distinguish chimpanzees from humans. **2 marks**

More exam questions are available in your learnON title.

10.3 Major trends in hominin evolution

KEY KNOWLEDGE

- Evidence for major trends in hominin evolution from the genus *Australopithecus* to the genus *Homo*: changes in brain size and limb structure

Source: VCE Biology Study Design (2022–2026) extracts © VCAA; reproduced by permission.

10.3.1 Major trends in hominin evolution

Timeline of hominin species

There is no universal agreement about the precise evolutionary history of our human species. However, there is agreement that small-brained hominins separated from the line that led to the gorillas and chimpanzees, possibly about 7 to 10 million years ago, and developed the ability to walk erect. It is agreed that these hominins gradually became less ape-like and more human-like as generations of hominins spent more time at ground level and were subjected to various selection pressures, including climate change. These first hominins were not members of the genus *Homo*.

The main genus' that are classified as hominins include:

- *Australopithecus*
- ***Paranthropus***
- *Homo*
- *Ardipithecus*.

Paranthropus a group of extinct hominins known as the robust hominins, who are not thought to be ancestors of modern humans

Figure 10.26 shows the timelines of the major hominin species.

FIGURE 10.26 The timeline and time frames in which the different hominin species were thought to be present. Each genus is shown with a different colour.

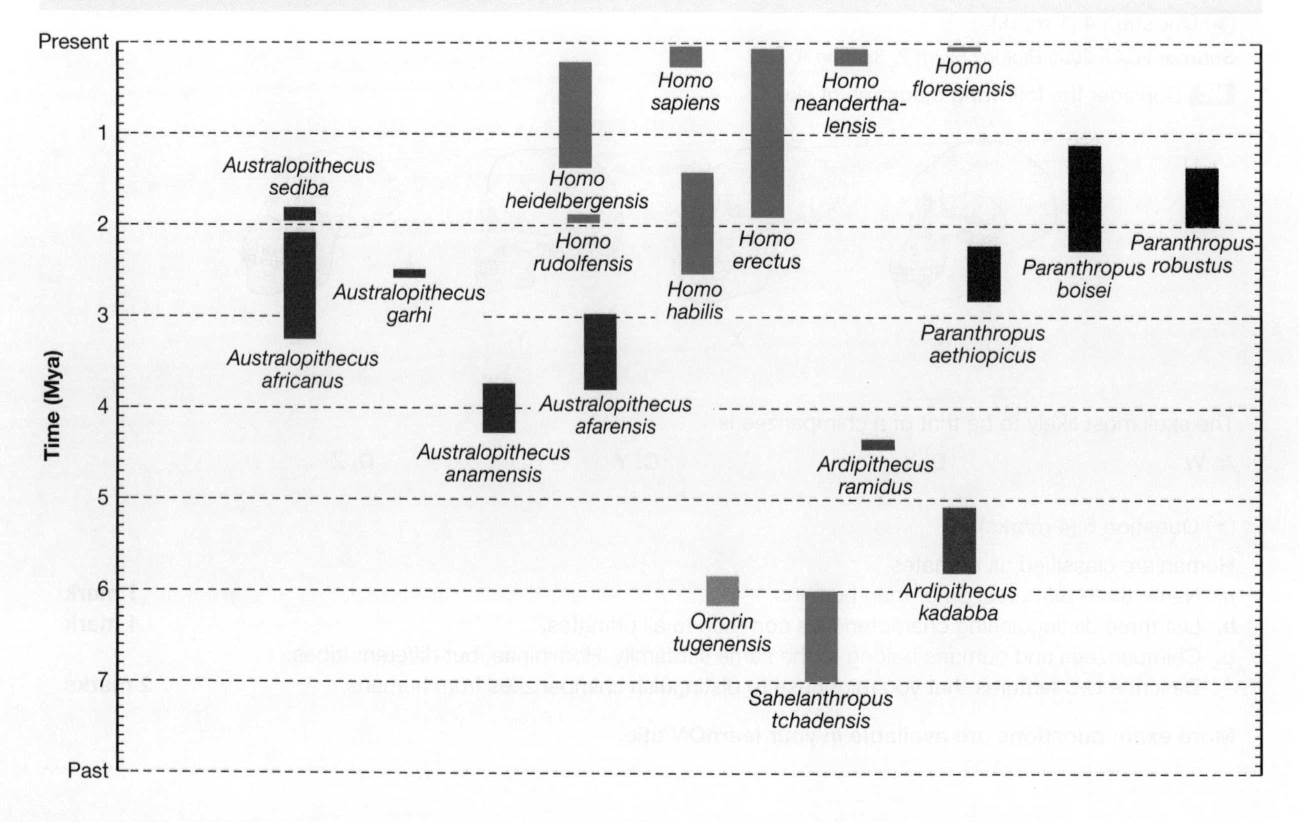

While it is thought that *Australopithecus* are our direct ancestors, the genus *Paranthropus* is not thought to be a direct ancestor of *Homo*.

One example of a possible phylogenetic tree for hominins is shown in figure 10.27. This is constantly up for debate and open for interpretation. For example, some scientists believe that some of the *Homo* species are actually the same species (i.e. *H. erectus* and *H. ergaster*), rather than different species. It is very difficult to determine this with certainty using fossil and DNA evidence alone. This will be further explored in subtopic 10.4.

FIGURE 10.27 A possible evolutionary pathway of hominin species

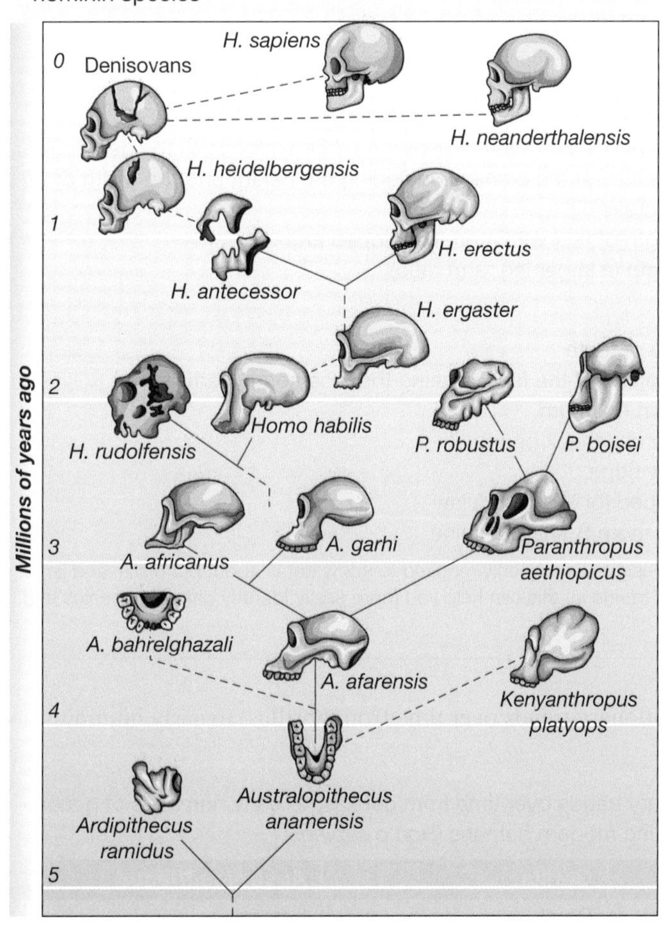

TIP: While you do not have to memorise all the names of the different species, you should know the general order of hominins from *Australopithecus* to *Homo*. For example, if you are asked to investigate the skulls of *A. afarensis, H. erectus* and *H. sapiens*, you should be able to note the trends and features in the skulls and determine which skull belongs to which species using a basic understanding of the order they appeared in geological time.

As explored in section 10.2.4, there are many features that help distinguish species as hominins. However, there is a slow progression of these different traits from early hominins to *Homo sapiens*. This progression is evident in the skulls shown in figure 10.28.

FIGURE 10.28 The development of primate skulls over time. From left to right: *Adapis* (a lemur-like animal that lived around 50 Mya); *Proconsul* (a primate from 23–15 Mya); *Australopithecus africanus* (3–1.8 Mya); *Homo habilis* (2.1–1.6 Mya); *Homo erectus* (1.8–0.3 Mya); a modern human (*Homo sapiens*, which is around 92 000 years old); and a French Cro-Magnon human from around 22 000 years ago. The last five skulls are all examples of hominins.

Trends in hominin evolution

The main trends observed over time from the genus *Australopithecus* to the genus *Homo* are:

- a marked increase the size of the cranium, including the height and the width of the skull, indicating that brain size also was increasing*
- an increase in the length of feet with more developed arches*
- an increase in leg length (a larger leg : arm ratio)*

Other trends include:

- a reduction in the size of teeth
- a flattening and shortening of the face, making the face nearly vertical
- a more central foramen magnum
- a smaller zygomatic arch (cheek bone)
- a less prominent brow ridge
- a more parabolic-shaped (or V-shaped) jaw
- the development of a more S-shaped spine.

**Note:* As part of the Study Design, you are only required to know detail about brain size and limb structure. However, it is useful to know some other trends as this can help you more easily identify different species and where they sit in the evolutionary line.

Table 10.5 shows the evolutionary trends over time from gorillas, to early hominins, to modern humans.

TABLE 10.5 The evolutionary trends over time from gorillas, to early hominins of genus *Australopithecus*, to those in the Homo genus, including modern humans (*Homo sapiens*)

Feature	Gorilla	*Australopithecus*	*Homo*
Brain size*	Small	Larger	Largest
Leg : arm ratio*	Small	Larger	Largest
Feet structure*	Short and flat	Longer, with arch	Longest, with most prominent arch
Size of canine teeth	Largest	Small	Smallest
Shape of dental arch	Box-shaped	More rounded	Parabolic
Position of foramen magnum	Towards back of skull	Further forward	Most forward
Zygomatic arch	Largest	Small	Smallest
Brow ridge	Large	Small	Very small
Spine shape	C-shaped	Between C and S-shaped	S-shaped

**Note:* Only information about brain size and limb structure are examinable for the Study Design.

10.3.2 Changes in brain size

Expansion of the human brain

One of the most important trends in hominin evolution relates to changes in brain size.

From the fossil record, it is clear that over time, the cranial capacity of hominins has increased. This indicates that over time, the size of the brain also increased. As we do not have brain specimens from earlier hominin species, we infer this from the size of the cranium.

Increased brain capacity, along with the greater cognitive capabilities it provides, distinguishes members of the genus *Homo* (humans) from members of *Australopithecus* (the australopithecines).

Figure 10.29 shows the brain volumes in various hominins, from *Australopithecus* to *Homo* species.

FIGURE 10.29 Graph showing the rate of increase in size of the hominin brain and cranial capacity over the 7 million years of hominin evolution

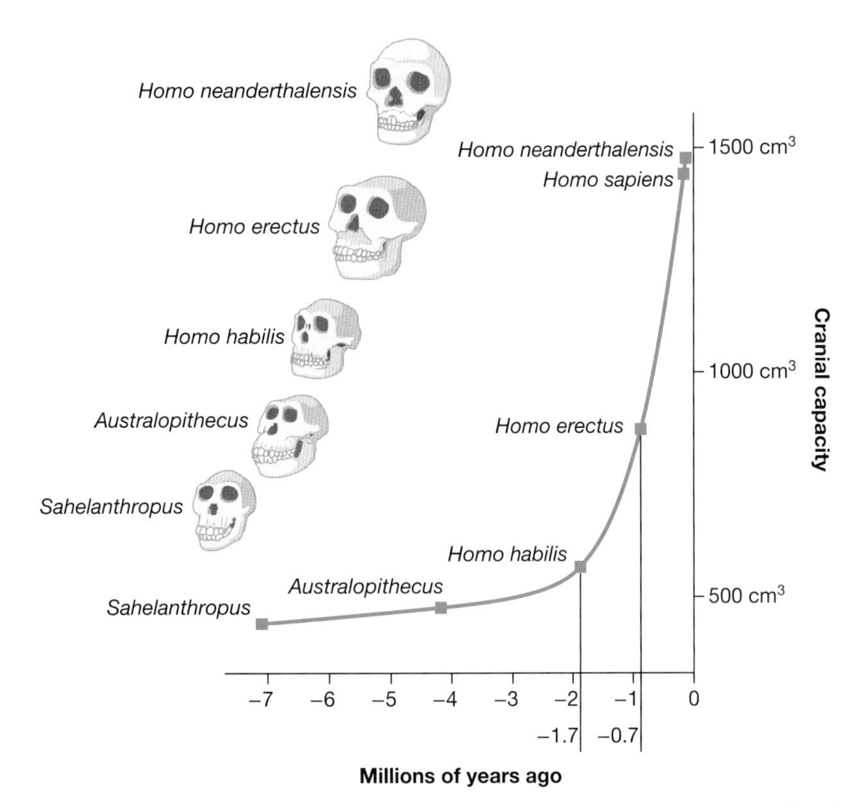

For about 2 million years (4.5 to 2.5 Mya), hominin brain volumes (based on measures of endocranial volume) remained at 400 to 500 millilitres, a little larger than that of the great apes.

However, after the appearance of *Homo habilis*, brain volume changed much more rapidly — it more than doubled in the next 2 million years, but this was not accompanied by a doubling of the sizes of other body parts. Research suggests a link between the increase in brain size and the use of fire and cooking by *Homo habilis*, which allowed for an increase in the digestion of nutrients. This indicates that a strong selection pressure for increased cognitive processing was acting on populations of early humans, with improved problem-solving skills and the ability to plan conferring a survival advantage. Speech was also to able become more complex and articulate with increased cognitive ability, allowing for richer communication and rituals (such as burial ceremonies).

The disproportionate size of the brain relative to average body mass in members of the genus *Homo* is seen even more clearly when average brain size is plotted against average body mass (see figure 10.30).

FIGURE 10.30 Graph of average brain volume against average body mass. Note the disproportionate expansion of brain volume in various human species as compared to the great apes and the early hominins

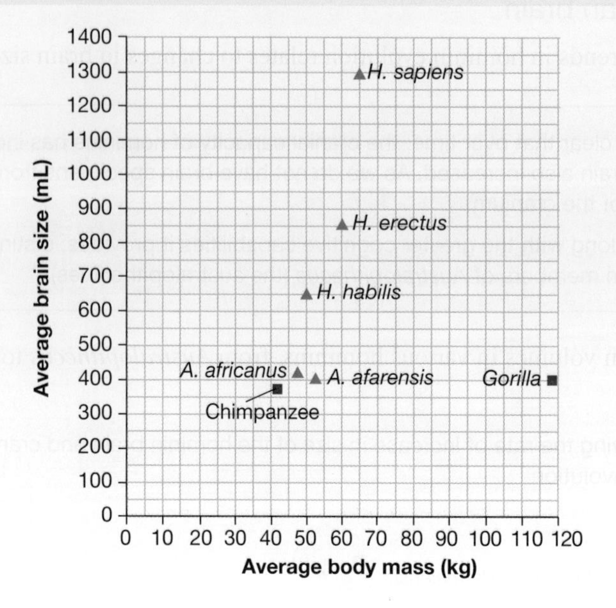

The expansion of the brain in *Homo* species did not occur evenly over all its parts. The cerebral cortex (neocortex) — the region of the brain concerned with cognitive abilities such as abstract reasoning, complex problem-solving skills, forward planning and social skills — expanded at a higher rate than most other brain regions, except for the cerebellum, which also expanded disproportionately (shown in figure 10.31).

FIGURE 10.31 The different components of the brain expanded at different rates

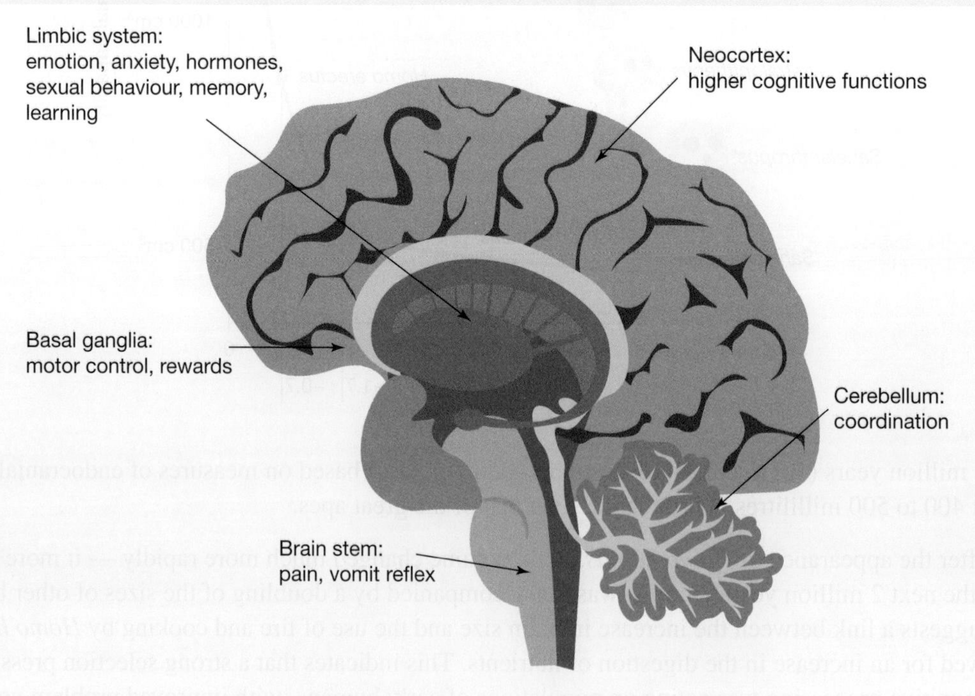

It should be noted that the power of an enlarged brain was only able to be used to advantage because of earlier evolutionary developments, such as the presence of an opposable thumb on the hands giving remarkable manual dexterity, and the development of sustained bipedal locomotion freeing the hands.

The changing skull size and structure

As the brain size of the hominins increased, the skull size and structure also changed.

Alongside a larger brain size, the following changes occured:

- The shape of the cranium became rounder.
- The slope of the forehead subsequently became more vertical (see table 10.5 and figure 10.33).
- The jaw shape become more parabolic and it reduced in size, as did the teeth (linked with changing diets due to the ability to use fire and cook food).
- The foramen magnum, which is the position where the vertebral column joins the skull, moved from the back to the middle of the skull.

Observations of skull shape related to brain size

You can see the result of the changing shape of the cranium in figure 10.32, which compares the skulls of an early hominin (*A. africanus*), an early *Homo* species (*H. erectus*) and a modern hominin (*H. sapiens*).

Note that as the human face became flatter and more vertical, the amount of the face that is visible from this perspective decreased. In early hominins the face tends to be concave and, because of their relatively much larger teeth and jaws, their faces project forward. A side view of these skulls can be seen in figure 10.33.

FIGURE 10.32 The change in cranium capacity over time correlates with the change in brain size.

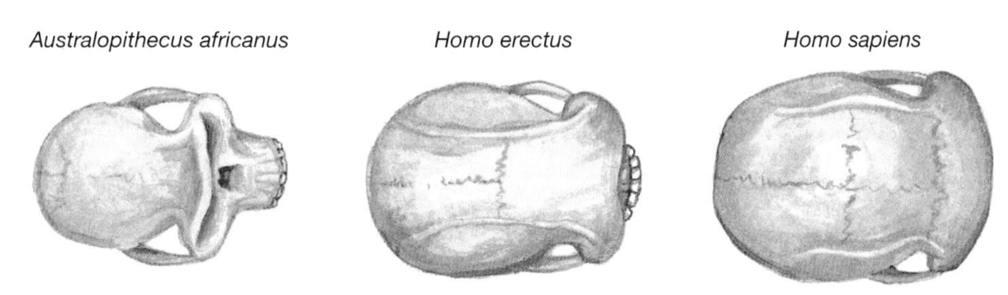

FIGURE 10.33 A side view of the skulls of different hominins, showing the flattening and shortening of the face

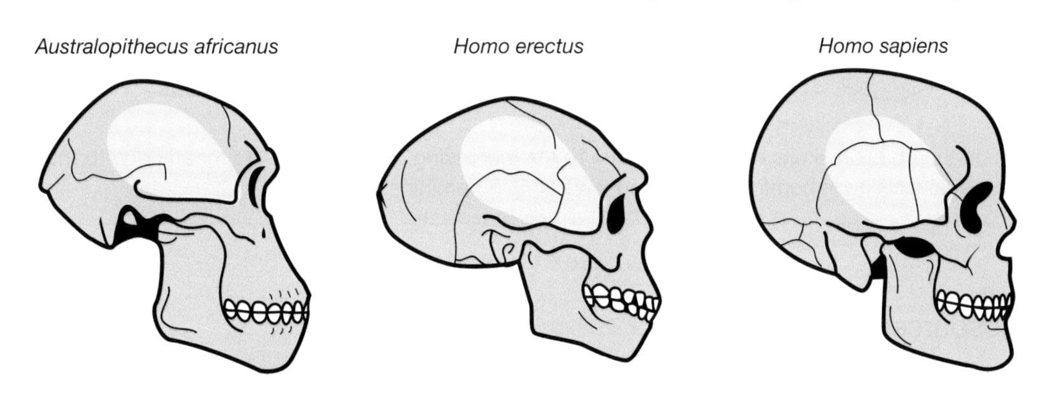

elog-0429

INVESTIGATION 10.2 online only

Investigating hominin fossils

Aim

To investigate different hominin skulls and determine trends in different features

CASE STUDY: The link between brain size and a changing diet

The brain is an energy-hungry organ. A large brain comes with a high energy cost.

- In modern humans, the brain represents about 2 per cent of body mass but uses about 20 per cent of the total energy intake.
- In apes, the brain uses about 13 per cent of total energy intake.
- In other mammals, the brain uses about 2 to 8 per cent of total energy intake.

The inclusion of meat in the human diet and the later use of fire for cooking produced the extra nutrients and energy for brain growth. Prior to this point, the main diet was herbivorous-based.

What were the selective pressures on early human populations for this brain growth? Some researchers propose that a global shift in climate (cooling and drying) about 3 to 2.4 million years ago reduced the area of lush forested land in Africa and gave a selective advantage to individuals with greater cognitive skills. These individuals were better equipped to exploit the resources in the new habitats arising from the climate change. For example, hominins may have followed game herds and obtained meat by scavenging the kills of predators.

BACKGROUND KNOWLEDGE: The position of the foramen magnum

While this does not directly link to skull size, it is clear that over time the position of the foramen magnum has become more centred on the head.

FIGURE 10.34 The movement of the foramen magnum over time. From left to right: chimpanzee, *A. africanus*, *H. erectus* and *H. sapiens*. While it can be seen that hominins have a much more central foramen magnum, this trend has continued as hominins have evolved.

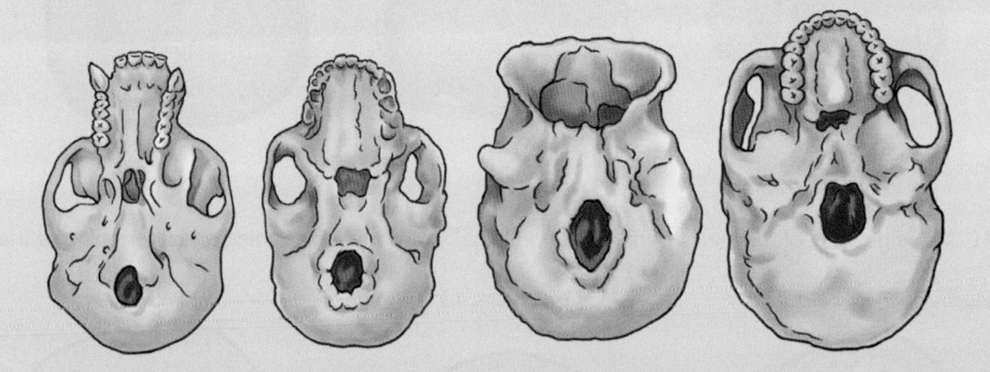

This change in the hominin skulls was not only due to the increasing brain size; it was mostly due to the change from quadrupedal walking in hominoids to bipedal walking in hominins.

on Resources

eWorkbook	Worksheet 10.3 Hominin relationships and evidence from skulls (ewbk-4109)
Video eLesson	Changes in hominin skulls (eles-4632)
Interactivity	Hominin skulls (int-0241)
Weblink	Australian Museum — Hominin skulls

10.3.3 Changes in limb structure

As primates transitioned from an arboreal life of moving through treetops to living and walking on land, their limbs adapted. The size, shape and angle of hominin limbs changed over time when particular characteristics gave populations a selective advantage.

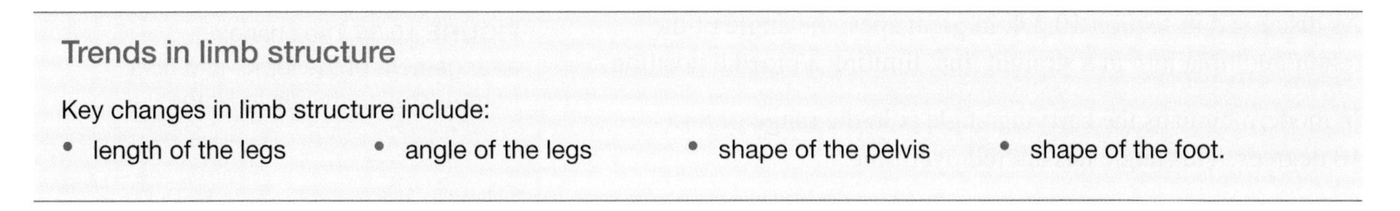

Trends in limb structure

Key changes in limb structure include:

- length of the legs
- angle of the legs
- shape of the pelvis
- shape of the foot.

Many of these changes were first introduced in section 10.2.4 when exploring the shared characteristics of hominins compared to hominoids such as the great apes. These features continued to change and further evolve through the hominin species.

Trends in the length of legs

Over time, the lengths of the legs in hominins have become longer in comparison to their arms. This better assists with bipedal motion, preventing knuckle dragging.

TABLE 10.6 The ratio of leg (hindlimb) to arm (forelimb) length for several primate species, including three hominins: *Australopithecus afarensis, Homo erectus* and modern humans

Species	Ratio of leg (hindlimb) to arm (forelimb) lengths
Gibbon, *Hylobates* sp.	0.76 : 1
Chimpanzee, *Pan troglodytes*	1.00 : 1
Lucy, *Australopithecus afarensis*	1.14 : 1
Homo erectus, an early human	1.43 : 1
Homo sapiens, modern human	1.43–1.47 : 1

FIGURE 10.35 From **a.** *Australopithecus afarensis* to **b.** *Homo erectus*, it can be seen that there has been a clear change in leg length. The leg length: arm length ratio is much greater in modern hominins.

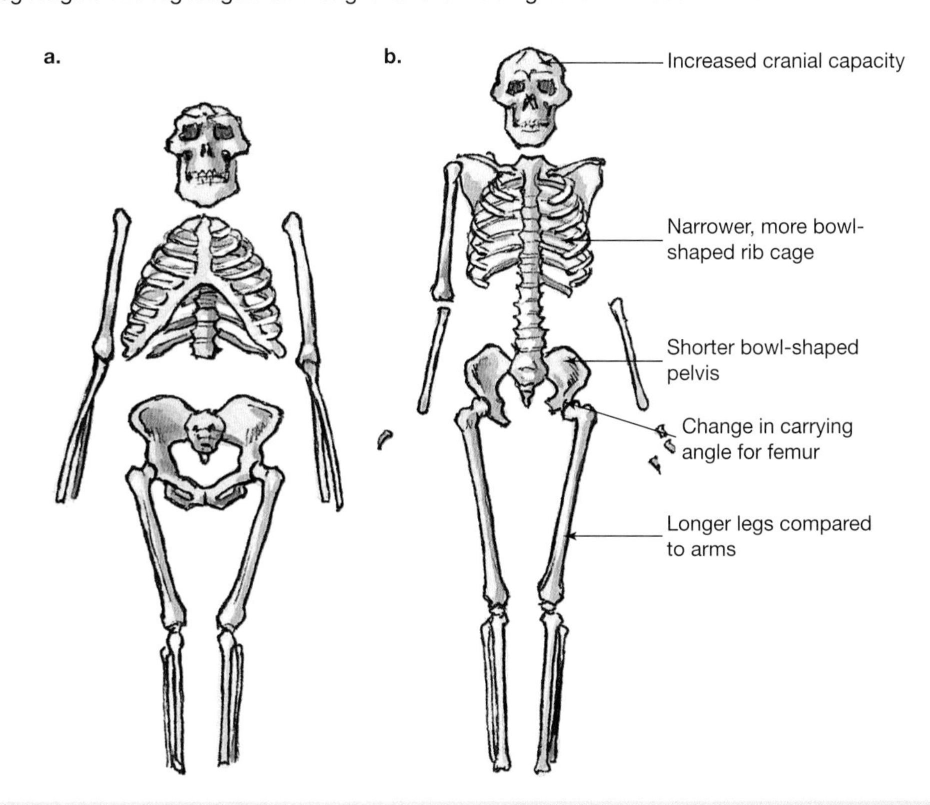

There are also many other features that can be observed in figure 10.35 to have changed from *Australopithecus* to *Homo*. The shape of the pelvis, the angle of the femur, the skull structure and the shape of the rib cage have all changed over time.

Trends in the angle of legs

As discussed in section 10.2.4, in great apes, the angle of the femur and tibia join in a straight line, limiting a bipedal position.

In modern humans the carrying angle is in the range of 8 to 10 degrees. This angle has altered over time.

In the *Australopithecus afarensis* skeleton, the femur is angled out from the line of the tibia and provides evidence that it walked erect. In *Australopithecus* species the carrying angle is larger — about 14 to 15 degrees. This angle is greater than in modern humans. This is the case because *Australopithecus* species were much shorter than modern humans and a greater angle is required to bring the centre of mass within the outline of the feet.

FIGURE 10.36 The changing arrangement of the femur and tibia from *Australopithecus* to *Homo*

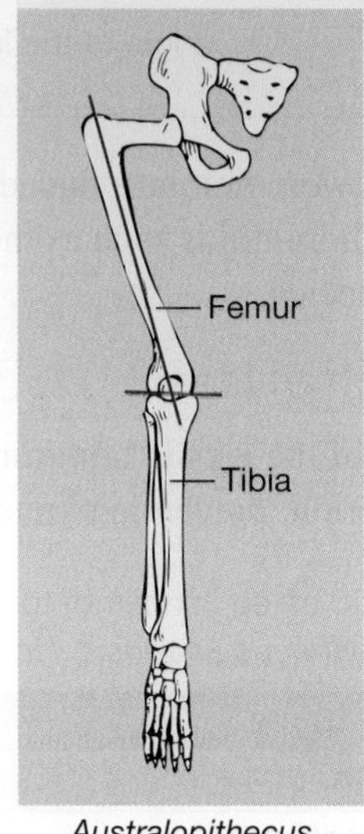

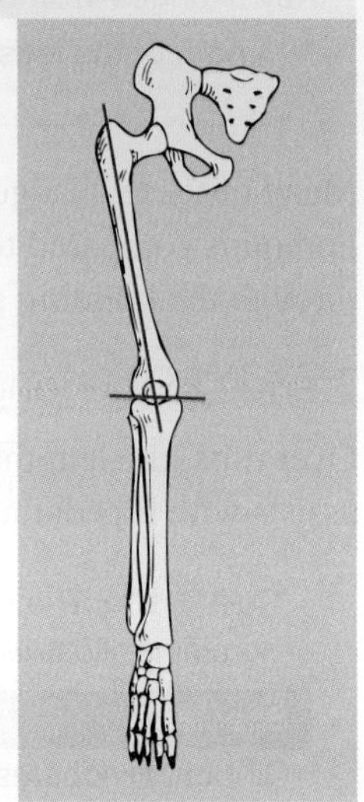

Trends in knee size

The size of the knees and kneecap (the patella) from *Australopithecus* to *Homo* also increased over time. When walking with a bipedal locomotion there is more body weight placed on two limbs rather than four. The increasing size of knees allows for greater muscle attachment and the ability to act as a shock absorber. However, this does not reduce all of the pressure on the knee joints, and this is why knee injuries can be commonplace when athletes assert a lot of pressure on the knees, as well as in old age.

Trends in pelvis shape

When the legs of hominins evolved to become angled inwards to walk erect, this also had an impact on their pelvic anatomy. To sustain walking upright, the pelvis needed to have a larger surface area for muscle attachment, but it also needed to be narrow enough to create the carrying angle for the legs. Evidence shows that the pelvis shape changed from a long and narrow shape, as seen in chimpanzees, to a bowl-shaped pelvis that is much shorter in hominins (see figure 10.37). The pelvis of *Australopithecus africanus* shows evidence of bipedal walking as it was shorter like a hominin pelvis, but it was wider than the pelvis of modern humans.

FIGURE 10.37 The changing pelvis shape from **a.** chimpanzees, to **b.** *Australopithecus*, to **c.** *Homo*

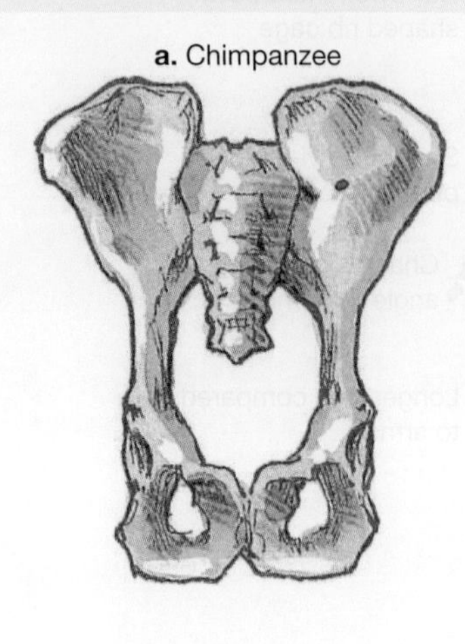

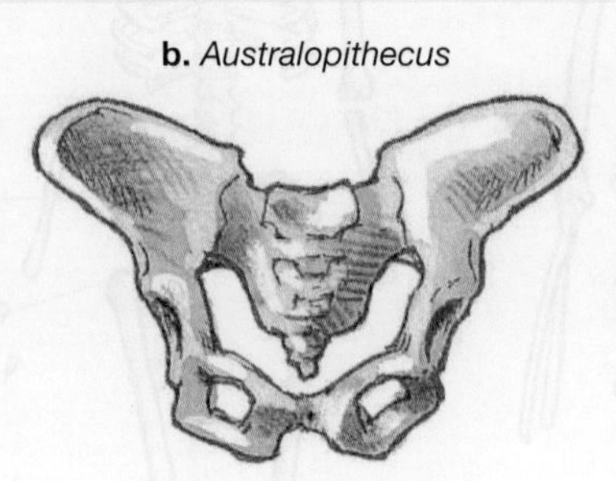

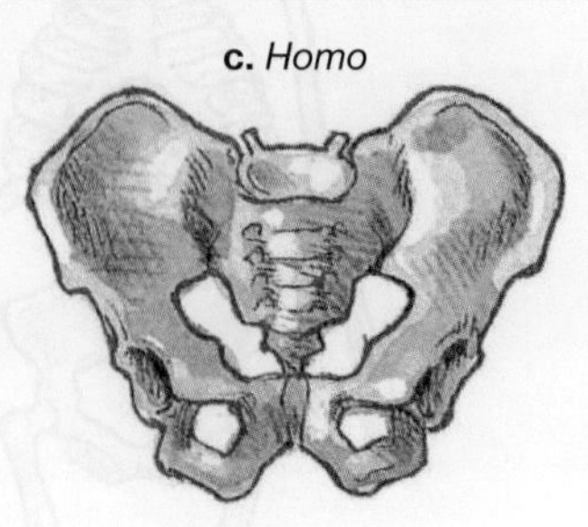

Trends in foot structure

The great apes predominantly live an arboreal life and this is reflected in their foot structure. Their feet are flat due to having straight bones, while their toes are curved (see figure 10.38). The curvature of their toes, along with a big toe that is opposable, allows them to grasp branches to move efficiently through trees.

FIGURE 10.38 The comparison of foot morphology between chimpanzees and modern humans. The foot of earlier hominins, such as the *Australopithecus* species, showed a transition between these traits.

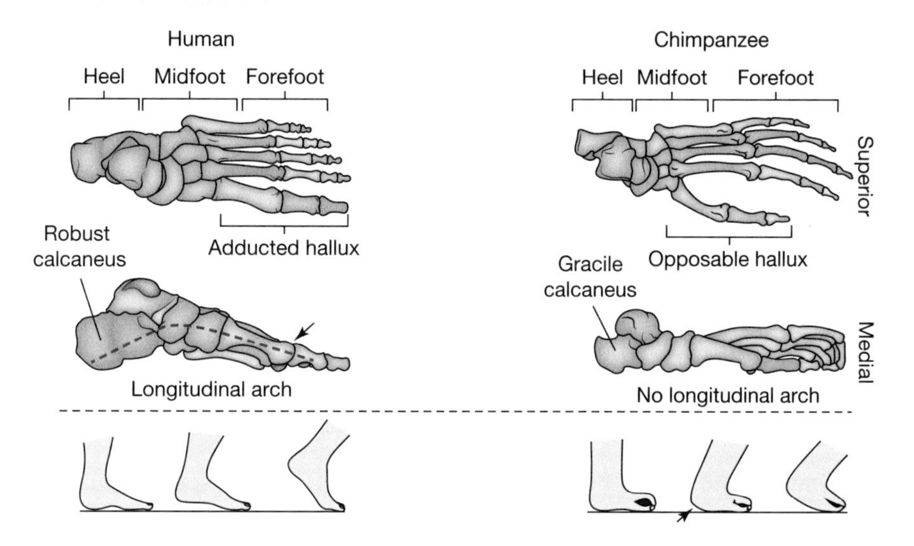

Evidence from the feet of *Australopithecus* species show they walked erect with their big toes oriented parallel to their other toes, as in the feet of modern humans. While *Australopithecus* species were capable of bipedal locomotion, it was not the long striding gait that is typical of modern humans.

The skeletons of *Australopithecus* species show features that are also adaptations for tree climbing, such as long, curved bones in the fingers and the toes, indicating that they still spent considerable time in trees.

Some scientists speculate that *Australopithecus* species could move by swinging from branch to branch with their arms, in a mode of locomotion known as **brachiation**. Their skeletal structure also allowed them to stand and walk erect when they descended to the ground, perhaps to move between clumps of trees or to seek food at ground level.

In modern humans and other species in the *Homo* genus, further modifications of the foot occurred from the foot structure of the australopithecines. The heel enlarged and this resulted in a longitudinal foot arch under the feet (see figure 10.38). This change allowed the heel to act as a shock absorber and distribute the pressure from the body across the whole foot. The foot structure also changed from having long, curved toes to ones that are short and straight. This, coupled with the big toe being positioned forward instead of being opposable, allowed the hominins to push off with their big toe when then walking with a bipedal locomotion.

brachiation a mode of locomotion involving swinging from one handhold to another

tlvd-1809

SAMPLE PROBLEM 2 Determining hominoid species from their limb structure

Images of the pelvises and lower limbs of three hominoids are shown. These are from a gorilla, *Australopithecus afarensis* and *Homo sapiens*.

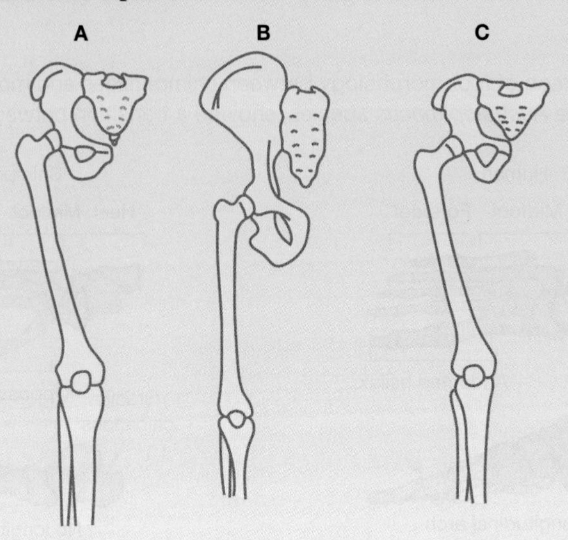

Examine the three images and match them to the appropriate hominoid species. Justify your response. (3 marks)

THINK	WRITE
1. Identify what the question is asking you to do. The first part of the question is asking you to *match*. In a match question, you must link two words, sentences or images to another.	
2. Consider the different features of the lower limbs of gorillas (great apes), *Australopithecus afarensis* and *Homo sapiens*. You should be looking at the pelvis and the angle of the limbs. Great apes have straight legs and a long pelvis, while *Homo sapiens* have a bowl-shaped pelvis and longer legs angled inwards.	• Gorilla — B • *Australopithecus afarensis* — C • *Homo sapiens* — A
3. The second part of the question asks you to *justify*. In a justify question, you must support your answer with evidence.	
4. Consider the impact that pelvis shape and leg angle has on how a species moves. A gorilla walks on all fours, while *Homo sapiens* and *Australopithecus afarensis* are bipedal. Use this to justify your answers.	B belongs to a gorilla as it has a long and narrow pelvis and a straight leg (1 mark). C belongs to *Australopithecus afarensis* as it has a shorter pelvis and a leg angled inwards (1 mark). A belongs to *Homo sapiens* as the pelvis is bowl-shaped and the leg is pointed inwards, and is longer than C (1 mark).

EXTENSION: Other trends observed in hominin evolution

Other than brain size and limb structure, many trends have been observed in the evolution of hominins.

Jaw and teeth

Throughout the evolution of hominins, the dental arch (jaw shape) has become more parabolic and shortened. The teeth have become much smaller and the diastema (the gap between teeth) has closed up.

Size of the zygomatic arch

The **zygomatic arch** (which forms the cheekbone) has decreased in size over time. The zygomatic arches not only protect the eyes, but allow muscles to attach that aid in chewing plant matter. As hominins have over time become more reliant on meat in their diet (compared to earlier hominins who where herbivores), less time is required to chew and break down plant product. Therefore the zygomatic arches are now smaller.

FIGURE 10.39 The shape of the dental arch has become more parabolic (V-shaped) rather than U-shaped and the teeth have become smaller throughout hominin evolution.

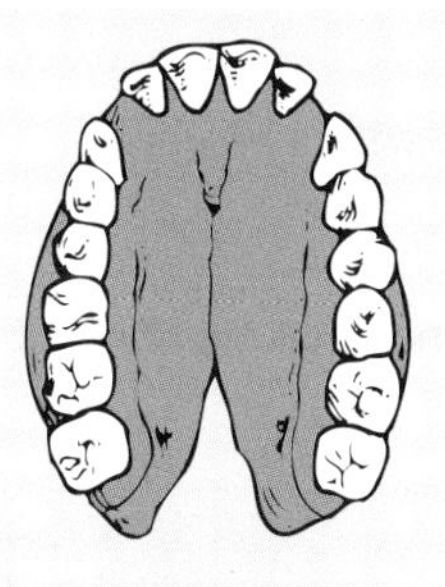

Australopithecus afarensis

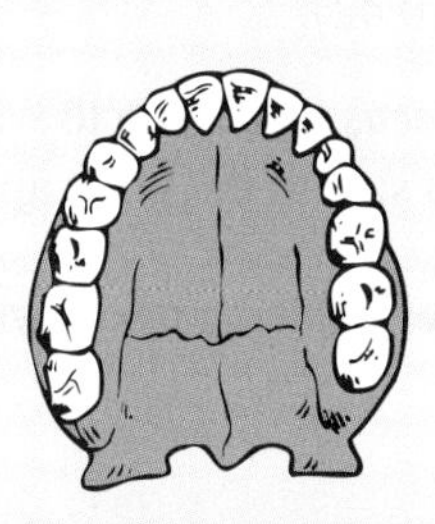

Homo sapiens (modern human)

FIGURE 10.40 The size of the zygomatic arch has decreased over time throughout hominin evolution.

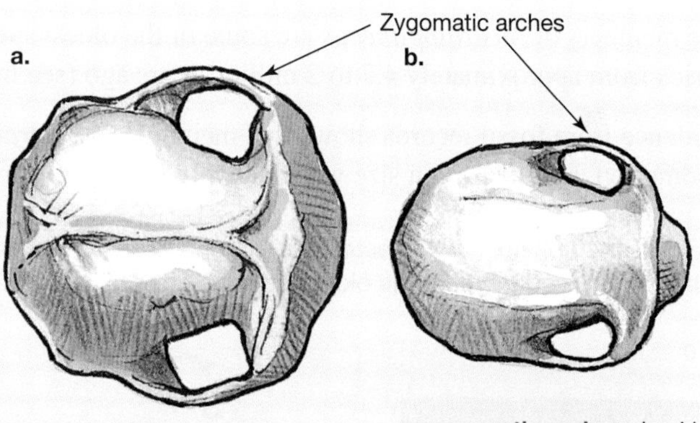

zygomatic arches cheekbones comprising horizontal bony ridges on a mammalian skull

Brow ridge

As hominins have evolved, their faces have become much flatter with a less prominent brow ridge. However, earlier hominins had a very prominent brow ridge. The brow ridge is located above the socket in primates and helps support the bones in the face. As our zygomatic arch got smaller, so too did the brow ridge.

FIGURE 10.41 a. Different skull specimens, showing the change in the brow ridge over time b. The brow ridge as seen in a model of a *Homo neanderthalensis*

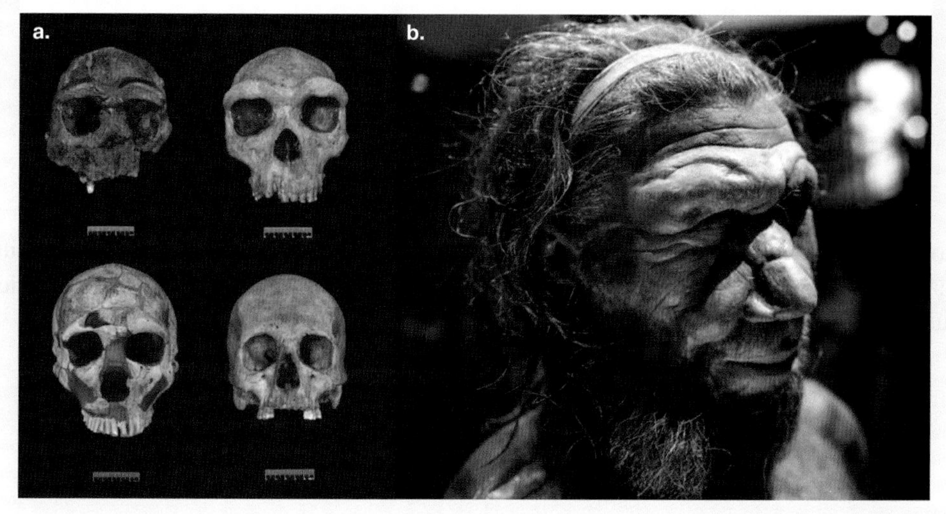

Size of the rib cage

Over time in hominin evolution, the rib cage has become much narrower and longer, with a bowl-shaped appearance. The rib cage of *Australopithecus* species was much wider at the lower end compared to the top end (refer back to figure 10.35). This is because early hominins were herbivores, and therefore had a longer digestive tract. Thus the rib cage structure allowed for the longer and larger digestive tracts to be accommodated.

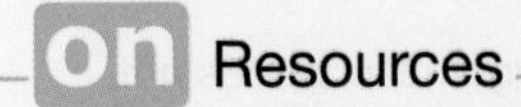

Resources

Interactivity Hominin limb structure (int-8330)

10.3.4 A detailed look at members of the early hominin tribe

The australopithecines

The members of *Australopithecus* are some of the oldest species of hominins. They lived in eastern and southern Africa from approximately 4.2 to 2 million years ago (see figure 10.42).

Evidence from fossil records shows that members of *Australopithecus* were able to walk with a bipedal locomotion, but their brain size was relatively small and similar in size to the great apes.

FIGURE 10.42 *Australopithecus* fossils found in eastern and southern Africa provide evidence that members of this genus lived there. Some older hominins are also shown.

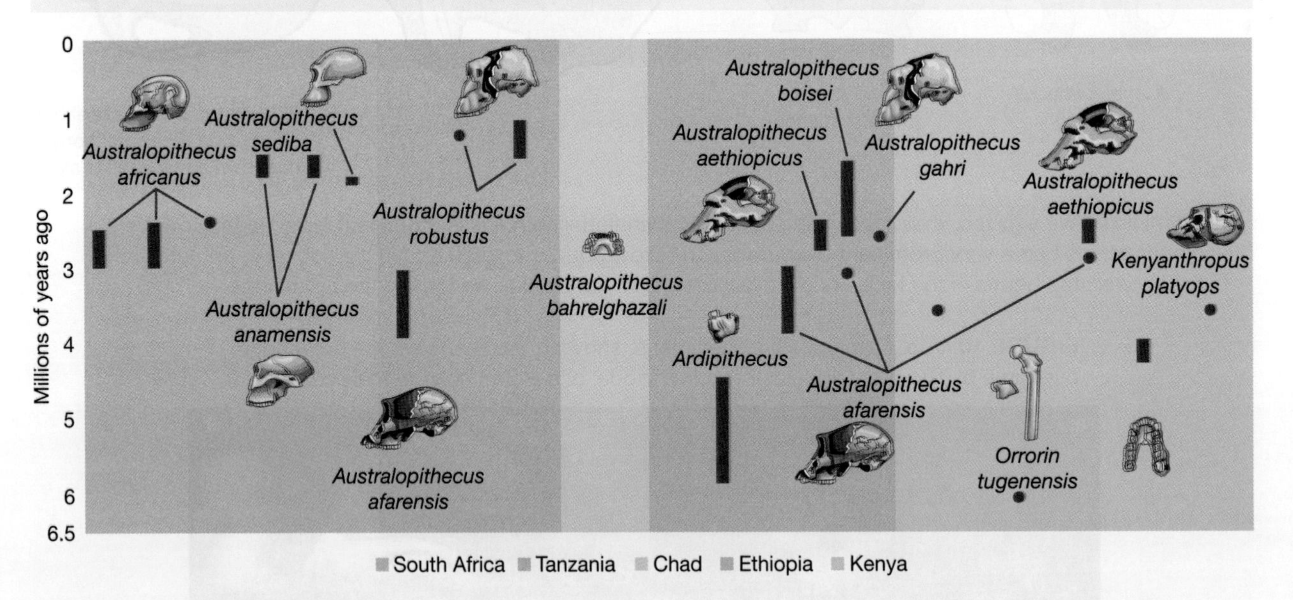

Two of the most famous members of the *Australopithecus* genus are *A. africanus*, which lived in southern Africa, and *A. afarenis* (as detailed in the provided Case study on Lucy), which lived in eastern Africa. Other members include *A. sediba* and *A. deyiremeda*.

CASE STUDY: Lucy from the Afar Triangle (*Australopithecus afarensis*)

It is 1974. A group of scientists are on a field trip in the Afar Triangle, a remote area of Ethiopia about 160 kilometres north-east of Addis Ababa. Donald Johanson, leader of the expedition, drives to an ancient lake bed to look for fossils. In a gully he finds part of an elbow joint, then part of a skull, a femur (thigh bone), several vertebrae, parts of a pelvis, some ribs, and other bones and bone fragments. He has found bones from the skeleton of a single individual.

Eventually, Johanson's team excavated several hundred pieces of bone that formed about 40 per cent of an entire skeleton (see figure 10.43). The skeleton was that of an ape-like creature who walked erect, was a little more than one metre tall, and who lived in this part of Africa about 3 million years ago. Because part of the pelvis was present, it was possible to identify the bones as being from a female. The condition of her teeth suggested she was 25 to 30 years old when she died.

This fossil skeleton was given the popular name Lucy. Lucy was not a member of the genus *Homo* but was a hominin, and is classified as a member of the species *Australopithecus afarensis*. Lucy is accepted by many palaeoanthropologists as being a possible early ancestor of the human species and on the direct line from which the first human species gradually evolved.

FIGURE 10.43 a. The hominin skull of Lucy (image courtesy of the Institute of Human Origins) b. A reconstruction of Lucy

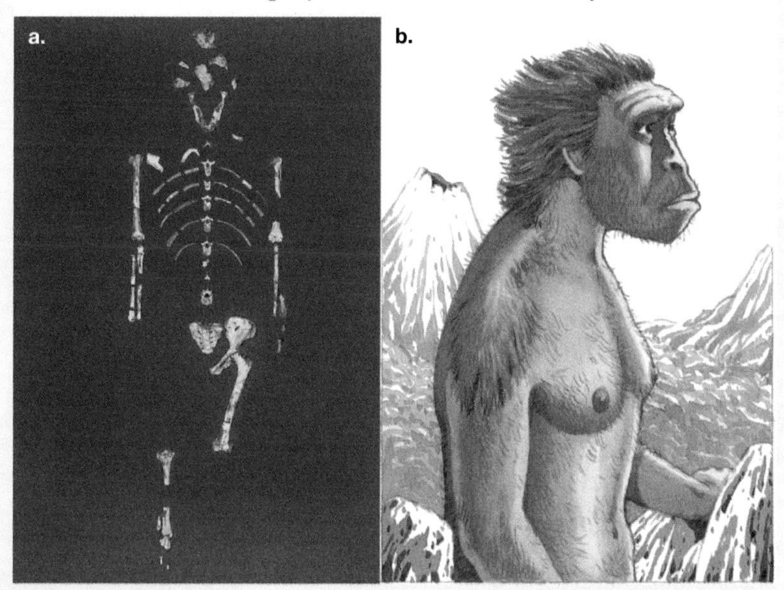

CASE STUDY: A detailed look at other early hominin species

Lucy was one of the most famous *Austalopithecus* discoveries. Another famous discovery was that of the 'Taung Child', a member of *Australopithecus africanus* that was uncovered in 1924.

Several more australopithecine species have been discovered in recent years, with one of the most recent being *Australopithecus deyiremeda*, found in 2015 in the Afar region of Ethiopia, just 35 kilometres north of where the Lucy fossil was found in 1974 (see figure 10.44). *Australopithecus sediba* (see figure 10.45) was another more recent discovery from this genus.

Australopithecus were not the only genus of early hominins. *Paranthropus* and *Kenyanthropus* were also part of these early hominins.

FIGURE 10.44 The left half of the lower jaw of *Australopithecus deyiremeda*

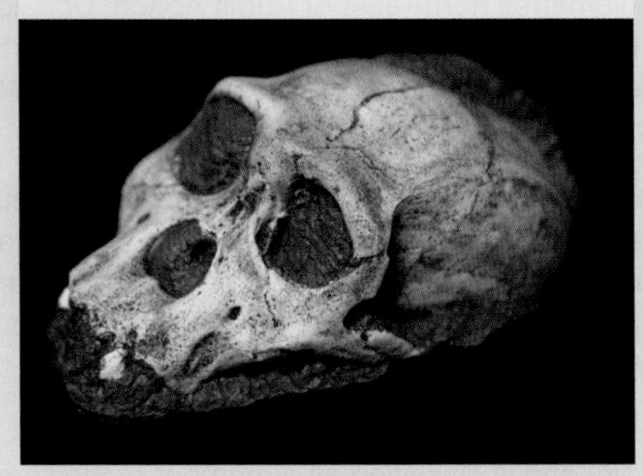

FIGURE 10.45 Cranium of *Australopithecus sediba* found in South Africa in 2008

To access more information about all of these early hominins, including the *Australopithecus* genus, please download the digital document.

Resources

Digital document Case study: A detailed look at early hominin species (doc-36149)

10.3.5 A detailed look at members of the genus *Homo*

The genus *Homo* is represented today by a single living species, *Homo sapiens* (meaning 'wise man'), which includes people of all racial groups. Human populations show phenotypic variation (see figure 10.46). Although variation exists between populations, humans share most features in common and are members of a single species, capable of interbreeding to produce viable, fertile offspring.

FIGURE 10.46 Variation exists in populations of *Homo sapiens.*

For extinct organisms, decisions about defining species are made using the evidence available, either directly or by inference from fossil bones and teeth. For this reason, there is no universal agreement on the number of different species within the genus *Homo* that have existed during its evolutionary history (over a period of about 2.5 Myr). Scientists who are 'splitters' tend to recognise more species, while those who are 'lumpers' tend to recognise fewer species.

Recent analysis has led to the recognition of more species in the genus *Homo*. Disagreements between scientists are reminders that, even when they examine the same fossil material, scientists may differ in their interpretations and conclusions. This will be further explored in subtopic 10.4.

What does it mean to be human?

'Being human' is not easy to define. No single feature defines humanity. The evolution of the genus *Homo* is linked to structural changes such as enlargement of the brain. Over the period starting from 2.8 million years ago — after the appearance of the first human species — the evolution of the genus *Homo* continued and is most significantly associated with the development of complex cognitive skills and behavioural changes, including:

- making and using fire
- cooperating in group activities
- caring for aged and ill members of the species

- burying the dead
- the development of art, language, mathematics and music
- the use of symbols
- an increase in the use of technology, which began with tools crudely fashioned from stone by *Homo habilis* and has extended to the present-day use of computers, lasers and spacecraft by *Homo sapiens*.

A summary of the main *Homo* species

Africa is the site where the first members of the genus *Homo* appeared at least 2.8 million years ago. Many different species within the genus *Homo* evolved and existed before the first appearance of modern humans (*Homo sapiens*) and some coexisted with our species. However, our species of *H. sapiens* is now the only living member of this genus.

Some of the main species of *Homo* (in order of approximate appearance) include:

- *Homo habilis*
- *Homo ergaster* (sometimes thought to be *Homo erectus ergaster)*
- *Homo erectus*
- *Homo heidelbergensis*
- ***Homo neanderthalensis***
- *Homo sapiens.*

Many trends observed from the evolution of *Australopithecus* to *Homo* (such as changes in brain size and limb structure) are also observed from early to modern *Homo* species. Some of these changes are summarised in table 10.7 and can be seen in figure 10.47.

Homo neanderthalensis an extinct *Homo* species known as Neanderthals, who were thought to have interbred with *Homo sapiens*

TABLE 10.7 Comparison of various *Homo* species. *H. erectus* is an older *Homo* species than *H. heidelbergensis* and *H. sapiens*.

Feature	*H. erectus*	*H. heidelbergensis*	*H. sapiens*
Prominent brow ridges	+	+	–
Rounded skull	–	–	+
Big teeth	+	+	–
Sloping forehead	+	+	–
Large brain (≥ 1200 mL)	–	+	+
Pointed chin	–	–	+

+ feature present

– feature absent

Source: Based on information from the Australian Museum.

FIGURE 10.47 Skulls of a. *Homo erectus* and b. a modern human, *Homo sapiens*

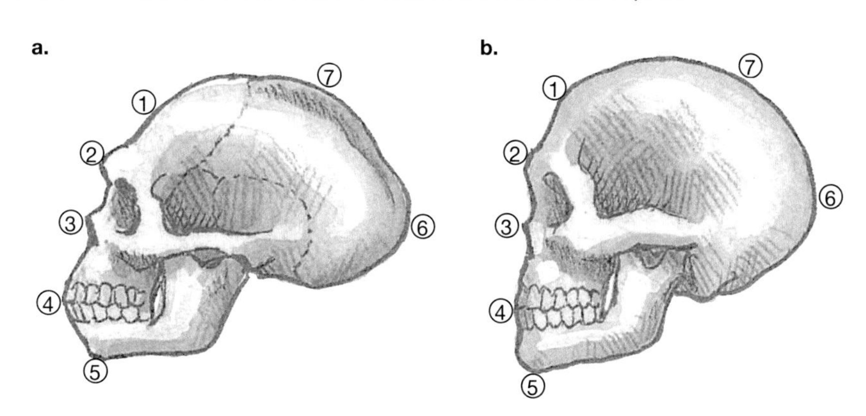

CASE STUDY: A detailed look at members of the genus *Homo*

Each *Homo* species has certain distinguishing features. These are summarised in table 10.8. Note that this table does not include all species within the genus *Homo*.

TABLE 10.8 A summary of species in the *Homo* genus

Species	Approximate time	Distinguishing feature	Comments
Homo habilis	Earliest fossils dated at 2.4 million years	The first toolmakers (used simple stone tools)	Fossils had a greater brain size (approximately 640 mL) and smaller teeth than *Australopithecus* so classed as *Homo*
Homo erectus	Earliest fossils dated at 1.8 million years. The oldest fossil was found in Kenya. Well-known fossil specimens are the Turkana Boy (found in Kenya), 'Peking Man' and 'Java Man'.	The first emigrants (*H. erectus* were thought to have migrated out of Africa to other continents)	Believed to be a direct ancestor of later human species As shown in figure 10.47, their skull structure has many similarities to *H. sapiens* (with a brain size of around 900 mL; around 75 per cent of modern humans). They were thought to use fire and more elaborate tools than *H. habilis*. Some scientists split into two separate species — *H. erectus* (later species that spread out of Africa) and *H. ergaster* (ancestral species only found Africa)
Homo floresiensis	Around 60 000 to 100 000 years old	The 'hobbit' — very small in stature	Found in Flores Island in Indonesia and was thought to have evolved from *H. erectus* Very small brain capacity compared to modern humans
Homo heidelbergensis	Around 700 000 to 300 000 years old	Intermediate features between *H. erectus* and *H. sapiens*	Larger brain case than *H. erectus* (1000–1200 mL); this is approximately 93 per cent of modern humans. Also has similar teeth to *H. sapiens* There is debate on the position of this species in the evolutionary line of hominins. Some research suggests it is an ancestor of *H. sapiens* and Neanderthals, while other findings place it as a separate branch.
Homo neanderthalensis	Mostly lived around 200 000 years to 30 000 years old (some fossils, however, are believed to be older — around 400 000 years old)	The Neanderthals (*H. sapiens*' 'cousins')	First reported in Neander Valley in Germany with most other fossils found in the Middle East and Europe Brain size around 1400 mL, with low and wide skulls, heavy brow ridges, protruding jaws, heavier skulls and low sloping foreheads

The location of some of the fossils for both *Homo erectus* (see figure 10.48) and *Homo heidelbergenis* (see figure 10.49) shows that these species are thought to have moved out of Africa. The evolution and appearance of *Homo sapiens* and their movement out of Africa will be further explored in section 10.5.1.

FIGURE 10.48 Map showing some sites where *H. erectus* fossils have been found. Approximate ages (Myr) of the fossils are given in parentheses.

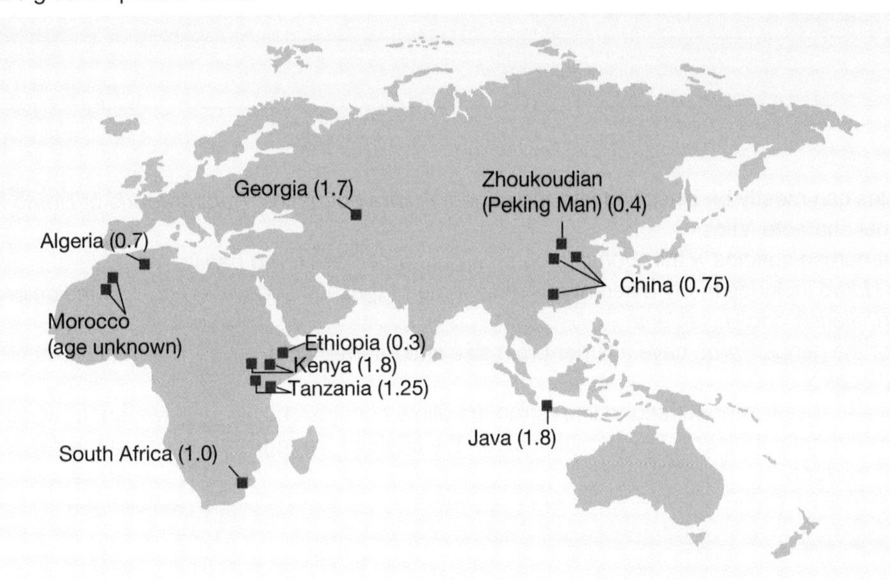

FIGURE 10.49 Map showing the location of some *H. heidelbergensis* fossil finds

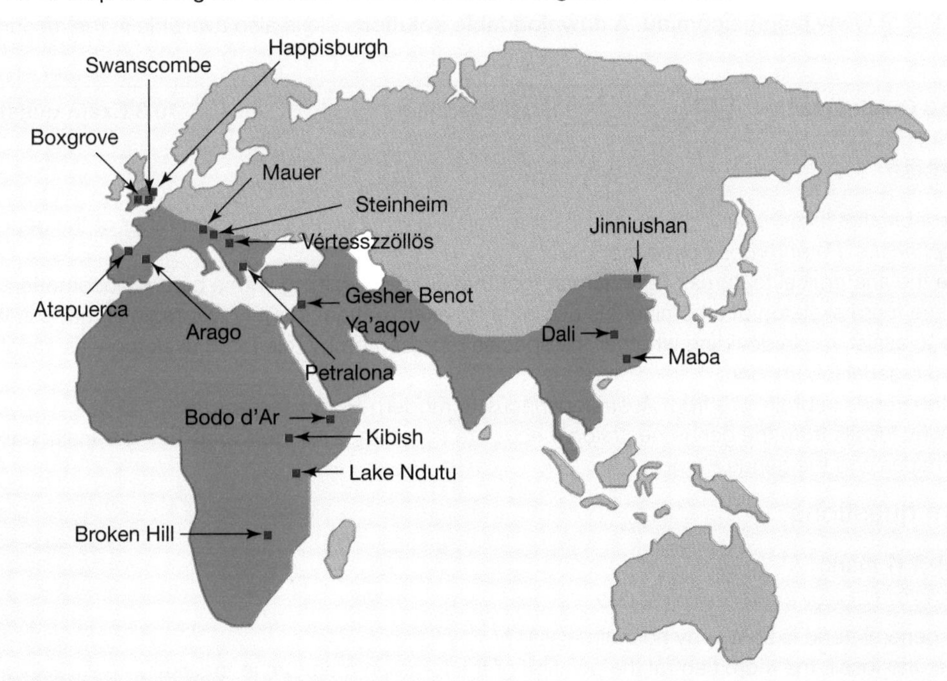

To access more information about these different species within the genus *Homo*, please download the digital document.

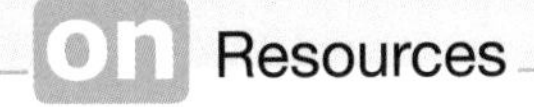

Resources

Digital document Case study: A detailed look at members of the genus *Homo* (doc-36150)

Resources

eWorkbook Worksheet 10.4 Summarising trends in hominin evolution (ewbk-8944)

KEY IDEAS

- The hominins can mostly be classified into the genera *Australopithecus, Paranthropus* or *Homo* based on their physical characteristics.
- The earliest human species within *Homo* is generally accepted as *Homo habilis*.
- The genus *Homo* includes many different species; some of these include *H. erectus*, *H. heidelbergensis*, *H. neanderthalensis* and *H. sapiens*.
- Members of the genus *Homo* have a larger brain size and inferred cognitive ability than members of the genus *Australopithecus*.
- Features of a fossil skeleton related to limb strucutre allow conclusions to be drawn about whether the species was capable of bipedal locomotion include the angle of the legs, the shape of the pelvis and the structure of the foot.

10.3 Activities

learnON

To answer questions online and to receive **immediate feedback** and **sample responses** for every question, go to your learnON title at **www.jacplus.com.au**. A **downloadable solutions** file is also available in the resources tab.

10.3 Quick quiz on	**10.3 Exercise**	**10.3 Exam questions**

10.3 Exercise

1. Create a table that lists the key characteristics that define a hominin as being a member of the genus *Australopithecus, Paranthropus* or *Homo*.
2. Describe the anatomical features of the human foot that support walking with a bipedal locomotion.
3. A complete fossil skeleton of a hominoid is available for examination. Identify two regions that you would examine to assist you in deciding whether the species represented by this fossil skeleton was capable of sustained bipedal locomotion.
4. Outline the trends in brain size from *Australopithecus* to modern *Homo sapiens.*
5. Explain why fossils from the genus *Australopithecus* are sometimes referred to as being transitional species of the great apes and the genus *Homo*.

10.3 Exam questions

Question 1 (1 mark)

Source: *VCAA 2018 Biology Exam, Section A, Q39*

MC Which general trend is shown by hominin fossils?

A. The older the fossil, the more central the position of the foramen magnum in the skull.
B. The older the fossil, the smaller the braincase that surrounds the cerebral cortex.
C. The more recent the fossil, the less bowl-shaped the pelvis.
D. The more recent the fossil, the larger the jaw bones.

Question 2 (1 mark)

Source: *VCAA 2019 Biology Exam, Section A, Q32*

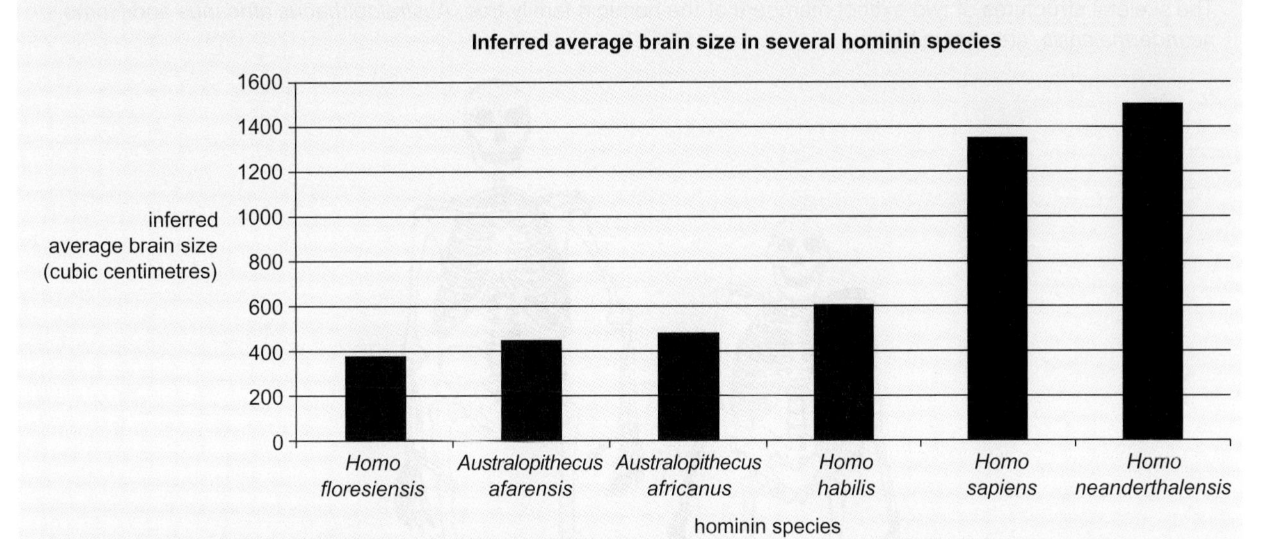

MC The graph above shows the inferred average brain size of a number of hominin species.

Based on your knowledge and using the information in the graph, which one of the following species does not follow the general trend in inferred brain size seen in hominin evolution?

A. *H. habilis*
B. *H. sapiens*
C. *A. afarensis*
D. *H. floresiensis*

Question 3 (1 mark)

Source: *VCAA 2019 Biology Exam, Section A, Q33*

MC Which row shows the group of characteristics that best reflects the trends in hominin evolution from the *Australopithecus* species to the *Homo* species?

	Characteristics
A.	Decreasing tooth size, increasing size of brow ridges, increasingly bowl-shaped pelvis, increasing size of zygomatic arch
B.	Decreasing tooth size, decreasing size of brow ridges, decreasing arch of feet, more-opposable big toe
C.	Increasing jaw size, decreasing size of zygomatic arch, increasing arch of feet, decreasing tooth size
D.	Decreasing size of canines, decreasing size of zygomatic arch, increasingly bowl-shaped pelvis, increasing arch of feet

Question 4 (1 mark)

MC A fossil human skeleton was found by paleontologists in East Africa.

What would identify it as belonging to the genus *Homo*?

A. Large teeth and parabolic jaw
B. Upright stance on two legs
C. Small cranial capacity
D. High arm to leg length ratio

Question 5 (4 marks)

Source: *Adapted from VCAA 2013 Biology Section B, Q11a*

The skeletal structures of two extinct members of the hominin family tree, *Australopithecus africanus* and *Homo neanderthalensis*, are shown below.

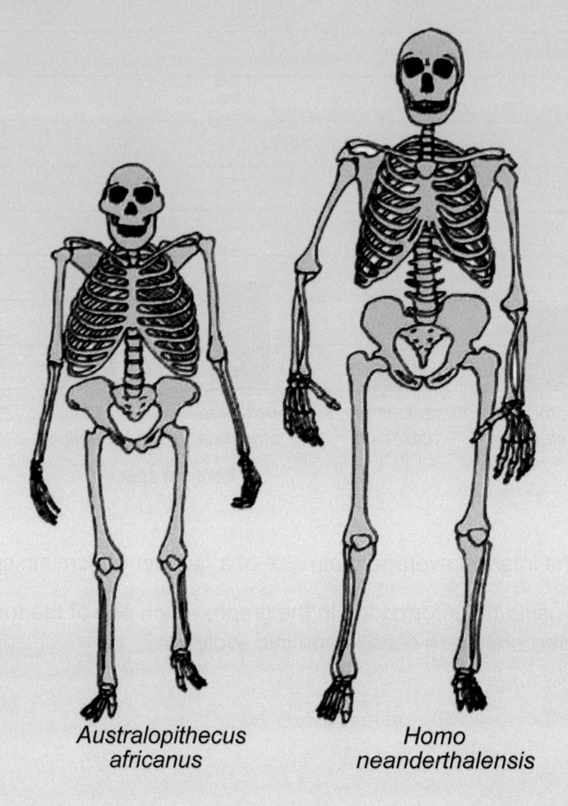

Examine the skeletal structures. For each of the features below, describe the difference between the two species and state the significance of the difference.

- pelvic structure
- arm to leg length ratio

More exam questions are available in your learnON title.

10.4 Interpretation of the human fossil record

KEY KNOWLEDGE

- The human fossil record as an example of a classification scheme that is open to differing interpretations that are contested, refined or replaced when challenged by new evidence, including evidence for interbreeding between *Homo sapiens* and *Homo neanderthalensis* and evidence of new putative *Homo* species

10.4.1 The changing classification schemes around evolution

A nineteenth-century concept was that there was a 'missing link' between ape-like ancestors and modern humans, as shown in figure 10.50a. The human family tree was seen as a straight line from ancestral forms, with each ancestor being more human-like and less ape-like over time.

Instead of a short, straight evolutionary line, a spate of hominin fossil discoveries during the last third of the twentieth century produced an evolutionary pathway leading to modern humans that is a bush with many branches, mostly with dead ends (see figure 10.50b). This will be explored further in section 10.4.2.

FIGURE 10.50 a. The nineteenth-century concept of a 'missing link' between the last common ancestor of modern African great apes and humans **b.** A late twentieth-century concept of an evolutionary 'bush' with many side branches

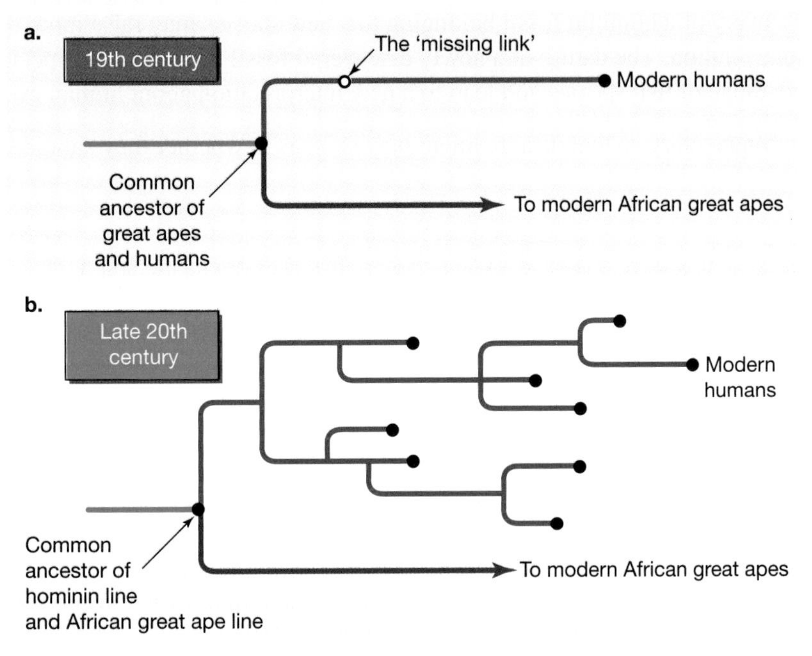

As new hominin species are found, the classification scheme can be changed and contested. Table 10.9 shows the major fossil species of the genus *Homo* and some members of *Australopithecus*, many of which have changed the ideas around evolutionary pathways.

TABLE 10.9 Table showing the major *Homo* fossils (as well as the recently discovered *Australopithecus* species), their year of discovery or public announcement, and the estimated age or age ranges for each species

Year	Species	Where discovered	Age range
2015	*Australopithecus deyiremeda*	Ethiopia	2.8 Mya
1964	*Homo habilis*	Olduvai Gorge, Tanzania	2.4–1.4 Mya
2016	*Homo naledi*	South Africa	1.98 Mya
1986	*Homo rudolfensis*	East Africa	~1.9–1.8 Mya
1891	*Homo erectus*	Trinil, Indonesia	1.89 Mya–143 000 years ago
1908	*Homo heidelbergensis*	Heidelberg, Germany	700 000–200 000 years ago
1829	*Homo neanderthalensis*	Neander Valley, Germany	400 000–40 000 years ago
2003	*Homo floresiensis*	Flores Island, Indonesia	100 000–60 000 years ago
2019	*Homo luzonensis*	Luzon Island, Philippines	50 000–67 000 years ago
2010	Denisovans*	Siberia	41 000 years ago

**Note:* As of 2020, there is no consensus around a formal species name. Proposed names have include *Homo denisova* or *Homo altaiensis*.

10.4.2 The human fossil record is always changing

Hominin fossils continue to be discovered and the pace of discovery has quickened in the last several decades.

Each discovery of a fossil hominin may provide new information that can be used to test currently accepted hypotheses. Often this results in clarification and increased understanding of hominin evolution.

In other cases, the discovery of a hominin fossil belonging to a new species may raise more questions about the pathway of human evolution. The dating of a newly discovered fossil of an existing species may extend the currently accepted age range of that species and lead to a new interpretation about human evolution.

Some fossils are very fragmentary so that disagreement may exist as to whether they should be identified as hominins or as apes.

Currently, it is not possible to identify the exact evolutionary pathway that links the first hominin species to the first *Homo* species. Much of the evidence we have is only based on fossils and DNA, and this evidence is largely incomplete. Thus, it is difficult to determine patterns, particularly when natural variation exists.

It is also hard to definitively prove whether extinct organisms were able to interbreed (and thus be classed as the same species) from fossils. Furthermore, it is also particularly important to note that there may be many species in which no fossilisation occurred (as the conditions did not favour this), or fossils are yet to be found.

It is reasonable to speculate that the earliest hominins were ancestral to the australopithecines. From the australopithecines, one evolutionary line gave rise to the robust *Paranthropus* species, and a second line gave rise to the first members of the genus *Homo* — the first humans. The *Paranthropus* species were an evolutionary 'dead-end'. One simple hypothetical pathway is shown in figure 10.51.

FIGURE 10.51 Possible relationships are shown by '?' symbols between the various hominin species.

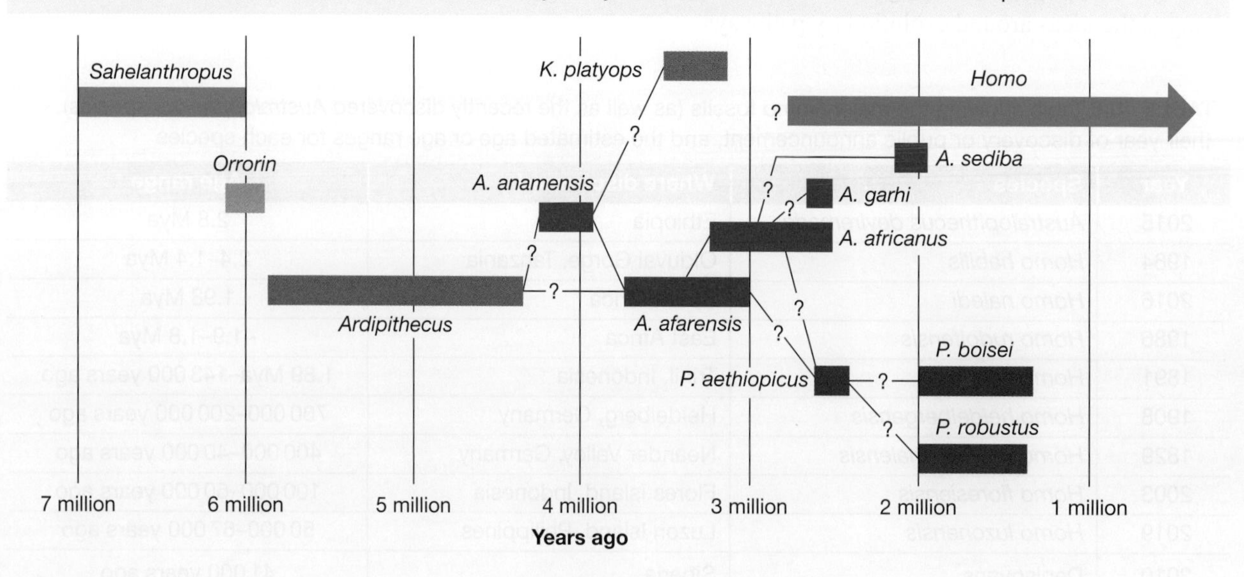

As new species are discovered (refer back to table 10.9 in section 10.4.1), such as the bones of *Homo luzonensis* in the Philippines in 2019, new additions are placed in the hominin classification scheme. Some of the relationships between the main hominin species, including various *Homo* species, are shown in figure 10.52. This is based on suggested evidence, but like figure 10.51, the relationships are only supported based on current evidence and may alter as new evidence comes to light.

FIGURE 10.52 Evolutionary links between major hominin groups

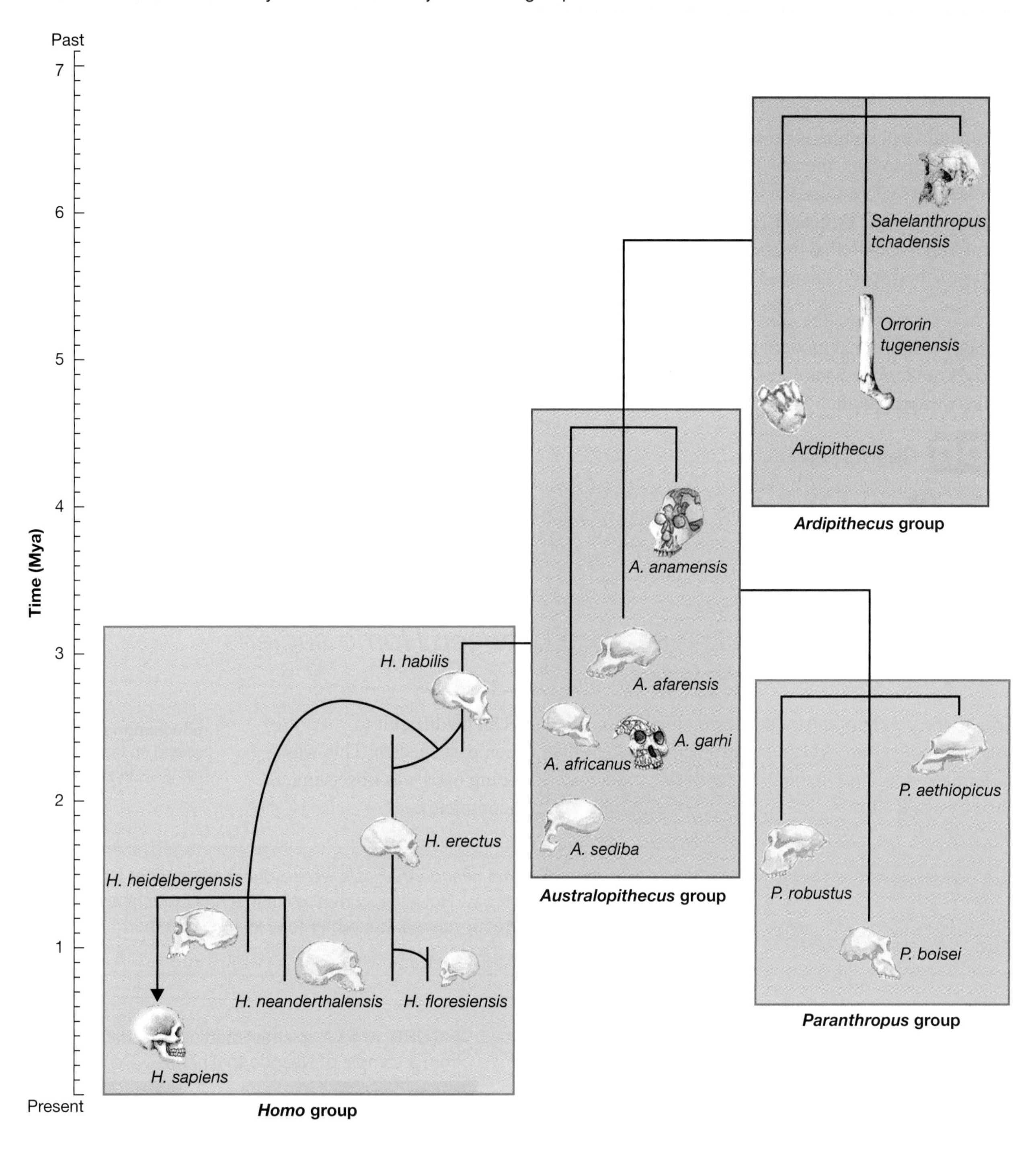

Classification schemes are not fixed but may change when new information becomes available or when new interpretations that provide better explanations are formulated.

Figure 10.52 shows a view of hominin evolution that identifies the major branches on the evolutionary 'bush'. However, it does not capture the most recent discoveries of new species, such as the Denisovans, which have made the hominin evolutionary bush even more tangled when interbreeding is brought into the equation. New fossil discoveries such as *H. naledi*, first described in 2015, and Denisovans, first described in 2010, raise questions about their places in hominin evolution.

Similar questions were asked when detailed descriptions of the fossil hominin *Australopithecus sediba* were first published in 2010, showing that it had some features similar to the australopithecines and others similar to early *Homo* species. As stated in the 9 September 2010 issue of the journal *Science*: 'Partial skeletons of 2-million-year-old hominin *Australopithecus sediba* leave researchers impressed by their completeness but scratching their heads over the implications for our family tree.'

Comparative genomics is providing new insights to hominin evolution. It was only through comparative genomics that the Denisovans, the new hominins in our family tree, were discovered. As of 2020, the Denisovans are only represented by five fossils: a tiny finger bone, three molar teeth and the parietal bone in the skull. The most recent Denisovan fossil, Denisova 13, was only found in 2019. Another fossil discovered in 2012 was found to be a Denisovan/Neanderthal hybrid, known as 'Denny'. Debate around the Denisovans still exists, and questions still surround their species name. Proposed names have included *Homo denisova* or *Homo altaiensis*.

Just a few years ago, the existence of interbreeding between modern humans and Neanderthals was not known. Today, we know that modern humans interbred with Neanderthals and Denisovans, and that various groups of today's modern humans carry that evidence in their genomes — most vividly the Melanesian people of islands of the western Pacific.

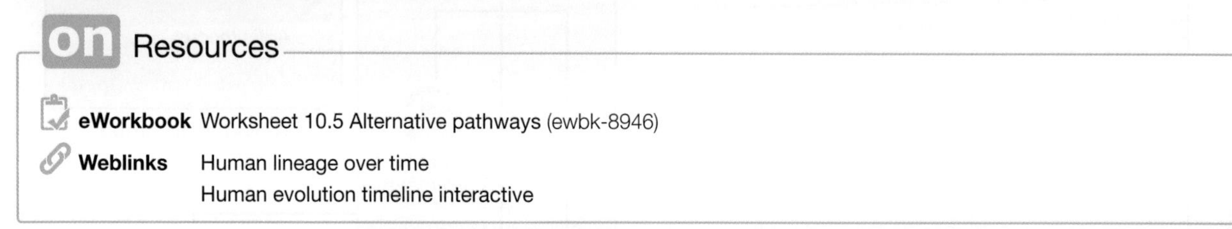

10.4.3 Evidence of interbreeding between *Homo sapiens* and *Homo neanderthalensis*

Due to the nature of the evidence obtained from fossils, it can be difficult to ascertain for certain whether interbreeding occurred between past species. This was particularly difficult in the past, when the main evidence being used was observing the physical characteristics of fossils (which were often incomplete).

mitochondrial DNA (mtDNA) DNA located on the mitochondria in cells that is usually inherited maternally

In the early days of fossil discovery it was thought that *Homo neanderthalensis* were a direct ancestor of *Homo sapiens*, a misconception that is still held by some people today. However, current evidence has brought about the idea that *H. neanderthalensis* and *H. sapiens* are not directly related, but rather they share a common ancestor.

Exploring the DNA from Neanderthal fossils

In 1997, a sample of **mitochondrial DNA (mtDNA)** was isolated from a 40 000-year-old piece of arm bone from a Neanderthal fossil. The mtDNA was partly degraded, but short DNA fragments were obtained. These many short fragments were amplified using the polymerase chain reaction (PCR). Overlapping regions were identified and, using these overlaps, a long sequence (379 bases) of Neanderthal mtDNA was obtained. The Neanderthal mtDNA sequence was compared with those of modern humans, and an average difference of 27 bases was noted. In contrast, the average difference between human mtDNA samples is eight bases. This finding strongly suggested that modern humans and Neanderthals are different species.

FIGURE 10.53 A scientist starting to extract DNA from a sample of Neanderthal bone

Later, in 1999, 2002 and 2004, longer segments of Neanderthal mtDNA were obtained from different Neanderthal fossils (see figure 10.53). All the Neanderthal sequences showed a high degree of similarity to one another, but were clearly distinct from the modern human samples.

In 2010, the first reconstruction of the entire Neanderthal nuclear genome obtained from the 38 000-year-old bones of several Neanderthals was achieved by Svante Pääbo's team at the Max Planck Institute for Evolutionary Anthropology in Leipzig, Germany. Comparison of the Neanderthal genome with the genome of modern humans revealed many single base differences, sufficient to identify them as two different species. (The differences between Neanderthals and modern humans are much greater than the differences between any two modern humans.)

Both the physical differences and the genetic differences support the conclusion that *H. neanderthalensis* and *H. sapiens* are different species, and that the former diverged at least 400 000 years ago from the line that later gave rise to *H. sapiens*. The two species coexisted in several areas of the Middle East and Europe for tens of thousands of years before Neanderthals became extinct.

FIGURE 10.54 Neanderthal fossil remains found in La Ferrassie in France

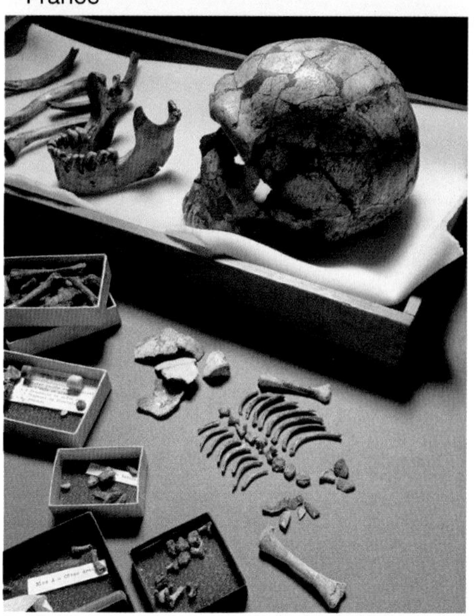

Did *Homo neanderthalensis* and *Homo sapiens* interbreed?

The Neanderthals coexisted with modern humans in some regions. Did Neanderthals and our modern human ancestors interbreed? Fossil evidence to definitively answer this question is not available, but genome sequencing has been crucial in providing an answer.

Comparisons of the nuclear genomes of modern humans with those of Neanderthals reveal that about one to four per cent of the DNA of all non-African modern humans living today came from Neanderthals. Only native Africans do not have any Neanderthal DNA.

This finding provides evidence that, after leaving Africa, modern humans interbred with Neanderthals. It is thought that:

- the major exodus of modern humans out of Africa occurred about 65 000 years ago
- this interbreeding most likely occurred in the Middle East, where both species coexisted for a long period before modern humans spread from there into Europe and Asia.

Evidence that modern humans and Neanderthals coexisted in the Middle East came with the discovery of a 55 000-year-old modern human skull in a cave in Israel. Neanderthals are known to have lived widely across this region.

Neanderthal genes identified in the genome of modern humans today include those:

- involved in features relating to skin and hair — thicker skin, paler skin and thicker hair
- related to the function of the innate immune system.

So, all modern humans today — except Africans — carry in their genomes evidence of gene flow between *H. neanderthalensis* and *H. sapiens*, particularly humans of European or Asian background. In all, about 30 per cent of the total Neanderthal genome is represented in the gene pool of non-African modern humans.

Figure 10.55 shows the two phases of interbreeding between Neanderthals and modern humans.

FIGURE 10.55 Diagram showing gene flow events between modern humans and Neanderthals

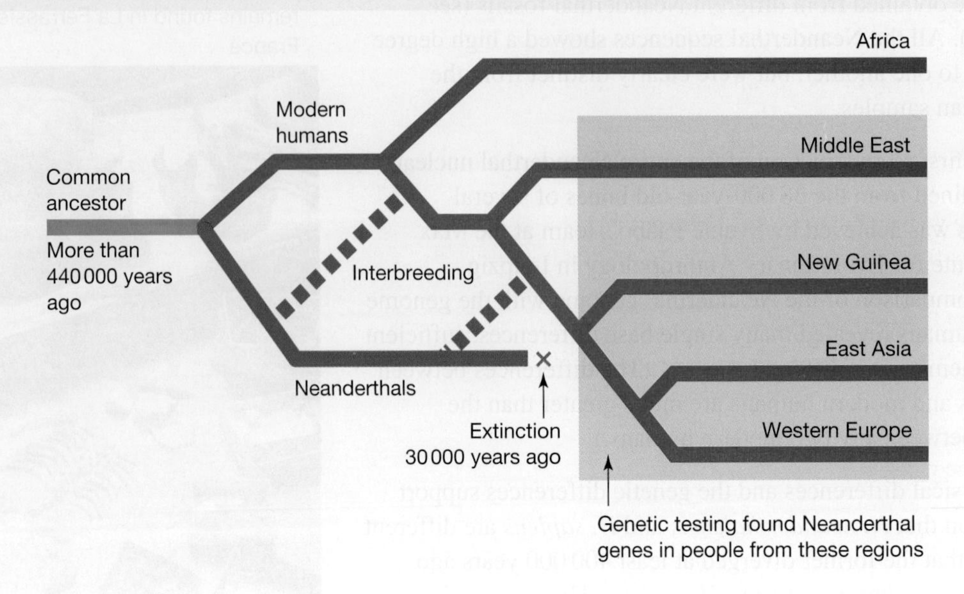

More recent evidence published in the science journal *Nature* in February 2016 indicates that some interbreeding between modern humans and Neanderthals occurred as early as 100 000 years ago.

This evidence comes from the genome sequence of the skeletal remains of a female Neanderthal found in the Altai mountains in Siberia and dated at 100 000 years ago. Her genomic DNA shows the presence of modern human DNA sequences, indicating gene flow from *H. sapiens* to *H. neanderthalensis*.

Not only does this provide evidence of much earlier interbreeding between Neanderthals and modern humans, it also provides evidence that some groups of modern humans migrated out of Africa thousands of years before the main 'Out of Africa' migration that occurred about 65 000 years ago. This migration will be explored in section 10.5.1.

If interbreeding occurred, should we still class Neanderthals and *Homo sapiens* as distinct species?

Generally, we still class Neanderthals and *Homo sapiens* as distinct species. However, some scientists, do not use the species name *H. neanderthalensis*. They instead refer to Neanderthals using the species name *H. sapiens neanderthalensis*. There is some evidence that there was reduced fertility or some sterility of hybrids between the two species.

Ideas around the Neanderthals and modern humans created challenges and the need for refinement to the term 'species', with debate around this continuing today.

10.4.4 Evidence of new putative *Homo* species

The hominin fossil record is incomplete, but palaeoarchaeologists are actively searching and making new discoveries of hominin fossils. These new fossils can be as small as a tooth, or can be a group of skeletal remains. With each new discovery, the fossils found need to be classified into a known species or classified as a new species. Specifically, new putative *Homo* species are constantly being discovered and are leading to contestation of our current classification scheme, leading to new interpretations.

Some relatively recent discoveries of hominins of the *Homo* genus are:

- *Homo naledi*
- *Homo luzonensis*
- **Denisovans**
- Cro-Magnon (early *Homo sapiens).*

Denisovans members of an extinct hominin species in the genus *Homo*

Homo naledi

In 2013, a large group of fossilised hominin remains were found in the Rising Star cave system in South Africa. It was the largest collection of fossilised bones ever found on the African continent and it comprised of 1500 bones from at least 15 different individuals (see figure 10.57). From this collection of bones, palaeoarchaeologists were able to classify them into a new species of the *Homo* genus, *Homo naledi*. However, classifying the bones into the genus *Homo* was not straightforward, as the bones exhibited characteristics from both *Homo* and the genus *Australopithecus*.

FIGURE 10.56 The right hand of *Homo naledi*

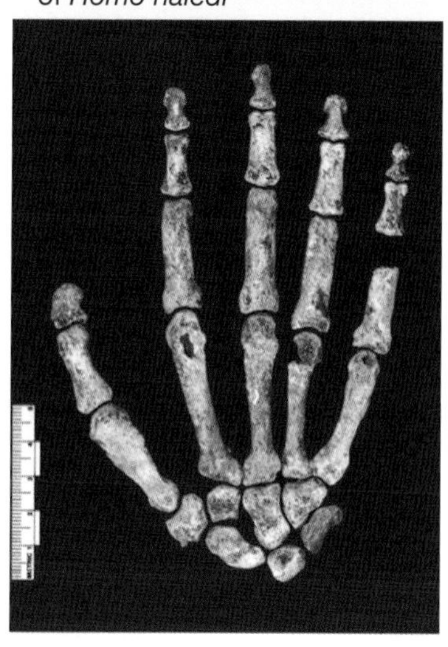

The characteristics of the bones that are similar to the australopithecines are:

- the species' small brain (based on cranial capacity
- their upper body (well-suited to climbing)
- their curved fingers.

In contrast, the characteristics similar to species in the *Homo* genus include:

- their feet
- their long legs (suited to walking long distances)
- their hands (well-suited to toolmaking as shown in figure 10.56).

The hand has a very similar shape to a modern human hand, but the curved fingers and the strong thumb are more primitive.

FIGURE 10.57 The skeletal remains of different individuals from *Homo naledi*

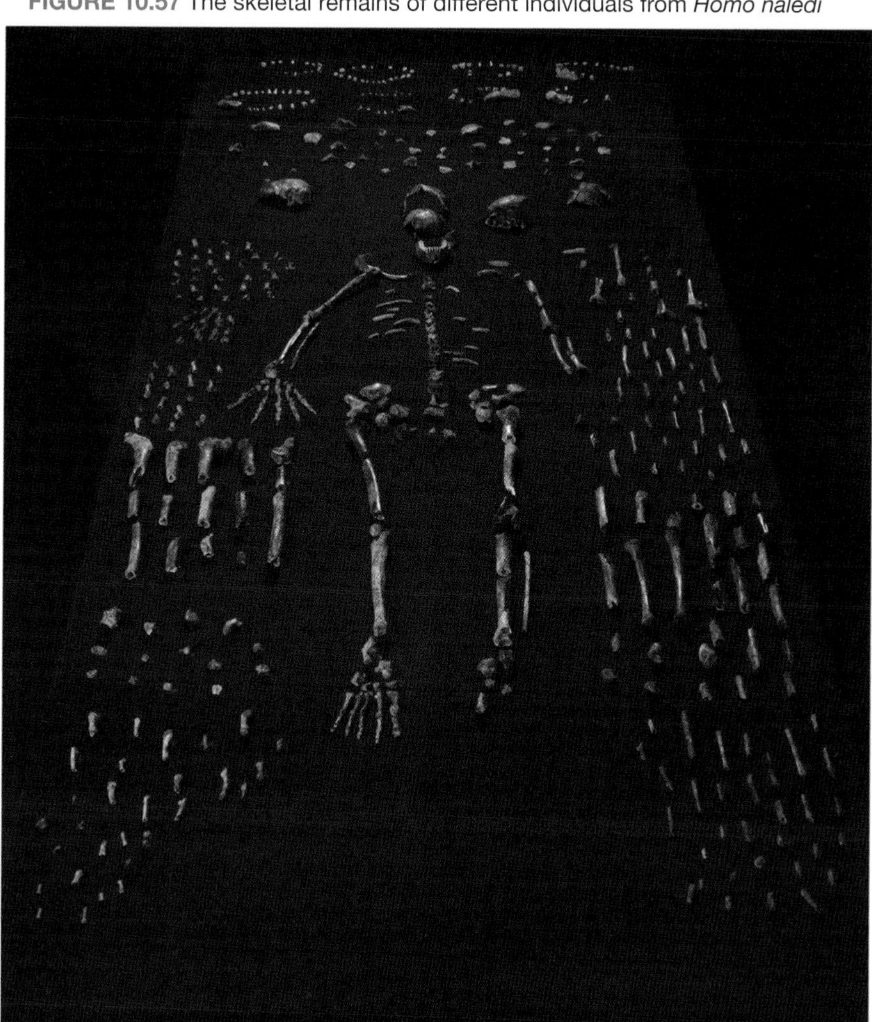

Homo luzonensis

On the island of Luzon, which is part of the Philippines archipelago, seven teeth and six small bones consisting of finger bones, part of a foot and part of a femur were discovered in the Callao Cave. This small collection of remains came from three separate individuals. From this evidence, palaeoarchaeologists were able to classify the fossils as belonging to a newly discovered small hominin, which was later named *Homo luzonensis.* This species lived on Luzon from at least 50 000 to 67 000 years ago — around the time many *Homo* genera, such as Denisovans and *H. neanderthalensis*, were thought to have gone extinct.

FIGURE 10.58 A bone from *Homo luzonensis* from Callao Cave in Luzon, an island in the Phillipines

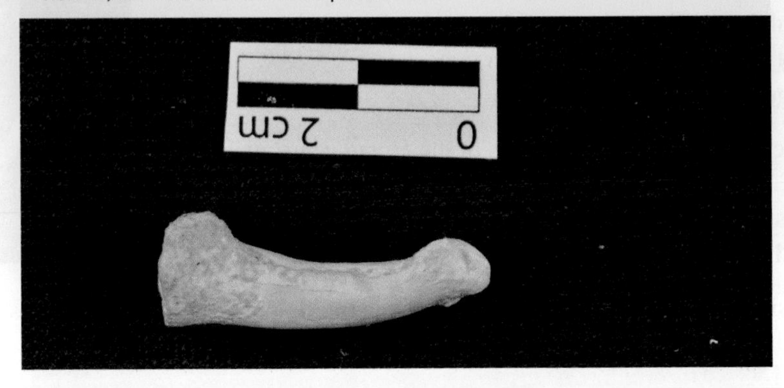

FIGURE 10.59 Fossilised teeth of *Homo luzonensis*

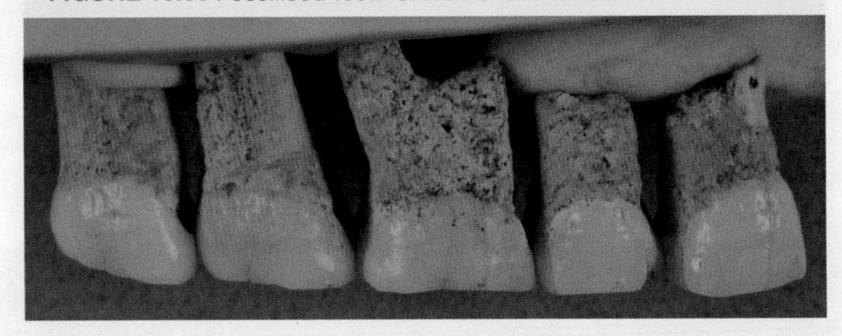

How was the genus and species of this new fossil determined?

For palaeoarchaeologists to classify the fossil remains found on Luzon into the *Homo* genus, they needed to examine the characteristics of the fossils and any other evidence of how the species lived. It was observed that:

- the fossilised teeth (figure 10.59) have structures in common with ancient hominin species, but other structures that are similar to modern humans. The teeth are similar in size and shape to modern humans but one of the premolars found has three roots, which is more similar to older hominin species.
- one of the foot bones was curved like an australopithecine rather than straight like species in the *Homo* genus.

So why were the fossils classified as *Homo*? The answer comes from evidence collected from the activity of the species rather than just from the bones themselves. Researchers also found deer bones in the Callao Cave that had signs of stone-tool marks on them. This points to *H. luzonensis* having the cognitive ability to make stone tools and to hunt with them. These are characteristics possessed by members of the *Homo* genus.

While the fossil remains from the Callao Cave are currently classified as *H. luzonensis*, not all researchers agree that they belong to the *Homo* genus. The small number of bones found makes this difficult to determine conclusively. If more fossils are found, they may either support or refute the classification possessed by this species.

Resources

Weblink New species of ancient human discovered in the Philippines

The Denisovans

A tiny finger bone that was found in 2008 in the Denisova Cave in south-western Siberia came from a human female who lived about 40 000 years ago.

The DNA extracted from this bone yielded a partial mitochondrial DNA sequence and, unexpectedly, this DNA did not match that of modern humans, Neanderthals or any other known hominin. The DNA was identified as that of a new human group that has been called the Denisovans.

So, another human group coexisted with modern humans and Neanderthals. This was the first time an extinct human group was recognised not by fossil bones, but by comparative DNA analysis.

FIGURE 10.60 Archaeologists working at an excavation site in the Denisova Cave in Siberia

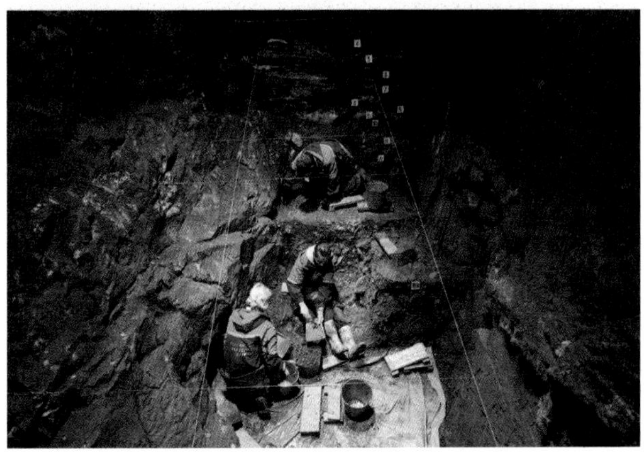

Although Denisovan fossil material has so far been found only in Siberia, it is probable that these people spread over Asia. In the case of the Denisovan female, her DNA included the gene variants (alleles) that would have made her brown-eyed, with brown hair and dark skin.

Interbreeding between Denisovans, Neanderthals and *Homo sapiens*

Teeth recovered in the same cave were also identified as Denisovan in origin. From these various fossils, the German research team constructed an entire Denisovan genome sequence. Comparative genomic studies revealed that the modern humans of that period interbred with Denisovans.

Traces of this interbreeding are seen today in the presence of Denisovan DNA in some human populations, such as present-day modern humans in the Oceania region and in mainland Asia. The highest percentages (3 to 5 per cent) of Denisovan DNA are found today in Melanesian populations, such as those of Papua New Guinea. This interbreeding probably occurred when the ancestors of these present-day humans were migrating across southern regions of Asia.

This putative species of *Homo* again led to the reconsideration of the classification of species, and new assessments of the relationship between *H. sapiens* and both related species and common ancestors.

Modern humans, Neanderthals and Denisovans all descended from a common ancestor that lived many hundreds of thousands of years ago. Some research suggests that this may have been *H. heidelbergensis.* Other research suggests that this was an ancestor of Denisovans only, and the common ancestor is another species that has yet to be determined. The ancestors of modern humans then diverged to form a separate branch that has lasted to the present time and produced *H. sapiens*, the modern human. The other branch was to give rise to the Neanderthals and the Denisovans. At a later time, the Neanderthals and the Denisovans split from each other to form two new evolutionary lines.

Figure 10.61 shows these divergences in the human evolutionary line and the interbreeding between the various human species that has been revealed by comparative genomic studies. These studies indicate that several gene flow (interbreeding) events occurred between Neanderthals, Denisovans and early modern humans. The possibility also exists of gene flow into Denisovans from an unknown archaic group.

FIGURE 10.61 Diagram showing gene flow or interbreeding (denoted by dotted arrows) between hominin species as revealed by comparative genomics. Neanderthals are shown as two geographically separated groups, but they are the one species. The fading of the lines for most groups denotes their extinction.

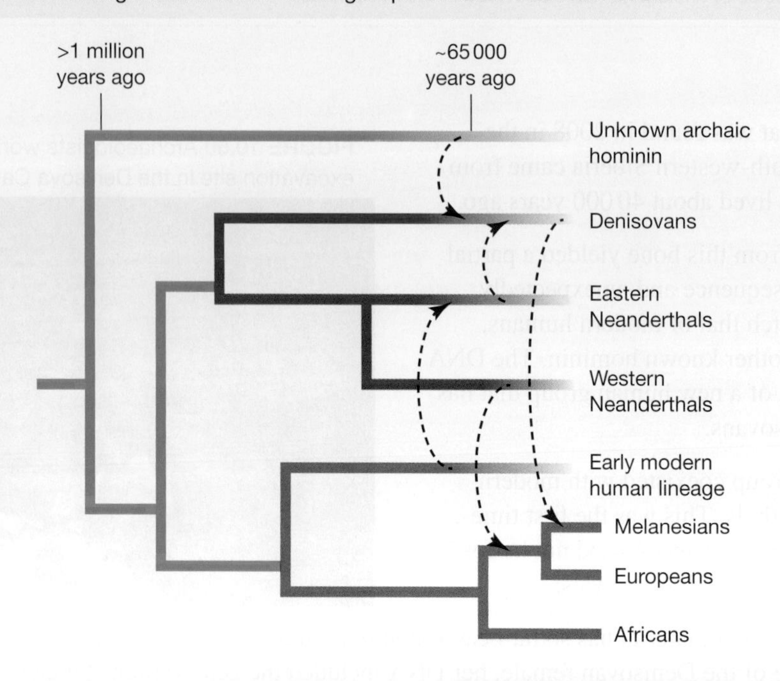

Based on comparisons of their genome sequences, the line leading to modern humans, and the Neanderthal–Denisovan line, most likely diverged about 600 000 years ago in Africa. The split of the Neanderthals from the Denisovans is estimated to have occurred at least 440 000 years ago. Following this split, the Denisovans migrated east into Asia and some Neanderthals migrated west into Europe.

FIGURE 10.62 The movement of different species of *Homo*

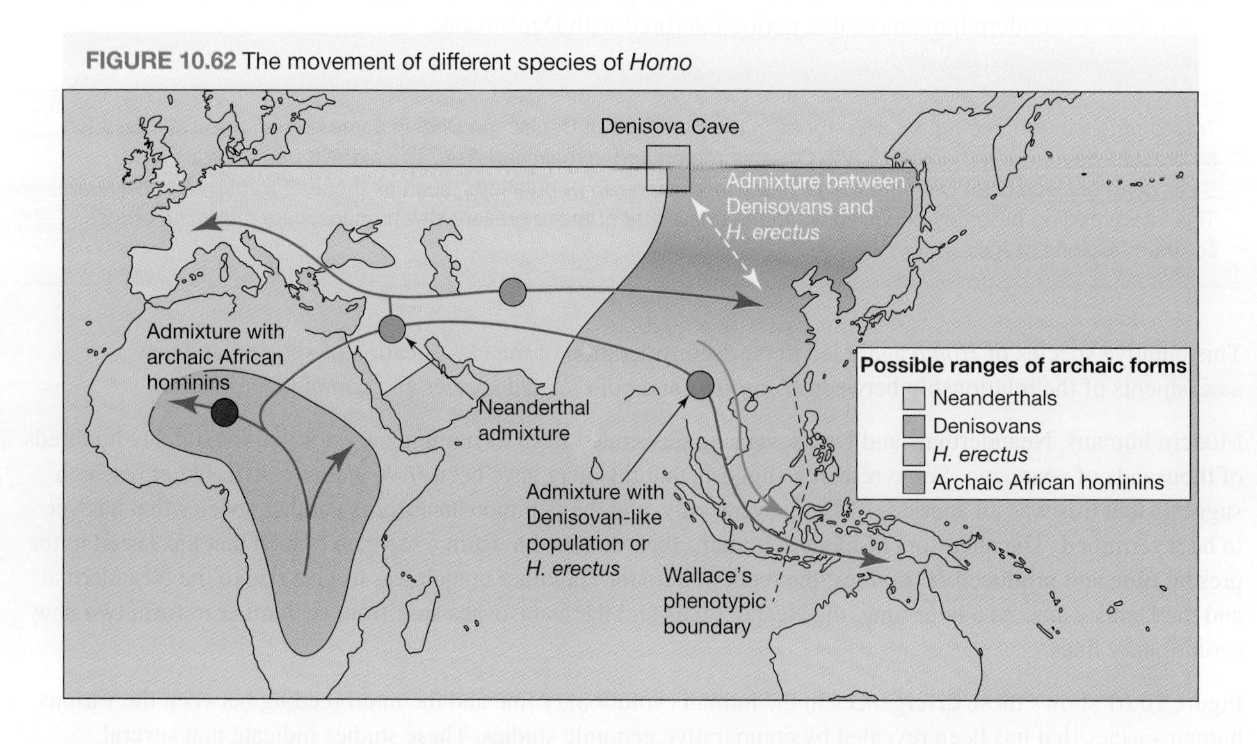

tlvd-1812

SAMPLE PROBLEM 3 Analysing *Homo* fossils

The skulls of two members of the *Homo* genus, *H. erectus* and *H. naledi*, are shown.

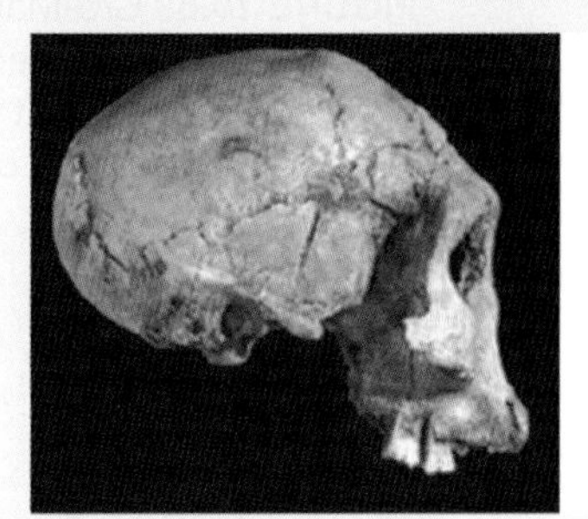

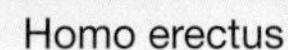

Homo erectus

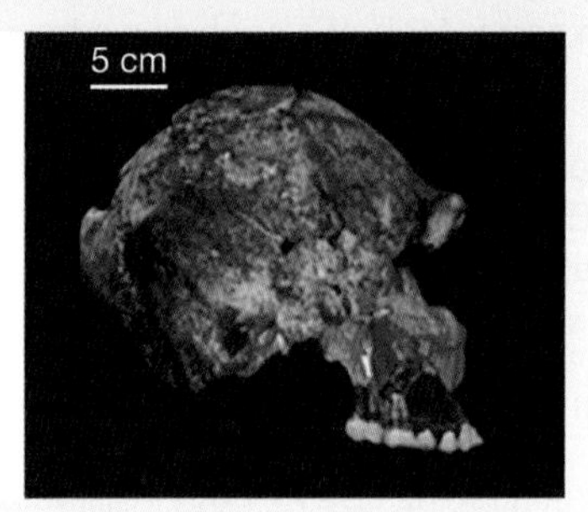

Homo naledi

a. **Examine the two skulls and describe two differences between them that enable them to be classified as either *H. erectus* or *H. naledi.*** **(2 marks)**

b. **When *H. naledi* was being classified, there was a debate over whether the species should be classified into the *Australopithecus* or *Homo* genus. Using data from the provided images, explain why this debate may have occurred.** **(2 marks)**

THINK	WRITE
a. 1. Identify what the question is asking you to do. In a *describe* question, you give a detailed account in sentence form.	
2. Consider the differences between the *H. erectus* and *H. naledi* skulls. Consider what features of hominin skulls are investigated for classification, such as brain size, jaw size and brow ridge. When writing your responses, ensure that you compare the two skulls.	Any two of the following: • The cranium capacity of *H. erectus* is much larger than *H. naledi*. • The brow ridge of *H. erectus* is less prominent than *H. naledi.* • The face slope of *H. erectus* is more vertical than *H. naledi.* (1 mark for each correct difference)
b. 1. Identify what the question is asking you to do. In an *explain* question, you need to account for the reason of why or how something occurs using your scientific knowledge.	
2. Consider the similarities between the *H. naledi* skull and an *Australopithecus* skull. What do they have in common for scientists to consider classifying them into this genus?	The small cranium capacity of *H. naledi* is more similar to species in the *Australopithecus* genus than to other *Homo* species (1 mark). This could lead to scientists wanting to classify this species as *Australopithecus* rather than *Homo* (1 mark).

CASE STUDY: Cro-Magnons — the first *Homo sapiens*

Sometimes the discovery of new fossils leads to not a new species, but rather an earlier form of a species we have already classified.

This occurred with the finding of Cro-Magnon fossils in a rock shelter at Les Eyzies, in southern France. It was determined that these fossils were not a new species, but rather the first *Homo sapiens*.

Indistinguishable from today's human beings, Cro-Magnon people had:

- large brains with an average size of 1350 millilitres
- high, steep foreheads
- short, high, narrow skulls
- small or absent brow ridges
- small eye sockets
- pointed chins
- no gap between the third molar and the jawbone.

FIGURE 10.63 Cro-Magnon art. Cro-Magnons were the earliest modern humans in Europe. Note the mammoth (at left) and the auroch (at right) painted in black outline.

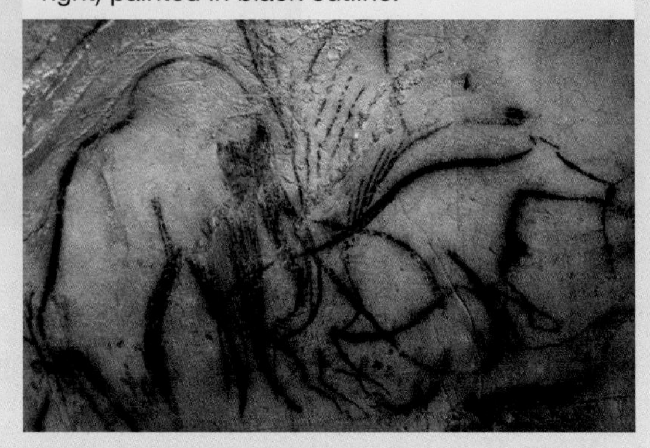

Cro-Magnons were very skilful tool-makers and produced a range of stone items that could be used to scrape, cut, chisel and pierce. They used their stone tools to craft other materials — such as bone and ivory — to produce a variety of instruments, including fish hooks and needles. Evidence of the art of Cro-Magnons exists in caves in France and Spain, where wall paintings depict animals that the Cro-Magnon people hunted (see figure 10.63).

Cro-Magnons buried their dead. Burials dated at 30 000 years ago have been found, where bodies were decorated with necklaces and surrounded by objects such as clay figurines and tools of bone and stone, which were buried with them.

Both the physical and the cultural findings led to Cro-Magnons being determined as *Homo sapiens.* However, evidence may come to light in the future that changes these findings.

Resources

 **eWorkbook** Worksheet 10.6 Interbreeding and new putative species (ewbk-8948)

 **Weblinks** Ancient girl's parents were two different species
DNA reveals first look at enigmatic human relative

KEY IDEAS

- Classification schemes are not fixed but may change when new information becomes available or when new interpretations that provide better explanations are formulated.
- Classification of newly discovered species can be contested.
- Comparative genomic evidence has identified the existence of a new *Homo* species called the Denisovans.
- Comparative genomics provides evidence that interbreeding occurred between modern humans, Neanderthals and Denisovans.
- The hominin evolutionary tree is highly branched and is subject to change as new hominin fossils are discovered.

10.4 Activities

To answer questions online and to receive **immediate feedback** and **sample responses** for every question, go to your learnON title at **www.jacplus.com.au**. A **downloadable solutions** file is also available in the resources tab.

10.4 Quick quiz	10.4 Exercise	10.4 Exam questions

10.4 Exercise

1. MC One to four per cent of DNA from many modern humans comes from Neanderthals, suggesting that interbreeding occurred between *Homo sapiens* and *Homo neanderthalensis*. This requires us to rethink our definition of species and challenges the previous model we have had in place. From the information in figure 10.55, it can be inferred that
   A. Neanderthal DNA would not be found in individuals of South African descent.
   B. Neanderthal DNA would only be found in individuals that live in New Guinea.
   C. Neanderthals must be ancestors of modern humans.
   D. individuals in Australia cannot have Neanderthal DNA in their genome.
2. State two types of evidence that can be used to identify a fossil as belonging to the *Homo* genus.
3. In 2013, two cavers found a large group of human-like bones sealed in a chamber of the Rising Star cave system in South Africa. There were approximately 1550 pieces of skeletal remains that scientists later determined to come from 15 different individuals. Some characteristics of the skeletons included curved fingers with opposable thumbs, pelvises with a flared shape and skulls with a cranium capacity of 40 per cent of the size of modern humans. Predict whether these skeletons were classified as hominoid or hominin as their highest level of classification.
4. Some palaeoarchaeologists consider the fossil remains of the species named *Homo luzonensis* to be incorrectly classified into the *Homo* genus and that it should be in the *Australopithecus* genus. Describe why these scientists could think this.
5. Explain why modern humans having Neanderthal DNA in their genome raises questions about whether we are the same species or distinct from each other.
6. Explain why the classification of hominins is open to interpretation.

10.4 Exam questions

Question 1 (1 mark)

Source: *VCAA 2017 Biology Exam, Section A, Q35*

MC Modern African *Homo sapiens* do not contain Neanderthal DNA. Modern non-African *H. sapiens* contain a small percentage of Neanderthal DNA because of interbreeding between Neanderthals and *H. sapiens*. This interbreeding is thought to have occurred within the time period 65 000 to 47 000 years ago. A recent study has found *H. sapiens* DNA in the genomes of 100 000-year-old Neanderthal remains.

From this new discovery, it would be reasonable to conclude that

A. modern Africans are the descendants of Neanderthals.
B. there was an early migration of *H. sapiens* out of Africa before 100 000 years ago.
C. the ancestors of modern Africans migrated from Europe to Africa between 65 000 and 47 000 years ago.
D. approximately 100 000 years ago, Neanderthals bred with *H. sapiens* in Africa before the Neanderthals spread to the rest of the world.

Question 2 (1 mark)

MC Pre-1980 classification schemes separated chimpanzees and humans into two different families. The scheme placed chimpanzees along with gorillas and orang-utans in family Pongidae, and humans alone were placed in family Hominidae. The modern classification scheme has separated chimpanzees from the other great apes and placed them in family Hominidae with humans.

It is reasonable to suggest that this revised classification was based on

A. molecular analysis showing the high degree of genetic relatedness between chimpanzees and humans.
B. anatomical studies showing similarities in structure between chimpanzees and humans.
C. physiological studies showing similarities in function between chimpanzees and humans.
D. studies of the brain size of early *Homo* fossils.

Question 3 (3 marks)

Source: *Adapted from VCAA 2013 Biology Section B, Q11b and c*

Neanderthals lived in the cold climate of Europe and Asia from 200 000 to 30 000 years ago. Modern humans and Neanderthals coexisted for around 10 000 years. Fossil evidence indicates Neanderthals shared much behaviour with modern humans. Scientists are undertaking research to find reasons why Neanderthals became extinct, but modern humans survived. Fossil evidence of the use of sewing needles and the division of labour between men and women was found only for modern humans, but not for Neanderthals.

a. Suggest one advantage each of the following would have had for the survival of populations of modern humans. **2 marks**

- use of sewing needles
- division of labour between men and women

b. People today with non-African heritage carry some Neanderthal DNA.
State a hypothesis to account for these findings. **1 mark**

Question 4 (3 marks)

Source: Adapted from *VCAA 2017 Biology Exam, Section B, Q7*

In 2013, about 1500 fossil bones of a hominin species were found in a cave in South Africa. From these bones, scientists have managed to construct an almost complete skeleton. The fossil bones have some features in common with those of the genus *Australopithecus*; however, they have enough similarities to the genus *Homo* that scientists have classified the fossil skeleton as belonging to a new species, *Homo naledi*.

a. What are two features that the fossil skeleton would need to have in order to be classified in the genus *Homo* and not in the genus *Australopithecus*? **2 marks**

Finding out the age of these *H. naledi* fossils has been both difficult and controversial. A group of scientists claims that the age of the fossils is more than 2 million years and suggests that *H. naledi* might be a 'link' between *Australopithecus* and *Homo*. A second group of scientists has calculated the age of the *H. naledi* fossils to be only about 900 000 years and claims that *H. naledi* cannot be the 'link' between *Australopithecus* and *Homo*. The diagram indicates the time periods for different *Australopithecus* and *Homo* species.

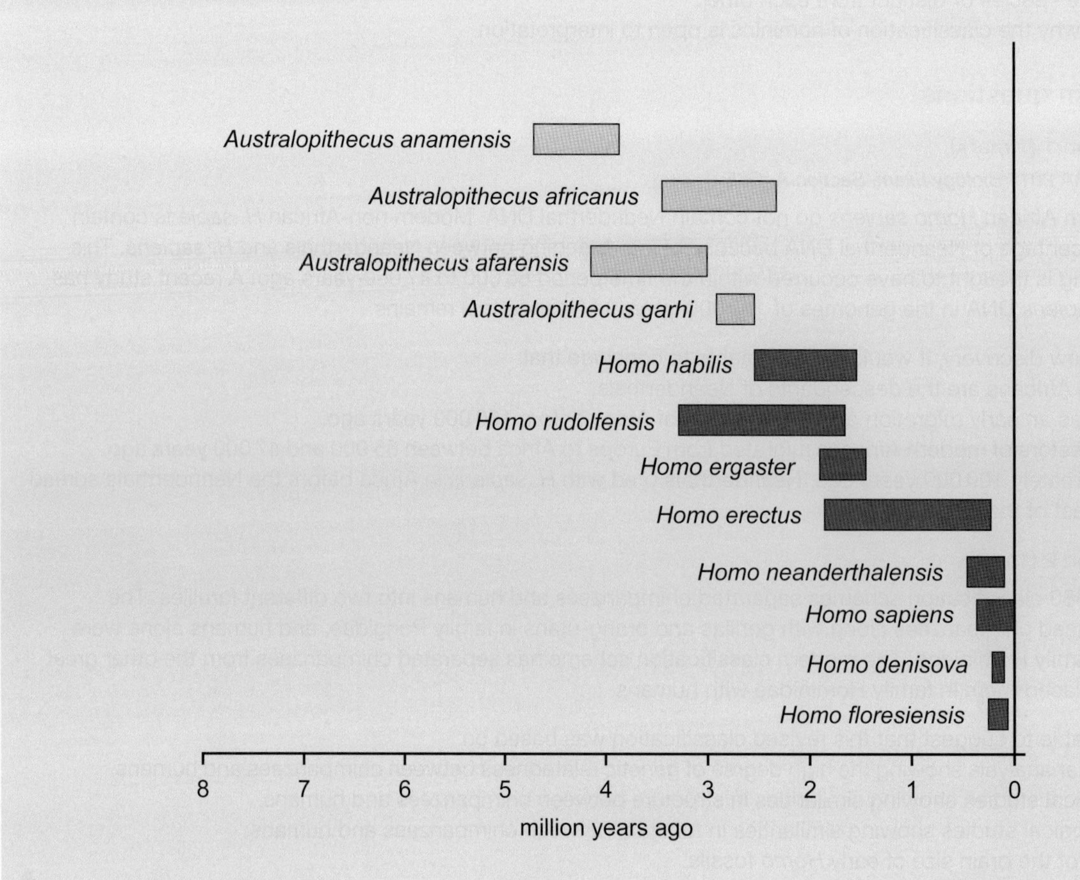

b. If the second group of scientists has correctly dated the *H. naledi* fossils, what evidence from the diagram supports this group's claim that *H. naledi* cannot be the 'link' between *Australopithecus* and *Homo*? **1 mark**

Question 5 (9 marks)

***Source:** Adapted from VCAA 2008 Biology Exam 2, Section B, Q8*

Two paleoanthropologists each used fossil data to draw a model of the human evolutionary tree. The two models they produced are shown.

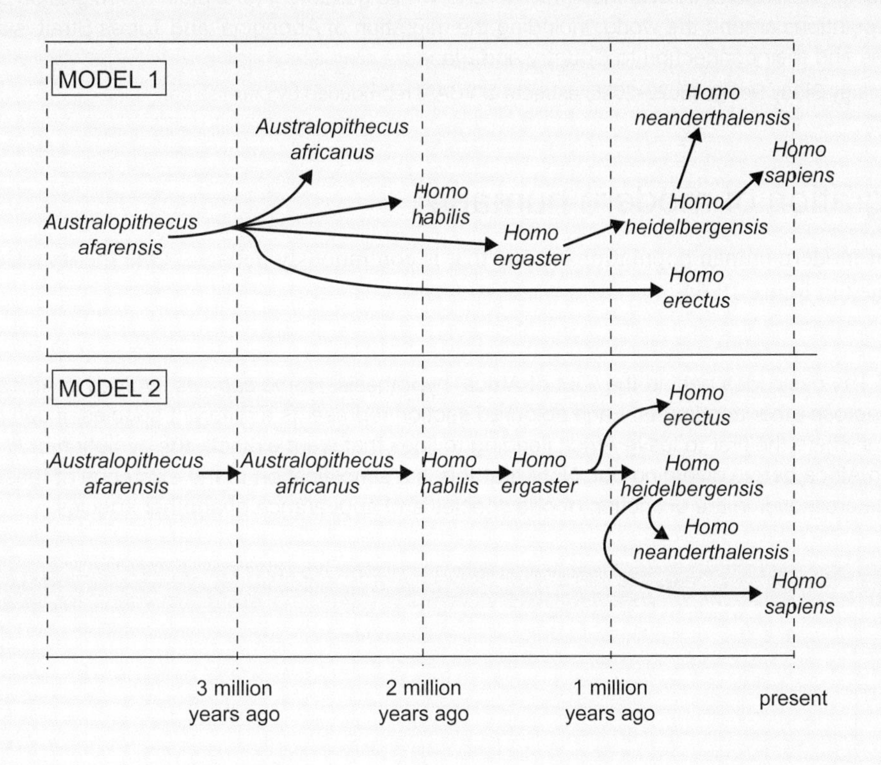

a. Explain how it is possible that the paleoanthropologists produced different models for the human evolutionary tree. **1 mark**

b. **i.** State one feature of agreement between the models. **1 mark**

ii. State one feature of conflict between the models. **1 mark**

c. Give **two** structural features that would distinguish between the fossils of *Homo erectus* and *Australopithecus afarensis.* **2 marks**

Evidence suggests that *Homo sapiens* and *Homo neanderthalensis* were living in the same areas some 30 000 years ago, but did not interbreed.

d. Give one reason why interbreeding might not have occurred. **1 mark**

Scientists have recently discovered a tiny fossil skull in Indonesia. It has been named *Homo floresiensis* (the hobbit) and dated to the time our own ancestors were colonising the world. Some scientists believe *Homo floresiensis* evolved from *Homo erectus*. Fossils of *Homo erectus* have also been discovered in Indonesia.

e. Modify **one** of the models shown to include *Homo floresiensis*. **1 mark**

There is still some debate about what the hobbit is. Two explanations have been proposed.

Explanation 1 — The hobbit belongs to a species of small-brained dwarf humans.

Explanation 2 — The hobbit is a Stone Age *Homo sapiens* with a disease that stunts brain development.

f. **i.** Suggest one piece of evidence that would support explanation 1. **1 mark**

ii. Suggest one piece of evidence that would support explanation 2. **1 mark**

More exam questions are available in your learnON title.

10.5 Evidence of the migration of modern human populations

KEY KNOWLEDGE

- Ways of using fossil and DNA evidence (mtDNA and whole genomes) to explain the migration of modern human populations around the world, including the migration of Aboriginal and Torres Strait Islander populations and their connection to Country and Place

10.5.1 Migration of modern humans

The migration of modern human populations is one that is continuously debated. One theory around this, 'Out of Africa', is outlined in figure 10.64.

FIGURE 10.64 The basic idea behind the 'Out of Africa' hypothesis. *Homo sapiens* evolved in Africa before spreading out across other continents. Each coloured section represents a different species. It can be seen that while *H. erectus* where the first to leave Africa (leading to lines that went extinct), the evolution of *H. sapiens* occurred within Africa before diverging out. Note that this is a simplification of the existence of these species, as there was some crossover in the extinction of one species and appearance of another species.

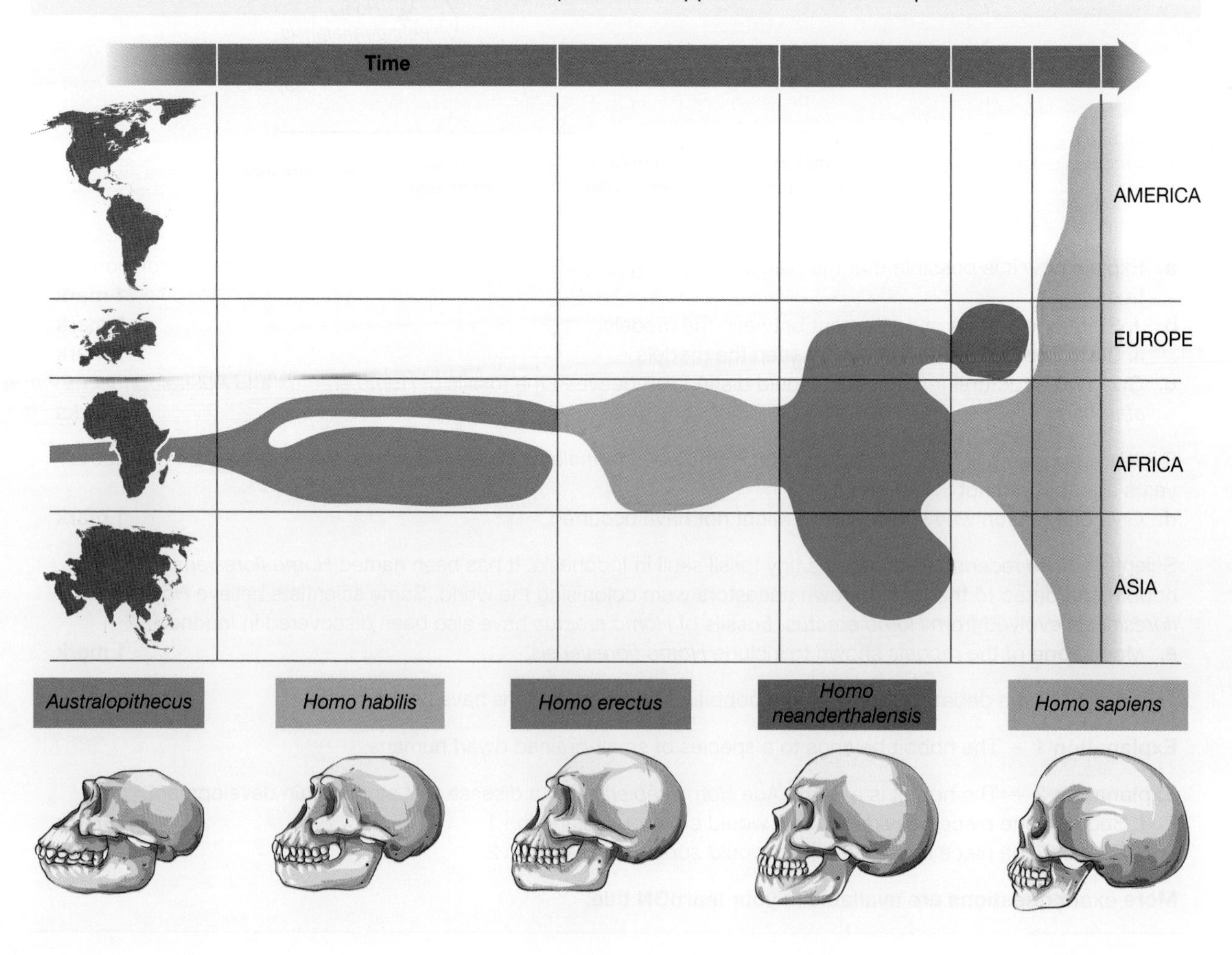

How and when modern humans migrated to different regions of the world has been debated and updated when new evidence is discovered, with multiple hypotheses being both supported and refuted by different scientists.

Some of the hypotheses that have been presented include the 'Out of Africa' hypothesis and the multiregional origin hypothesis.

The 'Out of Africa' hypothesis, shown in figure 10.64, suggests that *H. sapiens* first evolved in Africa. For hundreds of thousands of years, *H. sapiens* lived in Africa within a relatively small geographic area. From here, modern humans have explored and migrated to all parts of the Earth.

However, with the discovery of new fossils and molecular dating techniques, the story of modern human migration is being rewritten.

Updating 'Out of Africa'

The traditional 'Out of Africa' hypothesis (outlined in figure 10.64) suggests that all *Homo sapiens* evolved in Africa and then began to migrate out through North Africa about 60 000 years ago.

However, fossils of *H. sapiens* found in Israeli caves dating back to approximately 180 000 years old have made scientists rethink modern human migration. This latest hypothesis is referred to as 'Recent African origin' or 'Out of Africa II'. This theory proposes that the migration of modern humans occurred in two waves (see figure 10.65). These two waves of migration were:

1. small-scale migration, starting 120 000 years ago, to the Levant region, which makes up the current-day countries of Israel, Jordan, Lebanon, Syria and Iraq
2. large-scale migration, starting 60 000 years ago, to Europe and Asia, and then to all other regions.

FIGURE 10.65 The two waves of dispersal of *Homo sapiens* from Africa (ka refers to 1000 years)

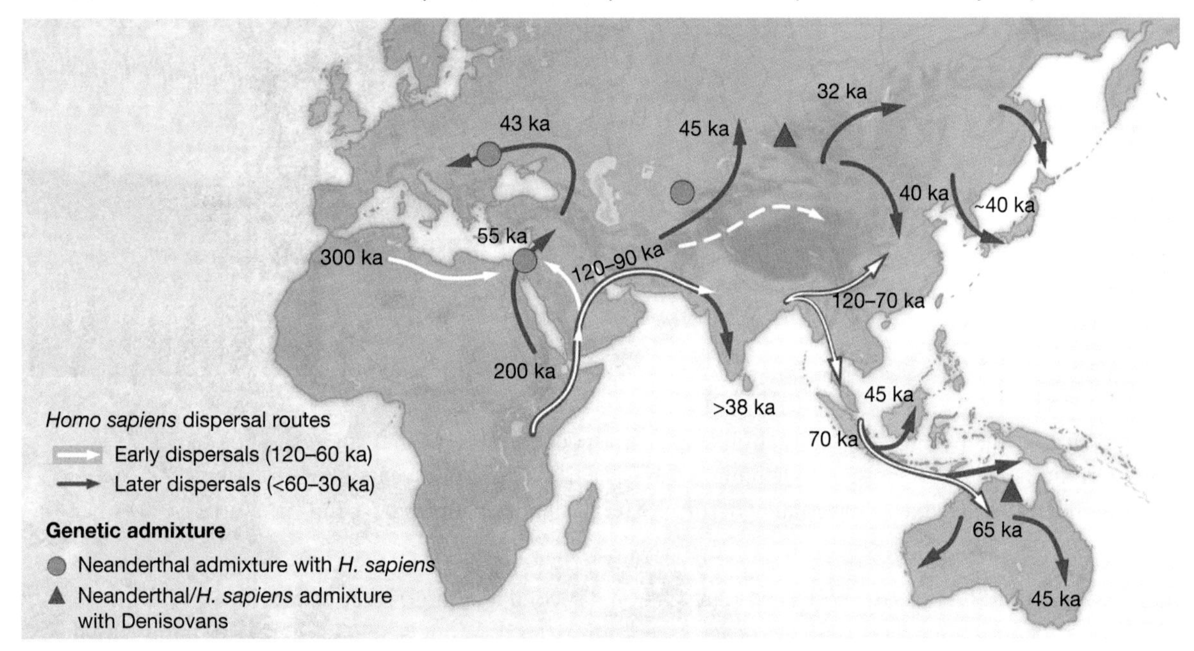

Fossil evidence of migrations through 'Out of Africa II'

The earlier migration of modern humans resulted in irregular occupation of the Levant region (Israel, Jordan, Lebanon, Syria and Iraq). The populations did not live in the region continuously, but their presence during this time did leave evidence.

The oldest fossils so far attributed to *H. sapiens* outside of Africa were found in Israel in the:

- Misliya cave, where an upper jawbone was discovered that was dated to approximately 180 000 years old
- Qafzeh cave, where the remains of five individuals were discovered and dated to approximately 120 000 years old
- Skhul cave, where the remains of seven adults and three children were discovered and dated to approximately 90 000 years old.

These fossils provide evidence that modern humans left Africa much earlier than 60 000 years ago.

FIGURE 10.66 The jawbone discovered in the Misliya cave and the skull found in the Qafzeh cave in Israel

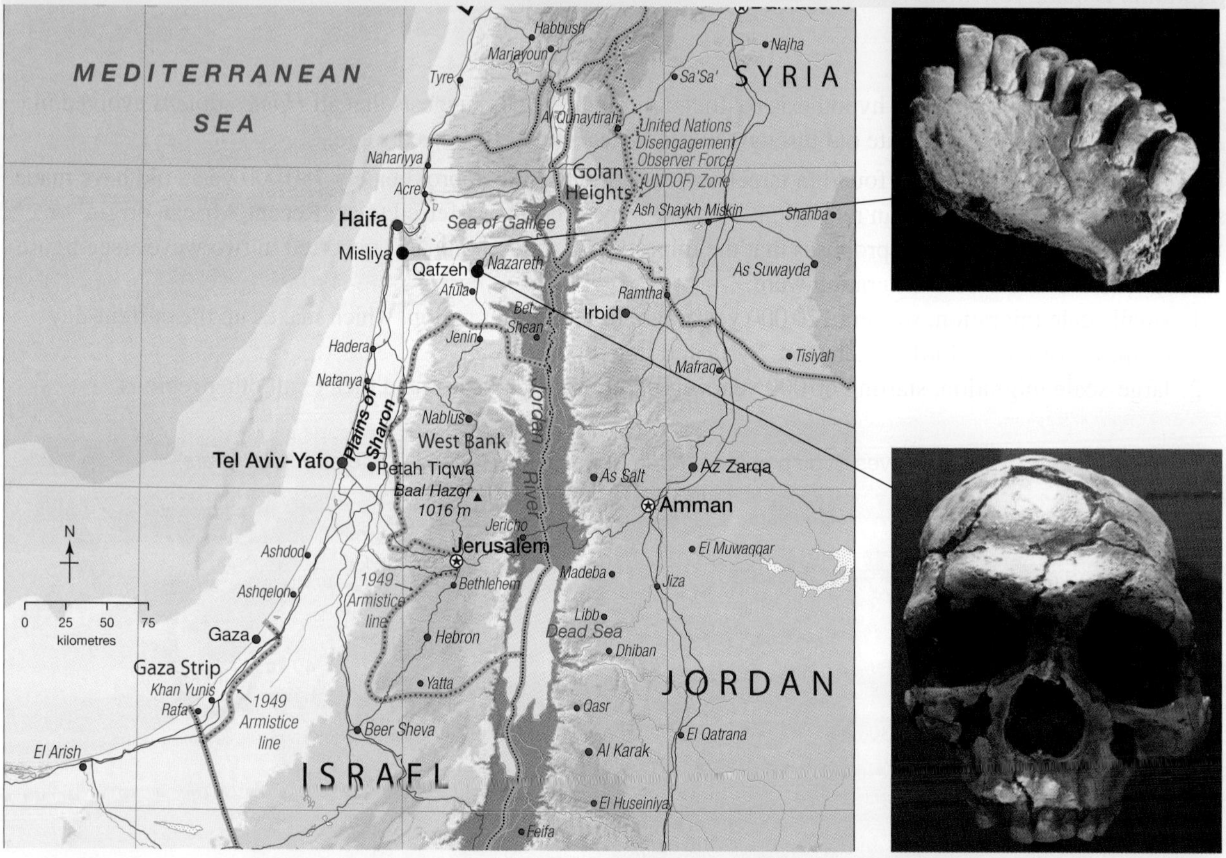

Other evidence of this first wave of migration comes from genetic studies that suggest that:

- modern humans may have started leaving Africa as early as 220 000 years ago
- the modern humans that Aboriginal and Torres Strait Islander peoples descend from arrived in Australia 65 000 years ago.

Research supporting 'Out of Africa II' also found that the second and large-scale migration of *H. sapiens* out of Africa began approximately 60 000 years ago (see figure 10.65). While scientists cannot be sure why this dispersal began, it is thought that environmental changes during this period were the driving factor for the movement.

The movement of modern humans has been tracked through fossils and DNA evidence. The evidence suggests modern humans started by moving through North Africa into Europe and Asia. Populations subsequently moved into the Indian subcontinent and then into Oceania, followed by Australia. The movement of modern humans to the American continent occurred later during the Pleistocene ice age. During this event, there was a land bridge connecting Asia to the American continent through Alaska. Groups of modern humans were able to travel across the bridge and then rapidly move south, populating the whole continent. The last places on Earth that modern humans populated were Micronesia and New Zealand, which occurred around 2500 years ago.

EXTENSION: What is the multiregional hypothesis?

The multiregional hypothesis suggests that *H. sapiens* evolved in various regions *after* they started moving out of Africa. This hypothesis states that *H. erectus* moved out of Africa and then evolved in the differing locations. However, during this time, there was still gene flow between these different populations.

This differs from the 'Out of Africa' hypothesis, which states that the evolution of *Homo sapiens* occurred in Africa, rather than in multiple regions worldwide.

The formation of the multiregional hypothesis shows the complexity of the fossil record and genomic data, highlighting how it is open to interpretation.

FIGURE 10.67 The multiregional hypothesis

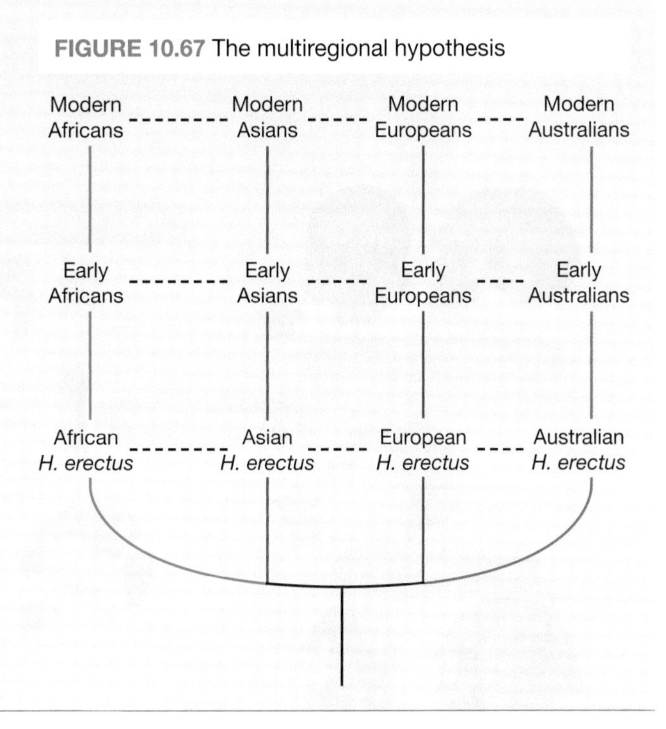

10.5.2 Evidence of migration

So how have scientists traced the migration of modern humans out of Africa?

It begins with the search and discovery of modern human fossils. This is not particularly easy, as it is rare for an individual to become fossilised and there are burial rituals that prevent this process from occurring incidentally. With this being the case, the fossils that are discovered are examined carefully and, if possible, DNA evidence is collected to analyse. This is usually focused on mitochondrial DNA (mtDNA).

Fossil evidence

The major finds of modern humans can be seen in figure 10.68. These fossils are some of the oldest ever to be found and provide evidence as to when modern humans were living in different regions of the world. Some of the important fossil discoveries are outlined in table 10.10.

TABLE 10.10 Notable early modern human fossils

Location of discovery	Fossil evidence	Age of fossil/s
Israel: Misliya cave	Upper jawbone	180 000 years ago
Israel: Qafzeh and Skhul caves	Skeletal remains of a number of individuals	90 000–120 000 years ago
India: Netankheri	Humerus and femur fragments	75 000 years ago
Afghanistan: Darra-i-Kur cave	Temporal bone from a skull	30 000 years ago
Sri Lanka: Fa Hien and Batadomba caves	Cranio-dental remains	25 000–33 000 years ago
Laos: Tam Pa Ling cave	Frontal bone from a skull	46 000 years ago
China: Luijiang	Skull and partial skeleton	Under 100 000 years ago
China: Luna cave	Teeth	70 000 years ago
Borneo: Niah cave	Partial skull	39 000–45 000 years ago
Australia: Lake Mungo	Skeleton	45 000 years ago

FIGURE 10.68 Fossil remains of modern humans found throughout the world

DNA evidence

To establish when modern humans migrated to different regions of the world, molecular genetic studies can be used to determine when lineages separated.

Traditionally, these studies have focused on:

- mitochondrial DNA (mtDNA)
- the non-recombining section of the Y-chromosome.

These genetic studies only focus on either the maternal or paternal lineage of the individual.

Recently, however, there has been an increase in investigating the full genome sequence of fossilised remains to determine the timing of migrations and the separation of populations from each other. This has challenged some of the dates for modern human populations and it has given a broader understanding, due to the ability to investigate more genes that are inherited from both parents.

A comparison of the evidence collected from mtDNA and the whole genome is outlined in table 10.11.

TABLE 10.11 Comparison of mtDNA and DNA of the whole genome

mtDNA	• Comes from the maternal lineage only in most cases (refer to Case study box later in this section), so no recombination occurs • Can be used to reconstruct the mutations and their subsequent effect in genes • Has a higher mutation rate because it does not have repair mechanisms • Contains a smaller amount of genetic material, with 37 genes and 16 569 nucleotide pairs • Does not provide evidence of interbreeding • Does not show the whole population's genetic history • Takes longer to degrade compared to nuclear DNA
Whole genome DNA	• Uses nuclear DNA, which provides evidence for both the maternal and paternal lineages due to recombination • Allows for larger amounts of DNA to be sequenced, with over 20 000 genes and 3.2 billion nucleotide pairs • Has a lower mutation rate because it has repair mechanisms • Provides evidence of interbreeding with another species

Why use mtDNA?

Mitochondrial DNA (mtDNA) is present in all individuals, but is maternally inherited (mothers pass on mtDNA to their children).

FIGURE 10.69 The structure and genes of mtDNA. The D-loop is mainly used for evolutionary studies.

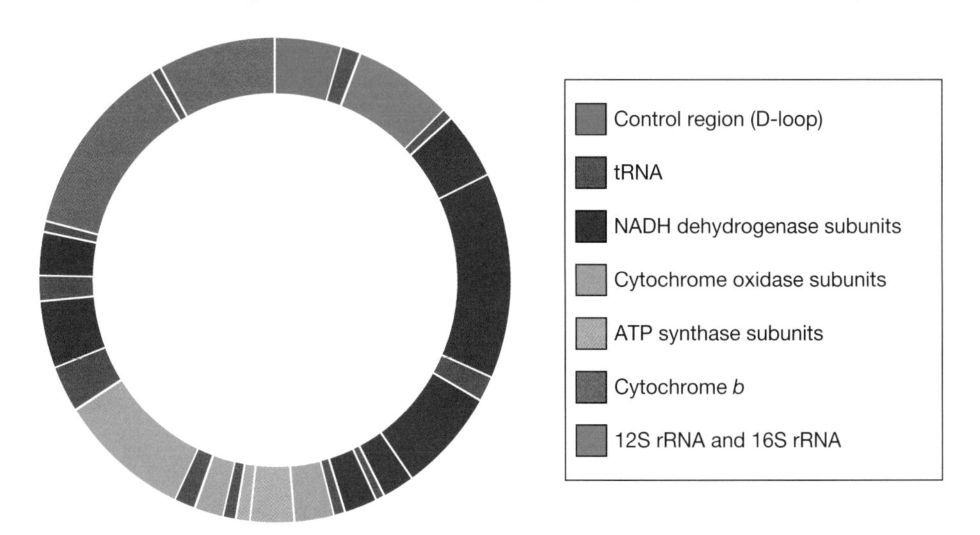

mtDNA contains a non-coding region known as the D-loop (see figure 10.69). Within the D-loop are two regions of DNA that undergo mutational change at a much higher rate than the rest of the molecule. These two regions within the non-coding region are denoted as HVR1 and HVR2 (HVR = hypervariable region). Over the 200 000 years or so of the existence of modern humans (*H. sapiens*), many mutations have occurred in these HVR regions. As a result, mtDNA sequences differ between populations and between individuals.

The particular mtDNA sequence of the D-loop of each person is called that person's **haplotype**. The various haplotypes worldwide fall into a number of large clusters known as **haplogroups**. Haplogroups are denoted by capital letters; for example, haplogroup H, haplogroup T and haplogroup U. Figure 10.70 shows the major human haplogroups. These can be used to track modern human migration across the world due to the lack of genetic recombination and high level of mutation in the D-loop.

haplotype a region of DNA in the D-loop of mtDNA that varies between individuals

haplogroup a group of people with similar haplotypes who share a common ancestor

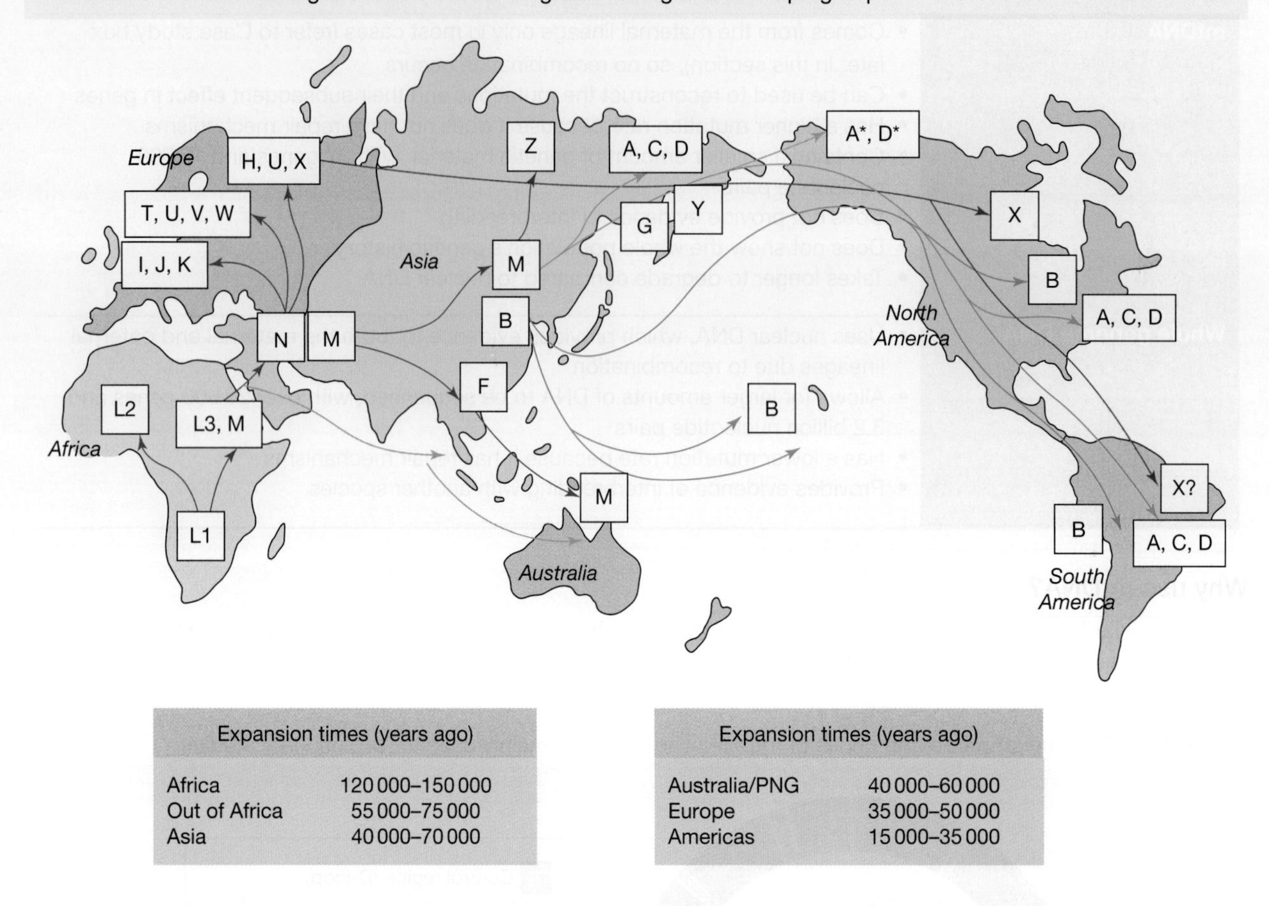

Expansion times (years ago)	
Africa	120 000–150 000
Out of Africa	55 000–75 000
Asia	40 000–70 000

Expansion times (years ago)	
Australia/PNG	40 000–60 000
Europe	35 000–50 000
Americas	15 000–35 000

FIGURE 10.70 The tracing of modern human migration using mtDNA haplogroups

The migration of individuals with certain haplogroups can be seen in figure 10.70. Each haplogroup represents significant differences in matrilineal mitochondrial DNA and allows for the migration of human populations to be tracked out of Africa. In Australia, haplogroups M, O, P and S are commonly found.

The examination of these haplotypes has also allowed an understanding to be gained about the matrilineal most recent common ancestor (MRCA) of *H. sapiens* in which other haplogroups have evolved. This ancestor is hypothetically referred to as Mitochondrial Eve (or mt-MRCA).

EXTENSION: mtDNA heteroplasmy

It has long been thought that mtDNA is inherited exclusively from the mitochondria in the maternal eggs and that no mitochondria from the paternal sperm cells is passed on to offspring. For the majority of people this is the case, with your mtDNA being the same as your mother's, her mother's and her mother's. However, a recent study has discovered that some individuals have mtDNA heteroplasmy. This is when an individual has mtDNA from both their mother and their father. This recent discovery has further complicated our understanding of human evolution.

mtDNA heteroplasmy has been found in families with diseases caused by mtDNA mutations. A study of a four-year-old boy who was displaying symptoms of a mitochondrial disorder found that he had mtDNA from both of his parents. Due to this being a highly unusual result, the scientists also tested the mtDNA of his close family members. They were shocked to discover that four of his relatives across multiple generations also had inherited their mtDNA from both parents (see figure 10.71).

At this early stage of mtDNA heteroplasmy research it is hard to determine how many families display this mode of mtDNA inheritance and why it occurs. It does however pose questions about how mtDNA is used to trace modern human migration.

FIGURE 10.71 Pedigree of a family showing mtDNA heteroplasmy

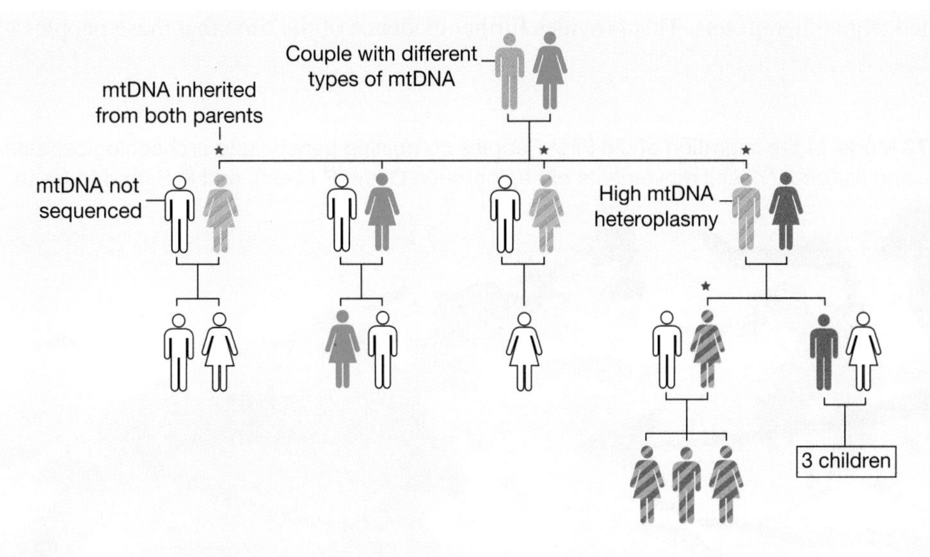

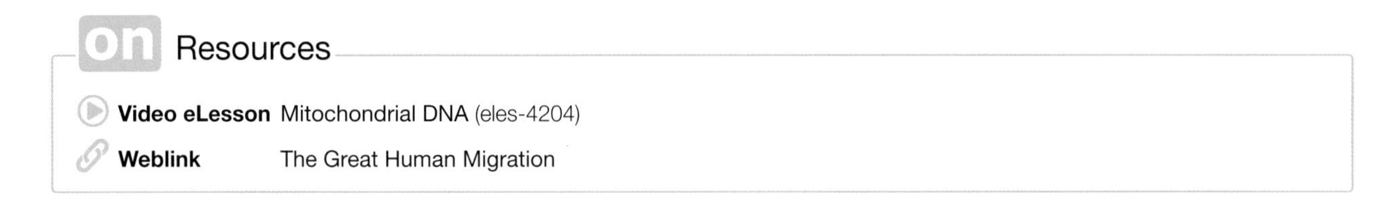

10.5.3 Migration of Aboriginal and Torres Strait Islander populations

As the original peoples of Australia, Aboriginal and Torres Strait Islander populations have lived on the continent for a significantly longer period than the first European colonisers who arrived at the end of the eighteenth century. They are the oldest living cultures in the world today, with a rich connection to Country and Place. How long these peoples have lived in Australia and how they migrated here has been tracked using the fossil and DNA evidence outlined in section 10.5.2.

The general consensus up until recently was that Aboriginal and Torres Strait Islander peoples have been living in Australia for around 45 000 years. However, new studies into the mtDNA of Aboriginal and Torres Strait Islander peoples has pushed this date back further to 50 000 years, with others speculating as far back as 60 000 years. It is proposed that their ancestors moved out of north-east Africa through the Middle East and Asia, and then travelled south to the landmass of Sahul (see figure 10.72). Sahul is the name of the landmass that previously existed where Australia and Papua New Guinea were joined with a land bridge. From here, different ancestors travelled both east and west, with evidence being provided from archaeological sites and through the investigation of DNA of Aboriginal and Torres Strait Islander peoples.

Haplotypes in Aboriginal and Torres Strait Islander peoples

Recently, scientists have researched the mtDNA haplogroups of Aboriginal and Torres Strait Islander peoples and found that there are haplogroups that are unique to these peoples. These groups are in the macrohaplogroups M, N and R and include N13, O, M42a, M14, M15, S and a number of P subtypes (see figures 10.72 and 10.73). The large variation in these haplogroups reflects the large time frame since Aboriginal and Torres Strait Islander peoples diverged from other groups. This provides further evidence of the time that these peoples have lived in Australia.

FIGURE 10.72 Model of the migration of the First Peoples combining genetic and archaeological data, showing approximate, and stylised, coastal movements of haplogroups O and R (west), and P, S, and M (east)

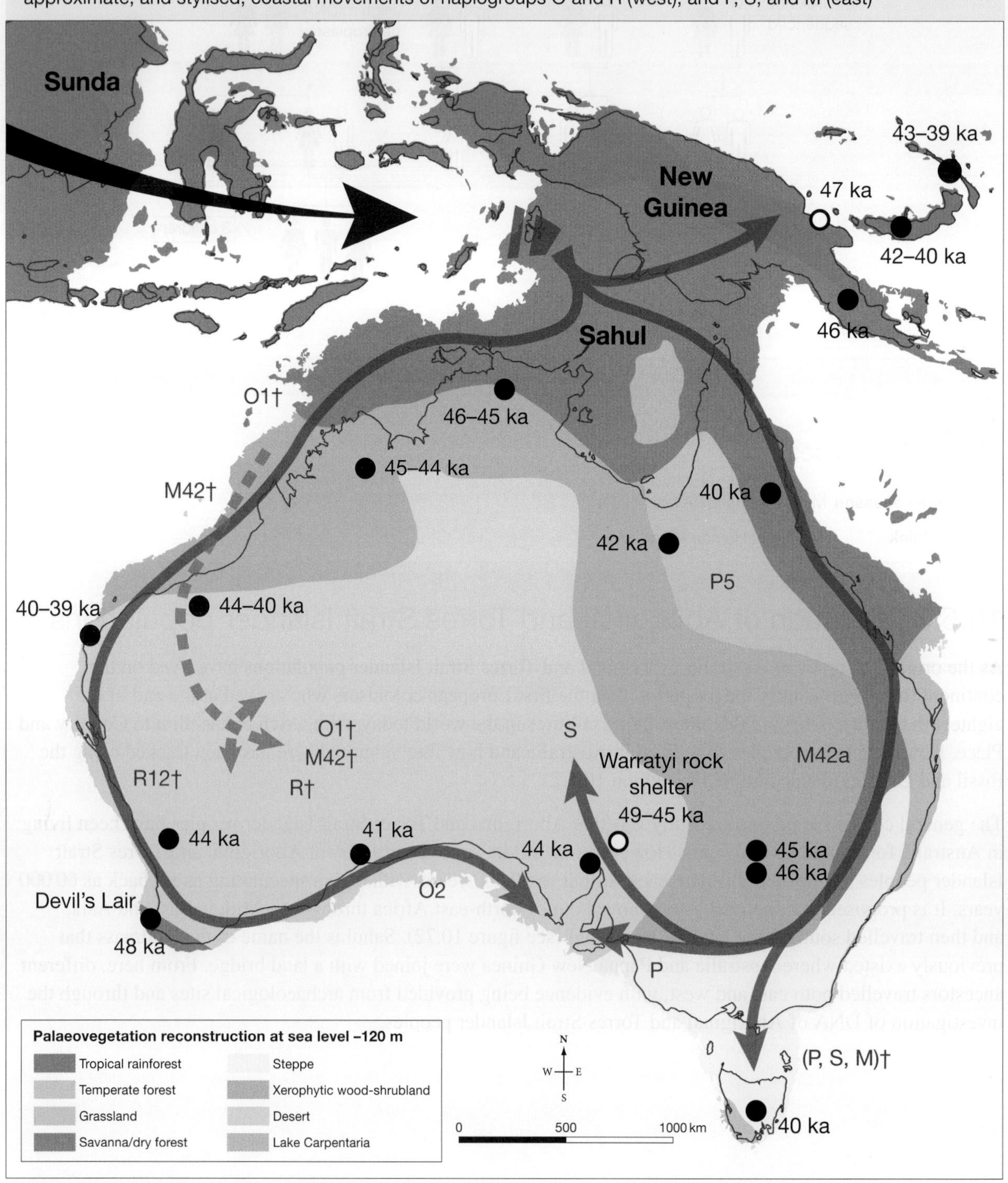

FIGURE 10.73 Phylogeny and geographical distributions of Aboriginal Australian mtDNA lineages and those of the surrounding regions of past and present studies. The length of branches is indicative of time since divergence. Coloured squares on the map indicate the presence of specific haplogroups depicted in the phylogeny. Note that the L3 haplogroup was found in Africa, associated with the 'Out of Africa' hypothesis.

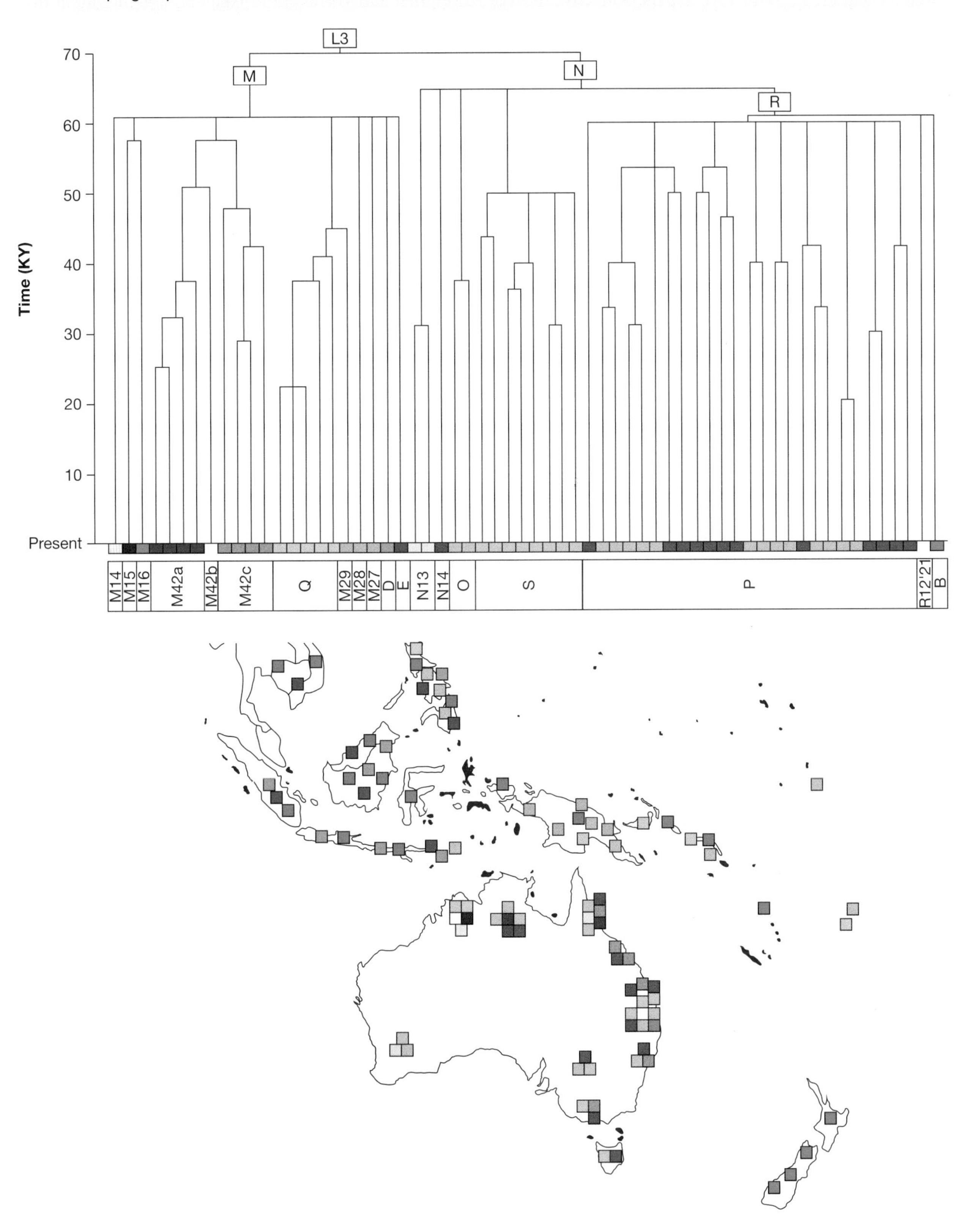

Connection to Country and Place

As outlined earlier, Aboriginal and Torres Strait Islander peoples are part of the longest living culture in the world. Due to length of time they have lived in Australia, they have created strong and rich connections to their Country and Place. Whilst we use the collective term of Aboriginal and Torres Strait Islander, it is important to remember that there are hundreds of different tribes (these also can be called clans or mobs) that have unique languages and connections to their land across various regions in mainland Australia, Tasmania and the Torres Strait. Some of the different language groups of the Aboriginal Australian peoples in Victoria can be seen in figure 10.74. The Torres Strait Islander peoples are a culturally distinct group to the Aboriginal Australian peoples. Kala Lagaw Ya, Merian Mir and Yumplatok are the three main languages spoken.

FIGURE 10.74 Different Aboriginal Australian language groups in Victoria. Note that this map is a simplification, and shows the general locations of language groups, and does not reflect the dynamic and overlapping nature of these groups. Note that some of these groups have alternate names.

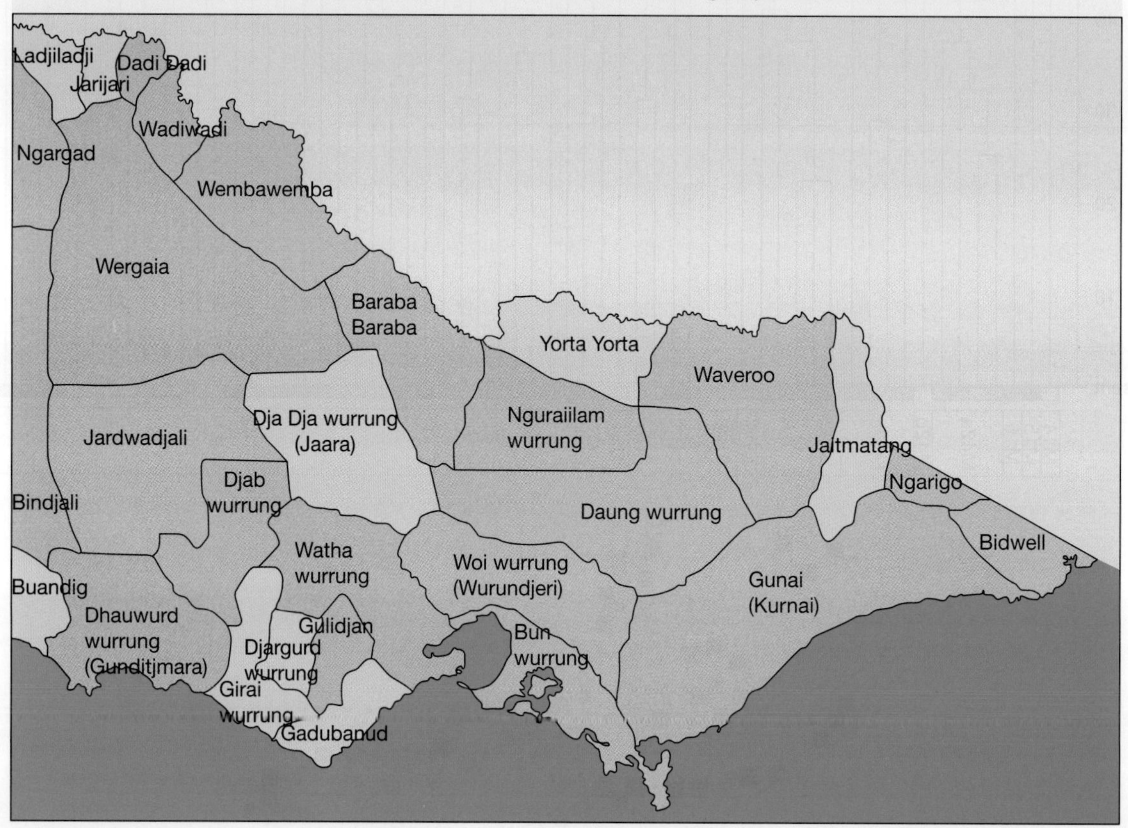

As the First Peoples of Australia, Aboriginal and Torres Strait Islander peoples have an indelible connection to the Country, which encompasses the land, sea, flora, fauna and its spirit. They have a long history of 'caring for Country' and this is exemplified by the traditional land management practices. These practices benefit the Australian ecosystems with fire management, conservation of biodiversity, and management of flora and fauna. All of these practices combined have allowed Aboriginal and Torres Strait Islander peoples to live sustainably in Australia for tens of thousands of years.

FIGURE 10.75 Aboriginal Australian rock art at the Bunjil Shelter

Some sites in Australia have more significant importance to the Aboriginal and Torres Strait Islander peoples and these are referred to as sacred sites. One significant sacred site in Victoria is Gariwerd, otherwise known as the Grampians National Park. This site has a high spiritual and cultural significance to the Djab Wurrang and Jardwadjali peoples. Gariwerd has the largest collection of Indigenous rock art in Victoria, which ranges in age from 1000 to 22 000 years old. This site provided food, water and shelter for the Djab Wurrang and Jardwadjali peoples. The rock art illustrates the peoples connection to the land and their important rituals (see figure 10.75).

tlvd-1813

SAMPLE PROBLEM 4 Exploring the migration of Aboriginal and Torres Strait Islander peoples

Aboriginal and Torres Strait Islander peoples have been living in Australia for over 50 000 years. The map depicts the movement of modern humans from East Africa to the rest of the world.

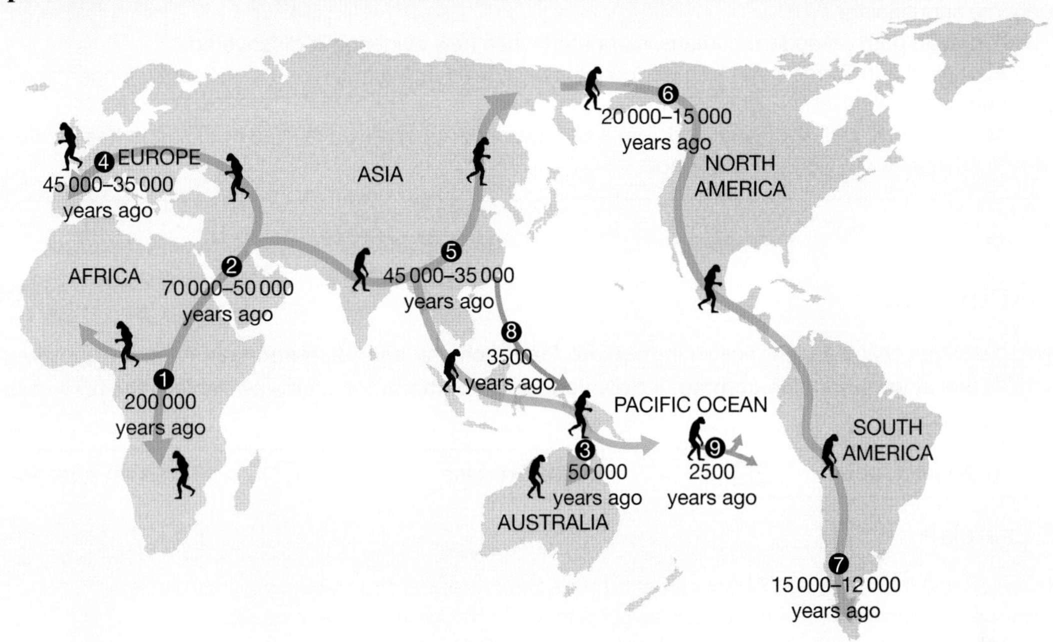

a. **Outline the migration path that the ancestors of Aboriginal and Torres Strait Islander peoples took to arrive in Australia.** **(1 mark)**

b. **Describe two different types of evidence scientists have used to identify the migration path and the timing of arrival of the Aboriginal and Torres Strait Islander peoples into Australia.** **(2 marks)**

THINK	WRITE
a. 1. In an *outline* question, you give a step-by-step account of what is occurring.	
2. Consider the start of the migration; the stem of the question states this is in East Africa. Trace the pathway from there to Australia. Name the major regions that were travelled.	East Africa → Middle East → Asia → Southeast Asia → Australia (1 mark)
b. 1. In a *describe* question, you give a detailed account in sentence form.	
2. The question is asking for two different types of evidence; these are fossils and DNA. These two types of evidence then need to be described with one sentence each.	Fossil evidence can be dated to discover how long a population was present (1 mark). DNA evidence can be used to compare the similarities and differences to other populations (1 mark).

on Resources

eWorkbook Worksheet 10.7 The migration of modern human populations (ewbk-8950)

Weblinks Aboriginal Australian mitochondrial genome variation
Australian Institute of Aboriginal and Torres Strait Islander Studies

KEY IDEAS

- The migration of modern humans out of Africa can be traced with fossil evidence.
- The migration of modern humans out of Africa can be traced with DNA evidence from mtDNA and whole genome sequencing.
- The migration path of modern humans is updated when new evidence is discovered.
- The Aboriginal and Torres Strait Islander peoples have lived in Australia for a significantly longer time than the first European colonisers.
- The length of time that Aboriginal and Torres Strait Islander peoples have lived in Australia has created a very strong connection to Country and Place.

10.5 Activities

learnon

To answer questions online and to receive **immediate feedback** and **sample responses** for every question, go to your learnON title at **www.jacplus.com.au**. A **downloadable solutions** file is also available in the resources tab.

10.5 Quick quiz on	10.5 Exercise	10.5 Exam questions

10.5 Exercise

1. Research and identify which Aboriginal and Torres Strait Islander tribe's land your school is on.
2. Describe the significance of the upper jawbone found in the Misliya cave in Israel.
3. Compare the evidence that mtDNA and whole genome DNA can give on the migration of modern humans.
4. Explain why the 'Out of Africa' model for human migration has been updated.
5. Explain how recent studies of mtDNA of Aboriginal and Torres Strait Islander peoples gives evidence for the length of time they have lived in Australia.

10.5 Exam questions

Question 1 (1 mark)

Source: *Adapted from VCAA 2009 Biology Exam 2, Section A, Q22*

MC Scientific opinion was once evenly divided regarding the geographical origin of the modern human. Two hypotheses were put forward – the 'Out of Africa' hypothesis and the multiregional hypothesis. In general, researchers now accept that the Out of Africa hypothesis is better supported by current information.

Findings from worldwide human fossil sites which would best support the Out-of-Africa hypothesis include

A. dating of fossils by radioactive uranium.
B. the degree of decomposition of remains.
C. the present-day climate of the region.
D. variations in mitochondrial DNA.

Question 2 (1 mark)

Source: *VCAA 2010 Biology Exam 2, Section A, Q14*

MC The 'Out of Africa' hypothesis

A. is also called the multiregional hypothesis.
B. proposes that mitochondrial DNA sequences are the same worldwide.
C. proposes that *Homo sapiens* first appeared in Africa and other continents at the same time.
D. proposes that *Homo erectus* evolved into *Homo sapiens* in Africa before migrating to other continents.

Question 3 (1 mark)

MC The 'Out of Africa' hypothesis state that modern humans evolved in Africa before migrating to other parts of the world. A piece of evidence that would better support the 'Out of Africa' hypothesis is

A. fossils of *Homo erectus* have been found in Northern Europe.
B. the proposal that genomes are the same worldwide.
C. proof that Neanderthals, Denisovans and *Homo sapiens* interbred.
D. more similarity in the DNA of Aboriginal Australian peoples when compared to Africans than when compared to Northern Europeans.

Question 4 (7 marks)

Source: *VCAA 2011 Biology Exam 2, Section B, Q7a–c*

The Mitochondrial Eve Hypothesis suggests that the mitochondrial DNA of all living people can be traced back to a few women in Africa.

a. **i.** Why is mitochondrial DNA useful for tracking human evolutionary history? **1 mark**
ii. Name another type of DNA that can be used to trace human ancestry. **1 mark**

It has been observed that

- samples of mitochondrial DNA taken from living humans are very similar to each other
- the greatest variation of mitochondrial DNA is observed within African populations.

b. Based on these observations, what are two inferences that can be made about human evolution? **2 marks**
c. **i.** Explain why DNA analysis cannot be used to trace very early hominin (previously called hominid) ancestry. **1 mark**
ii. Outline two types of evidence, other than DNA analysis, that can be used to determine the relatedness and age of early hominins. **2 marks**

Question 5 (9 marks)

Over time, there have been many fossil finds across the world, allowing us to better understand the migration of modern human populations around the world.

To establish when modern humans migrated to different regions of the world, molecular genetic studies can be used to determine when lineages separated. As part of these molecular genetic studies, both mitochondrial DNA (mtDNA) and whole genome analysis has been used.

a. Outline benefits of using mtDNA and benefits of using whole genomes for understanding migration patterns in modern human populations. **2 marks**
b. The particular mtDNA sequence within the D-loop of the mitochondria is referred to as a haplotype. These haplotypes are used to form haplogroups. Define the term *haplogroup* and outline why they are important. **2 marks**
c. Explain how haplogroups have been used to understand the migration of Aboriginal and Torres Strait Islander peoples. **2 marks**
d. The amount of Denisovan DNA in the genome of Aboriginal and Torres Strait Islander peoples is much greater when compared to those with European heritage. Outline the reason that this may be the case. **1 mark**
e. Recent research has found that the genetic diversity between Aboriginal Australian populations in the east of Australia and the west of Australia is much greater than expected. This genetic diversity is greater compared to populations in the Americas and Siberia. What does this suggest about the migration of Aboriginal Australian populations compared to populations in America and Siberia? **2 marks**

More exam questions are available in your learnON title.

10.6 Review

10.6.1 Topic summary

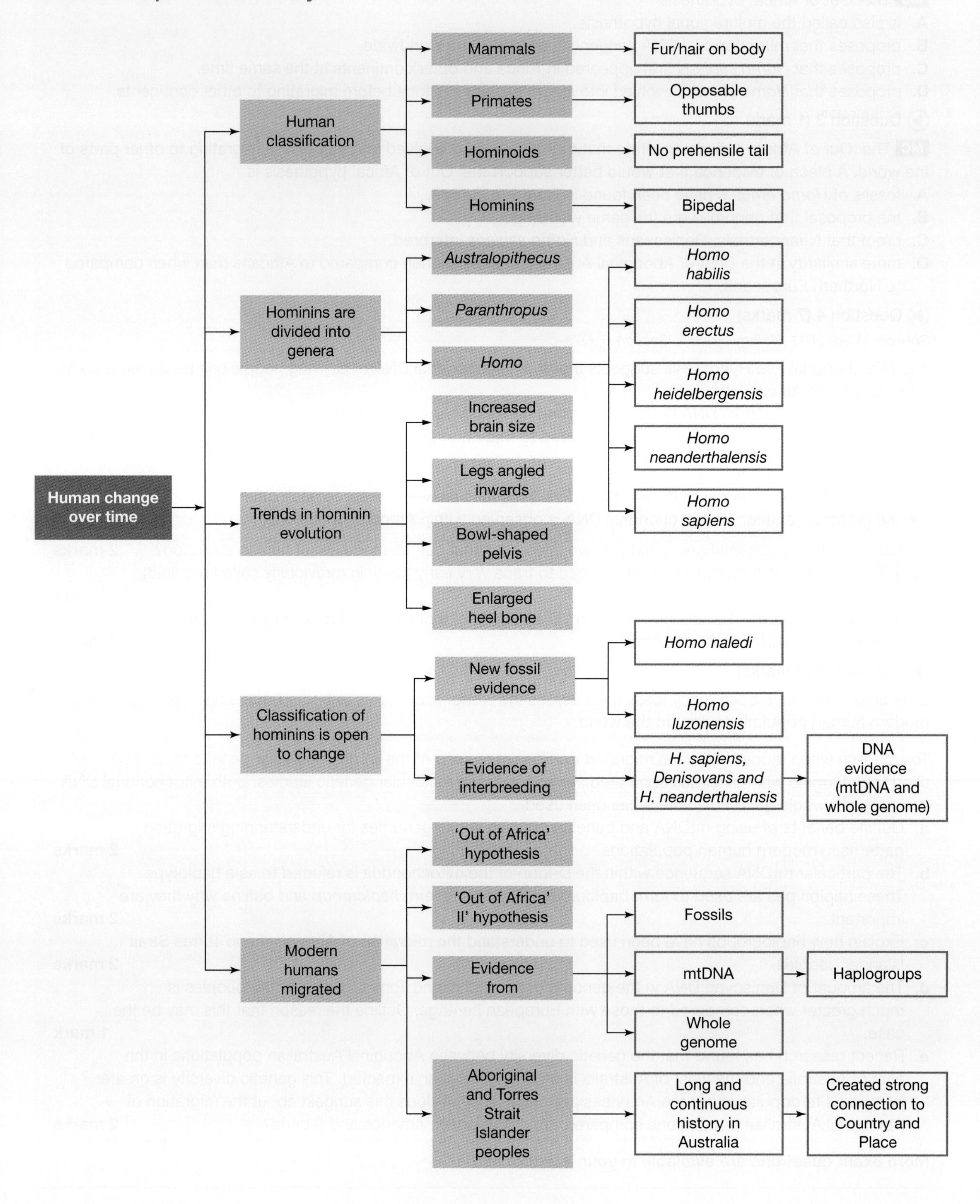

Resources

eWorkbook	Worksheet 10.8 Reflection — Topic 10 (ewbk-4550)
Practical investigation eLogbook	Practical investigation eLogbook — Topic 10 (elog-0197)
Digital documents	Key terms glossary — Topic 10 (doc-34624) Key ideas summary — Topic 10 (doc-34615)
Exam question booklet	Exam question booklet — Topic 10 (eqb-0021)

10.6 Exercises

learnON

To answer questions online and to receive **immediate feedback** and **sample responses** for every question, go to your learnON title at **www.jacplus.com.au**. A **downloadable solutions** file is also available in the resources tab.

10.6 Exercise 1: Review questions

1. Examine the figure showing the upper and lower jaws of an unnamed animal.

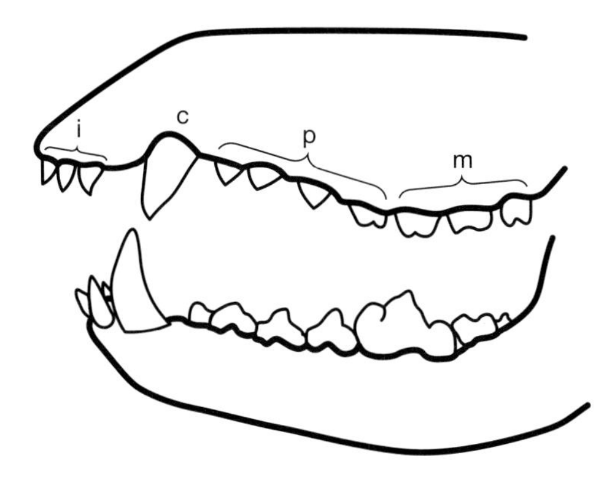

   a. On the basis of the information available, which of the following conclusions is reasonable?
      i. This jaw is from a reptile.
      ii. This jaw is from a mammal.
      iii. It is not possible to draw a conclusion from this figure.
   b. Identify the reason for your decision.
2. Examine the three fossil hominin skulls in the figure.

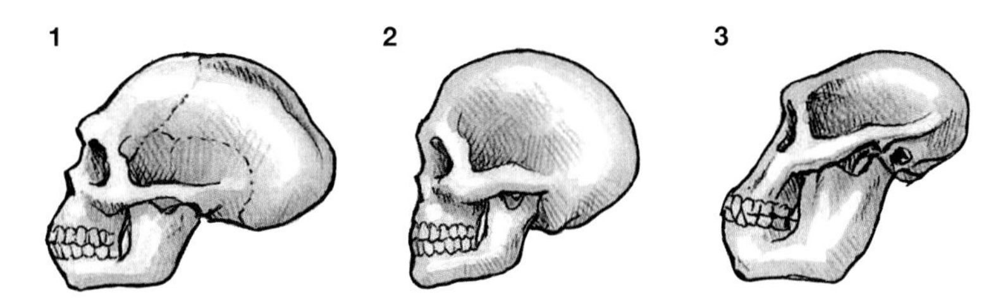

   Place the skulls in order from most ancient to most recent and justify your choice.
3. A skull is found at a dig site. One person says that it is the skull of a chimpanzee while a second person says that it is a hominin skull. Identify two key features that you might use to distinguish a chimpanzee skull from that of a hominin.

4. Discoveries of new putative *Homo* species can occur and this can rewrite the hominin classification schemes.
   a. What is the latest new putative *Homo* species discovery to be reported?
   b. Why was this species classified into the *Homo* genus?
5. Look at the provided figure, which shows layers of sediments and some fossil skulls.

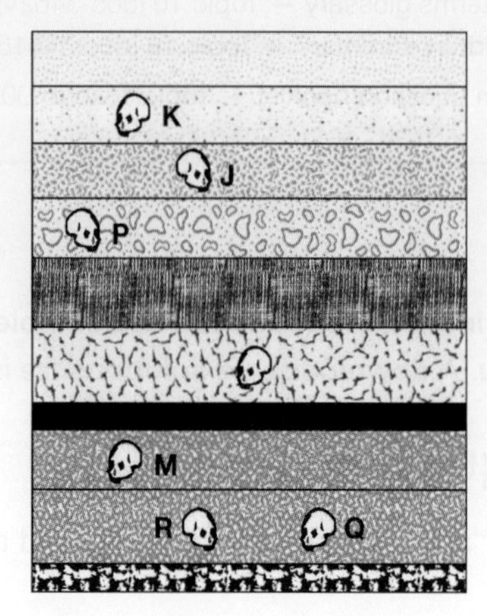

   a. Which fossil is older: M or P? Explain your choice.
   b. If fossil R were identified as *A. afarensis*, could fossil Q be *H. habilis*? Explain.
   c. Could fossils J and K be *H. sapiens* and *P. robustus* respectively? Explain.
6. Complete the table comparing the structural differences between hominoid species that walk with a quadrupedal versus bipedal locomotion.

Characteristic	Bipedal hominoid	Quadrupedal hominoid
Pelvis		
Angle of femur		
Size of heel bone		

7. Explain the advantages of using whole genome DNA testing when identifying the age of extinct hominin remains.
8. Describe how evidence from mtDNA haplogroups reveals the length of time Aboriginal and Torres Strait Islander peoples have lived in Australia.
9. Examine the feet of two different hominoids shown.
   a. Identify two other features of a skeleton that would enable it to be classified as a hominoid.
   b. Examine the feet to identify key features to match them to the hominoid species listed in the table. Justify your response.

Foot A

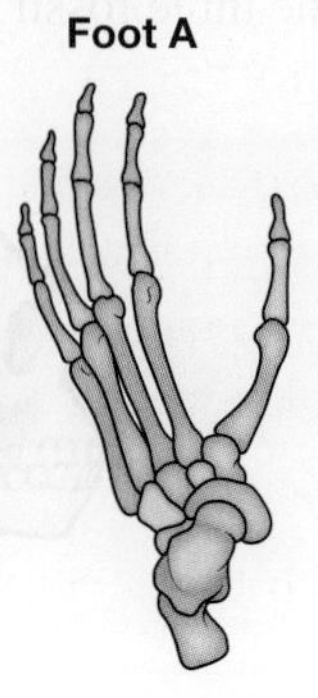

Foot B

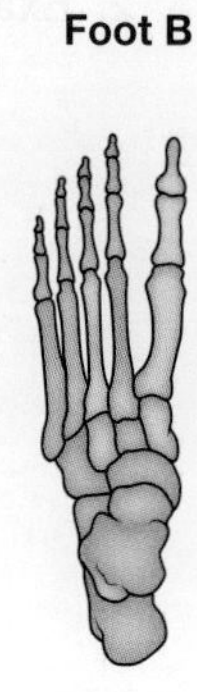

Hominoid species	Foot	Justification
Human (*Homo sapiens*)		
Chimpanzee (*Pan* sp.)		

10.6 Exercise 2: Exam questions

on Resources

Teacher-led videos Teacher-led videos for every exam question

Section A — Multiple choice questions

All correct answers are worth 1 mark each; an incorrect answer is worth 0.

Use the following information to answer Questions 1 and 2

A hominin species, *Homo floresiensis*, was identified from fossils found on an isolated Indonesian island. These fossils were dated to be 18 000 years old.

The adult skull of this upright bipedal hominin had a cranial volume less than one-third of the average cranial volume of a modern adult human. It had harder, thicker eyebrow ridges than *Homo sapiens*, a sharply sloping forehead and no chin.

H. floresiensis was just over one metre tall and their arm-to-leg ratio was slightly larger than modern humans. They weighed approximately 16 kg.

The fossils were found in sediment that also contained stone tools and fireplaces for cooking. The fireplaces contained the burnt bones of animals, each animal weighing more than 350 kg. The stone tools included blades, spearheads, and cutting and chopping tools.

Question 1

Source: *VCAA 2014 Biology Exam, Section A, Q39*

The above evidence and current theories of hominin evolution indicate that

A. the very small stature of *H. floresiensis* is unexpected in a species that is geographically isolated on a small island.
B. *H. floresiensis* was a species that showed considerable social cooperation and methods for passing on knowledge.
C. the sloping forehead and absence of chin of members of the *H. floresiensis* species suggest they were members of the first migration of hominins into Australia over 50 000 years ago.
D. the skull shape and size of *H. floresiensis* suggest a close relationship to the *Homo neanderthalensis* species.

Question 2

Source: *VCAA 2014 Biology Exam, Section A, Q40*

The fossils of *H. floresiensis* showed that they had opposable thumbs.

The development of an opposable thumb in primate evolution

A. is used to distinguish members of the genus *Homo* from the other great apes.
B. was a necessary step in the development of bipedalism in hominins.
C. was an important anatomical development that assisted tool-making in hominins.
D. is a significant factor in determining the arm-to-leg ratio of modern humans.

Question 3

Which of the following is **not** a feature of all primates?

A. Prehensile tail
B. Long gestation period
C. Stereoscopic 3-dimensional vision
D. Opposable thumbs

Question 4

Source: VCAA 2015 Biology Exam, Section A, Q38

Fossil remains of a number of individuals from the genus *Australopithecus* were found at various sites in the eastern half of Africa and have been dated to between 3–4 million years old.

These fossil remains

A. are descendants of *Homo erectus*.
B. represent the oldest evidence found of primates.
C. show early evidence that hominins were bipedal.
D. represent the earliest examples of the hominoid super-family.

Question 5

Source: Adapted from VCAA 2011 Biology Exam 2, Section A, Q25

Consider the following hominin species.

M *Homo erectus*
N *Homo neanderthalensis*
O *Australopithecus africanus*
P *Australopithecus afarensis*

The order of oldest to youngest hominin species is

A. M N O P.
B. P O M N.
C. N O P M.
D. O M P N.

Question 6

Source: VCAA 2017 Biology Exam, Section A, Q36

In recent years, scientists have discovered that Neanderthals took care of their elderly relatives, used burial rituals for their dead and gave symbolic meaning to natural objects. Studies have also shown that Neanderthals used complex methods to obtain sharp stone implements and produce glues to attach sharp stones to spears.

These discoveries suggest that Neanderthals

A. had bigger brains than previously thought.
B. had a highly developed culture.
C. lived a solitary lifestyle.
D. used a written language.

Question 7

Source: VCAA 2007 Biology Exam 2, Section A, Q24

Complex social activities require articulate speech. Evidence found with early *Homo* fossils suggests that this genus was the first to use articulate speech.

This suggestion would be best supported by evidence of

A. burial ceremonies.
B. use of stone tools.
C. living in groups.
D. organised hunting of prey.

Question 8

Fossil evidence reveals certain information about hominin evolution, such as the fact that

A. ape-like, erect-walking species appeared earlier than larger-brained human species.
B. just one hominin species existed at any one time.
C. hominin evolution followed a straight-line pathway.
D. hominin evolution originated at many points worldwide.

Question 9

A fossilised hominoid skeleton has legs angled inwards, curved fingers and a small brain size.

The skeleton would most likely be classified as

A. *Australopithecus afarensis.*
B. *Homo habilis.*
C. *Homo erectus.*
D. *Homo sapiens.*

Question 10

Current scientific theory states that Denisovans

A. are not members of the genus *Homo.*
B. lived around the same time as *Homo neanderthalensis.*
C. were isolated and did not breed with other groups.
D. are more closely related to *Homo sapiens* than to *Homo neanderthalensis.*

Section B — Short answer questions

Question 11 (3 marks)

Source: *VCAA 2009 Biology Exam 2, Section B, Q6a, b*

The press recently reported:

'Anthropologists have uncovered ancient fossil footprints in Kenya dating back 1.5 million years, the oldest evidence that indicates our ancestors walked like present-day humans ...'

a. Give one significant feature of the footprints that would have led anthropologists to this conclusion. **1 mark**

The pictures below show views of skulls from *Homo erectus* and *Homo sapiens*.

skull set 1

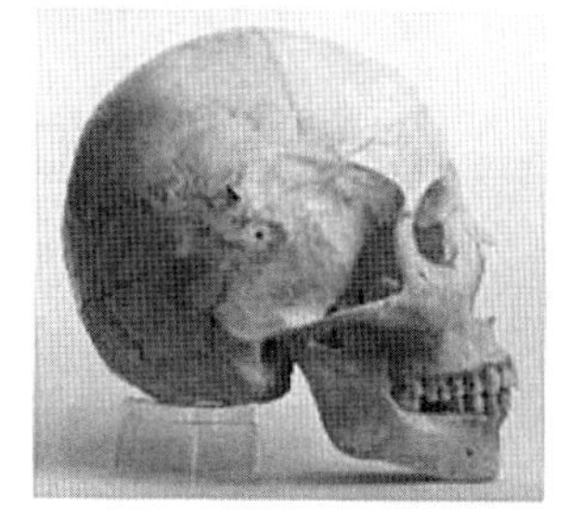

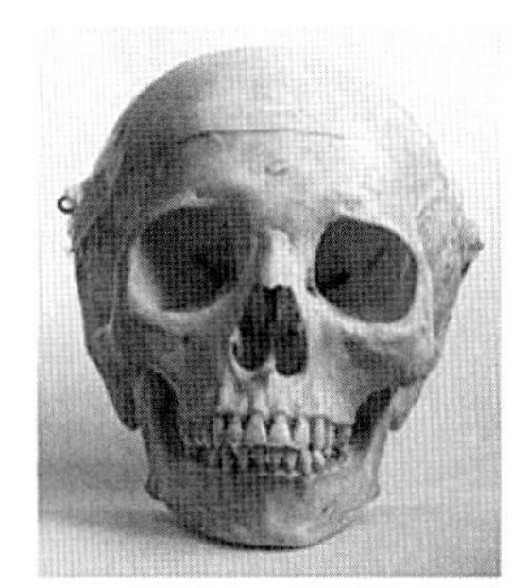

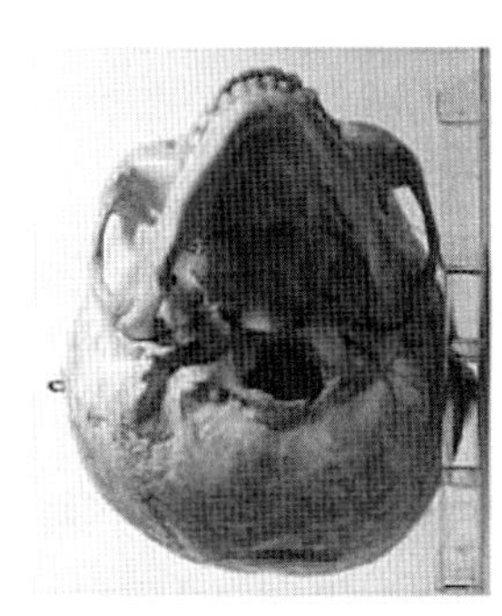

skull set 2

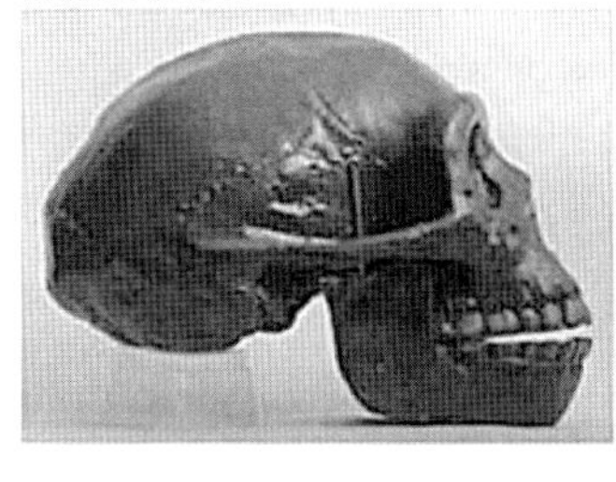

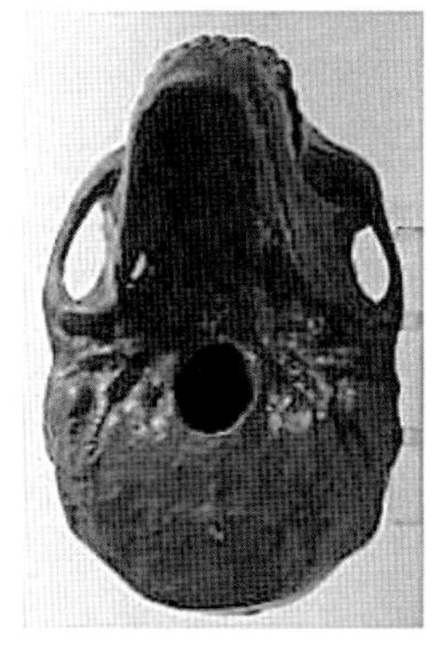

b. With reference to two structural features of the skull, which skull set represents *Homo erectus*? Justify your choice. **2 marks**

Question 12 (7 marks)

Source: *Adapted from VCAA 2020 Biology Exam, Section B, Q9*

In 1931 a team of Dutch archaeologists unearthed 12 skull and two leg bones at a site in Ngandong, Indonesia.

a. Describe **two** key structural features of these fossils that would suggest that the skull were from *Homo erectus* rather than from *Homo sapiens*. **2 marks**

Figure 1 and Figure 2 below show Homo evolutionary trees that were created in 1997 and 2013 respectively.

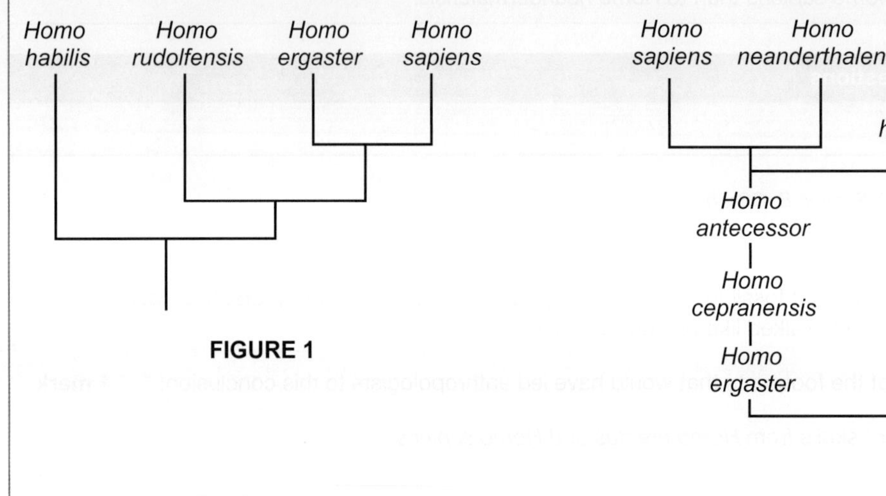

FIGURE 1

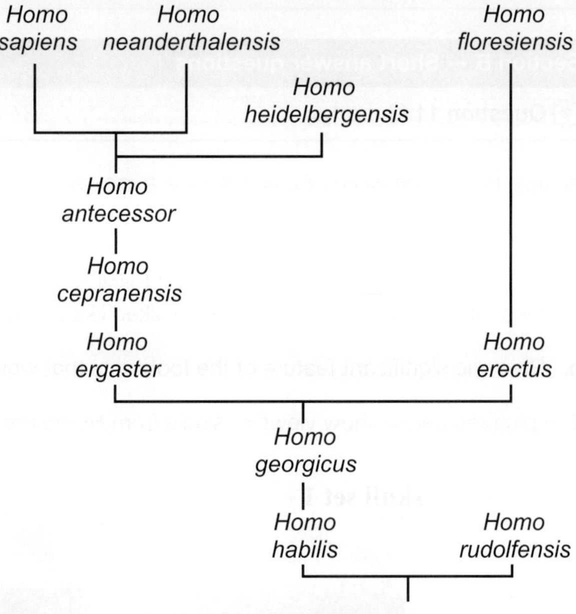

FIGURE 2

Sources (from left): adapted from DS Strait, FE Grine and MA Moniz, 'A reappraisal of early hominid phylogeny', *Journal of Human Evolution*, volume 32, issue 1, January 1997, pp. 17–82; C Schultz, '*Homo Sapiens*' family tree may be less complicated than we thought', *Smithsonian Magazine*, 18 October 2013, <www.smithsonianmag.com>

b. State **one** difference between Figure 1 and Figure 2 with respect to the evolutionary relationships between *Homo ergaster* and *Homo sapiens*. **1 mark**

Homo denisovans is one species that is missing from both evolutionary trees shown.

c. Draw a branch on Figure 2, the 2013 *Homo* evolutionary tree, that represents where you think *Homo denisovans* should be placed. Justify why you have put the branch in that position. **2 marks**

d. Scientists do not always agree on the position of each *Homo* species on evolutionary trees.

Explain why this is the case. **2 marks**

Question 13 (3 marks)

In 2002, an upper jawbone of an ancient *Homo sapiens* was discovered in the Misliya cave in Israel. The discovery of the human bone in this area was not surprising as many ancient bones had been found around there. What was surprising to scientists was the age of the fossil, which was approximately 180 000 years old. This was the oldest human fossil ever found outside of Africa and predates the earlier estimates of humans leaving Africa by 40 000 years. This left scientists with the need to rethink the theories on how modern humans migrated out of Africa and if there was more than one migration wave leaving Africa.

a. The upper jawbone was dated to approximately 180 000 years old using a radiometric dating technique with uranium isotopes.
Explain how scientists would have calculated the age of this upper jawbone. **2 marks**

b. Since the ancient upper jawbone was found in the Misliya cave, the theory of when modern humans left Africa has been revised.
Compare the hypotheses of the 'Out of Africa' and 'Out of Africa II' theories in relation to modern human migration. **2 marks**

Question 14 (5 marks)

Source: *VCAA 2012 Biology Exam 2, Section B, Q8a–c*

In 2008, two incomplete, fossilised skeletons were found in cave deposits in South Africa. The scientists compared the newly discovered bones with those of members of the genus *Australopithecus*, early *Homo*, modern humans and apes. The fossilised skeletons, named *Australopithecus sediba*, displayed an unusual mix of characteristics. They partly resembled primitive, ape-like animals.

Like the apes, *A. sediba* was of small stature. Scientists determined that *A. sediba*, like apes, was suited to climbing in trees.

a. What feature of the *A. sediba* skeleton allowed scientists to reach this conclusion? **1 mark**

A. sediba was found to have many characteristics in common both with the members of the genus *Australopithecus* and with the genus *Homo*. Two characteristics shared with the genus *Homo* included a projecting nose and hands with a precision grip.

b. Explain how each of these characteristics may have given *A. sediba* an advantage over other *Australopithecus* species. **2 marks**

Below is a drawing of the skull of a modern-day human.

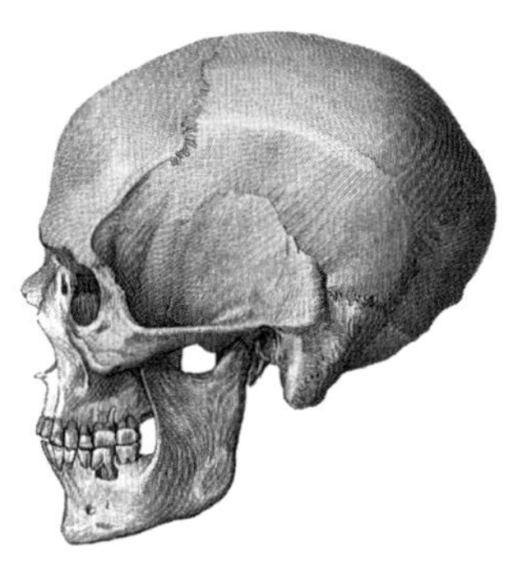

c. Using the drawing, suggest one feature of modern-day humans that makes humans more advanced than other *Homo* species. Explain the significance of this feature to the evolution of modern humans. **2 marks**

Question 15 (8 marks)

Source: *Adapted from VCAA 2015 Biology Exam, Section B, Q11*

Fossil evidence indicates that between 30 000–80 000 years ago, populations of the two hominin species – modern humans (*Homo sapiens*) and the extinct Neanderthals (*Homo neanderthalensis*) – lived close to one another in parts of the Middle East, Europe and Asia.

Researchers have constructed a theory about the relationships between ancient populations. This is represented in the following diagram.

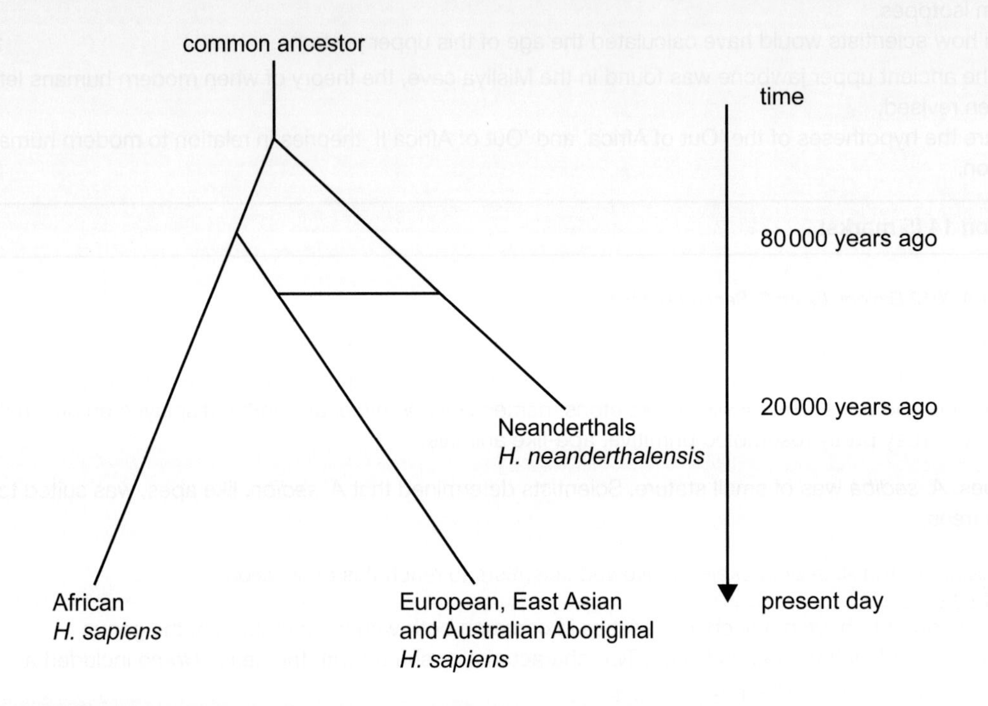

Recent DNA evidence has shown that:

- the genome of living humans of African descent does not contain Neanderthal DNA
- the genomes of living humans of European, East Asian and Australian Aboriginal descent all contain small amounts of Neanderthal DNA (1–4%).

a. **i.** Suggest how DNA from *H. neanderthalensis* entered the genome of present-day European, East Asian and Australian Aboriginal *H. sapiens*, and continues to be found in modern populations. **2 marks**

ii. What implication does this DNA evidence have for the classification of the two hominin species, *H. sapiens* and *H. neanderthalensis*, according to the common definition of a species? **1 mark**

b. There are several theories about the geographical origins of *H. sapiens*.
Scientists consider that the absence of Neanderthal DNA in present-day African *H. sapiens* lends support to one theory about the geographical origins of *H. sapiens*.

Name this theory and explain how the recent DNA evidence supports it. **3 marks**

Consider the map provided below.

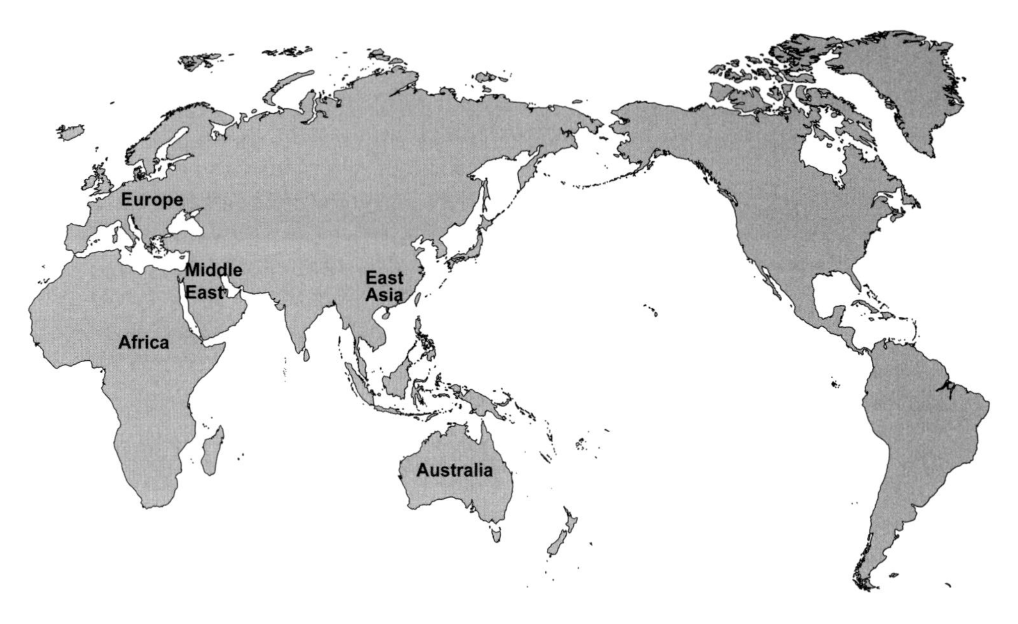

c What does the DNA evidence suggest about the route and timing of the migration of the first Australian Aboriginals to arrive in Australia? **2 marks**

10.6 Exercise 3: Biochallenge 

on Resources

eWorkbook Biochallenge — Topic 10 (ewbk-8091)

Solutions Solutions — Topic 10 (sol-0666)

Past VCAA examinations

Sit past VCAA examinations and receive immediate feedback, marking guides and examiner's report notes.
Access Course Content and select 'Past VCAA examinations' to sit the examinations online or offline.

teachon

Test maker
Create unique tests and exams from our extensive range of questions, including past VCAA questions.
Access the assignments section in learnON to begin creating and assigning assessments to students.

Online Resources

Below is a full list of **rich resources** available online for this topic. These resources are designed to bring ideas to life, to promote deep and lasting learning and to support the different learning needs of each individual.

eWorkbook

- **10.1** eWorkbook — Topic 10 (ewbk-1889) ☐
- **10.2** Worksheet 10.1 Mammals, primates and hominoids (ewbk-4105) ☐
  - Worksheet 10.2 Who's in the family (ewbk-4107) ☐
- **10.3** Worksheet 10.3 Hominin relationships and evidence from skulls (ewbk-4109) ☐
  - Worksheet 10.4 Summarising trends in hominin evolution (ewbk-8944) ☐
- **10.4** Worksheet 10.5 Alternative pathways (ewbk-8946) ☐
  - Worksheet 10.6 Interbreeding and new putative species (ewbk-8948) ☐
- **10.5** Worksheet 10.7 The migration of modern human populations (ewbk-8950) ☐
- **10.6** Worksheet 10.8 Reflection — Topic 10 (ewbk-4550) ☐
  - Biochallenge — Topic 10 (ewbk-8091) ☐

Solutions

- **10.6** Solutions — Topic 10 (sol-0666) ☐

Practical investigation eLogbook

- **10.1** Practical investigation eLogbook — Topic 10 (elog-0197) ☐
- **10.2** Investigation 10.1 Observing primates (elog-0427) ☐
- **10.3** Investigation 10.2 Investigating hominin fossils (elog-0429) ☐

Digital documents

- **10.1** Key science skills — VCE Units 1–4 (doc-34326) ☐
  - Key terms glossary — Topic 10 (doc-34624) ☐
  - Key ideas summary — Topic 10 (doc-34615) ☐
- **10.3** Case study: A detailed look at early hominin species (doc-36149) ☐
  - Case study: A detailed look at members of the genus *Homo* (doc-36150) ☐

Teacher-led videos

- Exam questions — Topic 10 ☐
- **10.2** Sample problem 1 Comparing skulls of hominins and hominoids (tlvd-1808) ☐
- **10.3** Sample problem 2 Determining hominoid species from their limb structure (tlvd-1809) ☐
- **10.3** Sample problem 3 Analysing *Homo* fossils (tlvd-1812) ☐
- **10.4** Sample problem 4 Exploring the migration of Aboriginal and Torres Strait Islander peoples (tlvd-1813) ☐

Video eLessons

- **10.3** Changes in hominin skulls (eles-4632) ☐
- **10.5** Mitochondrial DNA (eles-4204) ☐

Interactivities

- **10.3** Hominin skulls (int-0241) ☐
  - Hominin limb structure (int-8330) ☐

Weblinks

- **10.2** Humans are primates ☐
  - Hominid and hominin — what's the difference? ☐
  - Australian Museum Human Evolution resources ☐
- **10.3** Australian Museum — Hominin skulls ☐
- **10.4** Human lineage over time ☐
  - Human evolution timeline interactive ☐
  - New species of ancient human discovered in the Philippines ☐
  - Ancient girl's parents were two different species ☐
  - DNA reveals first look at enigmatic human relative ☐
- **10.5** The Great Human Migration ☐
  - Aboriginal Australian mitochondrial genome variation ☐
  - Australian Institute of Aboriginal and Torres Strait Islander Studies ☐

Exam question booklet

- **10.1** Exam question booklet — Topic 10 (eqb-0021) ☐

Teacher resources

There are many resources available exclusively for teachers online.

To access these online resources, log on to **www.jacplus.com.au**

AREA OF STUDY 2 How are species related over time?

OUTCOME 2

Analyse the evidence for genetic changes in populations and changes in species over time, analyse the evidence for relatedness between species, and evaluate the evidence for human change over time.

PRACTICE EXAMINATION

STRUCTURE OF PRACTICAL EXAMINATION		
Section	**Number of questions**	**Number of marks**
A	20	20
B	7	30
	Total	**50**

Duration: 50 minutes
Information:
- This practice examination consists of two parts. You must answer all question sections.
- Pens, pencils, highlighters, erasers and rulers are permitted.

SECTION A — Multiple choice questions

All correct answers are worth 1 mark each; an incorrect answer is worth 0.

1. The term 'gene pool' refers to the
A. genotype of an individual in a population.
B. genome of a species.
C. diploid chromosome number of a species.
D. genetic make-up of a population.

2. In a population, the frequency of allele *A* is 0.4 and that of allele *a* is 0.6. This statement describes a feature of
A. gene pool of a population.
B. genotype of an individual in a population.
C. gene pool of an individual in a population.
D. genome of a species.

3. The diagram shows the gene pool of a population over 20 generations.

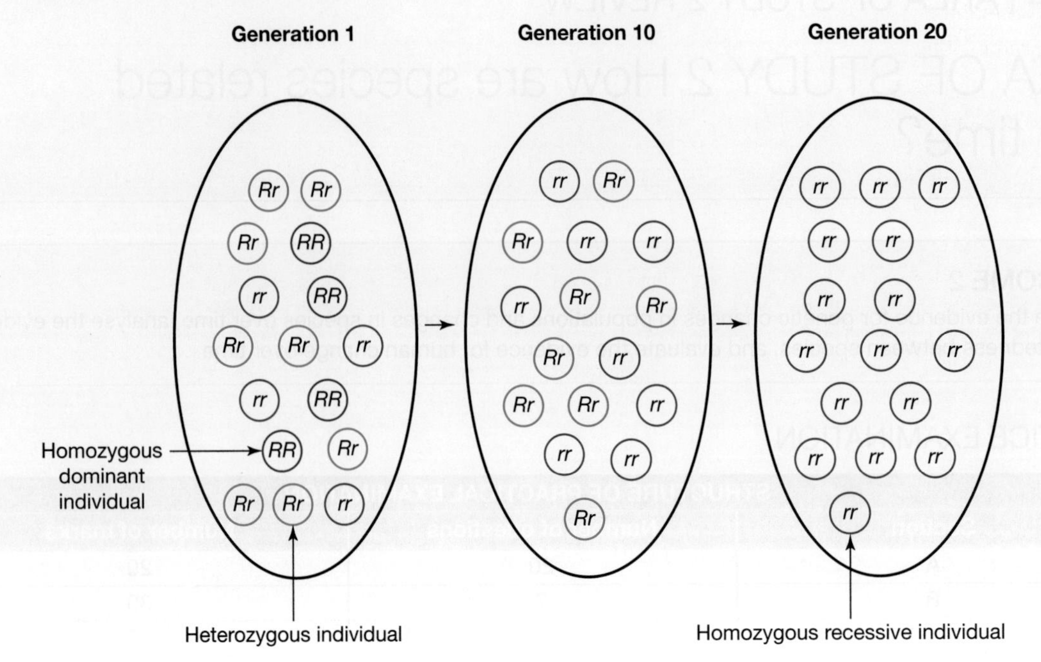

It is reasonable to say that

A. genetic diversity is increasing in the population.

B. genetic diversity is decreasing in the population.

C. the frequency of each allele remains unchanged in 20 generations.

D. individuals with genotype *RR* are at selective advantage.

4. The diagram shows change in the phenotype frequencies of an insect population. Figure **a** shows random mating in an insect population that includes pesticide-resistant (shown in red) and pesticide-sensitive insects. Figure **b** shows a sample of the insect population after several generations in the presence of pesticide.

It is reasonable to state that

A. the selecting agent is a pesticide.

B. the frequency of pesticide-sensitive insects increases after number of generations.

C. the selecting agent is a biological agent.

D. there is no change in the frequency of pesticide resistance over several generations in the two types of insects.

5. A DNA segment changes from GGT TAG to GGA TAG. What is this an example of?
   A. Nucleotide base insertion
   B. Frameshift mutation
   C. Nucleotide base deletion
   D. Nucleotide base substitution
6. X-rays, UV radiation and cigarette smoke can lead to errors being introduced in the DNA of a cell. What are X-rays, UV radiation and cigarette smoke examples of?
   A. Mutations
   B. Mutagenesis
   C. Mutagens
   D. Selection pressures
7. Through various actions, human beings are able to intervene in natural evolutionary processes. This intervention
   A. became possible only in the late twentieth century.
   B. can be achieved only by manipulation of the genotype.
   C. can be achieved only by manipulation at the phenotypic level.
   D. may be achieved through selective breeding.
8. The fossil record provides evidence that the types of organisms living on Earth changed over time. Which list places groups of organisms in the correct order according to their first appearance in the fossil record?
   A. Mammals, fish, reptiles, eukaryotes, prokaryotes, flowering plants, worms
   B. Prokaryotes, worms, fish, amphibians, ferns, flowering plants, birds
   C. Crustaceans, seaweeds, fish, reptiles, dinosaurs, horses, insects, flowering plants
   D. Fish, worms, unicellular organisms, multicellular organisms, worms, mammals
9. How is it possible to identify the relative age of a rock layer?
   A. Using a radiometric technique
   B. Using index fossils
   C. Identifying the rock type as sedimentary
   D. Examining the geologic time scale
10. It is reasonable to state that the absolute dating of the age of rocks
   A. could be done before relative dating was possible.
   B. requires a large sample of the rock to be dated.
   C. relies on the presence of fossils in the rocks to be dated.
   D. relies on the decay of radioactive elements in minerals in the rocks to be dated.

11. The image below shows a dinosaur footprint. What is it an example of?

A. Preserved remains
B. Trace fossil
C. Cast
D. Physical fossil

12. The saguaro cactus has features that equip it for survival in arid desert conditions. These cacti occur naturally only in American deserts. On other continents, such as Australia, other plant species are native to desert regions.
It is reasonable to conclude that saguaro cacti do not occur naturally in Australian deserts because
A. cacti were specially created for American deserts and different species were specially created for Australian deserts.
B. cacti were once present in Australian deserts but, at some time in the past, they became extinct.
C. the two desert regions were occupied by different ancestral species that evolved over time into species distinct to each region.
D. cacti migrated by seed dispersal from Australian deserts to American deserts.

13. Similar structures that appear in different species are said to be homologous. It is reasonable to predict that the similarity of such structures is due to the fact that
A. the structures have a similar function.
B. the species share a common ancestry.
C. the structures evolved independently.
D. homology is a result of chance.

14. Whales are found to have hip bones that have no current function but are evidence of evolution. Hip bones in whales are an example of
A. a homologous organ.
B. a biosignature.
C. an analogous structure.
D. a vestigial structure.

15. Divergent evolution occurs when populations of an ancestral species have been separated and exposed to different environmental selective pressures over a long time period. What is the result of divergent evolution?
A. At least one new species arises.
B. Only the ancestral species continues to survive.
C. All the separated populations become extinct.
D. Similar adaptations evolve in all the populations.

16. Human fossils are often difficult to interpret. One reason for this is because
  A. palaeontolgists lack sufficient training to make reliable interpretations.
  B. many fossils are incomplete due to disturbance or environmental factors.
  C. there are so many fossils that it is difficult to choose the best one to study.
  D. the technology to study fossils is inadequate.

17. Species A and species B are closely related by evolutionary descent. Another two species, M and N, are closely related to each other but are only distantly related to species A and B.
It may be confidently predicted that
  A. species M and N shared a common ancestor more recently than species A and B.
  B. the amino acid sequence of an enzyme from A would be more similar to the corresponding sequence from B than to that from N.
  C. a DNA sequence from species M would be more similar to the corresponding sequence from B than to that from N.
  D. the banding patterns of chromosomes from A and B would show more similarities than the banding patterns of M and N.

18. The genomes of six species were compared and the fractions of matching DNA sequences were identified. Based on this information, the following genome phylogeny was inferred.

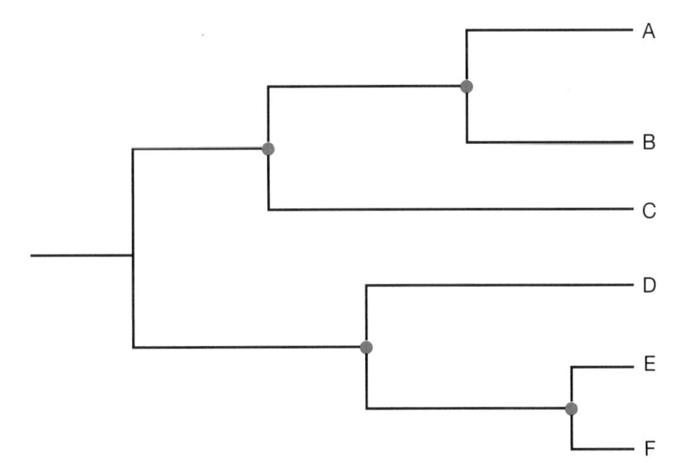

Using this inferred phylogeny, it is reasonable to conclude that the percentage of matching DNA sequences in the genome was
  A. greater in species C and D than in species A and C.
  B. greater in species D and E than in species E and F.
  C. greater in species B and C than in A and B.
  D. greater in species E and F than in species A and B.

19. Humans are classified as primates. What is a feature that is specific to all primates, including humans, but is not found in other mammals?
  A. Fur
  B. An embryonic tail
  C. Claws
  D. Stereoscopic vision

20. Evidence that the earliest hominins could walk erect comes from studies of their fossils. Confirmation that they walked erect came from examining
  A. a short femur (thigh bone) relative to the humerus (arm bone).
  B. the position of the junction of the lower jaw with the skull.
  C. patterns of teeth revealed by scanning electron microscopy.
  D. the position of the foramen magnum on the under surface of the skull, which is more central compared to other primates.

SECTION B — Short answer questions

Question 21 (3 marks)

The table shows three different species sharing a common ancestor.

Species	Tasmanian tiger	Dog (maltese terrier)	Blue whale
Mass	12–30 kg	3–4 kg	50 000–150 000 kg

a. Identify and describe the type of evolution shown, with reference to the provided table. **2 marks**

b. The death in 1936 of the last Tasmanian tiger (*Thylacinus cynocephalus*) marked the loss of the genus *Thylacinus* and the family Thylacinidae from this planet. Recent analysis of mitochondrial DNA suggested there was low genetic diversity in thylacine populations. Suggest a reason that this may have lead to their extinction. **1 mark**

Question 22 (4 marks)

At 10–12 weeks into pregnancy, a couple decided to get prenatal testing done to detect any mutations in a fetus.

The following codon chart can be used to determine the amino acids coded.

second letter

First letter	U	C	A	G	Third letter
U	UUU, UUC — Phe UUA, UUG — Leu	UCU, UCC, UCA, UCG — Ser	UAU, UAC — Tyr UAA Stop UAG Stop	UGU, UGC — Cys UGA Stop UGG Trp	U C A G
C	CUU, CUC, CUA, CUG — Leu	CCU, CCC, CCA, CCG — Pro	CAU, CAC — His CAA, CAG — Gln	CGU, CGC, CGA, CGG — Arg	U C A G
A	AUU, AUC, AUA — Ile AUG — Met	ACU, ACC, ACA, ACG — Thr	AAU, AAC — Asn AAA, AAG — Lys	AGU, AGC — Ser AGA, AGG — Arg	U C A G
G	GUU, GUC, GUA, GUG — Val	GCU, GCC, GCA, GCG — Ala	GAU, GAC — Asp GAA, GAG — Glu	GGU, GGC, GGA, GGG — Gly	U C A G

a. One type of mutation that can occur is known as a frameshift mutation. Describe what is meant by the term frameshift mutation. **1 mark**

b. Genetic abnormalities, such as substitution mutations, can be detected in prenatal screening. Name two potential outcomes of a substitution mutation. **2 marks**

c. The sequence below shows a section of gene without mutation.
AUG GCC UAU CGA GGG AUA CCA AAU
The section undergoes a mutation and the codon in the fifth position, GGG, is changed to UGG. Explain the consequence of the mutation. **1 mark**

Question 23 (4 marks)

A particular new allele appears in a closed animal population.

a. Contrast the source of new alleles in a closed population with the source of new alleles in an open population. **2 marks**

b. Two sources of changing alleles are gene flow and selection pressures. Suggest which of these is likely to act more quickly in altering allele frequencies. Justify your response. **2 marks**

Question 24 (4 marks)

The fossil record is the record of past life preserved in rock, and this record shows that the types of organisms living on Earth have changed over time.

a. Compare and contrast the absolute age of a rock layer and its relative age. **2 marks**

b. Give two reasons that explain why carbon-14 dating is not useful for calculating the absolute age of an igneous rock. **2 marks**

Question 25 (3 marks)

The diagram shows the life cycle of a frog.

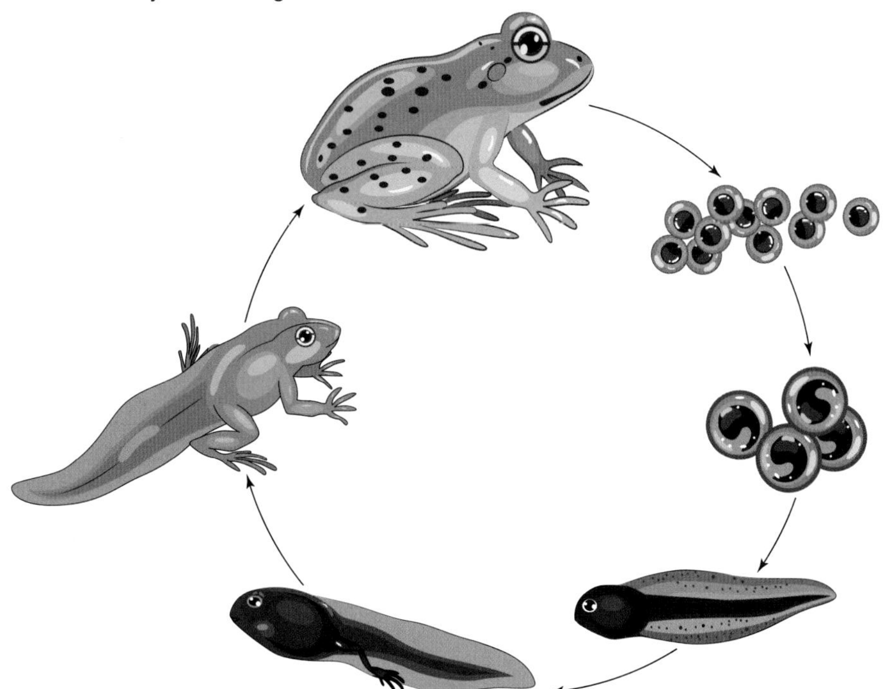

A frog develops from an embryo inside an egg, into a tadpole that has a tail and gills. The tadpole hatches from the egg to swim freely in its freshwater habitat. The tadpole survives by eating algae. Over time, it gradually loses its tail and gills while developing legs and lungs. It then emerges from the water to breathe air and to catch flying insects for food.

a. Name two structural similarities that exist in the early embryo of a human and in the early life stages of the frog. **2 marks**

b. What is the significance of your answer to part **a** in terms of the evolutionary relationship between frogs and humans? **1 mark**

Question 26 (5 marks)

The amino acid similarities of a conserved protein between three species compared to each other and humans are shown in the table.

	Humans	Species A	Species B	Species C
Humans		76%	87%	92%
Species A			70%	73%
Species B				90%
Species C				

a. Draw a clear phylogenetic tree showing the evolution of these species. **1 mark**

b. Which two species had the most recent common ancestor? Justify your response. **2 marks**

c. Explain why amino acid sequencing is considered a more accurate way of producing a phylogenetic tree than anatomical structures. **2 marks**

Question 27 (7 marks)

Consider the following statement: The line that gave rise to humans and the line that gave rise to the modern African apes diverged about 5 to 7 million years ago.

a. Which of the following comments correctly explains the statement above? **1 mark**

Comment 1: This statement means that humans evolved from modern African apes over a period of 5 to 7 million years.

Comment 2: This statement means that humans and modern African apes last shared a common ancestor about 5 to 7 million years ago.

b. Briefly explain your choice for part **a**. **1 mark**

c. A hominin skull has been unearthed at a historical burial site. List three features that would allow you to determine if it belonged to a modern human species or an earlier human species. **3 marks**

d. A nearly complete hominin skeleton was found at another location. This fossil was determined to be of the *Australopithecus* genus. With reference to limb structure, describe features that would have supported this idea. **2 marks**

END OF EXAMINATION

PRACTICE SCHOOL-ASSESSED COURSEWORK

ASSESSMENT TASK — Analysis and evaluation of collated secondary data

In this task, you will analyse the evidence for genetic changes in populations and changes in species over time, analyse the evidence for relatedness between species, and evaluate the evidence for human change over time.

- Students are permitted to use pens, pencils, highlighters, erasers, rulers and a scientific calculator.

Total time: 70 minutes (10 minutes reading, 60 minutes writing)
Total marks: 50 marks (to obtain score out of 40, divide raw score by 1.25)

TASK

Complete the series of questions shown. You should attempt to answer all questions.

Question 1 (25 marks)

Table 1 shows the first 60 bases of the nucleotide sequence from the cytochrome *b* gene in various species of the order Cetaeca (comprising marine mammals such as whales, dolphins and porpoises).

TABLE 1 First 60 nucleotides in the cytochrome *b* sequence of mtDNA of cetacean species

Species	Cytochrome *b* nucleotide sequence
Common bottlenose dolphin (*Tursiops truncatus*)	ATA CGA TAT GTC CTC CCG TTA GGC CAA ATA TCA TTC TGA GGG GCC ACA GTA ATT ACA AAC
Bowhead whale (*Balaena mysticetus*)	ATG GCA TAT GTC CAC CCG TGA GGC CAA ACA TCA TTC TGA CGG GCC GCA GTA ATT ACA TAC
Humpback whale (*Megaptera novaeangliae*)	ATG GCA TAT GTC CTC CCG TGA GGC CAA ACA TCA TTC TGA GGG GCC GCA GTA ATT ACA TAC
Common minke whale (*Balaenoptera acutorostrata*)	ATG GCA TAT GTC CTC CCG TGA GGC CAA ACA TCA TTC TGA GGG GCC GCA GTA ATT ACA TAC
Orca (*Orcinus orca*)	ATA CGA TAT GTC CTC CCG TTA GGC CAA ATA TCA TTC TGA GGG GCC ACA GTA ATT ACA AAC
Sperm whale (*Physeter macrocephalus*)	ATA GGA TAA CTC CTC CAG TGA GTT CAA ATA TCA GTC TGA GGG GGC ACA GCA ATT ACT ACC

a. Complete the following table to calculate the percentage sequence conservation shared between the pairs of species listed.

To calculate % sequence conservation, use: $\% \text{ sequence conservation} = \frac{\text{conserved nucleotides}}{\text{total nucleotides (60)}} \times 100.$

Pairs	Number of nucleotide differences	Number of nucleotides conserved (same)	% sequence conservation
Dolphin, orca			
Dolphin, humpback whale			
Dolphin, minke whale			
Dolphin, bowhead whale			
Dolphin, sperm whale			
Orca, humpback whale			
Orca, minke whale			
Orca, bowhead whale			
Orca, sperm whale			
Humpback whale, minke whale			
Humpback whale, bowhead whale			
Humpback whale, sperm whale			
Minke whale, bowhead whale			
Minke whale, sperm whale			
Bowhead whale, sperm whale			

5 marks

b. Draw a phylogenetic tree to illustrate the evolutionary relationships between the members of order Cetacea based on the percentage of sequence conservation in the cytochrome *b* gene. **5 marks**

c. Identify the species which is acting as an 'outgroup'. **1 mark**

d. Using the provided data, determine if the evolution of cetacean species represents divergent or convergent evolution. **2 marks**

Sperm whales are unique in this group of species in that they are one of the most sexually dimorphic species of cetaceans. Sexual dimorphism is shown where males and females of the same species exhibit different physical characteristics beyond their sexual organs. Female sperm whales mature between 11 and 16 metres long, whereas males mature between 16 and 20 metres long.

e. Identify the source of alleles controlling adult length in sperm whales. **1 mark**

f. Outline the process by which larger size in male sperm whales increases in phenotypic frequency. **3 marks**

It could be said that a 60 bp sequence of DNA is not considered sufficient evidence to definitively determine the evolutionary relationships between these species.

g. Discuss whether this statement has merit or not. Justify your response. **3 marks**

All cetacean species are part of class Mammalia, as well as humans and their ancestors.

h. List three features common to all mammals. **3 marks**

i. Determine if modern humans and cetaceans possess homologous structures and provide examples. **2 marks**

Question 2 (10 marks)

Lord Howe Island is a small, sub-tropical island 600 km off the eastern coast of Australia. It was formed by volcanic activity as part of a 1000-km-long island chain approximately 6.4 to 6.9 millions years ago. The island is unique due to its geographic isolation, small size (14.55 km^2), small human settlement and endemic species.

Two species of *Howea* palms are endemic to the island, *Howea belmoreana* and *Howea forsteriana.* They descend from a common ancestral species and grow in abundance across the island. The graph below shows data relating to the flowering periods of each species and each biological sex within the species.

FIGURE 1 Flowering phenology in *Howea* species. (Savolainen et al, 2006)

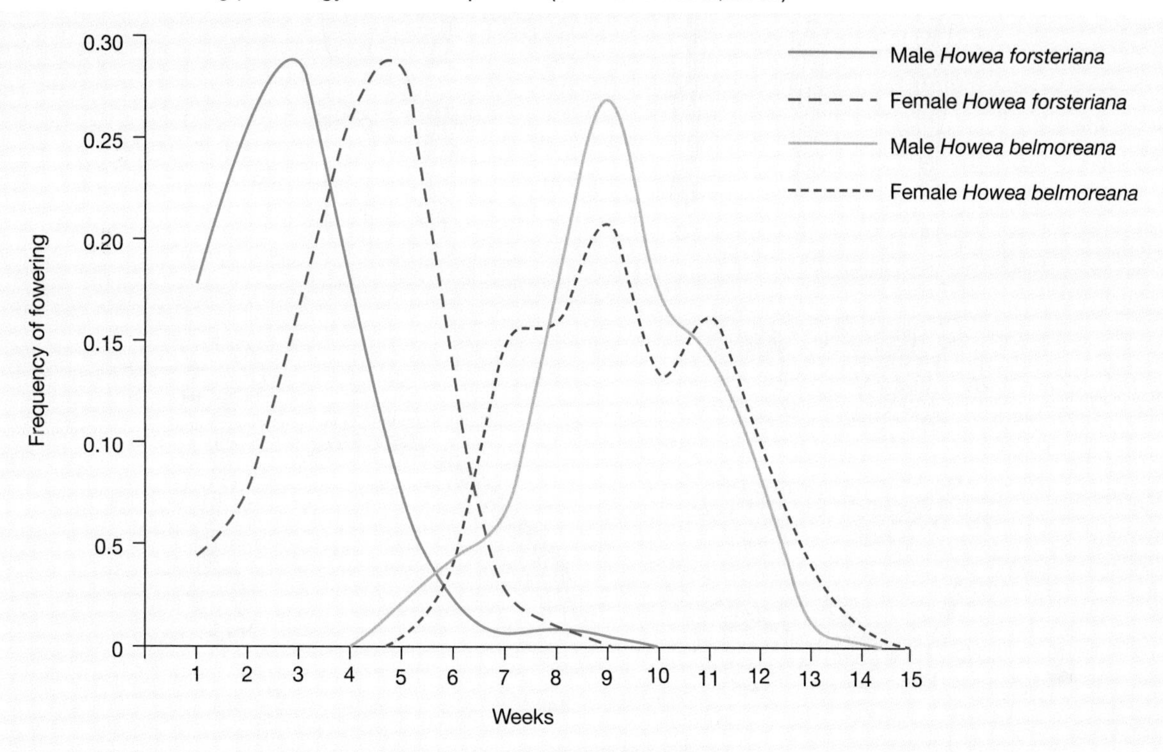

a. Describe the term 'sympatric speciation'. **1 mark**

b. Compare and contrast allopatric speciation with sympatric speciation. **2 marks**

c. i. With reference to figure 1, describe the difference in flowering trends between the two *Howea* species. **1 mark**

ii. Provide a possible explanation for the difference in these trends. **1 mark**

d. With reference to figure 1, explain how flowering times of the *Howea* species provide evidence of sympatric speciation. **2 marks**

e. Outline the steps that would have occurred in order for the descendent *Howea* species to arise from their common ancestral species. **3 marks**

Question 3 (8 marks)

The evolution of hominins over the course of geological history has been marked by trending changes in various aspects of skeletal structures that support bipedal locomotion.

FIGURE 2 Main anatomical adaptations to bipedal locomotion in apes, *Australopithecus afarensis* and *Homo sapiens* (Jankovic, 2015). Note that *H. sapiens* structures are shown at 1A, 2C, 3B and 4B. *A. afarensis* strucutres are shown in 1B and 2B. Ape structures are shown in 1C, 2A, 3A and 4A.

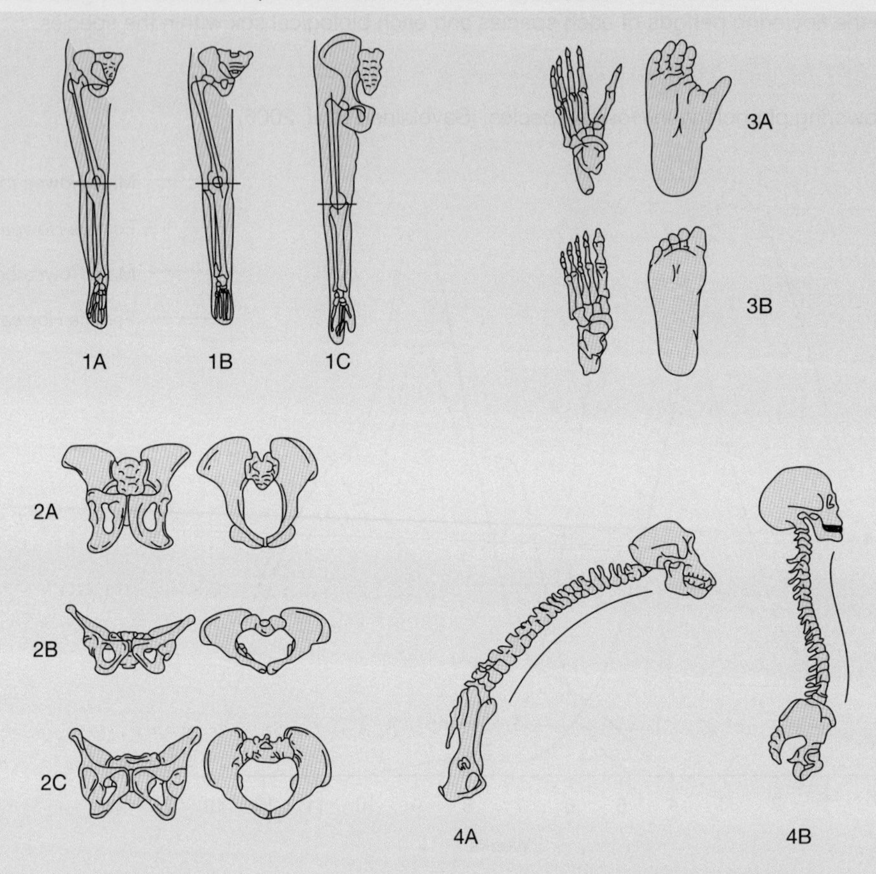

a. Describe how hominins are distinct from hominoids. **1 mark**

b. Some structural features shared by apes, *A. afarensis* and *H. sapiens* are functional in some species but vestigial in others. Using evidence from figure 2, describe a structure which is vestigial in one species but functional in another species. **2 marks**

c. Using evidence from the image, describe three trends in the changes of skeletons between *A. afarensis* and *H. sapiens*. Explain how these changes contributed to the success of *H. sapiens* when compared to other members of genus *Homo*. **5 marks**

Question 4 (7 marks)

Mitochondrial DNA (mtDNA) evidence is often used to explain the migration of various members of genus *Homo* through out the world. Human mtDNA is separated into types known as haplogroups, which are defined by the number of genetic differences from the original mtDNA sequence. Each haplogroup is named using a letter and possibly a number.

FIGURE 3 Human migrations and mitochondrial haplogroups.

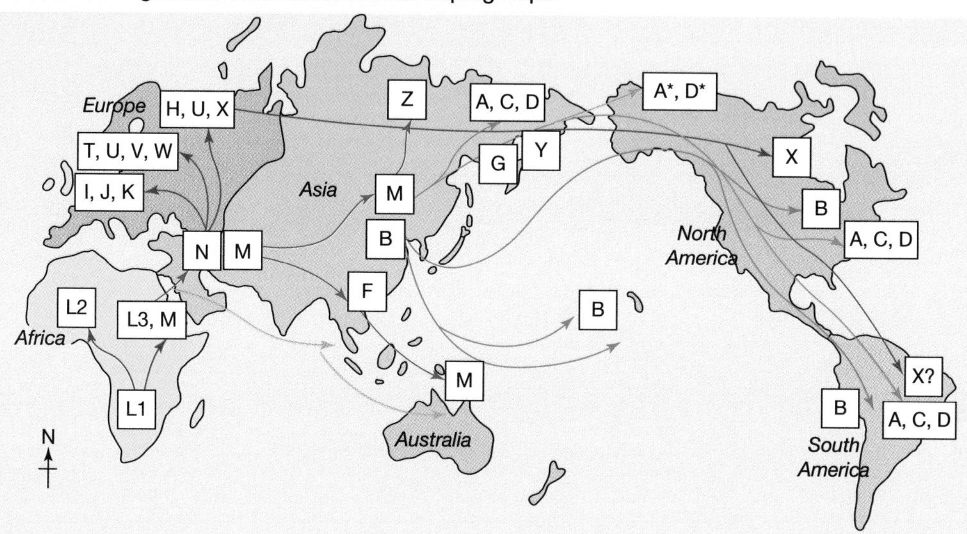

a. Outline why mtDNA evidence is useful for this type of analysis. **3 marks**

b. Using the data from figure 3 describe the migration patterns and times of members of genus *Homo* from 200 000 years ago to the modern day. **4 marks**

Resources

Digital document Unit 4 AOS 2 School-assessed coursework (doc-34819)

11 Scientific investigations

KEY KNOWLEDGE

online only

In this topic, you will investigate:

Investigation design

- biological concepts specific to the selected scientific investigation and their significance, including definitions of key terms
- characteristics of the selected scientific methodology and method, and appropriateness of the use of independent, dependent and controlled variables in the selected scientific investigation
- techniques of primary quantitative data generation relevant to the selected scientific investigation
- the accuracy, precision, reproducibility, repeatability and validity of measurements
- the health, safety and ethical guidelines relevant to the selected scientific investigation

Scientific evidence

- the nature of evidence that supports or refutes a hypothesis, model or theory
- ways of organising, analysing and evaluating primary data to identify patterns and relationships including sources of error and uncertainty
- authentication of generated primary data through the use of a logbook
- assumptions and limitations of investigation methodology and/or data generation and/or analysis methods

Science communication

- conventions of science communication: scientific terminology and representations, symbols, formulas, standard abbreviations and units of measurement
- conventions of scientific poster presentation, including succinct communication of the selected scientific investigation and acknowledgements and references
- the key findings and implications of the selected scientific investigation.

This topic is available online at **www.jacplus.com.au**

Online Resources

Below is a full list of **rich resources** available online for this topic. These resources are designed to bring ideas to life, to promote deep and lasting learning and to support the different learning needs of each individual.

eWorkbook

11.1 eWorkbook — Topic 11 (ewbk-1890) ☐
11.2 Worksheet 11.1 Reviewing variables (ewbk-9174) ☐
Worksheet 11.2 Writing aims, hypotheses and scientific questions (ewbk-9176) ☐
11.3 Worksheet 11.3 Designing investigations and scientific methodology (ewbk-9178) ☐
11.5 Worksheet 11.4 Validity, precision and accuracy (ewbk-9180) ☐
11.6 Worksheet 11.5 Displaying primary data (ewbk-9182) ☐
Worksheet 11.6 Analysing primary data (ewbk-9184) ☐
Worksheet 11.7 Identifying errors (ewbk-9186) ☐
11.7 Worksheet 11.8 Identifying strong and weak evidence (ewbk-9188) ☐
11.9 Worksheet 11.9 Acknowledgements and referencing (ewbk-9190) ☐
11.10 Worksheet 11.10 Scientific posters (ewbk-9192) ☐
11.11 Worksheet 11.11 Reflection — Topic 11 (ewbk-4552) ☐
Biochallenge — Topic 11 (ewbk-8092) ☐

Solutions

11.11 Solutions — Topic 11 (sol-0667) ☐

Digital documents

11.1 Key science skills — VCE Biology Units 1–4 (doc-34326) ☐
Key terms glossary — Topic 11 (doc-34625) ☐
Key ideas summary — Topic 11 (doc-34616) ☐
11.4 Blank risk assessment template (doc-36155) ☐

Teacher-led videos

Exam questions — Topic 11 ☐

11.2 Sample problem 1 Identifying types of variables (tlvd-1820) ☐
Sample problem 2 Developing questions, making aims and formulating hypotheses (tlvd-1821) ☐
11.5 Sample problem 3 Comparing accuracy and precision (tlvd-1822) ☐
11.6 Sample problem 4 Drawing a graph (tlvd-1823) ☐
11.9 Sample problem 5 Writing numbers in scientific notation (tlvd-1824) ☐
Sample problem 6 Converting units (tlvd-1825) ☐
11.10 Producing a scientific poster (tlvd-1908) ☐

Video eLessons

11.2 SkillBuilder — Controlled, dependent and independent variables (eles-4156) ☐
SkillBuilder — Writing an aim and forming a hypothesis (eles-4155) ☐

Interactivities

11.2 SkillBuilder — Controlled, dependent and independent variables (int-8090) ☐
SkillBuilder — Writing an aim and forming a hypothesis (int-8089) ☐
11.6 Selecting graphs (int-8331) ☐

Weblinks

11.4 Animal ethics ☐
The Tuskegee Study Timeline ☐
National Statement of Ethical Conduct in Human Research ☐
11.6 Microsoft Excel video training ☐
11.9 Creative commons for teachers and students ☐
Copyright — Australian Government ☐
Citing and referencing guides — Monash University ☐
Online citation generator ☐

Exam question booklet

11.1 Exam question booklet — Topic 11 (eqb-0022) ☐

Teacher resources

There are many resources available exclusively for teachers online.

To access these online resources, log on to **www.jacplus.com.au**

UNIT 4 | AREA OF STUDY 3 EXAM

How is scientific inquiry used to investigate cellular processes and/or biological change?

OUTCOME 3

Design and conduct a scientific investigation related to cellular processes and/or how life changes and responds to challenges, and present an aim, methodology and method, results, discussion and a conclusion in a scientific poster.

PRACTICE EXAMINATION

STRUCTURE OF PRACTICAL EXAMINATION		
Section	**Number of questions**	**Number of marks**
A	20	20
B	5	30
	Total	**50**

Duration: 50 minutes

Information:

- This practice examination consists of two parts. You must answer all question sections.
- Pens, pencils, highlighters, erasers and rulers are permitted.

SECTION A — Multiple choice questions

All correct answers are worth 1 mark each; an incorrect answer is worth 0.

1. A student conducts an investigation to determine the optimal temperature of the *Taq* polymerase enzyme. This is known to be around 72.0 °C.

The student repeated the investigation 6 times, and obtained the following results:

64.5 °C, 64.1 °C, 63.9 °C, 64.0 °C, 63.8 °C, 63.8 °C

It can be stated that

A. the student's results were neither accurate or precise.
B. the student's results were accurate and precise.
C. the student's results were accurate but not precise.
D. the student's results were precise but not accurate.

2. A student conducts an investigation to explore how the amount of salt in buffer solution affects the time for a fragment of 300 base pairs to move 5 centimetres down a gel during gel electrophoresis. The dependent variable in this investigation is

A. the time taken for the fragment to travel 5 cm.
B. the distance the fragment travelled.
C. the amount of salt in the buffer solution.
D. the length of the fragment.

3. The following data is collected by a student exploring the optimal pH for anaerobic fermentation.

pH	Amount of CO_2 produced (mL)
1	0.2
2	6.7
3	18.2
4	29.6
5	35.2
6	41.1
7	43.7
8	42.8
9	39.9
10	33.4
11	27.2
12	17.1
13	9.2
14	0.1

The best type of graph to show this data would be
A. a bar graph.
B. a line graph.
C. a pie chart.
D. a scatterplot with a straight line of best fit.

4. Which of the following should NOT be included in the introduction of a scientific report?
A. The aim
B. The purpose of the investigation
C. Suggestions for further points of investigation
D. Background information

5. Before the start of an investigation, a set of scales had not been calibrated, so all measurements were reading 4.3 grams higher than their actual value. This is an example of
A. a personal error.
B. a human error.
C. a systematic error.
D. a random error.

6. A student was using a micropipette. They drew up 125 μL of liquid in the micropipette. This amount is equivalent to
A. 0.0125 mL
B. 0.125 mL
C. 1.25 mL
D. 12.5 mL

7. A student is investigating the effect of changing carbon dioxide concentrations on the rate of photosynthesis. They use the same plant across all tests and keep all other environmental factors constant. This is an example of
A. fieldwork.
B. a case study.
C. a simulation.
D. a controlled experiment.

8. A risk assessment should NOT include
A. an outline of ethics.
B. a list of hazards.
C. recommended safety precautions.
D. information about the disposal of chemicals.

9. Which of the following is an example of an acceptable hypothesis for an investigation?
A. To determine how the size of fragments affects the movement through a gel
B. If a fragment is longer, it will travel a shorter distance.
C. If a DNA fragment is shorter, then it will travel further.
D. If a DNA fragment is of a shorter length, then it will travel further compared to a longer fragment when run through gel electrophoresis, as it can move easily move through agarose when gel electrophoresis is performed.

10. In a scientific poster, the centre section of the poster should
A. show a graph with data from your investigation.
B. have a one-sentence summary reporting the key finding of your investigation.
C. outline your scientific method.
D. show the title of your investigation.

11. Quantitative data

A. involves numerical data.

B. is more subject to bias than qualitative data.

C. can include the species of a plant used in an investigation.

D. always is continuous.

12. A student conducted some research for their investigation, and found a journal to assist them. Which of the following is not required when including the journal in the reference list?

A. All contributing authors

B. The location the article was obtained from or the DOI

C. A reference to bias in the article

D. The title of the journal

13. Images in the public domain

A. can be used freely without reference.

B. can only be used for non-commercial purposes.

C. must be fully referenced.

D. can be used freely, but adaptations or derivatives need to be referenced.

14. An appropriate standard abbreviation for adenosine triphosphate is

A. ATE **B.** ADT **C.** ATP **D.** ADTRI

15. A model is

A. a scientific drawing.

B. a measure of how close repeated measurements are to each other.

C. the same as reproducibility.

D. a measure of uncertainty in an investigation.

16. Validity is

A. the credibility of the research results from experiments or from observations.

B. a measure of how close repeated measurements are to each other.

C. the same as reproducibility.

D. a measure of uncertainty in an investigation.

Use the following information to answer Questions 17 and 18.

A student was exploring the number of individuals diagnosed with a specific disease over a three-month period.

They graphed the data and the results are shown.

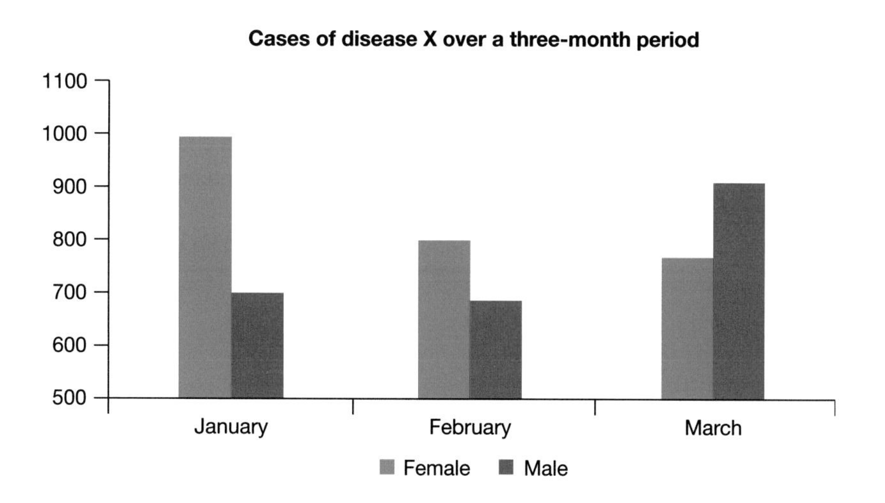

17. From this data, it can be concluded that
    A. in January, more than twice the number of females had disease X than males.
    B. there was a decrease in the number of cases over the 3-month period for both males and females.
    C. it can be predicted that the number of infected males in March will be 700.
    D. there were just over 100 more females infected in February than males.

18. Which of the following is correct regarding the data shown in the graph?
    A. Quantitative data is on the *y*-axis.
    B. All data collected is qualitative.
    C. The different colours represent differences in quantitative data.
    D. The data on the *x*-axis is numerical.

Use the following information to answer Questions 19 and 20.

A student conducts an investigation to observe how the temperature of the water during heat shock affects the uptake of plasmids by bacteria. They plated *E. coli* bacteria on nutrient agar. From this initial plate, they collected exactly 60 bacterial colonies and made 3 cultures, each with 20 identical bacterial colonies. All three cultures were placed in an ice bath at 0 °C, and 20 mL of plasmid was placed in each culture.

The first culture was then moved into a solution at 40 °C. The second culture was moved into a solution at 50 °C, and the final culture was placed in a solution at 60 °C. They remained at these temperatures for 50 seconds before being moved back into the ice bath for 2 minutes.

Each bacterial culture was then spread onto an agar plate with antibiotics and incubated overnight. The number of colonies that grew on the plate the next day were counted. (Only bacteria with the plasmid were able to grow.)

At the conclusion of the experiment, the results were recorded in the provided table. Each test was repeated three times.

The number of bacterial colonies after heat shock at different temperatures

Number of colonies	40 °C	50 °C	60 °C
Test 1	25	19	14
Test 2	29	20	12
Test 3	31	23	13

19. Which of the following is NOT a controlled variable in this investigation?
    A. The temperature of the ice bath
    B. The time the culture is in the hot solution
    C. The temperature of the hot solution
    D. The number of colonies initially plated

20. Which one of the following hypotheses is supported by the results?
    A. If the temperature of the hot solution used for heat shock is closer to the currently used temperature of 42 °C, then the success of bacterial transformation will increase.
    B. If the temperature of heat shock is changed, the number of bacteria will change.
    C. If the temperature of the hot solution of heat shock is increased, then the number of transformed bacteria will increase.
    D. If the temperature of heat shock decreases, bacteria will be unable to take up the plasmids.

SECTION B — Short answer questions

Question 21 (3 marks)

Describe the difference between:

a. repeatability and reproducibility. **1 mark**

b. precision and accuracy. **1 mark**

c. controlled variables and control groups. **1 mark**

Question 22 (7 marks)

Four students conducted an investigation to explore the size of a plasmid run through a gel. Each student used the same plasmid, which was known to be 2021 base pairs (bp) long. Each student ran three gels, each gel having a DNA ladder with fragments of known size to determine the size of the plasmid. The results for each student are shown:

The size of a plasmid run through a gel

Student	Test 1 (bp)	Test 2 (bp)	Test 3 (bp)
Student 1	2000	2500	2645
Student 2	2100	2002	2111
Student 3	2300	2200	2100
Student 4	2000	1700	2222

a. In relation to accuracy and precision, describe the data obtained by student 1. **2 marks**

b. Which student had the most accurate data? Provide reasons for your response. **1 mark**

c. Are the differences in results more likely due to random or systematic error for student 4? Justify your response and suggest a reason this may have occurred. **3 marks**

d. Convert the data of student 3 from base pairs (bp) to kilobase pairs (kbp). **1 mark**

Question 23 (7 marks)

Alex wanted to explore how temperature impacted anaerobic fermentation in yeast. As she was not using chemicals, she did not think that she needed any safety equipment. She collected 6 balloons and 6 conical flasks. She placed 100 mL of water in each flask. She placed 2 flasks at room temperature, 2 flasks at 40 °C and 2 flasks at 60 °C. In each flask she placed a packet of yeast and 10 grams of sugar. She quickly placed a balloon over the neck of each conical flask and measured the diameters of the balloons every minute for 5 minutes.

a. Outline any possible health or safety hazards in this investigation. **1 mark**

b. Suggest the safety precautions you would recommend Alex takes. **1 mark**

c. Identify the dependent and independent variable in this investigation. **2 marks**

d. Identify one piece of measuring equipment that would allow for the accurate quantitative collection of data in this investigation. **1 mark**

e. Suggest two improvements to the scientific method for this investigation. **2 marks**

Question 24 (6 marks)

Consider the table provided, which shows the effect of different concentrations of the antibiotic ampicillin on the growth of *E. coli*. This was measured by the zone of inhibition (the distance from the antibiotic disk to the edge of the area in which the bacteria do not grow).

The effects of ampicillin concentration on the growth of *E. coli*

Concentration of ampicillin (μg/mL)	Student results (zone of inhibition in mm)						Average zone of inhibition (mm)
	1	2	3	4	5	6	
0.00	0.00	0.12	0.00	0.08	0.01	0.02	0.04
31.25	1.20	1.27	1.18	1.15	1.76	1.24	1.30
62.50	2.61	3.00	2.10	2.50	2.71	2.86	2.63
125.00	5.48	5.50	5.00	5.42	5.89	5.06	5.39
250.00	8.18	7.80	7.95	8.53	8.43	8.23	8.19

a. Draw a graph of the average data, ensuring you add a title and labels. **3 marks**

b. The minimum inhibitory concentration is the concentration that is ideal in the treatment of bacterial infection (too high and you may harm the host, too low and you won't kill the bacteria). In this case, it is the point where the zone of inhibition is 6 mm. Using your graph from part **a**, determine the minimum inhibitory concentration of ampicillin. **1 mark**

c. Would you consider the method to be reproducible? Why or why not? **1 mark**

d. Based on the student's results, write a conclusion for this investigation. **1 mark**

Question 25 (7 marks)

A student wishes to investigate the following research question: 'Does an increase in glucose increase the rate of cellular respiration?'

In order to do this, they collect an *Elodea* plant, place it in darkness and measure the amount of carbon dioxide produced by examining the pH of the solution (as carbon dioxide forms carbonic acid in water, which causes the solution to become more acidic).

a. Identify the independent and dependent variable. **1 mark**

b. Write a clear aim for this investigation. **1 mark**

c. Write a suitable hypothesis. **1 mark**

d. Write a clear method for this investigation, ensuring the method is repeatable and valid. **4 marks**

END OF EXAMINATION

APPENDIX
Amino acid data

Amino acids: building blocks of proteins

Amino acids are the basic building blocks of peptides, polypeptides and proteins. Figure A.1 shows the general structure of amino acids. The common structure is a C atom to which are attached an amino group (NH_2), a carboxyl group (COOH) and an H atom. The fourth group, denoted R, is variable and distinguishes the different amino acids.

Table A.1 shows the R group of each of the 20 common amino acids. The R groups are organised according to their chemical nature. The simplest R group is an H atom — as seen in glycine. This amino acid forms part of a group of amino acids with aliphatic and non-polar R groups. Note that aromatic groups have a ring, whereas aliphatic groups do not.

FIGURE A.1 The general structure of an amino acid

Amino group
Carboxyl group
Variable group

TABLE A.1 The twenty common amino acids grouped according to the chemical properties of their R group.

Nonpolar, aliphatic R groups

Glycine, Alanine, Valine, Leucine, Methionine, Isoleucine

Polar, uncharged R groups

Serine, Threonine, Cysteine, Proline, Asparagine, Glutamine

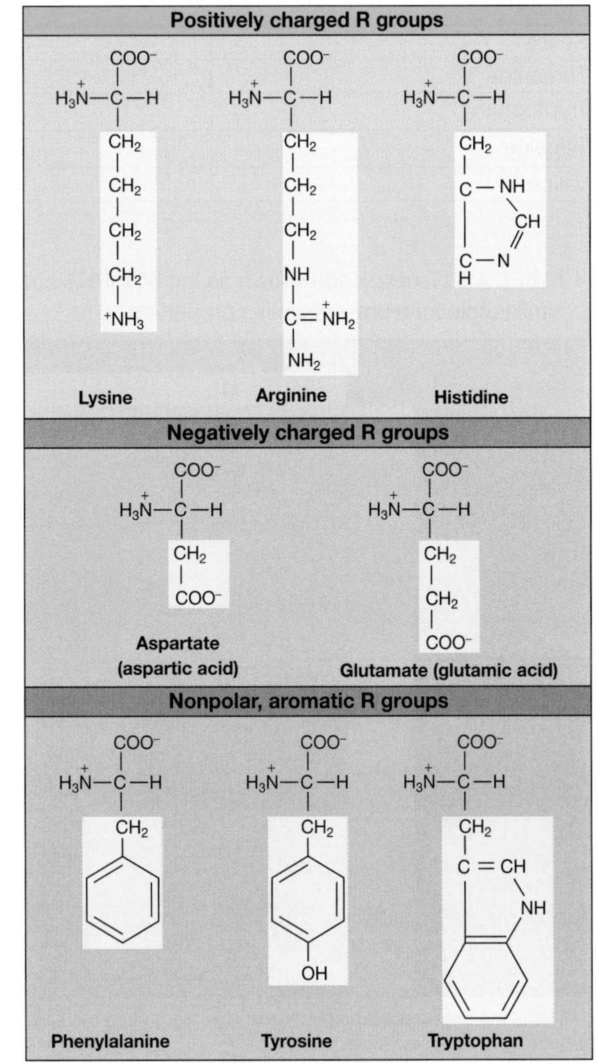

Table A.2 shows abbreviations for the twenty common amino acids. Two schemes are in use, a three-letter abbreviation and a one-letter system. Note that glutamic acid and aspartic acid can be referred to as glutamate and aspartate. This is based on if the amino acid is ionised. Either of these may be used in VCE Biology.

TABLE A.2 Amino acids and their abbreviations.

Amino acid	three-letter code	one-letter code
Alanine	*Ala*	A
Arginine	*Arg*	R
Asparagine	*Asn*	N
Aspartic acid	*Asp*	D
Cysteine	*Cys*	C
Glutamic acid	*Glu*	E
Glutamine	*Gln*	Q
Glycine	*Gly*	G
Histidine	*His*	H
Isoleucine	*Ile*	I
Leucine	*Leu*	L
Lysine	*Lys*	K
Methionine	*Met*	M
Phenylalanine	*Phe*	F
Proline	*Pro*	P
Serine	*Ser*	S
Threonine	*Thr*	T
Tryptophan	*Trp*	W
Tyrosine	*Tyr*	Y
Valine	*Val*	V

TABLE A.3 Genetic code shown as the 64 mRNA codons and the amino acids they specify. Note that similar colours represent similar properties.

FIRST BASE		SECOND BASE: U		C		A		G	
		UUU	*Phe*	UCU	*Ser*	UAU	*Tyr*	UGU	*Cys*
	U	UUC	*Phe*	UCC	*Ser*	UAC	*Tyr*	UGC	*Cys*
		UUA	*Leu*	UCA	*Ser*	UAA	*STOP*	UGA	*STOP*
		UUG	*Leu*	UCG	*Ser*	UAG	*STOP*	UGG	*Trp*
		CUU	*Leu*	CCU	*Pro*	CAU	*His*	CGU	*Arg*
	C	CUC	*Leu*	CCC	*Pro*	CAC	*His*	CGC	*Arg*
		CUA	*Leu*	CCA	*Pro*	CAA	*Gln*	CGA	*Arg*
		CUG	*Leu*	CCG	*Pro*	CAG	*Gln*	CGG	*Arg*
		AUU	*Ile*	ACU	*Thr*	AAU	*Asn*	AGU	*Ser*
	A	AUC	*Ile*	ACC	*Thr*	AAC	*Asn*	AGC	*Ser*
		AUA	*Ile*	ACA	*Thr*	AAA	*Lys*	AGA	*Arg*
		AUG	*Met**	ACG	*Thr*	AAG	*Lys*	AGG	*Arg*
		GUU	*Val*	GCU	*Ala*	GAU	*Asp*	GGU	*Gly*
	G	GUC	*Val*	GCC	*Ala*	GAC	*Asp*	GGC	*Gly*
		GUA	*Val*	GCA	*Ala*	GAA	*Glu*	GGA	*Gly*
		GUG	*Val*	GCG	*Ala*	GAG	*Glu*	GGG	*Gly*

* *Met* is the amino acid methionine and is the first amino acid added (the start instruction).

GLOSSARY

absolute age the age of an object, such as a rock or fossil, expressed in actual years
acceptor regeneration a process where acceptor molecules are reformed, allowing for them to be recycled in biochemical pathways
accuracy how close an experimental measurement is to a known value
acquired immunity an immunity that develops during a person's lifetime
activation energy minimum amount of energy required to initiate a chemical reaction
active immunity the production of antibodies by a person in response to exposure to a particular antigen
active site region of an enzyme that binds temporarily with the specific substrate of the enzyme
active transport net movement of dissolved substances across a cell membrane by an energy-requiring process that moves substances against a concentration gradient from a region of lower to higher concentration
adaptive immunity an immune response that is specific to a particular antigen and develops through contact with an antigen
ADP adenosine diphosphate; a coenzyme that accepts a phosphate group to form ATP
aerobic cellular respiration oxygen-requiring process that converts chemical energy into ATP
agarose gel a special type of porous gel used in gel electrophoresis, which allows for DNA to be separated by fragment size
aim a statement outlining the purpose of an investigation, linking the dependent and independent variables
alcohol fermentation a process that occurs without oxygen to produce two molecules of ATP and ethanol
allele frequency the proportion of a specific allele in a population
alleles alternate forms of a gene
allergen an antigen that elicits an allergic response
allergic response rapid immune response to normally harmless antigens such as dust or pollen; involves production of IgE antibodies by B lymphocytes and release of histamines by mast cells
allergy an abnormal immune response to a substance that is harmless for most people
allopatric speciation speciation that occurs between two populations that are geographically isolated
allosteric activators molecules that bind to the allosteric site of an enzyme and increase enzyme activity
allosteric inhibitors molecules that bind to the allosteric site of an enzyme and stop enzyme activity
allosteric regulation the control of the reaction rate of enzymes through conformational changes in enzymes
allosteric site location on an enzyme molecule where a compound can bind and alter the shape of the enzyme
alpha helix a type of secondary structure in proteins that appears as a tight twist
amino acids basic building blocks or sub-units of polypeptide chains and proteins
anabolic chemical reaction in a cell in which complex molecules are built from simple molecules
anaerobic cellular respiration a process that converts chemical energy into ATP in the absence of oxygen
anaphylaxis acute and potentially lethal allergic reaction to an allergen to which a person has become hypersensitive
annealing the second stage in PCR, in which primers attach to the single-stranded DNA
antibiotics a class of antimicrobial drug used in the treatment and prevention of bacterial infections that act either by killing pathogenic bacteria or by inhibiting their growth
antibodies proteins produced by plasma cells in response to antigens and which react specifically with the antigen that induced their formation; also called immunoglobulins
anticodon sequence of three bases in a transfer RNA molecule that can pair with the complementary codon of a messenger RNA molecule
antigen-binding sites regions of an antibody molecule to which an antigen binds; also called variable regions
antigen-presenting cells cells of the immune system that use MHC-II markers on their surface to present antigens to helper T cells to elicit an immune response
antigenic drift when a point mutation alters a virus's nucleic material, resulting in small changes so that it continues to be recognised and reacted to by the body's immune system
antigenic shift when two or more strains of a virus combine to form a new strain of the virus so that it is no longer recognised and reacted to by the body's immune system

antigens molecules or parts of molecules that stimulate an immune response
ape a large primate that lacks a tail
artificial active immunity the deliberate administration of disabled antigens to elicit the production of antibodies
artificial immunity immunity that is formed through deliberate exposure and intervention
artificial passive immunity the administration of antibodies to provide an immediate, specific immune response
artificial selection the process by which humans breed animals or plants in such a way to increase the proportion of chosen phenotypic traits
assumptions ideas that are accepted as true without evidence in order to overcome limitations in experiments
asymptomatic carrier person with an infectious disease showing no symptoms but able to infect others
ATP adenosine triphosphate; the common source of chemical energy for cells
attenuated pathogen a pathogen that, while still living, has been treated so that it is no longer able to cause disease
Australopithecus a genus of hominins from which the *Homo* genus is considered to be descended; sometimes referred to as australopithecines
autoimmune diseases diseases in which the immune system fails to identify 'self' material and makes antibodies against the body's own tissues
autonomy the opportunity to make informed decisions without coercion or unfair influence
autotrophs organisms that, when given a source of energy, produce their own food from simple inorganic substances
B lymphocytes also called B cells; white blood cells that recognise antigens or pathogens and produce a large number of antibodies specific to an antigen
bacterial transformation process in which bacterial cells take up segments of foreign DNA that become part of their genetic make-up
bar graph a graph in which data is represented as a series as bars; usually used when one variable is qualitative
beneficence the act of doing what is good and right
beta-pleated sheet a type of secondary structure in proteins that appears as folded sheets, with a change in direction of the polypeptide chain
bias intentional or unintentional influence on a research investigation
binocular vision a type of vision that allows for both peripheral vision and a perception of depth using two eyes
biochemical pathways a series of linked biochemical reactions
biochemical reactions reactions occurring in cells that lead to the formation of a product from a reactant
biofuel any fuel source derived from biomass
biogeography geographical distribution of species
biomass the organic material from plants and animals; it is a renewable source of energy
biosignatures chemical or physical traces that can be inferred to have resulted from the action of life forms
bipedalism a form of locomotion involving routine movement on two feet
blunt ends ends of a DNA fragment with no overhanging bases after being cut by an endonuclease
bone marrow fatty substance in the internal cavity of bones; the site of blood cell formation
bottleneck effect chance effects on allele frequencies in a population as a result of a major reduction in population size
brachiation a mode of locomotion involving swinging from one handhold to another
branch a section of the phylogenetic tree which indicates a speciation event
broad spectrum antibiotics that act against many different kinds of pathogens
C_3 plants plants that carry out the original Calvin cycle using Rubisco and are prone to photorespiration
C_4 plants plants that carry out an adapted Calvin cycle, in which carbon fixation and glucose production occur in different cells
Calvin cycle cycle of reactions occurring in the stroma of chloroplasts in the light-independent stage of photosynthesis
CAM plants plants that thrive in arid conditions and have their two stages of the Calvin cycle occurring at different times

cancer a disease in which cells divide in an uncontrolled manner, forming an abnormal mass of cells called a tumour
capsid protein shell enclosing the genetic material of a virus
carbohydrate groups molecules that are associated with the plasma membrane and are associated with cell to cell communication and signalling
carbon dating a form of absolute dating in which the amount of stable nitrogen-14 is compared to that of the radioactive parent isotope carbon-14
carbon fixation process by which atmospheric carbon dioxide is incorporated into organic molecules such as sugars
carrying angle the angle formed between the femur and the vertical; also known as the bicondular angle
cascade a multi-step process in which each step must occur in a set order, with each step triggering the next in the sequence
cast type of fossil formed when a buried structure decays leaving a cavity, which is later filled by different material that forms a model of the original structure
catabolic chemical reaction in a cell in which complex molecules are broken down into simple molecules
catalyst a factor that causes an increase in the rate of a reaction
causal pathogen a pathogen that is causing a specific disease under investigation
causation when one factor or variable directly influences the results of another factor or variable
cell surface markers proteins (or glycoproteins) present on the plasma membrane that distinguish various cell types and discriminate between self and non-self antigens
cell surface receptors regions of a trans-membrane molecule exposed at the surface of a cell that act in cell signalling by receiving and binding to extracellular molecules
cell-mediated adaptive response a specific response in which cytotoxic T cells destroy virus-infected cells using perforin and granzyme B
cell-mediated immunity immune response that is mediated by immune cells
cellular pathogens any disease-causing agent made up of cells that can reproduce independently without relying on the host machinery
cellular respiration process of converting chemical energy into a useable form by cells, typically ATP
chemical barriers innate barriers that use enzymes to kill pathogens and prevent invasion into a host
chemotaxis movement of a cell or organism in response to a chemical substance (such as complement or cytokines)
chlorophyll green pigment that traps the radiant energy of light
chloroplast chlorophyll-containing organelle that is the site of photosynthesis
cilia (singular = cilium) in eukaryote cells, fine hair-like outfoldings formed by extensions of the plasma membrane involved in synchronised movement
cladogram diagram, based on cladistic study, that shows the inferred relationship between different group of organisms
clonal expansion a process of multiple cycles of cell division of a lymphocyte specific to a particular antigen, resulting in the production of large numbers of identical lymphocytes
clonal selection an event occurring in lymph nodes in which those lymphocytes with receptors that can recognise a new antigen come into contact with that particular antigen
clones groups of cells, organisms or genes with identical genetic make-up
coding region part of a gene that contains the coded information for making a polypeptide chain
coding strand one strand of a DNA double helix that is complementary to the template strand
codons sequences of three bases in a messenger RNA molecule that contain information either to bring amino acids into place in a polypeptide chain or to start or stop this process
coenzyme an organic molecule that acts with an enzyme to alter the rate of a reaction
coenzyme A a coenzyme that aids pyruvate decarboxylase during cellular respiration, accepting an acetyl group
cofactor a non-protein molecule or ions that is essential for the normal functioning of some enzymes
common ancestor the species that two or more descendants have diverged from
comparative genomics the comparison of DNA or genes between different species to identify evolutionary relationships

competitive inhibition inhibition in which a molecule binds to the active site of a molecule instead of the usual substrate
complement proteins proteins that assist other innate immune cells and can destroy bacterial cells by lysis
complementary a molecule having a specific chemical structure that allows it to bond in a 'lock-and-key' fashion to another structure
complementary DNA (cDNA) a strand of DNA that has complementary bases to the opposite strand and is usually produced through reverse transcription
conclusion a section at the end of the report that relates back to the question, sums up key findings and states whether the hypothesis was supported or rejected
conservative missense mutation a type of missense mutation in which the substituted amino acid is similar in properties to the initial amino acid
constant region the section of an antibody that does not vary between antibodies of the same class
continuous data a type of quantitative data that can take on any continuous value
control group a group that is not affected by the independent variable and is used as a baseline for comparison
controlled variables variables that are kept constant across different experimental groups
correlation measure of a relationship between two or more variables
cranium the vertebrate skull minus the lower jaw
CRISPR-Cas9 a tool for precise and targeted genome editing that uses specific RNA sequences to guide an endonuclease, Cas9, to cut DNA at the required positions
cristae folds in the inner membrane of the mitochondria where the electron transport chain occurs
cytokines signalling molecules of the immune system
cytotoxic T cells T cells that are activated by cytokines to bind to antigen–MHC-I complexes on infected host cells and kill infected body cells
degenerate the property of the genetic code in which more than one triplet of bases can code or one amino acid
degranulation the process by which immune cells release various chemicals (such as histamine and antimicrobials) stored within secretory vesicles known as granules
denaturation the loss of enzyme structure due to the breaking of bonds upon heating, irreversibly changing the shape of the active site
denaturing the first stage in PCR, in which a double-stranded piece of DNA is heated and separated into single-stranded DNA
dendritic cells a type of antigen-presenting cell and phagocyte that can activate T lymphocytes in the adaptive immune response
Denisovans members of an extinct hominin species in the genus *Homo*
deoxyribonucleic acid (DNA) nucleic acid consisting of nucleotide sub-units that contain the sugar deoxyribose and the bases A, C, G and T; DNA forms the major component of chromosomes
dependent variable the variable that is influenced by the independent variable and is measured by an investigator
dermicidin an antimicrobial protein that acts as a chemical barrier is animals, which is found is secreted sweat
developmental biology the process of comparing embryos of different species to infer relatedness
direct ELISA a form of ELISA where a primary antibody with an enzyme indicator directly binds to an antigen attached to a surface
direct transmission mechanism of transmission of pathogenic agents that involves direct person-to-person contact, such as by kissing or sexual contact
discrete data a type of quantitative data that can only take on set values
discussion a detailed area of a report in which results are discussed, analysed and evaluated; relationships to concepts are made; errors, limitations and uncertainties are assessed; and suggestions for future improvements are made
disease a condition in a living animal or plant body that impairs the normal functioning of an organ, part, structure or system
DNA fingerprinting technique for identifying DNA from different individuals based on variable numbers of tandem repeats of short DNA segments near the ends of chromosomes

DNA profiling technique for identifying DNA from different individuals based on variable regions known as short tandem repeats (STRs) or microsatellites

DNA sequencing identification of the order or sequence of bases along a DNA strand

double-blind type of trial in which neither the participant nor the researcher is aware of which participants are in a control or experimental group

electron spin resonance a dating technique that determines the age of an object based on absorbed radiation and trapped high-energy electrons

electron transport chain third stage of aerobic respiration in which there is a high yield of ATP

electroporation a technique that uses brief exposure of host cells to an electric field to enable the entry of segments of foreign DNA into the cells

ELISA a technique known as enzyme linked immunosorbent assay, which can detect specific antigens or antibodies

emerging disease a disease caused by a newly identified or previously unknown agent

emigration the movement of individuals and their alleles out of a population, and thus out of a gene pool

end-product inhibition inhibition of the early stage of a multi-step pathway by the final product of that pathway

endergonic a chemical reaction that is energy-requiring

endocytosis an energy-requiring process of bulk transport, in which solids or liquids move into the cell by engulfment

endonucleases enzymes, also known as restriction enzymes, that cut at specific sites within DNA molecules

endoplasmic reticulum cell organelle consisting of a system of membrane-bound channels that transport substances within the cell

endotoxins toxic parts of the outer membrane of some Gram-negative bacteria that are released when the bacteria die

enveloped viruses viruses with an outer envelope composed of part of the plasma membranc of the host cell when the viral particles bud from the cell

environmental selection pressures external agents which influence the ability of an individual to survive in their environment

enzyme a protein that acts as a biological catalyst, speeding up reactions without being used up

enzyme–substrate complex transient compound produced by the bonding of an enzyme with its specific substrate

eosinophils a type of white blood cell that contain granules, enabling them to kill larger parasitic agents

epidemic the widespread occurrence of an infectious disease in a community or in a restricted geographic area at a particular time

ethics principles of acceptable and moral conduct determining what 'right' and what is 'wrong'

ethidium bromide a dye that binds to DNA and illuminates under UV light

eukaryotes any cells or organisms with a membrane-bound nucleus

eukaryotic cells cells within eukaryotes that have a membrane-bound nucleus and other membrane-bound organelles

exergonic a chemical reaction that is energy-releasing

exocytosis an energy-requiring process of bulk transport, in which solids or liquids move out of the cell via vesicles

exons parts of the coding region of a gene that are both transcribed and translated

exotoxins toxins that are secreted into the surrounding medium by a micro-organism as it grows

experimental group a test group that is exposed to the independent variable

extension the third stage in PCR, in which the *Taq* polymerase enzyme synthesises a new strand of DNA by adding free nucleotides

extracellular locations within the body that are outside cells, such as blood plasma and extracellular fluid

extrapolation estimation of a value outside the range of data points tested

facilitated diffusion form of diffusion involving a specific carrier molecule for the substance

FAD a coenzyme with a similar function to NAD, accepting hydrogen ions and electrons during cellular respiration

$FADH_2$ the loaded form of FAD, which can donate hydrogen ions and electrons during cellular respiration
fair testing investigations that have variables (outside the independent variables) carefully controlled
falsifiable able to be proven false using evidence
faunal succession the principle that fossilised fauna and flora in sedimentary rock strata are arranged vertically in a specific order
feedback inhibition inhibition occurs when the end product of a pathway inhibits an enzymes earlier in the pathway as a negative feedback mechanism; also known as end-product inhibition
fermentation a metabolic process that produces some ATP in the absence of oxygen
first line of defence part of the defence against pathogens provided by barriers of the innate immune system that prevent entry of pathogens into the body
fitness the ability to survive and pass genetic material on to the next generation
flanking regions regions located either downstream or upstream of the coding region of a gene
fluid mosaic model a model which proposes that the plasma membrane and other intracellular membranes should be considered as two-dimensional fluids in which proteins are embedded
foramen magnum a hole at the base of the skull in which the vertebral column attaches to
fossil evidence or remains of an organism that lived long ago
fossil impression a physical fossil formed when the organic matter has disappeared but the organism has left an impression of its structure
fossilisation process of preserving parts of organisms that lived in the geological past
founder effect chance effects on allele frequencies in a population that is formed from a small unrepresentative sample of a larger population
founder population a small group of organisms that starts a new population
frameshift mutation type of mutation in which, as a result of insertion or deletion of a base, all codons from that point are affected
gel electrophoresis a technique for sorting a mixture of DNA fragments (and other molecules with a net charge) through an electric field on the basis of different fragment lengths
gene a section of DNA that codes for a protein
gene flow the movement of individuals and their genetic material between populations
gene pool sum total of genetic information present in a population
genetic code representation of genetic information through a non-overlapping series of groups of three bases (triplets) in a DNA template chain
genetic divergence when two or more populations accumulate genetic changes, leading them to eventually be reproductively isolated
genetic diversity the amount of genetic variation there is within a population's gene pool
genetic drift random changes, unpredictable in direction, in allele frequencies from one generation to the next owing to the action of chance events
genetically modified organisms (GMOs) organisms whose genomes are altered through the use of genetic engineering technology
genome editing a process by which changes are made to the nucleic acid sequence of genes; also termed gene editing
genotype the genetic makeup of an organism; also refers to the combination of alleles at a particular gene locus
genotypic and molecular methods methods of identifying bacteria by examining its genetic material
genus a taxonomic rank that is less specific than species
geographically isolated the isolation of two or more populations due to a geographical barrier
geological time a system of chronological dating that relates geological strata to time
germline cells sex cells (sperm and egg) that are also known as gametes, that may pass on genetic information to the next generation
glycolysis the first stage of cellular respiration, in which glucose is broken down into pyruvate
Golgi apparatus organelle that packages material into vesicles for export from a cell (also known as Golgi complex or Golgi body)
grana stacks of flattened thylakoids; singular: granum

granzymes active protease enzymes present in granules that form part of the immune defences of NK cells and cytotoxic T cells

half-life the time taken for an amount of radioactive isotope to decay to half of its initial amount

haplogroup a group of people with similar haplotypes who share a common ancestor

haplotype a region of DNA in the D-loop of mtDNA that varies between individuals

heat shock a technique to transform bacteria in which cells are suspended in a ice cold solution and then moved into a warm solution to increase plasma membrane fluidity

helper T cells a class of T cell that bind to antigen–MHC-II complexes on antigen-presenting cells and activate B cells and cytotoxic T cells

herd immunity indirect protection, at the population level, against an infectious disease; the protection is created by the presence in the population of a high proportion of individuals who are vaccinated against the particular disease

histamine a substance involved in inflammation and allergic reactions that causes blood vessels to dilate and become more permeable to immune cells

histogram a graph in which the frequency of data is sorted into intervals is displayed

hominins modern human species and our extinct close relatives that could walk with a bipedal locomotion

hominoids a superfamily of primates that lack a prehensile tail, including apes and humans

Homo a genus of hominins that includes modern humans and closely related species with an increased brain size and a longer leg-to-arm ratio

Homo erectus an extinct human species appearing about two million years ago that was the first to migrate from Africa

Homo neanderthalensis an extinct *Homo* species known as Neanderthals, who were thought to have interbred with *Homo sapiens*

Homo sapiens modern humans

homologous structures structures of organisms that have evolved from a common ancestor to have possible differences in form and function but retain similarities in structure

host an organism that can get a disease

human leukocyte antigens antigens present on human cell surfaces that determine the 'self' status of a person's cells

humoral immunity immune response mediated by soluble molecules in the blood, lymph and interstitial fluid that disable pathogens

hydrophilic substances that dissolves easily in water; also termed polar

hydrophobic substances that tend to be insoluble in water; also termed nonpolar

hypervariable regions regions in DNA that are highly polymorphic

hypothesis a tentative, testable and falsifiable statement for an observed phenomenon that acts as a prediction for the investigation

immigration the movement of individuals and their alleles into a population, and thus into a gene pool

immune system the body system that helps resist infection and disease through specialised cells and proteins

immunoglobulin E (IgE) antibodies a type of antibody produced in response to exposure to a particular allergen; involved in allergic reactions

immunological memory ability of the adaptive immune response to remember antigens after primary exposure

immunological methods methods used to diagnose pathogens based on the presence of specific antigens or antibodies

immunotherapy a type of treatment that alters the immune response in an individual to combat diseases

incubation period the period between infection and the first appearance of the symptoms of a disease

independent variable the variable that is changed or manipulated by an investigator

index case the first individual known to have a case of an infectious disease

index fossils fossils of geologically short-lived species that are widely distributed but are found in a restricted depth of rock strata

indirect ELISA a form of ELISA where a primary antibody attached to a secondary antibody with an enzyme indicator binds to an antigen attached to a surface

indirect transmission mechanism of transmission of pathogenic agents that does not involve direct person-to-person contact, such as by airborne droplets or by ingestion of contaminated food
infectious diseases that can be transmitted between individuals and are caused by pathogens
inflammation an innate reaction by the immune response to foreign particles or injury resulting in redness and swelling
inflammatory response a reaction to an infection, typically associated with the reddening of the skin owing to an increased blood supply to that region
innate immunity the type of immunity that is present from birth, is fast acting but not long lasting, and produces non-specific (generic) responses against classes of pathogens
insulin a hormone that allows for glucose to enter cells, reducing blood glucose levels
integral proteins proteins that are embedded in the phospholipid bilayer
integrity possessing honesty and morality
interferons proteins secreted by some cells, in response to a virus infection, that helps uninfected cells resist infection by that virus
interpolation estimation of a value within the range of data points tested
intracellular anything that is within a cell
introns parts of the coding region of a gene that are transcribed but not translated
investigation question the focus of a scientific investigation in which experiments act to provide an answer
isolation the process in which populations are separated and gene flow is reduced or stopped
isotopes the different forms of an element which differ in the number of neutrons
knuckle walk a style of walking on four limbs in which the forelimbs are in contact with the ground through the knuckles of flexed hands
Krebs cycle second stage of aerobic respiration in which coenzymes are loaded and carbon dioxide is produced
lactic acid fermentation a process that occurs without oxygen to produce two molecules of ATP and lactic acid
leukocytes white blood cells that are involved in protecting the body from infectious disease
ligands any molecule that binds to a specific target to form an active complex
ligase an enzyme that catalyses the joining of two double-stranded DNA fragments
light saturation point the point in which increasing the light intensity no longer increases the rate of photosynthesis
light-dependent stage the first stage of photosynthesis where light energy is trapped by chlorophyll
light-independent stage the second stage of photosynthesis, in which glucose is produced
limitations factors that affect the interpretation and/or collection of findings in a practical investigation
limiting factor environmental condition that restricts the rate of biochemical reactions in an organism
line graph graph in which points of data are joined by a connecting line; these are used when both pieces of data are quantitative (numerical)
line of best fit a trend line added to a scatterplot that best expresses the data shown
loaded the form of coenzymes that can act as electron donors
logbook a record containing all the details of progress through the steps of a scientific investigation
lymph interstial fluid surrounding the tissues that is filtered through the tiny holes between the capillaries into the lymphatic system
lymph nodes organs of the lymphatic system where B cells and T cells are activated and adaptive immune responses occur
lymphatic system a network of tissues and organs that plays a key role in the immune response of the mammals
lymphocytes class of white blood cells found in all tissues including blood, lymph nodes and spleen, and which play a role in specific immunity
lysis destruction of cells by rupturing the membrane of the cell
lysosomes membrane-bound vesicles containing digestive enzymes
lysozyme an enzyme present in body secretions such as saliva and tears that helps in the first line of defence
macrophages phagocytic antigen-presenting cells derived from monocytes that may be found in various tissues throughout the body and can engulf foreign material
macroscopic of a size that is visible to the unaided eye
major histocompatibility complex receptor proteins on the surface of cells that identify the cells as 'self'

mammals a group of animals that are characterised by the presence of fur or hair and milk-producing mammary glands
mast cells immune cells containing histamine, which is involved in allergic responses and inflammation
matrix the gel-like solution within the mitochondria
measurement bias a type of influence on results in which an investigator either intentionally or unintentionally manipulates results to get a desired outcome
membrane-attack complex (MAC) one of the defence mechanisms resulting from activation of complement proteins that destroys pathogen cells by osmotic shock
memory cells long-lived cells specific to an antigen that are retained in lymph nodes and can respond to future reinfection
messenger RNA (mRNA) form of RNA synthesised by the transcription of a DNA template strand in the nucleus; mRNA carries a copy of the genetic information into the cytoplasm
metabolism total of all chemical reactions occurring in an organism
MHC-I a type of major histocompatibility complex found on nucleated cells
MHC-II a type of major histocompatibility complex found on specific white blood cells involved in the adaptive immune response
microbiological barriers innate barriers involving normal flora in the body
microfossils very small fossils of a size that can only be studied using a microscope
mineralised fossil a physical fossil formed when the organic material of the structure is replaced by minerals
missense mutation a type of mutation in which a single base change leads to the change in the amino acid translated in the protein chain
mitochondria organelles in eukaryotic cells that are the major site of ATP production; singular: mitochondrion
mitochondrial DNA (mtDNA) DNA located on the mitochondria in cells that is usually inherited maternally
models representations of ideas, phenomena or scientific processes; can be physical, mathematical or conceptual
molecular homology the process involved in comparing similarities in molecular structures to infer relatedness, with a particular focus on DNA and amino acid sequences
molecular techniques methods using DNA or RNA to identify a pathogen
monoclonal antibodies specifically designed antibodies used in the treatment of some diseases such as cancer and autoimmune disease
mould type of fossil that results when a buried organism decays, leaving an impression of the original organism
mucous membranes cellular linings of the inner spaces within the airways, the gut and the urogenital tract
mucus a gelatinous fluid secreted by cells of the mucous membranes
mutagen chemical or physical agent that can cause mutation in DNA
mutation a permanent alteration in the DNA sequence of a gene that leads to the formation of new alleles
NAD^+ the unloaded form of NADH, which can accept hydrogen ions and electrons during cellular respiration
NADH the loaded form of NAD^+ which can donate hydrogen ions and electrons during cellular respiration
$NADP^+$ the unloaded form of NADPH, which can accept hydrogen ions and electrons during photosynthesis
NADPH the loaded form of $NADP^+$, which can donate hydrogen and electrons during photosynthesis
narrow spectrum antibiotics that act against a limited variety of microorganisms
natural active immunity a type of immunity in which the body produces antibodies in response to a normal infection by a pathogen
natural immunity a form of specific immunity in which antibodies are produced or obtained through natural means
natural killer cells special white blood cells involved in the innate immune response that kill virus-infected cells
natural passive immunity a form of immunity in which an individual receives antibodies from a natural means, such as through breastfeeding
natural selection the process in which organisms better adapted for an environment are more likely to pass on their traits to the next generation
naïve refers to an immune cell that has not yet been activated

neutralisation the process of binding of an antibody to toxins or antigens on the surface of the pathogens that inhibits their action
neutrophils the most common type of white blood cell; one kind of phagocyte
node shows an ancestor of two or more descendants on a phylogenetic tree
nominal data a type of qualitative data that has no logical sequence
non-competitive inhibition inhibition in which a molecule binds to the allosteric site of an enzyme causing a conformation change in
non-conservative missense mutation a type of missense mutation in which the substituted amino acid is very different in properties to the initial amino acid
non-enveloped viruses viruses that lack an outer membrane; also referred to as naked
non-infectious diseases that cannot spread from affected people to healthy people via the environment
non-maleficence the act of avoiding harm
non-self antigens antigens that do not belong to the body's own cells
nonsense mutation a type of mutation in which a single base change leads to a STOP signal being received, resulting in a truncated protein
nucleotides basic building blocks or sub-units of DNA and RNA consisting of a phosphate group, a base and a five-carbon sugar
operator a region found in an operon where a repressor is able to bind
operon a cluster of adjacent structural genes in bacteria controlled by a single promoter and operating as a coordinated unit
opinion an idea based on personal beliefs
opposable thumb a thumb that allows for grasping and can be rotated and moved to touch other fingers
opsonisation the coating of the surface of pathogen cells by complement proteins, making the pathogens more susceptible to phagocytosis
optimum temperature the temperature at which the rate of reaction catalysed by an enzyme is at its highest
ordinal data a type of qualitative data that can be ordered or ranked
organic compound any carbon and hydrogen containing compound
origin of replication (ORI) a DNA base sequence in a plasmid in which DNA replication begins
orthologous gene or proteins derived from a shared ancestor in which the primary function is conserved
osmosis a specialised process of passive transport in which water molecules move across a partially permeable membrane from an area of high water (low solute) to an area of low water (high solute)
outlier result that is a long way from other results and seen as unusual
pandemic a situation when, over a relatively short time, many people worldwide contract a specific disease as it spreads from a region of origin
Paranthropus a group of extinct hominins known as the robust hominins, who are not thought to be ancestors of modern humans
passive immunity short-term immunity acquired from an external source of antibodies
pathogen-associated molecular patterns molecules that are found in pathogens but are not found in a host, allowing them to be recognised as foreign
pathogens agents that cause diseases in their hosts
pattern recognition receptor (PRR) protein receptors present on phagocytic cells of the innate immune system that enable these cells to recognise and bind to pathogens, with recognition being at a generic level
peer-reviewed a piece of work that has been reviewed by a board of independent experts who have no conflicts of interest and check for validity and bias
peptidoglycan a polymer consisting of sugars and amino acids that forms a major part of the cell wall of Gram-positive bacteria
perforin a protein, released by some immune cells, that produces a pore in the membrane of cells undergoing an immune attack
peripheral proteins proteins that are anchored to the exterior of the plasma membrane through bonding with either lipids or integral proteins
personal errors human errors or mistakes that can affect results but should not be included in analysis

phagocytes types of white blood cell, including neutrophils and macrophages, that can engulf and destroy foreign material
phagocytosis bulk movement of solid material into cells where the cell engulfs a particle to form a phagosome
phagosome a membrane-bound vesicle formed within a phagocytic cell that encloses the engulfed pathogen
phenotypic methods techniques to differentiate bacteria on the basis of microscopy, different media and biochemical tests
phospholipids major type of lipid found in plasma membranes and the main structural component of plasma membranes
photorespiration a process in which plants take up oxygen rather then carbon dioxide in the light, resulting in photosynthesis being less efficient
photosynthesis process by which plants use the radiant energy of sunlight trapped by chlorophyll to build carbohydrates from carbon dioxide and water
phylogenetic tree a branching diagram showing inferred evolutionary relationships between life forms based on their observed physical and genetic similarities and differences
physical barriers innate barriers that act to prevent the entry of pathogens into the body
physical fossil the remains of all or part of the structures of organisms, such as bones, teeth or leaves
physical methods methods of identifying pathogens based on size and shape
placebo substance or treatment that is designed to affect an individual (used as a control)
plasma cells B cells that are short-lived and secrete soluble antibodies against the specific antigen
plasma membrane partially permeable boundary of a cell controlling entry to and exit of substances from a cell
plasmid a small circular piece of double-stranded DNA that is able to reproduce independently and may be taken up by cells (usually bacteria) in addition to chromosomal DNA
polymerase enzyme involved in synthesising nucleic acids
polymerase chain reaction a technique used to amplify a segment of DNA
population members of one species living in one region at a particular time
post-transcription modification process occurring after transcription in which pre-mRNA is altered to become mature mRNA
post-zygotic isolating mechanisms barriers that prevent a fertile offspring developing after mating
pre-proinsulin an inactive precursor molecule of insulin that includes a signal peptide
pre-zygotic isolating mechanisms barriers that prevent an organism from finding and securing a mate
precision how close multiple measurements of the same investigation are to each other
preserved organism a physical fossil formed when an organism is completely preserved in a substance such as amber, ice or tar
primary antibody response production of antibodies induced in an individual by the first exposure to an antigen
primary data direct or firsthand evidence obtained from investigations or observations
primary lymphoid organs part of the lymphatic system that comprises the bone marrow and thymus
primary structure the specific linear sequence of amino acids in a protein
primates a group of mammals with opposable thumbs, flat nails and binocular vision
prions infectious particles made of protein that lack nucleic acids
probe single-stranded segment of DNA (or RNA) carrying a radioactive or fluorescent label with a base sequence complementary to that in a target strand of DNA
producer organism that can build organic matter from simple inorganic substances; also known as autotrophs
product the compound that is produced in a reaction
proinsulin a modified form of pre-proinsulin and the precursor to insulin, in which the signal peptide has been removed and disulfide bridges have formed between chains
prokaryotes any cells or organisms without a membrane-bound nucleus
prokaryotic cells cells within prokaryotes that lack a membrane-bound nucleus
promoter part of the upstream flanking region of a gene where RNA polymerase binds that contains base sequences that control the activity of that gene

proteins macromolecules built of amino acid sub-units and linked by peptide bonds to form a chain, sometimes termed a polypeptide
proteome the complete array of proteins produced by a single cell or an organism in a particular environment
proteomics the study of the proteome, the complete array of proteins produced by an organism
pyruvate a three-carbon molecule produced during glycolysis
qualitative data data that has labels or categories rather that numberical quantities; also known as categorical data
quantitative data numerical data that examines the quantity of something (e.g. length or time); also known as numerical data
quarantine the act of isolating infected individuals to prevent the spread of a disease
quaternary structure the final level of protein structure in which multiple polypeptides join together to form a protein complex
radiometric dating technique a technique for obtaining an absolute age that depends on the known rate of decay of a radioactive parent isotope to a stable daughter product
random coiling a type of secondary structure in proteins that does not fit in as either a alpha helix or beta-pleated sheet
random errors chance variations in measurements
randomised the assigning of individuals to an experiment or control group is random and not influenced by external means
rational drug design construction of a drug to fit the active site of a molecule so that the natural action of the molecule cannot occur
re-emerging disease reappearance of a known disease after a significant decline in incidence
reactant a substance that is changed during a chemical reaction
recent common ancestor an organism that two groups are descended from
recombinant plasmids plasmids that carry foreign DNA
redundant see 'degenerate'
regulator genes genes that produce proteins that control the activity of other genes
relatedness how recently species split from a common ancestor
relative age the age of an object expressed in relative terms so that it is identified as younger or older than another object
repeatability how close the results of successive measurements are to each other in the exact same conditions
repressor a protein produced by a regulatory gene that can bind to DNA and prevent transcription
reproducibility how close results are when the same variable is being measured but under different conditions
reproductively isolated the inability of species to breed and produce fertile offspring
reservoir the habitat in which a pathogen lives, grows and multiplies
resolution stage the final stage of inflammation, in which the normal state in restored
response bias a type of influence on results in which only certain members of a target population respond to an invitation to participate in a trial
results a section in a report in which all data obtained is recorded, usually in the form of tables and graphs
reverse transcriptase enzyme that directs the formation of copy DNA from a messenger RNA template through reverse transcription
ribonucleic acid (RNA) nucleic acid consisting of a single chain of nucleotide sub-units that contain the sugar ribose and the bases A, U, C and G; RNA
ribosomal RNA (rRNA) stable form of RNA found in ribosomes
ribosomes organelles that are major sites of protein production in cells in both eukaryotes and prokaryotes
risk assessment a document that examines the different hazards in an investigation and suggests safety precautions
RNA polymerase an enzyme that controls the synthesis of an RNA strand from a DNA template during transcription
RNA processing occurs after transcription and involves modifying pre-mRNA to form mature mRNA; also known as post-transcription modification
root the common ancestor of all taxa on a phylogenetic tree

rough endoplasmic reticulum endoplasmic reticulum with ribosomes attached
Rubisco an important enzyme involved in the process of carbon fixation
R_0 value the basic reproduction number that identifies the expected number of individuals a person with a certain disease will infect
sample size the number of trials or individuals being tested in an investigation
sampling bias a type of influence on results in which participants chosen for a study are not representative of the target population
sampling error differences that occur when a sample or subset is not representative of an initial population
sandwich ELISA a form of ELISA where an antigen attached to an antibody with an enzyme indicator binds to an antibody attached to a surface
scatterplots graphs in which two quantitative variables are plotted as a series of dots
scientific method the procedure that must be followed in scientific investigations, consisting of questioning, researching, predicting, observing, experimenting and analysing; also called scientific process
scientific methodology the type of investigation being conducted to answer a question and resolve a hypothesis
scientific poster a hard-copy or digital poster used to display the key findings from investigations conducted to answer a scientific question or hypothesis
sebum the oily secretion produced by sebaceous glands of the skin
second line of defence part of the defence provided by the immune cells and soluble proteins of the innate immune system against attacking pathogens that gain entry to the body
secondary antibody response the rapid production of high levels of specific antibodies to a foreign antigen that occurs in a person who was previously exposed to the same antigen
secondary data comments on or summaries and interpretations of primary data
secondary lymphoid organs part of the lymphatic system that comprises the lymph nodes and the spleen
secondary structure a type of protein structure where three different folds of alpha helices, beta-pleated sheets and random coils can occur in amino acid chains, depending on the R groups in the different amino acids
selectable marker genes carried by plasmids for certain traits, often for antibiotic resistance
selection the process in which a variant is more advantageous under certain selective pressures, enhancing its chance of survival and reproduction
selection bias a type of influence on results in which test subjects are not equally or randomly assigned to experimental or control groups
selective advantage relative higher genetic fitness of a phenotype compared with other phenotypes controlled by the same gene
selective breeding a process of mating that is not random, but uses parents chosen by the breeder on the basis of particular phenotypic characteristics that they display
selectively permeable another term for semi-permeable, where only particular molecules can pass through
self-antigens antigens on cells that are recognised by self-receptors as being part of the same body
self-tolerance inability of an adaptive immune system to respond to the body's own self-antigens
semi-permeable allows only certain molecules to cross by diffusion
sensitivity test test to establish the most effective drug to use for treatment against a particular bacterial infection
serotypes variants within a species of bacterium or virus that are distinguished by their surface antigens
short tandem repeats (STRs) chromosomal sites where many copies of a short DNA sequence are joined end-to-end; the number of repeats is variable between unrelated people
silent mutation a type of mutation in which a single base change does not change the amino acid and final protein expressed
simple diffusion the movement of substances from a region of higher concentration to one of lower concentration of that substance; that is, *down* its concentration gradient
sister taxa two groups with a common ancestor that is not shared with another taxa
somatic cells all body cells other than the reproductive cells (sperm and egg)
speciation process of formation of new species
species organisms that are able to interbreed and produce fertile offspring
specificity the ability to recognise and respond to a specific antigen

spliceosomes complex molecules present in the nucleus that remove introns from the pre-mRNA transcript
sticky ends ends of a DNA fragment with overhanging bases after being cut by an endonuclease
stratigraphic method the method of obtaining the relative age of an object by its position within a given sequence of rock strata
stroma in chloroplasts, the semi-fluid substance which contains enzymes for some of the reactions of photosynthesis
structural genes genes that produce proteins that contribute to the structure or functioning of an organism
structural morphology the process of comparing similarities in body structures to infer relatedness
substitution a gene mutation where one nucleotide is replaced with another
substrate a compound on which an enzyme acts
sympatric speciation speciation that occurs between two populations that have no geographical barrier between them
systematic errors errors that affect the accuracy of a measurement and cannot be improved by repeating an experiment. They are usually due to equipment or system errors.
T cell receptor a molecule found on the surface of T cells that is responsible for recognising fragments of antigen as peptides bound to major histocompatibility complex (MHC) molecules
T lymphocytes also called T cells; white blood cells that mature in the thymus and participate in the adaptive immune response
***Taq* polymerase** an enzyme used in PCR that adds free nucleotides to the single stranded DNA in order to synthesise a new strand
TATA box short base sequence consistently found in the upstream flanking region of the coding region of genes of many different species
template strand one strand of a DNA double helix that is used to produce a complementary mRNA strand during transcription; sometimes called the sense strand
tentative not fixed or certain, may be changed with new information
tertiary structure the total irregular 3D folding of a protein held together by various bonds forming a complex shape
testable able to be supported or proven false through the use of observations and investigation
theory a well-supported explanation of a phenomenon based on facts that have been obtained through investigations, research and observations
third line of defence part of the defence provided by the immune cells of the adaptive immune system through the various actions of T cells and B cells
thylakoid flattened membranous sacs in chloroplasts that contain chlorophyll
tip the terminal point of a phylogenetic tree, representing a particular species
toxoids inactivated toxins used for active immunisation
trace fossils preserved evidence of the activities of organisms, such as footprints, toothmarks, tracks, burrows and coprolites
trans-membrane proteins proteins that are embedded within and span the plasma membrane, allowing them to have parts exposed to both the intracellular and extracellular environment
transcription process of copying the genetic instructions present in DNA to messenger RNA
transfer RNA (tRNA) form of RNA that can attach to specific amino acids and carry them to a ribosome during translation
transgenic organisms organisms that carry in their genomes one or more genes artificially introduced from another species
transitional fossil the fossilised remains of a life form that exhibits traits common to both an ancestral group and its derived descendant group(s)
translation process of decoding the genetic instructions in mRNA into a protein (polypeptide chain) built of amino acids
transmission passing a pathogen on to another individual
triplet code the idea that the genetic code consists of triplets or three-base sequences
***trp* operon** a collection of adjacent genes in bacteria that code for the enzymes needed in the production of tryptophan

uncertainty a limit to the precision of data obtained; a range within which a measurement lies
universal the property of the genetic code in which the code is essentially the same across all organisms
unloaded a form of coenzymes that can act as electron acceptors
unrepresentative sample a sample or subset that is not representative of the allele frequencies in an initial population
vaccination an artificially active process in which an individual is injected with either antigens or weakened pathogens in order to produce their own antibodies and memory cells
vaccination programs mandated programs that set a schedule in which vaccinations against specific diseases should be administered
vaccines soluble antigens derived from the causative agents of diseases that are administered to individuals, providing them with protection
validity credibility of the research results from experiments or from observation; a measure of how accurately and precisely results measure what they are intending to measure through fair testing
variable any factor that can be changed in an investigation
variable regions the region that is unique to each antibody, where a specific antigen is able to bind
variation differences exhibited among members of a population owing to the action of genes
vector an agent or vehicle used to transfer pathogens or genes between cells and organisms
vectors organisms that carry pathogenic agents and spread them to other organisms
vesicles membrane-bound sacs found within a cell, such as secretory vesicles, which are involved in the export of proteins
vestigial structures non-functional structures that are remnants of functional structures in ancestral species
virion the extracellular form of a virus that can transfer between hosts
viroids simple forms of viruses that lack a capsid
viruses non-cellular pathogens that use the host cell in order to replicate their genetic material
water deficit when there is a limited amount of water
waterlogged when excess water has reached a plant
zoonotic diseases diseases that have been transferred from other animals
zygomatic arches cheekbones comprising horizontal bony ridges on a mammalian skull

INDEX

K

L

M

Q

R

S